Neural Functions of the Delta-Opioid Receptor

Ying Xia
Editor

Neural Functions of the Delta-Opioid Receptor

Springer

Editor
Ying Xia
University of Texas Medical School
Houston, TX, USA

ISBN 978-3-319-25493-7 ISBN 978-3-319-25495-1 (eBook)
DOI 10.1007/978-3-319-25495-1

Library of Congress Control Number: 2015960241

Springer Cham Heidelberg New York Dordrecht London

Printed on acid-free paper

Springer International Publishing AG Switzerland is part of Springer Science+Business Media (www.springer.com)

Preface

The early 1970s through the early 1990s was an important period in opioid research. From the discovery of endogenous receptor selective for opioids in 1973 to the cloning of three major opioid receptors, i.e., delta-, mu-, and kappa-opioid receptors (DOR, MOR, and KOR), in 1992–1993, this time period showed major advancements in our understanding of opioid receptors. During and after this period, many scientists and clinicians devoted their research efforts to understanding the biology of DOR and the development of various agonists and antagonists for DOR. In 2004, the book entitled *The Delta Receptor* edited by Chang, Porreca, and Woods comprehensively reviewed scientific accomplishments in DOR research from 1973 to 2003.

Since the cloning of DOR, many research efforts have been devoted to elucidating its function. However, most of the past studies focused on its role in pain relief as well as in opioid addiction because DOR was traditionally thought to be primarily involved in modulating the transmission of messages along pain pathways. Although there were scattered studies on other functions, inconsistent results and contradicting conclusions, due to the limitations of ligand specificity and experimental approaches, were pervasive in the past literature, especially in terms of DOR's role in response to hypoxic/ischemic stress. With the advancement of research techniques, more diversified studies on neural functions of DOR merged with traditional topics, especially in the past decade.

The distribution and density of the three major opioid receptors are vastly different among brain regions during development in mammals as well as across species, suggesting that opioid receptors are differentially involved in neural activities of the brain. Indeed, abundant recent data from independent laboratories at various institutions worldwide have shown that DOR, unlike MOR and KOR, plays a unique role in neuroprotection against hypoxic/ischemic injury and many other functions. Moreover, accumulating evidence suggests that DOR dysfunction participates in a variety of pathophysiological events in many diseases.

This book broadly reviews recent progress in the functional research of DOR. Written by currently active researchers in functional studies of DOR, this book discusses research on DOR's neural functions, excluding its roles in pain modulation and opioid addiction. Apart from the progress in ligand research and evolutional and developmental distribution of DOR, topics include ionic homeostasis, glutamate transportation, glutamatergic signaling, mitochondrial function, neurogenesis, hibernation for stroke therapy, Parkinson's disease, deep brain stimulation, electroacupuncture therapy for cerebral ischemia and epilepsy, and a comparative view on brain and heart protection. In the final chapter, the various functions of opioids in pathophysiological conditions are comprehensively addressed. There is some overlapping information among the different chapters in certain aspects, especially in the background of the topic. We chose to format the book in this style because in today's world, readers may download only a single chapter that is interesting to them, instead of the whole book. In this way, all readers may see a complete picture of any topic they are interested in. I believe that this unique book provides a broad perspective on functional research of DOR for neuroscientists, neurologists, and other clinicians. For medical students and undergraduate and graduate students majoring in biology, this book is a modern course for learning updated achievements in the research of opioid receptors.

I am very grateful to all the authors for their significant contributions that made this book happen. I also highly appreciate the efforts of the referees who carefully reviewed our proposal and made constructive comments and suggestions. Finally, I would like to thank Ms. Simina Calin and Mr. Jacob Rosati of Springer for their efforts to turn this book into a reality.

Houston, TX, USA Ying Xia

Contents

Contributors

Sandra Acosta, Ph.D. Department of Neurosurgery and Brain Repair, University of South Florida College of Medicine, Tampa, FL, USA

Tetsuya Asakawa, M.D., Ph.D. Department of Neurosurgery, Hamamatsu University School of Medicine, Hamamatsu, Shizuoka, Japan

Department of Psychiatry, Hamamatsu University School of Medicine, Hamamatsu, Shizuoka, Japan

Cesar V. Borlongan, Ph.D. Department of Neurosurgery and Brain Repair, University of South Florida College of Medicine, Tampa, FL, USA

Kwen-Jen Chang, Ph.D. Faculty of Life Science and Technology, Kunming University of Science and Technology, Kunming, Yunnan, China

Dongman Chao, Ph.D. Department of Neurosurgery, The University of Texas Medical School at Houston, Houston, TX, USA

Tao Chen, M.D., Ph.D. Department of Neurosurgery, The University of Texas Medical School at Houston, Houston, TX, USA

Department of Neurology, Hainan General Hospital, Haikou, Hainan, China

Daniel N. Deussen, M.S. VA San Diego Healthcare System, San Diego, CA, USA

Department of Anesthesiology, University of California, San Diego, CA, USA

Albert J. Fenoy, M.D. Department of Neurosurgery, Mischer Neuroscience Institute, The University of Texas Medical School at Houston, Houston, TX, USA

Huimin Gao, M.D., Ph.D. Department of Neurosurgery, The University of Texas Medical School at Houston, Houston, TX, USA

Department of Neurology, The Third Affiliated Hospital of Sun Yat-Sen University, Guangzhou, Guangdong, China

Jingchun Guo, Ph.D. State Key Laboratory of Medical Neurobiology, Department of Neurobiolgy, Fudan University, Shanghai, China

Taylor E. Hinchliffe, B.A. Department of Neurosurgery, The University of Texas Medical School at Houston, Houston, TX, USA

Yuji Kaneko, Ph.D. Department of Neurosurgery and Brain Repair, University of South Florida College of Medicine, Tampa, FL, USA

Naoko Kuzumaki, Ph.D. Department of Pharmacology, Hoshi University School of Pharmacy and Pharmaceutical Sciences, Tokyo, Japan

Jason Eli Maas, M.D., Ph.D. Department of Anesthesiology, University of Virginia, Charlottesville, VA, USA

Minoru Narita, Ph.D. Department of Pharmacology, Hoshi University School of Pharmacy and Pharmaceutical Sciences, Tokyo, Japan

Life Science Tokyo Advanced Research Center (L-StaR), Hoshi University School of Pharmacy and Pharmaceutical Sciences, Tokyo, Japan

Mibel Pabon, Ph.D. Department of Neurosurgery and Brain Repair, University of South Florida College of Medicine, Tampa, FL, USA

Matthew E. Pamenter, Ph.D. Department of Biology, University of Ottawa, Ottawa, ON, Canada

Hemal H. Patel, Ph.D. VA San Diego Healthcare System, San Diego, CA, USA

Department of Anesthesiology, University of California, San Diego, CA, USA

Lu Pu, Ph.D. School of Medicine, Case Western Reserve University, Cleveland, OH, USA

David M. Roth, M.D., Ph.D. VA San Diego Healthcare System, San Diego, CA, USA

Department of Anesthesiology, University of California, San Diego, CA, USA

Jan M. Schilling, M.D. VA San Diego Healthcare System, San Diego, CA, USA

Department of Anesthesiology, University of California, San Diego, CA, USA

Wolfgang Schwarz, Ph.D. Shanghai Research Center for Acupuncture and Meridians, Shanghai, China

Shanghai Key Laboratory of Acupuncture Mechanism and Acupoint Function, Fudan University, Shanghai, China

Institute for Biophysics, Goethe-University Frankfurt, Frankfurt am Main, Germany

Yuehai Shen, Ph.D. Faculty of Life Science and Technology, Kunming University of Science and Technology, Kunming, Yunnan, China

Meaghan Staples Department of Neurosurgery and Brain Repair, University of South Florida College of Medicine, Tampa, FL, USA

Naoki Tajiri, P.T., Ph.D. Department of Neurosurgery and Brain Repair, University of South Florida College of Medicine, Tampa, FL, USA

Sebastian J. Villarreal, M.D. Department of Neurosurgery, The University of Texas Medical School at Houston, Houston, TX, USA

Qing Wang, M.D., Ph.D. Department of Neurology, The Third Affiliated Hospital of Sun-Sen University, Guangzhou, Guangdong, China

Guoqiang Wen, M.D. Department of Neurology, Hainan General Hospital, Haikou, Hainan, China

Ying Xia, M.D. Ph.D. University of Texas Medical School, Houston, TX, USA

Yong Xia, M.D., Ph.D. Department of Neurosurgery, The University of Texas Medical School at Houston, Houston, TX, USA

Yinfang Xu, Ph.D. Shanghai Research Center for Acupuncture and Meridians, Shanghai, People's Republic of China

Shanghai Key Laboratory of Acupuncture Mechanism and Acupoint Function, Fudan University, Shanghai, People's Republic of China

Fei Yao, M.D., Ph.D. Department of Neurosurgery, The University of Texas Medical School at Houston, Houston, TX, USA

Shoupu Yi Faculty of Life Science and Technology, Kunming University of Science and Technology, Kunming, Yunnan, China

Daniel Yiu, B.A. Department of Neurosurgery, The University of Texas Medical School at Houston, Houston, TX, USA

Zhiyi Zuo, M.D., Ph.D. Departments of Anesthesiology, University of Virginia, Charlottesville, VA, USA

Introduction

Written by currently active researchers in functional studies of the delta-opioid receptor, this book discusses recent research on DOR's neural functions. The delta-opioid receptor was traditionally thought to be primarily involved in modulating the transmission of messages along pain pathways. New findings from different institutions worldwide have shown its unique role in neuroprotection and many other functions. Instead of traditional topics (e.g., pain modulation and opioid addition) on the delta-opioid receptor, this book focuses on newly explored fields. Besides the progress in ligand research and evolutional and developmental distribution of the receptor, topics include ionic homeostasis, glutamate transportation, glutamatergic signaling, mitochondrial function, neurogenesis, hibernation for stroke therapy, Parkinson's disease, deep brain stimulation, electroacupuncture therapy for cerebral ischemia and epilepsy, and a comparative view on brain and heart protection. Finally, the various functions of opioids in pathophysiological conditions are comprehensively addressed. This unique book provides a broad perspective on functional research of the delta-opioid receptor for neuroscientists, neurologists, and other clinicians. For medical students and undergraduate and graduate students majoring in biology, this book is a modern course for learning updated achievements in the research of opioid receptors.

Keywords Neuroscience, Medicine, Translational research, Opioid receptors, Neuroprotection, Excitotoxicity, Hypoxia, Ischemia, Parkinson's disease, Epilepsy

About the Editor

Ying Xia is Professor and Vice-Chairman for Research of the Department of Neurosurgery at the University of Texas Medical School at Houston and Professor of Neuroscience at the University of Texas at Houston Graduate School of Biomedical Sciences. He obtained his medical training at Soochow Medical College, China, and his Master Degree of Medicine and Doctoral Degree of Medical Science (Ph.D.) in Neurobiology and Integrative Medicine at Shanghai Medical University, China. Dr. Xia joined Yale University School of Medicine for his post-doctoral training and initiated his laboratory at Yale. During his 22-year academic career at Yale University School of Medicine, he was appointed as a faculty member in 1993 and then promoted to Associate Professor in 2002. He joined the University of Texas Medical School at Houston in May 2010.

Dr. Xia's research has been consecutively supported by NIH, America Heart Association, March of Dimes Foundation, Cerebral Palsy Foundation, and the Vivian L. Smith Neurologic Foundation since the estabolishment of his laboratory in 1994. He has been invited to deliver speeches by many academic institutions and international conferences. He has served as a reviewer for over 70 international journals and as a grant reviewer for NIH, US National Science Foundation, other American and European foundations, Hong Kong University Grants Committees, and Natural Science Foundation of China.

Chapter 1
Current Research on the δ-Opioid Receptor: From Neuroprotection Against Hypoxia/Ischemia to Broad Neural Functions

Ying Xia

Abstract The delta-opioid receptor (DOR) was traditionally thought to be primarily involved in modulating the transmission of messages along pain pathways. Therefore, most previous studies focused on its function in pain modulation. Although there were scattered studies on other functions, inconsistent results and contradicting conclusions were present in the past literature, especially in terms of DOR and neuroprotection against hypoxic/ischemic injury. Based on our observations on the evolutionary developmental distribution of DOR in the brain, and taking inspiration from the finding that the turtle brain exhibited a much higher DOR density and greater tolerance to hypoxic/ischemic insult, we clarified DOR's specific role in the brain against hypoxic/ischemic injury and reconciled previous controversies in this aspect. Furthermore, we expanded our work to other neural functions and achieved significant progress. Our first finding showed that DOR is an endogenous inhibitor of sodium channels, and DOR signaling is involved in the maintenance of ionic homeostasis stabilization in physiological conditions as well as under hypoxic/ischemic stress. DOR activation upregulates the expression of astrocytic glutamate transporters and increases astrocytic ability to clean up extracellular glutamate accumulation. DOR signals promote the function of Nrf2 in both neurons and glia. DOR can also exert a beneficial action against Parkinson's disease and its activation reduces α-synuclein overexpression and oligomer formation in in vitro models of Parkinson's disease. DOR activation prevents the collapse of mitochondrial membrane potential, thereby reducing cell injury. In addition to DOR neuroprotection against ischemic injury in the model of ischemic stroke, DOR acts as a major mediator in electroacupuncture-induced brain protection against cerebral ischemia. The DOR system may also serve as an inhibitor of epileptic hyperexcitability through the inhibition of sodium channel function, which would explain the positive effects electroacupuncture has on epilepsy. Apart from our research, recent studies from other independent laboratories consistently show a variety of functions

Y. Xia, M.D., Ph.D. (✉)
University of Texas Medical School, Houston, TX, USA
e-mail: ying.xia@uth.tmc.edu; y55738088@gmail.com

Y. Xia (ed.), *Neural Functions of the Delta-Opioid Receptor*,
DOI 10.1007/978-3-319-25495-1_1

for DOR that are mostly beneficial, including neurogenesis, in the brain and heart. On the other hand, dysfunction of DOR may be involved in the pathophysiological events of many diseases. An in-depth understanding of DOR's functions and the underlying mechanisms and development of more effective DOR ligands will eventually open a door for the prevention and treatment of many neurological and other diseases in the future.

Keywords δ-Opioid receptor • Brain • Heart • Neurons • Glia • Hypoxia • Ischemia • Ionic homeostasis • Na^+ channels • Nrf2 • Glutamate transporters • Mitochondria • Neurogenesis • Protection • Stroke • Epilepsy • Parkinson's disease

Abbreviations

DADLE	[D-ala2, D-leu5] enkephalin
DAMGO	[D-Ala2, N-MePhe4, Gly-ol]-enkephalin
DOR	δ-Opioid receptor
KOR	κ-Opioid receptor
MOR	μ-Opioid receptor
N/OFQ	Nociceptin/Orphanin FQ
NMDA	N-methyl-D-aspartate
NOP (ORL-1)	Nociceptin receptor
Nrf2	Nuclear factor (erythroid-derived 2)-like 2
PL017	Tyr-Pro-MePhe-D-Pro
UFP-512	H-Dmt-Tic-NH-CH(CH2-COOH)-Bid

1 Introduction

The delta-opioid receptor (DOR) is one of three classic opioid receptors in the opioid system. Leu- and met-enkephalins constitute the main endogenous agonists because they are produced in many types of cells and preferentially bind to DOR in physiological concentrations (Lord et al. 1977; Chang and Cuatrecasas 1979; Feng et al. 2012; refer to Chaps. 2 and 18).

Since DOR was cloned in the early 1990s (Evans et al. 1992; Kieffer et al. 1992), abundant research efforts have been paid to explore its function. However, most previous studies focused on pain modulation and addiction (Chang et al. 2004; Costantino et al. 2012; Lutz and Kieffer 2013; Giri and Hruby 2014). Because DOR was traditionally thought to be primarily involved in modulating the transmission of messages along pain pathways, much like MOR. Conversely, much more clinical and laboratory research attention had been paid to MOR than to DOR, because MOR agonists seem to play a more important role in pain control and opioid addiction. Indeed, in clinical settings, a majority of anti-pain drugs are MOR agonists like morphine and methadone.

Although scattered studies on other functions of DOR can be found in the literature, the outcomes and conclusions of these studies are inconsistent, and sometimes very contradictory. In fact, controversies exist in many functional studies of opioids and their receptors. For instance, some studies suggested that opioid agonists were neuroprotective against hypoxic or ischemic injury (Hayward et al. 1993a, b; Mayfield and D'Alecy 1994a, b; Mayfield et al. 1994), while other work showed that naloxone, a non-selective opioid antagonist, was neuroprotective against hypoxic or ischemic injury (Baskin and Hosobuchi 1981; Baskin et al. 1984; Hosobuchi et al. 1982; Phillis et al. 1985; Skarphedinsson and Thorén 1988). These controversies may be generated by many factors including inappropriate methodologies and nonspecific ligands.

My interest in opioid receptors began in the early 1980s when I was a graduate student working on the role of opioid receptors in the neural regulation of cardiovascular activities. I was deeply intrigued by the neural functions of DOR in the late 1980s. At that time, I was exploring differences in the expression of membrane functional proteins like DOR in the brain between mammals and turtles, an organism that is highly tolerant to hypoxic/ischemic stress. The findings of my research led me to seriously consider the neural functions of DOR. The distinctive distribution of DOR among brain regions and between species prompted me to believe that DOR has unique functions in the brain as well as in peripheral organs, and its functions may be very different from those of MOR and KOR, even though they share homologous sequences in molecular structure and certain pathways in signaling transduction.

Since then, I (along with my research team that was developed after 1993, when I became a faculty member at Yale Medical School and started building up my own laboratory) have conducted many studies with diversified approaches including biochemical, molecular, transgenic, and electrophysiological techniques to investigate the neuroprotective role of DOR against hypoxic/ischemic injuries. Through our work, we have made significant progress towards a better understanding of DOR's neural functions. Our findings have also been confirmed by many independent investigators. Moreover, over the past decade, many laboratories worldwide have explored new fields and have generated exciting and consistent results related to DOR neuroprotection and other functions (refer to Chaps. 2–18).

This chapter briefly recaps the major progress in the functional research of DOR and serves as an introductory guide to this book. For the purpose of comparison, studies on other opioid receptors are also discussed briefly.

2 Historical Recapitulation of the δ-Opioid Receptor

The opioid system serves as a neuromodulator in the nervous system and also exists in many peripheral organs/cells for other functions. There are three major subtypes of opioid receptors, i.e., mu-, kappa-, and delta-opioid receptors (MOR, KOR, and DOR) (Snyder and Pasternak 2003; Xia and Haddad 1991a, b, 2001; Waldhoer et al. 2004; Feng et al. 2012) with several endogenous agonists, including enkephalins, endorphins, and dynorphins (Hughes et al. 1975). Opioid receptors and endogenous opioids are widely and distinctly distributed in both the central nervous system and peripheral organs (Xia and Haddad 1991a, 2001; Hiller and Fan 1996; Peng et al. 2012; Feng et al. 2012; refer to Chaps. 3, 4 and 18).

2.1 Opiates and Opioids

Opiates have long been utilized by mankind for both their therapeutic and non-therapeutic effects (Brownstein 1993). In the early 1800s, the isolation of morphine, with an alkaloid as the active ingredient (Brownstein 1993), limited the use of the term opiate to all natural alkaloids of opium. Specifically, an opiate is a substance derived from the poppy plant (which contains opium). Morphine and codeine are two commonly used opiates, and both are isolated directly from the poppy plant. Since the active ingredient of opiates are produced by nature, rather than by chemical synthesis, opiates are sometimes referred to as "natural" drugs.

Acheson coined the term "opioids" for all synthetic and semi-synthetic morphine derivatives that demonstrate similar activity to morphine but distinct chemical structures (Martin 1967). Strictly speaking, an opioid is a substance (molecule) that is synthetic or partly synthetic, meaning the active ingredients (molecules) are manufactured via chemical synthesis. Opioids often act just like opiates in the human body, because the molecules are so similar. For brevity's and for convenience' sake, many references currently use opioid(s) to refer to all opium-like substances (both opiates and opioids), while limiting the use of "opiates" to only natural, poppy-derived drugs like morphine.

Opioids have been widely exploited for their potent analgesic and ecstasy-inducing effects in both medical and non-medical settings. By the 1960s, extensive research had been conducted to better understand the mechanisms of morphine/opioid-induced analgesia and pain modulation. This work later spawned many important concepts including the existence of opioid receptors. At around the same time, the existence of endogenous opioids was discovered (Hughes et al. 1975). This marked the beginning of later discoveries of various types of endogenous opioids–a family of chemically distinct endogenous compounds with properties like morphine, including enkephalins, endorphins, and dynorphins (Hughes et al. 1975; Cox et al. 1976; Goldstein 1976; Snyder and Pasternak 2003). In the early 1980s, opioid peptide precursor genes were identified with molecular approaches, proving the existence of the preproenkephalin gene that encoded several copies of enkephalins (Noda et al. 1982), the preprodynorphin gene encoding dynorphins (Civelli et al. 1985), and the proopiomelanocortin gene encoding β-endorphin (Nakanishi et al. 1979).

2.2 Opioid Receptors

Opioid receptors were first recognized in the 1960s during pharmacological studies that focused on the specific receptor sites at which opioids exerted their actions. Martin (1967) provided the first evidence of opioid receptors in vivo. In the early

1970s, the receptors were first identified, by the use of radio-binding assay, as specific molecules with distinct distribution patterns (Goldstein et al. 1971; Pert and Snyder 1973; Simon et al. 1973). Studies also delineated the specificity of opioid activity to certain areas of the brain that were blocked by opioid antagonists (Pert and Yaksh 1975; Snyder and Matthysse 1975). Martin and his co-workers first proposed the existence of μ- and κ-opioid receptors (MOR, KOR) by using morphine and ketacyclazocine as ligands (Martin et al. 1976). In the following year, the existence of DOR was proposed based on the preference of leu-enkephalin to bind to receptors in the mouse vas deferens (Lord et al. 1977).

In late 1992, two groups independently cloned DOR from the neuroblastoma × glioma cell line NG-108 by using an expression cloning strategy. Both reported the identification of a novel member of the 7-helix family of G-protein coupled receptors that had pharmacological properties typical of DOR (Evans et al. 1992; Kieffer et al. 1992). Almost one year later, MOR (Chen et al. 1993; Wang et al. 1993) and KOR (Meng et al. 1993; Minami et al. 1993) were cloned as well.

These opioid receptors are members of the rhodopsin subfamily in the superfamily of seven-transmembrane nucleotide binding regulatory G-protein coupled receptors (GPCRs) (Law and Loh 1999; Ma et al. 2005; Audet et al. 2008; Feng et al. 2012). Although homology is present, these receptors recognize structurally diverse exogenous/endogenous peptide and non-peptide ligands (Pasternak 2004; Ballet et al. 2008; Feng et al. 2012; refer to Chap. 2). These receptors are present not only in the nervous system, but also inhabit other organs such as the heart, kidneys, lungs, liver, and gastrointestinal and reproductive tracts (Peng et al. 2012; Feng et al. 2012; refer to Chap. 18).

Pharmacological evidence indicated the existence of two distinct DOR subtypes, DOR1 and DOR2 (Sofuoglu et al. 1991; Portoghese et al. 1992). These two subtype receptors differ in their selectivity to DOR agonists and antagonists, with no development of cross-tolerance. They can have opposing physiological effects when binding to subtype-selective DOR ligands (van Rijn et al. 2013). However, unlike MOR subtypes that are encoded by different genes, no DOR subtypes have been cloned, and there exists only a single DOR gene for distinct subtypes of DOR (Pasternak 2004). DOR subtypes likely originate from a single DOR gene, and differ due to multiple affinity states or to post-translational modifications at the protein level rather than at the gene level (van Rijn et al. 2013).

Although a significant amount of homology exists in the three major opioid receptors, there is a major difference in their density and distributional pattern in the brain (Fig. 1.1), suggesting that they are differentially involved in various neural functions in different regions.

In addition to the three classic opioid receptors, another non-classic opioid receptor, nociceptin receptor and its endogenous ligand, nociceptin/orphanin FQ (N/OFQ), were discovered in the middle of the 1990s (Mollereau et al. 1994; Meunier and Mollereau 1995). Both of them are widely distributed in the CNS. Nociceptin, a 17-amino acid neuropeptide, is a potent anti-analgesic, but does not act at DOR, MOR and KOR. Also, its actions are not antagonized by naloxone.

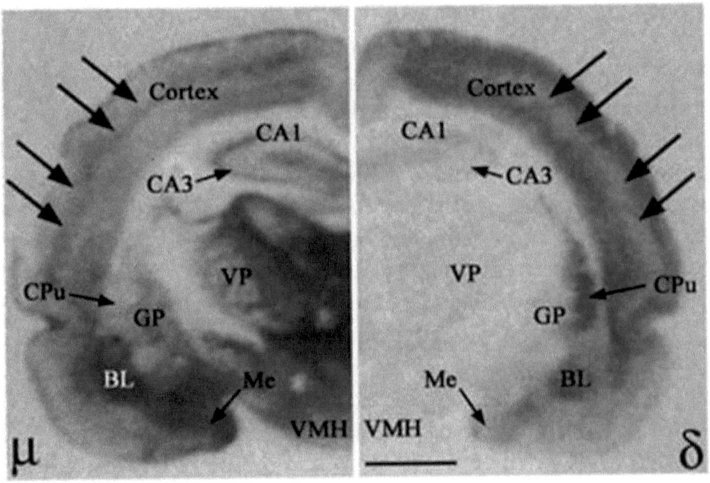

Fig. 1.1 Differential distribution of MOR (*left*) and DOR (*right*) in the rat rostral brain. *BL* basolateral amygdaloid nucleus, *CA1* field CA1 of hippocampus, *CA3* field CA3 of hippocampus, *CPu* caudate putamen, *GP* globus pallidus, *Me* medial amygdaloid nucleus, *VMH* ventromedial hypothalamic nucleus, *VP* ventral posterior thalamic nucleus. Note that the distributional patterns of MOR and DOR are totally different with MOR having a high density in all major regions including subcortical areas and DOR having a much higher density in the cortex, caudate putamen and amygdala than most of the subcortical regions. Also note that there is a higher density of MOR in the middle layer than the outer and inner layers of the cortex, while the opposite is true for the density of DOR. The other cortical layers have an opposite pattern of expression. Scale bar = 2 mm. Cited from Xia and Haddad (2001)

3 An Insightful Clue from the Turtle: High Density of DOR in the Brain

I started my research on opioid receptors in the early 1980s when I was a graduate student investigating the role of opioid receptors in neural regulation of cardiovascular activities at Shanghai Medical University (Xia et al. 1985, 1986, 1987, 1988, 1989) and became deeply fascinated by the neural functions of DOR in the late 1980s as a postdoctoral fellow at Yale Medical School.

My interest in DOR began once I turned my academic attention towards the major differences in hypoxic/ischemic tolerance between mammalian and freshwater turtle brains (Jiang et al. 1992; Xia et al. 1992; Nilsson and Lutz 2004; Milton and Prentice 2007). Indeed, the mammalian neurons are extremely sensitive to hypoxia/ischemia and many cellulr and moleculr events occur in response to hypoxic/ischemic stress (Jiang et al. 1992; Xia et al. 1992; Xia et al. 1993; Zhang et al. 2002; Zhao et al. 2005b, c; Chao et al. 2006, 2007a, b, 2008, 2009, 2012a, 2012b; Hong et al. 2007; Malthankar-Phatak et al. 2008; Sung et al. 2008; Kang et al. 2009; Feng et al. 2011; Qian et al. 2011; Ding et al. 2012; Wang et al. 2014; Yang et al. 2015). In sharp contrast, the turtle can survive without oxygen for days at room temperature and for weeks or months during winter hibernation (Milton and Prentice 2007). Several minutes of hypoxic stress can induce major disruptions of ionic homeostasis (e.g., increase in potassium leakage) in the rat brain slice, while hours of the same stress causes little change in ionic homeostasis in the turtle brain (Jiang et al. 1992; Xia et al. 1992). The cultured turtle cortical neurons could survive 5-min exposures to glutamate at concentrations up to 3 mM, doses 30 times the LD50 and 6 times the LD100 for mouse cortical neurons (Wilson and Kriegstein 1991). Apparently, a mechanistic understanding of this huge disparity would be an invaluable clue for new solutions to hypoxic/ischemic brain injury.

Since channels/exchangers/receptors are membrane proteins that are fundamental in regulating ionic homeostasis, and since their densities may directly affect neuronal functions, we started comparing distributional differences in many functional membrane proteins including Na^+ channels, K^+ channels, glutamate receptors, GABA receptors, opioid receptors, etc. At the time, DOR had not been cloned. The best approach to map the distribution of DOR in the brain was autoradiography with radiolabeled ligands. Therefore, we used a lower concentration (4 nM) of ^{3H}DADLE to label DOR sites, along with a higher concentration (1 μM) of non-radiolabel PL017, a selective MOR ligand (Hawkins et al. 1987; Feng et al. 2012) in order to prevent ^{3H}DADLE from binding to MOR in the experiments of DOR autoradiography. In early 1989, after several trial studies to ensure optimization of experimental conditions, the autoradiographic films showed that under the same experimental conditions, the density of DOR was much higher in the turtle brain than in the rat brain, while MOR showed an opposite distribution, i.e., lower density of MOR in the turtle than in the rat brain. Moreover, we found that in addition to a higher binding capacity, the binding affinity of DOR was also higher in the turtle than in the rat brain (Xia and Haddad 2001).

This finding immediately prompted me to hypothesize a unique role of DOR in brain protection against hypoxic/ischemic injury because none of the other membrane proteins, e.g., Na^+ channels (Xia and Haddad 1993) and sulfonylurea receptors (Xia and Haddad 1991b), we studied showed a similar distribution to that of DOR. This hypothesis was based not only on the high density and high affinity of DOR in the hypoxia/ischemia-resistant turtle brain, but also the unique distribution of DOR in the cortex and hippocampus in our observations (Xia and Haddad 1991a, 2001). The distribution of DOR varies in the rat cortex with the outer and inner layers having higher densities than the middle layer. Congruently, the outer and inner layers of the cortex are more tolerant to hypoxia/ischemia than the middle layer (Fig. 1.1; Kornblum et al. 1987; Nedergaard 1988; Lin et al. 1990; Xia and Haddad 1991a, 2001; Georges et al. 1998). Also, subcortical regions such as the hippocampus, thalamus, and brain stem show a lower density of DOR than the cortex (Xia and Haddad 1991a, 2001), and are more sensitive to hypoxia than the cortex (Lin et al. 1990; O'Reilly et al. 1995; Dunn et al. 1999; Buresh et al. 1999). This circumstantial evidence provided my initial support for the role of DOR in protection against hypoxic/ischemic stress.

Unfortunately, because of the lack of sufficient research resources, I could not validate the data and finish all the work required to publish and expand this side project to the next level at the time. Another reason I was unable to expand my research was that few people believed that DOR could be important in neuroprotection due to major controversies and significant uncertainty in past research (refer to Sect. 4.1). Even 10 years later when I presented this idea to my colleagues, there were still many doubts from all aspects.

Finally, in 1997 when I had enough funding from NIH and other funding agencies, I and my team members had the chance to validate my findings by repeating the experiments in the turtle vs. the rat/mouse. We then published this rediscovered finding in 2001.

As shown in Fig. 1.2, both DOR binding capacity and affinity are higher in the turtle than rat brain (Xia and Haddad 2001). This finding was exciting to me because it provided a clue for DOR neuroprotection. Although I do not intend to argue that the high density and high affinity of DOR is the sole reason for high tolerance of the turtle brain to hypoxia/ischemia, DOR is, at least, one of the factors that rescue the turtle brain from injury in hypoxic/ischemic conditions. Ultimately, this insightful clue gives us reason to explore a new mechanism of neuroprotection in the mammalian brain.

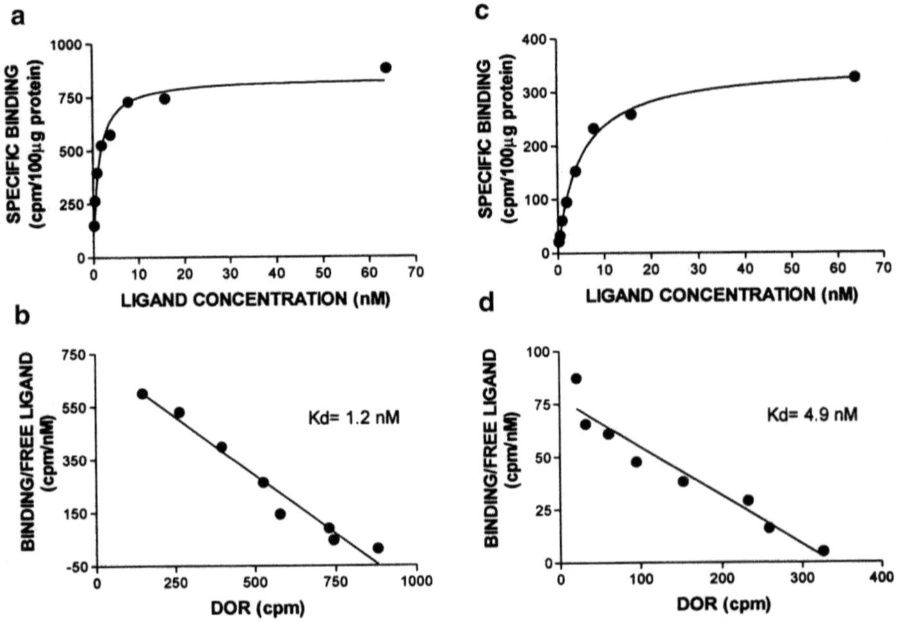

Fig. 1.2 DOR binding properties in turtle vs. rat brain. Experiments were performed with 0.1–32 nM of ^{3H}DADLE with addition of non-radiolabeled MOR ligand PL017 (1 μM). (**a**, **c**) Saturation curves. (**b**, **d**) Scatchard plots. Note that the binding sites are saturable and the Scatchard plots are linear in both turtle and rat brain tissues, suggesting the presence of one binding site. However, the dissociation constant (Kd) of turtle brains (1.2 nM) was more than four-fold lower than that of rat brain (4.9 nM) and the maximal binding capacity was more than double in turtle than in rat brains, suggesting that DOR has a higher binding affinity and density in the turtle brain than rat brain. Cited from Xia and Haddad (2001)

4 The δ-Opioid Receptor (DOR) and Neuroprotection against Hypoxic/Ischemic Injury

When I speculated on the role of DOR in brain protection in the later 1980s after I observed the unique distribution of DOR in the turtle brain, I broadly searched for the effects of opioid receptors on hypoxic/ischemic stress in the literature. Overall, the limited data in the literature led to confusion about the role of opioid receptors in neuroprotection against hypoxic/ischemic injury. Ambiguity, spawned from many conflicting views in this field, kept me from seeing a clear picture until the late 1990s when we produced our own data and compared it with that of others thereafter.

4.1 Historical Controversies: Uncertainty in Past Studies

There were many contradicting reports in the 1980s and 1990s.

4.1.1 In Vivo Studies on Neuroprotection: Opioid Receptor Activation or Inhibition?

As early as the 1980s, some studies (Baskin and Hosobuchi 1981; Baskin et al. 1984; Hosobuchi et al. 1982) had reported that naloxone, a nonspecific opioid antagonist, caused a reversal in ischemic neurological deficits in gerbils, baboons, and man. Baskin and Hosobuchi (1981) reported a naloxone-induced reversal of ischemic neurological deficits in man. Hosobuchi et al. (1982) used a gerbil ischemia model induced by occluding the right common carotid artery, and discovered that naloxone could reverse hemiparesis caused by cerebral ischemia, and continuous administration could reduce the associated mortality. Moreover, Baskin et al. (1984) further indicated that morphine, a MOR agonist, exacerbated neurological deficits in baboons secondary to focal cerebral ischemia. Some other researchers (Phillis et al. 1985; Skarphedinsson and Thorén 1988) also reported the beneficial effects of opioid antagonists on cerebral function in ischemic rats. Since naloxone binds non-specifically to DOR, MOR, and KOR depending on its concentration (Chang et al. 2004; Feng et al. 2012), these studies seemed to imply that *opioid receptor inhibition induces neuroprotection.*

In contrast, Hayward et al. (1993a) reported the neuroprotective efficacy of KOR agonist enadoline (CI-977) in two acute rat models of focal cerebral ischemia. Their data showed that enadoline at doses of 0.1, 0.3 and 1.0 mg/kg (s.c. 30 min prior to ischemia) produced a dose-dependent amelioration of cortical damage without any significant effect on any of the physiological parameters monitored such as blood pressure, blood gases, glucose, and pH. In the recovery model in left focal cerebral ischemia, enadoline produced a dose-dependent reduction in the volumes of infarction and brain swelling (Hayward et al. 1993a). These data suggest that this KOR agonist induces neuroprotection in both the non-recovery and recovery models of focal cerebral ischemia in the rat (Hayward et al. 1993a, b). However, Iwai et al.'s work did not support this notion (Iwai et al. 1992). They investigated the effect of opioids on delayed neuronal death in the hippocampus of male Mongolian gerbils subjected to transient forebrain ischemia. When hypothermia during and after ischemia was prevented, treatment with the KOR agonist U-50488H or MOR agonist morphine did not induce any significant protection. On the other hand, Mayfield et al. (Mayfield and D'Alecy 1994a, b; Mayfield et al. 1994) did several interesting studies to show that hypoxic conditioning increases survival time during subsequent lethal hypoxic conditions in mice, and this protective effect was blocked by naloxone, suggesting an opioid-dependent mechanism. Furthermore, they reported DOR-1 and DOR-2 selective antagonists, 7-benzylidenenaltrexone and naltrindole respectively, blocked the hypoxic conditioning-induced increase in survival time. The lowest effective 7-benzylidenenaltrexone dose was 3000-fold lower than the lowest effective naltrindole dose, indicating that the acute adaptation to hypoxia was predominantly sensitive to DOR-1 blockade. Neither the MOR antagonist (beta-funaltrexamine), nor KOR antagonist (norbinaltorphimine) significantly changed survival time in sham or hypoxic conditioned mice. They suggested that DOR-1 mediated the mechanism of hypoxic conditioning induced increases in animal survival time. All these data contradicted the conclusion that "opioid receptor inhibition induced neuroprotection" and implied that *opioid receptor activation induced neuroprotection.*

Closer examination of the literature caused even greater confusion. Adams et al. (1986) performed a clinical trial on the dose-dependent effect of naloxone on 27 patients with acute or progressive stroke for assessment of a safe and optimal dose to treat cerebral ischemia. They reported a transient or sustained improvement in only 13 patients with a worsening in 3 patients within 3 h of the discontinuation of the naloxone. Olinger et al. (1990) evaluated the safety and possible efficacy of high-dose (160 mg/m^2) naloxone for the treatment of acute cerebral ischemia. Among the 38 patients, only 12 of them showed early neurologic improvement (by completion of the naloxone loading dose). However, there was no correlation between such a loading dose response and clinical outcome at 3 months. They suggested that data for the efficacy are inconclusive. Therefore, these data do not support the view that opioid receptor inhibition induces neuroprotection.

Upon examination of the studies by Mayfield and D'Alecy (1994a, b; Mayfield et al. 1994), it was unclear if neuroprotection was involved in increased survival time because an increased animal survival time following systemic administration of opioid ligands does not necessarily indicate "neuroprotection". Since DOR is widely distributed in the central and peripheral systems including the heart and kidney (Feng et al. 2012; Peng et al. 2012; refer to Chaps. 3, 4 and 18), and since an intravenous ligand may elicit systemic effects through many complex and varied mechanisms, it is difficult to reach a conclusion in terms of the specificity of action and associate it with nerve cells per se. The extended survival may have resulted from the diverse effects of opioid agonists on various organs and organ systems, especially in the heart and kidney, e.g., preconditioning induced cardioprotection via DOR-1 (Schultz et al. 1997; Takasaki et al. 1999; Gross 2003; Wang et al. 2008; Huang et al. 2009; Fraessdorf et al. 2015; Headrick et al. 2015; Maslov et al. 2015), which are sensitive to hypoxic stress and are critical in animal survival under hypoxic or ischemic conditions. Moreover, it would be difficult to explain the contradicting observations between "naloxone neuroprotection" (Baskin and Hosobuchi 1981; Baskin et al. 1984; Hosobuchi et al. 1982) and "naloxone neurotoxicity" if neuroprotection was indeed involved in increased survival time (Mayfield and D'Alecy 1994a, b; Mayfield et al. 1994).

A later study caused further difficulty in interpreting the conclusions made by Mayfield et al. (Mayfield and D'Alecy 1994a, b; Mayfield et al. 1994) regarding the role of DOR in animal survival under hypoxia. Endoh et al. (1999) placed male ddY mice in an airtight plastic chamber with 5 % oxygen. They found that 1 and 5 mg/kg naloxone had no significant effect on the survival rate of the mice subjected to acute hypoxic hypoxia, whereas 10 mg/kg naloxone decreased the survival rate. On the other hand, morphine at 2 and 5 mg/kg was shown to have a protective action against acute hypoxic hypoxia. The protective effects of 5 mg/kg morphine against hypoxia was even antagonized by 5 mg/kg naloxone, which did not itself show any significant effect on the survival rate. The oxygen consumption in the morphine-treated (5 mg/kg) mice was significantly ($p < 0.05$) lower than that of the saline-treated animals. Therefore, the authors suggested that the endogenous opioid system did not play a significant role on the pathophysiology caused by acute hypoxic hypoxia and that the improved survival of the hypoxic animals by morphine is at least partly attributable to its depressant effect on oxygen consumption.

Apparently, these contradicting results complicated the issue rather than clarifying the role of opioids during hypoxia/ischemic stress. At this point, it was unclear as to which opioid receptors, if any, played a role in neuroprotection against hypoxic/ischemic injury.

4.1.2 In Vitro Studies on Neuroprotection: Opioid or Non-Opioid Effects?

I then began to wonder if in vitro studies would provide clearer information on this issue.

Unfortunately, there were very few studies in this field at the time.

It is well known that excitotoxicity critically mediates neuronal injury induced by hypoxia and/or ischemia (Albers et al. 1989; Sung et al. 2008; Puyal et al. 2013; Lai et al. 2014). In cultured murine cortical neurons subjected to excitotoxicity induced by N-methyl-D-aspartate (NMDA), Choi and Viseskul (1988) studied neuroprotective effects of several opioid agonists. They found that the addition of 1 μM to 1 mM methadone, a MOR agonist, to the culture medium produced a concentration-dependent reduction in the neurotoxicity of exogenously applied NMDA in the cortical culture with EC50 being about 100 μM, which is extremely high in terms of opioid concentration. Methadone also protected against exposure to quinolinate but not to quisqualate or kainate. At concentrations of 100 μM to 3 mM, several other opioids, e.g., agonists more selective towards MOR (morphine, fentanyl, codeine and meperidine), as well as agonist (dextropropoxyphene) and antagonist (naltrexone) selective towards both MOR and KOR, also induced concentration-dependent reductions in NMDA neurotoxicity. The authors believed that this neuroprotective effect of opioids was not mediated by conventional opioid receptors, but through a "non-opioid" mechanism because the non-opioid enantiomers of methadone and morphine both exhibited a potency equal to or greater than those of the opioid enantiomer, and because 1 mM naloxone did not reduce this potency.

With two different neuronal culture toxicity paradigms implicating either postsynaptic-mediated neurotoxicity (brief exposure of cultures to a low concentration of NMDA or kainate) or pre- and postsynaptic mechanisms (exposure to hypoxic/hypoglycemic conditions) in mixed neuronal/glial cultures, Lockhart et al. (1995) investigated the potential neuroprotective effect of KOR ligands. They observed that KOR agonists CI-977 and U-50488H failed to attenuate kainate-mediated neurotoxicity. On the other hand, KOR agonists were potent inhibitors of hypoxia/hypoglycemia-induced neurotoxicity. The authors suggested that KOR agonists are able to attenuate hypoxia/hypoglycemia-, but not NMDA- or kainate-induced toxicity (Lockhart et al. 1995).

Taken together, the limited in vitro data still did not clear up the confusion over the above-mentioned controversies, especially on the role of DOR. First, the results were not consistent with those of in vivo studies. Secondly, the limited studies contradicted each other - one claimed a "non-opioid" effect while the other argued for a KOR protective effect. Third and most importantly, there is no convincing data on the role of DOR in neuroprotection against hypoxic/ischemic injury.

4.2 A Unique Role of DOR in Neuroprotection against Hypoxic/Ischemic Injury

The above-mentioned controversies puzzled us in terms of the role of opioid receptors in neuroprotection. Was it opioid receptor activation or inhibition that ultimately protected the brain from hypoxic or ischemic injury? This fundamental question perplexed many clinicians and scientists. Partially due to this confusion, few people agreed with me when I first presented my idea about the role of DOR in neuroprotection against hypoxia/ischemia to my colleagues in the early 1990s.

Based on our observations on evolutionary (Refer to Chap. 3) and developmental (Refer to Chap. 4) distribution of DOR in the brain, and taking inspiration from the observations on the turtle brain that showed a much higher DOR density (Xia and Haddad 2001) and greater tolerance to hypoxic/ischemic insult than the rat brain (Wilson and Kriegstein 1991; Jiang et al. 1992; Xia et al. 1992; Milton and Prentice 2007; Nilsson and Lutz 2004), I strongly believed that DOR plays an important role against hypoxic/ischemic injury. Since it was difficult to derive any solid conclusions from the past studies at the time, I and my research team started our research

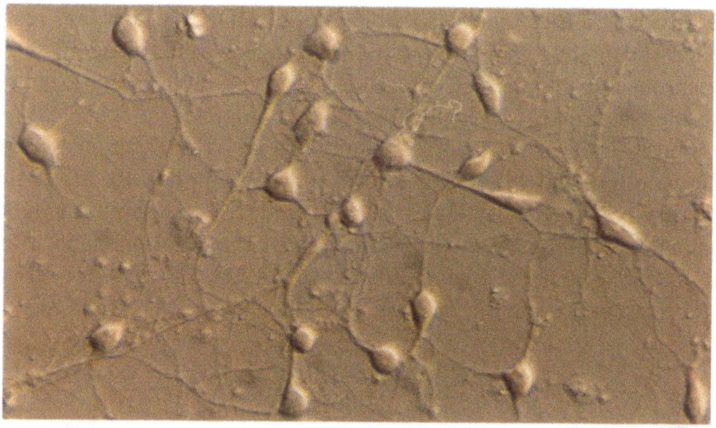

Fig. 1.3 Cultured cortical neurons for the evaluation of DOR neuroprotection. The cortical neurons were isolated from the cortex of embryonic day 17 rats and cultured in a neuron-defined culture medium, serum-free Neurobasal Medium. Cited from Zhang et al. (2000)

on the role of DOR in neural functions under hypoxic, ischemic, and excitatory conditions in 1997. Since then, we have demonstrated the neuroprotective role of DOR against hypoxic/ischemic insult with ample evidence.

As the first step, we attempted to determine if DOR is neuroprotective against hypoxic/ischemic injury with a reliable model. Since the cortex is rich in DOR expression compared to subcortical regions such as the hippocampus (Fig. 1.1; Xia and Haddad 1991a, 2001; also refer to Chaps. 3 and 4), we cultured cortical neurons (Fig. 1.3), exposed them to glutamate-induced excitotoxicity to induce neuronal injury, and applied [D-ala2, D-leu5] enkephalin (DADLE), a popular and commercially available DOR agonist at that time, to evaluate its effect on the neurons. In these neurons, the density of DOR underwent a rapid development and reached a plateau after ~10 days. To obtain objective results for DOR action on the cells, we used several approaches including "same-field morphology", LDH measurement, DNA defragmentation analysis, fluorescent staining techniques for live and dead neurons, etc. in our work. The "same-field morphology" we developed (Zhang et al. 2000) provided us a unique and objective tool to assess cell status after treatment and to compare it to cells before the treatment in the same field of the culture dish.

We found that DADLE indeed reduced excitotoxicity-induced injury (Zhang et al. 1999). At the same time, Borlongan et al. (1999) also showed that DADLE protects against ischemia reperfusion damage in the striatum and cerebral cortex.

However, it was extremely difficult to come to a reliable conclusion about the role of DOR in neuroprotection based on the data of DADLE because DADLE is not highly specific to DOR and might bind to other opioid receptors. For example, it binds to MOR sites and inhibits the binding of DMGO, a MOR ligand, in a competitive manner (Blurton et al. 1986). This was why we added a much higher concentration of non-radiolabled MOR agonist (1 μM PL017) in the binding buffer when detecting DOR sites by using ^{3H}DADLE (4 nM), (Xia and Haddad 1991a, 2001).

Therefore, we further clarified the actions of DOR on neuronal responses to excitotoxicity by directly applying various opioid agonists and antagonists to cultured cortical neurons to compare their effects on glutamate-induced neuroexcitotoxicity. We found that naltrindole, a DOR antagonist, completely blocked the protective effect offered by DADLE. In contrast, administration of MOR agonist DAMGO had no protective effect and MOR antagonism did not significantly affect glutamate-induced injury either. KOR agonist U50,488H induced a slight, but not significant, reduction in neuronal injury, and KOR antagonism had no appreciable effect on the glutamate-induced injury (Fig. 1.4; Zhang et al. 1999, 2000). After this work, we were inclined to believe that activation of DOR, but not MOR nor KOR, induced

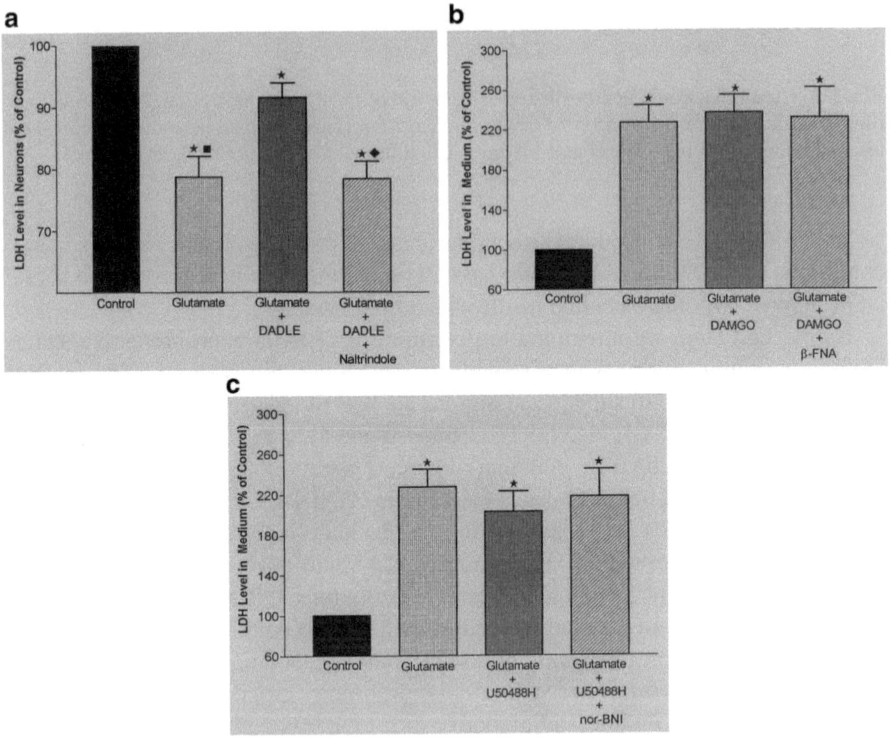

Fig. 1.4 Effects of opioid agonists and antagonists on LDH leakage from cortical neurons exposed to glutamate excitotoxicity. (**a**) Effects of DADLE and naltrindole. *Star,* $P<0.01$ vs. the control. *Closed diamond,* $P<0.05$ vs. glutamate excitotoxicity plus DADLE. *Closed square,* $P<0.01$ vs. glutamate excitotoxicity plus DADLE. Note that LDH leakage, an index of neuronal injury, increased with the treatment of glutamate and this increase was reduced by DOR activation with DADLE. The effect of DADLE was completely blocked by naltrindole, a DOR antagonist. (**b**) Effects of DAMGO and β-FNA. *Star,* $P<0.01$ vs. the control. Note that DAMGO, a MOR agonist, did not reduce the LDH leakage and the addition of a MOR antagonist, β-FNA, did not alter the leakage, either. (**c**) Effects of U50488H and nor-BNI. *Star,* $P<0.01$ vs. the control. Note that U50488H, a KOR agonist, slightly reduced the LD leakage, but the change was not statistically different as compared to that of glutamate alone. The addition of a KOR antagonist, BNI, also did not alter the leakage significantly. Cited from Zhang et al. (2000)

neuroprotection against glutamate excitotoxicity in cortical neurons. Soon after, we observed similar protection in the cortical neurons exposed to hypoxia (Zhang et al. 2002). Consistently, our electrophysiological studies obtained stronger evidence to show that DOR is relatively specific in the protection against anoxic disruption of ionic homeostasis because MOR activation was not shown to induce any protective effect (Chao et al. 2007a) (Fig. 1.5). In a rat model of focal cerebral ischemia, we found that electroacupuncture reduced ischemic infarct via a DOR mechanism, which was concluded when DOR antagonists greatly attenuated the electroacupuncture effect (Zhao et al. 2002). Furthermore, our transgenic studies showed that cortical DOR overexpression attenuated anoxia-induced disruption of ionic homeostasis, suggesting that an increase in DOR expression renders the cortex more tolerant to hypoxic stress (Chao et al. 2006).

To yield more valid data for DOR action, we collaborated with domestic and international scientists to use more potent and selective DOR ligands such as UFP-512 (Balboni et al. 2002) starting in the early 2000s.

Our strong evidence indicates that DOR is a unique neuroprotector against hypoxic/ischemic stress in the brain. Also, there have been abundant data thereafter showing DOR neuroprotection from many independent laboratories worldwide (Borlongan et al. 2004, 2009; Lim et al. 2004; Horiuchi et al. 2004, 2008; Narita et al. 2006; Iwata et al. 2007; Su et al. 2007; Xiong et al. 2007; Charron et al. 2008; Govindaswami et al. 2008; Pamenter and Buck 2008; Peng et al. 2009; Zhu et al. 2009; Johnson and Turner 2010; Nandhu et al. 2010; Duan et al. 2011; Gao et al. 2012; Turner and Johnson 2011; Yang et al. 2011; Wang et al. 2011; Zheng et al. 2012). One of the earliest reports came from Dr. Zuo's group and demonstrated that morphine preconditioning is neuroprotective against Purkinje cell injury induced by simulated ischemia-reperfusion via a DOR, but not MOR nor KOR, mediated mechanism (Lim et al. 2004). Our finding on the DOR-mediated electroacupuncture protection against cerebral ischmeia (Zhao et al. 2002) was soon confirmed by other investigators (Xiong et al. 2004). Moreover, we found that PKC-ERK pathway is a critical pathway in the signal transduction of DOR neuroprotection (Ma et al. 2005), which was also validated by other researchers. For example, Narita et al. (2006) demonstrated that DOR plays a crucial role in neurogenesis and neuroprotection and the signaling mechanism could be linked to PI3K, PKC, CaMKII and MEK (Narita et al. 2006).

Based on our consistent results and those of others, it is our belief that DOR plays a specific role in neuroprotection against hypoxic/ischemic stress, particularly in DOR-rich regions like the cerebral cortex.

4.3 *Reconciliation of Our Findings with Past Controversies*

All the evidence from our work provides compelling sources to reconcile the previous controversies that existed in the early literature.

When evaluating the effect of ligands on neuroprotection, the early results based on an intravenous injection of a ligand might not yield a reliable conclusion.

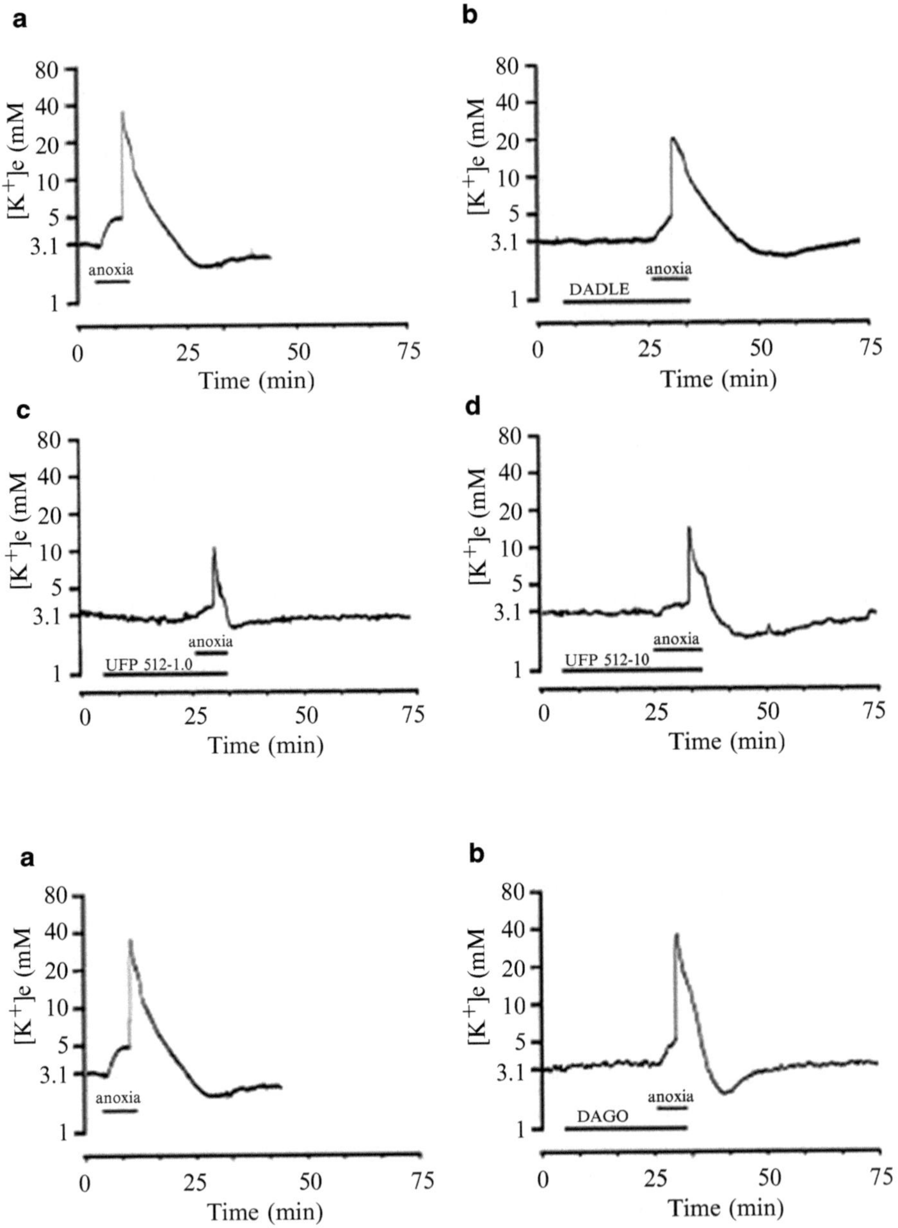

Fig. 1.5 Comparative effects of DOR and MOR agonists on anoxic K^+ leakage. *Top panel*: Remarkable reduction of anoxic K^+ leakage by DOR activation. (**a**) control. (**b**) DADLE (10 μM). (**c**) UFP-512 (1 μM). (**d**) UFP-512 (10 μM). Note that 10 μM DADLE largely reduced the peak increase in $[K^+]e$ and the rate of increase in $[K^+]e$ to peak in comparison to the control and delayed the response of the cortical slices to anoxia. UFP-512, a more potent and selective DOR agonist, at 1 μM induced an even larger reduction in the anoxic changes, while increasing concentrations of UFP-512 (up to 10 μM) did not further attenuate the anoxia-induced changes except for a slight increase in latency. *Bottom panel*: No appreciable effect of MOR activation on anoxic K^+ leakage. (**a**), Control. (**b**), DAMGO (10 μM). Note that MOR activation with DAMGO had no appreciable effect on the anoxia-induced K^+ leakage. Cited from Chao et al. (2007a)

First of all, non-selective ligands were used intravenously in previous studies and had limited receptor selectivity and specificity, thereby resulting in a diverse number of responses following their simultaneous action at multiple receptors. Moreover, a high dose of opioid ligands emphasizes their ability to cross-react with more than one type of opioid receptor and increases the likelihood of inducing non-specific effects. In addition, intravenous administration of opioid ligands, used in the majority of these earlier studies, could induce a multitude of systemic effects, affecting various organs including the heart, blood vessels, kidneys, etc. These diverse effects make it a herculean effort to correctly delegate the observed effect to a specific receptor action or non-opioid action in a specific organ or multiple organs. Therefore, the outcome of those experiments varied in all aspects depending on many factors including the conditions of subjects, dose of opioids used, time point of assessment, etc.

Another limitation of these early studies was the inappropriateness of the outcome measurement used. Some authors interpreted that DOR was neuroprotective based on extended animal survival during hypoxic exposure. In view of the complex interplay of mechanisms and the multiple effects involved in animal survival, such interpretations are likely to have been a mistake. Since the increase in animal survival could be a result of other associated protective/beneficial effects such as cardioprotection or renal protection (Schultz et al. 1997; Takasaki et al. 1999; Gross 2003; Peart et al. 2005; Huang et al. 2009; Headrick et al. 2015; Maslov et al. 2015; refer to Chap. 17), to report it as a result of neuroprotective effects seems to be a hasty and unsupported assumption.

In the in vivo studies, the lack of neuroprotective response to MOR ligands observed in the hippocampus by Iwai et al. (1992) could be explained based on the pattern of opioid receptor distribution in various regions of the brain. Our autoradiographic studies have shown a low density of DOR binding sites in the hippocampus, including the dentate gyrus, while MOR binding sites are present in abundance (Fig. 1.1; Xia and Haddad 1991a, 2001). A relatively low concentration of MOR or KOR agonists applied to the hippocampus can activate MOR or KOR, but cannot induce (or does not induce enough) DOR effects in this region. Since MOR and KOR activation do not induce a significant neuroprotective effect (Zhang et al. 2000; Chao et al. 2007a; Lim et al. 2004), it is not surprising that MOR or KOR ligands did not cause any major neuroprotection in the hippocampus (Iwai et al. 1992). Indeed, this notion reaffirmed that MOR and KOR do not play significant roles in neuroprotection. In fact, the MOR-preferable agonist morphine may be harmful to hippocampal neurons in hypoxia/ischemia (Ammon-Treiber et al. (2005, 2007). However, Biphalin, a dimeric opioid peptide that exhibits affinity for DOR, MOR and KOR, is neuroprotective against excitotoxic injury in hippocampal organotypic culture (Kawalec et al. 2011).

Since there is a higher density of DOR in the cortex, a high dose (up to 3 mM; Choi and Viseskul, 1988) of MOR or KOR agonists may activate DOR along with MOR and/or KOR, and thus induce DOR-mediated neuroprotection (Choi and Viseskul 1988). This might be the reason behind the observation that high doses of MOR or KOR agonists protect the cortical neurons against neuroexcitotoxic injury.

The reason why naloxone at 1 mM did not act as an antagonist to reduce the effects induced by these opioid agonists (Choi and Viseskul, 1988) may be complicated since the applied concentrations of opioid agonists were so high. Another possible explanation could be due to the fact that it is a non-specific opioid antagonist and was bound to all kinds of opioid receptors in the cultures, and therefore did not have enough capacity to fully block DOR's effect.

In mixed neuronal/glial cultures, KOR agonists could not attenuate NMDA- or kainate-induced neurotoxicity (Lockhart et al. 1995). In the "pure" neuronal culture, we observed that KOR agonist induced a slight but not significant reduction (Zhang et al. 2000). These data strongly support the unique role of DOR in neuroprotection since DOR activation effectively reduces excitatory neurotoxicity (Zhang et al. 2000). Because the same KOR agonists attenuated hypoxia/hypoglycemia induced injury in the mixed neuronal/glial cultures, Lockhart et al. (1995) proposed that KOR agonists may exert a neuroprotective role by a predominantly presynaptic mechanism, possibly by inhibiting ischemic-mediated glutamate release. Because glia also express DOR that increases glial glutamate transporter expression and enhances glial ability to reuptake released excitatory amino acids (Liang et al. 2014), glial DOR is likely activated by KOR agonists, at least partially, to display a neuroprotective effect. All these data provide further support for the uniqueness of DOR in neuroprotection because it acts against hypoxic/ischemic insults at multiple cellular/molecular levels (see Sect. 6).

Finally, to address the results regarding the critical role of DOR-1 in hypoxic conditioning-induced increase in survival time (Mayfield and D'Alecy 1994a, b; Mayfield et al. 1994), the argument that cardiac and/or other protection, rather than neuroprotection, is involved in the improved survival time (see Sect. 4.1.1) is very likely valid. This is because (1) preconditioning indeed causes cardioprotection via DOR-1 (Gross 2003; Wang et al. 2008; Huang et al. 2009; Fraessdorf et al. 2015; Headrick et al. 2015; Maslov et al. 2015) and (2) more importantly, most current studies show that DOR's neuroprotection is through a DOR-2 (if naltrindole is truly the antagonist of DOR-2) since naltrindole reverses the DOR neuroprotection (Zhang et al. 2000, 2002, 2006; Ma et al. 2005; Cao et al. 2015).

Based on our reconciliation with past data, I am more confident about the important role of DOR as a neuroprotector against hypoxia/ischemia in the brain.

However, the debates are still ongoing regarding the role of MOR in the response of the brain to hypoxic/ischemic stress. Feng et al. (2008) observed that MOR agonists morphine, endomorphin 1, and endomorphin 2 effectively protect brain mitochondria against oxidative stresses induced by in vitro anoxia-reoxygenation. In contrast, more studies showed the opposite results. For example, Ammon-Treiber et al. (2005) showed that morphine increases the hypoxia/hypoglycemia induced neurotoxicity in the hippocampus in a concentration dependent manner. One-hour morphine perfusion, immediately followed by a short hypoxic/hypoglycemic episode, resulted in an impaired restoration of evoked field potentials in the CA1 region as compared to untreated control brain slices undergoing the same hypoxia/hypoglycemia without any drug pretreatment (Ammon-Treiber et al. 2005). After comparing different MOR agonists, they found that receptor-internalizing opioids such as fentanyl or DAMGO may have fewer neurotoxic effects in hypoxia/hypoglycemia than the non-internalizing drug morphine (Ammon-Treiber et al. 2007). Other studies also suggested naloxone may act through a blockade of MOR activation, thus reducing brain infarction, neutrophil accumulation, and chemokine expression in the model of cerebral ischemia reperfusion insult (Liao et al. 2003).

There are different cases in the immature brain because of low density of MOR in the brain (Xia and Haddad 1991a; refer to Chap. 4). Kawaraguchi et al. (2008) observed that continuous administration of morphine does not worsen brain damage 7 days after hypoxic-ischemic insults in the neonatal brain, Moreover, neonatal stress plus morphine treatment might influence early gene expression and impair neurodevelopment and learning behavior, but did not affect cerebral vessel density or ischemic brain injury in the adult rodent brain (Hays et al. 2013). Another study also showed that morphine was ineffective as a neuroprotectant in rat pups with severe hypoxic-ischemic injury and may have even contributed to their decreased survival (Festekjian et al. 2011). More investigations are needed to clarify the role of MOR in the hypoxic/ischemic injury in the immature brain.

There are scattered studies on the role of nociceptin in the brain under hypoxic/ischemic conditions, especially its contribution to hypoxic/ischemic hypotensive cerebrovasodilation impairment (Jagolino and Armstead 2001; Armstead 2002, 2011; Philip and Armstead 2003; Zhang et al. 2013).

5 Recent Outcomes of Research on DOR's Neural Functions

After identifying the specific role DOR plays in neuroprotection during hypoxic/ischemic conditions and with improvements in experimental approaches, the studies of DOR's neural functions in the past decade have achieved promising and consistent results. Also, the advancement in research on opioid ligands in general (refer to Chaps. 2 and 18) further promotes the progress in this field. Moreover, the research progress in evolutional (see Chap. 3) and developmental (see Chap. 4) distribution of DOR led to better research design with various experimental models and reliable data interpretation. The below is a short list of DOR's neural functions that will be discussed in detail in the following chapters of this book.

Regulation of ionic homeostasis. We have conducted a series of studies involving the role of DOR in the regulation of ionic homeostasis in the brain (Chao et al. 2007a, b, 2008, 2009, 2012a, b; Chao and Xia 2010). Our data show that DOR functions as a stabilizer in the homeostasis of Na^+, K^+ and Ca^{++} in both physiological conditions (see Chap. 5) and hypoxic/ischemic circumstances (see Chap. 6) with an emphasized role under hypoxia/ischemia. DOR activation largely reduces hypoxia or simulated ischemia induced Na^+ (Fig. 1.6) and Ca^{++} influxes and K^+ leakage. More importantly, we observed for the first time with solid evidence that DOR expression and activation attenuate Na^+ channel currents (Figs. 1.7 and 1.8), suggesting that DOR may be an endogenous inhibitor of Na^+ channels (Kang et al. 2009). Na^+ channel distribution and functional properties are very similar between the rat and

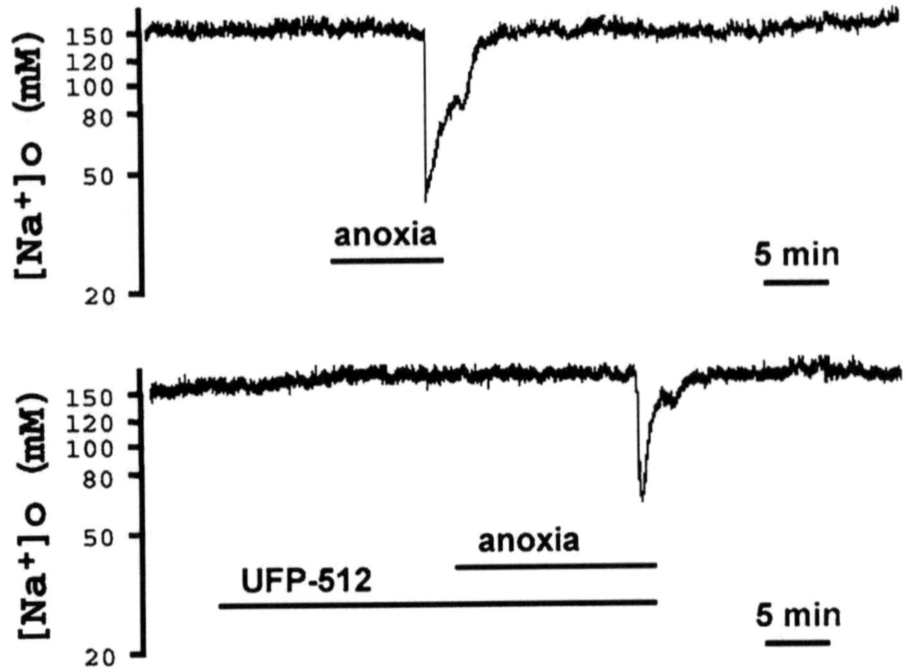

Fig. 1.6 Effect of DOR activation on anoxia-induced Na^+ influx. The cortical slices were exposed to anoxic stress to induce Na^+ influx. By using Na^+-sensitive microelectrodes, the drop of extracellular Na^+ concentration was measured as an index of Na^+ influx. Note that anoxia induced a sudden drop of extracellular $[Na^+]$ from its baseline of ~152 to 45.93±3.67 mM with 7.7±0.9 min for recovery from peak drop to the baseline after re-introducing oxygen (*upper trace*). Activation of DOR with UFP-512 (5 μM) greatly attenuated this drop (*lower trace*) and accelerated the recovery of the anoxia-induced drop of extracellular $[Na^+]$, suggesting that DOR activation reduced the anoxic Na^+ influx in the cortex. Cited from Kang et al. (2009)

human, especially in the cortex (Xia and Haddad 1993; Cummins et al. 1994). They function as a basic component for neuronal excitability and their dysfunction can lead to severe neurological disorders (Cummins et al. 1994; Xia et al. 2000, 2003; Zhao et al. 2005a; Catterall 2014). The discovery of this DOR-mediate inhibitory mechanism may provide a hint for new solutions to some neurological diseases.

Function of excitatory amino acid reuptake. Several lines of study suggest that DOR regulates the expression and function of glutamate transporters, and thus affects the reuptake of excitatory amino acids (see Chap. 7). For instance, we found that DOR activation with UFP-512 significantly upregulates the expression of EAAT1 and EAAT2 in astrocytes and increases astrocytic ability to uptake glutamate (Figs. 1.9 and 1.10), suggesting an increased ability to clean up extracellular excitatory amino acids (Liang et al. 2014).

Modulation of glutamatergic signaling. Besides the functional regulation of excitatory amino acid reuptake, DOR signals may also interact with glutamatergic synapses and directly inhibit neuronal glutamatergic signaling at post-synaptic

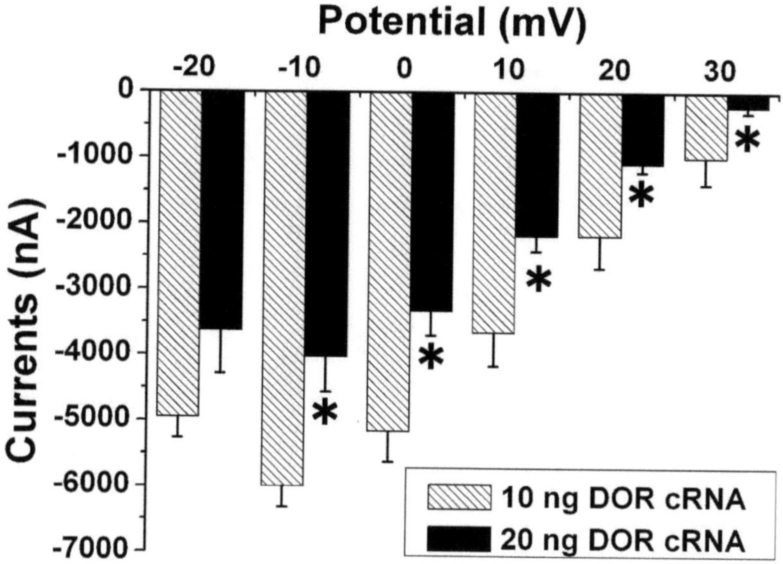

Fig. 1.7 Effect of DOR expression on Na^+ currents. Five ng of Nav1.2 cRNA with 10 or 20 ng of DOR cRNA were injected into the oocytes that had no expression of Na^+ channels and cultured for 72 h. *$p<0.05$ vs. 10 ng DOR cRNA. Note that the Na^+ current was significantly smaller in the oocytes with 20 ng DOR cRNA as compared to that of the oocytes with 10 ng DOR cRNA in the range of −10 to 30 mV of test potentials, suggesting that an increase in DOR expression leads to a decrease in Na^+ current amplitude. Cited from Kang et al. (2009)

terminals. This mechanism may significantly contribute to the hypoxia-tolerance of the turtle brain to hypoxia/ischemia (see Chap. 8).

Neurogenesis. DOR has been recognized to be involved in neural differentiation, maturation, and protection, and may contribute to the possible recovery of emotional dysfunction. Stimulation of the δ-opioid receptor in neural progenitor cells enhances neural differentiation, maturation, and protection, which could lead to anti-depressive and anti-anxiolytic effects (see Chap. 9). Since hypoxic preconditioning with relatively mild hypoxic stress can increase DOR activity (Ma et al. 2005; Zhang et al. 2006), hypoxia-promoted differentiation of stem cells (Lee et al. 2009; Park et al. 2011; Kim et al. 2011, 2012, 2013) may also be related, at least partially, to the activity of DOR.

Hibernation for stroke therapy. The DOR agonist DADLE was highly effective in inducing hibernation (Oeltgen et al. 1988). DOR and delta opioid peptides demonstrate promising potential for treatment of stroke and other degenerative neurological conditions (Borlongan et al. 2009; Yang et al. 2009). Recent studies present strong evidence in support of the therapeutic use of the delta-opioid receptor, and provide insight into the underlying mechanisms of action. DOR has been shown to confer protective effects by mediating ionic homeostasis and activating endogenous neuroprotective pathways, thereby decreasing apoptosis and promoting neuronal survival (see Chap. 10).

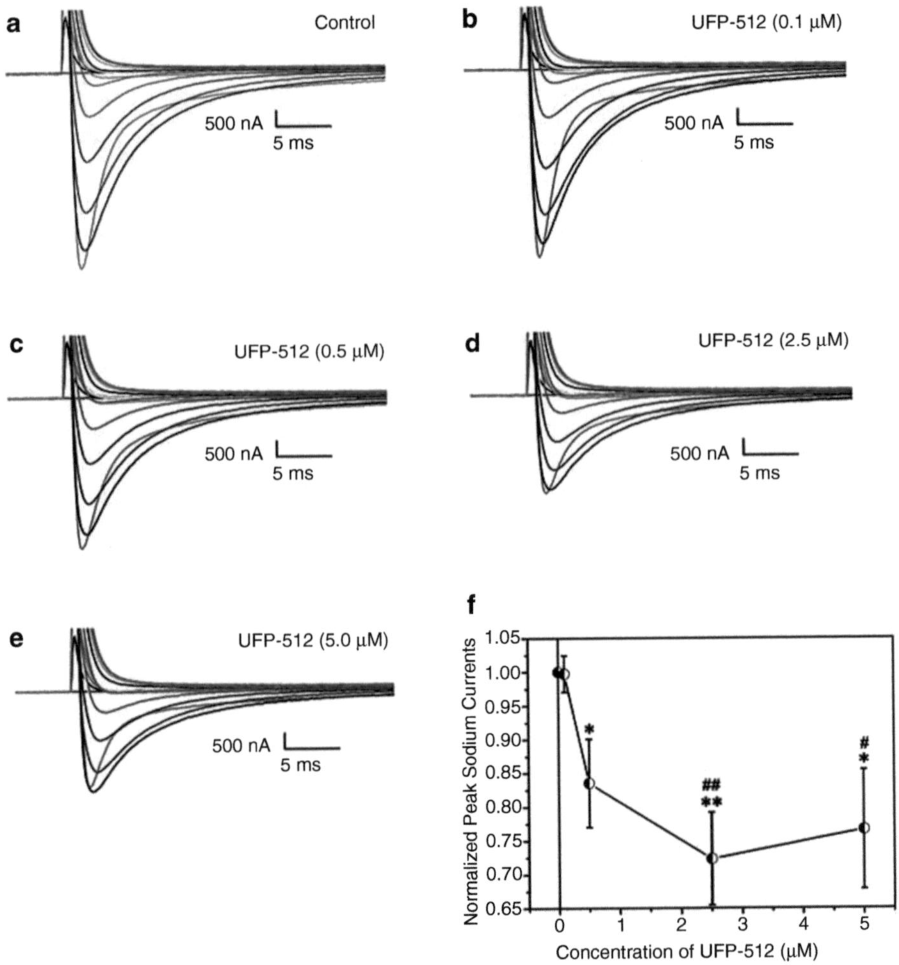

Fig. 1.8 DOR activation induced inhibition of Na^+ currents. The recordings were performed in the oocytes co-expressing Na^+ channels (5 ng of Nav1.2 cRNA) and DOR (10 ng of DOR cRNA). The families of representative currents traces show the inward currents recorded in the control (**a**), 0.1 μM UFP-512 (**b**), 0.5 μM UFP-512 (**c**), 2.5 μM UFP-512 (**d**), and 5.0 μM UFP-512 (**e**), respectively. Panel **f** shows the relationship curve of the current's size vs. concentration of UFP-512 (n=4–5). *$p<0.05$ and **$p<0.01$ vs. control. #$p<0.05$, ##$p<0.01$ vs. 0.1 μM UFP-512. Note that the Na^+ current amplitude was reduced by the DOR agonist in a dose–response manner and an increase in the concentration of UFP-512 to more than 2.5 μM could not further reduce the Na^+ current amplitude. Cited from Kang et al. (2009)

Electroacupuncture-induced brain protection against ischemic injury. Acupuncture is an important modality of traditional Chinese medicine and widely used as an alternative therapy for neurological diseases including stroke (Xia et al. 2010, 2012). The neural input induced by manual acupuncture and electro-acupuncture regulates the activities of various neurotransmitter/modulator systems

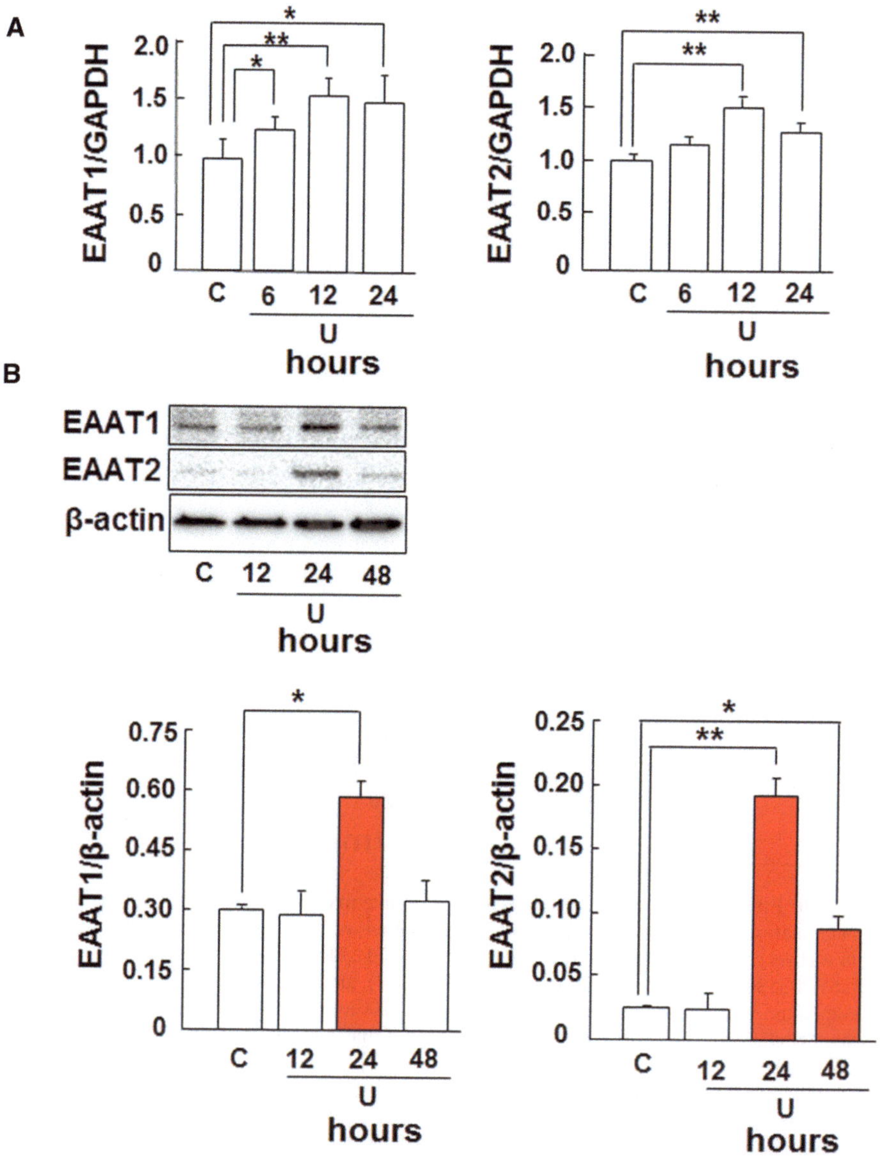

Fig. 1.9 DOR activation induced EAAT1 and EAAT2 expression in the astrocytes. (**a**) Quantitative PCR analysis showing that UFP-512 induced EAAT1 and EAAT2 mRNA expression in a time-dependent manner with expression reaching peak at 12 h. (**b**) Western blot showing that UFP-512 induced production of EAAT1 and EAAT2 proteins in a time-dependent manner with production reaching peak at 24 h. (**c**), non-treatment; U, DOR activation with 10 μM UFP-512. $^{*}P<0.05$; $^{**}P<0.01$. Note that UFP-512 increases EAAT1 and EAAT2 expression at both mRNA and protein levels in a time-dependent manner. Cited from Liang et al. (2014)

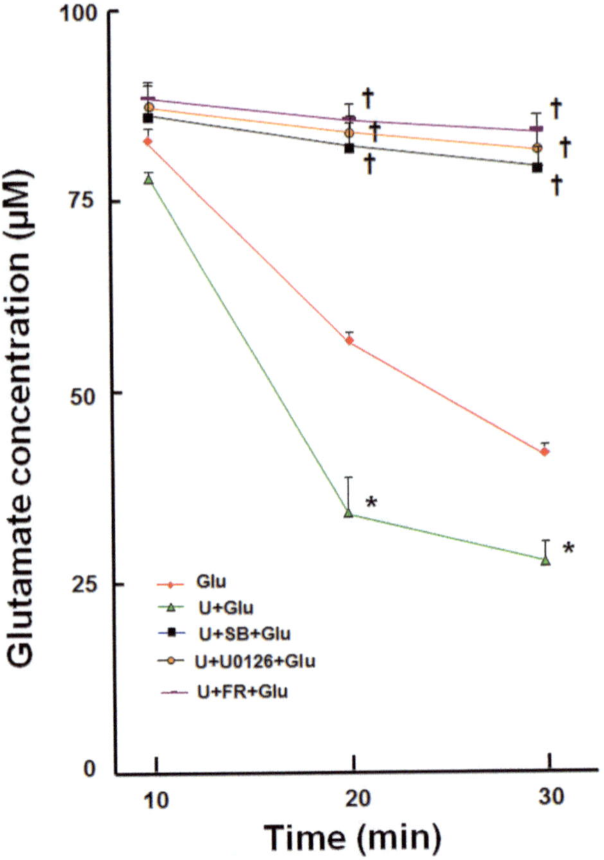

Fig. 1.10 Effect of DOR activation on glutamate uptake in cultured astrocytes. Extracellular glutamate analysis was conducted with the L-glutamate assay. The cultured astrocytes were treated with various chemicals: Glu, control (glutamate only); FR, FR180204 (500 nM), an ERK inhibitor; SB, SB203580 (10 µM), a p38 inhibitor; U0126 (5 µM), an MEK inhibitor. The astrocytic condition medium were collected from each group at 10-, 20- and 30-min time points for L-glutamate assay. $^{*}P<0.05$ vs. Glu. $^{\dagger}P<0.05$ vs. U+Glu. Note that UFP-512 greatly enhanced glutamate uptake in the astrocytes and this effect was almost completely abolished by inhibitors for MEK, ERK and p38, suggesting that MAP kinases were involved in the DOR-induced increase in glutamate uptake in the astrocytes. Also note that p38 may play a different role in astrocytes than in neurons. Cited from Liang et al. (2014)

including DOR system (Zhou et al. 2010; Wen et al. 2010a, b; Liang and Xia 2012). In the rat stroke model, an appropriate electroacupuncture stimulation reduces ischemic infarction (Xia et al. 2007a, b; Zhou et al. 2008, 2011a, 2013a, 2013b; Tian et al. 2008a, b, c) (Fig. 1.11). Strong evidence suggests that electroacupuncture increases endogenous opioid activity, e.g., increased met-enkephalin expression (Li et al. 2012), and increases the level of DOR in the brain (Tian et al. 2008b, c; Guo et al. 2010; Chen et al. 2014a). Also, electroacupuncture-induced brain protection is critically related to an increase in regional cerebral blood flow (Zhou et al. 2011a,

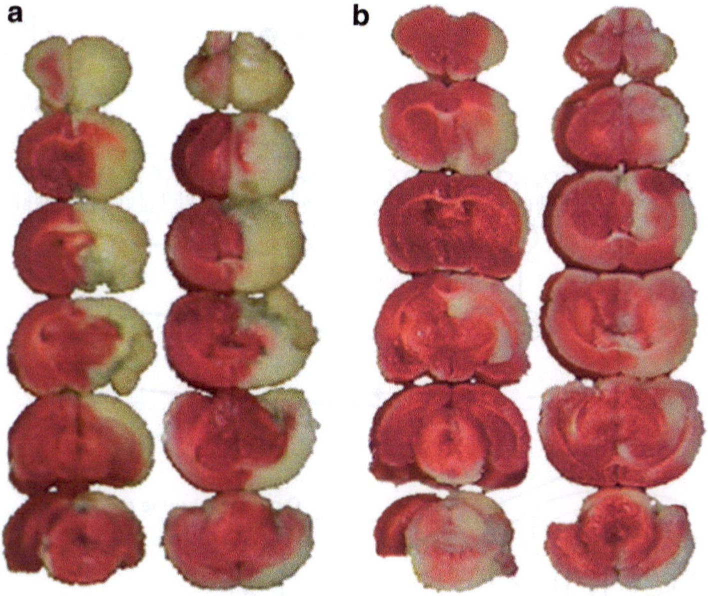

Fig. 1.11 Electroacupuncture induced reduction in ischemic brain infarction. The experiments were carried out in adult male rats. (**a**) Ischemia. (**b**) Ischemia + electroacupuncture. The model of focal cerebral ischemia was established by performing the middle cerebral artery occlusion (MCAO). Electroacupuncture was applied at contralateral acupoints DU20 and DU26 with optimal electrical stimulation (Zhou et al. 2011a, 2013a, b). The brain slices were stained with TTC (triphenyltetrazolium chloride) and subjected to computerized analysis of brain infarct volume. The slices on the right of each column show the backside of the contiguous slices. Note that appropriate electroacupuncture largely reduced the ischemic brain infarct. Cited from Zhou et al. (2013a)

2013a, b). Chapter 11 will recap the research progress in DOR-mediated mechanisms for electroacupuncture-induced neuroprotection against ischemic injury based on previous literature as well as our recent work.

Cell/brain preconditioning. Our studies have shown that delayed hypoxia preconditioning (HPC) renders cortical neurons more tolerant to subsequent severe hypoxia, an effect that was blocked by the DOR antagonist naltrindole (Ma et al. 2005). The DOR-mediated HPC protection depends on an increase in ERK and Bcl-2 activity, which counteracts severe hypoxia induced increase in p38 MAPK activity and cytochrome c release (Fig. 1.12). Also, rapid hypoxia preconditioning attenuates neuronal injury induced by glutamate excitotoxicity, and this protection was largely reversed by DOR antagonism (Zhang et al. 2006). More recently, we found that DOR activation increased Nrf2 expression and translocation (Fig. 1.13), thereby inducing cell protection (Cao et al. 2015). Indeed, Nrf2 signaling plays an important role for neuroprotection in ischemia-reperfusion injury and is involved in ischemia-preconditioning induced protection (Shokeir et al. 2015; Takagi et al. 2014; Xu et al. 2014). This DOR-Nrfs signaling may contribute, at least partially, to the mechanism of hypoxic/ischemic preconditioning in neurons/brains. Moreover,

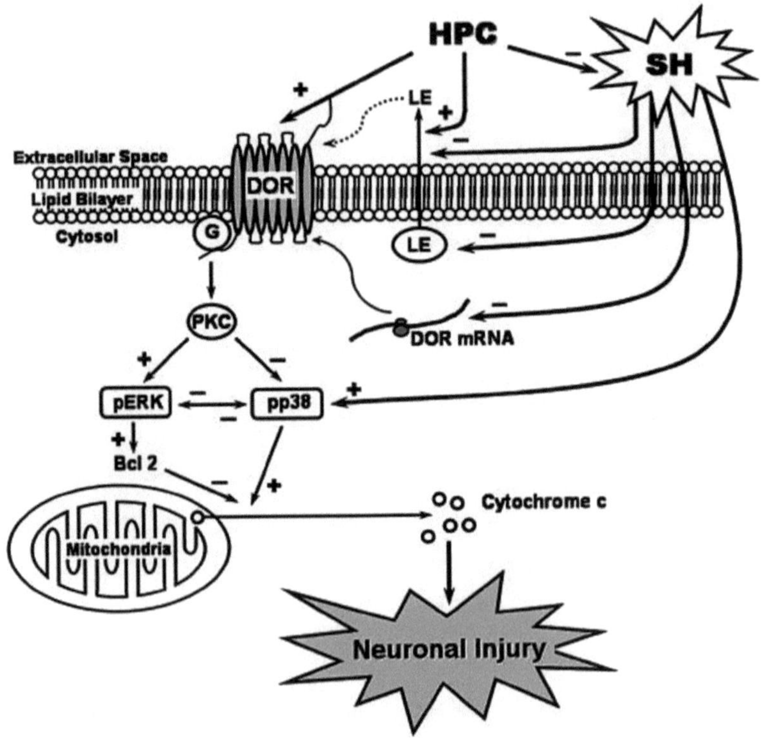

Fig. 1.12 Schematic representation of the pathways that mediate neuroprotection induced by delayed hypoxic preconditioning (HPC). HPC prevents severe-hypoxia (SH) induced reduction in DOR mRNA and protein levels and enkephalin (LE) release. HPC enhances DOR function to promote neuroprotection through the enkephalin release and the G protein (G)-PKC-pERK signal cascade, which, in turn, increases the level of the cellular protective protein, Bcl-2. This pathway may antagonize the p38-mediated cytochrome *c* release from mitochondria during severe hypoxia. Cited from Ma et al. (2005)

morphine can precondition neurons and induce neuroprotection via a DOR mechanism (Lim et al. 2004). Chapter 12 will discuss the role of DOR in preconditioning.

Maintenance of mitochondrial function. The dysfunction of mitochondria is involved in the pathogenesis of hypoxia/ischemia-related cardiac disorders, stroke, and neurodegenerative diseases. Accumulated evidence shows that DOR protection is associated with a better maintenance of mitochondrial function (Zhu et al. 2009, 2011; Chen et al. 2014b), suggesting that the DOR-mediated protection against myocardial and neuronal injury in hypoxia/ischemia is likely related to the regulation of mitochondrial pathways (see Chap. 13).

Beneficial actions against Parkinson's disease. Parkinson's disease results from the death of dopaminergic neurons in the substantia nigra pars compacta which innervate all of the nuclei of the basal ganglia. Cell loss in the substantia nigra pars compacta causes a consequent striatal dopamine deficiency, leading to a cascade of functional changes in basal ganglia circuitry and ultimately the development of the cardinal features of Parkinson's disease. In addition, many complex factors are

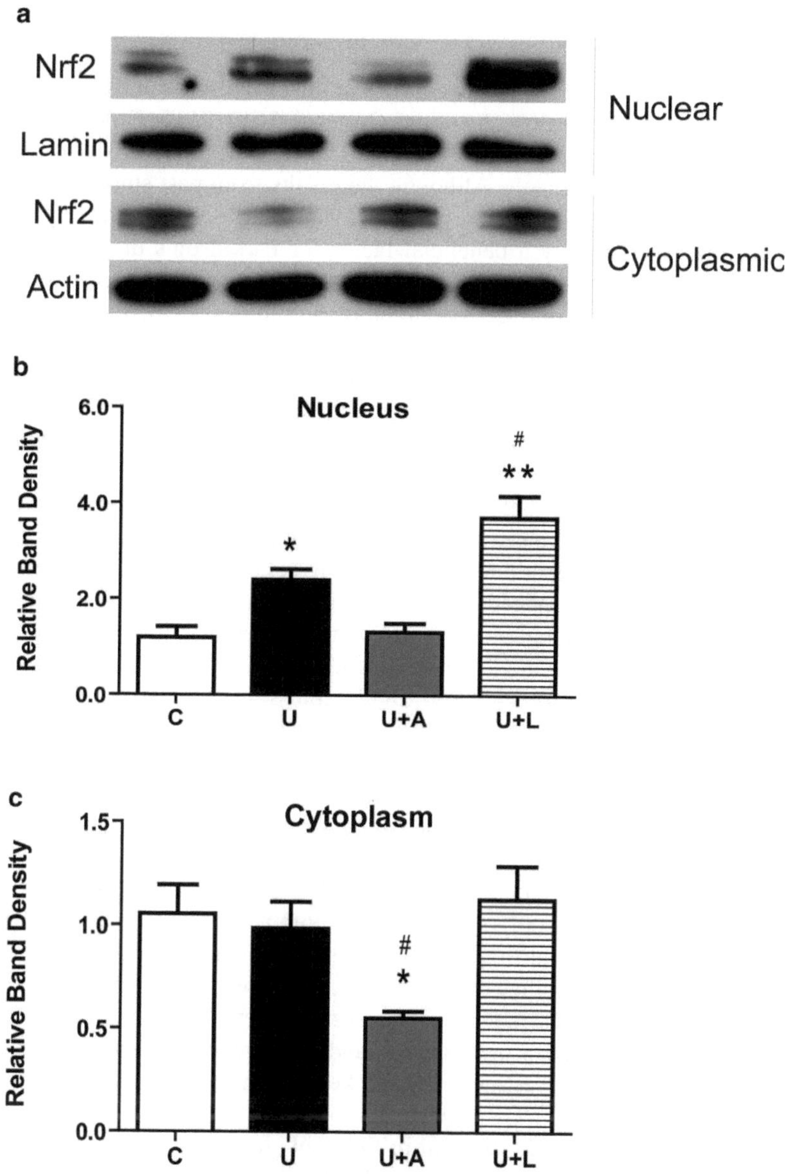

Fig. 1.13 DOR activation induced promotion of Nrf2 translocation from the cytoplasm to nucleus. (**a**) Representative Western blots of nuclear and cytoplasmic Nrf2 proteins. (**b**) Relative quantitation of Nrf2 protein in the nucleus. (**c**) Relative quantitation of Nrf2 protein in the cytoplasm. UFP-512 was added to the medium of cultured HEK293t cells to activate DOR. C, vehicle control. A, calphostin C (a PKC inhibitor, 100 nM). L, LY294002 (a PI3K inhibitor, 10 μM). U, DOR activation with UFP-512. $^{*}P<0.05$, $^{**}P<0.01$ vs. the control in normoxia. $\#P<0.05$ vs. UFP-512 alone. At least three independent experiments were carried out in each group. Note that DOR activation significantly increased nuclear Nrf2 protein without reducing the level of cytoplasmic Nrf2 protein, suggesting that DOR activation increases Nrf2 protein expression and translocation from the cytoplasm to nucleus. Also note that PKC inhibitor calphostin C lowered the Nrf2 protein in cytoplasm but had no effect on the level of Nrf2 protein in the nucleus, whereas the PI3K inhibitor LY294002 further elevated DOR-mediated increase in nuclear Nrf2 protein without affecting cytoplasmic levels, suggesting differential roles of PKC and PI3K pathways in DOR-mediated regulation of Nrf2 in cytoplasm and nucleus. Cited from Cao et al. (2015)

involved in the pathophysiological events of Parkinson's disease and cause a variety of premotor, motor and other dysfucntions (Xu et al. 2012, 2015; Fenoy et al. 2014; Gao et al. 2015; Wei et al. 2015; Zou et al. 2015). DOR is highly distributed in the caudate putamen and amygdala. Although the results from past studies on the role of DOR in Parkinson's disease were not conclusive, our recent studies suggest that DOR signaling is generally a beneficial factor for Parkinson's disease. More in-depth investigations may reveal useful clues for developing new therapeutic strategies against Parkinson's disease by targeting δ-opioid system. Chapter 14 will discuss this topic.

Potential role of DOR in deep brain stimulation. Despite its widespread use, the underlying mechanism of deep brain stimulation (DBS) remains unknown. There is now increasing evidence showing that DBS actually can both inhibit neurons and activate axons, generating a wide range of effects. This implies that the mechanisms that underlie DBS work not only locally but also at the network level. On the other hand, our recent studies showed that DOR activation reduces α-synuclein overexpression and oligomer formation, thus attenuating cellular injury after MPP+ exposure (Chen et al. 2014b). Indeed, DOR activation protects both glial and neuronal cells against injury according to our in-vitro studies. Furthermore, we observed that in a Parkin-deficient mouse model expressing a C-terminal truncated human mutant Parkin (Parkin-Q311X) (Lu et al. 2009), DOR expression decreased with an increase in the level of α-synuclein in the striatum, which occurs before behavioral deficits. This observation suggests that PD pathophysiology is potentially attributable to DOR impairment in the brain. Moreover, recent evidence suggests that DOR activation promotes the release of dopamine in the striatum. Since we have previously shown that brain stimulation increased the release of encephalin, an endogenous DOR agonist, in the brain, the mechanism of DBS may be, at least partially, attributed to DOR activity that confers protection in both neurons and astrocytes (Fenoy et al. 2014). Together, the plastic changes in these glial-neuronal network-wide interactions likely underlie the range of effects seen, from the variable temporal latencies to observed effects on global activation patterns. Chapter 15 provides a broad review of our understanding of DBS mechanisms, and then presents recent research progress in the literature on how astrocytes play a key role in DBS efficacy and our novel observations on a potential DOR mechanism for DBS.

Electroacupuncture-induced inhibition of epileptic hyperexcitability. Although the precise mechanism of its pathophysiological changes in the brain is unknown, epilepsy has been recognized as a disorder of brain excitability characterized by recurrent unprovoked seizures that result from abnormal, excessive, and synchronous activity of clusters of nerve cells in the brain. Currently available therapies, including medical, surgical, and other strategies such as ketogenic diet and vagus nerve stimulation, are all used for the control of epileptic symptoms with their own limitations and complications. Recent studies suggest that acupuncture may exert its antiepileptic effects by normalizing the disrupted neuronal and network excitability through several mechanisms, including lowering the overexcited neuronal activity, enhancing the inhibitory system, and attenuating the excitatory system in the brain via regulation of the interaction between δ-opioid receptors

(DOR) and Na^+ channels (Kang et al. 2009; Chao and Xia 2012b, 2013). Chapter 16 reviews the progress in this field and summarizes new knowledge based on our work and the work of others.

Cardiac protection. The δ-opioid receptor has actions in both the central and peripheral nervous systems to produce cell protection including cardiac protection against hypoxic/ischemic insults. Chapter 17 will review the role of DOR in the protection of the heart and compare it to that of the brain. The chapter will consider trigger, mediator, and end effectors involved in DOR-mediated protection in the heart and brain and will specifically focus on common and divergent mechanisms of DOR signaling and provide a description of novel and less well studied mechanisms of protection in the heart versus brain.

Various functions of opioids in pathophysiological conditions. Recent studies suggest that DOR, as well as other opioid receptors like MOR and KOR, participates in various physiological activities, such as the regulation of microRNAs in the brain (Yang et al. 2012), heart (Zhi et al. 2013) and kidney (He et al. 2013a) and protection against hypoxic/ischemic injury in the retina (Husain et al. 2012) and jejunum (Tubbs et al. 2002). There is accumulating evidence suggesting that opioid receptor dysfunction may play a role in numerous diseases, including hypoxic/ischemic injury in the brain, epileptic seizures, Parkinson's disease, Multiple Sclerosis, cardiovascular disease, respiratory depression, pulmonary artery hypertension, malignancies, diabetic cutaneous wounds, immune disease, chronic kidney disease, uremic pruritus, renal ischemia-reperfusion injury, and pre-eclampsia. There is also strong evidence showing that opioid receptors are involved in the regulation of stress, feeding, mood disorders and obesity, and play an essential role in hibernation in some species (Feng et al. 2012; Chu Sin Chung and Kieffer 2013). The last chapter of this book (Chap. 18) will update recent research on opioids and their various functions, particularly in pathophysiological conditions. This chapter aims to deliver an informative reference for better understanding the opioid system from a pathophysiological point of view.

6 The Mechanisms for DOR Function: Focusing on Neuroprotection

Over the past decade, many data have been produced, especially from our laboratory, that help us explore the mechanisms underlying neuroprotection. Our recent reviews (Feng et al. 2009; Chao and Xia 2010; He et al. 2013a) have summarized progress in diversified aspects. Indeed, recent data from our laboratory and those of others have presented some groundbreaking mechanisms involved in DOR neuroprotection against both acute and prolonged hypoxic/ischemic injuries. All these observations lead us to paint a picture depicting the workings of DOR that can present a helpful aid for understanding and narrowing potential therapeutic targets. Broadly speaking, the mechanisms of DOR neuroprotection can be mainly grouped as: (1) stabilization of the ionic homeostasis, (2) inhibition of excitatory transmitter release, (3) attenuation of disrupted neuronal transmission, (4) increase in

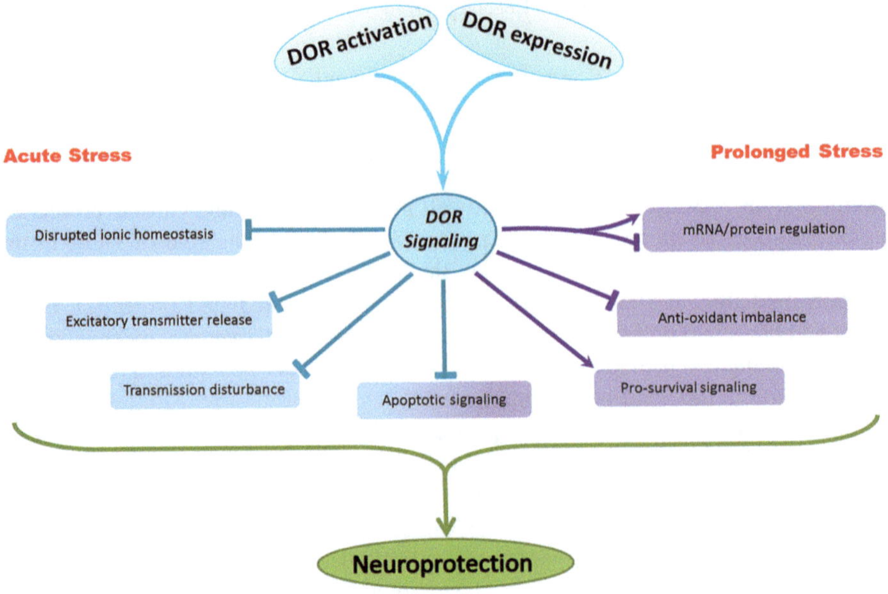

Fig. 1.14 Overview of the mechanisms underlying DOR neuroprotection. Cited from He et al. (2013a)

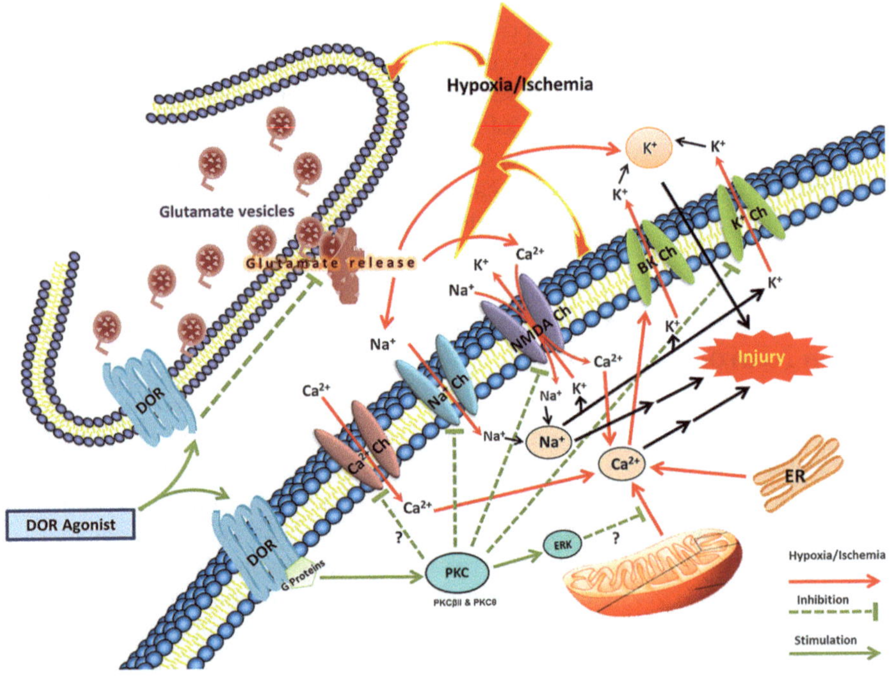

Fig. 1.15 DOR effect on ionic homeostasis under hypoxia/ischemia. *BK Ch* Calcium-activated potassium channels, *TRK* tyrosine kinase, *Na^+ Ch* sodium channels, *Ca^{2+} Ch,* calcium channels, *NMDA Ch,* NMDA channels, *K^+Ch* potassium channels. *Arrows* in *red* indicate the effect of hypoxic/ischemic stress. *Dashed lines* indicate inhibition. *Green arrows/lines* indicate the survival or neuroprotective mechanisms/changes. Cited from He et al. (2013a)

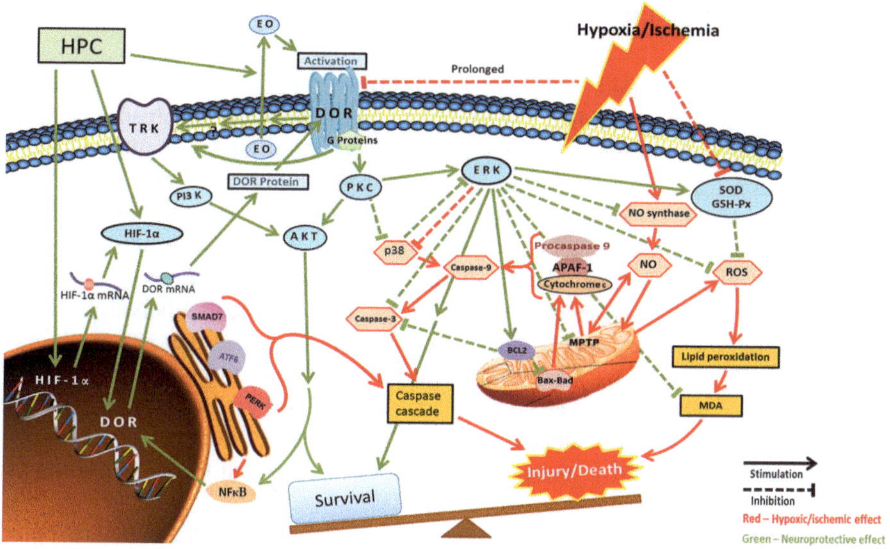

Fig. 1.16 DOR-mediated cellular and molecular regulation in prolonged hypoxia/ischemia. *ROS* reactive oxygen species or oxygen free radicals, *HPC* hypoxic pre-conditioning, *HIF* hypoxia inducible factor, *EO* endogenous opioids, *SOD* superoxide dismutase, *MPTP* mitochondrial permeability transition pore. *Arrows* in *red* indicate effect of hypoxic/ischemic stress. *Dashed lines* indicate inhibition. *Green arrows/lines* indicate the survival or neuroprotective mechanisms/changes. *Blue* ovals represent components involved in the neuroprotective pathways while *red hexagons* represent components playing role in the hypoxic/ischemic injury. Cited from He et al. (2013a)

antioxidant capacity, (5) regulation of intracellular pathways – inhibition of apoptotic signals and activation of pro-survival signaling (e.g., neurotrophic factors and TrkB) (Tian et al. 2013a, b; Liang et al. 2011), (6) regulation of specific gene and protein expression, and (7) up-regulation of endogenous opioid release and/or DOR expression (Fig. 1.14). Of these, some occur immediately in response to acute hypoxic stress (Fig. 1.15) while other changes appear gradually in response to a prolonged hypoxic/ischemic insult (Fig. 1.16). In response to acute and prolonged phases of hypoxic or ischemic stress, DOR signaling may use different strategies for neuroprotection against neuronal responses. In the acute phase, DOR may mainly target the disruption of ionic homeostasis and other immediate changes after the onset of stress that are mostly based on inherited functions of various molecules at the level of expression when encountering stress. In addition, DOR may activate/promote many protective mechanisms to prevent further injury. After stress is prolonged, transcriptional and translational changes can occur and cellular and molecular structures may be altered or even damaged or disrupted. DOR protection may work to attenuate some pathophysiological changes. However, our experience is that the sooner DOR is activated, the better the protection. It is difficult to reverse cellular injury at the late stage of hypoxic/ischemic stress with DOR signaling.

Please refer to our previous reviews (Feng et al. 2009; Chao and Xia 2010; He et al. 2013a) for more details.

7 Concluding Remarks

Recent studies have cleared up the confusion surrounding the functional research of DOR, and accumulating evidence shows that DOR is not only involved in physiological and pathophysiological events of pain modulation and addiction, but also plays critical roles in a variety of neural functions in the brain as well as the heart and other peripheral organs/cells – from the maintenance of ionic homeostasis and regulation of membrane protein expression to beneficial action against hypoxic/ischemic injury and other pathophysiological conditions. Our findings on the differences in DOR distribution among brain regions, across ages, and between species have prompted me to pay special attention to the brain regions that express higher levels of DOR, such as the cortex, caudate putamen, and amygdala. Indeed, our recent studies have shown that DOR exerts a beneficial action against ischemic stroke, Parkinson's' disease, and epilepsy in animal models. Interestingly, these diseases are closely relevant to the pathophysiological events in these brain regions, directly or indirectly.

In general, DOR's beneficial role in the brain as well as the heart is well accepted now, especially under hypoxic/ischemic conditions. KOR may have the same but much less effect in comparison to DOR (Cheng et al. 2007), but needs more investigations because of different voice (Lu et al. 1997; Tu et al. 2008). In contrast, most of studies suggest that activation of MOR is likely harmful in hypoxia/ischemia.

We have to realize that elucidating the function of DOR as well as other opioid receptors is a difficult task, especially if we attempt to distinguish the differences among these opioid receptors. This is simply due to the high homology in their genes. At present, it is extremely difficult, if not impossible, to have a "pure" ligand to interact with only one type of opioid receptor. It is hard to obtain valid data and make a reliable conclusion by solely applying an opioid ligand. By doing this, the outcomes of the research may not be consistent among the in vivo animals and the in vitro preparations, or even in a simple model like cell cultures. The status of the animals/organs/cells, interactions among various organs/regions/cells, the expression of various opioid receptors at the moment the ligand is added, the rate for the degradation of the ligand, and many other factors all influence the measurements at a given time point and lead to a complex outcome. This was the major cause of the unexplainable controversies in the past literature discussed in Sect. 4.1. As an example, our studies suggest that, at least in cortical neurons, MOR activation does not offer neuroprotection, and does not induce injury either in the concentrations we used (Zhang et al. 2000; Chao et al. 2007a). In some brain regions, MOR-preferable agent morphine is neuroprotective against ischemia-reperfusion via a DOR, but not MOR, mechanism (Lim et al. 2004). On the other hand, some studies suggest that MOR agonists can be harmful in hypoxic/ischemic condition (Ammon-Treiber et al. 2005, 2007; Festekjian et al. 2011), suggesting that MOR plays a detrimental role in certain regions/organs under certain conditions. In such a complex situation, a beneficial or detrimental outcome after intravenous administration of opioid agonists or antagonists is dependent on the ratio of protective (e.g., DOR-mediated) and harmful (e.g., MOR-mediated or blockade of DOR) powers at the time of assessment, which is also a major reason for the controversies in the past and should be taken into account in future studies.

To avoid pitfalls and flaws, therefore, a careful consideration in the selection of ligands and more importantly, using diversified approaches, e.g., siRNA and transgenic techniques (Chao et al. 2006; Shim and Kwon 2010; Cao et al. 2015), are essential for the functional research of DOR. Fortunately, current advancement of research techniques has greatly increased our power in research. In our research since the early 2000s, we have collaborated with domestic and international scientists and have used potent and selective DOR ligands (e.g., UFP-512 as a DOR agonist) in our work. More importantly, we have used biochemical, morphological, molecular, transgenic, and electrophysiological approaches to verify our findings through various methods for reliable conclusions.

An important goal regarding the clinical manipulation of DOR system is to develop potent and selective DOR ligands with minimal side effects in addiction that is a common problem for all opiates and opioids. However, this aim is still far from reach. On the other side, gene therapy is not yet practical, though we have shown DOR overexpression can mimic the "turtle effect" on the cortex of the brain and renders it more tolerant to hypoxic/ischemic stress. Nevertheless, our research has endorsed a relatively safe, cost-efficient and convenient strategy for the upregulation of DOR system - electroacupuncture that has been shown to promote DOR activity in the brain and display a beneficial effect against cerebral ischemia, epilepsy and other diseases (Guo et al. 2010; Chao and Xia 2010; Wen et al. 2010a, b; Kang and Xia 2010a, b, 2012; Xia et al. 2010, 2012; Zhou et al. 2011a, 2013a, b; Asakawa and Xia 2012a, b, c; Liang and Xia, 2012; Chao and Xia 2013; Chen et al. 2014a). It is worthwhile to expand research to the next level for better practice of electroacupuncture in clinical settings. In addition, if we cannot get an ideal DOR agonist for clinical use in a short amount of time, we may try to gain a protective effect by targeting a key point at DOR signal transduction pathway underneath the membrane where DOR sits. This may be possible once we fully understand the mechanisms for DOR's beneficial actions. On the other hand, hypoxic/ischemic preconditioning may be a strategy for brain protection against severe hypoxia or ischemia since we have shown that hypoxic precondition leads to DOR upregulation in neurons (Ma et al. 2005; Zhang et al. 2006). In fact, there is evidence showing that remote limb ischemic postconditioning protects against neonatal hypoxic-ischemic brain injury by the opioid receptor (Zhou et al. 2011b).

It is my belief that comprehensive and in-depth research on the functions of DOR will eventually lead to a better solution for many diseases, especially neurological disorders.

Acknowledgments I deeply thank my previous mentor, Dr. Haddad, an intelligent physician and scientist, for his permission for me to start DOR work as a side project in the first place, and for his dedicated and hard-working sprit that has inspired me along my scientific journey. I would also like to present special thanks to my students, postdoctoral follows, research assistants, visiting scientists, other team members in my laboratory, as well as my international collaborators whom I have worked with over the past 20 years. Without them, it would not have been possible to gain any major progress in our research. I am very grateful to Drs. Lazarus, Balboni and Salvadori for their collaboration with me in using selective and potent DOR ligands in order to gain reliable data on DOR research. I also attribute my research success, if any, to consecutive grant support from NIH, America Heart Association, March of Dimes Foundation, Cerebral Palsy Foundation and The Vivian L. Smith Neurologic Foundation since 1994.

References

Adams HP, Olinger CP, Barsan WG, Butler MJ, Graff-Radford NR, Brott TG, Biller J, Damasio H, Tomsick T, Goldberg M (1986) A dose-escalation study of large doses of naloxone for treatment of patients with acute cerebral ischemia. Stroke 17:404–409

Albers GW, Goldberg MP, Choi DW (1989) N-methyl-D-aspartate antagonists: ready for clinical trial in brain ischemia? Ann Neurol 25:398–403

Ammon-Treiber S, Stolze D, Schröder H, Loh H, Höllt V (2005) Effects of opioid antagonists and morphine in a hippocampal hypoxia/hypoglycemia model. Neuropharmacology 49:1160–1169

Ammon-Treiber S, Stolze D, Höllt V (2007) Differential effects of mu-opioid receptor agonists in a hippocampal hypoxia/hypoglycemia model. Brain Res 1183:60–65

Armstead WM (2002) Role of Nociceptin/Orphanin FQ in the physiologic and pathologic control of the cerebral circulation. Exp Biol Med (Maywood) 227:957–968

Armstead WM (2011) Nociceptin/orphanin phenylalanine glutamine (FQ) receptor and cardiovascular disease. Cardiovasc Ther 29:23–28

Asakawa T, Xia Y (2012a) Acupuncture treatment for Parkinson's disease. In: Xia (ed) Current research in acupuncture. Springer, New York, pp 215–254

Asakawa T, Xia Y (2012b) Can acupuncture treat Alzheimer's disease and other neurodegenerative disorders? In: Xia (ed) Current research in acupuncture (research monograph). Springer, New York, pp 255–302

Asakawa T, Xia Y (2012c) Future research in acupuncture—better design and analysis for novel and valid findings. In: Xia (ed) Current research in acupuncture (research monograph). Springer, New York, pp 687–726

Audet N, Galés C, Archer-Lahlou E, Vallières M, Schiller PW, Bouvier M, Pineyro G (2008) Bioluminescence resonance energy transfer assays reveal ligand-specific conformational changes within preformed signaling complexes containing deltaopioid receptors and heterotrimeric G proteins. J Biol Chem 283:15078–15088

Balboni G, Salvadori S, Guerrini R, Negri L, Giannini E, Jinsmaa Y, Bryant SD, Lazarus LH (2002) Potent d-opioid receptor agonists containing the Dmt-Tic pharmacophore. J Med Chem 45:5556–5563

Ballet S, Pietsch M, Abell AD (2008) Multiple ligands in opioid research. Protein Pept Lett 15:668–682

Baskin DS, Hosobuchi Y (1981) Naloxone reversal of ischaemic neurological deficits in man. Lancet 2:272–275

Baskin DS, Kieck CF, Hosobuchi Y (1984) Naloxone reversal and morphine exacerbation of neurologic deficits secondary to focal cerebral ischemia in baboons. Brain Res 290:289–296

Blurton PA, Broadhurst AM, Wood MD, Wyllie MG (1986) Is there a common, high-affinity opioid binding site in rat brain? J Recept Res 6:85–93

Borlongan CV, Oeltgen PR, Su TP, Wang Y (1999) Delta opioid peptide (DADLE) protects against ischemia reperfusion damage in the striatum and cerebral cortex. Soc Neurosci Abstr 24:979

Borlongan CV, Wang Y, Su T-P (2004) Delta opioid peptide (D-Ala 2, D-Leu 5) enkephalin: linking hibernation and neuroprotection. Front Biosci 9:3392–3398

Borlongan CV, Hayashi T, Oeltgen PR, Su TP, Wang Y (2009) Hibernation like state induced by an opioid peptide protects against experimental stroke. BMC Biol 7:31
Brownstein MJ (1993) A brief history of opiates, opioid peptides, and opioid receptors. Proc Natl Acad Sci U S A 90:5391–5393
Buresh Y, Koroleva VI, Korolev OS, Maresh V (1999) Changes in the constant potential in brain structures in rats during focal ischemia and systemic hypoxia. Neurosci Behav Physiol 29:569–579
Cao S, Chao DM, Zhou HH, Balboni G, Xia Y (2015) A novel mechanism for cytoprotection against hypoxic injury: DOR-mediated increase in Nrf2 translocation. Br J Pharmacol 172:1869–1881
Catterall WA (2014) Sodium channels, inherited epilepsy, and antiepileptic drugs. Annu Rev Pharmacol Toxicol 54:317–338
Chang KJ, Cuatrecasas P (1979) Multiple opiate receptors. Enkephalins and morphine bind to receptors of different specificity. J Biol Chem 254:2610–2618
Chang KJ, Porreca F, Woods JH (2004) The delta receptor. Marcel Dekker, Inc, New York
Chao D, Xia Y (2010) Ionic storm in hypoxic/ischemic stress: can opioid receptors subside it? Prog Neurobiol 90:439–470
Chao DM, Xia Y (2012) Acupuncture treatment of epilepsy. In: Xia Y (ed) Current Research in Acupuncture. Springer, New York, pp 129–214
Chao DM, Xia Y (2013) From acupuncture to interaction between delta-opioid receptors and Na^+ channels: a potential pathway to inhibit epileptic hyper-excitability. Evid Based Complement Alternat Med 2013:216016
Chao, D., Qian, H., Ghassemi, F., Chen, J., Xia, Y (2006) Transgenic overexpression of delta-opioid receptors protects the cortex from anoxic disruption of ionic homeostasis. Program No. 87.19/MM68, Neuroscience Meeting Planner. Society for Neuroscience, Atlanta
Chao DM, Bazzy-Asaad A, Balboni G, Xia Y (2007a) Delta-, but not μ-, opioid receptor stabilizes K^+ homeostasis by reducing Ca^{2+} influx in the cortex during acute hypoxia. J Cell Physiol 212:60–67
Chao DM, Donnelly D, Feng Y, Bazzy-Asaad A, Xia Y (2007b) Cortical delta-opioid receptors potentiate K^+ homeostasis during anoxia and oxygen-glucose deprivation. J Cereb Blood Flow Metab 27:356–368
Chao DM, Bazzy-Asaad A, Balboni G, Salvadori S, Xia Y (2008) Activation of DOR attenuates anoxic K^+ derangement via inhibition of Na^+ entry in mouse cortex. Cereb Cortex 18:2217–2227
Chao DM, Balboni G, Lazarus LH, Salvadori S, Xia Y (2009) Na^+ mechanism of delta-opioid receptor induced protection from anoxic K^+ leakage in the cortex. Cell Mol Life Sci 66:1105–1115
Chao DM, He XZ, Yang YL, Balboni G, Salvadori S, Dong KH, Xia Y (2012a) Hydrogen sulfide induced disruption of Na^+ homeostasis in the cortex. Toxicol Sci 128:198–208
Chao DM, He XZ, Yang YL, Bazzy-Asaad A, Lazarus LH, Balboni G, Kim DH, Xia Y (2012b) DOR activation inhibits anoxic/ischemic Na^+ influx through Na^+ channels via PKC mechanisms in the cortex. Exp Neurol 236:228–239
Charron C, Messier C, Plamondon H (2008) Neuroprotection and functional recovery conferred by administration of kappa and delta1-opioid agonists in a rat model of global ischemia. Physiol Behav 93:502–511
Chen Y, Mestek A, Liu J, Hurley JA, Yu L (1993) Molecular cloning and functional expression of a mu-opioid receptor from rat brain. Mol Pharmacol 44:8–12
Chen F, Qi ZF, Luo YM, Hinchliffe T, Xia Y, Ji XM (2014a) Non-pharmaceutical therapies for stroke: mechanisms and clinical implications. Prog Neurobiol 115:246–269
Chen T, Li J, Chao DM, Sandhu HK, Liao XP, Zhao JL, Wen GQ, Xia Y (2014b) δ-Opioid receptor activation reduces α-synuclein overexpression and oligomer formation induced by MPP+ and/or hypoxia. Exp Neurol 255:127–136
Cheng L, Ma S, Wei LX, Guo HT, Huang LY, Bi H, Fan R, Li J, Liu YL, Wang YM, Sun X, Zhang QY, Yu SQ, Yi DH, Ma XL, Pei JM (2007) Cardioprotective and antiarrhythmic effect of U50,488H in ischemia/reperfusion rat heart. Heart Vessels 22:335–344

Choi DW, Viseskul V (1988) Opioids and non-opioid enantiomers selectively attenuate N-methyl-D-aspartate neurotoxicity on cortical neurons. Eur J Pharmacol 155:27–35

Chu Sin Chung P, Kieffer BL (2013) Delta opioid receptors in brain function and diseases. Pharmacol Ther 140:112–120

Civelli O, Douglass J, Goldstein A, Herbert E (1985) Sequence and expression of the rat prodynorphin gene. Proc Natl Acad Sci U S A 82:4291–4295

Costantino CM, Gomes I, Stockton SD, Lim MP, Devi LA (2012) Opioid receptor heteromers in analgesia. Expert Rev Mol Med 14, e9

Cox BM, Goldstein A, Hi CH (1976) Opioid activity of a peptide, betalipotropin-(61–91), derived from etalipotropin. Proc Natl Acad Sci U S A 73:1821–1823

Cummins TR, Xia Y, Haddad GG (1994) Functional properties of rat and human neocortical voltage sensitive sodium currents. J Neurophysiol 71:1052–1064

Ding GH, Liu RQ, Chao DM, Xia Y (2012) Neuronal responses to hypoxic stress in mouse cognitive cortex. Circ Res 111:A338

Duan YL, Wang SY, Zeng QW, Su DS, Li W, Wang XR, Zhao Z (2011) Astroglial reaction to delta opioid peptide [D-Ala2, D-Leu5] enkephalin confers neuroprotection against global ischemia in the adult rat hippocampus. Neuroscience 192:81–90

Dunn JF, Zaim Wadghiri Y, Meyerand ME (1999) Regional heterogeneity in the brain's response to hypoxia measured using BOLD MR imaging. Magn Reson Med 41:850–854

Endoh H, Taga K, Yamakura T, Sato K, Watanabe I, Fukuda S, Shimoji K (1999) Effects of naloxone and morphine on acute hypoxic survival in mice. Crit Care Med 27:1929–1933

Evans CJ, Keith DE Jr, Morrison H, Magendzo K, Edwards RH (1992) Cloning of a delta opioid receptor by functional expression. Science 258:1952–1955

Feng Y, Lu Y, Lin X, Gao Y, Zhao Q, Li W, Wang R (2008) Endomorphins and morphine limit anoxia-reoxygenation-induced brain mitochondrial dysfunction in the mouse. Life Sci 82:752–763

Feng Y, Chao D, He X, Yang Y, Kang X, Lazarus LH, Xia Y (2009) A novel insight into neuroprotection against hypoxic/ischemic stress. Acta Physiol Sin 61:585–592

Feng Y, He X, Yang Y, Chen J, Yin K, Xia Y (2011) Effect of delta-opioid receptor over-expression on cortical expression of GABAA receptor alpha1-subunit in hypoxia. Chin J Physiol 54:118–123

Feng Y, He X, Yang Y, Chao D, Lazarus LH, Xia Y (2012) Current research on opioid receptor function. Curr Drug Targets 13:230–246

Fenoy AJ, Goetz L, Chabardès S, Xia Y (2014) Deep brain stimulation: are astrocytes a Key driver behind the scene? CNS Neurosci Ther 20:191–201

Festekjian A, Ashwal S, Obenaus A, Angeles DM, Denmark TK (2011) The role of morphine in a rat model of hypoxic-ischemic injury. Pediatr Neurol 45:77–82

Fraessdorf J, Hollmann MW, Hanschmann I, Heinen A, Weber NC, Preckel B, Huhn R (2015) Role of endogenous opioid system in ischemic-induced late preconditioning. PLoS One 10(7), e0134283

Gao CJ, Niu L, Ren PC, Wang W, Zhu C, Li YQ, Chai W, Sun XD (2012) Hypoxic preconditioning attenuates global cerebral ischemic injury following asphyxial cardiac arrest through regulation of delta opioid receptor system. Neuroscience 202:352–362

Gao H, Wei X, Liao J, Wang R, Xu J, Liu X, Pan X, Li Z, Li Z, Xia Y, Wang Q (2015) Lower bone mineral density in Patients with Parkinson's disease: a cross-sectional study from Chinese Mainland. *Frontiers in Aging Neuroscience*, http://dx.doi.org/10.3389/fnagi.2015.00203, October 27, 2015

Georges F, Normand E, Bloch B, Moine CL (1998) Opioid receptor gene expression in the rat brain during ontogeny, with special reference to the mesostriatal system: an in situ hybridization study. Develop Brain Res 109:187–199

Giri AK, Hruby VJ (2014) Investigational peptide and peptidomimetic μ and δ opioid receptor agonists in the relief of pain. Expert Opin Investig Drugs 23:227–241

Goldstein A (1976) Opioid peptides endorphins in pituitary and brain. Science 193:1081–1086

Goldstein A, Lowney LI, Pal BK (1971) Stereospecific and nonspecific interactions of the morphine congener levorphanol in subcellular fractions of mouse brain. Proc Natl Acad Sci U S A 68:1742–1747

Govindaswami M, Brown SA, Yu J, Zhu H, Bishop PD, Kindy MS, Oeltgen PR (2008) Delta 2-specific opioid receptor agonist and hibernating woodchuck plasma fraction provide ischemic neuroprotection. Acad Emerg Med 15:250–257

Gross GJ (2003) Role of opioids in acute and delayed preconditioning. J Mol Cel Cardiol 35:709–718

Guo JC, Cheng JS, Xia Y (2010) Acupuncture therapy for stroke. In: Acupuncture therapy for neurological diseases: a neurobiological view. Springer/Tsinghua, Heidelberg, pp 226–262

Hawkins KN, Morelli M, Gulya K, Chang KJ, Yamamura HI (1987) Autoradiographic localization of [3H] [MePhe3, D-Pro4] morphiceptin ([3H]PL017) to mu opioid receptors in rat brain. Eur J Pharmacol 133:351–352

Hays SL, Valieva OA, McPherson RJ, Juul SE, Gleason CA (2013) Adult responses to an ischemic stroke in a rat model of neonatal stress and morphine treatment. Int Dev J Neurosci 31:25–29

Hayward NJ, McKnight AT, Woodruff GN (1993a) Neuroprotective effect of the kappa-agonist enadoline (CI-977) in rat models of focal cerebral ischaemia. Eur J Neurosci 5:961–967

Hayward NJ, McKnight AT, Woodruff GN (1993b) Brain temperature and the neuroprotective action of enadoline and dizocilpine in the gerbil model of global ischaemia. Eur J Pharmacol 236:247–253

He X, Sandhu HK, Yang Y, Hua F, Belser N, Kim DH, Xia Y (2013a) Neuroprotection against hypoxia/ischemia: δ-opioid receptor-mediated cellular/molecular events. Cell Mol Life Sci 70:2291–2303

He X, Yang YL, Zhi F, Moore ML, Kang XZ, Chao DM, Wang R, Balboni G, Salvadori S, Kim DH, Xia Y (2013b) δ-Opioid receptor activation modified microRNA expression in the rat kidney under prolonged hypoxia. PLoS One 8(4):e61080

Headrick JP, See Hoe LE, Du Toit EF, Peart JN (2015) Opioid receptors and cardioprotection—'opioidergic conditioning' of the heart. Br J Pharmacol 172:2026–2050

Hiller JM, Fan LQ (1996) Laminar distribution of the multiple opioid receptors in the human cerebral cortex. Neurochem Res 21:1333–1345

Hong SS, Qian H, Zhao P, Bazzy-Asaad A, Xia Y (2007) Anisomycin protects cortical neurons from prolonged hypoxia with differential regulation of p38 and ERK. Brain Res 1149:76–86

Horiuchi T, Kawaguchi M, Sakamoto T, Kurita N, Inoue S, Nakamura M, Konishi N, Furuya H (2004) The effects of the delta-opioid agonist SNC80 on hind-limb motor function and neuronal injury after spinal cord ischemia in rats. Anesth Analg 99:235–240

Horiuchi T, Kawaguchi M, Kurita N, Inoue S, Sakamoto T, Nakamura M, Konishi N, Furuya H (2008) Effects of delta-opioid agonist SNC80 on white matter injury following spinal cord ischemia in normothermic and mildly hypothermic rats. J Anesth 22:32–37

Hosobuchi Y, Baskin DS, Woo SK (1982) Reversal of induced ischemic neurologic deficit in gerbils by the opiate antagonist naloxone. Science 215:69–71

Huang MH, Nguyen V, Wu Y, Rastogi S, Lui CY, Birnbaum Y, Wang HQ, Ware DL, Chauhan M, Garg N, Poh KK, Ye L, Omar AR, Tan HC, Uretsky BF, Fujise K (2009) Reducing ischaemia/reperfusion injury through delta-opioid-regulated intrinsic cardiac adrenergic cells: adrenopeptidergic co-signalling. Cardiovasc Res 84:452–460

Hughes J, Smith TW, Kosterlitz HW, Fothergill LA, Morgan BA, Morris HR (1975) Identification of two related pentapeptides from the brain with potent opiate agonist activity. Nature 258:577–580

Husain S, Abdul Y, Potter DE (2012) Non-analgesic effects of opioids: neuroprotection in the retina. Curr Pharm Des 18:6101–6108

Iwai T, Niwa M, Nakashima M, Kambara T, Yamada H, Tsurumi K, Nozaki M (1992) Effect of opioids on delayed neuronal death in the gerbil hippocampus. Life Sci 50:PL239–PL244

Iwata M, Inoue S, Kawaguchi M, Nakamura M, Konishi N, Furuya H (2007) Effects of delta-opioid receptor stimulation and inhibition on hippocampal survival in a rat model of forebrain ischaemia. Br J Anaesth 99:538–546

Jagolino A, Armstead WM (2001) Nociceptin/orphanin FQ contributes to hypoxic/ischemic impairment of hypercapnic cerebrovasodilation. Brain Res Bull 55:465–468

Jiang C, Xia Y, Haddad GG (1992) Role of ATP-sensitive K^+ channels during anoxia: major differences between rat (newborn and adult) and turtle neurons. J Physiol 448:599–612

Johnson SM, Turner SM (2010) Protecting motor networks during perinatal ischemia: the case for delta-opioid receptors. Ann N Y Acad Sci 1198:260–270

Kang XZ, Xia Y (2010a). Effects of acupuncture on arrhythmia and other cardiac diseases. In: Acupuncture therapy for neurological diseases: a neurobiological view. Springer/Tsinghua, Heidelberg, pp 263–288

Kang XZ, Xia Y (2010b) Acupuncture therapy for hypertension and hypotension. In: Acupuncture therapy for neurological diseases: a neurobiological view. Springer/Tsinghua, Heidelberg, pp 289–325

Kang XZ, Chao DM, Gu QB, Ding GH, Wang YW, Balboni G, Lazarus LH, Xia Y (2009) Delta-opioid receptor activation decreases anoxic Na^+ influx and K^+ leakage via sodium channel regulation. Cell Mol Life Sci 66:3505–3516

Kang XZ, Shen XY, Xia Y (2012) Electroacupuncture-induced attenuation of experimental epilepsy: A comparative evaluation of acupoints and stimulation parameters. Evid Based Complement Alternat Med 2012:149612

Kawalec M, Kowalczyk JE, Beresewicz M, Lipkowski AW, Zablocka B (2011) Neuroprotective potential of biphalin, multireceptor opioid peptide, against excitotoxic injury in hippocampal organotypic culture. Neurochem Res 36:2091–2095

Kawaraguchi Y, Sasaoka N, Kawaguchi M, Inoue S, Hayashi H, Konishi N, Furuya H (2008) Effect of continuous morphine infusion on hypoxic-ischaemic brain damage of neonatal rats. Acta Anaesthesiol Scand 52:1111–1115

Kieffer BL, Befort K, Gaveriaux-Ruff C, Hirth CG (1992) The delta-opioid receptor: isolation of a cDNA by expression cloning and pharmacological characterization. Proc Natl Acad Sci U S A 89:12048–12052

Kim JH, Park SH, Park SG, Choi JS, Xia Y, Sung JK (2011) The pivotal role of reactive oxygen species generation in the hypoxia-induced stimulation of adipose-derived stem cells. Stem Cells Dev 20:1753–1761

Kim JH, Song SY, Park SG, Song SU, Xia Y, Sung JH (2012) Primary involvement of NADPH oxidase 4 in hypoxia-induced generation of reactive oxygen species in adipose-derived stem cells. Stem Cells Dev 21:2212–2221

Kim JH, Kim SH, Song SY, Kim WS, Song SU, Yi TG, Jeon MS, Chung HM, Xia Y, Sung JH (2013) Hypoxia induces adipocyte differentiation of adipose-derived stem cells by triggering reactive oxygen species generation. Cell Biol Int 38:32–40

Kornblum HI, Hurlbut DE, Leslie FM (1987) Postnatal development of multiple opioid receptors in rat brain. Brain Res 465:21–41

Lai TW, Zhang S, Wang YT (2014) Excitotoxicity and stroke: identifying novel targets for neuroprotection. Prog Neurobiol 115:157–188

Law PY, Loh HH (1999) Regulation of opioid receptor activities. J Pharmacol Exp Ther 89:607–624

Lee EY, Xia Y, Kim WS, Kim MH, Kim TH, Kim KJ, Park BS, Sung JH (2009) Hypoxia-enhanced wound-healing function of adipose-derived stem cells: increase in stem cell proliferation and up-regulation of VEGF and bFGF. Wound Repair Regen 17:540–547

Li M, Tjen-A-Looi SC, Guo ZL, Longhurst JC (2012) Repetitive electroacupuncture causes prolonged increased Met-enkephalin expression in the rVLM of conscious rats. Auton Neurosci 170:30–35

Liang JF, Xia Y (2012) Acupuncture modulation of neural transmitters/modulators. In: Xia Y (ed) Current research in acupuncture. Springer, New York, pp 1–36

Liang JF, Chao DM, Sandhu HK, Yu Y, Zhang L, Balboni G, Kim DK, Xia Y (2014) δ-Opioid receptors up-regulate excitatory amino acid transporters of astrocytes in mouse astrocytes. Br J Pharmacol 171:5417–5430

Liang JF, Teng S, Chao DM, Kim DH, Xia Y (2011) Activation of the delta-opioid receptor up-regulates expression of neurotrophic factors in astrocytes. Program No. 673.22, 2011 Neuroscience Meeting Planner. Society for Neuroscience, Washington, DC

Liao SL, Chen WY, Raung SL, Chen CJ (2003) Neuroprotection of naloxone against ischemic injury in rats: role of mu receptor antagonism. Neurosci Lett 345:169–172

Lim YJ, Zheng S, Zuo Z (2004) Morphine preconditions Purkinje cells against cell death under in vitro simulated ischemia-reperfusion conditions. Anesthesiology 100:562–568

Lin CS, Polsky K, Nadler JV, Crain BJ (1990) Selective neocortical and thalamic cell death in the gerbil after transient ischemia. Neuroscience 35:289–299

Lockhart BP, Soulard P, Benicourt C, Privat A, Junien JL (1995) Distinct neuroprotective profiles for sigma ligands against N-methyl-D-aspartate (NMDA), and hypoxia-mediated neurotoxicity in neuronal culture toxicity studies. Brain Res 675:110–120

Lord JA, Waterfield AA, Hughes J, Kosterlitz HW (1977) Endogenous opioid peptides: multiple agonists and receptors. Nature 267:495–499

Lu X, Hong X, Wang C (1997) Effect of dynorphin A1–13 on hypoxia-ischemic brain injury in neonatal rats. Zhonghua Fu Chan Ke Za Zhi 32:198–201

Lu XH, Fleming SM, Meurers B, Ackerson LC, Mortazavi F, Lo V, Hernandez D, Sulzer D, Jackson GR, Maidment NT, Chesselet MF, Yang XW (2009) Bacterial artificial chromosome transgenic mice expressing a truncated mutant parkin exhibit age-depedent hypokinetic motor deficits, dopaminergic neuron degeneration, and accumulation of proteinase K-resistant alpha-synuclein. J Neurosci 29: 1962–1976

Lutz PE, Kieffer BL (2013) The multiple facets of opioid receptor function: implications for addiction. Curr Opin Neurobiol 23:473–479

Ma MC, Qian H, Ghassemi F, Zhao P, Xia Y (2005) Oxygen sensitive delta-opioid receptor-regulated survival and death signals: novel insights into neuronal preconditioning and protection. J Biol Chem 280:16208–16218

Malthankar-Phatak GH, Patel AB, Xia Y, Hong S, Chowdhury GM, Behar KL, Orina IA, Lai JC (2008) Effects of continuous hypoxia on energy metabolism in cultured cerebro-cortical neurons. Brain Res 1229:147–154

Martin W (1967) Opioid antagonists. Pharmacol Rev 19:283–323

Martin WR, Eades CG, Thompson JA, Huppler RE, Gilbert PE (1976) The effects of morphine- and nalorphine-like drugs in the nondependent and morphine-dependent chronic spinal dog. J Pharmacol Exp Ther 197:517–532

Maslov LN, Naryzhnaya NV, Prokudina ES, Kolar F, Gorbunov AS, Zhang Y, Wang H, Tsibulnikov SY, Portnichenko AG, Lasukova TV, Lishmanov YB (2015) Preserved cardiac mitochondrial function and reduced ischaemia/reperfusion injury afforded by chronic continuous hypoxia: role of opioid receptors. Clin Exp Pharmacol Physiol 42:496–501

Mayfield KP, D'Alecy LG (1994a) Delta-1 opioid agonist acutely increases hypoxic tolerance. J Pharmacol Exp Ther 268:683–688

Mayfield KP, D'Alecy LG (1994b) Delta-1 opioid receptor dependence of acute hypoxic adaptation. J Pharmacol Exp Ther 268:74–77

Mayfield KP, Hong EJ, Carney KM, D'Alecy LG (1994) Potential adaptations to acute hypoxia: Hct, stress proteins, and set point for temperature regulation. Am J Physiol 266:R1615–R1622

Meng F, Xie GX, Thompson RC, Mansour A, Goldstein A, Watson SJ, Akil H (1993) Cloning and pharmacological characterization of a rat kappa opioid receptor. Proc Natl Acad Sci U S A 90:9954–9958

Meunier JC, Mollereau C (1995) Isolation and structure of the endogenous agonist of opioid receptor-like ORL1 receptor. Nature 377(6549):532–535

Milton SL, Prentice HM (2007) Beyond anoxia: the physiology of metabolic downregulation and recovery in the anoxia-tolerant turtle. Comp Biochem Physiol A Mol Integr Physiol 147:277–290

Minami M, Toya T, Katao Y, Maekawa K, Nakamura S, Onogi T, Kaneko S, Satoh M (1993) Cloning and expression of a cDNA for the rat kappa-opioid receptor. FEBS Lett 329:291–295

Mollereau C, Parmentier M, Mailleux P, Butour JL, Moisand C, Chalon P, Caput D, Vassart G, Meunier JC (1994) ORL1, a novel member of the opioid receptor family. Cloning, functional expression and localization. FEBS Lett 341:33–38

Nakanishi S, Inoue A, Kita T, Nakamura M, Chang AC, Cohen SN, Numa S (1979) Nucleotide sequence of cloned cDNA for bovine corticotropin-beta-lipotropin precursor. Nature 278:423–427

Nandhu MS, Naijil G, Smijin S, Jayanarayanan S, Paulose CS (2010) Opioid system functional regulation in neurological disease management. J Neurosci Res 88:3215–3221
Narita M, Kuzumaki N, Miyatake M, Sato F, Wachi H, Seyama Y, Suzuki T (2006) Role of delta-opioid receptor function in neurogenesis and neuroprotection. J Neurochem 97:1494–1505
Nedergaard M (1988) Mechanisms of brain damage in focal cerebral ischemia. Acta Neurol Scand 77:81–101
Nilsson GE, Lutz PL (2004) Anoxia tolerant brains. J Cereb Blood Flow Metab 24:475–486
Noda M, Teranishi Y, Takahashi H, Toyosato M, Notake M, Nakanishi S, Numa S (1982) Isolation and structural organization of the human preproenkephalin gene. Nature 297:431–434
O'Reilly JP, Jiang C, Haddad GG (1995) Major differences in response to graded hypoxia between hypoglossal and neocortical neurons. Brain Res 683:179–186
Oeltgen PR, Nilekani SP, Nuchols PA, Spurrier WA, Su TP (1988) Further studies on opioid and hibernation: delta opioid receptor ligand selectively induced hibernation in summer-active ground squirrels. Life Sci 43:1565–1574
Olinger CP, Adams HP Jr, Brott TG, Biller J, Barsan WG, Toffol GJ, Eberle RW, Marler JR (1990) High-dose intravenous naloxone for the treatment of acute ischemic stroke. Stroke 21:721–725
Pamenter ME, Buck LT (2008) Delta-opioid receptor antagonism induces NMDA receptor-dependent excitotoxicity in anoxic turtle cortex. J Exp Biol 211:3512–3517
Park SG, Kim JH, Xia Y, Sung JH (2011) Generation of reactive oxygen species in adipose derived stem cells: friend or foe? Expert Opin Ther Targets 15:1297–1306
Pasternak GW (2004) Multiple opiate receptors: deja vu all over again. Neuropharmacology 47(Suppl 1):312–323
Peart JN, Gross ER, Gross GJ (2005) Opioid-induced preconditioning: recent advances and future perspectives. Vascul Pharmacol 42:211–218
Peng PH, Huang HS, Lee YJ, Chen YS, Ma MC (2009) Novel role for the delta-opioid receptor in hypoxic preconditioning in rat retinas. J Neurochem 108:741–754
Peng J, Sarkara S, Chang S (2012) Opioid receptor expression in human brain and peripheral tissues using absolute quantitative real-time RT-PCR. Drug Alcohol Depend 124:223–228
Pert CB, Snyder SH (1973) Opiate receptor: demonstration in nervous tissue. Science 179:1011–1014
Pert A, Yaksh T (1975) Localization of the antinociceptive action of morphine in primate brain. Pharmacol Biochem Behav 3:133–138
Philip S, Armstead WM (2003) Newborn pig nociceptin/orphanin FQ activates protein tyrosine kinase and mitogen activated protein kinase to impair NMDA cerebrovasodilation after ischemia. Neuroreport 14:201–203
Phillis JW, DeLong RE, Towner JK (1985) Naloxone enhances cerebral reactive hyperemia in the rat. Neurosurgery 17:596–599
Portoghese PS, Sultana M, Nelson WL, Klein P, Takemori AE (1992) Delta opioid antagonist activity and binding studies of regioisomeric isothiocyanate derivatives of naltrindole: evidence for delta receptor subtypes. J Med Chem 35:4086–4091
Puyal J, Ginet V, Clarke PG (2013) Multiple interacting cell death mechanisms in the mediation of excitotoxicity and ischemic brain damage: a challenge for neuroprotection. Prog Neurobiol 105:24–48
Qian H, Feng Y, He XZ, Yang YL, Sung JH, Xia Y (2011) Effects of inhibitory amino acids on expression of GABAA and glycine receptor in hypoxic cortical neurons during development. Brain Res 1425:1–12
Schultz JJ, Hsu AK, Gross GJ (1997) Ischemic preconditioning and morphine-induced cardioprotection involve the delta-opioid receptor in the intact rat heart. J Mol Cell Cardiol 29:2187–2195
Shim MS, Kwon YJ (2010) Efficient and targeted delivery of siRNA in vivo. FEBS J 277:4814–4482
Shokeir AA, Barakat N, Hussein AM, Awadalla A, Harraz A, Khater S, Hemmaid K, Kamal AI (2015) Activation of Nrf2 by ischemic preconditioning and sulforaphane in renal ischemia/reperfusion injury: a comparative experimental study. Physiol Res 64:313–323

Simon EJ, Hiller JM, Edelman I (1973) Stereospecific binding of the potent narcotic analgesic (3H) etorphine to rat-brain homogenate. Proc Natl Acad Sci U S A 70:1947–1949

Skarphedinsson JO, Thorén P (1988) Endorphin mechanisms are responsible for the beneficial effects of opioid antagonists on cerebral function during relative cerebral ischaemia in rats. Acta Physiol Scand 132:281–288

Snyder SH, Matthysse S (1975) Opiate receptor mechanisms. Neurosci Res Program Bull 13:1–166

Snyder SH, Pasternak GW (2003) Historical review: opioid receptors. Trends Pharmacol Sci 24:198–205

Sofuoglu M, Portoghese PS, Takemori AE (1991) Differential antagonism of delta opioid agonists by naltrindole and its benzofuran analog (NTB) in mice: evidence for delta opioid receptor subtypes. J Pharmacol Exp Ther 257:676–680

Su D, Wang Z, Zheng Y, Zhao Y, Wang X (2007) Dosedependent neuroprotection of delta opioid peptide [D-Ala2, D-Leu5] encephalin in neuronal death and retarded behavior induced by forebrain ischemia in rats. Neurosci Lett 423:113–117

Sung JH, Chao DM, Xia Y (2008) Neuronal responses to hypoxia. In: Ying W, Wang D (eds) New frontiers in neurological research. Research Signpost, Kerala, pp 73–153

Takagi T, Kitashoji A, Iwawaki T, Tsuruma K, Shimazawa M, Yoshimura S, Iwama T, Hara H (2014) Temporal activation of Nrf2 in the penumbra and Nrf2 activator-mediated neuroprotection in ischemia-reperfusion injury. Free Radic Biol Med 72:124–133

Takasaki Y, Wolff RA, Chien GL, van Winkle DM (1999) Met5-enkephalin protects isolated adult rabbit cardiomyocytes via delta-opioid receptors. Am J Physiol 277:H2442–H2450

Tian XS, Zhou F, Yang R, Xia Y, Wu GC, Guo JC (2008a) Effects of intracerebroventricular injection of delta-opioid receptor agonist TAN-67 or antagonist naltrindole on acute cerebral ischemia in rats. Acta Physiol Sinica 60:475–484

Tian XS, Zhou F, Yang R, Xia Y, Wu GC, Guo JC (2008b) Role of δ-opioid receptors in cumulative electro-acupuncture induced protection from ischemic injury in the rat brain. SH J TCM 42:71–74

Tian XS, Zhou F, Yang R, Xia Y, Wu GC, Guo JC (2008c) Electro-acupuncture protects the brain against ischemic injury via up-regulation of delta-opioid receptors. J Chin Integr Med 6:632–638

Tian XS, Guo JC, Zhu M, Li M, Wu G, Xia Y (2013a) δ-opioid receptor activation rescues the functional TrkB receptor and protects the brain from ischemia-reperfusion injury in the rat. PLoS One 8(7):e69252

Tian XS, Hua F, Sandhu HK, Chao DM, Balboni G, Salvadori S, He XZ, Xia Y (2013b) Effect of δ-opioid receptor activation on BDNF-TrkB vs. TNF-α in the mouse cortex exposed to prolonged hypoxia. Int J Mol Sci 14:15959–15976

Tu IH, Yen HT, Cheng HW, Chiu JH (2008) Baicalein protects chicken embryonic cardiomyocyte against hypoxia-reoxygenation injury via mu- and delta- but not kappa-opioid receptor signaling. Eur J Pharmacol 588:251–258

Tubbs RJ, Porcaro WA, Lee WJ, Blehar DJ, Carraway RE, Przyklenk K, Dickson EW (2002) Delta opiates increase ischemic tolerance in isolated rabbit jejunum. Acad Emerg Med 9:555–560

Turner SMF, Johnson SM (2011) Delta-opioid receptor activation prolongs respiratory motor output during oxygen-glucose deprivation in neonatal rat spinal cord in vitro. Neuroscience 187:70–83

van Rijn RM, Defriel JN, Whistler JL (2013) Pharmacological traits of delta opioid receptors: pitfalls or opportunities? Psychopharmacology (Berl) 228:1–18

Waldhoer M, Bartlett SE, Whistler JL (2004) Opioid receptors. Annu Rev Biochem 73:953–990

Wang JB, Imai Y, Eppler CM, Gregor P, Spivak CE, Uhl GR (1993) mu opiate receptor: cDNA cloning and expression. Proc Natl Acad Sci U S A 90:10230–10234

Wang J, Gao Q, Sun GQ, Zhou HO, Xia Q (2008) Delta-opioid receptor mediates the cardioprotective effect of ischemic postconditioning. Zhongguo Ying Yong Sheng Li Xue Za Zhi 24:184–189

Wang S, Duan Y, Su D, Li W, Tan J, Yang D, Wang W, Zhao Z, Wang X (2011) Delta opioid peptide [D-Ala2, D-Leu5] enkephalin (DADLE) triggers postconditioning against transient forebrain ischemia. Eur J Pharmacol 658:140–144

Wang QY, Chao DM, Xia Y (2014) δ-opioid receptors and inflammatory cytokines in hypoxia: differential regulation between neuron-like and glial cells. Transl Stroke Res 5:476–483

Wen GQ, Yang YL, Lu Y, Xia Y (2010a) Acupuncture-induced activation of endogenous opioid system. In: Acupuncture therapy for neurological diseases: a neurobiological view. Springer/Tsinghua, Heidelberg, pp 104–119

Wen GQ, He XZ, Lu Y, Xia Y (2010b) Effect of acupuncture on neurotransmitters/modulators. In: Acupuncture therapy for neurological diseases: a neurobiological view. Springer/Tsinghua, Heidelberg, pp 120–142

Wei X, Zou J, Liu X, Xu Y, Chen D, Liao J, Tang B, Zhang Z, Cai X, Jin K, Xia Y, Wang Q (2015) Contra-directional coupling of Nur77 and Nurr1 in neurodegeneration: A novel mechanism for Memantine-induced neuroprotection. Mol Neurobiology, http://dx.doi.org/10.3389/fnagi.2015.00203

Wilson AM, Kriegstein AR (1991) Turtle cortical neurons survive glutamate exposures that are lethal to mammalian neurons. Brain Res 540:297–301

Xia Y, Haddad GG (1991a) Ontogeny and distribution of opioid receptors in the rat brainstem. Brain Res 549:181–193

Xia Y, Haddad GG (1991b) Major differences in CNS sulfonylurea receptor distribution between the rat (newborn, adult) and turtle. J Comp Neurol 314:278–289

Xia Y, Haddad GG (1993) Neuroanatomical distribution and binding properties of saxitoxin sites in the rat and turtle CNS. J Comp Neurol 330:363–380

Xia Y, Haddad GG (2001) Major difference in the expression of delta- and mu-opioid receptors between turtle and rat brain. J Comp Neurol 436:202–210

Xia Y, Cao XD, Zhang AZ, Guo XQ, Li P (1985) Inhibitory effect of analogous electro-acupuncture on experimental arrhythmia. Acupuncture Electro-therapeutics Res Int J 10:13–34

Xia Y, Cao XD, Zhang AZ (1986) Roles of opioid peptides of PAG in analogous electroacupuncture inhibition of experimental arrhythmia: analyzed by specific antisera microinjection. Acupuncture Electro Therapeutics Res Int J 11:191–198

Xia Y, Cao XD, Zhang AZ, Tang QM, Xu XR (1987) Inhibitory effect of analogous electroacupuncture on sympathetic cardio-vascular response to stimulation of hypothalamic defense area in rabbits. J Tradit Chin Med 7:211–214

Xia Y, Cao XD, Zhang AZ (1988) Increased leu-enkephalin immunoreactivity in cerebrospinal fluid during stimulation of hypothalamic defense area in rabbits. Acta Physiologica Sin 40:365–367

Xia Y, Cao XD, Zhang AZ, Tang QM, Xu XR (1989) Changed contents of monoamines and their metabolites of cerebrospinal fluid during inhibition of defense pressor response by inputs of deep peroneal nerve in rabbits. Chin Sci Bulletin 34:1134–1139

Xia Y, Jiang C, Haddad GG (1992) Oxidative and glycolytic pathways in rat (newborn, adult) and turtle: role in anoxia. Am J Physiol 262:R595–R603

Xia Y, Eisenman D, Haddad GG (1993) Sulfonylurea receptor expression in rat brain: effect of chronic hypoxia during development. Pediatr Res 34:634–641

Xia Y, Fung ML, O'Reilly JP, Haddad GG (2000) Increased neuronal excitability after long-term O_2 deprivation is mediated mainly by sodium channels. Mol Brain Res 76:211–219

Xia Y, Zhao P, Xue J, Gu XQ, Sun XL, Yao H, Haddad GG (2003) Na^+ channel expression and neuronal function in Na^+/H^+ exchanger 1 null mutant mouse. J Neurophysiol 89:229–236

Xia Y, Zhou, F, Guo JC, Cheng JS, Wu GC (2007a) Optimal conditions for electro-acupuncture to protect the brain from experimental stroke. In: Scientific approaches to Chinese medicine. Heidelberg School Edition, Heidelberg, pp 75–78

Xia Y, Zhou, F, Guo JC, Chao DM, Tian XS, Lu Y, Kang XZ, Cheng JS, Wu GC (2007b) Electro-acupuncture, delta-opioid receptors and brain protection from hypoxic/ischemic stress. In: Scientific approaches to Chinese medicine. Heidelberg School Edition, Heidelberg, pp 79–82

Xia Y, Cao XD, Wu GC, Cheng JS (2010) Acupuncture therapy for neurological diseases: a neurobiological view. Springer/Tsinghua, Heidelberg, pp 1–480

Xia Y, Ding GH, Wu GC (2012) Current research in acupuncture. Springer, New York, pp 1–752

Xiong L, Yang J, Wang Q, Liu Y (2004) Cerebral ischemic tolerance induced by repeated electroacupuncture involved delta-opioid receptor and enkephalin in rats. Anesthesiology 101:A767

Xiong L, Yang J, Wang Q, Lu Z (2007) Involvement of delta and mu-opioid receptors in the delayed cerebral ischemic tolerance induced by repeated electroacupuncture preconditioning in rats. Chin Med J 120:394–399

Xu YQ, Yan JQ, Zhou P, Li JJ, Gao HM, Xia Y, Wang Q (2012) Neurotransmitter receptors and cognitive dysfunctions in Alzheimer's disease and Parkinson's disease. Prog Neurobiol 97:1–13

Xu J, Huang G, Zhang K, Sun J, Xu T, Li R, Tao H, Xu W (2014) Nrf2 activation in astrocytes contributes to spinal cord ischemic tolerance induced by hyperbaric oxygen preconditioning. J Neurotrauma 31:1343–1353

Xu Y, Wei X, Liu X, Liao J, Lin J, Zhou P, Zhu C, Meng X, Xie D, Chao DM, Fenoy AJ, Cheng M, Tang B, Xia Y, Wang Q (2015). Low cerebral glucose metabolism: A potential predictor for the severity of vascular Parkinsonism and Parkinson's disease. Aging and Disease, 10.14336/AD, http://www.aginganddisease.org/EN/10.14336/AD. 2015.0204

Yang YL, Xia XW, Zhang Y, Wang Q, Li L, Luo GH, Xia Y (2009) Delta-opioid receptor activation attenuates oxidative injury in the ischemic rat brain. BMC Biol 7:55

Yang L, Wang H, Shah K, Karamyan VT, Abbruscato TJ (2011) Opioid receptor agonists reduce brain edema in stroke. Brain Res 1383:307–316

Yang Y, Zhi F, He X, Moore ML, Kang X, Chao D, Wang R, Kim DH, Xia Y (2012) δ-opioid receptor activation and microRNA expression of the rat cortex in hypoxia. PLoS One 7:e51524

Yang YL, Sandhu HK, Zhi F, Hua F, Wu M, Xia Y (2015) Effects of hypoxia/ischemia on brain microRNAs. Curr Med Chem 22:1292–1301

Zhang JH, Xia Y, Haddad GG (1999) Activation of δ-opioid receptors protects cortical neurons from glutamate excitotoxic injury. Society for neuroscience online: SfN Abstract. Program no. 25:736.

Zhang JH, Haddad GG, Xia Y (2000) Delta, but not mu and kappa, opioid receptor activation protects neocortical neurons from glutamate-induced excitotoxic injury. Brain Res 885:143–153

Zhang J, Gibney GT, Zhao P, Xia Y (2002) Neuroprotective role of delta-opioid receptors in cortical neurons. Am J Physiol Cell Physiol 282:C1225–C1234

Zhang J, Qian H, Zhao P, Hong SS, Xia Y (2006) Rapid hypoxia preconditioning protects cortical neurons from glutamate toxicity through delta-Opioid Receptor. Stroke 37:1094–1099

Zhang J, Yin C, Zhang Y, Zhao L, Fu H, Feng J (2013) The role of OLR1 polymorphisms in determining the risk and prognosis of ischemic stroke in a Chinese population. NeuroRehabilitation 32:391–396

Zhao P, Guo J, Hong S, Bazzy-Asaad A, Cheng J, Xia Y (2002) Electro-acupuncture and brain protection from cerebral ischemia: the role of delta-opioid receptor. Society for Neuroscience: SfN Abstract. Program no. 490.13

Zhao P, Ma MC, Qian H, Xia Y (2005a) Decreased density of delta-opioid receptors in Na^+/H^+ Exchanger 1 null mutant mouse brain with epilepsy. Neurosci Res 53:442–446

Zhao P, Qian H, Xia Y (2005b) GABA and glycine are protective to mature and toxic to immature in cortical neurons under hypoxia. Eur J Neurosci 22:289–300

Zhao P, Xue J, Gu XQ, Haddad GG, Xia Y (2005c) Intermittent hypoxia modulates Na^+ channel expression in developing mouse brain. Int J Dev Neurosci 23:327–333

Zheng YJ, Wang XR, Chen HZ, Wu XJ, Zhao YH, Su DS (2012) Protective effects of the delta opioid peptide [D-Ala2, D-Leu5]enkephalin in an ex vivo model of ischemia/reperfusion in brain slices. CNS Neurosci Ther 18:762–766

Zhi F, Yang YL, He XZ, Sandhu HK, Kang XZ, Chao DM, Wang R, Kim DH, Xia Y (2013) Activation of δ-opioid receptors regulates cardiac microRNA expression under hypoxia. Circulation Res 113:A96

Zhou F, Guo J, Cheng J, Wu G, Xia Y (2008). Chinese herbal medicine or electroacupuncture: Which is better for brain protection from cerebral ischemia? In Proceedings of the 7th Meeting of the Consortium for Globalization of Chinese Medicine, Bio-II-104. p 149

Zhou F, Huang DK, Xia Y (2010) Neuroanatomic basis of acupuncture points. In: Acupuncture therapy for neurological diseases: a neurobiological view. Springer/Tsinghua, Heidelberg, pp 32–80

Zhou F, Guo JC, Cheng JS, Wu GC, Xia Y (2011a) Electroacupuncture increased cerebral blood flow and reduced ischemic brain injury: Dependence on stimulation intensity and frequency. J Appl Physiol 111:1877–1887

Zhou Y, Fathali N, Lekic T, Ostrowski RP, Chen C, Martin RD, Tang J, Zhang JH (2011b) Remote limb ischemic postconditioning protects against neonatal hypoxic-ischemic brain injury in rat pups by the opioid receptor/Akt pathway. Stroke 42:439–444

Zhou F, Guo JC, Cheng JS, Wu GC, Sun J, Xia Y (2013a) Electroacupuncture and brain protection against cerebral ischemia: Specific effects of acupoints. Evid Based Complement Alternat Med 2013:e804397

Zhou F, Guo JC, Cheng JS, Wu GC, Xia Y (2013b) Effect of electroacupuncture on rat ischemic brain injury: importance of stimulation duration. Evid Based Complement Alternat Med 2013:e878521

Zhu M, Li MW, Tian XS, Ou XM, Zhu CQ, Guo JC (2009) Neuroprotective role of delta-opioid receptors against mitochondrial respiratory chain injury. Brain Res 1252:183–191

Zhu M, Li M, Yang F, Ou X, Ren Q, Gao H, Zhu C, Guo J (2011) Mitochondrial ERK plays a key role in delta-opioid receptor neuroprotection against acute mitochondrial dysfunction. Neurochem Int 59:739–748

Zou J, Gao H, Chen D, Wei X, Xia Y, Wang Q (2015) Position Emission Tomography/Single Photon Emission Tomography Neuroimaging for Detection of Pre-Motor Parkinson's Disease. CNS Neuroscience & Therapeutics, in press, 2015.

Chapter 2
Recent History on Delta Opioid Receptors and Ligands: Biased Mechanisms, and Opioid Delta-Mu and Delta-Kappa Receptor Heteromers

Kwen-Jen Chang, Shoupu Yi, and Yuehai Shen

Abstract For the last decade, progresses in the understanding of opioid actions are enormous. The X-ray crystalline structures of all four receptor proteins were successfully obtained and analyzed in the antagonist bound stable conformation. Among all four opioid receptors cloned previously, we believe that MOR is the major player in mediating the good pain-killing effects of opioids as well as the bad adverse side effects of opioid analgesics. DOR, KOR and ORL-1 all seem to play modulating functions on MOR-mediating activities including good and bad effects in addition to their own independent activities. Recently, several new concepts have been evolved. These are biased mechanisms, opioid receptor heteromerization, signaling, internalization, trafficking and recycling. This Chapter will focus on the DOR related development during the last decade.

Keywords Opioid receptor heteromers • Delta opioid receptor • Analgesia • Biased mechanisms • Opioid receptor signaling and trafficking

Abbreviations

6′-GNTI	6′-Guanidinonaltrindole
ARM390	AR-M1000390, N,N-dietyl-4-(phenyl-piperidin-4-ylidenemethyl)-benzamide
BK	Bradykinin
BNTX	7-Benzylidenenaltrexone
CTAP	D-Phe-Cys-Tyr-D-Trp-Arg-Thr-Pen-Thr-NH_2
CTOP	H-D-Phe-Cys-Tyr-D-Trp-Orn-Thr-Pen-Thr-NH_2
DADLE	[D-Ala^2, D-Leu^5]enkephalin

K.-J. Chang, Ph.D. (✉) • S. Yi • Y. Shen, Ph.D.
Faculty of Life Science and Technology, Kunming University of Science and Technology, Kunming, Yunnan 650500, China
e-mail: kwenjenchang@yahoo.com

Y. Xia (ed.), *Neural Functions of the Delta-Opioid Receptor*,
DOI 10.1007/978-3-319-25495-1_2

DAMGO	[D-Ala^2, N-MePhe^4, Gly^5-ol]enkephalin
DMEM	Dulbecco's modified Eagle's medium
DOR	Delta opioid receptor
DPDPE	[D-Pen^2, D-Pen^5]-enkephalin
DPI-3290	[(+)-3-((α-R)-α-((2S,5R)-4-Allyl-2,5-dimethyl-1-piperazinyl)-3-hydroxybenzyl)-N-(3-fluorophenyl)-N-methylbenzamide
ERK1/2	Extracellular signal-regulated protein kinases 1 and 2
GPCR	G-protein-coupled receptors
GRK	GPCR receptor kinases
HEK	Human embryonic kidney
KDN21	Delta-kappa opioid antagonist bivalent ligand (K, KOR pharmacophore; D, DOR pharmacophore; N, antagonist; 21, number of atoms in the spacer)
KOR	Kappa opioid receptor
MAPK	Mitogen-activated protein kinase
MOR	mu opioid receptor
nor-BNI	Norbinaltorphimine
NTB	Naltriben
NTI	Naltrindole
ORL-1	Opioid receptor-like 1 receptor
PGE_2	Prostaglandin E_2
SNC80	(+)[α-Rα-((2S,5R)-4-allyl-2,5-dimethyl-1-piperazinyl)-3-methoxybenzyl]-N,N-diethylbenzamide
TG	Trigeminal ganglion
TIPP(ψ)	OH-Tyr-Tic(CH_2NH)-Phe-Phe-H
U69593	(+)-(5,7,8)-N-methyl-N-[7-(1-pyrrolidinyl)-1-oxaspiro[4.5]dec-8-yl] benzeneacetamide

1 Introduction

The book titled "The Delta Receptor" edited by Chang, Porreca and Woods was published in 2004 (Chang et al. 2004). This book was the first one devoted to comprehensively cover all scientific accomplishments during the three decades from year 1973 to 2003 by many distinguished scientists who had made contributions to our understanding of delta opioid receptor (DOR) biology. Molecular, biochemical, medicinal chemistry, pharmacological and medical aspects of the delta opioid receptor biology were thoroughly reviewed by these outstanding contributors to the book.

All four types of opioid receptor, mu opioid receptor (MOR), DOR, kappa opioid receptor (KOR) and opioid receptor-like 1 receptor (ORL-1) belong to the family of 7-transmembrane spanning (7-TM)/G-protein coupled receptors (GPCR). Recently, crystalline structures of all four opioid receptors have been elucidated for the antagonist bound stabilized conformation (Granier et al. 2012; Wu et al. 2012; Thompson et al. 2012; Manglik et al. 2012). Their corresponding endogenous peptide ligands and genes were also described two decades ago. Those are β-endorphin, enkephalins, dynorphins and nociceptin and their pro- and precursor polypeptides. Selective agonists and antagonists for all four opioid receptors have well been discovered and extensively studied both in vivo pharmacology and in vitro cellular/molecular biology (Chang et al. 2004). Even though these enormous improvements of our understanding of the biology of opioid receptor and endorphins, MOR agonists such as morphine, oxycodone, fentanyl are still mainstays for the treatment of moderate-to-severe pain including pre- and post-surgery, diagnosis procedures, chronic pain and cancer pain. Synthetic methadone and semisynthetic buprenorphine are used for the replacement therapy for heroin addicts.

Among these four opioid receptors, MOR is clearly the one primary responsible for mediating all analgesic activities and nearly all known unwanted side effects such as respiratory depression, tolerance and dependence, addiction, nausea and vomiting, as well as the constipation and urinary retention after chronic uses of opioid analgesics including morphine and oxycodone. Due to the aversive effects such as dysphoria and hallucination in human (Pfeiffer et al. 1986), KOR agonists have not been developed for any clinical utility. KOR agonists are often producing effects opposite to that produced by MOR agonists (Pan 1998). ORL-1 agonists are in general anti-analgesic. DOR agonists are attractive for their lacking of addictive properties (Negus 2004), but were not usefully for acute analgesic development due to their weak acute analgesic activity (Chang et al. 1993; Wild et al. 1993; Dykstra et al. 1993) and convulsion activity observed in rodents (Comer et al. 1993a, b).

During the last decade, many studies have demonstrated modulating functions of DOR on the MOR mediated analgesia, respiratory depression, as well as the development of tolerance and dependence. Therapeutic application potentials of DOR agonists have progressed to the areas including chronic inflammatory pain such as hyperalgesia, locomotor functions in improving Parkinson's disease (Hille et al. 2001; Mabrouk et al. 2008) neuronal protection against ischemia insults such as stroke (see review by He et al. 2013), anti-depression activities (Jutkiewicz and Woods 2004; Jutkiewicz 2006) cardioprotection against heart attack (Gross et al. 2004), and overactive bladder or urinary incontinence (Holt et al. 2005). These areas will be discussed in various chapters of the monograph by others. Many outstanding reviews were recently published to cover different aspects of opioid receptors, signaling, trafficking, opioid receptor heteromerization, biased mechanisms etc. (Law et al. 2013; Stockton and Devi 2012; Dang and Christie 2012; Al-Hasani and Bruchas 2011; van Rijn et al. 2010). This chapter will focus on recently developed general concepts regarding to pharmacology, cellular/molecular mechanisms of DOR and its ligands.

2 Overcome the Convulsive Effect of DOR Agonists

DOR agonists produce a unique pharmacological profile that results in activities differing from those of mu- and kappa-opioid agonists. DOR activation appears beneficial in several disease conditions that include chronic pain, emotional disorders such as depression and perhaps some aspects of dependence (Lee et al. 1993; Filliol et al. 2000), and to counteract the respiratory depression induced by mu-agonists (Su et al. 1998). However, delta opioid receptor activation may lead to convulsions (Jutkiewicz and Woods 2004). Delta opioid agonist-induced seizures are mild, closer to a model of absence seizures rather than grand mal, and are prevented by drug midazolam that attenuates absence seizure in humans. Convulsive effects were considered a hindrance to the development for delta drugs. Interestingly, however, convulsive effects of SNC80 can be separated from other effects (Jutkiewicz and Woods 2004; Jutkiewicz 2006) by slow intravenous (i.v.) infusion SNC80 and/or by subcutaneous (s.c.) administration of analogues with higher hydrophobicity such as DPI287 as compared to SNC80. In addition, tolerance has been shown to develop rapidly to the convulsive effects of SNC80 (Comer et al. 1993a) but not to its anti-depressant activity (Pradhan et al. 2010). Although there are ways to minimize or even eliminate convulsions, a nonpeptidic delta-opioid agonist without any convulsive properties would be preferred. Many of the recently developed delta opioid agonists showed no convulsive effects at therapeutic doses (Hudzik et al. 2011; Nozaki et al. 2012; Pradhan et al. 2010). Therefore, the pro-convulsion and convulsion activity of delta opioid agonists appears to be avoidable or surmountable. The DOR system has an interesting behavioral profile distinct from other opioids.

The mechanism of lacking convulsion effects of these new compounds has not been investigated. It has recently been suggested that β-arrestins may transduce the signal to produce adverse side effects such as the seizure effects of DOR agonists. G-proteins biased DOR agonists that avoid β-arrestins signaling pathway may not produce convulsive effects. This remains to be investigated.

3 Biased Mechanisms

The recent studies (see reviews by Raehal et al. 2011) have indicated that the nature of MOR signaling, regulation and responses are functions of both the agonist acting at the receptor and the cellular environment. Biochemical and electrophysiological studies demonstrated that ligand-directed signaling, regulation and responses of the opioid receptor could not be sole explained by the intrinsic efficacy of the agonist alone. Instead, these studies demonstrated that opioid ligands might biased the receptor toward or against interactions with specific intracellular proteins, such as G-proteins, GPCR receptor kinases (GRK), and β-arrestins, and these different interactions resulted in the differential signaling, phosphorylation, internalization, trafficking, receptor degradation and desensitization of the MOR. The *in vivo* studies suggest that ligand-directed signaling may have major implications for opioid receptor-mediated physiological responses, including the development of analgesic tolerance and producing adverse side effects. Although the regulatory events leading to tolerance may be complex, scaffolding molecules such as β-arrestins, as well as other adaptors involved in the phosphorylation and dephosphorylation of the receptors, ubiquitination and degradation of the receptors and subsequence receptor down-regulation are certain to affect the ability to restore sensitivity to further agonist activation. Although many of the studies published have focused on processes involved in MOR desensitization, it may be that facilitation of receptor resensitization, whether via an internalization process or through dephosphorylation at the membrane, is a key means to reverse or prevent the development of tolerance.

This biased mechanism was very recently explored for delta-agonists, SNC80 and ARM390 *in vivo* in DOR-eGFP knock-in mice (Pradhan et al. 2009, 2010) for their long-term consequences of ligand-biased trafficking of DOR, at both cellular and behavioral level. SNC80 and ARM390 have similar receptor binding and analgesic properties, SNC80 has high and ARM390 has low DOR-internalization potencies. A single SNC80 but not ARM390 administration triggered acute desensitization of analgesic response in mice. However, daily injections of either compounds over 5 days produced full analgesic tolerance. SNC80-tolerant mice showed widespread receptor down regulation, and tolerance to analgesic, locomotor and anxiolytic effects of the agonist. Hence, internalization-dependent tolerance developed, as a result of generalized receptor degradation. In contrast, ARM390-tolerant mice showed intact receptor expression, but DOR-coupling to Ca^{2+} channels was abolished in dorsal root ganglia, concomitantly, tolerance developed for agonist-induced analgesia, but not locomotor or anxiolytic responses. Therefore, internalization-independent tolerance was produced by anatomically restricted adaptation leading to pain-specific tolerance. This study revealed SNC80 and ARM390 produced chronic analgesic tolerance of different mechanisms. ARM390 retained its locomotor and anxiolytic activities despite of analgesic tolerance developed. DOR agonists like ARM390 might be suitable for the development for therapeutic agents of treating depression and Parkinson's disease as mentioned above because of its lacking tolerance development to these beneficial activities.

Our understanding of how different agonists induce distinct trafficking and signaling events and relating these events to the pharmacological effects of the drugs will be instrumental to the development of new therapeutics with enhanced desirable properties but limited adverse side effects. It will be essential to understand how the receptor is regulated in neurons that directly control each of the responses to DOR ligands. There is also suggestion that activation of β-arrestins pathway may be responsible for convulsive effects, therefore DOR agonists biased for the G-proteins signaling pathway may be able to avoid the convulsive activity (Comer et al. 1993a).

4 Opioid Receptor Heteromers: Delta-Mu Receptor Dimers or Tetramers

As drug targets, heteromeric receptors offer an additional level of selectivity and functionality or biased, because of allosteric interactions between protomers of the complexes. Mountain evidences have been generated during the last decade to indicate the unusualness of the actions of DOR ligands in *in vivo* pharmacology and *in vitro* cellular biochemistry and molecular biology. These data are often contradictory, complex and very difficult to comprehend. In this section, we will attempt to summarize these information according to *in vivo* pharmacological studies and *in vitro* studies with various opioid receptors expressed or co-expressed in cultured HEK293 or cos-7 cells or neuronal cell lines, and hopefully to integrate them for a unified hypothesis.

4.1. *In vivo* Pharmacology Data
4.2. Cellular and molecular mechanisms.
4.3. Hypothesis: To reasonably interpret above complicated results *in vivo* pharmacology and *in vitro* cellular/molecular studies, we propose the hypothesis as follows:

   4.1.1. SNC80 the delta-agonist potentiated and NTI the delta-antagonist antagonized acute analgesic effects of mu-agonists: morphine, fentanyl and methadone in rhesus monkeys (Stevenson et al. 2003; Yekkirala et al. 2010, 2012).
   4.1.2. SNC80 facilitated and NTI inhibited acute and chronic morphine tolerance in mice (Roy et al. 2005).
   4.1.3. SNC80 is an agonist of delta-mu receptor heteromers. The antinociception of SNC80 intrathecal (i.t.) in mice could be reduced by knocking out either mu-receptors or delta-receptors (Metcalf et al. 2012; Yekkirala et al. 2012).
   4.1.4. Both the potentiation and the inhibition effects of NTI on morphine antinociception have been reported by i.t. administration in mice. The reduction of delta-agonists, deltorphin I, SNC80, leu-enkephalin on morphine antinociception was only demonstrated after 30 min i.t. pretreatment in mice (Yekkirala et al. 2012; He et al. 2011; Gomes et al. 2004).
   4.1.5. The antinociception mediated by a delta-agonist DPDPE in spinal cord of mice could be blocked by NTI, a delta-antagonist, as well as CTAP, a mu-antagonist. DPDPE had a potentiating effect on DAMGO, a mu-agonist. This potentiation was lost in animals made tolerant to systemic morphine (He and Lee 1998). MOR knock-out mice also lost the DPDPE mediated analgesia (Matthes et al. 1996; Sora et al. 1997).
   4.1.6. Ultra low doses of delta and mu antagonists prevented the development of morphine tolerance, and reverse the tolerance developed after

morphine treatment in mice and rats (Powell et al. 2002; Abul-Husn et al. 2007).

4.1.7. Delta-agonists and antagonists reversed the respiratory depression induced by mu-agonists in rats. This inhibitory effect of delta-ligands could be antagonized by DPI2505 while it had no effect on respiratory depression induced by mu-agonist (Su et al. 1998).

4.1.8. In mice, BW373U86 produced convulsive side effects and fentanyl produced muscle rigidity, the Straub tail effects. The co-administration of BW373U86 and fentanyl, convulsive effects and Straub tail effects were reduced through the interactions between MOR and DOR in vivo (O'Neill et al. 1997).

4.1.9. The mixed mu-agonist/delta-agonist DPI3290 produced potent analgesic activity with reduced respiratory depression and physical dependence as compared with morphine in rodents but developed analgesic tolerance.(Gengo et al. 2003a, b; Gengo and Chang 2004).

4.1.10. Studies of co-administration of delta-antagonist with mu-agonist in rodents (Abdelhamid et al. 1991), DOR antisense knockdown (Sanchez-Blazquez et al. 1997), and DOR and preproenkephalin knockout mice (Zhu et al. 1999; Nitsche et al. 2002), had all indicated the possible role of DOR in development of morphine tolerance and dependence in the context of DOR/MOR heteromer formation in the same neuron.

4.1.11. Bivalent ligands with mu-agonist/delta-antagonist activities (Daniels et al. 2005; Lenard et al. 2007) showed potent analgesia without tolerance and dependence.

4.2.1. DOR and MOR form dimers or tetramers, or higher oligomers when co-expressed in a cell (George et al. 2000; Wang et al. 2005; Ferre et al. 2009; Rozenfeld and Devi 2010; Golebiewska et al. 2011). The evidences also suggest that symmetric 2delta-2mu tetramer may be the constitutive form of delta-mu receptor complex (Gomes et al. 2000).

4.2.2. Heteromerization of MOR and DOR leads to a constitutive recruitment of β-arrestin2 to the complex resulting in changes in the spatio-temporal regulation on DAMGO induced ERK1/2 signaling. The binding of both mu-antagonist CTOP at low concentration and delta-ligands deltorphin II or delta-antagonist TIPP(ψ) and mu-agonist DMAGO together to the heteromer expressed cells induced the dissociation of β-arrestin2 from the complexes and restored the normal signaling pattern of ERK1/2 phorphorylation (Rozenfeld and Devi 2007).

4.2.3. Mu-agonists, morphine, DAMGO, Fentanyl and methadone induced more robust responses (measured by the increase of intracellular Ca^{++} release) and [^{35}S]GTPrS binding as compared to cells individually expressing DOR, MOR or opioid mu-kappa receptors dimers (Yekkirala et al. 2010 and 2012).

4.2.4. SNC80 stimulated more robust responses in cells co-expressing delta-mu receptors as compared to cells expressing receptors of delta-, mu-, kappa, kappa-delta or mu-kappa (Metcalf et al. 2012).

4.2.5. Very low concentrations (~10 pM) of delta-ligands (both agonists and antagonists) increased the binding (B_{max} increase) of a mu-agonist [^{3}H] DAMGO in cells co-expressing DOR and MOR. Very low concentration (~1 pM) of mu-ligands (both agonists and antagonists) also increased the binding of a delta agonist [^{3}H]delt II (B_{max} increase) to the same cells (Gomes et al. 2000, 2004, 2011).

4.2.6. Both mu-agonists and delta-agonists increased the signaling in cells co-expressing DOR and MOR (Fan et al. 2005; O'Dowd et al. 2007, 2012).

4.2.7. The opioid delta-mu receptor heteromers have an anti-analgesic effect on morphine activities (Milan-Lobo and Whistler 2011; Milan-Lobo et al. 2013).

4.2.8. In cells co-expressed DOR and MOR, both delta-agonists and mu-agonist independently induced co-internalization of DOR and MOR, either one induced internalization could be blocked by either delta-antagonist NTI or mu-antagonist CTOP (Kabli et al. 2010; Milan-Lobo and Whistler 2011; He et al. 2011). Morphine could not induce endocytosis of 2delta-2mu tetramers (Milan-Lobo and Whistler 2011).

4.2.9. Chronic treatment of morphine increased the immnuoactivities in central nervous system to a monoclonal antibody specific to opioid delta-mu receptor heteromers (Gupta et al. 2010). These regions include cortex, medial nucleus of trapezoid body (MNTB) an auditory relay nucleus, the rostral ventral medulla (RVM), as well as in the hypothalamus, hippocampus, nucleus accumbens and ventral tegmental area which are know to be relevant to opioid addiction.

4.2.10. The delta-mu heteromers have shift the G-protein coupling from Gi/o to Gz and opioid mu-delta heterooliger-Gz signaling complexes were preassembled during the de novo biosynthesis in ER and/or Golgi and trafficked to the plasma membrane by a Golgi chaperone protein RTP4 (Hasbi et al. 2007; Decaillot et al. 2008).

4.3.1. Opioid delta-mu receptor heteromers formed constitutively as tetramers, 2delta-2mu receptor tetramers.

4.3.2. There are allosteric interactions between and among protomers of the 2delta-2mu tetramers.

4.3.3. The tetramer initially has one high affinity delta-binding site and one mu-binding site with affinities similar to individual expressed DOR and MOR. The other mu- and delta-binding sites are masked with very low affinity in the tetramers and these sites can be induced upon the binding of mu-agonists or delta-agonists to form extremely high affinity sites for delta-ligands or mu-ligands respectively. The binding of delta- or mu-ligands to these induced extremely high affinity sites can further trigger the allosteric interactions between DOR and MOR to uncover high binding affinity and increased signaling. Antagonists behave similarly to agonists in these uniquely induced extremely high binding pockets. These traditional antagonists are tetramer biased agonists and thus renamed as delta- or mu-antagonist/2delta-2mu agonist. Some agonists will perform similarly.

4.3.4. Many delta-antagonists and mu-antagonists acts as agonists at 2delta-2mu tetramers at low doses, and as antagonists at high doses in *in vivo* pharmacology.

4.3.5. The 2delta-2mu tetramers alter their interactions with cytoplasmic coupled proteins such as G-proteins and β-arrestins, thus may alter the signaling and trafficking.

4.3.6. The 2delta-2mu tetramers act as a unit to respond to delta-agonists, mu-agonist, or mixed dual delta-mu agonists in term of activation, signaling, internalization and degradation, recycling and tolerance. The outcome is dependent upon the type of agonists used (agonist biased).

4.3.7. The responses initiated by the tetramers will be greater or stronger than individual receptor expressed alone, and can be antagonized by either traditional delta-antagonists or mu-antagonists competitively or allosterically.

In summary (Fig. 2.1), in resting normal condition, 2delta-2mu tetramers are formed complexes with GDP-bound Gαβr and β-arrestin 2. Upon the activation by a mu- or delta-agonist and a delta- or mu-antagonist/2delta-2mu agonist and the complexes are dissociated into four proteins, ligand bound 2delta-2mu tetramers staying in the membrane, and activated β-arrestin 2, Gβr subunit, and GTP-bound Gαi/o/z which procced to initiate signaling as indicated in the figure such as activation of K^+ channels and inhibition of Ca^{2+} channels that are believed to be the mechanism of analgesia and other opioid actions, and inhibition of adenylyl cyclase, phospholipase C activation, MAPK activation and downstream pathways that may be responsible for side effects and/or other long-term effects such as dependence and addiction. Agonist bound 2delta-2mu tetramers are subjected to the phosporylation by GPCR receptor kinases (GRK) and lead to β-arrestin 1/2 binding, followed by endocytosis through clathrin coated vesicles. When the receptor bound ligands are delta-agonists, fate of teramers is dephosporylated, ubiquitinated, degraded and eventually receptor down regulated and tolerance developed.

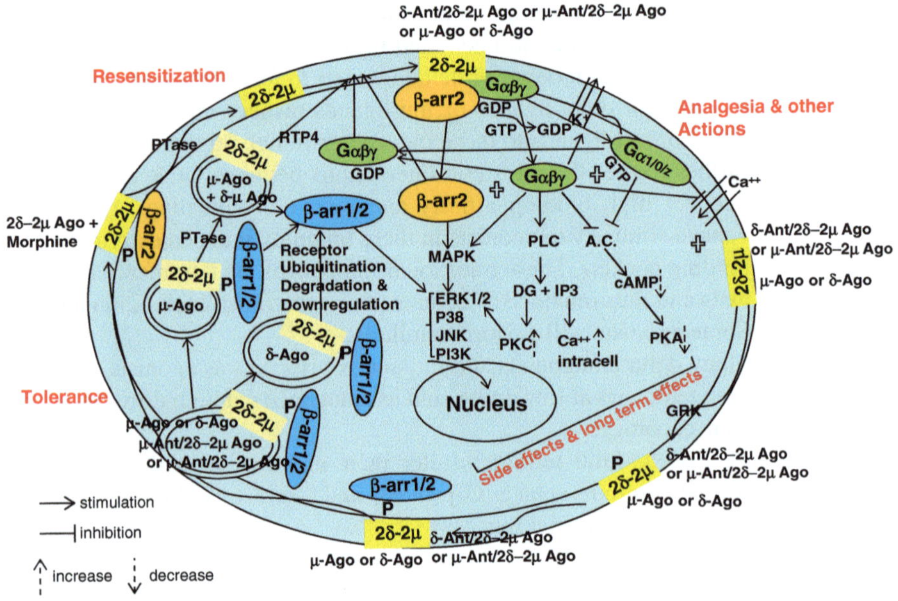

Fig. 2.1 Hypothetical opioid 2delta-2mu receptor tetramers model, intracelluar signaling and fate of tetramers stimulated by morphine, mu-agonists, delta-agonists, mixed mu-delta agonists, delta-antagonists/2delta-2mu agonists, and mu-antagonist/2delta-2mu agonist. *δ-ant/2δ-2 μ ago* delta-antagonist/2delta-2mu agonist, *μ-ant/2δ-2 μ ago* mu-antagonist/2delta-2mu agonist, *μ-ago* mu-agonist, *δ-ago* delta-agonist, *β-arrt2* β-arrestin2, *β-arr1/2* β-arrestin1/2, *A.C.* adenylyl cyclase, *PLC* phospholipase C, *MAPK* mitogen activation protein kinase, *IP3* inositol triphosphotate, *DG* diacyl glycerol, *ERK1/2* extracellular signal-regulated kinase ½, *p38* p38 protein kinase, *PTase* protein phosphotase, *JNK* JUN N-terminal kinase, *PI3K* phosphoinositide 3- kinase, *PKC* protein kinase C, *PKA* cAMP dependent protein kinase A, *P* phosphate, *GRK* GPCR receptor kinase, *RTP4* receptor transport protein 4. Xie et al. (2005)

When the receptor bound ligands are mu-agonists, the tetramers will be dephosphorylated, β-arrestin 1/2 dissociated and recycled back to plasma membrane perhaps through the transport by a receptor transport protein 4 (RTP4) and tetramers resensitized by the formtion of complex with other GDP-bound Gαβr and β-arrestin 2. However, when the tetramers bound ligands are morphine, the tetramers stay on the membrane and are not internalized. Tetramers remain desensitized until they are dephosphorylated, β-arrestin 2 dissociated and re-complexed with GDP-bound Gαβr and recruited β-arrestin 2 again.

5 Delta-Kappa Receptor Interactions: Opioid Delta-Kappa Receptor Dimers

There is also evidence that DOR can form heteromers with the KOR (Wang et al. 2005; Xie et al. 2005; Waldhoer et al. 2005). Evidences for delta-kappa heteromers are studies of DOR-KOR co-expressed in HEK293 cells with BRET (Bioluminescence Resonance Energy Transfer) studies, in peripheral sensory neurons included co-immunoprecipitation of DOR with KOR, a DOR-KOR heteromer selective antibody augmented the antinociceptive effect of DPDPE in vivo, and the delta-kappa heteromer agonist 6′-guanidinonaltrindole (6′-GNTI) inhibited PGE_2 stimulated adenylyl cyclase activity *in vitro* as well as PGE_2-stimulated thermal allodynia *in vivo* (Table 2.1) (Berg et al. 2012). Selective kappa-antagonists differentially altered the potency and/or efficacy of delta-agonists in primary cultures of adult rat peripheral sensory neurons and in a rat behavioral model of thermal allodynia (Berg et al. 2012). *In vitro*, the kappa-antagonist nor-binaltorphimine (nor-BNI) enhanced the potency of DPDPE, decreased the potency of [D-Ala^2,D-Leu^5]-enkephalin (DADLE), and decreased the potency and efficacy of SNC80 to inhibit prostaglandin E_2 (PGE_2)-stimulated adenylyl cyclase activity. *In vivo*, nor-BNI enhanced the effect of DPDPE and decreased the effect of SNC80 to inhibit PGE_2-stimulated thermal allodynia. In contrast to nor-BNI, the kappa-antagonist 5-guanidinonaltrindole (5′-GNTI) reduced the response of DPDPE both in cultured neurons and *in vivo*. Taken together, these data suggest that delta-kappa heteromers exist in rat primary sensory neurons and that kappa-antagonists can act as modulators of delta-agonist responses most likely through allosteric interactions between the protomers of the delta-kappa heteromers.

The opioid agonist ligand 6′-GNTI has the unique property of selectively activating only opioid delta-kappa receptor heterodimers but not homomers (Waldhoer et al. 2005). 6′-GNTI is a weak agonist in cells expressed KOR alone, and may be an antagonist in cells expressed DOR alone, but becomes a potent agonist when DOR and KOR are co-expressed in cells or neurons. This agonist effects could be blocked by either kappa-antagonist nor-BNI or delta-antagonist NTI. 6′-GNTI is an analgesic in mice, thereby demonstrating that opioid delta-kappa receptor heterodimers are indeed functionally relevant *in vivo* (Table 2.1). However, 6′-GNTI induces analgesia only when it is administered in the spinal cord but not in the brain, suggesting that the organization of heterodimers is tissue-specific only in spinal cord and peripheral neurons.

The profiles of interactions between KOR and DOR in the opioid delta-kappa receptor heteromers are different from the opioid 2delta-2mu receptor tetramers. The alterations of the receptor binding and responses are mainly in the affinities dependent upon the ligand examined and the relative potency of ligands (Table 2.2) (Xie et al. 2005). In the saturation binding of ^{3}H-labeled nor-BNI and NTI, the kappa-antagonist nor-BNI caused the left-shift of saturation binding curve of 3H-NTI with a decrease of K_d value of 27 folds, while the delta-antagonist NTI left-shifted the curve of ^{3}H-nor-BNI with a decrease of Kd value of 36 folds. These data suggest the allosteric interactions between DOR and KOR in the delta-kappa receptor dimers.

Table 2.1 Summary of Interaction of 6′-GNTI, Bivalent Ligand KDN21 and delta-, kappa-agonists and antagonist with heterodimeric delta-kappa opioid receptors in TG peripheral sensory neurons and Human Embryonic Kidney 293 Cells, and *in vivo* antinociception activities (Berg et al. 2012; Waldhoer et al. 2005)

Peripheral sensory			Rat	HEK293, δ-κ expressed		Mice
Neurons	BK primed		BK primed,			Tail flick assay
	PGE2 stimulated		PGE2 stimulated	Intracellular Ca^{++} release		Antinociception
	cAMP increase		Anti-allodynia			
Ligands	**EC50**	**Inh. Emax** (%)	**PW latency**	**EC50**	**Emax increase**	**ED50 i.t. only**
6′-GNTI	2 nM	67	Increase	40 nM	~400 RLU	0.45 nmoles
+ nor-BNI, 3 nM	Inhibition		Inhibition	Inh. IC50 ∼ 10-8 M		Inh. at 0.62 nmoles
+ NTI, 20 nM	Inhibition		Inhibition	Inh. IC50 ∼ 10-7 M		Inh. at 0.58 nmoles
+ KDN-21						Inh. at 0.067 nmoles
U69593				0.82 nM	~570 RLU	
+ nor-BNI				Inh. IC50 ∼ 10-8 M		
+ NTI				No effect		
DPDPE	0.5 nM	64	Increase	53 nM	~440 RLU	
+ nor-BNI, 3 nM	0.06 nM	63	Increase			
+5′-GNTI, 4 nM	Inhibition		Decrease			
δ-κ ab			Increase			
DADLE	0.06 nM	62				
+nor-BNI, 3 nM	1.8 nM	61				
SNC80	1.3 nM	63				
+nor-BNI, 3 nM	47 nM	30				
DAMGO	10 nM	72				
+nor-BNI, 3 nM	8.09 nM	66				

Table 2.2 Summary of the results on the binding assay with ^{3}H labeled nor-BNI and NTI to delta-kappa receptor co-expressed in HEK293 cells and MAPK activation by delta and kappa agonists, and inhibitory effects of KDN-21

HEK293 δ-κ receptor co-expressed				
	3H-nor-BNI		3H-NTI	
Saturation binding curves Kd	2 nM		0.1 nM	
+ norBNI, 10 nM			Left shift Kd decreases 27 folds	
+ NTI, 10 nM	Left shift Kd decreases 36 folds			
Competition curves				
KDN-21, Ki	0.16 nM		0.06 nM	
+ Bremazocine κ2. 100 nM			Right shift Ki increases 25 folds	
+ U69593 κ1, 100 nM			No effect	
+ BNTX δ1,100 nM	Right shift Ki increases 18 folds			
+ NBT. δ2, 100 nM	No effect			
MAPK activation				
Agonists	DPDPE δ1	Delt II δ2	U69593 κ1	Bremazocine κ2
Activation +	+	+	+	+
+Antagonist KDN-21, 100 nM				
Inhibition –	–	+	+	–

The evidences from the studies of competition binding and inhibition of MAPK activation with various delta- and kappa-ligands and KDN-21 a delta-kappa heteromer antagonist also suggest that this delta-kappa heteromers may be the previously proposed phenotypic delta1 and kappa2 receptors (Table 2.2) (Xie et al. 2005). The binding data and intrathecal studies have indicated that KDN21 is an antagonist selectively targeted to the delta1 and kappa2 phenotypic opioid receptors (Bhushan et al. 2004). KDN-21 could competitively inhibit the binding of 3H-nor-BNI (K_i of 0.16 nM) as well as the binding of ^{3}H-NTI (K_i of 0.06 nM). Furthermore, the KDN-21 inhibition curve against ^{3}H-NTI was shifted to right by the bremazocine but not by U69593. Similarly, KDN-21 inhibition curve against ^{3}H-nor-BNI was shifted to the right by BNTX a delta1-antagonist, but not NBT a delta2-antagonist.

Using different delta- and kappa-selective agonists to activate the phosphorylation of ERK1/2 and then observed the antagonism of KDN21 upon such activation. In DOR and KOR co-expressed cells, DPDPE (delta1 agonist), deltorphin II (delta2-agonist), U69593 (kappa1-agonist), and bremazocine (kappa2-agonist) could stimulate the phosphorylation of ERK1/2 (Table 2.2). 100 nM KDN21 could inhibit the activation of ERK1/2 induced by 100 nM DPDPE or 100 nM bremazocine, however, no inhibition on the activation induced by deltorphin II and U69593, providing direct evidence for the regulation of KDN21 on the function of delta1 and kappa2 opioid phenotypes.

Unlike the opioid 2delta-2mu tetramers, these binding studies and pharmacological studies can be simply interpreted by a opioid delta-kappa receptor heterodimer. The cell surface cross-link proteins, followed by KOR immunoprecipitation, 10 % SDS-PAGE and western blotting of DOR and KOR bands of 120 KD delta-kappa dimers was identified to support, indeed, that delta-kappa heteromers are dimer proteins of DOR and KOR (Berg et al. 2012).

Opioid receptors are expressed in peripheral primary sensory neurons that transduce pain information (nociceptors). It is noteworthy that peripherally restricted opioids generally do not elicit an analgesic response when administered to normal tissue, but can produce antinociception when administered to injured or inflamed tissue including a brief pre-treatment (priming) with bradykinin, arachidonic acid, protease-activated receptor-2 agonists, or direct activators of protein kinase C (PKC) are capable of inducing the functional competence of the opioid receptor system in cultures of primary sensory neurons *in vitro* (Rowan et al. 2009; see review by Bie and Pan 2007; Cahill et al. 2007). This finding suggests that some stimulus from the inflamed tissue interacts with opioid receptor systems in nociceptors to make them functionally competent to inhibit nociceptor signaling.

6′-GNTI was recently reported to be a potent kappa-agonist for G protein activation and fail to recruit arrestin and activate arrestin pathway. Thus, 6′-GNTI is a G-protein biased kappa-ligand (Rives et al. 2012). However, the co-expression of DOR and KOR in cells was previously reported to enhance the efficacy of 6′-GNTI for activation a chimeric G protein of Δ6-Gqi4-myc protein as described above (Waldhoer et al. 2005), Rives et al. (2012) did not observer an increase in the efficacy or potency of 6′-GNTI for $G\alpha o\beta\gamma$ activation or cAMP inhibition by KOR upon co-expression of DOR. It will be interesting to find out the reasons for the different responses in the opioid delta-kappa receptor heteromers. Finally, with the profiles of 6′-GNTI and targeting opioid delta-kappa heterodimers could provide an approach toward the design of analgesic drugs with reduced side effects.

6 New Ligands

Early mechanism studies revealed that small molecule delta-agonists SNC80 and ARM390 performed differently in terms of inducing receptor internalization (Marie et al. 2003). In antinociception experiments using knock-in mice expressing functional fluorescent delta-opioid receptors, Pradhan et al. (2009) discovered that although both of the agonists have antinociceptive effects initially, subsequent treatment of SNC80 showed no response due to delta-receptor internalization and degradation, while second dose of low internalizing ARM390 retained full analgesic efficacy. They later elucidated that ARM390 induced analgesic tolerance through different molecular events than SNC80 did (Pradhan et al. 2010). ARM390 did not appear to induce tolerance to its locomotor and anxiolytic effects, therefore it will be useful for its therapeutic applications of anti-depression and anti-Parkinsonism. It is worth mentioning that three delta-agonists currently in clinical trials, AZD2327

(Hudzik et al. 2011), ADL5747 and ADL5859 (Nozaki et al. 2012), also belong to the G-protein biased delta-agonists similar to ARM390. This single dose induced analgesic tolerance by SNC80 is a reminiscence of early report of single episode of convulsive response to BW373 and SNC80 (Comer et al. 1993a, b; Jutkiewicz and Woods 2004). They may share the same signaling pathway, β-arreestin2 pathway.

Using delta-mu heteromer-targeting monoclonal antibodies, Gomes et al. (2013) established a high throughput screening method and identified an opioid delta-mu receptor heteromers biased agonist CYM51010, which demonstrated antinociceptive effects similar to morphine with less tolerance tendency. Allosteric modulation of opioid receptors is a relatively new concept. It has the potential of diminishing opioid adverse effects by switching downstream signaling. In an effort towards allosteric modulation of MOR. Burford et al. (2013) employed a high throughput β-arrestin-recruitment assay to identify both of positive modulators, BMS-986121 and BMS-986122, and silent allosteric MOR modulators.

7 Concluding Remarks

The major concern for the clinical development for delta-agonists is their potential to induce seizure in human. Recent studies have indicated that the convulsion or pro-convulsion potential is avoidable or surmountable, thus, the major hurdle for delta-agonist was overcome for their development. Several delta-agonists lacking convulsion activities in animals have entered into clinical trials without seizure activity. The utilities of delta-agonists as anti-Parkinson's disease and antidepressant agents are currently under pre-clinical and/or clinical development.

DOR plays a role of modulating function on MOR as well as KOR in the complexes of delta-mu or delta-kappa heteromers in spinal cord or peripheral neurons (Wang et al. 2010) as described above. The interests on mu- and kappa-agonists are to develop analgesics without or reduced adverse side effects. However, both mu- and kappa-agonists can produce very serious side effects. Most studies on delta-mu receptor heteromers have focused on antinociceptive and analgesia tolerant effects. Studies on delta-kappa heterodimers have focused on peripheral and/or spinal analgesic effects to avoid central side effects. It was not yet explored whether the delta-kappa heterodimers could also induce dysphoria and aversive responses *in vivo* produced by traditional kappa-agonists.

The concept of opioid receptor heteromers in the supraspinal level of central nervous system remains elusive and needs more efforts devoted to it. Most evidences from *in vivo* studies in supporting the concept of the 2delta-2mu heterotetramers were also from spinal cord & dorsal root ganglia studies. In fact, there were reports of synergism between DOR and MOR in supraspinal studies on nucleus raphe magnus (NRM) (Zhang and Pan 2010) and rostral ventromedial medulla (RVM) (Sykes et al. 2007) by local microinjections of delta- and mu-agonists. However, the relationship of the observed synergism to delta-mu heteromers is currently unclear. The works reported by Su et al. (1998) indicated that delta-agonists as well as antagonists by intracelebral ventrical (i.c.v.) administration readily and potently reversed the respiratory depression induced by mu-agonists, suggesting that supraspinal 2delta-2mu tetramers might mediate the effects of delta ligands to counteract the mu-agonist induced respiratory depression.

Chronic treatment of morphine increased opioid delta-mu receptor heteromers (Gupta et al. 2010) in the rostral ventral medulla (RVM), and nucleus accumbens (NAc) and ventral tegmental area (VTA) and other brain regions. RVM is a brain stem nucleus associated with opioid induced respiratory depression (Koo and Eikemann 2011). The NAc and VTA are known to be regions relevant to opioid addiction (Robison and Nestler 2011). The role of opioid receptor heteromers in supraspinal level of brain is likely the focus for the future studies. Selective agonists and antagonist for 2delta-2mu tetramers and delta-kappa dimers will be extremely usefully for future studies and to explore clinical potentials.

Acknowledgement Financial support is obtained from Kunming University of Science and Technology, and Natural Science Foundation of China (81173029).

References

Abdelhamid EE, Sultana M, Portoghese PS, Takemori AE (1991) Selective blockage of delta opioid receptors prevents the development of morphine tolerance and dependence in mice. J Pharmacol Exp Ther 258:299–303

Abul-Husn NS, Sutak M, Milne B, Jhamandas K (2007) Augmentation of spinal morphine analgesia and inhibition of tolerance by low doses of mu- and delta-opioid receptor antagonists. Br J Pharmacol 151:877–887

Al-Hasani R, Bruchas MR (2011) Molecular mechanisms of opioid receptor-dependent signaling and behavior. Anesthesiology 115:1363–1381

Berg KA, Rowan MP, Gupta A, Sanchez TA, Silva M, Gomes I, McGuire BA, Portoghese PS, Hargreaves KM, Devi LA, Clarke WP (2012) Allosteric interactions between delta and kappa opioid receptors in peripheral sensory neurons. Mol Pharmacol 81:264–272

Bhushan RG, Sharma SK, Xie Z, Daniels DJ, Portoghese PS (2004) A bivalent ligand (KDN-21) reveals spinal delta and kappa opioid receptors are organized as heterodimers that give rise to delta(1) and kappa(2) phenotypes. Selective targeting of delta-kappa heterodimers. J Med Chem 47:2969–2972

Bie B, Pan ZZ (2007) Trafficking of central opioid receptors and descending pain inhibition. Mol Pain 3:37

Burford NT, Clark MJ, Wehrman TS, Gerritz SW, Banks M, O'Connell J, Traynor JR, Alt A (2013) Discovery of positive allosteric modulators and silent allosteric modulators of the mu-opioid receptor. Proc Natl Acad Sci U S A 110:10830–10835

Cahill CM, Holdridge SV, Morinville A (2007) Trafficking of delta-opioid receptors and other G-protein-coupled receptors: implications for pain and analgesia. Trends Pharmacol Sci 28:23–31

Chang KJ, Rigdon GC, Howard JL, McNutt RW (1993) A novel, potent and selective nonpeptidic delta opioid receptor agonist BW373U86. J Pharmacol Exp Ther 267:852–857

Chang KJ, Porreca F, Woods JH (2004) The delta receptor. Marcel Dekker, New York

Comer SD, Hoenicke EM, Sable AI, McNutt RW, Chang KJ, De Costa BR, Mosberg HI, Woods JH (1993a) Convulsive effects of systemic administration of the delta opioid agonist BW373U86 in mice. J Pharmacol Exp Ther 267:888–895

Comer SD, McNutt RW, Chang KJ, De Costa BR, Mosberg HI, Woods JH (1993b) Discriminative stimulus effects of BW373U86: a nonpeptide ligand with selectivity for delta opioid receptors. J Pharmacol Exp Ther 267:866–874

Dang VC, Christie MJ (2012) Mechanisms of rapid opioid receptor desensitization, resensitization and tolerance in brain neurons. Br J Pharmacol 165:1704–1716

Daniels DJ, Lenard NR, Etienne CL, Law PY, Roerig SC, Portoghese PS (2005) Opioid-induced tolerance and dependence in mice is modulated by the distance between pharmacophores in a bivalent ligand series. Proc Natl Acad Sci U S A 102:19208–19213

Decaillot FM, Rozenfeld R, Gupta A, Devi LA (2008) Cell surface targeting of mu-delta opioid receptor heterodimers by RTP4. Proc Natl Acad Sci U S A 105:16045–16050

Dykstra LA, Schoenbaum GM, Yarbrough J, McNutt R, Chang KJ (1993) A novel delta opioid agonist, BW373U86, in squirrel monkeys responding under a schedule of shock titration. J Pharmacol Exp Ther 267:875–882

Fan T, Varghese G, Nguyen T, Tse R, O'Dowd BF, George SR (2005) A role for the distal carboxyl tails in generating the novel pharmacology and G protein activation profile of mu and delta opioid receptor hetero-oligomers. J Biol Chem 280:38478–38488

Ferre S, Baler R, Bouvier M, Caron MG, Devi LA, Durroux T, Fuxe K, George SR, Javitch JA, Lohse MJ, Mackie K, Milligan G, Pfleger KD, Pin JP, Volkow ND, Waldhoer M, Woods AS, Franco R (2009) Building a new conceptual framework for receptor heteromers. Nat Chem Biol 5:131–134

Filliol D, Ghozland S, Chluba J, Martin M, Matthes HW, Simonin F, Befort K, Gaveriaux-Ruff C, Dierich A, LeMeur M, Valverde O, Maldonado R, Kieffer BL (2000) Mice deficient for delta- and mu-opioid receptors exhibit opposing alterations of emotional responses. Nat Genet 25:195–200

Gengo PJ, Chang KJ (2004) Mixed opioid receptor agonists as a new class of agents for the treatment of moderate to severe pain. In: Chang KJ, Porreca F, Woods JH (eds) The delta receptor. Marcel Dekker, New York, pp 231–244

Gengo PJ, Pettit HO, O'Neill SJ, Su YF, McNutt R, Chang KJ (2003a) DPI-3290 [(+)-3-((alpha-R)-alpha-((2S,5R)-4-Allyl-2,5-dimethyl-1-piperazinyl)-3-hydroxyben zyl)-N-(3-fluorophenyl)-N-methylbenzamide]. II A mixed opioid agonist with potent antinociceptive activity and limited effects on respiratory function. J Pharmacol Exp Ther 307:1227–1233

Gengo PJ, Pettit HO, O'Neill SJ, Wei K, McNutt R, Bishop MJ, Chang KJ (2003b) DPI-3290 [(+)-3-((alpha-R)-alpha-((2S,5R)-4-allyl-2,5-dimethyl-1-piperazinyl)-3-hydroxyben zyl)-N-(3-fluorophenyl)-N-methylbenzamide]. I A mixed opioid agonist with potent antinociceptive activity. J Pharmacol Exp Ther 307:1221–1226

George SR, Fan T, Xie Z, Tse R, Tam V, Varghese G, O'Dowd BF (2000) Oligomerization of mu- and delta-opioid receptors. Generation of novel functional properties. J Biol Chem 275:26128–26135

Golebiewska U, Johnston JM, Devi L, Filizola M, Scarlata S (2011) Differential response to morphine of the oligomeric state of mu-opioid in the presence of delta-opioid receptors. Biochemistry 50:2829–2837

Gomes I, Jordan BA, Gupta A, Trapaidze N, Nagy V, Devi LA (2000) Heterodimerization of mu and delta opioid receptors: a role in opiate synergy. J Neurosci 20:RC110

Gomes I, Gupta A, Filipovska J, Szeto HH, Pintar JE, Devi LA (2004) A role for heterodimerization of mu and delta opiate receptors in enhancing morphine analgesia. Proc Natl Acad Sci U S A 101:5135–5139

Gomes I, Ijzerman AP, Ye K, Maillet EL, Devi LA (2011) G protein-coupled receptor heteromerization: a role in allosteric modulation of ligand binding. Mol Pharmacol 79:1044–1052

Gomes I, Fujita W, Gupta A, Saldanha SA, Negri A, Pinello CE, Eberhart C, Roberts E, Filizola M, Hodder P, Devi LA (2013) Identification of a mu-delta opioid receptor heteromer-biased agonist with antinociceptive activity. Proc Natl Acad Sci U S A 110:12072–12077

Granier S, Manglik A, Kruse AC, Kobilka TS, Thian FS, Weis WI, Kobilka BK (2012) Structure of the delta-opioid receptor bound to naltrindole. Nature 485:400–404

Gross GJ, Fryer RM, Patel HH, Schultz JEJ (2004) Cardioprotection and delta opioid receptors. In: Chang KJ, Porreca F, Woods JH (eds) The delta receptor. Marcel Dekker, New York, pp 355–372

Gupta A, Mulder J, Gomes I, Rozenfeld R, Bushlin I, Ong E, Lim M, Maillet E, Junek M, Cahill CM, Harkany T, Devi LA (2010) Increased abundance of opioid receptor heteromers after chronic morphine administration. Sci Signal 3(131):ra54

Hasbi A, Nguyen T, Fan T, Cheng R, Rashid A, Alijaniaram M, Rasenick MM, O'Dowd BF, George SR (2007) Trafficking of preassembled opioid mu-delta heterooligomer-Gz signaling complexes to the plasma membrane: coregulation by agonists. Biochemistry 46:12997–13009

He L, Lee NM (1998) Delta opioid receptor enhancement of mu opioid receptor-induced antinociception in spinal cord. J Pharmacol Exp Ther 285:1181–1186

He SQ, Zhang ZN, Guan JS, Liu HR, Zhao B, Wang HB, Li Q, Yang H, Luo J, Li ZY, Wang Q, Lu YJ, Bao L, Zhang X (2011) Facilitation of mu-opioid receptor activity by preventing delta-opioid receptor-mediated codegradation. Neuron 69:120–131

He X, Sandhu HK, Yang Y, Hua F, Belser N, Kim DH, Xia Y (2013) Neuroprotection against hypoxia/ischemia: delta-opioid receptor-mediated cellular/molecular events. Cell Mol Life Sci 70:2291–2303

Hille CJ, Fox SH, Maneuf YP, Crossman AR, Brotchie JM (2001) Antiparkinsonian action of a delta opioid agonist in rodent and primate models of Parkinson's disease. Exp Neurol 172:189–198

Holt JD, Watson MJ, Chang JP, O'Neill SJ, Wei K, Pendergast W, Gengo PJ, Chang KJ (2005) DPI-221 [4-((alpha-s)-alpha-((2s,5r)-2,5-dimethyl-4-(3-fluorobenzyl)-1-piperazinyl)benzyl)-N, N-diethylbenzamide]: a novel nonpeptide delta receptor agonist producing increased micturition interval in normal rats. J Pharmacol Exp Ther 315:601–608

Hudzik TJ, Maciag C, Smith MA, Caccese R, Pietras MR, Bui KH, Coupal M, Adam L, Payza K, Griffin A, Smagin G, Song D, Swedberg MD, Brown W (2011) Preclinical pharmacology of AZD2327: a highly selective agonist of the delta-opioid receptor. J Pharmacol Exp Ther 338:195–204

Jutkiewicz EM (2006) The antidepressant -like effects of delta-opioid receptor agonists. Mol Interv 6:162–169

Jutkiewicz EM, Woods JH (2004) Antidepressant-like effects of delta opioid receptor agonists. In: Chang KJ, Porreca F, Woods JH (eds) The delta receptor. Marcel Dekker, New York, pp 355–372

Kabli N, Martin N, Fan T, Nguyen T, Hasbi A, Balboni G, O'Dowd BF, George SR (2010) Agonists at the delta-opioid receptor modify the binding of micro-receptor agonists to the micro-delta receptor hetero-oligomer. Br J Pharmacol 161:1122–1136

Koo CY, Eikemann M (2011) Respiratory effects of opioid in perioperative medicine. Open Anesthesiol J 5(Suppl M-6):23–34

Law PY, Reggio PH, Loh HH (2013) Opioid receptors: toward separation of analgesic from undesirable effects. Trends Biochem Sci 38:275–282

Lee PH, McNutt RW, Chang KJ (1993) A nonpeptidic delta opioid receptor agonist, BW373U86, attenuates the development and expression of morphine abstinence precipitated by naloxone in rat. J Pharmacol Exp Ther 267:883–887

Lenard NR, Daniels DJ, Portoghese PS, Roerig SC (2007) Absence of conditioned place preference or reinstatement with bivalent ligands containing mu-opioid receptor agonist and delta-opioid receptor antagonist pharmacophores. Eur J Pharmacol 566:75–82

Mabrouk OS, Volta M, Marti M, Morari M (2008) Stimulation of delta opioid receptors located in substantia nigra reticulata but not globus pallidus or striatum restores motor activity in 6-hydroxydopamine lesioned rats: new insights into the role of delta receptors in parkinsonism. J Neurochem 107:1647–1659

Manglik A, Kruse AC, Kobilka TS, Thian FS, Mathiesen JM, Sunahara RK, Pardo L, Weis WI, Kobilka BK, Granier S (2012) Crystal structure of the micro-opioid receptor bound to a morphinan antagonist. Nature 485:321–326

Marie N, Landemore G, Debout C, Jauzac P, Allouche S (2003) Pharmacological characterization of AR-M1000390 at human delta opioid receptors. Life Sci 73:1691–1704

Matthes HW, Maldonado R, Simonin F, Valverde O, Slowe S, Kitchen I, Befort K, Dierich A, Le Meur M, Dolle P, Tzavara E, Hanoune J, Roques BP, Kieffer BL (1996) Loss of morphine-induced analgesia, reward effect and withdrawal symptoms in mice lacking the mu-opioid-receptor gene. Nature 383:819–823

Metcalf MD, Yekkirala AS, Powers MD, Kitto KF, Fairbanks CA, Wilcox GL, Portoghese PS (2012) The delta opioid receptor agonist SNC80 selectively activates heteromeric mu-delta opioid receptors. ACS Chem Neurosci 3:505–509

Milan-Lobo L, Whistler JL (2011) Heteromerization of the mu- and delta-opioid receptors produces ligand-biased antagonism and alters mu-receptor trafficking. J Pharmacol Exp Ther 337:868–875

Milan-Lobo L, Enquist J, van Rijn RM, Whistler JL (2013) Anti-analgesic effect of the mu/delta opioid receptor heteromer revealed by ligand-biased antagonism. PLoS One 8, e58362

Negus SS (2004) Delta opioids and substance abuse. In: Chang KJ, Porreca F, Woods JH (eds) The delta receptor. Marcel Dekker, New York, pp 401–430

Nitsche JF, Schuller AG, King MA, Zengh M, Pasternak GW, Pintar JE (2002) Genetic dissociation of opiate tolerance and physical dependence in delta-opioid receptor-1 and preproenkephalin knock-out mice. J Neurosci 22:10906–10913

Nozaki C, Le Bourdonnec B, Reiss D, Windh RT, Little PJ, Dolle RE, Kieffer BL, Gaveriaux-Ruff C (2012) delta-Opioid mechanisms for ADL5747 and ADL5859 effects in mice: analgesia, locomotion, and receptor internalization. J Pharmacol Exp Ther 342:799–807

O'Dowd BF, Alijaniaram M, Ji X, Nguyen T, Eglen RM, George SR (2007) Using ligand-induced conformational change to screen for compounds targeting G-protein-coupled receptors. J Biomol Screen 12:175–185

O'Dowd BF, Ji X, O'Dowd PB, Nguyen T, George SR (2012) Disruption of the mu-delta opioid receptor heteromer. Biochem Biophys Res Commun 422:556–560

O'Neill SJ, Collins MA, Pettit HO, McNutt RW, Chang KJ (1997) Antagonistic modulation between the delta opioid agonist BW373U86 and the mu opioid agonist fentanyl in mice. J Pharmacol Exp Ther 282:271–277

Pan ZZ (1998) mu-Opposing actions of the kappa-opioid receptor. Trends Pharmacol Sci 19:94–98

Pfeiffer A, Brantl V, Herz A, Emrich HM (1986) Psychotomimesis mediated by kappa opiate receptors. Science 233:774–776

Powell KJ, Abul-Husn NS, Jhamandas A, Olmstead MC, Beninger RJ, Jhamandas K (2002) Paradoxical effects of the opioid antagonist naltrexone on morphine analgesia, tolerance, and reward in rats. J Pharmacol Exp Ther 300:588–596

Pradhan AA, Becker JA, Scherrer G, Tryoen-Toth P, Filliol D, Matifas A, Massotte D, Gaveriaux-Ruff C, Kieffer BL (2009) In vivo delta opioid receptor internalization controls behavioral effects of agonists. PLoS One 4, e5425

Pradhan AA, Walwyn W, Nozaki C, Filliol D, Erbs E, Matifas A, Evans C, Kieffer BL (2010) Ligand-directed trafficking of the delta-opioid receptor in vivo: two paths toward analgesic tolerance. J Neurosci 30:16459–16468

Raehal KM, Schmid CL, Groer CE, Bohn LM (2011) Functional selectivity at the mu-opioid receptor: implications for understanding opioid analgesia and tolerance. Pharmacol Rev 63:1001–1019

Rives ML, Rossillo M, Liu-Chen LY, Javitch JA (2012) 6'-Guanidinonaltrindole (6'-GNTI) is a G protein-biased kappa-opioid receptor agonist that inhibits arrestin recruitment. J Biol Chem 287:27050–27054

Robison AJ, Nestler EJ (2011) Transcriptional and epigenetic mechanisms of addiction. Nat Rev Neurosci 12:623–634

Rowan MP, Ruparel NB, Patwardhan AM, Berg KA, Clarke WP, Hargreaves KM (2009) Peripheral delta opioid receptors require priming for functional competence in vivo. Eur J Pharmacol 602:283–287

Roy S, Guo X, Kelschenbach J, Liu Y, Loh HH (2005) In vivo activation of a mutant mu-opioid receptor by naltrexone produces a potent analgesic effect but no tolerance: role of mu-receptor activation and delta-receptor blockade in morphine tolerance. J Neurosci 25:3229–3233

Rozenfeld R, Devi LA (2007) Receptor heterodimerization leads to a switch in signaling: beta-arrestin2-mediated ERK activation by mu-delta opioid receptor heterodimers. FASEB J 21:2455–2465

Rozenfeld R, Devi LA (2010) Receptor heteromerization and drug discovery. Trends Pharmacol Sci 31:124–130

Sanchez-Blazquez P, Garcia-Espana A, Garzon J (1997) Antisense oligodeoxynucleotides to opioid mu and delta receptors reduced morphine dependence in mice: role of delta-2 opioid receptors. J Pharmacol Exp Ther 280:1423–1431

Sora I, Funada M, Uhl GR (1997) The mu-opioid receptor is necessary for [D-Pen2, D-Pen5] enkephalin-induced analgesia. Eur J Pharmacol 324:R1–R2

Stevenson GW, Folk JE, Linsenmayer DC, Rice KC, Negus SS (2003) Opioid interactions in rhesus monkeys: effects of delta + mu and delta + kappa agonists on schedule-controlled responding and thermal nociception. J Pharmacol Exp Ther 307:1054–1064

Stockton SD Jr, Devi LA (2012) Functional relevance of mu-delta opioid receptor heteromerization: a role in novel signaling and implications for the treatment of addiction disorders: from a symposium on new concepts in mu-opioid pharmacology. Drug Alcohol Depend 121:167–172

Su YF, McNutt RW, Chang KJ (1998) Delta-opioid ligands reverse alfentanil-induced respiratory depression but not antinociception. J Pharmacol Exp Ther 287:815–823

Sykes KT, White SR, Hurley RW, Mizoguchi H, Tseng LF, Hammond DL (2007) Mechanisms responsible for the enhanced antinociceptive effects of micro-opioid receptor agonists in the rostral ventromedial medulla of male rats with persistent inflammatory pain. J Pharmacol Exp Ther 322:813–821

Thompson AA, Liu W, Chun E, Katritch V, Wu H, Vardy E, Huang XP, Trapella C, Guerrini R, Calo G, Roth BL, Cherezov V, Stevens RC (2012) Structure of the nociceptin/orphanin FQ receptor in complex with a peptide mimetic. Nature 485:395–399

van Rijn RM, Whistler JL, Waldhoer M (2010) Opioid-receptor-heteromer-specific trafficking and pharmacology. Curr Opin Pharmacol 10:73–79

Waldhoer M, Fong J, Jones RM, Lunzer MM, Sharma SK, Kostenis E, Portoghese PS, Whistler JL (2005) A heterodimer-selective agonist shows in vivo relevance of G protein-coupled receptor dimers. Proc Natl Acad Sci U S A 102:9050–9055

Wang D, Sun X, Bohn LM, Sadee W (2005) Opioid receptor homo- and heterodimerization in living cells by quantitative bioluminescence resonance energy transfer. Mol Pharmacol 67:2173–2184

Wang HB, Zhao B, Zhong YQ, Li KC, Li ZY, Wang Q, Lu YJ, Zhang ZN, He SQ, Zheng HC, Wu SX, Hokfelt TG, Bao L, Zhang X (2010) Coexpression of delta- and mu-opioid receptors in nociceptive sensory neurons. Proc Natl Acad Sci U S A 107:13117–13122

Wild KD, McCormick J, Bilsky EJ, Vanderah T, McNutt RW, Chang KJ, Porreca F (1993) Antinociceptive actions of BW373U86 in the mouse. J Pharmacol Exp Ther 267:858–865

Wu H, Wacker D, Mileni M, Katritch V, Han GW, Vardy E, Liu W, Thompson AA, Huang XP, Carroll FI, Mascarella SW, Westkaemper RB, Mosier PD, Roth BL, Cherezov V, Stevens RC (2012) Structure of the human kappa-opioid receptor in complex with JDTic. Nature 485:327–332

Xie Z, Bhushan RG, Daniels DJ, Portoghese PS (2005) Interaction of bivalent ligand KDN21 with heterodimeric delta-kappa opioid receptors in human embryonic kidney 293 cells. Mol Pharmacol 68:1079–1086

Yekkirala AS, Kalyuzhny AE, Portoghese PS (2010) Standard opioid agonists activate heteromeric opioid receptors: evidence for morphine and [d-Ala(2)-MePhe(4)-Glyol(5)]enkephalin as selective mu-delta agonists. ACS Chem Neurosci 1:146–154

Yekkirala AS, Banks ML, Lunzer MM, Negus SS, Rice KC, Portoghese PS (2012) Clinically employed opioid analgesics produce antinociception via mu-delta opioid receptor heteromers in Rhesus monkeys. ACS Chem Neurosci 3:720–727

Zhang Z, Pan ZZ (2010) Synaptic mechanism for functional synergism between delta- and mu-opioid receptors. J Neurosci 30:4735–4745

Zhu Y, King MA, Schuller AG, Nitsche JF, Reidl M, Elde RP, Unterwald E, Pasternak GW, Pintar JE (1999) Retention of supraspinal delta-like analgesia and loss of morphine tolerance in delta opioid receptor knockout mice. Neuron 24:243–252

Chapter 3
Evolutionary Distribution of the δ-Opioid Receptor: From Invertebrates to Humans

Taylor E. Hinchliffe and Ying Xia

Abstract The delta-opioid receptor (DOR), as well as other opioid receptors such as the mu-opioid receptor (MOR) and kappa-opioid receptor (KOR), hold evolutionary histories dating back hundreds of millions of years. DOR, in particular, has only experienced several small gene duplications throughout this expansive time period, and the receptor and its agonist like molecules can be found in an equally expansive range of species. Outside of the central nervous system (CNS), DOR is located on immune/inflammatory cells of both invertebrates and humans, and these shared immunoregulatory processes highlight a striking degree of evolutionary conservation. Upon examination within the CNS, a gradual pattern of increasingly light and heterogeneous DOR distribution and density over evolutionary time begins to emerge. When proceeding in the CNS from zebrafish, to turtle, to rat, and to human, DOR is comparatively denser at first and distributed in a more homologous manner in the zebrafish, with only a light rostral-to caudal pattern of DOR distribution apparent. This rostral-caudal pattern becomes increasingly pronounced and heterogeneous in turtles, more evident in rats, and even more prominent in humans. The clear heterogeneity of human DOR distribution is expressed via DOR levels that are highest in the cerebral cortex, caudate putamen and amygdala and dramatically drop off in density outside of the telencephalon. Also in the vertebrate CNS, not only does the regional distribution of DOR appear to become more selective through evolutionary time and hold various functions, but the binding properties of DOR also show a change in different species. Differential DOR distribution among species may correspond to a wide variety of functional differences. This chapter serves as a compilation and species-to-species comparison of DOR distribution, occasionally in comparison with MOR and KOR, as well as the primarily DOR-specific endogenous ligand, enkephalin. Focus is placed on the CNS, throughout the evolution of life and particularly in the context of organisms most often used for current research.

T.E. Hinchliffe, B.A.
Department of Neurosurgery, The University of Texas Medical School at Houston, Houston, TX 77030, USA

Y. Xia, M.D., Ph.D. (✉)
University of Texas Medical School, Houston, TX, USA
e-mail: ying.xia@uth.tmc.edu; y55738088@gmail.com

Y. Xia (ed.), *Neural Functions of the Delta-Opioid Receptor*,
DOI 10.1007/978-3-319-25495-1_3

Keywords Brain • Opioid receptors • Delta-opioid receptors • Distribution • Enkephalin • Evolution

Abbreviations

CNS	Central nervous system
DADLE	[D-ala2, D-leu5] enkephalin
DAMA	[D-Ala2, D-Met5] enkephalinamide
DELT-II	[D-Ala2, Glu4] deltorphin
DOR	Delta-opioid receptor
DPDPE	[D-Pen2, D-Pen5] enkephalin
KOR	Kappa-opioid receptor
MOR	Mu-opioid receptor
NEP	Neutral endopeptidase

1 Introduction

Many transmitters/modulators and their receptors have withstood tremendous tests of time, functioning as key components in the chemical processes of life forms today. The δ-opioid receptor (DOR) is no exception, with its mechanisms firmly in place at least 450 million years ago at the onset of gnathostome evolution, likely predating the existence of jawed vertebrates and extending to primitive jawless fishes (Dreborg et al. 2008; Larhammar et al. 2009). Even the hagfish, a member of the most primitive vertebrate class, agnathans, has been found to contain opioid-like receptor sequences with high homology to DOR at both the levels of amino acids and nucleotides (Li et al. 1996a, b). Additionally, opioid-like systems in less complex vertebrates such as jawless fishes and amphibians likewise exhibit less selective binding, coalescing in the proposal of a unireceptor in certain frog species capable of opioid mediation as if it were a system composed of DOR, MOR and KOR (Stevens and Newman 1999; McClendon et al. 2010). DOR and DOR-like binding, as well as the pattern of DOR distribution, appear to respectively become increasingly specific and heterogeneous upon examining increasingly complex organisms.

Despite having only undergone a small series of duplications from a single ancestral gene, DOR and similar, more ancient opioid systems can be found throughout a wide variety of species, all adapted to survive in radically different environments (Stevens et al. 2007). DOR adaptability and maintained physiological importance have resulted from a correlation between opioid receptor sequence divergence to the steady evolution of increasingly subsequent vertebrate species and, likewise, to increasing receptor type-selectivity (Stevens 2009). As might be expected, coming with this plethora of organisms and their distinct survivability adaptations are arrays of variant DOR distributional patterns and densities as well as correlating functional differences.

In studying DOR and its function, various species have been used in the past. The species most commonly used include the zebrafish (*Danio rerio*), house mouse (*Mus musculus*) and rat (*Rattus norvegicus*), with experimental models ranging from in vitro and in vivo mammalian model of ischemic stroke and cerebral hypoxia, to zebrafish models of morphine addiction (Ma et al. 2005; Zhang et al. 2006; Pamenter and Buck 2008a, b; Chao and Xia 2010; Rodriguez 2012; He et al. 2013). Thus, understanding how DOR has gradually come to be distributed within a myriad of bodily systems as well as studying marked differences between species, especially differences within the nervous system, not only offers up a more complete understanding of its functional versatility, but also grants valuable insight into the benefits and pitfalls of experimentation using animal models from less complex species for comparison to human physiology. Likewise, the discrepancies of DOR (and MOR/KOR) properties between humans and other species are best understood through the lens of evolution, of which ties all species together and closes the most gaps in knowledge. Examining DOR through the perspective of side-to-side species comparisons and evolutionary research forms the most complete picture of why DOR can differ in properties such as distribution and binding selectivity from fish, to rat, to human. This form of comparative examination also, from a wide-angle view, begins to form a DOR distributional pattern of increasingly honed heterogeneity through hundreds of millions of years of evolution. In turn, further macro-properties/functions of opioid receptors and their roles in humans may start to be deduced.

In this chapter, we will comprehensively review the distribution and density of DOR as well as its endogenous ligand, enkephalins, in the CNS between separate species and across evolutionary time. For the purpose of comparison, we will also occasionally review other opioid receptors such as MOR and KOR. Additionally, we will briefly discuss DOR density and distribution outside of the CNS, particularly in the context of less complex organisms where literature is scarce. Although only one DOR gene is cloned so far, many studies suggest that there are δ1 and δ2 subtypes of DOR in terms of their differences in pharmacological properties (Benyhe et al. 1990; Parkhill and Bidlack 2002; Bradford et al. 2006; van Rijn and Whistler 2009). We will adopt this concept in this review to keep consistency with that of the references we cited.

2 Ancient History of DOR and DOR-like Systems in Invertebrates

This section will begin a steady linkage between the great phylogenetic distance of invertebrates and mammals using various areas of common ground through DOR and DOR-like systems. Although studies offering highly detailed information at the intersection of invertebrates and opioid receptor function/distribution are quite rare, enough commonalities can be found to understand that DOR binding sites have been utilized by life forms for many millions of years. This grand timescale has therefore given DOR significant time and opportunity to evolve and integrate into various roles within the most complex organisms alive today. Due to the longstanding timeline and crossing points of DOR-related physiology between invertebrates and humans, it is possible to conclude that many surprises may be waiting to be uncovered in translational research with animal models (and thus, in humans). Finding instances of similarity in less complex organisms could offer clues of what regions and systems in mammals should be examined in greater depth. This is due to the fact that these overlapping regions and systems with DOR-like physiology in invertebrates and mammals have not been removed by natural selection through this time period, and may play even greater and more complex roles than currently known.

2.1 Differentiation of Invertebrate DOR

Beginning from the standpoint of older phylogenetic age, research pivoting around opioid systems in invertebrates, specifically DOR or similar molecules, remains sparse and limited. Yet it is possible to start piecing together scattered clues concerning not only the origins of DOR, but also how the localization of DOR and its ligands have fluctuated with time and adaptation, as well as DOR involvement in functions that are both highly conserved through evolution, and also utterly unique. The Asian scorpion *Buthus martensii*, for example, carries in its telson (stinger) a venom containing the enkephalin-like peptide BmK-YA that is selectively agonistic with mammalian DOR, 6.8 times more so than with MOR and 12 times more so than with KOR (Zhang et al. 2012). This offers up a glimpse at a potential δ-opioid-like system playing a notably different and distinct physiological function as compared to the roles of DOR in other organisms.

2.2 Prehistoric DOR, DOR Peptides and Immunoregulation: From Invertebrates to Humans

A less unique example of early DOR existence involves immunoregulation: an evolutionarily conserved commonality of a DOR-mediated activation of invertebrate immunocytes and human leukocytes that appears to be fairly specific to met-enkephalin (Stefano et al. 1989, 1992, 1995; Pryor and Elizee 2000). Although both invertebrate and human immunocytes can be stimulated by ligands binding to DOR, MOR and KOR, the DOR-specific ligand [D-Ala2, D-Met5] enkephalinamide (DAMA) was found to be the most effective by a considerable margin (Stefano et al. 1989). However, a highly similar DOR ligand [D-ala2, D-leu5] enkephalin (DADLE) did not display similar immunocyte stimulatory changes/activity (including cell flattening, elongation, pseudopodia formation and cell clustering) (Stefano et al. 1989). Such a notable difference in the results of DAMA vs. DADLE binding is thought to arise through met-enkephalin acting on pharmacologically described δ-2 receptors in immunocytes, instead of δ-1 receptors (Stefano et al. 1989, 1992, 1995). Indeed, it was later found that low concentrations of both DADLE and

[D-Pen2, D-Pen5] enkephalin (DPDPE) do *not* affect immunocyte activation (Stefano et al. 1995). Therefore, DOR has evolved into multiple, distinguishable micro-systems *outside* of those most commonly studied in the CNS, at least in binding properties at the pharmacological level.

The possible linkage of human and invertebrate immunoregulatory physiology through the high DOR affinity ligand met-enkephalin can additionally be examined outside of direct ligand-receptor binding. Neutral endopeptidase 24.11 (NEP), an enzyme that cleaves and inactivates enkephalins, can be detected in the cell membranes of invertebrate immunocytes (Stefano et al. 1991). The addition of the specific inhibitor to NEP, phosphoramidon, not only restricts NEP from inactivating met-enkephalin, but even induces a conformational change in *both* invertebrate and human granulocytes via increased met-enkephalin binding (Stefano et al. 1991). In addition to increased locomotor activity, met-enkephalin also increases the motility and velocity of responsive invertebrate and human immunocytes, but at a faster rate for all variables in humans (Stefano et al. 1991). The fact that the activation of immunocyte response can be potentiated in humans and invertebrates in sync by the same peptide further signifies the importance of DOR and DOR-like systems in immunoregulation.

The met-enkephalin and leu-enkephalin of invertebrates come in varied distributional proportions, being 3:1, respectively, for mussels, and 1:2 for leeches (Salzet and Stefano 1997). Despite these differences in enkephalin proportions, both share a similar mammalian-like proenkephalin molecule that contains an antibacterial peptide with a 98 % sequence identity to its counterpart in bovines (Salzet and Stefano 1997). Although the enkephalins are also capable of MOR and KOR binding, their high DOR affinity and presence in invertebrates offers strong clues pointing towards similar DOR-like systems between extremely different species. Likewise, both δ-1 and δ-2 subtypes are shared among the immune cells of leeches, mussels and humans, where it is thought that immune signaling may potentially lead to enhanced proenkephalin proteolytic processing (Salzet and Stefano 1997). The highly conserved and evolved DOR-like biochemical systems involved in immunoregulation may therefore likely be almost as complex as those encountered in the mammalian CNS.

DOR distribution is indeed varied throughout numerous cells involved in equally numerous immunoregulatory responses. DOR located on inflammatory cells, upon activation, play roles in the modulation and regulation of cellular and enzymatic degradation and regulation processes (Stefano et al. 1996). These roles are also shared between organisms that, likewise, range from invertebrates to humans (Stefano et al. 1996). DOR, as well as enkephalin, is also distributed to the gill (central axis and gill filaments), gonad, and hemocytes of the scallop *Chlamys farreri*, with hemocytes being the organism's most critical cells involved in its internal defense mechanisms (Liu 2007; Liu and Sun 2010). Within scallop hemocytes, increased enkephalin levels upregulate both intracellular and extracellular catalase and glutathione, and from this are correlated to an intricate involvement in invertebrate antioxidant defense systems (Liu 2007). Overall, in the case of comparative immunoregulatory processes, highly conserved systems may still be in place between invertebrates and humans. These primarily coalesce from the stimulatory actions of endogenous opioid peptides that result in activation, cellular motility and phagocytosis, and facilitate overall communication of information between the immune and nervous system (Liu and Sun 2010). Accumulating data therefore suggest that, in some situations, once DOR became entwined in certain regions and physiological functions, it stayed in place for hundreds of millions of years, but likewise evolved into different roles corresponding to different physiological systems.

2.3 DOR and Feeding Amongst Invertebrates and Mammals

Yet another function and correlating distribution of DOR with a longstanding evolutionary history is that of feeding and thus regions associated with feeding behavior, creating another potential evolutionary link between invertebrates and mammals (Kavaliers and Hirst 1987). Indeed, opioid systems have been linked to areas dictating the feeding behaviors of a variety of invertebrates, including cockroaches and snails (Kavaliers and Hirst 1987; Cooper et al. 2010; León-Olea et al. 2013). Additionally, radioimmunoanalysis shows that enkephalins are distributed broadly throughout snail peri*esophageal* ganglia in a circannual variation (León-Olea et al. 2013). Even DOR itself, as well as met-enkephalin, is extensively distributed within the digestive system of a far more complex invertebrate, the octopus (Sha et al. 2013). DOR and corresponding endogenous peptides are detectable in octopi in regions ranging from the bulbus oris, to the esophagus, crop, stomach, gastric cecum, intestine, anterior and posterior salivary glands, rectum, and digestive gland (Sha et al. 2013). The highest distributions and densities of both DOR and met-enkephalin are found in the inner epithelial cells of the digestive tract as compared to connective tissue (Sha et al. 2013).

It is believed that the wide distribution of octopi intestinal met-enkephalin and DOR, of varying densities, are involved in processes equally variable: food uptake and absorption, gastrointestinal smooth muscle movement, and digestive gland secretion (Sha et al. 2013). These examples, though limited in scope, coalesce to form the picture of a strikingly high degree of evolutionary conservation in both general distribution and function, as DOR also exists in the gastrointestinal tract and enteric nervous system of mammals such as rats, guinea pigs and humans (Lang et al. 1996; Nano et al. 2000; Sternini et al. 2004; Holzer 2009). In all three species, DOR distribution is particularly high in intestinal epithelial cells, and in both myenteric and submucosal plexuses: primarily in varicose fibers (Lang et al. 1996; Nano et al. 2000; Sternini et al. 2004; Holzer 2009). Thus, there are both commonalities between this broad range of species regarding epithelial DOR, as well as differences in more prominent DOR functional and ontological complexity in more complex organisms, correlated to increasingly sophisticated nervous systems through evolutionary history.

2.4 DOR in Other Systems of the Invertebrate

Apart from distribution of DOR and/or enkaphalin-like peptides throughout regions and cells involved in immunoregulation and feeding, there are clues showing the potential roles of DOR in other systems, including the nervous system, of the invertebrate. For example, leu- and met-enkephalin are detectable within the cerebral ganglia of land snails, which help to strengthen the possibility of a longstanding presence of DOR-like systems in invertebrates (León-Olea et al. 2013). DOR, both δ1- and δ2 subtypes, also carry an analgesic effect on snails, facilitated by the DOR-specific ligands DPDPE and [D-Ala2,Glu4] deltorphin (DELT-II) (Thomas et al. 1997).

Although sea sponges do not need, and therefore lack, nerve cells, muscle cells and internal organs of any kind, there is evidence showing that in *Amphimedon queenslandica*, one species of sea sponge, show that it synthesizes many of the proteins that are essential for the cell-to-cell communication that takes place within nervous systems (Sakarya et al. 2007). Carroll et al. (2010) found that Bastadins, the macrocyclic alkaloids produced by sea sponges, have a high affinity for DOR in guinea pigs. They are capable of inhibiting ^{3H}DPDPE binding to DOR in guinea pig forebrain membrane, but not KOR or MOR (Carroll et al. 2010). This is another example of a DOR-derived physiological linkage between vastly different species.

Still, current research over DOR distribution/function in invertebrates remains scattered, and the existence of any specific distributional similarities between octopi, other invertebrates and vertebrates has yet to be fully elucidated. Particularly in the case of the octopus CNS, as both MOR and KOR can also bind enkephalins at lower affinities, the detection of enkephalin-like substances shows promise and is a step forward, but only provides a clue hinting at the presence of a DOR system. Nevertheless, the existence of functional DOR/DOR-like systems in many of invertebrates, as summarized in Table 3.1, offers a great insight into the importance of DOR in the life of various organisms. On the other side, the relationship of DOR and its high affinity peptides to similar physiological functions between species separated by hundreds of millions of years of evolution hints at the existence of varying but often striking degrees of conserved distribution and function.

3 From Water to Land: DOR in Different Vertebrates

Steadily advancing the evolutionary tree from a chronological perspective and closer approaching ground between invertebrates and Homo sapiens marks the arrival at a gamut spanning fishes to small mammals. Here, DOR is still prone to degrees of evolutionary conservation, as one might expect, with certain, more generalized distributional patterns, functions and molecular properties holding their ground. Yet in other cases, these patterns disintegrate entirely or coalesce into something new.

Table 3.1 DOR/DOR-like systems and functions in some non-mammalian organisms

Species	Molecules	Function	References
Scorpion	BmK-YA (an enkephalin-like peptide)	Component of venom	Zhang et al. 2012
Snail	DOR-like systems	Feeding	Kavaliers and Hirst (1987), León-Olea et al. (2013)
Cockroach	DOR-like protein		Cooper et al. (2010)
Sea Sponge	Bastadins, with potent and selective DOR binding affinity	Cell-to-cell communication?	Sakarya et al. (2007), Carroll et al. (2010)
Octopus	DOR and met-enkephalin	Feeding and digestion	Sha et al. (2013)
Leeches, mussels	Pro- and met-/leu-enkephalin-like peptides	Immunoregulatory roles	Salzet and Stefano (1997)
Scallop	DOR-like proteins		Liu (2007), Liu and Sun (2010)
Carp	DOR-like proteins	Chemokine/chemokine receptor regulation and other immunoregulatory roles	Verburg-van Kemenade et al. (2009)
Freshwater turtle	DOR	Possible contribution to a high tolerance of the brain to hypoxia/ischemia	Xia and Haddad (2001), Pamenter and Buck (2008b)

3.1 *Fish*

A strong example of conservation through time comes from the carp, an aquatic vertebrate indeed falling into this *middle* ground that expresses DOR in its piscine leukocytes (Verburg-van Kemenade et al. 2009). DOR likewise facilitates immune and inflammatory responses in the carp, most notably the regulation of chemokine and chemokine receptors (Verburg-van Kemenade et al. 2009). Finding DOR participating in immunoregulation in the carp (or essentially any similar aquatic vertebrate) would be expected, noting its presence in the immunocytes of organisms both predating and postdating fish evolution, i.e. invertebrates and humans (Liu 2007; Liu and Sun 2010). This better completes the picture of some of the oldest physiological roles of DOR. Met- and/or leu-enkephalin are also detectable within the CNS of various aquatic species, such as the lamprey, hagfish, American eel, and coho salmon, as well as in the posterior pituitary of the bowfin (Dores et al. 1989; Dores and Gorbman 1990; McDonald and Dores 1991). Still, a more detailed mapping of DOR distribution and density in the CNS of these organisms has yet to be carried out.

Another aquatic vertebrate that has granted tremendous insight into the evolution of DOR and its chronologically earlier distribution throughout the history of chordate anatomy is the zebrafish. From *Danio rerio*, studies have identified two genes homologous to DOR: ZFOR1, with a 66 % homology to mammalian DOR, and

ZFOR4, with a 64 % identity to human DOR (Barrallo et al. 1998a, b; Rodriguez et al. 2000; Pinal-Seoane et al. 2006). The DOR-like proteins encoded by these two genes are also activated by the endogenous DOR peptides met- and leu-enkephalin (Barrallo et al. 1998a, b; Porteros et al. 1999). Although direct comparisons between the regions of the zebrafish CNS and those of mammals are not always feasible, a receptor distributional pattern begins to appear that will be found to further take shape in increasingly complex species. Aside from the main exception of high expression in the metencephalon, both ZFOR1 and ZFOR4 carry an overall CNS distribution that moves from high to low when advancing from rostral to caudal regions (Barrallo et al. 1998a, b; Porteros et al. 1999; Pinal-Seoane et al. 2006). ZFOR1 and ZFOR4 are, more specifically, distributed with a primarily moderate to low expression and density in the myelencephalon and spinal cord, as further compared to higher density the telencephalon, diencephalon, and mesencephalon (Barrallo et al. 1998a, b; Porteros et al. 1999; Pinal-Seoane et al. 2006).

It is noteworthy that multiple studies have found that opioid receptor binding in jawless fishes, such as the lamprey, is noticeably less selective than in mammalian counterparts (McClendon et al. 2010). McClendon et al. (2010) believe that this instance of subfunctionalization in a more primitive opioid system, where ligand selectivity overlaps, may also result in functional overlap that eventually became increasingly specific and uniquely selectively in mammals via neofunctionalization through evolutionary time (McClendon et al. 2010). Examining opioid systems in organisms such as the lamprey grants insight into opioid receptors at a less evolutionarily divergent state. Therefore, possible evolutionary precursors to mammalian DOR found in less physiologically complex aquatic vertebrates could indeed carry a higher number of overlapping functions, possibly correlating with their distribution in the less complex nervous systems of these organisms. This decreased selectivity in opioid binding in fish may also correspond to a more general and homogeneous distribution of opioid receptors.

3.2 *Amphibians*

Despite the existence of some conserved, broader similarities, in order to make way for more complex organisms, natural selection has gradually altered the properties of opioid receptors themselves. These alterations allow for more specified distributions, densities and functional roles. Changes in opioid receptor properties have occurred through a very gradual increase in the binding selectivity of near-identical opioid precursors, most notable when comparing organisms such as amphibians and jawless fishes to mammals. Research has indeed shown that opioid receptor subtypes are present in amphibian brains, including a DOR variant detected in rough skinned newt pooled brain samples, although the differences between these subtypes are more subtle than in mammals (Benyhe et al. 1990; Bradford et al. 2005, 2006). Still, such studies indicate the presence of a receptor with a high degree of similarity to mammalian DOR.

Yet at the same time, similarities from amphibians can also be drawn to the less selective opioid binding detected in more primitive jawless fishes (McClendon et al. 2010). Indeed, opioid receptor binding in some amphibian species has also been found to be considerably less selective than binding in mammals (Stevens et al. 2007). Some researchers have even proposed the existence of an additional, essentially nonselective *unireceptor* in the Northern grass frog, *Rana pipiens* (Stevens and Newman 1999). Three distinct opioid antagonists were used and found to not be selective in this frog model (Stevens and Newman 1999). In other words, in some amphibians, this possible unireceptor may act as a regulatory hub for physiological responses such as antinociception, capable of mediation through DOR-, MOR- and KOR-like systems (Stevens and Newman 1999; Stevens et al. 2007).

As compared to invertebrates, fish or mammals, another difference in the amphibian brain, at least in some species, is the proportionality of opioid receptors. The brain of the frog *Rana esculenta*, also often called the "edible frog" and characteristically quite similar to the Northern grass frog, appears to be an organ of kappa-like receptors (Simon et al. 1984). Specifically, membrane-bound opioid-like receptors yield a distribution of 70 % kappa, 17 % delta, and 13 % mu (Simon et al. 1984). As will be noted in the next section, the turtle brain carries an extremely high DOR density that decreases dramatically in the rat, seemingly in exchange for MOR (Xia and Haddad 2001). The prominence and high density of KOR, DOR, and then MOR in the brains of frogs, turtles, and rats, respectively, seems to correlate well with the hypothesis that opioid receptors evolved in the order of kappa, then delta, and most recently, mu (Stevens 2004). This offers an additional glimpse of insight into the roles and history of DOR alongside its opioid counterparts.

3.3 Turtle: Unique Distribution of DOR in the CNS

The rostral-caudal pattern of DOR distributional prominence observed in zebrafish (Barrallo et al. 1998a, b; Porteros et al. 1999; Pinal-Seoane et al. 2006) is far more pronounced and becomes increasingly heterogeneous when examining more complex vertebrates such as turtles and rodents, as well as pigeons (Reiner et al. 1989; Xia and Haddad 2001). In the brain of freshwater turtle, DOR binding density is indeed high, more so overall than in rat brain under the same autoradiographic condition, but displays a heterogeneous distribution pattern that is particularly high in the forebrain (Xia and Haddad 1991, 2001) (Fig. 3.1). Indeed, the turtle brain is actually an organ of DOR (Xia and Haddad 2001). Moreover, the dissociation constant (Kd) of turtle brain DOR is approximately four times lower than in rat brain, suggesting a higher DOR binding affinity in the turtle brain (Xia and Haddad 2001). Specifically, turtle forebrain DOR binding density is highest in the cortex and dorsal ventricular ridge (noted to be homologous to mammalian neocortex) while binding is low in the brainstem (Xia and Haddad 2001). DOR density levels in turtle cortex fall immediately behind levels in the dorsal ventricular ridge, being only slightly less (Xia and Haddad 2001). Yet these DOR density levels are still over

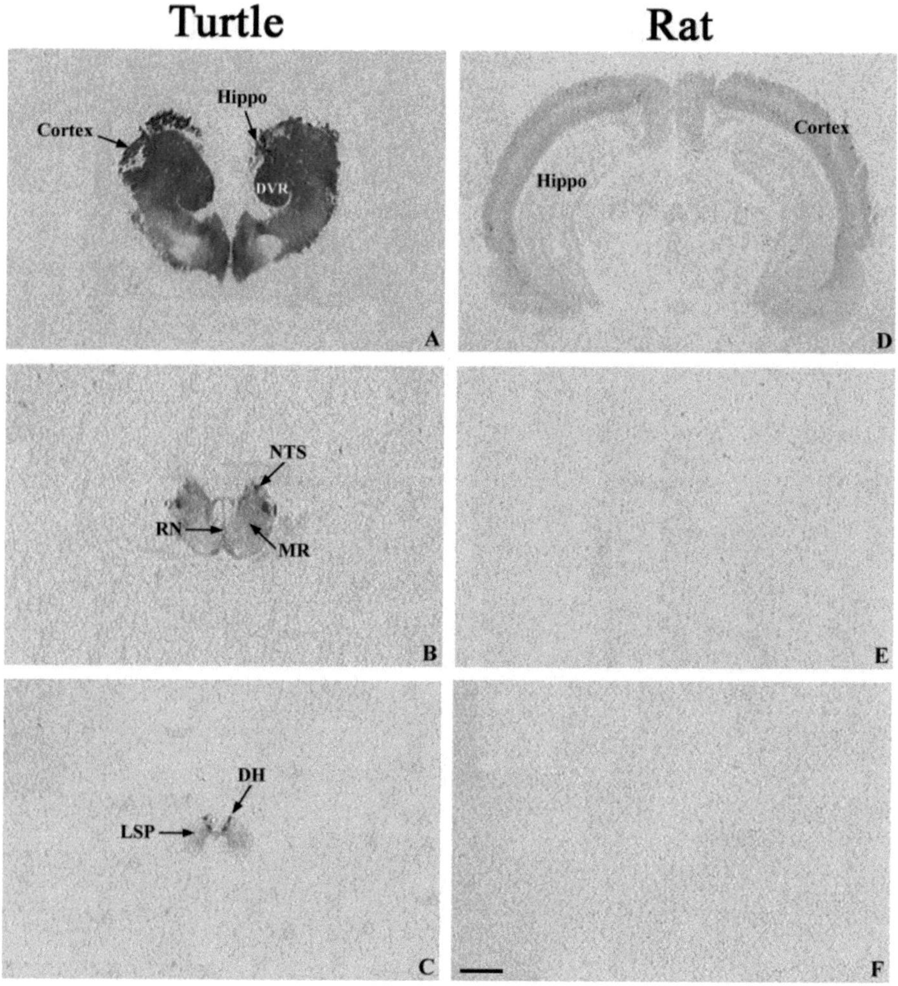

Fig. 3.1 Major difference in DOR distribution and density between turtle and rat brains. DOR binding was performed with 4 nM of ³H-DADLE with addition of non-radiolabled MOR ligand PL017 (1 μM). (**A-C**) Turtle brain, from forebrain to spinal cord. (**D-F**) Rat brain, from forebrain to spinal cord. Slices from both turtle and rat brains were studied simultaneously and all images were normalized to the same density scale. *DH* the dorsal horn of spinal cord, *DVR* dorsal ventricular ridge, *Hippo* Hippocampus, *LSP* lateral spinal nucleus, *MR* the reticular formation of medulla, *NTS* the nucleus of the solitary tract, *RN* raphe nucleus of medulla. Note that DOR density is much higher in the turtle than in rat brain. At a density scale that does not lead to DOR over-saturation in the images of the turtle brains, DOR density in rat medulla (**E**) and spinal cord (**F**) was too low to be appreciated. Scale bar = 2 mm in **F** (applies to **A–F**). Cited from Xia and Haddad (2001)

approximately 30–90 % higher than other regions in the turtle brain such as the reticular formation of the medulla, and lateral spinal nucleus, listed in order of descending binding density (Xia and Haddad 2001).

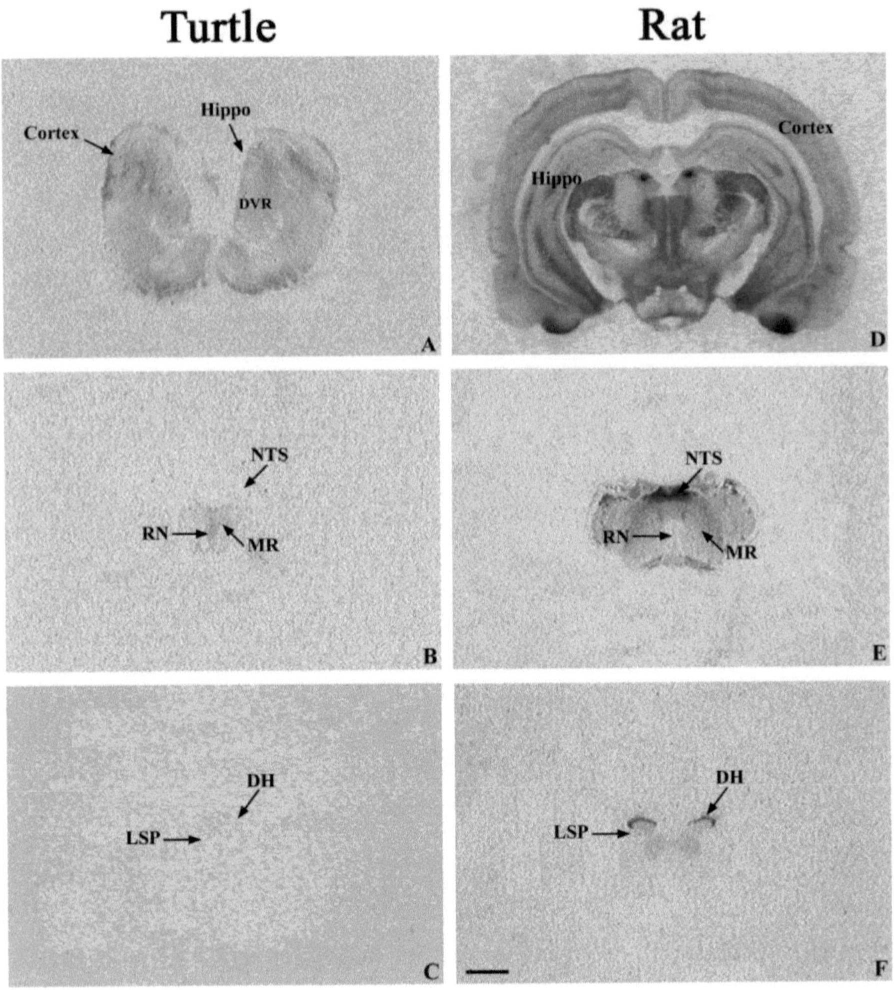

Fig. 3.2 Major differences in MOR distribution and density between turtle and rat brains. MOR binding was performed with 4 nM of ^{3H}-DAMGO. (**A-C**) Turtle brain, from forebrain to spinal cord. (**D-F**) Rat brain, from forebrain to spinal cord. Slices from both turtle and rat brains were studied simultaneously, and all images were normalized to the same density scale. Abbreviations are the same as in Fig. 3.1. Note that MOR density is lower in all turtle brain regions compared with the rat brain, especially in the caudal regions such as spinal cord. Scale bar=2 mm in **F** (applies to **A–F**). Cited from Xia and Haddad (2001)

Interestingly, when comparing turtle and rat brains, MOR binding sites are essentially expressed as the opposite to DOR distributional patterning (Fig. 3.2). In other words, MOR density is much lower throughout all turtle brain regions, as compared to the rat (Xia and Haddad 2001). MOR density in the turtle cortex is approximately 50 % of the levels of rat cortex MOR (Xia and Haddad 2001).

The only areas of the turtle brainstem that are comparable in DOR density to more rostral regions include the nuclei of the trigeminal nerve and solitary tract, of which contain densities similar to turtle hypothalamus (Xia and Haddad 2001). All other brainstem regions are otherwise much lower in DOR binding density than in the forebrain, but still higher than that of the rat brainstem (Xia and Haddad 1991, 2001). The spinal cord DOR binding is the least dense of all turtle CNS regions, though more DOR binding is found in the dorsal portion of spinal cord. As with the brainstem and turtle brain as a whole, turtle spinal cord DOR density is once again considerably higher than the density in rat spinal cord that is all-around low (Xia and Haddad 2001).

Since DOR is neuroprotective (Ma et al. 2005; Zhang et al. 2006; Chao and Xia 2010; He et al. 2013), higher DOR expression may render the brain more tolerant to neuronal lesion, such as hypoxic/ischemic injury. Indeed, turtle neurons do have the adaptive advantage of significantly higher tolerance to ischemic/hypoxic stress, as compared to rat neurons (Jiang et al. 1992; Xia et al. 1992; Pamenter and Buck 2008a, b; He et al. 2013). This adaptation may have arisen from lengthy periods of underwater submersion. Here we do not intend to say that the high density of DOR is solely responsible for the high tolerance of the turtle brain to hypoxic/ischemic stress. There are many factors (e.g., metabolism) determining the hypoxia/ischemia tolerance in the turtle brain. However, the DOR-mediated events certainly make a contribution to the neuroprotective mechanisms in the turtle CNS. For example, Pamenter and Buck (2008b) found that DOR plays an important role against NMDA receptor-dependent excitotoxicity in anoxic turtle cortex.

It is currently unknown as to how commonplace high DOR densities are in other reptiles, or if this is a case unique to certain aquatic turtle species.

3.4 *Rodent CNS DOR Distribution*

As compared to turtle brain, a strikingly more heterogeneous while simultaneously far less dense pattern of DOR distribution is observable in rat brain (Xia and Haddad 2001). DOR binding density within rat brain is highest in cortex and the caudate putamen, although DOR density in the middle layer of the cortex is much less than the inner and outer layers (Xia and Haddad 2001). DOR density in the rat brain then proceeds to decrease drastically when moving from rostral to caudal regions (Xia and Haddad 2001). Although the amygdala contains a high density of DOR binding sites, most subcortical regions such as the hippocampus, thalamus and hypothalamus, contain extremely low densities (Xia and Haddad 2001). In the brainstem, the DOR density is found to be the highest (although not comparable to cortex or caudate putamen) in the dorsal hypoglossal nucleus, but is otherwise low and distributed homogenously throughout the brainstem as well as spinal cord (Xia and Haddad 1991). The rodent caudal brain is thus thought of as a region of mu-receptors because of a much higher density of MOR than DOR (Xia and Haddad 1991, 2001). Therefore, upon progressing from zebrafish to turtle to rat, an overall increasingly honed and heterogeneous shape and pattern of CNS DOR distribution emerges, vaguely present in the zebrafish CNS yet suddenly far more apparent in that of the rat.

Although the overall CNS distributional patterns of DOR are similar between mice and rats, there are minor differences between the two rodents (Jenab et al. 1995). The caudate putamen of both mice and rats contains the highest overall levels of DOR mRNA, but mouse caudate putamen contains slightly more: 3.3 and 2.3 pg/μg of RNA for mice and rats, respectively (Jenab et al. 1995). Mid- and upper mid-level DOR mRNA distribution (1.5–2.6 pg/μg of RNA for mice and 0.9–1.5 pg/μg for rats) is most allocated to the frontal cortex, nucleus accumbens, whole brain tissue and the olfactory tubercle, and then descends and decreases in distribution and density in order of the spinal cord, periaqueductal gray and hippocampus for mice (Jenab et al. 1995). Similarly, mid-level DOR mRNA for rats is most detectable in the decreasing order of whole brain, cerebral cortex, olfactory bulb and brain stem (Jenab et al. 1995). The lowest amounts of DOR mRNA are found in the mouse medial thalamus and cerebellum, and in the thalamus, hippocampus, substantia nigra and cerebellum of rats (Jenab et al. 1995). Therefore, distributional differences in densities and in regionally specific mRNA expression must be taken into account for proper experimental design and model organism selection, as correlating and subtle DOR functional differences may have evolved between rats and mice, and such comparisons establish a framework in which to begin assessing these differences.

3.5 Guinea Pig DOR: Comparison to Other Rodents

The guinea pig, *Cavia porcellus*, has also been compared to the rat and other species in regards to DOR mechanisms and properties (Howard et al. 1986). A similar rostral-caudal distributional pattern is observable, with DOR density in both animals most concentrated in forebrain regions such as the striatum, nucleus accumbens and olfactory bulb, as well as cortex layers I–II and V–VI (Sharif and Hughes 1989). Also in the forebrain, comparative studies have given extra scrutiny to the rodent hippocampus, where pyramidal, dendate granular and dendate molecular layers contain low autoradiography labeled receptor binding in guinea pigs, but moderate to dense levels in rats (Foote and Maurer 1986). Additionally, myriad binding sites are detectable in the mossy fiber system of the hilus in guinea pigs, but not so in rats (Foote and Maurer 1986). The presence of high DOR density in the guinea pig mossy fiber system/hilus is further supported by intense enkephalin immunoreactivity in these regions (McLean et al. 1987). It has been observed that rat hippocampal DOR distribution more closely resembles that of rat MOR and KOR distribution, while in other rodents, hippocampal DOR distribution is often the inverse of MOR and KOR patterns (McLean et al. 1987). For example, rat hippocampal MOR and KOR are densest in the striatum and dendate molecular layers, while the inverse is true in guinea pigs (McLean et al. 1987).

Despite these similarities, DOR (and MOR) density is overall highest in rat brain, but guinea pig hindbrain regions, including the thalamus, geniculate bodies, inferior/superior colliculi and central grey contain higher DOR densities compared to these regions in rats (Sharif and Hughes 1989). On the other hand, DOR binding sites in both rat and guinea pig spinal cord are similar and fairly evenly distributed, but with the highest DOR densities in the cervical segment and lower levels in the lumbo-sacral region (Gouardères et al. 1985; Traynor and Wood 1987). Other studies have honed in on specific brain regions within multiple model rodents for comparative opioid receptor mapping, helping to fine-tune correlating future experiments, but such studies remain limited despite having the advantage of being subjected to the same experimental methods.

4 From Rodents to Humans: High Similarities with Minor Distinctions in DOR Distribution

For the purpose of comparison and application to human physiology and pathophysiology, experimental animal models within the bounds of mammalia are by far the most common test subjects. Therefore, past studies have granted tremendous insight into similarities and differences in DOR distribution and density. Even between two of the most commonplace model organisms; the mouse, *Mus musculus* and the rat, *Rattus norvegicus*, some differences in the location and density of DOR in the CNS have been uncovered (Moskowitz and Goodman 1984; Jenab et al. 1995; Bakota et al. 1998). These differences, in turn are key to note when fine tuning experimental design and discerning evolutionary distinctions.

The rostral-caudal pattern of DOR distribution is still observed in human brain. In human brain telencephalon, high DOR binding is found in the cortex, caudate putamen and amygdala, similar to DOR binding in rats, with minimal DOR binding in the globus pallidus of both species despite high quantities of detected enkephalin (Blackburn et al. 1988). In the medical and lateral segments of the human globus pallidus, no DOR binding is detectable (Blackburn et al. 1988). Concerning overall human cortical DOR density, DOR is highest in the superficial layers, laminae I, II and IIIa, as well as in deep cortical layers (Pilapil et al. 1987; Blackburn et al. 1988; Hiller and Fan 1996).

Although studies mapping DOR mRNA expression and density in human brain are limited, enough data exists to begin examining human DOR alongside other, well-studied mammals and shows that the distribution is even more heterogeneous than in rodents (Peckys and Landwehrmeyer 1999). The highest levels of DOR mRNA expression in the human brain are found in cerebral cortex, particularly in both prefrontal and occipital layers II–IV, and then proceed to decrease sharply upon leaving the telencephalon (Peckys and Landwehrmeyer 1999).

Down at the level of the spinal cord and dorsal root ganglia, notable differences have been uncovered in the distribution of DOR binding and mRNA between mice, rats, monkeys and humans via autoradiography and in situ hybridization (Mennicken et al. 2003). Such differences include how DOR binding is distributed throughout spinal cord gray matter in mice and rats, but is primarily restricted to laminae I–II in monkeys and humans (Mennicken et al. 2003). Additionally, DOR mRNA is

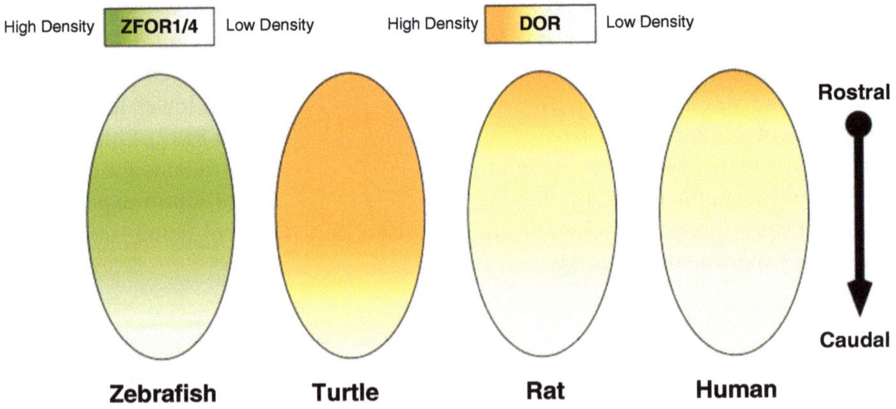

Fig. 3.3 Comparative DOR and DOR-like protein distribution among species. Note that Zebrafish ZFOR1 and ZFOR4 share a 66 % and 64 % homology, respectively, to mammalian/human DOR and their densities cannot be compared to those of the turtle, rat and human here because of no comparative study yet

highly expressed throughout the ventral and dorsal horns in rodent spinal cord, while expressed significantly less in primates—particularly so in humans, limited to Clarke's column and lamina IX (Mennicken et al. 2003). Finally, there are considerably high DOR mRNA levels in large ganglion cells in rodent dorsal root ganglia, while most DOR mRNA in primates is found in small or medium sized ganglion cells (Mennicken et al. 2003). The general pattern of DOR/DOR-like protein distribution between species is schematically shown in Fig. 3.3.

Overall, despite being taken from a fairly limited pool of studies, DOR appears to slightly increase in heterogeneous distribution when advancing from rodents to primates, though there is high similarity in the general pattern of DOR distribution in the brain. This, again, may correspond with increased functional specificity of DOR in the CNS of primates that could be advantageous and/or necessary in increasingly complex nervous systems. Therefore, although a high degree of similarities between commonly used rodent models and human DOR have started to be uncovered with currently available studies, much work is left unfinished in determining distributional/functional differences in the DOR of the human brain. When a more complete mapping of human CNS DOR with functional studies becomes available, continually placing such information in comparison with rodent data will help further elucidate potential differences between the animal and humans and improve the understanding of DOR function in translational research.

5 Concluding Remarks

Biologically, it is of high significance to understand the evolutionary changes of DOR in the brain and in other systems. More importantly, with the goal of translational research pivoting around the discovery of knowledge and solutions applicable to the humans, it is essential to be conscious of any potential pitfalls surrounding the differences between the physiology of model organisms and that of humans. And with another chronological, evolutionary leap from small mammals to primates, greater differences in DOR location and density emerge, all-the-while with a similar distributional framework still upheld. The even grander phylogenetic jump from invertebrates to primates/humans may seem like a stretch for translational research, but these less complex organisms hold keys of insight into which areas of opioid physiology have been essential to all life forms by staying in place through deep evolutionary time. Therefore, elucidating these longstanding areas of opioid physiology in such organisms may often warrant more detailed investigations in rodent studies.

It is an understanding of these key physiological differences between human opioid systems and those of model species, although often minor, that may act as a key component of efficient and successful translational research. One aspect of the framework that supports such insight is the thorough investigation and comparison of where DOR is distributed, particularly in the CNS, and how said distribution and density varies between organisms. The increasingly honed and heterogeneous pattern of DOR distribution in increasingly complex organisms is a product of hundreds of millions of years of evolution and underlies shifting and specifying functionality. Functional differences are likely less apparent between mammals and species separated by less evolutionary time, but still exist in varying degrees, and the initial blueprint of how DOR is allocated differently to particular CNS regions between organisms serves to facilitate the first step in the overall process of a greater understanding of human and model organism differences, that will in turn facilitate a smoother species-transitioning process of translational research.

Just how dramatic these functional differences may be, derived from the distributional differences of DOR and family opioid receptors as well as varied molecule selectivity, remains to be fully elucidated, but is key to the safe advancement of clinical trials. How DOR came to be distributed in its rostral to caudal manner through deeper evolutionary time, and what precisely this speaks of its current roles in human physiology is a further mystery, due to scarce information on the DOR-like systems of less complex organisms. Yet this is likewise capable of offering additional insight into broader views of human DOR (and MOR/KOR) functionality, as well as how these opioid systems may be altered for the sake of aiding human health and advancing medicine. Nevertheless, the high density of DOR in the cortex, caudate putamen and amygdale in human CNS in human CNS, after millions of years of evolution, reflects a specific importance of DOR in the function of these higher brain regions. On the other hand, the significant difference in DOR density and binding affinity between the rat and turtle brains results in a large, species-specific functional variation of DOR (high vs. low resistance to hypoxia/ischemia), and it is possible that equally striking and specific functional differences exist among other species.

At present, a sufficient evolutionary mapping and timeline of DOR distribution and density from invertebrates, to less complex vertebrates and to humans is far from complete, as available data in the literature remain scattered and often lack correlating detail on many key species in evolutionary history, of which is waiting to be further uncovered in future research.

Acknowledgements This work was supported by NIH (AT-004422, and HD-034852) and Vivian L Smith Neurologic Foundation.

References

Bakota L, Szikra J, Toth G, Gulya K (1998) Slide-binding characterization and autoradiographic localization of delta opioid receptors in rat and mouse brains with the tetrapeptide antagonist [3H]TIPP. Life Sci 63:1377–1385

Barrallo A, González-Sarmiento R, Porteros A, García-Isidoro M, Rodríguez RE (1998a) Cloning, molecular characterization, and distribution of a gene homologous to delta opioid receptor from zebrafish (Danio rerio). Biochem Biophys Res Commun 245:544–548

Barrallo A, Malvar FG, Gonzalez R, Rodriguez RE, Traynor JR (1998b) Cloning and characterization of a delta opioid receptor from zebrafish. Biochem Soc Trans 26:S360

Benyhe S, Varga E, Hepp J, Magyar A, Borsodi A, Wollemann M (1990) Characterization of kappa 1 and kappa 2 opioid binding sites in frog (Rana esculenta) brain membrane preparation. Neurochem Res 15:899–904

Blackburn TP, Cross AJ, Hille C, Slater P (1988) Autoradiographic localization of delta opiate receptors in rat and human brain. Neuroscience 27:497–506

Bradford CS, Walthers EA, Searcy BT, Moore FL (2005) Cloning, heterologous expression and pharmacological characterization of a kappa opioid receptor from the brain of the rough-skinned newt, Taricha granulosa. J Mol Endocrinol 34:809–823

Bradford CS, Walthers EA, Stanley DJ, Baugh MM, Moore FL (2006) Delta and mu opioid receptors from the brain of a urodele amphibian, the rough-skinned newt Taricha granulosa: cloning, heterologous expression, and pharmacological characterization. Gen Comp Endocrinol 146:275–290

Carroll AR, Kaiser SM, Davis RA, Moni RW, Hooper JN, Quinn RJ (2010) A bastadin with potent and selective delta-opioid receptor binding affinity from the Australian sponge Ianthella flabelliformis. J Nat Prod 73:1173–1176

Chao D, Xia Y (2010) Ionic storm in hypoxic/ischemic stress: can opioid receptors subside it? Prog Neurobiol 90:439–470

Cooper PD, Dennis SR, Woodman JD, Cowlings A, Donnelly C (2010) Effect of opioid compounds on feeding and activity of the cockroach, Periplaneta Americana. Comp Biochem Physiol C Toxicol Pharmacol 151:298–302

Dores RM, Gorbman A (1990) Detection of Met-enkephalin and Leu-enkephalin in the brain of the hagfish, Eptatretus stouti, and the lamprey, Petromyzon marinus. Gen Comp Endocrinol 77:489–499

Dores RM, McDonald LK, Crim JW (1989) Detection of Met-enkephalin and Leu-enkephalin in the posterior pituitary of the holostean fish, Amia calva. Peptides 10:951–956

Dreborg S, Sundström G, Larsson TA, Larhammar D (2008) Evolution of vertebrate opioid receptors. Proc Natl Acad Sci U S A 105:15487–15492

Foote RW, Maurer R (1986) Distribution of opioid binding sites in the guinea pig hippocampus as compared to the rat: a quantitative analysis. Neuroscience 19:847–856

Gouardères C, Cros J, Quirion R (1985) Autoradiographic localization of mu, delta and kappa opioid receptor binding sites in rat and guinea pig spinal cord. Neuropeptides 6:331–342

He X, Sandhu HK, Yang Y, Hua F, Belser N, Kim DH, Xia Y (2013) Neuroprotection against hypoxia/ischemia: delta-opioid receptor mediated cellular/molecular events. Cell Mol Life Sci 70:2291–2303

Hiller JM, Fan LQ (1996) Laminar distribution of the multiple opioid receptors in the human cerebral cortex. Neurochem Res 21:1333–1345

Holzer P (2009) Opioid receptors in the gastrointestinal tract. Regul Pept 155:11–17

Howard AD, Sarne Y, Gioannini TL, Hiller JM, Simon EJ (1986) Identification of distinct binding site subunits of mu and delta opioid receptors. Biochemistry 25:357–360
Jenab S, Kest B, Franklin SO, Inturrisi CE (1995) Quantitative distribution of the delta opioid receptor mRNA in the mouse and rat CNS. Life Sci 56:2343–2355
Jiang C, Xia Y, Haddad GG (1992) Role of ATP-sensitive K^+ channels during anoxia: major differences between rat (newborn and adult) and turtle neurons. J Physiol 448:599–612
Kavaliers M, Hirst M (1987) Slugs and snails and opiate tales: opioids and feeding behavior in invertebrates. Fed Proc 46:168–172
Lang ME, Davison JS, Bates SL, Meddings JB (1996) Opioid receptors on guinea-pig intestinal crypt epithelial cells. J Physiol 497:161–174
Larhammar D, Dreborg S, Larsson TA, Sundström G (2009) Early duplications of opioid receptor and Peptide genes in vertebrate evolution. Ann N Y Acad Sci 1163:451–453
León-Olea M, Miller-Pérez C, Sánchez-Islas E, Mendoza-Sotelo J, Garduño-Gutiérrez R, de Gortari P, Amaya MI (2013) The nociceptin/orphanin FQ-like opioid peptide in nervous periesophageal ganglia of land snail Helix aspersa. Brain Res 1505:22–46
Li X, Keith DE Jr, Evans CJ (1996a) Mu opioid receptor-like sequences are present throughout vertebrate evolution. J Mol Evol 43:179–184
Li X, Keith DE Jr, Evans CJ (1996b) Multiple opioid receptor-like genes are identified in diverse vertebrate phyla. FEBS Lett 397:25–29
Liu D (2007) Effects of leucine-enkephalin on catalase activity and glutathione level in haemolymph of the scallop Chlamys farreri. Int J Peptide Res Ther 14:16–27
Liu D, Sun H (2010) Immunohistological detection of mu, delta and kappa opioid-like receptors in the gill, gonad, and hemocytes of the scallop Chlamys farreri. Connect Tissue Res 51:67–70
Ma MC, Qian H, Ghassemi F, Zhao P, Xia Y (2005) Oxygen-sensitive {delta}-opioid receptor-regulated survival and death signals: novel insights into neuronal preconditioning and protection. J Biol Chem 280:16208–16218
McClendon J, Lecaude S, Dores AR, Dores RM (2010) Evolution of the opioid/ORL-1 receptor gene family. Ann N Y Acad Sci 1200:85–94
McDonald LK, Dores RM (1991) Detection of Met-enkephalin in the CNS of the teleosts, Anguilla rostrata and Oncorhynchus kisutch. Peptides 12:541–547
McLean S, Rothman RB, Jacobson AE, Rice KC, Herkenham M (1987) Distribution of opiate receptor subtypes and enkephalin and dynorphin immunoreactivity in the hippocampus of squirrel, guinea pig, rat, and hamster. J Comp Neurol 255:497–510
Mennicken F, Zhang J, Hoffert C, Ahmad S, Beaudet A, O'Donnell D (2003) Phylogenetic changes in the expression of delta opioid receptors in spinal cord and dorsal root ganglia. J Comp Neurol 465:349–360
Moskowitz AS, Goodman RR (1984) Light microscopic autoradiographic localization of mu and delta opioid binding sites in the mouse central nervous system. J Neurosci 4:1331–1342
Nano JL, Fournel S, Rampal P (2000) Characterization of delta-opioid receptors and effect of enkephalins on IRD 98 rat epithelial intestinal cell line. Pflugers Arch 439:547–554
Pamenter ME, Buck LT (2008a) Neuronal membrane potential is mildly depolarized in the anoxic turtle cortex. Comp Biochem Physiol A Mol Integr Physiol 150:410–414
Pamenter ME, Buck LT (2008b) Delta-Opioid receptor antagonism induces NMDA receptor-dependent excitotoxicity in anoxic turtle cortex. J Exp Biol 211:3512–3517
Parkhill AL, Bidlack JM (2002) Several delta-opioid receptor ligands display no subtype selectivity to the human delta-opioid receptor. Eur J Pharmacol 451:257–264
Peckys D, Landwehrmeyer GB (1999) Expression of mu, kappa, and delta opioid receptor messenger RNA in the human CNS: a 33P in situ hybridization study. Neuroscience 88:1093–1135
Pilapil C, Welner S, Magnan J, Gauthier S, Quirion R (1987) Autoradiographic distribution of multiple classes of opioid receptor binding sites in human forebrain. Brain Res Bull 19:611–615
Pinal-Seoane N, Martin IR, Gonzalez-Nuñez V, de Velasco EM, Alvarez FA, Sarmiento RG, Rodriguez RE (2006) Characterization of a new duplicate delta-opioid receptor from zebrafish. J Mol Endocrinol 37:391–403

Porteros A, García-Isidoro M, Barrallo A, González-Sarmiento R, Rodríguez RE (1999) Expression of ZFOR1, a delta-opioid receptor, in the central nervous system of the zebrafish (Danio rerio). J Comp Neurol 412:429–438
Pryor SC, Elizee R (2000) Evidence of opiates and opioid neuropeptides and their immune effects in parasitic invertebrates representing three different phyla: Schistosoma mansoni, Theromyzon tessulatum, Trichinella spiralis. Acta Biol Hung 51:331–341
Reiner A, Brauth SE, Kitt CA, Quirion R (1989) Distribution of mu, delta, and kappa opiate receptor types in the forebrain and midbrain of pigeons. J Comp Neurol 80:359–382
Rodriguez RE (2012) Morphine and microRNA activity: is there a relation with addiction? Front Genet 3:223
Rodriguez RE, Barrallo A, Garcia-Malvar F, McFadyen IJ, Gonzalez-Sarmiento R, Traynor JR (2000) Characterization of ZFOR1, a putative delta-opioid receptor from the teleost zebrafish (Danio rerio). Neurosci Lett 288:207–210
Sakarya O, Armstrong KA, Adamska M, Adamski M, Wang IF, Tidor B, Degnan BM, Oakley TH, Kosik KS (2007) A post-synaptic scaffold at the origin of the animal kingdom. PLoS One 2, e506
Salzet M, Stefano GB (1997) Invertebrate proenkephalin: delta opioid binding sites in leech ganglia and immunocytes. Brain Res 768:224–232
Sha A, Sun H, Wang Y (2013) Immunohistochemical observations of methionine-enkephalin and delta opioid receptor in the digestive system of Octopus ocellatus. Tissue Cell 45:83–87
Sharif NA, Hughes J (1989) Discrete mapping of brain Mu and delta opioid receptors using selective peptides: quantitative autoradiography, species differences and comparison with kappa receptors. Peptides 10:499–522
Simon J, Szücs M, Benyhe S, Borsodi A, Zeman P, Wollemann M (1984) Solubilization and characterization of opioid binding sites from frog (Rana esculenta) brain. J Neurochem 43:957–963
Stefano GB, Cadet P, Scharrer B (1989) Stimulatory effects of opioid neuropeptides on locomotory activity and conformational changes in invertebrate and human immunocytes: evidence for a subtype of delta receptor. Proc Natl Acad Sci U S A 86:6307–6311
Stefano GB, Casares F, Liu U (1995) Naltrindole sensitive delta 2 opioid receptor mediates invertebrate immunocyte activation. Acta Biol Hung 46:321–327
Stefano GB, Kream RM (2010) Opioid peptides and opiate alkaloids in immunoregulatory processes. Arch Med Sci 6:456–460
Stefano GB, Martin R (1983) Enkephalin-like immunoreactivity in the pedal ganglion of Mytilus edulis (Bivalvia) and its proximity to dopaminecontaining structures. Cell Tissue Res 230:147–153
Stefano GB, Melchiorri P, Negri L, Hughes TK Jr, Scharrer B (1992) [D-Ala2]deltorphin I binding and pharmacological evidence for a special subtype of delta opioid receptor on human and invertebrate immune cells. Proc Natl Acad Sci U S A 89:9316–9320
Stefano GB, Scharrer B, Smith EM, Hughes TK Jr, Magazine HI, Bilfinger TV, Hartman AR, Fricchione GL, Liu Y, Makman MH (1996) Opioid and opiate immunoregulatory processes. Crit Rev Immunol 16:109–144
Stefano GB, Shipp MA, Scharrer B (1991) A possible immunoregulatory function for [Met]-enkephalin-Arg6-Phe7 involving human and invertebrate granulocytes. J Neuroimmunol 31:97–103
Sternini C, Patierno S, Selmer IS, Kirchgessner A (2004) The opioid system in the gastrointestinal tract. Neurogastroenterol Motil 2:3–16
Stevens CW (2004) Opioid research in amphibians: an alternative pain model yielding insights on the evolution of opioid receptors. Brain Res Brain Res Rev 46:204–215
Stevens CW (2009) The evolution of vertebrate opioid receptors. Front Biosci (Landmark Ed) 14:1247–1269
Stevens CW, Brasel CM, Mohan S (2007) Cloning and bioinformatics of amphibian mu, delta, kappa, and nociceptin opioid receptors expressed in brain tissue: evidence for opioid receptor divergence in mammals. Neurosci Lett 419:189–194

Stevens CW, Newman LC (1999) Spinal administration of selective opioid antagonists in amphibians: evidence for an opioid unireceptor. Life Sci 64:PL125–PL130

Thomas AW, Kavaliers M, Prato FS, Ossenkopp KP (1997) Pulsed magnetic field induced "analgesia" in the land snail, Cepaea nemoralis, and the effects of mu, delta, and kappa opioid receptor agonists/antagonists. Peptides 1918:703–709

Traynor JR, Wood MS (1987) Distribution of opioid binding sites in spinal cord. Neuropeptides 10:313–320

van Rijn RM, Whistler JL (2009) The delta(1) opioid receptor is a heterodimer that opposes the actions of the delta(2) receptor on alcohol intake. Biol Psychiatry 66:777–784

Verburg-van Kemenade BM, Savelkoul HF, Chadzinska M (2009) Function of the opioid system during inflammation in carp. Ann N Y Acad Sci 1163:528–532

Xia Y, Haddad GG (1991) Ontogeny and distribution of opioid receptors in the rat brainstem. Brain Res 549:181–193

Xia Y, Haddad GG (2001) Major difference in the expression of delta- and mu-opioid receptors between turtle and rat brain. J Comp Neurol 436:202–210

Xia Y, Jiang C, Haddad GG (1992) Oxidative and glycolytic pathways in rat (newborn and adult) and turtle brain: role during anoxia. Am J Physiol 262:R595–R603

Zhang J, Qian H, Zhao P, Hong SS, Xia Y (2006) Rapid hypoxia preconditioning protects cortical neurons from glutamate toxicity through deltaopioid receptor. Stroke 37:1094–1099

Zhang Y, Xu J, Wang Z, Zhang X, Liang X, Civelli O (2012) BmK-YA, an enkephalin-like peptide in scorpion venom. PLoS One 7, e40417

Chapter 4
Developmental Distribution of the δ-Opioid Receptor in Mammalian Brains

Taylor E. Hinchliffe and Ying Xia

Abstract The delta-opioid receptor (DOR) is widely distributed within the mammalian body, especially in the brain, and plays critical roles in a variety of physiological functions during development and in the adult. As compared to other opioid receptors such as mu- and kappa-opioid receptors, DOR is the slowest to develop and thus overall initially lags behind in early developmental abundance. Yet later in development, specifically near the well-examined time of approximately of 2 weeks postnatal in rats, DOR undergoes a rapid developmental process that ends in a distinctly heterogeneous distribution within the central nervous system. Fairly consistent across adult mammals is the formation of a decreasing rostral to caudal pattern of DOR protein and mRNA density with the highest levels detected in the superficial and deep layers of the cortex, caudate putamen and amygdala and much lower levels in the subcortical regions, such as most regions of the hippocampus, thalamus and hypothalamus. Additionally, DOR is encountered outside of the central nervous system in areas of wide variance, such as the immune and musculoskeletal systems (e.g., bone/bone marrow and loose connective tissue), the heart (on both myocardial cells and sparse individual nerve fibers), kidney, the gastrointestinal tract (muscle, mucosa and varicose fibers in the plexuses), and even in reproductive organs (testis and ovary). The broad and heterogeneous patterns of DOR distribution may imply high functional versatility, and likewise, the developmental changes of DOR could also be correlated to functional changes in various organs and different regions/cells within the same organ (e.g., rostral vs. caudal brain) from early life to the adult. This chapter aims to serve as a collation of the developmental distribution of DOR as well as its putative endogenous ligand, enkephalin, with particular emphasis on the central nervous system and encompasses their changes in both adult mammals as well as the chronological patterns throughout the stages of embryonic and postnatal development.

T.E. Hinchliffe, B.A.
Department of Neurosurgery, The University of Texas Medical School at Houston, Houston, TX 77030, USA

Y. Xia, M.D., Ph.D. (✉)
University of Texas Medical School, Houston, TX, USA
e-mail: ying.xia@uth.tmc.edu; y55738088@gmail.com

Y. Xia (ed.), *Neural Functions of the Delta-Opioid Receptor*,
DOI 10.1007/978-3-319-25495-1_4

Keywords Opioid receptors • Delta-opioid receptor • Enkephalin • Development • Brain • Heart • Peripheral tissues • Immunoregulation • Neurological disorders

Abbreviations

CNS	Central nervous system
DADLE	[D-ala2, D-leu5] enkephalin
DAMGO	[D-Ala2, N-methyl-Phe4, Gly5-ol] enkephalin
DELT-I	[D-Ala2, Asp4] deltorphin
DELT-II	[D-Ala2, Glu4] deltorphin
DOR	Delta-opioid receptor
DPDPE	[D-Pen2, D-Pen5] enkephalin
KOR	Kappa-opioid receptor
MOR	Mu-opioid receptor

1 Introduction

The delta-opioid receptor (DOR), mu-opioid receptor (MOR) and kappa-opioid receptor (KOR), are major family members of opioid receptors. These G protein-coupled receptors with seven-transmembrane domains have been investigated with increasing intensity through previous decades, and the rate of discovery has only hastened with the elucidation of overwhelming functional versatility. DOR, in particular, has recently been implicated as a key player in a multitude of clinically significant mechanisms, e.g., neuroprotective roles against hypoxic and ischemic injury in the brain (Zhang et al. 2000, 2006; Ma et al. 2005; Chao and Xia 2010; Feng et al. 2011; Tian et al. 2013; He et al. 2013; Wang et al. 2014; Chen et al. 2014; Cao et al. 2015) and cardioprotection (Gross et al. 2009; Huang et al. 2009; Peart et al. 2011; Shen et al. 2012; Maslov et al. 2013; Surendra et al. 2013; Borges et al. 2014; Tanaka et al. 2014). Moreover, DOR activation has been shown to induce not only central but also peripheral antinociception (Pacheco Dda et al. 2012a, b; Zhang and Pan 2012). Furthermore, mammalian DOR is distributed broadly enough to be directly or indirectly involved in neural networks regulating pain sensitivity, drug reward and inhibitory controls, learning processes, and even food intake (Feng et al. 2012; Chu Sin Chung and Kieffer 2013; Kaneko et al. 2014).

There are major differences between ages in terms of DOR-relevant activities, being either central or peripheral events. Therefore, it is essential to learn these age-differences by mapping DOR distribution and density in the brain and other organs through development. This is crucial towards utilizing myriad cases of regionally specific functionality in different ages for accurately facilitating effective translational research pivoting around the roles of DOR in physiological and pathophysiological conditions.

In both the human and rat central nervous system (CNS), DOR is primarily distributed in a rostral to caudal pattern of which is highest in the cortex and caudate putamen and lowest in most regions of the spinal cord (Xia and Haddad 1991, 2001; Madar et al. 1996). DOR density also displays notable variance per layer of the cortex (Xia and Haddad 1991, 2001). DOR exists not only in neurons, but is additionally detectable in glial cells (Pilapil et al. 1987; Blackburn et al. 1988; Ruzicka et al. 1995; Hiller and Fan 1996; Liang et al. 2014). Besides the vast majority of DOR-related research focusing on the CNS, DOR is also found in a variety of peripheral organs and is involved in immunoregulatory functions (Hedner and Cassuto 1987; Wittert et al. 1996; Sternini et al. 2004). Its wide distribution grants even more importance to a thorough understanding of localization if DOR is to be utilized and manipulated effectively without unintended consequence. Such is crucial in order to avoid manipulator inhibitory or activating ligands spilling over to regions of potentially contrasting and/or separate physiological function.

The majority of literature, particularly those using rat models, note DOR to be the slowest developing opioid receptor that proceeds to rapidly increase in distribution and density not long after birth: in rats, approximately 2 weeks after birth (Wohltmann et al. 1982; Milligan et al. 1987; Szücs et al. 1987; Xia and Haddad 1991; Slamberová 2004). Much later in life, the levels of DOR in the CNS can display gradual drops with age, especially so in regions such as the caudate putamen, and these effects are notably exacerbated by diseases such as Alzheimer's (Hiller et al. 1987; Xia and Haddad 1991; Barg et al. 1993). Therefore, the effects of directly manipulating DOR-regulated processes with specific ligands or alterations in gene expression are capable of being blatantly distinct between different subject age groups.

Existing literature over DOR development remains scattered, and many gaps exist especially when transitioning from high specificity information at the cellular level up to mapped regional distinctions. Both inside and out of the CNS, the striking asynchrony of DOR developmental distribution is also not often factored in. In this chapter, we will comprehensively review the developmental distribution of DOR as well as its putative endogenous ligand, enkephalin, in the CNS. For the purpose of comparison, we will also review the developmental changes in other opioid receptors such as MOR and KOR. Moreover, we will briefly discuss the developmental changes and distribution of DOR and other opioid receptors outside of the central nervous system, with regions including a variety of peripheral organs and the immune and musculoskeletal systems.

2 Low Density of the δ-Opioid Receptor (DOR) in Fetal Life

At the earliest stages of mammalian development, the fetal and neonatal periods, the distribution of DOR is notably low, almost or completely undetectable in some instances, but increases rapidly within the first few weeks after birth (Milligan et al. 1987; Xia and Haddad 1991; Georges et al. 1998). In contrast, MOR density is much higher than that of DOR in the rat brainstem at birth (Xia and Haddad 1991). DOR therefore differs considerably in ontogeny as compared to other opioid receptors, as most studies have found it to be the slowest receptor to develop (Wohltmann et al. 1982; Milligan et al. 1987; Szücs et al. 1987; Slamberová 2004).

2.1 Rostral Brain

Specifically within the CNS of fetal rats, very little DOR mRNA is detectable overall in regions sporting neuronal proliferation, and DOR development is slow to act through and immediately after gestation, as compared to MOR and KOR (Leslie et al. 1998). As early as the 13th day of rat gestation, both MOR and KOR mRNAs are detectable in the telencephalon, while MOR mRNA is detected in the striatal neuroepithelium and cortical plate, and KOR mRNA, in the corroidal fissure (Georges et al. 1998). Also within 13 days of gestation in rat caudate putamen, MOR-G protein coupling first becomes detectable through the administration of [D-Ala2, N-methyl-Phe4, Gly5-ol] enkephalin (DAMGO), a MOR agonist (Nitsche and Pintar 2003). This activity then further increases 3 days later, extending to the midbrain, medial habenula and hypothalamus (Nitsche and Pintar 2003). It is not until approximately embryonic day 17.5 that extremely faint DOR activity is encountered in the hypothalamus and habenula (Nitsche and Pintar 2003). By week three of gestation, DOR mRNA begins to appear in small quantities in the nucleus accumbens and dorsal striatum, as well as the thalamus (Georges et al. 1998). Also at the third gestational week, MOR mRNA becomes visible as small patches in the striatum (Georges et al. 1998; Leslie et al. 1998). Upon the emergence of fetal DOR during rat gestation and through birth, regions containing DOR mRNA display homogenous DOR distribution throughout this time period (Georges et al. 1998).

It is noteworthy that there may be an exception in some nuclei/cells in terms of the general consensus of comparatively slow DOR development. In an investigation with immunohistochemistry into the development of DOR in human brain, for example, fluctuations in human DOR levels were detected in aforementioned ventricular and subventricular zone cells during the gestational period, overall highest in these regions during gestational weeks 11–16, and afterwards, are downregulated by the 20th gestation week (Tripathi et al. 2008). Yet despite the downregulation at the 20th gestational week, DOR levels here are still found to be *higher* than MOR levels (Tripathi et al. 2008). The authors believe that the gestational DOR fluctuations and downregulation may be explained by a decrease in DOR mRNA by immature neurons and developing glial cells as well as the concurrent decrease in stem cell number through the progression of development (Tripathi et al. 2008). It is known that the ventricular and subventricular zones are fairly unique in being significant sources of abundant self-renewing stem cells and, likewise, sources of cortical neurons in the developing mammalian brain (Tabata 2013). These zones acting as key developmental loci are thus even more pronounced in the large human brain (Tabata 2013), and it could be possible that DOR developmental ontogeny here is a special case. Likewise, concerning peak levels of DOR mRNA at gestational weeks 11–16, since cortical layers II and III are predicted to be formed during this stage of human fetal development, an upregulation of DOR in these zones at this time could indeed correspond to DOR regulatory/developmental involvement in human cortical formation (Tripathi et al. 2008).

DOR mRNA in the developing rat cortex emerges as two broad external and internal bands corresponding to cortical layers II and V, albeit with homogenous expression (Georges et al. 1998). Although fetal rat DOR distribution is observed to stay relatively low and homogeneous during rat gestation (Georges et al. 1998) and seems to lack the up- and down-regulation cycle as observed during human gestation (Tripathi et al. 2008), it is likely to play a role in cortical development in both species.

2.2 Cerebellum, Brainstem and Spinal Cord

Our early work showed that opioid receptors develop earlier in the brainstem than in the rostral regions such as the cortex (Xia and Haddad 1991). DOR becomes noticeable by embryonic day 17.5 in the pons and medulla, arising approximately 2 days later than MOR activity, of which is also notably higher during this period (Nitsche and Pintar 2003). Throughout these later embryonic developmental stages of rodents, DOR exists on both the cell membrane and within the cytoplasm of brainstem neurons, and these neurons often coexpress both DOR and MOR (Kivell et al. 2004a, b). The distribution of embryonic DOR in the brainstem differs considerably from embryonic forebrain DOR, at least among rat studies (Kivell et al. 2004a). DOR protein density is comparatively high in fetal and neonatal rat brainstem, but significantly lower in juveniles and adults (Xia and Haddad 1991; Kivell et al. 2004b). This developmental pattern contrasts with the homogeneity of DOR distribution upon its appearance in the fetal rodent forebrain (Georges et al. 1998). It is comparable to the cycles of DOR up- and downregulation through human embryonic forebrain development (Tripathi et al. 2008).

Not only is the brainstem a site of early DOR development, but some investigators believe that the brainstem may also serve as a regulatory hub of opioid receptors that controls or influences the development of DOR and/or MOR, which go on to further influence macro-developmental features ranging from breathing to analgesia (Kivell et al. 2004b). Kivell et al. (2004b) note that this control effect might be accomplished through actions such as changes in opioid receptor isoforms, as it has been found that MOR may undergo a postnatal isoform switch corresponding to the transition from fetal and neonatal developmental patterns to juvenile and adult patterns (Kivell et al. 2004b). They also observed that although an obvious postnatal DOR isoform switch in the brainstem was not observed, two distinct immunoreactive DOR bands (30 and 60 kDa) were detected that were never individually dominant throughout rat development (Kivell et al. 2004b).

Despite the overall scarcity of fetal DOR, studies have also investigated its presence (or lack thereof) in the cerebellum and spinal cord. KOR constitutes roughly 80 % of opioid receptors in the cerebellum during the earliest developmental stages of rodents and is widely expressed in all hindbrain germinal zones by the third gestational week (Barg and Simantov 1989; Leslie et al. 1998). Despite this, fairly abundant DOR and MOR mRNA expression is found in the external germinal layer of the immature cerebellum at embryonic day 16 in rats, offering up another example of opioid colocalization during early development (Mrkusich et al. 2004).

Colocalization of DOR and MOR is also detectable in fetal sheep cerebellum at gestational days 97–101, with putative ligand binding roughly 70 % of maternal values (Dunlap et al. 1986). In contrast to the cerebellum, embryonic spinal cord DOR in rodents is only detectable beginning at the first postnatal day (Attali et al. 1990). The discrepancies arise upon examining spinal cord in the humans. DOR and MOR binding is observable in human spinal cord at the fetal age of 14 weeks, primarily localized to gray matter and densest in the superficial layers of the dorsal horn (Salès et al. 1989). At 24 weeks, despite a lack of observable enkephalin-like immunoreactivity, DOR binding in human spinal cord is transiently localized within the fasciculus gracilis at the cervical level (Salès et al. 1989). Whether the noted variations in early spinal cord DOR between rats and humans should merely be attributed to differences at the species level, or are instead due to experimental methodologies requires further investigation.

2.3 Early Appearance of Endogenous Ligands for Opioid Receptors

Enkephalin is the major endogenous ligand of DOR, though it is not purely specific as it also binds to other opioid receptors at higher concentrations. The appearance of enkephalin seems earlier than that of DOR. During the second half of rat gestation near embryonic days 10 and 11, proenkephalin mRNA becomes widely expressed in forebrain germinal zones as well as the choroid plexus (Leslie et al. 1998). Met-enkephalin is detectable soon after as early as embryonic day 11.5 and continuously rises in immunoreactivity through prenatal brain development, but all the while no significant DOR quantities are found during this time period (Rius et al. 1991; Leslie et al. 1998). In addition to DOR, this appears to be true of MOR and KOR as well. Immunoreactivity of the highest affinity peptides of each receptor, including dynorphin to KOR and β-endorphin to MOR, is detectable before the appearance of their primary receptors for reasons still not fully understood (Rius et al. 1991). Furthermore, although dynorphin quantities continue to rise throughout prenatal rat brain development in a similar manner to met-enkephalin, levels of β-endorphin fluctuate during (and after) gestation, while met-enkephalin levels rise even more dramatically after birth (Rius et al. 1991). From this, it seems that the developmental patterns of these opioid peptides therefore act independently from opioid receptors and in an early and long-lasting manner.

2.4 Development of Peripheral Opioid Receptors

Studies mapping the existence and localization of prenatal DOR mRNA and protein outside of the CNS are scarce and a great deal of further research is required. One example of peripheral DOR protein distribution is in the developing mouse lung, of which indeed contains DOR, MOR and KOR (Gómez-Román et al. 2002).

More specifically, DOR mRNA arises in lung epithelial and mesenchymal cells as early as embryonic day 18, alongside MOR and KOR (Sunday et al. 2001). With such a varied DOR distribution even throughout the earliest stages of mammalian life, it is therefore clear that the receptor plays a wide variety of roles in helping to conduct the biochemical orchestra of development, and many of these roles are yet to be fully elucidated.

Other, less frequently studied regions in regards to opioid receptors and their peptides have also been found to contain DOR during the fetal and neonatal period, and even earlier. DOR may act in facilitating the maturation of human oocytes, as the receptor maintains a constant localization in peripheral tissue throughout the germinal vesicle stage to metaphase II (Agirregoitia et al. 2012). Although MOR and KOR mRNAs are also detectable peripherally and internally during the germinal vesicle, metaphase I and metaphase II stages, only MOR mRNA is found in mature human oocytes (Agirregoitia et al. 2012). Still, proteins of all three receptor types are found in the mature oocyte, and the disappearance of DOR and KOR mRNA is believed to be attributed to transcription silencing from meiosis resumption as well as selective mRNA degradation during the later stages of oocyte development (Agirregoitia et al. 2012). The developmental patterns of each opioid receptor are also distinct, as DOR maintains a homogenous peripheral localization throughout early oocyte development, while MOR and KOR display distinct peripheral and internal fluctuations in localization pattern (Agirregoitia et al. 2012). Therefore, it would appear that each receptor acts fairly independently very early on in human development, further differentiating the roles played by the three receptors during meiosis.

3 Rapid Development After Birth

Upon birth in mammals, DOR is generally similar in distribution and density as compared to the fetal period up until a certain point. It is only after birth when DOR in the neonatal mammalian brain undergoes a developmental change that have been well documented. Most noticeable is a rapid and significant increase in DOR density, while DOR binding affinity remains relatively unchanged (Tavani et al. 1985; McDowell and Kitchen 1986; Szücs and Coscia 1990; Xia and Haddad 1991). Although it is still unknown as to exactly which mechanisms accelerate the rapid ontogenesis of DOR in such a specific window of time, these mechanisms would be resultant of this developmental period's neuronal functional requirements.

3.1 Overview of Postnatal Changes in DOR Binding Capacity

By approximately the 14th postnatal day in rats, but varying by brain regions, DOR makes up for its overall late ontogenesis with notably rapid development at a level more comparable to MOR and KOR, despite its initial asynchrony (De Vries et al. 1990).

This stage of significant developmental progression lasts roughly 2 weeks, from the second postnatal week to the fourth, and can be noted as a marked increase in DOR binding capacity, but not so in binding affinity (to the highly selective DOR ligand DPDPE), throughout a range of rodent CNS regions (Tavani et al. 1985; McDowell and Kitchen 1986; Szücs and Coscia 1990). Over this period of approximately 2 weeks, beginning near postnatal days 10–14, studies using whole rat brain have found that DOR binding capacity can increase in ranges upwards of three- to eightfold, depending on the ligands used: DPDPE and ^{3H}[D-ala2, D-leu5] enkephalin, (DADLE), respectively (Spain et al. 1985; Kitchen and McDowell 1986). Over an even longer period of development, later results also using the ligand DPDPE found an overall DOR binding capacity increase of fivefold from the fourth postnatal day to adulthood (Szücs and Coscia 1990).

3.2 Rostral Brain

The evident display of heterogeneity between opioid receptors upon birth is not apparent in the rostral brain, including the cortex, as both cortical DOR and MOR are scarcely distributed during this time (Xia and Haddad 1991). More notably homogeneous distribution on the first postnatal day occurs in the brainstem (Xia and Haddad 1991). DOR undergoes a developmental delay followed by a relative rapidity in the rostral brain after birth. In the case of one such region, that of the rat basal forebrain, DOR density is very low at birth, and increases most markedly during the third postnatal week, instead of during the second, but the occurrence of rapid DOR development still overall lingers around the period of a few weeks postnatal (Kornblum et al. 1987). This pattern of developmental delay followed by rapidity is also observable down to both cellular and molecular levels. In rat caudate putamen, for example, DOR immunolabeling is undetectable at birth (P0), while DOR binding with DPDPE can be detected with autoradiography as early as the first postnatal day (Nitsche and Pintar 2003; Wang et al. 2003). By P5, DOR immunolabeling in caudate putamen is still light, but becomes dense by the tenth postnatal day, once again appearing later than MOR, of which is thought to be more directly related to synaptogenesis (Wang et al. 2003). Particularly during the period of postnatal days 5–10, DOR labeling is primarily localized to somatodendritic areas as well as axon terminals in the caudate putamen, while other labeled particles found in dendrite and terminal cytoplasm are relatively unaltered with rat age (Wang et al. 2003).

3.3 Cerebellum, Brainstem and Spinal Cord

DOR density is very low in the brainstem and spinal cord at birth (though higher than in the rostral brain) as it is hardly quantitated by autoradiographic approach (Fig. 4.1), while MOR density is much higher (Fig. 4.2) (Xia and Haddad 1991).

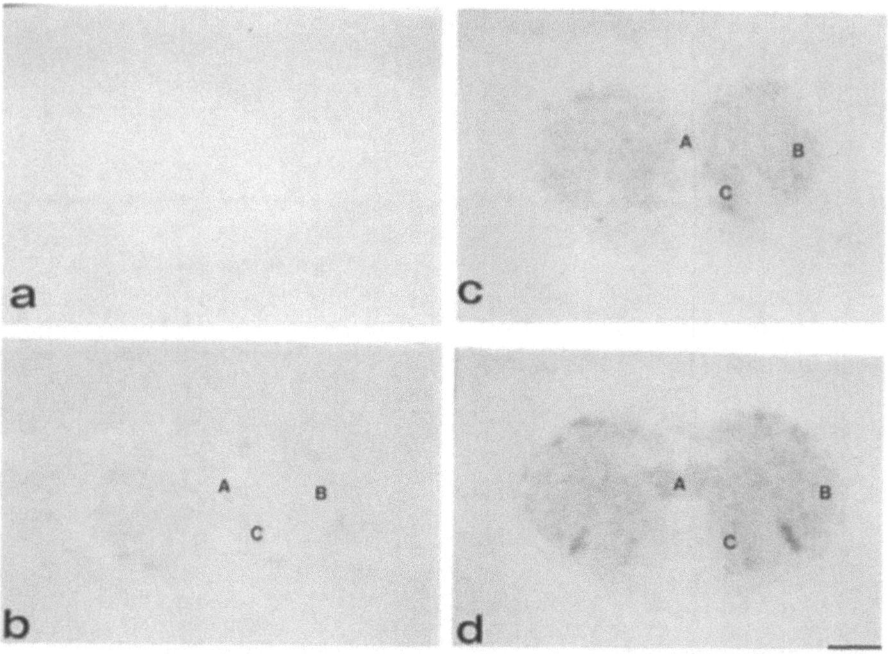

Fig. 4.1 Developmental time course of DOR binding sites in the medulla. DOR were labeled by 4 nM ^{3H}DADLE in the presence of 1 μM unlabeled PL017, a MOR ligand (to block DADLE binding to MOR). (**a**) P1; (**b**) P5; (**c**) P10; (**d**) P21. Bar scale = 1 mm. The labeled areas are as follows: *A* hypoglossal nucleus, *B*, caudal part of the spinal trigeminal nucleus, *C* ventral part of medullary reticular nucleus. Note that DOR density is very scarce at P1 and reaches peak level at P21. However, the density is still very low as compare to that of MOR (Fig. 4.2). Cited from Xia and Haddad (1991)

Likewise, the work of immunohistochemistry demonstrates similar results. Kivell et al. (2004a, b) observed that upon birth, neonatal DOR immunoreactivity is considered low overall in the brainstem and only slightly more intense than fetal DOR immunoreactivity, while for comparison, MOR displays a faster onset of development. Through postnatal days 5–10, DOR density in the brainstem, upon assessment with autoradiography, displays a pattern of relatively rapid development. However, the density was very low in all regions (Fig. 4.1) as compared to that of MOR (Fig. 4.2) (Xia and Haddad 1991). Peak DOR binding density is also reached sooner in the brainstem than the cortex of rats, occurring as early as the tenth postnatal day in some regions, compared to the 21st in the cortex (Xia and Haddad 1991). In most of the brainstem regions, DOR density reaches the highest level within 3 weeks after birth. As shown in our autoradiographic study, DOR density in most brainstem regions is lower at postnatal day 120 than at postnatal day 21 (Xia and Haddad 1991). Furthermore, the pattern of postnatal DOR development is more heterogeneous in the rat brainstem than in the cortex in the immature brain, which

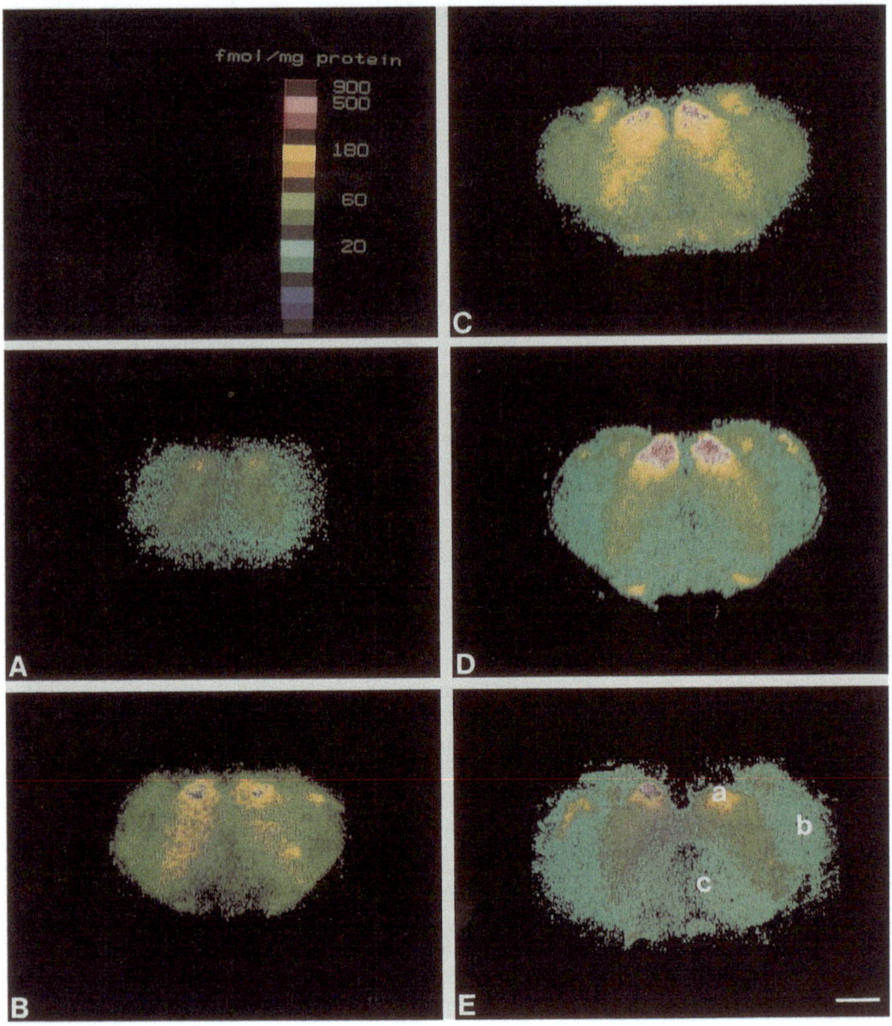

Fig. 4.2 Developmental time course of MOR binding sites in the medulla. MOR sites were labeled by 4 nM [3H]DAMGO. (**a**) P1; (**b**) P5; (**c**): Pl0; (**d**) P21; (**e**): adult. Bar scale = 1 mm. The labeled areas are as follows: *a* medial part of the nucleus of the solitary tract, *b* interpolar part of the spinal trigeminal nucleus, *c* paramedian reticular nucleus. Note that the nucleus of the solitary tract has a higher DAMGO-site density than other areas at postnatal day 1 and the binding density increases with age, peaking at P21. However, MOR sites in the reticular nucleus decrease in density after P10. Cited from Xia and Haddad (1991)

is very different from the distributional pattern in the adult brain that shows a highly heterogeneous pattern in the rostral regions (see Sect. 4).

Within rat spinal cord, DOR binding and immunoreactivity, intriguingly, decreases beginning from the end of the first postnatal week to the start of the third, specifically so in dorsal root ganglia larger primary sensory neurons, as well as in the superficial dorsal horn (Rahman et al. 1998; Beland and Fitzgerald 2001). Thus, not only is the rapid and transient development of DOR often not in parallel with other opioid receptors—fluctuations even occur in DOR binding and expression between different regions in the CNS, adding further degrees of complexity to its developmental functionality.

Zhao et al. (1998) observed that DOR binding on rat purkinje cells becomes detectable in abundance at parallel fiber synapses on postsynaptic membranes after the tenth postnatal day and on to adulthood, but are only found in high numbers at climbing fiber synapses from the 10th to 14th postnatal days (Zhao et al. 1998). Later, Mrkusich et al. (2004) showed that DOR as well as MOR mRNA and protein could be detected in the external germinal layer of the immature cerebellum at embryonic day 16 in rats. Then, in addition to the external germinal layer, DOR and MOR also become detectable within Purkinje cells (PK) as well as in the granular and molecular layers of 6-day old neonatal rats (Mrkusich et al. 2004). As Mrkusich et al. (2004) only examined DOR in embryonic, postnatal day 6 and adult rats, it remains unknown as to whether DOR in the rat cerebellum also exhibits a rapid quantitative increase at approximately 2 weeks postnatal as observed in other brain regions, particularly in the rostral brain.

On the other hand, an early study showed that in the human cerebellum, DOR binding capacities had *decreased* nearly 7.8-fold when comparing infant and adult brain (Zagon et al. 1990). Such a huge age-difference in DOR binding density is totally different from the pattern of DOR development in other brain regions and requires further investigation. More specifically, deeper investigation would aid in verifying whether or not this difference is specific to the human cerebellum or if it was related to other factors (e.g., the conditions of the measurements, tissue sampling and storage, etc.).

3.4 Continued Development of Peripheral Opioid Receptors

After birth, the period of rapid DOR development appears less observable in peripheral tissues, such as the lungs of rodents (Sunday et al. 2001). DOR development observed via immunohistochemical staining in rat lung peaks at birth and decreases several-fold by adulthood, instead of displaying the pattern of developmental delay followed by rapidity near the second postnatal week, as seen frequently in the CNS (Gómez-Román et al. 2002). Still, this peaking and decline following gestation implies an underlying series of developmental mechanisms facilitated or affected by DOR in the lungs that are yet to be clearly understood.

A rapid increase in postnatal DOR development (density) is also lacking in the heart, where DOR mRNA and protein overall only increase gradually from the first postnatal day into adulthood (Mousa et al. 2011). Early postnatal cardiac DOR is most significantly and specifically distributed to and detected in large-diameter principal neurons, catecholaminergic cells, and nerve terminals branching inside atrial myocardium and intracardiac ganglia (Mousa et al. 2011). The more gradual chronological increase in density of atrial myocardium nerve fibers with DOR co-expression parallels the growth of cardiac parasympathetic, sympathetic, and sensory innervation, of which suggests an important DOR regulatory role in heart development and function (Mousa et al. 2011).

Furthermore, a steadier DOR growth pattern also takes place in rat cochleae, with DOR mRNA levels rising gradually from birth to the beginning of the second postnatal week and ceasing increases in mRNA expression by the 16th postnatal day (Phansuwan-Pujito et al. 2003). At birth, DOR mRNA is detectable in the spiral ganglion on nerve cell soma, and appears in hair cells within the organ of Corti by postnatal day 4 (Phansuwan-Pujito et al. 2003). On the other hand, in some areas of the ear, DOR development is indeed delayed. Near the eighth postnatal day, DOR mRNA in the ear can first be found in the limbus on cells proximal to the tectorial membrane as well as in spiral ganglion nerve fibers (Phansuwan-Pujito et al. 2003). It is this transient and gradual development that hints at DOR also playing a role in the functional development of the inner ear, potentially including the development and onset of hearing (Phansuwan-Pujito et al. 2003). Although human data concerning early postnatal opioid receptor binding and distribution is scarce if not nonexistent, it is clear from studies on mammals that throughout a wide range of regions within the developing mammalian body, DOR is capable of contributing to an equally varied range of physiological functions, and many of these functions still require significantly more investigation.

4 Heterogeneous Distribution in the Adult

By adulthood, mammalian DOR ceases its rapid developmental patterns and can be detected heterogeneously throughout the brain and spinal cord as well as in peripheral nerves and other organs. Most information is currently derived from adult rodent studies with several cases of human research factored in, and distinct parallels can be drawn between the two over DOR distributional patterning. In general, the density of DOR is much higher in the rostral than caudal brain (Fig. 4.3).

4.1 General Distributional Pattern of Mammalian DOR in the CNS

The pattern in particular that is widely observable across mammals is a fairly prominent rostral to caudal DOR distribution (Xia and Haddad 1991, 2001; Madar et al. 1996). This rostral to caudal distribution in descending DOR density is upheld across species ranging from mice/rats and guinea pigs to humans, with only subtle distinctions in between (Sharif and Hughes 1989; Xia and Haddad 1991, 2001; Madar et al. 1996).

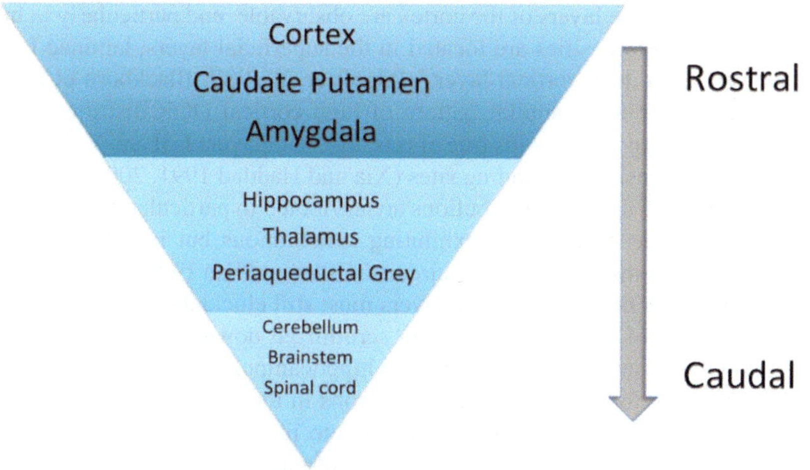

Fig. 4.3 Descending DOR density: Rostral to caudal rat brain

In the case of humans, density declines when proceeding from high levels in the neocortex (including insular, parietal, frontal, cingulate, and occipital cortices), caudate nucleus and putamen, to the amygdala with intermediate density, and to the cerebellum, thalamus and brain stem with the lowest densities (Xia and Haddad 1991, 2001; Madar et al. 1996). For comparison, this rostral to caudal distributional pattern is strikingly similar in the CNS of rats (Xia and Haddad 1991, 2001) and mice (Jenab et al. 1995). This pattern is also true in the detection of DOR mRNA. The highest levels of DOR mRNA are found in the frontal cortex (Jenab et al. 1995). Outside of the cortex, the most considerably high areas of rat and mouse DOR distribution that include significant levels of overlapping DOR mRNA expression and binding density are allocated to additional sensory processing areas such as the olfactory tubercle, as well as to caudate putamen, nucleus accumbens, amygdala and throughout the striatum (Moskowitz and Goodman 1984; Mansour et al. 1993, 1994), followed by intermediate to low levels in the hippocampus, periaqueductal grey, thalamus, cerebellum and spinal cord (Jenab et al. 1995). The regions in the rat CNS with a high-density DOR distributional pattern are consistent with studies analyzing human brain (Simonin et al. 1994).

4.2 *Rostral Brain*

With high DOR mRNA and protein expression particularly attributed to mammalian cortex, there are fewer notable distinctions between rats and humans in this region, as compared to other areas of the brain (Blackburn et al. 1988; George et al. 1994; Peckys and Landwehrmeyer 1999; Platzer et al. 2000). Still, distributional differences within the layers of the cortex are observable, and particularly in human brain, higher DOR densities are located in the superficial layers, laminae I, II and IIIa, as well as in deep cortical layers (Pilapil et al. 1987; Blackburn et al. 1988; Hiller and Fan 1996). A similar pattern of high cortical DOR distribution in the superficial and deep layers holds true of rats, too, with layers I–II and V–VI containing the highest densities of binding sites (Xia and Haddad 1991, 2001; Renda et al. 1993). Surely such specific distributions are attributed to particular neural functions that arise from the necessity of exhibiting synchronous but interdependent neurotransmitter systems in large neuro-circuits (Hiller and Fan 1996), though the exact roles played by DOR across cortical layers must still elucidated.

Other regions of the brain have been scrutinized down to the cellular level in regards to DOR distribution. Within mouse hippocampus, for example, DOR receptors are primarily found in GABAergic neurons in the Ammon's horn and dentate gyrus, and consequently have been concluded to likely play roles in hippocampal regulatory activity (Erbs et al. 2012). Glial cells from rat brain also display differential distribution of DOR mRNA expression, of which is highest in the cortex and hypothalamus (Ruzicka et al. 1995). Intriguingly, the expression of glial DOR mRNA is highest overall in the brain as compared to KOR and MOR mRNA: up to two and four times more, respectively (Ruzicka et al. 1995). Furthermore, DOR is not only allocated differently between neurons and glial cells, but also shows differential cellular and subcellular distributional patterns between its separate immunological variants, where C-terminus antibodies primarily label perikarya and proximal arbors, and N-terminus antibodies label dendritic and terminal arbors (Cahill et al. 2001).

4.3 Cerebellum, Brainstem and Spinal Cord

DOR mRNA and protein levels are low in the adult rodent cerebellum, but can still be observed. For example, cerebellar DOR mRNA in mice is only about 10 % of the level in the caudate putamen (0.4 pg/μg RNA in the cerebellum vs. 3.9 pg/μg RNA in the caudate-putamen) (Kest et al. 1994). Using DOR immunohistochemical localization and fluorescent in situ hybridization (FISH), DOR protein and mRNA were also detected within Purkinje cells (PK) and the granular layer of adult rat cerebellum (Mrkusich et al. 2004). Expression of DOR and MOR mRNA was also detected within cells of the molecular layer, but at lower levels than those seen within the PK cells (Mrkusich et al. 2004). Due to the lower levels of DOR and MOR mRNA in the molecular layer, Mrkusich et al. (2004) were unable to determine whether DOR and MOR are coexpressed here, or are resultant of two distinct cell types.

Further caudally in both the brainstem and spinal cord of the adult rat, DOR distribution is fairly homogenous in most regions, more homogenous than both MOR distribution in these regions as well as cortical DOR distribution (Xia and Haddad 1991, 2001) (Fig. 4.4). In rat spinal cord, for example, MOR is distributed heterogeneously with a high distribution in the dorsal region while DOR levels are low and homogeneous throughout the spinal cord (Xia and Haddad 2001). Indeed,

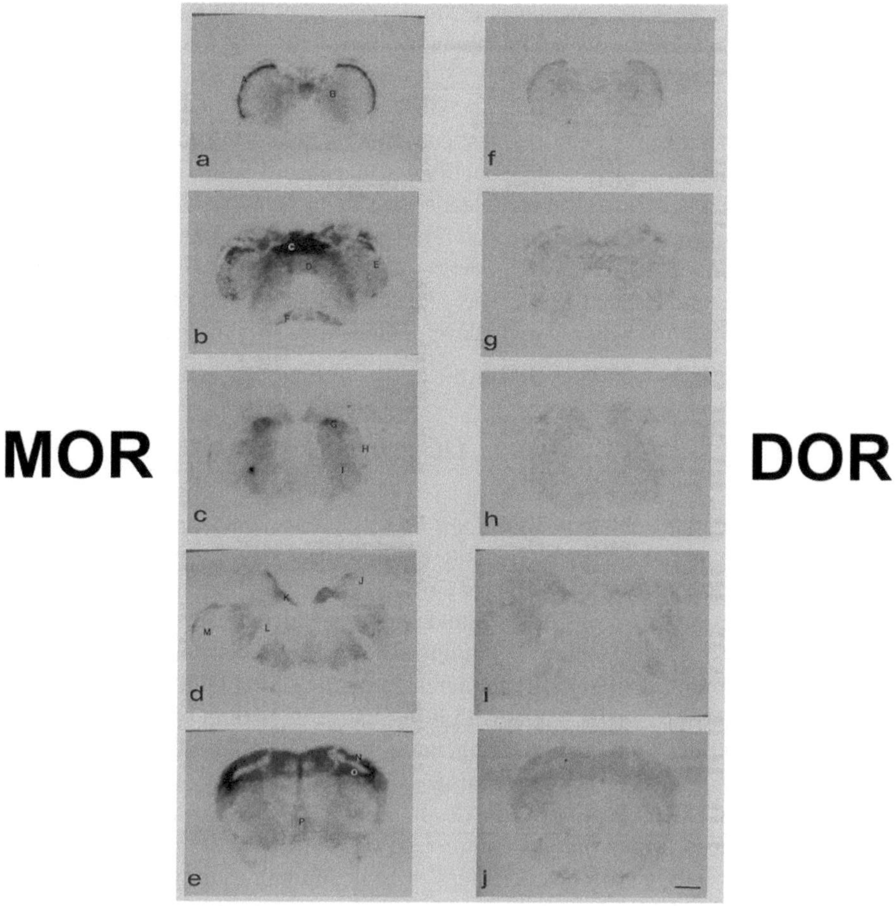

Fig. 4.4 Autoradiographic distribution of MOR and DOR in the brainstem. MOR sites (Left column, **a–e**) were labeled by 4 nM [3H]DAMGO and DOR sites (right column, **f–j**) were labeled by 4 nM [3H] DADLE in the presence of 1 μM unlabeled PL017, a MOR ligand. Film background and non-specific binding, which was diffuse and homogeneous all through a brainstem section, have been substrated from these images. *A* the spinal trigeminal nucleus (caudal part), *B* dorsal part of medullary reticular nucleus, *C* the nucleus of the solitary tract at the level of area postrema, *D* hypoglossal nucleus, *E* the spinal trigeminal nucleus (caudal part), *F* inferior olive, *G* the nucleus of the solitary tract (rostral part), *H* the spinal trigeminal nucleus (interpolar part), *I* ambiguus nucleus, *J* superior vestibular nucleus, *K* suprageniculate nucleus of the pons, *L* facial nerve, *M* the spinal trigeminal nucleus (oral part), *N* dorsal parabrachial area, *O* ventral parabrachial area, *P* raphe pontis nucleus. Note that the distribution of MOR sites is heterogeneous with high densities in the nucleus of the solitary tract, the ambiguus nucleus, dorsal motor nucleus of vagus, the parabrachial area, dorsal raphe nucleus, laterodorsal tegmental nucleus and the lateral margin of the spinal trigeminal nucleus, whereas DOR distribution is homogeneous with low densities all through the brainstem. Cited from Xia and Haddad (1991)

MOR binding is at a density several times higher than DOR binding (Xia and Haddad 1991) in the brainstem and spinal cord (Fig. 4.2). Therefore, the adult mammalian caudal brain is therefore generally thought of as a MOR region (Xia and Haddad 1991, 2001).

The brainstem also displays an overall similar pattern of DOR distribution as with the spinal cord. Yet despite overall homogenous distribution and low density of DOR in the brainstem (as compared to MOR), DOR density is highest in certain nuclei such as the dorsal hypoglossal nucleus (Xia and Haddad 1991). This instance of differential distribution to the dorsal hypoglossal nucleus strongly links DOR to possible regulatory involvement in developmental functions at the macro scale, such as sucking, swallowing, and respiration (Xia and Haddad 1991). In many areas of the rat brainstem, DOR density approaches peak/adult levels earlier (P10) than in the cortex (P21), and almost all brainstem regions experience a decrease in DOR binding sites from P21 to P120 (Xia and Haddad 1991). This indicates that in the brainstem, DOR might play more important roles during early development after birth than in adulthood.

4.4 Developmental Change in DOR Binding Properties

It is worth noting that some studies did *not* find a notable change in binding affinity from the neonatal rodent brain to adult brain using DPDPE as a DOR ligand (McDowell and Kitchen 1986; Szücs and Coscia 1990). Between postnatal day 5 and adult rats, Szücs and Coscia (1990) found a fivefold increase in DPDPE binding capacity, but no significant change in binding affinity. McDowell and Kitchen (1986) encountered a similar pattern: marked increases in DPDPE binding capacity during the second to fourth postnatal weeks with little change in binding affinity. It is worth noting, however, that whole brain homogenates were used in the work.

On the other side, several studies suggest that DOR displays notably different pharmacological characteristics during development. With DOR levels still lower in the neonatal stages as compared to adult rats, one difference is the chronological shift in DOR's binding affinities. Changes in DOR binding affinity are less noticeable during neonatal stages, such as the period of a rapid increase in DOR density, but become observable when neonatal and younger mammalian brain is placed in juxtaposition with older mammals. High affinity DOR binding sites are more abundant in the striatum and hippocampus of neonatal rat and guinea pig, as well as mouse whole brain, than those of 60-day adults, where binding sites with high affinity to DPDPE additionally decrease by approximately 35 % from the 25th postnatal day to the 60th (Barg and Simantov 1991; Negri et al. 1997). Furthermore, the large initial percentage of high to low affinity DPDPE binding sites is still apparent on the 15th postnatal day: 72 % and 28 %, respectively, and merely shifts to 65 % and 35 % by the 25th day (Negri et al. 1997). This produces a strong correlation with the onset of the rapid developmental stage by the 14th postnatal day, e.g., this period would consist of DOR both quickly reaching its peak, adult densities, while simultaneously yielding its highest binding affinities (De Vries et al. 1990; Negri et al. 1997).

Concerning DOR binding, DPDPE is known to be less potent than both of the ligands ^{3H}[D-Ala2, Asp4] deltorphin (DELT-I) and ^{3H}[D-Ala2Glu4] deltorphin (DELT-II), and these potencies also carry an ontological transience (Negri et al. 1997).

The efficacy of DELT-II increases at a rate three times faster than DELT-I and DPDPE during postnatal days 10–25, potentially due to DOR maturation progressing towards the delta-2 subtype (Negri et al. 1997). Such a progression seems further validated by both of the DOR ligands DPDPE and DADLE having limited to no interactive efficacy on δ2-subtypes (Vanderah et al. 1994; Stefano et al. 1995). In comparison, MOR also undergoes a change in peptide selectivity as a function of increasing age, including a gradual decrease in affinity to the mu-selective agonist DAMGO (Barg et al. 1992; Hoskins et al. 1998).

Elucidating the transient, developmental binding properties of DOR offers unique insight into its earliest physiological roles, and such developmental differences can also be viewed from varying CNS regions, and at various depths.

4.5 *Peripheral Opioid Receptors in Adult Mammals*

Although less frequently examined thus far, DOR is also present in varying, heterogeneous densities and is involved in many functions outside of the mammalian CNS, e.g., peripheral sensory neurons (Gooshe et al. 2015). DOR immunoreactivity, for example, has been detected alongside enkaphalin in the gastrointestinal tract, specifically within muscle, mucosa and varicose fibers in the plexuses in humans, as well as with a more general distribution throughout the large and small intestines of rats (Hedner and Cassuto 1987; Wittert et al. 1996; Sternini et al. 2004). Another muscular organ containing DOR is the mammalian heart, where it is present in both adult humans and rats in myocardial cells and on sparse individual nerve fibers, and even appears to be the predominant cardiac opioid receptor in the case of rodents, where little to no KOR or MOR binding is detectable (Wittert et al. 1996; Villemagne et al. 2002; Sobanski et al. 2014).

DOR transcripts have additionally been found in rat kidney, lung, spleen, liver, testis, ovary, and uterus, resulting in DOR holding additional dynamicity, playing regulatory roles at both central and peripheral levels (Wittert et al. 1996; Peng et al. 2012). Additional support of this comes from high affinity DOR peptides being detected throughout the musculoskeletal system, specifically in synovial membrane, periosteum, bone/bone marrow, the paratenon, the musculotendinous junction of the Achilles tendon, as well as in loose connective tissue (Spetea 2013). These examples of extensive peripheral distribution significantly strengthen the receptor's ties to regulatory roles related to pain (Spetea 2013). DOR mRNA can also be found in rat and macaque monkey skin (including both glabrous and hairy), as well as in corneas, eyelids, and the lip, and in close association with other, seemingly unlikely peripheral structures, including hair follicles, blood vessels and the glandular apparatus (Wenk and Honda 1999). And in human skin cells, DOR (and KOR) protein and mRNA is localized in mononuclear and fibroblast-like cells and found primarily on the cell membrane, with lower levels in the cytoplasm (Salemi et al. 2005). Therefore, with such a scattered and heterogeneous distribution throughout the mammalian body, the importance of drug targeting specificity as well as a continued attempt at fully uncovering the complexity of DOR is once again enunciated.

5 Development of Endogenous Opioid Ligands in the CNS

Despite the early appearance of prenatal proenkephalin and met-enkephalin mRNA without detectable DOR (Loughlin et al. 1985; Rius et al. 1991), met-enkephalin otherwise experiences a similar change in density as observed with rodent DOR. In mouse whole brain, the primary difference between met-enkephalin and DOR development is that met-enkephalin begins its rapid developmental period earlier, around embryonic day 12, and quickly increases in expression *in parallel* with the opioid peptides dynorphin and beta-endorphin, as well as with MOR and KOR (Rius et al. 1991). During this time frame in most examples, DOR development appears to be primarily dormant (Loughlin et al. 1985; Rius et al. 1991). Afterwards upon birth in rodents, met-enkephalin reaches the highest levels in the cerebellum at the first postnatal week, in the brainstem at the second week, and in the whole forebrain during the third (Tsang et al. 1982). This example of overall postnatal caudal-to-rostral peptide development appears to be in line with DOR development also occurring earlier/first in the brainstem (Tsang et al. 1982; Xia and Haddad 1991).

6 Decreased DOR Density in the Aged Brain and in Neurodegenerative Diseases

Accumulating data suggest an alteration in DOR density and distribution correlates with senescence, especially so when coupled with pathophysiology. Although research is currently limited and scattered to select brain regions of study, when taken together, a larger picture begins to form where the amount of DOR and endogenous DOR peptides sometimes decrease gradually and sometimes decrease dramatically throughout aging and at the intersection of aging and pathology. The amount of DOR decrease in these processes has, likewise, been found to vary greatly between brain regions.

6.1 *Aging and Decreased DOR Density*

The majority of research analyzing DOR density and distribution in adult rodents and through pre- and postnatal development cuts off at an adult age of a couple months, and is therefore not always capable of accurately offering a representation of opioid ontogeny and aging. Taken from this angle, rat DOR density would appear to only increase gradually in early adulthood throughout almost all brain regions after the onset of rapid postnatal development if sufficient time is not left for the physiology of aging to occur (Kornblum et al. 1987).

An early study observed that DOR distribution in aged (2-year-old) rats is similar to that of 3-month-old rats, despite observed differences in MOR and KOR binding

(Dondi et al. 1992). On the other hand, some investigators observed that aging exerts little influence on the number of KOR becuase KOR density and distribution in 19-month-old rats is comparable to 3-month-old rats, except for a notable KOR increase in the amygdala and thalamus of the older specimens (Maggi et al. 1989). However, our studies have found that after the rapid onset of DOR development following birth, DOR levels in adult rats reach a peak density and begin to slowly decline with age. In rat brainstem, for instance, DOR binding sites decreased from 21-day to 120-day-old rats with varying degrees in 28 out of 33 regions studied (Xia and Haddad 1991). For comparison, when moving from 21-day to 120-day-old rats, MOR binding sites in rat brainstem also decreased in approximately 80 % of regions analyzed (Xia and Haddad 1991). Overall, considerably more changes in DOR development occur during the first 3–4 months after birth, with the striking rise in density within 21 days after birth and a gradual decrease after the first month (Xia and Haddad 1991). Additionally, MOR levels are significantly lower in whole brain and hypothalamus of 22-month-old rats as compared to 2-month-old tissue samples (Piva et al. 1987).

During normal development, the gradual decline in DOR density from young rats to adults (such as those 4 months or older) is more likely to be attributed to overshoot in opioid density during earlier postnatal development, with a subsequent physiological adjustment phase as adult physiology/function is approached. Yet decreases in the aged brain, especially when pathophysiological changes are factored in, can be far more significant. However, a major decrease in DOR and MOR density in certain CNS regions through the aging process is likely a pattern found in pathologies related to neurodegenerative disorders such as Alzheimer's disease and Parkinson's disease (Hiller et al. 1987; Barg et al. 1993; Fernandez et al. 1994).

6.2 *DOR and Enkephalin in Alzheimer's Disease (AD)*

Shifts in DOR and enkephalin levels of older mammals are more pronounced when age-related diseases come into play. Most striking, and also most studied, is the case of AD.

Some studies found that in AD patients, DOR levels in certain regions of AD brains were indistinguishable from non-AD brains. These areas included the frontal cortex, caudate and hippocampus (Hiller et al. 1987), as well as the parahippocampal gyrus and occipitotemporal gyrus in a later study (Mathieu-Kia et al. 2001). However, more studies suggest a linkage between the changes in the DOR system and AD. Rinne et al. (1993) observed that primarily specific to DOR, both met- and leu-enkephalin binding is decreased considerably in hippocampus and temporal cortex, and met-enkephalin is additionally lowered in the internal globus pallidus of humans with AD. The Alzheimer's amygdala also displays a DOR decrease, but with no changes in binding affinity (Rinne et al. 1993). For comparison, patients with vascular dementia have no significant changes in enkephalin binding in the internal and external globus pallidus, amygdala, hippocampus, and temporal cortex

(Rinne et al. 1993). Other studies show that pathophysiological DOR alteration from AD is most notable in amygdala and putamen tissue (Mathieu-Kia et al. 2001). DOR sites decrease from roughly 102 to 69 fmol/mg of protein in AD putamen as compared to controls (Barg et al. 1993). DOR binding also decreases up to 55 % in the amygdala of AD patients (Hiller et al. 1987). On the other hand, MOR binding, too, decreases by approximately 44 % in AD amygdala (Hiller et al. 1987). Intriguingly, KOR binding actually increases in AD amygdala, as well as in the frontal cortex and hippocampus, and significantly so in the putamen and caudate: up 114 % and 53 %, respectively (Hiller et al. 1987). Furthermore, total opioid receptor binding in Alzheimer's putamen also increases by nearly 85 % (Hiller et al. 1987).

These studies suggest that AD is associated with a reduction in DOR density in the cortex, caudate putamen and amygdala where DOR is highly rich in quantity in healthy conditions.

6.3 DOR and Enkephalin in Parkinson's Disease (PD)

Although some instances of early research did not detect significant differences in the proportions and densities of opioid receptors throughout various brain regions of Parkinsonian patients (Delay-Goyet et al. 1987), it has still been observed for some time that altered opioid transmission is a part of the pathophysiology of PD (Taquet et al. 1983; Llorens-Cortes et al. 1984; Fernandez et al. 1992, 1994; Piccini et al. 1997). Furthermore, some studies indeed found altered and decreased opioid receptor binding and density, as also observed in Alzheimer's patients (Hiller et al. 1987; Fernandez et al. 1994; Piccini et al. 1997).

Opioid endogenous ligands, particularly the enkephalins with a high specificity to DOR, are also subject to substantial decreases in certain regions of the PD brain (Taquet et al. 1983; Llorens-Cortes et al. 1984; Fernandez et al. 1992, 1994; Piccini et al. 1997). Involvement of the substantia nigra in the pathology of PD, including notable dopamine depletion, has been established for some time (Llorens-Cortes et al. 1984). In this region, both met-enkephalin levels and all opiate receptor binding have been found to decrease by 72 % and 42 %, respectively (Llorens-Cortes et al. 1984). Even the activity of enkephalinase, the enzyme responsible for breaking down the endogenous enkephalin peptides, was observed to decrease by 39 % in PD substantia nigra (Llorens-Cortes et al. 1984). Other brain regions of PD patients also display altered ligand properties. In a subgroup of PD patients, met-enkephalin levels are reduced in the caudate nucleus (Fernandez et al. 1992). Both met- and leu-enkephalin are also significantly reduced in PD pallidum and putamen, and met-enkephalin levels are further decreased in the ventral tegmental area (as well as in the aforementioned substantia nigra) (Taquet et al. 1983). This case of differentially localized decreases in opioid peptides further complicates the scenario of PD but offers clues as to how met- and leu-enkephalin work as interacting yet distinguishably different systems, possibly within different neuronal populations (Taquet et al. 1983).

Therefore, it would appear that opioid physiology and particularly the systems likely linked to DOR go awry in this region with PD.

Regarding the opioid receptors, Parkinsonian posterior putamen also exhibits similar opioid behaviors to Alzheimer's putamen, as the densities of DOR, MOR and KOR are all reduced (Fernandez et al. 1994). MOR density is additionally decreased even more dramatically than the levels of DOR and KOR in the caudate nucleus and putamen (both anterior and posterior) of Parkinsonian patients (Fernandez et al. 1994). Thus, even from a fairly small pool of studies, it is likely that DOR is closely tied into the physiology of senescence alongside MOR and KOR, and most particularly so throughout the development of age-related diseases.

7 Concluding Remarks

DOR has a low density in prenatal life and undergoes a rapid developmental process that ends in a distinctly heterogeneous distribution within the central nervous system. Figure 4.5 roughly shows the rise of DOR density through the development of the CNS based on the data of rats. It reflects a general trend of DOR development in the brain, although differences exist in different brain regions (e.g., rostral vs. caudal regions).

The distributional variance of DOR throughout organ systems and throughout development, particularly so during the earliest and latest stages of life, indeed poses a good deal of inert complications. Still, this variation and uncertainty aims to become increasingly manageable as wider-scale compilations of research are drawn together. DOR cannot be viewed as a static receptor, as its distribution is too wide ranging, and likewise, is too functionally versatile. Taken together with its chronological and occasionally pathophysiological fluctuations, understanding DOR requires understanding its surroundings, its context and its dynamic changes during development. Such will only become possible through continued scrutiny and mapping at both macro and

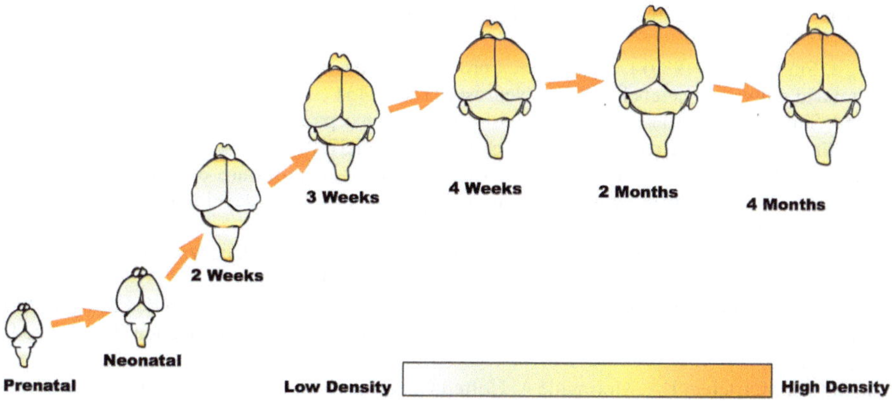

Fig. 4.5 General trend in the rise of DOR density through the development of the rat CNS

micro levels with diversified approaches, and eventually lead to more vivid visuals that will promote rapid, conceptual understanding and correspond to safer, more specialized, and more effective, methodologies of using the properties of DOR for clinical benefit.

The biggest issues at hand are gaps in research that often jump from the cellular level, to regional levels, and to vague, whole-organ levels. This makes the creation of a holistic, streamlined map of DOR distribution (both within the CNS and peripherally) from the cellular scale to the organ scale, coupled with the addition of time variances, very difficult to accomplish. Still, as more regions are scrutinized, more pieces are gradually filled in, and the complexity of DOR thus continues to be gradually understood. From today's data, it is very clear that after lengthy and complex development of DOR in pre- and post-natal life, DOR is highly distributed in the cortex, caudate putamen and amygdala in the mature brain. Likewise, high levels of DOR in these brain regions are well maintained throughout adulthood in healthy conditions. This phenomenon strongly suggests that DOR plays an extremely important role in the neural functions of these regions.

Acknowledgements This work was supported by supported by NIH HD-034852, NIH AT-004422 and Vivian L Smith Neurologic Foundation.

References

Agirregoitia E, Peralta L, Mendoza R, Expósito A, Ereño ED, Matorras R, Agirregoitia N (2012) Expression and localization of opioid receptors during the maturation of human oocytes. Reprod Biomed Online 24:550–557

Attali B, Saya D, Vogel Z (1990) Pre- and postnatal development of opiate receptor subtypes in rat spinal cord. Brain Res Dev Brain Res 53:97–102

Barg J, Belcheva M, Rowinski J, Ho A, Burke WJ, Chung HD, Schmidt CA, Coscia CJ (1993) Opioid receptor density changes in Alzheimer amygdala and putamen. Brain Res 632:209–215

Barg J, Rius RA, Bem WT, Belcheva MM, Loh YP, Coscia CJ (1992) Differential development of beta-endorphin and mu opioid binding sites in mouse brain. Brain Res Dev Brain Res 66:71–76

Barg J, Simantov R (1989) Developmental profile of kappa, mu and delta opioid receptors in the rat and guinea pig cerebellum. Dev Neurosci 11:428–434

Barg J, Simantov R (1991) Transient expression of opioid receptors in defined regions of developing brain: are embryonic receptors selective? J Neurochem 57:1978–1984

Beland B, Fitzgerald M (2001) Mu- and delta-opioid receptors are downregulated in the largest diameter primary sensory neurons during postnatal development in rats. Pain 90:143–150

Blackburn TP, Cross AJ, Hille C, Slater P (1988) Autoradiographic localization of delta opiate receptors in rat and human brain. Neuroscience 27:497–506

Borges JP, Verdoorn KS, Daliry A, Powers SK, Ortenzi VH, Fortunato RS, Tibiriçá E, Lessa MA (2014) Delta opioid receptors: the link between exercise and cardioprotection. PLoS One 9, e113541

Cahill CM, McClellan KA, Morinville A, Hoffert C, Hubatsch D, O'Donnell D, Beaudet A (2001) Immunohistochemical distribution of delta opioid receptors in the rat central nervous system: evidence for somatodendritic labeling and antigen-specific cellular compartmentalization. J Comp Neurol 440:65–84

Cao S, Chao D, Zhou H, Balboni G, Xia Y (2015) A novel mechanism for cytoprotection against hypoxic injury: δ opioid receptor mediated increase in Nrf2 translocation. British journal of pharmacology 172:1869–1881

Chao D, Xia Y (2010) Ionic storm in hypoxic/ischemic stress: can opioid receptors subside it? Prog Neurobiol 90:439–470

Chen T, Li J, Chao D, Sandhu HK, Liao X, Zhao J, Wen G, Xia Y (2014) δ-Opioid receptor activation reduces α-synuclein overexpression and oligomer formation induced by MPP and/or hypoxia. Exp Neurol 255:127–136

Chu Sin Chung P, Kieffer BL (2013) Delta opioid receptors in brain function and diseases. Pharmacol Ther 140:112–120

De Vries TJ, Hogenboom F, Mulder AH, Schoffelmeer AN (1990) Ontogeny of mu-, delta- and kappa-opioid receptors mediating inhibition of neurotransmitter release and adenylate cyclase activity in rat brain. Brain Res Dev Brain Res 54:63–69

Delay-Goyet P, Zajac JM, Javoy-Agid F, Agid Y, Roques BP (1987) Regional distribution of mu, delta and kappa opioid receptors in human brains from controls and parkinsonian subjects. Brain Res 414:8–14

Dondi D, Maggi R, Limonta P, Martini L, Piva F (1992) Binding characteristics of delta opioid receptors in different regions of the brain of young and old male rats as studied with the highly selective ligand [D-Pen2-D-Pen5] enkephalin. Aging (Milano) 4:69–75

Dunlap CE 3rd, Christ GJ, Rose JC (1986) Characterization of opioid receptor binding in adult and fetal sheep brain regions. Brain Res 389:279–285

Erbs E, Faget L, Scherrer G, Kessler P, Hentsch D, Vonesch JL, Matifas A, Kieffer BL, Massotte D (2012) Distribution of delta opioid receptor-expressing neurons in the mouse hippocampus. Neuroscience 221:203–213

Feng Y, He X, Yang Y, Chen J, Yin K, Xia Y (2011) Effect of delta-opioid receptor over-expression on cortical expression of GABAA receptor alpha1-subunit in hypoxia. Chin J Physiol 54:118–123

Feng Y, He X, Yang Y, Chao D, Lazarus LH, Xia Y (2012) Current research on opioid receptor function. Curr Drug Targets 13:230–246

Fernandez A, de Ceballos ML, Jenner P, Marsden CD (1992) Striatal neuropeptide levels in Parkinson's disease patients. Neurosci Lett 145:171–174

Fernandez A, de Ceballos ML, Jenner P, Marsden CD (1994) Neurotensin, substance P, delta and mu opioid receptors are decreased in basal ganglia of Parkinson's disease patients. Neuroscience 61:73–79

George SR, Zastawny RL, Briones-Urbina R, Cheng R, Nguyen T, Heiber M, Kouvelas A, Chan AS, O'Dowd BF (1994) Distinct distributions of mu, delta and kappa opioid receptor mRNA in rat brain. Biochem Biophys Res Commun 205:1438–1444

Georges F, Normand E, Bloch B, Le Moine C (1998) Opioid receptor gene expression in the rat brain during ontogeny, with special reference to the mesostriatal system: an in situ hybridization study. Brain Res Dev Brain Res 109:187–199

Gómez-Román JJ, Cifrián Martínez JM, Fernández Rozas S, Fernando Val-Bernal J (2002) Hormone expression and opioid receptors in fetal and adult lung. Arch Bronconeumol 38:362–366 (In Spanish with English abstract)

Gooshe M, Abdolghaffari AH, Aleyasin AR, Chabouk L, Tofigh S, Hassanzadeh GR, Payandemehr B, Partoazar A, Azizi Y, Dehpour AR (2015) Hypoxia/ischemia a key player in early post stroke seizures: modulation by opioidergic and nitrergic systems. Eur J Pharmacol 746:6–13

Gross ER, Hsu AK, Gross GJ (2009) Acute methadone treatment reduces myocardial infarct size via the delta-opioid receptor in rats during reperfusion. Anesth Analg 109:1395–1402

He X, Sandhu HK, Yang Y, Hua F, Belser N, Kim DH, Xia Y (2013) Neuroprotection against hypoxia/ischemia: δ-opioid receptor mediated cellular/molecular events. Cell Mol Life Sci 70:2291–2303

Hedner T, Cassuto J (1987) Opioids and opioid receptors in peripheral tissues. Scand J Gastroenterol Suppl 130:27–46

Hiller JM, Fan LQ (1996) Laminar distribution of the multiple opioid receptors in the human cerebral cortex. Neurochem Res 21:1333–1345

Hiller JM, Itzhak Y, Simon EJ (1987) Selective changes in mu, delta and kappa opioid receptor binding in certain limbic regions of the brain in Alzheimer's disease patients. Brain Res 406:17–23

Hoskins DL, Gordon TL, Crisp T (1998) The effects of aging on mu and delta opioid receptors in the spinal cord of Fischer-344 rats. Brain Res 791:299–302

Huang MH, Nguyen V, Wu Y, Rastogi S, Lui CY, Birnbaum Y, Wang HQ, Ware DL, Chauhan M, Garg N, Poh KK, Ye L, Omar AR, Tan HC, Uretsky BF, Fujise K (2009) Reducing ischaemia/reperfusion injury through delta-opioid-regulated intrinsic cardiac adrenergic cells: adrenopeptidergic co-signalling. Cardiovasc Res 84:452–460

Jenab S, Kest B, Franklin SO, Inturrisi CE (1995) Quantitative distribution of the delta opioid receptor mRNA in the mouse and rat CNS. Life Sci 56:2343–2355

Kaneko K, Mizushige T, Miyazaki Y, Lazarus M, Urade Y, Yoshikawa M, Kanamoto R, Ohinata K (2014) δ-opioid receptor activation stimulates normal diet intake but conversely suppresses high-fat diet intake in mice. Am J Physiol Regul Integr Comp Physiol 306:R265–R272

Kest B, Jenab S, Brodsky M, Elliott K, Inturrisi CE (1994) Supraspinal delta opioid receptor mRNA levels are not altered in [D-Ala2]deltorphin II tolerant mice. J Neurosci Res 39:674–679

Kitchen I, McDowell J (1986) Development of delta-opioid receptors in rat brain characterised by [3H]-[D-Pen2, D-Pen5] enkephalin binding. NIDA Res Monogr 75:67–68

Kivell BM, Day DJ, McDonald FJ, Miller JH (2004a) Mu and delta opioid receptor immunoreactivity and mu receptor regulation in brainstem cells cultured from late fetal and early postnatal rats. Brain Res Dev Brain Res 149:9–19

Kivell BM, Day DJ, McDonald FJ, Miller JH (2004b) Developmental expression of mu and delta opioid receptors in the rat brainstem: evidence for a postnatal switch in mu isoform expression. Brain Res Dev Brain Res 148:185–196

Kornblum HI, Hurlbut DE, Leslie FM (1987) Postnatal development of multiple opioid receptors in rat brain. Brain Res 465:21–41

Leslie FM, Chen Y, Winzer-Serhan UH (1998) Opioid receptor and peptide mRNA expression in proliferative zones of fetal rat central nervous system. Can J Physiol Pharmacol 76:284–293

Liang J, Chao D, Sandhu HK, Yu Y, Zhang L, Balboni G, Kim DH, Xia Y (2014) δ-opioid receptors up-regulate excitatory amino acid transporters in mouse astrocytes. Br J Pharmacol 171:5417–5430

Llorens-Cortes C, Javoy-Agid F, Agid Y, Taquet H, Schwartz JC (1984) Enkephalinergic markers in substantia nigra and caudate nucleus from Parkinsonian subjects. J Neurochem 43:874–877

Loughlin SE, Massamiri TR, Kornblum HI, Leslie FM (1985) Postnatal development of opioid systems in rat brain. Neuropeptides 5:469–472

Ma MC, Qian H, Ghassemi F, Zhao P, Xia Y (2005) Oxygen-sensitive delta-opioid receptor-regulated survival and death signals: novel insights into neuronal preconditioning and protection. J Biol Chem 280:16208–16218

Madar I, Lever JR, Kinter CM, Scheffel U, Ravert HT, Musachio JL, Mathews WB, Dannals RF, Frost JJ (1996) Imaging of delta opioid receptors in human brain by N1′-([11C]methyl)naltrindole and PET. Synapse 24:19–28

Maggi R, Limonta P, Dondi D, Martini L, Piva F (1989) Distribution of kappa opioid receptors in the brain of young and old male rats. Life Sci 45:2085–2092

Mansour A, Fox CA, Burke S, Meng F, Thompson RC, Akil H, Watson SJ (1994) Mu, delta, and kappa opioid receptor mRNA expression in the rat CNS: an in situ hybridization study. J Comp Neurol 350:412–438

Mansour A, Thompson RC, Akil H, Watson SJ (1993) Delta opioid receptor mRNA distribution in the brain: comparison to delta receptor binding and proenkephalin mRNA. J Chem Neuroanat 6:351–362

Maslov LN, Naryzhnaia NV, Tsibulnikov SY, Kolar F, Zhang Y, Wang H, Gusakova AM, Lishmanov YB (2013) Role of endogenous opioid peptides in the infarct size-limiting effect of adaptation to chronic continuous hypoxia. Life Sci 93:373–379

Mathieu-Kia AM, Fan LQ, Kreek MJ, Simon EJ, Hiller JM (2001) Mu-, delta- and kappa-opioid receptor populations are differentially altered in distinct areas of postmortem brains of Alzheimer's disease patients. Brain Res 893:121–134

McDowell J, Kitchen I (1986) Ontogenesis of delta-opioid receptors in rat brain using [3H][D-Pen2, D-Pen5]enkephalin as a binding ligand. Eur J Pharmacol 128:287–289

Milligan G, Streaty RA, Gierschik P, Spiegel AM, Klee WA (1987) Development of opiate receptors and GTP-binding regulatory proteins in neonatal rat brain. J Biol Chem 262:8626–8630

Moskowitz AS, Goodman RR (1984) Light microscopic autoradiographic localization of mu and delta opioid binding sites in the mouse central nervous system. J Neurosci 4:1331–1342

Mousa SA, Shaqura M, Schäper J, Treskatsch S, Habazettl H, Schäfer M, Abdul-Khaliq H (2011) Developmental expression of δ-opioid receptors during maturation of the parasympathetic, sympathetic, and sensory innervations of the neonatal heart: early targets for opioid regulation of autonomic control. J Comp Neurol 519:957–971

Mrkusich EM, Kivell BM, Miller JH, Day DJ (2004) Abundant expression of mu and delta opioid receptor mRNA and protein in the cerebellum of the fetal, neonatal, and adult rat. Brain Res Dev Brain Res 148:213–222

Negri L, Severini C, Lattanzi R, Potenza RL, Melchiorri P (1997) Postnatal development of delta-opioid receptor subtypes in mice. Br J Pharmacol 120:989–994

Nitsche JF, Pintar JE (2003) Opioid receptor-induced GTPgamma35S binding during mouse development. Dev Biol 253:99–108

Pacheco Dda F, Pacheco CM, Duarte ID (2012a) δ-opioid receptor agonist SNC80 induces central antinociception mediated by Ca^{2+}-activated Cl-channels. J Pharm Pharmacol 64:1084–1089

Pacheco Dda F, Pacheco CM, Duarte ID (2012b) Peripheral antinociception induced by δ-opioid receptors activation, but not μ- or κ-, is mediated by Ca^{2}-activated Cl channels. Eur J Pharmacol 674:255–259

Peart JN, Hoe LE, Gross GJ, Headrick JP (2011) Sustained ligand-activated preconditioning via δ-opioid receptors. J Pharmacol Exp Ther 336:274–281

Peckys D, Landwehrmeyer GB (1999) Expression of mu, kappa, and delta opioid receptor messenger RNA in the human CNS: a 33P in situ hybridization study. Neuroscience 88:1093–1135

Peng J, Sarkar S, Chang SL (2012) Opioid receptor expression in human brain and peripheral tissues using absolute quantitative real-time RT-PCR. Drug Alcohol Depend 124:223–228

Phansuwan-Pujito P, Saleema L, Mukda S, Tongjaroenbuangam W, Jutapakdeegul N, Casalotti SO, Forge A, Dodson H, Govitrapong P (2003) The opioid receptors in inner ear of different stages of postnatal rats. Hear Res 184:1–10

Piccini P, Weeks RA, Brooks DJ (1997) Alterations in opioid receptor binding in Parkinson's disease patients with levodopa-induced dyskinesias. Ann Neurol 42:720–726

Pilapil C, Welner S, Magnan J, Gauthier S, Quirion R (1987) Autoradiographic distribution of multiple classes of opioid receptor binding sites in human forebrain. Brain Res Bull 19:611–615

Piva F, Maggi R, Limonta P, Dondi D, Martini L (1987) Decrease of mu opioid receptors in the brain and in the hypothalamus of the aged male rat. Life Sci 40:391–398

Platzer S, Winkler A, Schadrack J, Dworzak D, Tölle TR, Zieglgänsberger W, Spanagel R (2000) Autoradiographic distribution of mu-, delta- and kappa 1-opioid stimulated [35S]guanylyl-5′-O-(gamma-thio)-triphosphate binding in human frontal cortex and cerebellum. Neurosci Lett 283:213–216

Rahman W, Dashwood MR, Fitzgerald M, Aynsley-Green A, Dickenson AH (1998) Postnatal development of multiple opioid receptors in the spinal cord and development of spinal morphine analgesia. Brain Res Dev Brain Res 108:239–254

Renda T, Negri L, Tooyama I, Casu C, Melchiorri P (1993) Autoradiographic study on [3H]-[D-Ala2]-deltorphin-I binding sites in the rat brain. Neuroreport 4:1143–1146

Rinne JO, Lönnberg P, Marjamäki P, Mölsä P, Säkö E, Paljärvi L (1993) Brain methionine- and leucine-enkephalin receptors in patients with dementia. Neurosci Lett 161:77–80

Rius RA, Barg J, Bem WT, Coscia CJ, Loh YP (1991) The prenatal development profile of expression of opioid peptides and receptors in the mouse brain. Brain Res Dev Brain Res 58:237–241

Ruzicka BB, Fox CA, Thompson RC, Meng F, Watson SJ, Akil H (1995) Primary astroglial cultures derived from several rat brain regions differentially express mu, delta and kappa opioid receptor mRNA. Brain Res Mol Brain Res 34:209–220

Salemi S, Aeschlimann A, Reisch N, Jüngel A, Gay RE, Heppner FL, Michel BA, Gay S, Sprott H (2005) Detection of kappa and delta opioid receptors in skin—outside the nervous system. Biochem Biophys Res Commun 338:1012–1017

Salès N, Charnay Y, Zajac JM, Dubois PM, Roques BP (1989) Ontogeny of mu and delta opioid receptors and of neutral endopeptidase in human spinal cord: an autoradiographic study. J Chem Neuroanat 2:179–188

Sharif NA, Hughes J (1989) Discrete mapping of brain Mu and delta opioid receptors using selective peptides: quantitative autoradiography, species differences and comparison with kappa receptors. Peptides 10:499–522

Shen H, Ben Q, Zhang Y, Jiang G, Yao C, Shi D, Sun Z, Tong C (2012) Role of δ2 opioid receptor in cardioprotection against hypoxia-reoxygenation injury. J Cardiovasc Pharmacol 60:253–261

Simonin F, Befort K, Gavériaux-Ruff C, Matthes H, Nappey V, Lannes B, Micheletti G, Kieffer B (1994) The human delta-opioid receptor: genomic organization, cDNA cloning, functional expression, and distribution in human brain. Mol Pharmacol 46:1015–1021

Slamberová R (2004) Opioid receptors of the CNS: function, structure and distribution. Cesk Fysiol 53:159–166 (In Czech with English Abstract)

Sobanski P, Krajnik M, Shaqura M, Bloch-Boguslawska E, Schäfer M, Mousa SA (2014) The presence of mu-, delta-, and kappa-opioid receptors in human heart tissue. Heart Vessels 29(6):855–63

Spain JW, Roth BL, Coscia CJ (1985) Differential ontogeny of multiple opioid receptors (mu, delta, and kappa). J Neurosci 5:584–588

Spetea M (2013) Opioid receptors and their ligands in the musculoskeletal system and relevance for pain control. Curr Pharm Des 19:7382–7390

Stefano GB, Casares F, Liu U (1995) Naltrindole sensitive delta 2 opioid receptor mediates invertebrate immunocyte activation. Acta Biol Hung 46:321–327

Sternini C, Patierno S, Selmer IS, Kirchgessner A (2004) The opioid system in the gastrointestinal tract. Neurogastroenterol Motil 2:3–16

Sunday ME, Haley KJ, Emanuel RL, Torday JS, Asokananthan N, Sikorski KA, Tooyama I, Kimura H, Renda T, Erspamer V (2001) Fetal alveolar epithelial cells contain [D-Ala(2)]-deltorphin I-like immunoreactivity: delta- and mu-opiate receptors mediate opposite effects in developing lung. Am J Respir Cell Mol Biol 25:447–456

Surendra H, Diaz RJ, Harvey K, Tropak M, Callahan J, Hinek A, Hossain T, Redington A, Wilson GJ (2013) Interaction of δ and κ opioid receptors with adenosine A1 receptors mediates cardioprotection by remote ischemic preconditioning. J Mol Cell Cardiol 60:142–150

Szücs M, Coscia CJ (1990) Evidence for delta-opioid binding and GTP-regulatory proteins in 5-day-old rat brain membranes. J Neurochem 54:1419–1425

Szücs M, Spain JW, Oetting GM, Moudy AM, Coscia CJ (1987) Guanine nucleotide and cation regulation of mu, delta, and kappa opioid receptor binding: evidence for differential postnatal development in rat brain. J Neurochem 48:1165–1170

Tabata H (2013) The behaviors of proliferative cells in the subventricular zone during cortical development. Nihon Shinkei Seishin Yakurigaku Zasshi 33:131–136, Review. (In Japanese with English Abstract)

Tanaka K, Kersten JR, Riess ML (2014) Opioid-induced cardioprotection. Curr Pharm Des 20:5696–5705

Taquet H, Javoy-Agid F, Hamon M, Legrand JC, Agid Y, Cesselin F (1983) Parkinson's disease affects differently Met5- and Leu5-enkephalin in the human brain. Brain Res 280:379–382

Tavani A, Robson LE, Kosterlitz HW (1985) Differential postnatal development of mu-, delta-and chi-opioid binding sites in mouse brain. Brain Res 355:306–309

Tian X, Hua F, Sandhu HK, Chao D, Balboni G, Salvadori S, He X, Xia Y (2013) Effect of δ-opioid receptor activation on BDNF-TrkB vs. TNF-α in the mouse cortex exposed to prolonged hypoxia. Int J Mol Sci 14:15959–15976

Tripathi A, Khurshid N, Kumar P, Iyengar S (2008) Expression of delta- and mu-opioid receptors in the ventricular and subventricular zones of the developing human neocortex. Neurosci Res 61:257–270

Tsang D, Ng SC, Ho KP, Ho WK (1982) Ontogenesis of opiate binding sites and radioimmunoassayable beta-endorphin and enkephalin in regions of rat brain. Brain Res 281:257–261

Vanderah T, Takemori AE, Sultana M, Portoghese PS, Mosberg HI, Hruby VJ, Haaseth RC, Matsunaga TO, Porreca F (1994) Interaction of [D-Pen2, D-Pen5]enkephalin and [D-Ala2, Glu4]deltorphin with delta-opioid receptor subtypes in vivo. Eur J Pharmacol 252:133–137

Villemagne PS, Dannals RF, Ravert HT, Frost JJ (2002) PET imaging of human cardiac opioid receptors. Eur J Nucl Med Mol Imaging 29:1385–1388

Wang Q, Chao D, Chen T, Sandhu H, Xia Y (2014) δ-opioid receptors and inflammatory cytokines in hypoxia: differential regulation between glial and neuron-like cells. Transl Stroke Res 5:476–483

Wang H, Cuzon VC, Pickel VM (2003) Ultrastructural localization of delta-opioid receptors in the rat caudate-putamen nucleus during postnatal development: relation to synaptogenesis. J Comp Neurol 467:343–353

Wenk HN, Honda CN (1999) Immunohistochemical localization of delta opioid receptors in peripheral tissues. J Comp Neurol 408:567–579

Wittert G, Hope P, Pyle D (1996) Tissue distribution of opioid receptor gene expression in the rat. Biochem Biophys Res Commun 218:877–881

Wohltmann M, Roth BL, Coscia CJ (1982) Differential postnatal development of mu and delta opiate receptors. Brain Res 255:679–684

Xia Y, Haddad GG (1991) Ontogeny and distribution of opioid receptors in the rat brainstem. Brain Res 549:181–193

Xia Y, Haddad GG (2001) Major difference in the expression of delta- and mu-opioid receptors between turtle and rat brain. J Comp Neurol 436:202–210

Zagon IS, Gibo DM, McLaughlin PJ (1990) Adult and developing human cerebella exhibit different profiles of opioid binding sites. Brain Res 523:62–68

Zhang J, Haddad GG, Xia Y (2000) Delta-, but not mu- and kappa-, opioid receptor activation protects neocortical neurons from glutamate-induced excitotoxic injury. Brain Res 885:143–153

Zhang Z, Pan ZZ (2012) Signaling cascades for δ-opioid receptor-mediated inhibition of GABA synaptic transmission and behavioral antinociception. Mol Pharmacol 81:375–383

Zhang J, Qian H, Zhao P, Hong SS, Xia Y (2006) Rapid hypoxia preconditioning protects cortical neurons from glutamate toxicity through delta-opioid receptor. Stroke 37:1094–1099

Zhao HM, Wenthold RJ, Petralia RS (1998) Glutamate receptor targeting to synaptic populations on Purkinje cells is developmentally regulated. J Neurosci 18:5517–5528

Chapter 5
The Role of δ-Opioid Receptors in Brain Ionic Homeostasis Under Physiological Condition

Dongman Chao and Ying Xia

Abstract Homeostasis is one of the most fundamental concepts in understanding the physiological functions of bodies as well as the pathophysiology of diseases, and in a global view of system biology, is also a vital guiding principle for medicine. Because of the particular importance of ionic environment in neuronal functions, ion homeostasis is the most fundamental mechanism among all the brain homeostasis regulations, and most of the functional activities of the brain are highly dependent on the dynamic and relatively steady state of ionic environment. The concentrations of ions are in dynamic balance under physiological conditions, which depend on the complicated, but elaborate, regulatory mechanisms. The δ-opioid receptor (DOR) regulates a diverse array of physiological functions. DOR and its cognate endogenous opioids are widely expressed throughout the central nervous system and have an extensive interaction with ion channels, receptors, and transporters. DOR can regulate the release of many neurotransmitters, modify neuronal electrical activities and synaptic plasticity, and thereby tightly regulates ion homeostasis in neural activities. In this chapter, we will focus on the regulation of homeostasis of the key cations such as Na^+, K^+, Ca^{2+} in the brain under physiological conditions and the role of DOR in such homeostatic regulation. Almost all the previous studies regarding DOR-mediated regulation of ionic homeostasis under normoxic condition focused on intracellular Ca^{2+} activity. Few studies highlighted the regulation of K^+ and Na^+ homeostasis despite the functional coupling of DOR with K^+ channels and Na^+ channels. Overall, the predominant effect of DOR activation on Ca^{2+} entry is inhibitory. However, some studies show either mobilization of intracellular Ca^{2+} or stimulation of Ca^{2+} entry with opioid activation. In general, DOR signaling is inhibitory to Na^+ influx and K^+ efflux. In regard to this issue, we have recently made

D. Chao, Ph.D.
Department of Neurosurgery, The University of Texas Medical School at Houston, Houston, TX 77030, USA
e-mail: chaodm21@yahoo.com

Y. Xia, M.D., Ph.D. (✉)
University of Texas Medical School, Houston, TX, USA
e-mail: ying.xia@uth.tmc.edu; y55738088@gmail.com

Y. Xia (ed.), *Neural Functions of the Delta-Opioid Receptor*,
DOI 10.1007/978-3-319-25495-1_5

the first finding that DOR activation and expression reduces Na^+ currents by targeting voltage-sensitive Na^+ channels. The DOR effects have been demonstrated to occur through activation of different kind of G proteins, including Gi, Go, and even Gs classes.

Keywords Ion homeostasis • δ-Opioid receptor • Brain • Ca^{2+} concentrations • K^+ concentrations • Na^+ concentrations • Receptors • Transporters • Channels • Na^+/K^+ ATPase • Exchangers • Endoplasmic reticulum • Mitochondria • Neuron • Glial cells

Abbreviations

$[Ca^{2+}]i$	Cytosolic free Ca^{2+} concentrations
$[Ca^{2+}]o$	Extracellular Ca^{2+} concentrations
$[K^+]e$	Extracellular K^+ concentrations
$[K^+]i$	Intracellular K^+ concentrations
$[Na^+]i$	Intracellular Na^+ concentrations
$[Na^+]o$	Extracellular Na^+ concentrations
ACSF	Artificial cerebrospinal fluid
AMPARs	AMPA receptors
BBB	Blood-brain barrier
CCCP	Carbonyl cyanide *m*-chlorophenyl hydrazone
CNS	Central nervous system
CSF	Cerebrospinal fluid
DA	Dopamine
DAG	Diacylglycerol
DAT	Dopamine transporter
DOR	δ-Opioid receptor
DRG	Dorsal root ganglion
EAAT	Excitatory amino acid transporter (glutamate transporter, also as GluT in Table 5.3)
eGFP	Enhanced green fluorescent protein
EPSPs/EPSCs	Excitatory postsynaptic potentials/currents
ER	Endoplasmic reticulum
GAT	GABA transporter
GIRK channels	G-protein-activated inwardly rectifying K^+ channels
GPCRs	G protein-coupled receptors
i.c.v	Intracerebroventricular
iGluRs	Ionotropic glutamate receptors
I_{Kir}	Inwardly rectifying K^+ currents
IMM	Inner mitochondrial membrane
IP_3	Inositol triphosphate
IP_3Rs	IP_3 receptors
IPSCs	Inhibitory postsynaptic currents
ISEs	Ion-selective electrodes

KOR	κ-Opioid receptor
LTD	Long-term depression
LTP	Long-term potentiation
MAM	Mitochondria-associated membrane
MAPK	Mitogen-activated kinase
MCU	Mitochondrial Ca^{2+} uniporter
mNCX	Mitochondrial Na^+/Ca^{2+} exchanger
MOR	μ-Opioid receptor
nAChRs	Nicotinic acetylcholine receptors
NCX	Na^+/Ca^{2+} exchanger
NKCC	Na^+-K^+-$2Cl^-$ cotransporter
NMDARs	NMDA receptors
$NMDG^+$	N-methyl-D-glucamine
NMR	Nuclear magnetic resonance
OMM	Outer mitochondrial membrane
PAG	Midbrain periaqueductal gray
PBFI	Potassium-binding benzofuran isophtalate
PIP_2	Phosphatidylinositol 4,5-bisphosphate
PLC	Phospholipase C
PMCA	Plasma membrane Ca^{2+}-ATPase
PTP	Permeability transition pores
ROCs	Receptor-operated channels
RYRs	Ryanodine receptors
SBFI	Sodium-binding benzofuran isophtalate
SERCA	Sarco-endoplasmic reticular Ca^{2+} ATPase
SNAP	Soluble N-ethylmaleimide sensitive factor Attachment Protein
SNARE	SNAP REceptor
SOCs	Store-operated Ca^{2+} entry channels
STIM	Stromal interaction molecules
TEA	Tetraethylammonium
TMA^+	Tetramethylammonium
TRPCs	Transient receptor potential channels
TTX	Tetrodotoxin
VGCCs	Voltage-gated Ca^{2+} channels
VGSCs	Voltage-gated sodium channels
WT	Wild type
$\Delta\Psi m$	Mitochondrial potential

1 Introduction

We know what we eat, but we may not know that what we eat needs to be balanced and kept stable after becoming part of us. Doctors always advise people to eat healthy and balanced food; too much sugar and fat in diet may lead to hyperglycemia and hyperlipemia, which increase the risks of diseases such as diabetes mellitus and heart attack. Similarly, our body and cells also need a balanced and stable ionic environment for our health. The idea of the maintenance of the stability of our internal environment was first presumably presented by French physiologist Claude Bernard in 1865 (Bernard 1865; Gross 1998) as the stability of "*milieu intérieur*" (internal environment, the extracellular fluids that provide stability to the organs), and later was coined as "homeostasis" [derived from the Greek words ***hómoios*** for "similar" and ***stásis*** for "standing still" (see review by Rose and Karus 2013)]. It was then further developed and popularized by American physiologist Walter Bradford Cannon in the 1920s (Cannon 1929, 1932) to describe the properties of a system that regulates its internal environment and tends to maintain a stable, relative constancy. One and a half centuries have passed since the first description of homeostasis, and it is still one of the most fundamental concepts in understanding the physiological functions of bodies as well as the pathophysiology of diseases, and in a global view of system biology, is also still a vital guiding principle for medicine (Ahn et al. 2006; Vodovotz et al. 2013). In this context, health (normal function) has been regarded as representing a living organism's ability to maintain stability in the face of changing environments (internal and external), while illness is defined as a failed homeostatic mechanism. Treatment, to a great extension, is an effort to help recover this failed mechanism by correcting deviations and placing parameters within the normal range (Ahn et al. 2006; Vodovotz et al. 2013). Indeed, living organisms have developed the remarkable ability to function effectively in a broad range of environmental conditions by homeostatic regulation of the vital activities such as body temperature, pH (acid-base balance), energy metabolism, etc. For the brain, homeostatic regulation of neuronal structures and functions such as development (Munakata et al. 2007), neurochemicals (Masino and Dulla 2005; Moulder et al. 2006), metabolism (Tsao et al. 1999a; Hamilton et al. 2003; Bienvenu et al. 2005; Zilberter et al. 2010), synaptogenesis (Slezak et al. 2006; Pfrieger 2010), neuronal plasticity (Yeung et al. 2004; Slezak et al. 2006; Gordon et al. 2009), and even gene expression (Nicot et al. 2003; McKinnon 2013) are of particular importance in normal brain activities and even brain diseases (also see the book edited by Walz 2002). Because of the particular importance of ionic environment in neuronal functions, ion homeostasis is the most fundamental mechanism among all the brain homeostasis regulations, and most of the functional activities of the brain are highly dependent on the dynamic and relatively steady state of the ionic environment.

The concentrations of ions are in dynamic balance under physiological conditions. With neuronal activity, the ionic concentrations fluctuate within a small range, and this fluctuation is often transient. Following ionic concentration fluctuation, a rapid increase in activity of ionic pumps (e.g. Na^+/K^+ ATPase, Ca^{2+} ATPase) and exchangers (e.g. Na^+/Ca^{2+} exchangers) as well as spatial ionic buffering mechanisms brings back the ionic gradients or levels towards their previous levels (Chao and Xia 2010). However, this ionic homeostasis may be disturbed under certain pathophysiological conditions such as hypoxia/ischemia and traumatic brain injury (Weber 2004; Stiefel et al. 2005; Chao and Xia 2010). Considering the critical importance of ionic homeostasis to normal brain function, therefore, keeping an appropriate balance of ions, especially the key ions such as Na^+, K^+, and Ca^{2+}, across the cell membrane is a critical and necessary condition for neuronal survival under pathophysiological conditions.

Our laboratory and others have proven activation of δ-opioid receptor (DOR) is neuroprotective against hypoxic/ischemic insult (See reviews by Chao and Xia 2010; He et al. 2013). One of the underlying mechanisms is related to the capability of DOR to stabilize the disrupted key ion homeostasis in the brain, which offers a potential therapeutic value of DOR in the early phase of insults.

We will focus on the regulation of homeostasis of the key cations (Na^+, K^+, Ca^{2+}) in the brain under physiological conditions and the role of DOR in their homeostatic regulation in this chapter. In the next chapter, we will discuss ionic dyshomeostasis and the underlying mechanism, and address the DOR regulation and stabilization of the disrupted ionic homeostasis under hypoxic/ischemic conditions. Maintenance of ionic homeostasis and its regulation under physiological conditions are very complicated processes, and are also fundamental to normal brain functions. For DOR regulation of ionic homeostasis in normal conditions, the currently available data mainly focus on the regulation of Ca^{2+} homeostasis in both neurons and glial cells. More studies are expected on DOR regulation of Na^+ and K^+ homeostasis and its functional implications under physiological conditions.

2 δ-Opioid Receptor (DOR)

Opioid receptors were first recognized in the mid-1960s in pharmacological studies on opiate drugs that were likely to exert their actions at specific receptor sites (Martin 1967). In the early 1970s, the receptors were first identified, by the use of radio-binding studies, as specific molecules with distinct distribution patterns (Goldstein et al. 1971; Pert and Snyder 1973; Simon et al. 1973), and the existence of multiple different opioid receptors was further supported by the isolation of endogenous opioids (Cox et al. 1976; Goldstein 1976). The existence of μ- and κ-opioid receptors (MOR, KOR) was first proposed by Martin and co-workers by using morphine and ketacyclazocine (Martin et al. 1976). Then, in the following year, δ-opioid receptor (DOR) was proposed based on the preference of leu-enkephalin to bind to receptors in the mouse vas deferens (Lord et al. 1977). In the early 1980s, opioid peptide precursor genes were identified with molecular approaches, proving the existence of preproenkephalin gene encoding several copies of enkephalins (Noda et al. 1982), preprodynorphin gene encoding dynorphins (Civelli et al. 1985), and the proopiomelanocortin gene encoding β-endorphin (Nakanishi et al. 1979) (also see review by Snyder and Pasternak 2003). In December 1992, two groups independently described the cloning of DOR from the neuroblastoma x glioma cell line NG-108 by using an expression cloning strategy, and reported the identification of a novel member of the 7-helix family of G-protein coupled receptors that had pharmacological properties typical of a DOR (Evans et al. 1992; Kieffer et al. 1992). Almost 1 year later, MOR (Chen et al. 1993; Wang et al. 1993a) and KOR (Meng et al. 1993; Minami et al. 1993) were cloned. The cloning of opioid receptors attracted more scientists to the research of opioid

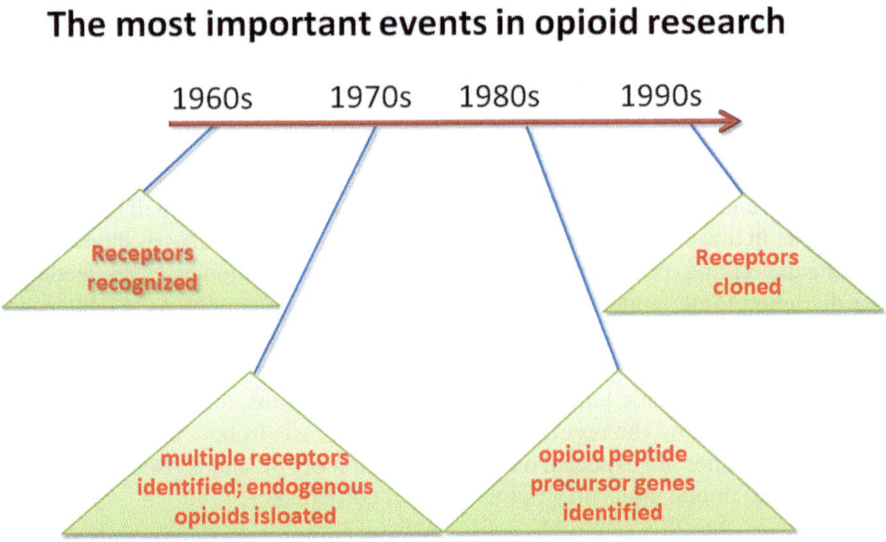

Fig. 5.1 The most important events in opioid research. The breakthrough studies occurred between the 1960s–1990s, which boomed the research on opioid receptors as well as their functions and boosted our knowledge on them

receptors as well as their functions, and greatly advanced our knowledge on them. The most important events in opioid research have been summarized in Fig. 5.1.

DOR, together with MOR and KOR, constitutes the classical opioid receptor family. Enkephalin, which preferentially binds to DOR in physiological levels, is the endogenous agonist of DOR (Lord et al. 1977; Chang and Cuatrecasas 1979).

The early studies with immunocytochemical, in situ hybridization and autoradiography methods revealed that DOR and its cognate endogenous opioids are widely expressed throughout the central nervous system (CNS) (Mansour et al. 1987, 1994; Xia and Haddad 1991, 2001; Bausch et al. 1995; Svingos et al. 1995). Lately, some investigators investigated the neuroanatomical distribution of DOR in the fluorescent knock-in mice that express a functional DOR fused at its carboxy terminus with the green fluorescent protein in place of the native receptor with the strategy of inserting the gene that encodes the enhanced green fluorescent protein (eGFP) into DOR-encoding gene, Oprd1, to express an approximately 80-kDa product (DOReGFP). Their data have provided further confirmation for the ubiquitous nature of DOR expression in the CNS observed in the early studies (Scherrer et al. 2006; Erbs et al. 2012; Rezaï et al. 2012).

Pharmacological evidence indicated that there are two distinct DOR subtypes, DOR1 and DOR2 (Sofuoglu et al. 1991; Portoghese et al. 1992). These two subtype receptors show distinct selectivity to DOR agonists and antagonists with no development of cross-tolerance. They can have opposing behavioral effects when binding subtype-selective DOR ligands (van Rijn et al. 2013). However, unlike MOR subtypes that are encoded by different genes, no DOR subtypes have been cloned, and there exists only a single DOR gene for distinct subtypes of DOR (Pasternak 2004).

DOR subtypes probably are a single DOR with multiple affinity states or appear due to post-translational modifications at the protein level rather than at the gene level (van Rijn et al. 2013).

Developmental studies have shown that DOR density is extremely low at birth in the brain, presents in only low levels during the first postnatal week, has a nearly linear increase from the second through the fourth week, and finally gradually reaches adult levels (Xia and Haddad 1991; Kornblum et al. 1987; Georges et al. 1998; Spain et al. 1985). Also, of all the three major opioid receptors, DOR has been proven more sensitive than others to stressful stimuli, such as hypoxia and ischemia (Zhang et al. 2006; Ma et al. 2005; Mayfield et al. 1996; Boutin et al. 1999).

DOR belongs to the Gi/Go-protein-coupled receptor family, characterized by a common three-dimensional structure that spans the cell membrane seven times. DOR activation inhibits adenylyl cyclase and regulates a number of other signalling intermediates, including Ca^{2+}, protein kinases (PKA, PKC, CaMK, GRK etc.), phospholipase C (PLC), mitogen-activated kinase (MAPK), and various ion channels (Law et al. 2000; Chao and Xia 2010; Al-Hasani and Bruchas 2011; Georgoussi et al. 2012; Bian et al. 2012). DOR also can physically interact with a variety of accessory proteins (novel opioid receptor interacting partners that link the receptors to alternative signalling pathways beyond G proteins), which include β-arrestin, calmodulin, calnexin, EAAC1, GASP, glycoprotein M6a, protachykinin, RGS4, Ribophorin I, SNX1, STAT5A/B, and ubiquitin. These interactions cause multifaceted activities of DOR such as receptor internalization and desensitization, interfering with G protein signaling, receptor lysosomal targeting and trafficking, possessing transcriptional activity, and affecting neurotransmitter release. This confirms that signal transduction of DOR is not restricted to heterotrimeric G protein activation (Georgoussi et al. 2012).

In line with its ubiquitous expression in the body, DOR is multifunctional, and has emerged as an attractive target in many aspects. In addition to the traditional roles in pain modulation and addiction, many novel functions of DOR have emerged with the development of highly selective delta opioid agonists, and rapid progress in mouse mutagenesis approaches targeting the Oprd1 gene. DOR is widely involved in various physiological and pathophysiological activities, including the regulation of membrane ionic homeostasis, cell proliferation, emotional response, epileptic seizures, immune function, feeding, obesity, respiratory and cardiovascular control, as well as some neurodegenerative disorders (see reviews by Chao and Xia 2010; Pradhan et al. 2011; Feng et al. 2012; He et al. 2013; Chao et al. 2013; Chu Sin Chung and Kieffer 2013).

3 Technical Considerations for the Measurement of Ion Activity

Since neuronal function is critically dependent on the ionic composition of extracellular fluid and ion movements across the cell membrane, electrophysiologists are generally interested in studying ion movements. Besides patch clamp technique that measures relatively weak (microscopic) transmembrane ion movement via ion channels (including voltage-gated and chemical-gated), and pumps/transporters (e.g. Na^+-K^+ pump, Ca^{2+} pump, Na^+/Ca^{2+} exchanger) in the plasma membrane, direct measurements of ionic activity inside and outside cells are undoubtedly of importance in understanding the roles of ionic activity in cellular functions. An ideal method for the measurement of ion activities would have the following properties: (1) it must be non-destructive (non-invasive) to cells/tissues; (2) high selectivity to discriminate the primary ions against others; (3) high spatial and temporal resolutions; (4) high sensitivity to the smaller changes of ion activities; (5) can continuously monitor event-related changes of ion activities (in real-time fashion); (6) insensitivity to pH and pharmacological inhibitors of ion transport; (7) determines an ion's activity rather than the total concentrations. For the last point, it is necessary to give some explanations. The activity of an ion species is essentially its effective concentration for explaining and predicting its equilibrium. It is the activity of an ion rather than the total or free concentration that determines the membrane potential, equilibrium potentials, or the thermodynamic conditions for ion transport mechanisms (Tsien 1983; Voipio et al. 1994), which makes ion activity especially attractive for clinical applications, since health and disorders are usually correlated to ion activity. The ion activity (effective concentration) is often quite different from the total concentration as measured by destructive chemical analysis. In a complex fluid system that contains multiple ionic constituents, for example, cytosol and cerebrospinal fluid (CSF), the mutual electrostatic repulsion between the similarly charged species and attraction between anions and cations reduce their mobility and freedom of ions, especially as the concentration of the solution becomes higher. Therefore, the ions exhibit non-ideal behavior, which is not due to incomplete ionization but to the existence of inter-ionic forces. Thus the activity of the ions in solution or their effective concentration is less than their total concentration (Tsien 1983; Voipio et al. 1994). The relationship between the activity (αi) and the concentration ([i]) of an ion follows the empirical formula: $\alpha i = [i]\gamma i$ in which γi refers to an appropriate activity coefficient that is closely related to the ionic strength of the solution and the valency of ion (Tsien 1983). However, for practical biological purposes (e.g. predicting electrophysiological reversal potential), because ions function at nearly constant ionic strength, the correction between ion activity and concentration are of less relevance to biological systems. Therefore the values are stated in concentrations due to the difficulties in determining activity coefficient in biological fluids (Tsien 1983; Hansen 1985). Therefore, in this review, we practically express the values in concentrations in accordance with the common usage.

Historically, scientists have been devoted to finding techniques to detect ion activities in biological tissues. Ion-selective electrodes (ISEs) and ion-sensitive fluorescent indicators (e.g. Na^+-, K^+-, and Ca^{2+}-sensitive fluorescent dyes) are the most important tools for detecting the intra- and extracellular ion activities. Some other methods have also been used for the determination of ion activity, such as flame photometry, electron probe X-ray microanalysis, and spectroscopy. In the following part of this section, with our knowledge, we will briefly retrospect the history of listed methods, and give some evaluations to these methods in the measurement of the intra- and extra-cellular ion activities.

Before the advent of ion-selective microelectrodes, no techniques were available for direct measurement of ion activities in the brain (Somjen 1979; Syková 1983; Hansen 1985). Based on the Noble Prize-winning ionic theory of membrane potential (including resting potential and action potential) (Bernstein 1902; Hodgkin and Katz 1949; Hodgkin and Huxley 1952a, b), electrophysiological techniques were used to indirectly determine ion concentrations by deducting the values from the Nernst equation since a Nernstian potential is created across the membrane when an electrochemical equilibrium is reached. For example, $[K^+]$ was measured by detecting glial cell membrane potential (since resting channels in glial cells are selective for K^+ only) or the magnitude of the undershoot after action potential, $[Na^+]$ was estimated from the height of action potential (which is the results of Na^+ influx), and $[Cl^-]$ was determined from the magnitude of the inhibitory postsynaptic potential (Syková 1983; Hansen 1985). This method was indeed helpful in exploring the ion activity of cells in the earlier studies, but it is too rough to reflect accurate activity.

Flame photometry is a type of emission photometry first invented in the 1920s in Germany (Caraway 1981), and was first introduced into clinical practice in 1947 by Pauline Hald at John Peters' laboratory at Yale University to measure the concentrations of Na^+, K^+, and Li^+ in serum and urine samples (Hald 1947; also see the website http://medicine.yale.edu/labmed/history.aspx, retrieved on July 22, 2015). This technique was later used to measure the ion concentrations in the brain (Nicholls and Kuffler 1965; Brown and Scholfield 1974; Patel et al. 1999; Raley-Susman et al. 2001). However, this method has some drawbacks such as low throughput, requirement of manual operation, as well as a time consuming procedure. Also, no standardization methods are available to date for this technique (Albert et al. 2011). This method lacks spatial resolution, meaning it cannot continuously monitor ionic changes. It is also tissue-destructive. For example, the method requires the brain tissues dried at high temperature (e.g. 85 °C), and then reconstituted in nitric acid for hours to days (Nicholls and Kuffler 1965; Brown and Scholfield 1974; Patel et al. 1999; Raley-Susman et al. 2001). The concentrations of ions are calculated by indirectly evaluating their contents in the water contents by detecting the difference between the wet and dry weight of brain tissues. This technique demands careful removal of extracellular fluid, and some ions may be lost from the intracellular space while washing out the extracellular space (Raley-Susman et al. 2001). Nevertheless, because this measurement cannot distinguish the free ions from those sequestered in intracellular organelles and those bound to intracellular buffers, the reported values are much higher than those detected using ISEs and ion-sensitive fluorescent sensors.

Electron probe X-ray microanalysis is a quantitative electron microscope technique that measures both water content (percentage water) and total (free plus bound) concentrations of biologically relevant elements (e.g., Na, K, S, P, Cl, Ca, and Mg) in cellular morphological compartments (Lopachin and Gaughan 1999). This technique was first proposed by Georg von Hevesy (the Noble Prize winner in chemistry in 1943 for characteristic X-ray spectra recordings and his key role in the development of radioactive tracers to study chemical processes such as in the metabolism of animals) in 1923 (Hevesy 1923) and applied by other workers in the following few years for the study of metal and mineral specimens. In 1981, Saubermann and his co-workers introduced this method to measure the contents of elements (Na, Mg, P, S, Cl, K, Ca, and Fe) in defined cellular compartments of the frozen-hydrated sections of mouse tissue specimens with little elemental and morphological disruption (Saubermann et al. 1981a, b). Electron probe X-ray microanalysis permits simultaneous determinations of multiple elements and allows optical differentiation of nervous tissue cell types and their processes (e.g., nerve and glial cell bodies, dendrites, axons) along with subsequent analyses of submembrane regions or organelles (e.g., axoplasm, mitochondria) (Lopachin and Gaughan 1999). Therefore, this method has been used for the determination of ion concentrations in neurons and glial cells in the brain (Pogorelov et al. 1997; Stys et al. 1997; Taylor et al. 1999; LoPachin et al. 2001). However, electron probe X-ray microanalysis requires the precision and accuracy of the specific specimen handling (Saubermann et al. 1981a, b; Taylor et al. 1999). Like flame photometry, this method is also low throughput and time consuming, does not distinguish free ions from bound element and those sequestered in intracellular organelles, but rather measures total elemental concentrations, and cannot detect event-related changes of ion concentrations in real time.

Spectroscopy is a collection of many spectroscopic techniques such as mass spectrometry, atomic absorption spectrometry (ultraviolet-visible and infrared spectroscopy), and nuclear magnetic resonance (NMR) spectroscopy, based on the concept of resonance and its corresponding resonant frequency of atoms, in which NMR spectroscopy has been widely used in medicine to help diagnosis of diseases. Spectroscopy was also used to detect the ion concentrations in the brain (Sussman et al. 1988; Gulya et al. 1991a, b, 1995; Boada et al. 2005; O'Donnell et al. 2013). For example, Sussman et al. utilized atomic absorption spectrometry to measure Na^+ and K^+ conventions of freeze-dried microscopic tissue samples (rat cerebellum, frog muscle, rabbit retinal tissue, and bovine liver), and proved the usefulness of this quantitative histochemical technique (Sussman et al. 1988). Gulya et al. used the same method to measure Na^+, K^+, Ca^{2+}, Mg^{2+} and Mn^{2+} levels in various brain regions including parietal cortex, hippocampus, striatum, medulla, and olfactory bulb (Gulya et al. 1991a). Magnetic resonance imaging diffusion-weighed imaging and magnetic resonance spectroscopy Na^+ spectroscopy were used to observe the changes in intracellular Na^+ contents during brain injury in animal stroke models (Boada et al. 2005; O'Donnell et al. 2013), and in patients with multiple sclerosis (Inglese et al. 2010). However, the destructive nature of atomic absorption spectrometry is obviously a drawback if time courses are desired, and this technique lacks spatial resolution, and cannot discriminate free ions from bound elements and those sequestered in intracellular organelles. NMR can quantify the amount of intracellular Na^+ and is nondestructive to tissues for the measurements; however, this technique requires relatively large amounts of tissue packed at high density in a magnet cavity, which limits other manipulations. For research purposes, these methods share some common drawbacks such as low throughput, the requirement of manual operation, as well as a time-consuming procedure.

Ion-selective microelectrode: An ISE is a transducer (or sensor) that converts the activity of a specific ion dissolved in a solution into an electrical potential, which can be measured by a potentiometer. The voltage is theoretically dependent on the logarithm of the ionic activity, according to the Nernst equation. ISEs are the oldest class of chemical sensors in the determination of the biologically relevant electrolytes in physiological fluids, and they provide a means of directly assessing the extracellular or intracellular activity of an ion and of making prolonged measurements of these ions (Voipio et al. 1994). The history of ISEs goes back all the way to 1906, when Max Cremer discovered pH-sensitive glasses, which led to the first commercial pH glass electrodes in the 1930s (Bühlmann and Chen 2012). In the 1950s–1960s, ISEs were introduced to detect the ionic activity in biological samples such as urine, bile, serum, cerebrospinal fluid, brain homogenate, and extremely large cells (e.g., squid giant axon, up to 1 mm in diameter) (Eisenman et al. 1957; Friedman et al. 1959; Isard 1959; Hinke 1961; Moore and Wilson 1963). There are mainly three classes of ISEs, i.e. glass electrodes (e.g. the earliest emerging electrodes), solid-membrane electrodes (Ag/AgCl electrode is the typical one), and liquid-membrane electrode (the most common version in recent decades). The earliest ISEs were fabricated from ion-selective glass and their usefulness for extra- and intracellular ion-measurements was greatly limited by relatively large tip diameter, slow response time, and expertise required to fabricate them (Lee et al. 2013). It was not until the emerging of miniaturized ISEs (i.e. ion selective microelectrodes) that made it possible for direct assessment of the extra- and intracellular activity of an ion. This has, to a large extent, been contributed to the progress of material chemistry and the invention of liquid selective ionophores (e.g., K^+ ionophore valinomysin, Na^+ ionophores I, VI, and X, Cl^- ionophore II, Ca^{2+} ionophore ETH 129, ionomycin etc.) which made the replacement of an ion-selective glass tip with an ionophore-doped liquid membrane that conferred ion selectivity to glass microelectrodes with a smaller (up to less than 1 μm) diameter tip and an electrode response time on the order of seconds (Lee et al. 2013). For example, the first K^+-selective liquid membrane microelectrodes were developed in 1971 (Walker 1971). After that, the electrophysiological studies on the CNS activities using this kind of K^+-selective microelectrodes boomed quickly (Krnjević and Morris 1972; Vyklicky et al. 1972; Vyskočil et al. 1972; Lux and Neher 1973; Prince et al. 1973; Syková et al. 1974; Mayevsky et al. 1974; Lothman and Somjen 1975). However, the earlier K^+-selective liquid membrane microelectrodes were poor in selectivity (30:1 for Na^+) with extremely high resistance (up to 10^{12} Ω, which greatly increases noise) and long response time (T_{95} up to 30 s). Then, a much-improved valinomycin-based K^+ selective microelectrode was introduced (Ammann et al. 1987b) with very high selectivity (5000:1 for Na^+), low resistance (10^{10} Ω), and short response time (0.6 s), which made it much better for K^+ study. Due to the development of K^+-selective microelectrodes there was a new surge of interest in the functional significance of K^+ activity in the nervous system in 1970s–1980s (see reviews by Syková 1983; Hansen 1985). Then, such studies largely gave way to those that used patch-clamp technique when the latter began to prevail in electrophysiological studies. Even so, K^+-selective microelectrode is still a very useful tool in studying the brain functions, especially under pathophysiological conditions (Jiang and Haddad 1991; Zetterström et al. 1995; Reid and Paterson 1996; Gido et al. 1997; Müller and Somjen 2000a, b; Mori et al. 2002; Stiefel et al. 2005; Chao et al. 2007a, b, 2008, 2009; Money et al. 2009; Shabala et al. 2010; Armstrong et al. 2011; Adámek and Vyskočil 2011; Stephan et al. 2012; Brocard et al. 2013). Ca^{2+}-selective liquid membrane microelectrodes based on either a synthetic neutral carrier (ETH 1001) or an organic phosphate were described in 1976 (Oehme et al. 1976; Brown et al. 1976). In 1977 they were first extracellularly used to observe Ca^{2+} changes during spreading depression or paroxysmal activity in the mammalian CNS (Heinemann et al. 1977; Nicholson et al. 1977), and in 1981 they were first used to detect intracellular Ca^{2+} activity in large snail neurons (Hofmeier and Lux 1981). In 1987 an optimized Ca^{2+}-selective liquid membrane microelectrode was introduced with much improved selectivity to allow it more realistic to measure very low Ca^{2+} levels such as in intracellular environments (Ammann et al. 1987a). Presently, the use of Ca^{2+}-selective microelectrodes has been largely eclipsed due to some practical difficulties and the introduction of a large number of fluorescent and other optical calcium probes with calcium sensitivities varying from the nanomolar to the millimolar range. However, it is still a complementary tool of fluorescent and other optical methods, and continues to be a good choice and one of the most straightforward ways to measure and quantify Ca^{2+} because it has some irreplaceable advantages including easy calibration, inexpensive cost for equipment, no side-effects in intracellular milieu (no photo-toxicity, don't change Ca^{2+} buffering), etc. (Thomas and Bers 2013). At almost the same time that the introduction of Ca^{2+}-selective liquid membrane microelectrodes occurred, Na^+-sensitive liquid membrane microelectrodes emerged and were then rapidly integrated and are still used for extracellular and intracellular recording in nerve cells (Thomas 1972; Kraig and Nicholson 1976; O'Doherty et al. 1979; Steiner et al. 1979; Grafe et al. 1982; Ballanyi et al. 1983; Müller and Somjen 2000a, b; Kang et al. 2009; Chao et al. 2012a, b) (also see Tables 5.1 and 5.2). The use of ISEs is critically dependent on four parameters: (1) high selectivity for the primary ion, (2) insensitivity to pH, (3) short response time, and (4) insensitivity to pharmacological inhibitors of ion transport (Messerli et al. 2008). The theoretical basis, practical operation, and applications of ISEs have been excellently reviewed for readers' reference (Voipio et al. 1994; Bühlmann and Chen 2012; Lee et al. 2013). ISEs determine an ion's activity rather than concentration because both electrodes and cellular processes respond to activity, not concentration. This method was widely used in the past to measure ion activities in the nerve cells with very minor destruction of brain tissue. Only the positioning of the electrode in the brain tissue may create a very limited dead space around the tip, but this region is functionally equilibrated with the surrounding undisturbed tissues. Therefore the influence on the measurement is negligible (Hansen 1985). Compared with the other methods mentioned above (flame photometry, electron probe X-ray microanalysis, and spectroscopy), the most striking advantages of ISEs are easy operation and its ability to continuously and long-lastingly monitor free ionic activity and event-related ionic changes in real time.

Optical imaging with ion-sensitive fluorescent indicators is a very popular technique and a very attractive alternative method for measuring intracellular ion activities today. The pioneering works of Roger Y Tsien (2008 Nobel Prize Winner

Table 5.1 Baseline $[Na^+]i$ of nerve cells

Species	Cell type	Preparation	Method	$[Na^+]i$ (mM)	References
Invertebrate preparations					
Snail	Neuron	Ganglia (In vivo)	ISME	4–6 mM	Thomas (1972) Alvarez-Leefmans et al. (1994)
Leech	Neuron	Ganglia	ISME	8 mM (De); 5.17 mM (Dö)	Deitmer and Schlue (1983) and Dörner et al. (1990)
	Glial cell		ISME	7 mM (M); 9.2 mM (Dö)	Dörner et al. (1990) and Munsch and Deitmer (1997)
Lobster	Stretch receptor neuron	Stretch receptor organ	ISME	20 mM	Edman et al. (1983)
Drone	Photoreceptor	Retinal slices	ISME	10.1 mM	Coles and Orkand (1985)
	Glia			37 mM	
Crayfish	Neuron	The slowly adapting stretch receptor organ	ISME (K); SBFI (Z)	7–10 mM	Kaila et al. (1987) and Zhong et al. (2001)
		Presynaptic terminal			
Vertebrate preparations					
Frog	Glia	Isolated spinal cord	ISME	44 mM	Bührle and Sonnhof (1983)
	Motoneuron			4–16 mM (G); 40 mM (B)	Grafe et al. (1982) and Bührle and Sonnhof (1983)
Rat	Neuron	Isolated superior cervical ganglia	ISME	7–11 mM	Ballanyi et al. (1983, 1984)

(continued)

Table 5.1 (continued)

Species	Cell type	Preparation	Method	[Na^+]i (mM)	References
Culture					
Rat	Neurons	Cultured cerebral cortical neurons	ISME (E); SBFI (F)	25–30 mM	Erecińska et al. (1991) and Friedman and Haddad (1994)
	Neuron	Cultured cerebellar granule cells	SBFI	4 mM (K); 7.5 mM (Co); 7.8 mM (Ch)	Kiedrowski et al. (1994a, b), Cousin et al. (1995), and Chen et al. (1995)
	Neurons	Cultured hippocampal neuron	SBFI	8.2 mM (P); 8.9 mM (R); 11 mM (S)	Pinelis et al. (1994), Rose and Ransom (1997b), and Sheldon et al. (2004)
	Astrocyte	Cultured hippocampal astrocytes	SBFI	About 15 mM (DIV10-21) (R-1996); 12 mM (DIV9-16) (K, R-1997); 19 mM (DIV20-34) (K)	Rose and Ransom (1996a, b, 1997a) and Kelly et al. (2009)
	Astrocyte	Cultured cortical astrocytes	CoroNa Green(R); SBFI (L)	15–17 mM	Reyes et al. (2012) and Longuemare et al. (1999)
Mouse	Astrocyte	Cultured cortical astrocytes	SBFI	12.7 mM (L); 9.4 mM (C); 6.6 mM (B-Hepes buffer), 13.2 mM (B); 19.0 mM (S)	Lenard et al. (2004), Chatton et al. (2000), Bernardinelli et al. (2004, 2006), and Su et al. (2002)
Fetal mouse	Neuron	culture	ISME	25 mM	Silver et al. (1997)
	Glia			20 mM	
Rat	Astrocyte	Cultured spinal cord astrocytes	SBFI	8–9 mM	Rose et al. (1997, 1998)
Embryonic mouse	Oligodendrocyte	Cultures spinal cord oligodendrocyte	ISME	15 mM	Ballanyi and Kettenmann (1990)

Slices					
Rat	Pyramidal neuron	Cortical slices	SBFI	9.9 mM	Pisani et al. (1998)
	Dopamine neuron	Substantia nigra pars compacta of midbrain slices	SBFI	13.5 mM (G); 11.7 mM (K)	Guatteo et al. (1998) and Knöpfel et al. (1998)
Mouse (P16–21)	Pyramidal neuron	Hippocampal slices	SBFI	12–15 mM	Kelly and Rose (2010) and Langer and Rose (2009)
Mouse	Astrocyte	Cortical slices	SBFI	17 mM	Unichenko et al. (2012)
Mouse (P12–21)	Astrocyte	Hippocampal slices	SBFI	11–13 mM	Langer et al. (2012), Kelly and Rose (2010), and Langer and Rose (2009)
Mouse (P20–30)	Bergmann glial cells	cerebellar slices	SBFI	5.2 mM	Kirischuk et al. (1997, 2007)
Guinea pig	Glial cells	Olfactory cortical slices	ISME	25.2 mM	Ballanyi et al. (1987)

Table 5.2 Baseline [Na^+]o in nerve cells

Species	Preparations	Method	[Na^+]o (mM)	References
Catfish	Cerebellum (in vivo)	ISME	149	Kraig and Nicholson (1978)
Adult cat	Sensorimotor cortex (in vivo)	ISME	146–150	Dietzel et al. (1980, 1982)
Rat (adult and neonate)	Brain stem slices	ISME	150	Jiang et al. (1992)
Adult rat	Parietal cortex (in vivo)	ISME	145	Stiefel and Marmarou (2002)
Rat	Hippocampal slices	ISME	152–155	Zanotto and Heinemann (1983), Fung et al. 1999, and Müller and Somjen (2000a, b)
Rat	In situ perfused rat brain	ISME	135–137	Xie et al. (1994)
Rat	Cerebral cortical slices	ISME	152	Kang et al. (2009) and Chao et al. (2012a, b)

in Chemistry) in the invention of Ca^{2+}, Na^+ and K^+-sensitive fluorescent indicator dyes in 1980s heralded a new era in the measurement of the key ions (Ca^{2+}, Na^+, K^+) by making available fluorescent probes that could be readily introduced into living cells. This led to the establishment of optical methods as a way of measuring the intracellular ion activities through a microscope where traditional microelectrode methods are not applicable (e.g., small cells, dendrites and spines of neurons). Tsien's laboratory designed and organically synthesized different series macrocyclic ligands, and ultimately demonstrated them as good quality intracellular Ca^{2+}, Na^+ and K^+ imaging (Tsien 1980, 1981; Tsien et al. 1982; Minta and Tsien 1989). Following Tsien's work, many other ion-sensitive fluorescent indicator dyes for the key ions have been invented (Simpson 1999; Schreiner and Rose 2012; Bootman et al. 2013). Ion-sensitive fluorescent indicator dyes, the key components for measurement of intracellular ion activities, constitute an ion-binding site and a fluorophore unit to translate the conformational change of the molecule on ion binding into an optical signal, changing their spectral properties in response to the binding of certain ions. A good fluorescent indicator dye would have the following properties: (1) highly specific for certain ions over other ions; (2) negligible sensitivity to change in pH; (3) its dissociation constant should be in the range of expected ion concentration; (4) the quantum yield (emission efficacy) of the indicator with primary ion binding should be high enough to allow easy detection as well as short exposure times and low indicator concentrations to minimize cellular stress; (5) the excitation and emission spectral should be easily distinguished, and have minimal overlaps (fluorescence shift) so that they can be easily detected without requiring very specialized detection equipment (e.g. ratiometric fluorescence measurements vs. single-wavelength measurements); (6) the indicators should have enough polar groups to render them highly water-soluble and impermeant through membranes to avoid the rapid lack out of cells; (7) the indicators could be modified by nonpolar groups (e.g., by acetoxymethyl esters) that is hydrolysable in the cytoplasm, to temporally mask the polar groups so that they can be loaded into cells; (8) the indicators should have minimal biological side-effects (e.g. changing intracellular pH, buffering etc.) and toxicities to cells; (9) the indicators should have good molecular stability (e.g., resistance to photo-blenching and extrusion by anion transporters) (Minta and Tsien 1989; Schreiner and Rose 2012). Optical imaging with ion-sensitive fluorescent indicators represents a most novel technique for measurement of intracellular ion activities with an apparently non-invasive nature and ease of use, especially in measuring the very rapid changes in relatively smaller cells or their parts (e.g., dendrites and spines of neurons, organelles).

In general, among all the aforementioned techniques and methods, both ISEs and optical imaging with ion-sensitive fluorescent indicators are the most widely used for extra- and intracellular ion activity studies in electrophysiology due to their advantages such as the non-invasive nature, continuous and real-time measurement, high spatial and temporal resolution, high throughput, measurement of only free ions, and ease of use over other techniques and methods. Though these two techniques are very different in their theoretical basis and operations, they play a considerably important role in studying ion activities in a complementary fashion. Compared to each other, both have their advantages and disadvantages (Voipio et al. 1994; Thomas and Bers 2013). For example, ISEs can measure both intracellular and extracellular ions, while optical imaging with ion-sensitive fluorescent indicators in most cases detect intracellular ions; for intracellular measurement, ISEs can measure two or more ions simultaneously for as long as one can maintain impalement of the cell (often many hours) even though it is difficult, though not impossible, to measures two ions simultaneously with ion-sensitive fluorescent indicators due to light interferences, and the measurement is only in a limited time (minutes) because of the photo-bleaching and photo-toxicity of the fluorescent indicators; ISEs themselves do not change intracellular fluids while fluorescent dyes may significantly increase ion buffering capacity of the cell; The calibration of ISEs is usually simple, and can be done separately, whereas it can be much more difficult with fluorescent indicators. Also ISEs measure only activities of cytoplasm because their tip size does not allow proper impalement of organelles without rupturing them, but incorporation of fluorescent indicators into cells can lead to the indicators being trapped in intracellular organelles, so it makes a mixed measurement in both cytosol and organelles; moreover, the equipment for ISE recording is simple and low cost (just basic electrophysiological apparatus), while the equipment for optical imaging with ion-sensitive fluorescent indicators can be much more expensive. However, optical imaging with ion-sensitive fluorescent indicators has some advantages ISE recording cannot surpass. For example, optical imaging with ion-sensitive fluorescent indicators is more advantageous for small cells, even dendrites and spines of neurons, while only large cells ($\geq$10 μm) are suitable for ISE recording; fluorescent indicators can detect the changes of intracellular ion activities in milliseconds while ISEs are only suitable to changes in the seconds timescale; the specificity is much better for fluorescent indicators than ISEs. Unlike ISE recording, electrical interference is never a problem for optical imaging with ion-sensitive fluorescent indicators. Therefore, it is better to complementarily use ISEs and optical imaging with ion-sensitive fluorescent indicators for the maximal advantages for measurement of intracellular ion activities.

Recently, Ca^{2+} imaging using green fluorescent protein-based genetically encoded Ca^{2+} indicators has been introduced as an alternative to using chemically synthesized fluorescent Ca^{2+} indicators (Palmer and Tsien 2006; Tian et al. 2009; Zhao et al. 2011; Gouriou et al. 2013). Genetically encoded Ca^{2+} indicators have several remarkable advantages over synthesized indicators: (1) they can be targeted to specific cell types and specific subcellular compartments by linking to cell-type selective promoters or sequences (Dreosti et al. 2009); (2) they are applicable to long-term expression (over months), allowing long-term recording from the same cellular structures (Mank et al. 2008; Tian et al. 2009; Ohkura et al. 2012; Gouriou et al. 2013); and (3) they allow reliable detection of neuronal activity with larger fluorescence signals and higher temporal resolution, and are capable of large-scale recording of Ca^{2+} dynamics from hundreds of postsynaptic spines and presynaptic boutons in a cell (Chen et al. 2013; Kobayashi et al. 2014). Therefore, combination of genetically encoded Ca^{2+} indicators with high-resolution, time-lapse in vivo as well as in vitro Ca^{2+} imaging (e.g. two-photo/multiphoton laser scanning microscopy) (Helmchen and Denk 2002, 2005; Helmchen et al. 2013) is a state-of-the-art tool and technique in the research of Ca^{2+} signaling.

4 Ion Homeostasis and Its Maintenance under Physiological Condition

The fulfillment of the fundamental functions of the nerve cells depends on a dynamically regulated ionic environment and ionic movement across the membrane. Nerve cells are surrounded by a very limited extracellular space (15–20 % of total brain volume) (Hrabìtová and Nicholson 2007), which is full of extracellular milieu containing various kinds of ions. Inside the cells, the cytosol also contains many kinds of ions. The extracellular ionic environment of the CNS is unique compared with most other tissues. Generally, in resting state, the extracellular fluids of mammalian nerve cells are abundant in Na^+ (concentrations up to 150 mM) and relatively poor in K^+ (around 3 mM). Conversely, the concentrations of these two ions are reversed inside the cells, with up to 150 mM of K^+ and 4–20 mM of Na^+. Extracellular Ca^{2+} concentration (1–2 mM) is about five orders of magnitude higher than cytosolic free Ca^{2+} concentrations (near 100 nM or below). Cl^- ions are also asymmetrically distributed, with high levels in the extracellular space (about 150 mM) and low intracellular levels in mature neurons (somata and dendrites) (around 5 mM) and astrocytes (30–60 mM). In immature CNS neurons $[Cl^-]i$ is relatively high, typically in the range of 25–40 mM (Blaesse et al. 2009). This difference between immature and mature neurons makes different responses of immature neurons in an increase in Cl^- conductance evoked by GABA that results in a membrane depolarization, not the conventionally regarded hyperpolarization in mature neurons (Blaesse et al. 2009). Moreover, $[Cl^-]i$ are uneven even in different parts of the same neurons. For example, measurement of $[Cl^-]i$ of cultured hippocampal neurons with a Cl^--sensitive fluorescent dye indicates the neuronal $[Cl^-]i$ is lower in perikaryonic than dendritic regions, which makes the estimated equilibrium potential of Cl^- (−35 mV) in neuronal processes more positive than the resting membrane potential (Hara et al. 1992), and therefore is responsible for the depolarization effect of activation of GABAergic receptors on pyramidal cell membranes at the dendritic regions in the hippocampal slices (Alger and Nicoll 1982a, b). Generally, the values of the key ions vary somewhat among multiple factors such as species, different CNS regions, cell types (neurons vs. glia), and maturation. For example, for squid giant axons, the cytoplasmic and extracellular concentrations of Na^+ (50 and 440 mM), K^+ (400 and 20 mM), Cl^- (51 and 560 mM) are very different from those in mammalian neurons. Na^+ imaging analysis with Na^+-selective fluorescent dye (e.g. SBFI) in rat neurons showed that $[Na^+]i$ is 4 mM in cultured cerebellar granular cells, approximately 9 mM in cultured hippocampal neurons, 11 mM in dopaminergic cells of substantia nigra pars compacta, and 10 mM in cortical pyramidal neurons; in contrast, glial $[Na^+]i$ is about twofold higher than in neurons (see review by Kirischuk et al. 2012). More detailed information about factors influencing the value of the key ions will be given in the later discussion in this chapter.

The maintenance of this steep electrochemical gradient across the membrane is determined by at least two major factors: selective permeability of the membrane to ions and active transport of ions. The lipid bilayer membrane functions as a permeability barrier selective to ions, in which some functional proteins (ionic channels/exchangers/transporters) act as the pathways in ionic movement. Active transport reversely moves the ions against the electrochemical gradients by directly/indirectly consuming the energy from hydrolysis of ATP. Electrogenic Na^+/K^+ ATPase (also called Na^+-K^+ pump or Na^+ pump), a transmembrane enzyme with the ability to bind and hydrolyze ATP, is the primary active transporter which functions to actively extrude three Na^+ outside the cells, and move two K^+ into the cells in each cycle of activity. Thus, the primary role of Na^+/K^+ ATPase is to maintain the high intracellular K^+ and low intracellular Na^+ (Sweadner 1989; Rose and Valdes 1994; Therien and Blostein 2000). An inhibition of Na^+ pumps induces an increase in $[Na^+]i$ and a fall in $[K^+]i$ in neurons (Alvarez-Leefmans et al. 1994). With $[K^+]e/[Na^+]i$ increase, Na^+/K^+ ATPase is expected to be activated and brings back $[K^+]e/[Na^+]i$ to resting levels (Sweadner 1989; Rose and Valdes 1994; Therien and Blostein 2000). Besides the key role of Na^+/K^+ ATPase, other machinery mechanisms are also involved in the steady-state maintenance of highly asymmetric concentrations of major inorganic cations and anions (Russell 2000; Kofuji and Newman 2004, 2009; Dipolo and Beaugé 2006; Giacomello et al. 2013).

The concentrations of ions are in dynamic balance under physiological conditions. With neuronal activity, the ionic concentrations fluctuate within a small range, and this fluctuation is often transient. For example, by using ISEs, it has been found that repetitive stimulation of the dorsal root nerves causes an increase in $[Na^+]i$ of 4–9 mM, a decrease in $[K^+]i$ of 8–20 mM and an increase in $[K^+]e$ of 2.0–3.5 mM to a so-called ceiling level followed by a post-stimulation K^+ undershoot, and an extracellular pH decrease after an alkaline shift in ganglions/spinal cords (Ballanyi et al. 1983; Syková and Chvátal 1993). By intracellular Na^+ imaging with Na^+-sensitive fluorescent dye sodium-binding benzofuran isophtalate (SBFI), spontaneous and transient $[Na^+]i$ increase (5 mM) was observed in 27 % of cultured hippocampal neurons (Rose and Ransom 1997b). An action potential-induced dendritic $[Na^+]i$ increase was accompanied by a rise in $[Ca^{2+}]i$ in hippocampal CA1 neurons (Spruston et al. 1995). Following ionic concentration fluctuation, a rapid increase in activity of ionic pumps (e.g. Na^+/K^+ ATPase, Ca^{2+} ATPase) and exchangers (e.g. Na^+/Ca^{2+} exchangers) as well as spatial ionic buffering mechanisms brings back the ionic gradients or levels towards their previous levels.

4.1 Potassium Ion (K^+)

K^+ is the most abundant cation in the cytoplasm. K^+-sensitive fluorescent indicator potassium-binding benzofuran isophtalate (PBFI) has been sometimes used to measure intracellular K^+ activity (Kiedrowski 1999; Liu et al. 2003; Kozoriz et al. 2010) since its invention (Minta and Tsien 1989). However, determination of $[K^+]i$ with PBFI is limited due to reasons such as its affinity for Na^+ ions, the relatively low selectivity for K^+ vs. Na^+, limited resolution for increases in $[K^+]i$ above the physiological range, and the technical difficulty of in situ calibration of PBFI (Minta and Tsien 1989). Recently, a new water-soluble, long-wavelength K^+-sensitive fluorescent indicator, TAC-Red, was invented for measuring K^+ activity. However, it is still in the initial stage and has not gotten popular (Padmawar et al. 2005; Namkung et al. 2008). Double-barrel K^+-selective microelectrodes are still the gold standard for determination of intracellular/extracellular K^+ activity (Amzica et al. 2002; Money et al. 2009; Xu et al. 2009; Shabala et al. 2010; Armstrong et al. 2011; Adámek and Vyskočil 2011; Stephan et al. 2012; Brocard et al. 2013; also see review by Syková 1983).

4.1.1 Extracellular and Intracellular K^+ Levels ($[K^+]e$ and $[K^+]i$) and Activity-Induced $[K^+]$ Changes

Although there are substantial differences in the results reported from the various laboratories and from various brain regions of various animal species, it is well accepted that in the resting state the basal level of K^+ is high in cytosol and low in the extracellular space of the brain. Measurement with K^+-selective microelectrodes shows a baseline of around 3 mM of $[K^+]e$ and 80–130 mM of $[K^+]i$ in nerve cells. Under the same conditions, the baseline $[K^+]i$ is much higher in glia (120–130 mM) than in neurons (75 mM) (Silver et al. 1997). Therefore, astrocytes generally have about a 20 mV more negative membrane potential than neighboring neurons (Walz 2000). The baseline $[K^+]e$ of the CNS seems higher in immature animals than in adults (Syková and Chvátal 1993). Most K^+ ions are in the free state in the cytoplasm, but they can be sequestered to mitochondria since direct measurement with electron probe X-ray microanalysis as well as fluorescence-sensitive K^+ dye, PBFI, indicates that intramitochondrial K^+ concentrations are higher than in the cytosol (Taylor et al. 1999; LoPachin et al. 2001; Liu et al. 2003; Kozoriz et al. 2010), and release of K^+ from the mitochondria into the cytoplasm can induce an increase in cytosolic K^+ concentrations (Liu et al. 2003; Kozoriz et al. 2010). The basal levels of $[K^+]$ in the brain are independent of the variations of plasma K^+ because of the excellent regulation by active transport across the blood-brain barriers (BBB) (Stummer et al. 1995; also see review by Somjen 2002). However, $[K^+]e$ can be easily increased during neuronal activity and the ensuing synaptic activity. For example, during spontaneous bursts of impulses of neurons in the reticular formation, a single action potential can increase 0.01–0.02 mM of $[K^+]e$ per cell (Syková 1983). The locomotor-like activity in the isolated neonatal rodent spinal interneurons increases $[K^+]e$ to 6 mM (Brocard et al. 2013). Repetitive stimulation of the nerve tissue or an afferent pathway can easily drive $[K^+]e$ from its resting level (around 3 mM) to 5–6 mM, and during seizures, $[K^+]e$ is further increased, but there seems to be a ceiling level of 10–12 mM (Heinemann and Lux 1977). Under normal conditions, the transient increases in $[K^+]e$ never exceed the ceiling level, even the stimuli are lasting or stimulus frequency/intensity is increased. Only under pathophysiological conditions such as hypoxia/ischemia, traumatic brain injury, and spreading depression, is this ceiling level disrupted and will $[K^+]e$ then increase to >25 mM, and sometimes even up to 100 mM (Hansen 1978; Dronne et al. 2007). After reaching the ceiling level following neuronal activity, $[K^+]e$ levels begin to recover towards baseline within seconds to minutes depending on the frequency/intensity of stimuli after discontinuing the stimulation, but do not stay in the basal level immediately. They continuously decrease to below baseline about 1 mM (undershoot phenomenon) and then gradually rise and stay to pre-stimulation levels (Heinemann and Lux 1975). Accompanying the increase in $[K^+]e$ during neuronal activity (Heinemann and Lux 1975, 1977; Křfž et al. 1975; Dietzel et al. 1980; Syková and Orkand 1980; Ballanyi et al. 1987; D'Ambrosio et al. 1999; David et al. 2009; Xu et al. 2009; Brocard et al. 2013; also see review by Syková 1983), intraneuronal K^+ decreases (Grafe et al. 1982; Ballanyi et al. 1983), while intraastrocytic K^+ increases in some degree (will be discussed next).

4.1.2 The Origination of Extracellular K^+ Accumulation during Neuronal Activity

The extracellular K^+ accumulation during neuronal activity largely originates from stimulated or spontaneously active neurons, unmyelinated fibers, and terminal arborization of axons (Syková 1983). Presynaptic terminals (e.g., activated primary afferent fibers) are responsible for 30–40 % of stimulation-evoked K^+ release, and the majority of K^+ ions are released postsynaptically (Syková and Chvátal 1993). Glial cells are not exactly known to give up their intracellular K^+ to the extracellular space. However, substantial evidence has suggested that glia cells, predominately astrocytes, largely contribute to the maintenance of K^+ homeostasis by uptake and/or spatial buffering of accumulated K^+ in the extracellular space (see reviews: Walz 2000; Kofuji and Newman 2004, 2009; MacAulay and Zeuthen 2012). Glial cell membrane is predominantly permeable to K^+ ions and sensitive to changes of $[K^+]e$ (Kuffler et al. 1966; Kimelberg et al. 1979; Ballanyi et al. 1987), and functions as a highly selective K^+ electrode in extracellular space. In addition, astrocytes generally have about a 20 mV more negative membrane potential than neighboring neurons (Walz 2000). Both properties provide a net driving force for K^+ influx in glial cells, particularly astrocytes. Therefore, surrounding glial cells, particularly astrocytes, take up extracellular K^+ into their cytoplasm and temporarily increase their $[K^+]i$ levels rather than release their intracellular K^+ to extracellular space, except for cases involving hypoxia/ischemia and other pathological conditions. Measurement of intracellular and extracellular K^+ concentrations of frog spinal motoneurons with double-barreled K^+-selective microelectrodes indicated that repetitive stimulation of a dorsal root yields an approximately equimolar concentration increase in $[K^+]e$ and decrease in $[K^+]i$ (Grafe et al. 1982). Based on a volume ratio of 4:1 for intracellular to extracellular space ratio, the expected $[K^+]i$ fall in motoneurons would be in the concentration range of only 1/4 of $[K^+]e$ rise. Therefore, the above data suggest that remarkable amounts of K^+ must be taken up into glial cells (Grafe et al. 1982). In addition, the existence of a ceiling level of ~12 mM under normal conditions (Heinemann and Lux 1977) and the K^+ undershoot below its resting levels after the transient increase in $[K^+]e$ (Heinemann and Lux 1975) are strong arguments in favor of the uptake of excess extracellular K^+ by glial cells (will be explained further next). Indeed, the surrounding glial cells act as "K^+-sinks" or "transient storage sites" and serve the function of restoring the extracellular K^+ concentration to its stable baseline (Walz 2000; MacAulay and Zeuthen 2012). Simultaneous measurement of intra- and extracellular K^+ concentrations with K^+-selective microelectrodes in guinea-pig olfactory cortical slices showed that neuronal stimulation induces about 6 mM of increase in $[K^+]e$, which is followed by a slow intracellular K^+ increase of about 16 mM in the glial cells (Ballanyi et al. 1987), accompanied with glial cell swelling (Walz 1992; MacAulay and Zeuthen 2012). Exposure of cultured astrocytes to saline containing high K^+ indeed induces reversibly the increase in intracellular K^+ concentrations and the subsequent cell swelling (Walz 1992). It is important to realize that nerve cells are surrounded by a very limited extracellular space (15–20 % of total brain volume) with low K^+ concentrations (around 3 mM in resting level), so even a very small amount of K^+ ions moving outside the neurons may cause a large increase in $[K^+]e$, and the subsequent glial cell swelling (Walz 1992; Kofuji and Newman 2004, 2009; MacAulay and Zeuthen 2012) as well as the resultant shrinkage of extracellular space (up to 50 % of its initial size) as demonstrated by 100 % increase in extracellular concentrations of membrane-impermeable tetramethylammonium (TMA^+) and choline (Ch^+) ions injected before stimulus-evoked $[K^+]e$ increase (Dietzel et al. 1980), which further enhances extracellular K^+ accumulation. In such a circumstance, a slow intracellular K^+ increase of about 16 mM in the glial cells (Ballanyi et al. 1987) is presumably underestimated because of the cell swelling.

4.1.3 The Pathways of K^+ Efflux During Neuronal Activity

During electrical activity, the repolarization and hyperpolarizing overshoot phase of each action potential or excitatory synaptic potential are the result of K^+ ions moving out of neurons and into extracellular space via a variety of K^+ channels including voltage-dependent K^+ channels, Ca^{2+}-activated K^+ channels, and Na^+-activated K^+ channels etc. (Franceschetti et al. 2003; Bhattacharjee and Kaczmarek 2005; Bean 2007), and contribute to increased $[K^+]e$ (Ballanyi et al. 1987). Measurement with K^+-selective microelectrodes has shown a single action potential can increase 0.01–0.02 mM of $[K^+]e$ per cell in the reticular formation (Syková 1983), while impulse activity induces a fast $[K^+]e$ increase of about 0.1 mM/pulse in the in vivo hippocampus (Chever et al. 2010), and an increase of $[K^+]e$ up to 0.3–0.5 mM/impulse of cerebellar Purkinje cells (Bruggencate et al. 1976). The relatively small increase in $[K^+]e$ per single impulse measured by K^+-selective microelectrodes is underestimated since a single impulse triggers K^+ release from point sources (vs. wide-spread or massive release), and K^+-selective microelectrodes with several μm in diameter cannot be placed sufficiently close to the surface of an active neuron or a nerve fiber which releases K^+ and makes $[K^+]e$ rise in the extracellular cleft (about 20–30 nm wide) (Syková 1983; Walz 2000). In addition to action potentials, synaptic currents may be another major reason for $[K^+]e$ increase. Repetitive stimulation of a lumbar dorsal root of frog spinal cord induced $[K^+]i$ decrease and $[K^+]e$ increase in spinal motoneurons (Syková and Orkand 1980; Grafe et al. 1982). In the hippocampal slices, when neuronal excitability is decreased and spontaneous and evoked impulse firing is blocked by tetrodotoxin (TTX) or lidocaine, spreading depression-related sharp $[K^+]e$ increase is depressed by 23 %, suggesting large amounts of $[K^+]e$ increase are dependent on other mechanisms, e.g. synaptic currents, rather than neuronal impulses in such a circumstance (Raley-Susman et al. 2001; Müller and Somjen 2000a, b). In addition, activation of NMDA, AMPA and kainate receptors (NMDARs, AMPARs, KARs) of mouse cortical neurons, even in a condition of decreased inward cation influx by lowering extracellular Na^+ and Ca^{2+}, induces a large outward K^+ current, which causes loss of about 50–80 % intraneuronal K^+ and great shrinkage of the cell body and consequently neuronal apoptosis (Yu et al. 1999; Xiao et al. 2001). High-frequency stimulation of stratum radiatum evokes a long-lasting GABA-mediated depolarization in pyramidal neurons of rat hippocampal slices, which is intimately associated with stimulation-induced $[K^+]e$ elevation (up to 7.4 mM) (Kaila et al. 1997). Activation of GABAergic receptors in the hippocampus induces a transient increase in $[K^+]e$, which involves the activation of $GABA_A$ receptors and $GABA_B$ receptor-activated K^+ conductance (Barolet and Morris 1991; Morris et al. 1996; Obrocea and Morris 1998).

4.1.4 The Clearance of Extracellular K^+

The active neurons lose their K^+ and induce extracellular K^+ accumulation. Excessive K^+ ions must be timely cleared to prevent the functional impairment of neurons (e.g. hyperexcitability and seizure-like discharges). Meanwhile, neurons must regain K^+ they lose to prevent the cease of their function. Diffusion of K^+ in the extracellular space is one of the ways for extracellular K^+ clearance; however, the capacity of pure diffusion is very limited and therefore has a minor role in clearing extracellular K^+ accumulation (Syková 1983; Kofuji and Newman 2009). Clearing of excess $[K^+]$

e is believed to involve both neurons and glial cells, and principally depends on both active and passive uptakes (Syková 1983; Walz 2000; Kofuji and Newman 2004, 2009; MacAulay and Zeuthen 2012). The occurrence of the ceiling level during continued neuronal activity suggests that $[K^+]e$ is actively cleared from the extracellular space (Xiong and Stringer 2000; Ransom et al. 2000; D'Ambrosio et al. 2002; Bay and Butt 2012). This process at least involves the active transport of neurons and glial cells (see reviews by Syková 1983; Walz 2000; Kofuji and Newman 2004, 2009; MacAulay and Zeuthen 2012).

Na^+/K^+ATPase (also known as Na^+-K^+ pump or Na^+ pump) plays a key role in the active transport of Na^+ and K^+ ions and the maintenance of their homeostasis in cells. Numerous studies have indicated that blockade of Na^+/K^+ ATPase increases the baseline $[K^+]e$, the amplitude of stimuli-induced $[K^+]e$ transient and $[K^+]i$ fall, and slows the rate of $[K^+]e$ recovery in the CNS (Cordingley and Somjen 1978; Grafe et al. 1982; Ballanyi et al. 1987; Alvarez-Leefmans et al. 1994; Ransom et al. 2000; Xiong and Stringer 2000; D'Ambrosio et al. 2002; Vaillend et al. 2002; Bay and Butt 2012; also see reviews by Syková 1983; MacAulay and Zeuthen 2012). Na^+/K^+ ATPase is an ubiquitous enzyme consisting of α, β and γ subunits, which couples ATP hydrolysis to active extrusion of Na^+ from the cells and uptake of K^+ at a stoichiometry of 3:2 for Na^+ and K^+ ions in each ATP-dependent turnover. The α-subunit contains the binding sites for Na^+, K^+ and ATP (and also the selective inhibitor ouabain binding site), while the β-subunit is required for the maturation, stability of the enzyme and the full enzymatic activity. The γ-subunit is not necessary for functional expression of the pump in some species, but it can alter pump activity by modulating Na^+ and K^+ affinity (Sweadner 1989; Rose and Valdes 1994; Therien and Blostein 2000). Na^+ binding from the cytoplasmic side triggers phosphorylation of the enzyme from ATP, whereas K^+ binding from the extracellular side leads to rapid dephosphorylation (Sweadner 1989; Rose and Valdes 1994). In the CNS, three isoforms of the catalytic α-subunit of Na^+/K^+ ATPase are expressed in a distinct pattern, i.e. $\alpha 1$, $\alpha 2$, and $\alpha 3$ isoform. Isoform $\alpha 1$ is ubiquitously distributed over the surfaces of the nerve cells (including neurons and glial cells) with very low affinity to ouabain (Na^+ pump inhibitor). High ouabain affinity isoform $\alpha 2$ is expressed predominately in glia, and $\alpha 3$ is expressed solely in neurons (McGrail et al. 1991; Cameron et al. 1994; Fink et al. 1996; Juhaszova and Blaustein 1997; Peng et al. 1997). Comparison of the substrate dependence properties of the rat Na^+/K^+ ATPase isoforms indicated that despite the variety of the reported Km values for K^+ and Na^+ affinity, $\alpha 1$ and $\alpha 2$ isoforms seem very similar in their affinity for extracellular K^+ and cytoplasmic Na^+, while $\alpha 3$ isoform has a two- to three-fold higher affinity for extracellular K^+ and two- to four-fold lower affinity for cytoplasmic Na^+ compared with the isoform $\alpha 1$ and $\alpha 2$ (Sweadner 1989; Jewell and Lingrel 1991; Munzer et al. 1994). Due to the properties of its ubiquitous distribution and kinetics, $\alpha 1$ isoform is believed to represent a "housekeeping" form of the Na^+ pump which is capable of responding to typical physiological demands (Munzer et al. 1994; MacAulay and Zeuthen 2012). In contrast, $\alpha 3$ isoform, which appears to be expressed predominantly in neurons with higher affinity for extracellular K^+ and lower affinity for cytoplasmic Na^+, and stimulated by increased neuronal $[Na^+]i$ and not by increased $[K^+]e$, would be ideally suited to run with large influxes of Na^+, such as those likely to occur during repeated action potentials, as well as during pathological processes such as seizures, ischemia, or excitotoxicity to recover Na^+-K^+ homeostasis (Munzer et al. 1994; Rose and Ransom 1997b), and the undershoot of rapid decrease in $[K^+]e$ in which the α3 isoform would be uniquely poised to continue pumping according to the demands of intracellular Na^+, without being significantly inhibited by this depletion of extracellular K^+ (Munzer et al. 1994). Glial Na^+/K^+ ATPase, which requires high K^+ (4–10 mM) stimulation (Sweadner 1989; Munzer et al. 1994), is suited to clear activity-dependent excess K^+ accumulation (Ransom et al. 2000; D'Ambrosio et al. 2002; Bay and Butt 2012; MacAulay and Zeuthen 2012). For example, small physiological relevant increases in $[K^+]e$ (4–8 mM) cause a decrease in $[Na^+]i$ in cultured hippocampal astrocytes (Rose and Ransom 1996a) but not in cultured hippocampal neurons (Rose and Ransom 1996b), indicating the activation of Na^+/K^+ ATPase in glial cells, but not in neurons, thus implying that glial isoform of Na^+/K^+ ATPase is well suited to play a significant role in the clearance of $[K^+]e$ (Rose and Ransom 1996a, 1997b; Ransom et al. 2000; D'Ambrosio et al. 2002; Bay and Butt 2012; MacAulay and Zeuthen 2012). It has been suggested that the ouabain-sensitive K^+ stimulation of Na^+/K^+ ATPase is several times less intense in neurons than in astrocytes (Rose and Ransom 1996a, 1997b; Hertz et al. 2013). In contrast, ouabain-sensitive K^+ stimulation of Na^+/K^+ ATPase in astrocytes is saturable only at a $[K^+]e$ of 10–12 mM (Walz and Hinks 1986). Therefore, under certain conditions, glial Na^+/K^+ ATPase can be modified to effectively become a K^+ pump, while neuronal Na^+/K^+ ATPase a Na^+ pump (Sweadner 1989), and the former contributes to the fast, initial phase of K^+ removal to K^+ active uptake, while the slower, sustained decline to K^+ uptake is via axonal Na^+ pumps during poststimulus recovery of $[K^+]e$ in rat optic nerve (Ransom et al. 2000). Therefore, glial and neuronal Na^+/K^+ ATPase are responsible largely for the activity-dependent ceiling levels and the undershoot of $[K^+]e$ respectively. It is worthy to point out that the undershoot phenomenon also involves the prominent role of glial cells. During repeated discharge firing, Na^+ ions also enter into neurons with K^+ efflux and accumulation in extracellular space. Because of the high efficiency of glial cells in clearing excess extracellular K^+ (see below discussion, Rose and Ransom 1996a), a large amount of K^+ ions are sequestered into glial cells. Even though Na^+/K^+ ATPase moves 3 Na^+ out to exchange 2 K^+ into cells in neurons, intracellular Na^+ ions still overwhelm the resting levels when extracellular K^+ ions have already recovered. The stimulation of neuronal Na^+/K^+ ATPase (especially α3 isoform, as discussed before) by high $[Na^+]i$ leads to the over-run of Na^+ pumps, thereby causing an excessive removal of Na^+ from the cells. Therefore, extracellular K^+ ions are continuously moved into neurons, resulting in the extracellular level below its basal level. Indeed, there is evidence showing pharmacological inhibition of Na^+/K^+ ATPase abolishes the K^+ recovery during higher frequency stimulation and the undershoot during the ensuing period (D'Ambrosio et al. 2002).

Neurons can replenish activity-dependent lost K^+ ions via active transportation by Na^+/K^+ ATPase. Several studies have provided evidence that K^+ clearance during neuronal activity is predominantly via glial and neuronal Na^+/K^+ ATPase (Xiong and Stringer 2000; Ransom et al. 2000; D'Ambrosio et al. 2002). However, the role of Na^+/K^+ ATPase in K^+ uptake was also argued by some investigators, i.e., the K^+ uptake via Na^+-K^+ pumps, particularly for neurons, is not thought to play a significant role in the rapid removal of K^+ from the extracellular space and the prevention of buildup of excess extracellular K^+ (Rose and Ransom 1997b; Walz 2000; Kofuji and Newman 2009; Hertz et al. 2013). In recent years, glial cells have been thought a key player, and play an essential role in clearing activity-dependent extracellular K^+ accumulation (Walz 2000; Kofuji and Newman 2004, 2009; MacAulay and Zeuthen 2012). For glia, in addition to the foregoing role of active transport, net K^+ uptake can occur passively by K^+ flux through transporters or K^+ channels, and can also be mediated by K^+ spatial buffering, in which both Na^+-K^+-$2Cl^-$ cotransporter 1 (NKCC1) and the inwardly rectifying K^+ channel subunit Kir4.1 have gained much attention (see reviews: Walz 2000; Kofuji and Newman 2004, 2009; MacAulay and Zeuthen 2012).

NKCC1: K^+ uptake via NKCC1 is strictly secondary active transport because it depends on the Na^+ gradients across the membrane built by the activity of Na^+-K^+ pumps by utilizing ATP. Na^+-K^+-Cl^- cotransporters are a class of membrane proteins that electroneutrally transport Na^+, K^+, and Cl^- ions into and out of a wide variety of cells almost always with stoichiometry of $1Na^+$:$1K^+$:$2Cl^-$. They are very sensitive to diuretics such as bumetanide and furosemide, and are essential in the regulation of intracellular Cl^- homeostasis along with K^+-Cl^- cotransporter 2 in the CNS (Haas and Forbush 1998; Blaesse et al. 2009). To date, two isoforms of NKCC have been confirmed, in which NKCC1 is present in a wide variety of tissues including the brain (while its counterpart NKCC2 is localized exclusively to the kidney) (Haas and Forbush 1998; Blaesse et al. 2009; MacAulay and Zeuthen 2012). Both neurons and glial cells express NKCC1 (Blaesse et al. 2009; MacAulay and Zeuthen 2012). Classic viewpoint regards that neuronal NKCC1 expression decreases during postnatal development (see review by Blaesse et al. 2009). However, this idea is challenged by some studies that show a developmental increase in the NKCC1 mRNA and protein levels in the brain (see review by Blaesse et al. 2009, and references cited therein). It has been demonstrated that NKCC1 in the adult CNS is absent from neuronal cell bodies despite its existence in dendrites in some neurons, while it is abundantly expressed in astrocytes (see review by Hertz et al. 2013, and references cited therein). Therefore, neuronal NKCC1 may contribute less for K^+ uptake by neurons. For glial cells, the value of NKCC1 for half-maximal K^+ activation is relatively low (2.7 mM), suggesting this transporter is constantly active even at a baseline $[K^+]e$ of about 3 mM (Tas et al. 1987). Because of its abundant expression, strong inward driving force, and high affinity for extracellular K^+ in glial cells, NKCC1 is well suited for K^+ uptake by glial cells (Walz 2000). With the measurement of K^+-selective microelectrodes and optical methods, it has been shown in astrocytes that intracellular K^+ accumulation and the consequent cell swelling following increased $[K^+]e$ can be partially prevented by bumetanide or furosemide or by removal of external Na^+ and Cl^-, suggesting the involvement of glial NKCC1 in K^+ buffering (Kofuji and Newman 2004, 2009). Physiologically, activity-dependent increases in $[K^+]e$ reach no more than the ceiling levels of 10–12 mM (Heinemann and Lux 1977). In this K^+ range, Na^+/K^+ ATPase activity is thought to be saturated (Walz and Hinks 1986), while NKCC1 is still running well. Therefore, the relative contribution of NKCC1 in K^+ uptake at higher extracellular K^+ concentrations is proposed (MacAulay and Zeuthen 2012). NKCC1 activity accumulates about 2 mM of astrocytic $[Na^+]i$ (Rose and Ransom 1996a; Kirischuk et al. 2012), which in turn stimulates Na^+/K^+ ATPase activity to extrude Na^+ out of astrocytes to complete a so-called "transmembrane Na^+ cycle" in astrocytes (Walz and Hinks 1986), and leave astrocytes with accumulated KCl and increased volume (glia swelling) (Walz 1992; MacAulay and Zeuthen 2012).

The spatial buffering mechanism: Besides the aforementioned mechanisms (simple diffusion, active transport via neuronal and glial Na^+-K^+ pumps as well as NKCC1), the spatial buffering mechanism, which largely depends on the K^+ influx via astrocyte Kir4.1 channels and gap junctions between astrocytes, is also contributable to the clearance of accumulated K^+. The concept of "K^+ spatial buffering" was first coined by Orkand et al. based on their experiments with giant glial cells in the leech CNS (Orkand et al. 1966), and has been well proven by numerous studies (see reviews by Walz 2000; Kofuji and Newman 2004, 2009; MacAulay and Zeuthen 2012). A specialized form of spatial buffering termed K^+ siphoning was delicately shown in Müller cells (the principal glial cell type) of the retina by Newman and coworkers (see reviews by Kofuji and Newman 2004, 2009). According to "K^+ spatial buffering hypothesis", neurons release K^+ ions during activity and increase $[K^+]e$, surrounding astrocytes take up excess extracellular K^+ ions, which makes their membrane potential more positive than that at the regions distant from high K^+ locations, and therefore generates a closed current loop between these regions to drive K^+ ions transferring via the gap-junction-coupled cell syncytium to astrocytes where $[K^+]e$ is low, and the net driving force causes K^+ to flow out of astrocytes at the sites where $[K^+]e$ is low (Orkand et al. 1966).

KIR 4.1 channels: Indeed, numerous studies have provided strong support to the K^+ spatial buffering hypothesis and demonstrated the critical roles of astrocytic Kir4.1 channels and gap junctions in this process (see reviews by Walz 2000; Kofuji and Newman 2004, 2009). Essentially, K^+ spatial buffering comprises the clearance and redistribution of extracellular K^+ ions, which is fulfilled by K^+ uptake via glial Kir4.1 channels and K^+ dissipation via glial gap junction-coupling and K^+ release to distant extracellular space with low $[K^+]e$. The classical property of Kir channels is that K^+ flows inwards when the resting membrane potential is negative compared to the equilibrium potential for K^+, but outward currents are inhibited at potentials that are more positive. This provides the driving force for glial uptake of K^+ that is released during neuronal activity (K^+ clearance) (Butt and Kalsi 2006). The first identified glia-associated Kir is the Ba^{2+}-sensitive weakly rectifying kir4.1 (Takumi et al. 1995). The Kir4.1 subunit is expressed almost exclusively in glial cells, particularly astrocytes, and to a much lesser degree in neurons (Takumi et al. 1995; Neusch et al. 2006; Djukic et al. 2007; Olsen et al. 2007; Tang et al. 2009; Seifert et al. 2009; David et al. 2009; and review by Olsen and Sontheimer 2008). Kir4.1 channels are critical in determining the K^+ selective membrane properties of astrocytes, and are largely responsible for the inward currents and for setting the strongly negative resting membrane potential of astrocytes (Neusch et al. 2006; Djukic et al. 2007; Olsen et al. 2007; Kucheryavykh et al. 2007; Tang et al. 2009; Seifert et al. 2009; Chever et al. 2010; Haj-Yasein et al. 2011; Bay and Butt 2012; Sibille et al. 2014). Both pharmacological blockade (D'Ambrosio et al. 1999) and gene knockout of Kir4.1 (Djukic et al. 2007; Sibille et al. 2014) result in a marked reduction of inward currents, several fold increases of membrane resistance, and severe membrane depolarization in astrocytes, and the consequently impaired ability to K^+ buffering and accompanying extracellular K^+ accumulation. Interestingly, Kir4.1 knockout does not affect membrane and action potential properties of hippocampal CA1 pyramidal neurons and basal synaptic transmission in the hippocampus, making a striking contrast with the changes of astrocytes (Djukic et al. 2007; Sibille et al. 2014). However, due to the impaired buffering ability and extracellular K^+ accumulation with Kir4.1 knockout, hippocampal spontaneous neuronal activity is reduced and high-frequency stimulation-induced short-term synaptic potentiation is greatly enhanced (Djukic et al. 2007; Sibille et al. 2014), providing further support of the critical role of astrocyte Kir4.1 channels in K^+ clearance. Most recently, it was found that a developmental upregulation of a Kir4.1 channels of adult oligodendrocyte precursor cells, which are the major source of myelinating oligodendrocytes during development, offers the ability of these cells to sense fine changes of local $[K^+]e$ increases induced by physiological neuronal activity, suggesting that these cells might play a role in K^+ uptake in the mature healthy brain, thus adding a further complexity, previously unsuspected, to the mechanisms used by these cells to detect neuronal activity and maintain K^+ homeostasis in mature gray matter (Maldonado et al. 2013).

Gap junctions: In addition to glial Kir4.1 channels, K^+ spatial buffering also involves the gap junctions between adjacent glial cells, particularly astrocytes. Gap junctions are critical to electrical coupling between astrocytes and are intimately associated with intrinsic properties of astrocytes (Adermark and Lovinger 2008; Meme et al. 2009). Multiple gap junction proteins (connexins) are exclusively expressed in astrocytes but not in neurons, which include Cx43, Cx30 and Cx26 (see reviews: Rouach et al. 2004; Sutor and Hagerty 2005). Among them, Cx43 is the principal gap junction protein that constitutes astrocytic gap junctions (Wallraff et al. 2006; Orthmann-Murphy et al. 2008). Tracer (e.g., biocytin, Lucifer Yellow) spread from single injected astrocytes indicates a widespread of astrocyte coupling, and an astrocyte can have coupling with up to 240 astrocytes via gap junction (Wallraff et al. 2006; Adermark and Lovinger 2008; Meme et al. 2009; Xu et al. 2009; Strohschein et al. 2011). The gap junction coupling makes astrocytes act as a functional syncytium, allowing rapid intercellular K^+ diffusion and release in the locations where $[K^+]e$ is low (K^+ redistribution) and thus preventing extracellular K^+ accumulation during neuronal firing (Kofuji and Newman 2004). Indeed, combined electrophysiological recordings of hippocampal and cortical glial cells and extracellular K^+ measurements with K^+-selective microelectrodes have suggested a crucial role of faster K^+ diffusion through the interglial syncytium in the homeostasis of extracellular K^+ in pathophysiological processes (D'Ambrosio et al. 1999; Amzica et al. 2002; Strohschein et al. 2011). Both pharmacological blockade of gap junctions (Xu et al. 2009; Spong and Robertson 2013) and downregulation of Cx43 expression in the hippocampal astrocytes (Xu et al. 2009) cause a progressive impairment in the ability to maintain baseline $[K^+]e$ levels and buffering of activity-dependent peak increase in $[K^+]e$, leading to spreading depression-like K^+ surging activity (Spong and Robertson 2013) and seizure-like activity (Uhlmann et al. 2002). Genetically-engineered mice deficient in Cx43 and Cx30 completely eliminated astrocyte coupling within the hippocampus, and these mice exhibited reduced K^+ buffering capabilities (substantially higher $[K^+]e$ peak and slower decay of elevated $[K^+]e$ during synchronized neuronal firing) and increased susceptibility to epileptiform activity (Wallraff et al. 2006). In fact, impaired astrocytic gap junction protein expression and/or interglia syncytium function are important causes for accumulation of K^+ as well as glutamate in the extracellular space and the subsequent pathophysiological changes for the generation of epileptiform events (D'Ambrosio et al. 1999; Amzica et al. 2002; Wallraff et al. 2006; Xu et al. 2009; Steinhäuser et al. 2012; Bedner and Steinhäuser 2013). In addition, oligodendrocyte gap junctions (consisting of Cx32, Cx29 and Cx47 subunits) also contribute to spatial buffering of K^+ released during neuronal activity in myelinated nerves such as optical nerves because mice lacking both Cx32 and Cx47 display a characteristic vacuolation of myelin surrounding axons (Odermatt et al. 2003; Menichella et al. 2006), and this phenotype is attributable to the extremely high peri-internodal K^+ concentrations (up to 100 mM) released after intensive nerve stimulation (Rash 2010). In such a scenario, and also considering the developmental upregulation of Kir4.1 channels of adult oligodendrocyte precursor cells to sense fine changes of local $[K^+]e$ increases induced by physiological neuronal activity (Maldonado et al. 2013), with the lack of Cx32 and Cx47, excess K^+ cannot diffuse among oligodendrocytes which form the myelin of axons (due to the elimination of oligodendrocyte gap junctions). Consequently, excess K^+ cannot access astrocyte cytosol via gap junctions between astrocytes and oligodendrocytes (panglial syncytium) or finally release K^+ to distal locations or into capillaries via astrocytic end feet (Kofuji and Newman 2004, 2009; Menichella et al. 2006; Rash 2010), leading to local pathological changes such as vacuolation of myelin surrounding axons. Taken together, K^+ uptake via glial Kir4.1 channels along with K^+ dissipation via glial gap junction-coupling and K^+ release to distant extracellular space with low $[K^+]e$ synergically constitute K^+ spatial buffering mechanisms to maintain K^+ homeostasis with other mechanisms.

For the convenience of our readers to understand K^+ activity and its regulation, a schematic cartoon has been presented to display the simplified toolkit for K^+ regulation (Fig. 5.2).

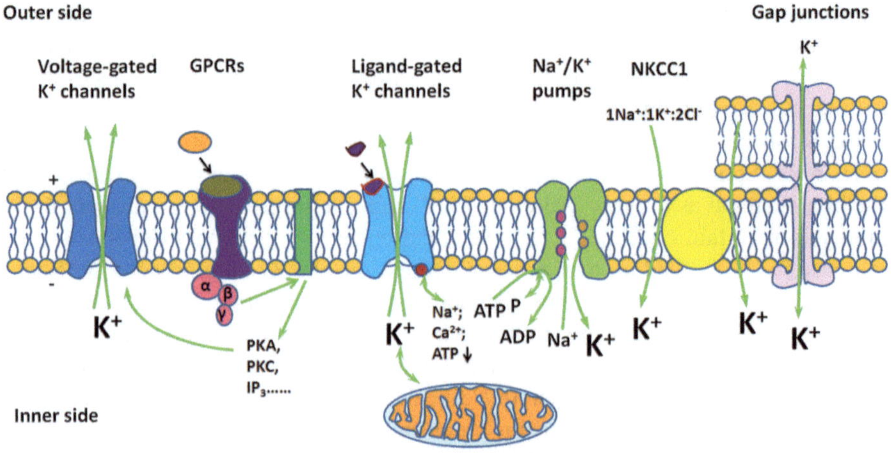

Fig. 5.2 Schematic cartoon displaying the simplified toolkit for K^+ regulation. The extracellular K^+ accumulation during neuronal activity largely originates from stimulated or spontaneously active neurons, unmyelinated fibers, and terminal arborization of axons via a variety of K^+ channels including voltage-dependent K^+ channels, Ca^{2+}-activated K^+ channels, and Na^+-activated K^+ channels etc. Some ligand-gated receptor channels such as ionotropic glutamatergic receptor channels also mediate K^+ efflux during excitatory neurotransmission. The surrounding glial cells act as "K^+-sinks" or "transient storage sites" and serve the function of restoring the extracellular K^+ concentration to its stable baseline. During electrical activity, repolarization and the hyperpolarizing overshoot phase of each action potential or excitatory synaptic potential occur as a result of K^+ ions moving out of neurons and into extracellular space via a variety of K^+ channels. Clearing excess $[K^+]e$ involves both neurons and glial cells, and principally depends on active uptake and passive uptake. This process involves at least primary active transport of neurons and glial cells by Na^+/K^+ ATPase, and secondary active transport by NKCC1. It also involves the spatial buffering mechanism that uses Kir4.1 channels and gap junctions, which is not fully included in this schematic demonstration. A specialized form of spatial buffering, termed K^+ siphoning, was delicately shown in Müller cells (the principal glial cell type) of the retina by Newman and coworkers. Readers who are interested in the spatial buffering mechanism are recommended to see the pioneer research work by Orkand et al. (1966) and the excellent reviews by Kofuji and Newman (2004, 2009)

4.1.5 The Influences of K^+ Ions on the Nerve Cells

The fundamental functions of a cell are based on the appropriate membrane potentials. For excitable cells such as neurons, in addition to the important role of the resting membrane potential, action potentials serve as the cellular language by which neurons communicate with one another. It has been well known that K^+ ions are the basis of resting membrane potential. For action potentials though, the explosive and regenerative Na^+ influx via voltage-dependent Na^+ channels, as well as the minor Ca^{2+} influx, contributes to the uprising stroke of the action potential, while K^+ efflux via a fraction of the voltage-dependent K^+ channels chiefly contributes to the repolarization and hyperpolarizing overshoot phase of the action potential. Thus K^+ ions largely influence the duration and firing rate of the action potentials. It is now clear that the duration and frequency of action potentials play a key role in information encoding. In this sense, K^+ ions act as a signaling molecule for neurons. In addition, the changes in the cellular K^+ levels exert multiple impacts on other ion activity. For example, as we will discuss in the next section, small changes in $[K^+]$e (e.g., increase from 3 to 4–13 mM to mimic $[K^+]$e changes that occur during variably intense neuronal activity or decrease to 2, and 1 mM) produce large effects in $[Na^+]$i in the nerve cells, in particular astrocytes (Rose and Ransom 1996a, 1997a), which are intimately related to astrocytic function in sensing neuronal activity via G protein-coupled receptors (e.g. type I metabotropic glutamate and ATP purinergic P_2Y1 receptors) in their membrane and sending feedback responses to neurons to modulate CNS signaling via the gliotrasmitters (e.g. glutamate, D-serine, and ATP) (Perea et al. 2009; Nedergaard and Verkhratsky 2012). Moreover, K^+-Na^+ interaction may influence the removal of synaptically released neurotransmitters such as glutamate and GABA in synaptic cleft by perisynaptic astrocytes to timely terminate their action and prevent their spillover (Kirischuk et al. 2012; Rose and Karus 2013), and might play a feedback mechanism to synaptic activity, thus influencing synaptic plasticity (Unichenko et al. 2012). The clearance of accumulated extracellular K^+ during neuronal activity also stimulates the activity of Na^+/K^+ ATPase and NKCC1 in the glia. The former plays a key role in bioenergetics via lactate genesis and astrocyte–neuron lactate shuttling between neurons (mainly oxidative) and astrocytes (mainly glycolytic) (Pellerin and Magistretti 2012), and the latter is critical in regulation of osmotic homeostasis and cellular volumes (MacAulay and Zeuthen 2012). K^+ activity also influences Ca^{2+} behavioral phenotype, which may exert multiple impacts on the cells such as neurotransmitter release, and gene regulation. Large and/or sustained changes in K^+ ions are a marker of the pathophysiology of the nerve cells, which result in cell death and brain diseases (Somjen 2002; Yu 2003; Chao and Xia 2010).

4.2 *Sodium Ion (Na^+)*

In the early 1970s, Na^+-selective liquid membrane-based microelectrodes emerged and were then rapidly used for the direct measurement of extracellular and intracellular Na^+ concentrations ($[Na^+]$o and $[Na^+]$i) of the nerve cells. Na^+-selective microelectrodes have proven a very useful tool to measure $[Na^+]$o. However, due to their large tip diameter, long response time, and a long time to reach steady-state, the earlier measurements of $[Na^+]$i with Na^+-selective microelectrodes were mainly limited to the nerve cells of invertebrates, such as snail, leech, lobster, drone, and crayfish in which nerve cells commonly have a large diameter >200 μm (see Table 5.1). With the improvement of the electrodes, they were also used in vertebrate nerve cells in later studies (Table 5.1). The application of Na^+-selective microelectrodes has yielded important information for understanding the regulation and the basic mechanisms of Na^+ homeostasis in the CNS. However, due to the obvious flaws (the damage of cells especially with small diameter caused by impalement, and lack of spatial resolution—allowing the measurement only in cell bodies not in small cellular microdomains such as axons, dendrites or spines), Na^+-selective microelectrodes are seldom used for $[Na^+]$i measurement nowadays (Rose 2002). Significant progress in the study of Na^+ signaling in vertebrate nerve cells has been made with the invention and application of the Na^+-sensitive fluorescent dye, sodium-binding benzofuran isophtalate (SBFI) (Minta and Tsien 1989). SBFI has been proven an ideal Na^+-sensitive fluorescent dye with optimum properties for measuring $[Na^+]$i (Minta and Tsien 1989; Rose and Ransom 1996a, b, 1997b; Schreiner and Rose 2012). SBFI combining with high-resolution imaging, particularly confocal and two-photon microscopy, has enabled the quantitative measurement of Na^+ signals with considerably high spatial and temporal resolution, even in cellular subdomains of both neurons and glial cells such as spines and fine dendrites of neurons and mitochondria of astrocytes (Jaffe et al. 1992; Rose et al. 1999; Rose and Konnerth 2001; Rose 2003; Bernardinelli et al. 2006; Langer and Rose 2009).

4.2.1 Basal Level of Extracellular and Intracellular Na^+ ($[Na^+]$o and $[Na^+]$ i)

Unlike high $[K^+]$i and low $[K^+]$e distribution, Na^+ ions are remarkably abundant in extracellular space but are relatively sparse in the cytosol. The reported $[Na^+]$o in the brain both in vivo and in vitro (different regions of a variety of species) is around 150 mM at rest, as measured with ion-selective electrodes (see Table 5.2). In contrast, the reported values of steady-state baseline $[Na^+]$i in normal resting state vary depending on the differences in species (invertebrates such as snail, leech, lobster, drone and crayfish, vs. vertebrates such as frog, rat, mouse, and guinea pig), animal/cell ages, regions of the nervous system (e.g., cortex, hippocampus, cerebellum, midbrain, spinal cord, retina, ganglia etc.), cell types (e.g., neurons vs. astrocytes), preparations (isolated tissue blocks, cultured cells, and brain slices), experimental conditions (e.g., using CO_2/HCO_3^- buffering solution vs. HEPES buffering one, and at (near) physiological temperature vs. room temperature), and even differences in measurement methods (Na^+-selective microelectrodes vs. Na^+ imaging with Na^+-sensitive fluorescent indicators such as SBFI and CoroNa Green) (see Table 5.1). For example, in invertebrates, the reported baseline $[Na^+]$i is as low as 4 mM (Thomas 1972; Alvarez-Leefmans et al. 1994), and as high as about 20 mM (Edman et al. 1983), as measured in snail ganglia neurons and lobster stretch receptor neurons respectively with Na^+-selective microelectrodes, but is mostly in the range of 7–10 mM (Table 5.1). For invertebrate glial cells, baseline $[Na^+]$i is reported in the range of 7–37 mM (Table 5.1). In contrast, the steady-state baseline $[Na^+]$i of vertebrate neurons is reported as the lowest concentration being 4 mM in motoneurons of isolated spinal cord of frogs as measured with Na^+-selective microelectrodes (Grafe et al. 1982) and cultured cerebellar granule cells of rats by SFBI fluorescence imaging (Kiedrowski et al. 1994a, b) and the highest as 40 mM in frog spinal cord motoneurons (Bührle and Sonnhof 1983), but the majority of reported values of steady-state baseline $[Na^+]$i of neurons fall in the range of 7–15 mM, whether measured with Na^+-selective microelectrodes or Na^+ imaging with SBFI, and in isolated tissue blocks or cultured cells or acute brain slices from different CNS regions in a variety of animals (Table 5.1). For vertebrate glial cells, the reported baseline $[Na^+]i$ values vary from the lowest concentration being 5.2 mM in Bergmann cells of mouse cerebellar slices as measured with SBFI fluorescence imaging (Kirischuk et al. 1997, 2007) to the highest being 44 mM in glia of isolated frog spinal cord as measured with Na^+-selective microelectrodes (Bührle and Sonnhof 1983), and mostly fall in the range of 10–20 mM in different CNS regions of different species (Table 5.1). Unlike some neurons that spontaneously fluctuate their baseline $[Na^+]i$ levels (Rose and Ransom 1997b), glial cells always show a very steady resting $[Na^+]i$ level. Moreover, for the same species, same types of cells from different regions of the CNS may have different resting baseline $[Na^+]i$ levels. For example, rat cerebral neurons (cultures and slices) have a relatively higher resting $[Na^+]i$ level (9.9 mM by Pisani et al. 1998, 25–30 mM by Erecińska et al. 1991 and Friedman and Haddad 1994) than that of rat cerebellar granule cells (4–8 mM) (Kiedrowski et al. 1994a, b; Cousin et al. 1995; Chen et al. 1999). The resting $[Na^+]$ i in the astrocytes of rat cortex (15–17 mM) (Reyes et al. 2012; Longuemare et al. 1999) and hippocampus (12–19 mM) (Rose and Ransom 1996a, b, 1997a; Kelly et al. 2009) is much higher than in the astrocytes of rat spinal cord (8–9 mM) (Rose et al. 1997, 1998). Even for the same cells, detection with different cell ages or in different buffer solution comes to the different baseline $[Na^+]i$. For instance, under control conditions with HCO_3^-/CO_2 buffered saline, the steady-state $[Na^+]i$ is 11.6 mM for cultured hippocampal astrocytes in DIV 9–16, while it is 19 mM in DIV 20–34 astrocytes. In contrast, in HEPES-buffered saline, the values are 8.1 and 15.5 mM for cultured astrocytes in DIV 9–16 and 20–34 respectively, a level of 3–4 mM lower than in HCO_3^-/CO_2 buffered saline (Kelly et al. 2009). SBFI fluorescence imaging indicated that cultured mouse cortical astrocytes have a resting $[Na^+]i$ of 6.6 mM in HEPES-buffered saline, but 13.2 mM in normal standard HCO_3^-/CO_2-buffered artificial cerebrospinal fluid (ACSF), two times higher than in HEPES-buffered ACSF (Bernardinelli et al. 2004, 2006). Similar observations were also found in cultured spinal cord astrocytes (DIV 6–10) (Rose et al. 1997), and this may be caused by inwardly directed Na^+-HCO_3^- contransport and age-dependent expression of Na^+-driven transports (e.g., NKCC) (Rose and Ransom 1997b; Kelly et al. 2009). The differences of steady-state baseline $[Na^+]i$ might reflect special nature and properties of certain cells and species, and could also be technique dependent, e.g., Na^+-selective microelectrode-related damages of small cells during impalement and the consequently artificially high $[Na^+]i$ (see Sect. 3). Nevertheless, the studies on baseline $[Na^+]o$ and $[Na^+]i$ illustrate a general phenomenon, i.e., there exists a very steep electrochemical gradient for Na^+ in both neurons and glial cells, which is estimated to be as high as +140 mV based on the reported baseline $[Na^+]o$ and $[Na^+]i$ (Rose and Karus 2013). Such a steep electrochemical gradient of Na^+ plays a central role in many fundamental functions of cells, such as generation of action potential and fast excitatory transmission in neurons, the homeostasis of other ions, osmoregulation, the uptake of nutrients, and uptake of transmitters etc. (see the following discussion).

Measurement of $[Na^+]i$ with the same techniques from the same or different laboratories showed resting baseline $[Na^+]i$ in astrocytes is generally higher than in neurons (Bührle and Sonnhof 1983; Coles and Orkand 1985; Dörner et al. 1990; Rose and Ransom 1996a, b, 1997b; also see review by Kirischuk et al. 2012 and Table 5.1). Bührle and Sonnhof (1983), by using Na^+-selective microelectrodes to measure the $[Na^+]i$ in frog spinal cord, found that the resting $[Na^+]i$ in motoneurons is 40 mM, while in glial cells it is 44 mM. In the retina of the drone, the resting $[Na^+]$ i, measured also with Na^+-selective microelectrodes, is 10 and 37 mM in photoreceptors and glia cells respectively (Coles and Orkand 1985). Dörner et al. (1990) found in leech ganglia that neurons have a resting $[Na^+]i$ of 5 mM, whereas it is 9 mM for glial cells. The $[Na^+]i$, as measured with SBFI fluorescence imaging, in cultured hippocampal neurons was reported about 9 mM, while it is 15 mM in cultured hippocampal astrocytes (Rose and Ransom 1996a, b, 1997b). This difference may be determined by essentially different roles, physiological functions, and different channel/receptor/transporter expression between neurons and glial cells (will be discussed). In addition, baseline $[Na^+]i$ can be kept stable very well in all the astrocytes during up to 2 h of experimental period (Rose and Ransom 1996a). In contrast, 27 % of cultured hippocampal neurons were observed to have a spontaneous, transient $[Na^+]i$ increase of about 5 mM, which was often synchronized among nearby neurons, and could be reversibly blocked by TTX or by high Mg^{2+} saline (Rose and Ransom 1997b). Therefore, in some neurons, baseline $[Na^+]i$ is influenced by periodic bursting activity of synaptically coupled cells and the opening of voltage-gated Na^+ channels (Rose and Ransom 1997b).

4.2.2 Regulation of Baseline [Na⁺]i in Neurons and Glial Cells

The simplified regulatory mechanisms for Na^+ activity have been shown in Fig. 5.3. More detailed information about the regulatory mechanisms will be discussed in the following sections.

The establishment of the extremely steep Na^+ electrochemical gradients across the membrane and the behavior phenotypes of baseline Na^+ ions are absolutely dependent on the permanent activity of Na^+/K^+ ATPase, which is responsible for at least 60 % of energy expenditure of nerve cells in the CNS (Schreiner and Rose 2012). Despite the different properties of its ubiquitous distribution and kinetics in neurons and glial cells (see Sect. 4.1), the primary role of Na^+/K^+ ATPase is to maintain the high intracellular K^+ and low intracellular Na^+ against their steep electrochemical gradients by transporting three Na^+ ions out of the cells in exchange of the uptake of two K^+ ions (Sweadner 1989; Munzer et al. 1994). Blockade of Na^+/K^+ ATPase activity by either removal of $[K^+]e$ (0 $[K^+]e$) or application of ouabain leads to an increase in $[Na^+]i$ in both neurons and glial cells (Ballanyi et al. 1983, 1984; Alvarez-Leefmans et al. 1994; Rose and Ransom 1996a, 1997b; Rose et al. 1998; Chatton et al. 2000). However, the responses of neurons and glial cells to the blockade of Na^+/K^+ ATPase activity are not exactly the same. For example, in neurons, removal of $[K^+]e$ causes an increase in $[Na^+]i$ of 5 mM/min, leading to $[Na^+]i$ values

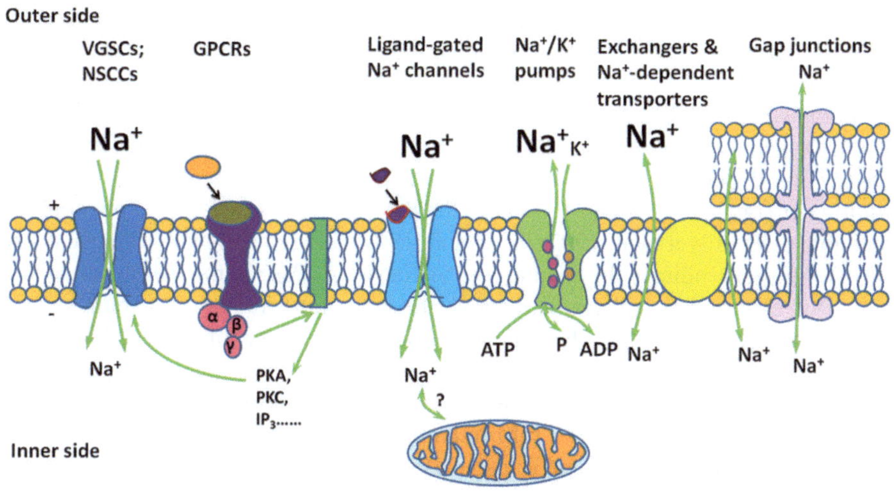

Fig. 5.3 Schematic demonstration of the simplified regulatory mechanisms for Na^+ activity in nerve cells. The baseline $[Na^+]i$ are absolutely dependent on the permanent activity of Na^+/K^+ ATPase as well as other molecules such as Na^+-dependent Cl^--HCO_3^- exchangers, Na^+-HCO_3^- cotransporter, or NKCC. Under normal conditions and during neuronal activity (e.g., action potential firing and synaptic activity), for neurons, voltage-gated Na^+ channels (VGSCs) and iGluR compose the most important passage for Na^+ entry. However, VGSCs play a minor role in glial cells. The most important pathways for Na^+ entry into glial cells are iGluR (particularly CNQX-sensitive AMPARs) and Na^+-dependent (glutamate, GABA, HCO_3^-) transporters. Non-selective cation channels, Na^+/Ca^{2+} exchangers (NCXs), $P2Y_{1-2}$ purinergic receptor channels, and Na^+-K^+-$2Cl^-$ cotransporters (NKCCs) also contribute to Na^+ entry in both neurons and glial cells, while gap junctions contribute to Na^+ entry in glial cells. Gap junction coupling can quickly equalize intracellular Na^+ concentration in astrocytes. More detailed information has been given in the text

of 49.7 mM after 10 min exposure to K^+-free saline, which are able to recover partly in >50 % of cells within 20 min after returning to normal $[K^+]e$ (Rose and Ransom 1997b). For astrocytes, removing $[K^+]e$ results in an increase in $[Na^+]i$ of 4 mM/min, and $[Na^+]i$ reaches to 60 mM after 10 min of 0 $[K^+]e$ exposure and only 32 % cells can recover from this Na^+ loading with external K^+ returned (Rose and Ransom 1996a). Ouabain at 0.5 mM **irreversibly** causes $[Na^+]i$ increase in cultured hippocampal neurons with a rate of 4.7 mM/min, while 1 mM of ouabain (which completely blocks glial Na^+/K^+ ATPase activity) causes $[Na^+]i$ increase in cultured hippocampal astrocytes by 2.5 mM/min, and this increase reaches 42 mM during a 10 min exposure to ouabain, and **reversibly** recovers in all the astrocytes observed after removal of the drug, indicating a different response between neurons and astrocytes (Rose and Ransom 1996a, 1997b). The different responses of neurons and glial cells to Na^+/K^+ ATPase blockade may be related to the differences of isoform expression and kinetics of Na^+/K^+ ATPase in these cells. Neurons predominantly express the α3 isoform of Na^+/K^+ ATPase, which has higher affinity for extracellular K^+ and lower affinity for cytoplasmic Na^+, and is stimulated by increased neuronal $[Na^+]i$ and not by increased $[K^+]e$, and can be modified to effectively become a Na^+ pump, and is ideally suited to run with large influxes of Na^+ (Sweadner 1989; Munzer et al. 1994). In contrast, glial α2 isoform requires high K^+ (4–10 mM) stimulation (Sweadner 1989; Munzer et al. 1994), and is functionally an effective K^+ pump suited to clear activity-dependent excess K^+ accumulation (Ransom et al. 2000; D'Ambrosio et al. 2002; Bay and Butt 2012; MacAulay and Zeuthen 2012). Due to the dependence of Na^+/K^+ ATPase activity on ATP, depletion of ATP by limiting energy synthesis can also cause the large increase in $[Na^+]i$ in both neurons and glial cells (Silver et al. 1997; Rose et al. 1997, also see review by Chao and Xia 2010). All these indicate the critical role of Na^+/K^+ ATPase in the maintenance of a steep Na^+ electrochemical gradient across the membrane. In addition, maintenance of baseline $[Na^+]i$ is strongly dependent on the presence of extracellular Na^+. Complete replacement of extracellular Na^+ with N-methyl-D-glucamine ($NMDG^+$) and choline causes exponential depletion of $[Na^+]i$ with a time constant of 2.7 min and a maximal efflux rate of 4.1 mM/min reaching close zero in 10 min in cultured hippocampal neurons (Rose and Ransom 1997b), whereas in cultured hippocampal astrocytes, it leads to a much slower $[Na^+]i$ decrease (0.15 mM/min) and a long period (20 min) to reach zero (Rose and Ransom 1996a). Changing $[Na^+]o$ by 20 mM results in small changes in $[Na^+]i$ of 2–3 mM in astrocytes (Rose and Ransom 1996a). Switching $[Na^+]o$ from 145.2 to 19.2 mM results in a slow (over 10–20 min) fall of resting $[Na^+]i$ from 11.7 to 7.5 mM in dopaminergic neurons of substantia nigra par compacta of rat midbrain slices (Knöpfel et al. 1998). After re-introducing $[Na^+]o$, neurons rapidly restore their baseline $[Na^+]i$ to the original level with a maximal Na^+ influx rate of 7.3 mM/min, while astrocytes rapidly increase their $[Na^+]i$ higher than the initial baseline and subsequently fall by 4.6 mM/min to reach a new steady-state $[Na^+]i$ significantly lower than the original levels (Rose and Ransom 1996a, 1997b).

In addition to the above-mentioned important factors, other molecules may also regulate baseline $[Na^+]i$, which include some transporters, channels, and exchangers (see review by Kirischuk et al. 2012), each with different contributions to $[Na^+]i$ between neurons and glial cells. It has been shown that Na^+-dependent Cl^--HCO_3^- exchange, Na^+-HCO_3^- cotransport, and NKCC do not contribute significantly to steady-state baseline $[Na^+]i$ in neurons since their blockade or activation does not produce obvious changes in steady-state baseline $[Na^+]i$ (Rose and Ransom 1997b). In contrast, both NKCC and Na^+-HCO_3^- cotransporters respectively have a 2 mM contribution to steady-state baseline $[Na^+]i$ in astrocytes, because blockade of NKCC with furosemide or bumetanide leads to a slow decrease in $[Na^+]i$, which indicates a steady inward transport of Na^+ via NKCC under resting condition; changing from CO_2/HCO_3^--free to CO_2/HCO_3^--buffered saline (to activate Na^+-HCO_3^- cotransport) induces a transient increase in $[Na^+]i$ of 5 mM and then falls to a new steady state of 2.1 mM higher than in CO_2/HCO_3^--free, and HEPES-buffered saline (Rose and Ransom 1996a, 1997b). Furthermore, the baseline $[Na^+]i$ can also be regulated by extracellular K^+ levels. For neurons, small changes in $[K^+]e$ (e.g., increase from 3 mM to 4–13 mM to mimic $[K^+]e$ changes occurred during variably intense neuronal activity, or decrease to 2, and 1 mM) produce little effects in $[Na^+]$ i in most of neurons during relatively brief alterations in $[K^+]e$, and <5 mM of changes in $[Na^+]i$ in a small part of neurons during relatively long time alterations in $[K^+]e$; even increased $[K^+]e$ from 3 mM to 20 and 40 mM for 2 min transiently increase $[Na^+]i$ only 3.0 and 4.7 mM respectively (Rose and Ransom 1997b), indicating neuronal $[Na^+]i$ is relatively insensitive to $[K^+]e$ changes, which is well consistent with the viewpoint that neuronal Na^+/K^+ ATPase is an effective Na^+ pump and stimulated by increased neuronal $[Na^+]i$ and not by increased $[K^+]e$ (Sweadner 1989; Munzer et al. 1994). For astrocytes, a brief increase (3 min) in $[K^+]e$ from 3 mM to 4, 5, 6, 7, and 8 mM causes reversible decreases of $[Na^+]i$ of 1–3 mM, and a 10 min-increase of $[K^+]e$ to 8 and 13 mM decreases $[Na^+]i$ of 5.8 and 7.1 mM respectively in all the tested cells. Reducing $[K^+]e$ from 3 to 2 mM (10 min) increases $[Na^+]i$ of 3 mM, and reduction of $[K^+]e$ to 1 mM causes a steep increase in $[Na^+]i$ with a rate of 1.9 mM/min, which is comparable with that of Na^+/K^+ ATPase inhibition (Rose and Ransom 1996a). This indicates that the glial Na^+/K^+ ATPase is functionally an effective K^+ pump suited to clear activity-dependent excess K^+ accumulation (Ransom et al. 2000; D'Ambrosio et al. 2002; Bay and Butt 2012; MacAulay and Zeuthen 2012).

4.2.3 Activity-Induced $[Na^+]i$ Signals in Neurons and Glia Cells

During neuronal activity (e.g., **action potential firing** and **synaptic activity**), the opening of voltage- and ligand-gated receptor channels can lead to the alterations in $[Na^+]o$ and $[Na^+]i$ of both neurons and glial cells, which transduce multiple important signals to regulate cellular functions. Action potentials are regenerative electrical activities that are mainly mediated by voltage-gated Na^+ channels and initiated following depolarization of the membrane potential to a threshold level. Both dendrites and axon initial segment of many neurons (e.g., pyramidal neurons of the neocortex and hippocampus) can generate Na^+ action potentials (Spruston et al. 1995; Stuart et al. 1997; Golding and Spruston 1998; Magee and Carruth 1999; Ariav et al. 2003). Dendritic spikes can amplify and sharpen the summed somatic potential as a trigger for action potential initiation in the axon, though they are strongly attenuated as they spread to the soma; reversely, somatic action potentials can back-propagate far into the dendritic tree where they open the voltage-gated Na^+ channels as well as Ca^{2+} channels to induce dendritic ion influx (Spruston et al. 1995; Stuart et al. 1997; Golding and Spruston 1998; Magee and Carruth 1999; Ariav et al. 2003), which may cause changes in $[Na^+]i$.

Action potential-induced $[Na^+]i$ changes in neurons. Indeed, action potential-induced rapid $[Na^+]i$ increase was first directly measured with SBFI fluorescence imaging in proximal dendrites of CA1 pyramidal neurons in rat hippocampal slices, which showed that a single Na^+ spike, elicited in the soma by injection of depolarization current pulse via a sharp intracellular microelectrode, causes a rapid rise in $[Na^+]i$ in proximal dendrites which can be completely blocked by voltage-gated Na^+ channel blocker TTX (Jaffe et al. 1992). With the aid of high-resolution two-photo Na^+ imaging with SBFI, Rose et al. (1999) found in mouse hippocampal slices that back-propagating action potentials evoke $[Na^+]i$ transients of about 4 mM for a train of 20 action potentials not only in the proximal part of the dendritic tree, but also the adjacent spines (diameter <1 μm) of CA1 pyramidal neurons. These [Na^+]i transients monotonically decayed with a time constant of several seconds, and are sensitive to TTX. Nevertheless, with the numbers of back-propagating action potential decreasing, no or only minor Na^+ changes were induced in the dendritic tree of CA1 pyramidal neurons (Rose and Konnerth 2001). In cultured rat hippocampal neurons, spontaneous, transient [Na^+]i increases of 5 mM were observed in 27 % of neurons, which could be reversibly blocked by TTX, suggesting an action potential-evoked, voltage-gated Na^+ channel-mediated, Na^+ entry and [Na^+]i increase (Rose and Ransom 1997b). In sharp contrast, cerebellar Purkinje neurons show a different [Na^+]i response to action potentials. When Purkinje cells are induced to fire a burst of action potentials by direct depolarizing current injection in the cell bodies, the antidromically activated, TTX-sensitive [Na^+]i increases are strictly confined to the region of the soma and axon of Purkinje cells (Lasser-Ross and Ross 1992; Callaway and Ross 1997; Knöpfel et al. 2000). [Na^+]i transients as observed in the dendrites and the spines of hippocampal pyramidal neurons (Jaffe et al. 1992; Rose et al. 1999) can never be detected in the dendrites and the spines of Purkinje cells (Lasser-Ross and Ross 1992; Callaway and Ross 1997; Knöpfel et al. 2000). Interestingly, dopaminergic neurons of the substantial nigra pars compacta of rat midbrain slices show a completely different [Na^+]i response to action potentials. It is hyperpolarization (not depolarization) of cell body that evokes a rise in somatic [Na^+]i with the amplitude positively increasing when the hyperpolarizing potential is more negative, while induction of action potential firing by depolarizing from subthreshold membrane potentials does not induce a measurable increase in [Na^+]i (Knöpfel et al. 1998). The hyperpolarization-induced [Na^+]i elevation was also observed in the dendrites of pyramidal neurons in mouse hippocampus (Tsubokawa et al. 1999). These differences in action potential-induced [Na^+]i changes may be a reflection of essential differences in the functions of different types of neurons in the brain.

Direct stimulation-induced [Na^+]i changes in glial cells. Unlike neurons, glial cells cannot generate action potentials due to the low densities of voltage-gated ion channels and high expression of K^+ channels that prevent substantial depolarization of glial membrane. However, glial cells can respond to mechanical and electrical stimulation to release gliotransmitters and induce intracellular Na^+ and Ca^{2+} transients to form "glial Ca^{2+} and Na^+ excitability" (see reviews by Kirischuk et al. 2012; Parpura and Verkhratsky 2012). In cultured mouse cortical astrocytes, both electrical and mechanical (that always occurs during vascular regulation in glial-neuronal vascular units in the brain) stimulation of a single astrocyte induces a [Na^+]i elevation in the stimulated cell that spreads to the surrounding cells at 14.4 μm/s (Bernardinelli et al. 2004). Similar observations were also made in situ astrocytes in mouse hippocampal slices (Langer et al. 2012). In this study, using SBFI fluorescence imaging, the investigators found direct electrical stimulation of a single SR101-positive astrocyte (see Table 5.3) via a fine glass pipette (tip diameter 1 μm) induces an increase of [Na^+]i up to 26 mM in the directly stimulated cell that spreads to neighboring astrocytes within a distance of about 100 μm with the

Table 5.3 Two distinct subpopulations of astrocyte-like glia in the hippocampus

	Passive cells	Complex cells
Electrophysiological properties	Large time- and voltage-independent leak currents	Time- and voltage-dependent K^+ and Na^+ currents
Cellular marker	GFAP and GLAST immunopositive, and can be stained by the red fluorescent dye SR101	S100β immunopositive with 20 % of cells also expressing NG2 proteoglycan
GluT/iGluR expression	Expressing GluT but not iGluR, also termed "GluT astrocytes"	Having functional iGluR of AMPA subtype, also called "GluR cells"
Gap junction coupling	Extensive	–
Percentage of cell numbers at postnatal 2–3 weeks	Comprising >90 % of cells with glial morphology	Few
Nature of the cells	Mature protoplasmic astrocytes	A mixed population of immature astrocytes and NG2-immunopositive multipotent precursors

References: Langer and Rose (2009) J Physiol 587:5859–5877; Djukic et al. (2007) J Neurosci 27:11354–11365.
Note that GluT in this form refers to glutamate transporter.

amplitude reducing as the distances increase and with variable propagation speeds, i.e., amplitude reaching 30 % of that of stimulated cell within a radius of 40–50 μm, and less than 10 % at 70–110 μm, and with the propagation speed ranging from >60 μm/s in the direct neighborhood of the stimulated cell to less than 10 μm/s at distances >60 μm (Langer et al. 2012). Based on their own mechanism analysis, these two groups gave different explanations for the stimulation-induced $[Na^+]i$ elevation and spreading to neighboring astrocytes (Bernardinelli et al. 2004; Langer et al. 2012) (see further discussion below).

Measurements of Na^+ activity in cat sensorimotor cortex with Na^+-selective microelectrodes indicate that burst stimulation of cortex itself or the corresponding thalamic projection nuclei leads to 4–8 mM decreases in $[Na^+]o$ in all cortical layers (Dietzel et al. 1980, 1982). Robust neuronal activity, mostly occurring in pathological conditions such as spreading depression (Kraig and Nicholson 1978) and cerebral hypoxia/ischemia (see review by Chao and Xia 2010, also will be discussed in the next chapter), results in more pronounced decrease in $[Na^+]o$, that can reach up to 100 mM falls from the basal level (~150 mM). Correspondingly, stimulus can induce the changes in $[Na^+]i$. For example, repetitive stimulation of a dorsal root of frog spinal cord causes an increase in $[Na^+]i$ of about 5 mM in motoneurons, as measured with Na^+-selective microelectrodes (Grafe et al. 1982). In guinea-pig olfactory cortical slices, electrical stimulation (20 Hz, 15 s) of afferent nerve fibers in the lateral olfactory tract results in an average **decrease** rather than an increase of $[Na^+]i$ of 6 mM in glial cells, and the authors proposed that this is the result of overrunning of glial Na^+/K^+ ATPase stimulated by excess extracellular K^+, and thus raised intracellular Na^+ is not essential in glial cells during stimulus-induced activity (Ballanyi et al. 1987). This observation is very different with those made by others on cerebellar Bergmann glial cells and hippocampal astrocytes evoked by afferent fibers by using SBFI imaging (Kirischuk et al. 2007; Bennay et al. 2008; Langer and Rose 2009; also see reviews by Deitmer and Rose 2010; Rose and Karus 2013, also see the following detailed discussion).

Synaptically induced sodium transients in glial cells: Since the inrooted conception of glial cells as "structural support and passive elements" of the nervous system was challenged by the discoveries of "glial Ca^{2+} excitability" (see review by Kirischuk et al. 2012), the investigation of the roles of glial cells (especially astrocytes) in synaptic events has been attracting more and more attention (Nedergaard and Verkhratsky 2012). The fine terminal processes of a single astrocyte can reach approximately 140,000 synapses simultaneously in rodents or up to two million synapses in humans (Bushong et al. 2002; Kirischuk et al. 2012) and thus form a "tripartite synapse" or "synaptic cradle" (Perea et al. 2009; Nedergaard and Verkhratsky 2012) to sense neuronal activity via G protein-coupled receptors (e.g. type I metabotropic glutamate and ATP purinergic P_2Y1 receptors) in their membrane and then send feedback responses to neurons to modulate CNS signaling via the gliotrasmitters (e.g. glutamate, D-serine, and ATP) (Perea et al. 2009; Nedergaard and Verkhratsky 2012). One of the most important responses of perisynaptic glia to synaptic activity is the changes in intracellular ions, and thus the generation of Na^+ signaling (Deitmer and Rose 2010; Kirischuk et al. 2012; Parpura and Verkhratsky 2012; Rose and Karus 2013). In mouse cerebellar slices, measurement of $[Na^+]i$ in Bergmann glial cells with SBFI imaging showed that short burst stimulation of **parallel fibers** results in **local** Na^+ transients of several mM in amplitudes that are dependent on the stimulus intensity and are largest in the processes close to the stimulation pipette (Kirischuk et al. 2007; Bennay et al. 2008). Burst stimulation of **climbing fibers** elicits **global** intracellular Na^+ transients with similar kinetics, amplitudes, and time courses throughout the entire tree of processes of Bergmann glial cells (Bennay et al. 2008). Further analysis of the kinetic properties of parallel fiber stimulation-evoked Na^+ transients uncovered two types of Na^+ signals. The "slow Na^+ transients" are characterized by smaller peak amplitude (<2 mM), slower rise time (10–30 s), and a broad peak with the decay not being fitted by a monoexponential decay time course; in contrast, the "fast Na^+ transients" have a large peak amplitude (average 5.6 mM), an average rise time of 4.8 s, a clearly defined peak, and strictly monoexponential decay (Bennay et al. 2008). Both kinds of Na^+ transients not only exist in different Bergmann glial cells, but also within the different main processes of a given cell (Bennay et al. 2008). Similar observations were made in the astrocytes of mouse hippocampal slices (Langer and Rose 2009). The hippocampus contains two distinct subpopulations of astrocyte-like glia, i.e., "complex" cells and "passive" cells, which are divided based on their electrophysiological properties and cellular marker expression (Djukic et al. 2007; Bennay et al. 2008) (see Table 5.3). Measurement of $[Na^+]i$ in SR101-positive passive astrocytes in mouse hippocampal slices with SBFI fluorescence imaging indicated that short-burst stimulation of Schaffer collaterals or direct soma stimulation of a single astrocyte induces Na^+ transients in the millimolar range which are restricted to astrocytes located within a radius of about 50 μm around the stimulation pipette (Langer and Rose 2009). The peak amplitudes and the slopes of the rising phase of Na^+ elevations of these Na^+ transients in surrounding cells decrease with increasing distance from the stimulated cell (a wave-like propagation) (Langer and Rose 2009). With low stimulation intensities, synaptic activity-evoked Na^+ signals have large differences in the amplitude and kinetics between the different cellular domains in single astrocytes. The largest signals in one or two cellular domains represent a primary branch with its fine processes, and only weakly invaded the soma and other cellular domains (Langer and Rose 2009).

Synaptically induced sodium transients in neurons: As done in glial cells, similar observations were also made in synaptic activation-induced Na^+ transients in cerebellar Purkinje cells (Lasser-Ross and Ross 1992; Callaway and Ross 1997; Knöpfel et al. 2000; Kuruma et al. 2003; Bennay et al. 2008) and hippocampal CA1 pyramidal neurons as well as granule cells of the dentate gyrus (Rose and Konnerth 2001; Langer and Rose 2009). In guinea-pig cerebellar slices, stimulation of climbing fibers causes $[Na^+]i$ rise, as displayed by the fluorescence changes of SBFI, in the main dendritic shafts of anatomically expected synaptic contact sites of Purkinje cells, which results from the Na^+ influx through AMPAR channels, and diffuses towards the fine dendrites (Lasser-Ross and Ross 1992; Callaway and Ross 1997). Climbing fiber stimulation also causes a large $[Na^+]i$ increase in the soma and axon hillock where there are no synaptic contacts, and this increase is entirely due to the Na^+ spike via voltage-gated channels (Lasser-Ross and Ross 1992; Callaway and Ross 1997). Stimulation of parallel fibers induces similar changes in $[Na^+]i$ in dendrites mediated by Na^+ entry via AMPAR channels as well as metabotropic glutamate receptor channels and in soma independent of the propagation generated at the dendrites in Purkinje cells (Callaway and Ross 1997; Knöpfel et al. 2000; Kuruma et al. 2003; Bennay et al. 2008). Calibration of SBFI signals indicated several millimolar increases in $[Na^+]i$ with the stimulation of parallel fibers (Kuruma et al. 2003; Bennay et al. 2008). In mouse hippocampal slices, measurements of $[Na^+]i$ of in situ CA1 pyramidal neurons using quantitative Na^+ imaging with SBFI indicated that stimulation of Schaffer collaterals (five 2 ms square pulses delivered at 50 Hz) via glass electrodes placed in close proximity to the dendrite (10–15 μm) synaptically induces Na^+ transients with the maximal response in the dendritic region near (10–50 μm) stimulation electrode and extending across a dendritic length of 20–30 μm (the site of synaptic input), and with small Na^+ signals just above the detection threshold in other dendritic branches, which was thought to be a result of short burst of back-propagating action potentials in these dendritic branches (Rose and Konnerth 2001; Langer and Rose 2009). These synaptically induced dendritic Na^+ signals can be detected only with suprathreshold synaptic stimulation, a typical "all-or-none" response with the amplitudes depending on the number of suprathreshold stimuli. For example, a single stimulus resulting in one action potential evokes a Na^+ rise of about 4.3 mM, three stimuli (resulting in 2–3 action potentials) evoking

8.6 mM, and five stimuli (resulting in 4–5 action potentials) evoking 13.4 mM, of dendritic Na^+ elevation (Rose and Konnerth 2001). The further analysis of spatial distribution indicated that synaptically induced Na^+ responses are also present in the spines only with suprathreshold synaptic stimulation, namely five suprastimuli evoke a $[Na^+]i$ transients of 35.7 mM with a biexponential decay in active spines (25 % of total spines that are thought the direct synaptic input sites), but 14.6 mM of $[Na^+]i$ rise with a monoexponential decay in passive spines (75 % of total spines in which Na^+ is diffused from active spines), which is similar to the response in the dendrite (13.4 mM increase) (Rose and Konnerth 2001). Furthermore, a synaptic stimulation with an long-term potentiation (LTP) induction protocol (100 Hz, 1 s) induces the same Na^+ elevation of 44.6 mM in both dendrites and passive spines, but of >100 mM in active spines (Rose and Konnerth 2001). These synaptically induced Na^+ transients were not only observed in CA1 pyramidal neurons, but also in granule cells of the dentate gyrus with the suprathreshold synaptic stimulation of the molecular layer of the dentate gyrus (Rose and Konnerth 2001).

It is very interesting to compare the $[Na^+]i$ responses to the same or similar synaptic stimulation between the glial cells and neurons from the same regions, and neurons or glial cells from different brain regions. In mouse hippocampal slices, burst stimulation of Schaffer collaterals (50 Hz, 200 ms) evokes Na^+ transients in both CA1 pyramidal neurons and astrocytes. However, the amplitude of $[Na^+]i$ increase is 2.5 times higher in neurons than in astrocytes, with the time taken to reach its peak at about two times slower in neurons than in astrocytes (Langer and Rose 2009). Burst stimulation of parallel fibers (50 Hz, 100 ms) in mouse cerebellar slices also induces Na^+ transients in both Purkinje cells and Bergmann glial cells. However, the amplitude of Na^+ transients is 3–4 times lower in both Purkinje cells and Bergmann glial cells with slow Na^+ transients than in Bergmann glial cells with fast Na^+ transients. The decay of Na^+ signals is strictly monoexpenential in both Purkinje cells and Bergmann glial cells with fast Na^+ transients, but does not fit the monoexpenential in Bergmann glial cells with slow Na^+ transients (Bennay et al. 2008). More interestingly, similar stimulation induced Na^+ transients are much larger in hippocampal CA1 pyramidal neurons (average 5.1 mM) than in cerebellar Purkinje cells (average 1.5 mM), but almost the same in hippocampal astrocytes (average 2.1 mM) and cerebellar Bergmann glial cells (average 2.7 mM) (Bennay et al. 2008; Langer and Rose 2009). As we mentioned before, these differences may be a reflection of essential differences in the functions of different types of neurons and glial cells in the brain.

4.2.4 Mechanisms of Activity-Induced $[Na^+]i$ Signals in Neurons and Glia Cells

The activity-induced $[Na^+]i$ changes are determined by the net output of the balance between Na^+ extrusion dominated by the activities of Na^+/K^+ ATPase and Na^+ entry via various channels (ligand- and voltage-gated), transporters (e.g., EAAT, GAT etc.), and exchangers (e.g., NCX, NHE etc.). Meanwhile, elevated cytosolic Na^+ concentrations can be taken into mitochondria, probably via mitochondrial Na^+/Ca^{2+}

exchangers or/and mitochondrial cation uniporters, and thus buffer the peak changes in cytosolic Na^+ concentrations (Bernardinelli et al. 2006; Azarias et al. 2008), and therefore, influence activity-induced Na^+ response. Because the roles and regulation of Na^+/K^+ ATPase have been discussed earlier, here we focus on the role of Na^+ entry in the regulation of activity-induced $[Na^+]i$ changes.

Na^+entry in neurons. ***Voltage-gated sodium channels*** (VGSCs) are the most important channels in response to the generation of action potentials. Spontaneous, transient $[Na^+]i$ increases in a small part of cultured hippocampal neurons can be reversibly blocked by 1 μM TTX. Opening of VGSCs by veratridine (a VGSC inactivation blocker), or depolarization of neurons by elevation of $[K^+]e$ (from 3 to 40 mM) induces significant increases in $[Na^+]i$ which are sensitive to TTX, suggesting a role of VGSCs in activity-induced Na^+ entry in neurons (Rose and Ransom 1997b). Indeed, even a single Na^+ spike in soma can actively invade the dendrites of hippocampal pyramidal neurons and induces Na^+ transients in the proximal dendrites and the adjacent spines by activation of the TTX-sensitive Na^+ channels existing on the dendrites (Jaffe et al. 1992; Rose et al. 1999). In the cerebellar Purkinje neurons, TTX-sensitive VGSC-mediated $[Na^+]i$ increases are only confined in the soma and axon, and not in the dendrites since somatic action potentials cannot back-propagate to the dendrites (Lasser-Ross and Ross 1992; Callaway and Ross 1997; Knöpfel et al. 2000). In contrast, hyperpolarization-evoked $[Na^+]i$ rise in the dopaminergic neurons of the substantia nigra par compacta (Knöpfel et al. 1998) and the dendrites of hippocampal pyramidal neurons (Tsubokawa et al. 1999) is related to Na^+ influx through the ***non-selective cation channels*** (I_h channels) since both of I_h currents and Na^+ transients can be depressed by Cs^+ (which is supposed to block I_h channels) but not affected by TTX. Synaptic activity triggers the releases of neurotransmitter, which may induce Na^+ influx by activation of ***ligand-gated receptor channels***. Indeed, Na^+-selective microelectrode recording in rat sympathetic neurons revealed that application of carbachol, a cholinergic agonist, to mimic sympathetic neuronal activity induces a dose-dependent $[Na^+]i$ elevation from 7 up to 40 mM. The authors thus attributed this to Na^+ entry via nicotinic acetylcholine receptor channels (Ballanyi et al. 1983, 1984). Dendritic spines represent the major postsynaptic input sites for excitatory transmission in the brain, and are the main targets of synaptically released excitatory amino acid transmitters (predominantly glutamate) (Harris and Kater 1994). Stimulation of Schaffer collaterals-induced Na^+ rise in the dendrites and adjacent spines of hippocampal CA1 pyramidal neurons can be largely depressed by the blockade of ionotropic glutamate receptors (iGluRs), particularly NMDARs, and pressure application of glutamate or D-aspartate (an agonist of NMDARs) evokes the similar Na^+ response with Schaffer collateral stimulation, which can also be strongly reduced by selective NMDAR blocker, indicating iGluR channels largely contribute to Na^+ influx and $[Na^+]i$ elevation in postsynaptic neurons during excitatory transmission (Rose and Konnerth 2001; Langer and Rose 2009). In the cerebellum, dendritic and somatic glutamate receptor channels in Purkinje cells appear to be functionally AMPA-, not NMDA-subtype receptors (Häusser and Roth 1997), and Na^+ transients in Purkinje cell dendrites induced by short burst stimulation of parallel or climbing fibers are largely blocked by AMPAR blocker

CNQX, and AMPA iontophoresis mimics stimulation-like responses, indicating Na^+ responses synaptically activated in Purkinje cell dendrites are predominantly caused by Na^+ influx via AMPAR channels (Lasser-Ross and Ross 1992; Callaway and Ross 1997; Kuruma et al. 2003; Bennay et al. 2008). In addition, metabotropic glutamate receptors in cerebellar Purkinje cells can also be activated by synaptically released glutamate, and can induce Na^+ transients in the dendrites of Purkinje cells (Knöpfel et al. 2000). ***Na^+/Ca^{2+}exchangers*** (NCX) are also suggested to contribute to activity-induced $[Na^+]i$ elevation in neurons (Kiedrowski et al. 1994a, b; Regehr 1997). NCX, which operates in a forward mode to maintain $[Ca^{2+}]i$ at physiological levels by extruding 1 Ca^{2+} out with 3 Na^+ in by consuming Na^+ electrochemical gradient generated by the ATP-driven Na^+ pumps under normal condition (Blaustein and Lederer 1999), can operate in either the reverse or forward mode depending on the changes in transmembrane potential and Na^+ and Ca^{2+} gradients. Neuronal activity can also induce Ca^{2+} transients in the dendrites and depolarization (Jaffe et al. 1992; Kuruma et al. 2003), which favors NCX running in the reverse mode to extrude Ca^{2+}, therefore increasing $[Na^+]i$ (Kiedrowski et al. 1994a, b; Regehr 1997). However, NCX is not a requisite for activity-induced increases in $[Na^+]i$. In the presence of the Ca^{2+} chelator BAPTA (10 mM) in the cytosol, tetanus stimulation of Schaffer collaterals still evokes considerable Na^+ increases in the dendrites (44.6 mM) and the active spines (>100 mM) of the hippocampal CA1 neurons. The authors thus argued that Na^+ increases can occur without a contribution of NCX (Rose and Konnerth 2001). Besides, other transporter/exchangers such as glutamate transporters, NKCC1, and NHE are also expressed by neurons (Leng et al. 2014), and might contribute to activity-induced Na^+ entry in neurons. However further studies are needed before these ideas become conclusive.

Na^+ entry in glial cells. Like neurons, glial cells are capable of expressing a large variety of ion channels, Na^+-dependent transporters, and exchangers (see reviews by Verkhratsky and Steinhäuser 2000; Kirischuk et al. 2012). Astrocytes also possess ionotropic receptors (iGluR and purinergic P_2X receptors) that mediate rapid influx of Na^+ and Ca^{2+} (Gallo and Ghiani 2000; Lalo et al. 2011). These molecules might enable perisynaptic glia to sense neuronal activity by changing their cytosic Na^+ dynamics, and to integrate Na^+ signaling within glial networks (see reviews by Deitmer and Rose 2010; Kirischuk et al. 2012; Parpura and Verkhratsky 2012; Rose and Karus 2013). ***VGSCs*** are found in astrocytes in the brain and spinal cord, with properities similar to neuronal Na^+ channels (see reviews by Sontheimer et al. 1996; Verkhratsky and Steinhäuser 2000). However, their low density and the general reluctance of astrocytes to depolarize make VGSC contribution to physiological Na^+ influx somewhat minor (Rose and Ransom 1996a; Rose et al. 1997; and review by Kirischuk et al. 2012). Activation of ***iGluRs, particularly CNQX-sensitive AMPARs,*** partially contributes to glial Na^+ transients (Rose and Ransom 1996a, b; Munsch and Deitmer 1997; Chatton et al. 2000; Kirischuk et al. 2007; Bennay et al. 2008; Langer and Rose 2009). Glutamatergic synaptic transmission-induced Na^+ transients in cerebellar Bergmann glial cells (Kirischuk et al. 2007; Bennay et al. 2008) and hippocampal astrocytes (Langer and Rose 2009), as well as glutamate-induced [Na^+]i elevation in cultured cortical astrocytes (Chatton et al. 2001), however, are largely mediated by activation of glial ***Na+-dependent glutamate transporters (EAATs)***. In addition, glutamatergic transmission and the subsequently released glutamate may cause perisynaptic glia acidification (Rose and Ransom 1996b; Munsch and Deitmer 1997), which leads to [Na^+]i increase by activation of ***$Na^+/HCO3^-$cotransporters*** in astrocytes (Rose and Ransom 1996a, 1997b). Na^+ influx through ***non-selective cation channels*** is also suggested to contribute to kainate acid-induced [Na^+]i transients in leech giant glial cells (Munsch and Deitmer 1997). For GABAergic synaptic transmission, released GABA stimulates presynaptic astrocytes to take up GABA by activation of astrocytic ***Na^+-dependent GABA transporters (GAT)***, which can elevate [Na^+]i several millimolar (Chatton et al. 2003; Doengi et al. 2009; Unichenko et al. 2012). Blockade of P_2Y1-2 purinergic receptors with suramin (100 μM) causes >70 % reduction of stimulation-evoked [Na^+]i in cultured astrocytes, suggesting ***P_2Y1-2 purinergic receptor channels*** also contribute to astrocytic Na^+ transients (Bernardinelli et al. 2004). Astrocytes ubiquitously express all three isoforms of ***plasmalemmal NCX*** (NCX1-3), and astrocytic NCX reversal potential is close to the resting potential. Therefore, NCX dynamatically fluctuates between forward and reverse modes in regulating astrocytic Ca^{2+} and Na^+ homeostasis (Juhaszova et al. 1996; Kirischuk et al. 2012). It has been shown in Bergmann cells of mouse cerebellar slices and cultured rat cortical astrocytes, that running of NCX in reverse mode increases [Na^+]i (Kirischuk et al. 1997; Reyes et al. 2012). NKCC1, as has been discussed earlier, is abundantly expressed in astrocytes (see review by Hertz et al. 2013), and is important in clearing extracellular K^+ accumulated by neuronal activity (see Sect. 4.1). Stimulation of ***NKCC1*** by high [K^+]e (or NH_4^+/NH_3 which is similar to K^+ in stimulation) can raise astrocytic [Na^+]i several millimolar (Rose and Ransom 1996a; Su et al. 2002; Kelly et al. 2009), which in turn stimulates Na^+/K^+ ATPase activity to extrude Na^+ out of astrocytes to complete a so-called "transmembrane Na^+ cycle" in astrocytes (Walz and Hinks 1986). It is important to note that Na^+ transients in the directly stimulated single astrocyte can spread to neighboring astrocytes and induce their Na^+ responses (Bernardinelli et al. 2004; Langer et al. 2012). Based on their study in cultured cortical astrocytes, Bernardinelli et al. (2004) explained that this is predominantly due to the activity of Na^+/glutamate cotransporters taking up astrocyte-released glutamate, which is triggered by the Ca^{2+} wave due to the action of released gliotransmitter ATP from stimulated astrocytes on P_2Y1-2 purinergic receptors. However, Langer et al. (2012) argued that in their in situ astrocytes in hippocampal slices, Ca^{2+} signaling and metabotropic glutamate receptors are only supportive of, but not prerequisites for the spread of Na^+ between hippocampal astrocytes in situ. The spread of Na^+ is mediated by the ***gap junctions*** formed by Cx30/Cx43, since gap junction coupling can quickly equalize intracellular Na^+ concentration in astrocytes, and the spread of Na^+ to neighboring astrocytes can be disturbed by pharmacological inhibition of gap junctions and completely omitted in Cx30/Cx43 double-deficient mice (Rose and Ransom 1997a; Langer et al. 2012). The reasons for this difference are complicated, but at least can be related to the different preparations (cultured astrocytes from mouse cortex vs. in situ astrocytes of mouse hippocampal slices) used in their experiments. Generally, in contrast to neurons, glial cells possess more machinery mechanisms in regulation of their cytosolic Na^+ dynamics. This reflects a central role of astrocytic Na^+ homeostasis to the homeostasis of other ions (Ca^{2+}, H^+, K^+, Cl^-), transmitters (glutamate, GABA), linking membrane signaling and metabolic pathways, and coupling glial metabolism to neuronal activity (see reviews by Deitmer and Rose 2010; Kirischuk et al. 2012; Parpura and Verkhratsky 2012; Rose and Karus 2013).

4.2.5 Physiological Implications of Na^+ Dynamics in Neurons and Glial Cells

The establishment and maintenance of the steep Na^+ electrochemical gradients across the membrane by the permanent activity of Na^+/K^+ ATPase cost at least 60 % of the energy expenditure of nerve cells in the CNS (Schreiner and Rose 2012). It thus regulates many physiological functions (see reviews by Rose 2002; Deitmer and Rose 2010; Kirischuk et al. 2012; Parpura and Verkhratsky 2012; Rose and Karus 2013). Because the physiological implications of Na^+ dynamics in nerve cells have been excellently elaborated in many review papers (Rose 2002; Bhattacharjee and Kaczmarek 2005; Deitmer and Rose 2010; Kirischuk et al. 2012; Parpura and Verkhratsky 2012; Rose and Karus 2013), here we only briefly outline them for both neurons and glial cells.

Neurons. Firstly, the transmembrane Na^+ gradients provide the critical and necessary conditions for the generation of action potentials, the cellular language by which neurons communicate with one another. Disturbance of these gradients may alter the properties of action potentials and neuronal excitability, and contribute to certain neurological symptoms (Arakaki et al. 2011). Secondly, Na^+ transients may affect intracellular Ca^{2+} and pH homeostasis via Na^+/Ca^{2+} exchange, Na^+/H^+ exchange, and Na^+/HCO_3^- cotransport. Thirdly, increases in $[Na^+]i$, especially in the dendrites and the spines (up to >100 mM with tetanic stimulation), might induce synaptic plasticity by regulating either glutamate receptor functions (Yu and Salter 1998; Steinberg et al. 2006) and firing properties via activation of Na^+-activated K^+ channels spatially localized along axons, dendrites and somata (Bhattacharjee and Kaczmarek 2005) or neurotransmitter release by increasing cytosolic Ca^{2+} concentrations through Ca^{2+} release from mitochondria via mitochondria Na^+/Ca^{2+} exchangers (Yang et al. 2003). This increase plays a role in the induction of LTP in the hippocampus (Rose and Konnerth 2001), and of long-term depression (LTD) in cerebellar Purkinje neurons (Linden et al. 1993). Finally, synaptic $[Na^+]i$ transients in the dendrites and active spines can reduce the reversal potential of glutamatergic currents to a more negative potential, eventually leading to dendritic saturation, which largely diminishes glutamatergic currents and thereby protects the postsynaptic input sites from excessive accumulation of Na^+ and Ca^{2+} that may induce cellular excitatoxicity (Rose and Konnerth 2001).

Glial cells. Glial cells, especially astrocytes, are very active participants and direct communication partners of neurons for normal brain functions.

Astrocytes provide metabolic support to neurons via intercellular networks through activity-dependent intercellular trafficking of lactate necessary to sustain glutamatergic synaptic activity (Takahashi et al. 1995; Voutsinos-Porche et al. 2003; Rouach et al. 2008). They are also vital for maintaining the homeostasis of neuronal microenvironments, including maintenance of low interstitial glutamate concentrations, control of the brain interstitial volume, and regulation of cellular pH, extracellular K^+ concentrations and local blood flow (Kofuji and Newman 2004, 2009; MacAulay and Zeuthen 2012). Astrocytes can sense neuronal activity via G protein-coupled receptors (e.g. type I metabotropic glutamate and ATP purinergic P_2Y1 receptors) in their membrane and send feedback responses to neurons to modulate CNS signaling via the gliotrasmitters (e.g. glutamate, D-serine, and ATP) (Perea et al. 2009; Nedergaard and Verkhratsky 2012). These functions are intimately related to cytosolic Na^+ dynamics in astrocytes (Deitmer and Rose 2010; Kirischuk et al. 2012; Parpura and Verkhratsky 2012; Rose and Karus 2013). On one hand, removal of synaptically released neurotransmitters such as glutamate and GABA in the synaptic cleft to timely terminate their action and prevent their spillover is predominantly carried out by perisynaptic astrocytes, and is largely dependent on the activities of Na^+-dependent transporters (EAAT, GAT) driven by the tremendous transmembrane Na^+ electrochemical gradients established by Na^+/K^+ ATPase, and causes increase in $[Na^+]i$; on the other hand, elevated $[Na^+]i$ in astrocytes ultimately reduces the driving force and activity of Na^+-dependent transporters, which reduces the capacity of glial glutamate uptake, and even causes efflux of GABA (because the reversal potential of GAT is close to the resting potential in glia) (Kirischuk et al. 2012; Rose and Karus 2013), and might work as a feedback mechanism to synaptic activity, thus influencing synaptic plasticity (Unichenko et al. 2012). In addition, stabilization of iGluR activation- or neuronal activity-induced pH changes in perisynaptic glia (Rose and Ransom 1996b; Munsch and Deitmer 1997) also are highly dependent on the Na^+/H^+ exchange and/or Na^+-HCO_3^- cotransport (Kirischuk et al. 2012; Rose and Karus 2013). The clearance of accumulated extracellular K^+ during neuronal activity also stimulates the activity of Na^+/K^+ ATPase and NKCC1 in glia, and the latter is highly dependent on the tremendous transmembrane Na^+ electrochemical gradients, and its activity is critical in regulation of osmotic homeostasis and cellular volumes (MacAulay and Zeuthen 2012). Dynamic changes in astrocytic $[Na^+]i$ modulate Ca^{2+} signaling via NCX, which can induce exocytotic release of gliotransmitters and send feedback responses to neurons to modulate CNS signaling (Perea et al. 2009; Nedergaard and Verkhratsky 2012; Kirischuk et al. 2012; Parpura and Verkhratsky 2012). Last but not least, glial glutamate transporter-mediated $[Na^+]i$ elevation results in an activation of Na^+/K^+ ATPase, causing an enhanced cellular glucose utilization and glycolysis not only in activated astrocytes but also the neighboring glia (Takahashi et al. 1995; Voutsinos-Porche et al. 2003; Bernardinelli et al. 2004; Rouach et al. 2008). The end product of glycolysis, lactate, then trafficks via gap junction coupling and is sent to active zone, released, and serves as a metabolic substrate for neurons (Rouach et al. 2008). This astrocyte–neuron lactate shuttle between neurons (mainly oxidative) and astrocytes (mainly glycolytic) (Pellerin and Magistretti 2012) ultimately provides a sustained guarantee for glutamatergic synaptic activity (Rouach et al. 2008). Thus, the Na^+ dynamics in glial cells represents a key signal for coupling membrane signaling, metabolic pathway, and neuronal activity (Deitmer and Rose 2010; Kirischuk et al. 2012; Parpura and Verkhratsky 2012; Rose and Karus 2013).

4.3 Calcium Ion (Ca^{2+})

Calcium ion is a universal and pluripotent intracellular messenger controlling a broad spectrum of intracellular events in all euraryotic cells. Ca^{2+} is of critical importance in neuronal physiology, such as regulation of the release of neurotransmitters, membrane excitability, dendrite development, synaptogenesis, synaptic transmission and plasticity, gene expression, as well as many other processes in neuronal activity (Berridge 1998; Kawamoto et al. 2012; Brini et al. 2014). The discovery of the Ca^{2+} response of glial cells leads to a concept of "glial Ca^{2+} excitability", which overturns the traditionally inrooted thought of glial cells as "structural support and passive elements" (Kirischuk et al. 2012; Verkhratsky et al. 2012; Parpura and Verkhratsky 2012), and thus upsurges the studies of neuron-glia interaction (Volterra et al. 2014). The invention and development of a large number of fluorescent Ca^{2+} indicators with diverse properties (e.g., spectral qualities, dissociation constants, chemical forms etc.) (see reviews by Takahashi et al. 1999; Paredes et al. 2008), which, combined with the development of confocal and two-photon microscopy (Denk et al. 1996; Helmchen and Denk 2002, 2005; Helmchen et al. 2013; Pérez-Alvarez et al. 2013), have allowed a close look at the Ca^{2+} signaling and its ubiquitous roles and underlying mechanisms in cellular processes. Recently, Ca^{2+} imaging using green fluorescent protein-based genetically encoded Ca^{2+} indicators has been introduced as an alternative to using chemically synthesized fluorescent Ca^{2+} indicators (Palmer and Tsien 2006; Tian et al. 2009; Zhao et al. 2011; Gouriou et al. 2013). Genetically encoded Ca^{2+} indicators have several remarkable advantages over synthesized indicators (also refer to Sect. 3). Therefore, they represent a new trend in Ca^{2+} signaling research. Currently, numerous reviews have been published to summarize Ca^{2+} signaling in various aspects, and readers can easily find these excellent review papers from websites. In this section, we focus on Ca^{2+} homeostasis and its regulation in neurons and glia. Under normal conditions, intracellular Ca^{2+} is strictly controlled by a Ca^{2+} toolkit to ensure the proper Ca^{2+} signaling. The Ca^{2+} toolkit includes extrusion by plasma membrane Ca^{2+} pumps and Na^+/Ca^{2+} exchanges, influx via ion channels, release and sequestration by intracellular organelles such as endoplasmic reticulum (ER) and mitochondria, and buffering by Ca^{2+} binding proteins (Burgoyne 2007; Zündorf and Reiser 2011; Kawamoto et al. 2012; Brini et al. 2014).

The regulation of Ca^{2+} activity in nerve cells has been summarized in a simplified cartoon in Fig. 5.4.

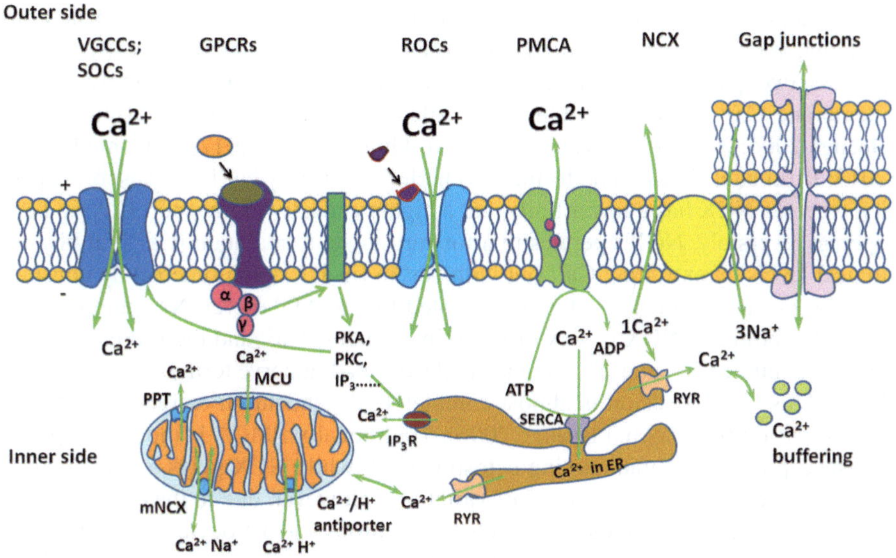

Fig. 5.4 Schematic representation of the toolkits regulating Ca^{2+} activity in the nerve cells. The homeostatic mechanisms include a vast array of processes such as the activation and inactivation of Ca^{2+} sources, the regulation of Ca^{2+} permeability of ion channels and intracellular compartments, the activities of endogenous Ca^{2+}-binding proteins (Ca^{2+} buffering proteins and sensors), efficient extrusion, and even the morphology of cellular structures, in particular that of the dendritic spines. The elevated $[Ca^{2+}]$i in both neurons and glial cells during neuronal activity originates from either the influx from extracellular fluid or the release from intracellular compartments, or the combination of both. The influx of Ca^{2+} is primarily mediated by a diversity of plasma membrane Ca^{2+}-permeable channels, which can be divided in three major groups based on their mechanism of opening: the voltage-gated Ca^{2+} channels (VGCCs), the receptor-operated channels (ROCs), and the store-operated Ca^{2+} entry channels (SOCs). The most important intracellular compartments of neuronal activity-induced cytoplasmic Ca^{2+} dynamics are ER and mitochondria. Ca^{2+}-binding proteins can reduce the amplitude of $[Ca^{2+}]$i, retard decay kinetics, and shape diffusional profiles with a very limited capacity. Two parallel, independent mechanisms (PMCA and NCX) extrude Ca^{2+}. For the more detailed information, please see the text

4.3.1 Basal Intracellular Ca^{2+} and Its Maintenance

Intracellular Ca^{2+} is determined by the interaction of membrane transporters (Ca^{2+} pump, NCX etc.) and cytoplasmic buffers (e.g., Ca^{2+} binding proteins and organelles). At rest, $[Ca^{2+}]$o is about 1.2–1.5 mM, while the cytosolic free Ca^{2+} concentration ($[Ca^{2+}]$i) is 10,000-fold lower than in extracellular fluid in both neurons (generally kept at a level between 40 and 100 nM) and glia (varies from 30–40 to 200–400 nM) (Hansen 1985; Verkhratsky et al. 1998, 2012; Burgoyne 2007), thus forming a very large concentration difference between the inside and outside of cells. The membrane potential of nerve cells (neurons and astrocytes) is relatively stable at 60–90 mV more negative inside the cell than outside, and together with the large transmembrane concentration difference, providing a very powerful electrochemical driving force for the influx of Ca^{2+} into cells. Nevertheless, cell membranes have very low permeability to Ca^{2+} for most cells, including neurons and glia, at rest. Nerve cells (both neurons and glia), like many other types of cells, possess two parallel, independent mechanisms to extrude Ca^{2+}, namely plasma membrane Ca^{2+}-ATPase (PMCA, also called Ca^{2+} pump) and NCX (Dipolo and Beaugé 2006; Zündorf and Reiser 2011; Verkhratsky et al. 2012; Giacomello et al. 2013; Brini et al. 2014). The former has high Ca^{2+} affinity but low transport capacity, whereas NCX has a low affinity, but a higher capacity to transport Ca^{2+} and can operate reversibly. NCX operates in a forward mode to maintain $[Ca^{2+}]i$ at physiological levels by extruding Ca^{2+} out with Na^+ in (in 1:3 stoichiometry) by consuming Na^+ electrochemical gradients generated by the ATP-driven Na^+ pumps under normal conditions. NCX is very efficient in extruding Ca^{2+} and therefore is considered a "genuine Ca^{2+} pump". Even a small Ca^{2+} load in nerve terminals can activate NCX. This Ca^{2+} extrusion is almost completely dependent on extracellular Na^+ in this condition (Dipolo and Beaugé 2006). Traditionally PMCA is thought to play a housekeeping role in maintaining cytosolic Ca^{2+}, while NCX plays a dynamic role of counteracting large cytosolic Ca^{2+} variations. However, due to the co-expression of these two systems in most cells including neurons and astrocytes, these two Ca^{2+} extrusion mechanisms not only work in parallel in the plasma membrane to keep basal $[Ca^{2+}]i$ at physiological levels under physiological conditions, but also may play different signaling roles (Dipolo and Beaugé 2006; Giacomello et al. 2013; Brini et al. 2014). In addition, two main intracellular compartments i.e., endoplasmic reticulum (ER) and mitochondria, acting either as a source or sink, sequester a fair amount of Ca^{2+} inside cells, although some other intracellular organelles such as Golgi complex, microsoma, synaptic vesicles and neurosecretory granules and cytoplasmic buffers such as Ca^{2+} binding proteins (also act as Ca^{2+} sensors, such as calmodulin, calbindin D-28k, calretinin, parvalbumin etc.) may also participate in intracellular Ca^{2+} homeostasis and signaling (Burgoyne 2007; Chao and Xia 2010; Verkhratsky et al. 2012; Brini et al. 2014). These homeostatic mechanisms together orchestrate the basal $[Ca^{2+}]i$ under physiological condition.

4.3.2 Activity-Induced Ca^{2+} Dynamics

Neuronal activity can lead to changes in $[Ca^{2+}]i$ ranging from highly localized and transient Ca^{2+} elevations within compartments such as the spines and dendrites to long-lasting and global changes through the neuron (Berridge 1998; Augustine et al. 2003) and the adjacent perisynaptic glial cells (Verkhratsky et al. 1998, 2012; Agulhon et al. 2008; Deitmer and Rose 2010; Volterra et al. 2014). For example, weak stimuli that activate only a few synapses to produce subthreshold postsynaptic potentials only lead to Ca^{2+} accumulation in an isolated spine via Ca^{2+} influx through a single Ca^{2+} channel, and thus form the unitary Ca^{2+} signal called a nanodomain, while the suprathreshold stimuli produce summed microdomain Ca^{2+} signals in the dendritic spines via Ca^{2+} influx through Ca^{2+} channel clusters and ligand-gated Ca^{2+} channels, and even a long-range type of local Ca^{2+} signal throughout the dendrites via Ca^{2+} influx and internal Ca^{2+} release from intracellular stores in neurons when stimuli are very strong (Berridge 1998; Sabatini et al. 2001; Augustine et al. 2003; Bloodgood and Sabatini 2007a). In contrast to synaptic-specific Ca^{2+} signaling produced by synaptic activity, back-propagating action potentials also produce Ca^{2+} response in neurons, and this response is always global, initiating from soma and invading into the dendrites and sometimes the adjacent spines (Sabatini et al. 2001; Augustine et al. 2003). Following synaptic activity or synergistic activity of metabotropic glutamate receptors and back-propagating action potentials, intracellularlly released Ca^{2+} in a single site of origin can propagate bidirectionally in the dendrites, and forms a second type of dendritic Ca^{2+} signal, the Ca^{2+} wave (Augustine et al. 2003). It is interesting to note that there are several different behavior phenotypes between activity-induced changes in localized Na^+ and Ca^{2+} transients in the dendrites. For Ca^{2+}, both subthreshold and suprathreshold stimulations can evoke $[Ca^{2+}]$ i rise, while only suprathreshold stimulation can induce $[Na^+]i$ (all-or-none response); Synaptically induced changes always exhibit a widespread elevation for Ca^{2+}, but are in defined dendritic regions for Na^+; back-propagating action potentials can lead to increases in both Na^+ and Ca^{2+} in the dendrites, but the underlying mechanisms are very different. For Ca^{2+}, it is largely mediated by activation of voltage-gated Ca^{2+} channels and the subsequent Ca^{2+} entry through them. Although there are voltage-gated Na^+ channels in the dendrites, Na^+ influx through these channels has minor contribution to $[Na^+]i$ elevation. Na^+ increase is primarily due to NMDAR channels in the dendrites. Finally, synaptical increases in $[Ca^{2+}]i$ in the dendrites are mediated by multiple ways (to be discussed next), while those in $[Na^+]i$ are largely mediated by NMDAR channels, that are evoked by back-propagating action potentials (Sabatini et al. 2001; Rose and Konnerth 2001; Bloodgood and Sabatini 2007a).

Neuronal activity can also evoke Ca^{2+} signaling in perisynaptic glial cells (Deitmer and Rose 2010; Volterra et al. 2014). Indeed, as early as in the 1990s, investigators have found that astrocytes in brain slices respond to synaptic activity by increasing $[Ca^{2+}]i$ (see review by Volterra et al. 2014). Astrocytes in barrel cortex, visual cortex, somatosensory cortex, the hippocampus, and the cerebellum exhibit Ca^{2+} response to neuronal activity induced by physiological stimuli (see reviews by Agulhon et al. 2008; Volterra et al. 2014; and reference herein). Astrocytic Ca^{2+} increases due to neuronal activity can be found either in the fine processes that wrap presynaptic boutons or in somata, and are qualitatively similar in response to different physiological stimuli (Agulhon et al. 2008). However, recent studies revealed intense and previously unappreciated local activity that is dissociated from activity in the cell body. For example, $[Ca^{2+}]i$ changes in the astrocytic cell body are significantly slower and less frequent than any of the transients in the processes. And even within an individual process of an astrocyte, Ca^{2+} activity is complex and may be comprised of different types of events and depends on the specific properties and organization of the underlying signaling components (Volterra et al. 2014). Astrocytic Ca^{2+} increases due to neuronal activity are generally thought to be mediated through activation of a variety of G-protein-coupled receptors expressed by astrocytes due to the spillover of a number of synaptically released neurotransmitters from neurons, and play a critical role in neurophysiology and especially in modulation of neuronal activity (Verkhratsky et al. 1998; Agulhon et al. 2008; Deitmer and Rose 2010; Volterra et al. 2014). In addition, synaptic activity can induce a rise in astrocytic [Ca^{2+}]i through the stimulation of Ca^{2+}-permeant ionotropic receptors in some regions, via reversal of NCX following neurotransmitter uptake or other pumping activities that lead to an increase in astrocytic [Ca^{2+}]i levels, or through transient receptor potential A1 channels (Volterra et al. 2014). Elevated [Ca^{2+}]i in a single astrocyte can propagate through the syncytium of cells within astrocytic networks, and this intercellular Ca^{2+} wave could be a mechanism for long-distance signaling between astrocytes under certain conditions (Verkhratsky et al. 1998; Agulhon et al. 2008; Leybaert and Sanderson 2012). However, due to some limitations in the current research, for example the lack of a clear-cut physiological relevance of the experimental approach, inadequate selection of the experimental parameters for Ca^{2+} imaging and analysis and Ca^{2+} indicators, and differences in circuit and synaptic phenomena under the same studies, the studies on astrocytic Ca^{2+} signaling linking neuronal information in different spatial-temporal dimensions are still full of discrepancies and unexpected complexities which highlight key insufficiency in our current understanding of astrocytic Ca^{2+} signaling (Volterra et al. 2014).

4.3.3 Homeostatic Mechanisms of Activity-Induced Ca^{2+} Dynamics

The activity-induced Ca^{2+} dynamics are tightly controlled by homeostatic mechanisms to ensure the precise spatial-temporal course and magnitude of Ca^{2+} signaling. The homeostatic mechanisms include a vast array of processes such as the activation and inactivation of Ca^{2+} sources, the regulation of Ca^{2+} permeability of ion channels and intracellular compartments, the activities of endogenous Ca^{2+}-binding proteins (Ca^{2+} buffering proteins and sensors), efficient extrusion, and even the morphology of cellular structures, in particular that of the dendritic spines.

Ca^{2+} Sources

The elevated [Ca^{2+}]i in both neurons and glial cells during neuronal activity originates from either the influx from extracellular fluid or the release from intracellular compartments, or the combination of both. Because of the extremely low resting [Ca^{2+}]i and the tremendous transmembrane electrochemical gradients of Ca^{2+}, neuronal activity can easily increase [Ca^{2+}]i 10- to 100-fold by influx of Ca^{2+} via activated Ca^{2+}-permeable ion channels in the membrane. Direct measurement with Ca^{2+}-selective microelectrodes or membrane-impermeant Ca^{2+}-sensitive dyes have demonstrated that modest action potential or synaptic activity can result in a significant reduction in [Ca^{2+}]o, and this extracellular Ca^{2+} depletion increases with the increase of stimulation frequency and spine length (see review by Cohen and Fields 2004), suggesting external Ca^{2+} as a source for the increase in [Ca^{2+}]i, especially in synaptic-contacted active spines and dendrites. Indeed, the primary source of Ca^{2+} signaling in presynaptic terminals is the influx through voltage-gated Ca^{2+} channels since these channels are highly enriched in presynaptic terminals compared to other subcellular compartments of a neuron (Augustine et al. 2003). Dendritic Ca^{2+} signals have more numerous sources, which in large part cause Ca^{2+} entry through voltage- and transmitter-gated channels. Blocking these entry pathways decreases, and can even abolish, activity-induced intracellular Ca^{2+} signaling (Sabatini et al. 2001; Augustine et al. 2003). In addition, Ca^{2+} release/sequestration by intracellular compartments also contributes to activity-induced Ca^{2+} dynamics in both neurons and glial cells. Two main intracellular compartments i.e., the ER and the mitochondria are the major players in regulation of Ca^{2+} signals (Berridge 1998; Verkhratsky et al. 1998, 2012; Kawamoto et al. 2012; Brini et al. 2014). Both ER and mitochondria can regulate the dynamics of Ca^{2+} signaling by acting either as a source or as a sink of signal Ca^{2+} (see the following discussion).

Regulation of the Dynamics of Ca^{2+} Signals by Ca^{2+}-Permeable Ion Channels and Intracellular Ca^{2+} Store

Ca^{2+}-permeable ion channels. The influx of Ca^{2+} is primarily mediated by a diversity of plasma membrane Ca^{2+}-permeable channels, which can be divided into three major groups based on their mechanism of opening: the voltage-gated Ca^{2+} channels (VGCCs), the receptor-operated channels (ROCs), and the store-operated Ca^{2+} entry channels (SOCs) (Verkhratsky et al. 1998, 2012; Kawamoto et al. 2012; Brini et al. 2014).

VGCCs represent the main Ca^{2+} entry pathways in many cell types and transduce electrical signals (membrane potential changes) into intracellular Ca^{2+} transients that initiate many physiological events such as the release of neurotransmitters, regulation of neuronal excitability, synaptic transmission and plasticity, gene expression, enzyme activation, and neurite outgrowth and retraction (Catterall and Few 2008; Catterall 2011; Atlas 2013). Based on the physiological and pharmacological properties of Ca^{2+} current types, VGCCs are divided into T (low-voltage activated) and P/Q, N, L, R (high-voltage activated) types. VGCCs, depending on the types of α1 pore-forming subunit, can also be divided into three subfamilies: Cav1 (Cav1.1-1.4: L-type Ca^{2+} currents), Cav2 (Cav2.1: P/Q-type Ca^{2+} currents; Cav2.2: N-type Ca^{2+} currents; Cav2.3: R-type Ca^{2+} currents), and Cav3 (Cav3.1-3.3: T-type Ca^{2+} currents) (Catterall 2011). VGCCs are expressed in both neurons and glial cells (Verkhratsky et al. 1998, 2012; Kawamoto et al. 2012; Brini et al. 2014). In neurons, Cav1.2 and Cav1.3 are localized in the cell bodies and proximal dendrites at both presynaptic and postsynaptic sites, forming small clusters on dendritic shafts and spines (Kawamoto et al. 2012), thus generating nanodomains or/and microdomains of Ca^{2+} signals in cell bodies and dendrites that in turn control processes like secretion, enzyme activation, and gene expression (Augustine et al. 2003; Catterall 2011; Kawamoto et al. 2012; Brini et al. 2014). Cav2 are localized at presynaptic terminals, dendrites and cell bodies (Catterall 2011; Kawamoto et al. 2012; Brini et al. 2014). During depolarization, Ca^{2+} entry through a single presynaptic Cav2 allows the formation of high Ca^{2+} concentration nanodomains or microdomains in the proximity of the pore, and thus triggers vesicular release by direct interaction with the soluble N-ethylmaleimide sensitive factor attachment protein (**SNAP**) **RE**ceptor (SNARE) complex, in which the vesicle-associated v-SNARE protein synaptobrevin interacts with two plasma-membrane-associated t-SNARE proteins, SNAP-25 and syntaxin-1 (Catterall and Few 2008; Catterall 2011; Kawamoto et al. 2012; Atlas 2013). Therefore, presynaptic VGCCs serve as the regulatory node in a dynamic, multilayered signaling network that exerts short-term control of neurotransmission in response to synaptic activity (Catterall and Few 2008; Catterall 2011). Cav3 channels are mainly distributed at cell bodies and dendrites of neurons in a variety of brain regions (Kawamoto et al. 2012). The Cav3 subfamily conducts T-type Ca^{2+} currents that are activated at comparatively negative membrane potentials and have fast voltage-dependent inactivation compared to other Ca^{2+} currents. This kind of Ca^{2+} channel is well-suited for generation of large Ca^{2+} transients because the driving force for Ca^{2+} entry is large at negative membrane potentials, and plays a critical role in the generation of repetitive and rhythmic firing of action potentials (Catterall 2011; Kawamoto et al. 2012). For glial cells, particularly astrocytes, it is complicated whether activation of VGCCs can lead to Ca^{2+} signaling, even though astrocytes express all types of VGCCs (Verkhratsky et al. 1998, 2012). In cultured or freshly isolated astrocytes, activation of VGCCs indeed evokes Ca^{2+} signaling (Verkhratsky et al. 2012), while astrocytic VGCCs seem to have nothing to do with astrocytic Ca^{2+} elevations in situ following stimulation of neuronal afferents, but may be important for initiating Ca^{2+} oscillations that occur independent of neuronal input (Agulhon et al. 2008). This discrepancy might be resulted from several differences between cultured astrocytes and the ones in situ. For example, the morphological characteristics of astrocytes in situ are lost in cultured astrocytes, and cultured astrocytes express genes that are not necessarily expressed in vivo (Agulhon et al. 2008).

ROCs, also called ligand-gated channels, are a group of transmembrane ion channel proteins, which open to allow ions such as Na^+, K^+, and Ca^{2+} to pass through the membrane in response to the binding of a chemical messenger such as neurotransmitters, neuromodulators, and neurohormones. ROCs are ionotropic receptors, and most of them are permeable to Ca^{2+} and represent an important mechanism for Ca^{2+} signaling. Ionotropic glutamate receptors (iGluRs) such as NMDAR and AMPAR are some of the most intensively investigated ROCs in regulation of activity-induced Ca^{2+} dynamics in both neurons and glial cells. In addition, some other ROCs such as purinergic ionotropic P_2X receptors, nicotinic acetylcholine receptors (nAChRs), serotoninergic 5-HT_3 receptors, etc. are also permeable to Ca^{2+}, and contribute to intracellular Ca^{2+} signaling (Verkhratsky et al. 2012; Kawamoto et al. 2012; Brini et al. 2014).

Ionotropic glutamate receptors are present at excitatory synapses, and mediate the majority of excitatory neurotransmission in the brain. Dendritic spines possess abundant iGluRs, and represent the major postsynaptic input site for excitatory transmission in the brain (Harris and Kater 1994). NMDARs are also found at extra-synaptic sites where their intensive or chronic activation may lead to neuronal death (Hardingham and Bading 2010), and in presynaptic compartments where they probably regulate glutamate release and affect synaptic transmission and plasticity (Corlew et al. 2008). Glial cells also express iGluRs that participate in the shaping of Na^+ and Ca^{2+} signaling with other molecules to influence the communication between neurons and glial cells (Gallo and Ghiani 2000; Deitmer and Rose 2010; Lalo et al. 2011; Parpura and Verkhratsky 2012). NMDARs act as non-specific cation channels, which are permeable to Na^+, Ca^{2+}, and K^+. These receptors are hetero-tetrameric transmembrane proteins containing two NR1 and two NR2 or NR3 subunits (Dingledine et al. 1999). NMDARs exhibit a voltage-dependent blockade with the variable strength by extracellular Mg^{2+} depending on the type of NR2 or NR3 subunit(s) expressed. Therefore, their activation not only requires the binding of glutamate but also the membrane depolarization caused mainly by AMPAR-mediated fast local Na^+ spikelets to relieve Mg^{2+} blockade of the receptor channels (Dingledine et al. 1999; Antic et al. 2010). In excitatory neurotransmission, NMDARs have a major role in spine Ca^{2+} dynamics since activation of NMDARs generates highly localized microdomains of Ca^{2+} signaling, while blocking NMDARs abolishes or largely attenuates activity-induced Ca^{2+} transients in the dendrites and spines of neurons (Sabatini et al. 2001; Augustine et al. 2003; Bloodgood and Sabatini 2007a). Adult cerebellar Purkinje neurons relatively lack functional NMDARs (Yuzaki et al. 1996; Piochon et al. 2007). Therefore, their NMDARs contribute less to cytosolic Ca^{2+} dynamics to the climbing/parallel fiber response. Astrocytic NMDARs have different gating properties with those of neurons. They demonstrate rather weak Mg^{2+} blockade, and have a substantially lower Ca^{2+} permeability (Palygin et al. 2010; Verkhratsky et al. 2012). Astrocytic NMDARs also produce inward currents and cytoplasmic Ca^{2+} transients in response to both exogenous agonists and synaptic stimulation in cortical and hippocampal astrocytes (Steinhäuser et al. 1994; Palygin et al. 2010). However, they cannot directly cause effects on membrane permeability and $[Ca^{2+}]i$ in cerebellar Bergmann glial cells (Müller et al. 1993; Shao and McCarthy 1997). AMPARs mediate fast excitatory synaptic transmission in the CNS and play an important role in synaptic plasticity (Liu and Zukin 2007). AMPARs are also hetero-tetrameric transmembrane proteins composed of subunits GluR1-4, and are primarily permeable to Na^+ and K^+, but also become permeable to Ca^{2+} only when AMPARs lack the GluR2 subunit (Dingledine et al. 1999). Aspiny dendrites throughout the CNS express Ca^{2+} permeable AMPARs, and activation of this type of receptors of single synapses on aspiny dendrites of neocortical fast spiking interneurons creates highly localized Ca^{2+} microdomains, often restricted to less than 1 μm of dendritic space (Goldberg et al. 2003). Activation of a single parallel fiber opens Ca^{2+} permeable AMPARs, generating long-lived Ca^{2+} transients that are confined to submicron dendritic stretches in cerebellar stellate cells and play a role in induction of LTD (Soler-Llavina and Sabatini 2006). Astrocytes also express Ca^{2+} permeable AMPARs, and activation of them leads to raised $[Ca^{2+}]i$ (Verkhratsky et al. 1998, 2012). However, rapid desensitization of AMPARs in physiological conditions (e.g., half decay times c.a. 30 ms for AMPARs vs. 250 ms for NMDARs in hippocampal cells) (Chao and Xia 2010) severely limits Ca^{2+} influx via Ca^{2+} permeable AMPARs (Verkhratsky et al. 2012).

P_2X receptors are a family of trimeric ATP-activated ion channels permeable to Na^+, K^+ and Ca^{2+}. They include seven homomeric or heteromeric subtypes (P_2X1-7). P_2X receptors are widely distributed in neuronal soma, presynaptic, and postsynaptic nerve terminals as well as extrasynaptic sites and glial cells throughout the central, peripheral, and autonomic nervous systems (Burnstock 2000; Khakh 2001; North 2002; Khakh and North 2012). Though P_2X receptor-mediated fractional Ca^{2+} currents are very small (3–12 %), the receptors display considerably high Ca^{2+} permeability (*PCa/PNa* varies from 1.2 to 4.8), thus can mediate substantial Ca^{2+} influx (Khakh and North 2012). Stimulation of neurons from various brain regions (e.g., hippocampus, hypothalamus, cerebellar Purkinje cells, supraoptic neurons, neurohypophysis) and cortical astrocytes, as well as cerebellar Bergmann cells, with ATP can cause cytosolic Ca^{2+} signals (North 2002; Verkhratsky et al. 2012). P_2X receptor-mediated Ca^{2+} signals have multiple effects on synaptic plasticity and mediate slow neuromodulatory functions (Khakh and North 2012). In addition, P_2X and Ca^{2+}-permeable nAChRs undergo cross-inhibition; however, the consequences of this cross-inhibition are still unknown (Khakh and North 2012). nAChRs are expressed throughout the brain at pre- and postsynaptic as well as extrasynaptic sites, and play important roles in neuronal development and synaptic plasticity. nAChRs are non-selective cation permeable with unique affinity to Ca^{2+}, and can generate Ca^{2+} signals by direct Ca^{2+} influx through the receptor channels and subsequent Ca^{2+} influx via VGCCs due to nAChR-associated postsynaptic depolarization as well as triggering Ca^{2+}-induced Ca^{2+} release from ryanodine-dependent stores (Kawamoto et al. 2012). Besides the above-mentioned ligand-gated channels, another ligand-gated channel, the 5-HT_3 receptor, is also attracting attention in generating Ca^{2+} signals during neuronal activity (Kawamoto et al. 2012). Among all the receptors for serotonin (5-HT_{1-7}), 5-HT_3 receptor is the only ligand-gated ion channel (rather than a metabotropic receptor for other 5-HT receptors). Functional receptors are either hetero- or homopentamers with at least two obligatory 5-HT_3A subunits, and are permeable to cations such as K^+, Na^+ and Ca^{2+} (see reviews by Chameau and van Hooft 2006; Kawamoto et al. 2012). 5-HT_3 receptors are expressed throughout the CNS, with high levels of expression in the cortex, the hippocampus, the amygdala, several brainstem nuclei and the spinal cord, and relatively low levels in the cerebellum (Tecott et al. 1993; Oostland et al. 2011; also see review by Chameau and van Hooft 2006). Presynaptic 5-HT_3 receptors display high permeability to Ca^{2+}, and are involved in mediating or modulating neurotransmitter release (e.g., DA and GABA). Postsynaptic 5-HT_3 receptors exhibit relatively lower permeability to Ca^{2+} than to Na^+ and K^+, and are preferentially expressed on interneurons, and their activation contributes to fast excitatory synaptic transmission (Kawamoto et al. 2012). Ca^{2+} imaging with fluorescent dye, combining with confocal microscopy, directly demonstrated that activation of 5-HT_3 receptors leads to a rapid Ca^{2+} rise to a peak followed by an apparent plateau phase in intra synaptosoma of several brain regions, with similar kinetics of Ca^{2+} changes in all the regions (Nayak et al. 1999).

SOCs, also named Ca^{2+} release-activated channels, are specialized plasma membrane Ca^{2+} ion channels activated by Ca^{2+} release and depletion from ER to allow extracellular Ca^{2+} entry into the cytoplasm. This Ca^{2+} influx across the plasma membrane has been called "capacitative calcium entry," or "store-operated Ca^{2+} entry", and is thought to be ubiquitous and important in most electrically non-excitable cells, as well as in many excitable cells. For cells, this Ca^{2+} influx exerts many effects such as promoting the refilling of the depleted ER Ca^{2+} stores and the generation of short-term Ca^{2+} signaling (e.g., regulation of exocytosis, enzymatic activity, Ca^{2+} oscillations, sperm chemotaxis and the acrosome reaction) or long-lasting Ca^{2+} signals in controlling gene transcription that support cell growth, apoptosis, differentiation or activation (Parekh and Putney 2005). In the CNS, store-operated Ca^{2+} entry is present in both neurons (Gemes et al. 2011; Steinbeck et al. 2011; Koss et al. 2013) and glial cells (Verkhratsky and Parpura 2014), and modifies neuronal excitability, network activity, and other Ca^{2+} signaling. It is now evident that store-operated Ca^{2+} entry requires members of two families of proteins: the stromal interaction molecules (STIM) STIM1 and STIM2, which function as Ca^{2+} sensors in the ER, and the Orais (Orai1, Orai2 and Orai3), which function as pore-forming subunits of store-operated Ca^{2+} entry channels (Cahalan 2009; Smyth et al. 2010; Verkhratsky and Parpura 2014). STIM is a non-channel intermediary of store-operated Ca^{2+} entry and Ca^{2+}-release-activated Ca^{2+} channel activation. An EF-hand motif near the N terminus of STIM protein is localized in the ER lumen, which contains negative charges, and can bind and sense ER Ca^{2+}, while the C terminus has a lengthy bipartite coiled-coil within a region containing an ezrin/radixin/moesin (ERM) domain, which physically contacts Orai proteins in the plasma membrane (Cahalan 2009; Smyth et al. 2010). Orai1 is a tetraspanning membrane protein with both terminals facing the intracellular space that assemble as tetramers to form the store-operated Ca^{2+} entry channel (Prakriya et al. 2006). STIM and corresponding Orai channel proteins are always present in clusters, termed the 'elementary unit' of store-operated Ca^{2+} entry and Ca^{2+}-release-activated Ca^{2+} channel activation, since Ca^{2+} influx through Ca^{2+}-release-activated Ca^{2+} channels occurs precisely at STIM–Orai clusters (Cahalan 2009). When the ER Ca^{2+} store is depleted, the EF-hand domain releases bound Ca^{2+}, and STIM oligomerization occurs within seconds and interacting with Orai to activate Orai currents which allows Ca^{2+} entry into cytoplasm (Cahalan 2009; Smyth et al. 2010). It is worth pointing out that transient receptor potential channels (TRPCs) are also involved in store-operated Ca^{2+} entry. TRPCs are a group of ion channels located mostly on the plasma membrane of numerous cell types including neurons and glial cells (Bollimuntha et al. 2011; Verkhratsky et al. 2012; Verkhratsky and Parpura 2014), and display permeability to K^+, Na^+, Cs^+, Ca^{2+}, Ba^{2+}, and other divalents to varying degrees (Cheng et al. 2013). The TRPC family contains seven members (TRPC1-7), in which TRPC1, TRPC4, and TRPC5 channels are activated primarily by store-depletion (Bollimuntha et al. 2011). TRPCs have intimate functional interactions with STIM and Orai proteins. For example, TRPCs (except TRPC7) can be activated by STIM1, and Orai1 can act as a regulatory unit that confers STIM1-mediated store depletion sensitivity to TRPCs (Cahalan 2009; Smyth et al. 2010; Bollimuntha et al. 2011; Cheng et al. 2013), and therefore are essential for replenishing and for maintaining ER Ca^{2+}. Indeed, TRPC1-mediated Ca^{2+} entry and cation currents in cells stimulated with agonist or thapsigargin are inhibited by low [Gd^{3+}] and 10–20 μM 2APB (conditions that block store-operated Ca^{2+} entry) (Cheng et al. 2013), absence of this pathway during chronic depletion of ER Ca^{2+} could influence ER-dependent processes such as protein folding and trafficking, the ER stress response, and apoptosis, and has been associated with many neurodegenerative diseases such as Parkinson's diseases, Alzheimer's disease, and Huntington's disease (Bollimuntha et al. 2011).

Intracellular Ca^{2+} store. Intracellular compartments also participate in the generation of the distinct types of spatio-temporally organized Ca^{2+} signals that also differ in amplitude during neuronal activity. The most important intracellular compartments of neuronal activity-induced cytoplasmic Ca^{2+} dynamics are ER and mitochondria. Both serve as a source or as a store of signal Ca^{2+}.

ER is a continuous membrane system that is extended from the nuclear envelope throughout the cell including the fine processes of the nerve cells (Berridge 1998; Kawamoto et al. 2012; Lam and Galione 2013). The ER network is the primary intracellular Ca^{2+} storage and release organelle in most cells, possessing two major families of Ca^{2+} release channels: the inositol triphosphate (IP_3) receptors (IP_3Rs) and ryanodine receptors (RYRs), both of which are capable of regenerative Ca^{2+} releases, which makes ER a central role in cellular Ca^{2+} homeostasis and signaling (Berridge 1998; Verkhratsky et al. 1998, 2012; Lam and Galione 2013; Brini et al. 2014). In neurons, regenerative propagating cytoplasmic Ca^{2+} waves by ER makes ER a "neuron-within-neuron" (Berridge 1998). In glial cells, ER regulation of cytosolic Ca^{2+} endows glial cell non-electrical excitability, termed "glial Ca^{2+} excitability" which triggers glial cells to release gliotransmitters and thus influences neuronal activity (Zorec et al. 2012). Recently it was suggested that ER also possesses physiological ER Ca^{2+} leak channels, presenilin 1 and 2 which form low conductance Ca^{2+} channels responsible for about 80 % of the passive Ca^{2+} leak from ER (Tu et al. 2006). The sarco-endoplasmic reticular Ca^{2+} ATPase (SERCA pump), which is similar to plasmalemmal Ca^{2+} ATPase, pumps back cytosolic Ca^{2+} in an energy-requiring process against a steep concentration gradient across ER membrane (Lam and Galione 2013). Recently, studies using electron tomography techniques revealed the presence of overlapping regions between ER and mitochondria known as "mitochondria-associated membrane" (MAM), which is separated by a minimum distance of 10–25 nm, and allows the direct physical association of ER proteins with components of the outer mitochondrial membrane (OMM). The tight vicinity of ER and mitochondria makes these zones "hotspots", where ER releases Ca^{2+} via IP_3Rs to form Ca^{2+} microdomains, and mitochondria are able to very efficiently take up released Ca^{2+} directly from activated IP_3Rs in ER to limit Ca^{2+} diffusion, and then gradually transfer Ca^{2+} to ER (Graier et al. 2007; Patergnani et al. 2011; Lam and Galione 2013). This process can also regulate spatiotemporal shapes of cytosolic Ca^{2+} signals. ER Ca^{2+} in lumen is in high micromolar levels or even in the millimolar range, and is mostly bound in Ca^{2+} binding proteins and divided into a number of subcompartments, such as cisternal organelle, spine apparatus, and annulate lamellae in nerve cells (Foskett et al. 2007; Lam and Galione 2013). The luminal Ca^{2+} can be released to cytoplasm. Depletion of ER Ca^{2+} leads to the activation of SOCs and Ca^{2+} entry into cytoplasma to replenish ER Ca^{2+} and also to generate Ca^{2+} signals (see above discussion).

Either a single or trains of action potential(s) can trigger Ca^{2+} release from ER in neurons of various brain regions (Berridge 1998). A large number of Gq protein- or tyrosine kinase-coupled receptors can also trigger ER Ca^{2+} release when they are activated by neurotransmitters, neuromodulators, or neurohormones (Foskett et al. 2007; Agulhon et al. 2008). Ca^{2+} efflux from ER is achieved though IP_3Rs and RYRs release as well as Ca^{2+} leakage from ER Ca^{2+} leak channels. The latter is responsible for about 80 % of the passive Ca^{2+} leak from ER at rest (Tu et al. 2006). The most important Ca^{2+} efflux from ER induced by neuronal activity is mediated by IP_3Rs and RYRs.

IP_3Rs are homo- or heterotetramic proteins encoded by three genes, with 60–80 % homology in full-length amino acid sequences of IP_3R1-3 (Foskett et al. 2007). They are widely distributed throughout the brain (Sharp et al. 1993; Pieper et al. 2001; Hertle and Yeckel 2007). IP_3Rs are enriched in cerebellar Purkinje cells and hippocampal CA1 pyramidal cells, and are primarily identified in dendritic shafts (mainly the phosphorylated IP_3R1) and cell bodies (predominantly the unphosphorylated receptors) throughout the brain (Sharp et al. 1993; Pieper et al. 2001). There are striking variations in regional and subcellular distributions of IP_3Rs. For example, in the hippocampus, IP_3R1 is observed in pyramidal cells and granule cells, IP_3R2 in perivascular astrocytes and endothelial cells, and IP_3R3 in axon terminals located in stratum pyramidal of CA1 and microvessels in stratum radiatum. Independent of age, IP_3R1 immunoreactivity is observed to form clusters within the primary apical dendrite and at dendritic bifurcations of pyramidal neurons (Hertle and Yeckel 2007). IP_3Rs can be activated by IP_3 that is generated through hydrolysis of phosphatidylinositol-4,5-bisphosphate by soluble phospholipase C (PLC) in response to the activation of various Gq protein- or tyrosine kinase-coupled receptors on the cell membrane. IP_3Rs are also regulated by Ca^{2+} itself, both in the cytosol and lumen (Berridge 1998; Foskett et al. 2007). Cytosolic Ca^{2+}, in modest elevation, activates IP_3Rs to induce Ca^{2+}-induced Ca^{2+} release, similar to that of RYR activation, and thus amplify Ca^{2+} signals, while in higher concentrations inhibits IP_3Rs to limit Ca^{2+} release via IP_3Rs (Berridge 1998; Foskett et al. 2007). Ca^{2+} inhibition is thought to serve as a negative-feedback mechanism to either terminate or prevent Ca^{2+} release as the local cytoplasmic Ca^{2+} concentration is raised by IP_3R-mediated Ca^{2+} release during Ca^{2+} signaling, even in the continuous presence of IP_3. This inhibition may play a significant role in the generation of Ca^{2+} spikes and oscillations, as well as the generation of highly localized Ca^{2+} signals, by preventing runaway Ca^{2+} release due to the positive feedback of Ca^{2+}-induced Ca^{2+} release (Foskett et al. 2007). IP_3R-mediated $[Ca^{2+}]i$ signals in single cells are spatio-temporally complex. Spatially, $[Ca^{2+}]i$ signals may initiate at specific locations and remain highly localized or propagate as waves; in the temporal domain, repetitive spikes or frequent oscillations of small Ca^{2+} signals often tuned to large stimulating levels to encode Ca^{2+} signaling (Foskett et al. 2007). Activation of IP_3Rs may generate different Ca^{2+} signals, which depend on the stimulus intensity and the number of IP_3Rs activated. Weak activation by low IP_3 levels evokes "fundamental" signals called "blips", resulted from the openings of individual IP_3R Ca^{2+} channels, with localized elevations of cytoplasmic $[Ca^{2+}]$ that arise stochastically and autonomously at discrete release sites; the "elementary" signals involve the opening of several IP_3Rs, and the large events called "puffs" involve the concerted opening of multiple IP_3R channels organized within a cluster (Foskett et al. 2007). If IP_3Rs are stimulated by higher level IP_3 associated with stronger extracellular agonist stimulation, Ca^{2+} released at one cluster site can trigger Ca^{2+} release at adjacent sites by Ca^{2+}-induced Ca^{2+} release, leading to the generation of Ca^{2+} waves that propagate in a saltatory manner by successive cycles of Ca^{2+} release, diffusion, and Ca^{2+}-induced Ca^{2+} release that generate global Ca^{2+} signals in single nerve cells and long-range signals in the networks. This leads to the naming of neuronal ER as a "neuron-within-neuron" (Berridge 1998; Foskett et al. 2007) and glial Ca^{2+} excitability (Zorec et al. 2012). IP_3R-mediated Ca^{2+} signals play a critical role in the brain functions such as neurotransmitter releases, neurite formation and extension during development, synaptic plasticity, learning, memory and behaviors (Kawamoto et al. 2012).

RYRs. Three mammalian RYR isoforms (RYR1-3) have been identified, and are encoded by three different genes on different chromosomes, sharing about 65 % identities (Fill and Copello 2002). They are homotetramers and the largest known ion channels with a molecular mass of about 2200 kDa. Their large mass challenges their structural analysis. RYRs are activated either by Ca^{2+} itself or by depolarization via coupling with voltage-gated Ca^{2+} channels. They can be modulated by numerous factors, including a number of physiological agents (e.g., Ca^{2+}, ATP, Mg^{2+}, and cyclic ADP ribose), various cellular processes (e.g., phosphorylation, oxidation, etc.), and several pharmacological agents (e.g., ryanodine, caffeine, and ruthenium) (Fill and Copello 2002). All the three RYR isoforms have high similarities in channel slope conductance, selectivity for Ca^{2+} vs. other mono- and divalent cations, the dependence of gating on membrane potential, and diagnostic pharmacology, but have different Ca^{2+}-sensitivities of activation and inactivation with a rank order to cytosolic Ca^{2+} of roughly RYR1 > RYR2 > RYR3 (Fill and Copello 2002; Bouchard et al. 2003). The RYR channels are activated by low Ca^{2+} concentrations (1–10 μM) but inhibited by high Ca^{2+} concentrations (1–10 mM). RYRs are expressed widely in both the neurons and glial cells in the nervous system (Sharp et al. 1993; Hertle and Yeckel 2007, also see reviews by Bouchard et al. 2003; Kawamoto et al. 2012). RYR1 is found in cerebellar Purkinje cells, moderate to strong levels of RYR2 and weak to moderate levels of RYR3 are also found in the cerebellum; In deep cerebellar nuclei, RYRs are located in cell bodies that appear devoid of IP_3R (Sharp et al. 1993; Bouchard et al. 2003; Kawamoto et al. 2012). The hippocampus contains high levels of RYRs, especially relatively high levels of RYR2 (Sharp et al. 1993; Hertle and Yeckel 2007; Bouchard et al. 2003; Kawamoto et al. 2012). In the hippocampus, RYR1 immunolabeling is enriched in CA1, RYR2 is most intense in CA3 and the dentate gyrus, and RYR3 immunolabeling is detected in all subfields of the hippocampus, but is the most intense in stratum lacunosum-moleculare. RYRs are located predominantly in axons, dendritic spines, and dendritic shafts near dendritic spines (Sharp et al. 1993; Hertle and Yeckel 2007). Cerebral cortex and other brain regions also have the expression of these three isoforms of RYRs (Bouchard et al. 2003). In synapses, RYRs are present in both pre- and postsynaptic terminals (Bouchard et al. 2003). Neurons are not the only cells in the brain that express RYRs. Astrocytes, which play key roles in extracellular ion balance, express RYR3 but not RYR1 or RYR2 (Matyash et al. 2002; Bouchard et al. 2003). The general idea for ER Ca^{2+} release via RYRs is that Ca^{2+} entering through either voltage-gated channels or receptor-operated channels provides the trigger Ca^{2+} to activate RYRs and subsequent Ca^{2+} release from ER, namely Ca^{2+}-induced Ca^{2+} release. Within the soma and the initial dendritic regions, portions of ER come into close contact with plasma membrane to form subsurface cisternae (Berridge 1998; Lam and Galione 2013), where RYR channels in ER are always closely apposed to L-type Ca^{2+} channels in the plasma membrane. Therefore, even a single action potential can induce Ca^{2+} entry through membrane Ca^{2+} channels to form highly localized Ca^{2+} in nanodomain or mircodomain, and trigger Ca^{2+} release from ER in neurons of various brain regions (Berridge 1998). The role of RYR channels in Ca^{2+} signaling in astrocytes remains debatable (Verkhratsky et al. 2012). Studies employing voltage-clamp methods in conjunction with ER-sensitive fluorescent Ca^{2+} indicators indicate that this Ca^{2+}-induced Ca^{2+} release in neurons can be either graded or regenerative (Bouchard et al. 2003), and the degrees of such Ca^{2+} signals are determined not only by the amplitude and duration of trigger Ca^{2+} stimulus, but also by some sort of negative-feedback mechanism that counters the inherent positive feedback of Ca^{2+}-induced Ca^{2+} release (Fill and Copello 2002). Typically, cell types displaying graded Ca^{2+}-induced Ca^{2+} release involve relatively low gain of Ca^{2+}-induced Ca^{2+} release. However, the high gain of Ca^{2+}-induced Ca^{2+} release process can also yield a graded Ca^{2+}-induced Ca^{2+} release by a low degree of coupling of functional release units, where Ca^{2+} release closely follows the open probability of membrane Ca^{2+} channels (Bouchard et al. 2003). This kind of Ca^{2+} signals always forms the fundamental Ca^{2+} release events at the level of individual RYRs and spontaneous Ca^{2+} release events at resting intracellular Ca^{2+} levels in cells, called Ca^{2+} sparks (Fill and Copello 2002; Bouchard et al. 2003). Regenerative Ca^{2+}-induced Ca^{2+} release in the Ca^{2+} release occurs in an all-or-none fashion once threshold levels of $[Ca^{2+}]i$ are reached. This release involves changes in bulk $[Ca^{2+}]$ i, and is responsible for amplifying Ca^{2+} signals coming from the outside and for forming Ca^{2+} oscillations or Ca^{2+} waves and global Ca^{2+} signal (Berridge 1998; Bouchard et al. 2003). RYR Ca^{2+} signals play important roles in the brain function. Presynaptic Ca^{2+}-induced Ca^{2+} release has been inferred to contribute 10–80 % of Ca^{2+} required for neurotransmission or Ca^{2+} transients, and plays an important role in neurotransmitter release in some neurons and in setting the level of tonic neuronal activity in the brain (Bouchard et al. 2003). RYR Ca^{2+} signals also regulate action potential hyperpolarization, axonal retrograde transport, synaptic plasticity and learning ability (Bouchard et al. 2003; Kawamoto et al. 2012).

SERCA pumps are located in the membranes of endo(sarco)plasmic reticulum and couple ATP hydrolysis to the transport of Ca^{2+} from cytoplasm to lumen. There are three major paralogs, SERCA1-3, which are encoded by a family of three highly homologous genes, and are expressed at various levels in different cell types. Four SERCA2 mRNAs express splice variants to generate further diversity (SERCA2a and SERCA2b). SERCA2b is the dominant neuronal form (Baba-Aissa et al. 1998). The cytoplasmic face of SERCA has three major domains: the phosphorylation and nucleotide-binding domains, which form the catalytic site, and the actuator domain, which is involved in the transmission of major conformational changes. Like PMCA, SERCA has two conformational states: E1 has high Ca^{2+} affinity on the cytosolic site, and E2 has the lower Ca^{2+} affinity on the luminal site. This favors the transport of Ca^{2+} from cytoplasm to lumen (Olesen et al. 2007; Toyoshima 2008). SERCA runs as long as ATP and Ca^{2+} are present in the cytoplasm. With the hydrolysis of ATP, ATP γ-phosphate is delivered to the phosphorylation domain leading to SERCA phosphorylation, which then causes a change in conformation from E1 to E2, leading to 2 Ca^{2+} ions being moved into lumen where Ca^{2+} ions are released due to the lower Ca^{2+} affinity. The operation of the pumps can establish a $>10^4$-fold concentration gradient across membranes (Toyoshima 2008). Therefore, SERCA pumps, together with PMCA and NCX, reset resting Ca^{2+} values after Ca^{2+} transient induced by cell stimulation.

Mitochondria are capable of uptake and release of Ca^{2+}. The resting level of cytosolic Ca^{2+} is partially dependent on the Ca^{2+} uptake by mitochondria (Lee et al. 2007; Kang et al. 2008). In cytosol, mitochondria act as a fixed buffer shaping the pattern of Ca^{2+} increase by rapidly taking up Ca^{2+} both entering from the plasma membrane and release from ER, and then releasing it in cytosol or shuttling into ER. These are also major components of Ca^{2+} signaling because they modulate both the amplitude and the spatiotemporal patterns of Ca^{2+} signals (Moreau et al. 2006). In glial cells such as astrocytes (Boitier et al. 1999; Fu et al. 2013) and oligodendrocyte progenitors (Haak et al. 2002), mitochondria are thought to participate in regulating the spontaneous rises of cytosolic Ca^{2+} and the spread of $[Ca^{2+}]i$ transients and modulating Ca^{2+}-dependent functions. In neurons, mitochondria are highly concentrated in pre- and postsynaptic terminals (Ly and Verstreken 2006), implying the importance of Ca^{2+} signaling and the role of mitochondria in these sites in synaptic transmission. Rat dorsal root ganglion (DRG) neurons show a temperature-dependent decrease in resting $[Ca^{2+}]i$, which cannot be affected by inhibition of either PMCA or ER Ca^{2+} uptake, but is largely reversed by inhibition of mitochondrial Ca^{2+} uptake, suggesting mitochondria Ca^{2+} uptake is necessary for temperature-dependent increase in Ca^{2+} buffering and contributes to the control of $[Ca^{2+}]i$ (Kang et al. 2008). Endogenous or stimuli-induced increases in $[Ca^{2+}]i$ accompany the increase in mitochondrial $[Ca^{2+}]$ in presynaptic terminals where mitochondria are densely packed (Ly and Verstreken 2006), and dissipation of the electrochemical gradient across inner mitochondrial membrane (IMM) with carbonyl cyanide *m*-chlorophenyl hydrazone (CCCP) rapidly and significantly increases $[Ca^{2+}]i$, but inhibits the increase in mitochondrial $[Ca^{2+}]$ through the stimulus train, suggesting mitochondrial uptake of Ca^{2+} contributes importantly to shaping presynaptic $[Ca^{2+}]$ i transients in response to brief trains of action potentials during neurotransmission (David et al. 1998; Talbot et al. 2003; Chouhan et al. 2012; Shutov et al. 2013). Evoked cytosolic Ca^{2+} transients have characteristic "peak and shoulder" features that are mediated by mitochondrial buffering and release (David et al. 1998; Murchison and Griffith 2000). In the hippocampus, presynaptic mitochondrial Ca^{2+} uptake limits the stimuli-induced $[Ca^{2+}]i$ build-up at large mossy fiber boutons but not at small ones (Lee et al. 2007). Postsynaptic mitochondria also show Ca^{2+}

responses similar to that in presynaptic mitochondria in response to stimuli (Young et al. 2008). Mitochondria are estimated to account for about 40 % of presynaptic Ca^{2+} clearance in dorsal root ganglia (Shutov et al. 2013). Developmental increases in mitochondrial activity of hippocampal granule cells show the increased ability in Ca^{2+} uptake and significantly contribute to Ca^{2+} clearance in mature granule cells compared to in young one (Lee et al. 2009). Hippocampal mitochondria are able to accumulate Ca^{2+} once the bulk cytosolic Ca^{2+} levels reach 500–600 nM (Young et al. 2008). Interestingly, the transient mitochondrial Ca^{2+} uptake in response to periods of synaptic activation exhibits regional difference in the same hippocampal neuron, with a greater sensitivity to synaptic activation in dendritic than somatic mitochondria, but are surprisingly similar in individual mitochondria located throughout the somatic compartment between the regions close to plasma membrane and the central region (Young et al. 2008). This regional difference may have high implications for spatial patterns of Ca^{2+} signals in defined cell portions. In addition, mitochondria can also release the accumulated Ca^{2+}. For example, high frequency stimulation can induce mitochondrial Ca^{2+} increase and then release at the motoneuron terminals, and the release can last a period of about 10 min after the cessation of the stimulation (David et al. 1998; Talbot et al. 2003; Verstreken et al. 2005). The modulation of cytosolic Ca^{2+} signals by mitochondria plays important roles in neuronal functions such as neurotransmitter release (Talbot et al. 2003; Yang et al. 2003; Verstreken et al. 2005), synaptic plasticity (Yang et al. 2003; Lee et al. 2007; Quinlan et al. 2008), energy metabolism (Chouhan et al. 2012), mitochondrial motility (Chang et al. 2011), and even cell fate (Barsukova et al. 2011a, b).

Isolated mitochondria have been shown to be capable of accumulating over 3 μmol Ca^{2+}/mg protein (equivalent to 3 mM at 1 mg/mL protein in the presence of ATP and Pi), most of which exists in the form of osmotically inactive $Ca_3(PO_4)_2$ in matrix to prevent a rise in the matrix osmolarity. Thus, mitochondria can function as a Ca^{2+} sink (Chao and Xia 2010). Under physiological conditions, Ca^{2+} uptake by mitochondria is mediated by a specific Ca^{2+} uniport system, which operates strictly according to the Ca^{2+} electrochemical equilibrium. Supposing the negative mitochondrial potential ($\Delta\Psi m$) is −180 mV, and the resting level of $[Ca^{2+}]i$ is about 100 nM (see previous discussion in Sect. 4.3.2), the theoretical Ca^{2+} concentration in matrix is up to 0.1–1 M (Bernardi 1999; Chao and Xia 2010; Santo-Domingo and Demaurex 2010). This concentration is at least 1,000-fold higher than the estimated value of 0.1–1 mM (Bernardi 1999) (the actual value can be as low as relatively lower μM, [David et al. 2003; Chalmers and Nicholls 2003]) due to the activity of mitochondrial extrusion system and the subsequent Ca^{2+} efflux from mitochondria. Ca^{2+} efflux from mitochondria is mediated primarily by the mitochondrial Na^+/Ca^{2+} exchangers (mNCX). In addition, under certain conditions, the Ca^{2+} uniporters (reverse mode) and permeability transition pores (PTP) may serve as pathways of Ca^{2+} release (Bernardi 1999; Graier et al. 2007).

Mitochondria process two membranes with different properties and roles in regulating Ca^{2+} uptake and release. The OMM has the abundant expression of voltage-dependent anion-selective channels acting as general diffusion pores for small hydrophilic molecules such as ATP, ADP, and cytochrome C. These channels are also highly permeable to Ca^{2+}, and can form Ca^{2+} tunnels with IP_3Rs at MAM via GRP75 (stress-70 protein) to tightly control ER Ca^{2+} signals into mitochondria (Betz et al. 2013). The IMM is impermeable and requires specific transporters for traffic between the inner mitochondrial matrix and cytosol. IMM Ca^{2+} uptake into mitochondrial matrix is predominantly mediated by low-affinity mitochondrial Ca^{2+} uniporter (MCU) that has the properties of a highly selective ion channel (Kirichok et al. 2004) and is composed of the channel forming protein MCU (Baughman et al. 2011; De Stefani et al. 2011) and the regulating accessory protein MICU1 (Perocchi et al. 2010). MCU has low affinity to cytosolic Ca^{2+} and thus had been thought to play few roles in Ca^{2+} homeostasis. This situation has been reversed by the seminal study by Rizzuto et al., in which they developed a novel mitochondria-targeting Ca^{2+} sensor, aequorin, for the first time to demonstrate that mitochondria are active players able to sense and target Ca^{2+} hotspots from ER release or influx across the plasma membrane, causing Ca^{2+} accumulation through MCU (Rizzuto et al. 1993, 1998). Indeed, inhibition of MCU increases not only the resting cytosolic Ca^{2+} level but also the stimulation-induced Ca^{2+} transients and the slow decline of Ca^{2+} signals (Lee et al. 2007; Kang et al. 2008; Chang et al. 2011); meanwhile the accumulation of Ca^{2+} in mitochondria is also abolished (Moreau et al. 2006). It is clear now that rapid Ca^{2+} uptake by MCU is driven by steep negative $\Delta\Psi m$ (approximately −180 mV) across IMM, and can generate a firewall that effectively delineates Ca^{2+} signaling domains to shape cytosolic Ca^{2+} signals (Santo-Domingo and Demaurex 2010; Raffaello et al. 2012). Mitochondrial MCU is gated by cytosolic Ca^{2+}, which first stimulates MCU activation and uptake and subsequently inactivates the uniporter to prevent further Ca^{2+} uptake (Moreau et al. 2006). The cytosolic Ca^{2+} concentrations that are required for half-maximal mitochondrial Ca^{2+} uptake and Ca^{2+}-dependent inactivation are approximately 10–20 μM, which can be reached at the locations where mitochondria are close to either Ca^{2+} influx channels in the plasma membrane or ER which can release Ca^{2+} via IP_3Rs (Moreau et al. 2006). MCU-mediated Ca^{2+} uptake allows mitochondria to act as mobile Ca^{2+} buffers that mitigate $[Ca^{2+}]i$ elevations. MCU only takes up Ca^{2+} in the micromolar range (due to its low Ca^{2+} affinity), while the nanomolar Ca^{2+} uptake also exists in mitochondria (Santo-Domingo and Demaurex 2010). The latter is found to be mediated by a high affinity mitochondrial Ca^{2+}/H^+ exchanger Letm1 (leucine-zipper-EF-hand-containing transmembrane region), a protein located in the IMM that catalyzes the 1:1 electrogenic exchange of Ca^{2+} and H^+ (Jiang et al. 2009), and has important implications for cell fate since this high affinity mode of Ca^{2+} uptake can sense slow and steady $[Ca^{2+}]i$ elevation and potentially transform small increase in $[Ca^{2+}]i$ into sustained increase in mitochondrial Ca^{2+} concentrations and initiate the signaling cascade of apoptosis (Santo-Domingo and Demaurex 2010). Most recently, a study with liposomes containing highly purified recombinant Letm1 indicated that Letm1-dependent Ca^{2+} influx was occurred when a minimum of 30 μM Ca^{2+} was added outside the liposomes, and such flux was stimulated only by a pH gradient but not by other monovalent cations, suggesting that Ca^{2+} flux occurs in exchange for H^+ (Tsai et al. 2014). However, such Letm1-mediated Ca^{2+} uptake is also questioned, because a Ca^{2+}/H^+ antiporter can catalyze Ca^{2+} uptake only if the H^+/Ca^{2+} stoichiometry is lower than 2, while an antiporter with a stoichiometry of $2H^+$/ Ca^{2+} can only catalyze Ca^{2+} efflux in energized mitochondria (Nowikovsky et al. 2012). Also, the Letm1 in the study by Jiang et al. (2009) seems not the same with that of Tsai et al. (2014), because the former is sensitive to ruthenium red, while the latter is not. It should be pointed out that Letm1 can contribute to both mitochondrial Ca^{2+} uptake and release depending on mitochondrial proton and Ca^{2+} chemical gradients (Nowikovsky et al. 2012). In addition, as discussed earlier, IP_3R-mediated Ca^{2+} release from ER in MAM forms Ca^{2+} microdomains, which can be very efficiently taken up by mitochondria to limit Ca^{2+} diffusion (Graier et al. 2007; Patergnani et al. 2011; Lam and Galione 2013). In isolated heart mitochondria, Ca^{2+} can be imported via mNCX operating in the reverse mode (Jung et al. 1995). However, it is not clear if this pathway runs in nerve cells. Under physiological conditions, cytosolic Ca^{2+} elevation can rapidly cause moderate Ca^{2+} elevation in the mitochondrial matrix via above pathways, where they increase the activity of enzymes of Krebs cycle and ATP synthase to increase ATP production (see review by Glancy and Balaban 2012; Tarasov et al. 2012) and thus energize Ca^{2+} pumps (both PMCA and SERCA) to extrude/sequestrate excess Ca^{2+}.

Stimulation or neurotoxin-induced Ca^{2+} release from neuronal mitochondria always requires an increase in $[Na^+]i$, suggesting that the mNCX plays a critical role in mitochondrial Ca^{2+} release (Tsang et al. 2000; Yang et al. 2003). The hunt for mNCX gene has been a difficult task with an inability to identify whether mNCX is the plasma membrane exchange isoform that is targeted to mitochondria and carry the exchange process or another different isoform that is specific for mitochondria (Palty and Sekler 2012). However, the molecular identity of this transporter most recently reveals that the Na^+/Ca^{2+} exchanger NCLX is enriched in mitochondria, where it is localized to the cristae (Palty et al. 2010). Indeed, RT-PCR analysis detected the expression of NCLX in cultured DRG neurons (Shutov et al. 2013). Western blotting analysis and in situ electron microscopy immunocytochemistry indicated that a conspicuous population of neuronal and astrocytic mitochondria in the cortex and the hippocampus express NCX, with distinct isoforms exhibiting differential patterns of mitochondrial expression, i.e., the majority of NCX-expressing mitochondria are found in the dendrites of neurons, often located beneath the plasmalemma and near postsynaptic sites, while most NCX-labelled mitochondria in astrocytes are situated close to the cellular surface (Gobbi et al. 2007). Mitochondiral NCX is electrogenic with a stoichiometry of 3–4 Na^+ ions/1 Ca^{2+} ion per transport cycle (Palty et al. 2010). In cultured or acutely isolated neurons, application of mNCX inhibitors (e.g., CGP37157, clonazepam) not only causes a modest but significant decrease in resting $[Ca^{2+}]i$ (Kang et al. 2008), but also slows the decay kinetics of mitochondrial response to neurotransmitters, resulting in a sustained elevation of mitochondrial Ca^{2+} levels (Young et al. 2008) and partially decreases mitochondrial depolarization-induced Ca^{2+} release and $[Ca^{2+}]i$ rise (Murchison and Griffith 2000). CGP37157 also attenuates tetanus-induced residual $[Ca^{2+}]i$ without altering the peak cytosolic Ca^{2+} in presynaptic terminals, and markedly reduces the effectiveness of tetanus to induce synaptic potentiation, indicating mNCX mediates tetanus-induced mitochondrial Ca^{2+} release and subsequent synaptic plasticity

(Yang et al. 2003; Lee et al. 2007). Nevertheless, Zhong et al. (2001) reported in crayfish motoneuron terminals that post-tetanic mitochondrial Ca^{2+} release and potentiation cannot be blocked by CGP37157, implying that Na^{+}-independent efflux of Ca^{2+} is an alternative route for mitochondrial Ca^{2+} release and synaptic plasticity. Indeed, as we discussed before, only if the H^{+}/Ca^{2+} stoichiometry is lower than two can a Ca^{2+}/H^{+} antiporter catalyze Ca^{2+} uptake. Otherwise, for $2H^{+}$-Ca^{2+} antiporters and $3H^{+}$-Ca^{2+} antiporters, direction of Ca^{2+} flux depends on the proton and Ca^{2+} chemical gradients, which under physiological conditions favor Ca^{2+} efflux (Nowikovsky et al. 2012). Therefore, mitochondrial Ca^{2+}/H^{+} antiporter can be a pathway for mitochondrial Ca^{2+} release. Generally, the activity of the mNCX is ubiquitously found in most cell types and tissues studied so far, and are particularly robust in excitable cells, whereas the activity of the Ca^{2+}/H^{+} exchanger is primarily found in non-excitable cells (e.g. glial cells) and plays an important role in mitochondrial Ca^{2+} release. In addition, the opening of the permeability transition pore (PTP) is also implicated as a Ca^{2+} release mechanism of mitochondria in physiological Ca^{2+} signaling (Murchison and Griffith 2000; Barsukova et al. 2011b). In some mitochondria, such as those in the CNS, high levels of Ca^{2+} within mitochondria can cause the PTP to open (Brustovetsky et al. 2002; Doczi et al. 2011). The dissipation of $\Delta\psi m$ can cause PTP to open too, but the contribution of $\Delta\psi m$ to the induction of PTP opening in neurons and astrocytes is complex (Doczi et al. 2011). The opening of PTP can allow Ca^{2+} efflux through the spontaneously irreversible megachannel conformation, which is widely believed to initiate cell death, and is associated with pathophysiological processes (Brustovetsky et al. 2002; Doczi et al. 2011; Barsukova et al. 2011a). However, recent studies demonstrated that physiological stimulations, which lead to non-toxic mitochondrial Ca^{2+} load, could cause Ca^{2+} release from mitochondria through PTP (Murchison and Griffith 2000; Barsukova et al. 2011b). For example, in acutely isolated rat basal forebrain neurons, mitochondria are able to buffer caffeine-evoked non-toxic Ca^{2+} loads, shaping a characteristic intracellular Ca^{2+} transient and shoulder, and PTP blocker cyclosporine A can reversibly block the Ca^{2+} shoulder without altering the Ca^{2+} transient, indicating the shoulder is resulted from Ca^{2+} release from loaded mitochondria through the opened PTP (Murchison and Griffith 2000). In cultured cerebral cortical neurons of wild type (WT) and cyclophilin D (a key regulator of the PTP)-knockout (CyPD-KO) mice, simultaneous stimulation of IP_3R-mediated ER release and plasma membrane Ca^{2+} channels evokes higher levels of $[Ca^{2+}]i$ in WT than CyPD-KO neurons and a higher level of mitochondrial Ca^{2+} accumulation in CyPD-KO than in WT neurons. Also, application of PTP blocker cyclosporine A to WT neurons can mimic responses observed in CyPD-KO neurons with a significant increase in mitochondrial Ca^{2+} levels following simultaneous stimulation of both pathways, confirming the role of PTP in mitochondrial Ca^{2+} release under physiological conditions (Barsukova et al. 2011b). Importantly, such a Ca^{2+} release through PTP does not lead to the activation of cell death pathways (Murchison and Griffith 2000; Barsukova et al. 2011b), suggesting PTP also mediates normal Ca^{2+} signals in the CNS. The released mitochondrial Ca^{2+} through pathways such as mNCX, Ca^{2+}/H^{+} antiporter and PTP in neurons plays a major role in modulating cytosolic Ca^{2+} response, because it generates a shoulder-like extension that shapes the magnitude and duration of cytosolic Ca^{2+} signals (David et al. 1998; Talbot et al. 2003; Verstreken et al. 2005).

Regulation of the Dynamics of Ca^{2+} Signals by the Activities of Endogenous Ca^{2+}-Binding Proteins (Ca^{2+} Buffering Proteins and Sensors)

There is a fair amount of Ca^{2+} ions inside cells, but most of them are either bound to specific proteins or sequestered in organelles, principally in the ER and mitochondria (for ER and mitochondria, see above discussion). Ca^{2+}-binding proteins are proteins that participate in Ca^{2+} signaling pathways by binding to Ca^{2+}. A number of Ca^{2+}-binding proteins have been identified, most of which share the typical EF-hand structure in their binding sites (see reviews by Ikura and Ames 2006; Burgoyne 2007; Schwaller 2009, 2010; also refer to the research topics edited by Kreutz et al, 2012 in *Frontiers in Molecular Neuroscience*, http://www.frontiersin.org/Molecular_Neuroscience/researchtopics/The_neuronal_functions_of_EF-h/585). In this section, we only summarized the general role of Ca^{2+}-binding proteins in regulation of cytosolic Ca^{2+} signals. For more detailed information, it is recommended to read the excellent reviews and research topics listed above.

More than 240 identified EF hand Ca^{2+}-binding proteins are encoded by the human genome (Schwaller 2009). Based on several characteristic features they can be divided into two groups: the Ca^{2+} buffers (e.g., parvalbumins, calbindin-D9k, calbindin-D28k, and calretinin) which act as chelators for Ca^{2+} ions and modulators of short-lived intracellular Ca^{2+} signals, affecting both the temporal and spatial aspects of these transient increases in $[Ca^{2+}]i$ (Schwaller 2009, 2010), and the Ca^{2+} sensors [e.g., calmodulin, neuronal calcium sensor (NCS) proteins] which are characterized by their ability to bind and release Ca^{2+} over the physiological range of Ca^{2+} concentrations, to undergo a significant conformational change on Ca^{2+} binding, and to consequently bind and regulate specific target proteins to modify their function (Ikura and Ames 2006; Burgoyne 2007). However, it is becoming hard to clearly distinguish between these two groups since Ca^{2+} sensor proteins are not the pure signal transducing molecules, and they obviously also buffer Ca^{2+} (Burgoyne 2007; Schwaller 2009). Calmodulin is the best-studied Ca^{2+}-binding protein that is ubiquitously expressed in almost all cell types (Ikura and Ames 2006). Calmodulin is highly concentrated in the brain (30–50 μM) with a considerable fraction (up to 100–200 μM) of the Ca^{2+} buffering capacity. It acts primarily to buffer cytosolic Ca^{2+}, but also functions as other signaling messengers (Ikura and Ames 2006; Burgoyne 2007). Ca^{2+}-binding proteins in neurons convert spatiotemporal Ca^{2+} concentration signals into enzymatic activity, but also shape Ca^{2+} transients themselves. Under physiological conditions, when Ca^{2+} enters neurons or releases from internal stores, Ca^{2+}-binding proteins promptly bind some Ca^{2+} during its diffusion to keep Ca^{2+} in a proper localization and level, and therefore, serve as a modulator of Ca^{2+} signaling (Burgoyne 2007). Undoubtedly, Ca^{2+}-binding proteins can reduce the amplitude of $[Ca^{2+}]i$, retard decay kinetics, and shape diffusional profiles. Although the process of Ca^{2+} binding to cytosolic Ca^{2+} binding proteins is rapid (several ms), the binding is very limited in capacity, and is easily saturated with higher [Ca^{2+}]i. The saturation kinetics, together with buffering kinetics, can result in complex features of Ca^{2+} signals. It is Ca^{2+}-binding proteins, together with organelle buffers and extrusion machinery, that determine the spatiotemporal aspect of Ca^{2+} signals, which allows concentration changes of a simple Ca^{2+} ion to modify neuronal function in a multitude of ways. This also enables the same ion to produce distinct outcomes over short, medium, or long distances and timescales in the same types of neurons (Burgoyne 2007).

Regulation of the Dynamics of Ca^{2+} Signals by Efficient Extrusion

As we have discussed in Sect. 4.3.1, both neurons and glial cells possess two parallel, independent mechanisms that cause extrusion of Ca^{2+}, namely PMCA and NCX (Dipolo and Beaugé 2006; Zündorf and Reiser 2011; Verkhratsky et al. 2012; Giacomello et al. 2013; Brini et al. 2014). Though they have their own distinctive properties in extrusing Ca^{2+} (see earlier discussion), these machineries not only work in parallel in the plasma membrane to keep basal [Ca^{2+}]i at physiological levels under physiological conditions, but also efficiently dissipate activity-induced Ca^{2+} elevations in cytoplasma (Dipolo and Beaugé 2006; Giacomello et al. 2013; Brini et al. 2014). These two extrusion mechanisms, together with other tools including the two main intracellular compartments, ER and mitochondria, some other intracellular organelles such as Golgi complex, microsoma, synaptic vesicles, and neurosecretory granules, and cytoplasmic buffers such as Ca^{2+} binding proteins, shape the spatiotemporal aspect of Ca^{2+} signals to regulate the brain functions (Dipolo and Beaugé 2006; Giacomello et al. 2013; Brini et al. 2014).

The Contribution of the Morphology of Dendritic Spines on the Regulation of the Dynamics of Ca^{2+} Signals

Spine Ca^{2+} dynamics are an indispensable part of activity-induced Ca^{2+} signals in neurons. Dendritic spines are tiny protrusions from neurons' dendrites, and represent the major postsynaptic input sites for excitatory transmission in the brain (Harris and Kater 1994; von Bohlen Und Halbach 2009). A typical dendritic spine consists of a small head (<1 μm diameter) that is connected to the dendritic shaft by a narrow neck (~0.2 μm diameter). The geometry of dendritic spines is diverse, which, based on the shape (relative size of the head and neck), can be divided into five groups: mushroom spine (which has a large head and a narrow neck), thin spine (which has a smaller head and a narrow neck), stubby spine (which does not have an obvious constriction between the head and the attachment to the shaft), filopodium spine (which has a hair-like morphology), and branched spine (which has multiple heads that emerge from a shared origin) (Harris and Kater 1994; von Bohlen Und Halbach 2009). Dendritic spines concentrate a density of spatially localized pumps and voltage- and ligand-operated channels such as VGCCs,

NMDAR- and AMPAR channels (Yuste et al. 2000; Matsuzaki et al. 2001; Yasuda et al. 2003; Noguchi et al. 2005; Bloodgood and Sabatini 2007a, b). Dendritic spines rarely contain mitochondria. However, about 50 % of spines contain smooth ER, a main intracellular Ca^{2+} store (Berridge 1998; von Bohlen Und Halbach 2009). Only in most of the large, mushroom-shaped dendritic spines do the ER become laminated spine apparatuses, whereas the smaller, thin spines have a thin tube of smooth ER (Berridge 1998; von Bohlen Und Halbach 2009).

The spatiotemporal Ca^{2+} dynamics of spines are controlled by (1) Ca^{2+} influx and release from internal store, (2) Ca^{2+} buffering, (3) Ca^{2+} diffusion, and (4) Ca^{2+} extrusion in spines (Yuste et al. 2000). Synaptic stimulation can induce Ca^{2+} influx and Ca^{2+} transients in the spines. Unlike Na^+ responses in the spines that require suprathreshold synaptic stimulation (Rose and Konnerth 2001), Ca^{2+} transients in the spines can be induced either by subthreshold or suprathreshold synaptic stimulation, or by back-propagating action potentials (Yuste et al. 2000; Augustine et al. 2003). Subthreshold stimulation-induced Ca^{2+} transients are graded and dependent on the strength of synaptic activity, and these Ca^{2+} transients are mediated by Ca^{2+} influx through NMDARs or Ca^{2+}-permeable AMPARs. During suprathreshold synaptic stimulation, VGCCs are also activated to allow Ca^{2+} influx. Both NMDARs and VGCCs in the spines can also be activated by back-propagating action potentials to shape Ca^{2+} transients (Yuste et al. 2000; Augustine et al. 2003; Bloodgood and Sabatini 2007a; Grunditz et al. 2008; Bloodgood et al. 2009). In addition to Ca^{2+} influx via the cell membrane, Ca^{2+} release from ER also exists in the spines. Ca^{2+} release is controlled by either IP_3Rs or RYRs in ER (Yuste et al. 2000). Despite the similar resting Ca^{2+} levels in the spines and their parent dendrites, stimulation-induced Ca^{2+} accumulations are larger in the spines than in their parent dendrites with a delay of several milliseconds for Ca^{2+} rise in the dendritic shaft, suggesting that Ca^{2+} accumulations are generated at the spines and can diffuse to dendrites via the spine neck (Yuste et al. 2000; Sabatini et al. 2002; Noguchi et al. 2005). The endogenous buffer capacity is low in spines (Sabatini et al. 2002; Holthoff et al. 2002; Noguchi et al. 2005), so Ca^{2+} transients in the spine head are very large (up to 20 μM with NMDAR activation) (Noguchi et al. 2005). The low endogenous buffer capacity also favors Ca^{2+} diffusion within the spine heads. The clearance of elevated Ca^{2+} in spines is mainly achieved by two pathways: diffusion of Ca^{2+} to the dendrites (diffusers) and active extrusion (pumpers) (Yuste et al. 2000; Majewska et al. 2000; Holthoff et al. 2002). Due to the high heterogeneity in spines, some spines (predominantly with large head and short/thick neck) show efflux dominated by diffusers, while others (mainly the small spines) show strong extrusion by pumpers, particularly SERCAs (Majewska et al. 2000; Holthoff et al. 2002; Noguchi et al. 2005). These two pathways, together with the diameter of the parent dendrites, control Ca^{2+} decay kinetics in spines (Majewska et al. 2000; Holthoff et al. 2002).

Spines vary from morphology, receptor composition and Ca^{2+} influx/efflux pathways. During synaptic activity, spine geometry/numbers can undergo rapid structural plasticity, e.g., transient or persistent expansion/shrinkage, generation of a new spine or complete loss of existing spines (Hayashi and Majewska 2005). For example, synaptic enhancement leads to an enlargement of thin spines (suggested as "learning spines") into mushroom spines (which are relatively stable and suggested as "memory spines") (Bourne and Harris 2007). The rapid expansion of spines can begin within 20s of initiation of stimulation and can last for at least an hour (Matsuzaki et al. 2004). In addition to the changes in size/numbers, spines such as filopodia and mature spines are motile and undergo constant remodeling (Dunaevsky and Mason 2003). These changes of spine morphology indispensably affect Ca^{2+} dynamics in spines. Furthermore, the expression of functional receptors/channels is closely related to the geometry of spines. For example, in hippocampal CA1 pyramidal neurons, AMPARs are abundant in mushroom spines (up to 150/spine) but sparse in thin spines and filopadia (Matsuzaki et al. 2001). The level of AMPAR expression increases with spine size (Noguchi et al. 2005). The spines that show a higher level of AMPAR expression also express a large number of NMDARs with an increased AMPAR/NMDAR ratio for the increased spine head volume, and the spines with a low level of AMPAR expression also express a substantial number of NMDARs (Noguchi et al. 2005). Such distributions of receptors in geometrically heterogenous spines may undoubtedly influence Ca^{2+} signals in spines, for example, a marked NMDAR-mediated Ca^{2+} increase can be detected in most small spines where AMPARs are very sparse and AMPAR-mediated currents are low (Noguchi et al. 2005). Though diffusion within the spine heads is fast due to the low endogenous buffer capacity in spines (Sabatini et al. 2002; Holthoff et al. 2002; Noguchi et al. 2005), diffusion between the spine and its parent dendrite can be very slow depending on the morphology of the spine neck (e.g., length, thickness). The thin necks separate dendritic spines from dendrites to form diffusional barriers that slow down the exchange of biochemical signals (e.g., signaling molecules such as IP_3 and Ca^{2+}) between spine heads and their parent dendrites and also control local depolarization of spine heads, making the spine heads act as a biochemical, even electrical compartment (Müller and Connor 1991; Svoboda et al. 1996; Sabatini et al. 2002; Noguchi et al. 2005; Bloodgood and Sabatini 2005; Santamaria et al. 2006; Grunditz et al. 2008; Bloodgood et al. 2009). Ca^{2+} decay in most spines have two components: the initial fast decay from a high amplitude followed by a slower decay with the time scale and amplitude similar to that of the dendrite. Also, long-necked spines have prolonged fast time constants of spine Ca^{2+} decay (Majewska et al. 2000; Holthoff et al. 2002). In addition, small spines tend to have narrow necks, which impede the efflux of Ca^{2+} into the dendritic shaft and contribute to the large isolated increases in $[Ca^{2+}]i$ in small spine heads, while larger spines have relatively thicker necks that have high Ca^{2+} conductance and permit great efflux into dendritic shaft, causing only moderate Ca^{2+} increases in large spine heads (Noguchi et al. 2005). Therefore, spine neck geometry may regulate the spread of spine Ca^{2+} signal, and such spine neck-regulated Ca^{2+} dynamics make the individual expression of LTP earlier in small spines than in large one (Matsuzaki et al. 2004). Holthoff et al. (2002) observed in layer 5 pyramidal neurons in mouse primary visual cortical slices that the peak spine Ca^{2+} transients are regulated intrinsically by the spine, and do not depend on dendritic Ca^{2+} dynamics since there is no clear relationship between peak spine Ca^{2+} transients and dendritic diameter, though Ca^{2+} dynamics in dendrites are determined by dendritic diameter (Holthoff et al. 2002). In the dendritic spines of hippocampal CA1 pyramidal neurons, Sabatini et al. (2002) observed back-propagating action potential-induced Ca^{2+} gradients across the spine neck that persist for the entire duration of action potential-evoked Ca^{2+} transients, and the estimated diffusional equilibration of Ca^{2+} across the spine neck is >1 s in unperturbed spines. The authors thus concluded that under physiological conditions, Ca^{2+} diffusion across the spine neck is negligible, and the spine head functions as a separate compartment on long time scales, allowing localized Ca^{2+} buildup during trains of synaptic stimuli (Sabatini et al. 2002). However, the barrier to diffusion posed by spine neck is dynamically regulated by neuronal activity (Bloodgood and Sabatini 2005; Grunditz et al. 2008). For example, Noguchi et al. (2005) demonstrated that following NMDAR-mediated potentiation in the spines of hippocampal CA1 pyramidal neurons, not only is the spine head volume enlarged, but also the morphology and function of spine neck changes with an increase in neck radius and Ca^{2+} conductance that allow fast diffusion between spines and dendrites, thus limiting the peak Ca^{2+} accumulation following NMDA stimulation.

Taken together, dendritic spines are highly heterogenous in morphology. Spine geometry is at least one determinant for the spatial and temporal pattern of the dynamics of Ca^{2+} signals in spines, which is closely related with the plasticity and functions of spines.

5 Role of DOR in Regulation of Ionic Homeostasis under Physiological Condition

The regulation of ionic homeostasis in nerve cells depends largely on the activities of functional proteins (e.g., ionic channels, receptors, transporters, and pumps) integrated in the membrane (see previous discussion). DOR is a member of the G protein-coupled receptor superfamily (Law et al. 2000), and may functionally couple with these functional proteins. Activation of DOR intensively influences a diversity of neural activities via the interaction with a large amount of membrane functional proteins, which also play a key role in regulation of ionic homeostasis in the CNS. Therefore, DOR may have some roles in ionic homeostasis.

5.1 DOR Regulation of Neural Activities

DOR regulates a diverse array of physiological functions (Feng et al. 2012). Studies with immunocytochemical, in situ hybridization and autoradiography methods as well as fluorescent knock-in mice that express a functional DOR, revealed that DOR and its cognate endogenous opioids are widely expressed throughout the CNS (Mansour et al. 1987, 1994; Xia and Haddad 1991, 2001; Bausch et al. 1995; Svingos et al. 1995; Abeyta et al. 2002; Scherrer et al. 2006; Erbs et al. 2012; Rezaï et al. 2012; Pierce et al. 2014), suggesting ubiquitous roles of DOR in the regulation of neural activities. Indeed, in the nervous system, DOR has extensive interactions with ion channels, receptors, and transporters. DOR can regulate the release of many neurotransmitters, and modify neuronal electrical activities and synaptic plasticity. These interactions and regulations might tightly relate to DOR regulation of ion phenotypes during neural activities.

5.1.1 Interaction between DOR and Channels, Receptors, and Transporters

DOR is functionally coupled with various ion channels, such as K^+ channels, Ca^{2+} channels, Na^+ channels, and Ca^{2+}-activated Cl^- channels (see the following discussion). A large amount of studies have demonstrated that DOR can regulate the activities of a variety of ***K⁺channels***, which include G-protein-activated inwardly rectifying K^+ (GIRK) channels (North et al. 1987; Ikeda et al. 1995; Piros et al. 2000; Shi et al. 2000; Marker et al. 2005; Richard-Lalonde et al. 2013), voltage-gated K^+ channels (Fan et al. 1991; Wild et al. 1991; Moore et al. 1994; Fan and Crain 1995; Shen and Johnson 2002; Jose et al. 2007), and K_{ATP} channels (Wild et al. 1991; Pacheco and Duarte 2005; Saloman et al. 2011). For example, in the slices of the midbrain periaqueductal gray (PAG) of normal mice, the opioid agonist Met-enkephalin produces an outward current in PAG neurons, which is sensitive to the blockade of GIRK channels (Bagley et al. 2005). GIRK channels, the homotetrameric and hetertetrametric complexes of GIRK/Kir3 subunits (Krapivinsky et al. 1995; Kunkel and Peralta 1995), include four subtypes (GIRK1-4), and GIRK1-3 are broadly distributed throughout the CNS (Kobayashi et al. 1995; Karschin et al. 1996; Marker et al. 2005). Early in vitro studies in guinea pig submucous plexus found that both DOR agonist DPDPE and DOR endogenous ligand Met-enkephalin increase the conductance of inwardly rectifying K^+ currents (I_{Kir}) and strongly hyperpolarize the membrane in a concentration-dependent manner, which can be mimicked by applying non-hydrolyzable GTP derivative, GTP[S] (to indicate G protein activation) (North et al. 1987). Stimulation of DOR with DPDPE in *Xenopus* oocytes co-expressing DOR and GIRK1 showed that DOR activation induced the I_{Kir} with an EC_{50} value of 45 nM for DPDPE, and the currents could be blocked by either opioid receptor antagonist naloxone or 300 μM Ba^{2+} (which is supposed to block GIRK channels), indicating DOR-induced I_{Kir} are mediated mainly by GIRK1 channels (Ikeda et al. 1995). In the rodent pituitary-derived GH_3 cell lines expressing G protein, voltage-activated L-type Ca^{2+} channels, and both DOR and MOR, DOR agonist DPDPE dose-dependently inhibits spontaneous Ca^{2+} oscillations and prolactin release, which is sensitive to pertussis toxin and can be inhibited by Kir channel inhibition. Also, DPDPE indeed activates Ba^{2+}-sensitive inward K^+ channel activity, suggesting a role of Gi/Go protein-coupled Kir channels in this process (Piros et al. 2000). In neuroblastomal x glioma NG108-15 cells, Shi et al. (2000) found that μM DPDPE increases the amplitude, delays the activation and inactivation, of I_{Kir} rapidly and reversibly, and these effects can be reversed by naloxone but are not sensitive to pertussis toxin or GDP-βs. The authors concluded that DPDPE-induced changes of I_{Kir} are mediated by DOR but do not involve G-protein (Shi et al. 2000). In contrast, in acutely dissociated rat brainstem neurons, 5-HT-induced GIRK channel currents can be reversibly inhibited by DOR antagonists, naltrindole and naltriben, and this inhibition cannot be reversed by DPDPE. The investigators suggested that naltrindole and naltriben inhibit GIRK channel currents through a non-opioid action (Shirasaki et al. 2004). Based on these studies, there are obvious contradictions in regards to the role of DOR and the involvement of G-protein. The reasons for these mixed results are complex. First, the observations were collected from different preparations and cell types; second, GIRKs can be activated by a wide variety of G-protein coupled receptors, including the M_2-muscarinic receptor, A_1-adenosine receptor, α_2-adrenergic receptor, D_2-dopaminergic receptor, MOR, DOR, KOR, 5-HT_1A receptor, somatostatin receptor, galanin receptor, metabotropic glutamate receptor, $GABA_B$ receptor, and sphingosine-1-phosphate receptor (North et al. 1987; Yamada et al. 1998; Marker et al. 2005). Under such a circumstance, DOR activation may utilize the same effector mechanisms (North et al. 1987). However, for different G-protein coupled receptors, the involved G-protein signaling molecules are not exactly the same. Therefore, these mixed results are not surprising (some of DOR's effects are sensitive and some are insensitive to G-protein interference, and some of GIRK channel currents are blocked by DOR antagonists while some are not). Further support for DOR regulation of GIRK channels comes from a recent study that explores the role of GIRK in opioid analgesia in the mouse spinal cord (Marker et al. 2005). In this study, immunohistochemistry and immunelectron microscopy demonstrated that GIRK1 and GIRK2 are concentrated in the outer layer of the substantia gelatinosa of the dorsal horns, almost exclusively in the postsynaptic membranes of putative excitatory synapses; similar to MOR activation, DOR activation with DPDPE dose-dependently produces analgesia effects, which can be blunted by genetic ablation or pharmacological inhibition of spinal GIRK (Marker et al. 2005), indicating DOR is analgesic by activation of postsynaptic GIRK1 and/or GIRK2-containing channels in the spinal cord. DOR-induced activation of GIRK can be changed in some cases. For example, in the neurons of the central nucleus of amygdala of chronic morphine-treated mice, DOR agonist activation of GIRK is increased (Chieng and Christie 2009). Most recently, Richard-Lalonde et al. (2013) used bioluminescence resonance energy transfer donor/acceptor pairs to investigate how ligand binding to DOR modulates constitutive interactions among DOR, $G\alpha o\beta 1\gamma 2$-G-protein, and neuronal Kir3.1/Kir3.2 channels. They found that DOR activation with different ligands (DPDPE, UFP 512, SNC80 etc.) induces conformational rearrangements between DOR-$G\alpha o$ and DOR-$G\beta\gamma$. The conformational changes of $G\beta\gamma$ subsequently induce the conformational changes at the $G\beta\gamma$/Kir3 interface to increase channel permeability and evoke channel currents (Richard-Lalonde et al. 2013). This study provides the direct support for the regulation of GIRK by DOR. In addition, DOR agonist, under low concentrations (nanomolar) decreases voltage-gated K^+ currents in neurons and neuron-like cell lines, and increases them when DOR agonist is in high concentrations (micromolar) (Fan et al. 1991; Fan and Crain 1995; Jose et al. 2007). In hippocampal CA3 pyramidal neurons, DOR agonists (DADLE and DPDPE) reduce the voltage-dependent K^+ current known as the M-current, which can be partially reversed by naloxone (Moore et al. 1994). Further analysis indicated that DOR inhibition of voltage-gated K^+ channels in some cases (e.g., presynaptic terminals of subthalamic nucleus and corticostriatal synapses) is 4-aminopyridine-sensitive transient K^+ conductance (Shen and Johnson 2002; Jose et al. 2007), while in others is tetraethylammonium (TEA)-sensitive (Wild et al. 1991). It has been shown that these two channels play a different role in action potential shaping and neuronal excitability. Namely, in the same sciatic nerve fibers, the 4-aminopyridine-sensitive potassium channels are important for rapid action potential repolarization and the TEA-sensitive potassium channels may serve to modulate axonal excitability during repetitive firing (Kocsis et al. 1987). Moreover, K_{ATP} channels are functionally associated with DOR function. Multiple measurements (in situ hybridization, single-cell PCR, and immunostaining) indicated that DORs are widely expressed in various peripheral sensory neurons (Wang et al. 2010), and peripheral antinociceptive actions of DOR have been causally linked with the activation of K_{ATP} channels (Pacheco and Duarte 2005; Saloman et al. 2011). In fact, intracerebroventricular (i.c.v) injection of DOR agonists (DPDPE, [D-Ala2]deltorpin II) also produces antinociceptive actions which can be blocked by i.c.v injection of K_{ATP} channel blocker glyblenclamide, suggesting that, in the CNS, DOR activation may also play its role via central K_{ATP} channels (Wild et al. 1991).

DOR can also interact with ***Na⁺channels***. For example, we observed that DOR down-regulation (Zhao et al. 2005) is associated with an up-regulation of voltage-gated Na^+ channels (Xia et al. 2003) in the brain. Activation of presynaptic DOR by enkephalin prevents the increase in neuronal Nav1.7 in the DRG, which relieves pain in diabetic neuropathy (Chattopadhyay et al. 2008). In a recent study, SNC80, a putative DOR agonist, was applied to the acutely isolated dentate granule cells of rat hippocampus, and the properties of voltage-gated Na^+ currents were investigated by whole cell patch-clamp recording (Remy et al. 2004). The authors observed that SNC80 (1–1,000 μM) reduced the maximal Na^+ current amplitude in a dose-dependent manner and selectively prolonged the time course of recovery from slow inactivation without effects on fast inactivation processes (Remy et al. 2004). However, the authors concluded that these effects are opioid receptor independent because the effects of SNC80 were not mimicked by another DOR agonist DPDPE (10 μM), and were not inhibited by high concentrations of opioid receptor antagonists, naloxone (50–300 μM) and naltrindole (10 and 100 μM) (Remy et al. 2004). This study may be complicated by several factors. For example, our autoradiographic work showed that the hippocampus, including the dentate gyrus, has a low density of DOR binding sites, while the cortex has a very high density (Xia and Haddad 1991, 2001). Applying low concentrations of DOR agonists to hippocampal neurons may induce an insignificant effect because of a scarcity in the density of DOR. On the other hand, applying DOR agonists at high concentrations may lead to non-selective or adverse effects because of the stimulation of multiple and non-specific membrane proteins. Furthermore, some experimental procedures such as mechanical trituration and chemical digestion (e.g., Pronase) (Remy et al. 2004) may affect the function of DOR and Na^+ channels and their interactions, causing complex changes in the effects of DOR agonists/antagonists on Na^+ channel currents. In addition, high concentrations of DOR ligands, either SNC80, naloxone or naltrindole, may stimulate a variety of membrane proteins and induce a complicated effect on cells since no "pure"

DOR agonists/antagonists exist at present. For instance, high concentrations of SCN80 can activate μ-opioid receptors (Knapp et al. 1996), which may inhibit the voltage-dependent Na^+ currents as shown in the cortex (Witkowski and Szulczyk 2006). Also, naltrindole at high concentrations generates adverse effects in neurons (Ma et al. 2005; Chao et al. 2007b). To avoid these pitfalls and to explore a "pure" interaction between DOR and Na^+ channels, we recently co-expressed DOR and Na^+ channels in *Xenopus* oocytes and made novel and exciting observations. We made the following observations: (1) Nav1.2 expression induced TTX-sensitive inward currents; (2) DOR expression reduced the inward currents; (3) Activation of DOR reduced the amplitude of the current and rightly shifted the activation curve of the current in the oocytes with both Nav1.2 and DOR, but not in oocytes with Nav1.2 alone; (4) The DOR agonist-induced inhibition of Nav1.2 currents was dose-dependent and saturable; and (5) The selective DOR agonist had no effect on naive oocytes. These data represent the first demonstration that activation of DOR inhibits Na^+ channel function by decreasing the amplitude of Na^+ currents and increasing its threshold of activation (Kang et al. 2009).

VGCCs are divided into T (low-voltage activated) and P/Q, N, L, R (high-voltage activated) types according to the physiological and pharmacological properties of Ca^{2+} current types. DOR regulation of VGCCs has been extensively studied in neuronal cell lines and DRG neurons as well as in other neurons. In *Xenopus* oocytes co-expressing DOR, G protein (Gi3α or Go1α), and N-type (α1B) Ca^{2+} channels, Ca^{2+} channel currents were reduced reversibly by 40–60 % in the presence of leucine-enkephalin, and were sensitive to pertussis toxin (Yoshii et al. 2004), suggesting functional coupling of DOR and Ca^{2+} channels via G protein. This is further supported by the study with mouse neuroblastoma x rat glioma NG 108-15 hybrid cells, in which it was found that DOR agonist DADLE inhibits Ca^{2+} currents, which is almost completely abolished by pretreatment with pertussis toxin, but is restored by intracellular application of Gi and Go (Hescheler et al. 1987). Indeed, in neuroblastoma cell lines (NG108-15, SH-SY5Y, F-11), DOR activation decreases high-voltage-activated, ω-conotoxin-sensitive N-type Ca^{2+} currents (Nah et al. 1997; Toselli et al. 1997, 1999), but does not affect nifedipine-sensitive L-type Ca^{2+} channels (Nah et al. 1997). After this inhibition is relieved by a large depolarizing prepulse, a potentiation of ω-conotoxin-sensitive N-type Ca^{2+} currents can be found in the presence of DOR agonist, which are sensitive to pertussis toxin and DOR antagonist, suggesting DOR exerts a dual action on N-type Ca^{2+} channels via pertussis toxin-sensitive G proteins (Morikawa et al. 1999). Such a dual action of DOR on N-type voltage-dependent Ca^{2+} channels is further supported by DOR-induced dual (inhibitory and excitatory) regulation of depolarization-evoked dopamine (DA) release in SK-N-SH human neuroblastoma cells by modulating these Ca^{2+} channels despite their insensitivity to pertussis toxin (Keren et al. 1997). DOR inhibition of Ca^{2+} currents in NG108-15 cells also shows a characteristic of partial desensitization with a time constant of 4.4 min. This inhibition is reversible after removal of agonist (Morikawa et al. 1998). In the rodent pituitary-derived GH_3 cell lines expressing G protein, voltage-activated L-type Ca^{2+} channels, and both DOR and MOR, DPDPE dose-dependently reversibly inhibits Ca^{2+} channel currents, which are sensitive to pertussis toxin and GDP βs (the non-hydrolyzable analogue of GDP) and are completely reversed by depolarizing prepulses; DPDPE also inhibits adenylyl cyclase activity, which is selectively attenuated by DOR antagonist TIPP-Ψ in GH_3 cell lines. The authors thus concluded that DOR in GH_3 cells voltage-dependently couple through Gi/Go proteins to L-type Ca^{2+} channels (Piros et al. 1996). For neurons, DRG neurons are the most extensively studied models regarding DOR regulation of Ca^{2+} channels since multiple measurements (in situ hybridization, single-cell PCR, and immunostaining) indicated that DORs are widely expressed not only in large DRG neurons but in small ones. They coexist with MORs in peptidergic small DRG neurons, with protachykinin-dependent localization in large dense-core vesicles. Also, because DRG is the most important afferent neurons and plays an important role in DOR analgesia in the spinal cords (Wang et al. 2010). However, the functional coupling of DOR to VGCCs in DRG neurons has remained uncertain. Some studies failed to show the effects of DOR on VGCCs in DRG neurons (Moises et al. 1994; Walwyn et al. 2005). Others, however, indicated that DOR activation with agonists (DADLE, deltorphin II) inhibits $VGCC_S$ (including L-, N-, P-, and Q-type currents) in DRG neurons in vitro. This inhibition can be blocked by DOR antagonist (naltrindole), but not by MOR antagonists (Acosta and López 1999). Recent studies demonstrated that all DRG neurons that are immunoreactive to DOR exhibit Cav3.2 immunoreactivity, and isolectin B4 (IB4)-negative small diameter DRG neurons are more sensitive to DOR agonists (Wu et al. 2008; Wang et al. 2010). Both DPDPE and deltorphin II, as well as SNC80 produce a greater inhibition of N-, P/Q-, and L-type Ca^{2+} currents in IB4-negative than in IB4-positive neurons. This inhibition can be blocked by the DOR antagonist naltrindole (Wu et al. 2008; Wang et al. 2010). DPDPE has no significant effects on R-type Ca^{2+} currents in both cells and T-type or high voltage-activated Ca^{2+} currents in all the DRG neurons with T-type Ca^{2+} currents (Wu et al. 2008). Interestingly, Ca^{2+} currents in a single small neuron can be inhibited by a DOR agonist (SNC80 or deltorphin II) as well as by the MOR agonist DAMGO, suggesting that co-expressed DOR and MOR in small DRG neurons mediate inhibitory effects on voltage-gated Ca^{2+} channels (Wang et al. 2010). Indeed, DOR are required for full inhibitory coupling of MOR to VGCCs in DRG neurons (Walwyn et al. 2009). Increases in coupling of DOR to VGCCs in certain condition (e.g., chronic inflammatory pain) can enhance the DOR regulation of VGCCs in DRG neurons, and thus enhance the analgesic effect of DOR in the spinal cord (Walwyn et al. 2005; Pradhan et al. 2013). In acutely dissociated neostriatal neurons, DOR activation can decrease postsynaptic whole-cell Ca^{2+} currents, and thus modulate the excitability of neostriatal neurons (Stefani et al. 1994). In contrast to DOR activation-induced elevation in $[Ca^{2+}]i$ in cultured cortical astroglial cells, cultured cortical neurons show a decrease in $[Ca^{2+}]$ i by inhibition of L-type Ca^{2+} channels with DOR activation in the same culture (Thorlin et al. 1998a). Overall, DOR activation can regulate the activities of VGCCs.

In addition to the above channels, some other channels have been implied to be involved in DOR function. For example, ***Ca^{2+}-activated Cl^- channels*** are found to be involved in DOR, but neither MOR nor KOR, antinociceptive action in rodent spinal cord, as well as in peripheral terminals (Yamazaki et al. 2000; Pacheco et al. 2012).

Ca^{2+}-activated K^+channels are involved in DOR regulation of K^+ homeostasis in the cortex during hypoxia (Chao et al. 2007a). ***Mitochondrial K_{ATP} channels*** are also linked to DOR neuroprotection against ischemia in Purkinje cells of rat cerebellar brain slices with morphine preconditioning (Lim et al. 2004), a similar action of DOR agonist SNC80 on mitochondrial K_{ATP} channels in the heart (Fischbach et al. 2003). Moreover, DOR and its subtypes have been implied to protect from myocardium injury via reducing ***gap junction channel*** permeability (Miura et al. 2007) or inhibiting the opening of ***mitochondrial permeability transition pore*** (mPTP) (Zeng et al. 2011) in the heart. However, it is unknown if such coupling of DOR to gap junction channels and mPTP exists in nerve cells.

DOR interplays not only with ion channels, but also with pumps and transporters. In a recent study, Deng et al. (2009) used a "*Xenopus* oocyte" model system to investigate the effect of heterologously expressed DOR on the endogenous ***Na^+pumps*** as well as heterologously expressed pumps. They found that the expression of DOR leads to reduction of the activity, rather than the numbers, of Na^+ pumps. Stimulation of DOR with DPDPE has no pronounced additional effect on pump activity. The authors suggested that there is a direct protein-protein interaction between DOR and Na^+ pumps since these two membrane proteins not only co-localize in oocyte model system but also in neuronal cells, and inhibition of Na^+ pumps, in turn, modulates agonist sensitivity of DOR (Deng et al. 2009). Similar protein-protein interactions between DOR and ***SERCA pumps*** were also observed recently in HEK-293i-derived cell lines expressing truncated human DOR (hDOR) and SERCA2b constructs (Tuusa et al. 2010). This study showed the tetracycline-inducible expression of the N-terminally *myc*-tagged and C-terminally *FLAG-tagged* hDOR in a ternary complex with SERCA2b and ER molecular chaperone calnexin. The interaction between SERCA2b and hDOR, which was regulated by Ca^{2+} and ATP, did not require calnexin, but hDOR was able to mediate co-immunoprecipitation of calnexin and C-terminally truncated SERCA2b (Tuusa et al. 2010). This ternary complex is important to the expression of mature hDOR; however, it is still not clear whether DOR regulates the pumping function of SERCA. In another recent study, these authors examined the influence of the polymorphism of hDOR (the cysteine variant $hDOR^{Cys27}$ and the phenylalanine variant $hDOR^{Phe27}$) on ER Ca^{2+} signaling (Tuusa and Petäjä-Repo 2011). They found that $HDOR^{Cys27}$ and $hDOR^{Phe27}$ variants have an intrinsic difference in SERCA2b association (easy dissociation of $hDOR^{Phe27}$ from the complex). PKC activation enhances the ability of SERCA2b to remove cytosolic Ca^{2+} liberated from ER and hDOR expression, probably via direct phosphorylation of the receptors, and over-expression of $hDOR^{Cys27}$ rather than $hDOR^{Phe27}$. This causes a significant suppression of ATP-induced Ca^{2+} release from the ER, which makes the rate of Ca^{2+} removal from cytosol not reliably measurable, therefore compromising normal Ca^{2+} signaling with intracellular Ca^{2+} stores (Tuusa and Petäjä-Repo 2011).

Synaptic transmissions lead to the release of neurotransmitters into synaptic cleft. Removal of synaptically released neurotransmitters such as glutamate and GABA in synaptic cleft, in order to timely terminate their action and prevent their spillover, is largely dependent on the activities of Na^+-dependent transporters.

To date, five ***glutamate transporters*** (EAAT1-5) (Maragakis and Rothstein 2004; Sheldon and Robinson 2007) and four ***GABA transporters*** (GAT1, GAT2, GAT3, and BGT1) (Borden 1996; Conti et al. 2004; Madsen et al. 2010) have been cloned from human and animal tissues. Of these, EAAT1 and EAAT2 are predominantly expressed in glial cells found in cerebellum, forebrain and hippocampus. EAAT3 and EAAT4 are typically localized in neurons of cerebellum, and EAAT5 is located mainly in retinal ganglion cells (Danbolt 2001). GAT1 and GAT3 have been found exclusively in astrocytes and neurons of the brain, whereas GAT2 and BGT1 have been found in multiple organs including the brain (Borden 1996; Conti et al. 2004; Madsen et al. 2010). Using a "*Xenopus* oocyte" model system, Xia et al. (2006) reported that DOR co-expressed with EAAT1 down-regulates EAAT1 function, which is manifested as the reduced rate of glutamate uptake and decreased EAAT1-mediated transport currents. This down-regulation does not occur when EAAT1 is coupled with MOR, nor with Na^+-K^+ pumps. Also, DPDPE stimulation of DOR can counteract the downregulation of EAAT1-mediated glutamate uptake. The investigators further demonstrated that DOR and EAAT1 are co-localized, not only in oocytes, but also in rat hippocampal neurons, and thus concluded that DOR can reduce EAAT1 function by direct protein-protein interaction, and that DOR activation releases the inhibitory interaction (Xia et al. 2006). In rat glioma cell lines (C6 cells), as well as in the prefrontal cortex of rats with morphine-induced conditioned place preference reinstatement, EAAT3 protein levels decrease significantly with chronic morphine exposure. DOR agonist DPDPE significantly increases EAAT3 expression in C6 cells, and even reverses decreased EAAT3 expression by chronic morphine exposure. This effect can be blocked by non-specific opioid receptor antagonist naloxone, but not by DOR inhibitor naltrindole, which only weakens conditioned place preference reinstatement (Wu et al. 2013). The authors thus concluded that DOR can affect the expression of EAATs but cannot mediate morphine-induced downregulation of EAAT3 in C6 cells and in the prefrontal cortex of rats (Wu et al. 2013). Reported that treatment of cultured astrocytes in cerebral cortex with DPDPE (10 μM) decreases EAAT2 mRNA expression and glutamate uptake. This finding is not consistent with our recent observation (Liang et al. 2014), in which we demonstrated that DOR activation, with a more potent and specific agonist UFP 512, induces astrocytic expression of EAAT1 and EAAT2, but not EAAT3, an effect that is completely reversed by DOR antagonist or siRNA knock-down. Furthermore, we observed increased glutamate uptake in the astrocytes treated with UFP 512, an effect completely blocked by an EAAT inhibitor or DOR knock-down. Our observations strongly suggest that DOR activation up-regulates astrocytic EAAT expression and function through a MEK-ERK-p38 signaling (Liang et al. 2014). There are several reasons for this discrepancy. The specificity of DPDPE is much lower compared to that of UFP 512. Since the EC_{50} of DPDPE for DOR is only 5.2 nM (Chandrakumar et al. 1992), 10 μM may also activate μ-opioid receptor (Han 2003) or other receptors, thus inducing a complex effect, or even an opposing outcome (Zhang et al. 2000, 2002; Chao et al. 2007a). Also, it is unclear as to whether EAAT2 protein levels changed in the study of. Furthermore, they observed glutamate uptake after 48 h of DPDPE treatment. Our study shows that after DOR activation, astrocytic glutamate transporter production reached a peak at 24 h and dramatically decreased by 48 h (Liang et al. 2014). This might be due to several reasons, such as desensitization of DOR and loss of activity of the added agonist. Therefore, comparing their data to ours may not be a fair comparison on many different levels. Moreover, we have used multiple approaches, including DOR antagonist and siRNA, to verify our results (Liang et al. 2014). Taking these observations together, it is clear that DOR can regulate EAAT activity. In addition to EAATs, some studies suggested a role of DOR in regulating the activities of GABA transporters (Bagley et al. 2005; Pu et al. 2012). In morphine-treated mouse PAG, the opioid agonist Met-enkephalin induces a non-reversing current in neurons, which is poorly fitted by the GIRK equation, but is abolished by inhibition of GAT1 (Bagley et al. 2005). This suggests a possible interaction between opioid receptor and GAT1. Chronic morphine exposure reduces GABA uptake in rat hippocampal synaptosomes, further indicating that GABA transport might be regulated by opioid receptors (Pu et al. 2012). Direct measurement of GAT1-mediated currents and GABA uptake in GAT1 and DOR, or MOR co-expressed *Xenopus* occytes, demonstrated that co-expression of DOR leads to a reduced number of fully functional GAT1, reduced substrate translocation as well as GAT1-mediated currents. Activation of DOR further reduces the rate of GABA uptake and GAT1-mediated currents, and these effects do not appear when there is co-expression of GAT1 and MOR and MOR activation (Pu et al. 2012). These investigators further confirmed the inhibition of GAT1-mediated currents by DOR activation in PAG with whole-cell patch-clamp experiments (Pu et al. 2012). These findings clearly indicate a regulation of GAT1 by DOR.

DOR also can interact with ***monoamine transporters*** (Tsao et al. 1998, 1999b; Svingos et al. 1999; Le Moine et al. 2002; Jutkiewicz et al. 2008). For example, unlike MOR and KOR mRNA levels that are not modified, the levels of DOR mRNA are increased significantly in caudate putamen and nucleus accumbens of the adult ***dopamine transporter*** (DAT) knockout mouse brain (Le Moine et al. 2002), indicating regulation of DOR expression by DAT. DOR also regulates DAT activity. Though a lack of all three opioid receptors (triple DOR-, MOR-, and KOR knockout) does not influence DAT expression in the mouse brain (Yoo et al. 2010), the ultrastructural evidence showed that in the nucleus accumbens shell, a place where opiate- and psychostimulant-induced modulation of dopamine transmission is thought to play a key role in potent addiction, desire, reward and reinforcement, and locomotor effects (Katsuura and Taha 2010; Laurent et al. 2012, 2014; Castro and Berridge 2014), DOR are present in axons and dendritic spines of nondopaminergic neurons that are contacted by DAT-containing terminals (Svingos et al. 1999), and δ-opioid peptide DADLE can reverse methamphetamine-induced long-term DAT loss in nucleus accumbens as well as in the striatum, an effect that can be blocked by opioid receptor antagonist naltrexone (Tsao et al. 1998, 1999b). Pharmacological evidence indicated that similar to other DOR agonists [(+) BW373U86, oxymorphindole], SNC80 produces robust, dose-dependent increases in locomotor activity mediated through DOR. SNC80 also enhances the stimulatory effects of dopamine/norepinephrine transporter ligands (cocaine, nomifensine), but not the direct dopamine receptor agonists (SKF81297, quinpirole), suggesting that DOR activation might alter the functional activity of monoamine transporters (Jutkiewicz et al. 2008).

As for the interactions of DOR with receptors, its interactions with ***NMDARs*** have been well documented (Dourmap and Costentin 1994; Bhargava and Zhao 1996; Bhargava et al. 1997; Cao et al. 1997; Cai et al. 1997; Fan et al. 1998; Suzuki et al. 2000; Zhao and Joo 2008; Pamenter and Buck 2008; Tanahashi et al. 2012). On one hand, activation of DOR and its subtypes upregulates the numbers of brain NMDARs (Bhargava et al. 1997) and enhances NMDAR currents of cultured rat dorsal horn neurons (Zhao and Joo 2008). It can also prevent the release of glutamate from presynaptic vesicles and depress the amplitudes of stimulus-evoked excitatory postsynaptic potentials/currents (EPSPs/EPSCs) of rat neocortical neurons (Tanaka and North 1994; Ostermeier et al. 2000) as well as NMDAR activities in rat trigeminothalamic neurons (Wang and Mokha 1996). Inhibition of DOR with naloxone or naltrindole potentiates glutamate release or NMDAR currents in cortical neurons (Vlaskovska et al. 1997; Pamenter and Buck 2008); on the other hand, incubation in glutamate upregulates the abundance of DOR mRNA in culture cortical astrocytes in a dose dependent way (Thorlin et al. 1997). Activation of NMDARs attenuates DOR-G protein coupling and DOR-mediated inhibition of adenylyl cyclase in both NG108-15 cells that functionally express both DOR and NMDAR as well as Giα2 proteins and cultured brain neurons (Cai et al. 1997; Fan et al. 1998). In addition, neurochemical evidence showed the DOR subtype DOR1 agonist-induced downregulation of brain DOR1, and DOR agonist DPDPE-induced dopamine release in the striatum can be blocked by NMDAR but not AMPAR or KAR antagonists (Dourmap and Costentin 1994; Cao et al. 1997). In median prefrontal cortex in vivo, DOR1 activation-induced GABA release is mediated by activation of NMDAR because it is inhibited by NMDAR antagonist MK801. MK801 does not affect DOR1 agonist-inhibited GABA release in the nucleus accumbens of rats, but dopamine D2 receptor antagonist sulpiride does block it (Tanahashi et al. 2012). Behavioral evidence also demonstrated that superspinal analgesic action of DOR and its subtypes is enhanced by NMDA but antagonized by NMDAR antagonists (Bhargava and Zhao 1996; Suzuki et al. 2000). All these clearly support the interaction of DOR and NMDAR. In addition, superspinal analgesic action of DOR also involves brain ***group I metabotropic glutamate receptors*** (Suzuki et al. 2000) and spinal ***$GABA_A$receptors*** (Rady and Fujimoto 1996). In isolated bullfrog DRG neurons, enkephalin and other opioid agonists, e.g., DOR agonist DPDPE, depress $GABA_A$ receptor-mediated currents in a concentration-dependent manner, and this effect can be blocked by non-selective opioid receptor antagonist naloxone and DOR antagonist naltrindole, suggesting DOR regulation of the activity of $GABA_A$ receptors in DRG (Xi and Akasu 1997). Also DPDPE inhibits glutamate-evoked endogenous acetylcholine release from rat neostriatal slices, and this effect can be completely abolished by either $GABA_A$ or $GABA_B$ receptor antagonist (Arenas et al. 1990). In addition to regulating the function, DOR also regulates $GABA_A$ receptor expression. In our recently developed tetracycline-inducible DOR transigenic mice, we found that $GABA_A$ receptor α1-subunit (the most abundant α-sununit of $GABA_A$ receptors in the adult brain) is lower in the cortex of DOR transigenic mice than in WT, suggesting that DOR over-expression induces an inhibitory effect on $GABA_A$ receptor expression (Feng et al. 2011). A recent study also indicated the regulation of ***dopamine D1 receptors*** by DOR in mouse medial prefrontal cortex (Olianas et al. 2012). In this study, activation of either DOR or MOR, but not KOR, potentiates dopamine (DA) D1-like receptor-induced stimulation of adenylyl cyclase activity through a G protein βγ subunit-dependent mechanism. Double immunofluorescence analysis indicates that a large percentage of dopamine D1 receptor positive cells express either DOR or MOR immunoreactivity, indicating in mouse medial prefrontal cortex that activation of DOR (and MOR) enhances dopamine D1-like receptor signaling (Olianas et al. 2012). Also, intra-accumbens infusion DOR agonist (D-Ala2)deltorphin II evokes dose-dependent motor stimulation (characterized by locomotion, sniffing, and oral stereotypies) and an increase of extracellular dopamine concentrations; these effects cannot be evoked from the caudate, nor by MOR agonist PL017 in the accumbens. Instead they are reduced by high doses of naloxone (1.0 mg/kg) and by DOR antagonist naltrindole; blockade of dopamine D1 receptors by SCH 23390 abolished (D-Ala2)deltorphin II effects, while blockade of dopamine D2 receptors by raclopride or by haloperidol had no effect, suggesting these effects of DOR activation are mediated by dopamine D1 receptors (Longoni et al. 1991). In in vivo rat nucleus accumbens, DOR1 inhibition of GABA release is largely abolished by dopamine D2 receptor antagonist sulpiride, suggesting a role of ***dopamine D2 receptors*** in mediating the stimulating effect of DOR in this region (Tanahashi et al. 2012). DOR also interacts with ***somatostatin receptors*** in the brain. For example, the inhibition of acetylcholine release from human neocortex by DPDPE (that can be blocked by DOR1 antagonist) is prevented by activation of type 2 somatostatin receptors (Feuerstein et al. 1998). In addition, there also exist interactions between DOR and ***other opioid receptors***. For example, as we indicated earlier, DOR is required for full inhibitory coupling of MOR to VGCCs in DRG neurons (Walwyn et al. 2009). In opioid-induced DA release in nucleus accumbens, stimulation of DOR1 or DOR2 subtype inhibits MOR that is involved in the slow onset increase in extracellular DA, while stimulation of DOR1 but not DOR2 activates MOR that is involved in the rapid increase in extracellular DA (Hirose et al. 2005). In contrast, in PAG neurons, induction of functional DOR signaling and DOR-mediated presynaptic inhibition of GABAergic synaptic currents requires MOR expression and chronic activation because in MOR knockout mice, no DOR responses are observed, and in WT mice, DOR effects cannot appear until several days of chronic stimulation of MOR with morphine treatment (Hack et al. 2005).

5.1.2 DOR Regulation of Neurotransmitter Release

DOR regulation of neurotransmitter release is an important aspect for DOR regulation of neuronal activity. Indeed, DOR can influence the release of a large amount of neurotransmitters as well as neuromodulators. It has been demonstrated that activation of DOR inhibits, while blockade of DOR potentiates, ***acetylcholine*** release in various brain regions such as rodent striatum (Arenas et al. 1990; Sandor et al. 1992), septal nuclei (Gazyakan et al. 2000), and human neocortex (Feuerstein et al. 1998). In contrast, DOR activation in the hypothalamic paraventricular nucleus does not influence the release of acetylcholine (vs. the inhibition of DA) from the functionally linked site, the nucleus accubens (Rada et al. 2010). DOR regulation of ***glutamate*** release is also observed in the CNS. In rat cerebral cortical primary cultures, naloxone increases glutamate and aspartate release, suggesting endogenous opioids partially inhibit glutamate and aspartate release (Vlaskovska et al. 1997). Electrophysiological study indicated that inhibition of DOR with naltrindole potentiates NMDAR currents in cortical neurons of turtle (Pamenter and Buck 2008), also suggesting DOR inhibits glutamate release from cortical neurons. This is also supported by other electrophysiological studies in which DOR activation prevents the release of glutamate from presynaptic vesicles and depresses the amplitudes of stimulus-evoked EPSPs/EPSCs of rat neocortical neurons (Tanaka and North 1994; Ostermeier et al. 2000). Direct measurement of glutamate release in rat spinal dorsal horn indicated that DOR agonist DPDPE (1–10 μM) does inhibit capsaicin-evoked release in a concentration-dependent manner, which can be antagonized by DOR antagonist naltrindole (1 μM) (Ueda et al. 1995). However, Rawls and McGinty (2000) found that in rat striatum, naltrindole (10–100 μM) decreases amphetamine-evoked glutamate release, which can be reversed by DOR agonist DPDPE (100, 500 μM). That means DOR activation enhances glutamate release in the striatum. A recent study also indicated that intraperitoneal administration of a novel DOR agonist KNT-127 increases the release of glutamate as well as DA in the striatum, nucleus accumbens, and median prefrontal cortex. This effect is abrogated by DOR1 antagonist 7-benzylidenenaltrexone, but not by DOR2 antagonist naltriben (Tanahashi et al. 2012). This enhancement of glutamate release in the striatum might be related to the maintenance of functional glutamatergic neurotransmission by glial cells since selective lesion of glial cells with gliotoxin prevents the increase in extracellular glutamate levels induced by DPDPE (Billet et al. 2007). However, it is also important to note that the study by Rawls and McGinty (2000) used an extremely high concentration of ligands, which may stimulate a variety of membrane proteins and induce a complicated effect on cells, and can even generate adverse effects in neurons (Ma et al. 2005; Witkowski and Szulczyk 2006; Chao et al. 2007b). Intracellular recording and whole-cell patch clamp studies demonstrated that DOR agonists inhibit ***GABA*** release in the brain (Cohen et al. 1992; Sugita and North 1993; Standford and Cooper 1999; Hack et al. 2005; Kang-Park et al. 2007; Margolis et al. 2008; Mitchell et al. 2014). For example, in amygdala slices, synaptic potentials mediated by GABA acting at $GABA_A$ receptors are inhibited up to 60 % by DPDPE, which can be blocked by DOR antagonist, but not MOR antagonist (Sugita and North 1993). DOR knockout enhances acute ethanol-induced $GABA_A$-mediated inhibitory postsynaptic currents (IPSCs), which can be mimicked by DOR inverse agonist (ICI174864) but inhibited by DOR agonist DPDPE in WT (Kang-Park et al. 2007). DOR action inhibits GABA release in the nucleus of the amygdala. In rat hippocampus, DOR agonist DADLE exerts an unknown inhibitory effect on the synaptic terminals of interneurons that terminate onto CA1 pyramidal neurons, resulting in a decrease in spontaneous action potential-independent GABA release, as evidenced by DADLE's ability to reduce miniature IPSC frequency without affecting amplitude. DOR also reduces the amplitude of action-potential-evoked, GABA-mediated IPSCs and decreases their frequencies (Cohen et al. 1992). In the ventral tegmental area, DOR activation decreases ethanol consumption, and this action depends upon DOR inhibition of GABA release in this area (Margolis et al. 2008; Mitchell et al. 2014). Electrophysiological studies also indicated that in rat globus pallidus, presynaptic DORs are preferentially located on terminals of quiescent globus pallidus cells. DOR activation serves to reduce GABA release in these cells to maintain the inhibitory function of the globus pallidus in basal ganglia circuitry (Standford and Cooper 1999). In PAG neurons, DOR stimulation also inhibits synaptic release, which requires MOR expression and chronic activation (Hack et al. 2005). These studies suggest that DOR presynaptically inhibits the release of GABA. However, DOR may also enhance GABA release. For example, though DOR1 indirectly inhibits GABA release in nucleus accumbens via activated dopaminergic transmission, it also enhances GABA release in median prefrontal cortex via activated glutamatergic transmission (Tanahashi et al. 2012).

The DOR regulation of ***DA*** release in the brain has been getting much attention due to its particular importance in psychoses (addiction, desire, reward and reinforcement) and motor control. Most of the studies indicated that in a variety of the brain regions such as the striatum (Longoni et al. 1991; Pentney and Gratton 1991; Acquas et al. 1993; Dourmap and Costentin 1994; Billet et al. 2007; Bosse et al. 2008; Tanahashi et al. 2012), the cortex (Tanahashi et al. 2012), and nucleus accumbens (Pentney and Gratton 1991; Acquas et al. 1993; Hirose et al. 2005; Hipólito et al. 2008; Rada et al. 2010; Tanahashi et al. 2012), activation of DOR and its subtypes increases DA release. For example, intraperitoneal administration of a novel DOR agonist KNT-127 increases the release of DA in the striatum, nucleus accumbens, and median prefrontal cortex. This effect is abrogated by DOR1 antagonist 7-benzylidenenaltrexone, but not by DOR2 antagonist naltriben (Tanahashi et al. 2012). Intra-accumbens infusion with DOR agonist (D-Ala2)deltorphin II evokes dose-dependent motor stimulation (characterized by locomotion, sniffing, and oral stereotypies) and an increase of extracellular DA concentrations. These effects cannot be evoked in the caudate. They also cannot be evoked in the accumbens by MOR agonist PL017. Instead, the effects of intra-accumbens infusion with DOR agonist are reduced by high doses of naloxone (1.0 mg/kg) and by DOR antagonist naltrindole (Longoni et al. 1991). Also, stimulation of DOR1 or DOR2 subtype in nucleus accumbens promotes MOR-associated initially rapid and then slow onset increase in extracellular DA (Hirose et al. 2005). It should be note that in nucleus accumbens, DOR-induced DA release is closely related to the subregions (Hipólito et al. 2008). In the core of nucleus accumbens, DOR agonist DPDPE significantly increases DA extracellular levels, whereas in the shell, DPDPE persistently reduces DA levels. In the transition zone of nucleus accumbens, DPDPE does not significantly modify DA levels (Hipólito et al. 2008). Another study also showed a subregional difference in DA overflow with DOR activation (Britt and McGehee 2008).

In the shell of nucleus accumbens, δ-opioids, similar to μ-opioids, inhibit DA overflow elicited with single-spike stimuli (which mimic midbrain DA neurons fire during quiescent period), while leaving DA overflow produced by burst stimuli (which mimic bursting firing patterns in response to rewarding or reward-predicting stimuli) unaffected. In the core, extracellular DA levels are even less affected (Britt and McGehee 2008). In the striatum, DPDPE significantly enhances gliotoxin L-α-aminoadipate-induced DA, as well as glutamate release, and DPDPE-induced DA release seems to require a glia-related initial glutamate release, since 2 days after an intra-striatal injection of L-α-aminoadipate, which destroys glial cells, DPDPE-induced increase in extracellular glutamate and DA levels are prevented, but can be restored by local infusion of glutamate (Billet et al. 2007). Taken together, these studies indicate that DOR activation in the brain predominantly increases DA release.

In addition to the above-mentioned neurotransmitters, studies have demonstrated DOR regulation of some other neurotransmitters/neuromodulators. For example, DOR activation inhibits ***substance P*** release from rat trigeminal nucleus (Suarez-Roca and Maixner 1992) and dorsal horn (Zachariou and Goldstein 1996), ***luteinizing hormone releasing hormone*** in SK-N-SH cells (Bennett and Ratka 2003), ***β-endorphin*** secretion from rat neurointermediate pituitary (Carr and Lovering 2000), and ***prolactin*** release from GH3 cells expressing recombinant DOR and MOR (Piros et al. 2000). Also, DOR2 activation may increase, while DOR1 activation may reduce, endogenous ***cholecystokinin*** release, respectively resulting in potentiation/attenuation of μ-opioid analgesic action (Noble et al. 1994, 1996). Nanomolar DOR1 and DOR2 agonists enhance MOR-induced release of ***adenosine*** from rat spinal cord synaptosomes (Cahill et al. 1996). In DRG neurons, DOR activation results in the release of ***calcitonin gene-related peptide*** (Bao et al. 2003). Moreover, DOR antagonist, but not MOR antagonist, can prevent plant-originated hallucinogenic neoclerodane diterpenes (e.g., Salvinorin A and Hardwichiic acid) (Shirota et al. 2006) facilitated ***noradrenalin*** release in the mouse hippocampus, suggesting a role of DOR at presynaptic terminals in the regulation of noradrenalin release and its neurotransmission (Pittaluga et al. 2013).

5.1.3 DOR Regulation of Neuronal Electrical Activity and Neurotransmission

As has been discussed above, DOR has been functionally coupled with various ion channels, and has been shown to interact with receptors, transporters and pumps. DOR also regulates the release of many kinds of neurotransmitters/neuromodulators. Therefore, DOR activation inevitably influences neuronal electrical activity and synaptic transmission. Indeed, there have been numerous studies for this. Due to the diversity of the investigations, and also because of the complexity of brain structure and function as well as diversity of cell types in the CNS, we will not extend our discussion in detail. Instead, we will give the general and well-established information in this section since this information is very helpful in understanding the role of DOR in the regulation of ion homeostasis.

In most cases, DOR activation inhibits neuronal electrical activity and synaptic transmission, but sometimes it also produces excitation of neurons. For example, in DRG neurons, low concentrations (nanomolar range) of opioids prolong the duration of Ca^{2+}-dependent component of the action potential, whereas higher concentrations (micromolar range) shorten it. These dual actions are mediated by different G proteins. The excitatory response is mediated via a Gs protein-dependent mechanism, while the inhibition is Gi/Go-dependent (Shen and Crain 1989, 1990). In the cortical neurons, DOR agonist suppresses spontaneous and evoked electrical discharges (Janiri et al. 1988). Parvalbumin-expressing interneuron transmission of the hippocampus is persistently depressed by DOR activation in CA2 area, which results in a novel form of LTD at these interneurons and makes them act as gates to prevent CA3 inputs from driving CA2 neurons (Piskorowski and Chevaleyre 2013). For the inhibition, DOR, in most cases, acts on the presynaptic receptors to limit the release of neurotransmitters, either through inhibition of Ca^{2+} influx via Ca^{2+} channels, or through hyperpolarizing presynaptic membrane by activation of K^+ channels, especially GIRK channels (see earlier discussion). Activation of DOR hyperpolarizes some pyramidal neurons in the cortex, and presynaptically inhibits the release of excitatory amino acids and GABA at synapses onto pyramidal neurons as manifested by the inhibition of glutamate receptor-mediated EPSPs/EPSCs (Tanaka and North 1994; Ostermeier et al. 2000). Such inhibition of excitatory transmission was also found in other regions such as trigeminothalamic neurons (Wang and Mokha 1996), solitary tract nucleus (Cui et al. 2012), spinal cord (Bardoni et al. 2014), and parasympathetic ganglia (Kennedy and Krier 1987), while inhibition of DOR with naltrindole potentiates NMDAR currents (Pamenter and Buck 2008). In amygdala slices, $GABA_A$-mediated synaptic potentials are inhibited up to 60 % by DPDPE, which can be blocked by DOR but not MOR antagonist (Sugita and North 1993), and DOR knockout enhances acute ethanol-induced $GABA_A$-mediated IPSCs, which can be mimicked by DOR inverse agonist (ICI174864) but inhibited by DOR agonist DPDPE in WT (Kang-Park et al. 2007). In rat hippocampus, DOR agonist DADLE reduces miniature IPSC frequency without affecting amplitude and the amplitude of action-potential-evoked, GABA-mediated, IPSCs and their frequency (Cohen et al. 1992). Similar inhibition of GABAergic transmission was also observed in many other regions such as ventral tegmental area (Margolis et al. 2008; Mitchell et al. 2014), globus pallidus (Standford and Cooper 1999), PAG neurons (Hack et al. 2005), brainstem nucleus raphe magnus (Zhang and Pan 2012) and DRG neurons (Xi and Akasu 1997).

DOR also produces excitation of neuronal activity and transmission (Watson and Lanthorn 1993; Zhao and Joo 2008). For example, DOR agonist-induced increases in the amplitude of population spikes, which were antagonized by DOR antagonist naltrindole, were observed in hippocampal neurons (Watson and Lanthorn 1993). DOR activation enhances NMDAR currents of cultured rat dorsal horn neurons (Zhao and Joo 2008). Another feature of DOR-induced excitation comes from the studies of DOR regulation of intracellular Ca^{2+} concentration, in which some studies showed that DOR activation increases intracellular Ca^{2+} concentrations (see the discussion in the next section). In addition, DOR activation in some cases is proconvulsive (Snead 1986; Negus et al. 1998; Grecksch et al. 1999; Yajima et al. 2000; Danielsson et al. 2006; Jutkiewicz et al. 2005, 2006), also suggesting an excitatory role of neurons. For this phenomenon, in addition to the factors like animal species, seizure types, the methods for drug administration, and dose used, the target neurons might play a key role in such excitation (Chao and Xia 2012; Chao et al. 2013). It is important to realize that it is receptors themselves, not opioids that determine the effects of opioid action. The action of δ-opioids depends on the site where the opioids act. In the hippocampus, inhibitory interneurons express DOR (Erbs et al. 2012; Rezaï et al. 2012), and DOR activation in these interneurons indeed produces inhibitory action on their own activity (Zieglgänsberger et al. 1979; Madison and Nicoll 1988; Neumaier et al. 1988; Piskorowski and Chevaleyre 2013; Rezai et al. 2013). However, this inhibition of inhibitory interneurons may lead to post-synaptic de-inhibition, i.e. facilitation, rather than suppression of seizures, thus making neurons in the network excited (Chao and Xia 2012; Chao et al. 2013). Therefore, the net output in neurotransmission within the network becomes excitatory.

5.2 DOR Regulation of Ionic Homeostasis

Endogenous opioids are generally regarded as neurotransmitters/modulators and mainly play an inhibitory role in cellular activities with three major classes of receptors (DOR, MOR, KOR). They are widely distributed throughout the CNS. DOR is a member of the G protein-coupled receptor superfamily (Law et al. 2000), and may functionally couple with ionic channels, receptors, transporters, and pumps. Activation of DOR can influence the release of neurotransmitters, neuronal electrical activity, and synaptic transmission through the interaction with a large amount of membrane functional proteins (e.g., channels, transporters, receptors and pumps) (see previous discussion), which also play a key role in regulation of ionic homeostasis in the CNS. Therefore, DOR may have some roles in ionic homeostasis. Indeed, a study, by using atomic absorption spectrophotometry, reported that Na^{+}, K^{+}, Ca^{2+}, Mg^{2+} and Mn^{2+} levels are lower in the medulla and spinal cord than in the cortices, the striatum, and the hippocampus, while the highest metal ion levels are detected in the olfactory bulb (Gulya et al. 1991a). Intracerebroventricular administration of DOR agonist DPDPE dose-dependently decreases endogenous Ca^{2+} and Mg^{2+} contents in the parietal cortex, with no changes in the levels of monovalent cations or Mn^{2+} in the hippocampus and the striatum, and these changes can be antagonized by naloxone (Gulya et al. 1991a). The investigators thus concluded that DOR may specifically be involved in the regulation of endogenous ion levels and their movements in the CNS (Gulya et al. 1991a). These investigators also found a similar decrease in the endogenous aluminum and zinc contents with DPDPE in the cortex, the hippocampus, and the striatum (Gulya et al. 1991b, 1995). Except for these cases, almost all other previous studies regarding the regulation of ionic homeostasis by DOR under normoxic condition focused on the activity of intracellular Ca^{2+}. Few studies highlighted the regulation of K^+ and Na^+ homeostasis despite the functional coupling of DOR with K^+ channels and Na^+ channels (North et al. 1987; Ikeda et al. 1995; Chattopadhyay et al. 2008; Kang et al. 2009). The imbalance of the studies on the DOR-mediated regulation between Ca^{2+} and K^+ or Na^+ is understandable because Ca^{2+} has been recognized long ago as a ubiquitous signaling molecule with complex and diverse functions, and thus has had many highlights at its function research (Petersen et al. 2005). In fact, the recognition of its importance can be traced back to 1880s when Sydney Ringer found that Ca^{2+} is crucially important for survival of fish, for the contraction of the heart, and skeletal muscle, for fertilization of eggs and development of the tadpole.

Ca^{2+} ions play a particular role in the regulation of neurotransmitter release (Miller 1990), and numerous neurochemical studies during the 1980s–1990s had indicated that DOR robustly regulates the release of a large amount of neurotransmitters/modulators (see previous discussion). Under such circumstances, it is natural to explore the role of DOR in the regulation of Ca^{2+} homeostasis in nerve cells, which composes a key step to uncover the mechanisms of DOR regulation of neurotransmitter release and DOR functions in the CNS. That's why a large amount of studies have emerged on DOR regulation of Ca^{2+} homeostasis during the 1990s–2000s. In contrast, K^+ and Na^+ ions used to be regarded as important elements, mainly in the maintenance of membrane potential and shaping action potentials, and in controlling neuronal excitability. However, studies in recent years began to realize that Na^+ and K^+, especially Na^+ ions, can act as key signaling molecules that couple membrane signaling, metabolic pathway, and neuronal activity and play a critical role in many aspects of neuronal functions (Walz 1992, 2000; Deitmer and Rose 2010; Kirischuk et al. 2012; Parpura and Verkhratsky 2012; Rose and Karus 2013). Therefore, for better understanding of DOR function, re-investigation on DOR regulation of homeostasis of K^+ and Na^+ with more sophisticated techniques is essential and necessary. Table 5.4 summarizes the studies of the effects of DOR agonists/antagonists on ionic homeostasis under normal condition.

5.2.1 The Role of DOR in Regulation of K^+ and Na^+ Homeostasis under Normal Condition

Though DOR functionally interacts with various channels, receptors, transporters and pumps, and extensively influences neuronal functions, little is known about the role of DOR in regulation of K^+ and Na^+ homeostasis under normal conditions. Recently, we assessed the effect of DOR activation/inhibition on K^+ and Na^+ homeostasis in mouse cortical slices by directly measuring the extracellular K^+ and Na^+ activities with ion-selective microelectrodes (Chao et al. 2007a, b, 2008, 2009, 2012a, b; Kang et al. 2009). Unfortunately, we did not find obvious changes in $[K^+]$ e and $[Na^+]$o in the cortical slices during 20 min of perfusion of DOR agonists (DADLE, UFP 512) and DOR antagonist (naltrindole) before hypoxia/ischemia. MOR agonist (DAGO) also did not change extracellular K^+ concentration under

Table 5.4 Effects of DOR activation/inhibition on ionic homeostasis under physiological condition

Preparation	Observation	References
Ca^{2+}		
Elevation in $[Ca^{2+}]i$		
Cultured neuronal cell line	Brief exposure (30–90 s) to DOR agonists in nanomolar concentrations induces Ca^{2+} release from intracellular stores via a process that appears to activate phospholipase C (PLC) and consequent IP_3 formation, which mainly occur in undifferentiated cells	Yoon et al. (1998, 1999), Smart and Lambert (1996), Jin et al. (1992), Spencer et al. (1997), Harrison et al. (1999), and Connor et al. (1997)
	DOR activation-induced $[Ca^{2+}]i$ elevation is dependent on the presence of extracellular Ca^{2+} in differentiated cells line	Jin et al. (1992) and Fields et al. (1995)
Cultured cortical astrocytes of the rat	Exposure to DOR agonist causes significant increase in $[Ca^{2+}]i$ or induces a biphasic Ca^{2+} response, with an initial peak of rapid increase in $[Ca^{2+}]i$ followed by a sustained level of elevated $[Ca^{2+}]i$ or oscillations in $[Ca^{2+}]$ i, and the first peak in $[Ca^{2+}]i$ is due to mobilization of intracellular Ca^{2+} store, whereas the second part of response is a result of Ca^{2+} influx through membrane channels	Hauser et al. (1998) and Thorlin et al. (1998a)
Cultured dorsal root ganglion neurons of the mouse	Activation of three opioid receptor subtypes (DOR, MOR, KOR)-induced $[Ca^{2+}]i$ rise can be greatly attenuated by removal of extracellular Ca^{2+} or blockade of dihydropyridine-sensitive Ca^{2+} channels	Tang et al. (1996) and Bao et al. (2003)
Inhibition of Ca^{2+} entry/$[Ca^{2+}]i$		
Primary cerebral cortical cultures of the rat	Naloxone treatment increases glutamate release and the $^{45}Ca^{2+}$ uptake induced by depolarization; opioid agonists (DOR, MOR, KOR) inhibit the $^{45}Ca^{2+}$ uptake induced by K^+ depolarization	Vlaskovska et al. (1997) and Thorlin et al. 1998a
	Stimulating neurons with DOR agonist DPDPE(1 μM) decreases $[Ca^{2+}]i$ by inhibition of L-type Ca^{2+} channels with DOR activation, which can be blocked by DOR antagonist, ICI 174868 (10 μM).	
In vitro neurons of turtle cortex	DOR inhibition with naltrindole (1–10 μM) greatly potentiates NMDA receptor activities and NMDA-mediated increase in $[Ca^{2+}]$ in neurons	Pamenter and Buck 2008

(continued)

Table 5.4 (continued)

Preparation	Observation	References
Neurons/neuroblastoma cell lines	DOR activation inhibits diverse calcium currents (e.g. L-, N-, P- and Q-type Ca^{2+} channel currents) which link to Ca^{2+} influx through different high-voltage-activated channels types	Nah et al. (1997), Toselli et al. (1997, 1999), Adams and Trequattrini (1998), Morikawa et al. (1998, 1999), Acosta and López (1999), and Piros et al. (2000)
Na^+		
Acutely isolated dentate granule cells of the rat hippocampus	DOR agonist, SNC80 (1–1000 μM) reduces the maximal Na^+ current amplitude in a dose-dependent manner, which is "opioid-receptor independent"	Remy et al. (2004)
Mouse cortical slices	During 20 min of perfusion of DOR agonist, UFP 512 (1–10 μM) before hypoxia/ischemia, no appreciable changes in $[Na^+]o$ were observed	Chao et al. (2012a, b)
Xenopus oocytes with co-expressed DOR and Na^+ channels	DOR activation inhibits Na^+ channel function by decreasing the amplitude of Na^+ currents and increasing its threshold of activation	Kang et al. (2009)
K^+		
Mouse cortical slices	During 20 min of perfusion of DOR agonists (DADLE, UFP 512) and antagonist (Naltroindole) before hypoxia/ischemia, no appreciable changes in $[K^+]e$ were observed, suggesting little effect of DOR on K^+ homeostasis under normoxic condition	Chao et al. (2007a, b, 2008, 2009)

normal conditions (Chao et al. 2007a). Our results suggest that there is little effect of DOR and MOR on basic K^+ and Na^+ homeostatic properties (vs. the activities of K^+ and Na^+ channels) during resting unstimulated and normoxic conditions (vs. hypoxia/ischemia, see next chapter by Chao and Xia) (Chao et al. 2007a, b, 2008, 2009, 2012a, b; Kang et al. 2009), though DOR has been shown to regulate the activities of K^+ and Na^+ channels. However, we cannot completely exclude the potential roles of DOR in extracellular K^+ and Na^+ activities under normoxic condition with stimulation-induced neuronal activity. Since DOR activation in most cases is inhibitory on neuronal activity, to show its effects on extracellular K^+ and Na^+ activity, pre-manipulation (e.g., pre-stimulation) or pre-excitatory input might be required to evoke activity-dependent extracellular K^+ rise/Na^+ drop before DOR activation. In support of this is the fact that DOR agonist itself, in some cases, has no effects on spontaneous activities (e.g., transmitter release, alterations in $[Ca^{2+}]i$ and locomotor), but largely influences evoked activities (Pentney and Gratton 1991; Connor and Henderson 1996; Bosse et al. 2008; Jutkiewicz et al. 2008). Also, DOR activation does attenuate hypoxia/ischemia-induced K^+ and Na^+ derangement in the cortex (Chao et al. 2007a, b, 2008, 2009, 2012a, b; Kang et al. 2009). Moreover, electrophysiological studies indeed demonstrated that activation of DOR presynaptically inhibits the release of excitatory amino acids and GABA at synapses, inhibits glutamate receptor-mediated EPSPs/EPSCs and neuronal discharges (Kennedy and Krier 1987; Tanaka and North 1994; Wang and Mokha 1996; Ostermeier et al. 2000; Cui et al. 2012; Bardoni et al. 2014), which to a certain degree, relieve the burden of Na^+-dependent transporters (e.g., EAATs and GATs) and the subsequent Na^+ loading in cells, especially astrocytes (see Sect. 4). In addition to Na^+-dependent transporters, excitation of neurons and an excitatory synaptic transmission and transmitter release also accompany the changes in the activities of channels, receptors, and pumps as well as the corresponding alterations of ion homeostasis (see Sect. 4). Therefore, DOR possibly influences ion homeostasis in such a condition regarding its interaction with these membrane proteins. Indeed, activation of presynaptic DOR by enkephalin prevents the increase in neuronal Nav1.7 in the DRG (Chattopadhyay et al. 2008). SNC80, a putative DOR agonist, reduces the maximal Na^+ current amplitude in a dose-dependent manner and selectively prolongs the time course of recovery from slow inactivation, but does not affect fast inactivation processes in acutely isolated hippocampal neurons (Remy et al. 2004). DOR expression and activation reduce the amplitude of the Nav1.2-mediated inward currents and rightly shift the activation curve of the currents in a dose-dependent manner in *Xenopus* oocytes with co-expressed DOR and Na^+ channels (Kang et al. 2009). These findings suggest a potential influence of DOR in intracellular Na^+ activity. SBFI imaging in guinea-pig adrenal chromaffin cells showed that activation of nAChRs induces an increase in $[Na^+]i$ (Ito et al. 1997), while activation of DOR inhibits, and blockade of DOR potentiates, acetylcholine release in various brain regions (Arenas et al. 1990; Sandor et al. 1992; Feuerstein et al. 1998; Gazyakan et al. 2000). Therefore, it is possible that DOR might inhibit Na^+ entry via nAChRs and Na^+ loading in the brain. In addition, in comparison to previous works that record the transient (several seconds) and relative small (several mM) changes of extracellular K^+ and Na^+ (see Sect. 4) during normal neuronal activity with ISEs, our recording lasted at least 60–75 min and sometimes up to 90 min, and mainly focused on very large changes in extracellular K^+ and Na^+ (>25 mM) (Chao et al. 2007a, b, 2008, 2009, 2012a, b; Kang et al. 2009). To identify relatively small $[K^+]$ e changes, special amplifiers are required with an input resistance $>10^{14}\ \Omega$ to reduce noise and increase S/N ratio, but it is not necessary for the relatively large K^+ signal as shown in our studies (Chao et al. 2007a, b, 2008, 2009, 2012a, b; Kang et al. 2009) in which a common intracellular electrometer with relatively low input resistance ($10^{11}\ \Omega$) (Model IE-210, Warner Instrument Co., Hamden, CT) works well enough with a much improved valinomycin-based K^+ selective microelectrode that is 200 times higher in selectivity to K^+ than to Na^+. This was two orders of amplitude lower in resistance and 30 times shorter in response time than that of earlier used K^+ selective microelectrodes (Ammann et al. 1987b). Therefore, it is also possible that the potential effects of DOR on extracellular K^+ and Na^+ (that would be transient and very small under normoxia), if not impossible, might be masked in noise with our experiment system and a very long-lasting recording. Furthermore,

DOR is expressed not only in neurons but in glial cells as well (Maderspach and Solomonia 1988; Thorlin et al. 1998a; Erbs et al. 2012; Rezaï et al. 2012; Liang et al. 2014). Under normal conditions, K^+ fluxes out of, and Na^+ enters into, neurons with neuronal activity. Besides the reuptake of K^+ and expelling of Na^+ by neurons via the activity of Na^+/K^+ ATPase due to the stimulation of increased $[Na^+]$i and $[K^+]$e, a large amount of K^+ and Na^+ is efficiently taken up by glial cells, particularly astrocytes, through multiple mechanisms as discussed earlier. DOR activation might also enhance the ability of K^+ uptake and Na^+ extrusion as well as inhibition of K^+ efflux and Na^+ influx, so the transient and small changes are not emerged due to the balance of efflux and uptake. These changes, if they really exist, can be easily obtained with techniques that require very high resolution and rapid response time (e.g., confocal- and two-photo microscopy with ion-sensitive fluorescence imaging) other than extracellular ion selective microelectrodes, and also require the identification of individual single cells (neuron, astrocyte) in the regions of interest to determine their individual changes in ionic activity. Investigation of these possibilities is interesting and essential in understanding the role of DOR in physiological regulation of ionic activity.

5.2.2 DOR Regulation of Ca^{2+} Homeostasis

Substantial evidence indicates that δ-opioids can affect Ca^{2+} homeostasis. Overall, the predominant effect of opioids on Ca^{2+} entry is inhibitory, which seems to match with classical recognition of opioid system as an inhibitory regulator in the brain. However, some studies show either mobilization of intracellular Ca^{2+} or stimulation of Ca^{2+} entry with opioid activation. These effects have been demonstrated to occur through activation of different kind of G proteins, including Gi, Go, and even Gs classes.

δ-Opioid receptor induced elevation of intracellular Ca^{2+}. Most of the results about δ-opioid-induced elevation of intracellular Ca^{2+} come from studies in neuron-like cell lines (e.g., NG108-15, SK-N-SH, NMB, SH-SY5Y, Neuro2a cell lines) (Jin et al. 1992, 1994; Tang et al. 1994; Fields et al. 1995; Allouche et al. 1996; Connor and Henderson 1996; Smart and Lambert 1996; Spencer et al. 1997; Connor et al. 1997; Harrison et al. 1999; Yoon et al. 1998, 1999; Chen et al. 2000). Most of these cell lines, including human neuroblastoma SH-SY5Y cells (Kazmi and Mishra 1986), SK-N-BE cells (Polastron et al. 1994), SK-N-SH cells (Yu et al. 1986), NMB cells (Baumhaker et al. 1993), mouse neuroblastoma×rat glioma NG 108-15 hybrid cells (Law et al. 1983), and neuroblastoma x DRG neuron hybrid ND8-47 cells (Tang et al. 1994) endogenously and uniquely express DOR, and represent a useful tool to study the molecular functions of opioid receptor. Some other cell lines such as GH_3 cells (Piros et al. 2000), murine neuroblastoma Neuro2a cells (Spencer et al. 1997), and CHO cells (Harrison et al. 1999), that seem to lack the endogenous DOR but express recombinant DOR as well as other opioid receptors, have also been used to study DOR effects on Ca^{2+} homeostasis. Though these cells may not be identical to native brain cells, the data yielded from them provide initial clues for understanding the roles of DOR in regulation of [Ca^{2+}]i and the relevant physiological implications (e.g., regulation of transmitter release) in early research.

Some studies indicated that DOR agonists themselves, in low nanomolar concentration (1–100 nM), are enough to induce significant increases in [Ca^{2+}]i in a concentration-dependent manner in neuron-like cell lines (Jin et al. 1992, 1994; Tang et al. 1994; Smart and Lambert 1996; Spencer et al. 1997; Yoon et al. 1999; Harrison et al. 1999), and the maximal effect is obtained at about 100 nM of DOR agonists. Increasing concentrations only induce sub-maximal elevation of [Ca^{2+}]i (Jin et al. 1992, 1994; Smart and Lambert 1996). In contrast, some other studies found that opioid agonists, even in micromolar range, cannot evelate [Ca^{2+}]i when applied alone, but require the continued presence of Gq-coupled receptor agonists such as carbacol (cholinergic M receptor agonist) and bradykinin for the induction of [Ca^{2+}]i elevation in cell line (Connor and Henderson 1996; Connor et al. 1997; Chen et al. 2000). This difference in response may be attributed to multiple factors. Indeed, it has been suggested that at least three factors determine the stimulatory opioid effect on [Ca^{2+}]i in neuron-like cell lines: synchronization of the division cycle in culture, differentiation of the cells, and synergism with other stimulatory agents (Sarne and Gafni 1996). Since these immortal cell lines show high diversity in the position of their division cycles, even if δ-opioid agonists are capable of inducing increases in [Ca^{2+}]i without the presence of other stimulatory agents (e.g. carbacol), they show a very low response rate (10–20 %) for δ-opioid agonist-induced increases in [Ca^{2+}]i when growing in standard CO_2-dependent media (Jin et al. 1992, 1994; Sarne and Gafni 1996). However, the synchronization of their position in the division cycle greatly increases the response rate and the amplitude of [Ca^{2+}]i rise (Sarne and Gafni 1996). In addition, cells, if differentiated or in the presence of other stimulatory agents, show high response rates and high amplitudes in δ-opioid-induced [Ca^{2+}]i elevation (Sarne and Gafni 1996). In these studies, opioid agonists are generally applied over a very short period (30–90 s), and the evoked increases in [Ca^{2+}]i are transient and peak in seconds and then rapidly or gradually return to basal level. Long-lasting presence of agonists (>90 s) does not increase the response, but makes cells desensitized (Connor et al. 1997; Yoon et al. 1998; Song and Chueh 1999). The repeated induction of [Ca^{2+}]i elevation requires at least a 5 min-interval for washout and recovery, while full recovery from the desensitization requires more time (approximately 25 min). Most of the studies indicated that opioid agonist-induced [Ca^{2+}]i elevations can be blocked by naloxone or naltrindole, and are sensitive to pertussis toxin, suggesting that they are mediated by Gi/Go protein-coupling DOR (Jin et al. 1994; Smart and Lambert 1996; Connor and Henderson 1996; Spencer et al. 1997; Harrison et al. 1999; Yoon et al. 1999; Chen et al. 2000). However, some studies also indicated that in some cases, opioid receptor-mediated [Ca^{2+}]i elevations are dependent on mechanisms other than Gi/Go protein mechanism, since the responses in these studies are not sensitive to pertussis toxin (Fields et al. 1995; Allouche et al. 1996; Chen et al. 2000). For example, in acutely isolated cultured SH-SY5Y cells, the response rate of DPDPE-induced increases in [Ca^{2+}]i is not affected by pretreatment with pertussis toxin, but great reduced by cholera toxin, suggesting a role of Gs, not Gi/Go in this process

(Chen et al. 2000). Besides neuron-like cell lines, DOR activation-induced increase in [Ca^{2+}]i has also been reported in cultured astrocytes (Hauser et al. 1998; Thorlin et al. 1997, 1998a) and DRG neurons (Tang et al. 1996; Bao et al. 2003). Similar to response rates in cell lines cultured in standard CO_2-dependent media (10–20 %) (Jin et al. 1992; Sarne and Gafni 1996; Chen et al. 2000), cultured neurons and astrocytes also show a low response rate (5–27 %) to DOR agonists (Tang et al. 1994; Thorlin et al. 1997, 1998a). However, in contrast to cell lines, astrocytes show less desensitization to DOR agonist (Thorlin et al. 1998a). For example, compared to the 30–90 s δ-opioid application-induced transient intracellular Ca^{2+} alteration (rise and then return to basal level within seconds to 1.5 min), long-lasting presence of δ-opioid agonists (>90 s) does not increase the response but makes cells desensitized, i.e., opioid-evoked intracellular Ca^{2+} elevation completely decays in the continued presence of opioids in cell lines (Connor et al. 1997; Yoon et al. 1998; Song and Chueh 1999). In astrocytes, a 20-min exposure to DPDPE induced an initial peak of rapid increase in [Ca^{2+}]i, followed by a sustained level of elevated [Ca^{2+}]i or oscillations in [Ca^{2+}]i, which persists in many cells for at least 20 min (Thorlin et al. 1998a). Indeed, it has been found that in small DRG neurons, 5 min of DOR agonist stimulation causes [Ca^{2+}]i elevation and a subsequent long-lasting plasma DOR exocytosis and insertion into the membrane and the release of calcitonin gene-related peptide (Bao et al. 2003), which seems not to result in desensitization but in enhancement of DOR response. In both NG108-15 cells and cultured astrocytes, DOR agonists induce [Ca^{2+}]i elevation in a concentration-dependent manner (Jin et al. 1992; Thorlin et al. 1998a). However, in cell lines, nanomolar concentrations of agonist obtains maximal Ca^{2+} response, and increasing agonist to the micromolar range only produces sub-maximal response (Jin et al. 1992); in astrocytes, in the tested concentration ranges (0.1 nM–100 μM), DOR agonist concentration-dependently increases [Ca^{2+}]i with only transient peaks in [Ca^{2+}]i at 1–10 nM of agonist, and rapid transient peak followed by oscillations or the sustained plateau of increased [Ca^{2+}]i at 1–100 μM of agonist (Thorlin et al. 1998a). The maximal response can be got in the highest tested concentration (100 μM) of agonist. These findings suggest a difference in response in cell lines and real nerve cells.

DOR-induced [Ca^{2+}]i rise may be caused by mobilization of intracellular Ca^{2+} stores, stimulation of Ca^{2+} entry across the membrane, or the inhibition of sequestration by Ca^{2+} pumps. An excellent review has been presented by Samways and Henderson (2006) for opioid-induced elevation of intracellular Ca^{2+} under normal condition. In some studies, it was found that neither the removal of extracellular Ca^{2+} nor blockade of Ca^{2+} channels influences δ-opioid-induced [Ca^{2+}]i elevation. This means that δ-opioid-induced increases in [Ca^{2+}]i in these studies do not require extracellular Ca^{2+}, suggesting the mobilization of intracellular Ca^{2+} store in δ-opioid-induced elevation of [Ca^{2+}]i (Jin et al. 1992, 1994; Fields et al. 1995; Connor and Henderson 1996; Smart and Lambert 1996; Spencer et al. 1997; Yoon et al. 1998; Harrison et al. 1999; Chen et al. 2000). This Ca^{2+} mobilization appears to mainly occur in undifferentiated NG108-15 cells (Jin et al. 1992, 1994; Smart and Lambert 1996; Chen et al. 2000), but also exists in DRG neurons (Bao et al. 2003). Further investigation indicated that either IP_3-sensitive (Jin et al. 1994; Smart and Lambert

1996; Song and Chueh 1999; Yoon et al. 1999; Bao et al. 2003) or ryanodine-sensitive stores (Allouche et al. 1996) are involved in intracellular Ca^{2+} mobilization by DOR. For G protein-coupled receptors (GPCRs), one of the most important signaling pathways is phosphatidylinositol pathway. In this pathway, the extracellular signal molecule binds with GPCRs and activates membrane phospholipase C (PLC), and activated PLC hydrolyzes phosphatidylinositol 4,5-bisphosphate (PIP_2) into two second messengers: IP_3 and diacylglycerol (DAG). IP_3 binds with the IP_3Rs in the membranes of the smooth ER and mitochondria to open Ca^{2+} channels (Gilman 1987). Such a signal pathway also exists for DOR-induced Ca^{2+} mobilization. In NG108-15 cells, PLC inhibitor U73122 completely blocks the DADLE-evoked $[Ca^{2+}]i$ increase, while an inactive analog, U73433 has no effect, and pretreatment of cells with pertussis toxin, but not cholera toxin, also blocks DADLE effect, suggesting Gi/Go-mediated activation of PLC in DOR-induced Ca^{2+} mobilization (Jin et al. 1994). PLC includes six isoforms including PLCβ (Kadamur and Ross 2013). PLCβ1 can be activated by Gq, but not by Gi/Go, via α subunit (Gilman 1987; Kadamur and Ross 2013), while PLCβ2 can be stimulated via βγ subunits of heterotrimeric G proteins (Camps et al. 1992; Katz et al. 1992). δ-opioid indeed liberates Gβγ in NG108-15 cells and mobilizes intracellular Ca^{2+} stores to increase $[Ca^{2+}]i$ (Yoon et al. 1999). Direct measurement of IP_3 generation with a specific radioreceptor mass assay indicated DOR agonist DPDPE causes a dose-dependent increase in IP_3 formation, as well as $[Ca^{2+}]i$ with the EC_{50} of 3.1 and 2.8 nM for IP_3 and $[Ca^{2+}]i$ rise respectively in NG108-15 cells (Smart and Lambert 1996). Obviously, DOR mobilizes IP_3-sensitive Ca^{2+} stores and increases $[Ca^{2+}]i$ via an increase in IP_3 generated by Gβγ-activated PLCβ2. Also, Allouche et al. (1996) found in SK-N-BE cells that DADLE-induced Ca^{2+} mobilization is not dependent on IP_3R-mediated Ca^{2+} release but on RYR-sensitive Ca^{2+} stores because both opioids (DADLE, etorphine) and carbocol induce a similar response of $[Ca^{2+}]i$ increase, and combination of them additively increases $[Ca^{2+}]i$, an effect independent of extracellular Ca^{2+}. However, carbacol can increase IP_3 levels while opioid has no effect on IP_3 levels. In contrast, ryanodine significantly decreases the opioid-induced elevation of $[Ca^{2+}]i$ but without effect on that of carbacol. These authors also found that opioid effects on $[Ca^{2+}]i$ are independent on G protein since pretreatment with cholera toxin and pertussis toxin has no influence on opioid-induced $[Ca^{2+}]i$ elevation, though pertussis toxin indeed inhibits the inhibitory effect of opioid on adenylyl cyclase. In any case, it is clear that DOR activation can increase $[Ca^{2+}]i$ by mobilization of intracellular Ca^{2+} store.

In contrast to the mobilization of intracellular Ca^{2+} stores, stimulation of Ca^{2+} entry also contributes to δ-opioid-induced $[Ca^{2+}]i$ elevation. For example, despite the contribution of mobilization of intracellular Ca^{2+} stores due to DOR activation-induced $[Ca^{2+}]i$ elevation in undifferentiated NG108-15 cells, DOR activation-induced $[Ca^{2+}]i$ elevation is abolished in differentiated NG108-15 cells in the absence of extracellular Ca^{2+} (Jin et al. 1992). A similar observation was reported in SK-N-SH and NMB cells with 30–90 s opiates application (Fields et al. 1995). In cultured cortical astroglial cells, DOR activation induces a biphasic Ca^{2+} response, with an initial peak of rapid increase in $[Ca^{2+}]i$ followed by a sustained level of elevated $[Ca^{2+}]i$ or oscillations in $[Ca^{2+}]i$ (Thorlin et al. 1998a). Removal of extracellular Ca^{2+} does not affect the first peak in $[Ca^{2+}]i$, but abolishes the sustained level of elevated $[Ca^{2+}]i$ or oscillations in $[Ca^{2+}]i$. This suggests that the first peak in $[Ca^{2+}]i$ is due to mobilization of intracellular Ca^{2+} store, whereas the second part of responses is a result of Ca^{2+} influx through membrane channels (Thorlin et al. 1998a). In DRG neurons as well as cell line, activation of DOR-induced $[Ca^{2+}]i$ rise can be greatly attenuated by removal of extracellular Ca^{2+} or blockade of dihydropyridine-sensitive Ca^{2+} channels with nifedipine (Tang et al. 1994, 1996; Bao et al. 2003). These results suggest a contribution of opioid-induced Ca^{2+} entry. This Ca^{2+} entry may be mediated by dihydropyridine-sensitive, voltage-gated Ca^{2+} channels (Jin et al. 1992; Tang et al. 1996). Interestingly, DOR activation is generally thought to produce inhibitory effects on Ca^{2+} channels (see earlier discussion). In neuroblastoma cell lines (NG108-15, SH-SY5Y, F-11), DOR activation decreases high-voltage-activated, ω-conotoxin-sensitive N-type calcium currents (Nah et al. 1997; Toselli et al. 1997, 1999), but does not affect nifedipine-sensitive L-type Ca^{2+} channels (Nah et al. 1997). Therefore, dihydropyridine-sensitive, voltage-gated Ca^{2+} channels are not expected to mediate DOR-induced Ca^{2+} entry and $[Ca^{2+}]i$ elevation in these cases. However, the study by Jin et al. (1992) in differentiated NG108-15 cells did demonstrate that DOR-induced $[Ca^{2+}]i$ increase can be completely blocked by nitrendipine. In this study, the used concentration of DADLE is in nanomolar range (Jin et al. 1992). This is in accordance with the observations in DRG neurons, in which low concentrations (nanomolar range) of opioids prolong, whereas higher concentrations (micromolar range) shorten, the duration of Ca^{2+}-dependent component of the action potential (Shen and Crain 1989, 1990). Jin et al. (1992) further found that Na^+-dependent depolarization facilitated δ-opioid-induced Ca^{2+} entry through dihydropyridine-sensitive, voltage-gated Ca^{2+} channels, since either replacement of extracellular Na^+ with the impermeant cation $NMDG^+$ or blockade of Na^+ channels with TTX, largely decreases DADLE-induced response. These authors suggested that DADLE produces a modest depolarization, which is then amplified by activation of voltage-gated Na^+ channels, and this larger depolarization then recruits these high-voltage-sensitive Ca^{2+} channels, leading to Ca^{2+} influx (Jin et al. 1992).

In addition to the contribution of mobilization of intracellular Ca^{2+} and stimulation of Ca^{2+} entry to DOR-induced $[Ca^{2+}]i$ elevation, the possibility of DOR-induced $[Ca^{2+}]i$ elevation by influencing Ca^{2+} extrusion across the plasma membrane or Ca^{2+} sequestration by organelles (e.g. ER) cannot completely be ruled out. So far, it is still unknown whether DOR interacts with plasma membrane NCX or Ca^{2+}-ATPase to influence Ca^{2+} extrusion across the plasma membrane under normal condition. However, there are some clues for DOR interaction with SERCA pumps (Tuusa et al. 2010; Tuusa and Petäjä-Repo 2011; also see the earlier discussion). Overexpression of $hDOR^{Cys27}$ (but not $hDOR^{Phe27}$) variation causes a significant suppression of ATP-induced Ca^{2+} release from ER, which makes the rate of Ca^{2+} removal from cytosol not reliably measurable, thus compromising normal Ca^{2+} signaling from intracellular Ca^{2+} store (Tuusa and Petäjä-Repo 2011).

It should be kept in mind that only a small number of cell types, but not all cells, show [Ca^{2+}]i elevation with opioid activation. Most of the knowledge regarding [Ca^{2+}]i elevation with opioid activation comes from the studies in neuron-like cell lines, but not in "real" neurons. Though these cells may have similar characteristics in some degree as native brain cells, they are not totally identical to each other. In fact, the majority of nerve cells showed an inhibition of Ca^{2+} entry with opioid activation (see the following discussion), and only a very small amount of nerve cells studied showed an increase in [Ca^{2+}]i when stimulated by DOR activation. The DOR-induced elevation of [Ca^{2+}]i is related to the excitatory effect of opioids on some presynaptic neurons and consequent regulation of neurotransmitter release as well as neuroplastic processes such as LTP (Shen and Crain 1990; Bao et al. 2003; Samways and Henderson 2006).

δ-Opioid inhibition of Ca^{2+} entry. Though some studies showed an excitatory effect of δ-opioids on Ca^{2+} entry (see above discussion), the predominant effect of DOR is inhibitory on Ca^{2+} entry. Indeed, in NG108-15 cells, despite 10 % cells showing a transient [Ca^{2+}]i increase, DADLE in μM significantly inhibits the depolarization-induced [Ca^{2+}]i transients in approximately 30 % cells (Jin et al. 1992). Direct measurement of $^{45}Ca^{2+}$ uptake in rat primary culture of cerebral cortical neurons indicated that opioid antagonist naloxone significantly increases depolarization-induced $^{45}Ca^{2+}$ uptake, suggesting that an endogenous opioid causes tonic inhibition of depolarization-induced $^{45}Ca^{2+}$ entry into cells (Vlaskovska et al. 1997). Application of DOR agonists indeed causes significant inhibition of K^+-induced $^{45}Ca^{2+}$ uptake in cultured neurons (Tang et al. 1996; Vlaskovska et al. 1997; Thorlin et al. 1998a) and neuron-like cell lines (Jin et al. 1992; Wang et al. 1993b; Fields et al. 1995). These findings indicate that activation of DOR decreases Ca^{2+} entry and [Ca^{2+}]i elevation.

Multiple factors determine [Ca^{2+}]i response to DOR activation. The nature of cells seems an important factor in the determination of DOR-induced changes in [Ca^{2+}]i. For example, in contrast to DOR activation-induced elevation in [Ca^{2+}]i in cortical astroglial cells, cortical neurons show a sustained decrease in [Ca^{2+}]i with DOR activation in the same culture (Thorlin et al. 1998a). For the basal [Ca^{2+}]i, opioid-induced response in both SK-N-SH and NG108-15 cell lines shows mainly the increase in basal [Ca^{2+}]i and increased $^{45}Ca^{2+}$ uptake, while in NMB cells, up to 42 % cells show a clear and sustained reduction in basal [Ca^{2+}]i with reduced $^{45}Ca^{2+}$ uptake. Only 17 % NMB cells increase their basal [Ca^{2+}]i with increased $^{45}Ca^{2+}$ uptake in response to opioids stimulation (Fields et al. 1995). In these cell lines, opioids have dual roles in high K^+-induced [Ca^{2+}]i increase via Ca^{2+} entry, i.e., opioid agonists decrease evoked [Ca^{2+}]i elevation in some cells but enhance it in others (Fields et al. 1995). Similar dual effects are also observed by others in cell line (Jin et al. 1992). In addition, different signal pathways may also contribute to the determination of DOR-induced changes in [Ca^{2+}]i. DOR is usually coupled to pertussis toxin-sensitive inhibitory G proteins (Gi/Go), and activation of DOR leads to the dissociation of heterotrimeric G proteins into α and βγ subunits. Gαi/Go inhibits adenylyl cyclase (McKenzie and Milligan 1990), while Gβγ activates K^+ channels (Clapham and Neer 1997) and inhibits VGCCs (Herlitze et al. 1996; Ikeda 1996).

These signaling processes generally produce inhibitory effects on nerve cells. However, opioids may also produce stimulatory effects by either direct activation of PLC via Gαi (Miyamae et al. 1993) or Gβγ (Yoon et al. 1999), or sensitize a Gαs-stimulated process by Gβγ (Tang and Gilman 1991), or by activation of the stimulatory process via cholera toxin-sensitive G protein (Shen and Crain 1990). In NG108-15 cells, DOR agonist DPDPE may inhibit high K^+-induced increase in $[Ca^{2+}]$i via cAMP dependent protein kinase mechanism (Wang et al. 1993b), and cAMP dependent protein kinase pathway is usually stimulated by Gαs (Hanoune and Defer 2001). In contrast, in GH_3 cells transfected with recombinant DOR and MOR, DOR activation inhibits spontaneous $[Ca^{2+}]$i oscillations by inhibition of Ca^{2+} entry via Ca^{2+} channels (not the release of intracellular Ca^{2+} store) due to the activation of pertussis toxin-sensitive inhibitory G proteins (Gi/Go)-coupled Kir channels (Piros et al. 2000).

A major piece of supporting evidence for DOR inhibition of Ca^{2+} entry is the blockade of Ca^{2+} entry through voltage-gated Ca^{2+} channels by δ-opioids (see earlier discussion). DOR activation inhibits diverse calcium currents (e.g. L-, N-, P- and Q-type Ca^{2+} channel currents) which link to Ca^{2+} influx through different high-voltage-activated channel types in neurons as well as neuroblastoma cell lines (Hescheler et al. 1987; Nah et al. 1997; Toselli et al. 1997, 1999; Adams and Trequattrini 1998; Morikawa et al. 1998, 1999; Acosta and López 1999; Piros et al. 2000). This decreases $[Ca^{2+}]$i levels in resting state as well as depolarization-induced Ca^{2+} uptake in neuron-like cell lines (Jin et al. 1992; Wang et al. 1993b; Fields et al. 1995) and cultured cortical neurons (Vlaskovska et al. 1997; Thorlin et al. 1998a). DOR inhibition with naltrindole greatly potentiates NMDAR activities and NMDAR-mediated increase in $[Ca^{2+}]$i in neurons (Pamenter and Buck 2008). Obviously, depolarization activates VGCCs and NMDAR channels, which results in Ca^{2+} entry, while opioids inhibit $[Ca^{2+}]$i rise by directly decreasing Ca^{2+} influx through NMDAR channels and VGCCs, especially dihydropyridine-sensitive Ca^{2+} channels (Jin et al. 1992; Wang et al. 1993b; Tang et al. 1994, 1996; Yang et al. 2004; Pamenter and Buck 2008). Therefore, activation of DOR inhibits the release of neurotransmitters from presynapes, EPSPs/EPSCs, and neuronal activities (Tanaka and North 1994; Wang and Mokha 1996; Ostermeier et al. 2000; Piros et al. 2000; Bao et al. 2003).

6 Concluding Remarks

Maintenance of ionic homeostasis is a vital feature for organisms, especially in the CNS. DOR is widely distributed throughout the CNS, and has a diversity of functions. It intensively regulates the activities of nerve cells, and mainly plays an inhibitory role. In this chapter, we first discussed the phenotype of several of the most important cations' behaviors under normal condition. K^+, Na^+ and Ca^{2+} ions show very different responses in neuronal and synaptic activity. There are very different regulatory mechanisms for the maintenance of homeostasis of these ions. DOR can influence a variety of neural activities including electrical activity, the activity of channels, receptors, transporters and pumps, and the releases of neurotransmitters/neuromodulators via very complicated and highly diverse signaling mechanisms. However, current information for DOR regulation of ionic homeostasis is still limited. Almost all the previous studies regarding the regulation of ionic homeostasis by DOR under normoxic condition focused on the activity of intracellular Ca^{2+}. Few studies focused on the regulation of K^+ and Na^+ homeostasis despite the functional coupling of DOR with K^+ channels and Na^+ channels studies. Due to the recognition of Ca^{2+} as a ubiquitous signaling molecule with complicated and diverse functions in the earlier studies, the DOR-mediated regulation of Ca^{2+} homoestasis attracted considerable research attention in the past. These studies demonstrated either a decrease or an increase of intracellular Ca^{2+} activity. In contrast, K^+ and Na^+ ions used to be regarded as important elements mainly in the maintenance of membrane potential and shaping action potentials and in controlling neuronal excitability. However, studies in recent years began to realize that K^+, and especially Na^+ ions, can act as key signaling molecules that couple membrane signaling, metabolic pathway, and neuronal activity, and playing a critical role in many aspects of neuronal functions. Therefore, for better understanding of DOR function, re-investigation of DOR regulation of homeostasis of theses ions with more sophisticated techniques is essential and necessary.

In the measurement of ion activity, scientists experienced the technology progress from indirect electrophysiological measurement based on the principle of Nernst potential, a simple and rough direct content measurement (which is tissue-destructive and time consuming) (e.g., flame photometry, X-ray microanalysis, spectroscopy), to real-time in situ high-resolution direct activity measurement (ion-selective microelectrode, optical imaging with ion-selective fluorescent indicators). Currently, genetically encoded ion indicators, especially for Ca^{2+} ions, bring several remarkable advantages to studies of ion activities in targeting specific cell types and specific subcellular compartments, allowing long-term recording from the same cellular structures, reliable detection of neuronal activity with larger fluorescence signals and higher temporal resolution, and large-scale recording of ion dynamics from hundreds of postsynaptic spines and presynaptic boutons in a cell (Dreosti et al. 2009; Mank et al. 2008; Tian et al. 2009; Ohkura et al. 2012; Gouriou et al. 2013; Chen et al. 2013; Kobayashi et al. 2014). Combining genetically encoded ion indicators with the powerful tools for high-resolution, time-lapse in vivo as well as in vitro ion imaging (e.g. two-photo/multiphoton laser scanning microscopy) (Helmchen and Denk 2002, 2005; Helmchen et al. 2013) gives perspective in ion activity research. However, contrary to its efficiency with Ca^{2+}, it is still lacking in adequate genetically encoded ion indicators for Na^+ and K^+. Also, unlike Ca^{2+} ions, Na^+ and K^+ ions, have emerged as key signaling molecules that couple membrane signaling, metabolic pathway, and neuronal activity and play a critical role in many aspects of neuronal functions. These ions require more attention in future studies.

It is important to note that it is opioid receptors, not opioids themselves that determine the effects of opioid action. In the CNS, endogenous DOR agonist enkephalin, which preferentially binds to DOR in physiological levels, can also bind to MOR (Lord et al. 1977; Chang and Cuatrecasas 1979) to have a similar effect, for example, on Ca^{2+} channels (Toselli et al. 1997; Stefani et al. 1994). Most of other opioid agonists also have similar characteristics in selectivity. Ionic responses to opioids largely depend on the distribution of specific opioid receptor(s) in the brain region/neuron studied. Therefore, we should be very cautious to explain the effects of opioid agonists and clarify the role of the effector receptors.

Acknowledgements This work was supported by NIH (AT-004422, and HD-034852) and Vivian L Smith Neurologic Foundation.

References

Abeyta A, Dettmer TS, Barnes A, Vega D, Carta M, Gallegos N, Raymond-Stintz M, Savage DD, Valenzuela CF, Saland LC (2002) Delta opioid receptor localization in the rat cerebellum. Brain Res 931:100–105

Acosta CG, López HS (1999) δ opioid receptor modulation of several voltage-dependent Ca^{2+} currents in rat sensory neurons. J Neurosci 19:8337–8348

Acquas E, Meloni M, Di Chiara G (1993) Blockade of δ-opioid receptors in the nucleus accumbens prevents ethanol-induced stimulation of dopamine release. Eur J Pharmacol 230:239–241

Adámek S, Vyskočil F (2011) Potassium-selective microelectrode revealed difference in threshold potassium concentration for cortical spreading depression in female and male rat brain. Brain Res 1370:215–219

Adams DJ, Trequattrini C (1998) Opioid receptor-mediated inhibition of omega-conotoxin GVIA-sensitive calcium channel currents in rat intracardiac neurons. J Neurophysiol 79:753–762

Adermark L, Lovinger DM (2008) Electrophysiological properties and gap junction coupling of striatal astrocytes. Neurochem Int 52:1365–1372

Agulhon C, Petravicz J, McMullen AB, Sweger EJ, Minton SK, Taves SR, Casper KB, Fiacco TA, McCarthy KD (2008) What is the role of astrocyte calcium in neurophysiology? Neuron 59:932–946

Ahn AC, Tewari M, Poon CS, Phillips RS (2006) The limits of reductionism in medicine: could systems biology offer an alternative? PLoS Med 3, e208

Albert V, Subramanian A, Rangarajan K, Pandey RM (2011) Agreement of two different laboratory methods used to measure electrolytes. J Lab Phys 3:104–109

Alger BE, Nicoll RA (1982a) Feed-forward dendritic inhibition in rat hippocampal pyramidal cells studied in vitro. J Physiol 328:105–123

Alger BE, Nicoll RA (1982b) Pharmacological evidence for two kinds of GABA receptor on rat hippocampal pyramidal cells studied in vitro. J Physiol 328:125–141

Al-Hasani R, Bruchas MR (2011) Molecular mechanisms of opioid receptor-dependent signaling and behavior. Anesthesiology 115:1363–1381

Allouche S, Polastron J, Jauzac P (1996) The δ-opioid receptor regulates activity of ryanodine receptor in the human neuroblastoma cell line SK-N-BE. J Neurochem 67:2461–2470

Alvarez-Leefmans FJ, Cruzblanca H, Gamino SM, Altamirano J, Nani A, Reuss L (1994) Transmembrane ion movements elicited by sodium pump inhibition in Helix aspersa neurons. J Neurophysiol 71:1787–1796

Ammann D, Bührer T, Schefer U, Müller M, Simon W (1987a) Intracellular neutral carrier-based Ca^{2+} microelectrode with subnanomolar detection limit. Pflugers Arch 409:223–228

Ammann D, Chao PS, Simon W (1987b) Valinomycin-based K^+ selective microelectrodes with low electrical membrane resistance. Neurosci Lett 74:221–226

Amzica F, Massimini M, Manfridi A (2002) Spatial buffering during slow and paroxysmal sleep oscillations in cortical networks of glial cells in vivo. J Neurosci 22:1042–1053

Antic SD, Zhou WL, Moore AR, Short SM, Ikonomu KD (2010) The decade of the dendritic NMDA spike. J Neurosci Res 88:2991–3001

Arakaki X, Foster H, Su L, Do H, Wain AJ, Fonteh AN, Zgou F, Harrington MG (2011) Extracellular sodium modulates the excitability of cultured hippocampal pyramidal cells. Brain Res 1401:85–94

Arenas E, Marsal J, Alberch J (1990) GABAA and GABAB antagonists prevent the opioid inhibition of endogenous acetylcholine release evoked by glutamate from rat neostriatal slices. Neurosci Lett 120:201–204

Ariav G, Polsky A, Schiller J (2003) Submillisecond precision of the input-output transformation function mediated by fast sodium dendritic spikes in basal dendrites of CA1 pyramidal neurons. J Neurosci 23:7750–7758

Armstrong GA, Xiao C, Krill JL, Seroude L, Dawson-Scully K, Robertson RM (2011) Glial Hsp70 protects K^+ homeostasis in the Drosophila brain during repetitive anoxic depolarization. PLoS One 6(12), e28994. doi:10.1371/journal.pone.0028994

Atlas D (2013) The voltage-gated calcium channel functions as the molecular switch of synaptic transmission. Annu Rev Biochem 82:607–635

Augustine GJ, Santamaria F, Tanaka K (2003) Local calcium signaling in neurons. Neuron 40:331–346

Azarias G, Van de Ville D, Unser M, Chatton JY (2008) Spontaneous NA^+ transients in individual mitochondria of intact astrocytes. Glia 56:342–353

Baba-Aissa F, Raeymaekers L, Wuytack F, Dode L, Casteels R (1998) Distribution and isoform diversity of the organellar Ca^{2+} pumps in the brain. Mol Chem Neuropathol 33:199–208

Bagley EE, Gerke MB, Vaughan CW, Hack SP, Christie MJ (2005) GABA transporter currents activated by protein kinase A excite midbrain neurons during opioid withdrawal. Neuron 45:433–445

Ballanyi K, Grafe P, Reddy MM, ten Bruggencate G (1984) Different types of potassium transport linked to carbachol and gamma-aminobutyric acid actions in rat sympathetic neurons. Neuroscience 12:917–927

Ballanyi K, Grafe P, ten Bruggencate G (1983) Intracellular free sodium and potassium, post-carbachol hyperpolarization, and extracellular potassium-undershoot in rat sympathetic neurones. Neurosci Lett 38:275–279

Ballanyi K, Grafe P, Ten Bruggencate G (1987) Ion activities and potassium uptake mechanisms of glial cells in guinea-pig olfactory cortex slices. J Physiol 382:159–174

Ballanyi K, Kettenmann H (1990) Intracellular Na^+ activity in cultured mouse oligodendrocytes. J Neurosci Res 26:455–460

Bao L, Jin SX, Zhang C, Wang LH, Xu ZZ, Zhang FX, Wang LC, Ning FS, Cai HJ, Guan JS, Xiao HS, Xu ZQ, He C, Hökfelt T, Zhou Z, Zhang X (2003) Activation of delta opioid receptors induces receptor insertion and neuropeptide secretion. Neuron 37:121–133

Bardoni R, Tawfik VL, Wang D, François A, Solorzano C, Shuster SA, Choudhury P, Betelli C, Cassidy C, Smith K, de Nooij JC, Mennicken F, O'Donnell D, Kieffer BL, Woodbury CJ, Basbaum AI, MacDermott AB, Scherrer G (2014) Delta opioid receptors presynaptically regulate cutaneous mechanosensory neuron input to the spinal cord dorsal horn. Neuron 81:1312–1327

Barolet AW, Morris ME (1991) Changes in extracellular K^+ evoked by GABA, THIP and baclofen in the guinea-pig hippocampal slices. Exp Brain Res 84:591–598

Barsukova A, Komarov A, Hajnóczky G, Bernardi P, Bourdette D, Forte M (2011a) Activation of the mitochondrial permeability transition pore modulates Ca^{2+} responses to physiological stimuli in adult neurons. Eur J Neurosci 33:831–842

Barsukova AG, Bourdette D, Forte M (2011b) Mitochondrial calcium and its regulation in neurodegeneration induced by oxidative stress. Eur J Neurosci 34:437–447

Baughman JM, Perocchi F, Girgis HS, Plovanich M, Belcher-Timme CA, Sancak Y, Bao XR, Strittmatter L, Goldberger O, Bogorad RL, Koteliansky V, Mootha VK (2011) Integrative genomics identifies MCU as an essential component of the mitochondrial calcium uniporter. Nature 476:341–345

Baumhaker Y, Wollman Y, Goldstein MN, Sarne Y (1993) Evidence for mu-, delta-, and kappa-opioid receptors in a human neuroblastoma cell line. Life Sci 52:PL205–PL210

Bausch SB, Patterson TA, Appleyard SM, Chavkin C (1995) Immunocytochemical localization of δ opioid receptors in mouse brain. J Chem Neuroanat 8:175–189

Bay V, Butt AM (2012) Relationship between glial potassium regulation and axon excitability: a role for glial Kir4.1 channels. Glia 60:651–660

Bean BP (2007) The action potential in mammalian central neurons. Nat Rev Neurosci 8:451–465

Bedner P, Steinhäuser C (2013) Altered Kir and gap junction channels in temporal lobe epilepsy. Neurochem Int 63:682–687

Bennay M, Langer J, Meier SD, Kafitz KW, Rose CR (2008) Sodium signals in cerebellar Purkinje neurons and Bergmann glial cells evoked by glutamatergic synaptic transmission. Glia 56:1138–1149

Bennett L, Ratka A (2003) Delta opioid receptors are involved in morphine-induced inhibition of luteinizing hormone releasing hormone in SK-N-SH cells. Neuropeptides 37:264–270

Bernard C (1865) Introduction à l'étude de la médecine expérimentale. Paris. English edition: Bernard C (1927) An introduction to the study of experimental medicine (trans: Greene HC). Macmillan, New York

Bernardi P (1999) Mitochondrial transport of cations: channels, exchangers, and permeability transition. Physiol Rev 79:1127–1155

Bernardinelli Y, Azarias G, Chatton JY (2006) In situ fluorescence imaging of glutamate-evoked mitochondrial Na^+ responses in astrocytes. Glia 54:460–470

Bernardinelli Y, Magistretti PJ, Chatton JY (2004) Astrocytes generate Na^+-mediated metabolic waves. Proc Natl Acad Sci U S A 101:14937–14942

Bernstein J (1902) Untersuchungen zur Thermodynamik der bioelektrischen Ströme. Pflugers Arch 92:521–562, English edition: Boylan JW (ed) (1971) Founders of experimental psychology. JF Lehmanns, München, pp 258–299

Berridge MJ (1998) Neuronal calcium signaling. Neuron 21:13–26

Betz C, Stracka D, Prescianotto-Baschong C, Frieden M, Demaurex N, Hall MN (2013) mTOR complex 2-Akt signaling at mitochondria-associated endoplasmic reticulum membranes (MAM) regulates mitochondrial physiology. Proc Natl Acad Sci U S A 110:12526–12534

Bhargava HN, Kumar S, Bian J-T (1997) Up-regulation of brain N-methyl-D-aspartate receptors following multiple intracerebroventricular injections of [D-Pen2, D-Pen5]encephalin and [D-Ala2, Glu4]deltorphin II in mice. Peptides 18:1609–1613

Bhargava HN, Zhao G-M (1996) Effects of competitive and noncompetitive antagonists of the N-methyl-D-aspartate receptor on the analgesic action of δ1- and δ2-opioid receptor agonists in mice. Br J Pharmacol 119:1586–1590

Bhattacharjee A, Kaczmarek LK (2005) For K^+ channels, Na^+ in the new Ca^{2+}. Trends Neurosci 28:422–428

Bian JM, Wu N, Su RB, Li J (2012) Opioid receptor trafficking and signaling: what happens after opioid receptor activation? Cell Mol Neurobiol 32:167–184

Bienvenu G, Seurin D, Le Bouc Y, Even P, Babajko S, Magnan C (2005) Dysregulation of energy homeostasis in mice overexpressing insulin-like growth factor-binding protein 6 in the brain. Diabetologia 48:1189–1197

Billet F, Costentin J, Dourmap N (2007) Influence of glial cells in the dopamine releasing effect resulting from the stimulation of striatal δ-opioid receptors. Neuroscience 150:131–143

Blaesse P, Airaksinen MS, Rivera C, Kaila K (2009) Cation-chloride cotransporters and neuronal function. Neuron 61:820–838

Blaustein MP, Lederer WJ (1999) Sodium/calcium exchange: its physiological implications. Physiol Rev 79:763–854

Bloodgood BL, Giessel AJ, Sabatini BL (2009) Biphasic synaptic Ca influx arising from compartmentalized electrical signals in dendritic spines. PLoS Biol 7(9), e1000190. doi:10.1371/journal.pbio.1000190

Bloodgood BL, Sabatini BL (2005) Neuronal activity regulates diffusion across the neck of dendritic spines. Science 310:866–869

Bloodgood BL, Sabatini BL (2007a) Ca^{2+} signaling in dendritic spines. Curr Opin Neurobiol 17:345–351

Bloodgood BL, Sabatini BL (2007b) Nonlinear regulation of unitary synaptic signals by Cav2.3 voltage-sensitive calcium channels located in dendritic spines. Neuron 53:249–260

Boada FE, LaVerde G, Jungreis C, Nemoto E, Tanase C, Hancu I (2005) Loss of cell ion homeostasis and cell viability in the brain: what sodium MRI can tell us. Curr Top Dev Biol 70:77–101

Boitier E, Rea R, Duchen MR (1999) Mitochondria exert a negative feedback on the propagation of intracellular Ca^{2+} waves in rat cortical astrocytes. J Cell Biol 145:795–808

Bollimuntha S, Selvaraj S, Singh BB (2011) Emerging roles of canonical TRP channels in neuronal function. Adv Exp Med Biol 704:573–593

Bootman MD, Rietdorf K, Collins T, Walker S, Sanderson M (2013) Ca^{2+}-sensitive fluorescent dyes and intracellular Ca^{2+} imaging. Cold Spring Harb Protoc 2013:83–99. doi:10.1101/pdb.top066050

Borden LA (1996) GABA transporter heterogeneity: pharmacology and cellular localization. Neurochem Int 29:335–356

Bosse KE, Jutkiewicz EM, Gnegy ME, Traynor JR (2008) The selective delta opioid agonist SNC80 enhances amphetamine-mediated efflux of dopamine from rat striatum. Neuropharmacology 55:755–762

Bouchard R, Pattarini R, Geiger JD (2003) Presence and functional significance of presynaptic ryanodine receptors. Prog Neurobiol 69:391–418

Bourne J, Harris KM (2007) Do thin spines learn to be mushroom spines that remember? Curr Opin Neurobiol 17:381–386

Boutin H, Dauphin F, MacKenzie ET, Jauzac P (1999) Differential time-course decreases in non-selective, μ-, δ-, κ-opioid receptors after focal cerebral ischemia in mice. Stroke 30:1271–1277

Brini M, Calì T, Ottolini D, Carafoli E (2014) Neuronal calcium signaling: function and dysfunction. Cell Mol Life Sci 71:2787–2814

Britt JP, McGehee DS (2008) Presynaptic opioid and nicotinic receptor modulation of dopamine overflow in the nucleus accumbens. J Neurosci 28:1672–1681

Brocard F, Shevtsova NA, Bouhadfane M, Tazerart S, Heinemann U, Rybak IA, Vinay L (2013) Activity-dependent changes in extracellular Ca^{2+} and K^{+} reveal pacemakers in the spinal locomotor-related network. Neuron 77:1047–1054

Brown DA, Scholfield CN (1974) Changes of intracellular sodium and potassium ion concentrations in isolated rat superior cervical ganglia induced by depolarizing agents. J Physiol 242:307–319

Brown HM, Pemberton JP, Owen JD (1976) A calcium-sensitive microelectrode suitable for intracellular measurement of calcium(II) activity. Anal Chim Acta 85:261–276

Bruggencate GT, Nicholson C, Stöckle H (1976) Climbing fiber evoked potassium release in cat cerebellum. Pflugers Arch 367:107–109

Brustovetsky N, Brustovetsky T, Jemmerson R, Dubinsky JM (2002) Calcium-induced cytochrome c release from CNS mitochondria is associated with the permeability transition and rupture of the outer membrane. J Neurochem 80:207–218

Bühlmann P, Chen LD (2012) Ion-selective electrodes with ionophore-doped sensing membranes. In: Gale PA, Steed JW (eds) Supramolecular chemistry: from molecules to nanomaterials. Wiley, New York, pp 2539–2579

Bührle CP, Sonnhof U (1983) Intracellular ion activities and equilibrium potentials in motoneurones and glia cells of the frog spinal cord. Pflugers Arch 396:144–153

Burgoyne RD (2007) Neuronal calcium sensor proteins: generating diversity in neuronal Ca^{2+} signaling. Nat Rev Neurosci 8:182–193

Burnstock G (2000) P2X receptors in sensory neurones. Br J Anaesth 84:476–488

Bushong EA, Martone ME, Jones YZ, Ellisman MH (2002) Protoplasmic astrocytes in CA1 stratum radiatum occupy separate anatomical domains. J Neurosci 22:183–192

Butt AM, Kalsi A (2006) Inwardly retifying potassium channels (Kir) in central nervous system glia: a special role for Kir4.1 in glial functions. J Cell Mol Med 10:33–44

Cahalan MD (2009) STIMulating store-operated Ca^{2+} entry. Nat Cell Biol 11:669–677

Cahill CM, White TD, Sawynok J (1996) Synergy between μ/δ-opioid receptors mediates adenosine release from spinal cord synaptosomes. Eur J Pharmacol 298:45–49

Cai Y-C, Ma L, Fan G-H, Zhao J, Jiang L-Z, Pei G (1997) Activation of N-methyl-D-aspartate receptor attenuates acute responsiveness of δ-opioid receptors. Mol Pharmacol 51:583–587

Callaway JC, Ross WN (1997) Spatial distribution of synaptically activated sodium concentration changes in cerebellar Purkinje neurons. J Neurophysiol 77:145–152

Cameron R, Klein L, Shyjan AW, Rakic P, Levenson R (1994) Neurons and astroglia express distinct subsets of Na, K-ATPase alpha and beta subunits. Brain Res Mol Brain Res 21:333–343

Camps M, Carozzi A, Schnabel P, Scheer A, Parker PJ, Gierschik P (1992) Isozyme-selective stimulation of phospholipase C-beta 2 by G protein beta gamma-subunits. Nature 360:684–686

Cannon WB (1929) Organization for physiological homeostasis. Physiol Rev 9:399–431

Cannon WB (1932) The wisdom of the body. W. W, Norton, New York

Cao Y-J, Bian J-T, Bhargava HN (1997) Effects of NMDA receptor antagonists on δ1- and δ2-opioid receptor agonists-induced changes in the mouse brain [3H]DPDPE binding. Eur J Pharmacol 335:161–166

Caraway WT (1981) Major developments in clinical chemical instrumentation. J Clin Chem Clin Biochem 19:491–496

Carr JA, Lovering AT (2000) Mu and delta opioid receptor regulation of pro-opiomelanocortin peptide secretion from the rat neurointermediate pituitary in vitro. Neuropeptides 34:69–75

Castro DC, Berridge KC (2014) Opioid hedonic hotspot in nucleus accumbens shell: mu, delta, and kappa maps for enhancement of sweetness "liking" and "wanting". J Neurosci 34:4239–4250

Catterall WA (2011) Voltage-gated calcium channels. Cold Spring Harb Perspect Biol 3:a003947. doi:10.1101/cshperspect.a003947

Catterall WA, Few AP (2008) Calcium channel regulation and presynaptic plasticity. Neuron 59:882–901

Chalmers S, Nicholls DG (2003) The relationship between free and total calcium concentrations in the matrix of liver and brain mitochondria. J Biol Chem 278:19062–19070

Chameau P, van Hooft JA (2006) Serotonin 5-HT3 receptors in the central nervous system. Cell Tissue Res 326:573–581

Chandrakumar NS, Stapelfeld A, Beardsley PM, Lopez OT, Drury B, Anthony E, Savage MA, Williamson LN, Reichman M (1992) Analogs of the delta opioid receptor selective cyclic peptide [2-D-penicillamine,5-D-penicillamine]-enkephalin: 2′,6′-dimethyltyrosine and Gly3-Phe4 amide bond isostere substitutions. J Med Chem 35:2928–2938

Chang KJ, Cuatrecasas P (1979) Multiple opiate receptors: enkephalins and morphine bind to receptors of different specificity. J Biol Chem 254:2610–2618

Chang KT, Niescier RF, Min KT (2011) Mitochondrial matrix Ca^{2+} as an intrinsic signal regulating mitochondrial motility in axons. Proc Natl Acad Sci U S A 108:15456–15461

Chao D, Balboni G, Lazerus LH, Salvadori S, Xia Y (2009) Na^+ mechanism of DOR protection against anoxic K^+ derangement in the cortex. Cell Mol Life Sci 66:1105–1115

Chao D, Bazzy-Asaad A, Balboni G, Salvadori S, Xia Y (2008) Activation of DOR attenuates anoxic K^+ derangement via inhibition of Na^+ entry in mouse cortex. Cereb Cortex 18:2217–2227

Chao D, Bazzy-Asaad A, Balboni G, Xia Y (2007a) δ-, but not μ-, opioid receptor stabilizes K^+ homeostasis by reducing Ca^{2+} influx in the cortex during acute hypoxia. J Cell Physiol 212:60–67

Chao D, Donnelly D, Feng Y, Asaad-Bazzy A, Xia Y (2007b) Cortical δ-opioid receptors potentiate K^+ homeostasis during anoxia and oxygen-glucose deprivation. J Cereb Blood Flow Metab 27:356–368

Chao D, He X, Yang Y, Balboni G, Salvadori S, Kim DH, Xia Y (2012a) Hydrogen sulfide induced disruption of Na^+ homeostasis in the cortex. Toxicol Sci 128:198–208

Chao D, He X, Yang Y, Bazzy-Asaad A, Lazarus LH, Balboni G, Kim DH, Xia Y (2012b) DOR activation inhibits anoxic/ischemic Na^+ influx through Na^+ channels via PKC mechanisms in the cortex. Exp Neurol 236:228–239

Chao D, Shen X, Xia Y (2013) From acupuncture to interaction between δ-opioid receptors and Na^+ channels: a potential pathway to inhibit epileptic hyperexcitability. Evid Based Complement Alternat Med 2013:216016. doi:10.1155/2013/216016

Chao D, Xia Y (2010) Ionic storm in hypoxic/ischemic stress: can opioid receptors subside it? Prog Neurobiol 90:439–470

Chao D, Xia Y (2012) Acupuncture treatment of epilepsy. In: Xia Y, Ding G, Wu G (eds) Current research in acupuncture. Springer, New York, pp 129–214. doi:10.1007/978-1-4614-3357-6_6

Chatton JY, Marquet P, Magistretti PJ (2000) A quantitative analysis of L-glutamate-regulated Na^+ dynamics in mouse cortical astrocytes: implications for cellular bioenergetics. Eur J Neurosci 12:3843–3853

Chatton JY, Pellerin L, Magistretti PJ (2003) GABA uptake into astrocytes is not associated with significant metabolic cost: implications for brain imaging of inhibitory transmission. Proc Natl Acad Sci U S A 100:12456–12461

Chatton JY, Shimamoto K, Magistretti PJ (2001) Effects of glial glutamate transporter inhibitors on intracellular Na^+ in mouse astrocytes. Brain Res 893:46–52

Chattopadhyay M, Mata M, Fink DJ (2008) Continuous δ-opioid receptor activation reduces neuronal voltage-gated sodium channel (Nav1.7) levels through activation of protein kinase C in painful diabetic neuropathy. J Neurosci 28:6625–6658

Chen L, Zou S, Lou X, Kang H-G (2000) Different stimulatory opioid effects on intracellular Ca^{2+} in SH-SY5Y cells. Brain Res 882:256–265

Chen TW, Wardill TJ, Sun Y, Pulver SR, Renninger SL, Baohan A, Schreiter ER, Kerr RA, Orger MB, Jayaraman V, Looger LL, Svoboda K, Kim DS (2013) Ultrasensitive fluorescent proteins for imaging neuronal activity. Nature 499:295–300

Chen WH, Chu KC, Wu SJ, Wu JC, Shui HA, Wu ML (1999) Early metabolic inhibit-induced intracellular sodium and calcium increase in rat cerebellar granule cells. J Physiol 515:133–146

Chen Y, Mestek A, Liu J, Hurley JA, Yu L (1993) Molecular cloning and functional expression of a mu-opioid receptor from rat brain. Mol Pharmacol 44:8–12

Cheng KT, Ong HL, Liu X, Ambudkar IS (2013) Contribution and regulation of TRPC channels in store-operated Ca^{2+} entry. Curr Top Membr 71:149–179

Chever O, Djukic B, McCarthy KD, Amzica F (2010) Implication of Kir 4.1 channel in excess potassium clearance: an in vivo study on anesthetized glial-conditional Kir 4.1 knock-out mice. J Neurosci 30:15769–15777

Chieng B, Christie MJ (2009) Chronic morphine treatment induces functional delta-opioid receptors in amygdala neurons that project to periaqueductal gray. Neuropharmacology 57:430–437

Chouhan AK, Ivannikov MV, Lu Z, Sugimori M, Llinas RR, Macleod GT (2012) Cytosolic calcium coordinates mitochondrial energy metabolism with presynaptic activity. J Neurosci 32:1233–1243

Chu Sin Chung P, Kieffer BL (2013) Delta opioid receptors in brain function and diseases. Pharmacol Ther 140:112–120

Civelli O, Douglass J, Goldstein A, Herbert E (1985) Sequence and expression of the rat prodynorphin gene. Proc Natl Acad Sci U S A 82:4291–4295

Clapham DE, Neer EJ (1997) G protein beta gamma subunits. Annu Rev Pharmacol Toxicol 37:167–203

Cohen GA, Doze VA, Madison DV (1992) Opioid inhibition of GABA release from presynaptic terminals of rat hippocampal interneurons. Neuron 9:325–335

Cohen JE, Fields RD (2004) Extracellular calcium depletion in synaptic transmission. Neuroscientist 10:12–17

Coles JA, Orkand RK (1985) Changes in sodium activity during light stimulation in photoreceptors, glia and extracellular space in drone retina. J Physiol 362:415–435

Connor M, Henderson G (1996) δ- and μ-opioid receptor mobilization of intracellular calcium in SH-SY5Y human neuroblastoma cells. Br J Pharmacol 117:333–340
Connor MA, Keir MJ, Henderson G (1997) δ-opioid receptor mobilization of intracellular calcium in SH-SY5Y cells: lack of evidence for δ-receptor subtypes. Neuropharmacology 36:125–133
Conti F, Minelli A, Melone M (2004) GABA transporters in the mammalian cerebral cortex: localization, development and pathological implications. Brain Res Brain Res Rev 45:196–212
Cordingley GE, Somjen GG (1978) The clearing of excess potassium from extracellular space in spinal cord and cerebral cortex. Brain Res 151:291–306
Corlew R, Brasier DJ, Feldman DE, Philpot BD (2008) Presynaptic NMDA receptors: newly appreciated roles in cortical synaptic function and plasticity. Neuroscientist 14:609–625
Cousin MA, Nicholls DG, Pocock JM (1995) Modulation of ion gradients and glutamate release in cultured cerebellar granule cells by ouabain. J Neurochem 64:2097–2104
Cox BM, Goldstein A, Hi CH (1976) Opioid activity of a peptide, betalipotropin-(61–91), derived from etalipotropin. Proc Natl Acad Sci U S A 73:1821–1823
Cui RJ, Roberts BL, Zhao H, Andresen MC, Appleyard SM (2012) Opioid inhibit visceral afferent activation of catecholamine neurons in the solitary tract nucleus. Neuroscience 222:181–190
D'Ambrosio R, Gordon DS, Winn HR (2002) Differential role of KIR channel and Na(+)/K(+)-pump in the regulation of extracellular K(+) in rat hippocampus. J Neurophysiol 87:87–102
D'Ambrosio R, Maris DO, Grady MS, Winn HR, Janigro D (1999) Impaired K^+ homeostasis and altered electrophysiological properties of post-traumatic hippocampal glia. J Neurosci 19:8152–8162
Danbolt NC (2001) Glutamate uptake. Prog Neurobiol 65:1–105
Danielsson I, Gasior M, Stevenson GW, Folk JE, Rice KC, Negus SS (2006) Electroencephalographic and convulsant effects of the delta opioid agonist SNC80 in rhesus monkeys. Pharmacol Biochem Behav 85:428–434
David G, Barrett JN, Barrett EF (1998) Evidence that mitochondria buffer physiological Ca^{2+} loads in lizard motor terminals. J Physiol 509:59–65
David G, Talbot J, Barrett EF (2003) Quantitative estimate of mitochondrial [Ca^{2+}] in stimulated motor nerve terminals. Cell Calcium 33:197–206
David Y, Cacheaux LP, Ivens S, Lapilover E, Heinemann U, Kaufer D, Friedman A (2009) Astrocytic dysfunction in epileptogenesis: consequence of altered potassium and glutamate homeostasis? J Neurosci 29:10588–10599
De Stefani D, Raffaello A, Teardo E, Szabò I, Rizzuto R (2011) A forty-kilodalton protein of the inner membrane is the mitochondrial calcium uniporter. Nature 476:336–340
Deitmer JW, Rose CR (2010) Ion changes and signaling in perisynaptic glia. Brain Res Rev 63:113–129
Deitmer JW, Schlue WR (1983) Intracellular Na^+ and Ca^{2+} in leech Retzius neurones during inhibition of the Na^+-K^+ pump. Pflugers Arch 397:195–201
Deng H, Yang Z, Li Y, Bao G, Friedrich T, Gu Q, Shen X, Schwarz W (2009) Interactions of Na^+, K^+-ATPase and co-expressed δ-opioid receptor. Neurosci Res 65:222–227
Denk W, Yuste R, Svoboda K, Tank DW (1996) Imaging calcium dynamics in dendritic spines. Curr Opin Neurobiol 6:372–378
Dietzel I, Heinemann U, Hofmeier G, Lux HD (1980) Transient changes in the size of the extracellular space in the sensorimotor cortex of cats in relation to stimulus-induced changes in potassium concentration. Exp Brain Res 40:432–439
Dietzel I, Heinemann U, Hofmeier G, Lux HD (1982) Stimulus-induced changes in extracellular Na^+ and Cl^- concentration in relation to changes in the size of the extracellular space. Exp Brain Res 46:73–84
Dingledine R, Borges K, Bowie D, Traynelis SF (1999) The glutamate receptor ion channels. Pharmacol Rev 51:7–61
Dipolo R, Beaugé L (2006) Sodium/calcium exchanger: influence of metabolic regulation on ion carrier interactions. Physiol Rev 86:155–203
Djukic B, Casper KB, Philpot BD, Chin LS, McCarthy KD (2007) Conditional knock-out of Kir4.1 leads to glial membrane depolarization, inhibition of potassium and glutamate uptake, and enhanced short-term synaptic potentiation. J Neurosci 27:11354–11365

Doczi J, Turiák L, Vajda S, Mándi M, Töröcsik B, Gerencser AA, Kiss G, Konràd C, Adam-Vizi V, Chinopoulos C (2011) Complex contribution of cyclophilin D to Ca^{2+}-induced permeability transition in brain mitochondria, with relation to the bioenergetic state. J Biol Chem 286:6345–6353

Doengi M, Hirnet D, Coulon P, Pape HC, Deitmer JW, Lohr C (2009) GABA uptake-dependent Ca(2+) signaling in developing olfactory bulb astrocytes. Proc Natl Acad Sci U S A 106:17570–17575

Dörner R, Ballanyi K, Schlue WR (1990) Glutaminergic responses of neuropile glial cells and Retzius neurones in the leech central nervous system. Brain Res 523:111–116

Dourmap N, Costentin J (1994) Involvement of glutamate receptors in the striatal encephalin-induced dopamine release. Eur J Pharmacol 253:R9–R11

Dreosti E, Odermatt B, Dorostkar MM, Lagnado L (2009) A genetically encoded reporter of synaptic activity in vivo. Nat Methods 6:883–889

Dronne MA, Grenier E, Dumont T, Hommel M, Boissel JP (2007) Role of astrocytes in grey matter during stroke: a modelling approach. Brain Res 1138:231–242

Dunaevsky A, Mason CA (2003) Spine motility: a means towards an end? Trends Neurosci 26:155–160

Edman A, Gestrelius S, Grampp W (1983) Intracellular ion control in lobster stretch receptor neurone. Acta Physiol Scand 118:241–252

Eisenman G, Rudin DO, Casby JU (1957) Glass electrode for measuring sodium ion. Science 126:831–834

Erbs E, Faget L, Scherrer G, Kessler P, Hentsch D, Vonesch JL, Matifas A, Kieffer BL, Massotte D (2012) Distribution of delta opioid receptor-expressing neurons in the mouse hippocampus. Neuroscience 221:203–213

Ereciñska M, Dagani F, Nelson D, Deas J, Silver IA (1991) Relations between intracellular ions and energy metabolism: a study with monensin in synaptosomes, neurons, and C6 glioma cells. J Neurosci 11:2410–2421

Evans CJ, Keith DE Jr, Morrison H, Magendzo K, Edwards RH (1992) Cloning of a delta opioid receptor by functional expression. Science 258:1952–1955

Fan G-H, Zhao J, Wu Y-L, Lou L-G, Zhang Z, Jing Q, Ma L, Pei G (1998) N-methyl-D-aspartate attenuates opioid receptor-mediated G protein activation and this process involves protein kinase C. Mol Pharmacol 53:684–690

Fan S-F, Crain SM (1995) Dual regulation by μ, δ, and κ opioid receptor agonists of K^+ conductance of DRG neurons and neuroblastoma x DRG neuron hybrid F11 cells. Brain Res 696:97–105

Fan S-F, Shen K-F, Crain SM (1991) Opioids at low concentration decrease openings of K^+ channels in sensory ganglion neurons. Brain Res 558:166–170

Feng Y, He X, Yang Y, Chao D, Lazarus LH, Xia Y (2012) Current research on opioid receptor function. Curr Drug Targets 13:230–246

Feng Y, He X, Yang Y, Chen J, Yin K, Xia Y (2011) Effect of delta-opioid receptor over-expression on cortical expression of GABAA receptor α1-subunit in hypoxia. Chin J Physiol 54:118–123

Feuerstein TJ, Albrecht C, Wessler I, Zentner J, Jackisch R (1998) δ1-opioid receptor-mediated control of acetylcholine (Ach) release in human neocortex slices. Int J Dev Neurosci 16:795–802

Fields A, Gafni M, Oron Y, Sarne Y (1995) Multiple effects of opiates on intracellular calcium level and on calcium uptake in three neuronal cell lines. Brain Res 687:94–102

Fill M, Copello JA (2002) Ryanodine receptor calcium release channels. Physiol Rev 82:893–922

Fink D, Knapp PE, Mata M (1996) Differential expression of Na, K-ATPase isoforms in oligodendrocytes and astrocytes. Dev Neurosci 18:319–326

Fischbach PS, Barrett TD, Reed NJ, Lucchesi BR (2003) SNC-80-induced preconditioning: selective activation of the mitochondrial adenosine triphosphate-gated potassium channel. J Cardiovasc Pharmacol 41:744–750

Foskett JK, White C, Cheung KH, Mak DO (2007) Inositol trisphosphate receptor Ca^{2+} release channels. Physiol Rev 87:593–658

Franceschetti S, Lavazza T, Curia G, Aracri P, Panzica F, Sancini G, Avanzini G, Magistretti J (2003) Na^+-activated K^+ current contributes to postexcitatory hyperpolarization in neocortical intrinsically bursting neurons. J Neurophysiol 89:2101–2111

Friedman JE, Haddad GG (1994) Anoxia induces an increase in intracellular sodium in rat central neurons in vitro. Brain Res 663:329–334

Friedman SM, Jamieson JD, Nakashima M, Friedman CL (1959) Sodium- and potassium-sensitive glass electrodes for biological use. Science 130:1252–1254

Fu W, Ruangkittisakul A, MacTavish D, Baker GB, Ballanyi K, Jhamandas JH (2013) Activity and metabolism-related Ca^{2+} and mitochondrial dynamics in co-cultured human fetal cortical neurons and astrocytes. Neuroscience 250:520–535

Fung ML, Croning MD, Haddad GG (1999) Sodium homeostasis in rat hippocampal slices during oxygen and glucose deprivation: role of voltage-sensitive sodium channels. Neurosci Lett 275:41–44

Gallo V, Ghiani CA (2000) Glutamate receptors in glia: new cells, new inputs and new functions. Trends Pharmacol Sci 21:252–258

Gazyakan E, Disko U, Haaf A, Heimrich B, Jackisch R (2000) Postnatal development of opioid receptors modulating acetylcholine release in hippocampus and septum of the rat. Brain Res Dev Brain Res 123:135–141

Gemes G, Bangaru ML, Wu HE, Tang Q, Weihrauch D, Koopmeiners AS, Cruikshank JM, Kwok WM, Hogan QH (2011) Store-operated Ca^{2+} entry in sensory neurons: functional role and the effect of painful nerve injury. J Neurosci 31:3536–3549

Georges F, Normand E, Bloch B, Moine CL (1998) Opioid receptor gene expression in the rat brain during ontogeny, with special reference to the mesostriatal system: an in situ hybridization study. Dev Brain Res 109:187–199

Georgoussi Z, Georganta EM, Milligan G (2012) The other side of opioid receptor signalling: regulation by protein-protein interaction. Curr Drug Targets 13:80–102

Giacomello M, De Mario A, Scarlatti C, Primerano S, Carafoli E (2013) Plasma membrane calcium ATPases and related disorders. Int J Biochem Cell Biol 45:753–762

Gido G, Kristian T, Siesjo BK (1997) Extracellular potassium in a neocortical core area after transient focal ischemia. Stroke 28:206–210

Gilman AG (1987) G proteins: transducers of receptor-generated signals. Annu Rev Biochem 56:615–649

Glancy B, Balaban RS (2012) Role of mitochondrial Ca^{2+} in the regulation of cellular energetics. Biochemistry 51:2959–2973

Gobbi P, Castaldo P, Minelli A, Salucci S, Magi S, Corcione E, Amoroso S (2007) Mitochondrial localization of Na^+/Ca^{2+} exchangers NCX1-3 in neurons and astrocytes of adult rat brain in situ. Pharmacol Res 56:556–565

Goldberg JH, Tamas G, Aronov D, Yuste R (2003) Calcium microdomains in aspiny dendrites. Neuron 40:807–821

Golding NL, Spruston N (1998) Dendritic sodium spikes are variable triggers of axonal action potentials in hippocampal CA1 pyrimidal neurons. Neuron 21:1189–1200

Goldstein A (1976) Opioid peptides endorphins in pituitary and brain. Science 193:1081–1086

Goldstein A, Lowney LI, Pal BK (1971) Stereospecific and nonspecific interactions of the morphine congener levorphanol in subcellular fractions of mouse brain. Proc Natl Acad Sci U S A 68:1742–1747

Gordon GRJ, Iremonger KJ, Kantevari S, Ellis-Davies GCR, MacVicar BA, Bains JS (2009) Astrocyte-mediated distributed plasticity at hypothalamic glutamate synapses. Neuron 64:391–403

Gouriou Y, Bijlenga P, Demaurex N (2013) Mitochondrial Ca^{2+} uptake from plasma membrane Cav3.2 protein channels contributes to ischemic toxicity in PC12 cells. J Biol Chem 288:12459–12468

Grafe P, Rimpel J, Reddy MM, ten Bruggencate G (1982) Changes of intracellular sodium and potassium ion concentrations in frog spinal motoneurons induced by repetitive synaptic stimulation. Neuroscience 7:3213–3220

Graier WF, Frieden M, Malli R (2007) Mitochondria and Ca(2+) signaling: old guests, new functions. Pflugers Arch 455:375–396

Grecksch G, Becker A, Schroeder H, Höllt V (1999) Involvement of delta-opioid receptors in pentylenetetrazol kindling development and kindling-related processes in rats. Naunyn Schmiedebergs Arch Pharmacol 360:151–156

Gross CG (1998) Claude Bernard and the constancy of the internal environment. Neuroscientist 4:380–385

Grunditz Å, Holbro N, Tian L, Zuo Y, Oertner TG (2008) Spine neck plasticity controls postsynaptic calcium signals through electrical compartmentalization. J Neurosci 28:13457–13466

Guatteo E, Mercuri NB, Bernardi G, Knöpfel T (1998) Intracellular sodium and calcium homeostasis during hypoxia in dopamine neurons of rat substantia nigra parscompacta. J Neurophysiol 80:2237–2243

Gulya K, Kovács GL, Kása P (1991a) Regulation of endogenous calcium and magnesium levels by delta opioid receptors in the rat brain. Brain Res 547:22–27

Gulya K, Kovács GL, Kása P (1991b) Partial depletion of endogenous zinc level by (D-Pen2, D-Pen5)enkephalin in the rat brain. Life Sci 48:PL57–PL62

Gulya K, Szikra J, Kása P (1995) [D-Pen2, D-Pen5]enkephalin, a delta opioid agonist reduces endogenous aluminum content in the rat central nervous system. Neuroscience 66:499–506

Haak LL, Grimaldi M, Smaili SS, Russell JT (2002) Mitochondria regulate Ca^{2+} wave initiation and inositol trisphosphate signal transduction in oligodendrocyte progenitors. J Neurochem 80:405–415

Haas M, Forbush B III (1998) The Na-K-Cl cotransporters. J Bioenerg Biomembr 30:161–172

Hack SP, Bagley EE, Chieng BCH, Christie M (2005) Induction of δ-opioid receptor function in the midbrain after chronic morphine treatment. J Neurosci 25:3192–3198

Haj-Yasein NN, Jensen V, Vindedal GF, Gundersen GA, Klungland A, Ottersen OP, Hvalby O, Nagelhus EA (2011) Evidence that compromised K^+ spatial buffering contributes to the epileptogenic effect of mutations in the human Kir4.1 gene (KCNJ10). Glia 59:1635–1642

Hald PM (1947) The flame photometer for the measurement of sodium and potassium in biological materials. J Biol Chem 167:499–510

Hamilton G, Mathur R, Allsop JM, Forton DM, Dhanjal NS, Shaw RJ, Taylor-Robinson SD (2003) Changes in brain intracellular pH and membrane phospholipids on oxygen therapy in hypoxic patients with chronic obstructive pulmonary disease. Metab Brain Dis 18:95–109

Han JS (2003) Acupuncture: neuropeptide release produced by electrical stimulation of different frequencies. Trends Neurosci 26:17–22

Hanoune J, Defer N (2001) Regulation and role of adenylyl cyclase isoforms. Annu Rev Pharmacol Toxicol 41:145–174

Hansen AJ (1978) The extracellular potassium concentration in brain cortex following ischemia in hypo- and hyperglycemic rats. Acta Physiol Scand 102:324–329

Hansen AJ (1985) Effect of anoxia on ion distribution in the brain. Physiol Rev 65:101–148

Hara M, Inous M, Yasukura T, Ohnishi S, Mikami Y, Inogaki C (1992) Uneven distribution of inctracellular Cl^- in rat hippocampal neurons. Neurosci Lett 143:135–138

Hardingham GE, Bading H (2010) Synaptic versus extrasynaptic NMDA receptor signalling: implications for neurodegenerative disorders. Nat Rev Neurosci 11:682–696

Harris KM, Kater SB (1994) Dendritic spines: cellular specializations imparting both stability and flexibility to synaptic function. Annu Rev Neurosci 17:341–371

Harrison C, Rowbotham DJ, Devi LA, Lambert DG (1999) The effect of C-terminal truncation of the recombinant δ-opioid receptor on Cai^{2+} signaling. Eur J Pharmacol 379:237–242

Hauser KF, Harris-White ME, Jackson JA, Opanashuk LA, Carney JM (1998) Opioids disrupt Ca^{2+} homeostasis and induce carbonyl oxyradical production in mouse astrocytes in vitro: transient increases and adaptation to sustained exposure. Exp Neurol 151:70–76

Häusser M, Roth A (1997) Dendritic and somatic glutamate receptor channels in rat cerebellar Purkinje cells. J Physiol 501:77–95

Hayashi Y, Majewska AK (2005) Dendritic spine geometry: functional implication and regulation. Neuron 46:529–532

He X, Sandhu HK, Yang Y, Hua F, Belser N, Kim DH, Xia Y (2013) Neuroprotection against hypoxia/ischemia: δ-opioid receptor-mediated cellular/molecular events. Cell Mol Life Sci 70:2291–2303

Heinemann U, Lux HD (1975) Undershoots following stimulus-induced rises of extracellular potassium concentration in cerebral cortex of cat. Brain Res 93:63–76

Heinemann U, Lux HD (1977) Ceiling of stimulus induced rises in extracellular potassium concentration in the cerebral cortex of cat. Brain Res 120:231–249

Heinemann U, Lux HD, Gutnick MJ (1977) Extracellular free calcium and potassium during paroxysmal activity in the cerebral cortex of the cat. Exp Brain Res 27:237–243

Helmchen F, Denk W (2002) New developments in multiphoton microscopy. Curr Opin Neurobiol 12:593–601

Helmchen F, Denk W (2005) Deep tissue two-photon microscopy. Nat Methods 2:932–940

Helmchen F, Denk W, Kerr JN (2013) Miniaturization of two-photon microscopy for imaging in freely moving animals. Cold Spring Harb Protoc 2013:904–913

Herlitze S, Garcia DE, Mackie K, Hille B, Scheuer T, Catterall WA (1996) Modulation of Ca^{2+} channels by G-protein beta gamma subunits. Nature 380:258–262

Hertle DN, Yeckel MF (2007) Distribution of inositol-1,4,5-trisphosphate receptor isotypes and ryanodine receptor isotypes during maturation of the rat hippocampus. Neuroscience 150:625–638

Hertz L, Xu J, Song D, Yan E, Gu L, Peng L (2013) Astrocytic and neuronal accumulation of elevated extracellular K(+) with a 2/3 K(+)/Na(+) flux ratio-consequences for energy metabolism, osmolarity and higher brain function. Front Comput Neurosci 7:114. doi:10.3389/fncom.2013.00114

Hescheler J, Rosenthal W, Trautwein W, Schultz G (1987) The GTP-binding protein, Go, regulates neuronal calcium channels. Nature 325:445–447

Hevesy G (1923) The absorption and translocation of lead by plants: a contribution to the application of the method of radioactive indicators in the investigation of the change of substance in plants. Biochem J 17:439–445

Hinke JA (1961) The measurement of sodium and potassium activities in the squid axon by means of cation-selective glass microelectrodes. J Physiol 156:314–335

Hipólito L, Sánchez-Catalán MJ, Zanolini I, Polache A, Granero L (2008) Shell/core differences in mu- and delta-opioid receptor modulation of dopamine efflux in nucleus accumbens. Neuropharmacology 55:183–189

Hirose N, Murakawa K, Takada K, Oi Y, Suzuki T, Nagase H, Cools AR, Koshikawa N (2005) Interactions among mu- and delta-opioid receptors, especially putative delta1- and delta2-opioid receptors, promote dopamine release in the nucleus accumbens. Neuroscience 135:213–225

Hodgkin AL, Huxley AF (1952a) Currents carried by sodium and potassium ions through the membrane of the giant axon of Loligo. J Physiol 116:449–472

Hodgkin AL, Huxley AF (1952b) The components of membrane conductance in the giant axon of Loligo. J Physiol 116:473–496

Hodgkin AL, Katz B (1949) The effect of sodium ions on the electrical activity of the giant axon of the squid. J Physiol 108:37–77

Hofmeier G, Lux HD (1981) The time courses of intracellular free calcium and related electrical effects after injection of CaCl2 into neurons of the snail, Helix pomatia. Pflugers Arch 391:242–251

Holthoff K, Tasy D, Yuste R (2002) Calcium dynamics of spines depend on their dendritic location. Neuron 33:425–437

Hrabìtová S, Nicholson C (2007) Biophysical properties of brain extracellular space explored with ion-selective microelectrodes, integrative optical imaging and related techniques. In: Michael

AC, Borland LM (eds) Electrochemical methods for neuroscience. CRC Press, Boca Raton, pp 167–204

Ikeda K, Kobayashi T, Ichikawa T, Usui H, Humanishi T (1995) Functional couplings of the δ- and κ-opioid receptors with the G-protein-activated K^+ channel. Biochem Biophys Res Commun 208:302–308

Ikeda SR (1996) Voltage-dependent modulation of N-type calcium channels by G-protein beta gamma subunits. Nature 380:255–258

Ikura M, Ames JB (2006) Genetic polymorphism and protein conformational plasticity in the calmodulin superfamily: two ways to promote multifunctionality. Proc Natl Acad Sci U S A 103:1159–1164

Inglese M, Madelin G, Oesingmann N, Babb JS, Wu W, Stoeckel B, Herbert J, Johnson G (2010) Brain tissue sodium concentration in multiple sclerosis: a sodium imaging study at 3 tesla. Brain 133:847–857

Isard JO (1959) Alkali ion determinations by means of glass electrodes. Nature 184:1616–1618

Ito S, Ohta T, Nakazato Y (1997) Changes in intracellular Na^+ concentration evoked by nicotinic receptor activation in the guinea-pig adrenal chromaffin cells. Neurosci Lett 238:111–114

Jaffe DB, Johnston D, Lasser-Ross N, Lisman J, Miyakawa H, Ross WN (1992) The spread of Na^+ spikes determines the pattern of dendritic Ca^{2+} entry into hippocampal neurons. Nature 357:244–246

Janiri L, d'Amato R, Zieglgänsberger W (1988) Dynorphin1-17 reduces the inhibitory actions of mu- and delta-selective opioid agonists in cortical neurons of the rat in vivo. Neurosci Lett 84:79–83

Jewell EA, Lingrel JB (1991) Comparison of the substrate dependence properties of the rat Na, K-ATPase alpha 1, alpha 2, and alpha 3 isoforms expressed in HeLa cells. J Biol Chem 266:16925–16930

Jiang C, Agulian S, Haddad GG (1992) Cl^- and Na^+ homeostasis during anoxia in rat hypoglossal neurons: interacellular and extracellular in vitro studies. J Physiol 448:697–708

Jiang C, Haddad GG (1991) Effect of anoxia in intracellular and extracellular potassium activity in hypoglossal neurons in vitro. J Neurophysiol 66:103–111

Jiang D, Zhao L, Clapham DE (2009) Genome-wide RNAi screen identifies Letm1 as a mitochondrial Ca^{2+}/H^+ antiporter. Science 326:144–147

Jin W, Lee NM, Loh HH, Thayer SA (1992) Dual excitatory and inhibitory effects of opioids on intracellular calcium in neuroblastoma x glioma hybrid NG108-15 cells. Mol Pharmacol 42:1083–1089

Jin W, Lee NM, Loh HH, Thayer SA (1994) Opioid mobilize calcium from inositol 1,4,5-trisphosphate-sensitive store in NG108-15 cells. J Neurosci 14:1920–1929

Jose X, Pineda JC, Rodriguez C, Mendoza E, Galarraga E, Bargas J, Barral J (2007) δ opioids reduce the neurotransmitter release probability by enhancing transient (Kv4) K^+-currents in corticostriatal synapses as evaluated by the paired pulse protocol. Neurosci Lett 414:150–154

Juhaszova M, Blaustein MP (1997) Na^+ pump low and high ouabain affinity alpha subunit isoforms are differently distributed in cells. Proc Natl Acad Sci U S A 94:1800–1805

Juhaszova M, Shimizu H, Borin ML, Yip RK, Santiago EM, Lindenmayer GE, Blaustein MP (1996) Localization of the Na^+-Ca^{2+} exchanger in vascular smooth muscle, and in neurons and astrocytes. Ann N Y Acad Sci 779:318–335

Jung DW, Baysal K, Brierley GP (1995) The sodium-calcium antiport of heart mitochondria is not electroneutral. J Biol Chem 270:672–678

Jutkiewicz EM, Baladi MG, Folk JE, Rice KC, Woods JH (2006) The convulsive and electroencephalographic changes produced by nonpeptidic delta-opioid agonists in rats: comparison with pentylenetetrazol. J Pharmacol Exp Ther 317:1337–1348

Jutkiewicz EM, Baladi MG, Folk JE, Rice KC, Woods JH (2008) The delta-opioid receptor agonist SNC80 [(+)-4-[alpha(R)-alpha-[(2S,5R)-4-allyl-2,5-dimethyl-1-piperazinyl]-(3-methoxybenzyl)-N, N-diethylbenzamide] synergistically enhances the locomotor-activating effects of some psychomotor stimulants, but not direct dopamine agonists, in rats. J Pharmacol Exp Ther 324:714–724

Jutkiewicz EM, Rice KC, Traynor JR, Woods JH (2005) Separation of the convulsions and antidepressant-like effects produced by the delta-opioid agonist SNC80 in rats. Psychopharmacology (Berl) 182:588–596

Kadamur G, Ross EM (2013) Mammalian phospholipase C. Annu Rev Physiol 75:127–154

Kaila K, Lamsa K, Smirnov S, Taira T, Voipio J (1997) Long-lasting GABA-mediated depolarization evoked by high-frequency stimulation in pyramidal neurons of rat hippocampal slices is attributable to a network-driven, bicarbonate-dependent K^+ transient. J Neurosci 17:7662–7672

Kaila K, Rydqvist B, Swerup C, Voipio J (1987) Stimulation-induced changes in the intracellular sodium activity of the crayfish stretch receptor. Neurosci Lett 74:53–57

Kang SH, Carl A, McHugh JM, Goff HR, Kenyon JL (2008) Role of mitochondria and temperature in the control of intracellular calcium in adult rat sensory neurons. Cell Calcium 43:388–404

Kang X, Chao D, Gu Q, Ding G, Wang Y, Balboni G, Lazarus LH, Xia Y (2009) Delta-opioid receptors protect from anoxic disruption of Na^+ homeostasis via Na^+ channel regulation. Cell Mol Life Sci 66:3505–3516

Kang-Park M-H, Kieffer BL, Roberts AJ, Siggins GR, Moore SD (2007) Presyanptic δ opioid receptors regulates ethanol actions in central amygdala. J Pharmacol Exp Ther 320:917–925

Karschin C, Dissmann E, Stühmer W, Karschin A (1996) GIRK(1-3) and GIRK(1-4) inwardly rectifying K^+ channel mRNAs are differentially expressed in the adult rat brain. J Neurosci 16:3559–3570

Katsuura Y, Taha SA (2010) Modulation of feeding and locomotion through mu and delta opioid receptor signaling in the nucleus accumbens. Neuropeptides 44:225–232

Katz A, Wu D, Simon MI (1992) Subunits beta gamma of heterotrimeric G protein activate beta 2 isoform of phospholipase C. Nature 360:686–689

Kawamoto EM, Vivar C, Camandola S (2012) Physiology and pathology of calcium signaling in the brain. Front Pharmacol 3:61. doi:10.3389/fphar.2012.00061

Kazmi SM, Mishra RK (1986) Opioid receptors in human neuroblastoma SH-SY5Y cells: evidence for distinct morphine (mu) and enkephalin (delta) binding sites. Biochem Biophys Res Commun 137:813–820

Kelly T, Kafitz KW, Roderigo C, Rose CR (2009) Ammonium-evoked alterations in intracellular sodium and pH reduce glial glutamate transport activity. Glia 57:921–934

Kelly T, Rose CR (2010) Ammonium influx pathways into astrocytes and neurones of hippocampal slices. J Neurochem 115:1123–1136

Kennedy C, Krier J (1987) δ-opioid receptors mediate inhibition of fast excitatory postsynaptic potentials in cat parasympathetic colonic ganglia. Br J Pharmacol 92:437–443

Keren O, Gafni M, Sarne Y (1997) Opioids potentiate transmitter release from SK-N-SH human neuroblastoma cells by modulating N-type calcium channels. Brain Res 764:277–282

Khakh BS (2001) Molecular physiology of P2X receptors and ATP signalling at synapses. Nat Rev Neurosci 2:165–174

Khakh BS, North RA (2012) Neuromodulation by extracellular ATP and P2X receptors in the CNS. Neuron 76:51–69

Kiedrowski L (1999) N-methyl-D-aspartate excitotoxicity: relationships among plasma membrane potential, Na(+)/Ca(2+) exchange, mitochondrial Ca(2+) overload, and cytoplasmic concentrations of Ca(2+), H(+), and K(+). Mol Pharmacol 56:619–632

Kiedrowski L, Brooker G, Costa E, Wroblewski JT (1994a) Glutamate impairs neuronal calcium extrusion while reducing sodium gradient. Neuron 12:295–300

Kiedrowski L, Wroblewski JT, Costa E (1994b) Intracellular sodium concentration in cultured cerebellar granule cells challenged with glutamate. Mol Pharmacol 45:1050–1054

Kieffer BL, Befort K, Gaveriaux-Ruff C, Hirth CG (1992) The delta-opioid receptor: isolation of a cDNA by expression cloning and pharmacological characterization. Proc Natl Acad Sci U S A 89:12048–12052

Kimelberg HK, Bowman C, Biddlecome S, Bourke RS (1979) Cation transport and membrane potential properties of primary astroglial cultures from neonatal rat brains. Brain Res 177:533–550

Kirichok Y, Krapivinsky G, Clapham DE (2004) The mitochondrial calcium uniporter is a highly selective ion channel. Nature 427:360–364

Kirischuk S, Kettenmann H, Verkhratsky A (1997) Na^+/Ca^{2+} exchanger modulates kainate-triggered Ca^{2+} signaling in Bergmann glial cells in situ. FASEB J 11:566–572

Kirischuk S, Kettenmann H, Verkhratsky A (2007) Membrane currents and cytoplasmic sodium transients generated by glutamate transport in Bergmann glial cells. Pflugers Arch 454:245–252

Kirischuk S, Parpura V, Verkhratsky A (2012) Sodium dynamics: another key to astroglial excitability? Trends Neurosci 35:497–506

Knapp RJ, Santoro G, De Leon IA, Lee KB, Edsall SA, Waite S, Porreca F, Roeske WR, Yamamura HI (1996) Structure-activity relationships for SNC80 and related compounds at cloned human delta and mu opioid receptors. J Pharmacol Exp Ther 277:1284–1291

Knöpfel T, Anchisi D, Alojado ME, Tempia F, Strata P (2000) Elevation of intracellular sodium concentration mediated by synaptic activation of metabotropic glutamate receptors in cerebellar Purkinje cells. Eur J Neurosci 12:2199–2204

Knöpfel T, Guatteo E, Bernardi G, Mercuri NB (1998) Hyperpolarization induces a rise in intracellular sodium concentration in dopamine cells of the substantia nigra pars compacta. Eur J Neurosci 10:1926–1929

Kobayashi C, Ohkura M, Nakai J, Matsuki N, Ikegaya Y, Sasaki T (2014) Large-scale imaging of subcellular calcium dynamics of cortical neurons with G-CaMP6-actin. Neuroreport 25(7):501–506

Kobayashi T, Ikeda K, Ichikawa T, Abe S, Togashi S, Kumanishi T (1995) Molecular cloning of a mouse G-protein-activated K^+ channel (mGIRK1) and distinct distributions of three GIRK (GIRK1, 2 and 3) mRNAs in mouse brain. Biochem Biophys Res Commun 208:1166–1173

Kocsis JD, Eng DL, Gordon TR, Waxman SG (1987) Functional differences between 4-aminopyridine and tetraethylammonium-sensitive potassium channels in myelinated axons. Neurosci Lett 75:193–198

Kofuji P, Newman EA (2004) Potassium buffering in the central nervous system. Neuroscience 129:1045–1056

Kofuji P, Newman EA (2009) Regulation of potassium by glial cells in the central nervous system. In: Parpura V, Haydon P (eds) Astrocytes in (patho)physiology of the nervous system, vol 6. Springer, New York, pp 151–175

Kornblum HI, Hurlbut DE, Leslie FM (1987) Postnatal development of multiple opioid receptors in rat brain. Brain Res 465:21–41

Koss DJ, Riedel G, Bence K, Platt B (2013) Store-operated Ca^{2+} entry in hippocampal neurons: regulation by protein tyrosine phosphatase PTP1B. Cell Calcium 53:125–138

Kozoriz MG, Church J, Ozog MA, Naus CC, Krebs C (2010) Temporary sequestration of potassium by mitochondria in astrocytes. J Biol Chem 285:31107–31119

Kraig PR, Nicholson C (1978) Extracellular ionic variations during spreading depression. Neuroscience 3:1045–1059

Kraig RP, Nicholson C (1976) Sodium liquid ion exchanger microelectrode used to measure large extracellular sodium transients. Science 194:725–726

Krapivinsky G, Gordon EA, Wickman K, Velimirović B, Krapivinsky L, Clapham DE (1995) The G-protein-gated atrial K^+ channel IKACh is a heteromultimer of two inwardly rectifying K(+)-channel proteins. Nature 374:135–141

Kreutz MR, Naranjo JR, Koch K, Schwaller B (2012) The neuronal functions of EF-hand Ca(2+)-binding proteins. Front Mol Neurosci 5:92, http://www.frontiersin.org/Molecular_Neuroscience/researchtopics/The_neuronal_functions_of_EF-h/585

Kříž N, Syková E, Vyklický L (1975) Extracellular potassium changes in the spinal cord of the cat and their relation to slow potentials, active transport and impulse transmission. J Physiol 249:167–182

Krnjević K, Morris ME (1972) Extracellular K^+ activity and slow potential changes in spinal cord and medulla. Can J Physiol Pharmacol 50:1214–1217

Kucheryavykh YV, Kucheryavykh LY, Nichols CG, Maldonado HM, Baksi K, Reichenbach A, Skatchkov SN, Eaton MJ (2007) Downregulation of Kir4.1 inward rectifying potassium channel subunits by RNAi impairs potassium transfer and glutamate uptake by cultured cortical astrocytes. Glia 55:274–281

Kuffler SW, Nicholls JG, Orkand RK (1966) Physiological properties of glial cells in the central nervous system of amphibian. J Neurophysiol 29:768–787

Kunkel MT, Peralta EG (1995) Identification of domains conferring G protein regulation on inward rectifier potassium channels. Cell 83:443–449

Kuruma A, Inoue T, Mikoshiba K (2003) Dynamics of Ca^{2+} and Na^{+} in the dendrites of mouse cerebral Purkinje cells evoked by parallel fibre stimulation. Eur J Neurosci 18:2677–2689

Lalo U, Pankratov Y, Parpura V, Verkhratsky A (2011) Ionotropic receptors in neuronal-astroglial signalling: what is the role of "excitable" molecules in non-excitable cells. Biochim Biophys Acta 1813:992–1002

Lam AK, Galione A (2013) The endoplasmic reticulum and junctional membranes communication during calcium signaling. Biochim Biophys Acta 1833:2542–2559

Langer J, Rose CR (2009) Synaptically induced sodium signals in hippocampal astrocytes in situ. J Physiol 587:5859–5877

Langer J, Stephan J, Theis M, Rose CR (2012) Gap junctions mediate intercellular spread of sodium between hippocampal astrocytes in situ. Glia 60:239–252

Lasser-Ross N, Ross WN (1992) Imaging voltage and synaptically activated sodium transients in cerebellar Purkinje cells. Proc Biol Soc 247:35–39

Laurent V, Bertran-Gonzalez J, Chieng BC, Balleine BW (2014) δ-opioid and dopaminergic processes in accumbens shell modulate the cholinergic control of predictive learning and choice. J Neurosci 34:1358–1369

Laurent V, Leung B, Maidment N, Balleine BW (2012) μ- and δ-opioid-related processes in the accumbens core and shell differentially mediate the influence of reward-guided and stimulus-guided decisions on choice. J Neurosci 32:1875–1883

Law PY, Hom DS, Loh HH (1983) Opiate regulation of adenosine 3′:5′-cyclic monophosphate level in neuroblastoma X glioma NG108-15 hybrid cells. Relationship between receptor occupancy and effect. Mol Pharmacol 23:26–35

Law PY, Wong YH, Loh HH (2000) Molecular mechanisms and regulation of opioid receptor signaling. Annu Rev Pharmacol Toxicol 40:389–430

Le Moine C, Fauchey V, Jaber M (2002) Opioid receptor gene expression in dopamine transporter knock-out mice in adult and during development. Neuroscience 112:131–139

Lee D, Lee KH, Ho WK, Lee SH (2007) Target cell-specific involvement of presynaptic mitochondria in post-tetanic potentiation at hippocampal mossy fiber synapses. J Neurosci 27:13603–13613

Lee SH, Ho WK, Lee SH (2009) Characterization of somatic Ca^{2+} clearance mechanisms in young and mature hippocampal granule cells. Cell Calcium 45:465–473

Lee SK, Boron WF, Parker MD (2013) Monitoring ion activities in and around cells using ion-selective liquid-membrane microelectrodes. Sensors (Basel) 13:984–1003

Lenard B, Kintner DB, Shull GE, Sun D (2004) Na-K-Cl cotransporter-mediated intracellular Na^{+} accumulation affects Ca^{2+} signaling in astrocytes in an in vitro ischemic model. J Neurosci 24:9585–9597

Leng T, Shi Y, Xiong ZG, Sun D (2014) Proton-sensitive cation channels and ion exchangers in ischemic brain injury: new therapeutic targets for stroke? Prog Neurobiol 115:189–209

Leybaert L, Sanderson MJ (2012) Intercellular Ca^{2+} waves: mechanisms and function. Physiol Rev 92:1359–1392

Liang J, Chao D, Sandhu HK, Yu Y, Zhang L, Balboni G, Xia Y (2014) δ-opioid receptors up-regulate excitatory amino acid transporters in mouse astrocytes. Br J Pharmacol 171(23):5417–5430. doi:10.1111/bph.12857

Lim YJ, Zheng S, Zuo Z (2004) Morphine preconditions purkinje cells against cell death under in vitro simulated ischemia-reperfusion conditions. Anesthesiology 100:562–568

Linden DJ, Smeyne M, Connor JA (1993) Induction of cerebellar long-term depression in culture requires postsynaptic action of sodium ions. Neuron 11:1093–1100

Liu D, Slevin JR, Lu C, Chan SL, Hansson M, Elmér E, Mattson MP (2003) Involvement of mitochondrial K^+ release and cellular efflux in ischemic and apoptotic neuronal death. J Neurochem 86:966–979

Liu SJ, Zukin RS (2007) Ca^{2+}-permeable AMPA receptors in synaptic plasticity and neuronal death. Trends Neurosci 30:126–134

Longoni R, Spina L, Mulas A, Carboni E, Garau L, Melchiorri P, Di Chiara G (1991) (D-Ala2) deltorphin II: D1-dependent stereotypies and stimulation of dopamine release in the nucleus accumbens. J Neurosci 11:1565–1576

Longuemare MC, Rose CR, Farrell K, Ransom BR, Waxman SG, Swanson RA (1999) K^+-induced reversal of astrocyte glutamate uptake is limited by compensatory changes in intracellular Na^+. Neuroscience 93:285–292

Lopachin RM, Gaughan CL (1999) Electron probe X-ray microanalysis: a quantitative electron microscopy technique for measurement of elements and water in nervous tissue cells. Methods Mol Med 22:289–299

LoPachin RM, Gaughan CL, Lehning EJ, Weber ML, Taylor CP (2001) Effects of ion channel blockade on the distribution of Na, K, Ca and other elements in oxygen-glucose deprived CA1 hippocampal neurons. Neuroscience 103:971–983

Lord JA, Waterfield AA, Hughes J, Kosterlitz HW (1977) Endogenous opioid peptides: multiple agonists and receptors. Nature 267:495–499

Lothman EW, Somjen GG (1975) Extracellular potassium activity, intracellular and extracellular potential responses in the spinal cord. J Physiol 252:115–136

Lux HD, Neher E (1973) The equilibration time course of $[K^+]0$ in cat cortex. Exp Brain Res 17:190–205

Ly CV, Verstreken P (2006) Mitochondria at synapse. Neuroscientist 12:291–299

Ma MC, Qian H, Ghassemi F, Zhao P, Xia Y (2005) Oxygen sensitive δ-opioid receptor-regulated survival and death signals: novel insights into neuronal preconditioning and protection. J Biol Chem 280:16208–16218

MacAulay N, Zeuthen T (2012) Glial K^+ clearance and cell swelling: key roles for cotransporters and pumps. Neurochem Res 37:2299–2309

Maderspach K, Solomonia R (1988) Glial and neuronal opioid receptors: apparent positive cooperativity observed in intact cultured cells. Brain Res 441:41–47

Madison DV, Nicoll RA (1988) Enkephalin hyperpolarizes interneurones in the rat hippocampus. J Physiol 398:123–130

Madsen KK, White HS, Schousboe A (2010) Neuronal and non-neuronal GABA transporters as targets for antiepileptic drugs. Pharmacol Ther 125:394–401

Magee JC, Carruth M (1999) Dendritic voltage-gated ion channels regulate the action potential firing model of hippocampal CA1 pyramidal neurons. J Neurophysiol 82:1895–1901

Majewska A, Brown E, Ross J, Yuste R (2000) Mechanisms of calcium decay kinetics in hippocampal spines: role of spine calcium pumps and calcium diffusion through the spine neck in biochemical compartmentalization. J Neurosci 20:1722–1734

Maldonado PP, Vélez-Fort M, Levavasseur F, Angulo MC (2013) Oligodendrocyte precursor cells are accurate sensors of local K^+ in mature gray matter. J Neurosci 33:2432–2442

Mank M, Santos AF, Direnberger S, Mrsic-Flogel TD, Hofer SB, Stein V, Hendel T, Reiff DF, Levelt C, Borst A, Bonhoeffer T, Hübener M, Griesbeck O (2008) A genetically encoded calcium indicator for chronic in vivo two-photon imaging. Nat Methods 5:805–811

Mansour A, Fox CA, Burke S, Meng F, Thompson RC, Akil H, Watson SJ (1994) Mu, delta, and kappa opioid receptor mRNA expression in the rat CNS: an in situ hybridization study. J Comp Neurol 350:412–438

Mansour A, Khachaturian H, Lewis ME, Akil H, Watson SJ (1987) Autoradiographic differentiation of mu, delta, and kappa opioid receptors in the rat forebrain and midbrain. J Neurosci 7:2445–2464

Maragakis NJ, Rothstein JD (2004) Glutamate transporters: animal models to neurologic disease. Neurobiol Dis 15:461–473

Margolis EB, Fields HL, Hjelmstad GO, Mitchell JM (2008) Delta-opioid receptor expression in the ventral tegmental area protects against elevated alcohol consumption. J Neurosci 28:12672–12681

Marker CL, Luján R, Loh HH, Wickman K (2005) Spinal G-protein-gated potassium channels contribute in a dose-dependent manner too the analgesic effect of μ- and δ- but not κ-opioids. J Neurosci 25:3551–3559

Martin WR (1967) Opioid antagonists. Pharmacol Rev 19:463–521

Martin WR, Eades CG, Thompson JA, Huppler RE, Gilbert PE (1976) The effects of morphine- and nalorphine-like drugs in the nondependent and morphine-dependent chronic spinal dog. J Pharmacol Exp Ther 197:517–532

Masino SA, Dulla CG (2005) Adenosine, glutamate and pH: interactions and implications. Neurol Res 27:149–152

Matsuzaki M, Ellis-Davies GC, Nemoto T, Miyashita Y, Iino M, Kasai H (2001) Dendritic spine geometry is critical for AMPA receptor expression in hippocampal CA1 pyramidal neurons. Nat Neurosci 4:1086–1092

Matsuzaki M, Honkura N, Ellis-Davies GC, Kasai H (2004) Structural basis of long-term potentiation in single dendritic spines. Nature 429:761–766

Matyash M, Matyash V, Nolte C, Sorrentino V, Kettenmann H (2002) Requirement of functional ryanodine receptor type 3 for astrocyte migration. FASEB J 16:84–86

Mayevsky A, Zeuthen T, Chance B (1974) Measurements of extracellular potassium, ECoG and pyridine nucleotide levels during cortical spreading depression in rats. Brain Res 76:347–349

Mayfield KP, Kozak W, Malvin GM, Porreca F (1996) Hypoxia decreases opioid delta receptor expression in mouse brain. Neuroscience 72:785–789

McGrail KM, Phillips JM, Sweadner KJ (1991) Immunofluorescent localization of three Na, K-ATPase isozymes in the rat central nervous system: both neurons and glia can express more than one Na, K-ATPase. J Neurosci 11:381–391

McKenzie FR, Milligan G (1990) Delta-opioid-receptor-mediated inhibition of adenylate cyclase is transduced specifically by the guanine-nucleotide-binding protein Gi2. Biochem J 267:391–398

McKinnon PJ (2013) Maintaining genome stability in the nervous system. Nat Neurosci 16:1523–1529

Meme W, Vandecasteele M, Giaume C, Venance L (2009) Electrical coupling between hippocampal astrocytes in rat brain slices. Neurosci Res 63:236–243

Meng F, Xie GX, Thompson RC, Mansour A, Goldstein A, Watson SJ, Akil H (1993) Cloning and pharmacological characterization of a rat kappa opioid receptor. Proc Natl Acad Sci U S A 90:9954–9958

Menichella DM, Majdan M, Awatramani R, Goodenough DA, Sirkowski E, Scherer SS, Paul DL (2006) Genetic and physiological evidence that oligodendrocyte gap junction contribute to spatial buffering of potassium released during neuronal activity. J Neurosci 26:10984–10991

Messerli MA, Kurtz I, Smith PJ (2008) Characterization of optimized Na^+ and Cl^- liquid membranes for use with extracellular, self-referencing microelectrodes. Anal Bioanal Chem 390:1355–1359

Miller RJ (1990) Receptor-mediated regulation of calcium channels and neurotransmitter release. FASEB J 4:3291–3299

Minami M, Toya T, Katao Y, Maekawa K, Nakamura S, Onogi T, Kaneko S, Satoh M (1993) Cloning and expression of a cDNA for the rat kappa-opioid receptor. FEBS Lett 329:291–295

Minta A, Tsien RY (1989) Fluorescent indicators for cytosolic sodium. J Biol Chem 264:19449–19457

Mitchell JM, Margolis EB, Coker AR, Allen DC, Fields HL (2014) Intra-VTA deltorphin, but not DPDPE, induces place preference in ethanol-drinking rats: distinct DOR-1 and DOR-2 mechanisms control ethanol consumption and reward. Alcohol Clin Exp Res 38:195–203

Miura T, Yano T, Naitoh K, Nishihara M, Miki T, Tanno M, Shimamoto K (2007) Delta-opioid receptor activation before ischemia reduces gap junction permeability in ischemic myocardium by PKC-epsilon-mediated phosphorylation of connexin 43. Am J Physiol Heart Circ Physiol 293:H1425–H1431

Miyamae T, Fukushima N, Misu Y, Ueda H (1993) Delta opioid receptor mediates phospholipase C activation via Gi in Xenopus oocytes. FEBS Lett 333:311–314

Moises HC, Rusin KI, Macdonald RL (1994) Mu- and kappa-opioid receptors selectively reduce the same transient components of high-threshold calcium current in rat dorsal root ganglion sensory neurons. J Neurosci 14:5903–5916

Money TG, Rodgers CI, McGregor SM, Robertson RM (2009) Loss of potassium homeostasis underlies hyperthermic conduction failure in control and preconditioned locusts. J Neurophysiol 102:285–293

Moore EW, Wilson DW (1963) The determination of sodium in body fluids by the glass electrode. J Clin Invest 42:293–304

Moore SD, Madamba SG, Schweitzer P, Siggins GB (1994) Voltage-dependent effects of opioid peptides on hippocampal CA3 pyramidal neuros in vitro. J Neurosci 14:809–820

Moreau B, Nelson C, Parekh AB (2006) Biphasic regulation of mitochondrial Ca^{2+} uptake by cytosolic Ca^{2+} concentration. Curr Biol 16:1672–1677

Mori K, Miyazaki M, Iwase H, Maeda M (2002) Temporal profile of changes in brain tissue extracellular space and extracellular ion (Na(+), K(+)) concentrations after cerebral ischemia and the effects of mild cerebral hypothermia. J Neurotrauma 19:1261–1270

Morikawa H, Fukuda K, Mima H, Shoda T, Kato S, Mori K (1998) Desensitization and resensitization of δ-opioid receptor-mediated Ca^{2+} channel inhibition in NG108-15 cells. Br J Pharmacol 123:1111–1118

Morikawa H, Mima H, Uga H, Shoda T, Fukuda K (1999) Opioid potentiation of N-type Ca^{2+} channel currents via pertussis-toxin-sensitive G proteins in NG108-15 cells. Pflugers Arch 438:423–426

Morris ME, Orbrocea GV, Avoli M (1996) Extracellular K^+ accumulations and synchronous GABA-mediated potentials evoked by 4-aminopyridine in the adult rat hippocampus. Exp Brain Res 109:71–82

Moulder KL, Meeks JP, Mennerick S (2006) Homeostatic regulation of glutamate release in response to depolarization. Mol Neurobiol 33:133–153

Müller M, Somjen GG (2000a) Na^+ and K^+ concentrations, extra- and intracellular voltages, and the effect of TTX in hypoxic rat hippocampal slices. J Neurophysiol 83:735–745

Müller M, Somjen GG (2000b) Na(+) dependence and the role of glutamate receptors and Na(+) channels in ion fluxes during hypoxia of rat hippocampal slices. J Neurophysiol 84:1869–1880

Müller T, Grosche J, Ohlemeyer C, Kettenmann H (1993) NMDA-activated currents in Bergmann glial cells. Neuroreport 4:671–674

Müller W, Connor JA (1991) Dendritic spines as individual neuronal compartments for synaptic Ca^{2+} responses. Nature 354:73–76

Munakata M, Watanabe M, Otsuki T, Nakama H, Arima K, Itoh M, Nabekura J, Iinuma K, Tsuchiya S (2007) Altered distribution of KCC2 in cortical dysplasia in patients with intractable epilepsy. Epilepsia 48:837–844

Munsch T, Deitmer JW (1997) Intracellular Ca^{2+}, Na^+ and H^+ transients evoked by kainate in the leech giant glial cells in situ. Neurosci Res 27:45–56

Munzer JS, Daly SE, Jewell-Motz EA, Lingrel JB, Blostein R (1994) Tissue- and isoform-specific kinetic behavior of the Na, K-ATPase. J Biol Chem 269(24):16668–16676

Murchison D, Griffith WH (2000) Mitochondria buffer non-toxic calcium loads and release calcium through the mitochondrial permeability transition pore and sodium/calcium exchanger in rat basal forebrain neurons. Brain Res 854:139–151

Nah S-Y, Unteutsch A, Bunzow JR, Cook SP, Beacham DW, Grandy DK (1997) μ and δ opioids but not κ opioid inhibit voltage-activated Ba^{2+} currents in neuronal F-11 cell. Brain Res 766:66–71

Nakanishi S, Inoue A, Kita T, Nakamura M, Chang AC, Cohen SN, Numa S (1979) Nucleotide sequence of cloned cDNA for bovine corticotropin-beta-lipotropin precursor. Nature 278:423–427

Namkung W, Padmawar P, Mills AD, Verkman AS (2008) Cell-based fluorescence screen for K^+ channels and transporters using an extracellular triazacryptand-based K^+ sensor. J Am Chem Soc 130:7794–7795

Nayak SV, Rondé P, Spier AD, Lummis SCR, Nichols RA (1999) Calcium changes induced by presynaptic 5-hydroxytryptamine-3 serotonin receptors on isolated terminals from various regions of the rat brain. Neuroscience 91:107–117

Nedergaard M, Verkhratsky A (2012) Artifact versus reality—how astrocytes contribute to synaptic events. Glia 60:1013–1023

Negus SS, Gatch MB, Mello NK, Zhang X, Rice K (1998) Behavioral effects of the delta-selective opioid agonist SNC80 and related compounds in rhesus monkeys. J Pharmacol Exp Ther 286:362–375

Neumaier JF, Mailheau S, Chavkin C (1988) Opioid receptor-mediated responses in the dentate gyrus and CA1 region of the rat hippocampus. J Pharmacol Exp Ther 244:564–570

Neusch C, Papadopoulos N, Müller M, Maletzki I, Winter SM, Hirrlinger J, Handschuh M, Bähr M, Richter DW, Kirchhoff F, Hülsmann S (2006) Lack of the Kir4.1 channel subunit abolishes K^+ buffering properties of astrocytes in the ventral respiratory group: impact on extracellular K^+ regulation. J Neurophysiol 95:1843–1852

Nicholls JG, Kuffler SW (1965) Na and K concent of glial cells and neurons determined by flame photometry in the central nervous system of the leech. J Neurophysiol 28:519–525

Nicholson C, Ten Bruggencate G, Steinberg R, Stockle H (1977) Calcium modulation in brain extracellular microenvironment demonstrated with ion-selective micropipette. Proc Natl Acad Sci U S A 74:1287–1290

Nicot A, Ratnakar PV, Ron Y, Chen CC, Elkabes S (2003) Regulation of gene expression in experimental autoimmune encephalomyelitis indicates early neuronal dysfunction. Brain 126:398–412

Noble F, Fournie-Zaluski MC, Roques BP (1996) Opposite role of delta 1- and delta 2-opioid receptors activated by endogenous or exogenous opioid agonists on the endogenous cholecystokinin system: further evidence for delta-opioid receptor heterogeneity. Neuroscience 75:917–926

Noble F, Smadja C, Roques BP (1994) Role of endogenous cholecystokinin in the facilitation of mu-mediated antinociception by delta-opioid agonists. J Pharmacol Exp Ther 271:1127–1134

Noda M, Teranishi Y, Takahashi H, Toyosato M, Notake M, Nakanishi S, Numa S (1982) Isolation and structural organization of the human preproenkephalin gene. Nature 297:431–434

Noguchi J, Matsuzaki M, Ellis-Davies GCR, Kasai H (2005) Spine-neck geometry determines NMDA receptor-dependent Ca^{2+} signaling in dendrites. Neuron 46:609–622

North RA (2002) Molecular physiology of P2X receptors. Physiol Rev 82:1013–1067

North RA, Williams JT, Surprenant A, Christie MJ (1987) μ and δ receptors belong to a family of receptors that are coupled to potassium channels. Proc Natl Acad Sci U S A 84:5487–5491

Nowikovsky K, Pozzan T, Rizzuto R, Scorrano L, Bernardi P (2012) The pathophysiology of LETM1. J Gen Physiol 139:445–454

O'Doherty J, Garcia-Diaz JF, Armstrong WM (1979) Sodium-selective liquid ion-exchanger microelectrodes for intracellular measurements. Science 203:1349–1351

O'Donnell ME, Chen YJ, Lam TI, Taylor KC, Walton JH, Anderson SE (2013) Intravenous HOE-642 reduces brain edema and Na uptake in the rat permanent middle cerebral artery occlusion model of stroke: evidence for participation of the blood-brain barrier Na/H exchanger. J Cereb Blood Flow Metab 33:225–234

Obrocea GV, Morris ME (1998) Changes in $[K^+]o$ evoked by baclofen in guinea pig hippocampus. Can J Physiol Pharmacol 76:148–154

Odermatt B, Wellershaus K, Wallraff A, Seifert G, Degen J, Euwens C, Fuss B, Bussow H, Schilling K, Steinhäuser C, Willecke K (2003) Connexin 47 (Cx47)-deficient mice with enhanced green fluorescent protein reporter gene reveal predominant oligodendrocytic expression of Cx47 and display vacuolized myelin in the CNS. J Neurosci 23:4549–4559

Oehme M, Kessler M, Simon W (1976) Neutral carrier Ca^{2+}- microelectrode. Chimia 30:204–206

Ohkura M, Sasaki T, Sadakari J, Gengyo-Ando K, Kagawa-Nagamura Y, Kobayashi C, Ikegaya Y, Nakai J (2012) Genetically encoded green fluorescent Ca^{2+} indicators with improved detectability for neuronal Ca^{2+} signals. PLoS One 7, e51286. doi:10.1371/journal.pone.0051286

Olesen C, Picard M, Winther AML, Gyrup C, Morth JP, Oxvig C, MØller JV, Nissen P (2007) The structural basis of calcium transport by the calcium pump. Nature 450:1036–1042

Olianas MC, Dedoni S, Onali P (2012) Potentiation of dopamine D1-like receptor signaling by concomitant activation of δ- and μ-opioid receptors in mouse medial prefrontal cortex. Neurochem Int 61:1404–1416

Olsen ML, Campbell SL, Sontheimer H (2007) Differential disruption of Kir4.1 in spinal cord astrocytes suggests regional differences in K^+ homeostasis. J Neurophysiol 98:786–793

Olsen ML, Sontheimer H (2008) Functional implications for Kir4.1 channels in glial biology: from K^+ buffering to cell differentiation. J Neurochem 107:589–601

Oostland M, Sellmeijer J, van Hooft JA (2011) Transient expression of functional serotonin 5-HT3 receptors by glutamatergic granule cells in the early postnatal mouse cerebellum. J Physiol 589:4837–4846

Orkand RK, Nicholls JG, Kuffler SW (1966) Effect of nerve impulses on the membrane potential of glial cells in the central nervous system of amphibia. J Neurophysiol 29:788–806

Orthmann-Murphy JL, Abrams CK, Scherer SS (2008) Gap junctions couple astrocytes and oligodendrocytes. J Mol Neurosci 35:101–116

Ostermeier AM, Schlosser B, Schwender D, Sutor B (2000) Activation of mu- and delta-opioid receptors causes presynaptic inhibition of glutamatergic excitation in neocortical neurons. Anesthesiology 93:1053–1063

Pacheco DF, Duarte DG (2005) δ-opioid receptor agonist SNC80 induces peripheral antinociception via activation of ATP-sensitive K^+ channels. Eur J Pharmacol 512:23–28

Pacheco DF, Pacheco CMF, Duarte IDG (2012) Peripheral antinociception induced by δ-opioid receptors activation, but not μ- or κ-, is mediated by Ca^{2+}-activated Cl^- channels. Eur J Pharmacol 674:255–259

Padmawar P, Yao X, Bloch O, Manley GT, Verkman AS (2005) K^+ waves in brain cortex visualized using a long-wavelength K^+-sensing fluorescent indicator. Nat Methods 2:825–827

Palmer AE, Tsien RY (2006) Measuring calcium signaling using genetically targetable fluorescent indicators. Nat Protoc 1:1057–1065

Palty R, Sekler I (2012) The mitochondrial Na^+/Ca^{2+} exchanger. Cell Calcium 52:9–15

Palty R, Silverman WF, Hershfinkel M, Caporale T, Sensi SL, Parnis J, Nolte C, Fishman D, Shoshan-Barmatz V, Herrmann S, Khananshvili D, Sekler I (2010) NCLX is an essential component of mitochondrial Na^+/Ca^{2+} exchange. Proc Natl Acad Sci U S A 107:436–441

Palygin O, Lalo U, Verkhratsky A, Pankratov Y (2010) Ionotropic NMDA and P2X1/5 receptors mediate synaptically induced Ca^{2+} signalling in cortical astrocytes. Cell Calcium 48:225–231

Pamenter ME, Buck LT (2008) δ-opioid receptor antagonism induces NMDA receptor-dependent excitotoxicity in anoxic turtle cortex. J Exp Biol 211:3512–3517

Paredes RM, Etzler JC, Watts LT, Zheng W, Lechleiter JD (2008) Chemical calcium indicators. Methods 46:143–151

Parekh AB, Putney JW Jr (2005) Store-operated calcium channels. Physiol Rev 85:757–810

Parpura V, Verkhratsky A (2012) Homeostatic function of astrocytes: Ca^{2+} and Na^+ signaling. Transl Neurosci 3:334–344

Pasternak GW (2004) Multiple opiate receptors: deja vu all over again. Neuropharmacology 47(suppl 1):312–323

Patel TR, Schielke GP, Hoff JT, Keep RF, Lorris Betz A (1999) Comparison of cerebral blood flow and injury following intracerebral and subdural hematoma in the rat. Brain Res 829:125–133

Patergnani S, Suski JM, Agnoletto C, Bononi A, Bonora M, De Marchi E, Giorgi C, Marchi S, Missiroli S, Poletti F, Rimessi A, Duszynski J, Wieckowski MR, Pinton P (2011) Calcium signaling around Mitochondria Associated Membranes (MAMs). Cell Commun Sgl 9:19. doi:10.1186/1478-811X-9-19

Pellerin L, Magistretti PJ (2012) Sweet sixteen for ANLS. J Cereb Blood Flow Metab 32:1152–1166

Peng L, Martin-Vasallo P, Sweadner KJ (1997) Isoforms of Na, K-ATPase alpha and beta subunits in the rat cerebellum and in granule cell cultures. J Neurosci 17:3488–3502

Pentney RJW, Gratton A (1991) Effects of local delta and mu opioid receptor activation on basal and stimulated dopamine release in striatum and nucleus accumbens of rat: an in vivo electrochemical study. Neuroscience 45:95–102

Perea G, Navarrete M, Araque A (2009) Tripartite synapses: astrocytes process and control synaptic information. Trends Neurosci 32:421–431

Pérez-Alvarez A, Araque A, Martín ED (2013) Confocal microscopy for astrocyte in vivo imaging: recycle and reuse in microscopy. Front Cell Neurosci 7:51. doi:10.3389/fncel.2013.00051

Perocchi F, Gohil VM, Girgis HS, Bao XR, McCombs JE, Palmer AE, Mootha VK (2010) MICU1 encodes a mitochondrial EF hand protein required for $Ca(^{2+})$ uptake. Nature 467:291–296

Pert CB, Snyder SH (1973) Opiate receptor: demonstration in nervous tissue. Science 179:1011–1014

Petersen OH, Michalak M, Verkhratsky A (2005) Calcium signaling: past, present and future. Cell Calcium 38:161–169

Pfrieger FW (2010) Role of glial cells in the formation and maintenance of synapses. Brain Res Rev 63:39–46

Pieper AA, Brat DJ, O'Hearn E, Krug DK, Kaplin AI, Takahashi K, Greenberg JH, Ginty D, Molliver ME, Snyder SH (2001) Differential neuronal localizations and dynamics of phosphorylated and unphosphorylated type 1 inositol 1,4,5-trisphosphate receptors. Neuroscience 102:433–444

Pierce JP, Kelter DT, McEwen BS, Waters EM, Milner TA (2014) Hippocampal mossy fiber leu-enkephalin immunoreactivity in female rats is significantly altered following both acute and chronic stress. J Chem Neuroanat 55:9–17

Pinelis VG, Segal M, Greenberger V, Khodorov BI (1994) Changes in cytosolic sodium caused by a toxic glutamate treatment of cultured hippocampal neurons. Biochem Mol Biol Int 32:475–482

Piochon C, Irinopoulou T, Brusciano D, Bailly Y, Mariani J, Levenes C (2007) NMDA receptor contribution to the climbing fiber response in the adult mouse Purkinje cell. J Neurosci 27:10797–10809

Piros E, Prather PL, Law PY, Evans CJ, Hales TG (1996) Voltage-dependent inhibition of Ca^{2+} channels in GH3 cells by cloned μ- and δ-opioid receptors. Mol Pharmacol 50:947–956

Piros ET, Charles A, Dong L, Evans C, Hales TG (2000) Cloned δ-opioid receptors in GH3 cells inhibit spontaneous Ca^{2+} oscillations and prolactin release through Kir channel activation. J Neurophysiol 83:2691–2698

Pisani A, Calabresi P, Tozzi A, Bernardi G, Knöpfel T (1998) Early sodium elevations induced by combined oxygen and glucose deprivation in pyramidal cortical neurons. Eur J Neurosci 10:3572–3574

Piskorowski RA, Chevaleyre V (2013) Delta-opioid receptors mediate unique plasticity onto parvalbumin-expressing interneurons in area CA2 of the hippocampus. J Neurosci 33:14567–14578

Pittaluga A, Olivero G, Di Prisco S, Merega E, Bisio A, Romussi G, Grilli M, Marchi M (2013) Effects of the neoclerodane Hardwickiic acid on the presynaptic opioid receptors which modulate noradrenaline and dopamine release in mouse central nervous system. Neurochem Int 62:354–359

Pogorelov A, Budantsev A, Pogorelova V, Mizin I (1997) Assay of acetylcholinesterase activity and elemental composition in brain compartments by electron probe microanalysis. Brain Res Brain Res Protoc 1:44–48

Polastron J, Mur M, Mazarguil H, Puget A, Meunier JC, Jauzac P (1994) SK-N-BE: a human neuroblastoma cell line containing two subtypes of delta-opioid receptors. J Neurochem 62:898–906

Portoghese PS, Sultana M, Nelson WL, Klein P, Takemori AE (1992) Delta opioid antagonist activity and binding studies of regioisomeric isothiocyanate derivatives of naltrindole: evidence for delta receptor subtypes. J Med Chem 35:4086–4091

Pradhan A, Smith M, McGuire B, Evans C, Walwyn W (2013) Chronic inflammatory injury results in increased coupling of delta opioid receptors to voltage-gated Ca^{2+} channels. Mol Pain 9:8. doi:10.1186/1744-8069-9-8

Pradhan AA, Befort K, Nozaki C, Gavériaux-Ruff C, Kieffer BL (2011) The delta opioid receptor: an evolving target for the treatment of brain disorders. Trends Pharmacol Sci 32:581–590

Prakriya M, Feske S, Gwack Y, Srikanth S, Rao A, Hogan PG (2006) Orai1 is an essential pore subunit of the CRAC channel. Nature 443:230–233

Prince DA, Lux HD, Neher E (1973) Measurement of extracellular potassium activity in cat cortex. Brain Res 50:489–495

Pu L, Xu N, Xia P, Gu Q, Ren S, Fucke T, Pei G, Schwarz W (2012) Inhibition of activity of GABA transporter GAT1 by δ-opioid receptor. Evid Based Complement Alternat Med 2012:818451. doi:10.1155/2012/818451

Quinlan ME, Alberto CO, Hirasawa M (2008) Short-term potentiation of mEPSCs requires N-, P/Q- and L-type Ca^{2+} channels and mitochondria in the supraoptic nucleus. J Physiol 586:3147–3161

Rada P, Barson JR, Sf L, Hoebel BG (2010) Opioids in the hypothalamus control dopamine and acetylcholine levels in the nucleus accumbens. Brain Res 1312:1–9

Rady JJ, Fujimoto JM (1996) Supraspinal delta2 opioid agonist analgesia in Swiss-Webster mice involves spinal GABAA receptors. Pharmacol Biochem Behav 54:363–369

Raffaello A, De Stefani D, Rizzuto R (2012) The mitochondrial Ca^{2+} uniporter. Cell Calcium 52:16–21

Raley-Susman KM, Kass IS, Cottrell JE, Newman RB, Chambers G, Wang J (2001) Sodium influx blockade and hypoxic damage to CA1 pyramidal neurons in rat hippocampal slices. J Neurophysiol 86:2715–2726

Ransom CB, Ransom BR, Sontheimer H (2000) Activity-dependent extracellular K^+ accumulation in rat optic nerve: the role of glial and axonal Na^+ pumps. J Physiol 522:427–442

Rash JE (2010) Molecular disruptions of the panglial syncytium block potassium siphoning and axonal salutatory conduction: pertinence to neuromyelitis optica and other demyelinating diseases of the central nervous system. Neuroscience 168:982–1008

Rawls SM, McGinty JF (2000) Delta opioid receptors regulate calcium-dependent, amphetamine-evoked glutamate levels in the rat striatum: an in vivo microdialysis study. Brain Res 861:296–304

Regehr W (1997) Interplay between sodium and calcium dynamics in granule cell presynaptic terminals. Biophys J 73:2476–2488

Reid JM, Paterson DJ (1996) Role of K^+ in regulating hypoxic cerebral blood flow in the rat: effect of glibenclamide and ouabain. Am J Physiol 270:H45–H52

Remy C, Remy S, Beck H, Swandulla D, Hans D (2004) Modulation of voltage-dependent sodium channels by the δ-agonist SNC80 in acutely isolated rat hippocampal neurons. Neuropharmocology 47:1102–1112

Reyes RC, Verkhratsky A, Parpura V (2012) Plasmalemmal Na^+/Ca^{2+} exchanger modulates Ca^{2+}-dependent exocytotic release of glutamate from rat cortical astrocytes. ASN Neuro 4, e00075. doi:10.1042/AN20110059

Rezaï X, Faget L, Bednarek E, Schwab Y, Kieffer BL, Massotte D (2012) Mouse δ opioid receptors are located on presynaptic afferents to hippocampal pyramidal cells. Cell Mol Neurobiol 32:509–516

Rezai X, Kieffer BL, Roux MJ, Massotte D (2013) Delta opioid receptors regulate temporoammonic-activated feedforward inhibition to the mouse CA1 hippocampus. PLoS One 8, e79081. doi:10.1371/journal.pone.0079081

Richard-Lalonde M, Nagi K, Audet N, Sleno R, Amraei M, Hogue M, Balboni G, Schiller PW, Bouvier M, Hébert TE, Pineyro G (2013) Conformational dynamics of Kir3.1/Kir3.2 channel activation via δ-opioid receptors. Mol Pharmacol 83:416–428

Rizzuto R, Brini M, Murgia M, Pozzan T (1993) Microdomains with high Ca^{2+} close to IP3-sensitive channels that are sensed by neighboring mitochondria. Science 262:744–747

Rizzuto R, Pinton P, Carrington W, Fay FS, Fogarty KE, Lifshitz LM, Tuft RA, Pozzan T (1998) Close contacts with the endoplasmic reticulum as determinants of mitochondrial Ca^{2+} responses. Science 280:1763–1766

Rose AM, Valdes R Jr (1994) Understanding the sodium pump and its relevance to disease. Clin Chem 40:1674–1685

Rose CR (2002) Na^+ signals at central synapses. Neuroscientist 8:532–539

Rose CR (2003) High-resolution Na^+ imaging in dendrites and spines. Pflugers Arch 446:317–321

Rose CR, Karus C (2013) Two sides of the same coin: sodium homeostasis and signaling in astrocytes under physiological and pathophysiological conditions. Glia 61:1191–1205

Rose CR, Konnerth A (2001) NMDA receptor-mediated Na^+ signals in spines and dendrites. J Neurosci 21:4207–4214

Rose CR, Kovaichuk Y, Eilers J (1999) Two-photon Na^+ imaging in spines and fine dendrites of central neurons. Pflugers Arch 439:201–207

Rose CR, Ransom BR (1996a) Intracellular sodium homeostasis in rat hippocampal astrocytes. J Physiol 491:291–305

Rose CR, Ransom BR (1996b) Mechanisms of H^+ and Na^+ changes induced by glutamate, kainate, and D-aspartate in rat hippocampal astrocytes. J Neurosci 16:5393–5404

Rose CR, Ransom BR (1997a) Gap junctions equalize intracellular Na^+ concentration in astrocytes. Glia 20:299–307

Rose CR, Ransom BR (1997b) Regulation of intracellular sodium in cultured rat hippocampal neurons. J Physiol 499:573–587

Rose CR, Ransom BR, Waxman SG (1997) Pharmacological characterization of Na^+ influx via voltage-gated Na^+ channels in spinal cord astrocytes. J Neurophysiol 78:3249–3258

Rose CR, Waxman SG, Ransom BR (1998) Effects of glucose deprivation, chemical hypoxia, and simulated ischemia on Na^+ homeostasis in rat spinal cord astrocytes. J Neurosci 18:3554–3562

Rouach N, Koulakoff A, Abudara V, Willecke K, Giaume C (2008) Astroglial metabolic networks sustain hippocampal synaptic transmission. Science 322:1551–1555

Rouach N, Koulakoff A, Giaume C (2004) Neurons set the tone of gap junctional communication in astrocytic networks. Neurochem Int 45:265–272

Russell JM (2000) sodium-potassium-chloride cotransport. Physiol Rev 80:211–276

Sabatini BL, Maravall M, Svoboda K (2001) Ca^{2+} signaling in dendritic spines. Curr Opin Neurobiol 11:349–356

Sabatini B, Oertner TG, Svoboda K (2002) The life cycle of Ca^{2+} ions in dendritic spines. Neuron 33:439–452

Saloman JL, Niu KY, Ro JY (2011) Activation of peripheral delta-opioid receptors leads to anti-hyperalgesic responses in the masseter muscle of male and female rats. Neuroscience 190:379–385

Samways DS, Henderson G (2006) Opioid elevation of intracellular free calcium: possible mechanisms and physiological relevance. Cell Signal 18:151–161

Sandor NT, Lendvai B, Vizi ES (1992) Effect of selective opiate antagonists on striatal acetylchonline and dopamine release. Brain Res Bull 29:369–373

Santamaria F, Wils S, De Schutter E, Augustine GJ (2006) Anomalous diffusion in Purkinje cell dendrites caused by spines. Neuron 52:635–648

Santo-Domingo J, Demaurex N (2010) Calcium uptake mechanisms of mitochondria. Biochim Biophys Acta 1797:907–912

Sarne Y, Gafni M (1996) Determinants of the stimulatory opioid effect on intracellular calcium in SK-N-SH and NG108-15 neuroblastoma. Brain Res 722:203–206

Saubermann AJ, Beeuwkes R, Peters PD (1981a) Application of scanning electron microscopy to X-ray analysis of frozen-hydrated sections. II. Analysis of standard solutions and artificial electrolyte gradients. J Cell Biol 88:268–273

Saubermann AJ, Echlin P, Peters PD, Beeuwkes R (1981b) Application of scanning electron microscopy to X-ray analysis of frozen-hydrated sections. I. Specimen handling techniques. J Cell Biol 88:257–267

Scherrer G, Tryoen-Tóth P, Filliol D, Matifas A, Laustriat D, Cao YQ, Basbaum AI, Dierich A, Vonesh JL, Gavériaux-Ruff C, Kieffer BL (2006) Knockin mice expressing fluorescent delta-opioid receptors uncover G protein-coupled receptor dynamics in vivo. Proc Natl Acad Sci U S A 103:9691–9696

Schreiner AE, Rose CR (2012) Quantitative imaging of intracellular sodium. In: Méndez-Vilas A (ed) Current microscopy contributions to advances in science and technology. Formatex, Badajoz, pp 119–129

Schwaller B (2009) The continuing disappearance of "pure" Ca^{2+} buffers. Cell Mol Life Sci 66:275–300

Schwaller B (2010) Cytosolic Ca^{2+} buffers. Cold Spring Harb Perspect Biol 2:a004051. doi:10.1101/cshperspect.a004051

Seifert G, Hüttmann K, Binder DK, Hartmann C, Wyczynski A, Neusch C, Steinhäuser C (2009) Analysis of astroglial K^+ channel expression in the developing hippocampus reveals a predominant role of the Kir4.1 subunit. J Neurosci 29:7474–7488

Shabala L, Howells C, West AK, Chung RS (2010) Prolonged Abeta treatment leads to impairment in the ability of primary cortical neurons to maintain K^+ and Ca^{2+} homeostasis. Mol Neurodegener 5:30. doi:10.1186/1750-1326-5-30

Shao Y, McCarthy KD (1997) Responses of Bergmann glia and granule neurons in situ to N-methyl-D-aspartate, norepinephrine, and high potassium. J Neurochem 68:2405–2411

Sharp AH, McPherson PS, Dawson TM, Aoki C, Campbell KP, Snyder SH (1993) Differential immunohistochemical localization of inositol 1,4,5-trisphosphate- and ryanodine-sensitive Ca^{2+} release channels in rat brain. J Neurosci 13:3051–3063

Sheldon AL, Robinson MB (2007) The role of glutamate transporters in neurodegenerative diseases and potential opportunities for intervention. Neurochem Int 51:333–355

Sheldon C, Diarra A, Cheng YM, Church J (2004) Sodium influx pathways during and after anoxia in rat hippocampal neurons. J Neurosci 24:11057–11069

Shen KF, Crain SM (1989) Dual opioid modulation of the action potential duration of mouse dorsal root ganglion neurons in culture. Brain Res 491:227–242

Shen KF, Crain SM (1990) Cholera toxin-B subunit blocks excitatory effects of opioids on sensory neuron action potentials indicating that GM1 ganglioside may regulate Gs-linked opioid receptor functions. Brain Res 531:1–7

Shen KZ, Johnson SW (2002) Presynaptic modulation of synaptic transmission by opioid receptor in rat subthalamic nucleus in vitro. J Physiol 541:219–230

Shi Y, Huang F, Chen W, Wu Y, Tang Y, Hu Q (2000) A DPDPE-induced enhancement of inward rectifier potassium current via opioid receptor in neuroblastoma x glioma NG108-15 cells. Neurosci Res 36:209–214

Shirasaki T, Abe K, Soeda F, Takahama K (2004) δ-Opioid receptor antagonists inhibit GIRK channel currents in acutely dissociated brainstem neurons of rat. Brain Res 1006:190–197

Shirota O, Nagamatsu K, Sekita S (2006) Neo-clerodane diterpenes from the hallucinogenic sage Salvia divinorum. J Nat Prod 69:1782–1786

Shutov LP, Kim MS, Houlihan PR, Medvedeva YV, Usachev YM (2013) Mitochondria and plasmamembrane Ca^{2+}-ATPase control presynaptic Ca^{2+} clearance in capsaicin-sensitive rat sensory neurons. J Physiol 591:2443–2462

Sibille J, Pannasch U, Rouach N (2014) Astroglial potassium clearance contributes to short-term plasticity of synaptically evoked currents at the tripartite synapse. J Physiol 592:87–102

Silver IA, Deas J, Erecińska M (1997) Ion homeostasis in brain cells: differences in intracellular ion responses to energy limation between cultured neurons and glial cells. Neuroscience 78:589–601

Simon EJ, Hiller JM, Edelman I (1973) Stereospecific binding of the potent narcotic analgesic (3H) etorphine to rat-brain homogenate. Proc Natl Acad Sci U S A 70:1947–1949

Simpson AWM (1999) Fluorescent measurement of $[Ca^{2+}]c$. In: Lambert DG (ed) Methods in molecular biology, vol 114, Calcium signaling protocols. Humana Press, Totowa, pp 3–30

Slezak M, Pfrieger FW, Soltys Z (2006) Synaptic plasticity, astrocytes and morphological homeostasis. J Physiol Paris 99:84–91

Smart D, Lambert DG (1996) δ-opioid stimulate inositol 1,4,5-trisphosphate formation, and so mobilize Ca^{2+} from intracellular stores, in undifferentiated NG108-15 cells. J Neurochem 66:1462–1467

Smyth JT, Hwang SY, Tomita T, DeHaven WI, Mercer JC, Putney JW (2010) Activation and regulation of store-operated calcium entry. J Cell Mol Med 14:2337–2349

Snead OC III (1986) Opiate-induced seizures: a study of mu and delta specific mechanisms. Exp Neurol 93:348–358

Snyder SH, Pasternak GW (2003) Historical review: opioid receptors. Trends Pharmacol Sci 24:198–205

Sofuoglu M, Portoghese PS, Takemori AE (1991) Differential antagonism of delta opioid agonists by naltrindole and its benzofuran analog (NTB) in mice: evidence for delta opioid receptor subtypes. J Pharmacol Exp Ther 257:676–680

Soler-Llavina GJ, Sabatini BL (2006) Synapse-specific plasticity and compartmentalized signaling in cerebellar stellate cells. Nat Neurosci 9:798–806

Somjen GG (1979) Extracellular potassium in the mammalian central nervous system. Annu Rev Physiol 41:159–177

Somjen GG (2002) Ion regulation in the brain: implications for pathophysiology. Neuroscientist 8:254–267

Song S-L, Chueh S-H (1999) Phosphorylation promotes the desensitization of the opioid-induced Ca^{2+} increase in NG108-15 cells. Brain Res 818(2):316–325

Sontheimer H, Black JA, Waxman SG (1996) Voltage-gated Na^{+} channels in glia: properties and possible functions. Trends Neurosci 19:325–331

Spain JW, Roth BL, Coscia CJ (1985) Differential ontogeny of multiple opioid receptors (mu, delta, and kappa). J Neurosci 5:584–588

Spencer RJ, Jin W, Thayer SA, Chakrabarti S, Law P-Y, Loh HH (1997) Mobilization of Ca^{2+} from intracellular store in transfected Neuro2a cells by activation of multiple opioid receptor subtypes. Biochem Pharmacol 54:809–818

Spong KE, Robertson RM (2013) Pharmacological blockade of gap junctions induces repetitive surging of extracellular potassium within the locust CNS. J Insect Physiol 59:1031–1040

Spruston N, Schiller Y, Stuart G, Sakmann B (1995) Activity-dependent action potential invasion and calcium influx into hippocampal CA1 dendrites. Science 268:297–300

Standford IM, Cooper AJ (1999) Presynaptic μ and δ opioid receptor modulation of GABAA IPSCs in the rat globus pallidus in vitro. J Neurosci 19:4796–4803

Stefani A, Surmeier DJ, Bernardi G (1994) Opioids decrease high-voltage activated calcium currents in acutely dissociated neostriatal neurons. Brain Res 642:339–343

Steinbeck JA, Henke N, Opatz J, Gruszczynska-Biegala J, Schneider L, Theiss S, Hamacher N, Steinfarz B, Golz S, Brüstle O, Kuznicki J, Methner A (2011) A store-operated calcium entry modulates neuronal network activity in a model of chronic epilepsy. Exp Neurol 232:185–194

Steinberg JP, Takamiya K, Shen Y, Xia J, Rubio ME, Yu S, Jin W, Thomas GM, Linden DJ, Huganir RL (2006) Targeted in vivo mutations of the AMPA receptor subunit GluR2 and its interacting protein PICK1 eliminate cerebellar long-term depression. Neuron 49:845–860

Steiner RA, Oehme M, Ammann D, Simon W (1979) Neutral carrier sodium ion-selective microelectrode for intracellular studies. Anal Chem 51:351–353

Steinhäuser C, Jabs R, Kettenmann H (1994) Properties of GABA and glutamate responses in identified glial cells of the mouse hippocampal slice. Hippocampus 4:19–35

Steinhäuser C, Selfert G, Bedber P (2012) Astrocyte dysfunction in temporal lobe epilepsy: K^{+} channels and gap junction coupling. Glia 60:1192–1202

Stephan J, Haack N, Kafitz KW, Durry S, Koch D, Hochstrate P, Seifert G, Steinhäuser C, Rose CR (2012) Kir4.1 channels mediate a depolarization of hippocampal astrocytes under hyperammonemic conditions in situ. Glia 60:965–978

Stiefel MF, Marmarou A (2002) Cation dysfunction associated with cerebral ischemia followed by reperfusion: a comparison of microdialysis and ion-selective electrode methods. J Neurosurg 97:97–103

Stiefel MF, Tomita Y, Marmarou A (2005) Secondary ischemia impairing the restoration of ion homeostasis following traumatic brain injury. J Neurosurg 103:707–714

Strohschein S, Hüttmann K, Gabriel S, Binder DK, Heinemann U, Steinhäuser C (2011) Impact of aquaporin-4 channels on K^+ buffering and gap junction coupling in the hippocampus. Glia 59:973–980

Stuart G, Schiller J, Sakmann B (1997) Action potential initiation and propagation in rat neocortical pyramidal neurons. J Physiol 505:617–632

Stummer W, Lorris Betz A, Keep RF (1995) Mechanisms of brain ion homeostasis during acute and chronic variations of plasma potassium. J Cereb Blood Flow Metab 15:336–344

Stys PK, Lehning E, Saubermann AJ, LoPachin RM Jr (1997) Intracellular concentrations of major ions in rat myelinated axons and glia: calculations based on electron probe X-ray microanalyses. J Neurochem 68:1920–1928

Su G, Kintner DB, Flagella M, Shull GE, Sun D (2002) Astrocytes from Na(+)-K(+)-Cl(-) cotransporter-null mice exhibit absence of swelling and decrease in EAA release. Am J Physiol Cell Physiol 282:C1147–C1160

Suarez-Roca H, Maixner W (1992) Delta-opioid-receptor activation by [D-Pen2, D-Pen5]enkephalin and morphine inhibits substance P release from trigeminal nucleus slices. Eur J Pharmacol 229:1–7

Sugita S, North RA (1993) Opioid action on neurons of rat lateral amygdala in vitro. Brain Res 612:151–155

Sussman IP, MacGregor LC, Masters BR, Matschinsky FM (1988) Quantitative histochemical determination of Na^+ and K^+ in microscopic samples using carbon furnace atomic absorption spectrometry. J Histochem Cytochem 36:237–244

Sutor B, Hagerty T (2005) Involvement of gap junctions in the development of neocortex. Biochim Biophys Acta 1719:59–68

Suzuki T, Aoki T, Ohnishi O, Nagase H, Narita M (2000) Different effects of NMDA/group I metabotropic glutamate receptor agents in δ- and μ-opioid receptor agonist-induced supraspinal antinociception. Eur J Pharmacol 396:23–28

Svingos AL, Cheng PY, Clarke CL, Pickel VM (1995) Ultrastructural localization of δ-opioid receptor and Met5-enkephalin immunoreactivity in rat insular cortex. Brain Res 700:25–39

Svingos AL, Clarke CL, Pickel VM (1999) Localization of the δ-opioid receptor and dopamine transporter in the nucleus accumbens shell: implications for opiate and psychostimulant cross-sensitization. Synapse 34:1–10

Svoboda K, Tank DW, Denk W (1996) Direct measurement of coupling between dendritic spines and shafts. Science 272:716–719

Sweadner KJ (1989) Isozymes of the Na^+/K^+-ATPase. Biochim Biophys Acta 988:185–220

Syková E (1983) Extracellular K^+ accumulation in the central nervous system. Prog Biophys Mol Biol 42:135–189

Syková E, Chvátal A (1993) Extracellulae ionic and volume changes: the role in glia-neuron interaction. J Chem Neuroanat 6:247–260

Syková E, Orkand RK (1980) Extracellular potassium accumulation and transmission in frog spinal cord. Neuroscience 5:1421–1428

Syková E, Rothenberg S, Krekule I (1974) Changes of extracellular potassium concentration during spontaneous activity in the mesencephalic reticular formation of the rat. Brain Res 79:333–337

Takahashi A, Camacho P, Lechleiter JD, Herman B (1999) Measurement of intracellular calcium. Physiol Rev 79:1089–1125

Takahashi S, Driscoll BF, Law MJ, Sokoloff L (1995) Role of sodium and potassium ions in regulation of glucose metabolism in cultured astroglia. Proc Natl Acad Sci U S A 92:4616–4620
Takumi T, Ishii T, Horio Y, Morishige K, Takahashi N, Yamada M, Yamashita T, Kiyama H, Sohmiya K, Nakanishi S, Kurachi Y (1995) A novel ATP-dependent inward rectifier potassium channel expressed predominantly in glial cells. J Biol Chem 270:16339–16346
Talbot JD, David G, Barrett EF (2003) Inhibition of mitochondrial Ca^{2+} Ca^{2+} uptake affects phasic release from motor terminals differently depending on external [Ca^{2+}]. J Neurophysiol 90:491–502
Tanahashi S, Ueda Y, Nakajima A, Yamamura S, Nagase H, Okada M (2012) Novel δ1-receptor agonist KNT-127 increases the release of dopamine and L-glutamate in the striatum, nucleus accumbens and median pre-frontal cortex. Neuropharmacology 62:2057–2067
Tanaka E, North RA (1994) Opioid actions on rat anterior cingulated cortex neurons in vitro. J Neurosci 14:1106–1113
Tang T, Kiang JG, Cox BM (1994) Opioids acting through delta receptors elicit a transient increase in the intracellular free calcium concentration in dorsal root ganglion-neuroblastoma hybrid ND8-47 cells. J Pharmacol Exp Ther 270:40–46
Tang T, Stevens BA, Cox BM (1996) Opioid regulation of intracellular free calcium in cultured mouse dorsal root ganglion neurons. J Neurosci Res 44:338–343
Tang WJ, Gilman AG (1991) Type-specific regulation of adenylyl cyclase by G protein beta gamma subunits. Science 254:1500–1503
Tang X, Taniguchi K, Kofuji P (2009) Heterogeneity of Kir4.1 channel expression in glia revealed by mouse transgenesis. Glia 57:1706–1715
Tarasov AI, Griffiths EJ, Rutter GA (2012) Regulation of ATP production by mitochondrial Ca($^{2+}$). Cell Calcium 52:28–35
Tas PWL, Kassa PT, Kress HG, Koschel K (1987) Characterization of an $Na^+/K^+/Cl^-$ cotransport in primary cultures of rat astrocytes. Biochim Biophys Acta 903:411–416
Taylor CP, Weber ML, Gaughan CL, Lehning EJ, LoPachin RM (1999) Oxygen/glucose deprivation in hippocampal slices: altered intraneuronal elemental composition predicts structural and functional damage. J Neurosci 19:619–629
Tecott LH, Maricq AV, Julius D (1993) Nervous system distribution of the serotonin 5-HT3 receptor mRNA. Proc Natl Acad Sci U S A 90:1430–1434
Therien AG, Blostein R (2000) Mechanisms of sodium regulation. Am J Physiol Cell Physiol 279:C541–C566
Thomas RC (1972) Intracellular sodium activity and the sodium pump in snail neurones. J Physiol 220:55–71
Thomas RC, Bers DM (2013) Calcium-sensitive mini- and microelectrodes. Cold Spring Harb Protoc 2013:305–309. doi:10.1101/pdb.top066290
Thorlin T, Eriksson PS, Hansson E, Ronnback L (1997) [D-Pen2,5]enkephalin and glutamate regulate the expression of delta-opioid receptors in rat cortical astrocytes. Neurosci Lett 232:67–70
Thorlin T, Eriksson PS, Persson PAI, Åberg ND, Hansson E, Rőnnbäck L (1998a) Delta-opioid receptors on astroglial cells in primary culture: mobilization of intracellular free calcium via a pertussis sensitive G protein. Neuropharmacology 37:299–311
Thorlin T, Roginski RS, Choudhury K, Nilsson M, Ronnback L, Hansson E, Eriksson PS (1998b) Regulation of the glial glutamate transporter GLT-1 by glutamate and delta-opioid receptor stimulation. FEBS Lett 425:453–459
Tian L, Hires SA, Mao T, Huber D, Chiappe ME, Chalasani SH, Petreanu L, Akerboom J, McKinney SA, Schreiter ER, Bargmann CI, Jayaraman V, Svoboda K, Looger LL (2009) Imaging neural activity in worms, flies and mice with improved GCaMP calcium indicators. Nat Methods 6:875–881
Toselli M, Tosetti P, Taglietti V (1997) μ and δ opioid receptor activation inhibits ω-conotoxin-sensitive calcium channels in a voltage- and time-dependent mode in the human neuroblastoma cell line SH-SY5Y. Pflugers Arch 433:587–596
Toselli M, Tosetti P, Taglietti V (1999) Kinetic study of N-type calcium current modulation by δ-opioid receptor activation in the mammalian cell line NG108-15. Biophys J 76:2560–2574

Toyoshima C (2008) Structural aspects of ion pumping by Ca^{2+}-ATPase of sarcoplasmic reticulum. Arch Biochem Biophys 476:3–11
Tsai MF, Jiang D, Zhao L, Clapham D, Miller C (2014) Functional reconstitution of the mitochondrial Ca^{2+}/H^{+} antiporter Letm1. J Gen Physiol 143:67–73
Tsang CW, Elrick DB, Charlton MP (2000) α-Latrotoxin releases calcium in from motor nerve terminals. J Neurosci 20:8685–8692
Tsao JW, Paramananthan N, Parkes HG, Dunn JF (1999a) Altered brain metabolism in the C57BL/Wld mouse strain detected by magnetic resonance spectroscopy: association with delayed Wallerian degeneration? J Neurol Sci 168:1–12
Tsao L-I, Cadet JL, Su T-P (1999b) Reversal by peptide [D-Ala2, D-Leu5]enkephalin of the dopamine transporter loss caused by methamphetamine. Eur J Pharmacol 372:R5–R7
Tsao LI, Ladenheim B, Andrews AM, Chiueh CC, Cadet JL, Su TP (1998) Delta opioid peptide [D-Ala2, D-Leu5]enkephalin blocks the long-term loss of dopamine transporters induced by multiple administration of methamphetamine: involvement of opioid receptors and reactive oxygen species. J Pharmacol Exp Ther 287:322–331
Tsien RY (1980) New calcium indicators and buffers with high selectivity against magnesium and protons: Design, synthesis, and properties of prototype structures. Biochemistry 19:2396–2404
Tsien RY (1981) A non-disruptive technique for loading calcium buffers and indicators into cells. Nature 290:527–528
Tsien RY (1983) Intracellular measurement of ion activities. Annu Rev Biophys Bioeng 12:91–116
Tsien RY, Pozzan T, Rink TJ (1982) Calcium homeostasis in intact lymphocytes: cytoplasmic free calcium monitored with a new intracellularly trapped fluorescent indicator. J Cell Biol 94:325–334
Tsubokawa H, Miura M, Kano M (1999) Elevation of intracellular Na^{+} induced by hyperpolarization at the dendrites of pyramidal neurons of mouse hippocampus. J Physiol 517:135–142
Tu H, Nelson O, Bezprozvanny A, Wang Z, Lee SF, Hao YH, Serneels L, De Strooper B, Yu G, Bezprozvanny I (2006) Presenilins form ER Ca^{2+} leak channels, a function disrupted by familial Alzheimer's disease-linked mutations. Cell 126:981–993
Tuusa JT, Leskelä TT, Petäjä-Repo UE (2010) Human delta opioid receptor biogenesis is regulated via interactions with SERCA2b and calnexin. FEBS J 277:2815–2829
Tuusa JT, Petäjä-Repo UE (2011) Phe27Cys polymorphism of the human delta opioid receptor predisposes cells to compromised calcium signaling. Mol Cell Biochem 351:173–181
Ueda M, Sugimoto K, Oyama T, Kuraishi Y, Satoh M (1995) Opioidergic inhibition of capsaicin-evoked release of glutamate from rat spinal dorsal horn slices. Neuropharmacology 34:303–308
Uhlmann EJ, Wong M, Baldwin RL, Bajenaru ML, Onda H, Kwiatkowski DJ, Yamada KA, Gutmann DH (2002) Astrocyte-specific TSC1 conditional knockout mice exhibit abnormal neuronal organization and seizures. Ann Neurol 52:285–296
Unichenko P, Myakhar O, Kirischuk S (2012) Intracellular Na^{+} concentration influences short-term plasticity of glutamate transporter-mediated currents in neocortical astrocytes. Glia 60:605–614
Vaillend C, Mason SE, Cuttle MF, Alger BE (2002) Mechanisms of neuronal hyperexcitability caused by partial inhibition of Na^{+}-K^{+}-ATPase in the rat CA1 hippocampal region. J Neurophysiol 88:2963–2978
van Rijn RM, Defriel JN, Whistler JL (2013) Pharmacological traits of delta opioid receptors: pitfalls or opportunities? Psychopharmacology (Berl) 228:1–18
Verkhratsky A, Orkand RK, Kettenmann H (1998) Glial calcium: homeostasis and signaling function. Physiol Rev 78:99–141
Verkhratsky A, Parpura V (2014) Store-operated calcium entry in neuroglia. Neurosci Bull 30:125–133
Verkhratsky A, Rodríguez JJ, Parpura V (2012) Calcium signalling in astroglia. Mol Cell Endocrinol 353:45–56

Verkhratsky A, Steinhäuser C (2000) Ion channels in glial cells. Brain Res Brain Res Rev 32:380–412
Verstreken P, Ly CV, Venken KJ, Koh TW, Zhou Y, Bellen HJ (2005) Synaptic mitochondria are critical for mobilization of reserve pool vesicles at Drosophila neuromuscular junctions. Neuron 47:365–378
Vlaskovska M, Schramm M, Nylander I, Kasakov L, You ZB, Herrera-Marschitz M, Terenius L (1997) Opioid effects on $45Ca^{2+}$ uptake and glutamate release in rat cerebral cortex in primary culture. J Neurochem 68:517–524
Vodovotz Y, An G, Androulakis IP (2013) A systems engineering perspective on homeostasis and disease. Front Bioeng Biotechnol 1:6. doi:10.3389/fbioe.2013.00006
Voipio J, Pasternack M, MacLeod K (1994) Ion-sensitive microelectrodes. In: Ogden D (ed) Microelectrode techniques, the Plymouth workshop handbook, 2nd edn. The Company of Biologists, Cambridge, pp 275–316
Volterra A, Liaudet N, Savtchouk I (2014) Astrocyte Ca^{2+} signaling: an unexpected complexity. Nat Neurosci 15:327–335
von Bohlen Und Halbach O (2009) Structure and function of dendritic spines within the hippocampus. Ann Anat 191:518–531
Voutsinos-Porche B, Bonvento G, Tanaka K, Steiner P, Welker E, Chatton JY, Magistretti P, Pellerin L (2003) Glial glutamate transporters mediate a functional metabolic crosstalk between neurons and astrocytes in the mouse developing cortex. Neuron 37:275–286
Vyklicky L, Sykova E, Kriz N, Ujec E (1972) Post-stimulation changes of extracellular potassium concentration in the spinal cord of the rat. Brain Res 45:608–611
Vyskočil F, Kritz N, Bures J (1972) Potassium-selective microelectrodes used for measuring the extracellular brain potassium during spreading depression and anoxic depolarization in rats. Brain Res 39:255–259
Walker JL (1971) Ion specific liquid ion exchanger microelectrodes. Anal Chem 43:89A–92A
Wallraff A, Köhling R, Heinemann U, Theis M, Willecke K, Steinhäuser C (2006) The impact of astrocytic gap junctional coupling on potassium buffering in the hippocampus. J Neurosci 26:5438–5447
Walwyn W, John S, Maga M, Evans CJ, Hales TG (2009) δ receptors are required for full inhibitory coupling of μ receptors to voltage-dependent Ca^{2+} channels in dorsal root ganglion neurons. Mol Pharmacol 76:134–143
Walwyn W, Maidment NT, Sanders M, Evans CJ, Kieffer BL, Hales TG (2005) Induction of δ opioid receptor function by up-regulation of membrane receptors in mouse primary afferent neurons. Mol Pharmacol 68:1688–1698
Walz W (1992) Mechanism of rapid K^{+}-induced swelling of mouse astrocytes. Neurosci Lett 135:243–246
Walz W (2000) Role of astrocytes in the clearance of excess extracellular potassium. Neurochem Int 36:291–300
Walz W (ed) (2002) The neuronal environment: brain homeostasis in health and disease. Humana, Totowa
Wang XM, Mokha SS (1996) Opioids modulate N-methyl-D-aspartic acid (NMDA)-evoked responses of trigeminothalamic neurons. J Neurophysiol 76:2093–2096
Walz W, Hinks EC (1986) A transmembrane sodium cycle in astrocytes. Brain Res 368:226–232
Wang HB, Zhao B, Zhong YQ, Li KC, Li ZY, Wang Q, Lu YJ, Zhang ZN, He SQ, Zheng HC, Wu SX, Hökfelt TG, Bao L, Zhang X (2010) Coexpression of delta- and mu-opioid receptors in nociceptive sensory neurons. Proc Natl Acad Sci U S A 107:13117–13122
Wang JB, Imai Y, Eppler CM, Gregor P, Spivak CE, Uhl GR (1993a) mu opiate receptor: cDNA cloning and expression. Proc Natl Acad Sci U S A 90:10230–10234
Wang JF, Shun XJ, Yang HF, Ren MF, Han JS (1993b) Suppression by [D-Pen2, D-Pen5]enkephalin on cyclic AMP dependent protein kinase-induced, but not protein kinase C-induced increment of intracellular free calcium in NG108-15 cells. Life Sci 52:1519–1525
Watson GB, Lanthorn TH (1993) Electrophysiological actions of delta opioids in CA1 of the rat hippocampal slices are mediated by one delta receptor subtype. Brain Res 601:129–135

Weber JT (2004) Calcium homeostasis following traumatic neuronal injury. Curr Neurovasc Res 1:151–171

Wild KD, Vanderah T, Mosberg HI, Porreca F (1991) Opioid δ receptor subtypes are associated with different potassium channels. Eur J Pharmacol 193:135–136

Witkowski G, Szulczyk P (2006) Opioid μ receptor activation inhibits sodium currents in prefrontal cortical neurons via a protein kinase A- and C-dependent mechanism. Brain Res 1094:92–106

Wu Q, Xia S, Lin J, Cao D, Chen W, Liu L, Fu Y, Liang J, Cao M (2013) Effects of the altered activity of δ-opioid receptor on the expression of glutamate transporter type 3 induced by chronic exposure to morphine. J Neurol Sci 335:174–181

Wu Z-Z, Chen S-R, Pan H-L (2008) Distinct inhibition of voltage-activated Ca^{2+} channels by δ-opioid agonists in dorsal root ganglion neurons devoid of functional T-type Ca^{2+} currents. Neuroscience 153:1256–1267

Xi ZX, Akasu T (1997) Opioid peptides modulate GABAA receptor responses in neurons of bullfrog dorsal root ganglia. Brain Res 758:163–168

Xia P, Pei G, Schwarz W (2006) Regulation of the glutamate transporter EAAC1 by expression and activation of δ-opioid receptor. Eur J Neurosci 24:87–93

Xia Y, Haddad GG (1991) Ontogeny and distribution of opioid receptors in the rat brainstem. Brain Res 549:181–193

Xia Y, Haddad GG (2001) Major difference in the expression of delta- and mu-opioid receptors between turtle and rat brain. J Comp Neurol 436:202–210

Xia Y, Zhao P, Xue J, Gu XQ, Sun X, Yao H, Haddad GG (2003) Na^+ channel expression and neuronal function in the Na^+/H^+ exchanger 1 null mutant mouse. J Neurophysiol 89:229–236

Xiao AY, Homma M, Wang XQ, Wang X, Yu SP (2001) Role of K^+ efflux in apoptosis induced by AMPA and kainite in mouse cortical neurons. Neuroscience 108:61–67

Xie Y, Dengler K, Zacharias E, Wilffert B, Tegtmeier F (1994) Effects of the sodium channel blocker tetrodotoxin (TTX) on cellular ion homeostasis in rat brain subjected to complete ischemia. Brain Res 652:216–224

Xiong ZQ, Stringer JL (2000) Sodium pump activity, not glial spatial buffering, clears potassium after epileptiform activity induced in the dentate gyrus. J Neurophysiol 83:1443–1451

Xu L, Zeng LH, Wong M (2009) Impaired astrocytic gap junction coupling and potassium buffering in a mouse model of tuberous sclerosis complex. Neurobiol Dis 34:291–299

Yajima Y, Narita M, Takahashi-Nakano Y, Misawa M, Nagase H, Mizoguchi H, Tseng LF, Suzuki T (2000) Effects of differential modulation of mu-, delta- and kappa-opioid systems on bicuculline-induced convulsions in the mouse. Brain Res 862:120–126

Yamada M, Inanobe A, Kurachi Y (1998) G protein regulation of potassium ion channels. Pharmacol Rev 50:723–760

Yamazaki M, Mizoguchi H, Ohsawa M, Tseng LF, Suzuki T, Narita M (2000) Implications of Ca^{2+}-activated Cl^- channels in the δ-opioid receptor-mediated antinociception in the mouse spinal cord. Neurosci Lett 295:113–115

Yang F, He X, Russell J, Lu B (2003) Ca^{2+} influx-independent synaptic potentiation mediated by mitochondrial Na^+-Ca^{2+} exchanger and protein kinase C. J Cell Biol 163:511–523

Yang TT, Hung CF, Lee YJ, Su MJ, Wang SJ (2004) Morphine inhibits glutamate exocytosis from rat cerebral cortex nerve terminals (synaptosomes) by reducing Ca^{2+} influx. Synapse 51:83–90

Yasuda R, Sabatini BL, Svoboda K (2003) Plasticity of calcium channels in dendritic spines. Nat Neurosci 9:948–955

Yeung LC, Shouval HZ, Blais BS, Cooper LN (2004) Synaptic homeostasis and input selectivity follow from a calcium-dependent plasticity model. Proc Natl Acad Sci U S A 101:14943–14948

Yoo JH, Bailey A, Ansonoff M, Pintar JE, Matifas A, Kieffer BL, Kitchen I (2010) Lack of genotype effect on D1, D2 receptors and dopamine transporter binding in triple MOP-, DOP-, and KOP-opioid receptor knockout mice of three different genetic backgrounds. Synapse 64:520–527

Yoon SH, Jin W, Spencer RJ, Loh HH, Thayer SA (1998) Desensitization of δ-opioid-induced mobilization of Ca^{2+} stores in NG108-15 cells. Brain Res 802:9–18

Yoon SH, Lo T-M, Loh HH, Thayer SA (1999) δ-Opioid-induced liberation of Gβγ mobilizes Ca^{2+} stores in NG108-15 cells. Mol Pharmacol 56:902–908

Yoshii M, Furukawa T, Ogihara Y, Watabe S, Shiotani T, Ishikawa Y, Nishimura M, Nukada T (2004) Negative regulation of opioid receptor-G protein-Ca^{2+} channel pathway by the nootropic nefiracetam. Ann N Y Acad Sci 1025:389–397

Young KW, Bampton ETW, Pinòn L, Bano D, Nicotera P (2008) Mitochondrial Ca^{2+} signaling in hippocampal neurons. Cell Calcium 43:296–306

Yu SP (2003) Regulation and critical role of potassium homeostasis in apoptosis. Prog Neurobiol 70:363–386

Yu SP, Yeh C-H, Strasser U, Tian M, Choi DW (1999) NMDA receptor-mediated K^+ efflux and neuronal apoptosis. Science 284:336–339

Yu VC, Richards ML, Sadée W (1986) A human neuroblastoma cell line expresses mu and delta opioid receptor sites. J Biol Chem 261:1065–1070

Yu XM, Salter MW (1998) Gain control of NMDA-receptor currents by intracellular sodium. Nature 396:469–474

Yuste R, Majewska A, Holthoff K (2000) From form to function: calcium compartmentalization in dendritic spines. Nat Neurosci 3:653–659

Yuzaki M, Forrest D, Verselis LM, Sun SC, Curran T, Connor JA (1996) Functional NMDA receptors are transiently active and support the survival of Purkinje cells in culture. J Neurosci 16:4651–4661

Zachariou V, Goldstein BD (1996) Delta-opioid receptor modulation of the release of substance P-like immunoreactivity in the dorsal horn of the rat following mechanical or thermal noxious stimulation. Brain Res 736:305–314

Zanotto L, Heinemann U (1983) Aspartate and glutamate induced reductions in extracellular free calcium and sodium concentration in area CA1 of 'in vitro' hippocampal slices of rats. Neurosci Lett 35:79–84

Zeng X, Zhao X, Yang Y, Kuai J, Cao C, Yu D, Zhao H, Chai W, Yao L (2011) Opioid δ1 and δ2 receptor agonist attenuate myocardial injury via mPTP in rats with acute hemorrhagic shock. J Surg Res 169:267–276

Zetterström TSC, Vaughan-Jones RD, Grahame-Smith DG (1995) A short period of hypoxia produces a rapid and transient rise in $[K^+]e$ in rat hippocampus in vivo which is inhibited by certain K^+-channel blocking agents. Neuroscience 67:815–821

Zhang J, Gibney GT, Zhao P, Xia Y (2002) Neuroprotective role of delta-opioid receptors in cortical neurons. Am J Physiol Cell Physiol 282:C1225–C1234

Zhang J, Haddad GG, Xia Y (2000) Delta-, but not mu- and kappa-, opioid receptor activation protects neocortical neurons from glutamate-induced excitotoxic injury. Brain Res 885:143–153

Zhang J, Qian H, Zhao P, Hong SS, Xia Y (2006) Rapid hypoxia preconditioning protects cortical neurons from glutamate toxicity through δ-opioid receptor. Stroke 37:1094–1099

Zhang Z, Pan ZZ (2012) Signaling cascades for δ-opioid receptor-mediated inhibition of GABA synaptic transmission and behavioral antinociception. Mol Pharmacol 81:375–383

Zhao M, Joo DT (2008) Enhancement of spinal N-methyl-D-aspartate receptor function by remifentanil action at α-opioid receptors as a mechanism for acute opioid-induced hyperalgesia or tolerance. Anesthesiology 109:308–317

Zhao P, Ma MC, Qian H, Xia Y (2005) Down-regulation of delta-opioid receptors in Na^+/H^+ exchanger 1 null mutant mouse brain with epilepsy. Neurosci Res 53:442–446

Zhao Y, Araki S, Wu J, Teramoto T, Chang YF, Nakano M, Abdelfattah AS, Fujiwara M, Ishihara T, Nagai T, Campbell RE (2011) An expanded palette of genetically encoded Ca^{2+} indicators. Science 333:1888–1891

Zhong N, Beaumont V, Zucker RS (2001) Role for mitochondrial and reverse model Na^+/Ca^{2+} exchange and the plasmalemma Ca^{2+} ATPase in post-tetanic potentiation at crayfish neuromuscular junctions. J Neurosci 21:9598–9607

Zieglgänsberger W, French ED, Siggins GR, Bloom FE (1979) Opioid peptides may excite hippocampal pyramidal neurons by inhibiting adjacent inhibitory interneurons. Science 205:415–417

Zilberter Y, Zilberter T, Bregestovski P (2010) Neuronal activity in vitro and the in vivo reality: the role of energy homeostasis. Trends Pharmacol Sci 31:394–401

Zorec R, Araque A, Carmignoto G, Haydon PG, Verkhratsky A, Parpura V (2012) Astroglial excitability and gliotransmission: an appraisal of Ca^{2+} as a signalling route. ASN Neuro 4(2), e00080. doi:10.1042/AN20110061

Zündorf G, Reiser G (2011) Calcium dysregulation and homeostasis of neural calcium in the molecular mechanisms of neurodegenerative diseases provide multiple tragets for neuroprotection. Antioxid Redox Signal 14:1275–1288

Chapter 6
The δ-Opioid Receptor and Stabilization of Brain Ionic Homeostasis in Hypoxia/Ischemia

Dongman Chao and Ying Xia

Abstract Neurons in the mammalian central nervous system are extremely vulnerable to oxygen deprivation and blood supply insufficiency. Hypoxia/ischemia can produce considerably longer or sustained changes in ionic concentrations that are characterized by enhanced K^+ efflux and Na^+-, Ca^{2+}- and Cl^- influx. The hypoxic/ischemic disruption of ionic homeostasis is an initial and key step for hypoxic/ischemic neuronal injury and death. Limiting hypoxic/ischemic ion dysregulation and stabilization of ionic homeostasis in the initial stage of hypoxia/ischemia is an efficient strategy in the treatment of stroke and ischemia-related neurodegenerative conditions. Substantial studies from our and many independent laboratories have demonstrated that activation of the δ-opioid receptor (DOR) is neuroprotective against hypoxic/ischemic insults. In this chapter, we first updated the recent progress in the studies of DOR neuroprotection against hypoxic/ischemic insults and the features and potential mechanisms of hypoxia/ischemia-induced disruption of major cation homeostasis. On the basis of this discussion, we addressed the important role of DOR in the homeostatic maintenance of these ions and the underlying mechanisms. Based on our recent work and current literature, we have drawn an initial picture of how DOR protects neurons against hypoxia/ischemia through the regulation of ionic homeostasis. In brief, DOR signals inhibit Na^+ influx and reduce the increase in intracellular Ca^{2+}, thus decreasing the excessive K^+ leakage. This DOR-mediated action involves a PKC-dependent and PKA-independent signaling pathway.

Keywords Ion homeostasis • δ-Opioid receptor • Brain • Hypoxia • Ischemia • Oxygen-glucose deprivation • Neuroprotection

---

D. Chao, Ph.D.
Department of Neurosurgery, The University of Texas Medical School at Houston, Houston, TX 77030, USA
e-mail: chaodm21@yahoo.com

Y. Xia, M.D., Ph.D. (✉)
University of Texas Medical School, Houston, TX, USA
e-mail: ying.xia@uth.tmc.edu; y55738088@gmail.com

Y. Xia (ed.), *Neural Functions of the Delta-Opioid Receptor*,
DOI 10.1007/978-3-319-25495-1_6

Abbreviation

$[Ca^{2+}]i$	Cytosolic free Ca^{2+} concentrations
$[Ca^{2+}]o$	Extracellular Ca^{2+} concentrations
$[K^+]e$	Extracellular K^+ concentrations
$[K^+]i$	Intracellular K^+ concentrations
$[Na^+]i$	Intracellular Na^+ concentrations
$[Na^+]o$	Extracellular Na^+ concentrations
AD	Anoxic depolarization
AMPAR	AMPA receptor
ASICs	Acid-sensing ion channels
BK	Ca^{2+}-activated K^+ channel
CCE	Capacitative Ca^{2+} entry
CNS	Central nervous system
CPP	3-(2-Carboxypiperazin-4-yl)propyl-1-phosphonic acid
DAG	Diacylglycerol
DOR	δ-Opioid receptor
EAATs	Excitatory amino-acid transporters
EPSPs/EPSCs	Excitatory postsynaptic potentials/currents
ER	Endoplasmic reticulum
GPCRs	G protein-coupled receptors
H_2S	Hydrogen sulfide
HPC	Hypoxia preconditioning
iGluRs	Ionotropic glutamate receptors (iGluRs)
IMM	Inner mitochondrial membrane (IMM)
IP_3	Inositol triphosphahate
IP_3Rs	IP_3 receptors
KAR	Kainate receptor
K_{Na}	Na^+-activated K^+ channel
KOR	κ-Opioid receptor
LDH	Lactate dehydrogenase
MAM	Mitochondria-associated membrane
MCU	Mitochondrial Ca^{2+} uniporter
MOR	μ-Opioid receptor
$NC_{Ca\text{-}ATP}$	Ca^{2+}-activated, ATP-sensitive nonselective cation channel
NCX	Na^+/Ca^{2+} exchanger
NHE	Na^+/H^+ exchanger
NKCC	Na^+-K^+-$2Cl^-$ cotransporter
NMDAR	NMDA receptor
$NMDG^+$	N-Methyl-D-glucamine
NOP	Nociceptin receptor
NSCCs	Non-selective cation channels
OGD	Oxygen-glucose deprivation
OMM	Outer mitochondrial membrane

PLC	Phospholipase C
PMCA	Membrane Ca^{2+}-ATPase
PTP	Permeability transition pores
ROS/RNS	Reactive oxygen/nitrogen species
RYRs	Ryanodine receptors
SERCA	Sarco-endoplasmic reticular Ca^{2+} ATPase
SOCCs	Store-operated Ca^{2+} channels
STIM	Stromal interaction molecule
TRP	Transient receptor potential channels
TTX	Tetrodotoxin
VGCCs	Voltage-gated Ca^{2+} channels
VGSCs	Voltage-gated Na^+ channels
$\Delta\Psi m$	Mitochondrial potential

1 Introduction

Life is thought to have originated 3.5–3.8 billion years ago. The first emerging life is presumed to be the simplest prokaryotic cells that are bathed in the salt water of the primitive ocean. During the long evolutionary process, eukaryotic cells emerged and gradually developed the ability to maintain an appropriate ionic environment for their survival and function. Thus, a fluid environment around cells was established, which is full of the asymmetrically distributed ions across the membrane. Advanced life is composed of a large number of highly diverse cells that further perfect their ability to maintain stability in the ions of the "*milieu intérieur*" (internal environment), i.e., ionic homeostasis. Now, it is clear that the extracellular fluid is abundant in Na^+ (up to 150 mM) and Ca^{2+} (1–2 mM), and is relatively poor in K^+ (around 3 mM). Conversely, the concentrations of these ions are reversed inside the cells (with up to 150 mM of K^+, 4–12 mM of Na^+, and Ca^{2+} of 10,000-fold lower than in extracellular fluid for neurons), thus forming a steep electrochemical gradient across the membrane. The concentrations of ions inside and outside the cells, especially in nerve cells, are in dynamic balance under physiological conditions (see the preceding chapter by Chao and Xia). With neuronal activity, the ionic concentrations fluctuate within a small range, and this fluctuation is often transient. Following ionic concentration fluctuation, a rapid increase in activity of ionic pumps (e.g. Na^+/K^+ ATPase, Ca^{2+} ATPase) and exchangers (e.g. Na^+/Ca^{2+} exchangers), as well as spatial ionic buffering mechanisms, brings back the ionic gradients or levels towards their previous levels (see the preceding chapter by Chao and Xia). The maintenance of ionic homeostasis is critical to the normal functioning of neurons. Cells cannot live without an appropriate balance of ions across their cell membrane. The transmembrane ionic gradients provide the critical and necessary conditions not only for the maintenance of membrane potential, generation of action potentials, synaptic transmission, and electrogenic transport of neurotransmitters and neuronal plasticity, but also for regulating osmotic balance, cell volume, proper pH and ionic microenvironment for the normal functioning of macromolecules including enzymes, and Na^+-coupled transport of nutrients into cells. Therefore, the changes in ionic concentrations or the blockade of transmembrane ionic movement can affect a wide variety of neuronal processes (see the preceding chapter by Chao and Xia).

Neurons in the mammalian central nervous system (CNS) are extremely vulnerable to hypoxic/ischemic stress due to their high demand for energy from oxidative phosphorylation and extremely limited capacity in glucose storage. Hypoxia/ischemia triggers a series of alterations in neuronal events, such as changes in membrane function, metabolism, intracellular transduction, and even cell morphology, which eventually leads to cell injury or death (Sung et al. 2008). Initially, the most important alteration is the disruption of ionic homeostasis, that is characterized by enhanced K^+ efflux and influx of Na^+ and Ca^{2+} (Hansen 1985; Chao and Xia 2010). Since neuronal function is critically dependent on the maintenance of Na^+, K^+, and Ca^{2+} transmembrane distribution, deregulated ionic imbalances as well as loss of ion gradients are considered to be key events in brain pathological consequences (Hansen 1985; Yu 2003b; Chao and Xia 2010). Over the past two decades, there has been vast and growing interest in the importance of Ca^{2+} and Na^+ in neuronal responses to hypoxic/ischemic stress and their roles in neuronal injury. Recently, disrupted K^+ homeostasis as a cause of neuronal death began to attract much attention. In comparison with the harmful role of Ca^{2+} and Na^+ overload in hypoxia/ischemia, the role of K^+ efflux appears to be complicated. The enhanced K^+ efflux used to be considered an adaptive mechanism by decreasing neuronal excitability during hypoxia/ischemia. However, recent evidence suggests that an excessive loss of neuronal K^+ is causally linked to neuronal damage and death in various conditions including hypoxia/ischemia (Yu et al. 1997, 1999; Huang et al. 2001; Wei et al. 2003; Liu et al. 2003; Yu 2003b; Zhao et al. 2006b; Nistico et al. 2007; Karki et al. 2007). Presently, the process and features of hypoxic/ischemic disruption of ionic homeostasis as well as ionic interactions during the process are not well understood yet, despite their importance in terms of understanding neuronal responses to hypoxia/ischemia. In fact, there is a major knowledge gap in this field because previous research attention was mainly paid to the resultant consequence of ionic homeostasis in pathophysiological conditions. Therefore, it is essential to update the information concerning the process and features of hypoxia/ischemia-induced disruption of ionic homeostasis to gain a better understanding of hypoxic/ischemic neuronal injury and death.

The δ-opioid receptor (DOR) is a member of the opioid receptor family that also includes classic μ-, and κ-opioid receptors (MOR, KOR) and the later recognized nociceptin receptor (NOP), which all belong to the classical G protein-coupled receptor superfamily, and mainly play an inhibitory role in cell activities (Feng et al. 2012). DOR is widely distributed throughout the CNS (Xia and Haddad 1991, 2001; Mansour et al. 1994; Hiller and Fan 1996; Law et al. 2000; Scherrer et al. 2006) and has diverse functions (Feng et al. 2012). Recent studies, especially those from our laboratory, have demonstrated that activation of DOR is neuroprotective against hypoxic/ischemic insults, and DOR activity may be one of the key factors in determining neuronal survival during hypoxic/ischemic stress as well as excitotoxic stress (see reviews by Chao and Xia 2010; He et al. 2013). Early evidence suggested that opioids may alter ionic transportation across the membrane in normoxic conditions (see the preceding chapter by Chao and Xia). Our recent data further indicate that an important aspect of the DOR-mediated neuroprotection is its action against disruption of ionic homeostasis during anoxia or ischemia (Chao et al. 2007a, b, 2008, 2009, 2012b; Kang et al. 2009, also see our reviews: Chao and Xia 2010; He et al. 2013). Therefore, it appears that opioids are involved in the regulation of ionic homeostasis at rest and in hypoxic/ischemic conditions. In this chapter, we will summarize the research progress in this field and discuss the underlying mechanism based on the current information after reviewing hypoxic/ischemic disruption of ionic homeostasis.

2 DOR-Induced Neuroprotection against Hypoxic/Ischemic Injury

Hypoxia and ischemia induce neuronal injury in the brain, which is a leading cause of neurological disability and death. However, the treatment and prevention of hypoxic/ischemic brain injury remain a major medical challenge. The vast majority of the current research directed to finding treatments for hypoxic/ischemic brain injuries has focused on the use of Ca^{2+} channel blockers, glutamate antagonists, antioxidants/free radical scavengers, and some agents that regulate cytokines and other intracellular molecules. However, many methods that were proved effective in animal models of stroke demonstrated minor or no efficacy in patients in clinical trials (Ginsberg 2008; Green 2008; Chen et al. 2014). Therefore, seeking novel approaches to protect neurons from hypoxic/ischemic injury still attracts much attention from both clinicians and scientists. Our laboratory initiated the study on DOR neuroprotection in the 1990s. Our serial studies since 1997 have proven that activation of DOR is neuroprotective and predisposes neurons to survive under severe stress such as excitatory, hypoxic and ischemic stress, while inhibition of DOR by administration of DOR antagonists causes cell death and further augments neuronal injury during severe hypoxia (see our reviews: Chao and Xia 2010; He et al. 2013). Our observations have been further supported by others (Chao and Xia 2010; He et al. 2013). More information will be discussed in detail in the following.

In past years, some clues from studies of our laboratory and those of others suggest the involvement of DOR in neuroprotection against hypoxic/ischemic stress. Firstly, the brain of the freshwater turtle has a much higher density of DOR than the rat brain, and is more resistant to hypoxic/ischemic stress compared to the rat brain (Xia and Haddad 1991, 2001; Xia et al. 1992). Secondly, the rat cortex shows higher density of DOR than subcortical regions (Xia and Haddad 1991), and is more tolerant to hypoxia than subcortical regions (i.e. thalamus, hippocampus and brain stem) (Lin et al. 1990; O'Reilly et al. 1995; Dunn et al. 1999; Buresh et al. 1999). Thirdly, the distribution of DOR is heterogeneous in the cortex with the outer and inner layers having higher density than in the middle layer. Correspondingly, the outer and inner layers (layer 2 and 5) of the cortex are more tolerant to hypoxia/ischemia than the middle layer (layer 3) (Kornblum et al. 1987; Nedergaard 1988; Lin et al. 1990; Georges et al. 1998; Gniel and Martin 2010). In addition, mammalian hibernation, a distinct energy-conserving state, is associated with depletion of energy stores, intracellular acidosis, and hypoxia. These effects are similar to those that occur during hypoxia/ischemia and this hypoxia-tolerant state can be triggered by DOR activation (Borlongan et al. 2009; Staples et al. 2013). Moreover, endogenous enkephalin levels are greatly increased in the brain during hypoxia (Yan et al. 1995; Armstead 1998; Chen and Du 2000). This increase may be one strategy used to counteract hypoxic stress by increasing activity of DOR. Of all the three major opioid receptors (DOR, MOR, KOR), DOR has been proven more sensitive to stressful stimuli (e.g., hypoxia and ischemia) than the others (Mayfield et al. 1996; Boutin et al. 2003, 2007; Ma et al. 2005; Zhang et al. 2006; Peng et al. 2009). Prolonged exposure to hypoxia (9 % O_2) continuously for 7 days selectively decreases DOR, but not MOR and KOR expression in mouse brain homogenates (Mayfield et al. 1996). DOR is also significantly decreased as early as 6 h post transient focal cerebral ischemia (vs. that of MOR after 24 h and no changes in KOR) (Boutin et al. 2003, 2007). All these notions suggest a crucial role of the DOR system in neuronal regulation under hypoxic/ischemic conditions.

Actually, concerns for the role of opioids in hypoxic/ischemic brain injury can be seen as early as the 1980s. However, earlier studies on the precise role of opioid receptors in neuroprotection against hypoxic/ischemic injury yielded conflicting results. For example, some studies suggested that opioid receptor activation protected the brain from ischemia or extended animal survival time in hypoxia (Hayward et al. 1993; Mayfield and D'Alecy 1994; Endoh et al. 1999; Summers et al. 2003; Oeltgen et al. 2006). However, some other works directly opposited this idea (Hosobuchi et al. 1982; Adams et al. 1986; Skarphedinsson and Thoren 1988; Olinger et al. 1990; Chen et al. 2001b). These in vivo studies indicated opioid receptor inhibition with opioid antagonists (e.g., naloxone that binds to DOR, MOR and KOR) protected the brain from ischemia-induced injury, rather than the activation of opioid receptors. In addition, treatment with morphine (preferential MOR-ligand), U-50488H (KOR ligand) or naloxone provided no significant protection on the hippocampus subjected to ischemia (Iwai et al. 1992), suggesting that MOR activation does not induce neuroprotection in the hippocampal neurons. Another study showed that a variety of MOR-ligands at high concentrations (0.1–3 mM) reduced neuroexcitotoxicity in cortical neurons, which could not be blocked by naloxone, suggesting a "non-opioid effect" (Choi and Viseskul 1988). The major controversies thus hindered the building of a principle consensus for the role of opioids in hypoxic/ischemic brain injury. It is very important to realize the methodological limitations in these studies. A majority of these earlier studies were mainly based on the application of low-selective ligands at high concentrations via intravenous injection. Firstly, a high dose of opioid ligands can cross-react with various opioid receptor subtypes or induce non-opioid effects, which may have complicated the interpretation of the results observed. Secondly, the improved neurological outcome or extended animal survival time following systemic administration of opioid ligands cannot be simply attributed to neuroprotection since systemic opioid ligands may induce multiple effects on various organs, e.g., heart, blood vessels, liver and gastroenteric tissue (Feng et al. 2012), which makes the results more complex.

To clarify the important issues about the role of DOR as well as other opioid receptors in hypoxic/ischemic neuronal injury, more specific techniques and ligands are required in the studies. Our serial studies since 1997 have proven that activation of DOR is neuroprotective and predisposes neurons to survive under severe stress such as glutamate-, hypoxic- and ischemic-stress, while inhibition of DOR by administration of DOR antagonists causes cell death and further augments neuronal injury during severe hypoxia (Zhang et al. 2000, 2002, 2006; Zhao et al. 2005; Ma et al. 2005; Chao et al. 2006, 2007a, b, 2008, 2009, 2012b; Tian et al. 2008, 2013a, b; Kang et al. 2009; Yang et al. 2009; Wang et al. 2014). Since the cortex is abundant in DOR expression as opposed to the hippocampus and other subcortical regions (Xia and Haddad 1991, 2001), we chose the cortex as our research target. We first tested the effects of DOR on glutamate-induced neuronal injury by using two independent methods, i.e., "same field quantification" of morphologic criteria and a biochemical assay of lactate dehydrogenase (LDH) release (an index of cellular injury) (Zhang et al. 2000) since excessive remarkably released glutamate in extracellular space is a well-known mediator of neuronal death that occurs during hypoxia/ischemia (Choi and Rothman 1990). Our results indicated that after 8–10 days in culture, cortical neurons exposed to a very low dose of glutamate (100 μM, 4 h) show substantial neuronal injury. Activation of DOR with DADLE reduces glutamate-induced injury by almost half, which can be completely blocked by DOR antagonist naltrindole, while MOR and KOR ligands (agonists and antagonists) have no appreciable effect on the glutamate-induced injury (Zhang et al. 2000). It is very interesting to note that glutamate, even at a very high concentration (10 mM), does not cause apparent injury to neurons at 4 days of culture (Zhang et al. 2000). This is because cultured rodent cortical neurons do not fully express glutamate receptors until cultured for 6–8 days (Cheung et al. 2007). This fact is, consistent with the observation of immature neurons being more resistant to hypoxic/ischemic injury than mature neurons (Zhang et al. 2002; Sung et al. 2008). Direct examination of DOR effects on hypoxia-induced neuronal injury with the same methods gets the same results as that in the glutamate-induced injury, confirming DOR plays a crucial role in neuroprotection in hypoxic environments (Zhang et al. 2002). To examine the universality of endogenous δ-opioids and DOR's neuroprotection, we further expanded our observation in both rapid and delayed neuronal preconditioning-induced neuroprotection to inspect the involvement of DOR in this process (Ma et al. 2005; Zhang et al. 2006). Our results indicated that rapid hypoxia preconditioning (HPC)-induced neuroprotection is mediated by increasing endogenous DOR ligand binding density (though no appreciable change in DOR mRNA was detected), and can be completely abolished by naltrindole (Zhang et al. 2006). In contrast, severe hypoxia caused a decrease in DOR expression and neuronal injury; the delayed effects of HPC include an increase in DOR mRNA and protein levels and reversed the reduction in levels of the endogenous DOR peptide, Leu-enkephalin, normally seen during severe hypoxia, thus protecting the neurons from severe hypoxia insult (Ma et al. 2005). Moreover, HPC-induced delayed neuroprotection could be blocked by DOR antagonists (Ma et al. 2005). The analysis for signaling pathways of DOR-mediated delayed neuroprotection revealed that DOR-mediated HPC protection depended on its enhancement of the intracellular activity of the G protein-PKC-pERK-Bcl2 pathway and suppression of the phosphorylated p38 and cytochrome c death signals during long-term serve hypoxia (Ma et al. 2005). To validate our observation in vivo, we used a model of cerebral ischemia with middle cerebral artery occlusion to examine the effects of DOR (Tian et al. 2008, 2013a; Yang et al. 2009). Our results indicated that DOR activation indeed protects the cerebral cortex against ischemia, and this protection is mediated by enhancing antioxidant ability, rescuing TrkB signaling, and inhibiting caspase activity (Tian et al. 2008, 2013a; Yang et al. 2009). Most recently we found that DOR activation can partially reverse the chronic hypoxia-induced reduction of BDNF expression and functional TrkB (Tian et al. 2013b), and can also modify (enhance or reverse) chronic hypoxia-induced changes in microRNAs that target HIF, ion transport, axonal guidance, free radical signaling, apoptosis and many other functions (Yang et al. 2012). DOR activation also inhibits TNF-α-mediated inflammatory processes following exposure to severe hypoxia with a different regulation between glial vs. neuronal-like cells (Wang et al. 2014). We also found that DOR exerts its neuroprotection against hypoxic/ischemic insults by its action against disruption of ionic homeostasis at early stage of hypoxia/ischemia (Chao et al. 2006, 2007a, b, 2008, 2009, 2012b; Kang et al. 2009; and the detailed discussion in the latter part of this chapter). All these findings clearly demonstrate that DOR activation plays a neuroprotective role through multiple mechanisms (Chao and Xia 2010; He et al. 2013). In fact, our novel data have clarified several controversies that existed in the literature. For example, the reason that μ-ligands had little protective effect on hippocampal neurons (Iwai et al. 1992) is likely because of the low density of DOR in the hippocampus, while the cause of high-dose μ-ligand induced reduction of neuroexcitotoxicity in cortical neurons (Choi and Viseskul 1988) might be attributed to activation of DOR that are highly expressed in the cortex.

Indeed, in the past few years, many independent laboratories have also demonstrated that DOR is neuroprotective against hypoxic/ischemic stress (Lim et al. 2004; Zhao et al. 2006a; Su et al. 2007; Iwato et al. 2007; Xiong et al. 2007; Charron et al. 2008; Horiuchi et al. 2008; Kao et al. 2008; Govindaswami et al. 2008; Pamenter and Buck 2008; Borlongan et al. 2009; Peng et al. 2009; Zhu et al. 2009, 2011; Gao et al. 2010, 2012; Johnson and Turner 2010; Duan et al. 2011; Yang et al. 2011, 2014; Turner and Johnson 2011; Wang et al. 2011; Kim et al. 2012; Abdul et al. 2013; Staples et al. 2013; Sen et al. 2013; Zheng et al. 2013b; Liu et al. 2014). In rodent models of stroke or asphyxial cardiac arrest-induced global cerebral ischemic injury, for example, up-regulation of DOR expression, and peripheral/central administration of DOR agonists (DADLE, DPDPE, TAN-67, Deltorphin-$D_{variant}$,

BW373U86) significantly reduce the infarct volume, attenuate neurological deficits and neuronal loss, and increase neuron survival in post-ischemia in the brain, particularly in the cortex and hippocampus. These effects can be abolished by DOR antagonist naltrindole (Su et al. 2007; Iwato et al. 2007; Xiong et al. 2007; Charron et al. 2008; Kao et al. 2008; Govindaswami et al. 2008; Borlongan et al. 2009; Gao et al. 2010; Johnson and Turner 2010; Duan et al. 2011; Yang et al. 2014). On the other hand, DOR antagonist (naltrindole) alone significantly worsens ischemic neuronal injury (Iwato et al. 2007; Pamenter and Buck 2008), suggesting a "tonic" role of endogenous DOR system in neuroprotection. Intrathecal treatment with DOR agonist SNC80 can attenuate hind-limb motor dysfunction as well as neuronal and ventral white matter injury following spinal cord ischemia in rats (Horiuchi et al. 2008). In vitro studies also reveal that activation of DOR significantly attenuates neuronal damage and death induced by hypoxia/ischemia (Lim et al. 2004; Zhao et al. 2006a; Peng et al. 2009; Zhu et al. 2009, 2011; Turner and Johnson 2011; Yang et al. 2011; Zheng et al. 2013b). Importantly, HPC in the in vivo model of asphyxial cardiac arrest-induced global cerebral ischemic injury improved the neurological deficit score, inhibited neuronal apoptosis, and increased the number of viable hippocampal CA1 neurons by upregulating the neuronal expression of DOR, and synchronously elevating the content of Leu-enkephalin in the rat brain. This could be attenuated by naltrindole (Gao et al. 2012), which validates our HPC observations in cultured neurons (Ma et al. 2005; Zhang et al. 2006). Moreover, the data from microarray analysis by Kim et al. (2012) show that DOR activation reduces neuronal injury after 24 h of oxygen deprivation by affecting the activities of 5 % genes (1917 out of 39,511 genes) that are involved in several important pathways such as apoptosis, intracellular ion homeostasis, molecular chaperones, and glucose metabolism. Furthermore, Yang et al. (2012) found that DOR activation significantly affect microRNA expression in the cortex exposed to hypoxic stress, suggesting the regulation of microRNAs may play a role in the DOR-induced neuroprotection against hypoxic/ischmic injury.

Some recent studies showed that DOR activation induces neuroprotection via the upregulation of nerve growth factor function and suppression of ischemia/reperfusion-induced production of TNF-α (Abdul et al. 2013; Sen et al. 2013). These observations are also in good agreement with our results (Tian et al. 2013b; Wang et al. 2014). We found that PKC-MAPK pathway plays an important role in DOR protection (Ma et al. 2005). Later, the results of other studies are well consistent with our finding (Narita et al. 2006; Sen et al. 2013).

Most interestingly, DOR activation during the reperfusion period after forebrain ischemia also confers neuroprotective ability manifested as increased neuronal survival and decreased neurodegeneration in the hippocampus at 3 days after ischemia, and cognitive benefits detected at 5–9 days after ischemia (Wang et al. 2011). In other words, DOR activation even offers a therapeutic opportunity of postconditioning neuroprotection against transient forebrain ischemia. This observation has essential relevance in transitional medicine. With so much evidence from our laboratory and others, we now confidently believe that an up-regulation of δ-opioid system, either in function or in expression, may have a therapeutic potential in management of hypoxic/ischemic injury.

3 Hypoxia/Ischemia-Induced Disruption of Ionic Homeostasis

We have noticed that an important aspect of DOR-mediated neuroprotection may be its action against disruption of ionic homeostasis, particularly K^+ and Na^+ homeostasis at early stage of anoxia or ischemia. To better understand DOR regulation of ionic homeostasis in hypoxia/ischemia, it is essential to know the changes of ionic activity during hypoxia/ischemia first. In this section, we will update the information concerning the process and features of hypoxia/ischemia-induced disruption of ionic homeostasis.

Generally, ion gradients are well maintained via multiple homeostatic mechanisms under normal conditions (see the preceding chapter by Chao and Xia). The homeostatic mechanisms of key cations (Na^+, K^+, Ca^{2+}) are, to a great extent, dependent directly or indirectly on the permanent activities of ion pumps, especially Na^+/K^+ ATPase (also known as Na^+-K^+ pump or Na^+ pump), which is responsible for at least 60 % of energy expenditure of nerve cells in the CNS (Schreiner and Rose 2012). Therefore, ion gradients are extremely sensitive to the impairment of energy metabolism (Silver et al. 1997; Hertz 2008), which always occurs in certain pathophysiological conditions such as hypoxia/ischemia. During hypoxia/ischemia, the principal homeostatic mechanisms for ions are dysfunctional due to the depletion of ATP, thus making the phenotype of ion activity very different from that under normally condition. Hypoxia, ischemia, or oxygen-glucose deprivation (OGD, simulated ischemia) produce considerably longer or sustained changes in ionic concentrations that are characterized by enhanced K^+ efflux and Na^+-, Ca^{2+}- and Cl^- influx (Hansen 1978, 1985; Sick et al. 1982, 1987; Jiang and Haddad 1991; Müller and Somjen 2000a, b; Galeffi et al. 2004; Martinez-Sánchez et al. 2004; Pond et al. 2006; Chao et al. 2007a, b, 2008, 2009, 2012b; Pamenter and Buck 2008; Kang et al. 2009; Armstrong et al. 2011). These changes of ions in the models of the intact brain (Hansen 1978, 1985; Sick et al. 1982; Zetterstrő̃m et al. 1995; Reid and Paterson 1996) as well as in vitro brain slices (Sick et al. 1987; Müller and Somjen 2000a, b; Chao et al. 2007a, b, 2008, 2009, 2012b; Kang et al. 2009) can be distinctly divided into two phases. Within the first minutes, hypoxia/ischemia only induces a slow and progressive rise in extracellular K^+ concentration ($[K^+]e$), with little changes in other ions. However, when $[K^+]e$ reaches a threshold level (the so-called "ceiling level" often no more than 10–12 mM), and as hypoxia/ischemia continues, a rapid, additional large increase in $[K^+]e$ bursts out with an accompanied negative shift in extracellular DC potential, and a substantial decrease in extracellular Na^+, Ca^{2+} and Cl^- concentrations (Hansen 1978, 1985; Zetterstrő̃m et al. 1995; Reid and Paterson 1996; Gido et al. 1997; Müller and Somjen 2000a, b; Chao et al. 2007a, b,

2008, 2009, 2012b). After re-introducing oxygen and/or glucose, these changes gradually recover.

The intensity of ion responses to hypoxia/ischemia is determined by multiple inherent and environmental factors including the duration and severity of insults, the species, the age of animals, the region of the brain, the type of cells, channel/receptor/transporter distribution and density etc. For example, hippocampal neurons in hypoxia-tolerant naked mole-rats show a significantly blunted intracellular calcium response to hypoxia when compared to neurons in mice, and slices from neonatal animals from both species show a weaker $[Ca^{2+}]i$ increase than slices from older animals, in response to hypoxia (Peterson et al. 2012). The hypoxic/ischemic increase in $[K^+]e$ is also observed in the brain of some anoxia-tolerant species such as turtles and frogs. However, the magnitude is very small (usually several mM) and the latency is very long (up to 4 h) in comparison to mammalian brains (Sick et al. 1982, 1987; Doll et al. 1991; Xia et al. 1992; Knickerbocker and Lutz 2001). Five minutes of OGD was associated with a significant increase in $[Ca^{2+}]i$ in both young and aged synapses, but a much higher level was reached in the aged synapses than in the younger synapses and was irreversible in the aged synapses (Tonkikh and Carlen 2009). The younger the animal, the longer the time until onset of K^+ leakage from cells and the lower the amount of K^+ leakage are. The rate of pumping of the K^+ ions into the cells during the recovery stage is higher in the oldest group (Yoles et al. 2000). Both mild and severe hypoxia influences the physiology of synaptosomal mitochondria; the modifications are reversible after mild hypoxia but are partly irreversible after severe hypoxia (Aldinucci et al. 2005). The same ischemic stress immediately and acutely damages anoxic depolarization (AD) propagating through the gray matter of the neocortex, the hippocampus, the striatum, the thalamus, and the cerebellar cortex (Brisson et al. 2014). This damage causes sudden membrane potential loss that does not recover in hippocampal and striatal neurons, and also causes non-recoverable swelling and dendritic beading in hippocampal neurons. In contrast, this ischemic stress causes little injury in AD propagation in adjacent brainstem nuclei (Brisson et al. 2014). With the same stress, brainstem neurons from locus ceruleus and mesencephalic nucleus as well as from sensory and motor nuclei are only slowly depolarized and then repolarized post-ischemia, and mesencephalic neurons in midbrain appear uninjured in morphology (Brisson et al. 2014). Respiratory neurons in the pre-Bötzinger complex of the mammalian ventrolateral medulla oblongata are more tolerant than other neurons to short-term hypoxia (Haddad and Mellins 1984), illustrated by increases in $[Na^+]i$ and $[Ca^{2+}]i$ of only half amplitude of that in non-respiratory neurons and also by prompt recovery (Mironov and Langohr 2005). Similar age- and region-dependent $[Ca^{2+}]i$ changes were also found in the same cortex (Takita et al. 2004). In addition, the distribution and density of the channels/receptors/transporters and their subtypes are heterogeneous in different brain regions (Kanyshkova et al. 2014), in the same cells (Lukyanetz et al. 2003), or even in the different parts of the same cells (dendrites vs. somatic regions) (Connor et al. 1988; Wadman and Connor 1992; Ladewig et al. 2003; Nicholson et al. 2006; Vander Jagt et al. 2008). Therefore, it is not strange that ions show heterogeneous response to the same stress in different regions or different part of the same cells. In this respect, Lukyanetz et al. (2003), for example, found in the same hippocample neurons, both N- and L-types of Ca^{2+} channels are sensitive to hypoxia, however, L-type is about 3.5 times more sensitive to oxygen reduction than N-type of Ca^{2+} channels. OGD-induced $[Ca^{2+}]i$ elevation was more pronounced in dendritic branches than in somatic region of neurons (Pisani et al. 1999; Zhang and Lipton 1999). Therefore, it is important to unscramble the results from different reports with caution.

3.1 Hypoxia/Ischemia-Induced K^+ Efflux

K^+ is the most abundant cation in the cytoplasm. The concentrations of K^+ ions in the nerve cells are around 3 mM in extracellular fluid and up to 150 mM in the cytosol (see the preceding chapter by Chao and Xia). The increase in $[K^+]e$ is a typical response of the brain tissues to hypoxia or ischemia. $[K^+]e$ increases up to 50 mM in mild hypoxia/ischemia and around 100 mM in severe hypoxia/ischemia (Hansen 1978; Dronne et al. 2007). This increase in $[K^+]e$ can be a result of increased K^+ efflux, decreased K^+ uptake, or both. Increased $[K^+]e$ always accompanies the loss of intracellular K^+ (Dronne et al. 2007). The cytoplasmic K^+ (approximate 100–150 mM) (Garlid 1996; Taylor et al. 1999; Lopachin et al. 2001) is the main source of accumulated extracellular K^+. The direct measurements show an intramitochondrial K^+ concentration of 180 mM or greater, which is higher than that of the cytosol (approximate 100–150 mM) (Garlid 1996; Taylor et al. 1999; Lopachin et al. 2001). Mitochondria can also contribute to the increase in extracellular K^+ by releasing K^+ into the cytosol, which then increases K^+ efflux into extracellular space during ischemia (Liu et al. 2003). K^+ uptake is principally dependent on glial cells, particularly astrocytes, via multiple mechanisms (see the preceding chapter by Chao and Xia). Glial cells are less sensitive to identical insults than neurons (e.g., a decrease in neuronal ion gradients after 10 min of energy limitation is comparable to the reduction seen in glial cells after 16 h) (Duffy and MacVicar 1996; Silver et al. 1997; Hertz 2008; Ding et al. 2009), and have high capacity of K^+ uptake and buffer (Kofuji and Newman 2004; Leis et al. 2005). In general, the main contribution of glial cells to K^+ redistribution during hypoxia/ischemia is buffering the increased extracellular K^+. Only when the capacity of K^+ uptake and buffering is severely damaged with prolonged/severe ischemia, do glial cells begin to contribute to extracellular K^+ rise by reducing K^+ uptake (Leis et al. 2005). Indeed, the role of astrocytes in K^+ movement during hypoxia/ischemia has been well mimicked in a mathematical model that is based on electrophysiological mechanisms and reproduces the movements of several ionic species (K^+, Na^+, Ca^{2+}, Cl^-, and glutamate ionized form) across neuron and astrocyte membranes during the excitotoxicity process and cytotoxic oedema development, and involves various channels, exchangers, transporters and pumps (Dronne et al. 2007). The model clearly shows that in conditions of mild ischemia, $[K^+]e$ increases up to 50 mM (which is consistent with the values measured in some experimental studies), and $[K^+]i$ in neurons decreases, whereas it slightly increases in astrocytes due to the important K^+ uptake by the astrocytic Na^+-K^+ pumps and Na^+-K^+-Cl^- cotransporters overwhelming K^+ release via astrocytic K^+ channels, resulting in a net K^+ influx during mild stress. In contrast, under severe ischemia (10 min after the vessel occlusion), $[K^+]e$ increases up to around 100 mM [also similar to experimental data by Hansen (1978) with 4 min brain ischemia], $[K^+]i$ in both neurons and astrocytes decreases, but at a relatively lower degree in astrocytes (Dronne et al. 2007). The decrease in astrocytes is due to the drop in Na^+/K^+ ATPase activity (which is not able to take up extracellular K^+ anymore in severe ischemia) and the activity of Na^+-K^+-Cl^- cotransporters is not enough to balance K^+ efflux via K^+ channels (such as Kir and BK channels), suggesting astrocytes contribute to increase in $[K^+]e$ in severe ischemia (Dronne et al. 2007). After ischemic insults, astrocytes have been found to decrease Kir4.1 protein expression and Kir currents with progressive depolarization in the gliotic tissue. This might profoundly influence the maintenance of K^+ homeostasis in the damaged tissue and the periinfarct area during recovery after ischemia, and contribute to the sustained damage during reperfusion period (Köller et al. 2000; Pivonkova et al. 2010). Therefore, glial cells contribute less to $[K^+]e$ increase than neurons during hypoxia/ischemia, but may play a key role in the maintenance of K^+ homeostasis after stress. The increase in $[K^+]e$ seen in adult brain tissue during anoxia is, to a great extent, secondary to neuronal loss of intracellular K^+ (Jiang and Haddad 1991).

The phenotype of K^+ activity in hypoxia/ischemia is a distinctive two-phase response when compared with the activities of Na^+ and Ca^{2+} ions (Hansen 1978, 1985; Jiang and Haddad 1991; Silver and Erecińska 1990; Xie et al. 1994; Reid and Paterson 1996; Gido et al. 1997; Müller and Somjen 2000a, b; Chao et al. 2007a, b, 2008, 2009). During hypoxia/ischemia, $[K^+]e$ first shows a relatively slow increase starting after 20–60 s of hypoxia/ischemia. At this time, $[Na^+]o$ and $[Ca^{2+}]o$ remain unchanged. Over several minutes of hypoxia/ischemia, $[K^+]e$ increase reaches a threshold level [no more than the so-called "K^+ ceiling" level of 10–12 mM (Heinemann and Lux 1977)]. Then, a rapid, abrupt, and large increase in $[K^+]e$, accompanied by a negative shift in extracellular DC potential and a sudden change in $[Na^+]o$ and $[Ca^{2+}]o$ occurs, and $[K^+]e$ peaks (range from 25 to 100 mM depending on animal ages, brain regions, neuronal types, severity and duration of insult, etc.) within about 20 s of onset of this rapid and large increase, and then gradually decreases to a relatively stable level when hypoxia/ischemia lasts. After reintroduction of oxygen/glucose, $[K^+]e$ falls rapidly, and often undershoots to below the basal levels before steadily returning to the pre-hypoxia/ischemia level. The increase in $[K^+]e$ in the intact brain is more profound than in vitro brain tissues/cells (Silver and Erecińska 1990; Xie et al. 1994; Zetterstrőm et al. 1995; Gido et al. 1997). The hypoxic/ischemic increase in $[K^+]e$ is also observed in the brain of some anoxia-tolerant species such as turtles and frogs. However, the magnitude is very small (usually several mM) and the latency is very long (up to 4 h) in comparison with mammalian brain (Sick et al. 1982, 1987; Doll et al. 1991; Xia et al. 1992; Knickerbocker and Lutz 2001).

The loss of intracellular K^+ and the accumulation of extracellular K^+ reflect an imbalance between the rate of active K^+ uptake via the Na^+/K^+ ATPase and the rate of K^+ efflux. Many mechanisms have been proposed for this imbalance during and after cerebral hypoxia/ischemia, including inhibition of Na^+/K^+ ATPase, increased leakage of K^+ due to an increase in excitability and action potential generation in neurons, and activation of K^+ channels (e.g. voltage-gated K^+ channels, ATP-sensitive K^+ channels, Ca^{2+}-activated K^+ channels and Na^+-activated K^+ channels, as well as ionotropic glutamate receptor channels) (Hansen 1985; Jiang and Haddad 1991; Haddad and Jiang 1993; Zetterstrőm et al. 1995; Reid and Paterson 1996; Silver et al. 1997; Erdemli et al. 1998; Yu et al. 1999; Xiao et al. 2001; Müller and Somjen 2000a; Lopachin et al. 2001; Yuan et al. 2003; Chao et al. 2007a, 2008) (also see Fig. 6.1).

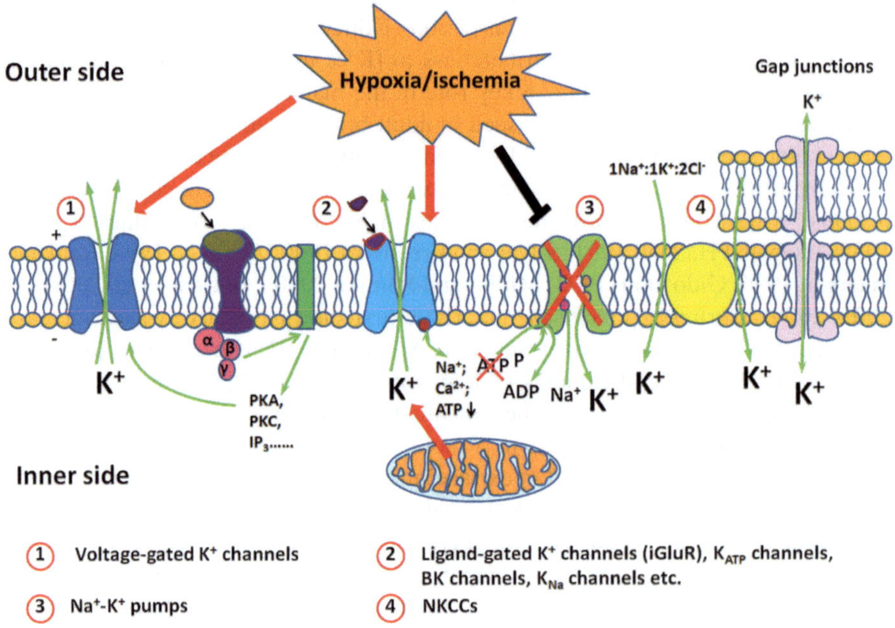

Fig. 6.1 Schematic demonstration of hypoxia/ischemia-induced loss of intracellular K^+ and accumulation of extracellular K^+ in the nerve cells. The loss of intracellular K^+ and the accumulation of extracellular K^+ reflect an imbalance between the rate of active K^+ uptake via the Na^+/K^+ ATPase and the rate of K^+ efflux. Many processes, including the inhibition of Na^+/K^+ ATPase, increased leakage of K^+ due to an increase in excitability and action potential generation in neurons. Also, activation of K^+ channels (e.g. voltage-gated K^+ channels, ATP-sensitive K^+ channels, Ca^{2+}-activated K^+ channels and Na^+-activated K^+ channels, as well as ionotropic glutamate receptor channels) may contribute to hypoxic/ischemic loss of intracellular K^+ and the accumulation of extracellular K^+ in the nerve cells. Na^+-K^+-$2Cl^-$ cotransporters (NKCC1) seem to contribute to K^+ and Na^+ entry (not efflux) during anoxia rather than immediately after anoxia (refer to the discussion of NKCC1 in the Sect. 3.2). Mitochondria can also contribute to the increase in extracellular K^+ by releasing K^+ into cytosol and then efflux into extracellular space during ischemia

3.1.1 Inhibition of Na^+/K^+ ATPase

Normal running of Na^+/K^+ ATPase constructs a large transmembrane electrochemical gradient of Na^+ and K^+ ions, which consumes at least 60 % of energy expenditure of nerve cells in the CNS (Schreiner and Rose 2012). Therefore, Na^+/K^+ ATPase is extremely sensitive to the impairment of energy metabolism. During hypoxia/ischemia, ATP generation is largely limited and ATP levels decrease rapidly (Silver et al. 1997; Raley-Susman et al. 2001; Martinez-Sánchez et al. 2004; Sheldon and Church 2004; Gouriou et al. 2013). These changes may cause the failure of Na^+/K^+ ATPase and the accumulation of extracellular K^+. Indeed, blockade of Na^+/K^+ ATPase with ouabain induces great intracellular K^+ loss, large increases in $[K^+]e$ and stimuli-induced $[K^+]e$ transient and $[K^+]i$ fall, and slows the rate of $[K^+]e$ recovery in both neurons and glial cells in the CNS (Alvarez-Leefmans et al. 1994; Balestrino et al. 1999; Xiao et al. 2002; D'Ambrosio et al. 2002; Vaillend et al. 2002; Bay and Butt 2012). However, after $[K^+]e$ rise in a steady-state level by Na^+-K^+ pump inhibition, anoxia/ischemia still induces a further significant increase in $[K^+]e$ in the brain neurons, suggesting inhibited activity of Na^+/K^+ ATPase only partially contributes to the loss of intracellular K^+ and the buildup of excess extracellular K^+ (Jiang and Haddad 1991; Reid and Paterson 1996). In fact, ATP levels are partially reserved during mild hypoxia/ischemia, which drives running Na^+-K^+ pumps to take up extracellular K^+ and decrease extracellular K^+ accumulation; only when ATP is completely exhausted during severe hypoxia/ischemia, does the drop of Na^+/K^+ ATPase activity make the pumps unable to take up extracellular K^+ any more, thus making extracellular K^+ accumulation from the K^+ release via other way (e.g., K^+ channels) (Dronne et al. 2007). In addition, Na^+/K^+ ATPase largely contributes to the phenomenon of K^+ undershoot during reoxygenation/reperfusion from hypoxia/ischemia as a result of compensatory over-transportation of K^+ from outside into inside stimulated by increased $[Na^+]i$ in the cells (Reid and Paterson 1996; also see the preceding chapter by Chao and Xia).

3.1.2 Activation of K^+ Channels

Hypoxia/ischemia can activate K^+ channels and enhance the outward K^+ currents in neurons (Jiang and Haddad 1993; Fujimura et al. 1997; Guatteo et al. 1998a; Erdemli et al. 1998; Mironov et al. 1998; Chi and Xu 2000, 2001; Kulik et al. 2002; Müller et al. 2002; Plant et al. 2002; Park et al. 2003; Chao et al. 2007a), suggesting a K^+ channel-mediated cellular K^+ loss/efflux. Diverse types of K^+ channels are involved, including K_{ATP} channels, voltage-gated K^+ channels, and two pore domain K^+ channels. In addition, Ca^{2+}-activated K^+ channels (BK channels) and Na^+-activated K^+ channels (K_{Na}) are also proposed to contribute to K^+ efflux and accumulation because they can be activated by largely increased $[Ca^{2+}]i$ and $[Na^+]i$ during hypoxia/ischemia (Murai et al. 1997; Tanabe et al. 1999; Sun et al. 2003; Yuan et al. 2003; Bhattacharjee and Kaczmarek 2005; Chao et al. 2007a, 2008). In the hippocampus, K^+ efflux through K^+ channels is responsible for at least 35 %

increase in hypoxia/ischemia-induced [K^+]e, because a completely non-selective K^+ channel blocker quinine attenuates the anoxia-induced rise in [K^+]e by 35 % (Zetterstrőm et al. 1995). During hypoxia/ischemia, K_{ATP} channels are activated as adaptive protection mechanism (Kulik et al. 2002; Ballanyi 2004; Huang et al. 2006), which may increase [K^+]e (Jiang and Haddad 1991; Chao et al. 2007a). Blockade of plasmalemmal and mitochondrial K_{ATP} channels (Jiang and Haddad 1991; Reid and Paterson 1996; Liu et al. 2003; Chao et al. 2007a) and voltage-gated K^+ channels (Zetterstrőm et al. 1995) attenuates hypoxic/ischemic [K^+]e rise in the brain. Though some studies using patch-clamp recording in neurons and in HEK 293 cells that stably co-express α- and β-subunits of human brain BK channels, showed an inhibition of BK channels by O_2 deprivation (Liu et al. 1999; Gao and Fung 2002; Lewis et al. 2002; Rodgers-Garlick et al. 2013), our work with K^+ selective microelectrode recording in mice cortical slices indicated a contribution of a 22 % increase in [K^+]e induced by anoxia because blockade of BK channels with paxilline (10 μM) attenuated anoxic [K^+]e rise by 22 % (Chao et al. 2007a). This result suggested that a part of BK channels in neurons might still run as an adaptive mechanism to counteract hypoxia-induced membrane depolarization by K^+ efflux. Indeed, it has been shown that activation of large-conductance Ca^{2+}-activated K^+ channels of the BK type by either sulfhydryl oxidation or by intracellular released Ca^{2+} induces a moderate hyperpolarization, blocks spontaneous spike discharges, and postpones massive hypoxic depolarization in hippocampal CA1 and locus coeruleus neurons, and thus reduces neuronal susceptibility to hypoxia-induced highly synchronized anoxic response (Murai et al. 1997; Erdemli et al. 1998; Hepp et al. 2005). Conversely, blockade of BK channels aggravates OGD-induced cell damage in hippocampal slices, suggesting that BK channels act as a kind of 'emergency brake' during and/or after ischemia (Rundén-Pran et al. 2002). Nevertheless, the BK channel-mediated [K^+]e increase (Chao et al. 2007a) may be a result of K^+ efflux via astrocytic Ca^{2+}-activated K^+ channels because in astrocytes (vs. neurons), hypoxia indeed enhances the activities of macroscopic outward Ca^{2+}-activated K^+ currents (K_{Ca}), and increases the open state probability of 71 pS and 161 pS single-channel K_{Ca} currents (Yamaura et al. 2006). In addition, we observed that the component of Na^+-based K^+ efflux and accumulation during anoxia is due to Na^+ entry through voltage-gated Na^+ channels and NMDA receptor channels (Chao et al. 2008, 2009), and the consequent activation of K_{Na} channels (Yuan et al. 2003; Bhattacharjee and Kaczmarek 2005; Ruffin et al. 2008). Therefore, K^+ channel-mediated K^+ efflux is one of the major routes for hypoxic/ischemic increase in [K^+]e. However, there is a study showing a blockade of a voltage-gated, dendrotoxin-sensitive homomeric type K^+ channels by moderate hypoxia in sensory neurons (Gruss et al. 2006), just like that of BK channels in some studies (Liu et al. 1999; Gao and Fung 2002; Lewis et al. 2002). It is unknown if this channel contributes to hypoxic/ischemic K^+ efflux and accumulation.

3.1.3 Ionotropic Glutamate Receptor Channels and Increased Leakage of K^+

Ionotropic glutamate receptors (iGluRs), which include NMDA receptor (NMDAR), AMPA receptor (AMPAR), and kainate receptor (KAR) (the latter two receptors are generally called non-NMDA receptors due to their similarity to functional properties), are present at excitatory synapses, and mediate the majority of excitatory neurotransmission in the brain. Dendritic spines possess abundant iGluRs, and represent the major postsynaptic input site for excitatory transmission in the brain (Harris and Kater 1994). Ionotropic glutamate receptor-gated channels act as non-specific cation channels that are permeable to Na^+, Ca^{2+}, and K^+ (Ozawa et al. 1998; Dingledine et al. 1999). However, in comparison with ionotropic receptor channel-mediated Ca^{2+} and Na^+ entry, iGluR channel-mediated K^+ efflux has received little attention.

Activation of iGluRs of mouse cortical neurons, even in a condition of decreased inward cation influx by lowering extracellular Na^+ and Ca^{2+}, induces a large outward K^+ current, which causes loss of approximately 50–80 % of intraneuronal K^+ and great shrinkage of cell body and consequently neuronal apoptosis (Yu et al. 1999; Xiao et al. 2001). Therefore, the iGluR-mediated K^+ loss may be another neuronal factor for hypoxia/ischemia-induced $[K^+]e$ increase (Croning et al. 1993; Lopachin et al. 2001; Chao et al. 2009). Indeed, blockade of iGluRs with NMDAR blockers (CPP, dizocilpine, (+)MK-801), or AMPAR blocker CNQX partially decreases OGD and hypoxia-induced intraneuronal K^+ loss in the hippocampus and cortex (Croning et al. 1993; Lopachin et al. 2001; Chao et al. 2009).

Ionotropic glutamate receptors mediate the majority of excitatory neurotransmission in the brain. An increase in excitability of neurons may also increase K^+ leakage in hypoxic/ischemic conditions. In hippocampal slices, when neuronal excitability is decreased and spontaneous and evoked impulse firing is blocked by tetrodotoxin (TTX) or lidocaine, hypoxia-induced sharp increase in $[K^+]e$ is significantly dampened (Müller and Somjen 2000a; Raley-Susman et al. 2001). Similar observations were made in hypoglossal neurons, and isolated in situ perfused rat brain (Jiang and Haddad 1991; Xie et al. 1994). These findings suggest a role of iGluR-mediated excitability in K^+ loss and extracellular K^+ accumulation during hypoxia. This is also supported by our recent study (Chao et al. 2008). We found in cortical slices that lowering external Na^+ concentration with substitution of impermeable N-methyl-D-glucamine ($NMDG^+$) greatly attenuates anoxic $[K^+]e$ rise, whereas lowering Na^+ concentration with substitution of permeable Li^+ tends to potentiate the anoxic disruption of K^+ homeostasis (Chao et al. 2008). Because $NMDG^+$ is a membrane impermeable organic cation, lowering external Na^+ with $NMDG^+$ replacement may reduce Na^+ influx and neuronal excitability because of a reduction of driving force for inward sodium current during anoxia (Calabresi et al. 1999; Sheldon et al. 2004). In contrast, Li^+, unlike $NMDG^+$, is permeable across the membrane by multiple transport ways, particularly by Na^+ channels, through which Li^+ passes nearly as well as Na^+ itself (Hemsworth et al. 1997). Li^+ entry gradually depolarizes membrane potential (Kiedrowski 1999; Franceschetti et al. 2003), broadens action potential (Colino et al. 1998), and positively shifts afterpotential (Franceschetti et al. 2003). In addition, it progressively potentiates high-frequency firing (Franceschetti et al. 2003), and presynaptically enhances excitatory synaptic transmission (Colino et al. 1998). Therefore, our observations (Chao et al. 2008) demonstrate that increased K^+ leakage due to an increase in excitability and action potential generation contributes to hypoxia/ischemia-induced increase in $[K^+]e$.

3.2 Hypoxia/Ischemia-Induced Increase in Na^+ Influx

Na^+ ions are remarkably abundant in extracellular space but are relatively sparse in cytosol in both neurons and glial cells, thus forming a very steep electrochemical gradient for Na^+ in both neurons and glial cells, which is estimated to be as high as +140 mV based on the reported baseline $[Na^+]o$ and $[Na^+]i$ (Rose and Karus 2013). Such a huge driving force for Na^+ can easily make large amounts of Na^+ entry in the conditions that enhance the membrane permeability, e.g., hypoxic/ischemic stress. Indeed, direct measurement with Na^+-selective microelectrodes have shown that during hypoxia, extracellular Na^+ (150–155 mM) sharply drops to about 60 mM in the hippocampus (Fung et al. 1999; Müller and Somjen 2000a, b) and 40 mM in hypoglossal neurons (Jiang et al. 1992). Meanwhile, an increase in $[Na^+]i$ from baseline level (about 8 mM) to 20–50 mM was observed in the nerve cells (neurons and astrocytes) of various regions, including the cortex (Friedman and Haddad 1994; Lynch et al. 1995; Pisani et al. 1998a; Banasiak et al. 2004; Lenart et al. 2004; Luo et al. 2005; Kintner et al. 2007; Chen et al. 2008), hippocampus (Taylor et al. 1999; Lopachin et al. 2001; Raley-Susman et al. 2001; Sheldon et al. 2004; Martinez-Sánchez et al. 2004; Cengiz et al. 2014), cerebellum (Chen et al. 1999), substantia nigra (Guatteo et al. 1998b), striatal neurons (Calabresi et al. 1999), hypoglossal cells (Jiang et al. 1992), medullary respiratory neurons (Mironov and Langohr 2005), optic nerve (Stys and Lopachin 1998; Nikolaeva et al. 2005), and spinal cord (Rose et al. 1998). This suggests that the hypoxic/ischemic increase in $[Na^+]i$ results from extracellular Na^+ entry. Indeed, chemical ischemia-induced increase in $[Na^+]i$ in cerebellar granule cells was reversed by replacement of external Na^+ ions with impermeable $NMDG^+$ (Chen et al. 1999). During the reoxygenation and reperfusion period, though Na^+ extrusion mechanisms (e.g. Na^+ pumps and reversely operated NCX) are rebuilt or are running, there is still Na^+ influx (Sheldon et al. 2004). Therefore, $[Na^+]i$ gradually recovers with varied durations or can have sustained increases for a period of time until it then recovers (Lenart et al. 2004; Luo et al. 2005). The type of recovery depends on the duration and severity of the insults and the balance between Na^+ extrusion and influx.

Extracellular Na^+ can enter cells through multiple ways. Under normal conditions, for neurons, voltage-gated Na^+ channels (VGSCs) and iGluRs compose the most important passage for Na^+ entry. However, VGSCs play only a minor role in glial cells. The most important pathways for Na^+ entry into glial cells are iGluRs (particularly CNQX-sensitive AMPARs) and Na^+-dependent (glutamate, GABA, HCO_3^-) transporters. Non-selective cation channels and Na^+/Ca^{2+} exchangers (NCXs) also contribute to Na^+ entry in both neurons and glial cells, while $P_2Y_{1\text{-}2}$ purinergic receptor channels, NKCC1, and gap junctions contribute to Na^+ entry

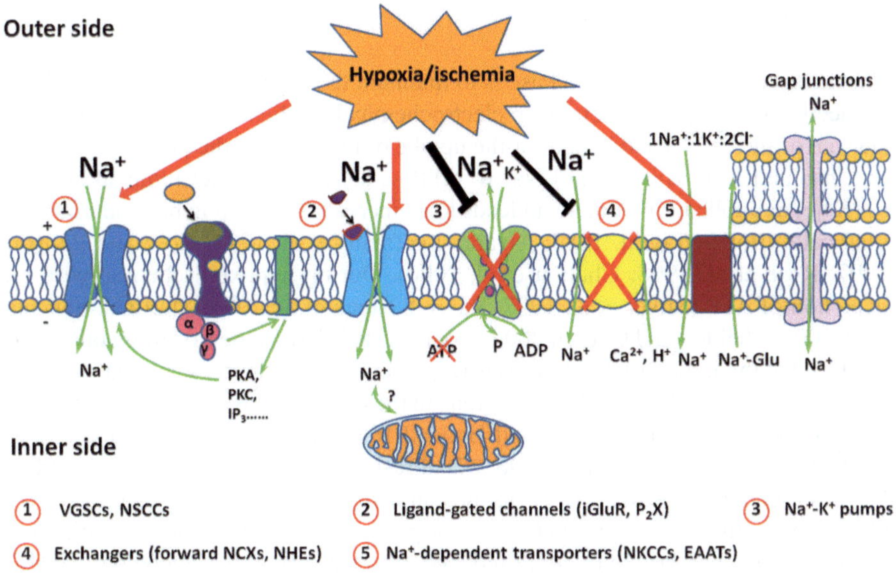

Fig. 6.2 Schematic demonstration of hypoxia/ischemia-induced Na^+ influx and accumulation in the nerve cells. Extracellular Na^+ can enter cells through multiple ways. Under normal conditions, for neurons, voltage-gated Na^+ channels (VGSCs) and iGluR compose the most important passage for Na^+ entry. However, VGSCs play a minor role for glial cells. The most important pathways for Na^+ entry into glial cells are iGluR (particularly CNQX-sensitive AMPARs) and Na^+-dependent (glutamate, GABA, HCO_3^-) transporters. Non-selective cation channels, Na^+/Ca^{2+} exchangers (NCXs), $P_2Y_{1\text{-}2}$ purinergic receptor channels, and Na^+-K^+-$2Cl^-$ cotransporters (NKCCs) also contribute to Na^+ entry in both neurons and glial cells, while gap junctions contribute to Na^+ entry in glial cells. Most importantly, the failure of Na^+ pumps also leads to accumulation of $[Na^+]i$. It is possible that all these pathways are involved in hypoxia/ischemia-induced Na^+ entry and accumulation. Several potential routes for Na+ entry during hypoxia/ischemia have been examined, which include the inhibition of Na^+/K^+ ATPase, activation of VGSCs, ligand-gated Na^+ channels (iGluR channels, ATP-activated P_2X receptor channels), NSCCs, running of Ca^{2+} extrusion/Na^+ entry mode of NCX etc. For NKCC, since inward NKCC may be limited by high $[Na^+]i$ and presumably $[Cl^-]i$ immediately after anoxia, NKCC-mediated Na^+ entry seems to be ascribed to $[Na^+]i$ rise during anoxia rather than immediately after anoxia. NHE is inhibited during hypoxia/ischemia but activated immediately after reoxygenation/reperfusion; therefore, it contributes to Na^+ influx in the period immediately after anoxia rather than during anoxia in neurons. Excitatory amino-acid transporters (EAATs) play a dual role in hypoxia/ischemia-induced changes in $[Na^+]o/[Na^+]i$ depending on their operating modes. Namely EAATs operate normally to eliminate released glutamate during the first minutes of ischemia (before energy exhaustion and Na^+ gradient collapse), and therefore may partially account for the limited $[Na^+]i$ increase in the early phase of hypoxia/ischemia. However, with hypoxia/ischemia being prolonged, largely increased $[Na^+]i$ and membrane depolarization make EAATs operate in reverse, thus releasing massive glutamate and moving intracellular Na^+ out

only in glial cells (see the preceding chapter by Chao and Xia). Most importantly, the failure of Na^+ pumps also leads to accumulation of $[Na^+]i$. It is possible that all these pathways are involved in hypoxia/ischemia-induced Na^+ entry and accumulation (Fig. 6.2). Several potential routes for Na^+ entry during hypoxia/ischemia have been examined, but the supporting evidence is still incomplete and even contradictory in some studies.

3.2.1 Na^+/K^+ ATPase

The primary role of Na^+/K^+ ATPase is to maintain the high intracellular K^+ and low intracellular Na^+ against their steep electrochemical gradients by transporting three Na^+ out of the cells in exchange for the uptake of two K^+ (see the preceding chapter by Chao and Xia). Blockade of Na^+/K^+ ATPase activity by either removal of $[K^+]e$ (0 $[K^+]e$) or application of ouabain leads to an increase in $[Na^+]i$ in both neurons and glial cells (Ballanyi et al. 1983; Alvarez-Leefmans et al. 1994; Rose and Ransom 1997; Rose et al. 1998; Chatton et al. 2000). During hypoxia/ischemia, with the depletion of ATP (Silver et al. 1997; Raley-Susman et al. 2001; Martinez-Sánchez et al. 2004; Sheldon and Church 2004; Gouriou et al. 2013), the first thing one might think is that the failure of Na^+ pumps is responsible for the increase in $[Na^+]i$ in both neurons and glial cells of different regions in the CNS such as cerebral cortex, striatum, and spinal cord (Friedman and Haddad 1994; Silver et al. 1997; Jiang et al. 1998; Rose et al. 1998; Stys 1998; Calabresi et al. 1999; Lenart et al. 2004). However, in dopamine neurons of rat substantia nigra pars compacta, supply of ATP (10 mM) via the recording pipette did not prevent the hypoxic increase in $[Na^+]i$, despite a great decrease in hypoxia-induced outward current (Guatteo et al. 1998a). Also, blockade of Na^+/K^+ ATPase with strophenthidin (100 μM) (an analog of ouabain) did not induce an increase in $[Na^+]i$ comparable to that produced by chemical ischemia in cerebellar granule cells (Chen et al. 1999). In the respiratory neurons of medullary slices, ouabain (10 μM) could not block a 2-min hypoxia-induced increase in $[Na^+]i$ (only several mM) (Mironov and Langohr 2005). These investigators thus concluded that ATP depletion-induced inability of Na^+/K^+ ATPase to extrude intracellular Na^+ is not the reason for the hypoxia/ischemia induced increase in $[Na^+]i$ (Guatteo et al. 1998a; Chen et al. 1999; Mironov and Langohr 2005). However, these studies cannot completely rule out the possibility for the potential contributions of the failure of Na^+/K^+ ATPase. Firstly, hypoxia/ischemia may release endogenous inhibitors of Na^+/K^+ ATPase, downregulating Na^+/K^+ ATPase activity (Yu 2003a). Under such a circumstance, supplement of ATP (Guatteo et al. 1998a) does not necessarily increase the efficiency of Na^+/K^+ ATPase, but may inhibit the activity of K_{ATP} channels that partially contribute to K^+ efflux and accumulation (Chao et al. 2007a; Mollajew et al. 2013). This would lead to the phenomenon observed by Guatteo et al. (1998a). Secondly, the binding sites for Na^+, K^+, and the ATP-containing isoform α1 (as a "housekeeping" form, see MacAulay and Zeuthen 2012) of Na^+/K^+ ATPase are ubiquitously distributed over the surfaces of the nerve cells (including neurons and glial cells) with very low affinity to ouabain (Na^+ pump inhibitor). High ouabain affinity isoforms α2 are expressed predominately in glia (McGrail et al. 1991; Cameron et al. 1994; Juhaszova and Blaustein 1997; Peng et al. 1997). Therefore, complete inhibition of neuronal Na^+/K^+ ATPase might require high concentration of strophenthidin, though the α3 isoform of Na^+/K^+ ATPase is also expressed in neurons. In this respect, though Chen et al. (1999) did not show an increase in $[Na^+]i$ in cerebellar granule cells with application of 100 μM of strophenthidin during ischemia, they did show that a K^+-free medium (which strongly inhibits Na^+/K^+ ATPase) produces a large increase in $[Na^+]i$ during ischemia. Thirdly, brief hypoxia (e.g. 2 min as did by Mironov and Langohr 2005) may not induce significant ATP depletion in relatively hypoxia-resistant neurons (e.g., respiratory neurons), and lower concentrations of ouabain (10 μM used by Mironov and Langohr 2005) may not completely block the activity of Na^+/K^+ ATPase. Under such a condition, Na^+/K^+ ATPase may still run efficiently to extrude intracellular Na^+ that has entered through other pathways (e.g., persistent Na^+ channels and non-selective cation channels as observed by Mironov and Langohr 2005), and to keep $[Na^+]i$ at lower levels. In addition, chemical ischemia is essentially not the same as real ischemia, and cultured single neurons can still obtain oxygen from environments where oxygen is not depleted and keep metabolism to a certain degree. Therefore, it is not impossible that the effects of both stimulations (chemical vs. real ischemia) on Na^+/K^+ ATPase activity are not exactly the same. Taking together, the failure of Na^+ pumps may at least be partially responsible for increase in $[Na^+]i$.

3.2.2 Voltage-Gated Sodium Channels (VGSCs)

VGSCs play a critical role in neuronal functions, and are involved in many pathophysiological conditions (Chao and Xia 2012; Chao et al. 2013). During hypoxia/ischemia, membrane depolarization (Calabresi et al. 1999; Müller and Somjen 2000a, b; Raley-Susman et al. 2001; Xia et al. 2003) may lead to the activation of and Na^+ entry through VGSCs and to a subsequent increase in $[Na^+]i$. This possibility has been extrapolated with a mathematical model of ion movements during stroke (Dronne et al. 2006). Experimental studies also provided supports for this possibility. For example, measurements with ion-selective microelectrodes revealed that blockade of VGSCs with TTX partially or completely prevented anoxia-induced $[Na^+]o$ drop in hippocampal slices (Fung et al. 1999; Müller and Somjen 2000a). Studies with other measurements (e.g., flame photometer, electron probe X-ray microanalysis, Na^+-sensitive fluorescent dye imaging) yielded similar findings in the hippocampus (Lopachin et al. 2001; Raley-Susman et al. 2001). Similar results were observed in neocortical neurons (Lynch et al. 1995; Banasiak et al. 2004), optic nerves (Stys and Lopachin 1998; Nikolaeva et al. 2005), and isolated, in situ perfused rat brains (Xie et al. 1994). All these reports support the idea that VGSCs are a major, though not sole, source of Na^+ influx during hypoxia/ischemia. Consistent with this idea is the observation that blockade of VGSCs with Na^+ channel blockers such as TTX and lidocaine attenuated/postponed anoxic depolarization, and prevented hypoxia/ischemia-induced neuronal injury and death (Probert et al. 1997; Takahashi et al. 1999; Müller and Somjen 2000a; Breder et al. 2000; Raley-Susman et al. 2001; Banasiak et al. 2004).

However, this viewpoint has also been challenged by some other studies (Guatteo et al. 1998b; Pisani et al. 1998a; Rose et al. 1998; Chen et al. 1999; Calabresi et al. 1999; Sheldon et al. 2004). Under conditions of voltage clamp mode with the membrane potential held at −60 mV, brief hypoxia still causes a reversible elevation in $[Na^+]i$ in dopamine neurons of the substantia nigra pars compacta. This elevation in

$[Na^+]i$ obviously is not mediated by VGSCs because VGSCs cannot be activated under such a condition (membrane potential in −60 mV) (Guatteo et al. 1998b). TTX (1 μM) or lidocaine (250 μM) failed to affect the magnitude of anoxia/OGD-induced increase in $[Na^+]i$ in hippocampal and cortical neurons (Pisani et al. 1998a; Sheldon et al. 2004). Similar results were also observed in cerebellar granule cells and spinal cord astrocytes (Rose et al. 1998; Chen et al. 1999). In striatal spiny neurons, OGD-induced membrane depolarization and increase in $[Na^+]i$ can be significantly reduced by lowering external Na^+, but not by blockade of VGSCs by either TTX or saxitoxin, which seems to imply that the ischemia-induced increase in $[Na^+]i$ is not due to Na^+ entry through TTX-sensitive Na^+ channels. However, these studies could not rule out the potential involvement of TTX-resistant voltage-gated Na^+ channels, which may carry the majority of inward currents in certain condition (Agostinho et al. 1996; Blair and Bean 2002). Indeed, brief hypoxia (2 min) induced $[Na^+]i$ elevation (several mM) can be completely blocked by persistent Na^+ channel inhibitor riluzole, indicating a contribution of TTX-resistant voltage-gated Na^+ channels to hypoxic increase in $[Na^+]i$ (Mironov and Langohr 2005).

The reasons for these conflicting results can be attributed to differences in regions, insult duration (acute vs. prolonged), and intensity (e.g. mild vs. severe hypoxia or ischemia). Moreover, observations made on isolated cells may be very different from those on brain slices that have an intact microenvironment. In the above-mentioned work, for example, most of the studies disproving VGSCs as an important pathway of hypoxia/ischemia-induced Na^+ influx were conducted in cultured cells (Rose et al. 1998; Chen et al. 1999; Sheldon et al. 2004) that lack excitatory neurotransmission, and are consequently relatively more silent than cells that are intact or in slices. In addition, it is important to realize that multiple pathways can interact with each other, but not all these pathways contribute equally. It is completely possible that under different conditions (e.g. cells from different regions, and different insult duration and intensity etc.), certain pathways predominate over others. Our recent observations made in the cortical slices directly or indirectly indicate that anoxia induces Na^+ influx through VGSCs (Chao et al. 2008, 2009, 2012b; Kang et al. 2009). For example, using $[K^+]e$ as an index, we found that blockade of VGSCs with TTX or partial replacement of extracellular Na^+ with impermeable $NMDG^+$ significantly reduces hypoxic response in the cortex, while Na^+ opener veratridine (which increases Na^+ channel permeability and prevent its inactivation) greatly enhances it (Chao et al. 2008, 2009), suggesting a role of VGSCs in hypoxic response. Direct measurement of $[Na^+]o$ with Na^+-selective microelectrodes indicated that perfusion of 1 μM of TTX completely abolishes $[Na^+]o$ response to anoxia in 64 % slices, while in the other 36 % slices, half show a reduced drop in $[Na^+]o$ induced by anoxia, and half do not have obvious changes in anoxia-induced $[Na^+]o$ drop (Kang et al. 2009; Chao et al. 2012b). Perfusion of 1 μM veratridine enhances anoxia-induced Na^+ influx by 9 % (Chao et al. 2012b). These findings prompt us to believe that VGSCs are a major route for Na^+ entry during hypoxia/ischemia.

3.2.3 Ligand-Gated Na^+ Channels

The most important ligand-gated Na^+ channels are iGluR channels. A common feature of iGluR channels is Na^+ permeability (Ozawa et al. 1998; Dingledine et al. 1999). During hypoxia/ischemia, massive amounts of glutamate are released into the synaptic cleft (Benveniste et al. 1984; Goldberg and Choi 1993; Camacho and Massieu 2006; Brongholi et al. 2006; Kirschner et al. 2009; Ye et al. 2010). Activation of iGluRs causes more than 45 mM reductions in $[Na^+]o$ in the hippocampus (Zanotto and Heinemann 1983), and particularly CNQX-sensitive AMPARs partially contribute to glial Na^+ transients (Chatton et al. 2000; Kirischuk et al. 2007; Bennay et al. 2008; Langer and Rose 2009). In the hippocampus, iGluR channels, especially NMDARs, largely contribute to Na^+ influx and $[Na^+]i$ elevation in postsynaptic neurons during excitatory transmission (Rose and Konnerth 2001; Vander Jagt et al. 2008; Langer and Rose 2009). In cerebellar Purkinje cells, synaptically activated Na^+ responses are predominantly caused by Na^+ influx via AMPAR channels (Lasser-Ross and Ross 1992; Kuruma et al. 2003; Bennay et al. 2008). In addition, metabotropic glutamate receptors in cerebellar Purkinje cells can also be activated by synaptically released glutamate, and induce Na^+ transients in the dendrites of Purkinje cells (Knöpfel et al. 2000). Ionotropic glutamate receptor-mediated increases in $[Na^+]i$ have been linked to hypoxic/ischemic neuronal injury (Goldman et al. 1994; Agostinho et al. 1996; Hasbani et al. 1998; Houzen et al. 1998; Mironov and Langohr 2005; Kiedrowski 2007; Rojas et al. 2007). Thus, it is conceivable that elevation of $[Na^+]i$ during hypoxia/ischemia is at least partially mediated by glutamate receptor channels (Lynch et al. 1995; Zhang and Lipton 1999; Lopachin et al. 2001; Mironov and Langohr 2005; Kiedrowski 2007; Vander Jagt et al. 2008; Chao et al. 2009). However, some studies indicate that activation of iGluR channels does not contribute to hypoxic/ischemic Na^+ influx in neurons and glial cells in various brain regions, because blockade of NMDARs, AMPARs/KARs, or even metabotropic glutamate receptors, has little effect on hypoxia/ischemia-induced increase in $[Na^+]i$ (Silver et al. 1997; Guatteo et al. 1998b; Pisani et al. 1998a; Calabresi et al. 1999; Müller and Somjen 2000b; Sheldon et al. 2004; Martinez-Sánchez et al. 2004). Therefore, the role of ligand-gated Na^+ channels, especially glutamate receptor channels, in hypoxic/ischemic influx and $[Na^+]i$ elevation is still full of controversy.

In addition to iGluR channels, ATP-activated P_2X receptor channels might also be a source of intracellular Na^+ rise via ligand-gated Na^+ channels. P_2X receptors are a family of cation-permeable (such as Na^+, Ca^{2+}, K^+ etc.) ligand gated ion channels of purinergic receptors that open in response to the binding of extracellular ATP (Liu and Adams 2001; Samways et al. 2014). P_2X receptors are widely distributed in excitable and non-excitable cells of vertebrates, including neurons and glial cells (North 2002; Bai and Li 2013). Correlation of voltage-clamp measurements of Na^+ entry with regional increases in Na^+-induced fluorescence demonstrated ATP-induced increases in $[Na^+]i$ exceeding 45 mM within 4 s in cochlear hair cells (Housley et al. 1998). Acidification (to pH 6.2) increased the ATP-induced current amplitude twofold via P_2X channels in dissociated neurons from rat parasympathetic ganglia (Liu and Adams 2001). During hypoxia/ischemia, excessive ATP release (Lin et al. 2008; Schock et al. 2008; Domercq et al. 2010; Bai and Li 2013) and acidification (Silver and Erecińska 1990; Diarra et al. 1999; Sheldon and Church 2004; Sheldon et al. 2004; Cengiz et al. 2014) may activate/enhance P_2X channel activity. Besides Ca^{2+} entry and overload via P_2X channels (see discussion in next section), Na^+ might also enter into intracellular cytosol and cause $[Na^+]i$ rise during hypoxia/ischemia. However, so far it is still unknown if P_2X channels are contributors to hypoxic/ischemic $[Na^+]i$ rise despite the fact that they are activated by hypoxia/ischemia (Bai and Li 2013) and their activation can induce large increases in $[Na^+]i$ (Housley et al. 1998). It has been shown that blockade of P_2X channels prevents ATP excitotoxicity in neurons and reduces brain damage after ischemia (Arbeloa et al. 2012; Bai and Li 2013). Therefore, investigating the role of P_2X channels in hypoxic/ischemic Na^+ overload is attractive and important in seeking new strategies for the treatment of glutamate-independent excitotoxicity and hypoxia/ischemia-induced brain damage and neuronal death.

3.2.4 Non-selective Cation Channels (NSCCs)

NSCCs superfamily consists of a large number of cation channels, such as transient receptor potential (TRP) channels, acid-sensing ion channels (ASICs), and Ca^{2+}-activated, ATP-sensitive nonselective cation channels ($NC_{Ca\text{-}ATP}$), which are mostly permeable to both monovalent and divalent cations, and are easily activated by Ca^{2+} and/or reactive oxygen/nitrogen species (ROS/RNS) which increase in response to hypoxia/ischemia (Chen and Simard 2001; Gao et al. 2005; Simard et al. 2006, 2007; Nilius et al. 2007; Leng et al. 2014). There is evidence showing that NSCCs' activation plays a critical role in hypoxic/ischemic neuronal death (Chen and Simard 2001; Aarts et al. 2003; Xiong et al. 2004; Gao et al. 2005; Simard et al. 2006; Leng et al. 2014). In rat cerebellar granule cells, chemical ischemia-induced large increases in $[Na^+]i$ cannot be attenuated by blocking TTX-sensitive Na^+ channels, NCX, NHE or NKCC, but can be attenuated by blocking ROS/RNS-activated NSCCs (Chen et al. 1999). Some other studies reveal that Na^+ influx during and after anoxia can be attenuated by blockade of NSCCs with Gd^{3+} and trolox (antioxidant) (Sheldon et al. 2004) or with flufenamic acid, a NSCC blocker (Mironov and Langohr 2005). Blockade of $NC_{Ca\text{-}ATP}$ channels is highly effective in preventing Na^+ influx, and reduces cerebral edema and infarct volume with a reduction of mortality by 50 % after ischemic stroke (Chen and Simard 2001; Simard et al. 2006). Marked accumulation of arachidonic acid seen during brain ischemia also induces cytosolic Na^+ and Ca^{2+} overload via a non-selective cation conductance, resulting in mitochondrial $[Na^+]$ and $[Ca^{2+}]$ overload (Fang et al. 2008). Under ischemic conditions, hypoxia enhances glycolysis, resulting in the buildup of lactic acid and subsequent tissue acidosis, and acidosis can activate homomeric ASIC1a, causing a large influx of Na^+ and Ca^{2+}. This leads to neuronal injury (Xiong et al. 2004; Leng et al. 2014). Therefore, NSCCs activation contributes to hypoxia/ischemia-induced increase in $[Na^+]i$.

3.2.5 Plasmalemmal Na^+/Ca^{2+} Exchanger (NCX)

NCX is a transmembrane antiporter protein that plays an essential role in the control of Ca^{2+} flux across the plasma membranes and intracellular compartments. NCX operates in a forward mode to maintain $[Ca^{2+}]i$ at physiological levels by extruding 1 Ca^{2+} out while bringing 3 Na^+ in by consuming the Na^+ electrochemical gradient generated by the ATP-driven Na^+ pumps under physiological conditions (Dipolo and Beaugé 2006). It can operate in either reverse or forward mode depending on the changes in transmembrane potential and on Na^+ and Ca^{2+} gradients. The roles of NCX following cerebral ischemia are complicated, and the exchanger may function as a double-edged sword (either beneficial or detrimental) (Jeffs et al. 2007). Though reverse operation of NCX has been demonstrated to be responsible for ischemic $[Ca^{2+}]i$ rise (Stys and Lopachin 1998; Tanaka et al. 2002; Martinez-Sánchez et al. 2004; Kiedrowski 2007), both forward and reverse Na^+/Ca^{2+} exchange can take place simultaneously in different areas of the same cell, depending on the combined effects of Na^+ and Ca^{2+} gradients (Yu and Choi 1997). Therefore, it is possible for NCX to contribute to hypoxic/ischemic Na^+ influx. In their mathematical model of ion movement in neurons, Dronne et al. (2006) extrapolated the involvement of NCX in Na^+ entry into neurons during severe ischemia by simulating the dynamics of Na^+. The data from some NCX knockout/knockdown models suggest that the Ca^{2+} extrusion/Na^+ entry mode of NCX possibly runs during hypoxia because NCX knockout/knockdown aggravates Ca^{2+} overload in neurons (Jeon et al. 2008; Jeffs et al. 2008; Molinaro et al. 2008). However, it seems that NCX does not contribute to $[Na^+]i$ rise in a major way in rat cortical (Pisani et al. 1998a) and cerebellar granule (Chen et al. 1999) neurons during chemical ischemia, and in the hippocampal neurons in brief anoxia (Sheldon et al. 2004), while it does contribute to Na^+ influx immediately after anoxia in hippocampal neurons (Sheldon et al. 2004). Therefore, NCX is one of the major routes in hypoxia/ischemia-induced $[Ca^{2+}]i$ rise rather than Na^+ influx during severe hypoxia/ischemia (particularly in the phase of the delayed uncontrolled $[Ca^{2+}]i$ rise) (see the discussion in the next section about Ca^{2+} homeostasis).

3.2.6 Na^+/H^+ Exchanger (NHE)

The NHE family of ion exchangers includes six isoforms (NHE1–NHE6) that function in an electroneutral exchange of intracellular H^+ for extracellular Na^+ (Pedersen et al. 2006). Type 1 NHE (NHE1) is the most abundant NHE isoform in the rat brain (Ma and Haddad 1997), and may be an important pathway for Na^+ entry in neurons during hypoxia/ischemia since oxygen deprivation or blood supply insufficiency evokes a significant fall in pHi (Silver and Erecińska 1990; Diarra et al. 1999; Sheldon and Church 2004; Sheldon et al. 2004; Cengiz et al. 2014), and activates NHE to extrude excessive intracellular H^+ (Diarra et al. 1999; Sheldon and Church 2004; Pedersen et al. 2006; Cengiz et al. 2014). However, it has been demonstrated that NHE is inhibited during hypoxia/ischemia, but is activated immediately after reoxygenation/reperfusion (Diarra et al. 1999; Sheldon and Church 2004; Sheldon et al. 2004). Therefore, NHE contributes to Na^+ influx in the period immediately after anoxia rather than during anoxia in neurons (Chen et al. 1999; Sheldon et al. 2004; Cengiz et al. 2014). This viewpoint is further supported by the results in both cultured cortical neurons and hippocampal astrocytes with genetic ablation of NHE1 or blockade of NHE1 with HOE 642 (Luo et al. 2005; Cengiz et al. 2014). In the study by Luo et al. (2005), OGD caused a two-fold increase in neuronal $[Na^+]i$, and during 60 min reperfusion, $[Na^+]i$ increased sevenfold. Genetic ablation of NHE1 or blockade of NHE1 had no effects on OGD-mediated initial $[Na^+]i$ rise, but reduced the second phase of $[Na^+]i$ by 40–50 %. Though 2 h of OGD does not increase basal $[Na^+]i$ in hippocampal astrocytes, $[Na^+]i$ significantly increases following reperfusion and stays at high levels for at least 24 h, and this increase in $[Na^+]i$ following reperfusion can be significantly decreased by NHE1 inhibitor HOE 642, indicating that NHE1 activity leads to sustained intracellular Na^+ overload after ischemia (Cengiz et al. 2014). Therefore, NHE activation after hypoxia/ischemia has a significant impact on the loss of Na^+ homeostasis and neuronal cell death (Luo et al. 2005; Xue et al. 2008; Wang et al. 2008; Leng et al. 2014; Cengiz et al. 2014).

3.2.7 Na^+-K^+-$2Cl^-$ Cotransporter (NKCC)

NKCCs are a class of membrane proteins that electroneutrally transport Na^+, K^+, and Cl^- ions into and out of a wide variety of cells, almost always with a stoichiometry of $1Na^+$:$1K^+$:$2Cl^-$. NKCCs are very sensitive to diuretics such as bumetanide and furosemide, and are essential in the regulation of intracellular Cl^- homeostasis in the CNS (Blaesse et al. 2009). To date, two isoforms of NKCC have been confirmed. NKCC1 is present in a wide variety of tissues including the brain, while its counterpart, NKCC2, is localized exclusively to the kidney (Blaesse et al. 2009; MacAulay and Zeuthen 2012). Both neurons and glial cells express NKCC1 (Blaesse et al. 2009; MacAulay and Zeuthen 2012). The classical viewpoint says that neuronal NKCC1 expression decreases during postnatal development (Blaesse et al. 2009). However, this idea is challenged by some studies that show a developmental increase in the NKCC1 mRNA and protein levels in the brain (Blaesse et al. 2009). NKCC1 in the adult CNS is absent from neuronal cell bodies despite its existence in the dendrites in some neurons, while it is abundantly expressed in astrocytes (Hertz et al. 2013).

In both cultured cortical neurons and astrocytes, an OGD-induced increase in $[Na^+]i$ can be significantly attenuated by inhibition of NKCC1 with bumetanide or genetic ablation of NKCC1 (Lenart et al. 2004; Kintner et al. 2007; Chen et al. 2008). Since inward NKCC may be limited by high $[Na^+]i$ and presumably $[Cl^-]i$ immediately after anoxia, NKCC-mediated Na^+ entry seems to be ascribed to $[Na^+]$ i rise during anoxia rather than immediately after anoxia. This is partially supported by the observation in cultured hippocampal neurons that 100 μM bumetanide caused a 40 % reduction in the rise in $[Na^+]i$ in 11–14 DIV neurons during anoxia, but failed to significantly reduce the increase in $[Na^+]i$ after anoxia (Sheldon et al. 2004). In cultured cerebellar granule cells, however, chemical ischemia-induced increase in $[Na^+]i$ cannot be attenuated by bumetanide. Thus, NKCC1 is not involved in ischemia-induced Na^+ influx in this kind of neuron (Chen et al. 1999). Taken together, NKCC may be one of the routes for hypoxia/ischemia-induced Na^+ influx in astrocytes and neurons in the early phase of stress.

3.2.8 Excitatory Amino-Acid Transporters (EAATs)

EAATs, also known as glutamate transporters, belong to the family of neurotransmitter transporters. Glutamate is the principal excitatory neurotransmitter in the vertebrate brain. EAATs serve to terminate the excitatory signal by removal (uptake) of glutamate from the neuronal synaptic cleft into glia and neurons (Camacho and Massieu 2006; Tzingounis and Wadiche 2007; Krzyżanowska et al. 2014). Downregulation/blockade or genetic deletion of EAAT2/GLT1 results in the elevation of residual glutamate in rodent brain as well as the development of epilepsy and exacerbated brain injury (Tanaka et al. 1997; Han et al. 2008). To date, five subtypes (EAAT1-5) have been identified from human and animal tissue. Subtypes EAAT1-2 are found in membranes of glial cells (astrocytes, microglia, and oligodendrocytes) as well as in endothelial cells, whereas subtypes EAAT3-4 are located on neurons (Anderson and Swanson 2000). EAAT5 is only found in the retina where it is principally localized to photoreceptors and bipolar neurons in the retina (Pow and Barnett 2000). In rodents, the orthologs for EAAT1-3 are named GLAST, GLT1, and EAAC1, respectively (Shigeri et al. 2004), whereas the acronyms EAAT4 and EAAT5 are conserved. Astrocyte-specific glutamate transporters (EAAT1/GLAST, EAAT2/GLT1) are largely responsible for glutamate clearance from the extracellular space (Chatton et al. 2000; Kirischuk et al. 2007; Bennay et al. 2008; Langer and Rose 2009). In astrocytes, glutamate is converted to glutamine via the astrocyte-specific enzyme glutamine synthetase, and glutamine is returned to neurons in order to replenish the presynaptic neurotransmitter pool and help maintain synaptic transmission (Eid et al. 2008; Kirischuk et al. 2012).

Glutamate uptake is driven by transmembrane Na^+ gradients, which dynamically control the functional activity of EAATs (Kirischuk et al. 2012). EAATs utilize the Na^+ gradient generated by membrane Na^+/K^+ ATPase with a complex stoichiometry of one glutamate molecule together with two or three Na^+ plus one H^+ cotransported into the cytoplasm and one K^+ out (Bouvier et al. 1992; Levy et al. 1998; Camacho and Massieu 2006). Thus, $[Na^+]i$ increases with EAAT activation (Chatton et al. 2000; Kirischuk et al. 2007; Rojas et al. 2007; Bennay et al. 2008; Langer and Rose 2009). It is believed that EAATs operate normally to eliminate released glutamate (Camacho and Massieu 2006) during the first minutes of ischemia (before energy exhaustion and Na^+ gradient collapse) (Yamaguchi et al. 1998), and therefore may partially account for the limited $[Na^+]i$ increase in the early phase of hypoxia/ischemia. However, with prolonged hypoxia/ischemia, largely increased $[Na^+]i$ (Friedman and Haddad 1994; Calabresi et al. 1999; Banasiak et al. 2004; Martinez-

Sánchez et al. 2004; Sheldon et al. 2004; Nikolaeva et al. 2005) along with membrane depolarization makes EAATs operate reversely (Rossi et al. 2000; Mitani and Tanaka 2003; Camacho and Massieu 2006; Chen et al. 2007a; Micu et al. 2007). This causes massive glutamate release and intracellular Na^+ to move out (Yamaguchi et al. 1998). In support of this model is the fact that blockade of the glutamate transporter during the early stages of hypoxia leads to accelerated accumulation of glutamate in the extracellular space, resulting in the inflated onset of the hypoxic response and the exaggerated peak magnitude of the hypoxic response; in the late stage of hypoxia, glutamate transporter begins to operate in the reverse direction, releasing glutamate into the extracellular space, and blockade of the glutamate transporter impedes the attenuating effects of high $[Na^+]o$ on the peak magnitude of hypoxic response (Yamaguchi et al. 1998). Therefore, EAATs play a dual role in hypoxia/ischemia-induced changes in $[Na^+]o/[Na^+]i$ depending on their operating modes.

3.3 Hypoxia/Ischemia-Induced Intracellular Ca^{2+} Accumulation

There is an elevation and accumulation of cytosolic Ca^{2+} in neurons during global or focal ischemia (Uematsu et al. 1988; Silver and Erecińska 1990, 1992; Kristián and Siesjő 1998; Nakamura et al. 1999; Vannucci et al. 2001) and in vitro hypoxia or ischemia (e.g., in the brain slices) (Mitani et al. 1993, 1995; Nabetani et al. 1997; Fukuda et al. 1998; Pisani et al. 1997, 1998b, 1999; Ballanyi and Kulik 1998; Connor et al. 1999; Zhang and Lipton 1999; Tjong et al. 2007), organotypic brain slice cultures (Martinez-Sánchez et al. 2004) and acutely isolated or cultured cells (Friedman and Haddad 1993; Goldberg and Choi 1993; Tymianski et al. 1993; Murai et al. 1997; Silver et al. 1997; Chen et al. 1999; Diarra et al. 1999; Smith et al. 2003, 2004; Kiedrowski 2007). During hypoxia/ischemia, most of the brain tissues/neurons show a progressive or transient small increase followed by a large sustained increase in cytosolic Ca^{2+} (Silver and Erecińska 1990, 1992; Mitani et al. 1993, 1995; Murai et al. 1997; Pisani et al. 1997; Ballanyi and Kulik 1998; Fukuda et al. 1998; Chen et al. 1999; Smith et al. 2004; Martinez-Sánchez et al. 2004). The increase in cytosolic Ca^{2+} varies from several hundred nanomolar to micromolar range in magnitude, and also varies in time required for the onset of such increase and recovery from stress after re-oxygentation/reperfusion (Silver and Erecińska 1990; Friedman and Haddad 1993; Nabetani et al. 1997; Zhang and Lipton 1999; Martinez-Sánchez et al. 2004).

The onset and magnitude of this increase depend on multiple inherent and environmental factors, including the duration (short vs. long) and severity (mild vs. severe) of insult, the age of animals (young vs. old), the region of the brain (hippocampus or cortex vs. other regions such as thalamus, striatum, dorsal vagal neurons, locus coeruleus, cerebellar granule cells), and even the type of cells (neurons vs. glial cells). For example, there is a minor or moderate increase in cytosolic Ca^{2+}

(tens to hundreds nanomolar) with short or transient hypoxia/ischemia, but a large sustained increase in [Ca^{2+}]i (up to 35 μM) in severe and prolonged hypoxia/ischemia (Uematsu et al. 1988; Silver and Erecińska 1990, 1992; Tymianski et al. 1993; Pisani et al. 1997; Silver et al. 1997; Chen et al. 1999; Zhang and Lipton 1999; Diarra et al. 1999; Nakamura et al. 1999; Martinez-Sánchez et al. 2004). Young animals show much less extension and longer latency (up to five times longer) of Ca^{2+} rise in neurons than those of mature animals during hypoxia/ischemia (Friedman and Haddad 1993; Nabetani et al. 1997; Vannucci et al. 2001) because young cells have a higher resistance to energy failure and adaptation capacity for disturbance of Ca^{2+} homeostasis (Hartmann et al. 1996; Nabetani et al. 1997). Compared to glial cells, neurons show a much faster increase in [Ca^{2+}]i during hypoxic/ischemic energy limitation (Goldberg and Choi 1993; Silver et al. 1997; Kahlert and Reiser 2004). This is because neurons lack glycogen storage (Silver et al. 1997), giving them less active energy generation, and have abundant ion pathways with marked suppression of Ca^{2+} extrusion during ischemic and excitotoxic stress (Kiedrowski et al. 1994; Duffy and MacVicar 1996).

Moreover, there exist regional differences in hypoxic/ischemic cytosolic Ca^{2+} accumulation. The magnitude of the rapid elevation of [Ca^{2+}]i is very large in many, but not all, hippocampal CA1 and cortical cells. In contrast to cortical and hippocampal CA1 cells, the change in [Ca^{2+}]i is smaller and relatively delayed in thalamic, dorsal vagal neurons, and cerebellar Purkinje cells during hypoxia/ischemia (Silver and Erecińska 1990; Mitani et al. 1995; Ballanyi and Kulik 1998). Striatal neurons are more vulnerable than cortical cells to hypoxia (Pisani et al. 1997). Even in the various parts of the same brain regions, hypoxia/ischemia-induced Ca^{2+} responses may also be different. For example, neocortical cells show a greater increase in [Ca^{2+}]i in layer II/III than in layer IV, V, or VI in more immature animals (postnatal day 7–14 rats) (Fukuda et al. 1998). They show a greater increase in layer V than layer II/III in relatively older animals (postnatal day 16–26 mice) (Gniel and Martin 2010) during OGD. In hippocampal CA1 neurons, ischemia induces an increase in [Ca^{2+}]i of up to 30 μM, and this change decreases rapidly to 73 nM during reperfusion. In contrast, the changes in [Ca^{2+}]i in CA3 area are much smaller (3 μM) and slightly slower than in CA1 area (Silver and Erecińska 1992). Similar observation of larger [Ca^{2+}]i elevation and mitochondrial Ca^{2+} overload in CA1 than CA3 neurons were also made with both chemical ischemia and NMDA challenge (Stanika et al. 2010). These notions sufficiently reflect a regional difference in the sensitivity to hypoxia/ischemia. In addition, Pisani et al. (1999) reported a very interesting phenomenon that OGD-induced [Ca^{2+}]i elevation was more pronounced in dendritic branches than in somatic regions in striatal neurons, and a similar result was also found in the hippocampus (Zhang and Lipton 1999). This interesting phenomenon can be explained by the larger surface area/volume ratios in dendrites than in soma, or by the variability of Ca^{2+} channel expression (Ladewig et al. 2003; Vander Jagt et al. 2008), or by higher Ca^{2+} conductance in dendrites than in soma due to more excitatory synaptic inputs and higher density of NMDA receptors in dendrites (Connor et al. 1988; Wadman and Connor 1992; Vander Jagt et al. 2008). Finally, not only are the onset and magnitude of Ca^{2+} accumulation different with differences in inherent and/or environmental conditions, but also posthypoxic/postischemic changes in $[Ca^{2+}]i$ show major variance depending on cellular status during hypoxia/ischemia. For instance, hypoxic/ischemic cytosolic Ca^{2+} accumulation gradually recovers with reperfusion or reoxygenation after short and/or mild hypoxia/ischemia, but exhibits a further increase during reperfusion period after prolonged and/or severe hypoxia/ischemia (Uematsu et al. 1988; Pisani et al. 1997; Vannucci et al. 2001). This may reflect damaged Ca^{2+} extrusion mechanisms (Schwab et al. 2002; Bano et al. 2005; Pászty et al. 2007; Amarjargal et al. 2008; Luo et al. 2008; Henrich and Buckler 2008, 2013), sequestering mechanisms of free cytosolic Ca^{2+} into organelles (Bergmann and Keller 2003; Ladewig et al. 2003), activation of non-selective cationic channels such as TRPM7 (a member of the melastatin branch of the transient receptor potential channel family), and calcium-permeable acid-sensing ion channels (Aarts et al. 2003; Xiong et al. 2004; Macdonald et al. 2006) in severe and prolonged insults. Thus, hypoxia/ischemia-induced intracellular Ca^{2+} accumulation is evident, but varies much in different cells and in various conditions.

Glial cells are the most abundant cell type, and astrocytes are the predominant glial-cell type in the CNS but little is known about the functional impact of ischemia on astrocytes. Research since the mid-1990s has shown that astrocytes propagate intercellular Ca^{2+} waves over long distances in response to stimulation, and similarly to neurons, release transmitters (called gliotransmitters) in a Ca^{2+}-dependent manner (Verkhratsky et al. 2012). The discovery of the Ca^{2+} response in glial cells led to a concept of "glial Ca^{2+} excitability", which overturns the traditional thought of glial cells as "structural support and passive elements" (Verkhratsky et al. 2012), and thus upsurges the studies of neuron-glia interaction (Volterra et al. 2014). Currently, hypoxic/ischemic response of astrocytic Ca^{2+} signaling is getting more and more attention and is increasingly becoming a target for seeking new therapeutic strategies for hypoxic/ischemic brain injury. Like neurons, glial cells also increase their $[Ca^{2+}]i$ in response to hypoxic/ischemic stress, as observed in cultures (Duffy and MacVicar 1996; Silver et al. 1997; Deng et al. 2003; Bondarenko et al. 2004; Smith et al. 2003, 2004; Aley et al. 2006; Iwabuchi and Kawahara 2009; Ouyang et al. 2011; Murugan et al. 2013; Cengiz et al. 2014), slices/in situ (Duffy and MacVicar 1996; Fern 1998; Thomas et al. 2004; Micu et al. 2006, 2007; Dong et al. 2013; Shinotsuka et al. 2014), and in vivo (Ding et al. 2009; Zheng et al. 2013a). However, there are dramatic differences in cytosolic Ca^{2+} dynamic kinetics and post-hypoxic/ischemic viability between glial cells and neurons. A few minutes of focal ischemia can trigger neuronal structural changes, dysfunction, and degeneration, whereas astrocytes remain structurally intact for up to 3 h following ischemia, and significant astrocytic death requires many hours of insults (Haun et al. 1992; Ding et al. 2009; Zheng et al. 2013a). In acutely isolated neurons and astrocytes, identical ischemic stimulation (1.5–6.5 min) causes large, irreversible $[Ca^{2+}]i$ responses in neurons which are not reversed by removal of extracellular Ca^{2+}, but has little effect on astrocytic $[Ca^{2+}]i$, and prolonged ischemia (7.5–34 min) only causes a slowly developing and modest increase in basal $[Ca^{2+}]i$ in astrocytes (Duffy and MacVicar 1996). Similar observations are also made in cultured glial cells and neurons (Silver et al. 1997). In cultured astrocytes, significant $[Ca^{2+}]i$ elevation and cellular dysfunction require many hours (up to 18 h) of continuous hypoxia and hypoglycemia (Haun et al. 1992). In contrast to the slow modest $[Ca^{2+}]i$ increase in acutely isolated or cultured astrocytes (Duffy and MacVicar 1996; Silver et al. 1997), astrocytes in brain slices can respond to a relatively short episode of OGD with a two-phase $[Ca^{2+}]i$ response: oscillating transients followed by large increases that occur simultaneously with a sudden and irreversible neuronal depolarization known as anoxic depolarization (Duffy and MacVicar 1996; Dong et al. 2013; Shinotsuka et al. 2014). This elevated astrocytic $[Ca^{2+}]i$ in brain slices can remain up to 60 min after reperfusion (Duffy and MacVicar 1996). In some cases, significant increases (20-fold) in oscillation frequency of spontaneous transient Ca^{2+} spikes can be observed in astrocytes (in cultures and hippocampal slices) even 7 days after hypoxia/ischemia-reperfusion (Butenko et al. 2012). It is interesting to note that 50 mM of K^+ (similar to $[K^+]e$ in response to hypoxia/ischemia) also triggers astrocytic $[Ca^{2+}]i$ responses similar to those in OGD in the brain slices, but 15 mM of K^+ (corresponding to the ceiling $[K^+]e$ increases during intense neuronal activity in situ) does not increase astrocytic $[Ca^{2+}]i$ in the brain slices (Duffy and MacVicar 1996). As a matter of fact, during anoxic depolarization and spreading depression, $[K^+]e$ can be elevated to levels of 50–80 mM. Therefore, the different $[Ca^{2+}]i$ increases in acutely isolated/cultured astrocytes and in astrocytes of brain slices during OGD may be due to the contributions of neuronal factors (such as large released K^+ as well as neurotransmitters' stimulation-induced membrane depolarization and the activation of Ca^{2+} channels and the subsequent Ca^{2+} influx through Ca^{2+} channels in astrocytes in slices). In vivo Ca^{2+} imaging also indicated an increase in astrocytic $[Ca^{2+}]i$ during hypoxia/ischemia (Ding et al. 2009; Mathiesen et al. 2013).

In addition, a striking feature of astrocytic Ca^{2+} signals is the Ca^{2+} wave, namely, that the localized elevation of $[Ca^{2+}]i$ spreads between neighboring astrocytes either via gap junctions between these cells or by extracellular ATP signaling-IP_3 stimulated regenerative $[Ca^{2+}]i$ of the neighboring astrocytes (Bernardinelli et al. 2004; Verkhratsky et al. 2012). Calcium waves occur not only under physiological, but also pathophysiological conditions such as hypoxia/ischemia. By using high-resolution time-lapse Ca^{2+} imaging (e.g. two-photo laser scanning microscopy), it has been shown in vivo (Ding et al. 2009; Mathiesen et al. 2013) and in brain slices (Dong et al. 2013; Shinotsuka et al. 2014) that hypoxia/ischemia not only increases the amplitude but also the frequency of astrocytic Ca^{2+} signals, and more astrocytes exhibit synchronous Ca^{2+} transients with ischemia, indicating enhanced intercellular Ca^{2+} waves during hypoxia/ischemia. It is generally thought that Ca^{2+} waves initiate from an initiation point in an ischemic core region where intracellular Ca^{2+} increases in single astrocytes propagate radially from the central initiation point (Ding et al. 2009; Mathiesen et al. 2013). This is believed because intra-astrocytic Ca^{2+} signals increase during ischemia in both ischemic core and the penumbra, but are more robust in the ischemic core, and exhibit a high degree of synchrony among astrocytes in the same imaging field (Ding et al. 2009), and a propagating spontaneous Ca^{2+} wave is refractory to a second wave beginning at the boarder of the previous wave (Mathiesen et al. 2013). In cultured cortical astrocytes, 2–8 h of OGD plus 1 h or 1 day of reperfusion not only significantly increase the propagating speed and distance of intercellular Ca^{2+} waves, but also the $[Ca^{2+}]$i of neighboring astrocytes (about 150 %), indicating that ischemia enhances the activity of Ca^{2+} waves (Iwabuchi and Kawahara 2009). The role of such glial Ca^{2+} signaling in hypoxic/ischemic neuronal damage is still unclear and the results are controversial. On one hand, some studies showed that astrocytic Ca^{2+} waves mediate the activation of extrasynaptic NMDAR to aggravate ischemic brain damage, and inhibition of astrocytic Ca^{2+} waves by either knockout of astrocytic IP_3R2 (activation of it enhaces Ca^{2+} wave activity) or intracellular loading of Ca^{2+} chelator BAPTA significantly decreases infarcted brain area, thus proposing a detrimental role of astrocytic Ca^{2+} waves (Dong et al. 2013; Ding et al. 2009). On the other hand, some studies suggested that the onset of the second phase of $[Ca^{2+}]$i dynamics is tightly synchronized with the appearance of the anoxic depolarization of neurons (the hallmark of ischemic neuronal responses causing glutamate release, cell swelling, and neuronal cell death), and pharmacological blockade of gap junction advances such events (Shinotsuka et al. 2014). Such astrocytic Ca^{2+} waves may drive a number of adaptive responses to hypoxia, including increases in local blood flow and changes in Purkinje cell bistability (Mathiesen et al. 2013). IP_3R2 knockout decreases the astrocytic P_2Y1R-IP_3-mediated Ca^{2+} signals and worsens ischemic neuronal lesions and dendritic damage (Zheng et al. 2013a). Therefore, these studies suggest a protective role of astrocytic Ca^{2+} signals (Zheng et al. 2013a; Mathiesen et al. 2013; Shinotsuka et al. 2014). Nevertheless, it is obvious that astrocytes increase their intracellular Ca^{2+} signals in response to hypoxia/ischemia with dramatic differences in cytosolic Ca^{2+} dynamic kinetics of neurons.

Other than astrocytes, oligodendrocytes also exhibit Ca^{2+} response to hypoxia/ischemia (Deng et al. 2003; Micu et al. 2006, 2007; Murugan et al. 2013). For example, OGD for 2 h results in robust $^{45}Ca^{2+}$ entry and intracellular $^{45}Ca^{2+}$ accumulation in cultured forebrain oligodendrocyte precursor cells of neonatal rats, which is sensitive to AMPAR/KAR antagonists, but not to the blockade of NMDAR, voltage-gated Ca^{2+} channels and NCX (Deng et al. 2003). Thirty minutes of chemical ischemia causes 50–60 % $[Ca^{2+}]$i elevations in both myelin (a specialized structure produced by oligodendrocytes that ensheaths axons) and its parent oligodendroglial cell body of adult rat optic nerves and dorsal columns (Micu et al. 2006, 2007). However, the sources of ischemic $[Ca^{2+}]$i rise in both compartments are completely different. Oligodendrocytes and their precursor cells had once been thought to express AMPARs/KARs, but not functional NMDARs (Patneau et al. 1994), and ischemia-induced $[Ca^{2+}]$i elevation has been believed to be mediated by Ca^{2+}-permeable AMPARs/KARs (Deng et al. 2003). However, recent studies indicated that NMDARs are present in the myelinating processes of oligodendrocytes and their precursor cells, and mediate NMDA-evoked currents (Káradóttir et al. 2005; Salter and Fern 2005; Micu et al. 2006). Ischemia leads to development of an inward current and process loss (but retaining of somata) in oligodendrocytes, which can be attenuated or abolished by the blockade of NMDARs (Káradóttir et al. 2005; Salter and Fern 2005). Direct Ca^{2+} imaging with two-photo microscopy indicated that the

AMPAR/KAR antagonist NBQX completely blocks the ischaemic Ca^{2+} increase in oligodendroglial cell bodies, but only modestly reduces the Ca^{2+} increase in myelin. In contrast, the Ca^{2+} increase in myelin is abolished by broad-spectrum NMDA receptor antagonists (MK-801, 7-chlorokynurenic acid, d-AP5), but not by more selective blockers of NR2A and NR2B subunit-containing receptors (NVP-AAM077 and ifenprodil) (Micu et al. 2006, 2007). These findings clearly indicate NMDARs mediate ischemic Ca^{2+} rise in oligodendroglial processes and myelin sheath, while AMPARs/KARs mediate this rise in oligodendroglial cell bodies. Since axons (in the absence of neurons and nerve terminals) can release glutamate during action potential propagation, which can be detected by ionotropic glutamate receptors on local ligodendrocytes (Kukley et al. 2007; Fu and Sretavan 2012; Frühbeis et al. 2013), and also because the small intracellular space in the myelinating processes of oligodendrocytes can lead to a large rise in intracellular ion concentrations in response to NMDAR activation (Micu et al. 2006, 2007), hypoxia/ischemia-induced dysregulation of glutamate homeostasis and NMDAR-mediated $[Ca^{2+}]i$ rise in oligodendroglial processes (Micu et al. 2006, 2007; Murugan et al. 2013) may have profound influences in the pathophysiology of oligodendrocytes, and contribute to white matter injury that leads to many forms of brain disorders such as cerebral palsy, spinal cord injury, multiple sclerosis, and stroke (Deng et al. 2003; Káradóttir et al. 2005; Salter and Fern 2005; Micu et al. 2006; Murugan et al. 2013). Therefore, it represents a potential therapeutic target for preventing white matter damage in a variety of diseases.

The hypoxia/ischemia-induced Ca^{2+} accumulation may originate from two sources: an influx of extracellular Ca^{2+}, and the release of Ca^{2+} from intracellular stores. At least three factors are involved in the regulation of Ca^{2+} homeostasis, including Ca^{2+} influx and efflux through the plasma membrane, Ca^{2+} uptake and release from intracellular stores, and intracellular Ca^{2+} buffering. In contrast to Ca^{2+} entry and release from Ca^{2+} stores, intracellular Ca^{2+} buffering of cytosolic Ca^{2+} binding proteins plays a very minor role in the regulation of $[Ca^{2+}]i$ overload during hypoxia/ischemia due to the limited capacity of Ca^{2+} of cytosolic Ca^{2+} binding proteins (Chao and Xia 2010), and therefore will not be discussed in this chapter. The most important regulation of Ca^{2+} dynamics during hypoxia/ischemia is Ca^{2+} entry and release from Ca^{2+} stores (Fig. 6.3).

3.3.1 Ca^{2+} Influx and Efflux Through the Plasma Membrane

Calcium ion is a universal and pluripotent intracellular messenger controlling a broad spectrum of intracellular events in all eukaryotic cells. Ca^{2+} is of critical importance in the excitability and interaction with neurons of electrically silent astrocytes (Kirischuk et al. 2012; Verkhratsky et al. 2012; Volterra et al. 2014) and neuronal physiology, such as regulation of the release of neurotransmitters, membrane excitability, dendrite development, synaptogenesis, synaptic transmission and plasticity, gene expression, as well as many other processes in neuronal activity (Kawamoto et al. 2012; Brini et al. 2014). At rest, $[Ca^{2+}]o$ concentration is about

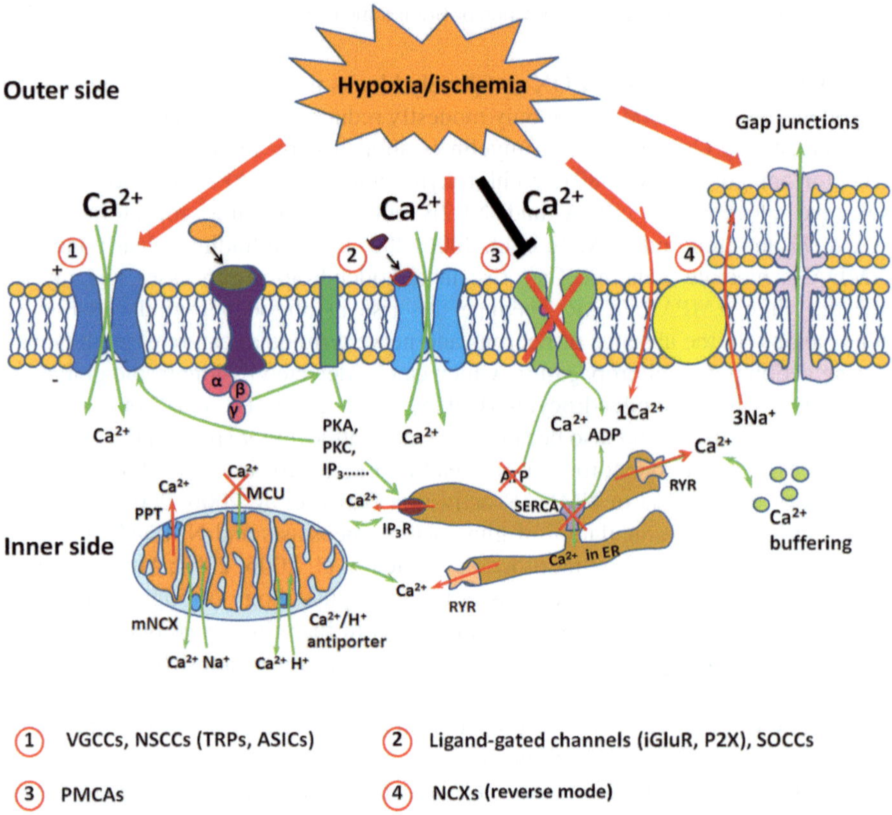

Fig. 6.3 Schematic demonstration of hypoxia/ischemia-induced intracellular Ca^{2+} accumulation in the nerve cells. The hypoxia/ischemia-induced intracellular Ca^{2+} accumulation may originate from two sources: an influx of extracellular Ca^{2+} and the release of Ca^{2+} from intracellular store. At least three factors are involved in the regulation of Ca^{2+} homeostasis, including Ca^{2+} influx and efflux through the plasma membrane, Ca^{2+} uptake and release from intracellular stores, and intracellular Ca^{2+} buffering. Neuronal Ca^{2+} influx leads to elevation of $[Ca^{2+}]i$ via ionotropic receptors (iGluRs, P_2X purinergic receptor) and/or voltage-gated Ca^{2+} channels. Plasmalamel Na^+/Ca^{2+} exchangers (NCXs), especially NCX1, can also greatly contribute to Ca^{2+} entry by operating in a reverse mode under hypoxic/ischemic condition. Some newly identified pathways, such as non-selective cation channels (e.g., transient receptor potential [TRP] channels, acid-sensing ion channels [ASICs]), gap junctions and hemichannels, and store-operated Ca^{2+} channels (SOC) also contribute to hypoxic/ischemic Ca^{2+} influx. Despite the substantial influx of extracellular Ca^{2+}, Ca^{2+} overload also reflects the failure of Ca^{2+} extruding systems (e.g., the inability and cleavage of plasma membrane Ca^{2+}-ATPase [PMCAs] and NCX3) to cope with excess Ca^{2+} that enter cells across the dysregulated plasma membrane channels. In addition to extracellular Ca^{2+} entry, intracellular Ca^{2+} stores [the most important stores are endoplasmic reticulum (ER) and mitochondria] are a source for cytosolic Ca^{2+} rise in hypoxic/ischemic neurons. The inositol triphosphahate receptors (IP_3Rs) and ryanodine receptors (RYRs) in the ER network are capable of regenerative Ca^{2+} release during hypoxia/ischemia, which together with the failure of sarco-endoplasmic reticular Ca^{2+} ATPase (SERCA pump) to sequester cytosolic Ca^{2+} into the ER due to ATP depletion, increase intracellular Ca^{2+} accumulation. Mitochondria are capable of storage and release of intracellular Ca^{2+}. During hypoxia/ischemia, Ca^{2+} buffering capacity of motochdria is largely damaged in both neurons and astrocytes. This may also contribute to hypoxia/ischemia-induced increase in $[Ca^{2+}]i$. Several pathways are responsible for hypoxia/ischemia-induced Ca^{2+} release from mitochondria, which include enhanced Ca^{2+} efflux via permeability transition pores (PTP), mitochondrial Na^+/Ca^{2+} exchanger (NCLXs), the decreased uptake due to the decrease of mitochondrial Ca^{2+} uniporter (MCU) activity, etc. Intracellular Ca^{2+} buffering of cytosolic Ca^{2+} binding proteins plays a very minor role in the regulation of $[Ca^{2+}]i$ overload during hypoxia/ischemia due to the limited capacity of Ca^{2+} of cytosolic Ca^{2+} binding proteins

1.2–1.5 mM, while the cytosolic free Ca^{2+} concentration is 10,000-fold lower than in extracellular fluid in both neurons (generally kept at a level between 40 and 100 nM) and glia (varies from 30–40 to 200–400 nM) (Hansen 1985; Verkhratsky et al. 2012), thus forming a very large concentration gradient between the inside and outside of cells. The membrane potential of neurons is relatively stable at 60–90 mV negative inside than outside, which, together with the large transmembrane concentration difference, provides a very powerful electrochemical driving force for the influx of Ca^{2+} into cells. Nevertheless, the cell membranes have very low permeability to Ca^{2+} for most cells, including neurons, at rest. Neurons, like many other types of cells, also possess two parallel, independent mechanisms to extrude Ca^{2+}, namely plasma membrane Ca^{2+}-ATPase (PMCA) (also called Ca^{2+} pump), and the Na^+/Ca^{2+} exchanger (NCX) (Dipolo and Beaugé 2006; Giacomello et al. 2013). PMCA has high Ca^{2+} affinity, but low transport capacity, whereas NCX has a low affinity, but a higher capacity to transport Ca^{2+} and can operate reversibly. NCXs operate in a forward mode to maintain $[Ca^{2+}]i$ at physiological levels by extruding Ca^{2+} out with Na^+ in (in 1:3 stoichiometry) by consuming Na^+ electrochemical gradients generated by the ATP-driven Na^+ pumps under normal conditions. NCXs are very efficient in extruding Ca^{2+} and are therefore considered as a "genuine Ca^{2+} pump". Even a small Ca^{2+} load in nerve terminals can activate Na^+/Ca^{2+} exchangers. This Ca^{2+} extrusion is almost completely dependent on extracellular Na^+ in this condition (Dipolo and Beaugé 2006). These two Ca^{2+} extrusion mechanisms, together with low permeability to Ca^{2+}, work in parallel in the plasma membrane to keep $[Ca^{2+}]i$ at physiological levels under normal conditions (Dipolo and Beaugé 2006; Giacomello et al. 2013).

Since neurons have a very powerful electrochemical driving force for the influx of Ca^{2+} into cells (though very low permeability to Ca^{2+} at rest), a very minor change in Ca^{2+} permeability of the plasma membrane may produce great increases in $[Ca^{2+}]$ i by the influx of Ca^{2+}. Indeed, such an event can happen under hypoxic/ischemic conditions (Silver and Erecińska 1990; Friedman and Haddad 1993; Mitani et al. 1993; Diarra et al. 1999; Nakamura et al. 1999).

There are several strategies to determine the contribution of Ca^{2+} influx to hypoxic/ischemic $[Ca^{2+}]i$ evelation, such as simultaneous measurement of the changes in $[Ca^{2+}]o$ and $[Ca^{2+}]i$, removal of extracellular Ca^{2+}, and blockade of the entry pathways across the plasma membrane. By using ion-selective microelectrodes, Silver and Erecińska (1990) directly evaluated the profiles of changes in $[Ca^{2+}]o$ and $[Ca^{2+}]i$ of neurons during anoxia and ischemia in different regions of the rat brain in vivo. They found that the major increase in $[Ca^{2+}]i$ (rising from ~70 nM to 24–35 μM) was associated with a large reduction in $[Ca^{2+}]o$ (dropping from ~1.5 to 0.5 mM or less) in many but not all neurons examined in the hippocampus, cortex and thalamus (Silver and Erecińska 1990). The drop of $[Ca^{2+}]o$ in this study might be greatly underestimated providing that the extracellular fluid space during hypoxia/ischemia decreases to approximately half of the control (Hansen 1985; Jiang and Haddad 1991), it can be approximately 0.1 mM during hypoxia/ischemia (Kristián and Siesjő 1998). Despite the real value, this study lends support to the idea that hypoxic/ischemic overload of intracellular Ca^{2+} in neurons is attributable primarily to Ca^{2+} influx from the extracellular space (Yao and Haddad 2004; Martinez-

Sánchez et al. 2004). Indeed, removal of extracellular Ca^{2+} also largely attenuates hypoxic/ischemic increase in [Ca^{2+}]i, further supporting the contribution of Ca^{2+} influx to hypoxic/ischemic [Ca^{2+}]i rise (Mitani et al. 1993, 1995; Friedman and Haddad 1993; Brown et al. 1998; Diarra et al. 1999; Nakamura et al. 1999). Obviously, neurons with very high calcium conductance, e.g., hippocampal CA1 and neocortical layer II/III neurons (with high density of NMDARs) (Kristián and Siesjő 1998; Fukuda et al. 1998), can suffer more severe Ca^{2+} overload due to an over-influx of extracellular Ca^{2+} (Silver and Erecińska 1990, 1992; Gniel and Martin 2010). In contrast to neuronal gray matter, measurement in the CNS white matter tract using ion-sensitive microelectrodes indicated that anoxia induces a two-phase change in [Ca^{2+}]o: an initial transient increase followed by a sustained reduction in [Ca^{2+}]o (Brown et al. 1998). The initial increase in [Ca^{2+}]o is supposed to be due to shrinkage of the extracellular space that results in a passive rise in [Ca^{2+}], and the following fall in [Ca^{2+}]o results from Ca^{2+} influx into an intracellular compartment, presumably via reverse operation of the Na^+/Ca^{2+} exchanger and depolarization induced activation of voltage-gated Ca^{2+} channels (Stys et al. 1992; Brown et al. 1998).

There are several pathways for the influx of Ca^{2+} during hypoxia/ischemia. A variety of reports using microfluorometric techniques have shown that neuronal Ca^{2+} influx leads to elevation of [Ca^{2+}]i via ionotropic receptors and/or voltage-gated Ca^{2+} channels. Plasmalemmal Na^+/Ca^{2+} exchangers can also greatly contribute to Ca^{2+} entry by operating in a reverse mode under hypoxic/ischemic condition. Further evidence also indicated that some newly identified pathways such as non-selective cation channels [e.g., transient receptor potential (TRP) channels, ASICs], gap junction and hemichannels, and store-operated Ca^{2+} channels also contribute to hypoxic/ischemic Ca^{2+} influx. Nevertheless, for certain cells, these pathways do not necessarily contribute equally to the influx of Ca^{2+} through the plasma membrane. They may function for different phases of [Ca^{2+}]i rise depending on inherent and environmental factors, including the duration and severity of insults, the species and age of animals, the region of the brain, the type of cells, etc., as we have discussed earlier.

Voltage-Gated Ca^{2+} Channels (VGCCs)

The activation of VGCCs is triggered by a depolarized change in membrane potential and plays a major role in shaping neuronal excitability. According to their activation properties and pharmacological profiles, VGCCs are divided into T (low-voltage activated) and P/Q, N, L, R (high-voltage activated) types based on the physiological and pharmacological properties of Ca^{2+} current types (Budde et al. 2002). Depending on the types of $\alpha 1$ pore-forming subunit, $VGCC_S$ can also be divided into three subfamilies: Cav1 (Cav1.1-1.4: L-type Ca^{2+} currents), Cav2 (Cav2.1: P/Q-type Ca^{2+} currents; Cav2.2: N-type Ca^{2+} currents; Cav2.3: R-type Ca^{2+} currents), and Cav3 (Cav3.1-3.3: T-type Ca^{2+} currents) (Catterall 2011). VGCCs are expressed in both neurons and glial cells (see preceding chapter by Chao and Xia).

Some reports indicate that VGCCs are a major pathway for Ca^{2+} influx and contribute to neurodegeneration following hypoxia/ischemia (Lukyanetz et al. 2003; Yao and Haddad 2004). During hypoxia/ischemia, neuronal membrane depolarizes (Calabresi et al. 1999; Müller and Somjen 2000b; Raley-Susman et al. 2001; Xia et al. 2003), which may activate VGCCs, especially L-type channels (Bruehl et al. 2000; Lukyanetz et al. 2003; Ouardouz et al. 2003; Mironov and Langohr 2005; Xiang et al. 2012; Gouriou et al. 2013), and results in a remarkable increase in $[Ca^{2+}]i$ (Duffy and MacVicar 1996; Pisani et al. 1998b; Bergmann and Keller 2003; Lukyanetz et al. 2003; Tjong et al. 2007; Tian et al. 2013c; Gouriou et al. 2013). Glutamate receptor-induced membrane depolarization can also activate VGCCs and result in a remarkable increase in Ca^{2+} entry through VGCCs and Ca^{2+} accumulation (Christie and Jahr 2008; Joshi et al. 2011). Different types of VGCCs may have different contributions to hypoxic/ischemic Ca^{2+} entry. It is generally believed that high-voltage activated VGCCs are responsible for Ca^{2+} influx induced by hypoxia/ischemia (Pisani et al. 1998b; Zhang and Lipton 1999; Lopachin et al. 2001; Bruehl et al. 2000; Lukyanetz et al. 2003; Yao and Haddad 2004; Tjong et al. 2007; Xiang et al. 2012). L-type Ca^{2+} channels, especially the dihydropyridine-sensitive channel (Cav1.2) (Szydlowska and Tymianski 2010), which open upon neuronal depolarization, can cause massive Ca^{2+} entry during ischemia (Pisani et al. 1998b; Lukyanetz et al. 2003; Ouardouz et al. 2003; Tjong et al. 2007). N-type (Lukyanetz et al. 2003) and P/Q- (e.g., Cav2.1) Ca^{2+} channels (Tian et al. 2013c) also make their contribution to hypoxic/ischemic Ca^{2+} entry, but their contributions are slightly smaller than L-type Ca^{2+} channels. Low-voltage activated T-type Ca^{2+} channels were once regarded to have minor contributions to hypoxic/ischemic Ca^{2+} entry and neuronal death because some studies showed that the blockade of low-voltage activated VGCCs cannot attenuate hypoxic/ischemic $[Ca^{2+}]i$ increase (Pisani et al. 1998b; Zhang and Lipton 1999; Lopachin et al. 2001; Bruehl et al. 2000; Lukyanetz et al. 2003; Yao and Haddad 2004; Tjong et al. 2007). However, more recent work with genetically encoded Ca^{2+} indicator in PC12 cells showed that the delayed, sustained phase of $[Ca^{2+}]i$ rise of OGD-induced prototypical biphasic $[Ca^{2+}]i$ elevations is remarkedly reduced by Cav3.1/Cav3.2 blockade and silencing, and is increased by Cav3.2 overexpression, indicating T-type Ca^{2+} channels contribute to Ca^{2+} entry during ischemia (Gouriou et al. 2013). T-type Ca^{2+} channel-mediated $[Ca^{2+}]i$ elevation can also evoke mitochondrial matrix Ca^{2+} evelation via Ca^{2+} uptake by recently identified low-affinity mitochondrial Ca^{2+} uniporter (Baughman et al. 2011; De Stefani et al. 2011), leading to cell death (Gouriou et al. 2013). Therefore, inhibition or knockout/knockdown of T-type Ca^{2+} channels has been shown to protect organotypic hippocampal slices against ischemia-induced delayed cell death (Nikonenko et al. 2005) and in vivo neuronal death in global brain ischemia (Bancila et al. 2011) or hypoxic insult (Kim et al. 2011). Nevertheless, there are also reports showing that VGCCs are not significantly involved in hypoxic/ischemic Ca^{2+} influx (Chen et al. 1999; Martinez-Sánchez et al. 2004). This discrepancy can be partially explained by the quick inactivation of VGCCs in some preparations during a maintained depolarization (Ozawa et al. 1989; Young and Somjen 1992), which often occurs during hypoxia/ischemia (Young and Somjen 1992; Calabresi et al. 1999).

Receptor-Operated Channels

Because VGCCs in cultured hippocampal neurons have been shown to be quickly inactivated during maintained depolarization (Ozawa et al. 1989), the major component of hypoxic/ischemic Ca^{2+} influx under this condition may be through receptor-operated channels, especially iGluR channels (Mitani et al. 1993). The receptor-operated channels in neurons are activated predominantly by the acidic amino acid transmitters (Silver and Erecińska 1990) and extracellular ATP (Bai and Li 2013).

Indeed, the contribution of Ca^{2+}-permeable glutamate receptor channels to hypoxic/ischemic Ca^{2+} influx has been proposed in numerous studies (Silver and Erecińska 1990; Goldberg and Choi 1993; McDonald et al. 1998; Zhang and Lipton 1999; Lopachin et al. 2001; Martinez-Sánchez et al. 2004; Yao and Haddad 2004; Kiedrowski 2007; Micu et al. 2007; Pamenter and Buck 2008). During hypoxic/ischemic conditions, glutamate is remarkably released to the extracellular space from presynaptic vesicles through Ca^{2+}-dependent mechanisms (involving N-type and P/Q-type voltage-dependent Ca^{2+} channels) or Ca^{2+}-independent mechanisms, such as by activating reversely operated glutamate-Na^+ transport system due to the reversal of Na^+ gradients (Benveniste et al. 1984; Goldberg and Choi 1993; Yamaguchi et al. 1998; Camacho and Massieu 2006; Micu et al. 2007), by gap-junction, hemichannel-mediated and P_2X channel-dependent release from astrocytes, release from neurons through exocytosis (Andrade and Rossi 2010; Krzyżanowska et al. 2014), or by a volume-sensitive pathway (Pérez-Domínguez et al. 2014). The massive accumulation of glutamate in extracellular space is causally linked to Ca^{2+} entry through iGluRs, because activation of iGluRs triggers more than 1 mM of $[Ca^{2+}]o$ reduction due to Ca^{2+} entry (Zanotto and Heinemann 1983) and subsequently leads to remarkable $[Ca^{2+}]i$ rise (Vlaskovska et al. 1997; Vander Jagt et al. 2008). This increase is similar to that seen in Ca^{2+} response to hypoxic/ischemic stress (Tymianski et al. 1993; Dugan et al. 1995; McDonald et al. 1998; Kiedrowski 1999, 2007; Zhang and Lipton 1999; Shen and Slaughter 2002; Micu et al. 2006, 2007; Stanika et al. 2010; Sivakumar et al. 2013). Pharmacologically blocking iGluRs largely eliminates anoxia/ischemia-induced $[Ca^{2+}]o$ falls (Macdonald et al. 2006) and $[Ca^{2+}]i$ rises (Silver and Erecińska 1990; Goldberg and Choi 1993; Zhang and Lipton 1999; Lopachin et al. 2001; Martinez-Sánchez et al. 2004; Yao and Haddad 2004; Micu et al. 2006; Stanika et al. 2010; Sivakumar et al. 2013). These findings strongly suggest that iGluR channels are one of the major components responsible for Ca^{2+} influx induced by severe hypoxia/ischemia.

Traditionally, iGluRs can be pharmacologically classified to three subtypes, namely NMDARs, AMPARs, and KARs. The latter two receptors are also generally called non-NMDA receptors due to their similarity to functional properties and clear indistinguishability by agonists or antagonists (Ozawa et al. 1998; Dingledine et al. 1999). Molecular cloning and expression studies have revealed that the diversity of iGluRs is much larger than expected from electrophysiological and pharmacological studies (Ozawa et al. 1998; Dingledine et al. 1999; Kumar and Mayer 2013). NMDARs are heterotetrameric and are coassembled by GluN1 (previously as NR1) with GluN2A-D (NR2A-D) and GluN3A-B (NR3A-B). They can allow the passage of Ca^{2+} and Na^+ into the cell and K^+ out of the cell through the channel pore (Ozawa et al. 1998; Dingledine et al. 1999; Paoletti and Neyton 2007). AMPARs and KARs are also heterotetrameric. AMPARs are composed of various combinations of four subunits, GluA1-4 (previously as GluR1-4), and only AMPARs that lack the GluR2 subunit are permeable to Ca^{2+}; KARs are formed by coassembly of GluK1-5 (GluK1-3 were previously denoted as GluR5-7; GluK4 and GluK5 as KA1 and KA2). The ion channel formed by KARs is permeable to Na^+ and K^+ ions. Their permeability to Ca^{2+} is usually very slight but varies with subunits and RNA editing at the tip of the p loop (Ozawa et al. 1998; Dingledine et al. 1999). Though all these iGluRs mediate Ca^{2+} influx contributing to intracellular Ca^{2+} accumulation, the relative contributions of different receptor subtypes are not equal in terms of their role in hypoxic/ischemic $[Ca^{2+}]i$ rise. In fact, the relative contributions are not well recognized yet because of some contradictory results. For example, in hippocampal slices, blockade of NMDARs with MK-801 or 3-(2-carboxypiperazin-4-yl) propyl-1-phosphonic acid (CPP) can partially (Zhang and Lipton 1999; Martinez-Sánchez et al. 2004) or completely (Lopachin et al. 2001) prevent ischemic Ca^{2+} accumulation. Broad-spectrum NMDAR antagonists (MK-801, 7-chlorokynurenic acid, D-AP5) abolish ischemia-induced Ca^{2+} increases in myelin (Micu et al. 2006). On the other hand, blocking AMPARs/KARs with CNQX or NBQX completely prevents Ca^{2+} accumulation (Lopachin et al. 2001; Micu et al. 2006), reduces approximately 50 % of the sustained $[Ca^{2+}]i$ rise (Zhang and Lipton 1999), or has no effect on ischemic $[Ca^{2+}]i$ rise (Martinez-Sánchez et al. 2004). By contrast, blockade of either NMDARs or AMPARs/KARs with MK-801 and CNQX respectively does not attenuate the increase in $[Ca^{2+}]i$ during anoxia/ischemia in acutely isolated hippocampal CA1 neurons and cultured cerebellular granular cells (Friedman and Haddad 1993; Chen et al. 1999). In the in vitro cortex, blocking NMDARs strongly suppresses intracellular Ca^{2+} accumulation induced by OGD (Goldberg and Choi 1993; Fukuda et al. 1998; Zanelli et al. 1999), whereas blockade of AMPARs/KARs with CNQX either greatly attenuates OGD-induced $[Ca^{2+}]i$ rise (Fukuda et al. 1998) or has no effects on $[Ca^{2+}]i$ rise (Goldberg and Choi 1993). These discrepant outcomes can be attributed to many factors, especially the differences in (1) tissue/cell preparation (acutely isolated/cultured neurons vs. brain slices which keep relatively intact microenvironments and neuronal connections) (Friedman and Haddad 1993; Chen et al. 1999; Zhang and Lipton 1999; Lopachin et al. 2001; Martinez-Sánchez et al. 2004); (2) methodology of $[Ca^{2+}]i$ determination [cryoultramicrotomy and electron probe X-ray microanalysis by Lopachin et al. (2001). vs. microfluorometric imaging by others]; (3) drugs used (e.g., MK-801 vs. CPP and AP5 for NMDARs; CNQX vs. NBQX for AMPARs/KARs); (4) drug doses adopted (higher vs. lower concentrations). Aside from these considerations, the following issues should also be taken into account:

1. *Ca^{2+} permeability of receptor subtypes*. NMDARs have a relatively higher permeability to Ca^{2+} than that of AMPARs/KARs (Ozawa et al. 1998; Dingledine et al. 1999; Paoletti and Neyton 2007). Activation/inhibition of these subtypes to the same extension could result in different changes in $[Ca^{2+}]i$ rise.

2. *Sensitivity to extracellular pH changes*. In contrast to NMDARs, AMPAR-mediated [Ca^{2+}]i rise is more sensitive to increased extracellular acidity (McDonald et al. 1998) which often occurs during ischemia/hypoxia (Silver and Erecińska 1992; Lipton 1999; Sheldon and Church 2004; Sheldon et al. 2004). Lowering extracellular pH significantly reduces the AMPAR-mediated peak elevation in intracellular free Ca^{2+}, but slows the recovery of intracellular homeostasis (McDonald et al. 1998).
3. *Rapid receptor desensitization*. AMPARs/KARs are easily desensitized after activation (half decay times c.a. 30 vs. 250 ms for NMDARs in hippocampal cells) (Vyklický et al. 1990; Moudy et al. 1994; Fedele and Raiteri 1996; Wilding and Huettner 1997). Therefore, AMPARs/KARs were not thought to be major contributors to glutamate-induced Ca^{2+} influx, but could participate in a Na^+-dependent depolarization and Ca^{2+} influx through voltage-activated Ca^{2+} channels and Na^+/Ca^{2+} exchangers (Goldman et al. 1994; Kiedrowski 2007) and Ca^{2+} release from intra-axonal Ca^{2+} store during ischemia (Ouardouz et al. 2006). However, recent studies indicated that Ca^{2+}-permeable AMPARs are also important in hypoxic/ischemic Ca^{2+} overload and neuronal death (Deng et al. 2003; Liu et al. 2004; Oguro et al. 2004; Noh et al. 2005; Ouardouz et al. 2006; Sivakumar et al. 2013). AMPARs may not only allow Ca^{2+} influx from the extracellular space through their own channels, but may also significantly influence Ca^{2+} release from intracellular Ca^{2+} store (Ouardouz et al. 2006). Moreover, Ca^{2+}-permeable AMPARs also allow Zn^{2+} influx and accumulation (Noh et al. 2005; Frazzini et al. 2007), along with Ca^{2+}, aggravating mitochondrial dysfunction, Ca^{2+} deregulation and cell death in ischemia (Medvedeva et al. 2009). In addition, global ischemia or OGD in cultured neurons triggers down-regulation of GluR2 gene expression and protein abundance (Opitz et al. 2000; Fernandes et al. 2014) and enhances AMPAR-mediated Ca^{2+} influx in vulnerable CA1 pyramidal neurons before the onset of neuronal death (Gorter et al. 1997), demonstrating the importance of GluR2-lacking Ca^{2+}-permeable AMPARs in hypoxic/ischemic Ca^{2+} dysregulation and cell death.
4. *Distribution of receptor subtypes*. The distributions of iGluRs are heterogeneous in dendrite and soma as well as pre-/post-synaptic membrane (Schiller et al. 1997; Dodt et al. 1998; Yin et al. 1999b; Nicholson et al. 2006; Groom et al. 2006; Micu et al. 2006). Therefore, different magnitudes of hypoxic/ischemic Ca^{2+} influx seen in different cells/tissues are likely attributed to different receptor subtypes and/or density in these preparations. In support of this hypothesis are the findings that hypoxic/ischemic Ca^{2+} entry is largely attributed to glutamate-operated channels in many (though not all) neurons examined in the hippocampus, cortex and thalamus (Silver and Erecińska 1990), as well as that OGD-induced [Ca^{2+}]i elevation is more pronounced in dendritic branches, where the main targets of synaptically released excitatory amino acid transmitters (predominantly glutamate) are located (Harris and Kater 1994) rather than in somatic regions (Pisani et al. 1999; Zhang and Lipton 1999). Central myelin-forming cells, oligodendrocytes, were once thought to express only AMPARs, but not functional NMDARs (Patneau et al. 1994), and ischemia-induced [Ca^{2+}]i elevation has been

believed to be mediated by Ca^{2+}-permeable AMPARs (Deng et al. 2003). However, recent studies indicated that both AMPARs and NMDARs are present in oligodendrocytes with Ca^{2+}-permeable AMPARs in the cell body and NMDARs in the myelinating processes (Káradóttir et al. 2005; Salter and Fern 2005; Micu et al. 2006, 2007). Therefore, AMPAR antagonist NBQX completely blocks ischemic $[Ca^{2+}]i$ increases in oligodendroglial cell bodies, only modestly reduces the $[Ca^{2+}]i$ increase in myelin, and prevents the loss of oligodendrocyte somata. However, it has no effect on the loss of processes, whereas the $[Ca^{2+}]i$ increases and damaging to processes are completely abolished by broad-spectrum NMDAR antagonists (MK-801, 7-chlorokynurenic acid, D-AP5) (Salter and Fern 2005; Micu et al. 2006, 2007). That is also the reason why previous studies that have focused on the somata have not detected a role for NMDARs in oligodendrocyte injury (Salter and Fern 2005).

5. *Animal/cell maturation (young vs. adult and old; immature vs. mature) and receptor subunit composition and brain region differences (e.g. hippocampal CA1 vs. CA3).* Increases of $[Ca^{2+}]i$ in hippocampal CA1 neurons after OGD is significantly lower in developing gerbils than in adults because GluR2 expression is significantly higher in developing than in adult brains, making AMPARs highly impermeable to Ca^{2+} in developing brains (Oguro et al. 2004). Similarly, NMDAR channel open time and NMDA-induced increase in $[Ca^{2+}]i$ of hippocampal CA1 neurons are largely inhibited by hypoxia in neonatal but not in adult rats, this is because NMDARs of rat hippocampal CA1 neurons are prevalently formed by NR1 plus NR2B or NR2D subunits in neonates, while by NR1 plus NR2A or NR2C in mature gerbils. Hypoxia is shown to reduce currents of NR1/NR2D receptors and to increase currents of NR1/NR2C (with no effect on NR1/NR2A or NR1/NR2B) (Bickler et al. 2003). This is one of the reasons why neonatal neurons are more tolerant to hypoxia/ischemia than mature ones. In organotypic slice cultures, NMDAR-mediated ischemia-induced $[Ca^{2+}]i$ elevations are much larger in vulnerable CA1 neurons than in resistant CA3, which makes CA1 mitochondria have stronger Ca^{2+} accumulation, more extensive swelling and damage, stronger depolarization of their membrane potential, and a significant increase in ROS/RNS generation than CA3 mitochondria (Stanika et al. 2010). In rat neocortical slices, OGD-induced NMDAR-mediated neuronal $[Ca^{2+}]i$ elevations are significantly greater in layers II/III than in layers IV, V, VI (Fukuda et al. 1998). These studies clearly indicate the animal/cell maturation and receptor subunit composition and brain region differences in iGluR-mediated hypoxia/ischemia-induced Ca^{2+} accumulation.

Although there is a well established idea that iGluR-mediated Ca^{2+} influx is one of the major sources of hypoxia/ischemia-induced increase in $[Ca^{2+}]i$, we should keep in mind that ischemia and hypoxia are not the same stimulus for iGluR activation and the subsequent increase in $[Ca^{2+}]i$. In addition, glutamate is not equal to hypoxic or ischemic stress. For example, although anoxia and glutamate induce a similar increase in $[Ca^{2+}]i$ in acutely isolated neurons (Friedman and Haddad 1993), they are actually different in the nature of the stress. In isolated hippocampal CA1 neurons, there are very large extracellular spaces and no synaptic input, which provides a microenvironment in the bathing fluid in the absence of glutamate accumulation. Anoxia indeed induces a significant increase in $[Ca^{2+}]i$. This increase could be completely eliminated by removal of extracellular Ca^{2+} or by replacement of it with EGTA, but could not be attenuated by glutamate receptor blockers (Friedman and Haddad 1993). The authors thus argue that glutamate release is not at the basis of the anoxia-induced increase in intracellular Ca^{2+} in such a condition (Friedman and Haddad 1993). Chen et al. (1999) observed similar results in cultured cerebellar granule cells. Taking all the results together, these findings suggest a complicated and delicate Ca^{2+} response to hypoxia/ischemia and diversity of Ca^{2+} signaling in stress.

In addition to iGluR, increasing evidence indicated that P_2X receptors are also important receptor-operated channels that contribute to Ca^{2+} overload in hypoxia/ischemia (Bai and Li 2013). P_2 receptors are ATP receptors, which include P_2X and P_2Y subtypes. P_2X receptors are a family of cation-permeable ligand gated ion channels that open in response to the binding of extracellular ATP. They belong to a larger family of receptors known as the purinergic receptors. P_2Y receptors are ATP activated G protein-coupled receptors (GPCRs) that initiate an intracellular chain of reactions. P_2 receptors are widely distributed in excitable and non-excitable cells of vertebrates (North 2002; Bai and Li 2013). Activation of both P_2X and P_2Y receptors can increase $[Ca^{2+}]i$ in nerve cells, however, metabotropic P_2Y purinoreceptors mobilize intracellular Ca^{2+} at physiological concentrations of ATP (by stimulating IP_3 production and IP_3R activation), whereas ionotropic P_2X purinoreceptors induce Ca^{2+} influx across the plasmalemma only at high concentrations of ATP, such as those that occur following CNS injury (James and Butt 2001). Hypoxia/ischemia can cause excessive ATP release into the extracellular space from damaged neurons, efflux of ATP through neuronal and glial (astrocytic and oligodendrocytic) hemichannels, and amplified autocrine from reactive microglia (Thompson et al. 2006; Lin et al. 2008; Schock et al. 2008; Domercq et al. 2010; Bai and Li 2013). Activation of P_2X receptors, especially P_2X7 subtype that has a permeability significantly higher to Ca^{2+} than to Na^+ and K^+, can cause Ca^{2+} influx and overload in both neurons and glial cells during ischemia and lead to brain damage (Domercq et al. 2010; Arbeloa et al. 2012; Bai and Li 2013). In addition, ATP, by binding to glial G protein-coupled purinergic (P_2Y) receptors, evokes glial Ca^{2+} increase and intercellular Ca^{2+} wave in glial network as well as reduction of brain oxygen tension, which may facilitate hypoxic Ca^{2+} signaling and propagation to Purkinje cells of cerebellar cortex in aging brain (James and Butt 2001; Hoogland et al. 2009; Mathiesen et al. 2013). Therefore, P_2 purinergic receptor channels also contribute to hypoxic/ischemic increase in $[Ca^{2+}]i$.

Plasmalemmal Na^+/Ca^{2+} Exchanger (NCX)

In certain cases, pronounced $[Ca^{2+}]i$ elevation in ischemia-like conditions greatly depends on an increase in intracellular Na^+, which results in the reversal of the operation of NCX (Itoh et al. 1998; Kiedrowski 1999, 2007; Rojas et al. 2007; Luo et al. 2008; Brittain et al. 2012). During hypoxia/ischemia, massive amounts of Na^+ enter into neurons (Friedman and Haddad 1994; Calabresi et al. 1999; Banasiak et al. 2004; Martinez-Sánchez et al. 2004; Sheldon et al. 2004; Nikolaeva et al. 2005). It has been demonstrated that hypoxia-induced membrane depolarization (Calabresi et al. 1999; Müller and Somjen 2000b; Raley-Susman et al. 2001; Xia et al. 2003) and large increase in $[Na^+]i$ (Friedman and Haddad 1994; Calabresi et al. 1999; Banasiak et al. 2004; Martinez-Sánchez et al. 2004; Sheldon et al. 2004; Nikolaeva et al. 2005) favor NCX to become reversed, i.e. Ca^{2+} entry/Na^+ extrusion mode (Kiedrowski 2007), though Ca^{2+} extrusion/Na^+ entry mode may also concomitantly run (Jeon et al. 2008; Molinaro et al. 2008). The net output of both modes thus favors Ca^{2+} entry during hypoxia/ischemia, and therefore further enhances the Ca^{2+} overload (Stys and Lopachin 1998; Tanaka et al. 2002; Kiedrowski 2007; Chen et al. 2008; Luo et al. 2008).

There exist three different but highly homologous isoforms of NCX (NCX1, NCX2, and NCX3) in mammals (Dipolo and Beaugé 2006). All of these isoforms are expressed throughout the brain in both neurons and glial cells with a distinct distribution in different brain regions and subcellular compartments (Thurneysen et al. 2002; Papa et al. 2003; Minelli et al. 2007). Following hypoxia or ischemia/reperfusion, the expression of these isoforms shows heterogenous changes depending on the brain regions, insult duration, and severity (Hwang et al. 2006; Annunziato et al. 2007; Bojarski et al. 2008; Sirabella et al. 2009). For example, after permanent occlusion of middle cerebral artery in rats, all three NCX proteins are downregulated in the ischemic core; NCX3 decreases whereas NCX1 and NCX2 are unchanged in penumbra area (Annunziato et al. 2007). A non-damaging 3 min duration of global ischemia (that just induces ischemic tolerance or ischemic preconditioning) significantly increases NCX1 proteins at 6–12 h after ischemia, but does not make significant changes in NCX2 and NCX3 in the hippocampus (Bojarski et al. 2008). In contrast, an 8-min hippocampal CA1 damaging duration of global ischemia down-regulates NCX2 proteins for up to 48 h after ischemia, but upregulates NCX3 at 18 h and keeps NCX1 protein level relatively constant (Bojarski et al. 2008). Nevertheless, these studies suggest that these isoforms may exert different roles during hypoxic/ischemic condition. Indeed, in cultured cortical neurons, 3 h of OGD transcriptionally increases NCX1 transcript and protein expression, but posttranscriptionally down-regulates NCX3 protein expression and leaves NCX2 transcript and protein expression unchanged (Sirabella et al. 2009). With these expression changes, NCX currents are significantly higher in both reverse and forward modes of operation, with more increases in reverse mode, at 1 and 3 h of OGD. Correspondingly, $[Ca^{2+}]i$ significantly increases at 1 h of OGD, but becomes lower due to Ca^{2+} refilling into ER, which is supposed to help neurons to prevent ER stress (Sirabella et al. 2009; Formisano et al. 2013) and to increase resistance to ischemia and reduce the infarct extension after stroke (Cross et al. 2012; Formisano et al. 2013) though some studies also suggest NCX1 may induce an intracellular Ca^{2+} overload and act preferentially (vs. NCX2 and NCX3) to exacerabate the cerebral damage (Hwang et al. 2006; Morimoto et al. 2012). The downregulation of NCX3 (Sirabella et al. 2009) might be a result of proteolytical cleavage of NCX3 during ischemia (Bano et al. 2005), which jeopardizes Ca^{2+} extrusion and contributes to a second uncontrolled, irreversible Ca^{2+} increase that leads to neuronal demise (Bano et al. 2005; Luo et al. 2008). Most importantly, cellular ATP depletion inhibits NCX1 and NCX2, but does not affect NCX3 activity (Annunziato et al. 2007). This property offers NCX3 isoform a key role in the regulation of Ca^{2+} and Na^{+} homeostasis during hypoxia/ischemia. NCX3 knockout/knockdown thereby aggravates Ca^{2+} overload in neurons and increases ischemia brain damage (Jeffs et al. 2008; Molinaro et al. 2008), while upregulation of NCX3 enables neurons to maintain $[Ca^{2+}]i$ at low levels and decreases cell death during hypoxic/ischemic insults (Secondo et al. 2007; Cross et al. 2012; Pignataro et al. 2012). Therefore, hypoxic/ischemic preconditioning-induced upregulation of NCX (NCX1 and 3) expression represents a reasonable strategy to reduce the infarct extension after stroke (Pignataro et al. 2012), probably via the better control of Ca^{2+} dysregulation in the later phase of severe and prolonged hypoxia/ischemia and during reoxygenation/reperfusion period. In addition, NCX2 operation in the mode of exporting Ca^{2+} may be neuroprotective because in NCX2 deficient mouse hippocampus, a significantly slower recovery in population spike amplitudes, a sustained elevation of $[Ca^{2+}]i$, and increased membrane depolarization are all developed with larger infarction and more cell death in response to a transient focal cerebral ischemia (Jeon et al. 2008). These findings clearly indicate a diverse role of different NCX isoforms in the regulation of hypoxic/ischemic Ca^{2+} dyshomeostasis and brain damage.

In addition, activation of the reverse NCX-induced increase in $[Ca^{2+}]i$ can trigger a great amount of Ca^{2+} release from the endoplasmic reticulum (Ca^{2+}-Ca^{2+} release) (Rojas et al. 2007; Chen et al. 2008; Wu et al. 2008). Therefore, reverse operation and cleavage of the plasmalemmal NCX3 are also responsible for hypoxic/ischemic $[Ca^{2+}]i$ rises, with reverse operation of exchangers contributing to $[Ca^{2+}]i$ rise in the earlier phase of hypoxia/ischemia. The cleavage of NCX accounts for a second, uncontrolled, irreversible Ca^{2+} increase in the later phase of severe and prolonged hypoxia/ischemia and during the reoxygenation/reperfusion period.

Non-selective Cation Channels (NSCCs)

Hypoxia/ischemia-induced neuronal death in large part contributes to glutamate-receptor-driven excitotoxicity due to iGluR-mediated Ca^{2+} overload. However, failure to treat stroke in numerous clinical trials using glutamate receptor antagonists unmasked another source of Ca^{2+} overload and non-excitotoxic cell death in hypoxic/ischemic stress. NSCCs underlie such a non-excitotoxic cell death (Simard et al. 2007; Besancon et al. 2008). NSCCs are also important sources for hypoxia/ischemia-induced Ca^{2+} influx and accumulation in nerve cells, especially in response to the prolonged hypoxic/ischemic stress. TRPM7, a membrane of transient receptor potential (TRP) channel superfamily, is such a channel (Aarts et al. 2003). Structurally, TRPM7 channel is composed of six putative transmembrane domains with a pore domain wedged between the fifth and sixth transmembrane domains

(Ramsey et al. 2006). TRPM7 and its closest homolog TRPM6, are unique channels that possess both ion channel and protein kinase activities (which contain a functional COOH-terminal α-kinase, an atypical serine/threonine kinase) (Ramsey et al. 2006). TRPM7 has a large permeability to divalent cations and a small permeability to monovalent cations with a permeability sequence of $Zn^{2+} \sim Ni^{2+} \gg Ba^{2+} > Co^{2+} > Mg^{2+} \geq Mn^{2+} \geq Sr^{2+} \geq Cd^{2+} \geq Ca^{2+}$ (Monteilh-Zoller et al. 2003), and is essential for embryonic development, is necessary for cell survival and trace ion homeostasis in vitro, and whose global deletion in mice is lethal (Wei et al. 2007; Jin et al. 2008). It can be activated by reactive species (Aarts et al. 2003; Simard et al. 2007). Under physiological condition, TRP channels, including TRPM7, can contribute to changes in $[Ca^{2+}]i$ in multiple ways (Gees et al. 2010). During prolonged hypoxia/ischemia, a large amount of ROS/RNS are generated (Carvalho et al. 2009; Niatsetskaya et al. 2012) due to Ca^{2+} entry through iGluRs and the resultant mitochondrial matrix Ca^{2+} overload (Stanika et al. 2010; Sivakumar et al. 2013), which induces further Ca^{2+} influx and accumulation (Perez Velazquez et al. 1997; Chen et al. 1999) by activating NSCCs (Simard et al. 2007). Indeed, in cultured cortical neurons, prolonged OGD activates a distinct cation conductance termed I_{OGD}, enhanced $^{45}Ca^{2+}$ uptake and a second large intracellular Ca^{2+} overload, as well as subsequent neuronal death. These effects are not responsive to agonists or antagonists of glutamate receptors, voltage-sensitive Ca^{2+} or Na^{+} channels, or ion exchangers, but can be blocked by Gd^{3+}, the non-selective cation inhibitor, and activated by ROS/RNS (Aarts et al. 2003). Small interfering RNA suppressing TRPM7 expression blocks these responses, indicating the critical role of TRPM7 in prolonged ischemia-induced Ca^{2+} influx and a second uncontrolled $[Ca^{2+}]i$ rise, as well as consequent neuronal death (Aarts et al. 2003; Sun et al. 2009). Hypoxia/ischemia can also increase the activity of some other TRP channel members, such as TRPM2, TRPV1, and TRPV4, in slices or cultures, leading to Ca^{2+} entry and overload in both neurons and astrocytes (Lipski et al. 2006; Henrich and Buckler 2008; Butenko et al. 2012). Given the factors of a vast amount of ROS/RNS generated during severe/prolonged hypoxia/ischemia (Lafon-Cazal et al. 1993; Dugan et al. 1995; Perez Velazquez et al. 1997; Carriedo et al. 2000; Stanika et al. 2010; Niatsetskaya et al. 2012; Sivakumar et al. 2013), ROS/RNS-sensitive, NSCC-mediated Ca^{2+} influx (e.g., via TRPM7) is a major contributor to a second, uncontrolled $[Ca^{2+}]i$ rise. The increase in $[Ca^{2+}]i$ through ROS/RNS-sensitive NSCCs often occurs in the late phase of severe hypoxia/ischemia and posthypoxic/postischemic reperfusion period, and triggers delayed neuronal death (Aarts et al. 2003; Macdonald et al. 2006). This is because during the period of post-ischemic reperfusion, Ca^{2+} signaling through voltage-dependent Ca^{2+} channels is inactive due to the sustained membrane depolarization (Connor et al. 1999; Aarts et al. 2003), and glutamate receptor channels are down-regulated in expression (Pellegrini-Giampietro et al. 1992) and inactive in function (Aarts et al. 2003). Thus, NSCC-mediated delayed Ca^{2+} overload cannot be arrested by blocking Ca^{2+} channels and NMDAR channels during this period (Tymianski et al. 1993; Aarts et al. 2003).

In addition to TRP channels, ASIC1a channels also contribute to intracellular Ca^{2+} overload during ischemia (Xiong et al. 2004; Mari et al. 2010). ASIC1a is a member of six isoforms (ASIC1a, ASIC1b, ASIC2a, ASIC2b, ASIC3, and ASIC4) of acid-sensing ion channels (ASICs) which are proton-gated, voltage-insensitive cationic channels, and are distributed throughout the central and peripheral nervous systems (Leng et al. 2014). ASICs belong to the degenerin/epithelial Na^+ channel superfamily, are Na^+-selective, and have a reversal potential near the Na^+ equilibrium potential (Waldmann et al. 1997; Xiong et al. 2004). For homomeric ASIC1a channels, acid activation also induces Ca^{2+} entry directly through these channels (Waldmann et al. 1997; Xiong et al. 2004; Yermolaieva et al. 2004). Therefore, we would like to put ASIC1a channels under the catalog of NSCCs in regard to Ca^{2+} regulation under hypoxia/ischemia. Following ischemia (especially severe ischemia), the brain switches its metabolism from aerobic to anaerobic glycolysis to produce ATP, resulting in the accumulation of byproduct (lactic acid), acidosis and pH fall (Silver and Erecińska 1990; Diarra et al. 1999; Bondarenko et al. 2004; Pedersen et al. 2006). As a result, Ca^{2+}-permeable ASIC1a channels are activated to allow Ca^{2+} influx and accumulation (Xiong et al. 2004; Mari et al. 2010). This is an important contributor, together with other NSCCs, to hypoxia/ischemia-induced glutamate receptor-independent neuronal injury, which can be inhibited by ASIC1a blockers or genetic knockout of ASIC1a (Xiong et al. 2004; Mari et al. 2010). In addition, it has been reported that ASIC1a channels are functionally coupled with NMDARs (Gao et al. 2005). Ischemia enhances ASIC1a currents through phosphorylation of the channels by CaMKII, which is activated in response to Ca^{2+} influx through NMDARs (Gao et al. 2005). Therefore, the coupling of ASIC1a with NMDARs may exacerbate Ca^{2+} overload and neuronal damage during ischemia (Xiong et al. 2004; Gao et al. 2005).

Taking all of the data into consideration, it is clear that Ca^{2+} entry through non-selective cation channels is a major route for hypoxic/ischemic $[Ca^{2+}]i$ accumulation, especially a second uncontrolled $[Ca^{2+}]i$ increase, in the late phase of severe hypoxia/ischemia and posthypoxic/postischemic reperfusion period.

Other Possible Pathways

Despite the substantial influx of extracellular Ca^{2+}, Ca^{2+} overload also reflects the failure of Ca^{2+} extruding systems to cope with excess Ca^{2+} that enter cells across the dysregulated plasma membrane channels. The inability and cleavage of PMCAs (Schwab et al. 2002; Pászty et al. 2007; Amarjargal et al. 2008; Henrich and Buckler 2008, 2013) and NCXs (Bano et al. 2005; Luo et al. 2008) may also contribute to intracellular Ca^{2+} accumulation. This is because it is very difficult, if not impossible, for PMCAs to remove the overloaded intracellular Ca^{2+} efficiently to extracellular spaces due to the decrease or failure in ATP generation during hypoxia/ischemia (Silver et al. 1997; Raley-Susman et al. 2001; Martinez-Sánchez et al. 2004; Sheldon and Church 2004; Gouriou et al. 2013). In addition, loss of NCX function following its proteolysis is critical for the delayed Ca^{2+} deregulation and excitotoxic demise in neurons (Bano et al. 2005; Luo et al. 2008). The contribution of the inability of Ca^{2+} extrusion mechanisms to intracellular Ca^{2+} dysregulation is further supported by the fact that Ca^{2+} recovery is slowed during hypoxia/ischemia, an effect observed in many studies (Silver and Erecińska 1990; Friedman and Haddad 1993; Nabetani et al. 1997; Zhang and Lipton 1999; Martinez-Sánchez et al. 2004).

Hypoxia/ischemia evokes a significant fall in pHi, which activates Na^+/H^+ exchangers (NHE) to extrude excessive intracellular H^+ in an electroneutral exchange of intracellular H^+ for extracellular Na^+ (Silver and Erecińska 1990; Diarra et al. 1999; Bondarenko et al. 2004; Pedersen et al. 2006). This is one of the factors leading to $[Na^+]i$ rise immediately after (not during) hypoxia/ischemia (reperfusion period) (Sheldon et al. 2004; Luo et al. 2005; Cengiz et al. 2014). Hypoxia/ischemia also activates electroneutral NKCC, leading to intracellular Na^+ overload (Sheldon et al. 2004; Lenart et al. 2004; Kintner et al. 2007; Chen et al. 2008; Cengiz et al. 2014). In addition, released glutamate, through stimulation of glutamate transporters (Yamaguchi et al. 1998; Krzyżanowska et al. 2014), results in Na^+ influx (Chatton et al. 2000; Kirischuk et al. 2007; Rojas et al. 2007; Bennay et al. 2008; Langer and Rose 2009). As a result, $[Ca^{2+}]i$ increases significantly due to the operation of plasmalemmal NCX in a reverse mode (Zhang and Lipton 1999; Bondarenko et al. 2004; Sheldon et al. 2004; Martinez-Sánchez et al. 2004; Luo et al. 2005, 2008; Nikolaeva et al. 2005; Pedersen et al. 2006; Rojas et al. 2007; Kiedrowski 2007; Chen et al. 2007b; Cengiz et al. 2014). The increase in $[Ca^{2+}]i$, in turn, triggers endoplasmic reticulum to release Ca^{2+} into the cytosol (a typical Ca^{2+}-Ca^{2+} release) (Rojas et al. 2007; Chen et al. 2008; Wu et al. 2008). Therefore, NHE, NKCC, and glutamate transporters can be additional contributors to secondary intracellular Ca^{2+} elevation by indirectly activating reverse NCX operation and Ca^{2+}-Ca^{2+} release from endoplasmic reticulum.

Recent evidence indicated that hypoxia/ischemia opens gap junction and hemichannels in neurons (Thompson et al. 2006; Madry et al. 2010; Bargiotas et al. 2011) and astrocytes (Iwabuchi and Kawahara 2011; Shinotsuka et al. 2014) by way of ROS/RNS (Zhang et al. 2008) and Src family kinases (Weilinger et al. 2012). This leads to anoxic depolarization and cellular injury. Gap junctions are intercellular connections composed by two half gap junction (hemichannel) proteins connexins (Cxs) between two cells, allowing the exchange of signaling molecules such as ions and second messengers smaller than ~1 kDa (Thompson and Macvicar 2008). At least 20 Cxs genes code Cxs, and several Cxs (such as Cx36, Cx30, Cx43, Cx45) are expressed in neurons and glial cells. Like gap junctions, Cxs-forming hemichannels also exist widely in the CNS (Thompson and Macvicar 2008). Pannexins (Pxs), the cousin of Cxs proteins but with no significant sequence homology vs. Cxs, are recently discovered gap junctions in vertebrates (Bruzzone et al. 2003), and appear to function as hemichannels (Thompson and Macvicar 2008). Pxs have three family members (Px1, Px2, and Px3). Px1 and Px2 are expressed in the nerve cells, whereas Px3 appears exclusively in skin and osteoblasts (Thompson and Macvicar 2008). It has been shown that inhibition of gap junctional intercellular coupling between astrocytes significantly advances the onset of the sustained $[Ca^{2+}]i$ increase and concurrent neuronal anoxic depolarization after OGD exposure, which initiates brain damage during the acute phase of ischemia (Shinotsuka et al. 2014). Cx hemichannel blockade improves outcomes in brain ischemia (Davidson et al. 2012). The opening of Px hemichannels in neurons and astrocytes during ischemia also induces cellular injury, and exaggerates ischemic brain damage (Thompson et al. 2006; Zhang et al. 2008; Madry et al. 2010; Bargiotas et al. 2011; Iwabuchi and Kawahara 2011; Weilinger et al. 2012), while genetic ablation of Px1 protects neurons from ischemic injury (Dvoriantchikova et al. 2012). Px hemichannel opening is proposed to contribute to the profound ionic dysregulation during stroke (Thompson et al. 2006). However, experimental support for this proposal is still lacking.

Capacitative Ca^{2+} entry (CCE), also called store-operated Ca^{2+} entry, may be another source of Ca^{2+} overload in hypoxia/ischemia (Nguyen et al. 2002; Smith et al. 2003, 2004; Berna-Erro et al. 2009; Zhang et al. 2014). Under physiological conditions, this kind of Ca^{2+} entry is mediated by store-operated Ca^{2+} channels (SOCCs), and is thought to be ubiquitous and important in most electrically non-excitable cells, as well as in many excitable cells, to promote the refilling of the depleted ER Ca^{2+} stores and the generation of short-term Ca^{2+} signaling or long-lasting Ca^{2+} signals that control gene transcription for supporting cell growth, apoptosis, differentiation or activation (Parekh and Putney 2005). In the CNS, CCE is present in both neurons (Gemes et al. 2011; Steinbeck et al. 2011; Koss et al. 2013) and glial cells (Verkhratsky and Parpura 2014) to modify neuronal excitability, network activity, and other Ca^{2+} signaling. A SOCC consists of two families of proteins: the stromal interaction molecule (STIM1 and STIM2), which function as Ca^{2+} sensors in the ER, and the Orais (Orai1, Orai2 and Orai3), which function as pore-forming subunits of store-operated Ca^{2+} entry channels. ER Ca^{2+} depletion causes STIM oligomerization and interaction with Orai, which causes activation of Orai currents to allow Ca^{2+} entry into the cytoplasm (see the preceding chapter by Chao and Xia). In cultured cortical astrocytes, acute hypoxia indeed always evokes CCE, and chronic hypoxia markedly enhances it (Smith et al. 2003, 2004). In neurons, ischemia also induces cytosolic Ca^{2+} accumulation via CCE (Berna-Erro et al. 2009; Zhang et al. 2014). However, the roles of STIM1 and STIM2 in hypoxic/ischemic store operated Ca^{2+} entry in neurons are still in debate. Berna-Erro et al. (2009) reported that STIM2 is the predominant STIM isoform in neurons, and STIM2, but not STIM1, is essential for CCE and ischemia-induced cytosolic Ca^{2+} accumulation in neurons. They found that neurons from *Stim2*$^{-/-}$ mice showed significantly increased survival under 1 or 2 h of OGD both in culture and in acute hippocampal slice preparations. In addition, *Stim2*$^{-/-}$ mice were markedly protected from neurological damage in a model of focal cerebral ischemia, whereas *Stim1*$^{-/-}$ itself had minor effect (Berna-Erro et al. 2009). In contrast, Zhang et al. (2014) found that STIM1 and Orai1 expression in rat hippocampus increased continuously after 15-min global ischemia and peaked on day 4, accompanying an increase in neuronal $[Ca^{2+}]i$. Silencing of STIM1 with *Stim1* siRNA suppresses STIM1 and Orai1 expression and SOCC-induced neuronal Ca^{2+} overload, and improves rat neurological function (Zhang et al. 2014). This contradiction might be contributed to multiple factors such as different models, animal ages, etc. It is worthy to point out that transient receptor potential channels (TRPCs) are also involved in CCE (Parekh and Putney 2005). TRPC family members, such as canonical TRPC1, TRPC4, and TRPC5, can be activated by STIM1, and Orai1 can act as a regulatory unit that confers STIM1-mediated store depletion sensitivity to TRPCs (Cahalan 2009; Smyth et al. 2010; Bollimuntha et al. 2011; Cheng et al. 2013). However, it is still unclear if these TRPCs also contribute to CCE and subsequent Ca^{2+} accumulation in hypoxic/ischemic stress.

3.3.2 Ca^{2+} Uptake and Release from Intracellular Stores

In addition to extracellular Ca^{2+} entry, intracellular Ca^{2+} stores are a source of cytosolic Ca^{2+} rise in hypoxic/ischemic neurons (Dubinsky and Rothman 1991; Mitani et al. 1993; Murai et al. 1997; Grøndahl et al. 1998; Chen et al. 1999; Zhang and Lipton 1999; Bergmann and Keller 2003; Nikolaeva et al. 2005; Aley et al. 2006; Andrade and Rossi 2010; Henrich and Buckler 2008, 2013). The measurements with Ca^{2+}-sensitive electrodes showed 50–200 % increases in $[Ca^{2+}]i$ with no accompanying decrease in $[Ca^{2+}]o$ in several brain regions within the first minute after oxygen depletion and prior to the reduction of ATP (Silver and Erecińska 1990). When extracellular Ca^{2+} is removed, synaptic transmission and neuronal activities are blocked (to eliminate Ca^{2+} influx due to possible activation of Ca^{2+}-permeable iGluR channels and VGCCs). In this circumstance, chemical ischemia or OGD still induces an increase in $[Ca^{2+}]i$, though relatively less than that in normal $[Ca^{2+}]o$ in neurons (Dubinsky and Rothman 1991; Mitani et al. 1993; Duffy and MacVicar 1996; Grøndahl et al. 1998; Chen et al. 1999; Bergmann and Keller 2003; Nikolaeva et al. 2005; Misonou et al. 2005; Andrade and Rossi 2010; Henrich and Buckler 2013). Depletion of intracellular Ca^{2+} stores may also reduce hypoxia/ischemia induced $[Ca^{2+}]$ rise (Aley et al. 2006; Andrade and Rossi 2010; Henrich and Buckler 2013). In hippocampal slices and isolated CA1 neurons, a combination of the removal of extracellular Ca^{2+} and thapsigargin-induced depletion of intracellular Ca^{2+} store, greatly decreases ischemic $[Ca^{2+}]i$ increase (from 25 μM to 630 nM) (Grøndahl et al. 1998; Larsen et al. 2005). These results strongly suggest the involvement of intracellular Ca^{2+} stores in hypoxia/ischemia-induced increase in $[Ca^{2+}]i$. The most important intracellular stores are endoplasmic reticulum and mitochondria.

Endoplasmic Reticulum (ER)

The ER network is the primary intracellular Ca^{2+} storage and release organelle in most cells, possessing two major families of Ca^{2+} release channels: the inositol triphosphahate (IP_3) receptors (IP_3Rs) and ryanodine receptors (RYRs), both of which are capable of regenerative Ca^{2+} releases. This storage gives ER a central role in cellular Ca^{2+} homeostasis and signaling (Paschen 2001; Verkhratsky 2005; Bodalia et al. 2013; also see the preceding chapter by Chao and Xia). ER also possesses physiological ER Ca^{2+} leak channels, presenilin1 and 2, which form low conductance Ca^{2+} channels responsible for about 80 % of the passive Ca^{2+} leak from ER, and the sarco-endoplasmic reticular Ca^{2+} ATPase (SERCA pump), which is similar to plasmalemmal Ca^{2+} ATPase and pumps back cytosolic Ca^{2+} in an energy-requiring process against a steep concentration gradient (see the preceding chapter by Chao and Xia). ER proteins can also directly associate with components of the outer mitochondrial membrane, known as "mitochondria-associated membrane" (MAM), to regulate Ca^{2+} signaling between them and shape sytosolic Ca^{2+} response (see the preceding chapter by Chao and Xia). In neurons, regenerative propagating cytoplasmic Ca^{2+} waves from the ER makes the ER a "neuron-within-neuron" (Berridge 1998). In glial cells, ER regulation of cytosolic Ca^{2+} endows glial cells with non-electrical excitability, termed "glial Ca^{2+} excitability", which triggers glial cells to release gliotransimitters and thus influence neuronal activity (Zorec et al. 2012). Under physiological conditions, ER Ca^{2+} is in high micromolar levels or even millimolar range, and is divided into a number of subcompartments (Paschen 2001).

Considerable evidence suggests that ER is involved in hypoxia/ischemia-induced cytosolic Ca^{2+} accumulation in the CNS (Dubinsky and Rothman 1991; Mitani et al. 1993; Murai et al. 1997; Wang et al. 2002; Meini et al. 2003; Ouardouz et al. 2003, 2006; Smith et al. 2004; Nikolaeva et al. 2005; Verkhratsky 2005; Aley et al. 2006; Henrich and Buckler 2013; also see review by Bodalia et al. 2013). In both neurons and astrocytes, ischemia induces a two-phase response of ER Ca^{2+}: accumulation of Ca^{2+} in ER stores and subsequent release of Ca^{2+} from ER following ischemia/reperfusion (Chen et al. 2008; Liu et al. 2010). The first Ca^{2+} accumulation in ER store might be a result of stimulation of SERCA pumps by elevated $[Ca^{2+}]i$ in the initial period before ATP is rapidly depleted. Subsequent release of Ca^{2+} from ER involves the failure of SERCA pumps (Larsen et al. 2005; Henrich and Buckler 2008, 2013) and IP_3Rs and RYRs activation (see the following), which induce large $[Ca^{2+}]i$ rise. ER Ca^{2+} depleption activates Ca^{2+} store-operated Ca^{2+} entry (see discussion in Sect. 3.3.1) and subsequent ER Ca^{2+} refilling to replenish ER Ca^{2+} (Nguyen et al. 2002; Chen et al. 2008; Sisalli et al. 2014) probably via the activities of STIM-Orais system, TPR channels, and NCX.

SERCA pumps run as long as ATP and Ca^{2+} are present in the cytoplasm to move Ca^{2+} ions into lumen of ER and establish a tremendous Ca^{2+} gradation between ER and cytosol by using the energy releases from hydrolysis of ATP. Therefore, keeping constant ATP supply to SERCA pumps is necessary for ER Ca^{2+} regulation (Jackson and Thayer 2006). During hypoxia/ischemia, with the depletion of ATP (Silver et al. 1997; Raley-Susman et al. 2001; Martinez-Sánchez et al. 2004; Sheldon and Church 2004), SERCA pumps are inhibited, leading to cytosolic Ca^{2+} accumulation (Larsen et al. 2005; Henrich and Buckler 2008, 2013). Indeed, inhibition of SERCA activity with thapsigargin or cyclopiazonic acid increases $[Ca^{2+}]i$ in neurons (Grøndahl et al. 1998; Nguyen et al. 2002; Larsen et al. 2005; Jackson and Thayer 2006; Henrich and Buckler 2008). In addition to the immediate effect on SERCA activity and $[Ca^{2+}]i$, hypoxia/ischemia can also chronically change the activity and expression of SERCA (Paschen et al. 1996; Behringer et al. 2009). Transient cerebral ischemia sharply decreases SERCA2b mRNA levels in the hippocampus and striatum after 24 h of reperfusion (Paschen et al. 1996). Long-term hypoxia also causes a reduction in SERCA function in fetal cerebrovascular sympathetic neurons (Behringer et al. 2009). These changes aggravate Ca^{2+} dysregulation in hypoxic/ischemia stress.

RYRs consist of three members (RYR1, RYR2, and RYR3), and are expressed widely in both neurons and glial cells in the nervous system. Neurons express all three isoforms, while astrocytes, which play key roles in extracellular ion balance, express RYR3 but not RYR1 or RYR2 (see the preceding chapter by Chao and Xia). The general idea for ER Ca^{2+} release via RYRs is that Ca^{2+} entering through either voltage-gated channels or receptor-operated channels provides the trigger Ca^{2+} that activates RYRs and, subsequently, Ca^{2+} release from ER, namely Ca^{2+}-induced Ca^{2+} release. Within the neuronal soma and the initial dendritic regions, portions of ER come into close contact with plasma membrane to form subsurface cisternae (Berridge 1998; Lam and Galione 2013), where RYR channels in ER are always closely apposed to L-type Ca^{2+} channels in the plasma membrane. Therefore, even a single action potential can induce Ca^{2+} entry through membrane Ca^{2+} channels to form highly localized Ca^{2+} in nanodomains or mircodomains, and trigger Ca^{2+} release from ER in neurons of various brain regions (Berridge 1998). In the absence of extracellular Ca^{2+}, depletion of ER Ca^{2+} store by application of ryanodine or caffeine before hypoxia/ischemia significantly decreases hypoxic/ischemic $[Ca^{2+}]i$ rise (Meini et al. 2003; Nikolaeva et al. 2005; Larsen et al. 2005; Henrich and Buckler 2008, 2013). Blockade of Ca^{2+}-induced Ca^{2+} release from RYR channels with dantrolene reduces hypoxia/ischemia-induced increases in $[Ca^{2+}]i$ in hippocampal neurons as well as in neuronal cell lines in the absence of extracellular Ca^{2+} (Dubinsky and Rothman 1991; Mitani et al. 1993; Wang et al. 2002; Nguyen et al. 2002). In dorsal columns, ischemia-induced axoplasmic Ca^{2+} rise can be blocked by ryanodine or blockers of L-type Ca^{2+} channel voltage sensors (nimodipine, diltiazem), but not by blocker of channel pores (Cd^{2+}) in Ca^{2+} free perfusate (Ouardouz et al. 2003). Immunoprecipitation and immunohistochemistry reveal a colocalization of Ca^{2+} channels and RYR clusters and an association between L-type Ca^{2+} channels and RYRs (Ouardouz et al. 2003). These findings suggest that Ca^{2+} release from RYR channels can also be evoked by depolarization sensed by L-type Ca^{2+} channels during ischemia, similar to "excitation-contraction coupling" in skeletal muscle (Ouardouz et al. 2003, 2006). Recently, cyclic ADP ribose-activated RYRs have been proposed to be a contributor to hypoxia-induced $[Ca^{2+}]i$ increase in cultured cortical astrocytes (Aley et al. 2006). All these observations are in support of the contribution of RYR channels to hypoxic/ischemic $[Ca^{2+}]i$ rise.

IP_3Rs also have three isoforms (IP_3R1-3), and are widely distributed throughout the brain, and activated by IP_3 that is generated from phosphatidylinositol 4,5-bisphosphate by phospholipase C (PLC) during activation of G protein-coupled receptors (see the preceding chapter by Chao and Xia). Hypoxia/ischemia has been shown to cause a dramatic increase in IP_3 levels associated with IP_3R-dependent Ca^{2+} release from ER in both neurons and astrocytes (Chen et al. 2008; Bickler et al. 2009). Therefore, IP_3R channels also contribute to hypoxic/ischemic Ca^{2+} response. For instance, depletion of Ca^{2+} store with thapsigargin or cyclopiazonic acid (inhibitors of SERCA) (Murai et al. 1997; Grøndahl et al. 1998; Larsen et al. 2005; Henrich and Buckler 2008, 2013), or with IP_3-generating agonist, bradykinin (Smith et al. 2004), reduces hypoxic/ischemic $[Ca^{2+}]i$ rise in the absence of external Ca^{2+} in both neurons and astrocytes. In optic nerves, application of U73122 (an irreversible antagonist of PLC), neomycin (rendering PLC substrates unrecognizable), or heparin (inhibiting IP_3-induced release), reduces ischemic peak Ca^{2+} response by 50 % in zero external Ca^{2+} conditions (Nikolaeva et al. 2005). A similar observation is made with IP_3R blockade in hippocampal neurons (Larsen et al. 2005; Bickler et al. 2009) in both axon and myelin of rat spinal cord white matter (Ouardouz et al. 2006). G protein-coupled receptors (e.g., mGluRs, P_2YRs) increase the production of IP_3, triggering release of Ca^{2+} from ER stores (Abdul-Ghani et al. 1996; Wu et al. 2007). During hypoxia/ischemia, release and extracellular accumulation of glutamate (Benveniste et al. 1984; Goldberg and Choi 1993; Yamaguchi et al. 1998; Krzyżanowska et al. 2014) and ATP (Lin et al. 2008; Schock et al. 2008; Domercq et al. 2010; Bai and Li 2013) may activate mGluRs and P_2YRs to stimulate IP_3 generation. Therefore, Ca^{2+} release from IP_3R could be induced by glutamate and ATP release and mGluR/P_2YR activation during hypoxic/ischemic stress (Iwabuchi and Kawahara 2009; Zheng et al. 2013a). Indeed, P_2Y1R activation-induced IP_3-mediated astrocytic Ca^{2+} release from ER and $[Ca^{2+}]i$ elevation are only present in control wild-type mice, but not in IP_3R type 2 (astrocyte type) knockout mice during ischemia (Zheng et al. 2013a).

The possible mechanisms of Ca^{2+} release from ER during hypoxia/ischemia can be as follows: released glutamate and ATP binds to glutamate receptors (both mGluR and iGluR) and P_2 purinergic receptors (both P_2YR and P_2XR) respectively, inducing IP_3-sensitive Ca^{2+} release from ER and Ca^{2+} influx from extracellular space through iGluR and P_2XR channels, and VGCCs, which subsequently results in $[Ca^{2+}]i$ rise. Elevated cytosolic Ca^{2+} stimulates more Ca^{2+} release from RYRs. Therefore, ER-based Ca^{2+} release contributes not only to the initial progressive increase in $[Ca^{2+}]i$, but also to the large $[Ca^{2+}]i$ increase at the later phase of hypoxic/ischemic stress. Due to the depletion of ATP during hypoxia/ischemia, SERCA is inhibited (Larsen et al. 2005; Henrich and Buckler 2008, 2013), and therefore cannot efficiently pump back cytosolic Ca^{2+} against a steep concentration gradient into ER, which also aggravates hypoxic/ischemic Ca^{2+} accumulation.

Mitochondria

Mitochondria are capable of storage and release of intracellular Ca^{2+}. Under physiological conditions, Ca^{2+} uptake by mitochondria is mediated by a specific Ca^{2+} uniport system, identified recently as mitochondrial Ca^{2+} uniporter (MCU) (Baughman et al. 2011; De Stefani et al. 2011), which is driven by steep negative $\Delta\Psi m$ across the inner mitochondrial membrane (IMM), and operates strictly according to the Ca^{2+} electrochemical equilibrium. MCU has the low affinity to cytosolic Ca^{2+}, thus only takes up Ca^{2+} in the microdomains of Ca^{2+} concentrations. It has been indicated that a significant fraction of mitochondria are located close to ER or plasma membrane Ca^{2+} entry channels and are exposed to high Ca^{2+} concentrations, thus promoting Ca^{2+} uptake by MCU (Gouriou et al. 2011, 2013, also see the preceding chapter by Chao and Xia). For example, IP_3R-mediated Ca^{2+} release from ER in MAM forms Ca^{2+} microdomains (Ca^{2+} hotspots) that can be sensed by neighboring mitochondria, causing mitochondrial Ca^{2+} accumulation through MCU (Rizzuto et al. 1993, 1998). In addition, a recently identified high affinity mitochondrial Ca^{2+}/H^+ exchanger, Letm1 (Jiang et al. 2009), can sense slow and steady $[Ca^{2+}]$ i elevation and potentially transform small increases in $[Ca^{2+}]$i into sustained increase in mitochondrial Ca^{2+} concentrations by catalyzing 1:1 electrogenic exchange of Ca^{2+} and H^+ (Gouriou et al. 2011, also see the preceding chapter by Chao and Xia). Ca^{2+} efflux from mitochondria is mediated primarily by the mitochondrial Na^+/Ca^{2+} exchanger that was recently identified as NCLX, and is localized to the cristae of mitochondria (Palty et al. 2010). Most recent studies reported that the NCX3 isoform of classical plasma membrane NCX family is located on the outer mitochondrial membrane (OMM) of neurons, extrudes Ca^{2+} from mitochondria through AKAP121 [a member of the protein kinase A (PKA) anchoring protein (AKAP) family] interaction in a PKA-mediated manner, both under normoxia and hypoxia, and improves cell survival when it works in the Ca^{2+} efflux mode at the level of the OMM (Scorziello et al. 2013; Sisalli et al. 2014). Na^+-independent efflux of Ca^{2+} is mediated by mitochondrial Ca^{2+}/H^+ antiporters, and Letm1 can contribute to both mitochondrial Ca^{2+} uptake and release depending on mitochondrial proton and Ca^{2+} chemical gradients (Nowikovsky et al. 2012). High levels of Ca^{2+} within mitochondria can cause the opening of permeability transition pores (PTP) to release Ca^{2+} from mitochondrial matrix (Gouriou et al. 2011, also see the preceding chapter by Chao and Xia).

It has been shown that an initial small increase in $[Ca^{2+}]$i is attributed to Ca^{2+} release from mitochondria in cerebellar granule cells and brainstem motoneurons during chemical ischemia (Chen et al. 1999; Bergmann and Keller 2003). In hippocampal neurons, chemical ischemia induces a large and sustained increase in $[Ca^{2+}]$i in the absence of extracellular Ca^{2+}, which is also attributed to Ca^{2+} release from mitochondria (Misonou et al. 2005). Based on the work by Zhang and Lipton (1999), mitochondrial Ca^{2+} release is responsible for approximately 50 % of ischemic increase in $[Ca^{2+}]$i in hippocampal slices. Similar results are also observed in optical axon during ischemia (Nikolaeva et al. 2005). These results suggest that mitochondria are important contributors to hypoxia/ischemia-induced Ca^{2+} accumulation.

By using electron probe X-ray microanalysis, Taylor et al. (1999) reported that despite Ca^{2+} release from mitochondria during OGD, cytosolic Ca^{2+} levels increase substantially in hippocampal CA1 cells during 42 min of OGD. They also showed that total Ca^{2+} contents in mitochondria increased significantly at 5 min and then decreased after 12 min of ischemia, but was still kept at a significantly higher level than control during all the periods of OGD. The functional and morphological impairment of brain mitochondria is causally linked with Ca^{2+} accumulation in matrix during hypoxic/ischemic stress (Frantseva et al. 2001; Schild et al. 2003; Kintner et al. 2007; Ouyang et al. 2011; Gouriou et al. 2013; Sisalli et al. 2014). These observations suggest that mitochondria can also take up Ca^{2+} during hypoxia/ischemia. In support of this conclusion are studies of direct measurements of mitochondrial Ca^{2+} with fluorescent dyes (Rhod-2 acetyl ester, X-Rhod1, dihydroRhod-2) or genetically encoded, mitochondria-targeted, Cameleon-based (4mtD3cpv) or aequorin-based, Ca^{2+} probes, in which ischemia indeed causes increases in mitochondrial Ca^{2+} (Frantseva et al. 2001; Ouyang et al. 2011; Gouriou et al. 2013; Scorziello et al. 2013; Sisalli et al. 2014). Actually, there may be a "set-point" for mitochondrial response to cytosolic [Ca^{2+}] (the value is 0.5–1 μM in neurons), at which Ca^{2+} uptake and efflux are equal but opposite in mitochondria (Nicholls 2005). Therefore, hypoxia/ischemia, as a stimulus, may stimulate light Ca^{2+} release from the mitochondria before the significantly elevated cytosolic [Ca^{2+}] is higher than the "set-point", as has been shown in cerebellar granule cells and brainstem motoneurons (Chen et al. 1999; Bergmann and Keller 2003). When the cytosolic Ca^{2+} elevation reaches a level higher than the "set-point", the activities of MCU greatly increase (Nicholls 2005). Then, large quantities of Ca^{2+} are taken up, and accumulate in the mitochondrial matrix. However, the rise in mitochondrial Ca^{2+} can cause the decline, or even the collapse of the steep negative $\Delta\Psi m$ across IMM (Bahar et al. 2000; Schild et al. 2003; Kintner et al. 2007; Tanaka et al. 2009; Ouyang et al. 2011; Hawrysh and Buck 2013; Henrich and Buckler 2013; Utkina-Sosunova et al. 2013), and stimulate ROS/RNS generation (Lafon-Cazal et al. 1993; Dugan et al. 1995; Perez Velazquez et al. 1997; Carriedo et al. 2000; Frantseva et al. 2001; Stanika et al. 2010; Niatsetskaya et al. 2012; Sivakumar et al. 2013), resulting in the decrease of MCU activity (due to the decrease or loss of driving force) and the opening of PTP and subsequent large Ca^{2+} efflux to the cytosol (Frantseva et al. 2001; Rytter et al. 2005; Hawrysh and Buck 2013). This is supported by the facts that glutamate challenge of neurons evokes first an increase in intramitochondrial Ca^{2+}, then a rapid drop back to prestimulation levels despite the continued presence of increased levels of [Ca^{2+}]i (Bano et al. 2005). FCCP, an uncoupler of oxidative phosphorylation that fully depolarizes mitochondria, evokes mitochondrial Ca^{2+} release similar to that induced by hypoxia/ischemia (Henrich and Buckler 2008, 2013; Abramov and Duchen 2010; Sisalli et al. 2014). Such Ca^{2+} efflux from the mitochondria is responsible for delayed Ca^{2+} deregulation (Bahar et al. 2000; Hajnóczky et al. 2006). Therefore, the results of Taylor et al. (1999) are not surprising.

In addition, mitochondria are able to accumulate Ca^{2+} once the bulk cytosolic Ca^{2+} levels reach the region of 500–600 nM. The accumulated Ca^{2+} is then gradually released into the cytosol to buffer the sharp increase in [Ca^{2+}]i (Verstreken et al. 2005; Young et al. 2008). In this sense, mitochondira act as mobile Ca^{2+} buffers in cytosol. It has been found that the Ca^{2+} buffering capacity of mitochondria is largely damaged in both neurons and astrocytes exposed to hypoxia/ischemia (Iijima et al. 2008; Tanaka et al. 2009; Ouyang et al. 2011; Utkina-Sosunova et al. 2013). This may also contribute to hypoxia/ischemia-induced increase in [Ca^{2+}]i.

Several pathways are responsible for hypoxia/ischemia-induced Ca^{2+} release from mitochondria. Besides PTP (Frantseva et al. 2001; Rytter et al. 2005; Hawrysh and Buck 2013), IMM NCLXs (Hernández-SanMiguel et al. 2006; Palty et al. 2010) and OMM NCX3 (Scorziello et al. 2013; Sisalli et al. 2014) directly modulate Ca^{2+} efflux. It has been known that intracellular Na^+ can promote the release of Ca^{2+} from mitochondria by coupling via the mitochondrial $2Na^+/Ca^{2+}$ exchangers (Ritter et al. 2003). Na^+ entry during ischemia activates Ca^{2+} release from mitochondria by the

IMM NCLXs. Therefore, blockade of NCLXs with CGP37157 greatly reduces ischemia-induced mitochondrial Ca^{2+} release and the subsequent increase in cytosolic [Ca^{2+}] in hippocampal neurons and optic axon (Zhang and Lipton 1999; Nikolaeva et al. 2005). In addition, disruption of mitochondrial NCX3 activity by the silencing of endogenous AKAP121 with siRNA reduces the Ca^{2+} efflux activity of the mitochondrial NCX3, as demonstrated by the intramitochodrial Ca^{2+} increase caused by chemical hypoxia (Scorziello et al. 2013; Sisalli et al. 2014). Furthermore, reverse Ca^{2+} uniporters are another pathway for Ca^{2+} release from mitochondria. As a result of depolarization of the inner membrane during hypoxia/ischemia (Bahar et al. 2000; Schild et al. 2003; Kintner et al. 2007; Tanaka et al. 2009; Ouyang et al. 2011; Hawrysh and Buck 2013; Henrich and Buckler 2013; Utkina-Sosunova et al. 2013), Ca^{2+} uniporters can operate in a reverse mode to cause a rapid release of Ca^{2+} from the mitochondria (Jurkowitz et al. 1983; Montero et al. 2001). For mitochondrial Ca^{2+}/H^+ antiporter (Letm1), no information is available regarding its contribution to mitochondrial Ca^{2+} efflux and subsequent [Ca^{2+}]i elevation during hypoxia/ischemia. In general, among these pathways, mitochondrial Na^+/Ca^{2+} exchangers are the most important route in hypoxic/ischemic Ca^{2+} release from mitochondria (Zhang and Lipton 1999; Nikolaeva et al. 2005).

4 DOR-Mediated Regulation of Ionic Homeostasis during Hypoxia/Ischemia

Neuronal responses to stress vary depending on the duration of oxygen deprivation and survival time following hypoxia (Lipton 1999). Short-term hypoxia to central neurons causes an immediate loss of ionic homeostasis, which is largely based on the functional response of inherent membrane proteins to the stress (Hansen 1985; Lipton 1999). On the other hand, long-term hypoxia leads to major alterations in gene expression of intracellular elements, especially death/survival signal systems (Lipton 1999). DOR signals may trigger, depending on stress duration, different mechanisms at multiple levels to protect neurons from short- or long-term stress. For example, it is possible that DOR signals maintain ionic homeostasis in the initial stage of hypoxia, while during prolonged hypoxia, DOR then signals up-regulation of the expression and function of benefit-signaling molecules and down-regulates those involved in cell injury or death (Chao and Xia 2010; He et al. 2013). Indeed, our recent studies show that DOR plays its neuroprotective role by enhancing the intracellular activity of the G protein-PKC-pERK-Bcl2 pathway and suppressing the phosphorylated p38 and cytochrome c death signals during long-term hypoxia (Ma et al. 2005).

Indeed, some anoxia-tolerant species such as turtles have extraordinary abilities to counteract hypoxia/ischemia-induced disruption of ionic homeostasis in their brain. Even up to 4 h of anoxia could not induce the abrupt rise in [K^+]e in the turtle brain, while it is very evident in the rat brain within several minutes of anoxia (Sick et al. 1982, 1987; Doll et al. 1991; Xia et al. 1992; Knickerbocker and Lutz 2001).

Although the mechanisms underlying hypoxia/ischemia tolerance are very complex in the turtle (Lutz and Nilsson 2004; Milton and Prentice 2007; Larson et al. 2014), the uniquely high density of DOR in the turtle brain (Xia and Haddad 2001) may be one of the potential mechanisms. To explore the possible role of opioid receptors in the regulation of ionic homeostasis during hypoxia/ischemia, we have recently conducted a series of experiments by direct measurement of extracellular ion (K^+, Na^+) activities with ion-sensitive microelectrodes in mouse cortical slices (Chao et al. 2007a, b, 2008, 2009, 2012b; Kang et al. 2009). We also established a mouse model with neuron-specific and tetracycline-regulated up-regulation of DOR in the cortex, and obtained some interesting results with this model regarding the effects of opioids on disruption of ionic homeostasis during hypoxia/ischemia (Chao et al. 2006).

Under physiological conditions, DOR activation largely influences $[Ca^{2+}]i$, either increasing or decreasing intracellular Ca^{2+} levels (Chao and Xia 2010; also see the preceding chapter by Chao and Xia). However, the role of DOR in hypoxia/ischemia-induced $[Ca^{2+}]i$ increase is largely unknown. Our recent work suggested that hypoxia-induced Ca^{2+} entry and subsequent BK channel-mediated K^+ efflux in cortical slices are largely attenuated by DOR activation (Chao et al. 2007a). Consistent with our results is a recent study that showed that in DOR-abundant (Xia and Haddad 2001), hypoxia/ischemia-resistant turtle cortex (Sick et al. 1982, 1987; Doll et al. 1991; Xia et al. 1992; Knickerbocker and Lutz 2001), DOR antagonism with naltrindole not only potentiates normoxic NMDAR currents and Ca^{2+} influx, but also boosts the large irreversible neuronal depolarization and increases in $[Ca^{2+}]i$ following anoxia (Pamenter and Buck 2008). These findings indicated that DOR activation attenuates Ca^{2+} dysregulation and benefits the maintenance of Ca^{2+} homeostasis in hypoxic/ischemic insults. The mechanisms of DOR attenuation of Ca^{2+} dysregulation under hypoxic/ischemic condition are largely unknown. Pamenter and Buck (2008) suggested that DORs mediate NMDAR activity in a Gi-dependent manner and prevent deleterious NMDAR-mediated $[Ca^{2+}]i$ influx during anoxic insults in the cortex. Besides, DOR can regulate the activities of various types of VGCCs as well as SERCA, and thus regulates intracellular Ca^{2+} levels of nerve cells under normal condition (Chao and Xia 2010; also see the preceding chapter by Chao and Xia). However, it is still unknown if similar regulatory mechanisms of DOR exist under hypoxic/ischemic condition.

Actually, in the mechanistic exploration of DOR neuroprotection, we noticed that an important aspect of DOR-mediated neuroprotection may be its early action against disruption of ionic homeostasis, particularly K^+ and Na^+ homeostasis during anoxia or ischemia, though DOR activation seems to have minor effects on Na^+ and K^+ activities under physiological condition (Chao and Xia 2010; also see the preceding chapter by Chao and Xia). Our studies first demonstrated DOR action on ionic homeostasis under hypoxic/ischemic stress (Chao et al. 2006, 2007a, b, 2008, 2009, 2012b; Kang et al. 2009). As discussed earlier, K^+ and Na^+ ions are the most abundant ions inside and outside cells respectively, and excessive K^+ efflux and Na^+ influx are believed to be a crucial factor leading to neuronal death under pathophysiological conditions such as anoxic/ischemic and excitotoxic stress (Probert et al. 1997; Yu et al. 1997, 1999; Takahashi et al. 1999; Breder et al. 2000; Raley-Susman et al. 2001; Liu et al. 2003; Wei et al. 2003; Yu 2003b; Banasiak et al. 2004; Karki et al. 2007; Hertz 2008). Therefore, DOR counteracting hypoxic/ischemic ionic dysregulation confers neuroprotection of DOR against hypoxic/ischemic insults.

We used mouse cortical slices as our model to assess the effects of DOR activation on K^+ homeostasis in both normoxic and anoxic perfusion because the cortex has a much higher density of DOR (Xia and Haddad 1991, 2001). Anoxia was induced by switching from the control superfusate to one continuously aerated with 95 % N_2 and 5 % CO_2. Under such a condition for at least 30 min aeration before perfusion, the O_2 level in ACSF is almost completely eliminated. Ischemia was mimicked by OGD, i.e., exposing the slices to D-glucose-deficient ACSF (D-glucose was substituted by equal molar mannitol) continuously aerated with 95 % N_2 and 5 % CO_2. Each slice was subjected to a single period of anoxia or ischemia that was halted at about 1.5 min after the onset of anoxic depolarization (as assessed by a rapid increase in $[K^+]e$ and negative shift of extracellular DC potential), or for a period of 20 min if anoxic depolarization did not occur. Then oxygen/glucose was re-introduced in the perfusion solution. Drugs (e.g., DOR agonists and antagonists) were applied to the slices starting 20 min before anoxia/ischemia and lasting for the whole period of anoxic/ischemic stress. The extracellular ion activities were recorded with ion-selective microelectrodes. All the recording lasted 60–90 min. To distinguish the drug effects from an artifact caused by bad viability for the slices that did not show anoxic depolarization to anoxia, we checked slice viability by OGD perfusion after at least 30 min recovery from the first recording. If the slice showed a response of anoxic depolarization to OGD, we considered the slice had good viability, and the response in the first recording (with no anoxic depolarization) reflects the effects of drug application. Otherwise, the data of the first recording were expelled from analysis. In some experiments, we focused just on the mechanistic investigation by analyzing anoxia/ischemia-induced disruption of ionic homeostasis as well as changes in extracellular DC potential since we and others (see reviews by Chao and Xia 2010; He et al. 2013) have demonstrated that activation of DOR is neuroprotective and predisposes neurons to survive under severe stress such as glutamate-, hypoxia- and ischemia-stress.

We showed that DOR agonist DADLE (1–10 μM) concentration-dependently attenuated the anoxia- and OGD-induced disruption of K^+ homeostasis as well as the sudden, sharp deflation of extracellular DC potential (Chao et al. 2007b), the hallmarker of anoxic depolarization and hypoxic/ischemic neuronal response causing glutamate and ATP release, cell swelling, and neuronal death (Asai et al. 2000; Müller and Ballanyi 2003; Anderson et al. 2005; Leichsenring et al. 2013; Brisson et al. 2014) in cortical slices. This attenuation was manifested by a prolonged latency of the response to anoxia/ischemia, largely reduced maximal increase in $[K^+]e$, and attenuated rate of rise of $[K^+]e$ from latency to peak (Chao et al. 2007a, b). The effect of DADLE was completely blocked by 1 μM naltrindole, a potent DOR antagonist, suggesting that DOR activation is involved in the effect (Chao et al. 2007a). We previously observed that naltrindole, if administrated at 10 μM in neuronal culture, increased hypoxic injury in cortical neurons (Zhang et al. 2002). At the same concentration in the cortical slices, naltrindole induced an extremely larger increase in $[K^+]e$ during anoxia (Chao et al. 2007b). Therefore, it is possible that in addition to severe Ca^{2+} overload (Pamenter and Buck 2008), excessive inhibition of DOR may cause neuronal injury through severe disruption of K^+ homeostasis. Application of UFP 512, a more specific, potent, and naturally distinct DOR agonist (Balboni et al. 2002) to cortical slices had similar effects on anoxic disruption of K^+ homeostasis, as seen with DADLE (Chao et al. 2007a, b).

Since almost all DOR agonists, including DADLE and UFP 512, may bind to MOR, another highly homologous opioid receptor (Pogozheva et al. 2005), in a low affinity (Knapp and Yamamura 1992; Balboni et al. 2002), we asked whether MOR also plays a role in the DOR agonists' effect on the anoxia-induced disruption of K^+ homeostasis. To distinguish the effects of MOR from DOR on anoxia-induced disruption of K^+ homeostasis, we examined the effect of a MOR agonist, DAGO, in the cortex. We found that perfusion of DAGO (10 μM) did not produce any significant change in anoxia-induced response latency, increase in peak $[K^+]e$, the rate of rise of $[K^+]e$ to peak, and the undershoot of $[K^+]e$ during reoxygenation in cortical slices. Therefore, activation of MOR has no effect on anoxic disruption of K^+ homeostasis (Chao et al. 2007a). The lack of a neuroprotective effect of MOR against anoxic disruption of K^+ homeostasis may not be surprising because of the different distribution patterns of these two receptors in the cortex, the major different functions, and the different thermogenesis and metabolic effects on the body from these receptors (see Discussion in Chao et al. 2007a). Indeed, some studies showed that MOR activation actually aggravates neurotoxic effects of hypoxia/hypoglycemia, exacerbates incomplete forebrain ischemia, induces neuronal apoptosis, and inhibits cell survival (Hauser et al. 1994; Yin et al. 1999a; Kofke et al. 1999; Hu et al. 2002; Ammon-Treiber et al. 2005; Harburg et al. 2007; Kolodziej et al. 2008), and blocking MOR activation is beneficial in cerebral ischemia/reperfusion injury (Liao et al. 2003; Chen et al. 2005). These results therefore strongly confirm that DOR, but not MOR, activation is protective against anoxia-induced disruption of K^+ homeostasis in the cortex (Chao et al. 2007a, b).

In our neuron-specific and tetracycline-regulated transgenic mice model with DOR over-expression in the cortex, we found that anoxia-induced disruption of K^+ homeostasis was significantly less in the transgenic cortex than in the wild-type counterparts. In the transgenic mice, DOR activation with DADLE further attenuated the anoxic response of disruption of K^+ homeostasis in the cortex. Chronic hypoxia (9.5 % O_2, up to 5 days) rendered the frontoparietal cortex of the wide-type mice more vulnerable to the subsequent anoxia-induced disruption of K^+ homeostasis, while the transgenic over-expression of DOR in the cortex significantly attenuated this vulnerability. These findings suggest that transgenic over-expression of DOR protects the cortex against anoxic disruption of ionic homeostasis, which supports our hypothesis that DOR is an inherent neuroprotector in the cortex (Chao et al. 2006).

However, a key issue may arise in the mind of readers: is it beneficial (as we believed) or detrimental to neurons regarding DOR inhibition of hypoxia/ischemia-evoked K^+ efflux and $[K^+]e$ elevation since K^+ efflux from activated channels (K_{ATP}, BK and K_{Na} channels, etc.) is traditionally thought to be an adaptive mechanism to limit excessive membrane depolarization, over-excitability, and energy consumption for potential damages to neurons? Here it is necessary and very important to clarify the role of K^+ efflux in neuroprotection.

Indeed, in certain circumstances, activation of K^+ channels protects neurons from hypoxia (Ballanyi 2004; Buck and Pamenter 2006; Huang et al. 2006; Sun et al. 2007). For example, although K_{ATP} channels are normally inhibited by physiological levels of ATP, a decrease in cytosolic ATP concentration, as occurs in hypoxic condition (Silver et al. 1997; Raley-Susman et al. 2001; Martinez-Sánchez et al. 2004; Sheldon and Church 2004), opens K_{ATP} channels, leading to membrane hyperpolarization (Ballanyi 2004; Miki and Seino 2005; Huang et al. 2006). Activation of BK channels shapes neuronal excitability and can be an adaptive mechanism in some central neurons because these K^+ channels can dampen or limit the depolarization by K^+ efflux, and reduce neurotransmitter release. BK channels can also be activated by the decrease in ATP and the increase in $[Ca^{2+}]i$ (Murai et al. 1997; Misonou et al. 2005; Chao et al. 2007a). In addition, K_{Na} channels are sensitive to intracellular Na^+, and are inhibited by intracellular ATP and decreased pH in low $[Na^+]i$, and are activated by higher $[Na^+]i$ even in the presence of decreased pH, a situation often seen in hypoxia/ischemia (Yuan et al. 2003; Bhattacharjee and Kaczmarek 2005; Ruffin et al. 2008). In some neurons, K_{Na} channels also appear to be activated by Na^+ influx accompanying a single spike. K_{Na} channels may contribute to K^+ efflux and play a role in adaptation of firing rate and slowing after-hyperpolarization that follows repetitive firing (Lawrence and Rodrigo 2001; Bhattacharjee and Kaczmarek 2005). Under the conditions of decreased ATP concentration as well as increased $[Na^+]i$ and $[Ca^{2+}]i$ (often occur in hypoxia/ischemia), these channels (K_{ATP}, BK, and K_{Na} etc.) can be activated to limit depolarization to a certain certain degree. As it is well known, limitation of depolarization can reduce neuronal excitability and energy consumption and potentially protect nerve cells against hypoxic/ischemic damage.

However, these adaptive mechanisms (K_{ATP}, BK and K_{Na} channels, etc.) are beneficial to neurons in counteracting hypoxic/ischemic stress only in a limited range of K^+ efflux. We would like to address that excessive intracellular K^+ loss is not protective, but harmful to neurons. The loss of ionic homeostasis is the initial and key step in hypoxic/ischemic neuronal injury. Excessive K^+ fluxes play important roles in the neuronal death that occurs in various conditions, including during hypoxia/ischemia (Bortner et al. 1997; Yu et al. 1997, 1999; D'Ambrosio et al. 1999; Wei et al. 2003; Liu et al. 2003; Yu 2003b; Lauritzen et al. 2003; Remillard and Yuan 2004; Karki et al. 2007). For example, activation of iGluRs of mouse cortical neurons, even when inward cation influx is decreased by lowering extracellular Na^+ and Ca^{2+}, induced a large outward K^+ current; this caused loss of about 50–80 % intraneuronal K^+ and led to great shrinkage of the cell body and, consequently, neuronal apoptosis (Yu et al. 1999; Xiao et al. 2001). In fact, blockade of K^+ efflux attenuates hypoxia-, ischemia- and glutamate-induced neuronal death (Huang et al. 2001; Wei et al. 2003; Liu et al. 2003; Zhao et al. 2006b; Nistico et al. 2007). The reduced expression of K_{ATP} channels protects dopaminergic neurons from degeneration in Parkinson's disease (Liss and Roeper 2001). These studies suggest that inhibition of massive K^+ fluxes and maintaining cellular K^+ homeostasis is beneficial to neurons under severe hypoxic/ischemic stress, and may be of therapeutic benefit in the treatment of stroke and related neurodegenerative conditions (Wei et al. 2003; Liu et al. 2003; Yu 2003b; Zhao et al. 2006b; Nistico et al. 2007). A myriad of studies, including those from our laboratory and other independent laboratories, have demonstrated that activation of DOR is neuroprotective and predisposes neurons to survive under severe stress such as glutamate-, hypoxia- and ischemia-stress (see reviews by Chao and Xia 2010; He et al. 2013; Staples et al. 2013). Therefore, we believe DOR attenuation of hypoxia/ischemia-induced ion dysregulation (including not only K^+, but also Ca^{2+} and Na^+ ions that will be discussed next) offers proactive ability to nerve cells against hypoxic/ischemic insults.

In mechanistic investigation of DOR's protection against hypoxic/ischemic disruption of K^+ homeostasis, we found that DOR activation also attenuates hypoxic/ischemic Na^+ influx (see next section for details).

5 Mechanisms for DOR-Mediated Regulation of Ionic Homeostasis

The DOR attenuation of anoxic disruption of ionic homeostasis might rely on multiple strategies involved in the regulation of ionic homeostasis in neurons although the precise mechanisms are still not well understood yet. There was little information available on the mechanisms underlying opioid-mediated regulation of ionic homeostasis under hypoxic/ischemic conditions in the past. A recent study reported that the naltrindole-mediated depolarization and increased $[Ca^{2+}]i$ are prevented by NMDAR antagonism or by perfusion of the Gi protein agonist mastoparan-7, which also reverse the naltrindole-mediated potentiation of NMDAR currents in both normoxia and anoxia (Pamenter and Buck 2008). The authors thus suggest that DOR mediates NMDAR activity in a Gi-dependent manner and prevents deleterious NMDAR-mediated $[Ca^{2+}]i$ influx during anoxic insults in the turtle cortex (Pamenter and Buck 2008). We have recently shed light on mechanisms of DOR-mediated regulation of ionic homeostasis in hypoxic/ischemic condition through several studies (Chao et al. 2007a, b, 2008, 2009, 2012b; Kang et al. 2009). We found that DOR signal may modulate the function of several membrane channel proteins and involve a PKC-dependent and PKA-independent pathway in the downstream of G proteins.

5.1 Ionic Mechanisms

The precise mechanisms underlying the DOR regulation of K^+ homeostasis could be very complex. As discussed in Sect. 3, the increase in extracellular K^+ during anoxia could be caused by multiple pathways in neurons. Therefore, it is possible that activation of DOR attenuates anoxic disruption of K^+ homeostasis by directly or indirectly decreasing neuronal K^+ loss and/or increasing its influx in multiple routes and levels.

Hypoxia/ischemia may increase $[Ca^{2+}]i$, and might lead to BK channel activation and the resultant K^+ efflux, as discussed above. To explore Ca^{2+}-based factors in DOR attenuation of hypoxic/ischemic disruption of K^+ homeostasis, we first used Ca^{2+}-free perfusate to observe anoxia-induced changes in K^+ activity and the effect of DOR agonist in cortical slices (Chao et al. 2007a). Our data indicate that the absence of extracellular Ca^{2+} greatly attenuates the anoxia-induced disruption of K^+ homeostasis. Blockade of either BK channels or K_{ATP} channels decreases anoxia-induced increase in peak $[K^+]e$ with prolonged response latency (Chao et al. 2007a). These results suggest a key role of extracellular Ca^{2+} in anoxic disruption of K^+ homeostasis, and an important role of BK and K_{ATP} channels in hypoxic K^+ leakage. DOR activation by DADLE (10 μM) did not further enhance the effect of a Ca^{2+} free perfusion on anoxic disruption of K^+ homeostasis, suggesting that the effect of DOR activation is mediated at least partially by targeting Ca^{2+} entry. This is supported by the study with DOR antagonist natrindole in the turtle cortex during anoxic insults (Pamenter and Buck 2008). Moreover, DOR activation did not further reduce the anoxic disruption of K^+ homeostasis in the presence of a BK channel blocker, paxilline. In contrast, the blockade of K_{ATP} channels does not affect the DOR protection against anoxic disruption of K^+ homeostasis. These results indicate that the inhibition of hypoxia-induced increase in Ca^{2+} entry-BK channel activities is one of the mechanisms for DOR attenuation of anoxic disruption of K^+ homeostasis in the cortex, while K_{ATP} channels are less likely a target of DOR in such a protective action (Chao et al. 2007a).

Besides the inhibition of anoxia-induced increase in Ca^{2+} entry-BK channel activity, we hypothesized that DOR may stabilize anoxic disruption of K^+ homeostasis in the cortex by acting on Na^+ based neuronal factor(s). This is because our results indicated that Ca^{2+}-based component of K^+ leakage is not the sole factor for anoxic disruption of K^+ homeostasis. Even the absence of external Ca^{2+} or blockade of BK channels could not completely abolish DOR effect (Chao et al. 2007a). Also, Na^+ is the predominant ion in the extracellular space, and the changes in $[Na^+]o$ widely affect many fundamental cellular functions, e.g. neuronal excitability, intracellular Ca^{2+} homeostasis, pH stability, glutamate and K^+ uptake, and even coupling membrane signaling and metabolic pathway (see the preceding chapter by Chao and Xia). We first used the strategy of Na^+ substitution to explore the role of Na^+ in DOR attenuation of hypoxic/ischemic K^+ derangement. We used equimolar $NMDG^+$ or Li^+ to replace extracellular Na^+ to a desired concentration (Chao et al. 2008). $NMDG^+$ is a membrane impermeable organic cation always used to substitute Na^+. Lowering external Na^+ with $NMDG^+$ replacement may reduce Na^+ influx and lower neuronal excitability because of a reduction of driving force for inward Na^+ current during anoxia (Calabresi et al. 1999; Sheldon et al. 2004). In contrast, Li^+ is a membrane permeable cation passing across Na^+ channels as well as Na^+ itself (Hille 1972). Because of a reduction in driving force for inward Na^+ current, lowering external Na^+ and substituting with Li^+ may reduce anoxia-induced Na^+ influx, similar to the effects of $NMDG^+$. However, unlike $NMDG^+$ which lowers neuronal excitability, Li^+ gradually depolarizes membrane potential (Kiedrowski 1999;

Franceschetti et al. 2003), broadens action potential (Colino et al. 1998), positively shifts afterpotential (Franceschetti et al. 2003), progressively potentiates the high-frequency firing (Franceschetti et al. 2003), and presynaptically enhances excitatory synaptic transmission (Colino et al. 1998). Therefore, $NMDG^+$ and Li^+ substitution may have different effects on hypoxic/ischemic disruption of K^+ homeostasis if it is determined mainly by membrane depolarization rather than Na^+ ions themselves. In such a circumstance (same lowering external Na^+ but different effects on K^+ homeostasis), if the output of DOR agonist is the same, then it indicates that DOR plays its role primarily by targeting Na^+ influx, but not to affect action potential-based factors. Indeed, our results demonstrated that lowering $[Na^+]o$ by substituting it with impermeable $NMDG^+$ and permeable Li^+ from 152 to 140–90 mM has an opposite effect on anoxic K^+ derangement. Lowering Na^+ concentration by substituting with $NMDG^+$ causes a concentration-dependent attenuation of anoxic K^+ derangement, while lowering Na^+ concentration by substituting with Li^+ tends to potentiate the anoxic K^+ derangement (Chao et al. 2008). Most importantly, the DOR-induced protection against the anoxic K^+ responses is largely abolished by low Na^+ perfusion, regardless of the substituted cations (Chao et al. 2008). Together with our recent study showing that DOR activation inhibits Na^+ currents in DOR and Nav1.2 channel-cotransfected *Xenopus* oocytes (Kang et al. 2009), we conclude that DOR-induced protection against the anoxic K^+ responses in cortical slices is at least partially due to Na^+-based factors, probably by Na^+ channel inhibition.

To further confirm our conclusion, we used specific blockers and openers of Na^+ channels/receptors/exchangers to explore which pathway(s) mediating the Na^+ influx is the target of DOR signals on Na^+-based anoxic K^+ efflux (Chao et al. 2009). We found that the DOR attenuation of anoxic disruption of K^+ homeostasis is largely dependent on the inhibition of anoxic Na^+ influx mediated by voltage-gated Na^+ channels. Also, DOR activation inhibits Na^+ influx mediated by ionotropic glutamatergic NMDARs, but not those by non-NMDARs, though both of them play a role in anoxic disruption of K^+ homeostasis (Chao et al. 2009). In addition, we found that DOR activation has little effect on the NCX-based ionic responses to anoxia (Chao et al. 2009). Therefore, we conclude that in the cortex, (1) DOR activation attenuates anoxic disruption of K^+ homeostasis by restricting Na^+ entry through voltage-gated Na^+ channels and NMDAR channels, and (2) Non-NMDAR channels and NCX, though involved in anoxic disruption of K^+ homeostasis in certain degrees, are less likely the targets of DOR signals.

The above results suggested that DOR activation directly reduces Na^+ influx and plays an inhibitory role in the regulation of Na^+ channels. Indeed, our previous studies imply a potential interaction between DOR and Na^+ channels in the cortex. For example, after being exposed to prolonged hypoxia during postnatal development, cortical neurons are more sensitive to subsequent stress, which is largely attenuated by tetrodotoxin (TTX), a Na^+ channel blocker (Xia et al. 2000). Interestingly, an increased Na^+ channel density (Xia et al. 2000) and decreased DOR density (Xia et al. 2001) occurred in the exposed brain. Furthermore, we observed that in a mutant brain with epileptic seizures, cortical neurons are hyper-excitable with an up-regulation of voltage-gated Na^+ channels (Xia et al. 2003) and down-regulation of DOR expression (Zhao et al. 2005). All these observations, though circumstantial, suggest that DOR may mediate an inhibitory regulation of Na^+ channels in the brain under pathological conditions, and then influence Na^+ and K^+ homeostasis in the brain. To explore this possibility directly, we performed a series of experiments in mouse cortical slices and *Xenopus* oocytes with expression of Nav1.2 channels that are highly expressed in the cortex (Xia et al. 2003; Jarnot and Corbett 2006). The measurements were done using Na^+-selective microelectrodes and two-electrode voltage-clamp recording. We found that in *Xenopus* oocytes with co-expressed DOR and Na^+ channels, (1) Nav1.2 expression induced TTX-sensitive inward currents; (2) DOR expression reduced the inward currents; (3) Activation of DOR reduced the amplitude of the current and rightly shifted the activation curve of the current in the oocytes with both Nav1.2 and DOR, but not in oocytes with Nav1.2 alone; (4) The DOR agonist-induced inhibition of Nav1.2 currents was dose-dependent and saturable; and (5) The selective DOR agonist had no effect on naive oocytes. These data represent the first demonstration that activation of DOR inhibits Na^+ channel function by decreasing the amplitude of Na^+ currents and increasing its threshold of activation (Kang et al. 2009). Furthermore, direct measurement of $[Na^+]o$ with Na^+-selective microelectrodes indicated that DOR activation inhibits anoxia/ischemia-induced Na^+ influx in the cortical slices (Kang et al. 2009; Chao et al. 2012b). Therefore, DOR activation directly attenuates Na^+ influx induced by anoxia in the cortical slices (Kang et al. 2009; Chao et al. 2012b), which, in turn, attenuates Na^+-triggered massive K^+ leakage (Chao et al. 2008, 2009).

To address the role of TTX-sensitive VGSCs in DOR attenuation of anoxic/ischemic Na^+ influx, we performed two sets of experiments with Na^+ channel blocker TTX and opener veratridine (Chao et al. 2012b). First, we perfused slices with ACSF containing TTX with/without DOR agonist UFP 512 in normoxia for 20 min, and then began anoxia treatment to induce the drop of $[Na^+]o$. We found that TTX (1 μM) largely abolishes anoxic decrease in $[Na^+]o$, strongly supporting a key role of voltage-gated Na^+ channels in anoxic/ischemic Na^+ influx into neurons in the cortex. DOR agonist UFP 512 (1–5 μM) concentration-dependently attenuates anoxia/OGD-induced decrease in $[Na^+]o$, and is abolished by DOR antagonist naltrindole (1 μM). This finding suggests that DOR activation inhibits anoxia/OGD-induced Na^+ influx in the cortex. To verify DOR inhibition of Na^+ influx via Na^+ channels, we also perfused slices with veratridine, and found that perfusion of 1 μM veratridine itself for 20 min did not produce any obvious changes in $[Na^+]o$ in normoxia, but greatly enhanced anoxia-induced Na^+ influx, which could be greatly attenuated by DOR activation with 5 μM UFP 512 (Chao et al. 2012b). Increasing the concentrations of veratridine to 5–10 mM induced a drop in $[Na^+]o$ to the same degree even under normoxia, but with a much shorter period of perfusion when using 10 μM veratridine, and activation of DOR with UFP 512 (5 μM) significantly attenuates Na^+ influx induced by 5 μM, but not 10 μM, of veratridine under normoxia (Chao et al. 2012b). These results together clearly indicate that DOR activation attenuates Na^+ influx via Na^+ channels in the cortex not only in the condition of anoxia but also in that of ischemia, a clinically relevant situation in stroke.

Interestingly, hydrogen sulfide (H_2S), a known toxic environmental gaseous pollutant and recently recognized gasotransmitter, also concentration-dependently increases Na^+ influx via NMDARs and Na^+ channels in super-physiological concentrations (Chao et al. 2012a), which leads to the CNS lesions sharing some common features with anoxia/ischemia (Hooser et al. 2000; Solnyshkova et al. 2004; Cheung et al. 2007). However, our data suggest that DOR activation cannot attenuate H_2S-evoked Na^+ influx owing to the difference in target and signaling molecules of DOR involved in hypoxia from those of H_2S exposure (Chao et al. 2012a). This result reaffirms that DOR activation specifically attenuates hypoxic/ischemic (vs. H_2S-evoked) disruption of ionic homeostasis (such as Na^+ influx and K^+ leakage) via PKC dependent, PKA-independent pathway (the protein kinase mechanisms will be discussed in the next section) as well as in part by inhibition of Ca^{2+} entry in the cortex (Chao et al. 2007a, b, 2012b; Pamenter and Buck 2008; Kang et al. 2009).

It should be mentioned that in our studies of Na^+-based factors, the indirect effect of DOR on Na^+/K^+ ATPase activity is not completely excluded. A recently study suggested a direct protein-protein interaction between DOR and Na^+ pumps since these two membrane proteins not only co-localize in oocyte model system but also in neuronal cells, and inhibitions of Na^+ pumps, in turn, modulates agonist sensitivity of DOR (Deng et al. 2009). DOR activation inhibits Na^+ currents and Na^+ influx during normoxia and anoxia (Kang et al. 2009; Chao et al. 2012b). A reduction of intracellular Na^+ accumulation decreases energy consumption of Na^+/K^+ ATPase that is responsible for at least 60 % of energy expenditure of nerve cells in the CNS (Schreiner and Rose 2012), and thus may increase K^+ uptake during hypoxia/ischemia and contribute to DOR attenuation of hypoxic/ischemic K^+ derangement.

The relationship between Na^+ and Ca^{2+} in terms of their role in DOR protection against hypoxic/ischemic disruption of K^+ homeostasis remains unclear. In the subsequent Na^+-triggered activities, it is unknown as to which cellular activity(ies) is affected by DOR activation during the attenuation of hypoxic/ischemic disruption of K^+ homeostasis. However, some clues showed that DOR activation may also reduce Ca^{2+} influx in order to attenuate anoxic disruption of K^+ homeostasis (Chao et al. 2007a; Pamenter and Buck 2008). Indeed, anoxia/ischemia triggers a great amount of Na^+ and Ca^{2+} influx, while DOR activation inhibits both Ca^{2+} current (Toselli et al. 1999; Acosta and López 1999) and Na^+ current (Kang et al. 2009). Since Na^+ influx plays a prominent role in ischemia-induced depolarization (Calabresi et al. 1999), and triggers cytosolic $[Ca^{2+}]$ elevations in neurons (Friedman and Haddad 1993; Zhang and Lipton 1999; Nikolaeva et al. 2005; Kiedrowski 2007; Vander Jagt et al. 2008), the anoxia-induced increase in Na^+ influx may serve as an up-stream signal for the elevation of cytosolic Ca^{2+} in the cortex. DOR activation may inhibit the ionic reactions in series, though we cannot rule out the independent actions of DOR on Na^+ and Ca^{2+} in the stabilization of ionic homeostasis in hypoxia/ischemia.

5.2 Protein Kinase Mechanism

DOR belongs to a family of G-protein-coupled receptors, whose effect is mediated by G proteins and G-protein-dependent cytoplasmic second messengers involving protein kinases (Law et al. 2000; Ma et al. 2005; Narita et al. 2006). PKA and PKC are important signaling molecules in a variety of cellular functions, including modulation of neurotransmitter release, regulation of ion channels and enzymes, control of growth and differentiation, and modification of neuronal plasticity (Majewski and Iannazzo 1998; Leenders and Sheng 2005). In addition to the involvement in normal physiological events, these two kinases have been shown to play an important role in pathophysiological events such as cellular responses to hypoxia/ischemia (Tanaka 2001; Selvatici et al. 2002; Libien et al. 2005). Our studies and those of others have shown that DOR regulates PKA and PKC activities under certain conditions (Lou and Pei 1997; Yao et al. 2003; Ma et al. 2005), suggesting the involvement of these protein kinases in DOR signaling. In consistence with our previous work showing that DOR-mediated neuroprotection induced by hypoxic preconditioning could be blocked by PKC inhibition, but not by PKA inhibition (Ma et al. 2005), we demonstrated that PKC plays an important role in signal transduction of the DOR-mediated regulation of K^+ and Na^+ homeostasis (Chao et al. 2007b, 2012b). In our work (Chao et al. 2007b), we could not demonstrate the involvement of PKA in the DOR protection against anoxic increase in $[K^+]e$ because blocking PKA with N-(2-[p-Bromocinnamylamino]-ethyl)-5-isoquinolinesulfonamide dihydrochloride did not result in any change in DOR effect. In sharp contrast, PKC inhibition with chelerythrine reversed the protective effect of DOR activation, while PKC activation by phorbol 12-myristate 13-acetate mimicked it. Collectively, these data suggest that the effect of DOR activation on anoxic disruption of K^+ homeostasis is mainly via a PKC-dependent, but PKA-independent, pathway. Phosphorylation of K^+ channels by PKC activation reduces potassium conductance of neurons (Grega et al. 1987), which may partially lead to the blockade of voltage-dependent, rapidly inactivating K^+ efflux (Chaki et al. 1994). Besides its direct regulation of Na^+ and K^+ homeostasis, PKC signaling resulting from DOR activation may decrease anoxia-induced glutamate release (Selvatici et al. 2002) and in turn reduce K^+ efflux (Yu et al. 1999; Lopachin et al. 2001; Xiao et al. 2001). The interaction between DOR and PKA is worthy of further investigation.

In addition, voltage-gated Na^+ channels are important targets modulated by metabotropic receptors via G protein/protein kinases, including PKC (West et al. 1991; Dascal and Lotan 1991; Chen et al. 2006). Therefore, it is possible that PKC functions as a "signal bridge" between DOR and Na^+ channels, and mediates the effects of DOR on Na^+ channels in hypoxia/ischemia, which has been shown to be a key target of DOR in the regulation of hypoxia/ischemia-induced K^+ derangement and Na^+ influx (Chao et al. 2009, 2012b; Kang et al. 2009). We found that DOR attenuation of anoxia-induced K^+ efflux and Na^+ influx are abolished by chelerythrine, a broad-spectrum PKC blocker (Chao et al. 2007b, 2012b). Intriguingly, activation of PKC has been demonstrated to modulate the activities of Na^+ channels by phosphorylation of certain sites on Na^+ channel subunits leading to the inhibition of Na^+ currents (West et al. 1991; Dascal and Lotan 1991; Chen et al. 2006), and preventing ischemia/hypoxia-associated neuronal loss and dendrite spine and synapse damage (Sun et al. 2008). Since DOR activation attenuates hypoxic/ischemic K^+ derangement primarily by targeting Na^+ channels to reduce Na^+ entry in the cortex (Chao et al. 2009, 2012b; Kang et al. 2009), our work suggests that PKC is very likely the link between DOR and Na^+ channels, regulating Na^+ entry through Na^+ channels during anoxia/ischemia. Indeed, Fazzina et al. (2010) reported that activation of PKC with phorbol ester significantly reduced brain water concentration in the infarction area (brain edema) in a rodent stroke model, which was associated with the reduced shift of Na^+ that normally accompanies cell swelling and brain edema (Churchwell et al. 1996; Ayata and Ropper 2002). In addition, a recent study indicated that activation of opioid receptors including DOR significantly decreases water contents (brain edema) in OGD-exposed hippocampal slices (Yang et al. 2011). Taken together, these findings support the idea that DOR activation attenuates anoxic Na^+ influx through Na^+ channels via a PKC-dependent pathway in the cortex.

The PKC family of serine/threonine kinases consists of at least ten different isozymes that exist in the brain (Bright and Mochly-Rosen 2005). Based on their structure and regulatory properties, these isozymes have been classified into three groups: conventional PKCs (α, βI, βII, γ) that are activated by diacylglycerol (DAG) and Ca^{2+}, novel PKCs (δ, ε, θ, η) that are activated by DAG but not Ca^{2+}, and atypical PKCs (ζ, λ/ι) that are insensitive to both DAG and Ca^{2+} (Chou and Messing 2005). Various PKC isozymes mediate different and sometimes opposing functions after activation by the same stimulus (Chen et al. 2001a; Wang et al. 2004). Although PKCβII and PKCθ isozymes belong to different groups, blockade of them with their specific inhibitors has similar effects on DOR signaling in inhibiting the functions of Na^+ channels and Na^+ entry, namely that DOR attenuation of anoxic Na^+ influx via Na^+ channels is abolished by blockade of PKCβII and PKCθ isozymes (Chao et al. 2012b).

Therefore, PKC plays a key role in DOR neuroprotection against hypoxic/ischemic disruption of ionic homeostasis.

5.3 *Decreased Excitability and Excitatory Transmission*

In addition, DOR-induced protection against the anoxic disruption of ionic homeostasis may involve other mechanisms. In most cases, DOR activation inhibits neuronal electrical activity and synaptic transmission (see the preceding chapter by Chao and Xia). DOR activation may decrease neuronal excitability and consequently reduce K^+ leakage, and may regulate iGluR channel-mediated K^+ flux. Indeed, in rat cerebral cortical primary cultures, naloxone increases glutamate and aspartate release, suggesting that endogenous opioids partially inhibit glutamate and aspartate release (Vlaskovska et al. 1997). Electrophysiological study indicated that inhibition of DOR with naltrindole potentiates NMDAR currents in cortical neurons (Pamenter and Buck 2008), and activation of DOR depresses both spontaneous and stimulus-evoked action potential discharges as well as the amplitudes of stimulus-evoked excitatory postsynaptic potentials/currents (EPSPs/EPSCs) of neocortical neurons (Stanzione et al. 1989; Tanaka and North 1994; Ostermeier et al. 2000). These findings suggest that DOR inhibits glutamate release and subsequent iGluR activation in cortical neurons. This depression at least partially reduces K^+ leakage induced by an increase in excitability and action potential generation in cortical neurons. In hippocampal slices, hypoxia-induced sharp $[K^+]e$ increase is significantly dampened by decreased neuronal excitability or by cease of spontaneous and evoked impulse firings with perfusion of TTX or lidocaine (Müller and Somjen 2000a, b; Raley-Susman et al. 2001). Another potential mechanism underlying the DOR attenuation of anoxic disruption of K^+ homeostasis involves the regulation of glutamate signaling. Immunocytochemical studies reveal that DOR is localized at presynaptic terminals in a variety of neurons, including mammalian cortical neurons (Bausch et al. 1995; Svingos et al. 1995). DOR activation can prevent the release of glutamate from presynaptic vesicles, thereby reducing glutamate excitability (Tanaka and North 1994; Ostermeier et al. 2000), and suppressing NMDA receptor activity in neurons (Wang and Mokha 1996; Pamenter and Buck 2008). It has been demonstrated that activation of iGluRs of mouse cortical neurons, even in a condition of decreased inward cation influx by lowering extracellular Na^+ and Ca^{2+}, induces a large outward K^+ current, which causes loss of approximately 50–80 % intraneuronal K^+ and great shrinkage of cell body and consequently neuronal apoptosis (Yu et al. 1999; Xiao et al. 2001). Therefore, DOR inhibition of glutamate release and subsequent iGluR activation may reduce hypoxic/ischemic K^+ efflux via iGluRs. In addition, DOR may increase EAAT expression and function, and increase glutamate uptake in astrocytes (Liang et al. 2014), thus limiting glutamate spillover and subsequent over-activation of extrasynaptic NMDARs (Dong et al. 2013) that is proven to play a major role in NMDAR toxicity (Parsons and Raymond 2014). DOR activation thus potentially decreases hyperexcitability-induced K^+ leakage through iGluRs in neurons. In addition, DORs mediate NMDAR activity in a Gi-dependent manner and prevent deleterious NMDAR-mediated $[Ca^{2+}]i$ influx during anoxic insults in the cortex (Pamenter and Buck 2008), which may indirectly decrease K^+ efflux via BK channels (Chao et al. 2007a).

The DOR attenuation of anoxic disruption of K^+ homeostasis might involve multiple strategies involved in ionic homeostasis in the neurons although the precise mechanisms are still not well understood yet at present. Based on today's information, the possible mechanisms of DOR-mediated regulation of K^+ homeostasis during hypoxia/ischemia are schematically summarized in Fig. 6.4.

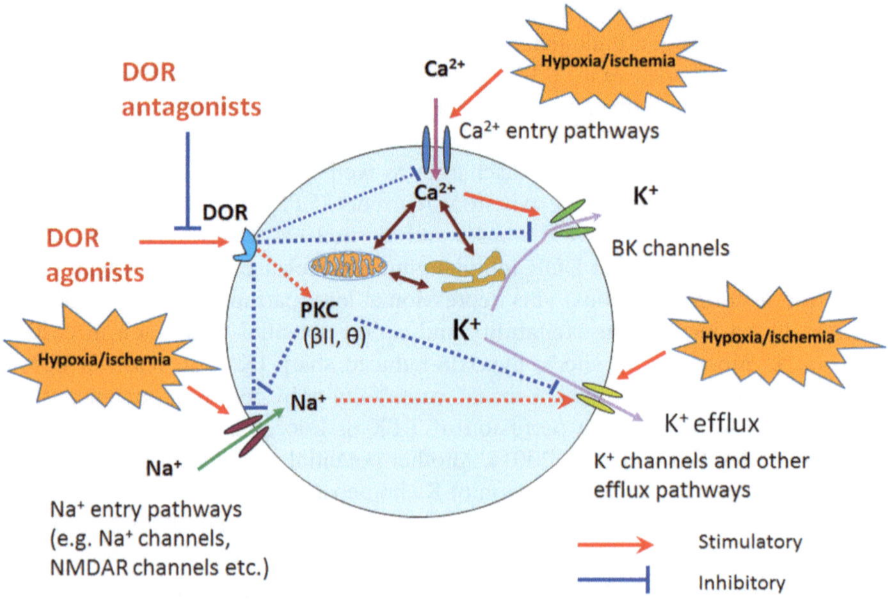

Fig. 6.4 Schematic demonstration of mechanisms of DOR attenuation of anoxic/ischemic disruption of K^+ and Na^+ homeostasis. Our data showed that the absence of extracellular Ca^{2+} and blockade of Ca^{2+}-activated K^+ channels (BK) greatly attenuated the anoxia-induced disruption of K^+ homeostasis, suggesting that the attenuation of DOR activation against anoxic disruption of K^+ homeostasis in the cortex is likely attributed to an inhibition of hypoxia-induced increase in Ca^{2+} entry-BK channel activities. Also, DOR-induced protection against the anoxic K^+ responses was largely abolished by low Na^+ perfusion with either impermeable N-methyl-D-glucamine or permeable Li^+, substitution, and blockade of voltage-gated Na^+ channels and NMDAR channels. While non-NMDAR channels and Na^+/Ca^{2+} exchangers are involved in anoxic disruption of K^+ homeostasis in certain degrees, they are less likely the targets of DOR signals. A PKC (especially PKCβII and PKCθ isozymes)-dependent, PKA-independent, pathway is also involved in the attenuation of anoxic disruption of K^+ and Na^+ homeostasis in the cortex. This was concluded because we could not demonstrate the involvement of PKA in the DOR protection against anoxic increase in extracellular K^+. Blocking PKA with H89 did not result in any changes in the DOR effect. In sharp contrast, PKC inhibition with chelerythrine as well as PKCβII and PKCθ isozymes inhibition reversed the protective effect of DOR activation against hypoxic/ischemic K^+ and Na^+ disarrangement, while PKC activation by phorbol 12-myristate 13-acetate mimicked it. Cited from Chao and Xia (2015) with permission of the journal

6 Implications of DOR Neuroprotection against Hypoxic/Ischemic Disruption of Ionic Homeostasis

Neuronal injury and death often occur as outcomes of hypoxic/ischemic stress (Lipton 1999; Zhang et al. 2004; Sung et al. 2008; Chen et al. 2009). Cell death was first classified as apoptosis and necrosis based on histologic criteria. Necrosis is generally a form of cell death that rapidly occurs in response to severe insults such as anoxia and cell trauma, and is characterized by swelling of the cell body and mitochondria, followed by vacuolization of the cytoplasm, random breakdown of DNA, rupture of the cell membrane, and cell lysis. In contrast, cells undergoing apoptosis share similar characteristic morphologies such as cell shrinkage, nuclear/chromatin condensation, internucleosomal cleavage of DNA, membrane blebbing, and formation of apoptotic bodies, which is controlled by an internally encoded death program executed by activation of endogenous proteases and endonucleases (Majno and Joris 1995; Kroemer et al. 1995; Banasiak et al. 2000). Distinguishing between apoptosis and necrosis is not always straightforward. It is sometimes very difficult to categorize an individual cell as apoptotic or necrotic using morphological or even biochemical parameters in the brain because necrosis, apoptosis, and even hybrid death always occur concurrently in stress (Tan et al. 1998; Xiao et al. 2002). The Nomenclature Committee on Cell Death (NCCD) recently proposed a set of recommendation for the definition of cell death and its different morphologies based on the biochemical and molecular criteria, rather than the morphological criteria (Kroemer et al. 2009; Galluzzi et al. 2012). According to NCCD, types of cell death can be classified as apoptosis, necrosis, autophagic cell death, mitotic catastrophe etc., with some subtypes (Kroemer et al. 2009; Galluzzi et al. 2012). For example, apoptosis includes "extrinsic apoptosis" and caspase-dependent and caspase-independent "intrinsic apoptosis". Necrosis has long been considered a merely accidental cell death mechanism with the absence of morphological traits of apoptosis or autophagy, but it is now clear that necrosis can occur in a regulated manner, which has been termed as "necroptosis" or "regulated necrosis" (Galluzzi et al. 2012). Despite the complex classifications, substantial evidence indicates that increased $[Na^+]i$ and $[Ca^{2+}]i$, and intracellular K^+ depletion induce both necrosis and apoptosis in various cell types (Friedman and Haddad 1993; Yu et al. 1997, 1999; Kristián and Siesjő 1998; Warny and Kelly 1999; Banasiak et al. 2000, 2004; Koike et al. 2000; Barros et al. 2002; Yu 2003b; Liu et al. 2003; Yamashima 2004; Wozniak et al. 2006; Petrat et al. 2006; Zhao et al. 2006b; Mongin 2007; Fang et al. 2008).

Indeed, excessive disruption of ionic homeostasis is the initial and key step in hypoxic/ischemic neuronal injury and death (see reviews by Lipton 1999; Banasiak et al. 2000; Yu 2003b; Sung et al. 2008). However, each ion (e.g., Na^+, K^+, Ca^{2+}) may play a different role in the processes. Large Na^+ influx induces cellular injury manifested by acute functional and morphological changes, e.g., loss of electrophysiological response to stimulus, cell swelling, bleb formation, and membrane injury (Friedman and Haddad 1993, 1994; Hasbani et al. 1998; Itoh et al. 1998; Calabresi et al. 1999; Shi et al. 2005; Petrat et al. 2006), which are characteristic of necrosis. Removal of extracellular Na^+ or blockade of Na^+ entry therefore prevents hypoxic/ischemic neuronal damage and death (Friedman and Haddad 1993, 1994; Lynch et al. 1995; Probert et al. 1997; Breder et al. 2000; Raley-Susman et al. 2001; Banasiak et al. 2004). Recently, the Na^+ dynamics in glial cells is considered a key signal for coupling membrane signaling, metabolic pathway, and neuronal activity (Kirischuk et al. 2012; Rose and Karus 2013). Thus, Na^+ overload-associated cell death has profound influence on brain function. In contrast, an increased level of intracellular Ca^{2+} is not necessary for anoxic/excitotoxic morphological changes as Na^+ does (Friedman and Haddad 1994; Hasbani et al. 1998). However, Ca^{2+} entry and overload, particularly uncontrolled entry in the later phase of hypoxic/ischemic stress, can induce severe damage to nerve cells and lead to delayed neuronal death (Friedman and Haddad 1994; Hasbani et al. 1998; Nakamura et al. 1999; Aarts et al. 2003; Xiong et al. 2004; Bano et al. 2005; Gao et al. 2005), which always involves mitochondrial dysfunction and ER stress (see reviews by Szydlowska and Tymianski 2010; Gouriou et al. 2011; Bodalia et al. 2013), and shares many biochemical and molecular characteristics intrinsic to apoptosis (Galluzzi et al. 2012). K^+ efflux is beneficial to neurons as an adaptive mechanism for counteracting hypoxic/ischemic stress in a very limited range, and is harmful to neurons when intracellular K^+ loss is excessive. K^+ depletion has been shown to be actively implicated in the changes in morphological features of necrosis in monocytic cells (Warny and Kelly 1999). In most cases, however, excessive K^+ efflux from neurons may induce cell shrinkage, cytochrome c release, and degradation of DNA and nuclear proteins—typical features of apoptosis (Bortner et al. 1997; Yu et al. 1997, 1999; D'Ambrosio et al. 1999; Xiao et al. 2001, 2002; Lauritzen et al. 2003; Yu 2003b; Remillard and Yuan 2004; Karki et al. 2007). The subsequent cellular K^+ uptake occurs mainly in astrocytes, which involves activation of Na^+/K^+ ATPase and NKCC1 (Walz 2000, 2004; Kofuji and Newman 2004; Wetherington et al. 2008). Activation of NKCC1 leads to intracellular Na^+ overload (Sheldon et al. 2004; Lenart et al. 2004; Kintner et al. 2007; Chen et al. 2008; Cengiz et al. 2014) and subsequently increases $[Ca^{2+}]i$ via the operation of plasmalemmal NCX in reverse (Zhang and Lipton 1999; Bondarenko et al. 2004; Sheldon et al. 2004; Martinez-Sánchez et al. 2004; Luo et al. 2005, 2008; Nikolaeva et al. 2005; Pedersen et al. 2006; Rojas et al. 2007; Kiedrowski 2007; Chen et al. 2007b; Cengiz et al. 2014). Activation of NKCC1 also causes intracellular accumulation of Na^+, K^+, and Cl^-, as well as osmotically driven water uptake and cell swelling, which contribute to ischemia-induced brain edema (Pond et al. 2006; Hertz 2008; Salter and Fern 2008). Therefore, blockade of K^+ efflux or other strategies for preventing excessive K^+ efflux as well as Na^+ and Ca^{2+} overload has been shown to prevent neuronal death in various pathological conditions including hypoxia/ischemia (Yu et al. 1997, 1999; Huang et al. 2001; Wei et al. 2003; Liu et al. 2003; Lauritzen et al. 2003; Zhao et al. 2006b; Karki et al. 2007; Nistico et al. 2007). It has been found that inhibition of excessive K^+ flux and maintaining cellular K^+ homeostasis is of therapeutic benefit in the treatment of stroke and ischemia-related neurodegenerative conditions (Wei et al. 2003; Liu et al. 2003; Zhao et al. 2006b; Karki et al. 2007; Nistico et al. 2007).

In general, among the major cations, $[Na^+]i$ rise mainly contributes to hypoxia/ischemia-induced rapid cell swelling and necrosis, often seen as brain edema (Churchwell et al. 1996; Ayata and Ropper 2002), and occurs at the core region of infarct. $[Ca^{2+}]i$ increase is mainly responsible for the delayed neuronal death, and K^+ efflux is beneficial to neurons in counteracting hypoxic stress in a limited range, and is harmful in the case of excessive K^+ loss during hypoxic/ischemic stress. No matter how these cations cause neuronal injury and death, stabilizing hypoxic/ischemic disruption of ionic homeostasis may be of therapeutic benefit in the treatment of stroke and hypoxia/ischemia related neurological disorders (Wei et al. 2003; Liu et al. 2003; Karki et al. 2007; Zhao et al. 2006b; Nistico et al. 2007). In this aspect,

DOR protection against the disruption of ionic homeostasis, particularly disruption of Na^+ and K^+ homeostasis, seems to have significant impact on prevention against hypoxia/ischemia-induced neuronal injury. Indeed, it has been shown that activation of opioid receptors, including DOR, significantly decreases water contents (brain edema) in OGD-exposed hippocampal slices (Yang et al. 2011), and protects both neurons and glial cells against hypoxic/ischemic injury and death (see Sect. 2).

Furthermore, we and others have demonstrated DOR-induced neuroprotection not only in hypoxia/ischemia (see Sect. 2), but also in other conditions including cytotoxic and oxidative stress (Zhang et al. 2000; Borlongan et al. 2000; Hayashi et al. 2002; Wallace et al. 2006; Narita et al. 2006), and even in drug and ethanol consumption (Margolis et al. 2008). Although multiple mechanisms exist (Chao and Xia 2010; He et al. 2013; Staples et al. 2013), DOR stabilization of ionic homeostasis plays an important role in DOR-mediated neuroprotection, especially in the early stage of environmental stress. Since neuronal responses to stress vary depending on the duration and severity of stress and survival time of neurons following hypoxia/ischemia (Lipton 1999), DOR-mediated regulation may occur at various levels to protect neurons from short- or long-term stress. In the initial stage of hypoxia, DOR signals maintain ionic homeostasis (Chao et al. 2007a, b, 2008, 2009, 2012b; Pamenter and Buck 2008; Kang et al. 2009), while during prolonged hypoxia, DOR signals up-regulate the expression and function of benefit-signaling molecules and down-regulate those involved in cell injury or death (Ma et al. 2005; Peng et al. 2009; He et al. 2013). Nevertheless, the stabilization of ionic homeostasis in the initial stage of hypoxia/ischemia is clearly an important aspect of DOR-mediated neuroprotection, which may have significant impacts on clinical strategies against hypoxic/ischemic injury in the brain.

7 Concluding Remarks

Maintenance of ionic homeostasis is critical to maintain the normal functions of neurons (see the preceding chapter by Chao and Xia), and the disruption of ionic homeostasis is a key trigger for neuronal injury and death. The brain possesses homeostatic mechanisms that maintain constant ionic concentrations in the cerebrospinal fluid and neurons (Hansen 1985; Kofuji and Newman 2004). However, such homeostatic mechanisms can be disrupted in certain pathophysiological conditions, particularly in energy depletion. For example, hypoxia/ischemia can produce considerably longer or sustained changes in ionic concentrations that are characterized by enhanced K^+ efflux and Na^+-, Ca^{2+}- and Cl^- influx. The hypoxic/ischemic disruption of ionic homeostasis is an initial and key step for hypoxic/ischemic neuronal injury and death. Thus limiting hypoxic/ischemic ion dysregulation, and stabilization of ionic homeostasis in the initial stage of hypoxia/ischemia are an efficient strategy in the treatment of stroke and ischemia-related neurodegenerative conditions (Kristián and Siesjő 1998). Substantial studies from our and many independent laboratories have demonstrated that DOR, a member of opioid receptor superfamily, is neuroprotective against hypoxic/ischemic insults. In the mechanistic exploration, we found that DOR activation and/or expression attenuate hypoxic/ ischemic disruption of ionic homeostasis.

In this chapter, we first updated the recent progress in the studies of DOR neuroprotection against hypoxic/ischemic insults and the features and potential mechanisms of hypoxia/ischemia-induced disruption of major cation homeostasis. On the basis of this discussion, we addressed the important role of DOR in the homeostatic maintenance of these ions and the underlying mechanisms. Based on our recent work and current literatures, we have drawn an initial picture of how DOR protects neurons against hypoxia/ischemia through the regulation of ionic homeostasis. In brief, DOR signals inhibit Na^+ influx and reduce the increase in intracellular Ca^{2+}, thus decreasing the excessive K^+ leakage, which involve a PKC-dependent and PKA-independent signaling pathway.

However, we are far away from the full understanding of DOR-mediated regulation of ionic homeostasis. Studies with immunocytochemical, in situ hybridization, autoradiography methods, and fluorescent knock-in mice that express a functional DOR, revealed that DOR and its cognate endogenous opioids are widely expressed throughout the CNS, suggesting an ubiquitous role of DOR in the regulation of neural activities. In the nerve system, DOR has extensive interaction with ion channels, receptors, and transporters. DOR can regulate the release of many neurotransmitters, and modify neuronal electrical activities and synaptic plasticity (see the preceding chapter by Chao and Xia). Hypoxic/ischemic disruption of ionic homeostasis is an integrative result, and involves functional membrane proteins (channels, receptors, transporters, pumps), neurochemicals (neurotransmitters, signaling molecules) and organelles (mitochondrial, ER, even nucleus) (see the previous discussion in this chapter). Therefore, DOR regulation of hypoxic/ischemic homeostasis might involve multifaceted mechanisms. However, it is still largely unkown with respect to this aspect. In addition, the brain composes a myriad of morphologically and functionally diverse cells, which have different sensitivities and vulnerabilities to hypoxic/ischemic stress. Therefore, distinguishing their role in hypoxic/ischemic disruption of ionic homeostasis and DOR action on them is of great importance in understanding the mechanisms of DOR neuroprotection. Fortunately, many novel and emerging technologies can help to complete our understanding. For example, targeted fluorescent-DOR-knockin (Scherrer et al. 2006), DOR-knockout (Zhu et al. 1999; Filliol et al. 2000), and transgenic DOR-overexpression mice (Chao et al. 2006) provide powerful tools for DOR research. Also, genetically encoded ion fluorescence indicators bring several remarkable advantages in studies of ion activities in targeting specific cell types and specific subcellular compartments, allowing long-term recording from the same cellular structures, reliable detection of neuronal activity with larger fluorescence signals and higher temporal resolution, and large-scale recording of ion dynamics from hundreds of postsynaptic spines and presynaptic boutons in a cell (Mank et al. 2008; Dreosti et al. 2009; Tian et al. 2009; Ohkura et al. 2012; Gouriou et al. 2013; Chen et al. 2013; Kobayashi et al. 2014). Combining the technique of genetically encoded ion fluorescence indicators with the powerful tools for high-resolution, time-lapse in vivo as well as in vitro ion imaging (e.g. two-photo/multiphoton laser scanning microscopy) (Helmchen and Denk 2002, 2005; Helmchen et al. 2013), draws a promising line of research in DOR regulation of ion homeostasis in hypoxia/ischemia as well as in normoxia.

Acknowledgements This work was supported by NIH (AT-004422, and HD-034852) and Vivian L Smith Neurologic Foundation.

References

Aarts M, Iihara K, Wei WL, Xiong ZG, Arundine M, Cerwinski W, MacDonald JF, Tymianski M (2003) A key role for TRPM7 channels in anoxic neuronal death. Cell 115:863–877

Abdul Y, Akhter N, Husain S (2013) Delta-opioid agonist SNC-121 protects retinal ganglion cell function in a chronic ocular hypertensive rat model. Invest Ophthalmol Vis Sci 54:1816–1828

Abdul-Ghani MA, Valiante TA, Carlen PL, Pennefather PS (1996) Metabotropic glutamate receptors coupled to IP3 production mediate inhibition of IAHP in rat dentate granule neurons. J Neurophysiol 76:2691–2700

Abramov AY, Duchen MR (2010) Impaired mitochondrial bioenergetics determines glutamate-induced delayed calcium deregulation in neurons. Biochim Biophys Acta 1800:297–304

Acosta CG, López HS (1999) δ-Opioid receptor modulation of several voltage-dependent Ca^{2+} currents in rat sensory neurons. J Neurosci 19:8337–8348

Adams HP Jr, Olinger CP, Barsan WG, Butler MJ, Graff-Radford NR, Brott TG, Biller J, Damasio H, Tomsick T, Goldberg M (1986) A dose-escalation study of large doses of naloxone for treatment of patients with acute cerebral ischemia. Stroke 17:404–409

Agostinho P, Duarte CB, Oliveira CR (1996) Intracellular free Na^+ concentration increases in cultured retinal cells under oxidative stress conditions. Neurosci Res 25:343–351

Aldinucci C, Carretta A, Pessina GP (2005) The effect of mild and severe hypoxia on rat cortical synaptosomes. Neurochem Res 30:981–987

Aley PK, Murray HJ, Boyle JP, Pearson HA, Peers C (2006) Hypoxia stimulates Ca^{2+} release from intracellular stores in astrocytes via cyclic ADP ribose-mediated activation of ryanodine receptors. Cell Calcium 39:95–100

Alvarez-Leefmans FJ, Cruzblanca H, Gamino SM, Altamirano J, Nanl A, Reuss L (1994) Transmembrane ion movements elicited by sodium pump inhibition in helix aspersa neurons. J Neurophysiol 71:1787–1796

Amarjargal N, Mazurek B, Haupt H, Andreeva N, Fuchs J, Gross J (2008) Effects of SERCA and PMCA inhibitors on the survival of rat cochlear hair cells during ischemia in vitro. Physiol Res 57:631–638

Ammon-Treiber S, Stolze D, Schröder H, Loh H, Höllt V (2005) Effects of opioid antagonists and morphine in a hippocampal hypoxia/hypoglycemia model. Neuropharmacology 49:1160–1169

Anderson CM, Swanson RA (2000) Astrocyte glutamate transport: review of properties, regulation, and physiological functions. Glia 32:1–14

Anderson TR, Jarvis CR, Biedermann AJ, Molnar C, Andrew RD (2005) Blocking the anoxic depolarization protects without functional compromise following simulated stroke in cortical brain slices. J Neurophysiol 93:963–979

Andrade AL, Rossi DJ (2010) Simulated ischaemia induces Ca^{2+}-independent glutamatergic vesicle release through actin filament depolymerization in area CA1 of the hippocampus. J Physiol 588:1499–1514

Annunziato L, Pignataro G, Boscia F, Sirabella R, Formisano L, Saggese M, Cuomo O, Gala R, Secondo A, Viggiano D, Molinaro P, Valsecchi V, Tortiglione A, Adornetto A, Scorziello A, Cataldi M, Di Renzo GF (2007) ncx1, ncx2, and ncx3 gene product expression and function in neuronal anoxia and brain ischemia. Ann N Y Acad Sci 1099:413–426

Arbeloa J, Pérez-Samartín A, Gottlieb M, Matute C (2012) P2X7 receptor blockade prevents ATP excitotoxicity in neurons and reduces brain damage after ischemia. Neurobiol Dis 45:954–961

Armstead WM (1998) Nitric oxide contributes to opioid release from glia during hypoxia. Brain Res 813:398–401

Armstrong GA, Xiao C, Krill JL, Seroude L, Dawson-Scully K, Robertson RM (2011) Glial Hsp70 protects K^+ homeostasis in the Drosophila brain during repetitive anoxic depolarization. PLoS One 6, e28994

Asai S, Kunimatsu T, Zhao H, Nagata T, Takahashi Y, Ishii Y, Kohno T, Ishikawa K (2000) Two distinct components of initial glutamate release synchronized with anoxic depolarization in rat global brain ischemia. Neuroreport 11:2947–2952

Ayata C, Ropper AH (2002) Ischaemic brain aedema. J Clin Neurosci 9:113–124

Bahar E, Fayuk D, Somjen GG, Aitken PG, Turner DA (2000) Mitochondrial and intrinsic optical signals imaged during hypoxia and spreading depression in rat hippocampal slices. J Neurophysiol 84:311–324

Bai HY, Li AP (2013) $P2X_7$ receptors in cerebral ischemia. Neurosci Bull 29:390–398

Balboni G, Salvadori S, Guerrini R, Negri L, Giannini E, Jinsmaa Y, Bryant SD, Lazarus LH (2002) Potent δ-opioid receptor agonists containing the Dmt-Tic pharmacophore. J Med Chem 45:5556–5563

Balestrino M, Young J, Aitken P (1999) Block of (Na^+, K^+)ATPase with ouabain induces spreading depression-like depolarization in hippocampal slices. Brain Res 838:37–44

Ballanyi K (2004) Protective role of neuronal KATP channels in brain hypoxia. J Exp Biol 207:3201–3212

Ballanyi K, Grafe P, ten Bruggencate G (1983) Intracellular free sodium and potassium, post-carbachol hyperpolarization, and extracellular potassium-undershoot in rat sympathetic neurons. Neurosci Lett 38:275–279

Ballanyi K, Kulik A (1998) Intracellular Ca^{2+} during metabolic activation of KATP channels in spontaneously active dorsal vagal neurons in medullary slices. Eur J Neurosci 10:2574–2585

Banasiak KJ, Burenkova O, Haddad GG (2004) Activation of voltage-sensitive sodium channels during oxygen deprivation leads to apoptotic neuronal death. Neuroscience 126:31–44

Banasiak KJ, Xia Y, Haddad GG (2000) Mechanisms underlying hypoxia-induced neuronal apoptosis. Prog Neurobiol 62:215–249

Bancila M, Copin JC, Daali Y, Schatlo B, Gasche Y, Bijlenga P (2011) Two structurally different T-type Ca^{2+} channel inhibitors, mibefradil and pimozide, protect CA1 neurons from delayed death after global ischemia in rats. Fundam Clin Pharmacol 25:469–478

Bano D, Young KW, Guerin CJ, Lefeuvre R, Rothwell NJ, Naldini L, Rizzuto R, Carafoli E, Nicotera P (2005) Cleavage of the plasma membrane Na+/Ca2+ exchanger in excitotoxicity. Cell 120:275–285

Bargiotas P, Krenz A, Hormuzdi SG, Ridder DA, Herb A, Barakat W, Penuela S, von Engelhardt J, Monyer H, Schwaninger M (2011) Pannexins in ischemia-induced neurodegeneration. Proc Natl Acad Sci U S A 108:20772–20777

Barros LF, Castro J, Bittner CX (2002) Ion movement in cell death: from protection to execution. Biol Res 35:209–214

Baughman JM, Perocchi F, Girgis HS, Plovanich M, Belcher-Timme CA, Sancak Y, Bao XR, Strittmatter L, Goldberger O, Bogorad RL, Koteliansky V, Mootha VK (2011) Integrative genomics identifies MCU as an essential component of the mitochondrial calcium uniporter. Nature 476:341–345

Bausch SB, Patterson TA, Appleyard SM, Chavkin C (1995) Immunocytochemical localization of δ opioid receptors in mouse brain. J Chem Neuroanat 8:175–189

Bay V, Butt AM (2012) Relationship between glial potassium regulation and axon excitability: a role for glial Kir4.1 channels. Glia 60:651–660

Behringer EJ, Leite LD, Buchholz NE, Keeney MG, Pearce WJ, Vanterpool CK, Wilson SM, Buchholz JN (2009) Maturation and long-term hypoxia alters Ca^{2+}-induced Ca^{2+} release in sheep cerebrovascular sympathetic neurons. J Appl Physiol 107:1223–1234

Bennay M, Langer J, Meier SD, Kafitz KW, Rose CR (2008) Sodium signals in cerebellar Purkinje neurons and Bergmann glial cells evoked by glutamatergic synaptic transmission. Glia 56:1138–1149

Benveniste H, Drejer J, Schousboe A, Diemer NH (1984) Elevation of the extracellular concentrations of glutamate and aspartate in rat hippocampus during transient cerebral ischemia monitored by intracerebral microdialysis. J Neurochem 43:1369–1374

Bergmann F, Keller BU (2003) Impact of mitochondrial inhibition on excitability and cytosolic Ca^{2+} levels in brainstem motoneurons from mouse. J Physiol 555:45–59

Berna-Erro A, Braun A, Kraft R, Kleinschnitz C, Schuhmann MK, Stegner D, Wultsch T, Eilers J, Meuth SG, Stoll G, Nieswandt B (2009) STIM2 regulates capacitive Ca^{2+} entry in neurons and plays a key role in hypoxic neuronal cell death. Sci Signal 2:ra67

Bernardinelli Y, Magistretti PJ, Chatton JY (2004) Astrocytes generate Na+-mediated metabolic waves. Proc Natl Acad Sci U S A 101:14937–14942

Berridge MJ (1998) Neuronal calcium signaling. Neuron 21:13–26

Besancon E, Guo S, Lok J, Tymianski M, Lo EH (2008) Beyond NMDA and AMPA glutamate receptors: emerging mechanisms for ionic imbalance and cell death in stroke. Trends Pharmacol Sci 29:268–275

Bhattacharjee A, Kaczmarek LK (2005) For K^+ channels, Na^+ is the new Ca^{2+}. Trends Neurosci 28:422–428

Bickler PE, Fahlman CS, Gray J, McKleroy W (2009) Inositol 1,4,5-triphosphate receptors and NAD(P)H mediate Ca^{2+} signaling required for hypoxic preconditioning of hippocampal neurons. Neuroscience 160:51–60

Bickler PE, Fahlman CS, Taylor DM (2003) Oxygen sensitivity of NMDA receptors: relationship to NR2 subunit composition and hypoxia tolerance of neonatal neurons. Neuroscience 118:25–35

Blaesse P, Airaksinen MS, Rivera C, Kaila K (2009) Cation-chloride cotransporters and neuronal function. Neuron 61:820–838

Blair NT, Bean BP (2002) Roles of tetrodotoxin (TTX)-sensitive Na^+ current, TTX-resistant Na^+ current, and Ca^{2+} current in the action potentials of nociceptive sensory neurons. J Neurosci 22:10277–100290

Bodalia A, Li H, Jackson MF (2013) Loss of endoplasmic reticulum Ca^{2+} homeostasis: contribution to neuronal cell death during cerebral ischemia. Acta Pharmacol Sin 34:49–59

Bojarski C, Meloni BP, Moore SR, Majda BT, Knuckey NW (2008) Na^+/Ca^{2+} exchanger subtype (NCX1, NCX2, NCX3) protein expression in the rat hippocampus following 3 min and 8 min durations of global cerebral ischemia. Brain Res 1189:198–202

Bollimuntha S, Selvaraj S, Singh BB (2011) Emerging roles of canonical TRP channels in neuronal function. Adv Exp Med Biol 704:573–593

Bondarenko A, Svichar N, Chesler M (2004) Role of Na^+-H^+ and Na^+-Ca^{2+} exchange in hypoxia-related acute astrocyte death. Glia 49:143–152

Borlongan CV, Hayashi T, Oeltgen PR, Su TP, Wang Y (2009) Hibernation-like state induced by an opioid peptide protects against experimental stroke. BMC Biol 7:31

Borlongan CV, Su TP, Wang Y (2000) Treatment with delta opioid peptide enhances in vitro and in vivo survival of rat dopaminergic neurons. Neuroreport 11:923–926

Bortner CD, Hughes FM Jr, Cidlowski JA (1997) A primary role for K^+ and Na^+ efflux in the activation of apoptosis. J Biol Chem 272:32436–32442

Boutin H, Catherine A, Mackenzie ET, Jauzac P, Dauphin F (2007) Long-term alteration in μ, δ and $\varkappa$ opioidergic receptors following middle cerebral artery occlusion in mice. Acta Neuropathol 114:491–500

Boutin H, Jauzac P, MacKenzie ET, Dauphin F (2003) Potential use of early alterations in mu and delta opioid receptors as a predictive index for delayed brain ischemic damage. Neurobiol Dis 13:63–73

Bouvier M, Szatkowski M, Amato A, Attwell D (1992) The glial cell glutamate uptake carrier countertransports pH-changing anions. Nature 360:471–474

Breder J, Sabelhaus CF, Opitz T, Reymann KG, Schröder UH (2000) Inhibition of different pathways influencing Na^+ homeostasis protects organotypic hippocampal slice cultures from hypoxic/hypoglycemic injury. Neuropharmacology 39:1779–1787

Bright R, Mochly-Rosen D (2005) The role of protein kinase C in cerebral ischemic and reperfusion injury. Stroke 36:2781–2790

Brini M, Calì T, Ottolini D, Carafoli E (2014) Neuronal calcium signaling: function and dysfunction. Cell Mol Life Sci 71:2787–2814

Brisson CD, Hsieh YT, Kim D, Jin AY, Andrew RD (2014) Brainstem neurons survive the identical ischemic stress that kills higher neurons: insight to the persistent vegetative state. PLoS One 9, e96585

Brittain MK, Brustovetsky T, Sheets PL, Brittain JM, Khanna R, Cummins TR, Brustovetsky V (2012) Delayed calcium dysregulation in neurons requires both the NMDA receptor and the reverse Na^+/Ca^{2+} exchanger. Neurobiol Dis 46:109–117

Brongholi K, Souza DG, Bainy AC, Dafre AL, Tasca CI (2006) Oxygen-glucose deprivation decreases glutathione levels and glutamate uptake in rat hippocampal slices. Brain Res 1083:211–218

Brown AM, Fern R, Jarvinen JP, Kaila K, Ransom BR (1998) Changes in $[Ca^{2+}]0$ during anoxia in CNS white matter. Neuroreport 9:1997–2000

Bruehl C, Neumann-Haefelin T, Witte OW (2000) Enhancement of whole cell calcium currents following transient MCAO. Brain Res 884:129–138

Bruzzone R, Hormuzdi SG, Barbe MT, Herb A, Monyer H (2003) Pannexins, a family of gap junction proteins expressed in brain. Proc Natl Acad Sci U S A 100:13644–13649

Buck LT, Pamenter ME (2006) Adaptive responses of vertebrate neurons to anoxia-matching supply to demand. Respir Physiol Neurobiol 154:226–240

Budde T, Meuth S, Pape HC (2002) Calcium-dependent inactivation of neuronal calcium channels. Nat Rev Neurosci 3:873–883

Buresh Y, Koroleva VI, Korolev OS, Maresh V (1999) Changes in the constant potential in brain structures in rats during focal ischemia and systemic hypoxia. Neurosci Behav Physiol 29:569–579

Butenko O, Dzamba D, Benesova J, Honsa P, Benfenati V, Rusnakova V, Ferroni S, Anderova M (2012) The increased activity of TRPV4 channel in the astrocytes of the adult rat hippocampus after cerebral hypoxia/ischemia. PLoS One 7, e39959

Cahalan MD (2009) STIMulating store-operated Ca^{2+} entry. Nat Cell Biol 11:669–677

Calabresi P, Marfia GA, Centonze D, Pisani A, Bernardi G (1999) Sodium influx plays a major role in the membrane depolarization induced by oxygen and glucose deprivation in rat striatal spiny neurons. Stroke 30:171–179

Camacho A, Massieu L (2006) Role of glutamate transporters in the clearance and release of glutamate during ischemia and its relation to neuronal death. Arch Med Res 37:11–18

Cameron R, Klein L, Shyjan AW, Rakic P, Levenson R (1994) Neurons and astroglia express distinct subsets of Na, K-ATPase alpha and beta subunits. Brain Res Mol Brain Res 21:333–343

Carriedo SG, Sensi SL, Yin HZ, Weiss JH (2000) AMPA exposures induce mitochondrial Ca^{2+} overload and ROS generation in spinal motor neurons in vitro. J Neurosci 20:240–250

Carvalho C, Correia SC, Santos RX, Cardoso S, Moreira PI, Clark TA, Zhu X, Smith MA, Perry G (2009) Role of mitochondrial-mediated signaling pathways in Alzheimer disease and hypoxia. J Bioenerg Biomembr 41:433–440

Catterall WA (2011) Voltage-gated calcium channels. Cold Spring Harb Perspect Biol 3:a003947

Cengiz P, Kintner DB, Chanana V, Yuan H, Akture E, Kendigelen P, Begum G, Fidan E, Uluc K, Ferrazzano P, Sun D (2014) Sustained Na^+/H^+ exchanger activation promotes gliotransmitter release from reactive hippocampal astrocytes following oxygen-glucose deprivation. PLoS One 9, e84294

Chaki S, Muramatsu M, Otomo S (1994) Involvement of protein kinase C activation in regulation of acetylcholine release from rat hippocampal slices by minaprine. Neurochem Int 24:37–41

Chao D, Balboni G, Lazarus LH, Salvadori S, Xia Y (2009) Na^+ mechanism of DOR protection against anoxic K^+ derangement in the cortex. Cell Mol Life Sci 66:1105–1115

Chao D, Bazzy-Asaad A, Balboni G, Salvadori S, Xia Y (2008) Activation of DOR attenuates anoxic K^+ derangement via inhibition of Na^+ entry in mouse cortex. Cereb Cortex 18:2217–2227

Chao D, Bazzy-Asaad A, Balboni G, Xia Y (2007a) Delta-, but not mu-, opioid receptor stabilizes K^+ homeostasis by reducing Ca^{2+} influx in the cortex during acute hypoxia. J Cell Physiol 212:60–67

Chao D, Donnelly DF, Feng Y, Bazzy-Asaad A, Xia Y (2007b) Cortical delta-opioid receptors potentiate K^+ homeostasis during anoxia and oxygen-glucose deprivation. J Cereb Blood Flow Metab 27:356–368

Chao D, He X, Yang Y, Balboni G, Salvadori S, Kim DH, Xia Y (2012a) Hydrogen sulfide induced disruption of Na^+ homeostasis in the cortex. Toxicol Sci 128:198–208

Chao D, He X, Yang Y, Bazzy-Asaad A, Lazarus LH, Balboni G, Kim DH, Xia Y (2012b) DOR activation inhibits anoxic/ischemic Na^+ influx through Na^+ channels via PKC mechanisms in the cortex. Exp Neurol 236:228–239

Chao D, Qian H, Ghassemi F, Chen J, Xia Y (2006) Transgenic overexpression of delta-opioid receptors protects the cortex from anoxic disruption of ionic homeostasis. Program No. 87.19/MM68, Neuroscience Meeting Planner. Society for Neuroscience, Atlanta

Chao D, Shen X, Xia Y (2013) From acupuncture to interaction between δ-opioid receptors and Na^+ channels: a potential pathway to inhibit epileptic hyperexcitability. Evid Based Complement Alternat Med 2013:216016

Chao D, Xia Y (2010) Ionic storm in hypoxic/ischemic stress: can opioid receptors subside it? Prog Neurobiol 90:439–470

Chao D, Xia Y (2012) Acupuncture treatment of epilepsy. In: Xia Y, Ding G, Wu G (eds) Current research in acupuncture. Springer, New York, pp 129–214. doi:10.1007/978-1-4614-3357-6_6

Chao D, Xia Y (2015) δ-Opioid receptor induced neuroprotection against hypoxic/ischemic injury: regulation of ionic homeostasis. J Anesth Perioper Med, in press

Charron C, Messier C, Plamondon H (2008) Neuroprotection and functional recovery conferred by administration of kappa- and delta-opioid agonists in a rat model of global ischemia. Physiol Behav 93:502–511

Chatton JY, Marquet P, Magistretti PJ (2000) A quantitative analysis of L-glutamate-regulated Na^+ dynamics in mouse cortical astrocytes: implications for cellular bioenergetics. Eur J Neurosci 12:3843–3853

Chen CJ, Liao SL, Chen WY, Hong JS, Kuo JS (2001a) Cerebral ischemia/reperfusion injury in rat brain: effects of naloxone. Neuroreport 12:1245–1249

Chen F, Qi Z, Luo Y, Hinchliffe T, Ding G, Xia Y, Ji X (2014) Non-pharmaceutical therapies for stroke: mechanisms and clinical implications. Prog Neurobiol 115:246–269

Chen L, Hahn H, Wu G, Chen CH, Liron T, Schechtman D, Cavallaro G, Banci L, Guo Y, Bolli R, Dorn GW II, Mochly-Rosen D (2001b) Opposing cardioprotective actions and parallel hypertropic effects of delta PKC and epsilon PKC. Proc Natl Acad Sci U S A 98:11114–11119

Chen M, Simard JM (2001) Cell swelling and a nonselective cation channel regulated by internal Ca^{2+} and ATP in native reactive astrocytes from adult rat brain. J Neurosci 21:6512–6521

Chen M, Tao YX, Gu JG (2007a) Inward currents induced by ischemia in rat spinal cord dorsal horn neurons. Mol Pain 3:10

Chen TW, Wardill TJ, Sun Y, Pulver SR, Renninger SL, Baohan A, Schreiter ER, Kerr RA, Orger MB, Jayaraman V, Looger LL, Svoboda K, Kim DS (2013) Ultrasensitive fluorescent proteins for imaging neuronal activity. Nature 499:295–300

Chen W, Ostrowski RP, Obenaus A, Zhang JH (2009) Prodeath or prosurvival: two facets of hypoxia inducible factor-1 in perinatal brain injury. Exp Neurol 216:7–15

Chen WH, Chu KC, Wu SJ, Wu JC, Shui HA, Wu ML (1999) Early metabolic inhibit-induced intracellular sodium and calcium increase in rat cerebellar granule cells. J Physiol 515:133–146

Chen X, Kintner DB, Jonse M, Matsuda T, Baba A, Kiedrowski L, Sun D (2007b) AMPA-mediated excitotoxicity in oligodendrocytes: role for Na^+-K^+-Cl^- cotransporter and reversal on Na^+/Ca^{2+} exchanger. J Neurochem 102:1783–1795

Chen X, Kintner DB, Luo J, Baba A, Matsuda T, Sun D (2008) Endoplasmic reticulum Ca^{2+} dysregulation and endoplasmic reticulum stress following in vitro neuronal ischemia: role of Na^+-K^+-Cl^- cotransporter. J Neurochem 106:1563–1576
Chen XQ, Du JZ (2000) Hypoxia influences enkephalin release in rats. Neuroreport 11:1555–1557
Chen Y, Yu FH, Surmeier DJ, Scheuer T, Catterall WA (2006) Neuromodulation of Na^+ channel slow inactivation via cAMP-dependent protein kinase and protein kinase C. Neuron 49:409–420
Chen ZC, Kuo JR, Huang YP, Lin MT (2005) μ-opioid receptor blockade protects against circulatory shock and cerebral ischemia during heatstroke. J Cardiovasc Pharmacol 46:754–760
Cheng KT, Ong HL, Liu X, Ambudkar IS (2013) Contribution and regulation of TRPC channels in store-operated Ca^{2+} entry. Curr Top Membr 71:149–179
Cheung NS, Peng ZF, Chen MJ, Moore PK, Whiteman M (2007) Hydrogen sulfide induced neuronal death occurs via glutamate receptor and is associated with calpain activation and lysosomal repture in mouse primary cortical neurons. Neuropharmacology 53:505–514
Chi XX, Xu ZC (2000) Potassium currents in CA1 neurons of rat hippocampus increase shortly after transient cerebral ischemia. Neurosci Lett 281:5–8
Chi XX, Xu ZC (2001) Alterations of single potassium channel activity in CA1 pyramidal neurons after transient forebrain ischemia. Neuroscience 108:535–540
Choi DW, Rothman SM (1990) The role of glutamate neurotoxicity in hypoxic-ischemic neuronal death. Annu Rev Neurosci 13:171–182
Choi DW, Viseskul V (1988) Opioids and non-opioid enantiomers selectively attenuate N-methyl-D-aspartate neurotoxicity on cortical neurons. Eur J Pharmacol 155:27–35
Chou WH, Messing RO (2005) Protein kinase C isozymes in stroke. Trends Cardiovasc Med 15:47–51
Christie JM, Jahr CE (2008) Dendritic NMDA receptors activate axonal calcium channels. Neuron 60:298–307
Churchwell KB, Wright SH, Emma F, Rosenberg PA, Strange K (1996) NMDA receptor activation inhibits neuronal volume regulation after swelling induced by veratridine-stimulated Na^+ influx in rat cortical cultures. J Neurosci 16:7447–7457
Colino A, García-Seoane JJ, Valentín A (1998) Action potential broadening induced by lithium may cause a presynaptic enhancement of excitatory synaptic transmission in neonatal rat hippocampus. Eur J Neurosci 10:2433–2443
Connor JA, Wadman WJ, Hockberger PE, Wong RK (1988) Sustained dendritic gradients of Ca^{2+} induced by excitatory amino acids in CA1 hippocampal neurons. Science 240:649–653
Connor JN, Bazani-Boroujerdi S, Greenwood AC, Cormier RJ, Petrozzino JF, Lin RCS (1999) Reduced voltage-dependent Ca^{2+} signaling in CA1 neurons after brief ischemia in gerbils. J Neurophysiol 81:299–306
Croning MDR, Grahame-Smith DG, Zetterström TSC (1993) Effect of (+)-MK-801 on the increase in extracellular [K^+] seen in rat dorsal hippocampal, in vivo, during transient hypoxia as measured by ion-selective microelectrodes. Br J Pharmacol 100:179
Cross JL, Boulos S, Shepherd KL, Craig AJ, Lee S, Bakker AJ, Knuckey NW, Meloni BP (2012) High level over-expression of different NCX isoforms in HEK293 cell lines and primary neuronal cultures is protective following oxygen glucose deprivation. Neurosci Res 73:191–198
D'Ambrosio R, Gordon DS, Winn HR (2002) Differential role of KIR channel and Na($^+$)/K($^+$)-pump in the regulation of extracellular K($^+$) in rat hippocampus. J Neurophysiol 87:87–102
D'Ambrosio R, Maris DO, Grady MS, Winn HR, Janigro D (1999) Impaired K^+ homeostasis and altered electrophysiological properties of post-traumatic hippocampal glia. J Neurosci 19:8152–8162
Dascal N, Lotan I (1991) Activation of protein kinase C alters voltage dependence of a Na^+ channel. Neuron 6:165–175
Davidson JO, Green CR, Nicholson LF, O'Carroll SJ, Fraser M, Bennet L, Gunn AJ (2012) Connexin hemichannel blockade improves outcomes in a model of fetal ischemia. Ann Neurol 71:121–132

De Stefani D, Raffaello A, Teardo E, Szabò I, Rizzuto R (2011) A forty-kilodalton protein of the inner membrane is the mitochondrial calcium uniporter. Nature 476:336–340
Deng H, Yang Z, Li Y, Bao G, Friedrich T, Gu Q, Shen X, Schwarz W (2009) Interactions of Na^+, K^+-ATPase and co-expressed δ-opioid receptor. Neurosci Res 65:222–227
Deng W, Rosenberg PA, Volpe JJ, Jensen FE (2003) Calcium-permeable AMPA/kainate receptors mediate toxicity and preconditioning by oxygen-glucose deprivation in oligodendrocyte precursors. Proc Natl Acad Sci U S A 100:6801–6806
Diarra A, Sheldon C, Brett CL, Baimbidge KG, Church J (1999) Anoxia-evoked intracellular pH and Ca^{2+} concentration changes in cultured postnatal rat hippocampal neurons. Neuroscience 93:1003–1016
Ding S, Wang T, Cui W, Haydon PG (2009) Photothrombosis ischemia stimulates a sustained astrocytic Ca^{2+} signaling in vivo. Glia 57:767–776
Dingledine R, Borges K, Bowle D, Traynelis SF (1999) The glutamate receptor ion channels. Pharmacol Rev 51:7–61
Dipolo R, Beaugé L (2006) Sodium/calcium exchanger: influence of metabolic regulation on ion carrier interactions. Physiol Rev 86:155–203
Dodt HU, Frick A, Kampe K, Zieglgänsberger W (1998) NMDA and AMPA receptors on neocortical neurons are differentially distributed. Eur J Neurosci 10:3351–3357
Doll CJ, Hochachka PW, Reiner PB (1991) Channel arrest: implications from membrane resistance in turtle neurons. Am J Physiol 261:R1321–R1324
Domercq M, Perez-Samartin A, Aparicio D, Alberdi E, Pampliega O, Matute C (2010) P2X7 receptors mediate ischemic damage to oligodendrocytes. Glia 58:730–740
Dong QP, He JQ, Chai Z (2013) Astrocytic $Ca(^{2+})$ waves mediate activation of extrasynaptic NMDA receptors in hippocampal neurons to aggravate brain damage during ischemia. Neurobiol Dis 58:68–75
Dreosti E, Odermatt B, Dorostkar MM, Lagnado L (2009) A genetically encoded reporter of synaptic activity in vivo. Nat Methods 6:883–889
Dronne MA, Boissel JP, Grenier E (2006) A mathematical model of ion movements in grey matter during a stroke. J Theor Biol 240:599–615
Dronne MA, Grenier E, Dumont T, Hommel M, Boissel JP (2007) Role of astrocytes in grey matter during stroke: a modelling approach. Brain Res 1138:231–242
Duan YL, Wang SY, Zeng QW, Su DS, Li W, Wang XR, Zhao Z (2011) Astroglial reaction to delta opioid peptide [D-Ala2, D-Leu5] enkephalin confers neuroprotection against global ischemia in the adult rat hippocampus. Neuroscience 192:81–90
Dubinsky JM, Rothman SM (1991) Intracellular calcium concentrations during "chemical hypoxia" and excitotoxic neuronal injury. J Neurosci 11:2545–2551
Duffy S, MacVicar BA (1996) In vitro ischemia promotes calcium influx and intracellular calcium release in hippocampal astrocytes. J Neurosci 16:71–81
Dugan LL, Sensl SL, Canzonlero LMT, Handran SD, Rothman SM, Lin TS, Goldberg MP, Choi DW (1995) Mitochondrial production of reactive oxygen species in cortical neurons following exposure to N-methyl-D-aspartate. J Neurosci 15:6377–6388
Dunn JF, Zaim Wadghiri Y, Meyerand ME (1999) Regional heterogeneity in the brain's response to hypoxia measured using BOLD MR imaging. Magn Reson Med 41:850–854
Dvoriantchikova G, Ivanov D, Barakat D, Grinberg A, Wen R, Slepak VZ, Shestopalov VI (2012) Genetic ablation of pannexin1 protects retinal neurons from ischemic injury. PLoS One 7, e31991
Eid T, Williamson A, Lee TSW, Petroff OA, de Lanerolle NC (2008) Glutamate and astrocytes—key players in human mesial temporal lobe epilepsy? Epilepsia 49(suppl 2):42–52
Endoh H, Taga K, Yamakura T, Sato K, Watanabe I, Fukuda S, Shimoji K (1999) Effects of naloxone and morphine on acute hypoxic survival in mice. Crit Care Med 27:1929–1933
Erdemli G, Xu YZ, Krnjević K (1998) Potassium conductance causing hyperpolarization of CA1 hippocampal neurons during hypoxia. J Neurophysiol 80:2378–2390

Fang KM, Chang WL, Wang SM, Su MJ, Wu ML (2008) Arachidonic acid induces both Na^+ and Ca^{2+} entry resulting in apoptosis. J Neurochem 104:1177–1189
Fazzina G, Amorini AM, Marmarou CR, Fukui S, Okuno K, Dunbar JG, Glisson R, Marmarou A, Kleindienst A (2010) The protein kinase C activator phorbol myristate acetate decreases brain edema by aquaporin 4 downregulation after middle cerebral artery occlusion in the rat. J Neurotrauma 27:453–461
Fedele E, Raiteri M (1996) Desensitization of AMPA receptors and AMPA-NMDA receptor interaction: an in vivo cyclic GMP microdialysis study in rat cerebellum. Br J Pharmacol 117:1133–1138
Feng Y, He X, Yang Y, Chao D, Lazarus LH, Xia Y (2012) Current research on opioid receptor function. Curr Drug Targets 13:230–246
Fern R (1998) Intracellular calcium and cell death during ischemia in neonatal rat white matter astrocytes in situ. J Neurosci 18:7232–7243
Fernandes J, Vieira M, Carreto L, Santos MA, Duarte CB, Carvalho AL, Santos AE (2014) In vitro ischemia triggers a transcriptional response to down-regulate synaptic proteins in hippocampal neurons. PLoS One 9, e99958
Filliol D, Ghozland S, Chluba J, Martin M, Matthes HW, Simonin F, Befort K, Gavériaux-Ruff C, Dierich A, LeMeur M, Valverde O, Maldonado R, Kieffer BL (2000) Mice deficient for delta- and mu-opioid receptors exhibit opposing alterations of emotional responses. Nat Genet 25:195–200
Formisano L, Guida N, Valsecchi V, Pignataro G, Vinciguerra A, Pannaccione A, Secondo A, Boscia F, Molinaro P, Sisalli MJ, Sirabella R, Casamassa A, Canzoniero LM, Di Renzo G, Annunziato L (2013) NCX1 is a new rest target gene: role in cerebral ischemia. Neurobiol Dis 50:76–85
Franceschetti S, Lavazza T, Curia G, Aracri P, Panzica F, Sancini G, Avanzini G, Magistretti J (2003) Na^+-activated K^+ current contributes to postexcitatory hyperpolarization in neocortical intrinsically bursting neurons. J Neurophysiol 89:2101–2111
Frantseva MV, Carlen PL, Perez Velazquez JL (2001) Dynamics of intracellular calcium and free radical production during ischemia in pyramidal neurons. Free Radic Biol Med 31:1216–1227
Frazzini V, Rapposelli IG, Corona C, Rockabrand E, Canzoniero LM, Sensi SL (2007) Mild acidosis enhances AMPA receptor-mediated intracellular zinc mobilization in cortical neurons. Mol Med 13:356–361
Friedman JE, Haddad GG (1993) Major differences in Ca2+i response to anoxia between neonatal and adult rat CA1 neurons: role of Ca2+o and Na+o. J Neurosci 13:63–72
Friedman JE, Haddad GG (1994) Anoxia induces an increase in intracellular sodium in rat central neurons in vitro. Brain Res 663:329–334
Frühbeis C, Fröhlich D, Kuo WP, Amphornrat J, Thilemann S, Saab AS, Kirchhoff F, Möbius W, Goebbels S, Nave KA, Schneider A, Simons M, Klugmann M, Trotter J, Krämer-Albers EM (2013) Neurotransmitter-triggered transfer of exosomes mediates oligodendrocyte-neuron communication. PLoS Biol 11, e1001604
Fu CT, Sretavan DW (2012) Ectopic vesicular glutamate release at the optic nerve head and axon loss in mouse experimental glaucoma. J Neurosci 32:15859–15876
Fujimura N, Tanaka E, Yamamoto S, Shigemori M, Higashi H (1997) Contribution of ATP-sensitive potassium channels to hypoxic hyperpolarization in rat hippocampal CA1 neurons in vitro. J Neurophysiol 77:378–385
Fukuda A, Muramatsu K, Okabe A, Shimano Y, Hida H, Fujimoto I, Nishino H (1998) NMDA receptor-mediated differential laminar susceptibility to the intracellular Ca^{2+} accumulation induced by oxygen-glucose deprivation in rat neocortical slices. J Neurophysiol 79:430–438
Fung ML, Croning MDR, Haddad GG (1999) Sodium homeostasis in rat hippocampal slices during oxygen and glucose deprivation: role of voltage-sensitive sodium channels. Neurosci Lett 275:41–44

Galeffi F, Sah R, Pond BB, George A, Schwartz-Bloom RD (2004) Changes in intracellular chloride after oxygen-glucose deprivation of the adult hippocampal slice: effect of diazepam. J Neurosci 24:4478–4488

Galluzzi L, Vitale I, Abrams JM, Alnemri ES, Baehrecke EH, Blagosklonny MV, Dawson TM, Dawson VL, El-Deiry WS, Fulda S, Gottlieb E, Green DR, Hengartner MO, Kepp O, Knight RA, Kumar S, Lipton SA, Lu X, Madeo F, Malorni W, Mehlen P, Nuñez G, Peter ME, Piacentini M, Rubinsztein DC, Shi Y, Simon HU, Vandenabeele P, White E, Yuan J, Zhivotovsky B, Melino G, Kroemer G (2012) Molecular definitions of cell death subroutines: recommendations of the Nomenclature Committee on Cell Death 2012. Cell Death Differ 19:107–120

Gao CJ, Li JP, Wang W, Lü BC, Niu L, Zhu C, Wei YY, Zhang T, Wu SX, Chai W, Li YQ (2010) Effects of intracerebroventricular application of the delta opioid receptor agonist [D-Ala2, D-Leu5] enkephalin on neurological recovery following asphyxial cardiac arrest in rats. Neuroscience 168:531–542

Gao CJ, Niu L, Ren PC, Wang W, Zhu C, Li YQ, Chai W, Sun XD (2012) Hypoxic preconditioning attenuates global cerebral ischemic injury following asphyxial cardiac arrest through regulation of delta opioid receptor system. Neuroscience 202:352–362

Gao J, Duan B, Wang DG, Deng XH, Zhang GY, Xu L, Xu TL (2005) Coupling between NMDA receptor and acid-sensing ion channel contributes to ischemic neuronal death. Neuron 48:635–646

Gao TM, Fung ML (2002) Decreased large conductance Ca^{2+}-activated K^+ channel activity in dissociated CA1 hippocampal neurons in rats exposed to perinatal and postnatal hypoxia. Neurosci Lett 332:163–166

Garlid KD (1996) Cation transport in mitochondria-the potassium cycle. Biochim Biophys Acta 1275:123–126

Gees M, Colsoul B, Nilius B (2010) The role of transient receptor potential cation channels in Ca^{2+} signaling. Cold Spring Harb Perspect Biol 2:a003962

Gemes G, Bangaru ML, Wu HE, Tang Q, Weihrauch D, Koopmeiners AS, Cruikshank JM, Kwok WM, Hogan QH (2011) Store-operated Ca^{2+} entry in sensory neurons: functional role and the effect of painful nerve injury. J Neurosci 31:3536–3549

Georges F, Normand E, Bloch B, Moine CL (1998) Opioid receptor gene expression in the rat brain during ontogeny, with special reference to the mesostriatal system: an in situ hybridization study. Brain Res Dev Brain Res 109:187–199

Giacomello M, De Mario A, Scarlatti C, Primerano S, Carafoli E (2013) Plasma membrane calcium ATPases and related disorders. Int J Biochem Cell Biol 45:753–762

Gido G, Kristian T, Siesjo KB (1997) Extracellular potassium in a neocortical core area after transient focal ischemia. Stroke 28:206–210

Ginsberg MD (2008) Neuroprotection for ischemic stroke: past, present and future. Neuropharmacology 55:363–389

Gniel HM, Martin RL (2010) Changes in membrane potential and the intracellular calcium concentration during CSD and OGD in layer V and layer II/III mouse cortical neurons. J Neurophysiol 104:3203–3212

Goldberg MP, Choi DW (1993) Combined oxygen and glucose deprivation in cortical cell culture: calcium-dependent and calcium-independent mechanisms of neuronal injury. J Neurosci 13:3510–3542

Goldman WF, Yarowsky PJ, Juhaszova M, Krueger BK, Blaustein MP (1994) Sodium-calcium exchange in rat cortical astrocytes. J Neurosci 14:5834–5843

Gorter JA, Petrozzino JJ, Aronica EM, Rosenbaum DM, Opitz T, Bennett MV, Connor JA, Zukin RS (1997) Global ischemia induces downregulation of Glur2 mRNA and increases AMPA receptor-mediated Ca^{2+} influx in hippocampal CA1 neurons of gerbil. J Neurosci 17:6179–6188

Gouriou Y, Bijlenga P, Demaurex N (2013) Mitochondrial Ca^{2+} uptake from plasma membrane Cav3.2 protein channels contributes to ischemic toxicity in PC12 cells. J Biol Chem 288:12459–12468

Gouriou Y, Demaurex N, Bijlenga P, De Marchi U (2011) Mitochondrial calcium handling during ischemia-induced cell death in neurons. Biochimie 93:2060–2067

Govindaswami M, Brown SA, Yu J, Zhu H, Bishop PD, Kindy MS, Oeltgen PR (2008) Delta 2-specific opioid receptor agonist and hibernating woodchuck plasma fraction provide ischemic neuroprotection. Acad Emerg Med 15:250–257

Green AR (2008) Pharmacological approaches to acute ischaemic stroke: reperfusion certainly, neuroprotection possibly. Br J Pharmacol 153:S325–S338

Grega DS, Werz MA, Macdonald RL (1987) Forskolin and phorbol ester reduce same potassium conductance of mouse neurons in culture. Science 235:345–348

Grøndahl TØ, Hablitz JJ, Langmoen IA (1998) Depletion of intracellular Ca^{2+} stores or lowering extracellular calcium alters intracellular Ca^{2+} changes during cerebral energy deprivation. Brain Res 796:125–131

Groom SY, Noh KM, Regis B, Bassell GJ, Bryan MK, Carroll RC, Zukin RS (2006) Activity bidirectionally regulates AMPA receptor mRNA abundance in dendrites of hippocampal neurons. J Neurosci 26:8339–8351

Gruss M, Ettorre G, Stehr AJ, Henrich M, Hempelmann G, Scholz A (2006) Moderate hypoxia influences excitability and blocks dendrotoxin sensitive K^+ currents in rat primary sensory neurons. Mol Pain 2:12

Guatteo E, Federici M, Siniscalchi A, Knőpfel T, Mercuri NB, Bernardi G (1998a) Whole cell patch-clamp recordings of rat midbrain dopaminergic neurons isolate a sulphonylurea- and ATP-sensitive component of potassium currents activated by hypoxia. J Neurophysiol 79:1239–1245

Guatteo E, Mercuri NB, Bernardi G, Knöpfel T (1998b) Intracellular sodium and calcium homeostasis during hypoxia in dopamine neurons of rat substantia nigra pars compacta. J Neurophysiol 80:2237–2243

Haddad GG, Jiang C (1993) O2 deprivation in the central nervous system: on mechanisms of neuronal response, differential sensitivity and injury. Prog Neurobiol 40:277–318

Haddad GG, Mellins RB (1984) Hypoxia and respiratory control in early life. Annu Rev Physiol 46:629–643

Hajnóczky G, Csordás G, Das S, Garcia-perez C, Saotome M, Roy SS, Yi M (2006) Mitochondrial calcium signaling and cell death: approaches for assessing the role of mitochondrial Ca^{2+} uptake in apoptosis. Cell Calcium 40:553–560

Han F, Shioda N, Moriguchi S, Qin ZH, Fukunaga K (2008) Downregulation of glutamate transporters is associated with elevation in extracellular glutamate concentration following rat microsphere embolism. Neurosci Lett 430:275–280

Hansen AJ (1978) The extracellular potassium concentration in brain cortex following ischemia in hypo- and hyperglycemic rats. Acta Physiol Scand 102:324–329

Hansen AJ (1985) Effect of anoxia on ion distribution in the brain. Physiol Rev 65:101–148

Harburg GC, Hall FS, Harrist AV, Sora I, Uhl GR, Eisch AJ (2007) Knockout of the mu opioid receptor enhances the survival of adult-generated hippocampal granule cell neurons. Neuroscience 144:77–87

Harris KM, Kater SB (1994) Dendritic spines: cellular specializations imparting both stability and flexibility to synaptic function. Annu Rev Neurosci 17:341–371

Hartmann H, Eckert A, Velbinger K, Rewsin M, Müller WE (1996) Down-regulation of free intracellular calcium in dissociated brain cells of aged mice and rats. Life Sci 59:435–449

Hasbani MJ, Hyrc KL, Faddi BT, Romano C, Goldberg MP (1998) Distinct role for sodium, chloride, and calcium in excitotoxic dendritic injury and recovery. Exp Neurol 154:241–258

Haun SE, Murphy EJ, Bates CM, Harrocks LA (1992) Extracellular calcium is a mediator of astroglial injury during combined glucose-oxygen deprivation. Brain Res 593:45–50

Hauser KF, Gurwell JA, Turbek CS (1994) Morphine inhibits purkinje cell survival and dendritic differentiation in organotypic cultures of the mouse cerebellum. Exp Neurol 130:95–105

Hawrysh PJ, Buck LT (2013) Anoxia-mediated calcium release through the mitochondrial permeability transition pore silences NMDA receptor currents in turtle neurons. J Exp Biol 216:4375–4387

Hayashi T, Tsal LI, Su TP (2002) Antiapoptotic and cytotoxic properties of delta opioid peptide [D-Ala2, D-Leu5]enkephalin in PC12 cells. Synapse 43:86–94
Hayward NJ, McKnight AT, Woodruff GN (1993) Neuroprotective effect of the kappa-agonist enadoline (CI-977) in rat models of focal cerebral ischemia. Eur J Neurosci 5:961–967
He X, Sandhu HK, Yang Y, Hua F, Belser N, Kim DH, Xia Y (2013) Neuroprotection against hypoxia/ischemia: δ-opioid receptor-mediated cellular/molecular events. Cell Mol Life Sci 70:2291–2303
Heinemann U, Lux HD (1977) Ceiling of stimulus induced rises in extracellular potassium concentration in the cerebral cortex of cat. Brain Res 120:231–249
Helmchen F, Denk W (2002) New developments in multiphoton microscopy. Curr Opin Neurobiol 12:593–601
Helmchen F, Denk W (2005) Deep tissue two-photon microscopy. Nat Methods 2:932–940
Helmchen F, Denk W, Kerr JN (2013) Miniaturization of two-photon microscopy for imaging in freely moving animals. Cold Spring Harb Protoc 2013:904–913
Hemsworth PD, Whalley DW, Rasmussen HH (1997) Electrogenic Li+/Li+ exchange mediated by the Na^+-K^+ pump in rabbit cardiac myocytes. Am J Physiol 272:C1186–C1192
Henrich M, Buckler KJ (2008) Effects of anoxia and aglycemia on cytosolic calcium regulation in rat sensory neurons. J Neurophysiol 100:456–473
Henrich M, Buckler KJ (2013) Cytosolic calcium regulation in rat afferent vagal neurons during anoxia. Cell Calcium 54:416–427
Hepp S, Gerich FJ, Müller M (2005) Sulfhydryl oxidation reduces hippocampal susceptibility to hypoxia-induced spreading depression by activating BK channels. J Neurophysiol 94:1091–1103
Hernández-SanMiguel E, Vay L, Santo-Domingo J, Lobatón CD, Moreno A, Montero M, Alvarez J (2006) The mitochondrial Na^+/Ca^{2+} exchanger plays a key role in the control of cytosolic Ca^{2+} oscillation. Cell Calcium 40:53–61
Hertz L (2008) Bioenergetics of cerebral ischemia: a cellular perspective. Neuropharmacology 55:289–309
Hertz L, Xu J, Song D, Yan E, Gu L, Peng L (2013) Astrocytic and neuronal accumulation of elevated extracellular K^+ with a 2/3 K^+/Na^+ flux ratio-consequences for energy metabolism, osmolarity and higher brain function. Front Comput Neurosci 7:114
Hille B (1972) The permeability of the sodium channel to metal cations in myelinated nerve. J Gen Physiol 59:637–658
Hiller JM, Fan LQ (1996) Laminar distribution of the multiple opioid receptors in the human cerebral cortex. Neurochem Res 21:1333–1345
Hoogland TM, Kuhn B, Göbel W, Huang W, Nakai J, Helmchen F, Flint J, Wang SS (2009) Radially expanding transglial calcium waves in the intact cerebellum. Proc Natl Acad Sci U S A 106:3496–3501
Hooser SB, van Alstine W, Kiupel M, Sojka J (2000) Acute pit gas (hydrogen sulfide) poisoning in confinement cattle. J Vet Diagn Invest 12:272–275
Horiuchi T, Kawaguchi M, Kurita N, Inous S, Sakamoto T, Nakamura M, Konishi N, Furuya H (2008) Effects of delta-opioid agonist SNC80 on white matter injury following spinal cord ischemia in normothermic and mildly hypothermic rats. J Anesth 22:32–37
Hosobuchi Y, Baskin DS, Woo SK (1982) Reversal of induced ischemic neurologic deficit in gerbils by the opiate antagonist naloxone. Science 215:69–71
Housley GD, Raybould NP, Thorne PR (1998) Fluorescence imaging of Na^+ influx via P2X receptors in cochlear hair cells. Hear Res 119:1–13
Houzen H, Kanno M, Kikuchi S (1998) AMPA/kainite receptor activation inhibits neuronal delayed rectifier K^+ current via Na^+ entry in rat cortical neurons. Biochem Biophys Res Commun 243:617–621
Hu S, Sheng WS, Lokensgard JR, Peterson PK (2002) Morphine induces apoptosis of human microglia and neurons. Neuropharmacology 42:829–836

Huang H, Gao TM, Gong LW, Zhuang ZY, Li X (2001) Potassium channel blocker TEA prevents CA1 hippocampal injury following transient forebrain ischemia in adult rats. Neurosci Lett 305:83–86

Huang L, Li W, Li B, Zou F (2006) Activation of ATP-sensitive K channels protects hippocampal CA1 neurons from hypoxia by suppressing p53 expression. Neurosci Lett 398:34–38

Hwang IK, Yoo KY, Kim DW, Kang TC, Choi SY, Kwon YG, Han BH, Kim JS, Won MH (2006) Na^+/Ca^{2+} exchanger 1 alters in pyramidal cells and expresses in astrocytes of the gerbil hippocampal CA1 region after ischemia. Brain Res 1086:181–190

Iijima T, Tanaka K, Matsubara S, Kawakami H, Mishima T, Suga K, Akagawa K, Iwao Y (2008) Calcium loading capacity and morphological changes in mitochondria in an ischemic preconditioned model. Neurosci Lett 448:268–272

Itoh T, Itoh A, Horiuchi K, Pleasure D (1998) AMPA receptor-mediated excitotoxicity in human NT2-N neurons results from loss of intracellular Ca^{2+} homeostasis following marked elevation of intracellular Na+. J Neurochem 71:112–124

Iwabuchi S, Kawahara K (2009) Oxygen-glucose deprivation-induced enhancement of extracellular ATP-P2Y purinoceptors signaling for the propagation of astrocytic calcium waves. Biosystems 96:35–43

Iwabuchi S, Kawahara K (2011) Functional significance of the negative-feedback regulation of ATP release via pannexin-1 hemichannels under ischemic stress in astrocytes. Neurochem Int 58:376–384

Iwai T, Niwa M, Nakashima M, Kambara T, Yamada H, Tsurumi K, Nozaki M (1992) Effect of opioids on delayed neuronal death in the gerbil hippocampus. Life Sci 50:PL239–PL244

Iwato M, Inoue S, Kawaguchi M, Nakamura M, Konishi N, Furuya H (2007) Effects of delta-opioid receptor stimulation and inhibition on hippocampal survival in a rat model of forebrain ischemia. Br J Anaesth 99:538–546

Jackson JG, Thayer SA (2006) Mitochondrial modulation of Ca^{2+}-induced Ca^{2+}-release in rat sensory neurons. J Neurophysiol 96:1093–1104

James G, Butt AM (2001) P2X and P2Y purinoreceptors mediate ATP-evoked calcium signalling in optic nerve glia in situ. Cell Calcium 30:251–259

Jarnot M, Corbett AM (2006) Immunolocalization of NaV1.2 channel subtypes in rat and cat brain and spinal cord with high affinity antibodies. Brain Res 1107:1–12

Jeffs G, Meloni BP, Bakker AJ, Knuckey NW (2007) The role of the Na^+/Ca^{2+} exchanger (NCX) in neurons following ischemia. J Clin Neurosci 14:507–514

Jeffs GJ, Meloni BP, Sokolow S, Herchuelz A, Schurmans S, Knuckey NW (2008) NCX3 knockout mice exhibit increased hippocampal CA1 and CA2 neuronal damage compared to wild-type mice following global cerebral ischemia. Exp Neurol 210:268–273

Jeon D, Chu K, Jung KH, Kim M, Yoon BW, Lee CJ, Oh U, Shin HS (2008) Na^+/Ca^{2+} exchanger 2 is neuroprotective by exporting Ca^{2+} during a transient focal cerebral ischemia in the mouse. Cell Calcium 43:482–491

Jiang C, Agulian S, Haddad GG (1992) Cl^- and Na^+ homeostasis during anoxia in rat hypoglossal neurons: intracellular and extracellular in vitro studies. J Physiol 448:697–708

Jiang C, Haddad GG (1991) Effect of anoxia on intracellular and extracellular potassium activity in hypoglossal neurons in vitro. J Neurophysiol 66:103–111

Jiang C, Haddad GG (1993) Short periods of hypoxia activate a K^+ current in central neurons. Brain Res 614:352–356

Jiang D, Zhao L, Clapham DE (2009) Genome-wide RNAi screen identifies Letm1 as a mitochondrial Ca^{2+}/H^+ antiporter. Science 326:144–147

Jiang RG, Zhang XQ, Eyzaguirre C (1998) Hypoxia induced by Na2S2O4 increases $[Na^+]i$ in mouse glomus cells, an effect depressed by cobalt. Experiments with Na^+-selective microelectrodes and voltage-clamping. Brain Res 797:197–208

Jin J, Desai BN, Navarro B, Donovan A, Andrews NC, Clapham DE (2008) Deletion of Trpm7 disrupts embryonic development and thymopoiesis without altering Mg^{2+} homeostasis. Science 322:756–760

Johnson SM, Turner SM (2010) Protecting motor networks during perinatal ischemia: the case for delta-opioid receptors. Ann N Y Acad Sci 1198:260–270

Joshi DC, Singh M, Krishnamurthy K, Joshi PG, Joshi NB (2011) AMPA induced Ca^{2+} influx in motor neurons occurs through voltage gated Ca^{2+} channel and Ca^{2+} permeable AMPA receptor. Neurochem Int 59:913–921

Juhaszova M, Blaustein MP (1997) Na^+ pump low and high ouabain affinity alpha subunit isoforms are differently distributed in cells. Proc Natl Acad Sci U S A 94:1800–1805

Jurkowitz MS, Geisbuhler T, Jung DW, Brierley GP (1983) Ruthenium red-sensitive and insensitive pathways for the release of Ca^{2+} from uncoupled heart mitochondria. Arch Biochem Biophys 223:120–128

Kahlert S, Reiser G (2004) Glial perspectives of metabolic states during cerebral hypoxia-calcium regulation and metabolic energy. Cell Calcium 36:295–302

Kang X, Chao D, Gu Q, Ding G, Wang Y, Balboni G, Lazarus LH, Xia Y (2009) δ-Opioid receptors protect from anoxic disruption of Na^+ homeostasis via Na^+ channel regulation. Cell Mol Life Sci 66:3505–3516

Kanyshkova T, Ehling P, Cerina M, Meuth P, Zobeiri M, Meuth SG, Pape HC, Budde T (2014) Regionally specific expression of high-voltage-activated calcium channels in thalamic nuclei of epileptic and non-epileptic rats. Mol Cell Neurosci 61C:110–122

Kao TK, Ou YC, Liao SL, Chen WY, Wang CC, Chen SY, Chiang AN, Chen CJ (2008) Opioids modulate post-ischemic progression in a rat model of stroke. Neurochem Int 52:1256–1265

Káradóttir R, Cavelier P, Bergersen LH, Attwell D (2005) NMDA receptors are expressed in oligodendrocytes and activated in ischaemia. Nature 438:1162–1166

Karki P, Seong C, Kim JE, Hur K, Shin SY, Lee JS, Cho B, Park IS (2007) Intracellular K^+ inhibits apoptosis by suppressing the Apaf-1 apoptosome formation and subsequent downstream pathways but not cytochrome c release. Cell Death Differ 14:2068–2075

Kawamoto EM, Vivar C, Camandola S (2012) Physiology and pathology of calcium signaling in the brain. Front Pharmacol 3:61

Kiedrowski L (1999) N-methyl-D-aspartate excitotoxicity: relationships among plasma membrane potential, Na^+/Ca^{2+} exchange, mitochondrial Ca^{2+} overload, and cytoplasmic concentrations of Ca^{2+}, H^+, and K^+. Mol Pharmacol 56:619–632

Kiedrowski L (2007) Critical role of sodium in cytosolic $[Ca^{2+}]$ elevations in cultured hippocampal CA1 neurons during anoxic depolarization. J Neurochem 100:915–923

Kiedrowski L, Brooker G, Costa E, Wroblewski JT (1994) Glutamate impairs neuronal calcium extrusion while reducing sodium gradient. Neuron 12:295–300

Kim H, Lee SW, Park JS, Min JH, Kim HK (2012) Genomic analysis of [d-Ala2, d-Leu5] enkephalin preconditioning in cortical neuron and glial cell injury after oxygen deprivation. Brain Res 1447:91–105

Kim J, Woo J, Park YG, Chae S, Jo S, Choi JW, Jun HY, Yeom YI, Park SH, Kim KH, Shin HS, Kim D (2011) Thalamic T-type Ca^{2+} channels mediate frontal lobe dysfunctions caused by a hypoxia-like damage in the prefrontal cortex. J Neurosci 31:4063–4073

Kintner DB, Luo J, Gerdts J, Ballard AJ, Shull GE, Sun D (2007) Role of Na^+-K^+-Cl^- cotransport and Na^+/Ca^{2+} exchange in mitochondrial dysfunction in astrocytes following in vitro ischemia. Am J Physiol Cell Physiol 292:C1113–C1122

Kirischuk S, Kattenmann H, Verkhratsky A (2007) Membrane currents and cytoplasmic sodium transients generated by glutamate transport in Bergmann glial cells. Pflugers Arch 454:245–252

Kirischuk S, Parpura V, Verkhratsky A (2012) Sodium dynamics: another key to astroglial excitability? Trends Neurosci 35:497–506

Kirschner DL, Wilson AL, Drew KL, Green TK (2009) Simultaneous efflux of endogenous D-ser and L-glu from single acute hippocampus slices during oxygen glucose deprivation. J Neurosci Res 87:2812–2820

Knapp RJ, Yamamura HI (1992) Delta opioid receptor radioligands. Biochem Pharmacol 44:1687–1695

Knickerbocker DL, Lutz PL (2001) Slow ATP loss and the defense of ion homeostasis in the anoxic frog brain. J Exp Biol 204:3547–3551

Knöpfel T, Anchisi D, Alojado ME, Tempia F, Strata P (2000) Elevation of intracellular sodium concentration mediated by synaptic activation of metabotropic glutamate receptors in cerebellar Purkinje cells. Eur J Neurosci 12:2199–2204

Kobayashi C, Ohkura M, Nakai J, Matsuki N, Ikegaya Y, Sasaki T (2014) Large-scale imaging of subcellular calcium dynamics of cortical neurons with G-CaMP6-actin. Neuroreport 25:501–506

Kofke WA, Garman RH, Garman R, Rose ME (1999) Opioid neurotoxicity: fentanyl-induced exacerbation of cerebral ischemia in rats. Brain Res 818:326–334

Kofuji P, Newman EA (2004) Potassium buffering in the central nervous system. Neuroscience 129:1045–1056

Koike T, Tanaka S, Oda T, Ninomiya T (2000) Sodium overload through voltage-dependent Na^+ channels induced necrosis and apoptosis of rat superior cervical ganglion cells in vitro. Brain Res Bull 51:345–355

Köller H, Schroeter M, Jander S, Stoll G, Siebler M (2000) Time course of inwardly rectifying $K(^+)$ current reduction in glial cells surrounding ischemic brain lesions. Brain Res 872:194–198

Kolodziej A, Stumm R, Becker A, Hollt V (2008) Endogenous opioid inhibit ischemia-induced generation of immature hippocampal neurons via the μ-opioid receptor. Eur J Neurosci 27:1311–1319

Kornblum HI, Hurlbut DE, Leslie FM (1987) Postnatal development of multiple opioid receptors in rat brain. Brain Res 465:21–41

Koss DJ, Riedel G, Bence K, Platt B (2013) Store-operated Ca^{2+} entry in hippocampal neurons: regulation by protein tyrosine phosphatase PTP1B. Cell Calcium 53:125–138

Kristián T, Siesjő BK (1998) Calcium in ischemic cell death. Stroke 29:705–718

Kroemer G, Galluzzi L, Vandenabeele P, Abrams J, Alnemri ES, Baehrecke EH, Blagosklonny MV, El-Deiry WS, Golstein P, Green DR, Hengartner M, Knight RA, Kumar S, Lipton SA, Malorni W, Nuñez G, Peter ME, Tschopp J, Yuan J, Piacentini M, Zhivotovsky B, Melino G (2009) Classification of cell death: recommendations of the Nomenclature Committee on Cell Death 2009. Cell Death Differ 16:3–11

Kroemer G, Petit P, Zamzami N, Vayssiere JL, Mignotte B (1995) The biochemistry of programmed cell death. FASEB J 9:1277–1287

Krzyżanowska W, Pomierny B, Filip M, Pera J (2014) Glutamate transporters in brain ischemia: to modulate or not? Acta Pharmacol Sin 35:444–462

Kukley M, Capetillo-Zarate E, Dietrich D (2007) Vesicular glutamate release from axons in white matter. Nat Neurosci 10:311–320

Kulik A, Brockhaus J, Pedarzani P, Ballanyi K (2002) Chemical anoxia activates ATP-sensitive and blocks Ca^{2+}-dependent K^+ channels in rat dorsal vagal neurons in situ. Neuroscience 110:541–554

Kumar J, Mayer ML (2013) Functional insights from glutamate receptor ion channel structures. Annu Rev Physiol 75:313–337

Kuruma A, Inoue T, Mikoshiba K (2003) Dynamics of Ca^{2+} and Na+ in the dendrites of mouse cerebral Purkinje cells evoked by parallel fibre stimulation. Eur J Neurosci 18:2677–2689

Ladewig T, Kloppenburg P, Lalley PM, Zipfel WR, Webb WW, Keller BU (2003) Spatial profiles of store-dependent calcium release in motoneurons of the nucleus hypoglossus from newborn mouse. J Physiol 547:775–787

Lafon-Cazal M, Pietri S, Culcasi M, Bockaert J (1993) NMDA-dependent superoxide production and neurotoxicity. Nature 364:535–537

Lam AK, Galione A (2013) The endoplasmic reticulum and junctional membrance communication during calcium signaling. Biochim Biophys Acta 1833:2542–2559

Langer J, Rose CR (2009) Synaptically induced sodium signals in hippocampal astrocytes in situ. J Physiol 587:5859–5877

Larsen GA, Skjellegrind HK, Moe MC, Vinje ML, Berg-Johnsen J (2005) Endoplasmic reticulum dysfunction and Ca^{2+} deregulation in isolated CA1 neurons during oxygen and glucose deprivation. Neurochem Res 30:651–659

Larson J, Drew KL, Folkow LP, Milton SL, Park TJ (2014) No oxygen? No problem! Intrinsic brain tolerance to hypoxia in vertebrates. J Exp Biol 217:1024–1039

Lasser-Ross N, Ross WN (1992) Imaging voltage and synaptically activated sodium transients in cerebellar Purkinje cells. Proc Biol Sci 247:35–39

Lauritzen I, Zanzouri M, Honoré E, Duprat F, Ehrengruber MU, Lazdunski M, Patel AJ (2003) K^+-dependent cerebellar granule neuron apoptosis: role of task leak K^+ channels. J Biol Chem 278:32068–32076

Law PY, Wong YH, Loh HH (2000) Molecular mechanisms and regulation of opioid receptor signaling. Annu Rev Pharmacol Toxicol 40:389–430

Lawrence CL, Rodrigo GC (2001) The Na^+-activated K^+ channel contributes to K^+ efflux in Na^+-loaded guinea-pig but not rat ventricular myocytes. Pflugers Arch 442:595–602

Leenders AGM, Sheng ZH (2005) Modulation of neurotransmitter release by the second messenger-activated protein kinases: implications for presynaptic plasticity. Pharmacol Ther 105:69–84

Leichsenring A, Riedel T, Qin Y, Rubini P, Illes P (2013) Anoxic depolarization of hippocampal astrocytes: possible modulation by P2X7 receptors. Neurochem Int 62:15–22

Leis JA, Bekar LK, Walz W (2005) Potassium homeostasis in the ischemia brain. Glia 50:407–416

Lenart B, Kintner DB, Shull GE, Sun D (2004) Na-K-Cl cotransporter-mediated intracellular Na^+ accumulation affects Ca^{2+} signaling in astrocytes in an in vitro ischemic model. J Neurosci 24:9585–9597

Leng T, Shi Y, Xiong ZG, Sun D (2014) Proton-sensitive cation channels and ion exchangers in ischemic brain injury: new therapeutic targets for stroke? Prog Neurobiol 115:189–209

Levy LM, Warr O, Attwell D (1998) Stoichiometry of the glial glutamate transporter GLT-1 expressed inducibly in a Chinese hamster ovary cell line selected for low endogenous Na^+-dependent glutamate uptake. J Neurosci 18:9620–9628

Lewis A, Peers C, Ashford ML, Kemp PJ (2002) Hypoxia inhibits human recombinant large conductance, Ca^{2+}-activated K^+ (maxi-K) channels by a mechanism which is membrane delimited and Ca^{2+} sensitive. J Physiol 540:771–780

Liang J, Chao D, Sandhu HK, Yu Y, Zhang L, Balboni G, Kim DH, Xia Y (2014) δ-Opioid receptors up-regulate excitatory amino acid transporters in mouse astrocytes. Br J Pharmacol 171(23):5417–5430. doi:10.1111/bph.12857

Liao SL, Chen WY, Raung SL, Chen CJ (2003) Neuroprotection of naloxone against ischemic injury in rats: role of mu receptor antagonism. Neurosci Lett 345:169–172

Libien J, Sacktor TC, Kass IS (2005) Magnesium blocks the loss of protein kinase C, leads to a transient translocation of PKCα and PKCε, and improves recovery after anoxia in rat hippocampal slices. Mol Brain Res 136:104–111

Lim YJ, Zheng S, Zuo Z (2004) Morphine preconditions purkinje cells against cell death under in vitro simulated ischemia-reperfusion conditions. Anesthesiology 100:562–568

Lin CS, Polsky K, Nadler JV, Crain BJ (1990) Selective neocortical and thalamic cell death in the gerbil after transient ischemia. Neuroscience 35:289–299

Lin JH, Lou N, Kang N, Takano T, Hu F, Han X, Xu Q, Lovatt D, Torres A, Willecke K, Yang J, Kang J, Nedergaard M (2008) A central role of connexin 43 in hypoxic preconditioning. J Neurosci 28:681–695

Lipski J, Park TI, Li D, Lee SC, Trevarton AJ, Chung KK, Freestone PS, Bai JZ (2006) Involvement of TRP-like channels in the acute ischemic response of hippocampal CA1 neurons in brain slices. Brain Res 1077:187–199

Lipton P (1999) Ischemic cell death in brain neurons. Physiol Rev 79:1431–1568

Liss B, Roeper J (2001) ATP-sensitive potassium channels in dopaminergic neurons: transducers of mitochondrial dysfunction. News Physiol Sci 16:214–217

Liu D, Slevin JR, Lu C, Chan SL, Hansson M, Elmer E, Mattson MP (2003) Involvement of mitochondrial K^+ release and cellular efflux in ischemic and apoptotic neuronal death. J Neurochem 86:966–979

Liu DM, Adams DJ (2001) Ionic selectivity of native ATP-activated (P2X) receptor channels in dissociated neurones from rat parasympathetic ganglia. J Physiol 534:423–435

Liu H, Chen B, Zhang Y, Qiu Y, Xia Y, Li S, Yao J (2014) Protective effect of delta opioid agonist [d-Ala2, d-Leu5] enkephalin on spinal cord ischemia reperfusion injury by regional perfusion into abdominal aorta in rabbits. Neurosci Lett 584:1–6. doi:10.1016/j.neulet.2014.09.048

Liu H, Moczydlowski E, Haddad GG (1999) O2 deprivation inhibits Ca^{2+}-activated K^+ channels via cytosolic factors in mice neocortical neurons. J Clin Invest 104:577–588

Liu S, Lau L, Wei J, Zhu D, Zou S, Sun HS, Fu Y, Liu F, Lu Y (2004) Expression of $Ca(^{2+})$-permeable AMPA receptor channels primes cell death in transient forebrain ischemia. Neuron 43:43–55

Liu Y, Kintner DB, Begum G, Algharabli J, Cengiz P, Shull GE, Liu XJ, Sun D (2010) Endoplasmic reticulum Ca^{2+} signaling and mitochondrial Cyt c release in astrocytes following oxygen and glucose deprivation. J Neurochem 114:1436–1446

Lopachin RM, Gaughan CL, Lehning EJ, Weber ML, Taylor CP (2001) Effects of ion channel blockade on the distribution of Na, K, Ca and other elements in oxygen-glucose deprived CA1 hippocampal neurons. Neuroscience 103:971–983

Lou LG, Pei G (1997) Modulation of protein kinase C and cAMP-dependent protein kinase by delta-opioid. Biochem Biophys Res Commun 236:626–629

Lukyanetz EA, Shkryl VM, Kravchuk OV, Kostyuk PG (2003) Action of hypoxia on different types of calcium channels in hippocampal neurons. Biochim Biophys Acta 1618:33–38

Luo J, Chen H, Kintner DB, Shull GE, Sun D (2005) Decreased neuronal death in Na^+/H^+ exchanger isoform 1-null mice after in vitro and in vivo ischemia. J Neurosci 25:11256–11268

Luo J, Wang Y, Chen H, Kintner DB, Cramer SW, Gerdts JK, Chen X, Shull GE, Philipson KD, Sun D (2008) A concerted role of Na^+-K^+-Cl^- cotransporter and Na^+/Ca^{2+} exchanger in ischemia damage. J Cereb Blood Flow Metab 28:737–746

Lutz PL, Nilsson GE (2004) Vertebrate brains at the pilot light. Respir Physiol Neurobiol 141:285–296

Lynch JJ III, Yu SP, Canzoniero LMT, Sensi SL, Choi DW (1995) Sodium channel blockers reduce oxygen-glucose deprivation-induced cortical neuronal injury when combined with glutamate receptor antagonists. J Pharmacol Exp Ther 273:554–560

Ma E, Haddad GG (1997) Expression and localization of Na^+/H^+ exchangers in rat central nervous system. Neuroscience 79:591–603

Ma MC, Qian H, Ghassemi F, Zhao P, Xia Y (2005) Oxygen sensitive δ-opioid receptor-regulated survival and death signals: novel insights into neuronal preconditioning and protection. J Biol Chem 280:16208–16218

MacAulay N, Zeuthen T (2012) Glial K^+ clearance and cell swelling: key roles for cotransporters and pumps. Neurochem Res 37:2299–2309

Macdonald JF, Xiong ZG, Jackson MF (2006) Paradox of Ca^{2+} signaling, cell death and stroke. Trends Neurosci 29:75–81

Madry C, Haglerød C, Attwell D (2010) The role of pannexin hemichannels in the anoxic depolarization of hippocampal pyramidal cells. Brain 133:3755–3763

Majewski H, Iannazzo L (1998) Protein kinase C: a physiological mediator of enhanced transmitter output. Prog Neurobiol 55:463–475

Majno G, Joris I (1995) Apoptosis, oncosis, and necrosis. An overview of cell death. Am J pathol 146:3–15

Mank M, Santos AF, Direnberger S, Mrsic-Flogel TD, Hofer SB, Stein V, Hendel T, Reiff DF, Levelt C, Borst A, Bonhoeffer T, Hübener M, Griesbeck O (2008) A genetically encoded calcium indicator for chronic in vivo two-photon imaging. Nat Methods 5:805–811

Mansour A, Fox CA, Burke S, Meng F, Thompson RC, Akil H, Watson SJ (1994) Mu, delta, and kappa opioid receptor mRNA expression in the rat CNS: an in situ hybridization study. J Comp Neurol 350:412–438

Margolis EB, Fields HL, Hjelmstad GO, Mitchell JM (2008) δ-opioid receptor expression in the ventral tegmental area protects against elevated alcohol consumption. J Neurosci 28:12672–12681
Mari Y, Katnik C, Cuevas J (2010) ASIC1a channels are activated by endogenous protons during ischemia and contribute to synergistic potentiation of intracellular Ca^{2+} overload during ischemia and acidosis. Cell Calcium 48:70–82
Martinez-Sánchez M, Striggow F, Schröder UH, Kahlert S, Reymann KG, Reiser G (2004) Na^+ and Ca^{2+} homeostasis pathways, cell death and protection after oxygen-glucose-deprivation in organotypic hippocampal slice cultures. Neuroscience 128:729–740
Mathiesen C, Brazhe A, Thomsen K, Lauritzen M (2013) Spontaneous calcium waves in Bergman glia increase with age and hypoxia and may reduce tissue oxygen. J Cereb Blood Flow Metab 33:161–169
Mayfield KP, D'Alecy LG (1994) Delta-1 opioid agonist acutely increases hypoxic tolerance. J Pharmacol Exp Ther 268:683–688
Mayfield KP, Kozak W, Malvin GM, Porreca F (1996) Hypoxia decreases opioid delta receptor expression in mouse brain. Neuroscience 72:785–789
McDonald JW, Bhattacharyya T, Sensi SL, Lobner D, Ying HS, Canzoniero LMT, Choi DW (1998) Extracellular acidity potentiates AMPA-mediated cortical neuronal death. J Neurosci 18:6290–6299
McGrail KM, Phillips JM, Sweadner KJ (1991) Immunofluorescent localization of three Na, K-ATPase isozymes in the rat central nervous system: both neurons and glia can express more than one Na, K-ATPase. J Neurosci 11:381–391
Medvedeva YV, Lin B, Shuttleworth CW, Weiss JH (2009) Intracellular Zn^{2+} accumulation contributes to synaptic failure, mitochondrial depolarization, and cell death in an acute slice oxygen-glucose deprivation model of ischemia. J Neurosci 29:1105–1114
Meini A, Benocci A, Frosini M, Sgaragli GP, Garcia JB, Pessina GP, Aldinucci C, Palmi M (2003) Potentiation of intracellular Ca^{2+} mobilization by hypoxia-induced NO generation in rat brain striatal slices and human astrocytoma U-373 MG cells and its involvement in tissue damage. Eur J Neurosci 17:692–700
Micu I, Jiang Q, Coderre E, Redsdale A, Zhang L, Woulfe J, Yin X, Trapp BD, McRory JE, Rehak R, Zamponi GW, Wang W, Sty PK (2006) NMDA receptors mediate calcium accumulation in myelin during chemical ischemia. Nature 439:988–992
Micu I, Ridsdale A, Zhang L, Woulfe J, McClintock J, Brantner CA, Brian Andrews S, Stys PK (2007) Real-time measurement of free Ca^{2+} changes in CNS myelin by two-photon microscopy. Nat Med 13:874–879
Miki T, Seino S (2005) Roles of KATP channels as metabolic sensors in acute metabolic changes. J Mol Cell Cardiol 38:917–925
Milton SL, Prentice HM (2007) Beyond anoxia: the physiology of metabolic downregulation and recovery in the anoxia-tolerant turtle. Comp Biochem Physiol A Mol Integr Physiol 147:277–290
Minelli A, Castaldo P, Gobbi P, Salucci S, Magi S, Amoroso S (2007) Cellular and subcellular localization of Na^+-Ca^{2+} exchanger protein isoforms, NCX1, NCX2, and NCX3 in cerebral cortex and hippocampus of adult rat. Cell Calcium 41:221–234
Mironov SL, Langohr K (2005) Mechanisms of Na^+ and Ca^{2+} influx into respiratory neurons during hypoxia. Neuropharmacology 48:1056–1065
Mironov SL, Langohr K, Haller M, Richter DW (1998) Hypoxia activates ATP-dependent potassium channels in inspiratory neurones of neonatal mice. J Physiol 509:755–766
Misonou H, Mohapatra DP, Menegola M, Trimmer JS (2005) Calcium- and metabolic state-dependent modulation of the voltage-dependent Kv2.1 channel regulates neuronal excitability in response to ischemia. J Neurosci 25:11184–11193
Mitani A, Tanaka K (2003) Functional changes of glial glutamate transporter GLT-1 during ischemia: an in vivo study in the hippocampal CA1 of normal mice and mutant mice lacking GLT-1. J Neurosci 23:7176–7182

Mitani A, Yanase H, Namba S, Shudo M, Kataoka K (1995) In vitro ischemia-induced intracellular Ca^{2+} elevation in cerebellar slices: a comparative study with the values found in hippocampal slices. Acta Neuropathol 89:2–7

Mitani A, Yanase H, Sakai K, Wake Y, Kataoka K (1993) Origin of intracellular Ca^{2+} elevation induced by in vitro ischemia-like condition in hippocampal slices. Brain Res 601:103–110

Molinaro P, Cuomo O, Pignataro G, Boscia F, Sirabella R, Pannaccione A, Secondo A, Scoriello A, Adornetto A, Gala R, Viggiano D, Sokolow S, Herchuelz A, Schurmans S, Di Renzo G, Annunziato L (2008) Targeted disruption of Na^+/Ca^{2+} exchanger 3 (NCX3) gene leads to a worsening of ischemic brain damage. J Neurosci 28:1179–1184

Mollajew R, Toloe J, Mironov SL (2013) Single KATP channel opening in response to stimulation of AMPA/kainate receptors is mediated by Na+ accumulation and submembrane ATP and ADP changes. J Physiol 591:2593–2609

Mongin AA (2007) Disruption of ionic and cell volume homeostasis in cerebral ischemia: the perfect storm. Pathophysiology 14:183–193

Monteilh-Zoller MK, Hermosura MC, Nadler MJ, Scharenberg AM, Penner R, Fleig A (2003) TRPM7 provides an ion channel mechanism for cellular entry of trace metal ions. J Gen Physiol 121:49–60

Montero M, Alonso MT, Albillos A, Garcia-Sancho J, Alvarez J (2001) mitochondrial Ca^{2+}-induced Ca^{2+} release mediated by the Ca^{2+} uniporter. Mol Biol Cell 12:63–71

Morimoto N, Kita S, Shimazawa M, Namimatsu H, Tsuruma K, Hayakawa K, Mishima K, Egashira N, Iyoda T, Horie I, Gotoh Y, Iwasaki K, Fujiwara M, Matsuda T, Baba A, Komuro I, Horie K, Takeda J, Iwamoto T, Hara H (2012) Preferential involvement of Na^+/Ca^{2+} exchanger type-1 in the brain damage caused by transient focal cerebral ischemia in mice. Biochem Biophys Res Commun 429:186–190

Moudy AM, Yamada KA, Rothman SM (1994) Rapid desensitization determines the pharmacology of glutamate neurotoxicity. Neuropharmacology 33:953–962

Müller M, Ballanyi K (2003) Dynamic recording of cell death in the in vitro dorsal vagal nucleus of rats in response to metabolic arrest. J Neurophysiol 89:551–561

Müller M, Brockhaus J, Ballanyi K (2002) ATP-independent anoxic activation of ATP-sensitive K^+ channels in dorsal vagal neurons of juvenile mice in situ. Neuroscience 109:313–328

Müller M, Somjen GG (2000a) Na^+ and K^+ concentrations, extra- and intracellular voltage, and the effect of TTX in hypoxic rat hippocampal slices. J Neurophysiol 83:735–745

Müller M, Somjen GG (2000b) Na^+ dependence and the role of glutamate receptors and Na^+ channels in ion fluxes during hypoxia of rat hippocampal slices. J Neurophysiol 84:1869–1880

Murai Y, Ishibashi H, Koyama S, Akaike N (1997) Ca^{2+}-activated K^+ currents in rat locus coeruleus neurons induced by experimental ischemia, anoxia, and hypoglycemia. J Neurophysiol 78:2674–2681

Murugan M, Ling EA, Kaur C (2013) Dysregulated glutamate uptake by astrocytes causes oligodendroglia death in hypoxic perventricular white matter damage. Mol Cell Neurosci 56:342–354

Nabetani M, Okada Y, Takata T, Takada S, Nakamura H (1997) Neural activity and intracellular Ca^{2+} mobilization in the CA1 area of hippocampal slices from immature and mature rats during ischemia or glucose deprivation. Brain Res 769:158–162

Nakamura T, Minamisawa H, Katayama Y, Ueda M, Terashi A, Nakamura K, Kudo Y (1999) Increased intracellular Ca^{2+} concentration in the hippocampal CA1 area during global ischemia and reperfusion in the rat: a possible cause of delayed neuronal death. Neuroscience 88:57–67

Narita M, Kuzumaki N, Miyatake M, Sato F, Wachi H, Seyama Y, Suzuki T (2006) Role of delta-opioid receptor function in neurogenesis and neuroprotection. J Neurochem 97:1494–1505

Nedergaard M (1988) Mechanisms of brain damage in focal cerebral ischemia. Acta Neurol Scand 77:81–101

Nguyen HN, Wang C, Perry DC (2002) Depletion of intracellular calcium stores is toxic to SH-SY5Y neuronal cells. Brain Res 924:159–166

Niatsetskaya ZV, Sosunov SA, Matsiukevich D, Utkina-Sosunova IV, Ratner VI, Starkov AA, Ten VS (2012) The oxygen free radicals originating from mitochondrial complex I contribute to oxidative brain injury following hypoxia-ischemia in neonatal mice. J Neurosci 32:3235–3244

Nicholls DG (2005) Mitochondria and calcium signaling. Cell Calcium 38:311–317

Nicholson DA, Trana R, Katz Y, Kath WL, Spruston N, Geinisman Y (2006) Distance-dependent differences in synapse number and AMPA receptor expression in hippocampal CA1 pyramidal neurons. Neuron 50:431–442

Nikolaeva MA, Mukherjee B, Stys PK (2005) Na^+-dependent sources of intra-axonal Ca^{2+} release in rat optic nerve during in vitro chemical ischemia. J Neurosci 25:9960–9967

Nikonenko I, Bancila M, Bloc A, Muller D, Bijlenga P (2005) Inhibition of T-type calcium channels protects neurons from delayed ischemia-induced damage. Mol Pharmacol 68:84–89

Nilius B, OWsianik G, Voets T, Peters JA (2007) Transient receptor potential cation channels in disease. Physiol Rev 87:165–217

Nistico R, Piccirilli S, Sebastianelli L, Nistico G, Bernardi G, Mercuri NB (2007) The blockade of K^+-ATP channels has neuroprotective effects in an in vitro model of brain ischemia. Int Rev Neurobiol 82:383–395

Noh KM, Yokota H, Mashiko T, Castillo PE, Zukin RS, Bennett MV (2005) Blockade of calcium-permeable AMPA receptors protects hippocampal neurons against global ischemia-induced death. Proc Natl Acad Sci U S A 102:12230–12235

North RA (2002) Molecular physiology of P2X receptors. Physiol Rev 82:1013–1067

Nowikovsky K, Pozzan T, Rizzuto R, Scorrano L, Bernardi P (2012) The pathophysiology of LETM1. J Gen Physiol 139:445–454

O'Reilly JP, Jiang C, Haddad GG (1995) Major differences in response to graded hypoxia between hypoglossal and neocortical neurons. Brain Res 683:179–186

Oeltgen PR, Govindaswami M, Witzke DB (2006) 24-hour pretreatment with δ opioid enhances survival from hemorrhagic shock. Acad Emerg Med 13:127–133

Oguro K, Miyawaki T, Yokota H, Kato K, Kamiya T, Katayama Y, Fukaya M, Watanabe M, Shimazaki K (2004) Upregulation of GluR2 decreases intracellular Ca^{2+} following ischemia in developing gerbils. Neurosci Lett 364:101–105

Ohkura M, Sasaki T, Sadakari J, Gengyo-Ando K, Kagawa-Nagamura Y, Kobayashi C, Ikegaya Y, Nakai J (2012) Genetically encoded green fluorescent Ca^{2+} indicators with improved detectability for neuronal Ca^{2+} signals. PLoS One 7, e51286

Olinger CP, Adams HP Jr, Brott TG, Biller J, Barsan WG, Toffol GJ, Eberle RW, Marler JR (1990) High-dose intravenous naloxone for the treatment of acute ischemic stroke. Stroke 21:721–725

Opitz T, Grooms SY, Bennett MV, Zukin RS (2000) Remodeling of alpha-amino-3-hydroxy-5-methyl-4-isoxazole-propionic acid receptor subunit composition in hippocampal neurons after global ischemia. Proc Natl Acad Sci U S A 97:13360–13365

Ostermeier AM, Schlosser B, Schwender D, Sutor B (2000) Activation of mu- and delta-opioid receptors causes presynaptic inhibition of glutamatergic excitation in neocortical neurons. Anesthesiology 93:1053–1063

Ouardouz M, Malek S, Coderre E, Stys PK (2006) Complex interplay between glutamate receptors and intracellular Ca^{2+} stores during ischemia in rat spinal cord white matter. J Physiol 577:191–204

Ouardouz M, Nikolaeva MA, Coderre E, Zamponi GW, McRory JE, Trapp BD, Yin X, Wang W, Woulfe J, Stys PK (2003) Depolarization-induced Ca^{2+} release in ischemic spinal cord white matter involves L-type Ca^{2+} channel activation of ryanodine receptors. Neuron 40:53–63

Ouyang YB, Xu LJ, Emery JF, Lee AS, Giffard RG (2011) Overexpressing GRP78 influences Ca^{2+} handling and function of mitochondria in astrocytes after ischemia-like stress. Mitochondrion 11:279–286

Ozawa S, Kamiya H, Tsuzuki K (1998) Glutamate receptors in the mammalian central nervous system. Prog Neurobiol 54:581–618

Ozawa S, Tsuzuki K, Iino M, Ogura A, Kudo Y (1989) Three types of voltage-dependent calcium current in cultured rat hippocampal neurons. Brain Res 495:329–336

Palty R, Silverman WF, Hershfinkel M, Caporale T, Sensi SL, Parnis J, Nolte C, Fishman D, Shoshan-Barmatz V, Herrmann S, Khananshvili D, Sekler I (2010) NCLX is an essential component of mitochondrial Na^+/Ca^{2+} exchange. Proc Natl Acad Sci U S A 107:436–441

Pamenter ME, Buck LT (2008) δ-opioid receptor antagonism induces NMDA receptor-dependent excitotoxicity in anoxic turtle cortex. J Exp Biol 211:3512–3517

Paoletti P, Neyton J (2007) NMDA receptor subunits: function and pharmacology. Curr Opin Pharmacol 7:39–47

Papa M, Canitano A, Boscia F, Castaldo P, Sellitti S, Porzig H, Taglialatela M, Annunziato L (2003) Differential expression of the Na^+-Ca^{2+} exchanger transcripts and proteins in rat brain regions. J Comp Neurol 461:31–48

Parekh AB, Putney JW Jr (2005) Store-operated calcium channels. Physiol Rev 85:757–810

Park YK, Jung SJ, Yoo JE, Kwak J, Lim W, Kim J (2003) Effect of acute hypoxia on ATP-sensitive potassium currents in substantia gelatinosa neurons of juvenile rats. Pflugers Arch 446:600–606

Parsons MP, Raymond LA (2014) Extrasynaptic NMDA receptor involvement in central nervous system disorders. Neuron 82:279–293

Paschen W (2001) Dependence of vital cell function on endoplasmic reticulum calcium levels: implications for the mechanisms underlying neuronal cell injury in different pathological states. Cell Calcium 29:1–11

Paschen W, Doutheil J, Uto A, Gissel C (1996) Changes in endoplasmic reticulum Ca^{2+}-ATPase mRNA levels in transient cerebral ischemia of rat: a quantitative polymerase chain reaction study. Neurosci Lett 217:41–44

Pászty K, Antalffy G, Hegedüs L, Padányi R, Penheiter AR, Filoteo AG, Penniston JT, Enyedi A (2007) Cleavage of the plasma membrane Ca^{2+}-ATPase during apoptosis. Ann N Y Acad Sci 1099:440–450

Patneau DK, Wright PW, Winters C, Mayer ML, Gallo V (1994) Glial cells of the oligodendrocyte lineage express both kainate- and AMPA-preferring subtypes of glutamate receptor. Neuron 12:357–371

Pedersen SF, O'Donnel ME, Anderson SE, Cala PM (2006) Physiology and pathophysiology of Na^+/H^+ exchange and Na^+-K^+-$2Cl^-$ cotransporter in heart, brain and blood. Am J Physiol Regul Integr Comp Physiol 291:R1–R25

Pellegrini-Giampietro DE, Zurin RS, Bennett MVL, Cho S, Palsinelli WA (1992) Switch in glutamate receptor subunit gene expression in CA1 subfield of hippocampus following global ischemia in rats. Proc Natl Acad Sci U S A 89:10499–10503

Peng L, Martin-Vasallo P, Sweadner KJ (1997) Isoforms of Na, K-ATPase alpha and beta subunits in the rat cerebellum and in granule cell cultures. J Neurosci 17:3488–3502

Peng PH, Huang HS, Lee YJ, Chen YS, Ma MC (2009) Novel role for the δ-opioid receptor in hypoxic preconditioning in rat retinas. J Neurochem 108:741–754

Perez Velazquez JL, Frantseva MV, Carlen PL (1997) In vitro ischemia promotes glutamate-mediated free radical generation and intracellular calcium accumulation in hippocampal pyramidal neurons. J Neurosci 17:9085–9094

Pérez-Domínguez M, Hernández-Benítez R, Peña Segura C, Pasantes-Morales H (2014) Thrombin-facilitated efflux of d-[(3)h]-aspartate from cultured astrocytes and neurons under hyponatremia and chemical ischemia. Neurochem Res 39:1219–1231

Peterson BL, Larson J, Buffenstein R, Park TJ, Fall CP (2012) Blunted neuronal calcium response to hypoxia in naked mole-rat hippocampus. PLoS One 7, e31568

Petrat F, Li T, Dehne N, de Groot H, Rauen U (2006) Sodium as the major mediator of NO-induced cell death in cultured hepatocytes. Life Sci 79:1606–1615

Pignataro G, Boscia F, Esposito E, Sirabella R, Cuomo O, Vinciguerra A, Di Renzo G, Annunziato L (2012) NCX1 and NCX3: two new effectors of delayed preconditioning in brain ischemia. Neurobiol Dis 45:616–623

Pisani A, Calabresi P, Bernardi G (1997) Hypoxia in striatal and cortical neurons: membrane potential and Ca^{2+} measurement. Neuroreport 8:1143–1147
Pisani A, Calabresi P, Centonze D, Marfia GA, Bernardi G (1999) Electrophysiological recordings and calcium measurements in striatal large aspiny interneurons in response to combined O2/ glucose deprivation. J Neurophysiol 81:2508–2516
Pisani A, Calabresi P, Tozzi A, Bernardi G, Knöpfel T (1998a) Early sodium elevations induced by combined oxygen and glucose deprivation in pyramidal cortical neurons. Eur J Neurosci 10:3572–3574
Pisani A, Calabresi P, Tozzi A, D'Angelo V, Bernardi G, Iadecol C (1998b) L-type Ca^{2+} channel blockers attenuate electrical changes and Ca^{2+} rise induced by oxygen/glucose deprivation in cortical neurons. Stroke 29:196–202
Pivonkova H, Benesova J, Butenko O, Chvatal A, Anderova M (2010) Impact of global cerebral ischemia on K^+ channel expression and membrane properties of glial cells in the rat hippocampus. Neurochem Int 57:783–794
Plant LD, Kemp PL, Peers C, Henderson Z, Pearson HA (2002) Hypoxic depolarization of cerebellar granule neurons by specific inhibition of TASK-1. Stroke 33:2324–2328
Pogozheva ID, Przydzial MJ, Mosberg HI (2005) Homology modeling of opioid receptor-ligand complexes using experimental constraints. AAPS J 7:E434–E448
Pond BB, Berglund K, Kuner T, Feng G, Augustine GJ, Schwartz-Bloom RD (2006) The chloride transporter Na^+-K^+-Cl^- cotransporter isoform-1 contributes to intracellular chloride increases after in vitro ischemia. J Neurosci 26:1396–1406
Pow DV, Barnett NL (2000) Developmental expression of excitatory amino acid transporter 5: a photoreceptor and bipolar cell glutamate transporter in rat retina. Neurosci Lett 280:21–24
Probert AW, Borosky S, Marcoux FW, Taylor P (1997) Sodium channel modulators prevent oxygen and glucose deprivation injury and glutamate release in rat neocortical cultures. Neuropharmacology 36:1031–1038
Raley-Susman KM, Kass IS, Cottrell JE, Newman RB, Chambers G, Wang J (2001) Sodium influx blockade and hypoxic damage to CA1 pyramidal neurons in rat hippocampal slices. J Neurophysiol 86:2715–2726
Ramsey IS, Delling M, Clapham DE (2006) An introduction to TRP channels. Annu Rev Physiol 68:619–647
Reid JM, Paterson DJ (1996) Role of K^+ in regulating hypoxic cerebral blood flow in the rat: effect of glibenclamide and ouabain. Am J Physiol 270:H45–H52
Remillard CV, Yuan JXJ (2004) Activation of K^+ channels: an essential pathway in programmed cell death. Am J Physiol 286:L49–L67
Ritter M, Sui Z, Philipson KD, Li F, Spitzer KW, Ishida H, Barry WH (2003) Ca^{2+} sparks induced by Na/Ca exchange. Cell Calcium 34:11–17
Rizzuto R, Brini M, Murgia M, Pozzan T (1993) Microdomains with high Ca^{2+} close to IP3-sensitive channels that are sensed by neighboring mitochondria. Science 262:744–747
Rizzuto R, Pinton P, Carrington W, Fay FS, Fogarty KE, Lifshitz LM, Tuft RA, Pozzan T (1998) Close contacts with the endoplasmic reticulum as determinants of mitochondrial Ca^{2+} responses. Science 280:1763–1766
Rodgers-Garlick CI, Hogg DW, Buck LT (2013) Oxygen-sensitive reduction in Ca^{2+}-activated K^+ channel open probability in turtle cerebrocortex. Neuroscience 237:243–254
Rojas H, Colina C, Ramos M, Benaim G, Jaffe EH, Caputo C, DiPolo R (2007) Na^+ entry via glutamate transporter activates the reverse Na^+/Ca^{2+} exchange and triggers Ca^{2+}i-induced Ca^{2+} release in rat cerebellar type-1 astrocytes. J Neurochem 100:1188–1202
Rose CR, Karus C (2013) Two sides of the same coin: sodium homeostasis and signaling in astrocytes under physiological and pathophysiological conditions. Glia 61:1191–1205
Rose CR, Konnerth A (2001) NMDA receptor-mediated Na^+ signals in spines and dendrites. J Neurosci 21:4207–4214
Rose CR, Ransom BR (1997) Regulation of intracellular sodium in cultured rat hippocampal neurons. J Physiol 499:573–587

Rose CR, Waxman SG, Ransom BR (1998) Effects of glucose deprivation, chemical hypoxia, and simulated ischemia on Na^+ homeostasis in rat spinal cord astrocytes. J Neurosci 18:3554–3562

Rossi DJ, Oshima T, Attwell D (2000) Glutamate release in severe brain ischemia is mainly by reversed uptake. Nature 403:316–321

Ruffin VA, Gu XQ, Zhou D, Douglas RM, Sun X, Trouth CO, Haddad GG (2008) The sodium-activated potassium channel Slack is modulated by hypercapnia and acidosis. Neuroscience 151:410–418

Rundén-Pran E, Haug FM, Storm JF, Ottersen OP (2002) BK channel activity determines the extent of cell degeneration after oxygen and glucose deprivation: a study in organotypical hippocampal slice cultures. Neuroscience 112:277–288

Rytter A, Cardoso CM, Johansson P, Cronberg T, Hansson MJ, Mattiasson G, Elmér E, Wieloch T (2005) The temperature dependence and involvement of mitochondria permeability transition and caspase activation in damage to organotypic hippocampal slices following in vitro ischemia. J Neurochem 95:1108–1117

Salter MG, Fern R (2005) NMDA receptors are expressed in developing oligodendrocyte processes and mediate injury. Nature 438:1167–1171

Salter MG, Fern R (2008) The mechanisms of acute ischemic injury in the cell processes of developing white matter astrocytes. J Cereb Blood Flow Metab 28:588–601

Samways DS, Li Z, Egan TM (2014) Principles and properties of ion flow in P2X receptors. Front Cell Neurosci 8:6

Scherrer G, Tryoen-Tóth P, Filliol D, Matifas A, Laustriat D, Cao YQ, Basbaum AI, Dierich A, Vonesh JL, Gavériaux-Ruff C, Kieffer BL (2006) Knockin mice expressing fluorescent delta-opioid receptors uncover G protein-coupled receptor dynamics in vivo. Proc Natl Acad Sci U S A 103:9691–9696

Schild L, Huppelsberg J, Kahlert S, Keilhoff G, Reiser G (2003) Brain mitochondria are primed by moderate Ca^{2+} rise upon hypoxia/reoxygenation for functional breakdown and morphological disintegration. J Biol Chem 278:25454–25460

Schiller J, Schiller Y, Stuart G, Sakmann B (1997) Calcium action potentials restricted to distal apical dendrites of rat neocortical pyramidal neurons. J Physiol 505:605–616

Schock SC, Leblanc D, Hakim AM, Thompson CS (2008) ATP release by way of connexin 36 hemichannels mediates ischemic tolerance in vitro. Biochem Biophys Res Commun 368:138–144

Schreiner AE, Rose CR (2012) Quantitative imaging of intracellular sodium. In: Méndez-Vilas A (ed) Current microscopy contributions to advances in science and technology. Formatex, Badajoz, pp 119–129

Schwab BL, Guerini D, Didszun C, Bano D, Ferrando-May E, Fava E, Tam J, Xu D, Xanthoudakis S, Nicholson DW, Carafoli E, Nicotera P (2002) Cleavage of plasma membrane calcium pumps by caspases: a link between apoptosis and necrosis. Cell Death Differ 9:818–831

Scorziello A, Savoia C, Sisalli MJ, Adornetto A, Secondo A, Boscia F, Esposito A, Polishchuk EV, Polishchuk RS, Molinaro P, Carlucci A, Lignitto L, Di Renzo G, Feliciello A, Annunziato L (2013) NCX3 regulates mitochondrial Ca^{2+} handling through the AKAP121-anchored signaling complex and prevents hypoxia-induced neuronal death. J Cell Sci 126:5566–5577

Secondo A, Staiano RI, Scorziello A, Sirabella R, Boscia F, Adornetto A, Valsecchi V, Molinaro P, Canzoniero LM, Di Renzo G, Annunziato L (2007) BHK cells transfected with NCX3 are more resistant to hypoxia followed by reoxygenation than those transfected with NCX1 and NCX2: possible relationship with mitochondrial membrane potential. Cell Calcium 42:521–535

Selvatici R, Marino S, Piubello C, Rodi D, Beani L, Gandini E, Siniscalchi A (2002) Protein kinase C activity, translocation, and selective isoform subcellular redistribution in the rat cerebral cortex after in vitro ischemia. J Neurosci Res 71:64–71

Sen D, Huchital M, Chen YL (2013) Crosstalk between delta opioid receptor and nerve growth factor signaling modulates neuroprotection and differentiation in rodent cell models. Int J Mol Sci 14:21114–21139

Sheldon C, Church J (2004) Reduced contribution from Na^+/H^+ exchange to acid extrusion during anoxia in adult rat hippocampal CA1 neurons. J Neurochem 88:594–603

Sheldon C, Diarra A, Cheng YM, Church J (2004) Sodium influx pathways during and after anoxia in rat hippocampal neurons. J Neurosci 24:11057–11069

Shen W, Slaughter MM (2002) A non-excitatory paradigm of glutamate toxicity. J Neurophysiol 87:1629–1634

Shi X, Gillespie PG, Nuttall AL (2005) Na^+ influx triggers bleb formation on inner hair cells. Am J Physiol Cell Physiol 288:C1332–C1341

Shigeri Y, Seal RP, Shimamoto K (2004) Molecular pharmacology of glutamate transporters, EAATs and VGLUTs. Brain Res Brain Res Rev 45:250–265

Shinotsuka T, Yasui M, Nuriya M (2014) Astrocytic gap junctional networks suppress cellular damage in an in vitro model of ischemia. Biochem Biophys Res Commun 444:171–176

Sick TJ, Rosenthal M, LaManna JC, Lutz PL (1982) Brain potassium ion homeostasis, anoxia, and metabolic inhibition in turtles and rats. Am J Physiol 243:R281–R288

Sick TJ, Solow EL, Roberts EL Jr (1987) Extracellular potassium ion activity and electrophysiology in the hippocampal slice: paradoxical recovery of synaptic transmission during anoxia. Brain Res 418:227–234

Silver IA, Deas J, Erecinska M (1997) Ion homeostasis in brain cells: differences in intracellular ion responses to energy limitation between cultured neurons and glial cells. Neuroscience 78:589–601

Silver IA, Erecińska M (1990) Intracellular and extracellular changes of $[Ca^{2+}]$ in hypoxia and ischemia in rat brain in vivo. J Gen Physiol 95:837–866

Silver IA, Erecińska M (1992) Ion homeostasis in rat brain in vivo: intra- and extracellular $[Ca^{2+}]$ and $[H^+]$ in hippocampus during recovery from short-term, transient ischemia. J Cereb Blood Flow Metab 12:759–772

Simard JM, Chen M, Tarasov KV, Bhatta S, Ivanova S, Melnitchenko L, Tsymbalyuk N, West GA, Gerzanich V (2006) Newly expressed SUR1-regulated NCCa-ATP channel mediates cerebral edema after ischemic stroke. Nat Med 12:433–440

Simard JM, Tarasov KV, Gerzanich V (2007) Non-selective cation channels, transient receptor potential channels and ischemic stroke. Biochim Biophys Acta 1772:947–957

Sirabella R, Secondo A, Pannaccione A, Scorziello A, Valsecchi V, Adornetto A, Bilo L, Di Renzo G, Annunziato L (2009) Anoxia-induced NF-kappaB-dependent upregulation of NCX1 contributes to Ca^{2+} refilling into endoplasmic reticulum in cortical neurons. Stroke 40:922–929

Sisalli MJ, Secondo A, Esposito A, Valsecchi V, Savoia C, Di Renzo GF, Annunziato L, Scorziello A (2014) Endoplasmic reticulum refilling and mitochondrial calcium extrusion promoted in neurons by NCX1 and NCX3 in ischemic preconditioning are determinant for neuroprotection. Cell Death Differ 21:1142–1149

Sivakumar V, Foulds WS, Luu CD, Ling EA, Kaur C (2013) Hypoxia-induced retinal ganglion cell damage through activation of AMPA receptors and the neuroprotective effects of DNQX. Exp Eye Res 109:83–97

Skarphedinsson JO, Thoren P (1988) Endorphin mechanisms are responsible for the beneficial effects of opioid antagonists on cerebral function during relative cerebral ischaemia in rats. Acta Physiol Scand 132:281–288

Smith IF, Boyle JP, Rome HA, Peers C (2004) Hypoxic regulation of Ca^{2+} signaling in cultured rat astrocytes. Glia 49:153–157

Smith IF, Plant LD, Boyle JP, Skinner RA, Pearson HA, Peers C (2003) Chronic hypoxia potentiates capacitative Ca^{2+} entry in type-I cortical astrocytes. J Neurochem 85:1109–1116

Smyth JT, Hwang SY, Tomita T, DeHaven WI, Mercer JC, Putney JW (2010) Activation and regulation of store-operated calcium entry. J Cell Mol Med 14:2337–2349

Solnyshkova TG, Shakhlamov VA, Volodina EP (2004) Cerebral cortex ultrastructure during exposure to hydrogen sulfide gas. Neurosci Behav Physiol 34:343–345

Stanika RI, Winters CA, Pivovarova NB, Andrews SB (2010) Differential NMDA receptor-dependent calcium loading and mitochondrial dysfunction in CA1 vs. CA3 hippocampal neurons. Neurobiol Dis 37:403–411

Stanzione P, Stefani A, Calabresi P, Mercuri NB, Bernardi G (1989) Met- and leu-enkephalins inhibit rat cortical neurons intracellularly recorded in vivo while morphine excites them: evidence for naloxone-sensitive and naloxone-insensitive effects. Exp Brain Res 77:302–308

Staples M, Acosta S, Tajiri N, Pabon M, Kaneko Y, Borlongan CV (2013) Delta opioid receptor and its peptide: a receptor-ligand neuroprotection. Int J Mol Sci 14:17410–17419

Steinbeck JA, Henke N, Opatz J, Gruszczynska-Biegala J, Schneider L, Theiss S, Hamacher N, Steinfarz B, Golz S, Brüstle O, Kuznicki J, Methner A (2011) A Store-operated calcium entry modulates neuronal network activity in a model of chronic epilepsy. Exp Neurol 232:185–194

Stys PK (1998) Anoxic and ischemic injury of myelinated axons in CNS white matter: from mechanistic concepts to therapeutics. J Cereb Blood Flow Metab 18:2–25

Stys PK, Lopachin RM (1998) Mechanisms of calcium and sodium fluxes in anoxic myelinated central nervous system axons. Neuroscience 82:21–32

Stys PK, Waxman SG, Ransom BR (1992) Ionic mechanisms of anoxic injury in mammalian CNS white matter: role of Na^+ channels and Na^+-Ca^{2+} exchanger. J Neurosci 12:430–439

Su DS, Wang ZH, Zheng YJ, Zhao YH, Wang XR (2007) Dose-dependent neuroprotection of delta opioid peptide [D-Ala2, D-Leu5]enkephalin in neuronal death and retarded behavior induced by forebrain ischemia in rats. Neurosci Lett 423:113–117

Summers RL, Li Z, Hildebrandt D (2003) Effect of a delta receptor agonist on duration of survival during hemorrhagic shock. Acad Emerg Med 10:587–593

Sun HS, Jackson MF, Martin LJ, Jansen K, Teves L, Cui H, Kiyonaka S, Mori Y, Jones M, Forder JP, Golde TE, Orser BA, Macdonald JF, Tymianski M (2009) Suppression of hippocampal TRPM7 protein prevents delayed neuronal death in brain ischemia. Nat Neurosci 12:1300–1307

Sun HS, Peng ZP, Barber PA, Buchan AM, French RJ (2007) Kir6.2-containing ATP-sensitive potassium channels protect cortical neurons from ischemic/anoxic injury in vitro and in vivo. Neuroscience 144:1509–1515

Sun MK, Hongpaisan J, Nelson TJ, Alkon DL (2008) Poststroke neuronal rescue and synaptogenesis mediated in vivo by protein kinase C in adult brains. Proc Natl Acad Sci U S A 105:13620–13625

Sun X, Gu XQ, Haddad GG (2003) Calcium influx via L- and N-type calcium channels activates a transient large-conductance Ca^{2+}-activated K^+ current in mouse neocortical pyramidal neurons. J Neurosci 23:3639–3648

Sung J, Chao D, Xia Y (2008) Neuronal response to hypoxia. In: Wang Q, Ying W (eds) New frontiers in neurological research. Research Signpost, Trivandrum, pp 73–153

Svingos AL, Cheng PY, Clarke CL, Pickel VM (1995) Ultrastructural localization of δ-opioid receptor and Met5-enkephalin immunoreactivity in rat insular cortex. Brain Res 700:25–39

Szydlowska K, Tymianski M (2010) Calcium, ischemia and excitotoxicity. Cell Calcium 47:122–129

Takahashi S, Shibata M, Fukunchi Y (1999) Role of sodium ion influx in depolarization-induced neuronal cell death by high KCl or veratridine. Eur J Pharmacol 372:297–304

Takita M, Puka-Sundvall M, Miyakawa A, Hagberg H (2004) In vivo calcium imaging of cerebral cortex in hypoxia-ischemia followed by developmental stage-specific injury in rats. Neurosci Res 48:169–173

Tan S, Wood M, Maher P (1998) Oxidative stress induces a form of programmed cell death with characteristics of both apoptosis and necrosis in neuronal cells. J Neurochem 71:95–105

Tanabe M, Mori M, Gähwiler BH, Gerber U (1999) Apamin-sensitive conductance mediates the $K(^+)$ current response during chemical ischemia in CA3 pyramidal cells. J Neurophysiol 82:2876–2882

Tanaka E, North RA (1994) Opioid actions on rat anterior cingulated cortex neurons in vitro. J Neurosci 14:1106–1113

Tanaka E, Uchikado H, Niiyama S, Uematsu K, Higashi H (2002) Extrusion of intracellular calcium ion after in vitro ischemia in the rat hippocampal CA1 region. J Neurophysiol 88:879–887

Tanaka K (2001) Alteration of second messengers during acute cerebral ischemia-adenylate cyclase, cyclic AMP-dependent protein kinase, and cyclic AMP response element binding protein. Prog Neurobiol 65:173–207

Tanaka K, Iijima T, Mishima T, Suga K, Akagawa K, Iwao Y (2009) Ca($^{2+}$) buffering capacity of mitochondria after oxygen-glucose deprivation in hippocampal neurons. Neurochem Res 34:221–226

Tanaka K, Watase K, Manabe T, Yamada K, Watanabe M, Takahashi K, Iwama H, Nishikawa T, Ichihara N, Kikuchi T, Okuyama S, Kawashima N, Hori S, Takimoto M, Wada K (1997) Epilepsy and exacerbation of brain injury in mice lacking the glutamate transporter GLT-1. Science 276:1699–1702

Taylor CP, Weber ML, Gaughan CL, Lehning EJ, LoPachin RM (1999) Oxygen/glucose deprivation in hippocampal slices: altered intraneuronal elemental composition predicts structural and functional damage. J Neurosci 19:619–629

Thomas R, Salter MG, Wilke S, Husen A, Allcock N, Nivison M, Nnoli AN, Fern R (2004) Acute ischemic injury of astrocytes is mediated by Na-K-Cl cotransport and not Ca^{2+} influx at a key point in white matter development. J Neuropathol Exp Neurol 63:856–871

Thompson RJ, Macvicar BA (2008) Connexin and pannexin hemichannels of neurons and astrocytes. Channels (Austin) 2:81–86

Thompson RJ, Zhou N, MacVicar BA (2006) Ischemia opens neuronal gap junction hemichannels. Science 312:924–927

Thurneysen T, Nicoll DA, Philipson KD, Porzig H (2002) Sodium/calcium exchanger subtypes NCX1, NCX2 and NCX3 show cell-specific expression in rat hippocampus cultures. Brain Res Mol Brain Res 107:145–156

Tian L, Hires SA, Mao T, Huber D, Chiappe ME, Chalasani SH, Petreanu L, Akerboom J, McKinney SA, Schreiter ER, Bargmann CI, Jayaraman V, Svoboda K, Looger LL (2009) Imaging neural activity in worms, flies and mice with improved GCaMP calcium indicators. Nat Methods 6:875–881

Tian X, Guo J, Zhu M, Li M, Wu G, Xia Y (2013a) δ-Opioid receptor activation rescues the functional TrkB receptor and protects the brain from ischemia-reperfusion injury in the rat. PLoS One 8, e69252

Tian X, Hua F, Sandhu HK, Chao D, Balboni G, Salvadori S, He X, Xia Y (2013b) Effect of δ-opioid receptor activation on BDNF-TrkB vs. TNF-α in the mouse cortex exposed to prolonged hypoxia. Int J Mol Sci 14:15959–15976

Tian X, Zhou Y, Gao L, He G, Jiang W, Li W, Takahashi E (2013c) Analysis of ischemic neuronal injury in Cav2.1 channel α1 subunit mutant mice. Biochem Biophys Res Commun 434:60–64

Tian XS, Zhou F, Yang R, Xia Y, Wu GC, Guo JC (2008) Effects of intracerebroventriclular injection of δ-opioid receptor agonist TAN-67 or antagonist naltrindole on acute cerebral ischemia in rat. Sheng Li Xue Bao 64:475–484

Tjong YW, Jian K, Li M, Chen M, Gao TM, Fung ML (2007) Elevated endogenous nitric oxide increases Ca^{2+} flux via L-type Ca^{2+} channels by S-nitrosylation in rat hippocampal neurons during severe hypoxia and in vitro ischemia. Free Radic Biol Med 42:52–63

Tonkikh AA, Carlen PL (2009) Impaired presynaptic cytosolic and mitochondrial calcium dynamics in aged compared to young adult hippocampal CA1 synapses ameliorated by calcium chelation. Neuroscience 159:1300–1308

Toselli M, Tosetti P, Taglietti V (1999) Kinetic study of N-type calcium current modulation by δ-opioid receptor activation in the mammalian cell line NG108-15. Biophys J 76:2560–2574

Turner SM, Johnson SM (2011) Delta-opioid receptor activation prolongs respiratory motor output during oxygen-glucose deprivation in neonatal rat spinal cord in vitro. Neuroscience 187:70–83

Tymianski M, Charlton MP, Carlen PL, Tator CH (1993) Source specificity of early calcium neurotoxicity in cultured embryonic spinal neurons. J Neurosci 13:2085–2104

Tzingounis AV, Wadiche JI (2007) Glutamate transporters: confining runaway excitation by shaping synaptic transmission. Nat Rev Neurosci 8:935–947

Uematsu D, Greenberg JH, Reivich M, Karp A (1988) In vivo measurement of cytosolic free calcium during cerebral ischemia and reperfusion. Ann Neurol 24:420–428

Utkina-Sosunova IV, Niatsetskaya ZV, Sosunov SA, Ratner VI, Matsiukevich D, Ten VS (2013) Nelfinavir inhibits intra-mitochondrial calcium influx and protects brain against hypoxic-ischemic injury in neonatal mice. PLoS One 8, e62448

Vaillend C, Mason SE, Cuttle MF, Alger BE (2002) Mechanisms of neuronal hyperexcitability caused by partial inhibition of Na^+-K^+-ATPase in the rat CA1 hippocampal region. J Neurophysiol 88:2963–2978

Vander Jagt TA, Connor JA, Shuttleworth CW (2008) Localized loss of Ca^{2+} homeostasis in neuronal dendrites is a downstream consequence of metabolic compromise during extended NMDA exposures. J Neurosci 28:5029–5039

Vannucci RC, Brucklacher RM, Vannucci SJ (2001) Intracellular calcium accumulation during the evolution of hypoxic-ischemic brain damage in the immature rat. Brain Res Dev Brain Res 126:117–120

Verkhratsky A (2005) Physiology and pathophysiology of the calcium store in the endoplasmic reticulum of neurons. Physiol Rev 85:201–279

Verkhratsky A, Parpura V (2014) Store-operated calcium entry in neuroglia. Neurosci Bull 30:125–133

Verkhratsky A, Rodríguez JJ, Parpura V (2012) Calcium signalling in astroglia. Mol Cell Endocrinol 353:45–56

Verstreken P, Ly CV, Venken KJ, Koh TW, Zhou Y, Bellen HJ (2005) Synaptic mitochondria are critical for mobilization of reserve pool vesicles at Drosophila neuromuscular junctions. Neuron 47:365–378

Vlaskovska M, Schramm M, Nylander I, Kasakov L, You ZB, Herrera-Marschitz M, Terenius L (1997) Opioid effects on $45Ca^{2+}$ uptake and glutamate release in rat cerebral cortex in primary culture. J Neurochem 68:517–524

Volterra A, Liaudet N, Savtchouk I (2014) Astrocyte Ca^{2+} signaling: an unexpected complexity. Nat Neurosci 15:327–335

Vyklický L Jr, Benveniste M, Mayer ML (1990) Modulation of N-methyl-D-aspartic acid receptor desensitization by glycine in mouse cultured hippocampal neurons. J Physiol 428:313–331

Wadman WJ, Connor JA (1992) Persisting modification of dendritic calcium influx by excitatory amino acid stimulation in isolated CA1 neurons. Neuroscience 48:293–305

Waldmann R, Champigny G, Bassilana F, Heurteaux C, Lazdunski M (1997) A proton-gated cation channel involved in acid-sensing. Nature 386:173–177

Wallace DR, Dodson SL, Nath A, Booze RM (2006) Delta opioid agonists attenuate TAT1-72-induced oxidative stress in SK-N-SH cells. Neurotoxicology 27:101–107

Walz W (2004) Potassium homeostasis in the brain at the organ and cell level. In: Hertz L (ed) Non-neuronal cells of the nervous system, function and dysfunction. Elsevier, Amsterdam, pp 595–609

Walz W (2000) Role of astrocytes in the clearance of excess extracellular potassium. Neurochem Int 36:291–300

Wang XM, Mokha SS (1996) Opioids modulate N-methyl-D-aspartic acid (NMDA)-evoked responses of trigeminothalamic neurons. J Neurophysiol 76:2093–2096

Wang C, Nguyen HN, Maguire JL, Perry DC (2002) Role of intracellular calcium stores in cell death from oxygen-glucose deprivation in a neuronal cell line. J Cereb Blood Flow Metab 22:206–214

Wang J, Bright R, Mochly-Rosen D, Giffard RG (2004) Cell-specific role for ε- and βI-protein kinase C isozymes in protecting cortical neurons and astrocytes from ischemia-like injury. Neuropharmacology 47:136–145

Wang Q, Chao D, Chen T, Sandhu H, Xia Y (2014) δ-opioid receptors and inflammatory cytokines in hypoxia: differential regulation between glial and neuron-like cells. Transl Stroke Res 5:476–483

Wang S, Duan Y, Su D, Li W, Tan J, Yang D, Wang W, Zhao Z, Wang X (2011) Delta opioid peptide [D-Ala2, D-Leu5] enkephalin (DADLE) triggers postconditioning against transient forebrain ischemia. Eur J Pharmacol 658:140–144

Wang Y, Luo J, Chen X, Chen H, Cramer SW, Sun D (2008) Gene inactivation of Na^+/H^+ exchanger isoform 1 attenuates apoptosis and mitochondrial damage following transient focal cerebral ischemia. Eur J Neurosci 28:51–61

Warny M, Kelly C (1999) Monocytic cell necrosis is mediated by potassium depletion and caspase-like proteases. Am J Physiol 276:C717–C724

Wei L, Yu SP, Gottron F, Snider BJ, Zipfei GJ, Choi DW (2003) Potassium channel blockers attenuate hypoxia- and ischemia-induced neuronal death in vitro and in vivo. Stroke 34:1281–1286

Wei WL, Sun HS, Olah ME, Sun X, Czerwinska E, Czerwinski W, Mori Y, Orser BA, Xiong ZG, Jackson MF, Tymianski M, MacDonald JF (2007) TRPM7 channels in hippocampal neurons detect levels of extracellular divalent cations. Proc Natl Acad Sci U S A 104:16323–16328

Weilinger NL, Tang PL, Thompson RJ (2012) Anoxia-induced NMDA receptor activation opens pannexin channels via Src family kinases. J Neurosci 32:12579–12588

West JW, Numann R, Murphy BJ, Scheuer T, Catterall WA (1991) A phosphorylation site in the Na^+ channel required for modulation by protein kinase C. Science 254:866–868

Wetherington J, Serrano G, Dingledine R (2008) Astrocytes in the epileptic brain. Neuron 58:168–178

Wilding TJ, Huettner JE (1997) Activation and desensitization of hippocampal kainate receptors. J Neurosci 17:2713–2721

Wozniak AL, Wang X, Stieren ES, Scarbrough SG, Elferink C, Boehning D (2006) Requirement of biphasic calcium release from the endoplasmic reticulum for Fas-mediated apoptosis. J Cell Biol 175:709–714

Wu J, Holstein JD, Upadhyay G, Lin DT, Conway S, Muller E, Lechleiter JD (2007) Purinergic receptor-stimulated IP3-mediated Ca^{2+} release enhances neuroprotection by increasing astrocyte mitochondrial metabolism during aging. J Neurosci 27:6510–6520

Wu MP, Kao LS, Liao HT, Pan CY (2008) Reverse mode Na^+/Ca^{2+} exchangers trigger the release of Ca^{2+} from intracellular Ca^{2+} stores in cultured rat embryonic cortical neurons. Brain Res 1201:41–51

Xia Y, Cao H, Zhang JH, Chen NY, Siegel K, Agulnik M, Haddad GG (2001) Effect of δ-opioid receptor activation on Na^+ channel expression in cortical neurons subjected to prolonged hypoxia in culture. Program No.740.6, SfN Abstract

Xia Y, Fung ML, O'Reilly JP, Haddad GG (2000) Increased neuronal excitability after long-term O2 deprivation is mediated mainly by sodium channels. Brain Res Mol Brain Res 76:211–219

Xia Y, Haddad GG (1991) Ontogeny and distribution of opioid receptors in the rat brainstem. Brain Res 549:181–193

Xia Y, Haddad GG (2001) Major difference in the expression of μ- and δ-opioid receptors between turtle and rat brain. J Comp Neurol 436:202–210

Xia Y, Jiang C, Haddad GG (1992) Oxidative and glycolytic pathways in rat (newborn and adult) and turtle brain: role during anoxia. Am J Physiol 262:R595–R603

Xia Y, Zhao P, Xue J, Gu XQ, Sun X, Yao H, Haddad GG (2003) Na^+ channel expression and neuronal function in the Na^+/H^+ exchanger 1 null mutant mouse. J Neurophysiol 89:229–236

Xiang K, Earl D, Dwyer T, Behrle BL, Tietz EI, Greenfield LJ Jr (2012) Hypoxia enhances high-voltage-activated calcium currents in rat primary cortical neurons via calcineurin. Epilepsy Res 99:293–305

Xiao AY, Homma M, Wang XQ, Wang X, Yu SP (2001) Role of K^+ efflux in apoptosis induced by AMPA and kainite in mouse cortical neurons. Neuroscience 108:61–67

Xiao AY, Wei L, Xia S, Rothman S, Yu SP (2002) Ionic mechanism of ouabain-induced concurrent apoptosis and necrosis in individual cultured cortical neurons. J Neurosci 22:1350–1362

Xie Y, Dengler K, Zacharias E, Wilffert B, Tegtmeier F (1994) Effects of the sodium channel blocker tetrodotoxin (TTX) on cellular ion homeostasis in rat brain subjected to complete ischemia. Brain Res 652:216–224

Xiong LZ, Yang J, Wang Q, Lu ZH (2007) Involvement of delta-and mu-opioid receptors in the delayed cerebral ischemic tolerance induced by repeated electroacupuncture preconditioning in rats. Chin Med J (Engl) 120:394–399

Xiong ZG, Zhu XM, Chu XP, Minami M, Hey J, Wei WL, MacDonald JF, Wemmie JA, Price MP, Welsh MJ, Simon RP (2004) Neuroprotection in ischemia: blocking calcium-permeable acid-sensing ion channels. Cell 118:687–698

Xue J, Zhou D, Yao H, Haddad GG (2008) Role of transporters and ion channels in neuronal injury under hypoxia. Am J Physiol Regul Integr Comp Physiol 294:R451–R457

Yamaguchi S, Endo K, Kitajima T, Ogata H, Hori Y (1998) Involvement of the glutamate transporter and the sodium-calcium exchanger in the hypoxia-induced increase in intracellular Ca^{2+} in rat hippocampal slices. Brain Res 813:351–358

Yamashima T (2004) Ca^{2+}-dependent proteases in ischemic neuronal death. A conserved 'calpain-cathepsin cascade' from nematodes to primates. Cell Calcium 36:285–293

Yamaura K, Gebremedhin D, Zhang C, Narayanan J, Hoefert K, Jacobs ER, Koehler RC, Harder DR (2006) Contribution of epoxyeicosatrienoic acids to the hypoxia-induced activation of Ca^{2+}-activated K^+ channel current in cultured rat hippocampal astrocytes. Neuroscience 143:703–716

Yan S, Zhang C, Laferrière A, Moss IR (1995) Met-enkephalin-like immunoreactivity in microdialysates from nucleus tractus solitarii in piglets during normoxia and hypoxia. Brain Res 687:217–220

Yang L, Wang H, Shah K, Karamyan VT, Abbruscato TJ (2011) Opioid receptor agonists reduce brain edema in stroke. Brain Res 1383:307–316

Yang L, Zhao X, Sun M, Sun X, Yao L, Yu D, Ding Q, Gao C, Chai W (2014) Delta opioid receptor agonist BW373U86 attenuates post-resuscitation brain injury in a rat model of asphyxial cardiac arrest. Resuscitation 85:299–305

Yang Y, Xia X, Zhang Y, Wang Q, Li L, Luo G, Xia Y (2009) delta-Opioid receptor activation attenuates oxidative injury in the ischemic rat brain. BMC Biol 7:55

Yang Y, Zhi F, He X, Moore ML, Kang X, Chao D, Wang R, Kim DH, Xia Y (2012) δ-Opioid receptor activation and microRNA expression of the rat cortex in hypoxia. PLoS One 7:e51524

Yao H, Haddad GG (2004) Calcium and pH homeostasis in neurons during hypoxia and ischemia. Cell Calcium 36:247–255

Yao L, Fan P, Jiang Z, Mailliard WS, Gordon AS, Diamond I (2003) Addicting drugs utilize a synergistic molecular mechanism in common requiring adenosine and Gi-βγdimers. Proc Natl Acad Sci U S A 100:14379–14384

Ye H, Jalini S, Zhang L, Charlton M, Carlen PL (2010) Early ischemia enhances action potential-dependent, spontaneous glutamatergic responses in CA1 neurons. J Cereb Blood Flow Metab 30:555–565

Yermolaieva O, Leonard AS, Schnizler MK, Abboud FM, Welsh MJ (2004) Extracellular acidosis increases neuronal cell calcium by activating acid-sensing ion channel 1a. Proc Natl Acad Sci U S A 101:6752–6757

Yin D, Mufson RA, Wang R, Shi Y (1999a) Fas-mediated cell death promoted by opioids. Nature 397:218

Yin HZ, Sensi SL, Carriedo SG, Weiss JH (1999b) Dendritic localization of Ca^{2+}-permeable AMPA/kainite channels in hippocampal pyramidal neurons. J Comp Neurol 109:250–260

Yoles E, Zarchin N, Zurovsky Y, Mayevsky A (2000) Metabolic and ionic responses to global brain ischemia in the newborn dog in vivo: II. Post-natal age aspects. Neurol Res 22:623–629

Young JN, Somjen GG (1992) Suppression of presynaptic calcium currents by hypoxia in hippocampal tissue slices. Brain Res 573:70–76

Young KW, Bampton ETW, Pinòn L, Bano D, Nicotera P (2008) Mitochondrial Ca^{2+} signaling in hippocampal neurons. Cell Calcium 43:296–306

Yu SP (2003a) Na^+, K^+-ATPase: the new face of an old player in pathgenesis and apoptotic/hybrid cell death. Biochem Pharmacol 66:1601–1609

Yu SP (2003b) Regulation and critical role of potassium homeostasis in apoptosis. Prog Neurobiol 70:363–386

Yu SP, Choi DW (1997) Na^+-Ca^{2+} exchange currents in cortical neurons: concomitant forward and reverse operation and effect of glutamate. Eur J Neurosci 9:1273–1281

Yu SP, Yeh CH, Sensi SL, Gwag BJ, Canzoniero LM, Farhangrazi ZS, Ying HS, Tian M, Dugan LL, Choi DW (1997) Mediation of neuronal apoptosis by enhancement of outward potassium current. Science 278:114–117

Yu SP, Yeh CH, Strasser U, Tian M, Choi DW (1999) NMDA receptor-mediated K^+ efflux and neuronal apoptosis. Science 284:336–339

Yuan A, Santi CM, Wei A, Wang ZW, Pollak K, Nonet M, Kaczmarek L, Crowder CM, Salkoff L (2003) The sodium-activated potassium channel is encoded by member of the slo gene family. Neuron 37:765–773

Zanelli SA, Numagami Y, McGowan JE, Mishra OP, Delivoria-Papadopoulos M (1999) NMDA receptor-mediated calcium influx in cerebral cortical synaptosomes of the hypoxic guinea pig fetus. Neurochem Res 24:437–446

Zanotto L, Heinemann U (1983) Aspartate and glutamate induced reductions in extracellular free calcium and sodium concentration in area CA1 of 'in vitro' hippocampal slices of rats. Neurosci Lett 35:79–84

Zetterstrőm TSC, Vaughan-Jones RD, Grahame-Smith DG (1995) A short period of hypoxia produces a rapid and transient rise in $[K^+]e$ in rat hippocampus in vivo which is inhibited by certain K^+-channel blocking agents. Neuroscience 67:815–821

Zhang F, Yin W, Chen J (2004) Apoptosis in cerebral ischemia: executional and regulatory signaling mechanisms. Neurol Res 26:835–845

Zhang J, Qian H, Zhao P, Hong SS, Xia Y (2006) Rapid hypoxia preconditioning protects cortical neurons from glutamate toxicity through δ-opioid receptor. Stroke 37:1094–1099

Zhang JH, Gibney GT, Zhao P, Xia Y (2002) Neuroprotective role of δ-opioid receptors in the cortical neurons. Am J Physiol Cell Physiol 282:C1225–C1234

Zhang JH, Haddad GG, Xia Y (2000) δ-, but not μ- and κ-, opioid receptor activation protects neocortical neurons from glutamate-induced excitotoxic injury. Brain Res 885:143–153

Zhang L, Deng T, Sun Y, Liu K, Yang Y, Zheng X (2008) Role for nitric oxide in permeability of hippocampal neuronal hemichannels during oxygen glucose deprivation. J Neurosci Res 86:2281–2291

Zhang M, Song JN, Wu Y, Zhao YL, Pang HG, Fu ZF, Zhang BF, Ma XD (2014) Suppression of STIM1 in the early stage after global ischemia attenuates the injury of delayed neuronal death by inhibiting store-operated calcium entry-induced apoptosis in rats. Neuroreport 25:507–513

Zhang Y, Lipton P (1999) Cytosolic Ca^{2+} changes during in vitro ischemia in rat hippocampal slices: major roles for glutamate and Na^+-dependent Ca^{2+} release from mitochondria. J Neurosci 19:3307–3315

Zhao P, Huang Y, Zuo Z (2006a) Opioid preconditioning induces opioid receptor-dependent delayed neuroprotection against ischemia in rats. J Neuropathol Exp Neurol 65:945–952

Zhao P, Ma MC, Qian H, Xia Y (2005) Down-regulation of delta-opioid receptors in Na^+/H^+ exchanger 1 null mutant mouse brain with epilepsy. Neurosci Res 53:442–446

Zhao YM, Sun LN, Zhou HY, Wang XL (2006b) Voltage-dependent potassium channels are involved in glutamate-induced apoptosis of rat hippocampal neurons. Neurosci Lett 398:22–27

Zheng W, Talley Watts L, Holstein DM, Wewer J, Lechleiter JD (2013a) P2Y1R-initiated, IP3R-dependent stimulation of astrocyte mitochondrial metabolism reduces and partially reverses ischemic neuronal damage in mouse. J Cereb Blood Flow Metab 33:600–611

Zheng YJ, Wang XR, Chen HZ, Wu XJ, Zhao YH, Su DS (2013b) Protective effects of the delta opioid peptide [D-Ala2, D-Leu5]enkephalin in an ex vivo model of ischemia/reperfusion in brain slices. CNS Neurosci Ther 18:762–766

Zhu M, Li M, Tian X, Ou X, Zhu C, Guo J (2009) Neuroprotective role of δ-opioid receptors against mitochondrial respiratory chain injury. Brain Res 1252:183–191

Zhu M, Li M, Yang F, Ou X, Ren Q, Gao H, Zhu C, Guo J (2011) Mitochondrial ERK plays a key role in δ-opioid receptor neuroprotection against acute mitochondrial dysfunction. Neurochem Int 59:739–748

Zhu Y, King MA, Schuller AG, Nitsche JF, Reidl M, Elde RP, Unterwald E, Pasternak GW, Pintar JE (1999) Retention of supraspinal delta-like analgesia and loss of morphine tolerance in delta opioid receptor knockout mice. Neuron 24:243–252

Zorec R, Araque A, Carmignoto G, Haydon PG, Verkhratsky A, Parpura V (2012) Astroglial excitability and gliotransmission: an appraisal of Ca^{2+} as a signalling route. ASN Neuro 4, e00080

Chapter 7
Regulation of Membrane Transporters by Delta-Opioid Receptors

Lu Pu, Yinfang Xu, and Wolfgang Schwarz

Abstract Neuronal excitability and synaptic transmissions are critical for both pain sensation and drug addiction. While the Na^+, K^+ ATPase contributes to maintaining the excitability of neurons, communications between neurons are mainly achieved by synaptic transmission via either glutamate or γ-aminobutyric acid (GABA), two main neurotransmitters that mediate excitatory postsynaptic potential (EPSP) and inhibitory postsynaptic potential (IPSP), respectively. Neurotransmitter transporters control the dwell time and concentration of the respective transmitters in the synaptic cleft; therefore, regulate dynamics of EPSP and IPSP. Opioid receptors are well known modulators in pain sensation and drug addiction, and are located abundantly in synaptic clefts as well. However, little is known whether and how opioid receptors interact regulatorily with Na^+, K^+ ATPase or neurotransmitter transporters. Here we review recent findings, which build the missing link between delta-opioid receptor (DOR) and glutamate/GABA transporters, and Na^+, K^+ ATPase. First, we show that excitatory amino-acid carrier 1 (EAAC1), a glutamate transporter, co-localizes with DOR; co-expression of DOR inhibits EAAC1 activity, and this inhibition is relieved by DOR activation. Second, we show that, similar to EAAC1, the activity of the major GABA transporter GAT1 is inhibited by DOR

L. Pu, Ph.D.
School of Medicine, Case Western Reserve University,
10900 Euclid Ave, Cleveland, OH 44106, USA

Y. Xu, Ph.D.
Shanghai Research Center for Acupuncture and Meridians,
Goushoujing Rd 199, Shanghai 201203, People's Republic of China

Shanghai Key Laboratory of Acupuncture Mechanism and Acupoint Function,
Fudan University, Shanghai 200433, People's Republic of China

W. Schwarz, Ph.D. (✉)
Shanghai Research Center for Acupuncture and Meridians,
Goushoujing Rd 199, Shanghai 201203, People's Republic of China

Shanghai Key Laboratory of Acupuncture Mechanism and Acupoint Function,
Fudan University, Shanghai 200433, People's Republic of China

Institute for Biophysics, Goethe-University Frankfurt,
Max-von-Laue-Str. 1, 60438 Frankfurt am Main, Germany
e-mail: schwarz@biophysik.org

Y. Xia (ed.), *Neural Functions of the Delta-Opioid Receptor*,
DOI 10.1007/978-3-319-25495-1_7

co-expression; but unlike EAAC1, activation of DOR further inhibits GAT1 activity. Third, similar to the findings with EAAC1 and GAT1, Na^+, K^+ pump activity is inhibited by co-expression with DOR, whereas DOR activation has no significant effect on Na^+, K^+ pump activity. On the other hand, pump stimulation reduces DOR sensitivity for the agonist. These studies provide novel regulatory roles of DOR in synaptic proteins that are essential in excitability, synaptic transmission, and synaptic plasticity, and shed light on how interactions between neurotransmitter transporters/Na^+, K^+ pumps and DOR may regulate analgesia and addictive drug use.

Keywords DOR • Neurotransmitter transporter • GAT1 • EAAC1 • Na^+ • K^+ pump

Abbreviations

CNS	Central nervous system
DAMGO	[D Ala^2,N-Me-Phe^4,Gly^5-ol]-enkephalin
DOR	δ-Opioid receptor (delta-opioid receptor)
DPDPE	[D-Pen^{2,5}]-enkephalin
EAAC1	Excitatory amino acid carrier 1
EPSP	Excitatory postsynaptic potential
GABA	Gamma-amino-butyric acid
GAT	GABA transporter
IPSP	Inhibitory postsynaptic potential
KOR	κ-Opioid receptor
MOR	μ-Opioid receptor

1 Introduction

Neurotransmitter transporters play a key role in the regulation of synaptic transmission (see Fig. 7.1). On the arrival of an action potential at the presynaptic nerve ending neurotransmitter is released in response to Ca^{2+} entry. The transmitter will subsequently bind to and activate transmitter-specific ionotropic and/or G-protein coupled receptors at the postsynaptic membrane. This will result in depolarisation or hyperpolarisation of the membrane, depending on whether an excitatory or inhibitory transmitter, respectively, is released at the particular synapse. To terminate synaptic transmission, the neurotransmitter needs to be removed, and this is achieved by high-efficient, Na^+-gradient-driven neurotransmitter transporters in the presynaptic neuron and the surrounding glia cells. The activity of these transporters determines to a large extent the concentration and dwell time of the respective transmitter in the synaptic cleft. The dominating excitatory and inhibitory transmitters in the

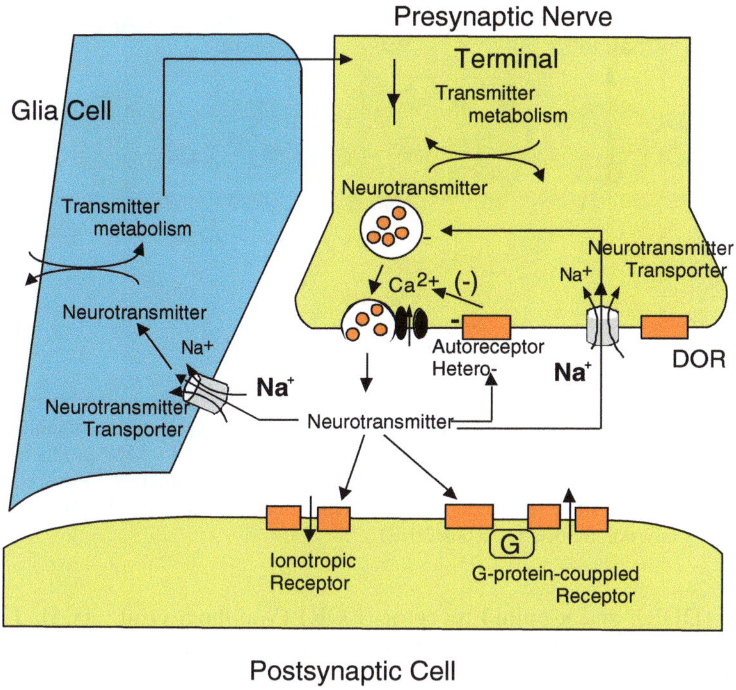

Fig. 7.1 Illustration of synaptic events. Release of neurotransmitters stored in vesicles in the presynaptic nerve terminal is initiated by the influx of Ca^{2+}. Subsequently, postsynaptic receptors become activated. Na^+-driven neurotransmitter transporters are involved in termination of neurotransmission (based on Schwarz and Gu, 2012)

mammalian central nervous system (CNS) are glutamate and gamma-amino-butyric acid (GABA).

In the brain, Na^+, K^+ pumps (also named as sodium pumps or Na^+, K^+-ATPases) consume half of the body's energy and are essential for maintaining membrane excitability and maintaining glutamate- or GABA-mediated postsynaptic potentials. During an action potential, Na^+ ions rush into the cell via voltage-gated Na^+ channels, leading to membrane depolarization. The membrane potential is then repolarized by Na^+-channel inactivation and K^+ ion efflux. The Na^+ ions flown in and K^+ ions flown out are continuously pumped back by Na^+, K^+-ATPase to maintain intracellularly low Na^+ and high K^+ activities, which is crucial for maintaining excitability. In addition, both glutamate transporters and GABA transporters are secondary-active transporters utilizing the inwardly directed gradient for Na^+ to transport neurotransmitter into the respective cell; therefore, proper Na^+, K^+-pump function is critical for neurotransmitter transporter functions, as well as efficacy of synaptic transmission.

Opiates including morphine have been used as analgesics for thousands of years. Both exogenous and endogenous opioids activate three main Gi/o-coupled opioid receptors located in both central and peripheral nervous systems: μ-opioid (MOR),

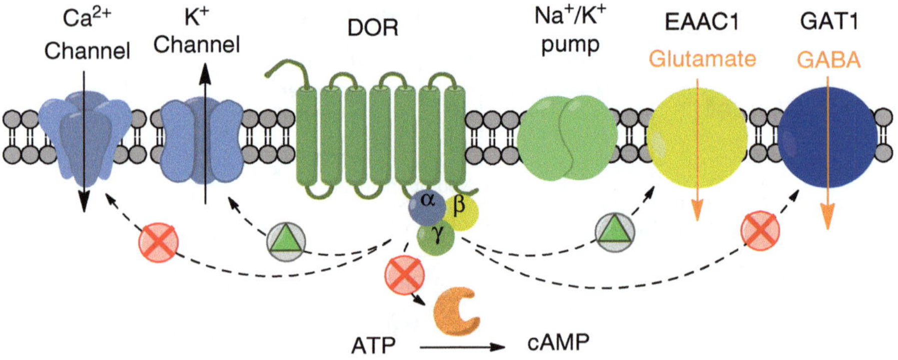

Fig. 7.2 Acute effect of DOR activation. Circles with *red cross* indicate inhibitory pathways; circles with *green triangle* indicate stimulation pathways. α, β, γ represent the subunits of G protein. Acute δ-opioid (DOR) activation inhibits adenylyl cyclase and cAMP production. It also reduces Ca^{2+} conductance and increases K^+ conductance (North et al. 1987; Piros et al. 2000). Recent studies further show that acute DOR activation leads to enhanced EAAC1 and reduced GAT1 activity (see e.g. Schwarz and Gu, 2012)

δ-opioid (DOR), and κ-opioid receptor (KOR) (Waldhoer et al. 2004). The three opioid receptors are quite similar at the protein level: for examples, the human DOR and MOR share 66 %, the human DOR and KOR share 58 %, and the human MOR and KOR share 61 % sequence identities, respectively (Knapp et al. 1994, 1995). The first endogenous opioid receptor agonist enkephalin was identified in 1974 (Terenius and Wahlström 1974, 1975; Hughes et al. 1975); endorphin and dynorphine were identified soon thereafter (Hughes et al. 1975; Simantov and Snyder 1976). Endorphin activates both MOR and DOR, but has higher affinity for MOR; enkephalin has a much higher affinity for DOR than that for MOR; and comparing to enkephalin and endorphine, the KOR agonist dynorphine has much weaker analgesia effect (Werz and Macdonald 1983; Ikeda et al. 2002).

Acute effects of DOR activation are illustrated in Fig. 7.2. When acutely activated, Gα subunits inhibit adenylyl cyclase and the cAMP production, which might contribute to the peripheral analgesia effects of opioids (Nestler and Aghajanian 1997). On the other hand, acute activation of opioid receptors reduces Ca^{2+} conductance and activates K^+ channels (North et al. 1987; Piros et al. 2000). In presynaptic neurons, the net effect of Ca^{2+} channel inhibition and K^+ channel activation lead to hyperpolarized membrane potential, reduced action potential firing, and reduced neurotransmitter release (Boehm and Huck 1997; Miller 1998). In the central nervous system, opioids inhibit ascending pain sensory pathway from the spinal cord and activate descending inhibitory signals from periaqueductal gray matter; the combination of these two effects leads to CNS analgesia (Yaksh and Luo 2011; Brenner and Stevens 2012).

Different from these acute effects, chronic activation of opioid receptors usually leads to tolerance—diminished effects of opioids; hence, higher drug doses are required to achieve the same pharmacological effects (Nestler and Aghajanian

1997; Yaksh and Luo 2011; Brenner and Stevens 2012). Chronic opioid exposure leads to up-regulation of cAMP-PKA pathway, down regulation of surface expression of the receptors, and increased expression of transcription factors including CREB and ΔFOSB; all these molecular changes lead to long term synaptic and behavioral changes and contribute to tolerance (Nestler and Aghajanian 1997; Nestler 2001).

Both MOR and DOR are localized at synapses and regulate neurotransmitter release or cellular excitability (Wang and Pickel 2001; Zhang and Pan 2010). As aforementioned, glutamate transporters and GABA transporters directly control the concentrations of glutamate and GABA at the synaptic cleft, respectively; while Na^+, K^+ pumps indirectly regulate transmitter concentrations via maintaining a vast difference of Na^+ and K^+ activities across the cellular membrane. Because their similar localizations and their common effects on regulating neurotransmitter levels and pain sensation, we were curious to know whether MOR/DOR could regulate neurotransmitter transporter and Na^+, K^+ pump functions.

In this review, we focus on three main studies (Pu et al. 2012; Deng et al. 2009; Xia et al. 2006), in which either MOR or DOR were co-expressed with EAAC1, GAT1, or Na^+, K^+ pump. The enkephalin analog [D-Pen^{2,5}]-enkephalin (DPDPE) was used as the DOR agonist; and the endorphin analog [D Ala^2,N-Me-Phe^4,Gly^5-ol]-enkephalin (DAMGO) was used as the MOR agonist. Recently, it was shown that stimulation of release of endorphins might affect expression of transporter protein (see e.g. Feng et al. 2011; Liang et al. 2014); here we demonstrate that activation of DOR resulted in immediate modulation of activity of the investigated transporters.

2 Regulation of EAAC1, GAT1, and Na^+, K^+ ATPase by Interaction with DOR

2.1 DOR and EAAC1

Glutamate transporters play an essential role in regulating temporal and spatial glutamate concentration in the synaptic cleft. The excitatory amino acid carrier 1 (EAAC1) is the dominating neuronal-specific isoform of glutamate transporters in nerve terminals (Rothstein et al. 1994; He et al. 2000; Levenson et al. 2002).

EAAC1 is a secondary active transporter; the uptake of one glutamate molecule is coupled to one proton and is driven by the inward movement of 3 Na^+ and outward movement of one K^+ (Fig. 7.3a). Hence the transporter is electrogenic transporting per glutamate 2 positive net charges cross the membrane. Transport activity can, therefore, be monitored by radioactive tracer fluxes, or by measuring the current mediated by EAAC1.

Opioid receptors are found to modulate the glutamatergic system in brain, but the regulatory mechanism remains to be investigated. Experiments on the model system

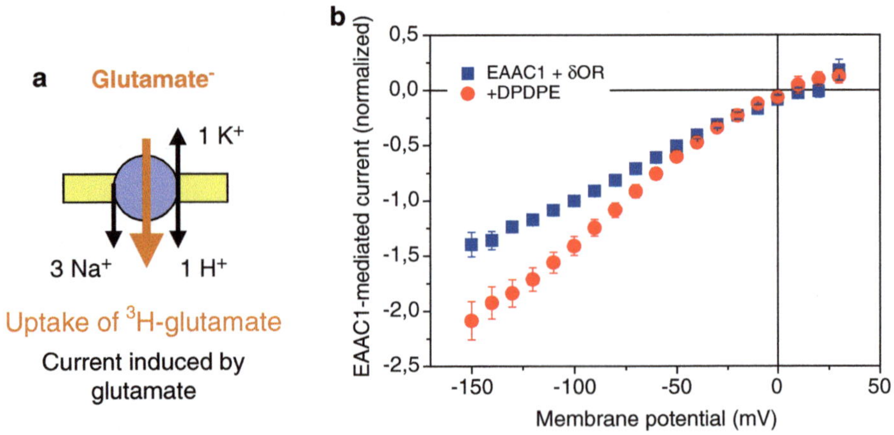

Fig. 7.3 (**a**) Schematic description of glutamate transporter EAAC1 that is driven by Na^+, K^+ and H^+ gradients. (**b**) Steady-state current-voltage dependence of EAAC1-mediated current in oocytes with co-expressed DOR; *squares* are data without DOR stimulation, *circles* with stimulation by 1 μM DPDPE (data based on Xia et al. 2006)

Table 7.1 Rate of [^{3}H]-glutamate uptake induced by DPDPE

	Rate of uptake		Transporter-mediated current (at –60 mV)	
	Control	+1 μM DPDPE	Control	+1 μM DPDPE
EAAC1	1±0.06	1.39±0.09	1±0.06	1.29±0.02
GAT1	1±0.10	0.56±0.17	1±0.13	0.61±0.13

Date are normalized averages (n=12–20)±SEM

Xenopus oocyte heterologously expressed membrane proteins allow us to investigate interaction between glutamate transporter and opioid receptor without interference with other components of the glutamatergic system. Co-expression of DOR with EAAC1 led to reduced transport activity, which had been attributed to direct protein-protein interaction (Xia et al. 2006). The total amount of EAAC1 expressed in the oocyte surface membrane was not affected by co-expression of DOR, but co-localization of the two proteins could be demonstrated not only in the *Xenopus* oocytes, but also in neuronal cells. Stimulation of DOR by 1 μM of the DOR-specific agonist DPDPE led to relief from the interaction and an increase in EAAC1-mediated current (Fig. 7.3b) and rate of [^{3}H]-glutamate uptake (see also Table 7.1).

2.2 *DOR and GAT1*

GABA is not only known as the chief inhibitory neurotransmitter in the mammalian CNS, but also as a natural pain-killer. There has been ample evidence showing that the GABAergic neurotransmission system plays a critical role in pain sensation.

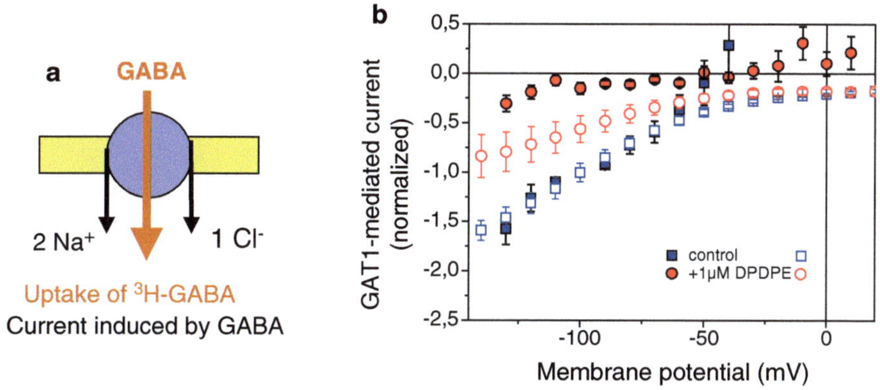

Fig. 7.4 (**a**) Schematic description of GABA transporter GAT1 that is driven by Na^+ and Cl^- gradients. (**b**) Steady-state current-voltage dependence of GAT1-mediated current in oocytes with co-expressed DOR (*open symbols*) and in PAG neurons (*filled symbols*); squares are data without DOR stimulation, circles with stimulation by 1 μM DPDPE (data based on Pu et al. 2012)

The GABA transporters (GATs) located at the presynaptic terminals and surrounding glia are responsible for maintaining low resting extracellular GABA level, thus the fine tuning of GABAergic synaptic transmission. Among the four GATs (GAT1-4) having been cloned so far, GAT1 is the most abundant one and contributes more than 75 % GABA reuptake in the CNS (Borden 1996; Jensen et al. 2003; Cai et al. 2006). Also GAT1 is a Na^+-driven electrogenic transporter; the inward movement of 2 Na^+ and 1 Cl^- drives the take up of one GABA molecule leading to the net transport of one positive charge into the cell (Fig. 7.4a).

It has been reported that elevated expression level of GAT1 in transgenic mice resulted in clear hyperalgesic effects (Hu et al. 2003). Experiments with the GAT1 knockout mice, on the contrary, showed that the deficiency of GAT1 can lead to a significant decrease of pain sensitivity (Xu et al. 2008). These data together suggested that GAT1 is involved in the regulation of pain threshold.

Since opioid receptors are well recognized in mediating analgesic effects as well, we investigated whether and how those two systems work together in pain sensation and modulation. *Xenopus* oocytes were again used as a model system to investigate the interaction between DOR and GAT1. Similar to co-expression of EAAC1 with DOR, GAT1-mediated current is reduced on co-expression of DOR (Pu et al. 2012). But in contrast to EAAC1, GAT1-mediated current (Fig. 7.4b) as well as rate of ^{3}H-GABA uptake becomes reduced on activation of DOR by DPDPE (see Table 7.1). To exclude that the reduced GAT1 activity is a result of overexpression in the oocytes, whole-cell patch-clamp experiments on rat brain slices of periaqueductal gray (PAG) were performed. Even in this native environment, the application of DOR-specific agonist DPDPE remarkably inhibited GAT1-mediated current. Synaptosomes of brain from rats chronically exposed to opiates also exhibited reduced GABA uptake, indicating that GABA transport can be regulated by opioid receptors in physiological states. Such inhibition of GAT1 function, consequently, will strengthen the inhibitory action of the GABAergic system and hence may contribute to pain modulation and other biological processes.

2.3 *DOR and Na^+, K^+ ATPase*

To investigate interference of DOR with the Na^+, K^+-ATPase in a simple model system, we used the *Xenopus* oocytes as an expression system. During the normal pump cycle one molecule of ATP is hydrolyzed, and the liberated energy is used to transport $3Na^+$ out of the cell and $2K^+$ into the cell (Fig. 7.5a). As a consequence of the $3Na^+$, $2K^+$ stoichiometry the pump is electrogenic, generating on outward current; this current can be detected under voltage-clamp and is a measure for the ATPase activity. Our results indicate that expression of the DOR results in reduction of sodium-pump activity. Stimulation of DOR by the DOR agonist DPDPE had no pronounced additional effect on pump activity. This we could demonstrate not only for the endogenous sodium pump, but also a variety of co-expressed exogenous sodium pumps (Deng et al. 2009). We suggest that reduced pump activity with DOR expression is brought about by an interaction of the pump with DOR, similarly to what was observed with co-expression of DOR and EAAC1 or GAT1. Direct interaction is also supported by co-immunoprecipitation, not only in the Xenopus oocytes but also in rat hippocampal neurons (Deng et al. 2009).

Xenopus oocytes express Ca^{2+}-activated Cl^- and K^+ channels (Weber 1999); since DOR is a Gi/o-coupled receptor, activation of DOR will stimulate these chan-

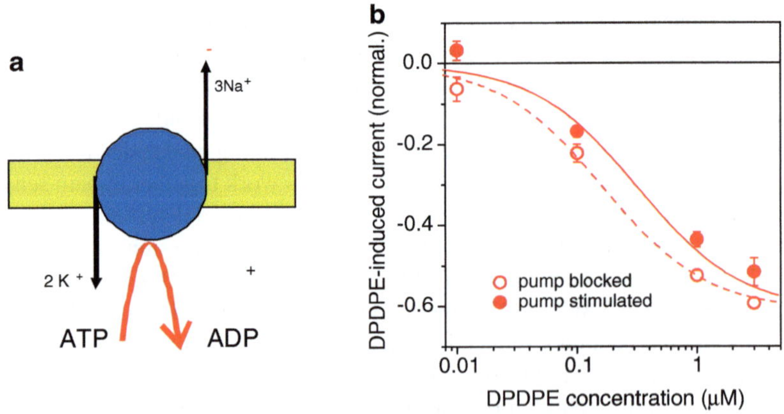

Fig. 7.5 (**a**) Schematic description of $3Na^+$, $2K^+$-ATPase. (**b**) Dependence of DPDPE-induced current (at −100 mV) on DPDPE concentration with sheep Na^+, K^+-ATPase blocked (by the absence of extracellular K^+, *open symbols*) and with pump stimulated by 10 mM K^+ (*filled symbols*). The *fitted line* indicate K1/2 values of 0.17 and 0.27 mM for current stimulation without and with sodium pump activity (data based on Deng et al. 2009)

nels. Figure 7.5 illustrates that the dependence of DPDPE-induced current can be modulated by the activity of the sodium pump The above mentioned DOR-ATPase interaction may be responsible for the altered agonist sensitivity of DOR. In conclusion, pump activity not only affects neural activity directly but our results also suggest that pump activity is affected through functional interaction with DOR that will modulate pain sensation.

3 Physical Significance of DOR Regulations of Transporters and ATPase

3.1 On Pain Regulation and Analgesia

Opioids, such as morphine, are among the oldest medications available for pain management. In spite of the powerful efficacy of these drugs, their usage is limited by the production of various undesirable side effects, especially tolerance and physical dependence. Endorphins, also known as endogenous opioid peptides, have been shown to produce analgesic effects in rodent models of both acute and chronic pain with less troublesome side effects. For instance, the involvement of endorphins in mediating acupuncture analgesia has been demonstrated in many research articles. Earlier studies have shown that acupuncture-induced analgesic effects can be blocked by opioid receptor specific antagonists both in human and rodents (Mayer et al. 1977; Cheng and Pomeranz 1979), indicating the involvement of endogenous opioid peptides. Han and his colleagues demonstrated that analgesia induced by electroacupuncture can be attributed to increased release of endorphins, including endomorphin, β-endorphin, enkephalin and dynorphin, in the CNS (Han 2004; Han et al. 1999). Among these endogenous opioid peptides, both β-endorphin and enkephalin can activate DOR, which may then trigger antinociception via interacting with neurotransmitter transporters and ATPase.

As mentioned above, activation of DOR can stimulate the function of EAAC1 and lower the concentration and dwell time of glutamate in the synaptic cleft, subsequently leading to decreased excitatory neurotransmission. In addition, DOR (in either resting or activated state) can inhibit the GABA reuptake activity of GAT1 causing the elevated extracellular GABA concentration and enhanced inhibitory synaptic transmission. Inhibition of excitatory synaptic transmission or enhancement of inhibitory transmission, either way could inhibit the propagation of pain signals in the CNS and result in analgesia.

The Na^+, K^+-ATPase was also recognized as a determinant of nociceptive sensitivities, spinal administration of ouabain, a sodium pump inhibitor, has been shown to produce analgesia in rats (LaCroix-Fralish et al. 2009; Zeng et al. 1999). According to our previous work (Yang et al. 2008; Deng et al. 2009), the functional interaction between DOR and sodium pumps cannot reduce pump activity, in contrast to EEAC1 and GAT1, but activation of sodium pump can in turn reduce the affinity of DOR to its agonist. Increased Na^+, K^+-ATPase activity was detected in spinal dorsal horn in rat models with inflammatory pain (Czaplinski et al. 2005), and likely be responsible for producing pain signals and be part of the body's self-defensive system.

3.2 Differences of Morphine and DOR Agonists on Tolerance and Dependence

As a golden standard for analgesia, morphine activates all three receptors, with the highest affinity for MOR (Harrison et al. 1998; Koyyalagunta and Waldman 2010). The clinical application of morphine is very much limited by respiratory depression, tolerance, and development of physical dependence; in addition, morphine administration for non-cancer patients is limited by the fear of developing psychological dependence or addiction (Koyyalagunta and Waldman 2010; McNicol et al. 2013). In contrast, endogenous DOR agonists such as enkephalin may serve as very promising clinical analgesics, partly because they have fewer side effects than morphine.

Enkephalins activate both MOR and DOR, with an affinity for DOR ten times higher than for MOR (Dhawan et al. 1996). In the peripheral nervous system, enkephalins are stored in vesicles in interneurons close to the synapses between the presynaptic primary afferent neuron and the postsynaptic second order spinal neurons; in response to noxious stimuli, enkephalins are released, activate opioid receptors, and inhibit the ascending pain pathway; these released enkephalins can either function locally or diffuse to nearby synapses and their actions can be quickly terminated by peptidases (Roques et al. 2012).

In clinical application, enkephalins have certain advantages over morphine. First, specific DOR agonists have been shown to relieve MOR-mediated respiratory depression (Su et al. 1998). Second, recent studies have shown proper receptor internalization is crucial in maintaining opioid receptor activity and in reducing tolerance (Whistler et al. 1999; Waldhoer et al. 2004). For example, morphine blocks receptor internalization, which leads to adaptation of the opioid receptor-mediate signaling and tolerance; quite the opposite, DPDPE induces receptor recycling; and mice receiving DPDPE develop decreased tolerance (Whistler et al. 1999).

In the previous sections, we have shown that DOR activation by DPDPE enhances EAAC1 activity and reduces GAT1 activity (Xia et al. 2006; Pu et al. 2012). These novel findings illustrate dual functions of DOR activation. For example, at the synapses from the primary afferent neurons to the secondary spinal neurons, DOR activation leads to inhibition of Ca^{2+} channels and activation of K^+ channels, which hyperpolarizes these neurons and reduces their excitability. Second, DOR activation of EAAC1 and GAT1 reduces synaptic glutamate concentration and increases GABA concentration and further reduces the excitability of these neurons. Thus, the analgesia effect of enkephalins including DPDPE is amplified by neurotransmitter transporters. A thorough understanding of the interaction between DOR and neurotransmitter transporters may provide new insight and improved methods for analgesics and pain management.

4 Concluding Remarks

Endogenous opioids such as enkephalin have been recognized as interesting targets in developing effective analgesics (Roques et al. 2012). However, no DOR agonists have been developed as analgesics yet. Recently, small molecules inhibiting enkephalin peptidases have been developed; clinical trials show that these enkephalin peptidase inhibitors do not lead to depression of respiratory system, tolerance, or other side effects (Roques et al. 2012). Our findings that a DOR selective agonist enhances EAAC1 (Xia et al. 2006) and reduces GAT1 function (Pu et al. 2012) promise new targets for pain management. We have shown that by promoting the association of DOR and EAAC1/GAT1, it is possible to enhance EAAC1 activity and to reduce GAT1 activity (Fig. 7.2); therefore, future therapeutics may exploit these dynamic interactions to reduce glutamate concentrations and enhances GABA concentrations at the synapses, which could enhance the analgesia effect of endogenously released enkephalins.

Ideally, the effect of these hypothetical drugs will be contingent upon noxious stimuli: when nerve terminals secret enkephalin and activate DOR, the drug can enhance their specific interaction and amplify the analgesic effect of enkephalin. Unlike morphine, which systematically activates opioid receptors in the body, the drugs targeting DOR-transporter interactions will only work at the locations close to enkephalin secretion. The localized effects can be understood because the diffusion of enkephalins is limited by enkephalin peptidases, and hence the effect of enkephalin is relatively close to the sites of enkephalin secretion. Because enkephalins are rapidly degraded, the over stimulation of opioid receptors can be avoided as well.

Acknowledgements This project was supported in part by the National Basic Research Program of China (973 program) and on the basis of an agreement between Max-Planck Society and Chinese Academy of Sciences. We also acknowledge the support from Green Valley Holding Co, Shanghai and from National Natural Science Foundation of China (Youth Program No. 81403489 to Y. Xu).

References

Boehm S, Huck S (1997) Recepptors controlling transmitter release from sympathetic neurons in vitro. Prog Neurobiol 51:225–242

Borden LA (1996) GABA transporter heterogeneity—pharmacology and cellular localization. Neurochem Int 29:335–356

Brenner GM, Stevens CW (2012) Opioid analgesics and antagonists. In: Brenner GM and Stevens CW (eds) Pharmacology. Elsevier Saunders, Philadelphia

Cai YQ, Cai GQ, Liu GX, Cai Q, Shi JH, Shi J, Ma SK, Sun X, Sheng ZJ, Mei ZT, Cui D, Guo L, Wang Z, Fei J (2006) Mice with genetically altered GABA transporter subtype I (GAT1) expression show altered behavioral responses to ethanol. J Neurosci Res 84:255–267

Cheng RS, Pomeranz B (1979) Electroacupuncture analgesia could be mediated by at least two pain-relieving mechanisms; endorphin and non-endorphin systems. Life Sci 25:1957–1962

Czaplinski M, Abad C, Eblen-Zajjur A (2005) Normal expression and inflammation-induced changes of Na and Na/K ATPase activity in spinal dorsal horn of the rat. Neurosci Lett 374:147–151

Deng HP, Yang ZJ, Li YT, Bao GB, Friedrich T, Gu QB, Schwarz W (2009) Interaction of Na^+, K^+-ATPase and co-expression of δ-opioid receptor. Neurosci Res 65:222–227

Dhawan BN, Cesselin F, Raghubir R, Reisine T, Bradley PB, Portoghese PS, Hamon M (1996) International Union of Pharmacology. XII. Classification of opioid receptors. Pharmacol Rev 48:567–592

Feng Y, He XZ, Yang YL, Chen JS, Yin KS, Xia Y (2011) Effect of delta-opioid receptor over-expression on cortical expression of $GABA_A$ Receptor α1-subunit in hypoxia. Chin J Physiol 54:118–123

Han J-S (2004) Acupuncture and endorphins. Neurosci Lett 361:258–261

Han Z, Jiang YH, Wan Y, Wang Y, Chang JK, Han JS (1999) Endomorphin-1 mediates 2 Hz but not 100 Hz electroacupuncture analgesia in the rat. Neurosci Lett 274:75–78

Harrison LM, Kastin AJ, Zadina JE (1998) Opiate tolerance and dependence: receptors, G-proteins, and antiopiates. Peptides 19:1603–1630

He Y, Janssen WG, Rothstein JD, Morrison JH (2000) Differential synaptic localization of the glutamate transporter EAAC1 and glutamate receptor subunit GluR2 in the rat hippocampus. J Comp Neurol 418:255–269

Hu JH, Yang N, Ma YH, Zhou XG, Jiang J, Duan SH, Mei ZT, Fei J, Guo LH (2003) Hyperalgesic effects of gamma-aminobutyric acid transporter I in mice. J Neurosci Res 73:565–572

Hughes J, Smith TW, Kosterlitz HW, Fothergill LA, Morgan BA, Morris HR (1975) Identification of two related pentapeptides from the brain with potent opiate agonist activity. Nature 258:577–580

Ikeda K, Kobayashi T, Kumanishi T, Yano R, Sora I, Niki H (2002) Molecular mechanisms of analgesia induced by opioids and ethanol: is the GIRK channel one of the keys? Neurosci Res 44:121–131

Jensen K, Chiu CS, Sokolova I, Lester HA, Mody I (2003) GABA transporter-1 (Gat1)-deficient mice: Differential tonic activation of $Gaba_A$ versus $GABA_B$ receptors in the hippocampus. J Neurophysiol 90:2690–2701

Knapp RJ, Malatynska E, Fang L, Li X, Abin E, Guyen M, Antoro G, Arga EV, Ruby VJ, Oeske WR, Amamura HI (1994) Identification of a human delta opioid receptor: cloning and expression. Life Sci 54:L463–L469

Knapp RJ, Malatynska E, Collins N, Fang L, Wang JY, Hruby VJ, Roeske WR, Yamamura HI (1995) Molecular biology and pharmacology of cloned opioid receptors. FASEB J 9:516–525

Koyyalagunta D, Waldman SD (2010) Opioid analgesics. In: Waldman SD (ed) Pain managemen. Elsevier, Philadelphia, pp 891–912

LaCroix-Fralish ML, Mo G, Smith SB, Sotocinal SG, Ritchie J, Austin JS, Melmed K, Schorscher-Petcu A, Laferriere AC, Lee TH, Romanovsky D, Liao G, Behlke MA, Clark DJ, Peltz G, Seguela P, Dobretsov M, Mogil JS (2009) The beta3 subunit of the Na^+, K^+-ATPase mediates variable nociceptive sensitivity in the formalin test. Pain 144:294–302

Levenson J, Weeber E, Selcher JC, Kategaya LS, Sweatt JD, Eskin A (2002) Long-term potentiation and contextual fear conditioning increase neuronal glutamate uptake. Nat Neurosci 5:155–161

Liang JF, Chao DM, Sandhu HK, Yu YB, Zhang L, Balbon G, Dong H, Kim DH, Xia Y (2014) δ-Opioid receptors up-regulate excitatory amino acid transporters in mouse astrocytes. Br J Pharmacol 171:5417–5430

Mayer DJ, Price DD, Rafii A (1977) Antagonism of acupuncture analgesia in man by the narcotic antagonist naloxone. Brain Res 121:368–372

McNicol ED, Midbari A, Eisenberg E (2013) Opioids for neuropathic pain. Cochrane Database Syst Rev 8, CD006146
Miller RJ (1998) Presynaptic receptors. Annu Rec Pharmacol Toxicol 38:201–227
Nestler EJ (2001) Molecular basis of long-term plasticity underlying addiction. Nat Rev Neurosci 2:119–128
Nestler EJ, Aghajanian GK (1997) Molecular and cellular basis of addiction. Science 278:58–63
North RA, Williams JT, Surprenant A, Christie MJ (1987) Mu and delta receptors belong to a family of receptors that are coupled to potassium channels. Proc Natl Acad Sci U S A 84:5487–5491
Piros ET, Charles RC, Song L, Evans CJ, Hales TG (2000) Cloned delta-opioid receptors in GH(3) cells inhibit spontaneous Ca(2+) oscillations and prolactin release through K(IR) channel activation. J Neurophysiol 83:2691–2698
Pu L, Xu NJ, Xia P, Gu QB, Ren SL, Fucke T, Pei G, Schwarz, W (2012). Inhibition of activity of GABA transporter GAT1 by delta-opioid receptor. eCAM ID818451-12 pages.
Roques BP, Fournie-Zaluski MC, Wurm M (2012) Inhibiting the breakdown of endogenous opioids and cannabinoids to alleviate pain. Nat Rev Drug Discov 11:292–310
Rothstein JD, Martin L, Levey AI, Dykes-Hoberg M, Jin L, Wu D, Nash N, Kuncl RW (1994) Localization of neuronal and glial glutamate transporters. Neuron 13:713–725
Schwarz W, Gu QB (2012) Cellular mechanisms in acupuncture points and affected sites. In: Xia Y, Ding GH, Wu G-C (eds) Current research in acupuncture. Springer, New York, pp 37–51
Simantov R, Snyder SH (1976) Morphine-like peptides in mammalian brain: isolation, structure elucidation, and interactions with the opiate receptor. Proc Natl Acad Sci U S A 73:2515–2519
Su YF, McNutt RW, Chang KJ (1998) Delta-opioid ligands reverse alfentanil-induced respiratory depression but not antinociception. J Pharmacol Exp Ther 287:815–823
Terenius L, Wahlström A (1974) Inhibitors of narcotic receptor binding in brain extracts and cerebrospinal fluid. Acta Pharmacol 35:55
Terenius L, Wahlström A (1975) Search for an endogenous ligand for the opiate receptor. Acta Physiol Scand 94:74–81
Waldhoer M, Bartlett SE, Whistler JL (2004) Opioid receptors. Annu Rev Biochem 73:953–990
Wang H, Pickel VM (2001) Preferential cytoplasmic localization of δ-opioid receptors in rat striatal patches: Comparison with plasmalemmal μ-opioid receptors. J Neurosci 21:3242–3250
Weber W-M (1999) Ion currents of *Xenopus laevis* oocytes: state of the art. Biochem Biophys Acta 1421:213–233
Werz MA, Macdonald RL (1983) Opioid peptides with differential affinity for mu and delta receptors decrease sensory neuron calcium-dependent action potentials. J Pharmacol Exp Ther 227:394–402
Whistler JL, Chuang HH, Chu P, Jan LY, von Zastrow M (1999) Functional dissociation of mu opioid receptor signaling and endocytosis: implications for the biology of opiate tolerance and addiction. Neuron 23:737–746
Xia P, Pei G, Schwarz W (2006) Regulation of the glutamate transporter EAAC1 by expression and activation of δ-opioid receptor. Eur J Neurosci 24:87–93
Xu YF, Cai YQ, Cai GQ, Jiang J, Sheng ZJ, Wang ZG, Fei J (2008) Hypoalgesia in mice lacking GABA transporter subtype 1. J Neurosci Res 86:465–470
Yaksh TL, Luo ZD (2011) Anatomy of the pain processing system. In: Waldman SD (ed) Pain management, 2nd edn. Elsevier, Philadelphia
Yang ZJ, Bao GB, Deng HP, Du HM, Gu QB, Pei G, Pu L, Schwarz W, Xia P (2008) Interaction of δ-opioid receptor with membrane transporters: possible mechanism in pain suppression by acupuncture. J Acupunct Tuina Sci 6:298–300
Zeng W, Dohi S, Shimonaka H, Asano T (1999) Spinal antinociceptive action of Na^{+}+K^{+}+ pump inhibitor ouabain and its interaction with morphine and lidocaine in rats. Anesthesiology 90:500–508
Zhang Z, Pan ZZ (2010) Synaptic mechanism for functional synergism between delta- and mu-opioid receptors. J Neurosci 30:4735–4745

Chapter 8
Neuroprotective Interactions Between Delta-Opioid Receptors and Glutamatergic Signaling Mediate Hypoxia-Tolerance in Brain

Matthew E. Pamenter

Abstract Delta-opioid receptors are a class of membrane proteins found throughout the nervous system. They have traditionally been associated with the transmission of sensations related to pain via neuromodulation of excitatory glutamatergic synaptic signaling. Research examining these interactions in nociception and related anesthesia applications has demonstrated that delta-opioid receptors are capable of mediating glutamatergic signaling via both pre-synaptic and post-synaptic mechanisms. In addition to normal neurotransmission functions, derangements in glutamatergic signaling are also associated with pathological brain damage due to low oxygen stresses, such as hypoxia or ischemic stroke; delta-opioid receptors are capable of mediating neuroprotective responses to such stresses via the inhibition of deleterious excitatory glutamatergic signaling. Specifically, studies of the mechanisms of hypoxic or ischemic preconditioning have demonstrated that delta-opioid receptors are central triggers that mediate inducible neuroprotective mechanisms against acute hypoxic, ischemic, and glutamatergic stresses in mammalian brain, and recent evidence points to the modulation of glutamate receptors as a critical component of this neuroprotective mechanism. In addition, a recent study has demonstrated that endogenous activation of similar mechanisms contributes to the innate anoxia-tolerance of the brain of one of the most hypoxia-tolerant vertebrates identified—the Western Painted turtle, and that the mechanism of neuroprotection in this organism involves the direct inhibition of neuronal glutamatergic signaling at the post-synapse. This chapter will focus on the putative neuroprotective effects of delta-opioid receptor signaling in models of hypoxic preconditioning in mammal brain and also in glutamatergic channel arrest in turtle brain. Similarities between the underlying neuroprotective mechanisms against hypoxia and the mechanistic interactions in nociception and analgesia will also be discussed.

Keywords δ-Opioid receptor • Glutamatergic signaling • Hypoxia • Neuroprotection • Turtle

M.E. Pamenter, Ph.D. (✉)
Department of Biology, University of Ottawa, Ottawa, ON, Canada, K1N 6N5
e-mail: mpamenter@uottawa.ca

Y. Xia (ed.), *Neural Functions of the Delta-Opioid Receptor*,
DOI 10.1007/978-3-319-25495-1_8

1 Introduction

Delta-opioid receptors are a class of inhibitory G protein (G_i) coupled receptors that are found throughout the central and peripheral nervous systems, although their expression varies considerably between brain and brainstem regions (Sharif and Hughes 1989). The natural endogenous ligands of delta-opioid receptors are enkephalins and these receptors were first identified as a unique subclass of opioid receptor in studies examining the relative efficacy of morphine compared to enkephalins on electrical contractions in the murine *vas deferens* (Lord et al. 1977), for which the delta (δ) designation was applied. The primary functions of delta-opioid receptors in neurotransmission are related to nocireception (i.e. pain sensations) and anesthesia, although their activity is also associated with antidepressant effects, convulsions related to seizures, and physical dependency and addiction to narcotics (e.g. to morpine; Quock et al. 1999). The initial studies of delta-opioid receptor function were focused upon the examination of pharmacological interactions associated with pain (Quock et al. 1999); however, more recent studies have demonstrated a central role for delta-opioid receptors in mediating neuronal tolerance to hypoxic or ischemic stresses, such that delta-opioid receptor activation (1) is neuroprotective against acute hypoxic, ischemic, or glutamate toxicity-mediated insults in mammalian neurons (Zhang et al. 2000, 2002; Chao et al. 2007b), (2) triggers neuroprotective pathways involved in mechanisms of hypoxia or ischemic preconditioning in mammalian brain (HPC/IPC; Zhang et al. 2006a), and (3) plays a key role in triggering endogenous neuroprotective mechanisms during prolonged environmental hypoxia or anoxia in the brain of a naturally hypoxia-tolerant model organism (i.e. the Western painted turtle; Buck and Pamenter 2006; Pamenter and Buck 2008a).

Unlike in Western painted turtle brain, the brains of most mammals are acutely sensitive to hypoxia or ischemia and rapidly undergo excitotoxic cell death (ECD) when deprived of oxygen or when exposed to high levels of glutamate. ECD is characterized by increased glutamate release, which leads to over-activation of glutamatergic N-methyl-D-aspartate receptors (NMDARs), toxic accumulation of intracellular Ca^{2+}, and a severe loss of membrane potential (Lundberg and Oscarsson 1953; Choi 1994). ECD can be avoided and mammalian neurons can be made relatively hypoxia-tolerant via inducible neuroprotective mechanisms known as ischemic or hypoxic preconditioning (IPC or HPC), whereby pretreatment with a tolerable ischemic or hypoxic insult confers neuroprotection against subsequent, otherwise neurotoxic insults (Murry et al. 1986; Schurr et al. 1986). The specific mechanisms of this inducible neuroprotection are poorly understood but recent studies have implicated delta-opioid receptors as key initiators of neuroprotective pathways that limit glutamatergic hyperexcitability and neural cell death during such low oxygen stresses. For example, in the brain of HPC-treated animals, delta-opioid receptor mRNA and protein expression are elevated concomitantly with increased neuroprotection, while a delta-opioid receptor antagonist prevents this increase in receptor expression and abolishes neuroprotection. In addition, delta-opioid receptor antagonism induces cell death in normoxic rat cortical neurons and accelerates anoxia-induced cell death (Zhang et al. 2002). This finding is particularly important since it suggests that a tonic background level of delta-opioid receptor activation is critical to neuronal survival in non-pathological environments as well as during pathological events.

Nonetheless, despite strong evidence supporting a role for delta-opioid receptors as neuroprotective against hypoxic and ischemic insults, the underlying intracellular mechanisms and molecular pathways of neuroprotection remain poorly understood; although recent studies suggest that limiting neuronal cationic (i.e. Na^+ and Ca^{2+}) influx during low oxygen stress is critical to such protective mechanisms. The primary entry point of cations during ECD is via excitatory glutamatergic receptors, with the majority of ions moving through NMDARs. Interestingly, research into the effects of NMDAR pharmacological agents on the mechanism of opioid drug dependency provided early evidence for a cross-talk mechanism between delta-opioid receptors and NMDARs such that activation of delta-opioid receptors reduces NMDAR activity and vice-versa (Wang and Mokha 1996; Cao et al. 1997). Since over-activation of NMDARs is central to ECD and prolonged neuronal depolarization following hypoxic or ischemic insults, and since delta-opioid receptor activation provides neuroprotection against such insults, it is perhaps not surprising that recent studies have begun to elucidate beneficial interactions between delta-opioid receptors and the regulation of glutamatergic activity in various models of low oxygen stress. This chapter will discuss what has been learned regarding the role of delta-opioid receptors in neuroprotection against low oxygen stress as it pertains to the regulation of excitatory glutamatergic signaling. Signaling mechanisms between glutamatergic receptors and delta-opioid receptors elucidated in pain and drug dependency studies will be discussed. These mechanisms will also be related to recent examination of the role of delta-opioid receptors in modulating excitatory glutamatergic activity during anoxia in the anoxia-tolerant turtle brain and the many similarities between this neuroprotective mechanism in turtle brain and neuroprotective mechanisms activated by preconditioning in the brains of hypoxia-intolerant species.

2 Glutamatergic Excitoxicity during Low Oxygen Stress in the Brain

Glutamate receptors play a central role in the synaptic transmission that mediates most of the body's systems, and are also critical to memory formation. However, deranged glutamatergic signaling also plays a key role in the pathology of brain cell death during low oxygen stress. Indeed, mammalian neurons are acutely sensitive to hypoxia and rapidly undergo ECD when deprived of oxygen. Specifically, in the brain of hypoxia-intolerant organisms, low oxygen stress induces elevations in the excitatory amino acid glutamate (Bosley et al. 1983; Abele et al. 1990; Andine et al.

1991), primarily by the reversed operation of glutamate transporters (Rossi et al. 2000). Accumulation of glutamate in the synaptic cleft chronically activates glutamatergic α-amino-3-hydroxy-5-methyl-4-isoxazolepropionic acid receptors (AMPARs) and NMDARs, which permit excessive Na^+ and Ca^{2+} influx that leads to neuronal depolarization and electrical hyper-excitability (Abele et al. 1990; Michaels and Rothman 1990; Andine et al. 1991; Crepel et al. 1993; Lyubkin et al. 1997). Often, seizure-like bursts of synaptic activity are observed and these excitatory events permit significant ion movement that requires compensation by ATP-dependent pump activity to restore ionic gradients and neuronal homeostasis; however, the hypoxic or ischemic cell is typically unable to utilize oxidative phosphorylation to generate ATP in the absence of oxygen and instead must rely on anaerobic glycolytic pathways of energy production during hypoxia. Glycolytic pathways typically yield only 1/10th of the energy produced by aerobic pathways, and thus the hypoxic or ischemic brain cell suffers an ~90 % reduction in ATP availability. This energy production deficit, combined with greatly increased ATP demand due to heightened neuronal excitability, results in a rapid depletion of cellular ATP stores (Kopp et al. 1984; Santos et al. 1996). ATP depletion results in the abolishment of ATP-dependent pump activity, triggering a further depolarization of neuronal membrane potential, which rapidly reaches a point of no return, becoming irreversible, even upon reoxygenation (Lundberg and Oscarsson 1953; Anderson et al. 2005). Extended neuronal depolarization chronically over-activates voltage-sensitive channels and deleterious concentrations of Ca^{2+} and Na^+ continue to enter the cell, leading to further depolarization and acceleration of excitatory events. As the cytosolic $[Ca^{2+}]$ rises, mitochondria take up free Ca^{2+} and mitochondrial $[Ca^{2+}]$ increases concomitantly, opposed by the activity of the mitochondrial Ca^{2+}/H^+ or Ca^{2+}/Na^+ exchangers (Pizzo et al. 2012). When cytosolic $[Ca^{2+}]$ reaches ~500 nM, the ability of these exchangers to oppose ion gradient-dependant mitochondrial Ca^{2+} uptake is overwhelmed and the mitochondrial $[Ca^{2+}]$ begins to rise rapidly. This is termed the 'set point' and the mitochondrial $[Ca^{2+}]$ becomes overloaded at 1–3 μM cytosolic $[Ca^{2+}]$ (Di Lisa and Bernardi 2009).

Excessive uptake of Ca^{2+} into mitochondria induces the formation of the mitochondrial permeability transition pore (MPTP), which is a junctional complex that permits ions and solutes up to 1500 Da in size to readily pass out of the mitochondrion and which also enables the release of mitochondrial apoptotic factors that trigger local cell death pathways in neighboring cells (Kannurpatti et al. 2004; Wang and Qin 2010). Many labs have shown that prevention of MPTP formation is critical to avoiding neuronal apoptosis and necrosis following hypoxic or ischemic damage. For example, in neonatal rat myocytes, ischemia-reperfusion results in apoptotic events that are abolished by cyclosporine A, an inhibitor of the MPTP (Xu et al. 2001). The MPTP is currently believed to form between the adenine nucleotide translocase protein of the inner mitochondrial membrane and an as yet unidentified protein of the outer mitochondrial membranes (for review see Crompton 2000; Halestrap 2010; Suh et al. 2013). However, there remains considerable debate regarding the identity of the core components of the MPTP, as genetic mutants lacking each of the individual proposed components of the MPTP (including adenine nucleotide translocase) still exhibit pore formation (Rasola and Bernardi 2007; Brenner and Moulin 2012). There is also some variability in the stage at which a given stressor induces MPTP formation. In some pathologies and tissues MPTP formation is induced upon reoxygenation (e.g. in cardiomyocytes, Assaly et al. 2012), while in others MPTP formation can occur within minutes of stress onset (e.g. in somatosensory cortex, Liu and Murphy 2009).

In response to hypoxia or ischemia, most mammalian neurons undergo intrinsic (mitochondrial) apoptotic cell death, mediated by cytochrome c release from mitochondria following mitochondrial Ca^{2+} overload (McClintock et al. 2002). Briefly, MPTP formation leads to mitochondrial outer membrane permeabilization (MOMP) via alterations in the balance of the pro- and anti-apoptotic members of the B cell lymphoma 2 (Bcl-2) protein family (Chipuk and Green 2008). MOMP permits several mitochondrial components to be released into the cytoplasm where they act as pro-apoptotic intracellular signaling molecules to execute programmed cell death. For example, cytochrome C, a key component of the electron transport chain, also activates caspase 9-dependant cell death when released into the cytoplasm (Liu et al. 1996). Similarly, apoptosis inducing factor, which is a component of complex I of the electron transport chain, acts directly on the nucleus to induce chromatin condensation and fragmentation of the nuclear envelope in caspase-independent cell death (Susin et al. 1999). Beyond the local cell, release of these factors into the surrounding perfusate also induces apoptotic events in neighboring cells, which is a key component of penumbral spread in stroke pathology (Lo 2008). Considerable research has extensively described cell death pathways related to hypoxic and ischemic cell death and this topic has been expertly reviewed elsewhere (Kroemer et al. 2007; Tait and Green 2010).

Conversely, preventing excitotoxicity and the execution of cell death pathways are hallmarks of the brains of hypoxia-tolerant organisms. This phenotype is made possible by a wide array of cytoprotective mechanisms that are likely initiated by a few cellular sensors that detect early changes in cellular oxygen or energy availability and trigger broader neuroprotective molecular and cellular responses to hypoxia (Pamenter 2014). At the cellular level, the underlying architecture of the nervous system is highly conserved across nature and differences in the expression or activation of the same proteins are often the difference between a brain that is tolerant to hypoxia and one that is not. The study of the role of delta-opioid receptors as oxygen sensors and triggers of neuroprotection against low oxygen stress is in its infancy but studies to date suggest that these receptors can function in a neuroprotective capacity in both hypoxia-tolerant and –intolerant brains, with the key difference that delta-opioid receptors are more broadly expressed throughout turtle brain and that they are naturally activated by hypoxia to trigger neuroprotection, whereas in hypoxia-intolerant mammal brain they have to be activated to confer protection. Therefore, differences in the expression and activation of delta opioid receptors underlie enormous differences in the hypoxia-tolerance of neurons. The underlying mechanisms of this protection as well as differences between the induction of neuroprotection between tolerant and intolerant species will be examined in the ensuing sections.

3 Delta-Opioid Receptors Trigger Glutamatergic Channel Arrest and Underlie Hypoxia-Tolerance in Facultative Anaerobe Brain

As discussed above, neurons in most mammals are exquisitely sensitive to hypoxic or ischemic stresses and rapidly exhibit severe derangements of ion gradients and ECD. In particular, excessive Ca^{2+} influx into the cytosol followed by the rapid overloading of mitochondria with Ca^{2+} is a critical point of no return in the initiation of cell death pathways in most mammalian neurons during hypoxia or ischemia. Conversely, many organisms experience daily or seasonal hypoxia or anoxia in their natural environment and these creatures have evolved a variety of cellular and molecular mechanisms that protect their brain cells during hypoxic episodes. Unlike in the brain of hypoxia-intolerant animals, limiting neuronal Ca^{2+} influx during periods of low oxygen stress is a hallmark of the brain in these hypoxia-tolerant facultative anaerobes. Although no measurements of changes in mitochondrial [Ca^{2+}] during hypoxia have been made to date in a hypoxia-tolerant species, many groups have demonstrated blunted increases in neuronal [Ca^{2+}] in such species relative to in hypoxia-intolerant species. For example, slower cytosolic [Ca^{2+}] accumulation during hypoxic or anoxic stresses have been reported in the brain of anoxia-tolerant freshwater turtles (Bickler 1992), in hypoxia-tolerant neonatal rat neurons (Bickler et al. 2003), and in juvenile and adult naked mole rat brains (Peterson et al. 2012b). Of these models, the mechanism mediating reduced Ca^{2+} accumulation during low oxygen stress is best understood in the brain of the most anoxia-tolerant vertebrate identified to date—the freshwater Western painted turtle, *Chrysemys picta bellii*. In cortical slices from *C. picta*, moderate elevations of 10–30 % in brain cytosolic [Ca^{2+}] have been reported during prolonged anoxic exposure (Bickler et al. 2000; Pamenter et al. 2007b, 2008c; Pamenter and Buck 2008a), and this increase has been linked to channel arrest of glutamatergic AMPARs and NMDARs (Bickler et al. 2000; Zivkovic and Buck 2010).

Channel arrest is a mechanism that occurs in the brain and other excitable cells of numerous hypoxia-tolerant organisms wherein the activity and/or expression of ion channels is reduced during periods of low oxygen stress to limit passive ion leak and the associated workload on ion pumps, thus reducing ATP demand during hypoxia (Hochachka 1986). In particular, several studies support the occurrence of channel arrest in the anoxic turtle brain, including observations of 50–65 % decreases in whole-cell AMPAR and NMDAR currents, NMDAR open probability, NMDAR-mediated Ca^{2+} influx, K^+ conductance, and whole-cell K^+ leakage (Chih et al. 1989; Buck and Bickler 1995, 1998; Shin and Buck 2003; Pamenter et al. 2008a; Rodgers-Garlick et al. 2013). In addition to decreased ion conductance, both NMDAR and Na^+ channel density decrease during prolonged anoxia in turtle brain (Perez-Pinzon et al. 1992; Bickler et al. 2000). Beyond turtles, channel arrest of NMDARs has also been demonstrated in hypoxia-tolerant goldfish brain (Wilkie et al. 2008), NMDAR NR1 subunit gene expression in anoxia-tolerant crucian carp brain decreases following 1 or 7 days of anoxic exposure (Ellefsen et al. 2008),

NMDAR expression and activity decrease in the brain of hypoxia-tolerant hibernating arctic ground squirrels (Ross et al. 2006; Zhao et al. 2006), and the expression of NMDAR isoforms that are associated with reduced receptor permeability to Ca^{2+} are upregulated in the brain of naked mole rats following prolonged hypoxia (Peterson et al. 2012a). Of these examples, the mechanism of glutamate receptor channel arrest in turtle brain has received the most attention and appears to be regulated by a mitochondrial-mediated change in cytosolic [Ca^{2+}] during low oxygen stress, which is initiated by delta-opioid receptor activation.

In turtle brain, the mitochondrial H^+ gradient is partially dissipated during anoxia due to the activation of mitochondrial ATP-sensitive K^+ (mK_{ATP}) channels (Pamenter et al. 2008d; Hawrysh and Buck 2013). Mitochondrial K_{ATP} channels are located on the inner membrane of the mitochondria and although their specific structure is unknown, it is thought to be similar to plasmalemmal K_{ATP} channels, which are composed of four pore-forming inward-rectifying K^+ channel subunits (KIR6.1, 6.2) and four modulatory sulfonylurea receptors (SUR-1, 2) (Karschin et al. 1998; Aguilar-Bryan and Bryan 1999). Mild mitochondrial uncoupling mediated by the activation of mK_{ATP} channels reduces the driving force on mitochondrial Ca^{2+}-uniporter-mediated mitochondrial Ca^{2+} uptake, and thereby reduces the rate of Ca^{2+} uptake into the mitochondria (Pamenter et al. 2008d). As mitochondrial Ca^{2+} buffering decreases, cytosolic [Ca^{2+}] rises and Ca^{2+} bind to calmodulin, which leads to the dephosphorylation of NMDARs and presumably AMPARs (Fig. 8.1a) (Bickler et al. 2000; Shin et al. 2005). This mechanism thus reduces excitatory Na^+ and Ca^{2+} influx and hyperexcitability during anoxia.

Evidence in support of this mechanism includes the observations that anoxic increases in cytosolic [Ca^{2+}] and channel arrest of AMPARs and NMDARs are prevented by: (1) intracellular Ca^{2+} chelation but not by chelation of extracellular Ca^{2+} or antagonism of endoplasmic reticulum-dependent Ca^{2+} cycling, and (2) agonists of mK_{ATP} channels; while (3) calmodulin inhibitors prevent channel arrest of NMDARs but presumably do not effect the anoxic increase in cytosolic [Ca^{2+}] (Bickler et al. 2000; Shin et al. 2005; Pamenter et al. 2008d; Zivkovic and Buck 2010). In addition, channel arrest and the anoxic increase in cytosolic [Ca^{2+}] can be mimicked in normoxia or restored in anoxia by agonists of (4) mK_{ATP} channels, or (5) the mitochondrial Ca^{2+} uniporter, which would reduce mitochondrial buffering of Ca^{2+}; and importantly, (6) these effects are all abolished by chelation of intracellular Ca^{2+} (Pamenter et al. 2008d; Zivkovic and Buck 2010).

While the signaling pathways connecting mitochondrial uncoupling to the induction of glutamatergic channel arrest have been described downstream of the mitochondria, the upstream extra-mitochondrial oxygen sensor in this system is poorly understood. Physiologically, several cellular messengers including protein kinase C (PKC), adenosine, superoxide ($O_2^{\bullet}$), and nitric oxide can activate mK_{ATP} channels (Sasaki et al. 2000; Korge et al. 2002), and the oxygen sensor in this system appears to involve cell membrane-based proteins. A recent study aimed at determining the cellular oxygen sensor supports a central role for delta-opioid receptors in this mechanism (Pamenter and Buck 2008a), while previous studies have also implicated a role for adenosine receptors (Buck and Bickler 1995, 1998; Pamenter et al.

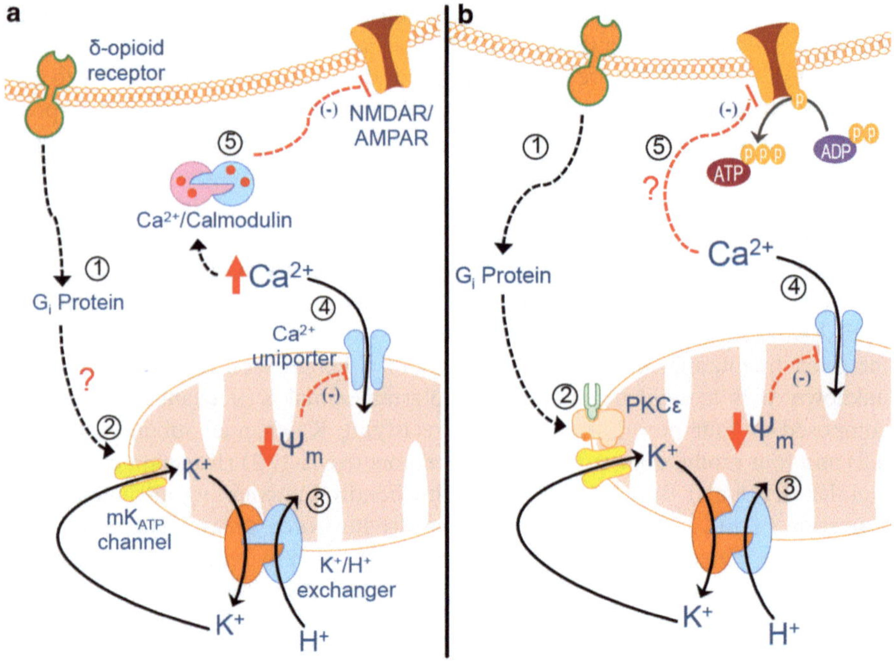

Fig. 8.1 Delta-opioid receptor activation initiates glutamatergic receptor channel arrest and hypoxic preconditioning via related cellular pathways. (**a**) In anoxia-tolerant turtle (*Chrysemys picta*) brain, (1) low oxygen stress activates delta-opioid receptors, which (2) activates mK_{ATP} channels via a G_i protein-dependent mechanism. (3) Mitochondrial K_{ATP} channel opening permits a futile cycling of K^+ ions across the mitochondrial membrane and reduces the mitochondrial H^+ gradient. This leads to "mild uncoupling" of the mitochondria, which is manifest as decreased mitochondrial membrane potential ($\Delta\Psi_m$). (4) The activity of the mitochondrial Ca^{2+}-uniporter is dependent on $\Delta\Psi_m$ and uncoupling reduces the driving force powering this pump, which decreases the rate of Ca^{2+} uptake into the mitochondria, causing (5) cytosolic Ca^{2+} to rise ~10–30 % and bind to calmodulin to dephosphorylate glutamate receptors. A similar mechanism acts to inhibit AMPARs, which further reduces neuronal electrical excitability by reducing excitatory Na^+ influx. (**b**) In hypoxia-intolerant species, a similar mechanism can be upregulated by preconditioning protocols. In this paradigm, preconditioning (1) activates delta-opioid receptors, which (2) activate mK_{ATP} channels via a G-protein mediated mechanism that modulates protein kinase C (PKC), (3) activating a similar futile cycling of K^+ ions across the mitochondrial membrane that reduces the mitochondrial H^+ gradient and thereby (4) reduces the driving force on the mitochondrial Ca^{2+} uniporter and prevents toxic accumulation of Ca^{2+} in the mitochondria and (5) related formation of the mitochondrial permeability transition pore (MPTP). *Black arrows* indicate upregulated pathways. *Red lines* indicated downregulated pathways

2008b). Pharmacological activation of either receptor during normoxia has a similar effect as anoxia on NMDAR currents (i.e. they mimic channel arrest).

Delta-opioid receptors are purported oxygen sensors in mammalian brain (He et al. 2013) and adenosine has been proposed as a detector of reduced energy availability during low oxygen stress due to breakdown of ATP (Buck 2004). The mechanism via which delta-opioid or adenosine receptors and NMDARs interact is poorly understood, but since both receptors are coupled to G_i proteins, the pathway likely involves G_i protein activation. G proteins are well suited to act as hypoxic messengers since they mediate numerous downstream mechanisms in neurons. There is also some indirect evidence to support a role for G protein-mediated responses in the turtle's neuronal anoxia-tolerance. For example, whole-brain [cAMP] decreases significantly in the anoxic cortex, and since cAMP is directly mediated by G_i protein activity this data suggests that anoxia-induced changes in G_i signaling occur in turtle brain (Pamenter et al. 2007a). Furthermore, anoxic channel arrest of turtle NMDAR activity is blocked by pertussis toxin, a specific G_i protein inhibitor (Pamenter et al. 2008b). Conversely, the G_i protein activator MP7 mimics channel arrest and depresses NMDAR currents in normoxic turtle brain (Pamenter and Buck 2008a).

Further upstream, there is considerable evidence implicating delta-opioid receptors as the triggers that initiate neuroprotective responses to hypoxia via G_i protein activation. For example, turtle neurons treated with the delta-opioid receptor antagonist naltrindole during anoxia lose their ability to tolerate low oxygen stress and exhibit a phenotype that is consistent with that of mammalian neurons exposed to glutamate toxicity or hypoxia/ischemia (Fig. 8.2a). In such mammalian neurons exposed to low oxygen stress excessive NMDAR-mediated Ca^{2+} influx is observed along with severe neuronal membrane potential depolarization (Garthwaite and Garthwaite 1986; Garthwaite et al. 1986; Coulter et al. 1992; Pamenter and Buck 2008a). Furthermore, upon reoxygenation, mammalian neuronal membrane potential remains depolarized, and all these effects can be prevented by perfusion of an NMDAR antagonist prior to the onset of severe depolarization (Limbrick et al. 2003). Similarly, in anoxic turtle neurons treated with naltrindole, terminal depolarization of neuronal membrane potential occurs in 31 % of experiments and electrical hyperexcitability and a large depolarization of membrane potential is observed in all neurons that is not reversed by reperfusion of normoxic naltrindole-free ACSF. Concomitantly, anoxic elevations in cytosolic [Ca^{2+}] are greatly increased and are further elevated following reoxygenation (Pamenter and Buck 2008a). This is contrary to control anoxia-alone experiments in which both the neuronal membrane potential depolarization and the increase in cytosolic [Ca^{2+}] during anoxia are significantly smaller in magnitude and reverse back to baseline following reoxygenation (Bickler 1992; Bickler et al. 2000; Pamenter and Buck 2008b; Pamenter et al. 2008d). As in mammalian brain, these deleterious events in anoxic turtle brain are likely dependent on NMDAR-mediated Ca^{2+} influx since naltrindole-mediated electrical hyper-excitability, severe prolonged depolarization, and enhanced Ca^{2+} influx are abrogated in cortical slices pre-treated with the specific NMDAR antagonist (2R)-amino-5-phosphonovaleric acid (APV; Fig. 8.2a) (Pamenter and Buck 2008a). Importantly, during anoxia, pharmacological activation of G_i proteins downstream of delta-opioid receptor antagonism prevents this extended neuronal depolarization and deleterious accumulations of cytosolic [Ca^{2+}] following anoxic/naltrindole perfusion in the anoxic turtle cortex (Pamenter and Buck 2008a), indicating anoxic channel arrest has been restored and that delta-opioid receptors activate channel arrest of NMDARs via G_i protein activity. Finally, delta-opioid receptor antagonism

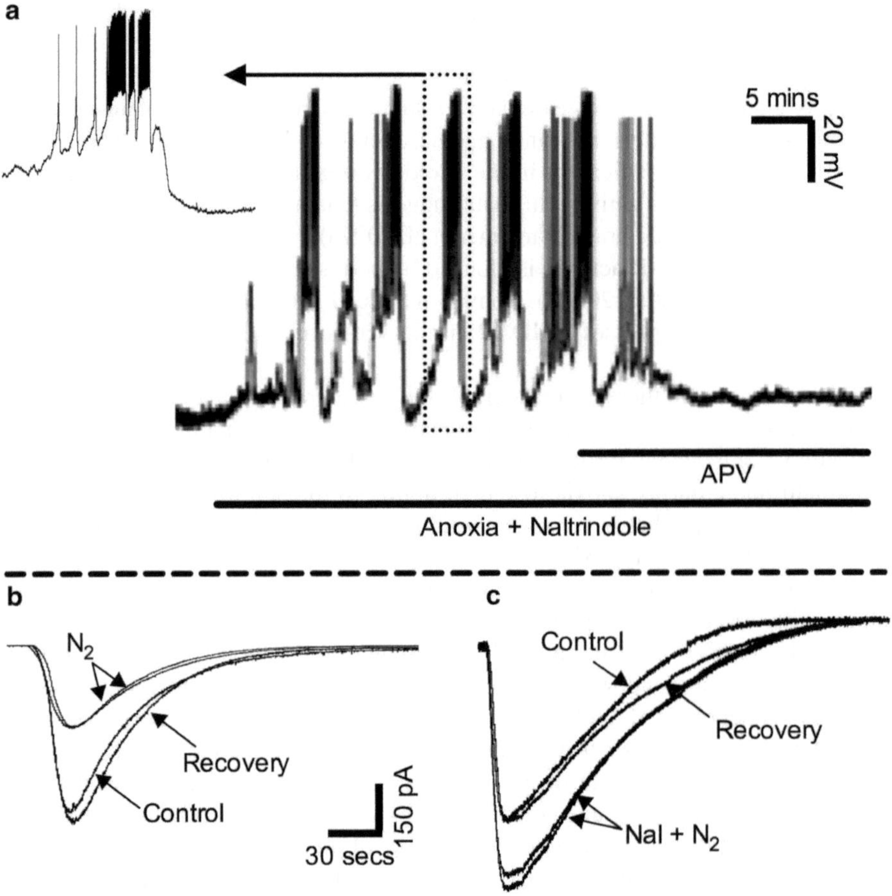

Fig. 8.2 Delta-opioid receptor antagonism potentiates NMDAR currents and induces electrical hyperexcitability. (**a**) Raw membrane potential recording from a turtle cortical neuron. Naltrindole in the presence of anoxia results in electrical hyperexcitability and seizure-like electrical events that were reversed by co-treatment with the specific NMDAR antagonist (2R)-amino-5-phosphonovaleric acid (APV). (**b**, **c**) Raw sample recordings of whole-cell NMDAR currents recorded from a single turtle cortical neuron treated with (**b**) control (normoxia), anoxia, and normoxic recovery, or (**c**) normoxia, followed by anoxia in the presence of the delta-opioid receptor antagonist naltrindole. NMDAR currents are normally reduced during anoxia (i.e. channel arrest) but potentiated when delta-opioid receptors are blocked in anoxia. (**a**) Is reprinted with permission from Pamenter and Buck (2008a)

with naltrindole also potentiates normoxic NMDAR currents (Fig. 8.2b) and Ca^{2+} influx, suggesting that a basal level of delta-opioid receptor activity is critical to regulating NMDARs and cytosolic [Ca^{2+}] during normoxia as well.

A key difference between turtle brain and mammalian brain that likely contributes to the enhanced anoxia-tolerance of the later tissue are relative differences in protein expression in turtle versus mammal brain. Indeed, delta-opioid receptor protein expression in the turtle brain and brainstem is greater and more homogenous in all regions relative to homologous areas of rat brain and brainstem, and receptor binding affinity is also markedly greater in turtle brain than in rat (Xia and Haddad 2001). These differences in receptor expression and binding affinity are in stark contrast to most studies of protein expression differences between turtle and mammal brain, in which turtle brains typically express an ~10-fold lower quantity of any given membrane protein (e.g. Xia and Haddad 1991). Such differences could partly explain the turtle's remarkable innate anoxia-tolerance relative to mammalian brain, despite their apparent reliance on similar neuroprotective mechanisms. In turtle brain, greater receptor expression and binding affinity would confer a higher sensitivity to opioid-mediated signaling, allowing turtle neurons to rapidly decrease glutamate receptor activity in response to decreasing environmental oxygen while long-term mechanisms of metabolic depression are being activated (e.g. removal of channels from neuronal membranes; Perez-Pinzon et al. 1992; Bickler et al. 2000). Conversely, mammalian brain requires a preconditioning pre-treatment to upregulate delta-opioid receptor protein expression and 'prime' the brain against subsequent low-oxygen insults (see next section). Thus turtle brain is better able to respond quickly to low-oxygen insults than mammalian brain, while activation of delta-opioid receptors in mammal brain imbues neuroprotection that mimics that of the turtle.

Conversely, the role of delta-opioid receptor-mediated signaling in long-term anoxia-tolerance in the turtle remains undetermined, however NMDAR activity remains depressed during at least the first 6 weeks of anoxia in turtle brain and delta-opioid receptor-mediated pathways may regulate this depression (Bickler 1998). Given the inverse relationship between delta-opioid receptor and glutamate receptor expression and activity in mammal brain, it would be fascinating to explore the relationship between the relative expression levels of these two receptors during prolonged anoxia in turtle brain when glutamate receptor protein expression is known to decrease (Bickler et al. 2000). Furthermore, while neuroprotective interactions between delta-opioid receptors and channel arrest of glutamate receptors in turtle brain are reasonably well mapped out, considerable work remains to determine whether or not delta-opioid receptors also mediate channel arrest in the brain of other hypoxia-tolerant organisms. Finally, interactions between delta-opioid receptors and glutamate receptors are only recently beginning to be explored in mammalian models of hypoxic and ischemic stress in the brain; however, early studies are supportive of similar links in mammalian brain, and in particular in models of hypoxic preconditioning. These links will be discussed in the next section.

4 Delta-Opioid Receptors Trigger Hypoxic/Ischemic Preconditioning in Hypoxia-Intolerant Mammalian Brain

Hypoxic and ischemic preconditioning confers neuro- or cardioprotection against subsequent, otherwise lethal insults in the brain and heart of hypoxia-intolerant organisms (Murry et al. 1986), including humans (Koch and Gonzalez 2013).

Beyond low oxygen stress, preconditioning-like cytoprotection can be induced by a wide variety of stressors, including ionic imbalances (Ashraf et al. 1994), spreading depression (Matsushima et al. 1996), hyperbaric oxygen (Kim et al. 2001), hyperoxia (Petrosillo et al. 2011), inhaled nitric oxide (Shinbo et al. 2013), hypothermia or heat shock (Ning et al. 1998), and inflammation (Deplanque and Bordet 2000), among others. Furthermore, in addition to the brain and heart, preconditioning has been observed in a variety of other tissues as diverse as lungs and testes (Gidday 2006). Remarkably, the underlying mechanisms of most of these protective pathways are largely conserved. For example, a period of exposure to high levels of Ca^{2+} will confer protection against a subsequent hypoxic insult (Xu et al. 2001), and pathways that employ common mechanistic components will typically confer this protection. The field of preconditioning research has ballooned since IPC was first described in dog myocardium challenged with ischemic/reperfusion injury (Murry et al. 1986); a full discussion of the pathways and mechanisms involved is beyond the scope of the present chapter, but these have been expertly reviewed elsewhere (Gidday 2006; Downey et al. 2007; Heusch et al. 2008).

The underlying signaling lynchpin of preconditioning in brain is generally accepted to be mitochondria, and appropriately, mitochondria have been dubbed the "gatekeepers of preconditioning" (Dirnagl and Meisel 2008). Given the similarly central role of mitochondria in mediating endogenous channel arrest mechanisms of hypoxia-tolerance in facultative anaerobe brain, it is perhaps not surprising that research has begun to elucidate remarkable similarities between the pathways that mediate preconditioning in hypoxia-intolerant organisms and those that underlie endogenous tolerance to low oxygen stress in the brain of facultative anaerobes. In particular, there is strong evidence supporting a central role for mild mitochondrial uncoupling and downstream Ca^{2+}-mediated signaling in the induction of preconditioned protection in the brain of hypoxia-intolerant species. Similar to the regulation of endogenous neuroprotective mechanisms in turtle brain, mK_{ATP} channels are presently favored as the mitochondrial uncoupling mechanism that underlies hypoxic or ischemic preconditioning in mammalian brain (Kis et al. 2004). As in turtle brain, activation of mK_{ATP} channels partially dissipates the mitochondrial H^+ gradient, reducing the driving force of the mitochondrial Ca^{2+}-uniporter and subsequently decreasing mitochondrial accumulation of Ca^{2+} during hypoxia, ischemia, or related stresses, along with subsequent MPTP formation and cytochrome C loss from the mitochondria (Fig. 8.1b). Such effects have been demonstrated in rat cortical neurons exposed to glutamate toxicity (Kis et al. 2003, 2004), following focal ischemia in rat brain (Shimizu et al. 2002), and following anoxia/reperfusion injuring in adult rat hippocampus and cortex (Heurteaux et al. 1995; Semenov et al. 2000), and in juvenile mouse brainstem (Muller et al. 2002). Conversely, blockade of mK_{ATP} channels abolishes preconditioning-mediated neuroprotection ubiquitously (Grover 1997; Takashi et al. 1999; Murata et al. 2001; Korge et al. 2002; Yoshida et al. 2004).

Notably, a similar mechanism has been described to underlie IPC-mediated cardioprotection in the hearts of hypoxia-intolerant species, and here again, mitochondrial K^+ channels are leading candidates to mildly uncouple mitochondria and provide preconditioning-based cytoprotection. In rat heart, mildly uncoupling mitochondria by activating mK_{ATP} channels is cardioprotective against subsequent hypoxic or ischemic insults and prevents MPTP formation (Holmuhamedov et al. 1999; Cao et al. 2005; Cuong et al. 2006). Mild uncoupling decreases the rate of Ca^{2+} uptake into isolated rat heart mitochondria and also increases the rate of Ca^{2+} release from isolated mitochondria that have been pre-loaded with Ca^{2+} (Holmuhamedov et al. 1998), and both of these responses to uncoupling have been linked to decreases in mitochondrial membrane potential. Similarly, in intact cardiomyocytes, activation of mitochondrial K^+ channels decreases mitochondrial membrane potential and reduces mitochondrial Ca^{2+} accumulation during ischemia (Murata et al. 2001; Wang et al. 2001; Sato et al. 2005). Importantly, pharmacological stimulation of MPTP formation abolishes the protective effects of preconditioning; suggesting that cytoprotection induced by preconditioning prevents MPTP formation (Cao et al. 2005). Taken all together, these data suggest preconditioning-mediated cytoprotection against hypoxic and ischemic stresses in both the brain and heart is due to prevention of mitochondrial Ca^{2+} accumulation resulting from mild mitochondrial uncoupling. As discussed above, the prevention of Ca^{2+} overload in the mitochondria theoretically prevents the rupture of mitochondrial membranes and release of pro-apoptotic signals via the MPTP. Whether or not this change in Ca^{2+} handling also modulates additional cytoprotective mechanisms—such as glutamatergic channel arrest in the brain—as a signaling intermediate, remains to be determined.

Mitochondrial K_{ATP} channels are not the only mitochondrial ion channels that provide neuroprotection against ischemic insults upon activation. Similar to mK_{ATP} channel activation, Ca^{2+}-mediated increases in mitochondrial Ca^{2+}-sensitive K^+ (mK_{Ca}) channel activity dissipates the mitochondrial H^+ gradient, partially uncouples mitochondria, reduces mitochondrial uniporter activity, and slows the cytotoxic accumulation of mitochondrial [Ca^{2+}]. Mitochondrial Ca^{2+}-sensitive K^+ channels are similar to plasmalemmal large-conductance K^+ channels: they are multi-conductance state channels with an open probability that is both voltage and [Ca^{2+}] dependent such that cellular membrane potential depolarization (or in the case of mK_{Ca} channels, mitochondrial membrane potential depolarization) potentiates channel activity (Siemen et al. 1999). In mitoplast-attached patches, mK_{Ca} currents increase when [Ca^{2+}] outside the pipette is increased, suggesting that the Ca^{2+} sensor of the channel is located on the matrix side of the mitochondrial membrane (Xu et al. 2002; Gu et al. 2014). Therefore, channel activity increases as mitochondrial [Ca^{2+}] rises due to sequestration of cytosolic Ca^{2+}. Such an accumulation of Ca^{2+} occurs during ischemia and mK_{Ca} channels are also activated by hypoxia (Gu et al. 2007, 2014). Since activation of either mK_{Ca} or mK_{ATP} channels should have the same effect on cells (increased mitochondrial K^+ conductance), it is not surprising that activation of mK_{Ca} channels in cardiac myocytes confers protection during global ischemia and reperfusion experiments that is similar in magnitude to the protection afforded by activation of mK_{ATP} channels or IPC (Xu et al. 2002; Cao et al. 2005). Furthermore, cytoprotection due to the activation of mK_{ATP} channels is not impaired by blockade of mK_{Ca} channels, or vice versa, suggesting that these two channels function independently, although their mechanism of action is similar—increasing mitochondrial K^+ conductance. Together, these data independently confirm the central role of K^+ influx into the mitochondrial matrix in IPC-mediated protection against ischemic injury that has been suggested by mK_{ATP} channel experiments. Interestingly, channel arrest of AMPARs and NMDARs in the anoxic turtle cortex can also be induced by mK_{Ca} channel activation, although blockade of mK_{Ca} channels does not prevent channel arrest, indicating they are not the primary mediators of this phenomenon (Pamenter et al. 2008d; Zivkovic and Buck 2010).

The mechanism via which mK_{ATP} channels are activated has not been directly elucidated in turtle brain and varies between systems and tissues in hypoxia-tolerant mammals; however, in mammals, this mechanism appears to commonly involve the activation of PKCε, which is functionally associated with mK_{ATP} in the mitochondrial inner membrane (Jaburek et al. 2006), and whose activation leads to opening of mK_{ATP} channels (Raval et al. 2003; Costa and Garlid 2008). In mammalian brain and heart, PKCε activation is mediated primarily through the activation of adenosine A_1 receptors (Liu et al. 1991; Perez-Pinzon et al. 1996), while redundant pathways mediated by bradykinins and opioids also triggers IPC in brain (Wall et al. 1994; Schultz et al. 1995). All three of these pathways are G_i-protein coupled and converge on PKCε activation. There is considerable evidence in particular for a central role for delta-opioid receptors as a key trigger in mediating hypoxia-tolerance and also hypoxic and ischemic preconditioning and reducing glutamate-associating neurotoxicity in mammalian brain. For example, delta-opioid receptor activation protects rat cortical neurons from hypoxic or glutamate-mediated stress (Zhang et al. 2002; Kawalec et al. 2011; Fig. 8.3), while morphine (which acts via opioid receptors) protects Purkinje cells from oxygen-glucose deprivation via a delta-opioid receptor-dependant mechanism (Lim et al. 2004).

Specific to models of preconditioning, in rat cortical cultures, hypoxic preconditioning increases delta-opioid receptor binding affinity, (but not mRNA expression) and attenuates glutamate-induced neurotoxicity, while delta-opioid receptor antagonism prevents preconditioning-mediated neuroprotection (Zhang et al. 2006a). Furthermore, hypoxic preconditioning increases delta-opioid receptor protein expression in a time-dependent fashion in rat cortical neuronal cultures and delta-opioid receptors protect these neurons against a subsequent severe hypoxia insult

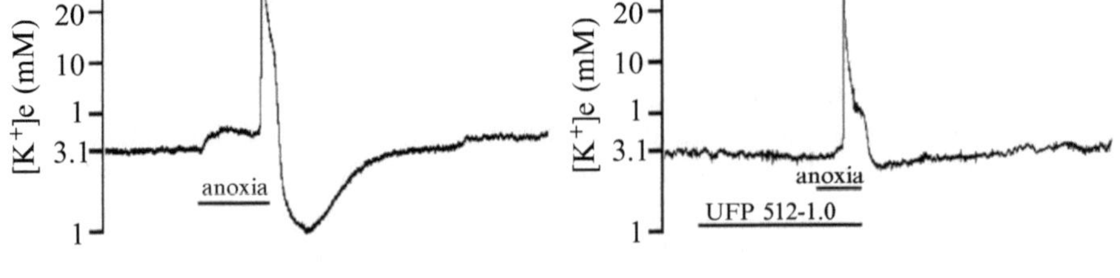

Fig. 8.3 Delta-opioid receptor activation with UFP 512 preserves K^+ homeostasis in anoxic rat cortical neurons. Reproduced from Chao and Xia (2010)

via a mechanism that involves G proteins and PKC, and which results in a decrease in the activation of p38 MAPK cell death pathways and prevents pro-apoptotic cytochrome C release from the mitochondria (Ma et al. 2005). Delta-opioid receptor-mediated preconditioning also prevents hypoxia- and ischemia-mediated loss of K^+ homeostasis, via a mechanism that is dependent on preventing cytotoxic increases in cytosolic [Ca^{2+}] (Chao et al. 2007a, b), similar to in turtle brain. These protective effects may be mediated by preventing glutamate-dependent excitotoxicity, which as discussed above, prevents cytotoxic cytosolic [Ca^{2+}] accumulation and protects mitochondria from Ca^{2+} overload and prevents MPTP formation and cytochrome c release. Furthermore, recent studies examining potential protective effects of delta-opioid receptor activation on neuronal Na^+ flux have provided more direct evidence of a role for delta-opioid receptors in mediating excitatory glutamatergic neurotransmission in mammal brain during low oxygen stress. In murine cortical slices, delta-opioid receptor activation reduces anoxia-dependent deleterious derangement of K^+ gradients via by limiting anoxic neuronal Na^+ entry through NMDAR but not AMPARs via a mechanism that involves PKC activation (Chao et al. 2008, 2009, 2012). Additional inhibition of Na^+ entry and prevention of the deleterious loss of K^+ gradients is also mediated through neuroprotective interactions between delta-opioid receptors and voltage-gated Na^+ channels in these studies, suggesting that delta-opioid receptors mediate neuroprotective pathways via the modulation of both glutamate receptor-dependent and -independent excitatory signaling in mammalian neurons.

It is fascinating to note that overall, the parallels between the delta-opioid receptor-dependent pathway that mediate the mild uncoupling mechanism that endogenously activates neuroprotective channel arrest in anoxic turtle brain and the delta-opioid receptor-dependent pathway that mediates inducible hypoxic or ischemic preconditioning in mammalian brain are numerous and clear. Specifically, (1) activation of mK_{ATP} or mK_{Ca} channels or inhibition of the mitochondrial Ca^{2+}-uniporter induces channel arrest in turtle brain and HPC/IPC in mammalian brain; (2) blockade of mK_{ATP} channels (but not mK_{Ca} channels: Gaspar et al. 2008, 2009; Pamenter et al. 2008d; Zivkovic and Buck 2010), or activation of the mitochondrial Ca^{2+}-uniporter prevents channel arrest in turtle brain and IPC in rat heart (Zhang et al. 2006b; Pamenter et al. 2008d; Zivkovic and Buck 2010); (3) mitochondria from turtles and mammals are similarly uncoupled 10–20 % by mK_{ATP} channel activation (Holmuhamedov et al. 1999; Murata et al. 2001; Pamenter et al. 2008d); (4) in both the turtle and preconditioned rat brain, small increases in cytoplasmic Ca^{2+} are associated with subsequent neuroprotection (Bickler et al. 2000; Bickler and Fahlman 2004); (5) pharmaceutical activation of mK_{ATP} channels increases cytosolic Ca^{2+} mildly by ~10 % in turtle brain and 50 % in rat heart, and these increases correlate with an ~80 % reduction in mitochondrial Ca^{2+} accumulation during a subsequent ischemic stress in rat and also with cell survival in both species (Bickler et al. 2000; Wang et al. 2001; Pamenter et al. 2008d), and most pertinent to this chapter, (6) the activation of delta-opioid receptors, upstream of the mitochondrial induction of cytoprotective events, acts as a trigger of mK_{ATP}-dependent neuroprotective mechanisms in both mammals and turtles (Schultz et al. 1995; Pamenter and Buck 2008a).

The mechanism through which preconditioning provides cytoprotection downstream of mitochondrial uncoupling in hypoxia-intolerant organisms is poorly understood but is thought to involve preservation of neuronal ion gradients, prevention of mitochondrial Ca^{2+} overload, subsequent MPTP formation, and the activation of related cell death pathways during deleterious stresses. Glutamatergic channel arrest in turtle neurons provides clear neuroprotection against low oxygen stress or glutamatergic excitotoxicity, and examination of the effect of IPC protocols on AMPAR and NMDAR activity in mammalian neurons may reveal a related neuroprotective mechanism that limits glutamatergic contributions to ECD, particularly given the myriad similarities between mitochondrial regulation of channel arrest and preconditioning in brain, and the important role of glutamate toxicity in ischemic or hypoxic mammalian brain.

In addition to a role in direct local preconditioning, opioid receptors have recently been implicated as key mediators of remote preconditioning. Remote preconditioning is a phenomenon whereby a stressor applied to a remote region of the body provides neuroprotection against subsequent hypoxic or ischemic insults. Common examples of these preconditioning stimuli include ischemic stress applied to a limb, or abdominal surgical trauma. For example, stimulation of spinal opioid receptors may confer cardioprotection against subsequent ischemic insults (Groban et al. 2004; Ling Ling et al. 2010; Wong et al. 2010); however, the specific opioid receptor subtype that mediates such protective effects are poorly understood and further research is required to determine whether or not the delta subtype of opioid receptor is involved in this peripheral form of preconditioning. It is also currently unknown whether delta-opioid receptors play a role in modes of preconditioned protection outside of the brain via such interactions in the peripheral nervous system, and further research is required in this area as well.

Importantly, preconditioning is not unique to hypoxia-intolerant organisms and was in fact first identified in anoxia-tolerant goldfish (Prosser et al. 1957). Preconditioning mechanisms appear to be a general protective response to stress and the observation that such pathways are endogenously upregulated in hypoxia-tolerant species suggests that such organisms are naturally “primed” to respond to a variety of stresses. This remarkable overlap highlights the utility of comparative model organisms in the study of stress responses that are relatable to clinical insults. Indeed the basic mechanism of mK_{ATP} channel-mediated cytoprotection (induced or endogenous) appears to be conserved in all animals studied, and hypoxic preconditioning mediated by mK_{ATP} channels has recently even been reported in *c elegans* (Wojtovich et al. 2012). The oxygen sensor and trigger of cytoprotective preconditioning mechanisms in these species have not been investigated but the high degree of overlap between the induction of channel arrest and hypoxic/ischemic preconditioning, along with their mutual reliance on delta-opioid receptors to initiate protective responses to low oxygen stress, suggest that delta-opioid receptors may play a key role in other species and models of preconditioning. Further research is required to investigate the role of delta-opioid receptors in these fascinating and non-traditional models.

5 Additional Evidence for Functional Coupling between Delta-Opioid and Glutamate Receptors in Nociception

Beyond investigations related to hypoxia and ischemia, considerable insights into the mechanisms of interaction between delta-opioid receptors and glutamatergic signaling have been revealed in studies of the neural transmission of pain sensations and related anesthetic interactions. In particular, a handful of studies examining the interactions between delta-opioid receptor and glutamate receptor signaling in the dorsal horn of the medulla provide considerable indirect support for the occurrence of delta-opioid receptor-mediated modulation of glutamate receptor currents in mammal brain that closely resembles the relationship between these receptor types reported previously in anoxic turtle brain. The dorsal horn region of the medulla, also known as the trigeminal nucleus caudalis, is a key relay circuit of pain and temperature sensation from the oro-facial region, and glutamate is the primary excitatory neurotransmitter in this circuit (Salt and Hill 1981, 1983; Sessle 1987; Clements et al. 1991). In an *in vivo* preparation in anesthetized rats, direct microinjection of NMDA activates trigeminothalamic neurons, and application of specific delta-opioid receptor antagonists reduces NMDA-evoked responses (Wang and Mokha 1996; Zhang et al. 1996). These pharmacologically-mediated responses in rat brain are remarkably similar to the endogenous response that occurs in turtle cortical neurons, in which pharmacological delta-opioid receptor activation decreases NMDA-evoked currents in normoxia, and persistent endogenous delta-opioid receptor activation during anoxia reduces stimulus-evoked NMDAR currents (Pamenter and Buck 2008a).

In addition to such post-synaptic modification of NMDAR currents, which presumably occur via phosphorylation of glutamate receptors in the post-synaptic neuron, delta-opioid receptors also modulate glutamate receptor activity via alterations of pre-synaptic glutamate release. In experiments aimed at understanding the cellular mechanisms of the effects of morphine by recording from layer V pyramidal neurons in slices of rat anterior singulate cortex, application of the delta-opioid receptor ligand enkephalin attenuates glutamatergic excitatory postsynaptic potentials (EPSPs) by ~60 % and this effect is prevented by delta-opioid receptor antagonists (Fig. 8.4) (Tanaka and North 1994). Importantly, the inhibitory effect of delta-opioid receptor antagonism is manifest on both the NMDAR- and AMPAR-dependent components of the EPSP, which supports modulation of presynaptic glutamate release as the underlying mechanism and which would limit the impact of glutamatergic excitability considerably beyond the impact of limiting NMDAR activity alone. Similar inhibitory effects of delta-opioid receptor antagonism on pre-synaptic glutamate release and post-synaptic NMDAR activity have also been reported in rat striatum (Jiang and North 1992), nucleus accumbens (Yuan et al. 1992), and in rat neocortex (Ostermeier et al. 2000). To my knowledge, the roll of delta-opioid receptor-dependent inhibition of glutamate release has not been investigated in any neuronal model of hypoxia/ischemia tolerance or HPC/IPC, and such

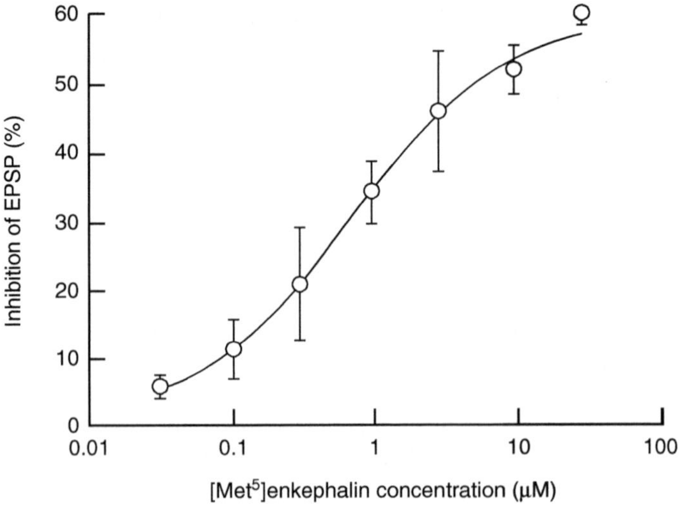

Fig. 8.4 Agonism of the delta-opioid receptor inhibits glutamatergic EPSP activity in a dose dependent manner in rat brain slices. Reproduced from Tanaka and North (1994)

investigations will likely provide a fascinating addition to our understanding of the ability of delta-opioid receptors to limit excitatory glutamatergic signaling in the brain during periods of low oxygen stress.

6 Concluding Remarks

Over-excitation of glutamatergic signaling is a central catalyst of neuronal cell death during low oxygen stresses, including those related to systemic hypoxia and ischemic stroke. Blocking glutamate receptors is an effective treatment for such stresses, but this approach also results in significant side effects, including psychomimetic effects, the impairment of memory processes, or coma. Indirect mechanisms to partially impair glutamatergic signaling during low oxygen stresses may circumvent these unwanted side effects, and as such, the study of natural upstream modulators of glutamatergic signaling are of pressing clinical interest. Inducible preconditioning mechanisms in hypoxia-intolerant mammal brain and innate neuroprotective mechanisms observed in the brains of endogenously hypoxia-tolerant model organisms are two naturally occurring phenomenon that offer promising avenues of exploration in the search for such indirect neuroprotective mechanisms of glutamate receptor modulation. Recent studies have demonstrated that similar pathways may mediate both of these neuroprotective mechanisms and that a common trigger– the delta-opioid receptor, may activate these pathways.

In anoxia-tolerant turtle brain, glutamatergic excitotoxicity is prevented in part by the channel arrest of NMDARs, which reduces the excitatory influx of Na^+ and

Ca^{2+} ions during anoxia. NMDAR channel arrest in turtle brain can be mimicked by activating delta-opioid receptors and prevented by inhibiting them. Preventing channel arrest by blocking delta-opioid receptors reverses the anoxia-tolerance of turtle neurons and they exhibit a phenotype consistent with ECD in mammalian neurons: electrical hyperexcitability, seizure-like electrical events, large scale Ca^{2+} influx, and irreversible depolarization of neuronal membrane potential. It is noteworthy that in mammalian models of pain transmission, delta-opioid receptor agonists also reduce NMDAR activity, and the degree to which NMDAR currents are depressed is similar between anoxic turtle brain and mammal brain in which delta-opioid receptors have been stimulated (~60 % suppression). Based on these studies, it is reasonable to hypothesize that delta-opioid receptors may be capable of inducing similar neuroprotective responses in mammal brain during low oxygen stress via related mechanisms.

The investigation of the putative role of delta-opioid receptors in hypoxic or ischemic mammal brain has largely been focused on preservation of neuronal ion homeostasis via a wide variety of ions and ion channels. Recently, a few studies have demonstrated neuroprotective interactions between delta-opioid receptor activation and the inhibition of glutamatergic signaling, and specifically the inhibition of NMDARs. However, direct electrophysiological examination of whole-cell or single-channel NMDAR currents in mammalian brain slice models of hypoxia and ischemia are required to fully determine the extent of these interactions and the degree to which they overlap with the mechanisms described previously in turtle brain.

References

Abele AE, Scholz KP, Scholz WK, Miller RJ (1990) Excitotoxicity induced by enhanced excitatory neurotransmission in cultured hippocampal pyramidal neurons. Neuron 4:413–419

Aguilar-Bryan L, Bryan J (1999) Molecular biology of adenosine triphosphate-sensitive potassium channels. Endocr Rev 20:101–135

Anderson TR, Jarvis CR, Biedermann AJ, Molnar C, Andrew RD (2005) Blocking the anoxic depolarization protects without functional compromise following simulated stroke in cortical brain slices. J Neurophysiol 93:963–979

Andine P, Orwar O, Jacobson I, Sandberg M, Hagberg H (1991) Changes in extracellular amino acids and spontaneous neuronal activity during ischemia and extended reflow in the CA1 of the rat hippocampus. J Neurochem 57:222–229

Ashraf M, Suleiman J, Ahmad M (1994) Ca^{2+} preconditioning elicits a unique protection against the Ca^{2+} paradox injury in rat heart. Role of adenosine. Fixed. Circ Res 74:360–367

Assaly R, de Tassigny A, Paradis S, Jacquin S, Berdeaux A, Morin D (2012) Oxidative stress, mitochondrial permeability transition pore opening and cell death during hypoxia-reoxygenation in adult cardiomyocytes. Eur J Pharmacol 675:6–14

Bickler PE (1992) Cerebral anoxia tolerance in turtles: regulation of intracellular calcium and pH. Am J Physiol 263:R1298–R1302

Bickler PE (1998) Reduction of NMDA receptor activity in cerebrocortex of turtles (Chrysemys picta) during 6 wk of anoxia. Am J Physiol 275:R86–R91

Bickler PE, Fahlman CS (2004) Moderate increases in intracellular calcium activate neuroprotective signals in hippocampal neurons. Neuroscience 127:673–683
Bickler PE, Donohoe PH, Buck LT (2000) Hypoxia-induced silencing of NMDA receptors in turtle neurons. J Neurosci 20:3522–3528
Bickler PE, Fahlman CS, Taylor DM (2003) Oxygen sensitivity of NMDA receptors: relationship to NR2 subunit composition and hypoxia tolerance of neonatal neurons. Neuroscience 118:25–35
Bosley TM, Woodhams PL, Gordon RD, Balazs R (1983) Effects of anoxia on the stimulated release of amino acid neurotransmitters in the cerebellum in vitro. J Neurochem 40:189–201
Brenner C, Moulin M (2012) Physiological roles of the permeability transition pore. Circ Res 111:1237–1247
Buck LT (2004) Adenosine as a signal for ion channel arrest in anoxia-tolerant organisms. Comp Biochem Physiol B Biochem Mol Biol 139:401–414
Buck LT, Bickler PE (1995) Role of adenosine in NMDA receptor modulation in the cerebral-cortex of an anoxia-tolerant turtle (Chrysemys-picta belli). J Exp Biol 198:1621–1628
Buck LT, Bickler P (1998) Adenosine and anoxia reduce N-methyl-D-aspartate receptor open probability in turtle cerebrocortex. J Exp Biol 201:289–297
Buck LT, Pamenter ME (2006) Adaptive responses of vertebrate neurons to anoxia—matching supply to demand. Respir Physiol Neurobiol 154:226–240
Cao YJ, Bian JT, Bhargava HN (1997) Effects of NMDA receptor antagonists on delta1- and delta2-opioid receptor agonists-induced changes in the mouse brain [3H]DPDPE binding. Eur J Pharmacol 335:161–166
Cao CM, Xia Q, Gao Q, Chen M, Wong TM (2005) Calcium-activated potassium channel triggers cardioprotection of ischemic preconditioning. J Pharmacol Exp Ther 312:644–650
Chao D, Xia Y (2010) Ionic storm in hypoxic/ischemic stress: can opioid receptors subside it? Prog Neurobiol 90:439–470
Chao D, Bazzy-Asaad A, Balboni G, Xia Y (2007a) delta-, But not mu-, opioid receptor stabilizes $K^{(+)}$ homeostasis by reducing $Ca^{(2+)}$ influx in the cortex during acute hypoxia. J Cell Physiol 212:60–67
Chao D, Donnelly DF, Feng Y, Bazzy-Asaad A, Xia Y (2007b) Cortical delta-opioid receptors potentiate K^+ homeostasis during anoxia and oxygen-glucose deprivation. J Cereb Blood Flow Metab 27:356–368
Chao D, Bazzy-Asaad A, Balboni G, Salvadori S, Xia Y (2008) Activation of DOR attenuates anoxic K^+ derangement via inhibition of Na^+ entry in mouse cortex. Cereb Cortex 18:2217–2227
Chao D, Balboni G, Lazarus LH, Salvadori S, Xia Y (2009) $Na^{(+)}$ mechanism of delta-opioid receptor induced protection from anoxic $K^{(+)}$ leakage in the cortex. Cell Mol Life Sci 66(6):1105–1115
Chao D, He X, Yang Y, Balboni G, Salvadori S, Kim DH, Xia Y (2012) Hydrogen sulfide induced disruption of Na^+ homeostasis in the cortex. Toxicol Sci 128:198–208
Chih CP, Rosenthal M, Sick TJ (1989) Ion leakage is reduced during anoxia in turtle brain: a potential survival strategy. Am J Physiol 257:R1562–R1564
Chipuk JE, Green DR (2008) How do BCL-2 proteins induce mitochondrial outer membrane permeabilization? Trends Cell Biol 18:157–164
Choi DW (1994) Calcium and excitotoxic neuronal injury. Ann N Y Acad Sci 747:162–171
Clements JR, Magnusson KR, Hautman J, Beitz AJ (1991) Rat tooth pulp projections to spinal trigeminal subnucleus caudalis are glutamate-like immunoreactive. J Comp Neurol 309:281–288
Costa AD, Garlid KD (2008) Intramitochondrial signaling: interactions among mitoKATP, PKCepsilon, ROS, and MPT. Am J Physiol Heart Circ Physiol 295:H874–H882
Coulter DA, Sombati S, DeLorenzo RJ (1992) Electrophysiology of glutamate neurotoxicity in vitro: induction of a calcium-dependent extended neuronal depolarization. J Neurophysiol 68:362–373

Crepel V, Hammond C, Chinestra P, Diabira D, Ben-Ari Y (1993) A selective LTP of NMDA receptor-mediated currents induced by anoxia in CA1 hippocampal neurons. J Neurophysiol 70:2045–2055

Crompton M (2000) Mitochondrial intermembrane junctional complexes and their role in cell death. J Physiol 529(pt 1):11–21

Cuong DV, Kim N, Youm JB, Joo H, Warda M, Lee JW, Park WS, Kim T, Kang S, Kim H et al (2006) Nitric oxide-cGMP-protein kinase G signaling pathway induces anoxic preconditioning through activation of ATP-sensitive K^+ channels in rat hearts. Am J Physiol Heart Circ Physiol 290:H1808–H1817

Deplanque D, Bordet R (2000) Pharmacological preconditioning with lipopolysaccharide in the brain. Stroke 31:1465–1466

Di Lisa F, Bernardi P (2009) A CaPful of mechanisms regulating the mitochondrial permeability transition. J Mol Cell Cardiol 46:775–780

Dirnagl U, Meisel A (2008) Endogenous neuroprotection: mitochondria as gateways to cerebral preconditioning? Neuropharmacology 55:334–344

Downey JM, Davis AM, Cohen MV (2007) Signaling pathways in ischemic preconditioning. Heart Fail Rev 12:181–188

Ellefsen S, Sandvik GK, Larsen HK, Stenslokken KO, Hov DAS, Kristensen TA, Nilsson GE (2008) Expression of genes involved in excitatory neurotransmission in anoxic crucian carp (Carassius carassius) brain. Physiol Genomics 35:5–17

Garthwaite G, Garthwaite J (1986) Amino acid neurotoxicity: intracellular sites of calcium accumulation associated with the onset of irreversible damage to rat cerebellar neurones in vitro. Neurosci Lett 71:53–58

Garthwaite J, Garthwaite G, Hajos F (1986) Amino acid neurotoxicity: relationship to neuronal depolarization in rat cerebellar slices. Neuroscience 18:449–460

Gaspar T, Katakam P, Snipes JA, Kis B, Domoki F, Bari F, Busija DW (2008) Delayed neuronal preconditioning by NS1619 is independent of calcium activated potassium channels. J Neurochem 105:1115–1128

Gaspar T, Domoki F, Lenti L, Katakam PV, Snipes JA, Bari F, Busija DW (2009) Immediate neuronal preconditioning by NS1619. Brain Res 1285:196–207

Gidday JM (2006) Cerebral preconditioning and ischaemic tolerance. Nat Rev Neurosci 7:437–448

Groban L, Vernon JC, Butterworth J (2004) Intrathecal morphine reduces infarct size in a rat model of ischemia-reperfusion injury. Anesth Analg 98:903–909, Table of contents

Grover GJ (1997) Pharmacology of ATP-sensitive potassium channel (KATP) openers in models of myocardial ischemia and reperfusion. Can J Physiol Pharmacol 75:309–315

Gu XQ, Siemen D, Parvez S, Cheng Y, Xue J, Zhou D, Sun X, Jonas EA, Haddad GG (2007) Hypoxia increases BK channel activity in the inner mitochondrial membrane. Biochem Biophys Res Commun 358:311–316

Gu XQ, Pamenter ME, Siemen D, Sun X, Haddad GG (2014) Mitochondrial but not plasmalemmal BK channels are hypoxia-sensitive in human glioma. Glia 62(4):504–513

Halestrap AP (2010) A pore way to die: the role of mitochondria in reperfusion injury and cardioprotection. Biochem Soc Trans 38:841–860

Hawrysh PJ, Buck LT (2013) Anoxia-mediated calcium release through the mitochondrial permeability transition pore silences NMDA receptor currents in turtle neurons. J Exp Biol 216:4375–4387

He X, Sandhu HK, Yang Y, Hua F, Belser N, Kim DH, Xia Y (2013) Neuroprotection against hypoxia/ischemia: delta-opioid receptor-mediated cellular/molecular events. Cell Mol Life Sci 70:2291–2303

Heurteaux C, Lauritzen I, Widmann C, Lazdunski M (1995) Essential role of adenosine, adenosine A1 receptors, and ATP-sensitive K^+ channels in cerebral ischemic preconditioning. Proc Natl Acad Sci U S A 92:4666–4670

Heusch G, Boengler K, Schulz R (2008) Cardioprotection: nitric oxide, protein kinases, and mitochondria. Circulation 118:1915–1919

Hochachka PW (1986) Defense strategies against hypoxia and hypothermia. Science 231:234–241
Holmuhamedov EL, Jovanovic S, Dzeja PP, Jovanovic A, Terzic A (1998) Mitochondrial ATP-sensitive K^+ channels modulate cardiac mitochondrial function. Am J Physiol 275:H1567–H1576
Holmuhamedov EL, Wang L, Terzic A (1999) ATP-sensitive K^+ channel openers prevent Ca^{2+} overload in rat cardiac mitochondria. J Physiol 519(2):347–360
Jaburek M, Costa AD, Burton JR, Costa CL, Garlid KD (2006) Mitochondrial PKC epsilon and mitochondrial ATP-sensitive K^+ channel copurify and coreconstitute to form a functioning signaling module in proteoliposomes. Circ Res 99:878–883
Jiang ZG, North RA (1992) Pre- and postsynaptic inhibition by opioids in rat striatum. J Neurosci 12:356–361
Kannurpatti SS, Sanganahalli BG, Mishra S, Joshi PG, Joshi NB (2004) Glutamate-induced differential mitochondrial response in young and adult rats. Neurochem Int 44:361–369
Karschin A, Brockhaus J, Ballanyi K (1998) KATP channel formation by the sulphonylurea receptors SUR1 with Kir6.2 subunits in rat dorsal vagal neurons in situ. J Physiol 509(pt 2):339–346
Kawalec M, Kowalczyk JE, Beresewicz M, Lipkowski AW, Zablocka B (2011) Neuroprotective potential of biphalin, multireceptor opioid peptide, against excitotoxic injury in hippocampal organotypic culture. Neurochem Res 36:2091–2095
Kim CH, Choi H, Chun YS, Kim GT, Park JW, Kim MS (2001) Hyperbaric oxygenation pretreatment induces catalase and reduces infarct size in ischemic rat myocardium. Pflugers Arch 442:519–525
Kis B, Rajapakse NC, Snipes JA, Nagy K, Horiguchi T, Busija DW (2003) Diazoxide induces delayed pre-conditioning in cultured rat cortical neurons. J Neurochem 87:969–980
Kis B, Nagy K, Snipes JA, Rajapakse NC, Horiguchi T, Grover GJ, Busija DW (2004) The mitochondrial K(ATP) channel opener BMS-191095 induces neuronal preconditioning. Neuroreport 15:345–349
Koch S, Gonzalez N (2013) Preconditioning the human brain: proving the principle in subarachnoid hemorrhage. Stroke 44:1748–1753
Kopp SJ, Krieglstein J, Freidank A, Rachman A, Seibert A, Cohen MM (1984) P-31 nuclear magnetic resonance analysis of brain: II. Effects of oxygen deprivation on isolated perfused and nonperfused rat brain. J Neurochem 43:1716–1731
Korge P, Honda HM, Weiss JN (2002) Protection of cardiac mitochondria by diazoxide and protein kinase C: implications for ischemic preconditioning. Proc Natl Acad Sci U S A 99:3312–3317
Kroemer G, Galluzzi L, Brenner C (2007) Mitochondrial membrane permeabilization in cell death. Physiol Rev 87:99–163
Lim YJ, Zheng S, Zuo Z (2004) Morphine preconditions Purkinje cells against cell death under in vitro simulated ischemia-reperfusion conditions. Anesthesiology 100:562–568
Limbrick DD Jr, Sombati S, DeLorenzo RJ (2003) Calcium influx constitutes the ionic basis for the maintenance of glutamate-induced extended neuronal depolarization associated with hippocampal neuronal death. Cell Calcium 33:69–81
Ling Ling J, Wong GT, Yao L, Xia Z, Irwin MG (2010) Remote pharmacological post-conditioning by intrathecal morphine: cardiac protection from spinal opioid receptor activation. Acta Anaesthesiol Scand 54:1097–1104
Liu RR, Murphy TH (2009) Reversible cyclosporin A-sensitive mitochondrial depolarization occurs within minutes of stroke onset in mouse somatosensory cortex in vivo: a two-photon imaging study. J Biol Chem 284:36109–36117
Liu GS, Thornton J, Van Winkle DM, Stanley AW, Olsson RA, Downey JM (1991) Protection against infarction afforded by preconditioning is mediated by A1 adenosine receptors in rabbit heart. Circulation 84:350–356
Liu X, Kim CN, Yang J, Jemmerson R, Wang X (1996) Induction of apoptotic program in cell-free extracts: requirement for dATP and cytochrome c. Cell 86:147–157

Lo EH (2008) A new penumbra: transitioning from injury into repair after stroke. Nat Med 14:497–500

Lord JA, Waterfield AA, Hughes J, Kosterlitz HW (1977) Endogenous opioid peptides: multiple agonists and receptors. Nature 267:495–499

Lundberg A, Oscarsson O (1953) Anoxic depolarization of mammalian nerve fibres. Acta Physiol Scand Suppl 111:99–110

Lyubkin M, Durand DM, Haxhiu MA (1997) Interaction between tetanus long-term potentiation and hypoxia-induced potentiation in the rat hippocampus. J Neurophysiol 78:2475–2482

Ma MC, Qian H, Ghassemi F, Zhao P, Xia Y (2005) Oxygen-sensitive {delta}-opioid receptor-regulated survival and death signals: novel insights into neuronal preconditioning and protection. J Biol Chem 280:16208–16218

Matsushima K, Hogan MJ, Hakim AM (1996) Cortical spreading depression protects against subsequent focal cerebral ischemia in rats. J Cereb Blood Flow Metab 16:221–226

McClintock DS, Santore MT, Lee VY, Brunelle J, Budinger GR, Zong WX, Thompson CB, Hay N, Chandel NS (2002) Bcl-2 family members and functional electron transport chain regulate oxygen deprivation-induced cell death. Mol Cell Biol 22:94–104

Michaels RL, Rothman SM (1990) Glutamate neurotoxicity in vitro: antagonist pharmacology and intracellular calcium concentrations. J Neurosci 10:283–292

Muller M, Brockhaus J, Ballanyi K (2002) ATP-independent anoxic activation of ATP-sensitive K^+ channels in dorsal vagal neurons of juvenile mice in situ. Neuroscience 109:313–328

Murata M, Akao M, O'Rourke B, Marban E (2001) Mitochondrial ATP-sensitive potassium channels attenuate matrix Ca^{2+} overload during stimulated ischemia and reperfusion. Circ Res 89:891–898

Murry CE, Jennings RB, Reimer KA (1986) Preconditioning with ischemia: a delay of lethal cell injury in ischemic myocardium. Circulation 74:1124–1136

Ning XH, Xu CS, Song YC, Xiao Y, Hu YJ, Lupinetti FM, Portman MA (1998) Hypothermia preserves function and signaling for mitochondrial biogenesis during subsequent ischemia. Am J Physiol 274:H786–H793

Ostermeier AM, Schlosser B, Schwender D, Sutor B (2000) Activation of mu- and delta-opioid receptors causes presynaptic inhibition of glutamatergic excitation in neocortical neurons. Anesthesiology 93:1053–1063

Pamenter ME (2014) Mitochondria: a multimodal hub of hypoxia tolerance. Can J Zool 92(7):569–589

Pamenter ME, Buck LT (2008a) delta-Opioid receptor antagonism induces NMDA receptor-dependent excitotoxicity in anoxic turtle cortex. J Exp Biol 211:3512–3517

Pamenter ME, Buck LT (2008b) Neuronal membrane potential is mildly depolarized in the anoxic turtle cortex. Comp Biochem Physiol A Mol Integr Physiol 150:410–414

Pamenter ME, Richards MD, Buck LT (2007a) Anoxia-induced changes in reactive oxygen species and cyclic nucleotides in the painted turtle. J Comp Physiol B 177:473–481

Pamenter ME, Cooray M, Buck LT (2007b) Mitochondrial KATP channels attenuate anoxic NMDAR currents via regulation of the mitochondrial Ca^{2+} uniporter in the western painted turtle. FASEB J 21:A594

Pamenter ME, Shin DS, Buck LT (2008a) AMPA receptors undergo channel arrest in the anoxic turtle cortex. Am J Physiol Regul Integr Comp Physiol 294:R606–R613

Pamenter ME, Shin DS, Buck LT (2008b) Adenosine A1 receptor activation mediates NMDA receptor activity in a pertussis toxin-sensitive manner during normoxia but not anoxia in turtle cortical neurons. Brain Res 1213:27–34

Pamenter ME, Hogg DW, Buck LT (2008c) Endogenous reductions in N-methyl-d-aspartate receptor activity inhibit nitric oxide production in the anoxic freshwater turtle cortex. FEBS Lett 582:1738–1742

Pamenter ME, Shin DS, Cooray M, Buck LT (2008d) Mitochondrial ATP-sensitive K^+ channels regulate NMDAR activity in the cortex of the anoxic western painted turtle. J Physiol 586:1043–1058

Perez-Pinzon MA, Rosenthal M, Sick TJ, Lutz PL, Pablo J, Mash D (1992) Downregulation of sodium channels during anoxia: a putative survival strategy of turtle brain. Am J Physiol Regul Integr Comp Physiol 262:R712–R715

Perez-Pinzon MA, Mumford PL, Rosenthal M, Sick TJ (1996) Anoxic preconditioning in hippocampal slices: role of adenosine. Neuroscience 75:687–694

Peterson BL, Park TJ, Larson J (2012a) Adult naked mole-rat brain retains the NMDA receptor subunit GluN2D associated with hypoxia tolerance in neonatal mammals. Neurosci Lett 506:342–345

Peterson BL, Larson J, Buffenstein R, Park TJ, Fall CP (2012b) Blunted neuronal calcium response to hypoxia in naked mole-rat hippocampus. PLoS One 7, e31568

Petrosillo G, Di Venosa N, Moro N, Colantuono G, Paradies V, Tiravanti E, Federici A, Ruggiero FM, Paradies G (2011) In vivo hyperoxic preconditioning protects against rat-heart ischemia/reperfusion injury by inhibiting mitochondrial permeability transition pore opening and cytochrome c release. Free Radic Bio Med 50:477–483

Pizzo P, Drago I, Filadi R, Pozzan T (2012) Mitochondrial Ca(2)(+) homeostasis: mechanism, role, and tissue specificities. Pflugers Arch 464:3–17

Prosser CL, Barr LM, Pinc RD, Lauer CY (1957) Acclimation of goldfish to low concentrations of oxygen. Physiol Zool 30:137–141

Quock RM, Burkey TH, Varga E, Hosohata Y, Hosohata K, Cowell SM, Slate CA, Ehlert FJ, Roeske WR, Yamamura HI (1999) The delta-opioid receptor: molecular pharmacology, signal transduction, and the determination of drug efficacy. Pharmacol Rev 51:503–532

Rasola A, Bernardi P (2007) The mitochondrial permeability transition pore and its involvement in cell death and in disease pathogenesis. Apoptosis 12:815–833

Raval AP, Dave KR, Mochly-Rosen D, Sick TJ, Perez-Pinzon MA (2003) Epsilon PKC is required for the induction of tolerance by ischemic and NMDA-mediated preconditioning in the organotypic hippocampal slice. J Neurosci 23:384–391

Rodgers-Garlick CI, Hogg DW, Buck LT (2013) Oxygen-sensitive reduction in Ca(2)(+)-activated K(+) channel open probability in turtle cerebrocortex. Neuroscience 237:243–254

Ross AP, Christian SL, Zhao HW, Drew KL (2006) Persistent tolerance to oxygen and nutrient deprivation and N-methyl-D-aspartate in cultured hippocampal slices from hibernating Arctic ground squirrel. J Cereb Blood Flow Metab 26:1148–1156

Rossi DJ, Oshima T, Attwell D (2000) Glutamate release in severe brain ischaemia is mainly by reversed uptake. Nature 403:316–321

Salt TE, Hill RG (1981) Excitatory amino acids as transmitter candidates of vibrissae afferent fibres to the rat trigeminal nucleus caudalis. Neurosci Lett 22:183–187

Salt TE, Hill RG (1983) Pharmacological differentiation between responses of rat medullary dorsal horn neurons to noxious mechanical and noxious thermal cutaneous stimuli. Brain Res 263:167–171

Santos MS, Moreno AJ, Carvalho AP (1996) Relationships between ATP depletion, membrane potential, and the release of neurotransmitters in rat nerve terminals. An in vitro study under conditions that mimic anoxia, hypoglycemia, and ischemia. Stroke 27:941–950

Sasaki N, Sato T, Ohler A, O'Rourke B, Marban E (2000) Activation of mitochondrial ATP-dependent potassium channels by nitric oxide. Circulation 101:439–445

Sato T, Saito T, Saegusa N, Nakaya H (2005) Mitochondrial Ca^{2+}-activated K^{+} channels in cardiac myocytes: a mechanism of the cardioprotective effect and modulation by protein kinase A. Circulation 111:198–203

Schultz JE, Rose E, Yao Z, Gross GJ (1995) Evidence for involvement of opioid receptors in ischemic preconditioning in rat hearts. Am J Physiol 268:H2157–H2161

Schurr A, Reid KH, Tseng MT, West C, Rigor BM (1986) Adaptation of adult brain tissue to anoxia and hypoxia in vitro. Brain Res 374:244–248

Semenov DG, Samoilov MO, Zielonka P, Lazarewicz JW (2000) Responses to reversible anoxia of intracellular free and bound $Ca^{(2+)}$ in rat cortical slices. Resuscitation 44:207–214

Sessle BJ (1987) The neurobiology of facial and dental pain: present knowledge, future directions. J Dent Res 66:962–981

Sharif NA, Hughes J (1989) Discrete mapping of brain Mu and delta opioid receptors using selective peptides: quantitative autoradiography, species differences and comparison with kappa receptors. Peptides 10:499–522

Shimizu K, Lacza Z, Rajapakse N, Horiguchi T, Snipes J, Busija DW (2002) MitoK(ATP) opener, diazoxide, reduces neuronal damage after middle cerebral artery occlusion in the rat. Am J Physiol Heart Circ Physiol 283:H1005–H1011

Shin DS, Buck LT (2003) Effect of anoxia and pharmacological anoxia on whole-cell NMDA receptor currents in cortical neurons from the western painted turtle. Physiol Biochem Zool 76:41–51

Shin DS, Wilkie MP, Pamenter ME, Buck LT (2005) Calcium and protein phosphatase 1/2A attenuate N-methyl-D-aspartate receptor activity in the anoxic turtle cortex. Comp Biochem Physiol A Mol Integr Physiol 142:50–57

Shinbo T, Kokubo K, Sato Y, Hagiri S, Hataishi R, Hirose M, Kobayashi H (2013) Breathing nitric oxide plus hydrogen gas reduces ischemia-reperfusion injury and nitrotyrosine production in murine heart. Am J Physiol Heart Circ Physiol 305:H542–H550

Siemen D, Loupatatzis C, Borecky J, Gulbins E, Lang F (1999) Ca^{2+}-activated K Channel of the BK-type in the inner mitochondrial membrane of a human glioma cell line. Biochem Biophys Res Commun 257:549–554

Suh DH, Kim MK, Kim HS, Chung HH, Song YS (2013) Mitochondrial permeability transition pore as a selective target for anti-cancer therapy. Front Oncol 3:41

Susin SA, Lorenzo HK, Zamzami N, Marzo I, Snow BE, Brothers GM, Mangion J, Jacotot E, Costantini P, Loeffler M et al (1999) Molecular characterization of mitochondrial apoptosis-inducing factor. Nature 397:441–446

Tait SW, Green DR (2010) Mitochondria and cell death: outer membrane permeabilization and beyond. Nat Rev Mol Cell Biol 11:621–632

Takashi E, Wang Y, Ashraf M (1999) Activation of mitochondrial K(ATP) channel elicits late preconditioning against myocardial infarction via protein kinase C signaling pathway. Circ Res 85:1146–1153

Tanaka E, North RA (1994) Opioid actions on rat anterior cingulate cortex neurons in vitro. J Neurosci 14:1106–1113

Wall TM, Sheehy R, Hartman JC (1994) Role of bradykinin in myocardial preconditioning. J Pharmacol Exp Ther 270:681–689

Wang XM, Mokha SS (1996) Opioids modulate N-methyl-D-aspartic acid (NMDA)-evoked responses of trigeminothalamic neurons. J Neurophysiol 76:2093–2096

Wang Y, Qin ZH (2010) Molecular and cellular mechanisms of excitotoxic neuronal death. Apoptosis 15:1382–1402

Wang L, Cherednichenko G, Hernandez L, Halow J, Camacho SA, Figueredo V, Schaefer S (2001) Preconditioning limits mitochondrial $Ca^{(2+)}$ during ischemia in rat hearts: role of K(ATP) channels. Am J Physiol Heart Circ Physiol 280:H2321–H2328

Wilkie MP, Pamenter ME, Alkabie S, Carapic D, Shin DS, Buck LT (2008) Evidence of anoxia-induced channel arrest in the brain of the goldfish (Carassius auratus). Comp Biochem Physiol C Toxicol Pharmacol 148(4):355–362

Wojtovich AP, DiStefano P, Sherman T, Brookes PS, Nehrke K (2012) Mitochondrial ATP-sensitive potassium channel activity and hypoxic preconditioning are independent of an inwardly rectifying potassium channel subunit in Caenorhabditis elegans. FEBS Lett 586:428–434

Wong GT, Ling Ling J, Irwin MG (2010) Activation of central opioid receptors induces cardioprotection against ischemia-reperfusion injury. Anesth Analg 111:24–28

Xia Y, Haddad GG (1991) Major differences in CNS sulfonylurea receptor distribution between the rat (newborn, adult) and turtle. J Comp Neurol 314:278–289

Xia Y, Haddad GG (2001) Major difference in the expression of delta- and mu-opioid receptors between turtle and rat brain. J Comp Neurol 436:202–210

Xu M, Wang Y, Ayub A, Ashraf M (2001) Mitochondrial K(ATP) channel activation reduces anoxic injury by restoring mitochondrial membrane potential. Am J Physiol Heart Circ Physiol 281:H1295–H1303

Xu W, Liu Y, Wang S, McDonald T, Van Eyk J, Sidor A, O'Rourke B (2002) Cytoprotective Role of Ca^{2+}-activated K^+ channels in the cardiac inner mitochondrial membrane. Science 298:1029–1033

Yoshida M, Nakakimura K, Cui YJ, Matsumoto M, Sakabe T (2004) Adenosine A(1) receptor antagonist and mitochondrial ATP-sensitive potassium channel blocker attenuate the tolerance to focal cerebral ischemia in rats. J Cereb Blood Flow Metab 24:771–779

Yuan XR, Madamba S, Siggins GR (1992) Opioid peptides reduce synaptic transmission in the nucleus accumbens. Neurosci Lett 134:223–228

Zhang KM, Wang XM, Mokha SS (1996) Opioids modulate N-methyl-D-aspartic acid (NMDA)-evoked responses of neurons in the superficial and deeper dorsal horn of the medulla (trigeminal nucleus caudalis). Brain Res 719:229–233

Zhang J, Haddad GG, Xia Y (2000) delta-, but not mu- and kappa-, opioid receptor activation protects neocortical neurons from glutamate-induced excitotoxic injury. Brain Res 885:143–153

Zhang J, Gibney GT, Zhao P, Xia Y (2002) Neuroprotective role of delta-opioid receptors in cortical neurons. Am J Physiol Cell Physiol 282:C1225–C1234

Zhang J, Qian H, Zhao P, Hong SS, Xia Y (2006a) Rapid hypoxia preconditioning protects cortical neurons from glutamate toxicity through delta-opioid receptor. Stroke 37:1094–1099

Zhang SZ, Gao Q, Cao CM, Bruce IC, Xia Q (2006b) Involvement of the mitochondrial calcium uniporter in cardioprotection by ischemic preconditioning. Life Sci 78:738–745

Zhao HW, Ross AP, Christian SL, Buchholz JN, Drew KL (2006) Decreased NR1 phosphorylation and decreased NMDAR function in hibernating Arctic ground squirrels. J Neurosci Res 84:291–298

Zivkovic G, Buck LT (2010) Regulation of AMPA receptor currents by mitochondrial ATP-sensitive K^+ channels in anoxic turtle neurons. J Neurophysiol 104:1913–1922

Chapter 9
δ-Opioids and Neurogenesis

Naoko Kuzumaki and Minoru Narita

Abstract Opioid analgesics and endogenous opioid peptides have a wide range of physiological and behavioral effects on pain perception, mood, motor control and autonomic function. The δ-opioidergic system has been recognized as a diagnostic neuropeptide that may be directly involved in neural differentiation, maturation and protection, and may contribute to the possible recovery of emotional dysfunction. Stimulation of δ-opioid receptors of neural progenitor cells enhanced neural differentiation, maturation and protection, which could lead to anti-depressive and anti-anxiolytic effects. We suggest that the δ-opioids may be a good target for the treatment of neuropsychiatric disease.

Keywords δ-Opioids • Neurogenesis • Neuroprotection • Neural maturation • iPS cells

Abbreviations

BDNF	Brain-derived neurotrophic factor
CNTF	Ciliary neurotrophic factor
DAMGO	[D-Ala2, N-Me- Phe4, Gly5-ol]-enkephalin
EAATs	Excitatory amino acid transporters
FGFs	Fibroblast growth factors

N. Kuzumaki, Ph.D. (✉)
Associate professor, Department of Pharmacology, Hoshi University School of Pharmacy and Pharmaceutical Sciences, 2-4-41 Ebara, Shinagawa-ku, Tokyo 142-8501, Japan
e-mail: n-kuzumaki@hoshi.ac.jp

M. Narita, Ph.D. (✉)
Professor and chair, Department of Pharmacology, Hoshi University School of Pharmacy and Pharmaceutical Sciences, 2-4-41 Ebara, Shinagawa-ku, Tokyo 142-8501, Japan

Life Science Tokyo Advanced Research Center (L-StaR), Hoshi University School of Pharmacy and Pharmaceutical Sciences, 2-4-41 Ebara, Shinagawa-ku, Tokyo 142-8501, Japan
e-mail: narita@hoshi.ac.jp

Y. Xia (ed.), *Neural Functions of the Delta-Opioid Receptor*,
DOI 10.1007/978-3-319-25495-1_9

GDNF	Glial-derived neurotrophic factor
GPCRs	G-protein-coupled receptors
NGF	Nerve growth factor
NT-3/4	Neurotrophin-3/4
PDYN	Prodynorphin
PENK	Proenkephalin
POMC	Proopiomelanocortin
SNC80	[(+)-4-[(aR)-a-((2S,5R)-4-allyl-2,5-dimethyl-1-piperazinyl)-3-methoxybenzyl] -N,N-diethylbenzamide]
U50,488H	(–)-Trans-(1S,2S)-U-50488 hydrochloride

1 Introduction

Opiate drugs hijack a complex neuromodulatory system composed of three receptors, μ, δ, and κ, which interact with the endogenous opioid peptides β-endorphin, enkephalins, and dynorphins. Among these, δ-opioid receptors form a subfamily of G-protein-coupled receptors (GPCRs) that are expressed throughout the peripheral and central nervous systems (Lutz and Kieffer 2013). δ-Opioids have a wide range of physiologic and behavioral effects on pain perception, mood, motor control and autonomic function (Narita et al. 2001; Cordonnier et al. 2005). However, the effects that exposure to δ-opioids have on the birth of neural progenitors in the mature brain are still unclear. In this review, we focus on recent experimental findings that δ-opioids induce neurogenesis and lead to a specific cell fate that has been implicated in both the pathogenesis of depression and the effects of anti-depressants.

2 Neural Stem Cell Differentiation *In Vitro*

Neural stem cells capable of producing new neurons and astrocytes are maintained in a self-replicating state by cytokines and neurotrophic factors (Eriksson et al. 1998). Neurotrophic factors, such as brain-derived neurotrophic factor (BDNF), nerve growth factor (NGF), neurotrophin-3/4 (NT-3/4), fibroblast growth factors (FGFs), glial-derived neurotrophic factor (GDNF) and ciliary neurotrophic factor (CNTF), have been shown to promote the survival of specific populations of brain neurons under experimental conditions relevant to brain aging and neurodegenerative disorders (Quintero et al. 2004). It has been reported that a stable δ-opioid peptide blocks apoptosis induced by the protein Bax, a member of the Bcl-2 family of proteins, through activation of the BDNF pathway (Yuan and Yankner 2000; Tsao and Su 2001). BDNF has been shown to support neural growth and survival during development and in the adult nervous system, and to regulate synaptic plasticity

(McAllister et al. 1999). Notably, δ-opioid receptor agonists increase BDNF gene expression in several brain regions (Torregrossa et al. 2004). In our study, the stimulation of δ-opioid receptors by the selective δ-opioid receptor agonist SNC80 [(+)-4-[(aR)-a-((2S,5R)-4-allyl-2,5-dimethyl-1-piperazinyl)-3-methoxybenzyl]-N, N-diethylbenzamide] (10 nM) promoted neural differentiation from multipotent neural stem cells obtained from embryonic C3H mouse forebrains. In contrast, neither a selective μ-opioid receptor agonist, [D-Ala2, N-Me- Phe4, Gly5-ol]-enkephalin (DAMGO), nor a specific κ-opioid receptor agonist, (–)-trans-(1S,2S)-U-50488 hydrochloride (U50,488H), had a similar effect. Furthermore, the stimulation of δ-opioid receptors promoted neurogenesis through a Trk-dependent tyrosine kinase pathway (Narita et al. 2006a).

3 Neural Protection of the δ-Opioid System

Chronic *in vitro* treatment of adult hippocampal progenitors with specific μ-(β-funaltrexamine) and δ- (naltrindole) opioid receptor antagonists decreased their differentiation into astrocytes and oligodendrocytes, whereas the κ-opioid receptor antagonist nor-BNI had no effect (Persson et al. 2003). It has been reported that STAT5B plays an important role in δ-opioid receptor signaling and function in Neuro-2A cells. It would be interesting to identify more proteins that are regulated by STAT5B activation during neurite outgrowth and determine if they are also involved in neuronal development (Georganta et al. 2013). Liang et al. (2014) showed that δ-opioid receptor activation up-regulates astrocytic excitatory amino acid transporters (EAATs), which in turn regulate the extracellular glutamate concentration and play a role in preventing neuroexcitotoxicity, via MEK-ERK-p38 signaling, suggesting that the δ-opioid receptor plays a critical role in the neuroprotective effect against excitotoxic injury.

4 Hippocampal Neurogenesis

One aspect of the mammalian hippocampus that has recently received considerable attention is the birth of new neurons in the dentate gyrus throughout the lifetime of the animal (Gould et al. 1999; Kornack and Rakic 1999; Van Praag et al. 1999). This phenomenon has been described in rodents, non-human primates, and, most recently, humans (Altman and Das 1965; Kaplan and Hinds 1977; Eriksson et al. 1998; Gould et al. 1998). Research suggests that cells are born in the subgranular zone of the dentate gyrus, migrate into the granule cell layer and express neuronal markers (Kaplan and Bell 1984; Cameron et al. 1993; Kornack and Rakic 1999), extend processes to CA3 pyramidal neurons (Stanfield and Trice 1988; Markakis and Gage 1999), receive synaptic connections (Kaplan and Hinds 1977; Kaplan and

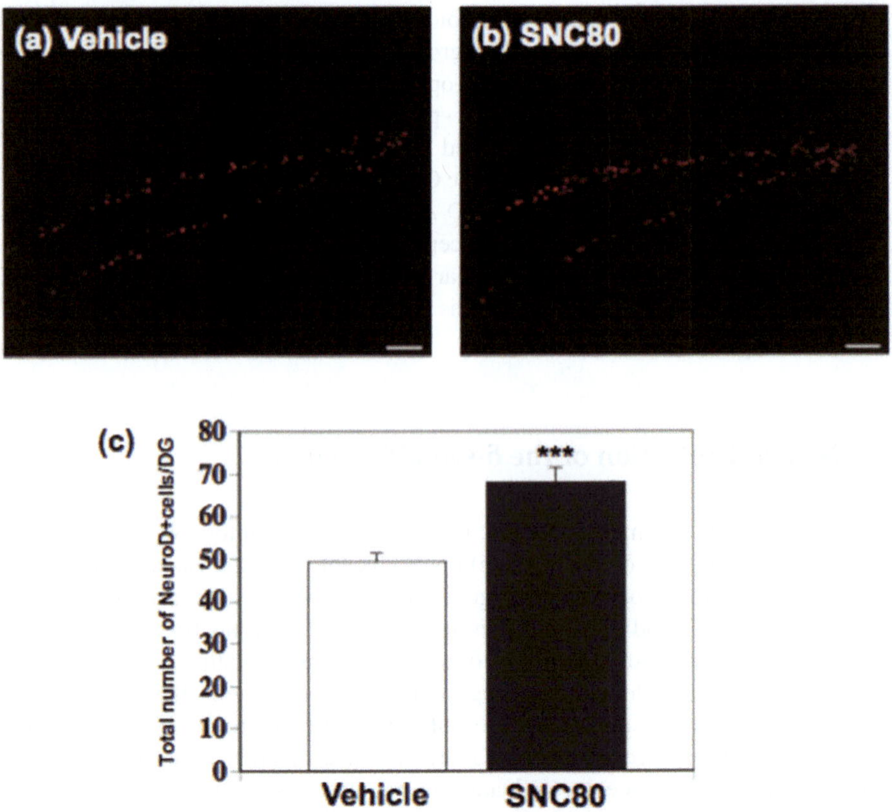

Fig. 9.1 Immunofluorescent staining for NeuroD in the dentate gyrus (**a, b**). Chronic *in vivo* treatment with the selective δ-opioid receptor agonist SNC80 (3 mg/kg, s.c.) twice a day for 1 week caused a significant increase in NeuroD-like immunoreactivity. (**c**) Results of the stereological counting of NeuroD-positive cells in the dentate gyrus of standard and enriched mice. The total number of NeuroD-positive cells was significantly increased in SNC80-treated groups compared to vehicle-treated groups. Each column represents the mean ± S.E.M. ***p < 0.001 vs. vehicle-treated group. Scale bar: 50 μm

Bell 1984; Markakis and Gage 1999) and demonstrate long-term potentiation (Wang et al. 2000). It has been proposed that the thousands of new neurons born each day in the adult rodent hippocampus may contribute to a variety of hippocampal-related functions, including learning and memory (Gould et al. 1999; Van Praag et al. 1999).

We previously reported that chronic treatment with SNC80 significantly increased the total number of NeuroD-positive cells (i.e., neural precursor cells) in the dentate gyrus (Fig. 9.1). In contrast, it has been reported that the μ-opioid receptor agonist morphine, but not the μ-opioid receptor agonist fentanyl, decreased NeuroD activity via the reduced activity of CaMKII, which is upstream of NeuroD (Eisch et al. 2000; Zheng et al. 2012), indicating that μ-opioids might suppress neurogenesis, which may depend on the ligands. On the other hand, an enriched environment, such as when groups of animals are housed together in a complex environment with various toys to provide more opportunity for learning and social interaction than standard laboratory living conditions, induces the proliferation of hippocampal progenitor cells and neurogenesis (Nilsson et al. 1999; van Praag et al. 1999; Kuzumaki et al. 2011). We reported that an enriched environment increases BDNF expression with notably sustained chromatin regulation in the mouse hippocampus (Kuzumaki et al. 2011). Under these conditions, our data showed that the expression of proenkephalin (PENK) mRNA in the hippocampus was significantly elevated after exposure to an enriched environment for both 3 and 4 weeks (Fig. 9.2). In contrast, mRNA levels of proopiomelanocortin (POMC) or prodynorphin (PDYN) in the hippocampus were not altered under the same conditions.

5 Anti-Depressive Effect of δ-Opioid Receptors

The degeneration of new cells in the adult mammalian brain may significantly modify pathophysiologic processes in neuropsychiatric disorders. Notably, neurogenesis and gliogenesis can be induced in the adult brain by environmental manipulations and various pharmacologic treatments (Kempermann et al. 1997; Gould et al. 1999; Nilsson et al. 1999; Yang et al. 2002; Ehringer and Kempermann 2003). In the hippocampus, synaptic plasticity and neurogenesis (in the subgranular zone of the dentate gyrus) have been implicated in both the pathogenesis of depression and the effects of antidepressants (Hanson et al. 2011; Snyder et al. 2011). Treatment with various anti-depressants promotes neurogenesis (Duman et al. 1997; Malberg and Duman 2003; Wang et al. 2004).

The δ-opioidergic system has also been recognized as a neurotransmitter system that could be directly involved in anxiety and depression (Broom et al. 2002). We found that the long-term blockade of δ-opioid receptor function using the selective δ-opioid receptor antagonist naltrindole in the adult forebrain of mice induced anxiety with astrogliosis in the cingulate cortex region (Narita et al. 2006a). We also demonstrated that the microinjection of cultured astrocytes, activated astrocytes or astrocyte-conditioned medium into the cingulate cortex of adult mice enhanced the expression of anxiety-like behavior (Narita et al. 2005, 2006b). Taken together, these findings raise the intriguing possibility that the δ-opioid system helps to regulate the differentiation of neural stem cells in the adult forebrain, leading to the possible modification of emotionality.

BDNF is one of the most studied molecular substrates of hippocampal synaptic adaptations and neurogenesis. BDNF promotes proper spine density in the hippocampus, mediates antidepressant responses, and regulates depressive-like behaviors (Castrén and Rantamäki 2010). Interestingly, opioid receptors regulate BDNF activity. Systemic administration of peptidic (Torregrossa et al. 2006)

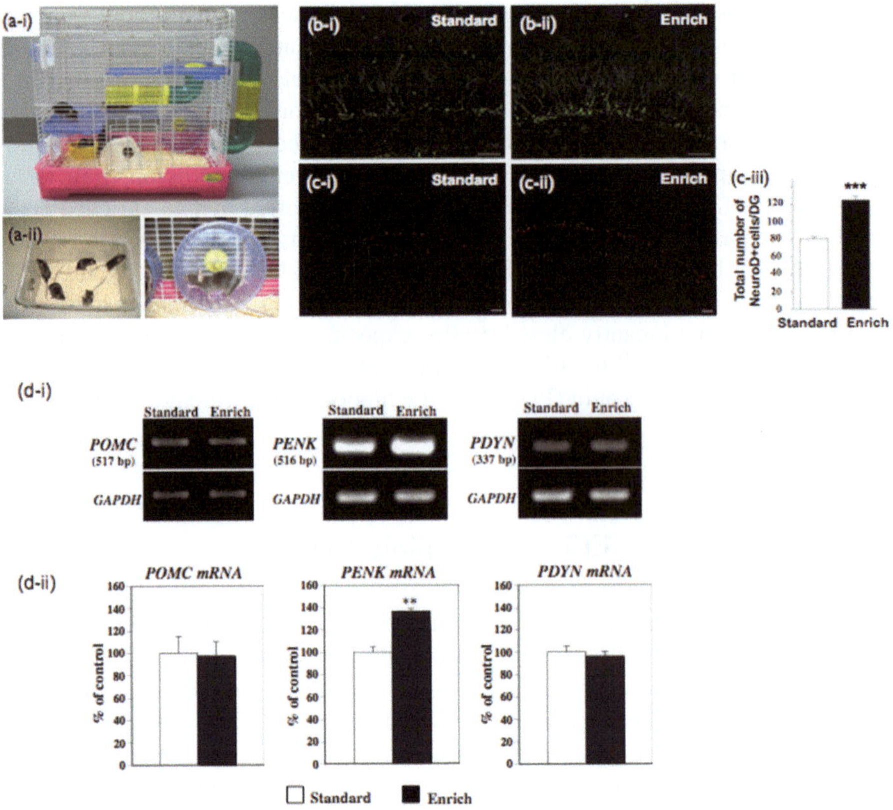

Fig. 9.2 (**a**) Cage equipped with an enriched environment. Enrichment consisted of greater social interaction (8 mice in a large cage vs. 4 mice in a standard cage), greater potential for exploratory behavior with objects such as toys and a rearrangeable set of tunnels, and physical activity on a running wheel. (*a-ii*) Standard cage. (**b**) Immunofluorescent staining for doublecortin in the dentate gyrus in mice housed under their respective experimental conditions for 4 weeks. Doublecortin-like IR in the dentate gyrus of enriched mice (*b-ii*) was increased compared to that in standard mice (*b-i*). Scale bar: 50 μm. (**c**) Immunofluorescent staining for NeuroD in the dentate gyrus in mice housed under standard or enriched conditions for 4 weeks. NeuroD-like IR in the dentate gyrus of enriched mice (*c-ii*) was increased compared to that of mice in standard cages (*c-i*). (*c-iii*) Results of the stereological counting of NeuroD-positive cells in the dentate gyrus of mice exposed to standard and enriched conditions. The total number of NeuroD-positive cells was significantly increased in the enriched group compared to the standard group. Each column represents the mean ± S.E.M. ***$p < 0.001$ vs. the standard group. Scale bar: 50 μm. These data (figure) were adopted from Kuzumaki et al. (2011), the author's original work. (*d-i*) Representative RT-PCR for proopiomelanocortin (POMC), preproenkephalin (PENK) and preprodynorphin (PDYN) mRNAs in the hypothalamus obtained from standard or enriched mice. (*d-ii*) The intensity of the bands was semi-quantified using NIH Image software. The value for each mRNA was normalized by that for the internal standard glyceraldehyde-3-phosphate dehydrogenase (GAPDH). The value for enriched mice is expressed as a percentage of that in standard mice. Each column represents the mean ± S.E.M. of six samples. **$p < 0.01$ vs. standard group

and non-peptidic (Torregrossa et al. 2004) δ-opioid receptor agonists had anti-depressant-like effects in rats and increased *BDNF* mRNA in the hippocampus, as well as other structures (e.g., frontal cortex, amygdala, hippocampus, endopiriform cortex). In addition, endogenous δ-opioids (enkephalins) upregulated *BDNF* mRNA in the hippocampus (Zhang et al. 2006), suggesting that δ-opioid receptors control the expression of this major neurotrophic factor. These results indicate that δ-opioids have a potentially beneficial effect (increased BDNF) on hippocampal mood substrates, yet their overall behavioral relevance remains to be determined.

6 Role of δ-Opioid Receptors in the Neural Maturation of Human iPSc-Derived Neurons

With the development of new techniques, we are now able to generate iPS cells from diseased patients, such as those with a neuropsychiatric disease. Disease-specific iPS cells are useful tools that may allow researchers to observe the development of diseases in the lab, outside of patients. We now have good reason to believe that this will make it possible to identify new treatments, and eventually drugs, to slow or even stop the course of several diseases. In a recent study, we investigated the effect of the stimulation of δ-opioid receptor on the maturation of human iPS cell-derived neurons using a synapsin::dsRed reporter system. The treatment of human iPS cell-derived neurons with SNC80 showed the enhancement of synapsin::dsRed intensity, indicating that the stimulation of δ-opioid receptor could affect neural maturation (Fig. 9.3). These results suggest that the δ-opioid system plays an important role in neural maturation in the central nervous system of humans.

7 Concluding Remarks

Proper control of the δ-opioidergic system in the central nervous system has a beneficial effect on the modulation of neural function. Stimulation of δ-opioid receptors on neural progenitor cells enhanced neural differentiation, maturation and protection, which may lead to anti-depressive and anti-anxiolytic effects. Most of these effects are related to the production of BDNF. Taken together, these findings suggest that the activated δ-opioidergic system has a wide variety of functions in controlling brain function, indicating that δ-ligands might be able to fine-tune mood disorders.

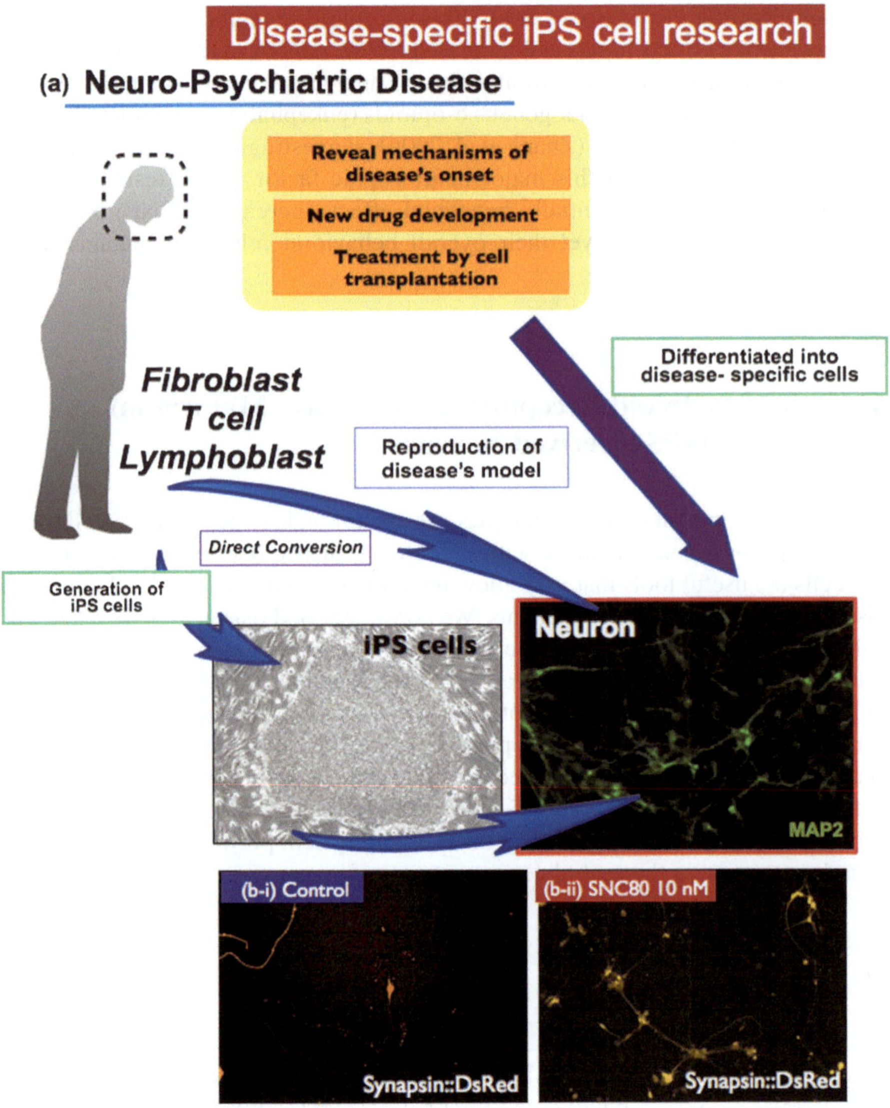

Fig. 9.3 Effect of a δ-opioid receptor agonist on the neural maturation of human iPS cell-derived neurons using synapsin::DsRed reporter. (**a**) Schematic model of research using disease-specific iPS cells. Disease-specific iPS cells are useful tools that may allow researchers to observe the development of diseases in the lab, outside of patients. (**b**) The treatment of control human iPS cell-derived neurons that had differentiated for 20 days with SNC80 (10 nM) enhanced the intensity of synapsin::dsRed. 201B7 iPSCs were provided by RIKEN Cell Bank. All of the experimental procedures for iPS culture were approved by the Hoshi University (H26-4) and Keio University (20-16-28)

References

Altman J, Das GD (1965) Autoradiographic and histological evidence of postnatal hippocampal neurogenesis in rats. J Comp Neurol 124:319–335

Broom DC, Jutkiewicz EM, Folk JE, Traynor JR, Rice KC, Woods JH (2002) Nonpeptidic δ-opioid receptor agonists reduce immobility in the forced swim assay in rats. Neuropsychopharmacology 26:744–755

Cameron HA, Woolley CS, McEwen BS, Gould E (1993) Differentiation of newly born neurons and glia in the dentate gyrus of the adult rat. Neuroscience 56:337–344

Castrén E, Rantamäki T (2010) The role of BDNF and its receptors in depression and antidepressant drug action: reactivation of developmental plasticity. Dev Neurobiol 70:289–297

Cordonnier L, Sanchez M, Roques BP, Noble F (2005) Facilitation of enkephalins-induced delta-opioid behavioral responses by chronic amisulpride treatment. Neuroscience 135:1–10

Duman RS, Heninger GR, Nestler EJ (1997) A molecular and cellular theory of depression. Arch Gen Psychiatry 54:597–606

Ehringer D, Kempermann G (2003) Regional effects of wheel running and environmental enrichment on cell genesis and microglia proliferation in the adult murine neocortex. Cereb Cortex 13:845–851

Eisch AJ, Barrot M, Schad CA, Self DW, Nestler EJ (2000) Opiates inhibit neurogenesis in the adult rat hippocampus. Proc Natl Acad Sci U S A 97:7579–7584

Eriksson PS, Perfilieva E, Bjork-Eriksson T, Alborn AM, Nordborg C, Peterson DA, Gage FH (1998) Neurogenesis in the adult human hippocampus. Nat Med 4:1313–1317

Georganta EM, Tsoutsi L, Gaitanou M, Georgoussi Z (2013) δ-opioid receptor activation leads to neurite outgrowth and neuronal differentiation via a STAT5B-Gαi/o pathway. J Neurochem 127:329–341

Gould E, Tanapat P, McEwen BS, Flügge G, Fuchs E (1998) Proliferation of granule cell precursors in the dentate gyrus of adult monkeys is diminished by stress. Proc Natl Acad Sci U S A 95:3168–3171

Gould E, Beylin A, Tanapat P, Reeves A, Shors TJ (1999) Learning enhances adult neurogenesis in the hippocampal formation. Nat Neurosci 2:260–265

Hanson ND, Owens MJ, Nemeroff CB (2011) Depression, antidepressants, and neurogenesis: a critical reappraisal. Neuropsychopharmacology 36:2589–2602

Kaplan MS, Bell DH (1984) Mitotic neuroblasts in the 9-day-old and 11-month-old rodent hippocampus. J Neurosci 4:1429–1441

Kaplan MS, Hinds JW (1977) Neurogenesis in the adult rat: electron microscopic analysis of light radioautographs. Science 197:1092–1094

Kempermann G, Kuhn HG, Gage FH (1997) More hippocampal neurons in adult mice living in an enriched environment. Nature 386:493–495

Kornack DR, Rakic P (1999) Continuation of neurogenesis in the hippocampus of the adult macaque monkey. Proc Natl Acad Sci U S A 96:5768–5773

Kuzumaki N, Ikegami D, Tamura R, Hareyama N, Imai S, Narita M, Torigoe K, Niikura K, Takeshima H, Ando T, Igarashi K, Kanno J, Ushijima T, Suzuki T, Narita M (2011) Hippocampal epigenetic modification at the brain-derived neurotrophic factor gene induced by an enriched environment. Hippocampus 21:127–132

Liang J, Chao D, Sandhu HK, Yu Y, Zhang L, Balboni G, Kim DH, Xia Y (2014) δ-Opioid receptors up-regulate excitatory amino acid transporters in mouse astrocytes. Br J Pharmacol 171:5417–5430

Lutz PE, Kieffer BL (2013) Opioid receptors: distinct roles in mood disorders. Trends Neurosci 36:195–206

Malberg JE, Duman RS (2003) Cell proliferation in adult hippocampus is decreased by inescapable stress: reversal by fluoxetine treatment. Neuropsychopharmacology 28:1562–1571

Markakis EA, Gage FH (1999) Adult-generated neurons in the dentate gyrus send axonal projections to field CA3 and are surrounded by synaptic vesicles. J Comp Neurol 406:449–460

McAllister AK, Katz LC, Lo DC (1999) Neurotrophins and synaptic plasticity. Annu Rev Neurosci 22:295–318
Narita M, Funada M, Suzuki T (2001) Regulations of opioid dependence by opioid receptor types. Pharmacol Ther 89:1–15
Narita M, Kuzumaki N, Narita M, Kaneko C, Tamai E, Khotib J, Miyatake M, Shindo K, Nagumo Y, Tanaka S, Suzuki T (2005) Age-related emotionality is associated with cortical δ-opioid receptor dysfunction-dependent astrogliosis. Neuroscience 137:1359–1367
Narita M, Kuzumaki N, Miyatake M, Sato F, Wachi H, Seyama Y, Suzuki T (2006a) Role of delta-opioid receptor function in neurogenesis and neuroprotection. J Neurochem 97:1494–1505
Narita M, Kuzumaki N, Narita M, Kaneko C, Hareyama N, Miyatake M, Shindo K, Miyoshi K, Nakajima M, Nagumo Y, Sato F, Wachi H, Seyama Y, Suzuki T (2006b) Chronic pain-induced emotional dysfunction is associated with astrogliosis due to cortical delta-opioid receptor dysfunction. J Neurochem 97:1369–1378
Nilsson M, Perfilieva E, Johansson U, Orwar O, Eriksson PS (1999) Enriched enviroment increases neurogenesis in the adult rat dentate gyrus and improves spatial memory. J Neurobiol 39:569–578
Persson AI, Thorlin T, Bull C, Zarnegar P, Ekman R, Terenius L, Eriksson PS (2003) Mu- and delta-opioid receptor antagonists decrease proliferation and increase neurogenesis in cultures of rat adult hippocampal progenitors. Eur J Neurosci 17:1159–1172
Quintero EM, Willis LM, Zaman V, Lee J, Boger HA, Tomac A, Hoffer BJ, Strömberg I, Granholm AC (2004) Glial cell line-derived neurotrophic factor is essential for neuronal survival in the locus coeruleus-hippocampal noradrenergic pathway. Neuroscience 124:137–146
Snyder JS, Soumier A, Brewer M, Pickel J, Cameron HA (2011) Adult hippocampal neurogenesis buffers stress responses and depressive behavior. Nature 476:458–461
Stanfield BB, Trice JE (1988) Evidence that granule cells generated in the dentate gyrus of adult rats extend axonal projections. Exp Brain Res 72:399–406
Torregrossa MM, Isgor C, Folk JE, Rice KC, Watson SJ, Woods JH (2004) The delta-opioid receptor agonist (+) BW373U86 regulates BDNF mRNA expression in rats. Neuropsychopharmacology 29:649–659
Torregrossa MM, Jutkiewicz EM, Mosberg HI, Balboni G, Watson SJ, Woods JH (2006) Peptidic delta opioid receptor agonists produce antidepressant-like effects in the forced swim test and regulate BDNF mRNA expression in rats. Brain Res 1069:172–181
Tsao LI, Su TP (2001) Hibernation-induction peptide and cell death: [-Ala2,-Leu5]enkephalin blocks Bax-related apoptotic processes. Eur J Pharmacol 428:149–151
Van Praag H, Kempermann G, Gage FH (1999) Running increases cell proliferation and neurogenesis in the adult mouse dentate gyrus. Nat Neurosci 2:266–270
Wang S, Scott BW, Wojtowicz JM (2000) Heterogenous properties of dentate granule neurons in the adult rat. J Neurobiol 42:248–257
Wang HD, Dunnavant FD, Jarman T, Deutch AY (2004) Effects of antipsychotic drugs on neurogenesis in the forebrain of the adult rat. Neuropsychopharmacology 29:1230–1238
Yang M, Stull ND, Berk MA, Snyder EY, Iacovitti L (2002) Neural stem cells spontaneously express dopaminergic traits after transplantation into the intact or 6-hydroxydopamine-lesioned rat. Exp Neurol 177:50–60
Yuan J, Yankner BA (2000) Apoptosis in the nervous system. Nature 407:802–809
Zhang H, Torregrossa MM, Jutkiewicz EM, Shi YG, Rice KC, Woods JH, Watson SJ, Ko MC (2006) Endogenous opioids upregulate brain-derived neurotrophic factor mRNA through delta- and micro-opioid receptors independent of antidepressant-like effects. Eur J Neurosci 23:984–994
Zheng H, Law PY, Loh HH (2012) Non-coding RNAs regulating morphine function: with emphasis on the in vivo and in vitro functions of miR-190. Front Genet 3:1–9

Chapter 10
Delta Opioid Receptor and Peptide: Hibernation for Stroke Therapy

Meaghan Staples, Sandra Acosta, Naoki Tajiri, Mibel Pabon, Yuji Kaneko, and Cesar V. Borlongan

Abstract In pursuit of neurological therapies, the opioid system, specifically delta opioids and delta opioid peptides, demonstrates promising potential for stroke, Parkinson's Disease, and other degenerative neurological conditions. Recent studies present strong evidence in support of the therapeutic use of delta opioid receptors, and provide insight into the underlying mechanisms of action. Delta opioid receptors have been shown to confer protective effects by mediating ionic homeostasis and activating endogenous neuroprotective pathways. Additionally, delta opioid agonists such as (D-Ala 2, D-Leu 5) enkephalin (DADLE) have been shown to decrease apoptosis and promote neuronal survival. In its entirety, the delta opioid system represents a promising target for neural therapies.

Keywords Ischemia • Delta opioids • Neuroprotection • DADLE • Hibernation

Abbreviations

DADLE	(D-Ala 2, D-Leu 5) enkephalin
DAT	Dopamine transporter
DOR	Delta (δ) opioid receptor
HIT	Hibernation Induction Trigger
KOR	Kappa (κ) opioid receptor
MAPK	Mitogen-activated protein kinase
MCAO	Middle Cerebral Artery Occlusion
METH	Methamphetamine
MOR	Mu (μ) opioid receptor
PKC	Protein kinase c

M. Staples • S. Acosta, Ph.D. • N. Tajiri, P.T., Ph.D. • M. Pabon, Ph.D.
Y. Kaneko, Ph.D. • C.V. Borlongan, Ph.D. (✉)
Department of Neurosurgery and Brain Repair, University of South Florida
College of Medicine, Tampa, FL 33612, USA
e-mail: cborlong@health.usf.edu

Y. Xia (ed.), *Neural Functions of the Delta-Opioid Receptor*,
DOI 10.1007/978-3-319-25495-1_10

1 Introduction

The opioid system is composed of various opioid peptides and their corresponding receptors. Opioids are a group of inhibitory neurotransmitters that are involved in a variety of functions including pain regulation and respiratory rate control. The effects exhibited by various opioid peptides are mediated by opioid receptors which each recognize a unique group of opioid and non-opioid ligands. The family of classical Gi-protein coupled receptors which inhibit adenylyl cyclase, is divided into three primary subgroups: μ- (MOR), κ- (KOR), and δ-opioid receptors (DOR). Endogenous opioid peptides including the endorphins, dynorphins, and enkephalins, associate with the MOR, KOR, or DOR respectively. Opioid receptors elicit diverse effects depending on which opioid class they belong to. These receptors are present throughout the central and peripheral nervous systems, as well as various peripheral organs including the heart, lungs, liver, and gastrointestinal tract (Mansour et al. 1987; Xia and Haddad 1991; Hiller and Fan 1996; Xiang et al. 1996; Lim et al. 2004; Feng et al. 2012). Accumulating evidence suggests that the opioid system may confer protection against degenerative neurological diseases characterized by oxygen-, blood-, and energy depleting states (Borlongan et al. 2004; Peart et al. 2005). A study conducted by Mayfield and colleagues demonstrated extended survival during hypoxia when animals were pretreated with an opioid receptor agonist (D'Alecy 1994). Additionally, it was shown that opioid-induced protection could be inhibited by MOR and KOR antagonists but not DOR antagonists (Bofetiado et al. 1996). These data suggest that the opioid system is involved in neuroprotection against hypoxic and ischemic events, and is likely mediated primarily by DOR and delta opioid peptides.

2 DOR: The Receptor and Neuroprotection

Accumulating evidence has demonstrated that DOR activation in response to hypoxic/ischemic stress, may offer neuroprotective effects. In the late 1980s, Xia and colleagues, observed that the turtle brain has a higher density of DOR than the rat brain (Xia and Haddad 2001). Furthermore, the turtle brain demonstrates a higher tolerance to hypoxic/ischemic conditions than the rat brain (Sick et al. 1983; Xia et al. 1992). Indeed, the potential relationship between these two phenomena becomes of interest. To investigate this proposed connection, researchers added glutamate to cultured cortical neurons in order to mimic neuroexcitotoxicity (Zhang et al. 1999). Neurons which had been exposed to 100 μmol/L glutamate for 4 h daily over a period of 8–10 days showed significant neuronal injury (Zhang et al. 1999). However, activation of the DOR, via administration of (D-Ala 2, D-Leu 5) enkephalin (DADLE), decreased glutamate-induced injury by almost half (Zhang et al. 1999). Additionally, activation of MOR and KOR did not elicit any significant protective effects (Zhang et al. 1999, 2000). This suggests that the delta opioid receptors, not mu or kappa opioid receptors, are responsible for the observed neuroprotective effects.

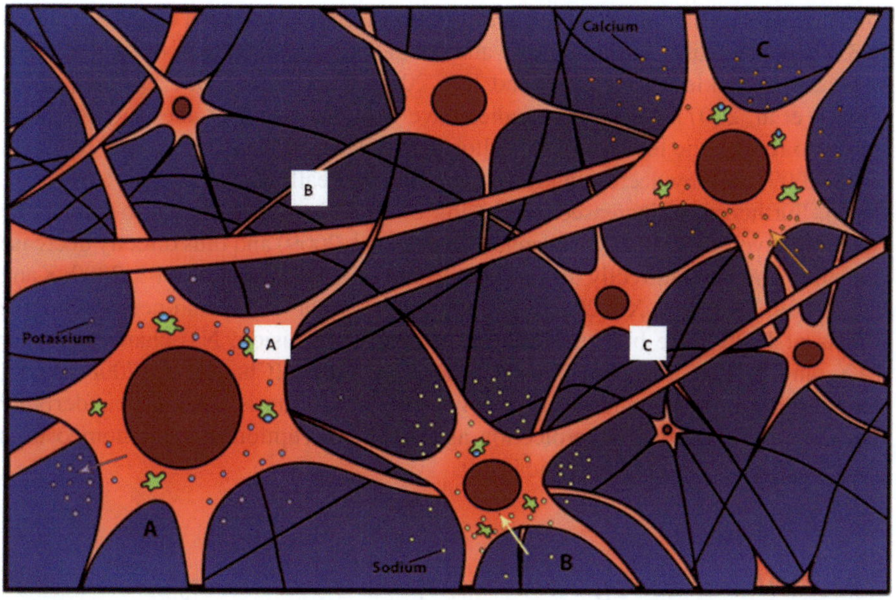

Fig. 10.1 DOR-Induced Neuroprotection via Maintenance of Ionic Homeostasis Following hypoxic/ischemic stress, there is an immediate loss of ionic homeostasis which leads to neuronal injury and death. Activation of DOR can prevent the disturbance of ionic homeostasis and attenuate injury. (**A**) Activation of DOR prevents efflux of potassium. (**B**) Activation of DOR prevents influx of sodium. (**C**) Activation of DOR prevents influx of calcium. Delta opioid receptor agonists, such as (D-Ala 2, D-Leu 5) encephalin, can bind to and activate DOR in the brain. The activation of the DOR results in decreased neuronal injury and death following an ischemic event

Recent studies have also implicated delta opioid receptors in ischemia (Frerichs and Hallenbeck 1998; Boutin et al. 1999; Kevelaitis et al. 1999). Following middle cerebral artery occlusion (MCAO) in mice, delta binding sites decreased prior to reductions in kappa or mu binding sites, concomitant with infarct core extension (Mayfield et al. 1996; Boutin et al. 1999). Reduction in DORs occurs prior to any observable brain damage, suggesting that DORs are more sensitive to brain insults than MORs and KORs. Consequently, it follows that stimulation of DORs may elicit anti-ischemic effects (Frerichs and Hallenbeck 1998; Kevelaitis et al. 1999) (Fig 10.1).

2.1 DOR and Ionic Homeostasis

Although the acting mechanism of DORs in neuroprotection is not fully known, recent advances have been made in our understanding of their role. Acute hypoxic/ischemic stress causes an immediate loss of ionic homeostasis characterized by an efflux of K^+, and an influx of Cl^-, Na^+, and Ca^{+2} (Sung et al. 2008; Kang et al. 2009).

This increased efflux of K^+ is typical of hypoxia/ischemia (Sung et al. 2008; Hansen 1985) and can cause neuronal injury and death (Yu et al. 1997; Liu et al. 2003; Wei et al. 2003; Bickler 2004; Chao and Xia, 2010; Karki et al. 2007; Mongin 2007; Nistico et al. 2007). Numerous studies have demonstrated that activation of DOR serves to reduce the leakage of K^+ following ischemia (Chao et al. 2006, 2007a, b, 2008, 2009) thereby decreasing resulting neuronal death (Liu et al. 2003; Wei et al. 2003). Furthermore, increased expression of DOR has been shown to inhibit the function of voltage-gated Na^+ channels (Kang et al. 2009) and thus directly decrease the influx Na^+ and indirectly decrease the efflux of K^+ (Chao et al. 2008, 2009; Kang et al. 2009; Chao and Xia 2010). The inhibition of Na^+ influx and neuroprotective effects of DOR were blocked following low Na^+ perfusion, Na^+ channel blocker TTX, and NMDA receptor channel blocker MK 801 (Chao et al. 2008, 2009). Together these data suggest that inhibition of Na^+ influx is the major underlying mechanism for DOR neuroprotection against the disruption of ionic homeostasis associated with hypoxic/ischemic insult.

2.2 DOR and Endogenous Neuroprotective Pathways

Another proposed mechanism underlying the neuroprotective effects of DOR is its mediation of neuroprotective pathways. Activation of DORs has been shown to prevent cell death by preventing the phosphorylation of p38 via stimulation of protein kinase C (PKC) and mitogen-activated protein kinase (MAPK)-ERK1/2 (Ma et al. 2005; Narita et al. 2006; Feng et al. 2009; Peng et al. 2009; Sun et al. 2009). In support, treatment with a PKC inhibitor has been shown to diminish DOR mediated neuroprotection against ischemia/hypoxia (Ma et al. 2005). This PKC-dependent pathway has also been implicated in DOR attenuation of K^+ efflux and maintenance of ionic homeostasis (Chao et al. 2007b). Peng demonstrated that an up-regulation of hypoxia inducible transcription factor occurs following hypoxic preconditioning (Peng et al. 2009). This up-regulation of hypoxia inducible transcription factor increases the expression of DOR which induced neuroprotection via the ERK signaling pathway (Peng et al. 2009). However, these DOR-induced neuroprotective effects are inhibited by treatment with an ERK inhibitor (Ma et al. 2005). Together these data demonstrate the pivotal role of DOR mediation of PKC and ERK signaling pathways for neuroprotection.

3 DADLE: The Ligand and Neuroprotection

Hibernation is a unique natural model that allows animals to survive typically detrimental oxygen-, blood-, and energy-deprived conditions. For this reason, hibernation has been a point of focus for many researchers in search of potential neurotherapies for disease states with similar conditions. A search for further understanding of the molecular components involved in hibernation lead to the discovery that plasma from thirteen-lined ground squirrels could induce hibernation when injected into summer active ground squirrels (Dawe and Spurrier 1969). The responsible hibernation inducing trigger (HIT) was identified as a protein factor that co-migrates with albumin (Bruce et al. 1987; Oeltgen et al. 1988). Due to the ability of opioids to elicit physiological conditions similar to hibernation, it has been speculated that HIT may function as an opioid, however recent evidence suggests that HIT releases endogenous opioids rather than function as one itself (Bruce et al. 1987). Additionally, this suggests that the opioid system is likely responsible for achieving hibernation. Investigation into the hibernation-inducing ability of opioids and opioid receptors demonstrated that each class of opioid receptors, μ, κ, and δ, has varying potency for inducing hibernation. MOR and KOR selective antagonists such as morphine and dynorphin, were ineffective in inducing hibernation in summer active ground squirrels (Oeltgen et al. 1987, 1988). However, the DOR agonist (D-Ala 2, D-Leu 5) enkephalin (DADLE) was highly effective in inducing hibernation (Oeltgen et al. 1988). As a result, several studies investigating the neuroprotective potential of the opioid system has focused on DORs and DOR selective ligands such as DADLE.

DADLE is an opioid peptide that binds primarily to delta opioid receptors and it is therefore of interest when pursuing potential neuroprotective therapies. In a study conducted by Tsao and colleagues investigated the effects of DADLE against the dopamine neurotoxicity of methamphetamine (METH). A high single dose of METH, or prolonged use at a medium dosage, generated long-term loss of striatal dopaminergic terminals (Tsao et al. 1998). When DADLE was administered 2 weeks after the delivery of METH, dopamine transporter (DAT) levels were restored from a loss of 30%, to normal levels (Tsao et al. 1999). Furthermore, administration of DALDE prior to exposure to METH completely inhibited, and even reversed METH-induced DAT loss (Hayashi et al. 1999; Tsao et al. 1999). Successive studies have demonstrated that the free radical scavenging nature of DADLE, and mediation by the DOR, are responsible for the protective effects of DADLE against METH-induced DAT loss (Tsao et al. 1998).

3.1 DADLE and Parkinson's Disease

The capacity of DADLE to protect the brain against METH-induced DAT loss has sparked investigation into DADLE's potential for neuroprotective therapy for other neurological diseases. Specifically, DADLE has been suggested as a potential therapy for Parkinson's Disease, a neurological disease characterized by dopamine depletion. When pretreated with DADLE, adult male rats treated with 6-hydroxydopamine lesion, a dopamine depleting neurotoxin, exhibit increased survival of tyrosine hydroxylase immunoreactive cells (Borlongan et al. 1999a, b, 2000). Similarly, pretreatment with DADLE has been shown to increase cell viability of cultured primary rat fetal mesencephalic cells in a dose dependent manner (Borlongan et al. 1999a, b, 2000). In an additional in vivo study, DADLE was shown to enhance the survival of serum deprived PC12 cells (Hayashi et al. 2002). While this suggests that DADLE may also involve a trophic factor mechanism, the primary pathway for DADLE's neuroprotective action remains the participation of opioid receptors.

3.2 *DADLE and Cerebral Ischemia*

The neuroprotective effects of DADLE have also been extended stroke. Studies have demonstrated that DADLE exhibits neuroprotection against ischemia reperfusion-induced brain damage following transient MCAO (Borlongan et al. 1999a, b). Rats subjected to a 60 min unilateral MCAO, followed by either a 24 or 72 h reperfusion, exhibited extensive striatum infarction which was completely inhibited when DADLE was administered prior to the MCAO (Borlongan et al. 1999a, b). Furthermore, treatment with the universal opioid receptor antagonist naltrexone transiently blocked the early phase of DADLE-induced protection but was ineffective in blocking the prolonged effects (Borlongan et al. 1999a, b) suggesting that opioid receptors are highly involved in the initial protective phase of DADLE, and the latter phase on an alternative mechanism of action. A recent study by Borlongan and colleagues further demonstrates the effectiveness of DADLE in protection against stroke (Borlongan et al. 2009). Animals that were pretreated with DADLE prior to exposure to MCAor surgery demonstrated decreased behavioral deficits when compared to animals who received saline exclusively (Hayashi et al. 2000). Furthermore, treatment with DADLE or DALDE and an opioid blocker, had almost no detectable dehydrogenase deficient tissue (necrotic infarction) in the ischemic core (Borlongan et al. 2009). In addition, DADLE has been shown to suppress p-53 mRNA expression, a marker for apoptosis characteristically associated with MCAO and stroke models (Hayashi et al. 2000; Wang et al. 2001; Borlongan et al. 2009). Together, these data suggest that DADLE's protective effects are promoted centrally. It has been suggested that DADLE may exert its protective effects by increasing the expression of GDNF, a highly selective dopamine neuron survival agent (Lin et al. 1993; Borlongan et al. 2001) that has been shown to protect against cerebral ischemia (Kearns et al. 1997; Wang et al. 1997; Fox et al. 2001). Accordingly, the increased levels of striatal GDNF following treatment with DADLE suggest that the striatal dopaminergic system may be a suitable target for DADLE in the treatment of ischemia (Borlongan et al. 2009).

4 Concluding Remarks

Recent investigations into potential neuroprotective therapies have lead to the discovery of numerous novel possibilities including DOR induced neuroprotection. As a result, there presently exists substantial evidence demonstrating the clinical potential of DOR in treating cytotoxic, hypoxic, and ischemic neurological stress. Recent investigations of DOR underlying neuroprotective mechanism have revealed several different mechanisms of action. Activation of DORs has been shown to stabilize ionic homeostasis and prevent ischemia-induced neuronal damage. Additionally, DORs have been shown exert neuroprotection by inducing endogenous neuroprotective pathways. Furthermore, DADLE, a DOR ligand, has been shown to be effective in preventing neuronal injury and death in ischemia, Parkinson's disease, and drug-induced stress. The opioid system, specifically DOR and DOR ligands, represent a strong potential for neuroprotective therapy. However, while our understanding of DORs potential for neuroprotective therapy has expanded, there are still many questions that remain unanswered. Further development of translational therapies into the clinical setting is needed. Additionally, because of the therapeutic benefits obtained with up-regulation of DOR expression, potential mechanisms that induce DOR up-regulation should be explored.

Acknowledgements This study was supported by the National Institutes of Health (1RO1NS071956-01A1) and the James and Easter King Biomedical Research Program (1KG01-33966).

References

Bickler PE (2004) Clinical perspectives: neuroprotection lessons from hypoxia-tolerant organisms. J Exp Biol 207:3243–3249

Bofetiado DM, Mayfield KP, D'Alecy LG (1996) Alkaloid d agonist BW 373U86 increases hypoxia tolerance. Anesth Analg 82:1237–1241

Borlongan CV, Oeltgen PR, Su TP, Wang Y (1999a) Delta opioid peptide (DADLE) protects against ischemia reperfusion damage in the striatum and cerebral cortex. Soc Neurosci Abstr 24:979

Borlongan CV, Wu JN, Su TP, Wang Y (1999b) Delta opioid peptide (DADLE) enhances survival of cultured fetal cells. In: Committee on problems of drug dependence 61st meeting, Abstract 13

Borlongan CV, Su TP, Wang Y (2000) Treatment with delta opioid peptide enhances in vitro and in vivo survival of rat dopaminergic neurons. Neuroreport 11:923–992

Borlongan CV, Zhou FC, Hayashi T, Su TP, Hoffer BJ, Wang Y (2001) Involvement of GDNF in neuronal protection against 6- OHDA-induced parkinsonism following intracerebral transplantation of fetal kidney tissues in adult rats. Neurobiol Dis 8:636–646

Borlongan CV, Wang Y, Su TP (2004) Delta opioid peptide (D-Ala 2, D-Leu 5) enkephalin: linking hibernation and neuroprotection. Front Biosci 9(11):3392–3398

Borlongan CV, Hayashi T, Oeltgen PR, Su TP, Wang Y (2009) Hibernation like state induced by an opioid peptide protects against experimental stroke. BMC Biol 7:31

Boutin H, Dauphin F, MacKenzie ET, Jauzac P (1999) Differential time-course decreases in non-selective, mu-, delta-, and kappa-opioid receptors after focal cerebral ischemia in mice. Stroke 30:1271–1277

Bruce DS, Cope GW, Elam TR, Ruit SK, Oeltgen PR, Su TP (1987) Opioids and hibernation. I. Effects of naloxone on bear HIT's depression of guinea-pig ileum contractility and on induction of summer hibernation in the ground squirrel. Life Sci 41:2107–2113

Chao D, Xia Y (2010) Ionic storm in hypoxic/ischemic stress: can opioid receptors subside it? Prog Neurobiol 90:439–470

Chao D, Qian H, Ghassemi F, Chen JS, Xia Y (2006) Transgenic over-expression of δ-opioid receptors protects the cortex from anoxic disruption of ionic homeostasis. Soc Neurosci Abstr Program No. 87.19/MM68.

Chao D, Bazzy-Asaad A, Balboni G, Xia Y (2007a) δ-, but not μ-, opioid receptor stabilizes K^+ homeostasis by reducing Ca^{2+} influx in the cortex during acute hypoxia. J Cell Physiol 212(1):60–67, PubMed: 17373650

Chao D, Donnelly DF, Feng Y, Bazzy-Asaad A, Xia Y (2007b) Cortical δ-opioid receptors potentiate K^+ homeostasis during anoxia and oxygen-glucose deprivation. J Cereb Blood Flow Metab 27(2):356–368, PubMed: 16773140

Chao D, Bazzy-Asaad A, Balboni G, Salvadori S, Xia Y (2008) Activation of DOR attenuates anoxic K^+ derangement via inhibition of Na^+ entry in mouse cortex. Cereb Cortex 18(9):2217–2227, PubMed: 18203692

Chao D, Balboni G, Lazarus LH, Salvadori S, Xia Y (2009) Na+ mechanism of δ-opioid receptor induced protection from anoxic K^+ leakage in the cortex. Cell Mol Life Sci 66(6):1105–1115, PubMed: 19189047

D'Alecy LG (1994) Delta-1 opioid agonist acutely increases hypoxic tolerance. J Pharmacol Exp Ther 268:683–688

Dawe AR, Spurrier WA (1969) Hibernation induced in ground squirrels by blood transfusion. Science 163:298–299

Feng Y, Chao D, He X, Yang Y, Kang X, Lazarus LH, Xia Y (2009) A novel insight into neuroprotection against hypoxic/ischemic stress. Sheng Li Xue Bao 61:585–592

Feng Y, He X, Yang Y, Chao D, Lazarus LH, Xia Y (2012) Current research on opioid receptor function. Curr Drug Targets 13:230–246

Fox CM, Gash DM, Smoot MK, Cass WA (2001) Neuroprotective effects of GDNF against 6-OHDA in young and aged rats. Brain Res 896:56–63

Frerichs KU, Hallenbeck JM (1998) Hibernation in ground squirrels induces state and species-specific tolerance to hypoxia and aglycemia: an in vitro study in hippocampal slices. J Cereb Blood Flow Metab 18:168–175

Hansen AJ (1985) Effect of anoxia on ion distribution in the brain. Physiol Rev 65(1):101–148, PubMed: 3880896

Hayashi T, Tsao LI, Cadet JL, Su TP (1999) [D-Ala2, D-Leu5] enkephalin blocks the methamphetamine-induced c-fos mRNA increase in mouse striatum. Eur J Pharmacol 366:7–8

Hayashi T, Sakai K, Sasaki C, Itoyama Y, Abe K (2000) Loss of bag-1 immunoreactivity in rat brain after transient middle cerebral artery occlusion. Brain Res 852:496–500

Hayashi T, Tsao LI, Su TP (2002) Antiapoptotic and cytotoxic properties of delta opioid peptide [D-Ala2, D-Leu5] enkephalin in PC12 cells. Synapse 43(1):86–94

Hiller JM, Fan LQ (1996) Laminar distribution of the multiple opioid receptors in the human cerebral cortex. Neurochem Res 21:1333–1345

Kang X, Chao D, Gu Q, Ding G, Wang Y, Balboni G, Lazarus LH, Xia Y (2009) δ-opioid receptors protect from anoxic disruption of Na^+ homeostasis via Na+ channel regulation. Cell Mol Life Sci 66(21):3505–3516

Karki P, Seong C, Kim JE, Hur K, Shin SY, Lee JS, Cho B, Park IS (2007) Intracellular K+ inhibits apoptosis by suppressing the Apaf-1 apoptosome formation and subsequent downstream pathways but not cytochrome c release. Cell Death Differ 14(12):2068–2075, PubMed: 17885667

Kearns CM, Cass WA, Smoot K, Kryscio R, Gash DM (1997) GDNF protection against 6-OHDA: time dependence and requirement for protein synthesis. J Neurosci 17:7111–7118

Kevelaitis E, Peynet J, Mouas C, Launay JM, Menasche P (1999) Opening of potassium channels: the common cardioprotective link between preconditioning and natural hibernation? Circulation 99:3079–3085

Lim YJ, Zheng S, Zuo Z (2004) Morphine preconditions Purkinje cells against cell death under in vitro simulated ischemia-reperfusion conditions. Anesthesiology 100:562–568

Lin LF, Doherty DH, Lile JD, Bektesh S, Collins F (1993) GDNF: a glial cell line-derived neurotrophic factor for midbrain dopaminergic neurons. Science 260:1130–1132

Liu D, Slevin JR, Lu C, Chan SL, Hansson M, Elmer E, Mattson MP (2003) Involvement of mitochondrial K+ release and cellular efflux in ischemic and apoptotic neuronal death. J Neurochem 86(4):966–979, PubMed: 12887694

Ma MC, Qian H, Ghassemi F, Zhao P, Xia Y (2005) Oxygen sensitive δ-opioid receptor-regulated survival and death signals: novel insights into neuronal preconditioning and protection. J Biol Chem 280(16):16208–16218, PubMed: 15687501

Mansour A, Khachaturian H, Lewis ME, Akil H, Watson SJ (1987) Autoradiographic differentiation of mu, delta and kappa opioid in the rat forebrain and midbrain. J Neurosci 7(8):2445–2464

Mayfield KP, Kozak W, Malvin GM, Porreca F (1996) Hypoxia decreases opioid delta receptor expression in mouse brain. Neuroscience 72:785–789, PubMed: 9157323

Mongin AA (2007) Disruption of ionic and cell volume homeostasis in cerebral ischemia: the perfectstorm. Pathophysiology 14(3–4):183–193, PubMed: 17961999

Narita M, Kuzumaki N, Miyatake M, Sato F, Wachi H, Seyama Y, Suzuki T (2006) Role of delta-opioid receptor function in neurogenesis and neuroprotection. J Neurochem 97:1494–1505

Nistico R, Piccirilli S, Sebastianelli L, Nistico G, Bernardi G, Mercuri N (2007) The blockade of K+- ATP channels has neuroprotective effects in an in vitro model of brain ischemia. Int Rev Neurobiol 82:383–395, PubMed: 17678973

Oeltgen PR, Welborn JR, Nuchols PA, Spurrier WA, Bruce DS, Su TP (1987) Opioids and hibernation. II. Effects of kappa opioid U69593 on induction of hibernation in summer-active ground squirrels by 'hibernation induction trigger' (HIT). Life Sci 41:2115–2120

Oeltgen PR, Nilekani SP, Nuchols PA, Spurrier WA, Su TP (1988) Further studies on opioid and hibernation: Delta opioid receptor ligand selectively induced hibernation in summer-active ground squirrels. Life Sci 43:1565–1574

Peart JN, Gross ER, Gross GJ (2005) Opioid-induced preconditioning: recent advances and future perspectives. Vascul Pharmacol 42(5):211–218

Peng P-HH, Huang H-SS, Lee Y-JJ, Chen Y-SS, Ma M-C (2009) Novel role for the delta-opioid receptor in hypoxic preconditioning in rat retinas. J Neurochem 108:741–754

Sick TJ, Rosenthal M, LaManna JC, Lutz PL (1983) Brain potassium ion homeostasis, anoxia, and metabolic inhibition in turtles and rats. Am J Physiol 243(3):R281–R288, PubMed: 6287869

Sun K, Su DS, Wang XR (2009) Delta opioid agonist [D-Ala2, D-Leu5] enkephalin (DADLE) reduced oxygen-glucose deprivation caused neuronal injury through the MAPK pathway. Brain Res 1292:100–106

Sung JH, Chao DM, Xia Y (2008) Neuronal responses to hypoxia. In: Wang Y (ed) New frontiers in neurological research. Research Signpost, Kerala, pp 73–153

Tsao LI, Ladenheim B, Andrews A, Chiueh CC, Cadet JL, Su TP (1998) Delta opioid peptide (D-Ala 2, D-Leu 5) enkephalin blocks the long-term loss of dopamine transporter induced by multiple administrations of methamphetamine: involvement of opioid receptors and reactive oxygen species. J Pharmacol Exp Ther 287:322–331

Tsao L, Cadet JL, Su TP (1999) Reversal by [D-Ala2, D-Leu5]enkephalin of the dopamine transporter loss caused by methamphetamine. Eur J Pharmacol 372:5–7

Wang Y, Lin SZ, Chiou AL, Williams LR, Hoffer BJ (1997) Glial cell line derived neurotrophic factor protects against ischemia-induced injury in the cerebral cortex. J Neurosci 17:4341–4348

Wang Y, Chang CF, Morales M, Chou J, Chen HL, Chiang YH, Lin SZ, Cadet JL, Deng X, Wang JY, Chen SY, Kaplan PL, Hoffer BJ (2001) Bone morphogenetic protein-6 reduces ischemia-induced brain damage in rats. Stroke 32:2170–2178

Wei L, Yu SP, Gottron F, Snider BJ, Zipfei GJ, Choi DW (2003) Potassium channel blockers attenuate hypoxia- and ischemia-induced neuronal death in vitro and in vivo. Stroke 34(5):1281–1286, PubMed: 12677023

Xia Y, Haddad GG (1991) Ontogeny and distribution of opioid receptors in the rat brainstem. Brain Res 549:181–193

Xia Y, Haddad GG (2001) Major difference in the expression of δ- and μ-opioid receptors between turtle and rat brain. J Comp Neurol 436(2):202–210, PubMed: 11438924

Xia Y, Jiang C, Haddad GG (1992) Oxidative and glycolytic pathways in rat (newborn, adult) and turtle: Role in anoxia. Am J Physiol 262(4 pt 2):R595–R603, PubMed: 1314516

Xiang H, Hochman DW, Saya H, Fujiwara T, Schwartzkroin PA, Morrison RS (1996) Evidence for p53-mediated modulation of neuronal viability. J Neurosci 16:6753–6765

Yu SP, Yeh CH, Sensi SL, Gwag BJ, Canzoniero LM, Farhangrazi ZS, Ying HS, Tian M, Dugan LL, Choi DW (1997) Mediation of neuronal apoptosis by enhancement of outward potassium current. Science 278(5335):114–117, PubMed: 9311914

Zhang JH, Xia Y, Haddad GG (1999) Activation of δ-opioid receptors protects cortical neurons from glutamate excitotoxic injury. Soc Neurosci Abstr 28:736

Zhang JH, Haddad GG, Xia Y (2000) δ-, but not μ- and κ-opioid receptor activation protects neocortical neurons from glutamate-induced excitotoxic injury. Brain Res 885(2):143–153, PubMed: 11102568

Chapter 11
An Important Role of the δ-Opioid Receptor in Electroacupuncture-Induced Protection Against Ischemic Brain Injury

Tetsuya Asakawa and Ying Xia

Abstract Acupuncture is an important modality of traditional Chinese medicine and widely used as an alternative therapy for neurological diseases including stroke. Strong evidence suggests that acupuncture-induced brain protection is critically related to the activity of the delta-opioid receptor (DOR) in addition to an increase in regional cerebral blood flow. This chapter will recap the cellular and molecular mechanisms for electroacupuncture (EA) induced neuroprotection against ischemic injury based on previous literature as well as our recent work. We will discuss EA-induced changes in endogenous opioids and opioid receptors, especially DOR, under ischemic conditions and their responses to EA treatment, the role of DOR in the EA-induced brain protection against ischemic injury, and the potential signaling pathways.

Keywords Acupuncture • Electroacupuncture • Endogenous opioids • Opioid receptors • Delta-opioid receptor • Ischemia • Stroke • Neuroprotection

Abbreviations

BBB	Blood-brain barrier
BDNF	Brain-derived neurotrophic factor
CaMKII	Ca^{++}/calmodulin-dependent protein kinase II
CNS	Central nervous system

T. Asakawa, M.D., Ph.D.
Department of Neurosurgery, Hamamatsu University School of Medicine, Hamamatsu, Shizuoka, Japan

Department of Psychiatry, Hamamatsu University School of Medicine, Hamamatsu, Shizuoka, Japan

Y. Xia, M.D., Ph.D. (✉)
University of Texas Medical School, Houston, TX, USA
e-mail: ying.xia@uth.tmc.edu; y55738088@gmail.com

Y. Xia (ed.), *Neural Functions of the Delta-Opioid Receptor*,
DOI 10.1007/978-3-319-25495-1_11

CREB	cAMP-response element binding protein
CT	Computed tomography
DADLE	[D-Ala2, D-Leu5]-enkephalin
DOR	Delta-opioid receptor
EA	Electroacupuncture
EOP	Endogenous opioid peptides
ERK	Extracellular signal-regulated kinase
GPx	Glutathione peroxidase
HMGB1	High-mobility group box 1 protein
HPC	Hypoxic preconditioning
IL-1β	Interleukin-1β
KOR	Kappa-opioid receptor
MAPKs	Mitogen-activated protein kinases
MCAO	Middle cerebral artery occlusion
MDA	Malondialdehyde
MOR	Mu-opioid receptor
NDMC	N-desmethylclozapine
NO	Nitric oxide
PI3K	Phosphatidylinositol 3-kinase
PKC	Protein kinase C
rCBF	Regional cerebral blood flow
ROS	Reactive oxygen species
SOD	Superoxide dismutase
TNF-α	Necrosis factor-alpha
VEGF	Vascular endothelial growth factor

1 Introduction

Ischemic/hypoxic injury is caused by an insufficient supply of cerebral oxygen and/or blood flow, leading to neuronal dysfunctions and subsequent apoptosis through a multitude of pathophysiological processes (Myles et al. 2009; Chao and Xia 2010; Guo et al. 2010; He et al. 2013; Ghosh et al. 2014). The most significant disease associated with ischemic/hypoxic pathophysiology is stroke, a profound health problem in senior populations and the third leading cause of morbidity and mortality in the world. Except for thrombolytic agents (e.g., the recombinant tissue plasminogen activator) within a short therapeutic window, there are few promising options for treating ischemic stroke. Finding new approaches against ischemic injury and seeking alternative solutions for stroke have been long-term battles and have attracted much attention from both scientists and clinicians. One such approach is aimed at complementary and alternative medicine. For example, acupuncture has been used to treat patients with stroke in both acute phase (Gao et al. 2013; Chen et al. 2014b) and chronic phase (Chai et al. 2014; Huang et al. 2014; Zhu et al. 2014).

Acupuncture is an alternative therapy based on the theories of traditional Chinese medicine. The basic technology of acupuncture is to simulate some specific "acupoints" on the body surface using fine needles to achieve therapeutic effects. Electroacupuncture (EA), developed from traditional manual acupuncture, stimulates acupoints by using electric currents generated through acupuncture needles. The major advantage of EA is that the stimulating parameters of EA are objective and detectable, and therefore more reproducible without dependence on personal experience of the manipulators (Asakawa and Xia 2012a; Xia et al. 2010, 2012). Therefore, some clinicians are now using EA more often than manual acupuncture. Substantial evidence shows that acupuncture or EA can markedly regulate the neurotransmitter systems in the central nervous system (CNS) (Wen et al. 2010a, b; Liang and Xia 2012) by stimulating peripheral nerves (Zhou et al. 2010). In fact, acupuncture is employed in the treatment of not only stroke (Guo et al. 2010), but also of Parkinson's disease (Asakawa and Xia 2012a), Alzheimer's disease, and other neurodegenerative diseases (Asakawa and Xia 2012b), especially in oriental countries, even though the mechanisms are not yet fully understood. Indeed, our studies and those of others have shown that acupuncture/EA induces a beneficial effect on patients with strokes (Gao et al. 2013; Chai et al. 2014; Chen et al. 2014a; Huang et al. 2014; Zhu et al. 2014), and reduces ischemic infarction and ameliorates the neurological deficits in animal models (Tian et al. 2008a, b; Zhou et al. 2011a, b, 2013a, b). The mechanisms may lie in the enhancement of regional cerebral blood flow (rCBF) in the ischemic brain (Zhou et al. 2007, 2011a, 2013a, b), stabilization of ionic homeostasis, and upregulation of the neuroprotector signaling (Chao and Xia 2010; Guo et al. 2010; He et al. 2013; Tian et al. 2013a, b). In particular, delta-opioid receptor (DOR) has been found to play an important role in EA-induced brain protection against ischemic injury. In this chapter, we will perform a systematic review concerning the role of DOR in EA treatment for ischemic brain injury and its cellular and molecular mechanisms.

2 Clinical Application of Acupuncture for Stroke

Acupuncture has been used to treat patients with stroke for a long time, especially in China and other oriental countries, though there are still some controversies in terms of efficacy (Wang et al. 2014a; Park et al. 2014; Zhang et al. 2014). Several reviews/chapters have provided general information regarding acupuncture therapy against stroke (Guo et al. 2010; Gao et al. 2013; Chai et al. 2014; Chen et al. 2014a; Huang et al. 2014; Zhu et al. 2014).

2.1 *Efficacy of Acupuncture/EA in Patients with Stroke*

Manual- and electro-acupuncture have been used to treat patients with stroke at different disease stages, including acute (Naeser et al. 1992, 1993; Hu et al. 1993; Chang et al. 2014; Man et al. 2014) and sub-acute stages (Chen et al. 2014a) as well as the convalescent period (Chai et al. 2014; Huang et al. 2014; Zhu et al. 2014).

In general, EA reduces ischemic injury and promotes the recovery of neural function based on the results of CT images and neurophysiologic assessments in patients with acute stroke (Naeser et al. 1992, 1993; Hu et al. 1993) and subacute stroke (Johansson et al. 1993; Chen et al. 2014a). Some studies showed that EA played a particular role in rehabilitative treatment for patients who suffered from motor deficits (such as hemiplegia), pain, and contracture (Naeser 1997; Chai et al. 2014; Zhu et al. 2014) after stroke attack, thereby improving quality of life (Huang et al. 2014).

More progresses have been recently made in this field, especially for acupuncture-induced recovery after stroke. Firstly, more attention has been paid to motor symptoms than to conventional hemiplegia. For example, the clinicians/acupuncturists have used acupuncture/EA to treat dysphagia after stroke (Chang et al. 2014; Cheng and Chen 2014; She and Ge 2014). Several studies claimed that EA had a beneficial effect on the recovery from speech and swallowing dysfunction caused by medulla oblongata palsy after brainstem infarction. Moreover, EA was also used to treat a wide-range of motor symptoms including central facial nerve paralysis (Li et al. 2014), Broca's aphasia (Tao et al. 2014), balance function (Huang et al. 2014) and constipation (Yang et al. 2014a) in stroke patients. Secondly, more efforts have been devoted to acupuncture treatment for non-motor symptoms. Some recent studies suggest the efficacy of acupuncture for several neuropsychological symptoms such as cognitive impairment (Liu et al. 2014; Yang et al. 2014b) and depression (Man et al. 2014).

2.2 *Different Voices and Potential Reasons*

There are still conflicting voices concerning the efficacy of acupuncture treatment for stroke and the quality and conclusions of previous clinical studies.

Zhang et al. (2014) investigated the quality of clinical reports on randomized controlled trials of acupuncture for stroke rehabilitation. They concluded that although the overall quality of the studies was generally good, there was a lack of key information in some studies. For example, five out of the 15 studies (33 %) did not report any of three key methodological items in detail.

Although substantial data support acupuncture as an efficacious and satisfactory treatment for stroke (Gao et al. 2013; Chai et al. 2014; Chen et al. 2014a; Huang et al. 2014; Zhu et al. 2014), some investigators questioned this efficacy (Gosman-Hedström et al. 1998; Johansson et al. 2001; Wang et al. 2014a; Park et al. 2014). For example, Zhang et al. (2013) pointed out that acupuncture might be effective for treating neurological deficits after stroke; however, it is uncertain if acupuncture is helpful in preventing post-stroke death or disability or in improving other aspects of stroke recovery, such as motor ability after stroke. Park et al. (2014) reviewed eight randomized trials covering 399 patients undergoing acupuncture for the treatment of spasticity after stroke and found that acupuncture had no effect on improving clinical and physiologic outcomes. However, because of the poor quality of available studies, they could not make a definitive conclusion about the effect of EA for spasticity after stroke.

There are several potential pitfalls and flaws behind the controversies. Firstly, it is difficult to overcome the systematic weaknesses of an acupunctural study in patients with serious diseases like stroke, especially in setting "double-blinded" controls (Asakawa and Xia 2012c). Secondly, non-systematic flaws such as selection bias and problems of sample size were found in the previous studies. Thirdly, the outcome of EA treatment may greatly depend on EA conditions, including the selection of acupoints, depth of the needles under the skin, current frequency and intensity, wave pattern, stimulation duration, times of treatment, applying windows, etc. Indeed, our experimental studies (Zhou et al. 2011a, 2013a, b) demonstrate that EA protection against ischemic injury is greatly dependent on EA conditions. More recently, a clinical report (Chang et al. 2014) also showed that different EA parameters led to different outcomes. Unfortunately, there is still no standard clinical guideline based on systematic research for acupuncture treatment of stroke. The methods of acupuncture/EA were not comparable among the different studies in the past. Therefore, previous controversies among different studies are largely due to methodological deficits in some clinical investigations. More research with optimal EA strategies are needed for better clinical practice.

2.3 Limitations of Clinical Research

There is a stumbling block on the way of clinical research because it is sometimes difficult, if not impossible, to make a randomized, double-blind placebo-controlled design for any acupuncture study in patients with stroke, a serious neurological disorder. Research involving human participants must obey several ethical principles. For example, it is impractical to perform rigorous control studies on severe patients. Due to the ethical limitations and technical difficulties in human studies, it is extremely hard to generate a clinical guideline for appropriate application of acupuncture/EA in patients with stroke through human research. Therefore, bench studies on animal stroke models, with a well-designed protocol are indispensable for acupunctural research. In this way, we have optimized the EA conditions for EA-induced brain protection against ischemic injury, including optimal acupoints, current intensity and frequency, stimulation duration, etc. (Guo et al. 2010; Zhou et al. 2007, 2011a, b, 2013a, b), which provide a useful guide for appropriate application of EA in clinical settings (refer to Sect. 3.4 for more details).

3 EA Effects on Ischemic Injury: Laboratory Research

Because of the above-mentioned limitations and difficulties in clinical research, experimental study is particularly important for the effects of acupuncture on ischemic brain injury. Without some of the ethical limitations that must be obeyed in human studies, rigorously randomized, controlled, and double-blinded experimental designs can be adopted in animal studies. This is very helpful for us to obtain solid evidence for the efficacy of EA protection against brain ischemia.

Ischemic brain injury is caused by a reduction of blood supply, causing glucose and oxygen deprivation in the brain, and by subsequent reperfusion injury if blood supply is restored. The pathophysiology of ischemic injury involves complex events in the brain, including energetic failure, perturbation of ion homeostasis, and consequent intracellular Ca^{2+} accumulation, excitotoxicity, free-radical injury, cytokine-mediated cytotoxicity, blood-brain barrier (BBB) damage, and infiltration of leukocytes (Ghosh et al. 2014). The major outcomes of such ischemic insults are brain infarction, neurological deficits, and even death.

Our studies and those of others have shown that EA is effective in reducing the infarct size, ameliorating the neural damage and neurological deficits, repairing the damage of BBB, and reducing the death rate (Guo et al. 2010; Zhou et al. 2011a, 2013a, b; Feng and Zhang 2014). Optimal EA conditions are critical for inducing brain protection (Guo et al. 2010; Zhou et al. 2011a, 2013a, b).

3.1 EA-Induced Reduction in Cerebral Ischemic Infarct Volume and Death Rate

The reduction of ischemic volume is an important index for treating ischemic brain injury, which can be easily detected by CT or MRI scanning in clinical settings. It is even more accurate and easier in the animal middle cerebral artery occlusion (MCAO) models to measure the volume of triphenyltetrazolium chloride staining (Zhou et al. 2011a, 2013a, b). Our studies have proven that optimal EA stimulation reduces the volume of ischemic infarction when EA was performed at Shuigou (GV 26) – Neiguan (PC 6) or Shuigou (GV 26) – Baihui (GV 20) (Tian et al. 2008a, b; Zhou et al. 2007, 2011a, 2013a, b). Also, the death rate after MCAO was significantly reduced in the group treated with optimal EA (Zhou et al. 2011a, 2013a, b). Xu et al. (2014) also confirmed that EA attenuated the infarct size in the rat brain with MCAO.

3.2 EA Attenuation of BBB Damage

Wu et al. (2001) performed EA in the rat ischemic model to investigate the influence of EA on permeabilization of BBB. They found that EA significantly reduced the level of the dye Evans blue, an index of BBB permeability, indicating that EA contributed to protecting the integrity of BBB in ischemic state. A later study also showed that EA at Baihui (GV 20) and Yamen (GV 15) protected BBB from damage (Lin et al. 2009). Yu et al. (2011) obtained similar results in mice with MCAO. The mechanism may lie in the inhibition of P-glycoprotein (Yu et al. 2011).

Recently, Peng et al. (2012) performed EA at Shuigou (GV 26) and Baihui (GV 20) and found that EA might regulate aquaporin-4 expression and relieve the damage to the blood-brain barrier in a rat model of cerebral ischemia/reperfusion injury. Xu et al. (2014) applied EA (2 Hz, 1 mA, 20 min) at Zusanli (ST 36) and Baihui (GV 20) in the MCAO rats and found that EA decreased aquaporin-4 at both protein and mRNA levels that is relevant to permeabilization of BBB and modulation of inflammatory cell infiltration.

3.3 EA Improvement of the Neurological Deficits After Cerebral Ischemia

Our serial studies showed that EA improved neurological symptoms caused by cerebral ischemia (Zhou et al. 2007, 2011a, 2013a, b; Tian et al. 2008a, b). Such improvement of behavioral performance is always related to the reduction of cerebral ischemic volume (Zhou et al. 2011a, 2013a, b). Kim et al. (2014) applied EA with a 2 Hz/2 V current intensity at Baihui (GV 20) and Dazhui (GV 14) in the mouse MACO model and found that EA improved neuromotor function as well as cognitive ability after stroke. It also helped increase the number of proliferated cells and differentiated cells in the hippocampus and subventricular zone. Xu et al. (2014) also confirmed that EA ameliorated neurological deficits and caused a reduction in ischemic infarction in rats with MCAO. All these studies show that EA improves neurological deficits after cerebral ischemia.

3.4 Importance of EA Conditions for EA Protection Against Ischemic Injury

We have recently systematically investigated the importance of EA conditions, especially acupoint selection along with EA parameters in the EA-induced neuroprotection against ischemic injury and found that appropriate EA stimulation is crucial to achieve effective brain protection.

We compared different effects of several acupoints in the EA-induced changes in rCBF and brain ischemic injury with a battery of "optimal" parameters, namely 5/20 Hz parse-dense wave at 1.0 mA (Zhou et al. 2007, 2011a, b, 2013a, b). We found that Baihui (GV 20) and Shuigou (GV 26) induced the maximum efficacy for brain protection, while stimulating Quchi (LI 11) and Neiguan (PC 6) induced a weaker but significant protective effect. In contrast, the stimulation of Neiguan (SP 6) and Yanglingquan (GB 34) had no appreciable effect. Other studies also demonstrated the importance of acupoints selection. Feng and Zhang (2014) pointed out that 2/15 Hz EA at bilateral Zusanli (ST 36) acupoints could increase rCBF and achieve good efficacy in brain protection against acute cerebral ischemia. Since the nerve distribution and afferent/efferent pathways are quite different among acupoints, it seems reasonable that stimulating various acupoints with different types of currents leads to different outcomes of the ischemic brain.

Besides acupoints and stimulation parameters, an appropriate duration of EA stimulation is also important for the EA-induced outcomes. We found that EA stimulation for 5–30 min at Baihui (GV 20) and Shuigou (GV 26) with sparse-dense wave (5/20 Hz) at 1.0 mA significantly increased the rCBF, attenuated cerebral ischemia volume, and improved neurological deficits. However, over-length stimulation, e.g., EA for 45 min, did not reduce the ischemic volume nor improve neurological deficits, although the rCBF still increased during the period of EA stimulation. This detached result between rCBF and brain protection suggests that the improvement of blood flow is not the sole factor for the brain protection against cerebral ischemia. Other mechanisms, e.g., rebalance of survival and death signals (He et al. 2013) also play a key role in brain protection.

3.5 An Insight into the Controversies in Clinical Reports

Our studies and those of others gave strong evidence for the efficacy of EA neuroprotection against ischemic injury through rigorous experimental design. Importantly, we found "divergent" outcomes of animal ischemia models after EA treatment due to differences in EA conditions, which help us achieve a better understanding regarding previous controversies in clinical reports.

Regardless, the problems of experimental design, the selection of acupoints, and acupuncture manipulations, including directions, angles, reinforcement and reduction, and quantity of stimulus, will surely affect the outcome of acupuncture. An important issue is the selection of EA parameters. Obviously, selection of "optimal" parameters will achieve the highest efficacy, whereas selection of "inappropriate parameters" may produce low efficacy. Because there is no "standard protocol" for acupuncture treatment of stroke, the manner of manual acupuncture and condition of EA were very different among various studies in the past, which is, at least partially, the cause of different outcomes of acupuncture treatment for stroke in clinical settings. More clinical studies are needed to gather more data of the "optimal" acupuncture conditions, which are essential to formulate an objective and optimal standard of acupuncture/EA treatment.

4 Mechanisms of EA Protection Against Ischemic Brain Injury

There have been substantial studies investigating the mechanisms of EA protection against ischemic brain injury. Several lines of evidence have suggested that EA presents its neuroprotection by enhancing cerebral blood flow and upregulating the DOR system.

4.1 EA Enhances Cerebral Blood Flow in the Ischemic Brain

Our studies (Zhou et al. 2007, 2011a, 2013a, b) and those from other laboratories (Byeon et al. 2011; Zhou et al. 2011b; Feng and Zhang 2014; Uchida and Kagitani 2014; Zhang et al. 2014) have shown that effective EA enhances the rCBF in the ischemic brain, though the mechanism is still not fully understood. EA conditions are crucial for the outcomes of blood flow in acupuncture treatment. The differences in stimulation parameters and acupoints may lead to very different outcomes (see Sect. 3.4).

As indicated in Sect. 3.4, however, the improvement of blood flow is not the sole factor for brain protection against cerebral ischemia because an "inappropriate" EA stimulation, though increasing rCBF, did not induce brain protection against ischemic injury (Zhou et al. 2013b). On the other hand, we found that DOR antagonists largely attenuated the EA-induced brain protection against cerebral ischemia (Zhao et al. 2002; Tian et al. 2008a, b; Zhou et al. 2013b), suggesting a crucial role of DOR signaling in the EA-induced neuroprotection (Zhou et al. 2013b). Therefore, we will mainly focus on the DOR mechanism for EA neuroprotection in the below sections.

4.2 Roles of DOR in EA Neuroprotection Against Ischemic Injury

Brain opioid peptide systems include endogenous opioid peptides (EOP) and opioid receptors. The EOP systems have been proven to play important roles in a wide range of regulatory functions for neuropsychological behaviors including emotion, motivation, pain and analgesia, attachment behavior, stress and social status, tolerance and dependence, learning and memory, control of food intake, alcohol and drugs of abuse, sexual activity and hormones, pregnancy, development and endocrinology, mental illness and mood, etc. (Bodnar 2014; also refer to Chap. 18). Opioid peptides achieve these complicated functions by binding with different opioid receptors including delta-, mu-, and kappa-opioid receptors (DOR, MOR and KOR) (Wen et al. 2010b; Liang and Xia 2012; also refer to Chaps. 2 and 18).

It has been well documented that acupuncture upregulates opioidergic activity in the brain (Sjölund et al. 1977; Yang et al. 2006; Zhou et al. 2013b; Wen et al. 2010b; Liang and Xia 2012; da Silva et al. 2013; Wang 2013; Xia et al. 2010, 2012). Several experiments in our laboratory verified the roles of DOR involved in EA treatment.

4.2.1 EA Upregulates Endogenous Opioid System

We have previously reviewed acupuncture and EA induced upregulation of EOP and opioid receptors in the CNS (Wen et al. 2010b; Liang and Xia 2012). More recently, several studies provide strong evidence for EA-induced increase in EOP release and expression at both CNS and peripheral levels. Such promotional effects are greatly affected by EA parameters, especially frequency, similarly seen in our studies on the EA-induced neuroprotection against ischemic injury (see Sect. 3.4). Here we briefly update the research progress in this field.

EOP Expression and Release

Cheng et al. (2013) investigated the dynamic variations of mRNA expression of several precursors of EOPs (proenkephalin, proopiomelanocortin, prodynorphin) in the CNS. They performed EA (60 Hz for 0.5 h) in a succession of acupoints, namely, Baihui (GV 20), Santai (EX-HN), Ergen (AH 1), and Sanyangluo (SJ 8) in 108 normal goats and found that effective EA significantly increased the mRNA expression of proenkephalin, proopiomelanocortin, and prodynorphin in the brain and spinal cord dorsal horn, although the dynamic variations of each precursor were different. Moreover, abundant evidence suggests that appropriate stimulation frequency is essential for EA-induced release of EOP. Han and his colleagues published a series of papers to clarify the relationship between EA frequency and potential EOP release (Han et al. 1991; Han 2003, 2004). Briefly, low frequency stimulation (1–2 Hz) contributes to the release of the β-endorphins and enkephalins, while high frequency (100 Hz) causes dynorphin release and does not influence endorphins or enkephalins. EA at 12–15 Hz can affect all three EOPs. Recently, Liu et al. (2013) performed EA at Zusanli (ST 36) in rats with experimental autoimmune encephalomyelitis and found that “effective” EA significantly elevated β-endorphin concentrations in both the hypothalamus and in plasma. Da Silva et al. (2013) performed EA (2, 100, and 2/100 Hz alternatively; 1.4–1.5 mA; 20 min; square waves with a 5-ms width) at Zusanli (ST36) and Sanyinjiao (SP 6) in normal rats. They found only EA at ST 36 and SP 6 with 2/100 Hz alternatively could increase the EOP release in the spinal cord. They pointed out that in regard to the antinociceptive effect of 2/100 Hz alternative EA, the mechanism is quite different from that of solely performing alternating stimulation at 2 and 100 Hz (da Silva et al. 2013). More recently, Moon et al. (2014) compared efficacy of low frequency EA (2 Hz) and high frequency EA (100 Hz) (the remaining parameters were similar, i.e., 0.3-ms pulse duration and 0.2–0.3 mA) at Zusanli (ST 36) in a rat model of oxaliplatin-induced neuropathic pain. They achieved good efficacy with EA at 2 Hz and found effective EA (2 Hz) increased the plasma β-endorphin right after the end of EA, and such efficacy could be completely blocked by naloxone, which was not seen when applying EA with noneffective stimulation (100 Hz).

Interestingly, a study showed that effective EA (2 Hz, 1 mA, 0.1 ms duration time) at Huantiao (GB 30) and Yanglingquan (GB 34) upregulated the expression of β-endorphin even in inflamed skin tissues (Su et al. 2011).

All these studies show that effective EA in low frequency or low/high-alternative frequency may increase release of EOP in CNS, and probably in peripheral tissues.

Opioid Receptors

Opioid receptors mainly include DOR, KOR, MOR and Nocientin receptor (Stevens 2009). They bind with endogenous opioids and play important roles in many functions. Recent studies further demonstrated that EA may upregulate the expression of opioid receptors, including DOR (Tian et al. 2008a, b; Kuo et al. 2013; Seo et al. 2013; Yi et al. 2013a, b; Xiang et al. 2014), MOR (Chen et al. 2013; Kim et al. 2013; Kuo et al. 2013; Silva et al. 2013; Seo et al. 2013; Yi et al. 2013a, b), KOR (Cheng et al. 2012; Yi et al. 2013a, b) in various organs under different conditions. The effect of EA on opioid receptors is likely dependent on stimulation parameters. Kuo et al. (2013) observed that MOR was activated by low frequency stimulation (2 Hz), while DOR was activated by high frequency EA (100 Hz). Silva performed EA (2, 100 Hz; 5- or 10-min duration time) at Zusanli (ST36) and Sanyinjiao (SP 6) to study the EA-induced analgesia in normal rats. They found that only low frequency EA (2 Hz) activated MOR in dorsal and ventral anterior pretectal nucleus (Silva et al. 2013). Cheng et al. (2013) performed EA (60 Hz for 0.5 h) in a succession of acupoints, namely, Baihui (GV 20), Santai (EX-HN), Ergen (AH 1), and Sanyangluo (SJ 8) in normal goats and found that effective EA significantly increased the mRNA expression of DOR, MOR and KOR in the brain and spinal cord dorsal horn. Ren et al. (2012) performed EA (2/100 Hz, 0.1–0.3 mA, 20 min) at Sanyinjiao (SP 6) and Xuanzhong (GB 39) to observe the analgesic effects of EA in dysmenorrhea rats and found that KOR expression is upregulated by effective EA in both acupoints, but EA at Sanyinjiao (SP 6) has a stronger effect. In contrast, Feng et al. (2012) performed low frequency EA (2 Hz, 0.8–1.4 mA) at the bilateral points of rectus abdominis and triceps surae muscles in the rat model with polycystic ovary syndrome and observed that effective EA at low frequency (2 Hz) downregulated the expression of mRNA of KOR in the hypothalamic arcuate nucleus. Therefore, more investigations are needed to verify the appropriate parameters for EA-induced upregulation of opioid receptors.

There is also evidence suggesting that EA might increase the expression of β-endorphin of peripheral CD4+ cells in multiple sclerosis rat models (Liu et al. 2013).

Taken together, EA generally upregulates EOP activity by increasing opioid receptor expression and promoting EOP expression and release, which is greatly dependent on appropriate EA conditions (Fig. 11.1).

4.2.2 EA Upregulates Opioid Receptors in the Ischemic Brain

We have previously found that DOR is sensitive to hypoxic and ischemic stress (Ma et al. 2005; Tian et al. 2008b). Indeed, DOR density is largely decreased in the ischemic brain (Tian et al. 2008b; Boutin et al. 1999). Excitingly, our study shows that EA stimulation with low-frequency and low-intensity current significantly increases the level of DOR in the ischemic brain (Tian et al. 2008b; Fig. 11.3). Interestingly, the EA-induced increase in DOR expression is also observed in the ischemic heart.

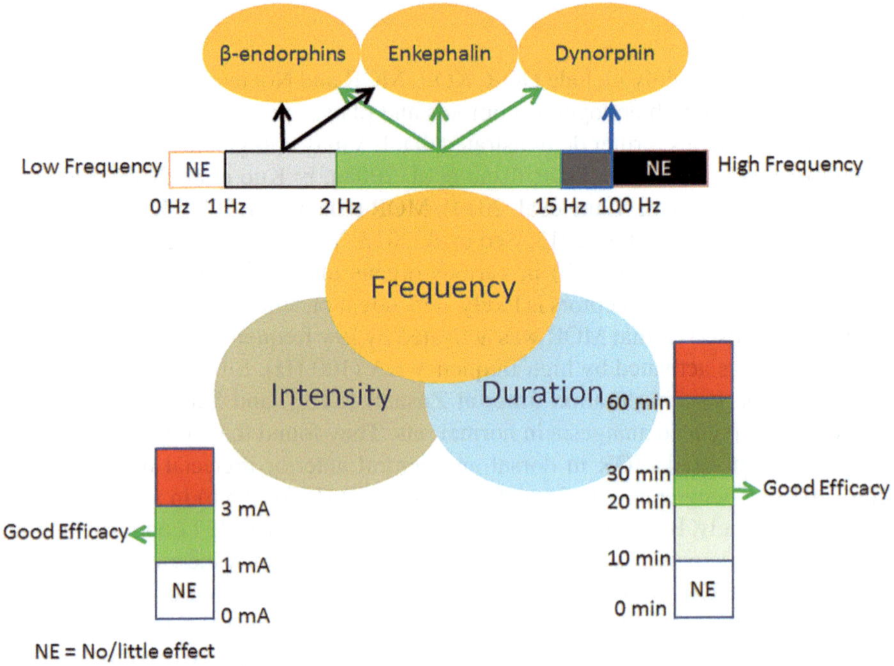

Fig. 11.1 EA parameter and EA effects

Li et al. (2011) performed EA (2 Hz, 20 min) at Neiguan (PC 6) in the rat model of myocardial ischemia and found that effective EA upregulated the expression of the mRNA of DOR.

Therefore, it is possible to increase the level of DOR in ischemic organs, which is of high significance in the treatment of hypoxic/ischemic encephalopathy, since DOR has been shown to be neuroprotective.

4.2.3 DOR Antagonists Block EA Protection Against Ischemic Brain Injury

As early as 2002, our laboratory firstly found the role of DOR in EA-induced protection against ischemic brain injury (Zhao et al. 2002). We found that pretreatment with naltrindole, a DOR antagonist, greatly attenuated EA-induced brain protection against ischemic injury (Zhao et al. 2002). Later, Yang et al. (2006) performed EA (2/15 Hz, 30 min, 1 mA in sparse-dense wave) at Baihui (GV 20) in MCAO rats and observed that naltrindole administrated before EA reversed EA-induced ischemic tolerance. Later studies from our laboratory and others (Xiong et al. 2004, 2007) consistently demonstrated that blockade of DOR greatly attenuates EA-induced neuroprotection. For example, Tian et al. (2008b) performed cumulative EA with a disperse-dense wave (disperse wave =3.85 Hz, duration time 1.28 s, 0.8 mA; dense wave = 6.25 Hz, duration time 2.08 s, 0.8 mA) at Shuigou (GV 26) and Neiguan

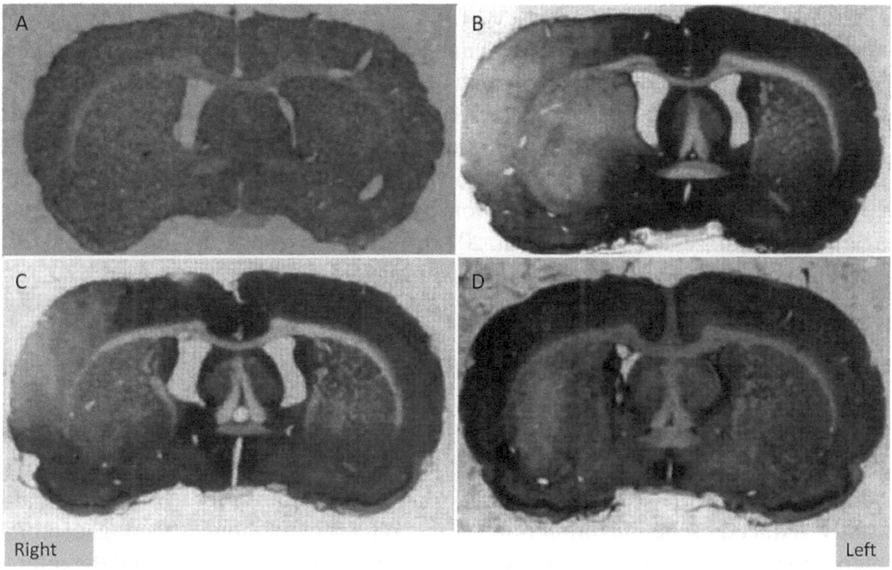

Fig. 11.2 Effect of EA on ischemic brain infarction. (**a**) Sham control of MCAO. No visible infarction. (**b**) MCAO, Major infarction. (**c**) MCAO plus sham EA. Major infarction similar to that of (**b**). (**d**) MCAO plus true EA. Reduced infarction volume. Cited from Tian et al. (2008a) with permission of the journal

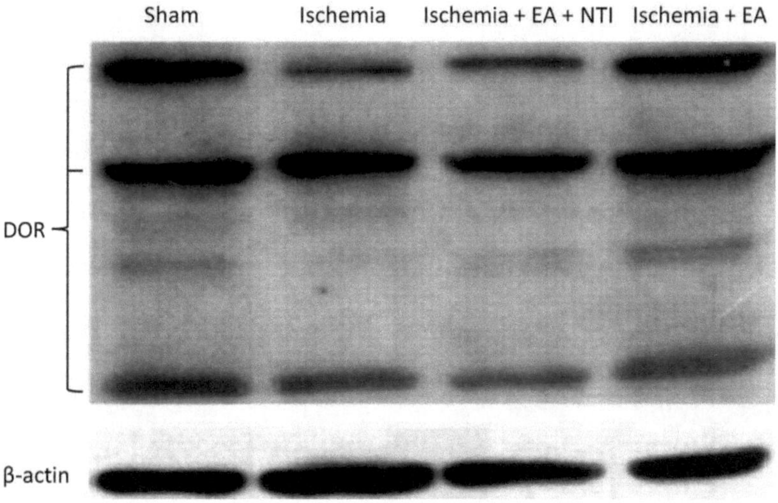

Fig. 11.3 EA-induced upregulation of DOR in the ischemia-exposed striatum. The ischemic tissues, with appropriate control, were sampled at 24 h after MCAO. DOR protein was detected by western blot analysis. β-Actin was employed as an internal standard. Note that ischemia reduced, while EA significantly increased the level of DOR, which could be reversed by naltrindole (NTI), a DOR antagonist. Cited from Tian et al. (2008b) with permission of the journal

(PC 6) in MCAO rats. The efficacy of EA was confirmed by the rating scale of neurological scales and infarction volume. The administration of DOR antagonist, naltrindole, largely reduced the EA-induced reduction of infarct volume and amelioration of behavioral performance (Tian et al. 2008a). It also attenuated the EA-induced increase in DOR expression (Fig. 11.3) (Tian et al. 2008b). These results confirmed the roles of DOR involved in EA therapy. Our later study further confirmed the fact that DOR inhibition by naltrindole attenuates EA protection against cerebral ischemia (Zhou et al. 2013b). All these results suggest a crucial role of DOR in the EA neuroprotection against ischemic injury.

4.2.4 DOR Activation Attenuates Ischemic/Hypoxic Injury

Indeed, DOR activation remarkably relieves ischemic brain injury. In one of our recent studies (Tian et al. 2013a), we found that DOR activation with TAN-67, administrated to the ipsilateral lateral ventricle 30 min before MCAO, significantly reduced ischemic infarction of the rat brain without affecting cerebral blood flow. In contrast, 100 nmoL of naltrindole significantly increased the infarct volume in the ischemic brain (Fig. 11.4) without any change in cerebral blood flow.

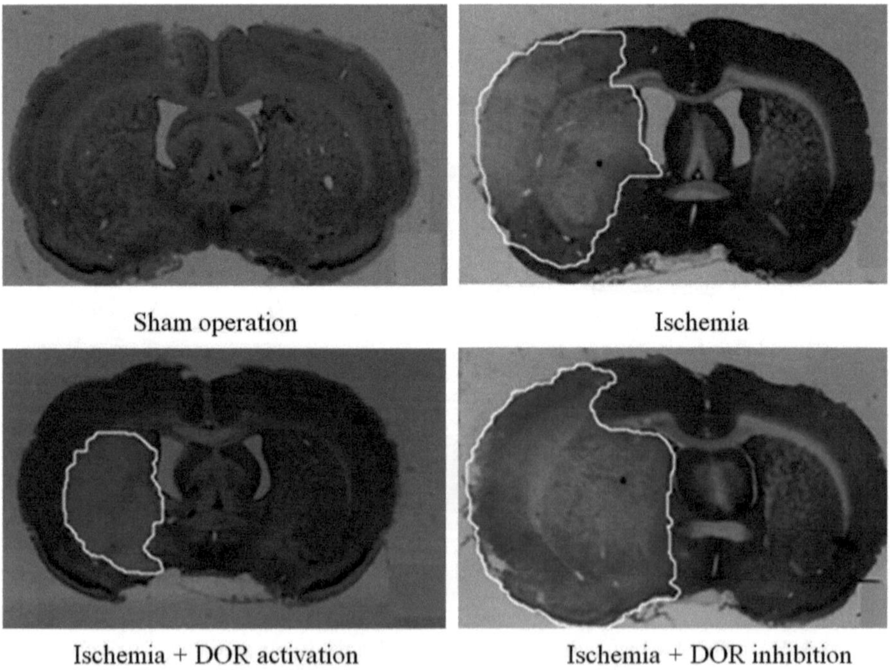

Fig. 11.4 DOR activation induced reduction of ischemic brain infarction in the rats with MCAO. Note that DOR activation with TAN-67, significantly reduced ischemic infarct volume, whereas administration of naltrindole (NTI), a DOR antagonist, significantly enlarged the infarction size. Cited from Tian et al. (2013a) with permission of the journal

These observations further confirmed the neuroprotective role of DOR in the ischemic brain and provide strong evidence that DOR is involved in the EA-induced neuroprotection against ischemic/hypoxic injury, since EA can upregulate DOR in the ischemic brain.

4.2.5 Cellular and Molecular Mechanisms for DOR Protection Against Ischemic/Hypoxic Injury

Accumulating data from our laboratory and those of others have suggested that DOR provides neuroprotection through multiple pathways at cellular and molecular levels.

DOR Stabilizes the Ionic Homeostasis in Hypoxic/Ischemic States

Ionic homeostasis is a crucial condition to maintain normal brain function. Appropriate transmembrane ionic gradients are important in maintaining membrane potentials, generating new action potentials, synaptic transmission, neuronal plasticity, and regulation of osmotic balance, pH, volume, and micro environment of ions (Chao and Xia 2010; refer to Chap. 5). In hypoxic/ischemic states in the brain, disturbance of ionic modulation will occur with disruption of Na^+, Ca^{2+}, K^+ and Cl^- homeostasis.

We firstly pointed out that activation of DOR attenuates the disruption of K^+ homeostasis undergoing anoxia or oxygen-glucose deprivation (Chao et al. 2006, 2007, 2008, 2009). We found that activation of DOR may inhibit Na^+ entry, and then ameliorate K^+ homeostasis (Chao et al. 2008). Moreover, We found that activation of DOR or upregulation of DOR expression attenuates the amplitude of Na^+ currents with increasing activation threshold of Na^+ channels (Kang et al. 2009; Chao et al. 2013). The DOR inhibition of Na^+ channels is the initial process that causes a reduction in K^+ leakage under hypoxic conditions (Kang et al. 2009). All the above reports suggest that stabilization of ionic homeostasis may play a role in the mechanism of DOR protection against cerebral ischemia (refer to Chap. 6).

DOR Increases Anti-oxidative Capacity

Oxidative stress plays a peculiar role in the pathophysiology of cerebral ischemia. Hypoxia/ischemia reperfusion can lead to oxidative stress, which contributes to neuron death, and causes neurological deficits in cerebral ischemia. Our previous study (Yang et al. 2009) showed that DOR activation decreased oxidative injury in the rat cerebral ischemia model, partially reversed the ischemia-induced decrease in superoxide dismutase (SOD) and glutathione peroxidase (GPx), and decreased the levels of malondialdehyde (MDA) and nitric oxide (NO) in the ischemic brain. These data suggest that anti-oxidative stress may play a role in the mechanisms for DOR protection.

Recently, several studies have provided evidence that EA activates respiratory enzymes, decreases the generation of reactive oxygen species (ROS), ameliorates dysfunctions of respiratory chain, and increases anti-oxidative capability in the infarct penumbra area (Siu et al. 2004; Zhong et al. 2006). In Siu's study, EA (0.7 V, 2 Hz, 0.5 ms) was performed on Fengchi (GB 20) of MCAO rats. They found that EA reduced lipid peroxidation in ischemia-reperfused rat brains, which may be related to activation of SOD and GPx (Siu et al. 2004). This conclusion was verified by a more recent study (Zhang et al. 2014) that showed that effective acupuncture ameliorated the cognitive impairment in multi-infarct dementia rats and attenuated mitochondrial deficits. Taken together, effective acupuncture elevated the activities of SOD, CuZnSOD and MnSOD, decreased the level of MDA and superoxide anion and regulated the ratio of reduced glutathione and oxidized glutathione in mitochondria.

In addition, there are several studies showing that in organs besides the brain, DOR also shows protective effects by activation of anti-oxidative mechanisms (McPherson and Yao 2001; Zhang et al. 2002; Husain et al. 2014).

DOR Downregulates the Inflammatory Cytokines

Several lines of evidence show that DOR downregulates the inflammatory cytokines. Husain et al. (2011) reported that DOR activation reduces the production of tumor necrosis factor-alpha (TNF-α) induced by ischemia/reperfusion in the retina. Tang et al. (2011) administrated a DOR agonist [D-Ala2, D-Leu5]-enkephalin (DADLE) in a sepsis model and reached an analogous conclusion. They observed that activation of DOR achieved better survival rate associated with a decrease in TNF-α, interleukin-1β (IL-1β) and high-mobility group box 1 protein (HMGB1). Our recent study showed that DOR activation reduces the production of inflammatory cytokines in the brain exposed to prolonged hypoxia (Tian et al. 2013b). Furthermore, we found that DOR activation reversed the hypoxia-induced changes in TNF-α levels in both astrocytes and neuro-like cells with different regulatory mechanisms between neurons and glia (Wang et al. 2014b). Our data suggest that downregulation of the inflammatory cytokines may be a potential mechanism of DOR protection. Therefore, EA-increased DOR activity may protect the brain against hypoxic/ischemic injury through the down regulation of inflammatory cytokines.

DOR Reduces Extracellular Accumulation of Excitatory Transmitters

As mentioned above, the release of excitatory transmitter plays a role in the pathophysiology of cerebral ischemia. As early as 1999, we found that DOR activation reduced glutamate-induced excitotoxic injury in rat neurons, and this protection can be reversed by DOR antagonist naltrindole. Yet when we performed the experiments using KOR and MOR agonists and antagonists, no/little neuroprotective effects were found. We therefore concluded that only activation of DOR protects neurons from glutamate-induced excitotoxicity (Zhang et al. 2000). Our recent study found that DOR activation increases the expression of astrocytic excitatory amino acid transporters (EAATs) via MAPK signaling (Liang et al. 2014). We compared the mRNA and protein expression of EAAT1, EAAT2, and EAAT3 in mouse astrocytes pretreated with a specific DOR receptor agonist (UFP-512) with and without a DOR antagonist, DOR siRNA or inhibitors of PKC, PKA, PI3K, p38, MAPK, MEK and ERK. We also evaluated the function of EAATs by measuring glutamate uptake. Our data showed that DOR activation upregulated mRNA and protein expression of EAAT1 and EAAT2, but not EAAT3, thereby enhancing glutamate uptake of astrocytes. DOR-induced EAAT1 and EAAT2 expression were remarkably reversed by DOR antagonist, naltrindole or by DOR siRNA knockdown, and suppressed by inhibitors of MEK, ERK and p38. DOR-accelerated glutamate uptake was inhibited by EAAT blockers, DOR siRNA knockdown, or inhibitors of MEK, ERK or p38. In contrast, inhibitors of PKA, PKC or PI3K had no significant effect on DOR-induced EAAT expression. Our results suggest a critical role of DOR in the regulation of astrocytic EAATs (Liang et al. 2014). Since glial EAATs play an important role in the removal of extracellular excitatory amino acids, DOR may protect neurons from ischemic/hypoxic injury by increasing astrocytes' abilities to reduce excitotoxicity.

DOR Regulates PKC-MAPK and the Downstream Pathways

Several studies suggested that DOR regulates some intracellular pathways by inhibition of apoptotic signals and activation of pro-survival signals (He et al. 2013). Our previous work (Ma et al. 2005) investigated the role of DOR in the regulation of survival and death signaling pathways including MAPK PKC pathway. We found that DOR-mediated hypoxic preconditioning (HPC) protection is dependent on an increase in extracellular signal-regulated kinase (ERK) and B cell lymphoma protein-2 (Bcl 2). Their effects counteracted the hypoxia-induced increase in p38 activities and cytochrome c release. The cross-talk between ERK and p38 showed a "yin-yang" antagonism under the control of the DOR-G protein-PKC pathway (Ma et al. 2005). A later report by Narita et al. (2006) furthermore pointed out that the protective effects of DOR were mediated through the activation of Trk-dependent tyrosine kinase, which could be linked to phosphatidylinositol 3-kinase (PI3K), PKC, Ca^{++}/calmodulin-dependent protein kinase II (CaMKII) and MARK (Narita et al. 2006). Olianas et al. (2011) investigated the regulating effects of N-desmethylclozapine (NDMC) on protein kinase B/Akt (Akt) signaling in the different cellular systems. They found that NDMC could stimulate Akt signaling and regulate PI3Kα by activating Gi/Go-coupled DORs. Another study confirmed the protective effects of DOR activation on acute mitochondrial dysfunction in neurons. They found that this protective effect was conducted through PKC-ERK pathway, and mitochondrial ERK activation was required for DOR neuroprotection against acute mitochondrial dysfunction (Zhu et al. 2011). Moreover, a recent report indicated an inhibition of DOR induced apoptosis of the brain glioma cells, which is closely related to the mitochondrial and PKC pathways (Zhou et al. 2013c). All these studies support our finding that PKC-MAPK plays an important role in the signaling of DOR protection.

DOR Upregulates Brain-Derived Neurotrophic Factor and Functional TrkB Receptors

Brain-derived neurotrophic factor (BDNF), a member of the "neurotrophin" family, plays a crucial role in supporting neuron survival, encouraging the growth as well as in the differentiation of new neurons and synapses in the hippocampus, cortex and basal forebrain (Yamada and Nabeshima 2003). BDNF conducts its functions by binding to at least two receptors on the cell surface, one of them being TrkB receptor.

Recently, Kim et al. (2014) reported effective EA significantly upregulated mRNA and protein expression of BDNF and vascular endothelial growth factor (VEGF) in mice, implying that an increased expression of growth factors may play a role in EA's effects.

Narita et al. found that DOR stimulation may mediate neurogenesis and neuroprotection through the release of BDNF, while these effects can be inhibited by inhibitors of a Trk-dependent tyrosine kinase inhibitor, K-252a (Narita et al. 2006). Another study found that administration of DOR agonist resulted in enhancement of BDNF mRNA levels in the frontal cortex, hippocampus, and amygdala, while most of these effects were inhibited by administration of DOR antagonist Naltrindole (Zhang et al. 2006a). Our study further demonstrated that DOR activation attenuates hypoxic/ischemic brain injury through a BDNF-TrkB pathway. We measured the expression of BDNF, TrkB, total CREB, p-CREB, p-ATF and CD11b in MCAO rats. DOR agonist TAN-67, with or without DOR antagonist naltrindole, was administrated before MCAO process. DOR activation effectively reduced the infarct volume of the ischemic brain. MCAO significantly reduced the level of full length TrkB (functional TrkB), while DOR activation largely reversed this reduction in spite of an unchanged expression level of BDNF. Therefore, DOR activation may rescue TrkB signaling by reversing the decrease in full-length TrkB caused by ischemia/reperfusion processes (Tian et al. 2013a). Moreover, we found that prolonged hypoxia downregulated the BDNF-TrkB signaling, leading to an increase in TNF-α in the cortex. DOR activation resulted in an upregulation of BDNF-TrkB and a decrease in TNF-α in the cortex exposed to prolonged hypoxia (Tian et al. 2013b). All these results support that activation of DOR upregulates the BDNF-TrkB signaling, which may play a key role in the DOR neuroprotection.

DOR Signaling Affects Other Mechanisms

Our recent work demonstrated that the protective effect of DOR against hypoxic/ischemic injury may be associated with modulations of α-synuclein. We found that both hypoxia and MPP+ stress enhanced the expression of α-synuclein, which formed oligomers that caused cell injury. DOR activation with UFP-512 significantly attenuated the cell injury and overexpression of α-synuclein, which could be reversed by DOR antagonist. Administration of the DOR agonist enhanced cAMP-response element binding protein (CREB) phosphorylation and prevented the collapse of mitochondrial membrane potential. These findings indicate that downregulation of the expression/aggregation of α-synuclein via a CREB pathway may play a role in the mechanisms of DOR protection (Chen et al. 2014b).

More recently, we reported that DOR activation upregulates Nuclear factor (erythroid-derived 2)-like 2 (Nrf2) expression and promotes its nuclear translocation, which may also contribute to DOR-induced neuroprotection (Cao et al. 2014, 2015). We found that exogenous activation of DOR upregulated Nrf2 protein expression and nuclear translocation and induced a strong cytoprotective effect against hypoxic injury. The DOR effects on Nrf2 nuclear translocation could be blocked by naltrindole, and could be remarkably reversed by transfection with Nrf2 siRNA, suggesting that DOR-mediated Nrf2 signaling is involved in cytoprotection.

Duan et al. (2011) investigated the effect of DOR activation in rat global ischemia models. The ischemic injury resulted in a significant neuronal loss and an increase in the dying astrocytes in the hippocampal CA1 region as compared with those in the sham control animals. DOR activation (intracerebroventricular injection of DADLE) improved neuronal survival and reduced dying astrocytes. However, DADLE also increased active caspase-3 positive cells in the ischemic group. Most of the active caspase-3-expressing cells were co-labeled with an astrocyte marker called glial fibrillary acidic protein (GFAP). They therefore concluded that DOR activation promotes beneficial astrocyte activation and induces astroglial apoptosis after reperfusion, which may be good for neuron survival. Since the roles of astrocytes in neuroprotection are very complex (Takuma et al. 2004; Giffard and Swanson; 2005 and Ouyang et al. 2007; Rossi et al. 2007; Takano et al. 2009 and Szydlowska et al. 2010), more studies are needed to elucidate the precise mechanism for the role of astrocytes in DOR neuroprotection. Toward this goal, we have recently performed several studies and found in the astrocytes that DOR activation down-regulates inflammatory cytokines induced by hypoxic stress (see Sect. 4.2.5.3) and increases the capacity of removing extracellular excitatory amino acids (see Sect. 4.2.5.4). If EA induces a DOR upregulation in the brain, these DOR-mediated events are certainly beneficial to the brain exposed to ischemia.

5 EA Preconditioning Protects from Hypoxic/Ischemic Injury via a DOR Mechanism

5.1 Hypoxia/Ischemia Preconditioning and Ischemic Tolerance

Hypoxia/ischemia preconditioning is an endogenous neuroprotective mechanism, activated by brief sublethal hypoxia or ischemia, which functions to protect against subsequent lethal hypoxic/ischemic insult. That is to say, "preconditioning hypoxic/ ischemic stimuli trigger a neuroprotective state against hypoxic/ischemic injury" (Gidday 2006). Recent reports indicated that such ischemic tolerance can be transiently induced by preexposure to a non-injurious preconditioning stimulus (Rosenstein and Fernandez 2014). The preconditioning mechanism has been found in many other organs including the heart (Rana et al. 2014) and retina (Rosenstein and Fernandez 2014). Two types of ischemic tolerance in the brain were reported, namely rapid and delayed ischemic tolerance.

5.2 EA Renders the Brain More Tolerant to Subsequent Ischemic Injury via a DOR Mechanism

We previously found that rapid or delayed HPC can increase enkephalin release and activate the DOR mechanism, thus protecting neurons from subsequent severe hypoxia and excitotoxicity (Ma et al. 2005; Zhang et al. 2006b). Since EA can promote DOR activity, EA pretreatment may render the brain more tolerant to hypoxic/ischemic stress via a DOR mechanism. This viewpoint was supported by several lines of evidence. Yang et al. performed 30-min EA (1 mA, 2/15 Hz sparse-dense wave) at Baihui (GV 20) as a preconditioning treatment in MCAO rats and found that EA significantly reduced the infarct volume. The EA-induced ischemic tolerance was abolished when naltrindole, a DOR antagonist was give before EA preconditioning. These data indicate that DOR plays a critical role in improving the ischemic tolerance induced by EA (Yang et al. 2006). Xiong et al. (2007) also reported that repeated EA preconditioning improved enkephalin release, which may bind δ- and μ-opioid receptors to induce the delayed cerebral ischemic tolerance. On the other hand, Wang et al. (2005) employed EA (1 mA, frequency 15 Hz) at acupoint Baihui (GV 20) for 30 min in MCAO rats and found that EA preconditioning at 2 h before MCAO improved neurologic outcomes and reduced the infarct volume at 24 h after reperfusion. They found that EA preconditioning induced rapid tolerance to focal cerebral ischemia is possibly mediated through an adenosine A1 receptor-related mechanism. Recently, Zhao et al. (2012) reported that EA pretreatment conducted the tolerance against cerebral ischemia through activation of canonical Notch pathway. Jin et al. (2013) found that EA pretreatment induced delayed ischemic tolerance may be related to the function of monocyte chemotactic protein-induced protein 1. Further research is needed to elucidate if there is an interaction between DOR signaling and other mechanisms in EA preconditioning induced ischemia tolerance.

6 Concluding Remarks

This chapter focuses on the roles of DOR in the EA-induced neuroprotection against ischemic injury. Although the data are still limited, the available literature shows that EA, if applied in an appropriate condition, can upregulate DOR signaling and regulate many cellular/molecular events at multiple levels, thereby rebalancing survival and death signals with beneficial outcomes. However, future studies are required to further investigate several important issues, for example, the roles of other memberane proteins in the EA neuroprotection and their interaction with

DOR. We believe that an in-depth study on DOR-mediated EA neuroprotection will guide us to a better understanding of EA neuroprotection against ischemic injury and eventuallly lead to better practice in clinical acupuncture.

Acknowledgements TA was supported by grants from the Japan Society for the Promotion of Science (Grant-in-Aid for Young Scientists, Type B, No. 20791025 and Grant-in-Aid for Scientific Research C, General, No. 24592157 and 15k10358). YX was supported by NIH grants (AT-004422 and HD-03582) and Vivian L Smith Neurologic Foundation.

References

Asakawa T, Xia Y (2012a) Acupuncture treatment for Parkinson's disease. In: Xia Y, Ding GH, Wu GC (eds) Current research in acupuncture. Springer, New York, pp 215–253

Asakawa T, Xia Y (2012b) Can acupuncture treat Alzheimer's disease and other neurodegenerative disorders? In: Xia Y, Ding GH, Wu GC (eds) Current research in acupuncture. Springer, New York, pp 255–301

Asakawa T, Xia Y (2012c) Future research in acupuncture: better design and analysis for novel and valid findings. In: Xia Y, Ding GH, Wu GC (eds) Current research in acupuncture. Springer, New York, pp 687–725

Bodnar RJ (2014) Endogenous opiates and behavior: 2013. Peptides 62:67–136

Boutin H, Dauphin F, MacKenzie ET, Jauzac P (1999) Differential time-course decreases in non-selective, μ-, δ-, and κ-opioid receptors after focal cerebral ischemia in mice. Stroke 30(6):1271–1278

Byeon HS, Moon SK, Park SU, Jung WS, Park JM, Ko CN, Cho KH, Kim YS, Bae HS (2011) Effects of GV20 acupuncture on cerebral blood flow velocity of middle cerebral artery and anterior cerebral artery territories, and CO_2 reactivity during hypocapnia in normal subjects. J Altern Complement Med 17(3):219–224

Cao S, Chen T, Chao D, Wen G, Balboni G, Xia Y (2014) δ-opioid receptor activation increases Nrf2 nuclear translocation and stabilizes mitochondrial membrane potential of hypoxic cells. Circ Res 115(Suppl 1):A147

Cao S, Chao D, Zhou H, Balboni G, Xia Y (2015) A novel mechanism for cytoprotection against hypoxic injury: δ-opioid receptor-mediated increase in Nrf2 translocation. Br J Pharmacol 172(7):1869–1881

Chai Y, Zhang R, Xue C, Liu Y, Wang X (2014) Stroke-induced acroparalysis treated with xunjing-cuiqi needling technique: a randomized controlled trial. Zhongguo Zhen Jiu 34(6):534–538, in Chinese with English abstract

Chang L, He P, Zhou Z, Li Y (2014) Efficacy observation of dysphagia after acute stroke treated with acupuncture and functional electric stimulation. Zhongguo Zhen Jiu 34(8):737–740, in Chinese with English abstract

Chao D, Bazzy-Asaad A, Balboni G, Xia Y (2007) δ-, but not μ-, opioid receptor stabilizes K^+ homeostasis by reducing Ca^{++} influx in the cortex during acute hypoxia. J Cell Physiol 212(1):60–67

Chao D, Bazzy-Asaad A, Balboni G, Salvadori S, Xia Y (2008) Activation of DOR attenuates anoxic K^+ derangement via inhibition of Na^+ entry in mouse cortex. Cereb Cortex 18(9):2217–2227

Chao D, Balboni G, Lazarus LH, Salvadori S, Xia Y (2009) Na^+ mechanism of δ-opioid receptor induced protection from anoxic K^+ leakage in the cortex. Cell Mol Life Sci 66(6):1105–1115.

Chao D, Donnelly DF, Feng Y, Bazzy-Asaad A, Xia Y (2006) Cortical δ-opioid receptors potentiate K^+ homeostasis during anoxia and oxygen–glucose deprivation. J Cereb Blood Flow Metab 27(2):356–368

Chao D, Shen X, and Xia Y (2013) From acupuncture to interaction between-opioid receptors and Na. Evid Based Complement Altern Med 2013:216016 (17 pages)
Chao D, Xia Y (2010) Ionic storm in hypoxic/ischemic stress: can opioid receptors subside it? Prog Neurobiol 90(4):439–470
Chen L, Fang J, Wu Y, Ma R, Xu S, Shen L, Luo K, Gao F, Bao Y, Ni K (2014a) Motor dysfunction in stroke of subacute stage treated with acupuncture: multi-central randomized controlled study. Zhongguo Zhen Jiu 34(4):313–318, in Chinese with English abstract
Chen T, Li J, Chao D, Sandhu HK, Liao X, Zhao J, Wen G, Xia Y (2014b) δ-Opioid receptor activation reduces α-synuclein overexpression and oligomer formation induced by MPP^+ and/or hypoxia. Exp Neurol 255:127–136
Chen YH, Yang HY, Lin CH, Dun NJ, Lin JG (2013) Electroacupuncture attenuates 5-guanidinonaltrindole-evoked scratching and spinal c-Fos expression in the mouse. Evid Based Complement Alternat Med 2013:319124, 12 pages
Cheng CH, Yi PL, Chang HH, Tsai YF and Chang FC (2012) Kappa-opioid receptors in the caudal nucleus tractus solitarius mediate 100 Hz electroacupuncture-induced sleep activities in rats. Evid Based Complement Altern Med 2012:715024 (10 pages)
Cheng F, Chen T (2014) Efficacy observation of post-stroke dysphagia treated with acupuncture at Lianquan (CV 23). Zhongguo Zhen Jiu 34(7):627–630, in Chinese with English abstract
Cheng LL, Ding MX, Wei J, Wu YQ, Qiu ZY, Chen JG, Liu DM, Hu CM, Hu ML, Shah Z (2013) Electroacupuncture-induced dynamic processes of gene expression levels of endogenous opioid peptide precursors and opioid receptors in the CNS of goats. Evid Based Complement Altern Med 2013:257682 (10 pages)
da Silva JRT, da Silva ML and Prado WA (2013) Electroacupuncture at 2/100 Hz activates antinociceptive spinal mechanisms different from those activated by electroacupuncture at 2 and 100 Hz in responder rats. Evid Based Complement Altern Med 2013:205316 (14 pages)
Duan YL, Wang SY, Zeng QW, Su DS, Li W, Wang XR, Zhao Z (2011) Astroglial reaction to delta opioid peptide [D-Ala2, D-Leu5] enkephalin confers neuroprotection against global ischemia in the adult rat hippocampus. Neuroscience 192:81–90
Feng R, Zhang F (2014) The neuroprotective effect of electro-acupuncture against ischemic stroke in animal model: a review. Afr J Tradit Complement Altern Med 11(3):25–29
Feng Y, Johansson J, Shao R, Holm LM, Billig H, Stener-Victorin E (2012) Electrical and manual acupuncture stimulation affect oestrous cyclicity and neuroendocrine function in an 5α-dihydrotestosterone-induced rat polycystic ovary syndrome model. Exp Physiol 97(5):651–662
Gao H, Li X, Gao X, Ma B (2013) Contralateral needling at unblocked collaterals for hemiplegia following acute ischemic stroke. Neural Regen Res 8(31):2914–2922
Ghosh N, Ghosh R, Bhat ZA, Mandal V, Bachar SC, Nima ND, Sunday OO, Mandal SC (2014) Advances in herbal medicine for treatment of ischemic brain injury. Nat Prod Commun 9(7):1045–1055
Gidday JM (2006) Cerebral preconditioning and ischaemic tolerance. Nat Rev Neurosci 7(6):437–448
Giffard RG, Swanson RA (2005) Ischemia–induced programmed cell death in astrocytes. Glia 50(4):299–306
Gosman-Hedström G, Claesson L, Klingenstierna U, Carlsson J, Olausson B, Frizell M, Fagerberg B, Blomstrand C (1998) Effects of acupuncture treatment on daily life activities and quality of life a controlled, prospective, and randomized study of acute stroke patients. Stroke 29(10):2100–2108
Guo JC, Cheng JS, Xia Y (2010) Acupuncture therapy for stroke. In: Xia Y, Cao X, Wu GC, Cheng J (eds) Acupuncture therapy for neurological diseases: a neurobiological view. Springer, Heidelberg, pp 226–262
Han J-S (2003) Acupuncture: neuropeptide release produced by electrical stimulation of different frequencies. Trends Neurosci 26(1):17–22
Han J-S (2004) Acupuncture and endorphins. Neurosci Lett 361(1):258–261

Han J, Chen X, Sun S, Xu X, Yuan Y, Yan S, Hao J, Terenius L (1991) Effect of low-and high-frequency TENS on Met-enkephalin-Arg-Phe and dynorphin A immunoreactivity in human lumbar CSF. Pain 47(3):295–298

He X, Sandhu HK, Yang Y, Hua F, Belser N, Kim DH, Xia Y (2013) Neuroprotection against hypoxia/ischemia: delta-opioid receptor-mediated cellular/molecular events. Cell Mol Life Sci 70(13):2291–2303

Hu HH, Chung C, Liu TJ, Chen RC, Chen CH, Chou P, Huang WS, Lin JC, Tsuei J (1993) A randomized controlled trial on the treatment for acute partial ischemic stroke with acupuncture. Neuroepidemiology 12(2):106–113

Huang SW, Wang WT, Yang TH, Liou TH, Chen GY, Lin LF (2014) The balance effect of acupuncture therapy among stroke patients. J Altern Complement Med 20(8):618–622

Husain S, Abdul Y, Singh S, Ahmad A, Husain M (2014) Regulation of nitric oxide production by δ-opioid receptors during glaucomatous injury. PLoS One 9(10), e110397

Husain S, Liou GI, Crosson CE (2011) Opioid receptor activation: suppression of ischemia/reperfusion-induced production of TNF-α in the retina. Invest Ophthalmol Vis Sci 52(5): 2577–2583

Jin Z, Liang J, Wang J, Kolattukudy PE (2013) Delayed brain ischemia tolerance induced by electroacupuncture pretreatment is mediated via MCP-induced protein 1. J Neuroinflammation 10:63

Johansson BB, Haker E, von Arbin M, Britton M, Långström G, Terént A, Ursing D, Asplund K (2001) Acupuncture and transcutaneous nerve stimulation in stroke rehabilitation a randomized, controlled trial. Stroke 32(3):707–713

Johansson K, Lindgren I, Widner H, Wiklund I, Johansson B (1993) Can sensory stimulation improve the functional outcome in stroke patients? Neurology 43(11):2189–2192

Kang X, Chao D, Gu Q, Ding G, Wang Y, Balboni G, Lazarus LH, Xia Y (2009) δ-Opioid receptors protect from anoxic disruption of Na^+ homeostasis via Na^+ channel regulation. Cell Mol Life Sci 66(21):3505–3516

Kim W, Kim SK, Min B-I (2013) Mechanisms of electroacupuncture-induced analgesia on neuropathic pain in animal model. Evid Based Complement Altern Med 2013:436913 (11 pages)

Kim YR, Kim HN, Ahn SM, Choi YH, Shin HK, Choi BT (2014) Electroacupuncture promotes post-stroke functional recovery via enhancing endogenous neurogenesis in mouse focal cerebral ischemia. PLoS One 9(2):e90000

Kuo C-C, Tsai H-Y, Lin J-G, Su H-L, Chen Y-F (2013) Spinal serotonergic and opioid receptors are involved in electroacupuncture-induced antinociception at different frequencies on ZuSanLi (ST 36) acupoint. Evid Based Complement Altern Med 2013:291972 (10 pages)

Li L, Tian G, Meng Z, Fan X, Zhang C, Shi X (2014) Different stimulation intensities of acupuncture at Hegu (LI 4) for central facial nerve paralysis after ischemic stroke: a randomized controlled trial. Zhongguo Zhen Jiu 34(7):669–674, in Chinese with English abstract

Li W, Zhong M, Yang J, Zhao W (2011) Effects of electroacupuncture preconditioning at "Neiguan" (PC 6) on gene expression of myocardial opioid receptors in rats with myocardial ischemia-reperfusion injury. Zhongguo Zhen Jiu 31(5):441–445

Liang J, Chao D, Sandhu HK, Yu Y, Zhang L, Balboni G, Kim DH, Xia Y (2014) δ-Opioid receptors up-regulate excitatory amino acid transporters in mouse astrocytes. Br J Pharmacol 171(23):5417–5430

Liang JF, Xia Y (2012) Acupuncture modulation of neural transmitters/modulators. In: Xia Y et al (eds) Current research in acupuncture. Springer, New York, pp 1–36

Lin XM, Tan KP, Zhang AJ, He LH (2009) Effect of electroacupuncture on the permeability of blood-brain barrier for nerve growth factor and its relevance to PKC pathway in cerebral ischemia rats. Zhen Ci Yan Jiu 34(2):110–113, in Chinese with English abstract

Liu Y, Wang H, Wang X, Mu L, Kong Q, Wang D, Wang J, Zhang Y, Yang J, Zhou M (2013) The mechanism of effective electroacupuncture on T cell response in rats with experimental autoimmune encephalomyelitis. PLoS One 8(1):e51573

Liu F, Li Z-M, Jiang Y-J, Chen L-D (2014) A meta-analysis of acupuncture use in the treatment of cognitive impairment after stroke. J Altern Complement Med 20(7):535–544

Ma MC, Qian H, Ghassemi F, Zhao P, Xia Y (2005) Oxygen-sensitive δ-opioid receptor-regulated survival and death signals novel insights into neuronal preconditioning and protection. J Biol Chem 280(16):16208–16218

Man SC, Hung BH, Ng RM, Yu XC, Cheung H, Fung MP, Li LS, Leung KP, Leung KP, Tsang KW (2014) A pilot controlled trial of a combination of dense cranial electroacupuncture stimulation and body acupuncture for post-stroke depression. BMC Complement Altern Med 14(1):255

McPherson BC, Yao Z (2001) Signal transduction of opioid-induced cardioprotection in ischemia-reperfusion. Anesthesiology 94(6):1082–1088

Moon HJ, Lim BS, Lee DI, Ye MS, Lee G, Min BI, Bae H, Na HS, Kim SK (2014) Effects of electroacupuncture on oxaliplatin-induced neuropathic cold hypersensitivity in rats. J Physiol Sci 64(2):151–156

Myles PS, Daly D, Silvers A, Cairo S (2009) Prediction of neurological outcome using bispectral index monitoring in patients with severe ischemic-hypoxic brain injury undergoing emergency surgery. Anesthesiology 110(5):1106–1115

Naeser MA (1997) Neurological rehabilitation: acupuncture and laser acupuncture to treat paralysis in stroke, other paralytic conditions, and pain in carpal tunnel syndrome. J Altern Complement Med 3(4):425–428

Naeser MA, Alexander MP, Stiassny-Eder D, Galler V, Hobbs J, Bachman D (1992) Real versus sham acupuncture in the treatment of paralysis in acute stroke patients: a CT scan lesion site study. Neurorehabil Neural Repair 6(4):163–174

Naeser M, Alexander M, Stiassny-Eder D, Galler V, Hobbs J, Bachman D (1993) Acupuncture in the treatment of paralysis in chronic and acute stroke patients–improvement correlated with specific CT scan lesion sites. Acupunct Electrother Res 19(4):227–249

Narita M, Kuzumaki N, Miyatake M, Sato F, Wachi H, Seyama Y, Suzuki T (2006) Role of δ-opioid receptor function in neurogenesis and neuroprotection. J Neurochem 97(5):1494–1505

Olianas MC, Dedoni S, Onali P (2011) Regulation of PI3K/Akt signaling by N-desmethylclozapine through activation of δ-opioid receptor. Eur J Pharmacol 660(2):341–350

Ouyang YB, Voloboueva LA, Xu LJ, Giffard RG (2007) Selective dysfunction of hippocampal CA1 astrocytes contributes to delayed neuronal damage after transient forebrain ischemia. J Neurosci 27(16):4253–4260

Park S-W, Yi S-H, Lee JA, Hwang PW, Yoo HC, Kang KS (2014) Acupuncture for the treatment of spasticity after stroke: a meta-analysis of randomized controlled trials. J Altern Complement Med 20(9):672–682

Peng Y, Wang H, Sun J, Chen L, Xu M, Chu J (2012) Electroacupuncture reduces injury to the blood-brain barrier following cerebral ischemia/ reperfusion injury. Neural Regen Res 7(36):2901–2906

Rana A, Goyal N, Ahlawat A, Jamwal S, Reddy BVK, Sharma S (2014) Mechanisms involved in attenuated cardio-protective role of ischemic preconditioning in metabolic disorders. Perfusion 30:94–105

Ren X, Guo M, Zhao Y, Ding X, Li C, Ji B, Zhu J, Zhang L (2012) Effects of electroacupuncture on pain reactions, expression of spinal kappa-opioid receptor and contents of enkephalin and beta-endorphin in periaqueductal gray of midbrain in dysmenorrhea model rats. Zhen Ci Yan Jiu 37(1):1–7, in Chinese with English abstract

Rosenstein RE, Fernandez DC (2014) Induction of ischemic tolerance as a promising treatment against diabetic retinopathy. Neural Regen Res 9(17):1581

Rossi DJ, Brady JD, Mohr C (2007) Astrocyte metabolism and signaling during brain ischemia. Nat Neurosci 10(11):1377–1386

Seo BK, Park DS, Baek YH (2013) The analgesic effect of electroacupuncture on inflammatory pain in the rat model of collagenase-induced arthritis: mediation by opioidergic receptors. Rheumatol Int 33(5):1177–1183

She R, Ge C (2014) Clinical observation on medulla oblongata palsy after brainstem infarction treated with electroacupuncture at eight-neck-occiput points. Zhongguo Zhen Jiu 34(6):539–542, in Chinese with English abstract

Silva ML, Silva JR, Prado WA (2013) Analgesia induced by 2-or 100-Hz electroacupuncture in the rat tail-flick test depends on the anterior pretectal nucleus. Life Sci 93(20):742–754

Siu FK, Lo SC, Leung MC (2004) Electroacupuncture reduces the extent of lipid peroxidation by increasing superoxide dismutase and glutathione peroxidase activities in ischemic-reperfused rat brains. Neurosci Lett 354(2):158–162

Sjölund B, Terenius L, Eriksson M (1977) Increased cerebrospinal fluid levels of endorphins after electro-acupuncture. Acta Physiol Scand 100(3):382–384

Stevens CW (2009) The evolution of vertebrate opioid receptors. Front Biosci (Landmark Ed) 14:1247

Su T, Zhang L, Peng M, Wu C, Pan W, Tian B, Shi J, Pan H-l, Li M (2011) Cannabinoid CB2 receptors contribute to upregulation of b-endorphin in inflamed skin tissues by electroacupuncture. Mol Pain 7:98

Szydlowska K, Gozdz A, Dabrowski M, Zawadzka M, Kaminska B (2010) Prolonged activation of ERK triggers glutamate-induced apoptosis of astrocytes: neuroprotective effect of FK506. J Neurochem 113(4):904–918

Takano T, Oberheim N, Cotrina ML, Nedergaard M (2009) Astrocytes and ischemic injury. Stroke 40(3 suppl 1):S8–S12

Takuma K, Baba A, Matsuda T (2004) Astrocyte apoptosis: implications for neuroprotection. Prog Neurobiol 72(2):111–127

Tang CW, Feng WM, Du HM, Bao Y, Zhu M (2011) Delayed administration of D-Ala2-D-Leu5-enkephalin, a delta-opioid receptor agonist, improves survival in a rat model of sepsis. Tohoku J Exp Med 224(1):69–76

Tao J, Fang Y, Wu Z, Rao T, Su Y, Lin L, Liu W, Wu J, Yang S, Zheng G (2014) Community-applied research of a traditional Chinese medicine rehabilitation scheme on Broca's aphasia after stroke: study protocol for a randomized controlled trial. Trials 15(1):290

Tian X, Guo J, Zhu M, Li M, Wu G, Xia Y (2013a) δ-Opioid receptor activation rescues the functional TrkB receptor and protects the brain from ischemia-reperfusion injury in the rat. PLoS One 8(7):e69252

Tian X, Hua F, Sandhu HK, Chao D, Balboni G, Salvadori S, He X, Xia Y (2013b) Effect of delta-opioid receptor activation on BDNF-TrkB vs. TNF-alpha in the mouse cortex exposed to prolonged hypoxia. Int J Mol Sci 14(8):15959–15976

Tian X, Zhou F, Yang R, Xia Y, Wu G, Guo J (2008a) Role of δ-opioid receptors in cumulative electro-acupuncture induced protection from ischemic injury in the rat brain. Shanghai J TCM 42:71–74

Tian XS, Zhou F, Yang R, Xia Y, Wu GC, Guo JC (2008b) Electro-acupuncture protects the brain against ischemic injury via up-regulation of deltaopioid receptors. Chin J Integr Med 6(6):632–638

Uchida S, Kagitani F (2014) Effect of acupuncture-like stimulation on cortical cerebral blood flow in aged rats. J Physiol Sci 2014:1–9

Wang BH, Lin CL, Li TM, Lin SD, Lin JG, Chou LW (2014a) Selection of acupoints for managing upper-extremity spasticity in chronic stroke patients. Clin Interv Aging 9:147–156

Wang Q, Chao D, Chen T, Sandhu H, Xia Y (2014b) δ-Opioid receptors and inflammatory cytokines in hypoxia: differential regulation between glial and neuron-like cells. Transl Stroke Res 5(4):476–483

Wang Q, Xiong L, Chen S, Liu Y, Zhu X (2005) Rapid tolerance to focal cerebral ischemia in rats is induced by preconditioning with electroacupuncture: window of protection and the role of adenosine. Neurosci Lett 381(1–2):158–162

Wang T (2013) Endogenous opiates in the parabrachial nucleus mediate the electroacupuncture–induce sleep activities in rats. Sleep Med 14:e301

Wen GQ, He XZ, Lu Y, Xia Y (2010a) Effect of acupuncture on neurotransmitters/modulators. In: Xia Y, Cao X, Wu GC, Cheng J (eds) Acupuncture therapy for neurological diseases: a neurobiological view. Springer, Heidelberg, pp 120–142

Wen GQ, Yang YL, Lu Y, Xia Y (2010b) Acupuncture-induced activation of endogenous opioid system. In: Xia Y, Cao X, Wu GC, Cheng J (eds) Acupuncture therapy for neurological diseases. Springer, New York, pp 104–119

Wu XD, Du LN, Wu GC, Cao XD (2001) Effects of electroacupuncture on blood-brain barrier after cerebral ischemia-reperfusion in rat. Acupunct Electrother Res 26(1–2):1–9
Xia Y, Cao XD, Wu GC, Cheng JS (eds) (2010) Acupuncture therapy for neurological diseases: a neurobiological view. Springer, Heidelberg, pp 1–480
Xia Y, Ding GH, Wu GC (eds) (2012) Current research in acupuncture. Springer, New York, pp 1–752
Xia Y, Haddad GG (2001) Major difference in the expression of δ- and μ-opioid receptors between turtle and rat brain. J Comp Neurol 436(2):202–210
Xiang XH, Chen YM, Zhang JM, Tian JH, Han JS, Cui CL (2014) Low- and high-frequency transcutaneous electrical acupoint stimulation induces different effects on cerebral μ-opioid receptor availability in rhesus monkeys. J Neurosci Res 92(5):555–563
Xiong L, Yang J, Wang Q, Liu Y (2004) Cerebral ischemic tolerance induced by repeated electroacupuncture involved delta-opioid receptor and enkephalin in rats. Anesthesiology 101:A767
Xiong LZ, Yang J, Wang Q, Lu ZH (2007) Involvement of delta-and mu-opioid receptors in the delayed cerebral ischemic tolerance induced by repeated electroacupuncture preconditioning in rats. Chin Med J (Engl) 120(5):394–399
Xu H, Zhang Y, Sun H, Chen S, Wang F (2014) Effects of acupuncture at GV20 and ST36 on the expression of matrix metalloproteinase 2, aquaporin 4, and aquaporin 9 in rats subjected to cerebral ischemia/reperfusion injury. PLoS One 9(5), e97488
Yamada K, Nabeshima T (2003) Brain-derived neurotrophic factor/TrkB signaling in memory processes. J Pharmacol Sci 91(4):267–270
Yang J, Xiong L, Wang Q (2006) Role of opioid receptors in cerebral ischemic tolerance induced by electroacupuncture preconditioning in rats. Chin J Multi Organ Dis Elder 5(3):218–291
Yang J, Liu J, Gu H, Lv W, Zhao H, Li G (2014a) Randomized controlled trials of acupuncture and moxibustion for post-stroke constipation: a meta analysis. Zhongguo Zhen Jiu 34(8):833–836, in Chinese with English abstract
Yang S, Ye H, Huang J, Tao J, Jiang C, Lin Z, Zheng G, Chen L (2014b) The synergistic effect of acupuncture and computer-based cognitive training on post-stroke cognitive dysfunction: a study protocol for a randomized controlled trial of 2 × 2 factorial design. BMC Complement Altern Med 14(1):290
Yang Y, Xia X, Zhang Y, Wang Q, Li L, Luo G, Xia Y (2009) δ-Opioid receptor activation attenuates oxidative injury in the ischemic rat brain. BMC Biol 7(1):55
Yang Z, Shen D, Guo X, Sun F (2004) Electroacupuncture enhances striatal neurogenesis in adult rat brains after a transient cerebral middle artery occlusion. Acupunct Electrother Res 30(3–4):185–199
Yi PL, Lu CY, Cheng CH, Tsai YF, Lin CT, Chang FC (2013a) Activation of amygdala opioid receptors by electroacupuncture of Feng-Chi (GB20) acupoints exacerbates focal epilepsy. BMC Complement Altern Med 13(1):290
Yi PL, Lu CY, Cheng CH, Tsai YF, Lin CT, Chang FC (2013b) Amygdala opioid receptors mediate the electroacupuncture-induced deterioration of sleep disruptions in epilepsy rats. J Biomed Sci 20(1):85
Yu B, Lu GH, Sun Y, Lin X, Fang TH (2011) [Effect of electroacupuncture combined with intragastric administration of borneol on the permeability of blood-brain barrier in the mouse]. Zhen Ci Yan Jiu 36(5):335–340, in Chinese with English abstract
Zhang HY, McPherson BC, Liu H, Baman T, McPherson SS, Rock P, Yao Z (2002) Role of nitric-oxide synthase, free radicals, and protein kinase C δ in opioid-induced cardioprotection. J Pharmacol Exp Ther 301(3):1012–1019
Zhang J, Haddad GG, Xia Y (2000) δ-, But not μ-and κ-, opioid receptor activation protects neocortical neurons from glutamate-induced excitotoxic injury. Brain Res 885(2):143–153
Zhang H, Torregrossa MM, Jutkiewicz EM, Shi YG, Rice KC, Woods JH, Watson SJ, Ko H (2006a) Endogenous opioids upregulate brain-derived neurotrophic factor mRNA through δ- and μ-opioid receptors independent of antidepressant-like effects. Eur J Neurosci 23(4):984–994

Zhang J, Qian H, Zhao P, Hong SS, Xia Y (2006b) Rapid hypoxia preconditioning protects cortical neurons from glutamate toxicity through δ–opioid receptor. Stroke 37(4):1094–1099

Zhang J-h, Wang D, Liu M (2013) Overview of systematic reviews and meta-analyses of acupuncture for stroke. Neuroepidemiology 42(1):50–58

Zhang X, Wu B, Nie K, Jia Y, Yu J (2014) Effects of acupuncture on declined cerebral blood flow, impaired mitochondrial respiratory function and oxidative stress in multi-infarct dementia rats. Neurochem Int 65:23–29

Zhao P, Cheng J (1996) Effects of electroacupuncture on extracellular contents of amino acid neurotransmitters in rat striatum following transient focal cerebral ischemia. Acupunct Electrother Res 22(2):119–126

Zhao P, Guo J, Xia Y, Hong S, Bazzy-Asaad A, Cheng J, Xia Y (2002) Electro-acupuncture and brain protection from cerebral ischemia: the role of delta-opioid receptor. Soc Neurosci Abstr 28:736

Zhao Y, Chen X, Ma L, Zuo Z, Zhu Z, Zhu X, Wang Q, He E, Xiong L, Pei J, Xu L, Hou L, Chen S (2012) Electroacupuncture pretreatment induces tolerance against focal cerebral ischemia through activation of canonical Notch pathway. BMC Neurosci 13:111

Zhong S, Li Z, Wang C-X, Chen B-Y (2006) Protective effect of electroacupuncture on mitochondrial function in rats with temporal cerebral ischemia. Acupunct Res 31(6):337–341

Zhou F, Guo J, Cheng J, Wu G, Xia Y (2007) Electro-acupuncture induced protection from cerebral ischemia is dependent on stimulation intensity, and frequency. In: Proceedings of the third Cell Stress Society International Congress on Stress Responses in Biology and Medicine and second World Conference of Stress, Budapet, p 250

Zhou F, Huang DK, Xia Y (2010) Neuroanatomic basis of acupuncture points. In: Xia Y, Cao X, Wu GC, Cheng J (eds) Acupuncture therapy for neurological diseases: a neurobiological view. Springer, Heidelberg, pp 32–80

Zhou F, Guo J, Cheng J, Wu G, Xia Y (2011a) Electroacupuncture increased cerebral blood flow and reduced ischemic brain injury: dependence on stimulation intensity and frequency. J Appl Physiol 111(6):1877–1887

Zhou HF, Wang J, Cao TJ, Ju QB, Huang CY, Feng Y, Li YM, Li XN, Qu F, Dou WB (2011b) Effects of eye acupuncture on SEPCT-determined cerebral blood flow in patients with cerebral infarction. Zhongguo Zhen Jiu 31(5):391–394, in Chinese with English abstract

Zhou F, Guo J, Cheng J, Wu G, Sun J, Xia Y (2013a) Electroacupuncture and brain protection against cerebral ischemia: specific effects of acupoints. Evid Based Complement Alternat Med 2013:804397, 14 pages

Zhou F, Guo J, Cheng J, Wu G, Xia Y (2013b) Effect of electroacupuncture on rat ischemic brain injury: importance of stimulation duration. Evid Based Complement Alternat Med 2013:878521

Zhou L, Guo X, Chen M, Fu S, Zhou J, Ren G, Yang Z, Fan W (2013c) Inhibition of δ-opioid receptors induces brain glioma cell apoptosis through the mitochondrial and protein kinase C pathways. Oncol Lett 6(5):1351–1357

Zhu H, Fan G, Zhao Y (2014) Status que of acupuncture for strephenopodia after stroke. Zhongguo Zhen Jiu 34(3):309–312, in Chinese with English abstract

Zhu M, Li M, Yang F, Ou X, Ren Q, Gao H, Zhu C, Guo J (2011) Mitochondrial ERK plays a key role in delta-opioid receptor neuroprotection against acute mitochondrial dysfunction. Neurochem Int 59(6):739–748

Zhuang L, He J, Zhuang X, Lu L (2014) Quality of reporting on randomized controlled trials of acupuncture for stroke rehabilitation. BMC Complement Altern Med 14(1):151

Chapter 12
Delta Opioid Receptor in Cerebral Preconditioning

Jason Eli Maas and Zhiyi Zuo

Abstract Cerebral infarction is a leading cause of morbidity and mortality in the United States. Activation of the delta-opioid receptor has been shown to decrease damage in many models of cerebral protection. In this chapter we gather the work that has been done on the topic, including work to establish the efficacy of the delta-opioid receptor in protection in animals, and the work done to elucidate potential mechanisms. However, most of the previous work was performed in rodents. Testing whether activation of the delta-opioid receptor can induce neuroprotection in large animals and then in humans is needed to fully establish its usefulness in clinical practice.

Keywords Delta opioid receptor • Cerebral preconditioning • Ischemic preconditioning • Hypoxic preconditioning • Superoxide • Ion homeostasis

Abbreviations

AMPA	α-Amino-3-hydroxy-5-methyl-4-isoxazolepropionic acid
ATP	Adenosine triphosphate
Bcl2	B-cell lymphoma-2
BNTX	7-Benzylidenenaltrexone
DADLE	[D-Ala^2, D-Leu^5]-enkephalinamide
DOR	Delta opioid receptor
ERK	Extracellular signal-regulated kinases
GABA	Gamma-aminobutyric acid
HIF-1 α	Hypoxia-inducible factor-1α
HPC	Hypoxic preconditioning
$mitoK_{ATP}$	Mitochondrial ATP-activated potassium channel
NMDA	N-methyl-D-aspartate
OGD	Oxygen-glucose deprivation
PKC	Protein kinase C
Trk	Tropomyosine receptor kinase

J.E. Maas, M.D., Ph.D. (✉) • Z. Zuo, M.D., Ph.D.
Department of Anesthesiology, University of Virginia, Charlottesville, VA 22903, USA
e-mail: jmaas@virginia.edu; zz3c@virginia.edu

Y. Xia (ed.), *Neural Functions of the Delta-Opioid Receptor*,
DOI 10.1007/978-3-319-25495-1_12

1 Introduction

Ischemia refers to a phenomenon of interruption of blood supply to an organ or tissue. Cerebral ischemia is termed stroke. Strokes are medical emergencies. Prolonged cerebral ischemia can progress to cell death and brain tissue infarction, and may result in death. In 2009, stroke accounted for one in every 19 deaths, approximately one death every 4 min in the United States, and is currently the fourth leading cause of death behind heart disease, cancer, and chronic respiratory disease. For those who survive their stroke, permanent impairment may ensue. Among ischemic stroke survivors over 65 years of age at 6 months after their stroke, 50 % had some hemiparesis, 30 % were unable to walk without assistance, 46 % had cognitive deficits, 35 % had depressive symptoms, 19 % had aphasia, 26 % were dependent in their activities of daily living, and 26 % were institutionalized in a nursing home (Go et al. 2013).

The central goal in acute ischemic stroke is to prevent permanent injury. This can be performed by many mechanisms, including limiting the duration of the ischemic insult by restoring blood flow to the compromised area and limiting the severity of the insult by cellular protection (Bhuiyan and Kim 2010). While efforts to limit the duration of ischemia are primarily directed at public awareness initiatives and revascularization techniques, such as invasive neurovascular interventions and pharmacological thrombolysis, methods to protect brain cells (neuroprotection) are primarily pharmacologic in nature and many of them are developed from the discovery and characterization of ischemic preconditioning and the subsequent pharmacological preconditioning.

Ischemic preconditioning in general refers to the process whereby sublethal stress induces a protective response to subsequent lethal stress. The preconditioning phenomenon has been described in a variety of organ systems including the heart, liver, intestines, lung, skeletal muscle, kidney, and bladder (Murry et al. 1986; Kume et al. 1996; Toosy et al. 1999; Pasupathy and Homer-Vanniasinkam 2005), however it was in the brain that the phenomenon was first reported in 1964 by Dahl and Balfour (1964). Since that time, the ability to precondition the brain has been confirmed in various models including global (Kitagawa et al. 1991) and focal (Stagliano et al. 1999) ischemia, in vitro brain slices (Xu et al. 2002), primary neuron cultures (Bruer et al. 1997), and in humans in the form of transient ischemic attacks (Moncayo et al. 2000).

In parallel to the attempt to elucidate the mechanism of ischemic preconditioning, multiple pharmacologic agents, such as inflammatory cytokines (Ginis et al. 2002), anesthetics (Kapinya et al. 2002), and metabolic inhibitors (Weih et al. 1999), have been found also to induce a preconditioned state. Multiple intracellular and extracellular messengers have been found to play a vital role in the ischemic preconditioning phenomenon. Like in cardiac preconditioning, adenosine, adenosine A1 receptors, and ATP-sensitive potassium channels may play a role in neuronal preconditioning (Heurteaux et al. 1995). N-methyl-D-aspartate (NMDA) (Grabb and Choi 1999; Marini et al. 2008), α-Amino-3-hydroxy-5-methyl-4-isoxazolepropionic acid (AMPA) (Sommer and Kiessling 2002; Tanaka et al. 2002), and gamma-aminobutyric acid (GABA) (Schwartz-Bloom and Sah 2001; Sommer et al. 2003) receptors may participate in the mechanisms of ischemic preconditioning, however recent evidence has brought the δ opioid receptor (DOR) to the forefront and an increasing number of papers have found evidence for its vital role in cerebral preconditioning.

2 DOR in Cerebral Protection

2.1 DOR in Hypoxic Preconditioning

In isolated cell cultures two models of ischemia-like insults are primarily utilized, hypoxia and oxygen-glucose deprivation (OGD). As such, ischemic preconditioning is typically referred to as hypoxic preconditioning (HPC) because it is brought about by hypoxia and not by ischemia.

In 2005, Ma, et al. published their early findings showing that HPC induced neuronal protection could be blocked by the DOR-specific antagonist, naltrindole (Ma et al. 2005) (Fig. 12.1). Furthermore, while severe hypoxia decreased DOR expression and reduced levels of the endogenous DOR peptide leucine enkephalin, HPC induced an increase in DOR mRNA and protein levels and reversed the reduction in leucine enkephalin. DOR specific antagonism with naltrindole completely abolished the HPC effect. This DOR-mediated HPC was also shown to depend on extracellular signal-regulated kinases (ERK) (Fig. 12.2) and B-cell lymphoma-2 (Bcl2) activity, counteracting severe hypoxia-induced p38 mitogen-activated protein kinase activity and cytochrome c release from the mitochondria.

HPC is completely abolished by DOR antagonism, but the exact mechanism required further elucidation. Work by Peng et al. suggests a role for hypoxia-inducible factor-1α (HIF-1α) in the HPC-induced ischemic tolerance. In their study, HPC increased HIF-1α, which then caused an increase in DOR expression in rat retinas (Peng et al. 2009). HPC also upregulates DOR expression in the rat brain via HIF-1α (Gao et al. 2012).

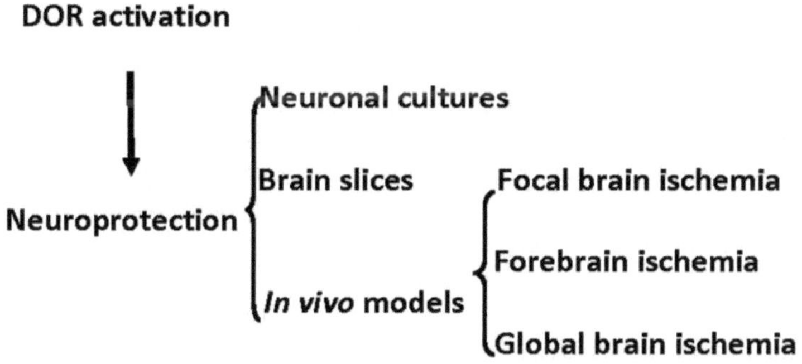

Fig. 12.1 Models used to show δ opioid receptor (DOR) activation-induced neuroprotection

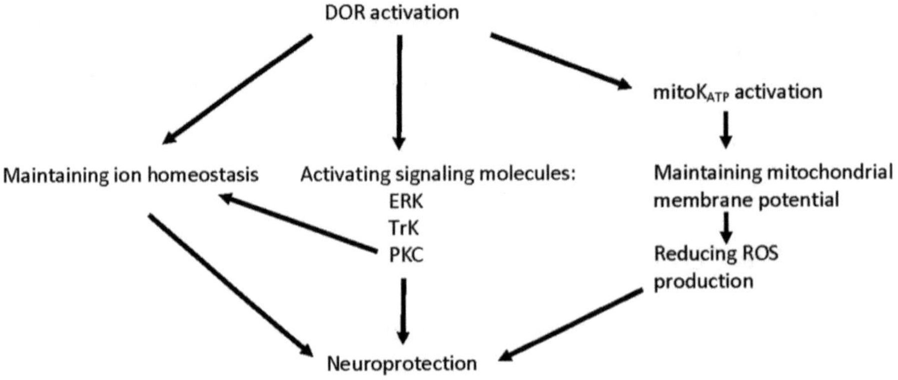

Fig. 12.2 Possible mechanisms for opioid receptor (DOR) activation-induced neuroprotection. *ERK* extracellular signal-regulated kinases, *mito$_{KATP}$* mitochondrial ATP-activated potassium channel, *PKC* protein kinase C, *ROS* reactive oxygen species, *Trk* tropomyosine receptor kinase

2.2 *Pharmacologic Neuroprotection with DOR Activation*

While there is limited evidence that the DOR is involved in the intrinsic neuroprotective mechanism of HPC, large amounts of data have shown that the DOR plays a crucial role in neuroprotection independent of HPC. Activation of the DOR alone is neuroprotective against both hypoxic and ischemic stress. In 2002, Zhang et al. published their findings showing that the DOR was protective during hypoxia (Zhang et al. 2002). In their *in vitro* study, cortical neurons were isolated in culture. Immature (day 4 and 6) and mature (day 8) neurons were subjected to a hypoxic environment (1 % O_2). Immature neurons were not significantly injured until 72-h exposure to hypoxic conditions, and mature neurons were injured after only 24 h. However, in both instances, activation of the DOR with [D-Ala^2, D-Leu^5]-enkephalinamide (DADLE) attenuated the amount of hypoxic injury. Stimulation of mu- and kappa-opioid receptors did not reduce hypoxic injury. Furthermore, this neuronal protection was inhibited by the DOR-specific inhibitor, naltrindole. In a similar study seeking to further elucidate the mechanism of this process, the selective DOR agonist (+)-4-[(α*R*)-α-((2*S*,5*R*)-4-allyl-2,5-dimethyl-1-piperazinyl)-3-methoxybenzyl]-*N*,*N*-diethylbenzamide prevented the increase in cleaved caspase 3-like immunoreactivity induced by H_2O_2 in cortical neuron/glia co-cultures (Narita et al. 2006). While this was clearly a different process, with peroxide stress rather than that of hypoxia, it is clear that activation of the DOR functions results in neuroprotection presented as an anti-apoptotic activity. In this study, the effect of the DOR was completely abolished by a tropomyosine receptor kinase (Trk)-dependent tyrosine kinase inhibitor, suggesting a role for the Trk-dependent tyrosine kinases in the signaling of this cascade (Narita et al. 2006).

In vivo studies have also suggested a role of the DOR in cerebral conditioning. Pre-administration of DADLE ([D-Ala^2, D-Leu^5]-enkephalinamide) to Sprague-Dawley rats 30 min before cerebral ischemia (bilateral carotid artery occlusion and hypotension) improved neuronal survival in the hippocampal CA3 and dentate gyrus but not in the CA1 and this effects was reversed by naltrindole (Iwata et al. 2007) (Fig. 12.1). Additionally, administration of naltrindole without DADLE significantly worsened neuronal injury in the CA1 region after mild ischemia, suggesting a role for an endogenous DOR ligand in cerebral protection. Additional studies have confirmed this finding with intracerebroventricular administration of DADLE leading to both a significant protective effect against CA1 neuronal loss and behavioral retardation revealed by the Morris water maze and motor score testing (Su et al. 2007).

In a subsequent study, Sprague Dawley rats were subjected to cardiac arrest induced by 8 min of asphyxiation followed by resuscitation (Gao et al. 2010) (Fig. 12.1). DOR protein expression was increased in the hippocampus 1 h after restoration of spontaneous circulation, and decreased from that point. Pre-administration of DADLE through intracerebroventricular administration at 30 min prior to cardiac arrest led to improved neurological functional recovery as assessed by the neurological deficit score and Morris water maze trials. DADLE also significantly increased the number of surviving hippocampal CA1 neurons, and was antagonized by the administration of naltrindole.

A potential role for the DOR in clinically used drugs-induced neuroprotection was suggested by Lim et al. (2004). In their study, cerebellar brain slices were preconditioned with morphine in the presence or absence of naloxone (nonspecific opioid receptor antagonist), β-funaltrexamine (mu-selective), nor-binaltorphimine (kappa-selective), naltrindole, 7-benzylidenenaltrexone (BNTX; delta1-selective), naltriben (delta2-selective), 5-hydroxydecanoate [mitochondrial ATP-activated potassium channel (mitoK_{ATP})-selective] or myxothiazol (mitochondrial electron transport inhibitor). The slices were then subjected to simulated ischemia (OGD) for 20 min and 5 h of recovery in an artificial cerebrospinal fluid with oxygen. It was found that Purkinje cells had greater survival when pre-treated with morphine, and that this increased survival was abolished by naloxone, naltrindole, and BNTX, but not by the selective mu-, kappa-, delta2-opioid receptor antagonists. These effects were partially blocked by 5-hydroxydecanoate or myxothiazol, suggesting a partial role for the mitoK_{ATP} channel and mitochondrial electron transport chain in the process (Lim et al. 2004). Later, delayed neuroprotection was shown in the same model and *in vivo* focal brain ischemia model through DOR agonism with morphine and Tan-67, a selective delta-1 opioid receptor agonist (Zhao et al. 2006).

While much work is required to bring DOR-agonists into clinical practice, the evidence is clear that DOR activation would provide neuroprotection.

3 Mechanism of DOR Involvement in Cerebral Preconditioning

3.1 DOR and Ion Homeostasis

Significant work has been contributed to elucidating the mechanism of the DOR in providing neuroprotection. One area that has been identified is ion homeostasis (Fig. 12.2). Cellular and extracellular ion concentrations are maintained in an extremely strict balance and in no organ is this more closely maintained than in the neuron and the cerebrospinal fluid. During hypoxia or OGD sustained changes in ionic concentrations, such as potassium efflux and/or sodium, calcium and chloride influx, can occur. The loss of these ionic gradients across the neuronal cell membranes is thought to mediate anoxia-induced depolarization, which can lead to neuronal cell death (Chao et al. 2006).

Steep increases in extracellular potassium have been associated with anoxia-induced depolarization (Hansen 1985). The blockade of potassium efflux has also been shown to attenuate hypoxia- and ischemia-induced neuronal cell death (Huang et al. 2001; Liu et al. 2003; Wei et al. 2003). Therefore, inhibition of potassium shifts has been hypothesized to be an important mediator in brain ischemic preconditioning. Interestingly, activation of the DOR by DADLE mimicked direct potassium blockade by preventing anoxia- and OGD-induced increases in extracellular potassium concentration and decreasing the direct current potential in cortical slices, the effect of which was blocked by the DOR-specific antagonist, naltrindole (Chao et al. 2006). Subsequent work further confirmed the role of the DOR by showing that activation of the mu-opioid receptor did not produce the same effect, and that the more specific and potent DOR agonist H-Dmt-Tic-NH-CH (CH_2-COOH)-Bid also significantly attenuated potassium derangement under anoxic and hypoxic conditions. Additionally, it was shown that the potassium stabilization occurred by reduction of calcium influx, leading to decreased calcium-activated potassium channel activation (Chao et al. 2007).

Later work to elucidate the mechanism for the potassium leakage targeted sodium entry into the mouse cortex. Substituting N-methyl-D-glucamine for sodium ions extracellularly in mouse cortical slices caused attenuation of anoxic potassium derangement as measured by potassium-sensitive microelectrodes. Permeable lithium ions potentiated the anoxic potassium derangements. However, the DOR-induced protection noted previously was abolished by low-sodium perfusion regardless of the substituted cation, leading to the conclusion that sodium was responsible for the potassium derangement (Chao et al. 2008). Likewise, the sodium influx was found to occur via tetrodotoxin-sensitive sodium channels, facilitated by veratridine (a sodium channel opener), and abolished by DOR activation. This appears to occur via protein kinase C (PKC, possibly subtypes PKCβII or PKCθ), as specific inhibitors were able to attenuate DOR-mediated inhibition of sodium influx during anoxia (Chao et al. 2012). Together, it appears that the DOR may inhibit sodium entry via PKCβII and PKCθ activation, thereby decreasing calcium entry and promoting potassium homeostasis during anoxia.

The role of the DOR in maintaining sodium ion homeostasis during anoxia was further confirmed in isolated mouse cortex subjected to anoxia (Kang et al. 2009). Activation of the DOR in this system reduced anoxic sodium influx in the cortex, an effect which was reversed by naltrindole. Furthermore, when sodium channels were co-expressed on xenopus oocytes with the DOR, activation of the DOR did inhibit sodium channel opening, decreasing the amplitude of their current and increasing the activation threshold when measured via two electrode voltage clamp technique.

3.2 DOR, Superoxide and Apoptosis

The DOR activation has been associated with decreasing pro-apoptotic factors. As mentioned earlier, severe hypoxia-induced p38 activation and cytochrome c release from the mitochondria were inhibited by DOR activation (Ma et al. 2005). Further work found that in a model of cerebral ischemia with middle cerebral artery occlusion, activity of the antioxidant enzymes, glutathione peroxidase and superoxide dismutase, were increased by DOR activation (Yang et al. 2009). Similarly, malondialdehyde and nitric oxide levels were reduced after ischemia reperfusion with DOR stimulation (Fig. 12.2). DOR activation also reduced caspase 3 expression but not interleukin-1β or tumor necrosis factor-α expression at the same timepoint. In another study, the DOR effect was investigated in the rat retina. In this study, while ischemia increased caspase 3 expression and superoxide production, and hypoxic-preconditioning reduced these increases, antagonism of the DOR was able to abrogate this response, suggesting a role for the DOR in HPC-reduced superoxide production (Peng et al. 2009).

The role of DOR signaling on the mitochondrial respiratory chain was also investigated at length (Zhu et al. 2009). Cultured cortical neurons were subjected to inhibition of the mitochondrial respiratory chain on day 8 of culture with sodium azide, and neuronal injury was assessed by quantification of the release of lactate dehydrogenase. It was found that DOR activation but not mu-opioid receptor activation decreased azide induced neuronal damage and that the decrease was abolished by naltrindole. Treatment of neurons in culture with naltrindole alone increased the neuronal damage by azide. Azide downregulates DOR expression and the downregulation of the DOR was attenuated by DOR activation with DADLE. Taken together, native DOR stimulation is important to inhibit the neuronal damage with interruption of the mitochondrial respiratory chain, but further stimulation by exogenous agonists can maintain its expression and function (Zhu et al. 2009).

3.3 Intracellular Signaling Associated with DOR-Mediated Cerebral Protection

The intracellular mechanism for cerebral protection via the DOR has not been clearly established. Hwever, multiple participating members have been elucidated (Fig. 12.2). Antagonists for Trk-dependent tyrosine kinases and ERK/Bcl2 signaling have been implicated (Ma et al. 2005; Narita et al. 2006; Tian et al. 2013). Work in cardioprotection also confirmed the role of ERK upon DOR activation (Ikeda et al. 2006). As discussed above, the effect of DOR activation to inhibit sodium entry appears via activation of PKCβII and PKCθ (Chao et al. 2012) (Fig. 12.2). Despite these and other advances, much work remains for elucidating the exact mechanism whereby the DOR mediates cerebral protection.

4 Concluding Remarks

While much work has been done for DOR activation-induced neuroprotection, many critical issues need to be addressed. Clear promise awaits as *in vivo* studies have revealed positive findings, leading to the belief that a clear clinical use is within reach. Nevertheless, there is a lack of data from large animals including non-human primates to demonstrate this protection. Clinical trials to test DOR activation-induced neuroprotection may then occur. Also, it would be difficult for patients to take a DOR-specific antagonist 30 min prior to having a stroke. However, there are clear instances in clinical situations, where the probability of having a stroke increases, that DOR stimulation would be beneficial. The perioperative arena is one of those moments. Patients are at increased risk of stroke, most likely hypoxic or ischemic events, during high risk surgeries, such as cardiac, liver and brain surgery. Vascular surgeries place the patient at risk of embolic stroke. Could a DOR-specific agonist be administered intraoperatively to patients just prior to going on bypass, entering the artery or dropping the patient's blood pressure? Despite the need for additional basic science and clinical research, the DOR is a clear favorite in the drug-development arena for neuroprotection.

Acknowledgements We would like to thank the National Institutes of Health, American Heart Association-Mid-Atlantic Affiliate and International Anesthesia Research Society for the grants supporting research in Dr. Zhiyi Zuo's laboratory. The support of the Department of Anesthesiology, University of Virginia, is also greatly appreciated.

References

Bhuiyan MIH, Kim YJ (2010) Mechanisms and prospects of ischemic tolerance induced by cerebral preconditioning. Int Neurourol J 14:203–212

Bruer U, Weih MK, Isaev NK, Meisel A, Ruscher K, Bergk A, Trendelenburg G, Wiegand F, Victorov IV, Dirnagl U (1997) Induction of tolerance in rat cortical neurons: hypoxic preconditioning. FEBS Lett 414:117–121

Chao D, Donnelly DF, Feng Y, Bazzy-Asaad A, Xia Y (2006) Cortical [delta]-opioid receptors potentiate K^+ homeostasis during anoxia and oxygen-glucose deprivation. J Cereb Blood Flow Metab 27:356–368

Chao D, Bazzy-Asaad A, Balboni G, Xia Y (2007) δ-, But not μ-, opioid receptor stabilizes K^+ homeostasis by reducing Ca^{2+} influx in the cortex during acute hypoxia. J Cell Physiol 212:60–67

Chao D, Bazzy-Asaad A, Balboni G, Salvadori S, Xia Y (2008) Activation of DOR attenuates anoxic K^+ derangement via inhibition of Na+entry in mouse cortex. Cereb Cortex 18:2217–2227

Chao D, He X, Yang Y, Bazzy-Asaad A, Lazarus LH, Balboni G, Kim DH, Xia Y (2012) DOR activation inhibits anoxic/ischemic Na+influx through Na+channels via PKC mechanisms in the cortex. Exp Neurol 236:228–239

Dahl N, Balfour W (1964) Prolonged anoxic survival due to anoxia pre-exposure: brain atp, lactate, and pyruvate. Am J Physiology 207:452–456

Gao CJ, Li JP, Wang W, Lu BC, Niu L, Zhu C, Wei YY, Zhang T, Wu SX, Chai W, Li YQ (2010) Effects of intracerebroventricular application of the delta opioid receptor agonist [d-Ala2, d-Leu5] enkephalin on neurological recovery following asphyxial cardiac arrest in rats. Neuroscience 168:531–542

Gao CJ, Niu L, Ren PC, Wang W, Zhu C, Li YQ, Chai W, Sun XD (2012) Hypoxic preconditioning attenuates global cerebral ischemic injury following asphyxial cardiac arrest through regulation of delta opioid receptor system. Neuroscience 202:352–362

Ginis I, Jaiswal R, Klimanis D, Liu J, Greenspon J, Hallenbeck JM (2002) TNF-[agr]-induced tolerance to ischemic injury involves differential control of NF-[kgr]B transactivation[colon] the role of NF-[kgr]B association with p300 adaptor. J Cereb Blood Flow Metab 22:142–152

Go AS, Mozaffarian D, Roger VL, Benjamin EJ, Berry JD, Borden WB, Bravata DM, Dai S, Ford ES, Fox CS, Franco S, Fullerton HJ, Gillespie C, Hailpern SM, Heit JA, Howard VJ, Huffman MD, Kissela BM, Kittner SJ, Lackland DT, Lichtman JH, Lisabeth LD, Magid D, Marcus GM, Marelli A, Matchar DB, McGuire DK, Mohler ER, Moy CS, Mussolino ME, Nichol G, Paynter NP, Schreiner PJ, Sorlie PD, Stein J, Turan TN, Virani SS, Wong ND, Woo D, Turner MB (2013) Heart disease and stroke statistics: 2013 update: a report from the American Heart Association. Circulation 127:e6–e245

Grabb MC, Choi DW (1999) Ischemic tolerance in murine cortical cell culture: critical role for NMDA receptors. J Neurosci 19:1657–1662

Hansen A (1985) Effect of anoxia on ion distribution in the brain. Physiol Rev 65:101–148

Heurteaux C, Lauritzen I, Widmann C, Lazdunski M (1995) Essential role of adenosine, adenosine A1 receptors, and ATP-sensitive K^+ channels in cerebral ischemic preconditioning. Proc Natl Acad Sci U S A 92:4666–4670

Huang H, Gao TM, Gong L, Zhuang Z, Li X (2001) Potassium channel blocker TEA prevents CA1 hippocampal injury following transient forebrain ischemia in adult rats. Neurosci Lett 305: 83–86

Ikeda Y, Miura T, Sakamoto J, Miki T, Tanno M, Kobayashi H, Ohori K, Takahashi A, Shimamoto K (2006) Activation of ERK and suppression of calcineurin are interacting mechanisms of cardioprotection afforded by δ-opioid receptor activation. Basic Res Cardiol 101:418–426

Iwata M, Inoue S, Kawaguchi M, Nakamura M, Konishi N, Furuya H (2007) Effects of delta-opioid receptor stimulation and inhibition on hippocampal survival in a rat model of forebrain ischaemia. Br J Anaesth 99:538–546

Kang X, Chao D, Gu Q, Ding G, Wang Y, Balboni G, Lazarus L, Xia Y (2009) Delta-opioid receptors protect from anoxic disruption of Na^+ homeostasis via Na^+ channel regulation. Cell Mol Life Sci 66:3505–3516

Kapinya KJ, Lowl D, Futterer C, Maurer M, Waschke KF, Isaev NK, Dirnagl U (2002) Tolerance against ischemic neuronal injury can be induced by volatile anesthetics and is inducible NO synthase dependent. Stroke 33:1889–1898

Kitagawa K, Matsumoto M, Kuwabara K, Tagaya M, Ohtsuki T, Hata R, Ueda H, Handa N, Kimura K, Kamada T (1991) 'Ischemic tolerance' phenomenon detected in various brain regions. Brain Res 561:203–211

Kume M, Yamamoto Y, Saad S, Gomi T, Kimoto S, Shimabukuro T, Yagi T, Nakagami M, Takada Y, Morimoto T, Yamaoka Y (1996) Ischemic preconditioning of the liver in rats: implications of heat shock protein induction to increase tolerance of ischemia-reperfusion injury. J Lab Clin Med 128:251–258

Lim YJ, Zheng S, Zuo Z (2004) Morphine preconditions purkinje cells against cell death under in vitro simulated ischemia-reperfusion conditions. Anesthesiology 100:562–568

Liu D, Slevin JR, Lu C, Chan SL, Hansson M, Elmér E, Mattson MP (2003) Involvement of mitochondrial K^+ release and cellular efflux in ischemic and apoptotic neuronal death. J Neurochem 86:966–979

Ma M-C, Qian H, Ghassemi F, Zhao P, Xia Y (2005) Oxygen-sensitive delta-opioid receptor-regulated survival and death signals: novel insights into neuronal preconditioning and protection. J Biol Chem 280:16208–16218

Marini A, Popolo M, Pan H, Blondeau N, Lipsky R (2008) Brain adaptation to stressful stimuli: a new perspective on potential therapeutic approaches based on BDNF and NMDA receptors. CNS Neurol Disord Drug Targets 7:382–390

Moncayo J, de Freitas G, Bogousslavsky J, Altieri M, van Melle G (2000) Do transient ischemic attacks have a neuroprotective effect? Neurology 54:2089–2094

Murry CE, Jennings RB, Reimer KA (1986) Preconditioning with ischemia: a delay of lethal cell injury in ischemic myocardium. Circulation 74:1124-1136

Narita M, Kuzumaki N, Miyatake M, Sato F, Wachi H, Seyama Y, Suzuki T (2006) Role of δ-opioid receptor function in neurogenesis and neuroprotection. J Neurochem 97:1494–1505

Pasupathy S, Homer-Vanniasinkam S (2005) Ischaemic preconditioning protects against ischaemia/reperfusion injury: emerging concepts. Eur J Vasc Endovasc Surg 29:106–115

Peng P-H, Huang H-S, Lee Y-J, Chen Y-S, Ma M-C (2009) Novel role for the δ-opioid receptor in hypoxic preconditioning in rat retinas. J Neurochem 108:741–754

Schwartz-Bloom RD, Sah R (2001) gamma-Aminobutyric acid(A) neurotransmission and cerebral ischemia. J Neurochem 77:353–371

Sommer C, Kiessling M (2002) Ischemia and ischemic tolerance induction differentially regulate protein expression of GluR1, GluR2, and AMPA receptor binding protein in the gerbil hippocampus: GluR2 (GluR-B) reduction does not predict neuronal death. Stroke 33:1093–1100

Sommer C, Fahrner A, Kiessling M (2003) Postischemic neuroprotection in the ischemia-tolerant state gerbil hippocampus is associated with increased ligand binding to inhibitory GABA(A) receptors. Acta Neuropathol 105:197–202

Stagliano NE, Perez-Pinzon MA, Moskowitz MA, Huang PL (1999) Focal ischemic preconditioning induces rapid tolerance to middle cerebral artery occlusion in mice. J Cereb Blood Flow Metab 19:757–761

Su D-S, Wang Z-H, Zheng Y-J, Zhao Y-H, Wang X-R (2007) Dose-dependent neuroprotection of delta opioid peptide [d-Ala2, d-Leu5] enkephalin in neuronal death and retarded behavior induced by forebrain ischemia in rats. Neurosci Lett 423:113–117

Tanaka H, Calderone A , Jover T, Grooms SY, Yokota H, Zukin RS, Bennett MV (2002) Ischemic preconditioning acts upstream of GluR2 down-regulation to afford neuroprotection in the hippocampal CA1. Proc Natl Acad Sci USA 99:2362–2367

Tian X, Guo J, Zhu M, Li M, Wu G, Xia Y (2013) δ-Opioid receptor activation rescues the functional TrkB receptor and protects the brain from ischemia-reperfusion injury in the rat. PLoS One 8:e69252

Toosy N, McMorris EL, Grace PA, Mathie RT (1999) Ischaemic preconditioning protects the rat kidney from reperfusion injury. BJU Int. 84:489–494

Wei L, Yu SP, Gottron F, Snider BJ, Zipfel GJ, Choi DW (2003) Potassium channel blockers attenuate hypoxia- and ischemia-induced neuronal death in vitro and in vivo. Stroke 34: 1281–1286

Weih M, Bergk A, Isaev NK, Ruscher K, Megow D, Riepe M, Meisel A, Victorov IV, Dirnagi U (1999) Induction of ischemic tolerance in rat cortical neurons by 3-nitropropionic acid: chemical preconditioning. Neurosci Lett 272:207–210

Xu G-P, Dave KR, Vivero R, Schmidt-Kastner R, Sick TJ, Perez-Pinzon MA (2002) Improvement in neuronal survival after ischemic preconditioning in hippocampal slice cultures. Brain Res 952:153–158

Yang Y, Xia X, Zhang Y, Wang Q, Li L, Luo G, Xia Y (2009) δ-Opioid receptor activation attenuates oxidative injury in the ischemic rat brain. BMC Biol 7:55

Zhang J, Gibney GT, Zhao P, Xia Y (2002) Neuroprotective role of delta-opioid receptors in cortical neurons. Am J Physiol Cell Physiol 282:C1225–C1234

Zhao P, Huang Y, Zuo Z (2006) Opioid preconditioning induces opioid receptor-dependent delayed neuroprotection against ischemia in rats. J Neuropathol Exp Neurol 65:945–952

Zhu M, Li M-W, Tian X-S, Ou X-M, Zhu C-Q, Guo J-C (2009) Neuroprotective role of delta-opioid receptors against mitochondrial respiratory chain injury. Brain Res 1252:183–191

Chapter 13
Delta-Opioid Receptor-Mediated Protection and Mitochondria

Jingchun Guo

Abstract Delta opioid receptor (DOR) is widely expressed in peripheral and central systems including the brain, heart, liver and kidney. The concept that DOR is cardioprotectant and neuroprotectant has been demonstrated and strengthened in recent years. However, the underlying mechanisms of DOR protection remain unclear. Mitochondria play essential roles in mediating cellular survival and death. The dysfunction of mitochondria is broadly involved in the pathogenesis of hypoxia/ischemia-related heart disorders, stroke, and neurodegenerative diseases. Accumulating evidence showed that DOR protection is associated with maintaining mitochondrial function, suggesting that DOR-mediated protection may be through mitochondria pathways.

Keywords Delta-opioid receptor • Mitochondria • Cardioprotection • Neuroprotection • Hypoxia • Ischemia • Heart

Abbreviations

Cyt c	Cytochrome *c*
DOR	Delta opioid receptor
DOR1	δ1-Opioid receptor
DOR2	δ2-Opioid receptor
ER	Endoplasmic reticulum
KOR	Kappa opioid receptor
Mito-K_{ATP}	Mitochondrial ATP-sensitive potassium channel
$\Delta\psi m$	Mitochondrial membrane potential
MOR	Mu opioid receptor

J. Guo, Ph.D. (✉)
State Key Laboratory of Medical Neurobiology, Department of Neurobiology, Fudan University, Shanghai 200032, China
e-mail: jingchunguo@shmu.edu.cn

Y. Xia (ed.), *Neural Functions of the Delta-Opioid Receptor*,
DOI 10.1007/978-3-319-25495-1_13

mPTP	Mitochondrial permeability transition pore
mROS	Mitochondrial reactive oxygen species
ROS	Reactive oxygen species
SOD	Superoxide dismutase

1 Introduction

Hypoxia/ischemia stress is one of the common risk factors leading to cardiovascular- and cerebro-vascular-related disorders, including heart diseases, stroke, brain trauma, and neurodegenerative diseases. Due to acute or chronic insufficient supply of oxygen and blood flow to the organ, hypoxia/ischemic injury ultimately induces cell death through multiple cellular and molecular signaling pathways. Patients with these diseases suffer greatly from high mortality and disability rate, and require long-term health care by professionals and families. Although being extensively studied, effective therapies against hypoxia/ischemia injury are still limited.

Delta opioid receptor (DOR) is classified as members of rhodopsin family of G protein coupled receptors. It belongs to opioid receptors, and is highly sequence homologous to other two opioid receptors: mu opioid receptor (MOR) and kappa opioid receptor (KOR). All of these opioid receptors inhibit adenylyl cyclase activity by coupling Gi/Go proteins. Together with MOR and KOR, DOR and the endogenous opioid peptides play important roles in physiological and pathological processes in peripheral and central system, such as mood control, pain/analgesia, addiction, hibernation, stress, and ischemic/hypoxic injury.

Recent evidence accumulates on neuroprotection or cytoprotection effects of DOR against ischemia/hypoxia injury. Mechanisms underlying these protections are not clearly clarified so far. Studies reported that the distributions of DOR are overlapped with MOR and KOR, and the functional interactions are across receptors of DOR/MOR or DOR/KOR. Despite of the complex crosstalk between opioid receptors, recent studies based on genetic and pharmacologic progresses have accumulated preclinical data supporting the role and potential signaling pathways of DOR in ischemic/hypoxia protection.

Ischemic/hypoxia injury is triggered by reduced oxygen and/or blood flow, and followed with energy disability and mitochondrial dysfunction. The increased mitochondrial damage/dysfunction is one of the key events in ischemia/hypoxia induced injury cascades (Borutaite et al. 2013; Walters et al. 2012). Our previous studies revealed that DOR activation protects neurons against mitochondrial respiratory chain injury (Zhu et al. 2009, 2011), suggesting a role for DOR to counteract mitochondrial damage and modulate mitochondria-related signaling. Here we will briefly summarize efforts in targeting DORs in mitochondrial abnormality, and outline possible signaling pathways in this process.

2 Mitochondrial Function

Mitochondria are one of the key intracellular organelles that play critical roles in cellular survival or death in response to physiological and pathological stresses. The outer membrane of mitochondria is permeable to most of small molecules and ions, while the inner membrane is a unique assembly of protein complexes that is selective for specific substrates permeation, such as adenosine monophosphate. Enzymes of tricarboxylic acid cycle and fatty acids/amino acids distribute in mitochondrial matrix, producing reducing equivalents from enzyme reaction. Normally, mitochondria mechanically active cells and maintain cellular functions by generation of ATP via oxidative phosphorylation in intact inner membrane of mitochondria (Mitchell 1961). After substrates oxidation and tricarboxylic acid cycle, reducing equivalents NADH and $FADH_2$ are yielding from coenzymes nicotinamide adenine dinucleotide (NAD) and flavin adenine dinucleotide (FAD), and used by electron transport chain to generate high-energy electron transferring and mitochondrial transmembrane potential across the inner membrane. With the proton gradient released energy, ATP is produced from ADP phosphorylation by inner membrane ATP synthase.

Beyond ATP production, mitochondria plays essential role in the process of reactive oxygen species (ROS) generation. As intermediate product of energy metabolism, superoxide radical anion O_2^- is constantly generated during electron transportation, due to the incomplete reduction of molecular oxygen. After production, superoxide anion could be rapidly converted to H_2O_2 by superoxide dismutase (SOD), and subsequently transformed to other ROS such as hydroxyl radical and peroxynitrite. ROS are highly reactive and can initiate attacks to nucleic acids, proteins, and lipids in both mitochondria and cytoplasm. Oxidative stress damage is induced by elevated mitochondrial ROS (mROS) production and leads to mitochondrial fragmentation, which is associated with neurodegenerative diseases, cancer and ischemia/reperfusion injury. Upon physiological conditions, ROS is maintained at extremely low level by antioxidant enzymes, including SOD, glutathione peroxidase (GPX), and peroxiredoxins (Prx). Accumulating evidence showed that mitochondria are highly resembled redox organelles. With both ROS generators and antioxidative enzymes, mitochondria are critical for cellular redox homeostasis (Greenlund et al. 1995; Murphy 2009, 2012). Recent findings also imply that mitochondrial ROS could mediate signaling transduction (Finkel 2012).

In addition, mitochondria are dominant organelle for the cellular calcium homeostasis. Under normal conditions, cytosolic calcium concentrations are precisely regulated and maintained at low levels. Increases of cytosolic calcium may eventually result in cell death via excessive activation of calcium-dependent signalings. Early from 1960s, mitochondria have been found to be capable of calcium uptake (Deluca and Engstrom 1961; Rizzuto et al. 1993). Increased cytosolic calcium could be uptaken by mitochondria via mitochondrial calcium uniporter to buffer calcium concentration alterations in the cytosol (Rizzuto et al. 2012; Perocchi et al. 2010; Baughman et al. 2011; De Stefani et al. 2011). Within mitochondria, free calcium ion further stimulates electron transport and ATP production (McCormack et al. 1990; Brini et al. 1999). Recently, studies have revealed that mitochondria are physically and functionally communicated with endoplasmic reticulum (ER), which is another cellular organelle for calcium storage (Kornmann et al. 2009; Dedkova and Blatter 2008). Upon stresses, calcium is released from ER and leads to calcium overloading in the cytosol. Mitochondria could absorb most of the excessive calcium and protects cells against the cytotoxic effect of calcium.

More importantly, mitochondria are actively responsible for cell death. Mitochondria-mediated apoptosis is initiated with mitochondria outer membrane disruption by the formation of permeable channels, or inner membrane permeabilization by the opening of mitochondrial permeability transition pore (mPTP). Bcl-2 family proteins Bax and Bak directly induce the pore formation in outer membrane which permit the release of proapoptotic molecules cytochrome *c* (Cyt *c*), apoptosis-inducing factor (AIF), and second mitochondria-derived activator caspase (SMAC)/direct IAP-binding protein with low pI (DIABLO) from the mitochondrial intermembrane space into the cytosol, and subsequently activate caspase-dependent/independent apoptosis. Other Bcl-2 family proteins such as Bid, Bim, Bad and p53 upregulated modulator of apoptosis (PUMA), also facilitate the function of Bax and Bak, while anti-apoptotic members such as Bcl-2 and Bcl-X_L, antagonize the pro-apoptotic function of Bax and Bak on the mitochondrial outer membrane. Bax/Bak activation causes slow loss of mitochondrial membrane potential ($\Delta\psi m$). The inner membrane transition pore mPTP opens upon several stresses and causes influx of solutes and water, following with disruption of proton gradient in electron transport chain, swelling of the inner membrane, destruction of the outer membrane, and the release of cytotoxic proteins. Cytosol Cyt *c* released from mitochondria could eventually initiates caspases activation and apoptosis program, while AIF induces DNA fragmentation in the nucleus. Cyt *c* is also reported to be able to bind the inositol reiphosphate receptors of ER and further induce calcium release from ER (Boehning et al. 2003). Over-uptake of calcium into mitochondria causes mitochondria swelling, mPTP opening and leads to necrotic cell death. In normal cells, these aberrant mitochondria could be cleared via a self-defense mechanism of autophagy, namely mitophagy, before it initiates cell death signaling pathway (Dieter and Gustafsson 2012).

Hypoxia/ischemia and reperfusion injury in heart, brain, liver, kidney, and other organs/tissues is tightly associated with mitochondrial dysfunction, which includes uncoupling in mitochondia electron transport chain, enhanced permeability in mitochondria membrane, cytochrome *c* released, and mitochondria-dependent apoptosis pathway. During hypoxia/ischemia, deficiency of oxygen/glucose supply leads to impaired oxidative phosphorylation, collapse of mitochondrial membrane potentials, and subsequently failure of ATP generation (Folbergrova et al. 1995). Apart from the fact that the reduced ATP level is accompanied with decreased capacity of mitochondrial electron transport upon focal cerebral ischemia (Anderson and Sims 1999), studies also demonstrated series of mitochondrial abnormality rapidly occurs upon hypoxia/ischemia stimuli (Schulze-Osthoff et al. 1992; Cadenas and Davies 2000; Gouriou et al. 2011; Borutaite et al. 2013), including increased generation of

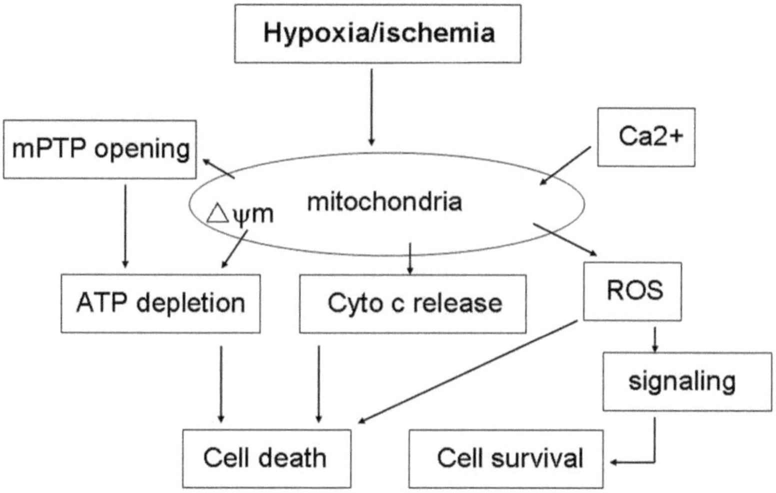

Fig. 13.1 Mitochondrial abnormality, including over-production of ROS, mPTP opening and Cyto c releasing, rapidly occurs upon hypoxia/ischemia, and eventually leads to cell death

ROS, excessive uptake of calcium, impairment of respiratory complexes, opening of mPTP, release of pro-death proteins such as Cyt *c* and AIF, activation of cell death pathways, and the induction of mitophagy. These hypoxia/ischemia-induced mitochondrial events are summarized in Fig. 13.1. Mitochondrial dysfunction contributes greatly to the progression of hypoxic/ischemic injury, leading to acute or chronic cellular damage. Increasing studies are focusing on exploring effective mitochondrial therapeutics to preserve mitochondrial function or reverse mitochondrial dysfunction during hypoxia/ischemia stresses.

3 Roles and Mechanisms of DOR in Relation to Mitochondrial Dysfunction

DOR is widely distributed in peripheral and central systems. To date, DOR has been pharmacologically classified into at least two subtypes, δ1-opioid receptor (DOR1) and δ2-opioid receptor (DOR2) (Zaki et al. 1996). The structural differences between these two subtypes have not been described so far. Accumulating evidence has demonstrated an interaction of DOR and mitochondria in the heart (Suo et al. 2014; Panneerselvam et al. 2013; Kim 2011; Ishikawa et al. 2012; Zeng et al. 2011; Maslov et al. 2010; Yao et al. 2007), brain (Zhu et al. 2009, 2011) and other organ systems or cell lines (Tang et al. 2011; Zhou et al. 2013; Chen et al. 2014; Plotnikov et al. 2013). For example, a recent study (Panneerselvam et al. 2013) reported that flavonoid epicatechin could directly improve the mitochondria tolerance to calcium overloading, increase the capacity of mitochondrial respiration, and enhance mitochondrial ROS production. These effects would be attenuated by blocking DOR with its antagonist naltrindole, suggesting a DOR-dependent manner for epicatechin effects in mouse myocardium. The activation of DOR has also been found to be essential for mediating neuroprotection, cardioprotection, cytoprotection and preconditioning in the situation of mitochondrial dysfunction induced by hypoxia/ischemia (He et al. 2013; Maslov et al. 2010; Tian et al. 2013), MPP^{+} treatment (Chen et al. 2014), NaN_3 stimulation (Zhu et al. 2009, 2011), serum-deprivation (Suo et al. 2014; Tang et al. 2011), and other mitochondria-related stresses (Plotnikov et al. 2013; Ishikawa et al. 2012). These studies provide support for the concept that DOR protection is tightly associated with mitochondrial pathway.

3.1 DOR Activation Is Protective Against Mitochondrial Injury

As mentioned above, the activation of DOR has been extensively demonstrated to be protective against mitochondrial injury (Zhu et al. 2009, 2011; Ishikawa et al. 2012; Chen et al. 2014). Studies revealed that DOR could modulate mitochondrial structure and function to attenuate or block mitochondria-induced damage, suggesting that mitochondria may be directly related to DOR protection.

3.1.1 mPTP Opening, Membrane Depolarization and ATP Depletion

Mitochondrial dysfunction is defined as mitochondrial membrane potential depolarization, associated with mPTP opening, and ATP depletion. As mentioned above, the opening of mPTP initiates mitochondrial apoptosis. Therefore, inhibition of the mPTP is considered as one of the targets for mitochondrial protection. Studies have observed a close relationship between DOR activation and mPTP inhibition (Jang et al. 2008). For example, Kim et al. (2011) reported that the mPTP opener atractyloside could totally block morphin-induced preconditioning in ischemia/reperfusion rat heart, which is similar to the effects of naloxon (nonspecific antagonist of opioid receptors), naltrindole (DOR antagonist) and 7-benzylidenenaltrexone (specific DOR1 antagonist). Similar results were found in rats with acute hemorrhagic shock and resuscitation (Zeng et al. 2011). The mPTP opening in myocardial mitochondria is decreased after administration of DOR1 agonist TAN-67 or DOR2 agonist deltorphin II, but increased under DOR1 antagonist BNTX or DOR2 antagonist NTB. Meanwhile, the mitochondrial and myocardial damage were found to be attenuated in TAN-67 and deltorphin II groups. These results indicated that DOR activation (including DOR1 and DOR2) possibly mediates cardioprotection via inhibition of mPTP.

DOR activation is also accompanied with the prevention of mitochondrial membrane potential ($\Delta\psi m$) collapse. In cortical neurons (Zhu et al. 2011), we observed that exposure to 10 mM NaN_3 for 30 min induced $\Delta\psi m$ decrease, which is represented by the reduction of tetramethylrhodomine ethyl ester (TMRE) fluorescence. Treatment with DADLE, a potent DOR agonist, attenuated the mitochondrial membrane depolarization. These effects of DADLE were abolished by administration of selective DOR antagonist naltrindole. UFP-512, another DOR agonist, is also found to activate DOR in HEK293 cells, and suppress the loss of $\Delta\psi m$ under hypoxia and/or MPP^+ stress (Chen et al. 2014). Similar results were reported in hepatic cells (Tang et al. 2011) and myocytes (Forster et al. 2010; Jang et al. 2008) that DOR activation is benefit for stabilization of mitochondrial membrane potential under serum deprivation or peroxide stimuli. ATP production is closely connected with mitochondrial membrane potential. In our previous study (Zhu et al. 2011), we do find that ATP concentration is remarkably reduced by NaN_3, which was reversed by DOR agonist DADLE. This effect of DADLE was abolished by DOR antagonist naltrindole. All of these data suggested that DOR functions to prevent against mitochondrial damage in targets of mPTP and $\Delta\psi m$.

3.1.2 ROS Production, Mitochondrial Calcium Overloading and Cyto c Releasing

There is increasing evidence that cellular ROS level plays a delicate role in mediating cellular surviving or death. Low concentrations of ROS in physiological conditions are reported to activate prosurvival signaling, whereas high ROS levels induce nonspecific permeability of the mitochondrial membrane, and activate signaling of cell death. In mitochondria, the mitochondrial ROS (mROS) is shown to regulate series of biological processes, such as stabilization of hypoxia inducible factors (HIFs) in physiological conditions (Chandel et al. 2000; Sena and Chandel 2012). Recent study also reported that hypoxia promotes mitochondrial clustering in perinuclear region, and mROS is released into the nucleus to induce hypoxia-triggered gene transcription (Al-Mehdi et al. 2012), further indicating that mROS regulates critical signaling.

Upon DOR protection, it is known that DOR activation alters both intracellular and mitochondrial ROS levels (Plotnikov et al. 2013; Zhu et al. 2011; Cohen et al. 2007; Ishikawa et al. 2012). In cultured cortical neurons, NaN_3 induced significantly elevation in intracellular ROS levels and mitochondrial calcium overloading (Zhu et al. 2011). The abnormalities were reversed by DADLE and suppressed at the presence of naltrindole, which suggests direct links to the function of DOR. In DADLE-triggered preconditioning, an increased ROS production was induced in cardiomyocytes (Cohen et al. 2007). Some studies revealed that mitochondrial ROS (mROS) generation is modulated under DOR activation as well. In H9c2 cardiomyocytes, DADLE is found to protect cells against peroxide- or antimycin A-induced necrosis, in association with significant ROS production in mitochondria (Ishikawa et al. 2012). Although recent evidence suggests that the increasing of mROS can activate cellular stress pathways in cellular survival/death (Ristow and Schmeisser 2011; Hekimi et al. 2011), the exact mechanisms in DOR modulation on mROS levels remain unknown and need further investigation.

Mitochondrial dysfunction would lead to Cyto *c* releasing from mitochondrial into cytosol, which cleaves pro-caspases-3 to form active caspase-3, and eventually, causes apoptosis. Several studies showed that DOR activation prevents Cyto *c* release. In serum-deprivation-induced apoptosis, activation of delta receptors protects hepatic cells against apoptosis, with a reduction in cytosolic Cyto *c* and Bax expression (Tang et al. 2011). In cortical neurons, we also observed that DADLE prevents the NaN_3-induced cytosolic Cyto *c* elevation, and subsequently caspase-3 activation (Zhu et al. 2011). These studies implied that the activation of DOR is able to prevent the initiation of apoptosis pathways induced by mitochondrial injury.

3.2 Mechanisms of DOR Protection Against Mitochondrial Dysfunction

Since the identification of this receptor, studies have revealed that DOR triggers alterations of cellular and molecular events at multiple levels to exhibit protective effects. Although it is believed that DOR-mediated protection is dependent on selective intracellular signaling cascades (He et al. 2013), it still remains uncertain how signaling transduction connects the membrane receptor specifically to mitochondria.

3.2.1 Mitochondrial ATP-Sensitive Potassium Channels

Some evidence of DOR-mitochondrial interactions relies on the facts that DOR activation induces the opening of mitochondrial ATP-sensitive potassium (Mito-KAT) channels in cardiac myocytes (Cao et al. 2003; Fang et al. 2006; Maslov et al. 2009; Zhang et al. 2005; Wu et al. 2004; Lim et al. 2004; Seymour et al. 2003; Fischbach et al. 2003; Fryer et al. 1999, 2001; Huh et al. 2001; Peart et al. 2003; Patel et al. 2002; McPherson and Yao 2001; Pateliya et al. 2008), and that DOR-mediated cardiac protection is suppressed by Mito-K_{ATP} channels blocker (Panneerselvam et al. 2013; Maslov et al. 2010). The opening of Mito-K_{ATP} channels eventually elicits cytoprotection by decreasing cytosolic and mitochondrial calcium overload. Although this concept is confirmed in several preconditioning conditions, and Mito-K_{ATP} channels were once considered as the end effector of preconditioning, recent investigations implied several other mediators in preconditioning signaling pathways (Busija et al. 2008; Downey et al. 2007). Therefore, Mito-K_{ATP} channels may be not the only target in mitochondria upon DOR protection. In fact, in NaN_3 treated cortical neurons, we found that blocking Mito-K_{ATP} channels with 5-HD did not affect DADLE neuroprotection, suggesting that Mito-K_{ATP} channels play an inessential role in DADLE neuroprotection against acute NaN_3 insults.

3.2.2 Putative Signaling Pathways

Protein kinase C (PKC) signaling pathway is presented in a variety of cellular activities and is reported to be involved in DOR protection against mitochondrial dysfunction (Tang et al. 2011; Maslov et al. 2009, 2010; Zhang et al. 2005; Wu et al. 2004;

Lim et al. 2004; Seymour et al. 2003; Huh et al. 2001). It is reported that during apoptosis, activation of DOR increases the levels of PKC, while blocking the PKC pathway led to increased apoptosis of liver cells (Tang et al. 2011). Inhibition of PKC with chelerythrine could abolish the cardioprotection of deltorphin II, a selective DOR2 agonist (Maslov et al. 2009, 2010). Since studies confirmed that opioid receptors act through the PTX-sensitive G protein signal pathway (Schultz et al. 1997), thus, the intracellular signaling transduction pathways activated by DOR may include the PTX-sensitive G protein-PKC pathway.

PKC-ERK pathway in DOR-mediated neuroprotection exists in hypoxia/ischemia stresses (He et al. 2013), and acute mitochondrial injury (Zhu et al. 2011) as well. We found that the activation of DOR with DADLE protects neurons from acute NaN_3 insults mainly through PKC-ERK pathway, and that mitochondrial ERK plays a key role in DOR protection against mitochondrial dysfunction. JNK and p38 MAPK signaling are not involved in this process (Zhu et al. 2011). ERK activation in DOR protection is also observed by other studies in cardiomyocytes (Yao et al. 2007; Cohen et al. 2007; Philipp et al. 2006), implying a possibly common signaling pathway mediating DOR protective effects.

Another important signaling pathway including phosphoinositol 3-kinase (PI3K)-Akt is also reported to be activated in DOR protection (Yao et al. 2007; Ishikawa et al. 2012; Cohen et al. 2007). In addition, glycogen synthase kinase 3β (GSK-3β) is considered as one of the downstream signals of PI3k/Akt pathway. Previous studies also demonstrated that GSK-3β phosphorylation is enhanced after treatment with DOR agonist dalargin or DADLE (Forster et al. 2010; Ishikawa et al. 2012; Plotnikov et al. 2013), suggesting that PI3K/Akt/GSK-3β signaling cascade is involved in DOR protection against mitochondria injury.

By activating DOR, a recent study showed that DOR-mediated morphine postconditioning protects the heart from ischemia/reperfusion, in association with nitric oxide production in cardiomyocytes, which is abolished by NOS inhibitor or PKG inhibitor (Jang et al. 2008). Therefore, the NO-cGMP-PKG pathway is also possibly involved in DOR protection.

Above all the signal transduction pathways are proposed to be activated by the stimulation of DOR to counteract mitochondria dysfunction. The signaling selectivity may be dependent on different types of cells/organs and animal/cellular models, and the severity of mitochondrial injuries. The potential mechanisms underlying DOR protection against mitochondrial dysfunction is summarized in Fig. 13.2.

3.3 Changes in DOR Protein Expression upon Mitochondrial Impairment

Several studies reported that the expression of DOR protein is sensitive in response to mitochondrial dysfunction. In some hypoxia/ischemia models, DOR expression is found to be up-regulated in a short-term hypoxic stress in crotical neurons (Zhang

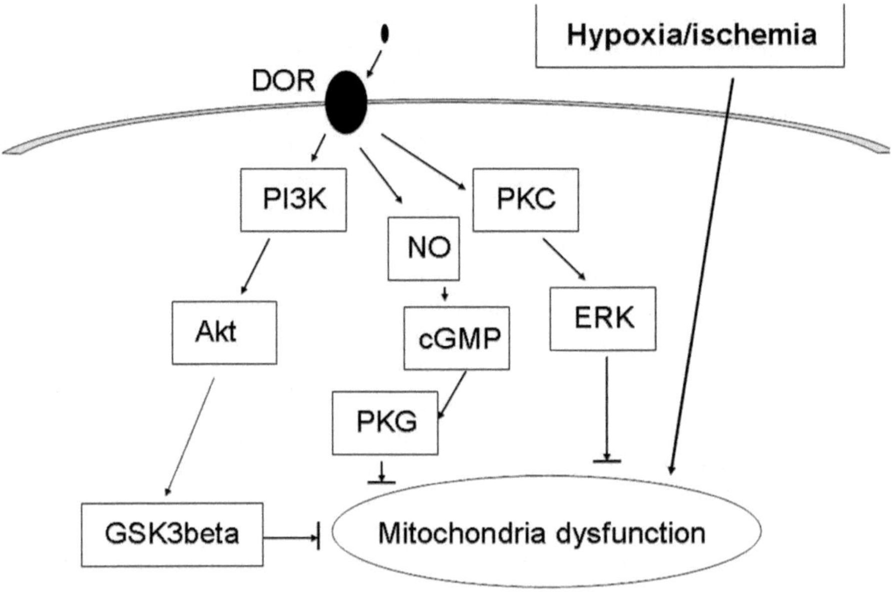

Fig. 13.2 DOR activation protects against hypoxia/ischemia-induced mitochondrial dysfunction. There are several putative signaling pathways underlying, including PI3K/Akt/GSK3 beta, NO/PKG, and PKC/ERK

et al. 2006), whereas down-regulated under prolonged and severely hyoxia/ischemia brain injury (Mayfield et al. 1996; Ma et al. 2005; Tian et al. 2008). Similar results were observed on cultured neurons with mitochondrial damage. After exposure to 3-nitropropionic acid for 2 h, no change in DOR protein levels are detected in differentiated SK-N-SH cells (Raut et al. 2007). However, in a NaN_3-induced prolonged mitochondrial respiratory chain injury model (Zhu et al. 2009), we found that DOR expression is reduced in cortical neurons upon NaN_3 treatment for 4 days, and the 60 and 32 kDa DOR protein variants are particularly sensitive. Since these two variants of DOR represent c-terminus truncated DORs, which is reported to be more mature in function than the full-length receptors (Persson et al. 2005), their reduction implies that the process of DOR maturation rather than DOR expression is far more affected by NaN_3. Mechanisms underlying the alteration of DOR protein have not been clearly studied so far. Possible explanations may relate to modulation of protein synthesis, maturation, and degradation as well.

4 Concluding Remarks

With the data obtained over the past decades of research on DOR protection against mitochondrial impairment, we summarized that the activation of DOR with selective DOR agonist could mediate protection via directly preserving mitochondrial function in targeting of mPTP, mitochondrial membrane potential, ROS production, and mitochondria-mediated apoptosis. Although these studies have implied possible involvement of Mito-K_{ATP} channel, PKC, ERK, PI3K, Akt, GSK-3β, and NO-cGMP-PKG in the process, the detailed molecular interactions linking DOR to mitochondria remain to be further elucidated.

Acknowledgements Our own relevant research is funded by the 973 program of China (No. 2005CB523306), and the Science and Technology Commission Program of Shanghai (No. 06DZ19734).

References

Al-Mehdi AB, Pastukh VM, Swiger BM, Reed DJ, Patel MR, Bardwell GC, Pastukh W, Alexeyev MF, Gillespie MN (2012) Perinuclear mitochondrial clustering creates an oxidant-rich nuclear domain required for hypoxia-induced transcription. Sci Signal 5(231):5369–5378

Anderson MF, Sims NR (1999) Mitochondrial respiratory function and cell death in focal cerebral ischemia. J Neurochem 73(3):1189–1199

Baughman JM, Perocchi F, Girgis HS, Plovanich M, Belcher-Timme CA, Sancak Y, Bao XR, Strittmatter L, Goldberger O, Bogorad RL, Koteliansky V, Mootha VK (2011) Integrative genomics identifies MCU as an essential component of the mitochondrial calcium uniporter. Nature 467:341–345

Boehning D, Patterson RL, Sedaghat L, Glebova NO, Kurosaki T, Snyder SH (2003) Cytochrome c binds to inositol (1,4,5) trisphosphate receptors, amplifying calcium-dependent apoptosis. Nat Cell Biol 5(12):1051–1061

Borutaite V, Toleikis A, Brown GC (2013) In the eye of the storm: mitochondrial damage during heart and brain ischaemia. FEBS J 280(20):4999–5014

Brini M, Pinton P, King MP, Davidson M, Schon EA, Rizzuto R (1999) A calcium signaling defect in the pathogenesis of a mitochondrial DNA inherited oxidative phosphorylation deficiency. Nat Med 5:951–954

Busija DW, Gaspar T, Domoki F, Katakam PV, Bari F (2008) Mitochondrial-mediated suppression of ROS production upon exposure of neurons to lethal stress: mitochondrial targeted preconditioning. Adv Drug Deliv Rev 60:1471–1477

Cadenas E, Davies KJA (2000) Mitochondrial free radical generation, oxidative stress, and aging. Free Radic Biol Med 29(3–4):222–230

Cao Z, Liu L, Van Winkle DM (2003) Activation of delta- and kappa-opioid receptors by opioid peptides protects cardiomyocytes via KATP channels. Am J Physiol Heart Circ Physiol 285:H1032–H1039

Chandel NS, McClintock DS, Feliciano CE, Wood TM, Melendez JA, Rodriguez AM, Schumacker PT (2000) Reactive oxygen species generated at mitochondrial complex III stabilize hypoxia-inducible factor-1 alpha during hypoxia: a mechanism of O_2 sensing. J Biol Chem 275: 25130–25138

Chen T, Li J, Chao D, Sandhu HK, Liao X, Zhao J, Wen G, Xia Y (2014) δ-Opioid receptor activation reduces α-synuclein overexpression and oligomer formation induced by MPP(+) and/or hypoxia. Exp Neurol 255:127–136

Cohen MV, Philipp S, Krieg T, Cui L, Kuno A, Solodushko V, Downey JM (2007) Preconditioning-mimetics bradykinin and DADLE activates PI3-kinase through divergent pathways. J Mol Cell Cardiol 42(4):842–851

De Stefani D, Raffaello A, Teardo E, Szabo I, Rizzuto R (2011) A forty-kilodalton protein of the inner membrane is the mitochondrial calcium uniporter. Nature 476:336–340

Dedkova EN, Blatter LA (2008) Mitochondrial Ca^{2+} and the heart. Cell Calcium 44:77–91
Deluca HF, Engstrom GW (1961) Calcium uptake by rat kidney mitochondria. Proc Natl Acad Sci U S A 47:1744–1750
Dieter AK, Gustafsson AB (2012) Mitochondria and mitophagy: the Yin and Yang of cell death control. Circ Res 111:1208–1221
Downey JM, Davis AM, Cohen MV (2007) Signaling pathways in ischemic preconditioning. Heart Fail Rev 12:181–188
Fang X, Tang W, Sun S, Weil MH (2006) Delta-opioid-induced pharmacologic myocardial hibernation during cardiopulmonary resuscitation. Crit Care Med 34:S486–S489
Finkel T (2012) Signal transduction by mitochondrial oxidants. J Biol Chem 287:4434–4440
Fischbach PS, Barrett TD, Reed NJ, Lucchesi BR (2003) SNC-80-induced preconditioning: selective activation of the mitochondrial adenosine triphosphate-gated potassium channel. J Cardiovasc Pharmacol 41(5):744–750
Folbergrova J, Zhao Q, Katsura K, Siesjo BK (1995) N-tertbutyl-alpha-phenylnitrone improves recovery of brain energy state in rats following transient focal ischemia. Proc Natl Acad Sci U S A 92(11):5057–5061
Forster K, Richter H, Alexeyev MF, Rosskopf D, Felix SB, Krieg T (2010) Inhibition of glycogen synthase kinase 3 beta prevents peroxide-induced collapse of mitochondrial membrane potential in rat ventricular myocytes. Clin Exp Pharmacol Physiol 37(7):684–688
Fryer RM, Hsu AK, Eells JT, Nagase H, Gross GJ (1999) Opioid-induced second window of cardioprotection: potential role of mitochondrial KATP channels. Circ Res 84(7):846–851
Fryer RM, Wang Y, Hsu AK, Gross GJ (2001) Essential activation of PKC-delta in opioid-initiated cardioprotection. Am J Physiol Heart Circ Physiol 280(3):H1346–H1353
Gouriou Y, Demaurex N, Bijlenga P, De Marchi U (2011) Mitochondrial calcium handling during ischemia-induced cell death in neurons. Biochimie 93(12):2060–2067
Greenlund LJS, Deckwerth TI, Johnson EM (1995) Superoxide dismutase delays neuronal apoptosis: a role for reactive oxygen species in programmed neuronal death. Neuron 14(2): 303–315
He X, Sandhu H, Yang Y, Hua F, Belser N, Kim DH, Xia Y (2013) Neuroprotection against hypoxia/ischemia: δ-opioid receptor-mediated cellular/molecular events. Cell Mol Life Sci 70:2291–2303
Hekimi S, Lapointe J, Wen Y (2011) Taking a 'good' look at free radicals in the aging process. Trends Cell Biol 21:569–576
Huh J, Gross GJ, Nagase H, Liang BT (2001) Protection of cardiac myocytes via delta(1)-opioid receptors, protein kinase C, and mitochondrial K(ATP) channels. Am J Physiol Heart Circ Physiol 280(1):H377–H383
Ishikawa S, Kuno A, Tanno M, Miki T, Kouzu H, Itoh T, Sato T, Sunaga D, Murase H, Miura T (2012) Role of connexin-43 in protective PI3K-Akt-GSK-3β signaling in cardiomyocytes. Am J Physiol Heart Circ Physiol 302(12):H2536–H2544
Jang Y, Xi J, Wang H, Mueller RA, Norfleet EA, Xu Z (2008) Postconditioning prevents reperfusion injury by activating delta-opioid receptors. Anesthesiology 108(2):243–250
Kim JH, Chun KJ, Park YH, Kim J, Kim JS, Jang YH, Lee MY, Park JH (2011) Morphine-induced postconditioning modulates mitochondrial permeability transition pore opening via delta-1 opioid receptors activation in isolated rat hearts. Korean J Anesthesiol 61(1):69–74
Kornmann B, Currie E, Collins SR, Schuldiner M, Nunnari J, Weissman JS, Walter P (2009) An ER-mitochondria tethering complex revealed by a synthetic biology screen. Science 325:477–481
Lim YJ, Zheng S, Zuo Z (2004) Morphine preconditions Purkinje cells against cell death under in vitro stimulated ischemia-reperfusion conditions. Anesthesiology 100(3):562–568
Ma MC, Qian H, Ghassemi F, Zhao P, Xia Y (2005) Oxygen sensitive delta opioid receptor-regulated survival and death signals: novel insights into neuronal preconditioning and protection. J Biol Chem 280:16208–16218
Maslov LN, Lishmanov YB, Oeltgen PR, Barzakh EI, Krylatov AV, Govindaswami M, Brown SA (2009) Activation of peripheral delta2 opioid receptors increases cardiac tolerance to ischemia/

reperfusion injury involvement of protein kinase C, NO-synthase, KATP channels and the autonomic nervous system. Life Sci 84(19–20):657–663

Maslov LN, Barzakh EI, Krylatov AV, Chernysheva GA, Krieg T, Solenkova NV, Lishmanov AY, Cybulnikov SY, Zhang Y (2010) Opioid peptide deltorphin II stimulates the cardioprotective effect of ischemic preconditioning: role of δ2-opioid receptors, protein kinase C, and K(ATP) channels. Bull Exp Biol Med 149(5):591–593

Mayfield KP, Kozak W, Malvin GM, Porreca F (1996) Hypoxia decreases opioid delta receptor expression in mouse brain. Neuroscience 72:785–789

McCormack JG, Halestrap AP, Denton RM (1990) Role of calcium ions in regulation of mammalian intramitochondrial metabolism. Physiol Rev 70:391–425

McPherson BC, Yao Z (2001) Morphine mimics preconditioning via free radical signals and mitochondrial K(ATP) channels in myocytes. Circulation 103(2):290–295

Mitchell P (1961) Coupling of phosphorylation to electron and hydrogen transfer by a chemiosmotic type of mechanism. Nature 191:144–148

Murphy MP (2009) How mitochondria produce reactive oxygen species. Biochem J 417(1):1–13

Murphy MP (2012) Mitochondrial thiols in antioxidant protection and redox signaling: distinct roles for glutathionylation and other thiol modifications. Antioxid Redox Signal 16:476–495

Panneerselvam M, Ali SS, Finley JC, Kellerhals SE, Migita MY, Head BP, Patel PM, Roth DM, Patel HH (2013) Epicatechin regulation of mitochondrial structure and function is opioid receptor dependent. Mol Nutr Food Res 57(6):1007–1014

Patel HH, Ludwig LM, Fryer RM, Hsu AK, Warltier DC, Gross GJ (2002) Delta opioid agonists and volatile anesthetics facilitate cardioprotection via potentiation of K(ATP) channel opening. FASEB J 16(11):1468–1470

Pateliya BB, Singh N, Jaggi AS (2008) Possible role of opioids and KATP channels in neuroprotective effect of postconditioning in mice. Biol Pharm Bull 31:1755–1760

Peart JN, Patel HH, Gross GJ (2003) Delta-opioid induced receptor activation mimics ischemic preconditioning in the canine heart. J Cardiovasc Pharmacol 42(1):78–81

Perocchi F, Gohil VM, Girgis HS, Bao XR, McCombs JE, Palmer AE, Mootha VK (2010) MICUI encodes a mitochondrial EF hand protein required for Ca^{2+} uptake. Nature 467:291–296

Persson AI, Thorlin T, Eriksson PS (2005) Comparison of immunoblotted delta opioid receptor proteins expressed in the adult rat brain and their regulation by growth hormone. Neurosci Res 51:1–9

Philipp S, Critz SD, Cui L, Solodushko V, Cohen MV, Downey JM (2006) Localization extracellular signal-regulated kinase (ERK) in pharmacological preconditioning's trigger pathway. Basic Res Cardiol 101(2):159–167

Plotnikov EY, Grebenchikov OA, Babenko VA, Pevzner IB, Zorova LD, Likhvantsev VV, Zorov DB (2013) Nephroprotective effect of GSK-3β inhibition by lithium ions and δ-opioid receptor agonist dalargin on gentamicin-induced nephrotoxicity. Toxicol Lett 220(3):303–308

Raut A, Rao VR, Ratka A (2007) Changes in opioid receptor proteins during mitochondrial impairment in differentiated SK-N-SH cells. Neurosci Lett 422:187–192

Ristow M, Schmeisser S (2011) Extending life span by increasing oxidative stress. Free Radic Biol Med 51:327–336

Rizzuto R, Brini M, Mk M, Pozzan T (1993) Microdomains with high Ca^{2+} close to IP3-sensitive channels that are sensed by neighboring mitochondria. Science 262:744–747

Rizzuto R, De Stefani D, Raffaello A, Mammucari C (2012) Mitochondria as sensors and regulators of calcium signalling. Nat Rev Mol Cell Biol 13:566–578

Schultz JJ, Hsu AK, Gross GJ (1997) Ischemic preconditioning and morphine-induced cardioprotection involve the delta (delta)-opioid receptor in the intact rat heart. J Mol Cell Cardiol 29:2187–2195

Schulze-Osthoff K, Bakker AC, Vanhaesebroeck B, Beyaert R, Jacob WA, Fiers W (1992) Cytotoxic activity of tumor necrosis factor is mediated by early damage of mitochondrial functions. Evidence for the involvement of mitochondrial radical generation. J Biol Chem 267(8):5317–5323

Sena LA, Chandel N (2012) Physiological roles of mitochondrial reactive oxygen species. Mol Cell 48:158–167

Seymour EM, Wu SY, Kovach MA, Romano MA, Traynor JR, Claycomb WC, Bolling SF (2003) HL-1 myocytes exhibit PKC and K(ATP) channel-dependent delta opioid preconditioning. J Surg Res 114(2):187–194

Suo C, Sun L, Yang S (2014) Alpinetin activates the δ receptor instead of the κ and μ receptor pathways to protect against rat myocardial cell apoptosis. Exp Ther Med 7(1):109–116

Tang B, Zhang Y, Liang R, Yuan P, Du J, Wang H, Wang L (2011) Activation of the δ-opioid receptor inhibits serum deprivation-induced apoptosis of human liver cells via the activation of PKC and the mitochondrial pathway. Int J Mol Med 28(6):1077–1085

Tian XS, Zhou F, Yang R, Xia Y, Wu GC, Guo JC (2008) [Effects of intracerebroventricular injection of delta-opioid receptor agonist TAN-67 or antagonist naltrindole on acute cerebral ischemia in rat]. Sheng Li Xue Bao 60(4):475–484

Tian X, Guo J, Zhu M, Li M, Wu G, Xia Y (2013) δ-Opioid receptor activation rescues the functional TrkB receptor and protects the brain from ischemia-reperfusion injury in the rat. PLoS One 8(7):e69252

Walters AM, Porter GA, Brookes JPS (2012) Mitochondria as a drug target in ischemic heart disease and cardiomyopathy. Circ Res 111:1222–1236

Wu S, Wong MC, Chen M, Cho CH, Wong TM (2004) Role of opioid receptors in cardioprotection of cold-restraint stress and morphine. J Biomed Sci 11(6):726–731

Yao LL, Wang YG, Cai WJ, Yao T, Zhu YC (2007) Survivin mediates the anti-apoptotic effect of delta-opioid receptor stimulation in cardiomyocytes. J Cell Sci 120(pt5):895–907

Zaki PA, Bilsky EJ, Vanderah TW, Lai J, Evans CJ, Porreca F (1996) Opioid receptor types and subtypes: the delta receptor as a model. Annu Rev Pharmacol Toxicol 36:379–401

Zeng X, Zhao X, Yang Y, Kuai J, Gao C, Yu D, Zhao H, Chai W, Yao L (2011) Opioid δ(1) and δ(2) receptor agonist attenuate myocardial injury via mPTP in rats with acute hemorrhagic shock. J Surg Res 169(2):267–276

Zhang Y, Irwin MG, Wong TM, Chen M, Cao CM (2005) Remifentanil preconditioning confers cardioprotection via cardiac kappa- and delta-opioid receptors. Anesthesiology 102(2): 371–378

Zhang J, Qian H, Zhao P, Hong S-S, Xia Y (2006) Rapid hypoxia preconditioning protects cortical neurons from glutamate toxicity through delta-opioid receptor. Stroke 37:1094–1099

Zhou L, Guo X, Chen M, Fu S, Zhou J, Ren G, Yang Z, Fan W (2013) Inhibition of δ-opioid receptors induces brain glioma cell apoptosis through the mitochondrial and protein kinase C pathways. Oncol Lett 6(5):1351–1357

Zhu M, Li M, Tian X, Ou X, Zhu C, Guo J (2009) Neuroprotective role of δ-opioid receptors against mitochondrial respiratory chain injury. Brain Res 1252:183–191

Zhu M, Li M, Yang F, Ou X, Ren Q, Gao H, Zhu C, Guo J (2011) Mitochondrial ERK plays a key role in δ-opioid receptor neuroprotection against acute mitochondrial dysfunction. Neurochem Int 59(6):739–748

Chapter 14
The Delta-Opioid System in the Brain: A Neglected Element in Parkinson's Disease?

Dongman Chao and Ying Xia

Abstract Parkinson's disease (PD) is the second most common age-related neurodegenerative disease resulting from the death of dopaminergic neurons in the substantia nigra pars compacta (SNc) which innervate all of the neuclei of the basal ganglia. Cell loss in the SNc causes a consequent striatal dopamine (DA) deficiency, leading to a cascade of functional changes in basal ganglia circuitry and ultimately the development of the cardinal features of PD, including resting tremor, bradykinesia, postural instability and rigidity, along with non-motor symptoms including cognitive, affective, psychiatric, and autonomic problems. DA replacement with the Levodopa (L-DOPA) is still a gold standard for symptomatic therapy of PD. However, the side effects of long-term use of L-DOPA are obvious. The emerging non-dopaminergic treatments for PD are becoming a scheduling program for the pipe-lines of drug development and clinical trials. The abundant expression of opioid receptors and their endogenous ligands, especially those for the δ-opioid system within the basal ganglia, has attracted much attention for its contribution to neurodegenerative diseases such as PD. In this article, we first discussed the relevant functional neuroanatomy and circuitry of the basal ganglia, and the evolving PD models of the basal ganglia. We then retrospected the studies on the location and expression of the δ-opioid system within the basal ganglia in health and PD, opioid-induced regulation of neurotransmitter release in basal ganglia, effects of pharmacological δ-opioid receptor (DOR) manipulation on PD and levodopa-induced dyskinesia (LID), and the effects of DOR activation on dopaminergic neuron injury/cellular model of PD. Based on these analyses, we believe that δ-opioid signaling is generally a beneficial, not detrimental factor for motor complications in PD. However, more in-depth investigations are expected to provide more solid and direct evidence for this benefit and to develop therapeutic strategies against PD by targeting δ-opioid system.

D. Chao, Ph.D.
Department of Neurosurgery, The University of Texas Medical School at Houston, Houston, TX 77030, USA
e-mail: chaodm21@yahoo.com

Y. Xia M.D., Ph.D. (✉)
University of Texas Medical School, Houston, TX, USA
e-mail: ying.xia@uth.tmc.edu; y55738088@gmail.com

Y. Xia (ed.), *Neural Functions of the Delta-Opioid Receptor*,
DOI 10.1007/978-3-319-25495-1_14

Keywords Opioids • δ-opioid receptors • Parkinson's disease • Motor complications • Neurodegenerative diseases

Abbreviations

BAC	Bacterial artificial chromosome
CM-pf	Centromedian-parafascicular nuclei
CNS	Central nervous system
DA	Dopamine
DBS	Deep brain stimulation
DOR	δ-Opioid receptor
GPe	Globus pallidus external segment
GPi	Globus pallidus internal segment
KOR	κ-Opioid receptor
L-DOPA	Levodopa
LID	Levodopa-induced dyskinesia
MOR	μ-Opioid receptor
MPPP	1-Methyl-4-phenyl-4-propionoxypiperidine
MPTP	1-Methyl-4-phenyl-1,2,3,6-tetrahydropyridine
MSNs	Medium spiny neurons
PD	Parkinson's disease
PENK	Proenkephalin
PPE-A	Preproenkephalin-A
PPN	Pedunculopontine nuclei
SNc	Substantia nigra pars compacta
SNr	Substantia nigra pars reticulate
STN	Subthalamic nucleus
VTA	Ventral tegmental area

1 Introduction

Parkinson's disease (PD) is the second most common age-related neurodegenerative disease after Alzheimer's disease, and is characterized by the cardinal clinical syndrome of motor impairments involving resting tremor, bradykinesia, postural instability and rigidity, along with non-motor symptoms including cognitive, affective, psychiatric, and autonomic problems (Samii et al. 2004; Lees et al. 2009; Goldman and Postuma 2014). The prevalence of PD steadily increases with age across all regions of the world, rising from 0.5–1 % in those over 60 years of age to 2–4 % in those over 80 (de Lau and Breteler 2006; Pringsheim et al. 2014). The incidence of PD rises steeply with age, from 17.4 in 100,000 person years for people between 50 and 59 years of age to 93.1 in 100,000 person years in people between 70 and 79 years, with a lifetime risk of developing the disease of 1.5 % (Lees et al. 2009). In the US, up to one million people have PD, with 50,000–60,000 new cases diagnosed annually, causing the predicted annual cost to US society to be over $50 billion by the year of 2040 (Goldman 2014). The mean age of onset is around 60 years. However, 5–10 % of cases, classified as young onset, begin between the ages of 20 and 50 (Samii et al. 2004). Aging is the biggest risk factor for the disease because with advancing age, the accumulation of many defects (significant generation of ROS by DA metabolism, a decline of mtDNA copy number and subsequent decreased ATP production, and a decline in the ability of cells to handle misfolded proteins, Ca^{2+} and iron levels) renders the neurons of substantia nigra (SN) pars compacta (SNc) more vulnerable to the additional insults of mitochondrial deficiencies and toxic misfolded proteins (Reeve et al. 2014). The most prevalent hypothesis about the etiology and pathogenesis of PD is that the environmental stress such as infection and toxic agents commonly used as herbicides and pesticides like 1-methyl-4-phenyl-1,2,3,6-tetrahydropyridine (MPTP), rotenone, paraquat, etc., are dopaminergic neurotoxins that cause Parkinsonian neurodegeneration (Dauer and Przedborski 2003; Goldman 2014). The emergent suffering of Parkinsonism in intravenous drug [1-methyl-4-phenyl-4-propionoxypiperidine (MPPP), an analog of the narcotic meperidine] users in San Jose, California in 1982 and the subsequent identification of the contaminating MPTP as the culprit (Langston et al. 1983) is a prototypic example frequently used to support the environmental hypothesis (Dauer and Przedborski 2003; Goldman 2014). This situation lasted for a long time until the discovery of PD-pathogenic mutations in the gene encoding α-synuclein (*SNCA*) (Polymeropoulos et al. 1997) and the demonstration of α-synuclein as the major component of filamentous Lewy bodies, a major pathophysiological hallmark in most PD cases (Spillantini et al. 1997). Since then, the genetic hypothesis has emerged and has shed light on the situation (Corti et al. 2011). This hypothesis addresses the roles of gene mutation in PD pathogenesis. In support of this hypothesis is the identification of at least 16 loci (designated as PARK1 to PARK18) and 11 genes thus far that have been associated with inherited forms of Parkinsonism (Corti et al. 2011; Singleton et al. 2013; Trinh and Farrer 2013). Indeed, it is now clear that mutations in *SNCA* (α-synuclein) and *LRRK2* cause autosomal dominant PD, while those in *PINK1*, *Park7* (*DJ-1*), *Park9* (*ATP13A2*), and *Park2* (*parkin*) cause autosomal recessive PD (Corti et al. 2011; Singleton et al. 2013; Trinh and Farrer 2013). Among all the PD cases, about 90–95 % are idiopathic (regarded as a sporadic disorder), and others are identified as genetically inherited. However, it has been realized that the previous supposition of either an environmental or a genetic cause of PD is a false dichotomy (Singleton et al. 2013). The environmental and genetic factors are not mutually exclusive; both probably play a role, albeit in varying degrees. Indeed, the disorder may result from a genetic predisposition to an environmental toxin, or to a combination of several genes that each increases the risk of the disease to only a modest extent (Corti et al. 2011). An intimate connection between genetic and environmental causes of PD neurodegeneration has been increasingly revealed (Vance et al. 2010; Gao and Hong 2011; Cannon and Greenamyre 2013; Dardiotis et al. 2013). As has been suggested, environmental risk factors and PD-associated gene mutations act in series and/or in parallel pathways, likely sharing some common molecular mechanisms to cause PD pathogenesis (Gao and Hong 2011).

The motor symptoms of PD result from the death of dopaminergic neurons in SNc. The cause of this cell death is unknown, but is presumably resulted from extremely high vulnerability to multiple insults. As a result of interplay between genetic and environmental factors, dopaminergic neurons in SNc are easy to suffer lysosomal and proteasomal dysfunction, which lead to the consequent neurotoxicity to misfolded proteins (α-synuclein), mitochondrial dysfunction, the consequent oxidative stress, and probably neuroinflammation (Dauer and Przedborski 2003; Subramaniam and Chesselet 2013; Antony et al. 2013; Stayte and Vissel 2014),. Dopaminergic neurons in SNc innervate all of the nuclei of the basal ganglia (brain structures that are situated at the base of the forebrain, comprise multiple subcortical nuclei of varied origin, and are critical to motor control), but project preferentially to the dorsolateral domains of the striatum, particularly the putamen (thus forming the nigrostriatal system), having a more defined role in the organization of motor plans (Smith and Villalba 2008; Benazzouz et al. 2014). Cell loss in SNc causes a consequent striatal dopamine (DA) deficiency, leading to a cascade of functional changes in basal ganglia circuitry and ultimately the development of the cardinal features of PD (Obeso et al. 2008; Ellens and Leventhal 2013). At the onset of symptoms, about 60 % of SNc dopaminergic neurons have been lost, accompanying the depletion of approximately 80 % of DA in the putamen (Dauer and Przedborski 2003). This situation will be aggravated in late advanced PD and can reach 80–90 % cell loss and >95 % DA depletion. Though the loss of dopaminergic neurons within SNc results in striatal DA deficiency and ultimately PD symptom, dysfunction and/or dysregulation of other neurotransmitter systems (e.g., GABAergic, peptidergic, glutamatergic, cholinergic, serotonergic, adenosinergic, adrenergic etc.) and ion homeostasis (e.g., disruption of Ca^{2+} and iron homeostasis) in the basal ganglia as well as other brain regions may also be involved in the pathogenesis of PD (Graybiel 1990; Brichta et al. 2013; Stayte and Vissel 2014; Rivero-Ríos et al. 2014; Calì et al. 2014). These changes, together with dopaminergic deficiency, are instrumental in the debilitating motor and/or non-motor features of PD.

Opioid receptors, including the δ-opioid receptor (DOR), and its endogenous opioids (enkephalins) are widely distributed throughout the central nervous system (CNS) (Mansour et al. 1993, 1994; Xia and Haddad 1991, 2001; Bausch et al. 1995; Svingos et al. 1998, 1999; Scherrer et al. 2006; Erbs et al. 2012), and are extensively involved in various physiological and pathophysiological activities (Chao and Xia 2010; Pradhan et al. 2011; Feng et al. 2012; He et al. 2013; Chao et al. 2013; Chu Sin Chung and Kieffer 2013; also refer to Chap. 18). The abundant expression of opioid receptors and their endogenous ligands, especially the δ-opioid system within the basal ganglia has attracted much attention for its contribution to neurodegenerative diseases such as PD (Samadi et al. 2006). Indeed, as a co-neurotransmitter, enkephalin exists in the terminals of dopaminergic and GABAergic neurons within basal ganglia circuitry. These systems regulate their activities and *vice versa*. The changes in the activities of DOR and its endogenous agonists (as well as its precursors) in the basal ganglia have been intensively demonstrated in PD patients as well as in animal models. Activation of DOR also alters the symptoms of PD. However, conflicting results limit in-depth research and its clinical usefulness (see the detailed discussion in this chapter). Establishing a framework for the relatively clear role of DOR in the PD pathophysiology is absolutely necessary and relevant in PD research and treatment strategy.

In addition, since the establishment of pharmacological treatment of PD (Cotzias et al. 1969), DA replacement with Levodopa (L-DOPA) is still a gold standard for symptomatic therapy of PD. However, the side-effects of long-term use of L-DOPA are obvious, which include progressive loss of efficacy, the development of dyskinesia (termed levodopa-induced dyskinesia, LID) and aggravated "on-off" effect for the standard treatment regimes in most of patients within 5 years of therapy, and multiple non-motor manifestations (Olanow et al. 2006; Brichta et al. 2013; Stayte and Vissel 2014). Other dopaminergic medicines (such as pramipexole, ropinirole, or pergolide) for PD treatments also produce some side-effects in PD patients (Vo et al. 2014). The emerging non-dopaminergic treatments for PD are becoming a scheduling program for the pipe-lines of drug development and clinical trials (Olanow and Schapira 2013; Brichta et al. 2013; Stayte and Vissel 2014; Connolly and Lang 2014). DOR and its endogenous agonists have been demonstrated to be involved in motor complications and LID, though the effects are conflicting; this also increases the urgent need for a framework for the relatively clear role of DOR in the PD pathophysiology and its potential treatment strategy.

In this article, we attempted to establish such a framework for the relatively clear role of DOR in PD pathophysiology. We reviewed the studies on the location and expression alteration of δ-opioid system within the basal ganglia in health and PD, opioid regulation of neurotransmitter release in basal ganglia, effects of pharmacological DOR manipulation on PD and LID, and the effects of DOR activation on dopaminergic neuron injury/cellular model of PD. In retrospect, it was found that some other neuropeptides are also involved in the comparison with δ-opioid system. Based on these studies, we believe that δ-opioid system is a beneficial, not detrimental factor for motor complications in PD. However, more in-depth investigations are expected to provide more solid and direct evidence, and the development of treatment strategies targeting the δ-opioid system is also expected.

2 Basal Ganglia and PD

The motor symptoms of PD originate from the death of dopaminergic neurons in SNc. The functional changes in basal ganglia circuitry due to SNc cell loss are the direct triggers for the development of the cardinal features of PD (Obeso et al. 2008; Ellens and Leventhal 2013). In this section, we presented necessary discussions of the relevant functional neuroanatomy and circuitry of the basal ganglia, and the evolving PD models of basal ganglia.

2.1 Organizations of the Basal Ganglia

The basal ganglia are a group of interconnected subcortical nuclei consisting of the striatum (caudate nucleus and putamen), the globus pallidus (internal and external segments, GPi in primates or its homolog the entopeduncular nucleus in cats and rodents and GPe), the substantia nigra (pars compacta and pars reticulate, SNc and SNr), and the subthalamic nucleus (STN) (Albin et al. 1989; Utter and Basso 2008; Ellens and Leventhal 2013; Nelson and Kreitzer 2014). Based on their functional organization, the basal ganglia nuclei can be generally sorted as three subgroups: two principal input nuclei (striatum and STN), two principal output nuclei (SNr and GPi), and the intrinsic nuclei (GPe). The canonical model of basal ganglia organization, known as the direct- and indirect-pathway model, was presented about 25 years ago (Albin et al. 1989; Delong 1990; Alexander and Crutcher 1990). This model, though somewhat rigid and simple regarding its speculation for basal ganglia function and connectivity, is proven still the most influential and also extremely valuable model of base ganglia circuit function. According to this model, the inputs to the striatal projection neurons, known as medium spiny neurons (MSNs) characterized by their high spine density, negative resting potential, and low firing rates in vivo (Kreitzer and Malenka 2008), are segregated into two information processing pathways: direct and indirect pathways, which exert in opposing ways to control movement (Albin et al. 1989; Delong 1990; Alexander and Crutcher 1990). The direct pathway arises from GABAergic MSNs containing substance P, and projects monosynaptically to GPi/SNr, while the indirect pathway arises from GABAergic MSNs containing enkephalin, and projects polysynaptically to GPi/SNr via a sequential relay of GPe and STN (Albin et al. 1989; Delong 1990; Alexander and Crutcher 1990). Current viewpoints about the anatomical and functional organizations of the basal ganglia indicate the basal ganglia is neither a simple component of extrapyramidal systems in the first generation of the basal ganglia model, nor merely part of a cortical closed loop as "the direct and indirect pathway model" speculated. Rather it is an even more complex map that contains many reciprocal and loop connections (Goldberg and Bergman 2011). The basal ganglia nuclei have extensive connections with other brain regions, including the cerebral cortex, thalamus, and several brain stem nuclei, thus forming even more complicated networks with both open- and closed-loop circuitry (Kreitzer 2009; Nelson and Kreitzer 2014). These circuits exert not only the well-known roles in motor control but also the newly-appreciated functions such as attention and time estimation, implicit learning and habit formation, and reward-related behaviors and emotions (Albin et al. 1989; Delong 1990; Kreitzer and Malenka 2008; Obeso et al. 2008; Nelson and Kreitzer 2014; Calabresi et al. 2014), which involve three functional units (motor, associative, and limbic regions) of the basal ganglia. Motor, associative, and limbic cortical regions correspondingly innervate motor, associative, and limbic regions of the striatum, palladium, and SNr in certain fashions of topographic and somatotopic distribution (Parent and Hazrati 1995a, b; Hamani et al. 2004). In general, the basal ganglia have two major features regarding their anatomic and functional organization. The first major feature of its connectivity is its unidirectional flow of information. More specifically, cortical and thalamic information comes down the basal ganglia network from the input structure (the striatum and STN), through the central networks of GPe (the intrinsic nucleus), to the basal ganglia outputs (GPi and SNr); the other key feature that affects information flow is the dramatic decrease in tissue volume (for example 9700 mm^3 of dorsal striatum, 800 mm^3 of GPe, and 480 and 410 mm^3 of GPi and SNr respectively in humans) and the number of neurons (e.g., ~10,000,000 projection neurons in the striatum, 150,000 in GPe, and 50,000 in GPi in the primate; 1,700,000–2,800,000 neurons in the rodent striatum, 46,000 and 3200 neurons in the rodent homologues of the GPe and GPi respectively) down the striatum-GPe-GPi axis (funneling organization), suggesting robust information integration of the basal ganglia (Goldberg and Bergman 2011).

A schematic representation summarizing the connectivity within the basal ganglia nuclei has been presented in Fig. 14.1. More detailed discussion will be given in the following section.

2.1.1 Input to the Basal Ganglia

The striatum is the largest and also the primary input nucleus of the basal ganglia. It, in most mammals, consists of two functionally similar nuclei in dorsal striatum (the medial caudate and the lateral putamen separated by the fibers of internal capsule) and nucleus accumbens in the ventral portion (Albin et al. 1989; Ubeda-Bañon et al. 2007; Utter and Basso 2008). The dorsal striatum is closely involved in motor control, while the ventral striatum in motivation, reward, and addiction (Kreitzer and Malenka 2008). The striatal MSNs receive multiply originated inputs, including that from the cortex (corticostriatal projection), thalamus (thalamostriatal projection), SNc (nigrostriatal projection), dorsal raphe nuclei, pedunculopontine nuclei (PPN), the amygdala and so on. According to the results Wall et al. (2013) obtained with a monosynaptic rabies virus system-generated brain-wide map of neurons that form synapses with direct- or indirect-pathway MSNs, corticostriatal neurons comprised 61.1 % of total inputs onto the direct pathway and 69.9 % onto the indirect pathway; the thalamus provides 22.0 % of total inputs onto the direct pathway, 25.5 % onto the indirect pathway; MSNs are the only projection neurons in the striatum. The dorsal raphe nuclei and PPN provide very small amount of inputs (<1 %) to striatal MSNs; the amygdala preferentially innervates direct pathway MSNs (4.3 % vs. 0.1 % of total inputs onto indirect pathway MSNs). Surprisingly, this method only shows a very small proportion of total labeled inputs (0.8 %) arising from SNc, similarly innervating both pathway MSNs (Wall et al. 2013). By using a similar rabies virus that functions as a traditional retrograde tracer, the same investigators found a much higher proportion of total inputs (7.6 %) arising from SNc (Wall et al. 2013), similar to that observed by Pan et al. (2010) using other retrograde tracers (red or green Lumafluor beads). This might be attributable to the limitation of monosynaptic spread of rabies virus by some factors (Wall et al. 2013).

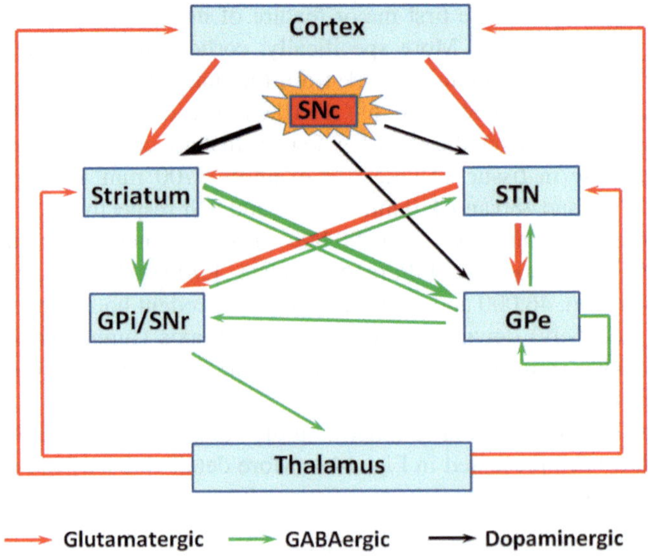

Fig. 14.1 Schematic demonstration of the connectivity within the basal ganglia nuclei and that of cortex and thamalus regulating motor control. The canonical direct- and indirect-pathway model of the basal ganglia proposes a dual organization of the striatum and basal ganglia outputs. According to this model, the inputs to the striatal projection neurons, known as medium spiny neurons (MSNs), are segregated into two information processing pathways: direct and indirect pathways, which are exerted in opposing ways to control movement (Albin et al. 1989; Delong 1990; Alexander and Crutcher 1990). The direct pathway arises from GABAergic MSNs containing substance P, and projects monosynaptically to GPi/SNr, while the indirect pathway arises from GABAergic MSNs containing enkephalin, and projects polysynaptically to GPi/SNr via a sequential relay of GPe and STN (Albin et al. 1989; Delong 1990; Alexander and Crutcher 1990). Current viewpoints about the anatomic and functional organizations of the basal ganglia indicate the basal ganglia is an even more complex map that contains many reciprocal and loop connections. The basal ganglia nuclei have extensive connections with other brain regions, including the cerebral cortex, thalamus, and several brain stem nuclei, thus forming even more complicated networks with both open- and closed-loop circuitry. These circuits perform not only the well-known roles in motor control but also in newly-appreciated functions such as attention and time estimation, implicit learning and habit formation, and reward-related behaviors and emotions, which involve three functional units (motor, associative, and limbic regions) of the basal ganglia. Motor, associative, and limbic cortical regions correspondingly innervate motor, associative, and limbic regions of the striatum, palladium, and SNr in certain fashions of topographic and somatotopic distribution. The major feature of basal ganglia connectivity is its unidirectional flow of information. Cortical and thalamic information comes down the basal ganglia network from the input structure (the striatum and STN), through the central networks of GPe (the intrinsic nucleus), to the basal ganglia outputs (GPi and SNr)

The striatal MSNs monosynaptically receive huge numbers of glutamatergic inputs from nearly every cortical area (corticostriatal projection) via distinct channels or with an overlapping fashion in subregions of the striatum to integrate cognitive and motor information (including action selection, motor control, sequence learning and habit formation etc.) from almost every cortical area (Albin et al. 1989;

Parent and Hazrati 1995a; Reiner et al. 2010; Shepherd 2013; Nelson and Kreitzer 2014). The inputs, primarily arising from the somatosensory and motor cortices mainly project to dorsolateral striatum, while significant inputs from prefrontal and limbic cortices predominantly give rise to dorsomedial striatum (Pan et al. 2010; Wall et al. 2013). Studies, by using electron microscopy (Lei et al. 2004) and genetic tool of Cre-dependent, cell-type-specific, monosynaptic rabies virus (Wall et al. 2013), demonstrated biased synaptic inputs of the cortical fibers onto MSNs of direct/indirect pathway. Namely, except prefrontal cortical structures (insular and orbitofrontal cortices, as well as the frontal association area) that equally innervate direct- and indirect-pathway MSNs, other cortical areas provide considerably biased synaptic input to one of these two pathways. For example, motor (primary and secondary) cortices provide significantly higher proportions of input to indirect way, whereas somatosensory (primary and secondary) and limbic (prelimbic, retrosplenial, cingulate, perirhinal, and entorhinal) cortices tend to provide a stronger proportion of input to the direct pathway; more posterior cortical structures (somatosensory and limbic cortices) provide more input to the direct pathway, while more rostral cortices (primary motor cortex) provide stronger input to indirect pathway (Wall et al. 2013). Based on their axon anatomy, there are at least two types of corticostriatal projection neurons in the cortex (Lei et al. 2004; Reiner et al. 2010). Pyramidal tract (PT-type) neurons project only ipsilaterally to the striatum and convey an efference copy of motor commands (for motor cortex at least), while intratelencephalically projecting (IT-type) neurons project ipsi- or bilaterally to the striatum (and the cortex) to convey sensory and motor planning information to striatum (Reiner et al. 2010). PT-type corticostriatal neurons are typically larger than IT-type corticostriatal neurons and mainly found in lower cortical layer 5, whereas IT-type corticostriatal neurons are mainly found in layer 3 and upper layer 5 (Lei et al. 2004; Reiner et al. 2010). It is thought that these two cortical neuron types differ in their targeting of the two main types of striatal projection neurons, with the IT-type input preferentially innervating direct pathway MSNs and the PT-type input preferentially innervating indirect pathway MSNs (Lei et al. 2004; Reiner et al. 2010). That means there is a laminar preference of the cortex to send its output to striatal direct- and indirect-pathway MSNs. In other words, lower cortical layer 5 (mainly IT-type neurons) preferentially sends input to direct- while layer 3 and upper layer 5 to indirect-pathway MSNs. However, a recent study performed by using the genetic tool of Cre-dependent, cell-type-specific, monosynaptic rabies virus found that direct- and indirect-pathway MSNs receive similar proportional levels of input from each cortical layer (Wall et al. 2013). This finding is further supported by an in vivo electrophysiological study that also shows direct- and indirect-pathway MSNs receive similar amounts of IT- and PT-type synaptic input (Ballion et al. 2008). An optogenetic study also provides direct evidence for this (Kress et al. 2013). The optogenetic approach is done by the use of a novel non-invasive tool that uses light at multiple wavelengths and locations to read out and control neural activity. This approach relies on genetically encoded proteins (probes) targeted to genetically defined cell types and sub-cellular compartments with exquisite spatial and temporal precision to allow the probes to be used for investigating multiple levels of nervous system function (Yizhar et al. 2011; Häusser 2014). By using the optogenetic approach combined with electrophysiological measurement of the responses of mouse striatal neurons to optogenetic activation of IT- and PT-type axons, Kress et al. (2013) demonstrated that direct and indirect pathway striatal MSNs respond to both IT- and PT-tract activation, arguing that these cortical networks innervate both striatal projection neurons. Thus the crucial aspect of corticostriatal organization remains to be established. In addition to the direct corticostriatal projection, the striatum may also receive an indirect cortical input via the relay of the intralaminar thalamic nuclei, especially the centromedian-parafascicular (CM-pf) nuclei (Albin et al. 1989).

The striatum also receives extensive inputs from the thalamus nuclei, including the medial dorsal, ventromedial, anterior medial, and anterior lateral nuclei, as well as CM-pf (thalamostriatal projection) (Pan et al. 2010; Wall et al. 2013; Smith et al. 2014), which are topographically organized into parallel but overlapping motor, cognitive, and limbic circuits (Obeso et al. 2008; Nelson and Kreitzer 2014). Although the thalamostriatal system originates from most thalamic nuclei, the CM/Pf (or the Pf in rodents) is the main source of thalamostriatal projections in both the primates and non-primates, CM/Pf neurons send massive and topographically organized projections to specific regions of the dorsal striatum, and are involved in the regulation of activities such as behavioral switching, attentional set-shifting and reinforcement as well as movement disorders in PD, while other (sensorimotor-, associative- and limbic-related) thalamic nuclei innervate functionally corresponding regions of the dorsal and ventral striatum (smith et al. 2014). Thalamic input into the striatum does not show biased connectivity to direct- and indirect-pathway MSNs (Wall et al. 2013). The thalamus is also a massive source of extrinsic glutamatergic inputs to the striatum that originates either in the CM/Pf nuclear complex, or in the numerous non-CM/Pf thalamic nuclei (Smith et al. 2014). The glutamatergic inputs from both cortex and thalamus converge upon single MSNs, and predominantly form asymmetric contacts on the heads of dendritic spines of MSNs (Parent and Hazrati 1995a; Smith and Villalba 2008; Obeso et al. 2008; Smith et al. 2014). It is estimated that approximately 5,000 glutamatergic afferents project onto each striatal MSN, and 100 MSNs project onto each pallidal neuron (Obeso et al. 2008), which demonstrates a highly convergent projection in the basal ganglia. These excitatory inputs get intensive dopaminergic modulation of afferents from the SNc via DA acting through a wide variety of pre- and/or post-synaptic mechanisms dependent on the type and localization of DA receptors and the physiological state of striatal MSNs (Surmeier et al. 2007; Tritsch and Sabatini 2012; Smith et al. 2014).

The other striatal inputs include a dense dopaminergic projection from SNc (nigrostriatal projection), and some non-canonical projections such as serotonergic afferents from dorsal raphe nuclei, a mixture of cholinergic, glutamatergic, and GABAergic fibers from PPN and inputs from the amygdala (Albin et al. 1989; Pan et al. 2010; Wall et al. 2013). These inputs to MSNs tend to form thousands of symmetric synaptic contacts on the spine and dendritic shafts of MSNs (Smith and Villalba 2008; Obeso et al. 2008). Though the dopaminergic nigrostriatal projection processes only a small proportion of total inputs to the striatum in contrast to corticostriatal and thalamostriatal inputs (Wall et al. 2013), it plays a critical role in PD pathogenesis. Nigrostriatal projection is topographically organized along a medial–lateral axis and includes projections from throughout the anterior–posterior extent of the dopaminergic cell bodies in the ventral midbrain (Pan et al. 2010). The ventral midbrain dopaminergic neurons process 75 % of all dopaminergic neurons, and are subdivided into three main groups: SNc, ventral tegmental area (VTA), and retrorubral field, with 400,000–600,000 dopaminergic neurons found in the human and 20,000–30,000 in the mouse (Smith and Villalba 2008; Hegarty et al. 2013). Neurons in VTA and the retrorubal field innervate the ventral striatum (and the prefrontal cortex) via the mesocorticolimbic system, and are involved in the regulation of emotion and reward (Smith and Villalba 2008; Hegarty et al. 2013). Dopaminergic neurons in SNc project to the dorsal striatum via the nigrostriatal pathway, and are required for the control of voluntary movement, and the loss of these neurons is the pathological hallmark of PD (Smith and Villalba 2008; Hegarty et al. 2013). Dopaminergic neurons in SNc are more sensitive to neurodegenerative insults than in other subgroups; this is likely due to subtle developmental differences during their ontogeny, the sparseness of some endogenous neuroprotective mechanisms (e.g., lack of the Ca^{2+} binding protein calbindin D28K), and greater expression of DA transporter which largely increases neurotoxicity of the environmental toxic agents such as MPTP to dopaminergic neurons (Dauer and Przedborski 2003; Smith and Villalba 2008; Hegarty et al. 2013).

Neurons in the striatum include morphologically distinct MSNs and aspiny neurons. Aspiny neurons, comprising only ~5 % of all striatal neurons, are the interneurons, which include large aspiny neurons or tonically active cholinergic interneurons (1–2 %) and several different subtypes (on the basis of their expression of the Ca^{2+} binding proteins parvalbumin or calretinin, or their expression of the neuropeptides somatostatin, or neuropeptide Y or nitric oxide synthase) of aspiny GABAergic interneurons (3–4 %) electrophysiologically distinguished as fast-spiking interneurons or persistent and low-threshold spike interneurons. They are indirect players and crucial regulators of striatal output due to their ability to form microcircuity within the striatum (Kreitzer 2009; Tepper et al. 2010; Gittis and Kreitzer 2012). MSNs make up approximately 95 % of all neurons and are the only projecting neurons integrating glutamatergic inputs from the cortex and thalamus and sending GABAergic projections to neurons in downstream basal ganglia nuclei. These neurons have typically been considered as homogeneous populations regarding their somatodendritic morphology. However, based on anatomical projection patterns and biochemical differences, MSNs can be divided into two classes. D1 DA receptor-expressing MSNs express substance P, dynorphin, and directly project to basal ganglia output nuclei (SNr and GPi in the primates or the entopeduncular nucleus in cats and rodents) (striatonigral projection). These are termed "direct-pathway MSNs", whereas D2 DA receptor-expressing MSNs are known as "indirect-pathway MSNs." They express enkephalin, and project indirectly to basal ganglia output nuclei via the GPe and STN (striatopallidal projection) (Gerfen et al. 1990). Such dipartition of striatal D1 and D2 MSNs is confirmed by the studies with

D1 and D2 transgenic mice and optogenetic identification (Bertran-Gonzalez et al. 2008; Kravitz et al. 2010; Freeze et al. 2013; Thibault et al. 2013; Calabresi et al. 2014). These neurons also have different electrophysiological characteristics, with D2 MSNs consistently firing at higher frequencies and more excitably than D1 MSNs (Calabresi et al. 2014). Approximately 80–90 % of striatal MSNs in the dorsolateral striatum belong to the direct/indirect pathway classification system, and the remaining 10–20 % are found in neurochemically distinct patches (defined by the intensity of histochemical staining for acetycholinesterase) throughout the striatum, and are termed striosomes or patches (Graybiel 1990; Gittis and Kreitzer 2012). These striosome neurons typically express D1 receptors, contain mainly substance P and GABA, and project to a subset of dopaminergic neurons in SNc (Albin et al. 1989; Gittis and Kreitzer 2012). It is thought that MSNs project to direct- and indirect- pathway with a ratio of 1:1 (Nelson and Kreitzer 2014). However, single-cell studies in rats revealed that 37 % of MSNs project exclusively to the GPe ("pure" indirect pathway), whereas only 3 % project only to the SNr or entopeduncular nucleus ("pure" direct pathway), and 60 % of labeled neurons project to the SNr/ entopeduncular nucleus and possess collateral terminal fields in the GPe termed "bridging collaterals" to bridge the direct pathway with the functionally opposing indirect pathway (see Cazorla et al. 2014 and the herein references). The bridging collaterals can be upregulated by enhanced neuronal excitability in indirect D2 MSNs, which in turn increases inhibition of pallidal neurons in vivo and disrupted locomotor activation after optogenetic stimulation of the direct pathway (Cazorla et al. 2014). In addition, MSNs, though only a small proportion, are proven to co-express D1 and D2 DA receptors in a single neuron (Aizman et al. 2000; Hasbi et al. 2009; Perreault et al. 2011). In vivo recordings using Cre-dependent viral expression of the genetically encoded Ca^{2+} indicator GCaMP3 in the dorsal striatum of D1-Cre (direct-pathway-specific) and A_{2A}-Cre (indirect-pathway-specific) mice combined with an innovative approach of fibre optics and time-correlated single-photon counting in mice performing an operant task indicate transient increases in neural activity in both direct- and indirect-pathway MSNs when animals initiated actions (Cui et al. 2013). These findings directly challenge the canonical direct- and indirect-pathway model of the basal ganglia which presumes these two pathways are not only functionally opposing, but also anatomically segregated (Albin et al. 1989; Delong 1990; Alexander and Crutcher 1990).

Following DA depletion, MSNs change their activities. D1- and D2- receptors couple to the different signaling systems in the MSNs. Gs-coupled D1 receptors are proposed to facilitate MSN output, whereas Gi-coupled D2 receptors inhibit MSN firing (Surmeier et al. 2007; Kreitzer and Malenka 2008). Thus, after DA depletion, D1-expression MSNs are largely inhibited, while D2-expressing MSNs become less inhibited (de-inhibition), which make the direct downstream indirect-pathway neurons in GPe hypoactive (due to the enhanced GABAergic inhibition by D2-expressing MSNs), and direct-pathway nuclei (GPi and SNr) hyperactive, resulting in the over-inhibition of the thalamus and the cortex, and ultimately the PD.

The subthalamic nucleus, first described by French neurologist Jules Bernard Luys in 1865, is a small biconvex-shaped obliquely oriented nucleus, located at the diencephalon-mesencephalic junction with the borders defined by the zona incerta superiorly and postero-medially, prelemniscal radiations and postero-lateral hypothalamus anteromedially and cerebral peduncle laterally. It has an average size of ~240 mm^3 and ~560,000 neurons in the human, 2.7 mm^3 and 35,000 neurons in marmosets, 34 mm^3 and 155,000 neurons in macaques, 50 mm^3 and 230,000 neurons in baboons, and 0.8 mm^3 and 25,000 cells in the rodent (Hamani et al. 2004; Castrioto et al. 2014). STN, presumed to be only a key node (as a relay nucleus) of the indirect pathway serving as a "gate" for cortico-basal ganglia-thalamocortical circuits in the canonical direct- and indirect-pathway model of the basal ganglia (Albin et al. 1989; Delong 1990; Alexander and Crutcher 1990), is now appreciated also as an important input nucleus of the basal ganglia (Nambu et al. 2002), and a critical component of complex networks controlling not only motor functions (Hamani et al. 2004; Baunez et al. 2011) but also cognition, emotion, behavior, and thalamocortical excitability (Temel et al. 2005; Baunez et al. 2011; Castrioto et al. 2014). STN receives extensive inputs from multiple brain regions (Parent and Hazrati 1995b; Hamani et al. 2004). Most of the cortical afferents to STN (corticosubthalamic projection) arising from the primary motor cortex (mainly layer 5), supplementary motor area, and premotor cortices, project predominantly to the dorsal part of STN and form motor loops. These loops are those that arise from the frontal and supplementary frontal eye movement cortices to the ventromedial portion of STN, and the prelimbic-medial orbital area of the prefrontal cortex to the medial STN as part of limbic loop (Parent and Hazrati 1995b; Nambu et al. 2002; Hamani et al. 2004). STN also receives inputs from other cortical areas (cingulate, somatosensory, insular cortex), but the function is unclear (Hamani et al. 2004). The other major afferents to STN derive from GPe or GPi in non-primates (pallidosubthalamic projection) with a complex topographical organization. In general, motor and limbic portions of GPe innervate their corresponding counterparts in STN, associative pallidal afferents innervate both the associative and motor portions of STN (Parent and Hazrati 1995b; Hamani et al. 2004). STN also receives the projections arising from the thalamus (mainly CM/pf, also the reticular nucleus) (thalamo-subthalamic projection), the brainstem nuclei (SNc, PPN, dorsal raphe nucleus), and several other regions such as zona incerta, the locus coeruleus, the hypothalamus, the amygdala, the parabrachial nuclear complex, and the dorsolateral tegmental nucleus (Parent and Hazrati 1995b; Hamani et al. 2004). Thus, STN cells integrate both excitatory glutamatergic inputs mainly from the cortex and the thalamus with inhibitory GABAergic inputs predominantly from GPe in order to modify their firing rate (increase or decrease) and pattern (tonic or bursting, oscillation or synchronization), which is critical to the physiology and pathophysiology of the basal ganglia (Parent and Hazrati 1995b; Hamani et al. 2004; Montgomery 2007; Obeso et al. 2008). The STN projects massively to the pallidal complex (both GPe and GPi) (subthalamo-pallidal projection) and SNr (subthalamo-nigral projection), but much less profusely to the striatum (subthalamo-striatal projection) and SNc. The other outputs of STN include the projection to PPN and the ventral tegmental nucleus (Parent and Hazrati 1995b; Hamani et al. 2004). Therefore, multiple-source inputs and multiple-target outputs of STN, even many reciprocal connectivities, form many large-scaled loops and microcircuitries, and make STN a center in the basal ganglia, which stimulates the proposal of a "center-surround" model of basal ganglia function (Mink 1996; Nambu et al. 2002). Recently, the "cortico-subthalamic-pallidal" projection has been of increasing interest in motor control and basal ganglia diseases (Nambu et al. 2002). This pathway conveys powerful excitatory effects from the motor-related cortical areas to GP, bypassing the striatum, and is faster in signal conduction from the cerebral cortex than the direct and indirect pathways, making it thus termed the "hyperdirect pathway" (Nambu et al. 2002). Based on this pathway and extensive connectivity of STN, a dynamic model of basal ganglia function, "center-surround model", was proposed and expanded (Mink 1996; Nambu et al. 2002), challenging the canonical direct- and indirect-pathway model of the basal ganglia (Albin et al. 1989; Delong 1990; Alexander and Crutcher 1990) regarding the regulation of voluntary movement. The key points of the center-surround model are as follows:

Before initiating a voluntary movement, corollary signals of the motor cortex are sent simultaneously to GPi via direct, indirect, and hyperdirect pathways, respectively. Because signal conduction is faster in hyperdirect pathway than in direct and indirect pathways, the signal in hyperdirect pathway first reaches GPi and activates GPi neurons extensively (a widespread facilitation of the surround), thus resulting in the inhibition of large areas of the thalamus and cortex that are related to both the selected motor program and other competing programs. Then the signal in the direct pathway reaches the GPi to inhibit a specific population of pallidal neurons in the center area, which causes disinhibition of pallidal targets and releases only the selected motor program. Finally, another signal, possibly through the indirect pathway, reaches the GPi to activate neurons, therein, suppressing their targets extensively. Through this sequential information processing, only the selected motor program is initiated, executed, and terminated at the appropriate timing, whereas other competing programs mediated by pallidal neurons in the surrounding area are canceled (Nambu et al. 2002). According to this model, activity of STN neurons during voluntary limb movements is likely to be mediated by the excitatory cortico–STN projection rather than by the inhibitory GPe–STN projection. Such increased STN activity may contribute to increased activity of pallidal neurons. Thus, it is most likely that the increased pallidal activity during voluntary limb movements is mediated by the net excitatory, faster cortico–STN–pallidal 'hyperdirect' pathway, while the decreased pallidal activity is mediated by the net inhibitory, slower direct pathway (Nambu et al. 2002). This model has gained supports from some recent studies (Sano et al. 2013; Schmidt et al. 2013), which address the involvement of the hyperdirect pathway in the development of action canceling or stop signals for motor and cognitive programs.

The average firing rate of STN neurons is 13–18 Hz in normal rats and 18–25 Hz in non-human primates, which provides tonic excitatory input to both the GPe and GPi (Hamani et al. 2004; Obeso et al. 2008; Ellens and Leventhal 2013). Following DA depletion in PD, the firing rate of STN increases, STN neurons become hyperactive, and the STN-GP circuit also becomes more reactive which provokes a surge of independent rhythms (circuit oscillation) in the STN-GP network (Hamani et al.

2004; Obeso et al. 2008). The circuit oscillations are highly related to the characteristic parkinsonian tremor in relative low frequency (<10 Hz), and bradykinesia and akinesia in higher frequency (15–30 Hz) in PD patients (Hamani et al. 2004; Obeso et al. 2008). To "silence" the hyperactive STN, chronic stereotactic deep brain stimulation (DBS) of STN has been established and proven an effective treatment in advanced PD since DBS is believed to cause a reduction of neuronal activity by means of a depolarization block (Temel et al. 2005). On the other hand, some investigators postulated that early denervation of the STN following PD may act as a compensatory mechanism that maintains the basal ganglia output within normal limits despite a progressive loss of striatal DA, which may be the reason why clinical features do not emerge until a 70–80 % reduction of striatal DA levels are developed (Obeso et al. 2008).

2.1.2 Intrinsic Nucleus

The globus pallidus is a component of the basal ganglia involved in the regulation of voluntary movement. In primates, it is separated into two segments, GPe and GPi, by small fibers of passage and pallidal border cells. In rodents, the GPi is embedded in the fibers of the internal capsule, and thus called the entopeduncular nucleus, and the GPe is referred to simply as the globus pallidus. To simplify the designation, we use GPi for the generic terms of GPi in the primates and its rodent homolog entopeduncular nucleus.

Though pallidal neurons in both GPe and GPi express the same morphological and chemical phenotype, their afferent and efferent connections clearly show that the two segments of pallidal complex are distinct functional entities (Parent and Hazrati 1995b; Goldberg and Bergman 2011; Schwab et al. 2013). The GPi is considered one of the output nuclei and the principal gateways of the basal ganglia (will be discussed next), while the GPe is part of the indirect pathway and the principal intrinsic nucleus of the basal ganglia since it receives inputs mainly from basal ganglia structures (striatum, STN, and GPe itself) and projects to all basal ganglia structures, but not to structures outside of the basal ganglia (Goldberg and Bergman 2011). The GPe establishes the reciprocal interconnections with both the striatum and the STN, both of which provide the major afferents of GPe, and also receive the back projections from GPe (Parent and Hazrati 1995b; Goldberg and Bergman 2011). The striatal MSNs provide the major GABAergic synapses onto the pallidal neurons, while the major glutamatergic and excitatory input to the pallidal neurons is derived from the STN, and there are many more GABAergic synapses (~90 %) from the striatum than STN glutamatergic synapses (5–10 %) on the pallidal neurons (Jaeger and Kita 2011; Goldberg and Bergman 2011). Single-cell studies in rats revealed that 37 % of MSNs project exclusively to the GPe ("pure" indirect pathway), and 60 % of labeled neurons project to the SNr/GPi and possess collateral terminal fields in the GPe termed "bridging collaterals" to bridge the direct pathway with the functionally opposing indirect pathway (see Cazorla et al. 2014 and the herein references). In addition to the afferent inputs from the striatum and STN,

GPe neurons also receive collateral innervation from other GPe neurons (Goldberg and Bergman 2011), and such projection comprises about 10 % of the GABAergic synapses in the GPe itself (Schwab et al. 2013).

The GPe neurons project to virtually every other basal ganglia nucleus, including itself (Obeso et al. 2006; Jaeger and Kita 2011; Goldberg and Bergman 2011). The projection to STN and GPi and the inhibition of these nuclei are frequently considered to be the primary model of action of GPe. The GPe to GPi projection that was neglected in the canonical direct- and indirect-pathway model of the basal ganglia (Albin et al. 1989; Delong 1990; Alexander and Crutcher 1990) actually processes a large percentile (~48 % in the monkey) of terminals arriving to the perikarya of GPi neurons (Obeso et al. 2006; Goldberg and Bergman 2011). The GPe and its reciprocally interconnected STN show a very precise anatomical projection to the GPi, namely the reciprocally interconnected areas of both nuclei converge on the same region in the GPi (Obeso et al. 2006). The feedback projection of GPe to the striatum is another output pathway of GPe that was neglected in the past (Obeso et al. 2006). Approximately 40 % of the GPe neurons of this projection predominantly express preprokephalin and form synapses with striatal interneurons (Obeso et al. 2006). In addition, GPe also sends fibers to SNr. Thus, the projections from the GPe to basal ganglia output stations provide an additional pathway that parallels the direct inhibitory striatal-GPi/SNr projection (Parent and Hazrati 1995b; Deniau et al. 2007; Zhou and Lee 2011). Taken together, these findings strongly suggest that the GPe is not a mere relay station in the indirect pathway of the basal ganglia, but rather, like STN (Nambu et al. 2002), occupies a central position in the basal ganglia circuitry, which is reciprocally connected to the striatum and the STN, and is a major source of innervation to the GPi and the SNr (Parent and Hazrati 1995b; Goldberg and Bergman 2011). Thus, GPe, together with GPi, act as bi-stable resonators in the information processing in the basal ganglia (Parent and Hazrati 1995a, b; Goldberg and Bergman 2011).

The pallidal neurons have unique cellular morphology and electrophysiology. Their cellular morphology is more uniform than that of most other areas in the CNS, with relatively large (20–60 μm) triangular or polygonal cell bodies and thick, sparsely spined, poorly branching dendrites (Nambu 2007; Goldberg and Bergman 2011). Most GPe cells fire at high rates with occasional pauses, and others tend to fire sparsely with intermittent bursts (Goldberg and Bergman 2011; Ellens and Leventhal 2013). GPe neurons generally show no spike correlation with each other (desynchronized activity) in normal awake animals (Jaeger and Kita 2011; Goldberg and Bergman 2011; Schwab et al. 2013). Following DA depletion in PD, GP neurons undergo plastic changes in their synaptic and cellular structure. For example, the STN-GP circuit becomes more reactive to the influence of the activity of cortical inputs after DA depletion (Hamani et al. 2004; Obeso et al. 2008). Their electrophysiological function also changes. With DA depletion, there is a decrease in average GPe discharge rate in the rodents, and increase in burst discharges and highly synchronized activity (pattern changes) (Jaeger and Kita 2011; Goldberg and Bergman 2011; Schwab et al. 2013). These changes exert powerful impacts on basal ganglia information processing and output in PD state.

2.1.3 Output of the Basal Ganglia

The GPi and SNr are the two major output structures and the smallest nuclei with relatively fewer neurons in the basal ganglia (Goldberg and Bergman 2011). Though both of the nuclei are separated by the dense fibers of the internal capsule in most mammals, they are considered as a continuum since they contain cytologically similar neurons, receive similar inputs from the striatum, GPe, and STN, and project to the cerebral cortex via the relay of the thalamus. Thus, they have similar functional role in principle (Albin et al. 1989; Parent and Hazrati 1995a; Nambu 2007). Their differences merely reflect their topographical organization, namely, the GPi has strong interactions with the motor cortices and control somato-motor behaviors (especially in movement initiation, sequential organization and movement and action selection), while the SNr has interactions with the prefrontal cortex and mainly control associative function (Nambu 2007). Both nuclei play a central role in the transmission of the basal ganglia output to the thalamus and ultimately to the motor cortex.

The cellular morphology of GPi neurons is similar to that of GPe neurons. However, unlike GPe neurons, GPi neurons do not have extensive local axon collaterals (Nambu 2007). In contrast to GPe neurons, GPi neurons also show a different firing pattern, displaying continuous high rate discharges of ~60 Hz with rapid firing rate fluctuations ranging from 10 to 107 Hz, but without displaying pause or prolonged silences that GPe neurons do in primates (DeLong 1971; Nambu 2007). The rodent homologous entopeduncular nucleus neurons also display a high firing rate with an average of 26 Hz (ranging from 2 to 76 Hz) in freely moving state, which can be significantly decreased (to 15 Hz ranging from 0.2 to 74 Hz) in anaesthetic state (Benhamou and Cohen 2014). The GPi activity either increases or decreases during voluntary limbic activity, which is probably influenced by phasic glutamatergic and GABAergic inputs from direct, indirect, and hyperdirect pathways (Nambu 2007; Sims et al. 2008; Tachibana et al. 2008; Gross et al. 2011). GPi neurons are enriched with GABA receptors, including the ionotropic $GABA_A$ and metabotripic $GABA_B$ receptors (Kita et al. 2006; Nambu 2007). $GABA_A$ receptors located at symmetric synapses are consistent with their role in fast inhibitory synaptic transmission, and the extrasynaptic distribution of both $GABA_A$ and $GABA_B$ receptors provides a substrate for complex modulatory functions that rely predominantly on the spillover of GABA (Nambu 2007). Under STN blockade, single stimulation of the putamen in awake monkeys induces $GABA_A$ receptor-mediated responses, while burst stimulation evokes $GABA_B$ receptor-mediated responses in GPi (Kita et al. 2006). Glutamate receptors, including ionotropic AMPA- and NMDA- and metabotropic receptors, are also present in GPi neurons (Nambu 2007; Tachibana et al. 2008). In addition, there are also D1 DA receptors in GPi neurons (Smith and Villalba 2008; Benazzouz et al. 2014). GPi neurons are thought to be essential for voluntary movements, and GPi lesion (Turner and Anderson 2005) or ablation (Obeso et al. 2009) may cause motor disturbances. In addition to motor control, GPi neurons may initiate reward related signals (Hong and Hikosaka 2008).

The GPi neurons receive a combination of major excitatory glutamatergic input from the STN (~10 %) and major inhibitory GABAergic inputs from the striatum (~70 %) and GPe (~15 %) (Parent and Hazrati 1995a; Nambu 2007; Nelson and Kreitzer 2014). The GPi neurons also receive glutamatergic inputs from the intralaminar nuclei of the thalamus, serotonergic inputs from the dorsal raphe nuclei, glutamatergic and cholinergic inputs from the PPN, and dopaminergic inputs from the SNc (Parent and Hazrati 1995a; Nambu 2007; Nelson and Kreitzer 2014). Synaptic terminals of these inputs differ in their localization on the GPi neurons, as well as in their density. For example, glutamatergic afferents from the STN are homogeneously distributed on the soma and dendrites of GPi neurons, GABAergic terminals from the striatum terminate more heavily on the distal dendrites than on the soma and proximal dendrites, while GABAergic terminals from the GPe terminate almost exclusively on the soma and proximal dendrites (Nambu 2007). These inputs are the main factors governing the spontaneous activity of GPi neurons. For example, stimulation of motor cortex induces a typical response of GPi neurons: an early excitation, an inhibition, and a late excitation, in which the early and late excitations are glutamatergic and induced by the cortico-STN-GPi and cortico-striato-GPe-STN-GPi pathways, respectively, while the inhibition is GABAergic and induced by the cortico-striato-GPi pathway (Tachibana et al. 2008).

The GPi nucleus projects primarily to motor thalamus (the ventral anterior/ventral lateral nuclei), which in turn projects back to the cerebral cortex as well as the striatum, thus forming loop circuits for motor control; other projections of GPi include CM/pf thalamic nucleus, lateral habenula, and the PPN (Parent and Hazrati 1995a; Nambu 2007; Nelson and Kreitzer 2014). The primate GPi consists of two concentric zones according to the organization pattern: a cenral core and a peripheral zone (Parent and Hazrati 1995a). The pallido-thalamic projection occurs in the core of the GPi along its entire rostrocaudal extent, and is topographically organized with the fibers arborizing in the form of numerous typical glomerule-like plexuses distribution (Parent and Hazrati 1995a). GPi neurons projecting to the PPN are also confined to the core of the GPi; whereas those projecting to the habenula are located more peripherally (in the dorsal and ventral borders of the GPi), and this projection is involved in the initiation of reward related signals (Parent and Hazrati 1995a; Hong and Hikosaka 2008). The GPi projections use GABA as their neurotransmitter, and exert an inhibitory impact on thalamo-cortical relay neurons.

The SNr is located ventral and adjacent to the SNc, and has a low density of neurons scattering within a dense neuropile of SN neurons in contrast to SNc. It is comprised primarily of GABAergic neurons as well as some intermixed dopaminergic and cholinergic neurons (Deniau et al. 2007; Zhou and Lee 2011). SNr GABAergic neurons are the essential projection cells with medium to large sized, and pleomorphic (from triangular, fusiform, ovoid to polygonal) shaping soma and 2–6 primary dendrites frequently branching in their distal parts. These neurons electrophysiologically show sustained spontaneous short duration (~1 ms) spike firing at very high frequency (e.g., 25–30 Hz in awake rodents, ~65 Hz in awake primates), providing a default tonic inhibitory signal to their targets (Deniau et al. 2007; Zhou and Lee 2011). In contrast, SNr dopaminergic neurons fire low frequency

(1–4 Hz) and long duration (~2.5 ms) spikes. In addition, SNr GABAergic neurons show no or little adaptation in both firing frequency and spike amplitude, whereas SNr dopaminergic neurons are adaptive in both parameters (Deniau et al. 2007; Zhou and Lee 2011). Some dopaminergic neurons also contain GABA. SNr neurons are inhibitory to their target structures.

Similar to GPi, SNr neurons also receive a combination of the primary inhibitory GABAergic inputs from the striatum and GPe and the primary excitatory glutamatergic input from the STN (Deniau et al. 2007). In addition, SNr neurons receive dopaminergic input from SNc, serotonergic input from raphe nuclei, and a mixture of cholinergic, glutamatergic, and GABAergic fibers from PPN (Zhou and Lee 2011). The striatonigral input originates from the MSN GABAergic neurons. In most cases, electrical stimulation of the striatum evokes $GABA_A$/$GABA_B$ receptor-mediated short duration inhibitory synaptic events in SNr neurons. However, excitatory events can also be elicited by striatal stimulation, which likely result from a disinhibition of the subthalamonigral neurons via the indirect striato-nigro-subthalamic circuit or via local synaptic interactions between SNr cells via their local axon collateral network, or a spurious axon reflex in the glutamatergic subthalamonigral projections elicited by antidromic activation of the subthalamostriatal neurons (Deniau et al. 2007; Zhou and Lee 2011). Recent studies using selective optogenetic control of the direct and indirect pathways demonstrate that optogenetic activation of each pathway elicits both excitation and inhibition of SNr neurons, and direct pathway-mediated movement initiation selectively correlates with the subpopulation of inhibited SNr neurons, while indirect pathway-mediated motor suppression is most strongly influenced by excited SNr neurons (Kravitz et al. 2010; Freeze et al. 2013). However, in another recent study (Sano et al. 2013), it is shown that the ablation of D2 receptor-expressing striatal MSN neurons has little effect on spontaneous activity in the SNr, but induces the reduction of the cortically evoked late excitation in the SNr and motor hyperactivity, which is not well consistent with the above observations with optogenetic control of indirect pathway (Kravitz et al. 2010; Freeze et al. 2013). Sano's results suggest the striatopallidal indirect pathway-mediated change of phasic signals, rather than the spontaneous firing rates and patterns in the SNr, seems to be responsible for motor control, especially in stopping movements (Sano et al. 2013). This is a paradox of prediction in the direct- and indirect-pathway model, in which changes of the mean firing rate of the output nuclei in the basal ganglia play a key role in motor control (Albin et al. 1989; Delong 1990; Alexander and Crutcher 1990). It has been demonstrated that the striatonigral axon terminal strongly expresses D1 DA receptors and cannabinoid CB1 receptors (Smith and Villalba 2008; Zhou and Lee 2011). Activation of D1 receptors presynaptically facilitates GABA release, while activation of CB1 receptors presynaptically inhibits it, thus acting as tonic modulators of SNr neuronal firing activity and the basal ganglia output (Zhou and Lee 2011). The GPe also sends GABAergical fibers to SNr (pallidonigral input). Its afferent terminals make large symmetrical synapses preferentially on the soma and proximal dendrites, forming perisomatic baskets on GABAergic neurons (Parent and Hazrati 1995b; Deniau et al. 2007; Zhou and Lee 2011). Because GPe occupies a central position in the basal ganglia circuitry and has powerful GPe-SNr projection, activation of indirect pathway significantly increases SNr GABAergic activity because of the de-inhibition caused by the inhibition of GPe neurons (Albin et al. 1989; Kravitz et al. 2010; Freeze et al. 2013). The subthalamonigral projection provides excitatory glutamatergic input to SNr cells. The STN projects massively to SNr (subthalamonigral projection), forming excitatory, asymmetric synapses with SNr GABAergic neurons which also receive inhibitory GABAergic inputs in the same cells from striatal D1-expressing MSNs and GPe neurons (note GPe also receives glutamatergic input from STN) (Deniau et al. 2007; Zhou and Lee 2011). Electrophysiological studies have demonstrated that high-frequency stimulation of STN induces a concomitant synaptic excitation–inhibition in SNr neurons resulting in a net increase of the spontaneous spiking activity in half of SNr neurons and a decreased activity in the remaining neurons (Bosch et al. 2011). It also results in large compound glutamatergic and GABAergic currents in SNr neurons with massive depression of glutamatergic currents but persistence of GABAergic currents during repeated and/or prolonged STN stimulation (Dvorzhak et al. 2013). Such dual excitatory and inhibitory responses to STN stimulation among SNr neurons arise probably from the concomitant activation of glutamatergic subthalamo-nigral and GABAergic GPe-nigral fibres passing through the STN (Bosch et al. 2011; Dvorzhak et al. 2013). In addition, STN activation stimulates GABA release from GPe-nigral fibres probably via STN-GPe pathway; the released GABA not only causes inhibitory responses in SNr neurons via $GABA_A$ receptors, but also suppresses glutamate release from subthalamonigral terminals by acting at presynaptic $GABA_B$ receptors (Zhou and Lee 2011; Dvorzhak et al. 2013), thus leading to massive depression of glutamatergic currents (Dvorzhak et al. 2013). Also, there might be presynaptic D1 and D2 receptors in subthalamonigral terminals (Smith and Villalba 2008). Activation of D1 receptors facilitates, while D2 inhibits glutamate vesicle release at subthalamonigral axon terminals. These effects also exert an impact on the intensity and pattern of glutamatergic subthalamonigral input to SNr GABA neurons (Zhou and Lee 2011). In addition, the dopaminergic SNc-SNr and raphe nuclei-SNr serotonergic input may exert a tonic excitatory influence on SNr GABAergic neurons; PPN-SNr cholinergic input preferentially innervates SNr dopaminergic neurons but also reaches to SNr GABAergic neurons. Stimulation of this pathway most often produces short latency excitation of SNr GABAergic neurons (Zhou and Lee 2011).

Like GPi, an output nucleus of the basal ganglia, the SNr sends massive inhibitory projections to specific thalamic relay nuclei (nigrothalamic projection), namely the ventromedial and intralaminar thalamic nuclei in rodents and the ventral anterior and ventral lateral nuclei in primates, where GABAergic terminals end at the proximal dendrites and soma of relay neurons to form synapses. These relay nuclei in turn project back to the motor cortex (Parent and Hazrati 1995a; Deniau et al. 2007; Zhou and Lee 2011). Antidromic and orthodromic stimulations clearly demonstrated that the caudolateral inhibitory SNr neurons in the monkeys monosynaptically project to the mediodorsal and ventral anterior thalamic nuclei, as well as the ventral lateral nucleus pars medialis, where thalamocortical neurons send projections to the frontal cortex, and the SNr neurons for this projection pathway are topographically organized in the SNr (Tanibuchi et al. 2009a). Caudolateral SNr neurons also project to thalamic mediodorsal pars multiformis and mediodorsal pars parvocellularis neurons, where the relay neurons send projection to the prefrontal area ventral to the principal sulcus (Tanibuchi et al. 2009a). This pathway seems to play more important roles in task performance (Tanibuchi et al. 2009b). There is no significant overlap of SNr and GPi terminal fields at thalamic levels in the primates (Parent and Hazrati 1995a). The SNr also projects to superior colliculus (nigrocollicular projection) which plays an important role in the control of saccadic eye movements and orienting behaviors, and to mesopontine tegmentum (nigrotegmental projection) which participates in both motor and attentional control mechanisms (Parent and Hazrati 1995a; Deniau et al. 2007; Zhou and Lee 2011). SNr projection neurons are topographically organized along the same onion-like architectural plan (Parent and Hazrati 1995a; Deniau et al. 2007). Nigrocollicular neurons located ventrolaterally in the peripeduncular lamina of SNr project to the intermediate gray layer of the superior colliculus with a mediolateral topography, while those located in more dorsal laminae and in the central core of SNr innervate the intermediate white layers of the superior colliculus in a series of patches organized with a precise topography (Deniau et al. 2007). Moreover, SNr projection neurons possess a recurrent axon collateral network with multiple targets and to SNr itself in order to exert mutual inhibitory interactions (Parent and Hazrati 1995a; Deniau et al. 2007; Zhou and Lee 2011). These interactions may desynchronize firing of adjacent SNr neurons innervating the same targets, and such desynchronization involves surrounding mechanisms of contrast acting in synergy of distant SNr neurons that partially share the same input (Deniau et al. 2007).

The motor control function of the basal ganglia is critically determined by the output of GPi/SNr. Indeed, many motor or motor-related disorders such as PD, Huntington's disease, dystonia, chorea, Tourette's disorder, and Wilson's disease, thought etiologically different, are resulted from the altered output signals of the basal ganglia. For example, PD symptoms are a result of increased inhibitory output of GPi/SNr and the consequent excessive inhibition of thalamocortical and/or brainstem motor systems. Following DA depletion, increased firing rates (Filion and Tremblay 1991; Hutchison et al. 1998; Wichmann et al. 1999; Kliem et al. 2010; Bosch et al. 2011) and enhanced bursting (Wichmann and Soares 2006; Tsiokos et al. 2013; Nambu and Tachibana 2014) in GPi/SNr of some MPTP-treated primate models, 6-hydroxydopamine-treated rodent models, and PD patients, as well as increased synchronization of neuronal activity throughout the basal ganglia in 6-hydroxydopamine treated rats (Dejean et al. 2008, 2012; Ge et al. 2012) have been demonstrated (also see review by Rubin et al. 2012; Ellens and Leventhal 2013). Indeed, stimulation of SNr neurons with single current pulses (0.5 mA strength, 0.3 ms duration) at a low frequency of 0.7 Hz evokes powerful inhibition in thalamic neurons in the monkeys (Tanibuchi et al. 2009a), supporting the prediction that the increased activity in GPi/SNr neurons decreases the activity in thalamic and cortical neurons, which result in akinesia in the direct- and indirect- pathway model (Albin et al. 1989; Delong 1990; Alexander and Crutcher 1990). Therapeutic DBS has been an established symptomatic treatment modality for Parkinson's disease and other movement disorders (Kalia et al. 2013). One of the potential therapeutic mechanisms of DBS involves the modification of these disrupted pathological firing patterns of the GPi/SNr and increased synchronization of neuronal activity throughout the basal ganglia (Cleary et al. 2013; McCracken and Kiss 2014). Indeed, high-frequency stimulation (therapeutic stimulation) of GPi has been shown to completely inhibit spontaneous firing of neighboring GPi neurons as well as both cortically evoked responses and spontaneous discharges by strong GABAergic inhibition in normal awake monkeys (Chiken and Nambu 2013). It also reduces or silences the mean neuronal firing rate, post-stimulus burst firing time in most (but not all) GPi neurons, and loosely entrains neuronal firing (namely, a period of silence in the seconds after the start of stimulation followed by a slow increase in firing rate over the course of stimulation) in some neurons of PD patients (Cleary et al. 2013) as well as dystonia patients (Liu et al. 2012). Finally, high-frequency stimulation of GPi widely alters the spontaneous and stimulus-induced oscillations of local field potential along a motor cortical-basal ganglia-thalamic circuit including the GPi (the entopeduncular nucleus) in the rodents and PD patients (Tsiokos et al. 2013; McCracken and Kiss 2014). These changes of DBS on GPi activity, as the direct- and indirect-pathway model predicted, are supposed to result in reduced inhibition and consequently, increased activity by de-inhibition in thalamic neurons (Albin et al. 1989; Delong 1990; Alexander and Crutcher 1990). However, it is strange to find that the therapeutically effective GPi DBS predominately produces inhibition rather than excitation of neurons in the thalamus (Anderson et al. 2003; Montgomery 2006; Liu et al. 2012) and the primary motor cortex (Johnson et al. 2009) in MPTP-treated monkeys and PD patients. Obviously, this finding is a paradox against the notion that decreased GPi output increases the activity of thalamic and cortical neurons as predicted by the direct- and indirect-pathway model. However, it is also important to note that even though GPi DBS predominately inhibits thalamic neurons, many neurons have a post-inhibitory rebound increased activity that results in net excitation in thalamic neurons (Montgomery 2006). This seems consistent with the prediction of the direct- and indirect-pathway model. In a computational model of the basal ganglia thalamic circuit, periodic high frequency stimulation of the STN reduced the incidence of thalamic cell errors from the high error rates seen in the Parkinsonian case back to the low error rates seen in the normal-healthy case (Dorval et al. 2009). Though the firing rate of GPi cells may play some minor role, disordered patterns of GPi activity lead to irregular synaptic inhibition of the thalamus while ordered patterns of GPi activity lead to regular synaptic inhibition of the thalamus, thus regularizing the pathological patterns of GPi activity improving parkinsonian motor symptoms (Dorval et al. 2009). In addition, during therapeutic stimulation of spanning GPe and GPi for rigidity and bradykinesia, neurons in primary motor cortex exhibit an inhibitory phase-locking to the stimulus and an overall decrease in mean discharge rate (another paradox of prediction in the direct- and indirect-pathway model). However, they also show an increase in response specificity to passive limb movement (that means a decreased error rate as seen in the thalamus) (Johnson et al. 2009). These findings support the viewpoint that DBS-induced changes of GPi activity do not necessarily facilitate thalamic (and cortical) discharges, but may act, instead, to interrupt abnormal patterns of thalamic (and cortical) discharges associated with parkinsonian symptoms, thus producing therapeutic effects on parkinsonian symptoms (Anderson et al. 2003; Chiken and Nambu 2014).

2.2 *The Evolving Basal Ganglia Models of PD*

The canonical direct- and indirect-pathway model of the basal ganglia proposed a dual organization of the striatum and basal ganglia outputs; these two pathways are not only functionally opposing, but also anatomically segregated (Albin et al. 1989; Delong 1990; Alexander and Crutcher 1990). Its explanation for the pathophysiology of movement disorders is that activity imbalance between the direct and indirect pathways changes the mean firing rate of the output nuclei (GPi/SNr) of the basal ganglia and induces hypokinesia or hyperkinesia, thus it is also called the firing rate model in some documents (Montgomery 2007; Nambu 2008; Nelson and Kreitzer 2014). Specifically for PD, DA depletion reduces tonic excitation to striatal neurons in the direct pathway through D1 receptors and tonic inhibition to striatal neurons in the indirect pathway through D2 receptors disorders. These effects result in increased activity in GPi/SNr neurons and decreased activity in thalamic and cortical neurons, which ultimately results in akinesia (Albin et al. 1989; Delong 1990; Alexander and Crutcher 1990). The firing rate model is a revolution in basal ganglia research history, and has been proven the most influential model for PD. It has been widely used to generate testable hypotheses regarding the physiology and connectivity of basal ganglia nuclei in health and disease, explain experimental findings, build models of basal ganglia disorders, and explain therapeutic effects of both pharmacological and neurosurgical treatments. It has also provided a conceptual framework and a remarkable foundation for better understanding the pathophysiology of PD and other movement disorders (Obeso et al. 2008; Calabresi et al. 2014; Nelson and Kreitzer 2014). Indeed, this model has been gaining much support from the studies of PD animal models and patients (Obeso et al. 2008; Calabresi et al. 2014; Nelson and Kreitzer 2014). In particular, as a consequence of the pathophysiological theory of the firing rate model, pallidotomy, subthalamotomy, and STN DBS in PD patients have been demonstrated to markedly improve the cardinal motor signs of PD, including tremor, rigidity, bradykinesia, abnormal gait, and balance (Obeso et al. 2008; Calabresi et al. 2014; Nelson and Kreitzer 2014). It is the simplicity of the model that made it extraordinarily attractive for clinicians and basic scientists alike (Nelson and Kreitzer 2014). However, with the numerous advances in our understanding of the anatomic and physiological state of the basal ganglia in the normal and dopamine depleted state, its rigidness and oversimplicity have made the model questionable. Many observations made in the laboratory and in PD patients over the years have highlighted some limitations of the firing rate model (see the scattered contents in the above discussion, and the reviews by Montgomery 2007; Nambu 2008; Obeso et al. 2008; Calabresi et al. 2014; Nelson and Kreitzer 2014). For example, although the firing rate model has successfully explained the bradykinesia and the alterations of automatic movements with reduced blinking rate, decreased arm swing, and short stride length when walking, it provides neither an adequate explanation for other essential characteristic motor abnormalities in PD, including difficulty in performing simultaneous and sequential movements, a progressive reduction in amplitude while performing a repetitive movement, and the striking improvement in gait freezing induced by visual cues. Neither does it explain the two other cardinal features of PD, that is, rigidity and tremor (Nambu 2008; Obeso et al. 2008). It also cannot explain several other key findings, for example, the finding that lesions of the motor thalamus or GPe do not result in bradykinesia or akinesia and that GPi lesions in parkinsonian patients do not result in dyskinesias (DeLong and Wichmann 2009). Also, this model assumes a clear anatomical segregation and functional opposition between the direct and indirect pathway. However, recent studies have shown these two pathways inextricably linked (Calabresi et al. 2014), and the cortico-STN-GPi/SNr hyperdirect pathway also plays an important role in motor control (Nambu et al. 2002; Nambu 2008). In addition, interneurons in the striatum play an important role in the regulation of neurotransmitter release (Threlfell et al. 2012; Nelson et al. 2014) and movement disorders via microcircuits (Gittis and Kreitzer 2012). The emerging non-dopaminergic treatments for PD, such as anticholinergics (e.g., trihexyphenidyl, benzatropine, orphenadrine, procyclidine, biperiden) and A_{2A} adenosine receptor antagonist (e.g., istradefyllineare, preladenant, SYN-115) which are mainly targeted at interneurons, are becoming a scheduling program for the pipe-lines of drug development and clinical trials (Brichta et al. 2013; Stayte and Vissel 2014; Connolly and Lang 2014). This is also an unnoticed part in the classical model. Indeed, it has been clear that the basal ganglia encode information using a more sophisticated algorithm than firing rate alone (Nelson and Kreitzer 2014). Generally in the firing rate model, its oversimplicity expels many new anatomical and functional findings, and its anatomical rigidness makes it a static rather than dynamic functional model. Thus, some other models have been proposed either as a supplement or a substitute for the classical model to explain the physiology and pathophysiology of the basal ganglia (Montgomery 2007; Nambu 2008; Obeso et al. 2008; Calabresi et al. 2014; Nelson and Kreitzer 2014), such as the center-surround model (Mink 1996; Nambu et al. 2002), system oscillators model (Montgomery 2007), and the firing pattern model (Rivlin-Etzion et al. 2006; Hammond et al. 2007; Rubin et al. 2012).

Neurons code information not only by firing rate but also by the amplitude and the pattern of firing. The firing rate model has been supported by many studies that found elevated firing rates in the GPi and STN, and depressed rates in the GPe, thalamus, and motor cortex in PD animal models and patients. However, many other studies also reported that firing rates in the GP, STN, thalamus, or cortex do not change in the way that the rate model would predict with the induction of parkinsonism (Hammond et al. 2007; Obeso et al. 2008; Rubin et al. 2012; Nelson and Kreitzer 2014). It is now becoming clear that abnormalities in the neuronal discharge pattern, beyond firing rate alone, play central roles in the genesis of parkinsonian signs. The firing pattern model emphasizes the presence of abnormal firing patterns such as burst and oscillatory patterns in the basal ganglia and abnormal synchronization throughout the motor circuit in PD (Rivlin-Etzion et al. 2006; Hammond et al. 2007; Rubin et al. 2012). Indeed, studies in PD animal models and patients have found significant increases in the prevalence of burst firing in GPe, STN, and GPi/SNr, which appear early during the induction of experimental parkinsonism, and therapy with L-DOPA or GPe/STN DBS reduces burst firing (Hammond et al. 2007; Rubin et al. 2012; Ellens and Leventhal 2013). Some studies also showed that DA depletion increases bursting firing in the thalamus, the primary motor cortex, the striatum, and the SNc (Rubin et al. 2012; Ellens and Leventhal 2013; Wang et al. 2014). Oscillatory activity can be recorded in single unit discharges or local field potentials; the latter reflects the activity of networks and is generally easier to measure than single unit activity. In PD animal models and patients, the STN, GPe, and GPi can produce regularly recurring fluctuations in firing, and striatal cholinergic interneurons produce tonic activity that coordinates synaptic inputs from both the SNc and intralaminar thalamic nuclei by pause activity (Kreitzer 2009; Rubin et al. 2012; Ellens and Leventhal 2013). The GPe-STN interconnections form a mutual inhibitory network, and may function as a GPe-STN oscillator. Following DA depletion, bursting GPe neurons may hyperpolarize STN neurons, and STN neurons, in turn, fire rebound bursts that stimulate GPe neurons, thus forming a regenerative loop to stimulate oscillation (Ellens and Leventhal 2013). However, independent mechanisms may underlie the burst discharges and oscillatory firing of most neurons in the parkinsonian GPi. Therefore, bursts and oscillations in the parkinsonian GPi represent independent properties of neural activity (Chan et al. 2011). Frequency analysis of oscillations in local field potentials reveals low frequency band oscillations (4–10) corresponding to resting tremor and higher frequency beta-band oscillations (10–30 Hz), which are correlated with rigidity and bradykinesia (but not tremor), in humans and rodent models of PD. These higher frequency oscillations can be suppressed by dopaminergic replacement therapies (Rubin et al. 2012; Ellens and Leventhal 2013). Synchronous firing is rare in different neurons of the basal ganglia under physiological condition, but becomes very prevalent in single unit spikes in the GPe, STN and GPi and among oscillating units after DA depletion in PD animal models and patients (Rivlin-Etzion et al. 2006; Hammond et al. 2007; Rubin et al. 2012; Ellens and Leventhal 2013). Synchrony is supposed to impair the selection and compression of cortical information (action selection) by the basal ganglia, which leads to the loss of specificity of the basal ganglia neurons for passive and active movement, thus causing bradykinesia (Hammond et al. 2007; Ellens and Leventhal 2013). Treatment with DA receptor agonists, L-DOPA, and DBS is able to reduce such synchrony, consequently improving the motor symptoms in PD (Hammond et al. 2007; Rubin et al. 2012; Ellens and Leventhal 2013; McCracken and Kiss 2014). The firing pattern model is attractive and is gaining support from experimental and clinical studies. However, there are also many unsolved problems of the firing pattern model. For example, a study of the dynamic relation between the emergence of abnormal synchronous oscillatory activity and the onset of motor symptoms during a progressive dopamine depletion process in monkeys has indicated that oscillatory firing patterns may emerge at a relatively later stage of DA depletion after the development of parkinsonian motor symptoms. Thus, a causality between the emergence of synchronous oscillations in the pallidum and main parkinsonian motor symptoms seems unlikely (Leblois et al. 2007). This study suggests that oscillatory activities may be present in advanced Parkinsonism, but are not necessarily the primary cause of akinesia or bradykinesia (DeLong and Wichmann 2009). There are also some studies showing neither L-DOPA treatment in animal models affecting burst firing within the basal ganglia, nor therapeutic STN and GPi DBS producing decreased synchronized regular firing pattern (which actually induces a significant increase in mean discharge rate and a stimulus-synchronized regular firing pattern) in both GPe and GPi of PD monkeys (see review by Montgomery 2007 and the references herein). The firing pattern model assumes that failure of action selection leads to bradykinesia in the context of PD; however, there is still a lack of studies on establishing the causal link between excessive synchronization and bradykinesia (Hammond et al. 2007). Moreover, this model does not provide full explanations regarding the muscle tone disorder (rigidity in PD) and tremor. The model suggests that low frequency band oscillations (4–10 Hz) correspond to resting tremor and higher frequency beta-band oscillations (10–30 Hz) are correlated with rigidity and bradykinesia (but not tremor). However, tremor-related activity involves not only the rhythmic activity in the basal ganglia but also the thalamus, the cortex, and the cerebellum. Tremor-related activity even primarily originates outside the basal ganglia (i.e., brainstem motor nuclei or spinal cord) (Rivlin-Etzion et al. 2006; Obeso et al. 2008). In addition, a recent study in PD patients indicated that though the power of beta band oscillations of contralateral primary sensori-motor cortex increases proportionally to the severity of the affected forearm during rest as well as during isometric contraction, cortico-muscular coherence does not differ among the groups of drug–naive, de novo patients, early-stage patients with chronic medication, and age-matched control subjects (Pollok et al. 2012). Thus, studies are needed to establish the causal link between increased beta band oscillations and muscle tone disorder.

The center-surround model of the basal ganglia shares the common assumption with the firing rate model of neuronal firing rate coding information of the basal ganglia output. However, as discussed earlier, the center-surround model addresses the involvement of the hyperdirect pathway in the development of action canceling or stop signals for motor and cognitive programs (Mink 1996; Nambu et al. 2002), which is neglected in the firing rate model (Albin et al. 1989; Delong 1990; Alexander and Crutcher 1990). This model assumes that the role of the striatal input is to provide a focused inhibition of basal ganglia output nuclei (center), while the cortico-STN hyperdirect pathway provides a widespread facilitation of the output nuclei (surround), and the integrated output on the target structures is to disinhibit wanted movements and inhibit undesirable movements (a signal filtering process) (Mink 1996; Nambu et al. 2002). A recent study using optically activated proteins targeting STN neurons showed that in freely moving hemiparkinsonian rats, direct STN cellular interventions by either silencing or exciting STN itself has no effects on amphetamine-induced behavioral rotation. Only selective stimulation of cortico-STN afferent axons can robustly stop the behavioral rotation in these rats, suggesting that the cortico-STN hyperdirect pathway, not STN itself plays a key role in DBS therapeutic effects on PD (Gradinaru et al. 2009). This finding favors the center-surround model. The center-surround model is also successful in the explanation of dystonia and tics, two hyperkinetic disorders that are difficult to explain with the firing rate model, but its predictions for some other basal ganglia movement disorders (e.g., Tourette syndrome) remain to be tested in animal models (Utter and Basso 2008).

The system oscillators model can be partially regarded as an expansion of the firing pattern model in a system viewpoint from the changes within the basal ganglia to the basal ganglia-thalamus-cortical system. This model proposes that the basal ganglia-thalamus-cortical system is a large set of multiple loosely coupled oscillators. Oscillators of different frequencies of the basal ganglia-thalamus-cortical could combine in the manner of an inverse Fourier transform to produce any complex signal necessary to encode the appropriate patterns for movement (Montgomery 2007). Accordingly, DA depletion in PD could affect synaptic efficiency within the ganglia-thalamus-cortical system, thus changing the oscillatory dynamics. Consequently, there may be a reduction of resonance modulation of the ventrolateral thalamus–motor cortex circuit with corresponding problems in positively and negatively modulating electromyographic activity. This results in a delay in reaching sufficient force to produce movement, which, rather than slowed responses to initiating cueing, ultimately leads to the symptoms of akinesia and bradykinesia (Montgomery 2007). The system oscillators model accounts for phenomena such as the similar efficacy of brain stimulation across multiple structures within the ventrolateral thalamus–motor cortex circuit, and the production of Parkinsonism by lesion of multiple different structures such as the GPe and supplementary motor area (Montgomery 2007). However, the changes of oscillations within the system are highly variable due to factors such as short-term plasticity in basal ganglia networks, noise and fluctuating inputs from other parts of the nervous system, and the intrinsic properties of the system. For example, even the synchronous oscillations can be easily formed in PD, not only in the appearance of motor symptoms, but at rest as well, and there have some healthy functions for the synchronous oscillations at rest (Rubchinsky et al. 2012; Tsiokos et al. 2013). In addition, this model is largely based on the deduction of information processed in sets of multiple nested, loosely coupled, synaptic, non-linear, re-entrant, polysynaptic oscillators. It seems too abstract and lacking in specific and detailed explanation of the cardinal symptoms of movement disorders in the basal ganglia. Indeed, increased beta band oscillations prevail not only in the basal ganglia, but also in the thalamus and the cortex (Ellens and Leventhal 2013), suggesting that they may have different roles and are not coherent in motor impairment of PD (Pollok et al. 2012). Thus, this model remains to be tested by more studies.

In general, the current models, just like two sides of the same coin, provide explanation only for some, but not all phenomena. Our knowledge of the basal ganglia has been evolving throughout the history of basal ganglia research, from a component of extrapyramidal subsystem of movement control in the first generation, to a dichotomous-component part of a cortical closed motor loop in the second generation, and an even more complex map of connectivity and functions for motor control, cognition, and emotion in the present (Goldberg and Bergman 2011). Our understanding of the physiology and pathophysiology of the basal ganglia has been changing too, thanks to the introduction and use of D1 and D2 bacterial artificial chromosome (BAC) transgenic mice (Valjent et al. 2009), and the development of new techniques such as optogenetic identification (Bertran-Gonzalez et al. 2008; Kravitz et al. 2010; Freeze et al. 2013; Thibault et al. 2013) and quantitative anatomical studies (e.g., single cell tracing study, monosynaptic rabies virus system-generated brain-wide maps study) (Wall et al. 2013; Cazorla et al. 2014), which provide powerful tools for the insight into basal ganglia functions. Now, it is clear that the basal ganglia are an even more complex map of connectivity, and encode information using a more sophisticated algorithm. It is time to produce new models for the basal ganglia. An ideal model of the basal ganglia should at least have such features: it is a multidisciplinary hybrid other than single subject (e.g., an interdiscipline comprehending subjects such as anatomy, physiology, molecular biology, medicine, even large scale system cybernetics and theory of information processing); it integrates new anatomical and functional discoveries of the basal ganglia; it is compatible to the logical and rational components of current models; finally, it can provide testable hypotheses to explain more related phenomena, and efficiently direct the treatments for basal ganglia movement disorders, especially PD, the most common disease in the basal ganglia. Fortunately, such attempts have been on the way. The findings that show the simultaneous activation of both direct and indirect pathway striatal MSNs during a motor task (Cui et al. 2013) and cooperative activity of these two pathways during voluntary movement (Isomura et al. 2013), at first glance, contradict the rate model, but upon further examination, support an action selection form of the model (Nelson and Kreitzer 2014). Also, the finding that rapid indirect pathway striatal MSN firing may drive pause-and-burst activity in GPe conceives the hypothesis that information may be translated from a rate code in the striatum to a pattern code in GPe (Ellens and Leventhal 2013). This hypothesis is interesting because it is an attempt to explain how selectively driving indirect pathway striatal MSNs can cause akinesia and excessive basal ganglia bursting by connecting both the firing rate and pattern models. Most recently, after the retrospection of the classical model and the experimental and clinical consequences of the model as well as the new findings beyond the model, Calabresi et al. (2014) proposed a new model of the basal ganglia. Here we'd like to call it "the coordinated action model". In this model, intrastriatal connections (via the possible molecular, biochemical, and synaptic cross-talks between direct and indirect pathway) are critical and the two pathways are structurally and functionally intertwined, and all MSNs might either facilitate or inhibit movement depending on the form of synaptic plasticity expressed at a certain moment. The authors further supposed that in PD, alterations of dopamine-dependent synaptic plasticity could alter this coordinated activity between the direct and indirect pathway via oscillatory activity. The lack of coordinated activity between these two pathways will result in the inability of the PD patient to physiologically start and stop action sequences, and in a global slowing of motor function, causes bradykinesia, freezing and gait festination (Calabresi et al. 2014). We are full of expectations for further experimental and clinical studies to validate this model.

3 δ-Opioid System and PD

Similar to other modulators within the basal ganglia, δ-opioid system is one of the components that may involve in the physiological and pathophysiological processes of the basal ganglia, particularly in PD (Steiner and Gerfen 1998; Samadi et al. 2006). The abundant expression of opioid receptors and their endogenous ligands, especially δ-opioid system within the basal ganglia, has attracted much attention for its contribution to such processes. However, there has been a relatively sparse number of highly influencing investigations on the role of δ-opioid system in pathophysiological functions, especially PD, within the basal ganglia over the past decades. In contrast to some other modulating systems such as dopaminergic, GABAergic, cholinergic, and glutamatergic systems in the basal ganglia physiology and pathophysiology, studies on δ-opioid system are basically at the margin position of investigation. One can image that nobody willingly dedicates himself to an inconspicuous thing that is not a remarkable hot spot issue. This is manifested by the fact that in sharp contrast to the rapid development and clinical trial of some new non-dopaminergic medicines for PD treatments (Brichta et al. 2013; Stayte and Vissel 2014; Connolly and Lang 2014), no opiodergic drugs are available for pharmacological PD treatments, a problem ascribable to our limited knowledge of δ-opioid system in the basal ganglia in PD. The reasons for such a situation are multiple. First, the extremely high homology of opioid receptor subtypes (DOR, MOR, KOR) and their signaling molecules and the lack of highly selective ligands make it hard to clearly differentiate the effect of a subtype without cross reactions. Thus, the contradictory results in the past studies reduced the enthusiasm of investigation. This is a common problem regarding opioid studies in the past. Second, in contrast to the rapid technical development and the consequent novel insight in some other modulating systems and specific neurons [e.g., the introduction and use of D1 and D2 BAC transgenic mice (Valjent et al. 2009) and optogenetic identification in striatal MSN study (Bertran-Gonzalez et al. 2008; Kravitz et al. 2010; Freeze et al. 2013; Thibault et al. 2013); optogenetic proteins selectively targeted to the input terminals or output fibers of the STN (Gradinaru et al. 2009)], the lack of the cell-specific targeting optogenetic proteins and animal models for δ-opioid system limits the in-depth investigation and insight of the role of δ-opioid system. These reasons make δ-opioid system largely a neglected element in PD research. However, it has gradually been realized that opioid system may contribute to neurodegenerative diseases such as PD and Alzheimer's disease (Samadi et al. 2006; Cai and Ratka 2012). In this section, we attempted to get the ideas of δ-opioid system into shape and establish a framework for the relatively clear role of DOR in the PD pathophysiology by discussing the issues of δ-opioid distribution within the basal ganglia: expression alteration of δ-opioid within basal ganglia in PD, δ-opioid regulation of neurotransmitter release in basal ganglia, effects of pharmacological DOR manipulation on PD and Levodopa-induced dyskinesia, and effects of DOR activation on dopaminergic neuron injury/cellular model of PD.

3.1 The Distribution of δ-Opioids and Receptors Within the Basal Ganglia

Both δ-opioid peptide precursor proenkephalin (PENK) and DOR are distributed all over the basal ganglia (Aubert et al. 2007) (see the schematic showing in Fig. 14.2). The striatum is the well-studied structure of δ-opioid distribution within the basal ganglia. In addition, in the GP (Tempel and Zukin 1987; Mansour et al.

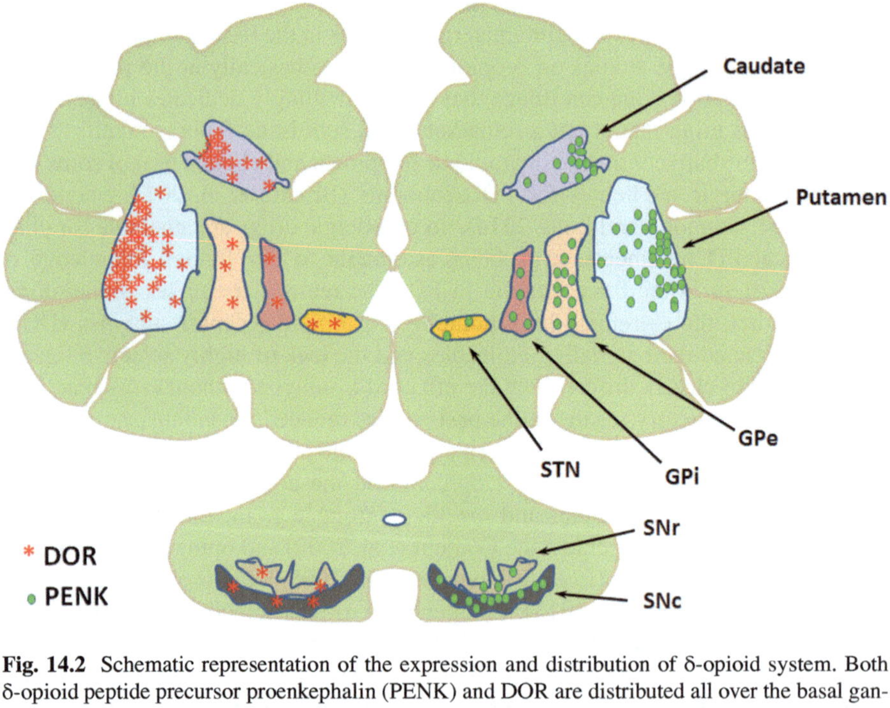

Fig. 14.2 Schematic representation of the expression and distribution of δ-opioid system. Both δ-opioid peptide precursor proenkephalin (PENK) and DOR are distributed all over the basal ganglia. DORs are highly abundant in the striatum. In the GP, both the SNr and SNc, and the entopeduncular nucleus (in rodents), show that there is also a DOR distribution despite there being a lower density than in the striatum. In the human brain, PENK mRNA expression is very high in the striatum (caudate-putamen and nucleus accumbens) and the SN, and scattered PENK mRNA hybridization signals (low and moderate to high) are also evident in the GP. In animals, PENK expression in the striatum is generally consistent with that in the human striatum, and scattered PENK-expressing cells are found in the GP. However, no PENK mRNA hybridization signals were found in the rat SN. Both DOR and PENK mRNA display a similar medial to lateral expression gradient in the caudate-putamen, with lateral levels at the maximum

1993, 1994; Bausch et al. 1995; Olive et al. 1997; Madar et al. 1996; Peckys and Landwehrmeyer 1999; Johansson et al. 2001; Aubert et al. 2007), both SNr and SNc (Tempel and Zukin 1987; Mansour et al. 1994; Jenab et al. 1995; Gray et al. 2006; Aubert et al. 2007; Léna et al. 2008), and entopeduncular nucleus (in the rodents) (Kornblum et al. 1987; Mansour et al. 1993), δ-opioid distributions exist despite a lower density than in the striatum. For example, multiple measurements have shown DOR expression is comparatively low in the globus pallidus and ventral pallidum, with a few scattered cells detected in both regions (Tempel and Zukin 1987; Mansour et al. 1993, 1994; Bausch et al. 1995; Olive et al. 1997; Madar et al. 1996; Aubert et al. 2007). In contrast to MOR presynaptic and postsynaptic localization in the pallidum, DOR immunostaining patterns reveals primarily a postsynaptic localization in pallidal cell bodies adjacent to enkephalin- or synaptophysin-positive fiber terminals, and the inhibitory GABAergic pallidostriatal feedback neurons project to the striatum (Olive et al. 1997). Thus it has been suggested that pallidal enkephalin release may be modulated by MOR located presynaptically on striatopallidal enkephalinergic neurons and by DOR located postsynaptically on pallidostriatal feedback neurons (Olive et al. 1997). In addition, quantitative receptor autoradiography combining a unilateral striatal lesion study suggested presynaptic terminals of striatal efferent fibers in the GP and SN also have DOR distribution, which might control GABA release from striatal efferent fibers in the GP and SNr (Abou-Khalil et al. 1984; Johansson et al. 2001). In the ventral thalamus where the STN is closely located, there are low levels of DOR binding (Kornblum et al. 1987; Mansour et al. 1993). However, it is not exactly clear whether the STN has DOR distribution. A study using in situ hybridization histochemistry reported that no hybridization signal is detected in the human STN, suggesting there is no DOR expression (Peckys and Landwehrmeyer 1999). However, another study using the same method showed a higher density of DOR in the STN, as well as in the striatum in comparison with the pallidal complex in monkeys (Aubert et al. 2007). A functional study with electrophysiological recording in the STN indicated that DOR is present in the presynaptic terminals where its activation inhibits inhibitory GABAergic synaptic transmission (Shen and Johnson 2002). The inhibitory GABAergic inputs to the rat STN are predominantly from GPe to modify their firing rate (increase or decrease) and pattern (tonic or bursting, oscillation or synchronization), which is critical to the physiology and pathophysiology of the basal ganglia (Parent and Hazrati 1995b; Hamani et al. 2004; Obeso et al. 2008). Therefore, the presynaptic inhibition of inhibitory GABAergic synaptic transmission by DOR might have certain relevance to PD.

It has been demonstrated with multiple measurements including in situ hybridization, immunocytochemistry, autoradiography, radioligand binding, and PET imaging in adult rodents and human beings that the striatum is one of the regions with the highest densities of DOR expression in the brain (Tempel and Zukin 1987; Mansour et al. 1993, 1994; Jenab et al. 1995; Bausch et al. 1995; Madar et al. 1996; Svingos et al. 1998, 1999; Peckys and Landwehrmeyer 1999; Johansson et al. 2001; Wang and Pickel 2001; Xia and Haddad 2001; Ambrose et al. 2006; Gray et al. 2006; Aubert et al. 2007; Léna et al. 2008). In addition, immunofluorescence imaging with confocal microscopy in DOR-eGFP knock-in transgenic mice has clearly shown the enrichment of DOR in the striatum (Scherrer et al. 2006; Bertran-Gonzalez et al. 2013). Developmental studies indicated that DOR develops considerably later than MOR and KOR in the CNS (Kornblum et al. 1987; Zhu et al. 1998; Georges et al. 1998). For example, DOR mRNA is first detected in the basal ganglia, especially the caudate-putamen, at embryonic day 14 (E14) (Zhu et al. 1998) or E17 (Winzer-Serhan et al. 2003) or E21 (Georges et al. 1998) in different reports in comparison with E11-13 for MOR and KOR mRNA (Zhu et al. 1998; Georges et al. 1998). Receptor binding analysis showed that MOR and KOR are extremely dense prior to birth, while DOR binding sites are almost completely absent at birth with a marked increase occurring between postnatal day 13 (P13) and P17 in the rat striatum (Kornblum et al. 1987; Wang et al. 2003). Cells expressing DOR have a more homogeneous distribution in the dorsolateral striatum (caudate-putamen) and the ventral striatum (nucleus accumbens), which is distinguishable from the patch- and cluster-like distribution of cells expressing MOR (Mansour et al. 1994; Ambrose et al. 2006). In the caudate-putamen, DOR expressing and binding show a medial to lateral gradient with the highest levels in the lateral portion of this structure (Kornblum et al. 1987; Mansour et al. 1993, 1994; Johansson et al. 2001; Wang and Pickel 2001; Ambrose et al. 2006) (Fig. 14.2). DOR is present in both patch and matrix compartments of the dorsolateral striatum in a mosaic pattern (Wang and Pickel 2001; Ambrose et al. 2006; Koizumi et al. 2013). In the ventral striatum, the highest levels of DOR expression and binding are seen in the ventral portions of the anterior shell with comparatively lower levers in the core portions (Mansour et al. 1993, 1994). In general, within the striatum, DOR is expressed both presynaptically and postsynaptically on glutamatergic, GABAergic, dopaminergic, and cholinergic afferents. Most of the DOR-immunoreactive somata are GABAergic MSN projecting neurons in the striatum (Svingos et al. 1998, 1999; Wang and Pickel 2001; Wang et al. 2003). A small portion of DOR-immunoreactive large cells are aspiny neurons co-localized with choline acetyltransferase (Le Moine et al. 1994; Wang and Pickel 2001; Winzer-Serhan et al. 2003), and are presumably the cholinergic interneurons in the striatum (Mansour et al. 1994; Le Moine et al. 1994; Winzer-Serhan et al. 2003; Wang et al. 2003). Indeed, Immunofluorescence imaging in DOR-eGFP knock-in mice has shown that most striatal cholinergic neurons expressed DOR-eGFP in the striatum (Scherrer et al. 2006; Bertran-Gonzalez et al. 2013). A functional study also suggested the existence of DOR in striatal cholinergic neurons (Britt and McGehee 2008). Ultrastructural evidence revealed that about half of D1 receptor-expressing neurons in the dorsal striatum also co-express DOR, suggesting a possible mechanism of D1 receptor modulation of DOR function (Ambrose et al. 2006). Immunocytochemical and immunoelectron microscopic labeling for DOR indicated that DOR has a preferential cytoplasmic distribution in spiny dendrites (mainly the site of postsynapse) and a more prominent presynaptic location (axonal terminals) in both dorsal and ventral striatum. Therefore, DOR can presynaptically regulate neurotransmitter releases and postsynaptic activity (Svingos et al. 1998, 1999; Wang and Pickel 2001; Wang et al. 2003). The axonal terminals immunoreactive for DOR contact either dendritic shafts or spines, forming both symmetric (with dendritic shafts and spine neck) and asymmetric (with spine head) synapses (Wang and Pickel 2001; Wang et al. 2003). Axon terminals forming asymmetric synapses on spine heads are mostly in the cerebral cortex (corticostriatal trace), and those on dendrites are mainly in the thalamus (thalamostriatal projection) (see Wang and Pickel 2001; Wang et al. 2003 and the reference herein). These kinds of synapses are predominantly the glutamatergic inputs from both cortex and thalamus converging upon single MSNs (Parent and Hazrati 1995a; Smith and Villalba 2008; Obeso et al. 2008; Smith et al. 2014). Thus, presynaptic DORs in these axonal terminals may play a role in the modulation of corticostriatal synapse formation as well as glutamate release (Wang et al. 2003). Indeed, electrophysiological studies have shown that in rat neocortical and corticostriatal brain slices, activation of DOR depresses glutamatergic excitatory transmission by presynaptic inhibition (Jiang and North 1992; Tanaka and North 1994; Ostermeier et al. 2000; Jose et al. 2007), though some studies with microdialysis also showed increased glutamate release from corticostriatal glutamatergic terminals by DOR stimulation (Billet et al. 2004, 2007). Axon terminals forming symmetric synapses with spine necks (Smith and Villalba 2008; Obeso et al. 2008) are probably from the dopaminergic nigrostriatal neurons and cholinergic interneurons in the striatum (Wang and Pickel 2001) since DOR-immunoreactive large cells are aspiny neurons colocalized with choline acetyltransferase in striatal neurons with typical characteristics of interneurons (Le Moine et al. 1994; Mansour et al. 1994; Wang and Pickel 2001; Winzer-Serhan et al. 2003; Wang et al. 2003). Electron microscopic imaging in the ventral striatum shows a colocalization of DOR- and DA transporter-immunoreactivity (Svingos et al. 1999). DOR in this kind of synapse may control neurotransmitter release from nigrostriatal terminals and striatal cholinergic interneurons. Indeed, it has been shown that activation of DOR inhibits, while blockade of DOR potentiates, acetylcholine release in the striatum (Beani et al. 1982; Mulder et al. 1984; Arenas et al. 1990; Sandor et al. 1992), and endogenous striatal acetylcholine release by synchronized activity in the cholinergic interneurons is sufficient to evoke DA release (Threlfell et al. 2012). Thus through this way, DOR may indirectly regulate DA release in the striatum. Moreover, most of the studies indicated that in the striatum (including both the dorsal and ventral portions), activation of DOR and its subtypes increases DA release (Chesselet et al. 1981, 1983; Longoni et al. 1991; Pentney and Gratton 1991; Sandor et al. 1992; Dourmap and Costentin 1994; Billet et al. 2004, 2007; Bosse et al. 2008; Hipólito et al. 2008; Tanahashi et al. 2012).

It has been shown that interactions between receptors can regulate DOR expression and distribution in the basal ganglia. Immunogold histochemical analysis indicated that prolonged treatment with morphine (48 h) specifically induces a translocation of DORs from intracellular to subplasmalemmal and membrane compartments in dendrites from both the ventral and dorsal striatum but not from the frontal cortex, suggesting a MOR-DOR interaction (Lucido et al. 2005). Similar observations were also made with enkaphalin in the striatum (Steece et al. 1986). In addition, dopaminergic D2 receptors, cannabinoid CB1 receptors, and DA transporters also regulate DOR expression and function in the basal ganglia, including in the caudate-putamen, the ventral striatum, the GP, and the ventral pallidum (Le Moine et al. 2002; Urigüen et al. 2005; Léna et al. 2008; Blume et al. 2013). For example, in DA transporter knockout mice, the levels of DOR mRNA, but not MOR and KOR mRNA, are largely increased in adult caudate putamen and nucleus accumbens at a time when these receptors are believed to be functional, suggesting a stimulation of DOR expression with chronic constitutive hyperdopaminergia (Le Moine et al. 2002). In contrast, CB1 receptor knockout mice exhibit increased expression and function of DOR in the caudate-putamen (Urigüen et al. 2005; Blume et al. 2013), and D2 receptor knockout mice show a down-regulation of DORs as well as MORs in the ventral striatum, the caudate-putamen, and the ventral pallidum and/or the GP (Léna et al. 2008), or an increase in DOR but not MOR expression due to a down-regulation of opioid peptide precursor proenkephalin in the striatum, mainly in the indirect pathway MSNs (Blume et al. 2013). These regulations of DOR expression and function are specific, not general changes, since complete knockout of the nociception/orphanin FQ receptor does not induce changes in DOR, MOR, nor KOR expression (Homberg et al. 2009). It is worthy to note CB1 receptors play a key role in biochemical cross-talk between direct and indirect pathways (Calabresi et al. 2014), thus its regulation of DOR expression might have some relevant contributions in PD pathology.

Proenkephalin (PENK), formerly known as proenkephalin A (now changed since proenkephalin B was renamed prodynorphin), is an endogenous opioid polypeptide which, via proteolyic cleavage, produces the pentapeptide enkephalin with 6 copies of Met-enkephalin, and 1 copy of Leu-enkephalin for each PENK (Noda et al. 1982). Met-enkephalin peptide sequence is coded by the enkephalin gene; Leu-enkephalin peptide sequence is coded by both the enkephalin gene and the dynorphin gene (Comb et al. 1982; Noda et al. 1982). DOR is considered the receptor for endogenous Met-enkephalin and Leu-enkephalin (Lord et al. 1977; Waldhoer et al. 2004). Similar to DOR, PENK is also widely expressed throughout the basal ganglia (Fig. 14.2). In situ hybridization showed that in the basal ganglia of the normal human brain, PENK mRNA expression is very high in the striatum (caudate-putamen and nucleus accumbens) and the SN, and scattered PENK mRNA hybridization signals (low and moderate to high) are also evident in the GP (Noda et al. 1982; Nisbet et al. 1995; Brana et al. 1995; Hurd 1996; Wang et al. 2007). Correspondingly, the high levels of the PENK-derived opioid peptides (Met-enkephalin and Leu-enkephalin) are found in the striatum, the GP, and the SN of human beings (Pioro et al. 1984; Haber and Watson 1985; Fernandez et al. 1994, 1996). In animals, PENK expression in the striatum is generally consistent with that in the human striatum (Shivers et al. 1986; Harlan et al. 1987; Reimer et al. 1992; Mansour et al. 1993; Svingos et al. 1998; Winzer-Serhan et al. 2003). In rats, scattered PENK-expressing cells are found in the GP, particularly in the ventral pallidum (Wamsley et al. 1980; Harlan et al. 1987; Mansour et al. 1993), but the highest concentrations of enkephalins were found within the GP of the mammalian brain (Hong et al. 1977; Palkovits et al. 1981). The enkephalinergic fibers are thought to originate primarily in the GABAergic MSN neurons of caudate-putamen nucleus and to project topographically upon the GP (Brann and Emson 1980; Del Fiacco et al. 1982), since GABA and enkephalin may coexist within individual caudate neurons of the type that are known to project to the GP (Aronin et al. 1984, 1986). No PENK mRNA hybridization signals were found in the rat SN (Wamsley et al. 1980; Harlan et al. 1987; Mansour et al. 1993). Immunohistochemical and in situ hybridization studies showed a moderate number of enkephalinergic terminals in the rat STN (Wamsley et al. 1980; Peckys and Landwehrmeyer 1999). Consistent with DOR distribution, PENK mRNA displays a similar medial to lateral expression gradient in the caudate-putamen, with the highest levels laterally (Mansour et al. 1993; Winzer-Serhan et al. 2003). The first detectable PENK mRNA is at E14 (Schwartz and Simantov 1988) and enkephalin at E16 (Bayon et al. 1979), which is earlier than DOR mRNA. PENK mRNA expression first appears in the ventrolateral striatum, spreading dorsally and medially throughout the caudate-putamen by the end of gestation (Winzer-Serhan et al. 2003). The distribution pattern of PENK mRNA seems changing with development. Specifically, PENK is mainly localized in matrix neurons with lower level of expression in the patch compartment in embryonic caudate-putamen of human beings and rats (Brana et al. 1995; Winzer-Serhan et al. 2003), but is more abundant in the patch than the matrix in adult rats (Gerfen and Young 1988; Wang et al. 2007). More than half of the neurons in the caudate-putamen express PENK (Shivers et al. 1986; Gerfen and Young 1988; Gerfen et al. 1991). In contrast to the prodynorphin-derived opioid peptides (endorphins) that are present predominantly in the D1 receptor-expressing striatonigral pathway (the direct pathway), PENK-derived opioid peptides are contained primarily in the D2 receptor-expressing striatopallidal pathway (the indirect pathway) (Cuello and Paxinos 1978; Del Fiacco et al. 1982; Gerfen and Young 1988; Gerfen et al. 1990).

3.2 *δ-Opioid Regulation of Neurotransmitter Release in the Basal Ganglia*

The basal ganglia contain a remarkable diversity of neuroactive substances (e.g., GABAergic, peptidergic, glutamatergic, cholinergic, serotonergic, adenosinergic, adrenergic etc.), organized into functional subsystems that have unique developmental histories and vulnerabilities in neurodegenerative diseases (Graybiel 1990). We will not attempt to discuss the effects of δ-opioid system on each neurotransmitter in this section, which will be exhausted for both the authors and the readers. Several main neurotransmitters that are critical to the basal ganglia functions will be mentioned. These include DA, glutamate, acetylcholine, and GABA.

It has been shown that the activation of DOR inhibits, while the blockade of DOR potentiates, acetylcholine release in the striatum (Beani et al. 1982; Mulder et al. 1984; Arenas et al. 1990; Sandor et al. 1992). The inhibition mediated by DOR is detectable at postnatal day 15, earlier than that of MOR (at P21) (Pérez-Navarro et al. 1993). In addition, most studies indicated that in the striatum (including both the dorsal and ventral portions), activation of DOR and its subtypes increases DA release (Chesselet et al. 1981, 1983; Longoni et al. 1991; Pentney and Gratton 1991; Sandor et al. 1992; Dourmap and Costentin 1994; Billet et al. 2004, 2007; Bosse et al. 2008; Hipólito et al. 2008; Tanahashi et al. 2012), although DOR inhibition of single-spike stimuli-induced DA release in the nucleus accumbens has also been observed (Britt and McGehee 2008). In addition, DOR-regulation of acetylcholine release from the striatal cholinergic interneurons (Beani et al. 1982; Mulder et al. 1984; Arenas et al. 1990; Sandor et al. 1992) may indirectly regulate DA release in the striatum, since endogenous striatal acetylcholine release by synchronized activity in the cholinergic interneurons has been proven sufficient to evoke DA release by acting at nAChRs on DA axons which act independently of cell body firing in the midbrain DA neurons (Threlfell et al. 2012; Cachope and Cheer 2014). Intraperitoneal administration of a novel DOR agonist KNT-127 increases the release of DA in the striatum, nucleus accumbens, and median prefrontal cortex, which is abrogated by DOR1 antagonist 7-benzylidenenaltrexone but not by DOR2 antagonist naltriben (Tanahashi et al. 2012). Intra-accumbens infusion of DOR agonist (D-Ala2)deltorphin II evokes dose-dependent motor stimulation (characterized by locomotion, sniffing, and oral stereotypes) and increases of extracellular DA concentrations, which cannot be evoked from the caudate, nor by MOR agonist PL017 in the accumbens, which instead are reduced by high doses of naloxone (1.0 mg/kg) and by DOR antagonist naltrindole (Longoni et al. 1991). In the striatum, DPDPE significantly enhances gliotoxin L-α-aminoadipate-induced DA and glutamate release. At the same time, DPDPE-induced DA release seems to require a glia-related initial glutamate release. This observation was made due to the fact that 2 days after an intrastriatal injection of L-α-aminoadipate, which destroys glial cells, DPDPE-induced increase in extracellular glutamate and DA levels is prevented, but can be restored by local infusion of glutamate (Billet et al. 2007). DOR-stimulated DA release may be a result of either direct action of DOR on the dopaminergic terminals or indirect action on DA transporters as well as cholinergic interneurons. As to the action on DA transporters, the ultrastructural evidence showed that in the ventral striatum, DOR are present in axons and dendritic spines of nondopaminergic neurons that are contacted by DA transporter-containing terminals (Svingos et al. 1999), and δ-opioid peptide DADLE can reverse methamphetamine-induced long-term DA transporter loss in the striatum, an effect that can be blocked by opioid receptor antagonist naltrexone (Tsao et al. 1998, 1999).

Presynaptic DORs in striatal axon terminals may also play a role in modulation of corticostriatal glutamate release (Wang et al. 2003). Both stimulation and inhibition of glutamate release by DOR are suggested. For example, microdialysis studies revealed that either systemic administration or local intranigral perfusion of DOR agonist SNC80 led to a reduction of nigral glutamate (Mabrouk et al. 2014). Electrophysiological studies have shown in rat neocortical and corticostriatal brain slices, that activation of DOR depresses glutamatergic excitatory transmission by presynaptic inhibition (Jiang and North 1992; Tanaka and North 1994; Ostermeier et al. 2000; Jose et al. 2007). In contrast, the inhibition in electrophysiological studies and studies with microdialysis showed increased glutamate release from corticostriatal glutamatergic terminal by DOR stimulation (Billet et al. 2004, 2007). The novel DOR agonist UFP 512, at an antiakinetic dose (10 μg/kg) decreases glutamate release in the SNr, while at a pro-akinetic dose (1000 μg/kg) increases nigral glutamate release (Mabrouk et al. 2009). Intrastriatal perfusion of an antiepileptic drug zonisamide (100 and 1000 μM) decreased the extracellular levels of glutamate in SNr via activation of DOR1 (Yamamura et al. 2009). In addition, Rawls and McGinty (2000) found that in rat striatum, naltrindole (10–100 μM) decreases amphetamine-evoked glutamate release, which can be reversed by DOR agonist DPDPE (100, 500 μM). That means DOR activation enhances glutamate release in the striatum. In consistence with this observation, a recent study with microdialysis also indicated that intraperitoneal administration of a novel DOR agonist KNT-127 increases the release of glutamate as well as DA in the striatum, which is abrogated by DOR1 antagonist 7-benzylidenenaltrexone but not by DOR2 antagonist naltriben (Tanahashi et al. 2012). The enhancement of glutamate release in the striatium might be related to the maintenance of functional glutamatergic neurotransmission by glial cells since selective lesion of glial cells with gliotoxin prevents the increase in extracellular glutamate levels induced by DPDPE (Billet et al. 2004, 2007). However, it is also important to note that the study by Rawls and McGinty (2000) used an extremely high concentration of ligands, which may stimulate a variety of membrane proteins and induce a complicated effect on cells, and can even generate adverse effects to neurons.

In the case of GABA, it has been shown that GABA and enkephalin may coexist within individual caudate neurons of the type which are known to project to the GP (Aronin et al. 1984, 1986). Met-enkephalin has also been shown to inhibit the K^+-stimulated release of GABA from rat brain synaptosomes (Brennan et al. 1980). Microdialysis studies revealed that either systemic administration or local intranigral perfusion of DOR agonist SNC80 led to the enhancement of nigral GABA and reduction of thalamic GABA levels. This is consistent with the viewpoint that DOR stimulation causes overinhibition of nigro-thalamic GABA neurons (Mabrouk et al. 2014). The novel DOR agonist UFP 512, at an antiakinetic dose (10 μg/kg) decreases GABA in GP as well as GABA and glutamate release in the SNr, while at a pro-akinetic dose (1000 μg/kg) increases pallidal GABA, simultaneously producing a decrease in GABA and an increase in nigral glutamate release (Mabrouk et al. 2009). Intrastriatal perfusion of an antiepileptic drug zonisamide (100 and 1000 μM) decreased the extracellular levels of GABA in STN via activation of DOR1 (Yamamura et al. 2009). Electrophysiological studies indicated that presynaptic DORs are preferentially located on terminals of quiescent GP cells and GABAergic nigral neurons, and DOR activation serves to reduce GABA release and IPSCs via presynaptic mechanisms in these cells to maintain the inhibitory function in basal ganglia circuitry (Standford and Cooper 1999; Mabrouk et al. 2014). In the ventral striatum, DOR indirectly inhibits GABA release via activated dopaminergic transmission; it also enhances GABA release in median prefrontal cortex via activated glutamate transmission (Tanahashi et al. 2012).

3.3 Alteration of δ-Opioid Expression Within the Basal Ganglia in PD and L-DOPA-Induced Dyskinesia (LID)

DOR and its endogenous agonist enkephalin (and its precursor PENK) are widely distributed within the basal ganglia. Their activities in the basal ganglia are largely influenced by DA depletion, which occurs in PD. In LID with L-DOPA treatment of PD, δ-opioid system is also affected within the basal ganglia. The changes in δ-opioid system in PD and LID imply certain relevance contributing to their pathophysiology. Though numerous studies indicated the changes of δ-opioid expression in the basal ganglia in PD and LID, the results are not consistent, sometime even opposing.

A study using non-selective opioid ligand [^{3}H]diprenorphine to label opioid sites within the basal ganglia of the 6-OHDA-lesioned hemiparkinsonian rats showed a significant reduction of opioid receptors in the ipsilateral caudate-putamen, the GP, and the SN, in contrast to no changes in the premotor and motor cortices (Johansson et al. 2001). Conversely, another study with the same method in MPTP-lesioned parkinsonian monkeys showed no changes of the [^{3}H]diprenorphine binding in the caudate-putamen, the GPe, and the GPi (Aubert et al. 2007). These two groups also reported different changes in DOR binding. Namely, Johansson et al. (2001) observed a 17–26 % decrease in [^{3}H]DPDPE binding in the DA-denervated caudate-putamen and ~50 % reduction in the GP in the rat model. In contrast, Aubert et al. (2007) found that [^{3}H]DPDPE binding does not affected by DA depletion as well as L-DOPA treatment in the parkinsonian monkey. These observations might reflect a difference in the species and models. Despite this, these two groups consistently revealed that there is no correlation between DOR density and the severity of PD or LID (Johansson et al. 2001; Aubert et al. 2007). The function analysis in another study with MPTP-lesioned parkinsonian monkeys, however, indicated that δ-agonist-stimulated GTPγS binding does not change in parkinsonian monkeys without LID but significantly elevates in the caudate and the dorsal putamen of animals with LID, and the degree of elevation in the latter correlates positively with the monkey dyskinesia (Chen et al. 2005a). These contradictions make the role of DOR in PD and LID more complicated. Indeed, these contradictions have been widely observed by others with various methods. For example, in PD patients, no change of δ-opioid binding (with [^{3}H]DPDPE or [^{3}H]DTLET) in all the brain structures including the basal ganglia (Delay-Goyet et al. 1987) or a significant decrease of binding in the posterior putamen (Fernandez et al. 1994) was present. In reserpine-treated rats, specific δ-opioid binding (by [^{3}H]naltrindole) was increased in the premotor cortex (+30 %), sensorimotor striatum (+20 %), and associative striatum (+17 %) rostrally, but was not changed in caudal forebrain (Hallett and Brotchie 2007). This finding directly opposes the observations of 20–50 % loss of DOR binding (by [^{3}H]DPDLE) most apparent in the central striatum in 6-OHDA-lesioned rat PD model (Eghbali et al. 1987; Smith et al. 1993). It also opposes the significant reduction of striatal immunohistochemistry for DOR in the patch but not in the matrix compartment in 6-OHDA-lesioned rat PD model (Koizumi et al. 2013). Finally, it opposes the fact that DOR binding is not significantly altered at any region analyzed, in either nondyskinetic or dyskinetic, MPTP-lesioned macaques (Hallett and Brotchie 2007). Thus the changes and their relevance to DOR within the basal ganglia in PD and LID is still a puzzle.

In contrast to the vagueness of the changes in DOR expression and function activity, the changes of endogenous enkephalins and their precursors PENK and preproenkephalin-A (PPE-A) are relatively explicit in PD. In rodents (Gerfen et al. 1990, 1991; Reimer et al. 1992; Smith et al. 1993; Nisenbaum et al. 1994, 1996; Henry et al. 1999; Westin et al. 2001; Ravenscroft et al. 2004), non-human primates (Quik et al. 2002; Tel et al. 2002; Aubert et al. 2007; Tamim et al. 2010; Bourdenx et al. 2014), and humans (Nisbet et al. 1995; Calon et al. 2002), parkinsonism is associated with an increased expression of PPE-A and PENK mRNA in striatal MSNs projecting to the GPe and a decreased PPE-B (which ultimately produces predominant dynorphin) mRNA expression in striatal MSNs projecting to the GPi. The onset of these changes in the striatum does not occur until three days following neurotoxic lesion of SNc DA neurons, and the elevation is gradually developed and becomes apparent within seven days and is maintained up to 24 weeks to more than one year post-lesion (Smith et al. 1993; Nisenbaum et al. 1994; Westin et al. 2001). Strangely, direct measurement of enkephalin levels with HPLC/radioimmunoassay in the postmortem brain of the parkinsonian patients revealed no difference (Fernandez et al. 1992) or a decrease in enkephalin contents (Taquet et al. 1983; Llorens-Cortes et al. 1984; Fernandez et al. 1996) in the caudate-putamen and the SN in comparison with the age-matched control subjects without any neurological diseases. Even so, the potential pseudomorphism is not completely excluded due to some methodological issues. For example, in these studies, the average time between death, refrigeration of the body, and brain removal is ~3 h and 15–22 h respectively (Taquet et al. 1983; Llorens-Cortes et al. 1984; Fernandez et al. 1992, 1996). It is possible that the some endogenous enkephalins are degraded or destroyed during such delays. However, it is also possible that due to the difference in the brain states between PD and control subjects, e.g., the activity of enzymes, the degree of enkephalin destruction may be different, thus resulting in different changes between the parkinsonian patients and the control subjects.

In LID, the reported changes of endogenous enkephalin precursors (PENK and PPE-A) are complicated. The reported results include unchanged, further increased, or even decreased levels in comparison with parkinsonism only. For example, Quik et al. (2002) reported that though MPTP-lesion results in significant increases in PPE mRNA in the striatum of the monkeys, L-DOPA treated animals do not show any changes in PPE mRNA even LID is development when compared with MPTP-lesion parkinsonism animals without LID. This result is identical in the animals killed 3–4 h after the last dose of L-DOPA and those after 3 days. The authors thus suggested that while elevations in striatal PPE mRNA levels are associated with nigrostriatal damage, they are not linked to the development of LID (Quik et al. 2002). This conclusion is consistent with others in non-human primate model (Tel et al. 2002) and the rodent model of PD (westin et al. 2001). In contrast, some studies have also demonstrated that in LID PPE-A mRNA is further increased in the striatum, in particular, the lateral part of the putamen in PD patients suffering from LID (Calon et al. 2002), the medial part of caudate nucleus of the parkinsonian monkeys with LID (Tamim et al. 2010), and the dorsolateral areas of the rostral caudate-putamen of 6-OHDA-lesioned rats with LID (Henry et al. 1999), where a significant positive correlation is observed between maximum dyskinesias scores and the levels of PPE mRNA. On the other hand, decreases in the PPE-A mRNA and the derived peptide levels were also observed in dyskinetic MPTP-treated monkeys, and this decrease achieved an apparent normalization of PPE mRNA levels (Aubert et al. 2007; Bourdenx et al. 2014), similar to the changes of PPE-A mRNA levels in the striatum in the PD animals treated with non-dyskinetia-developing DA receptor agonists bromocriptine and lisuride (Henry et al. 1999; Tel et al. 2002). Thus, these findings suggest a lack of correlation between the PPE-A mRNA/the derived peptide levels and LID severity (Aubert et al. 2007; Bourdenx et al. 2014). The studies of topographical organization of striatal PPE-A mRNA revealed a regional difference in PPE-A mRNA changes with LID (Henry et al. 1999; Calon et al. 2002; Ravenscroft et al. 2004). For example, in L-DOPA treated dyskinetic rats, PPE-A is increased in the caudal striatum, while decreased in the rostral striatum and the shell region of the nucleus accumbens (the ventral striatum) (Ravenscroft et al. 2004). In PD patients, LID is positively correlated with the increase of PPE expression in the lateral part of the putamen, but no alteration of PPE mRNA expression was observed in the caudate and in the medial putamen of dyskinetic patients (Calon et al. 2002). In L-DOPA treated 6-OHDA-lesioned rats, the increased PPE-A mRNA is only observed in the areas of the striatum associated with the control of motor activity, namely, the dorsolateral areas of the rostral caudate-putamen (Henry et al. 1999). Taken together, these findings reflect a very complicated situation for the changes of endogenous enkephalin precursors in LID. Thus, more investigations are required to clarify these changes and their relation with LID.

3.4 *Effects of Pharmacological DOR Manipulation on PD and L-DOPA-Induced Dyskinesia*

Activation of DOR in the basal ganglia may exert dual effects (either anti-akinetic or pro-akinetic) or even have no effects on motor activity of PD. These effects seem to largely depend on the agonists, dose, and the regions where DOR is acting. For example, in 6-OHDA-lesioned hemiparkinsonian rats, the novel DOR agonist UFP 512 causes opposite changes in motor activity, namely, UFP 512, at low dose, produces motor facilitation, while at high dose causes inhibition (Mabrouk et al. 2009). These effects respectively correspond to the decreased GABA releases in GP and

SNr as well as reduced glutamate release in SNr, and the subsequent reduced GABA and increased glutamate release in nigrothalamic output targets (the thalamus) at low dose of UFP 512. At high doses of UFP 512, there is increased pallidal GABA, simultaneously producing a decrease in GABA and an increase in nigral glutamate release, and subsequently increased GABA without affecting thalamic glutamate release (Mabrouk et al. 2009). Further study with microinjection revealed that UFP 512 in GP increases akinesia whereas in SNr reduced akinesia in this model (Mabrouk et al. 2009). These findings suggest a dose- and regional-related effect of DOR on PD motor activity. In rats, chronic neuroleptic administration (e.g., fluphenazine) results in a robust increase in vacuous chewing movements, which can be significantly attenuated by infusion of naloxone, or naltrindole into the GP, suggesting the increase in δ-opioid transmission in the indirect striatopallidal pathway following chronic neuroleptic administration is likely to contribute to tardive dyskinesia (McCormick and Stoessl 2002). A study in unilaterally 6-OHDA lesioned rats reported that the selective non-peptide DOR agonists BW 373U86 and SNC80 fail to induce motor asymmetries, but elicit contralateral turning after pretreatment with a dose of SKF 38393, a D1 agonist, or quinpirole, a D2 agonist, which per se did not induce contralateral turning, suggesting a pro-parkinsonian role of DOR (Pinna and Di Chiara 1998). On the other hand, SNC80 was also shown to cause pronounced reversal of motor deficits in MPTP-treated marmosets and reserpinized or 6-OHDA-hemilesioned (hemiparkinsonian) rats, suggesting an antiparkinsonian role of DOR (Hill et al. 2000; Hille et al. 2001; Mabrouk et al. 2008). This antiparkinsonian role of SNC80 seems related to the action of DOR in the SNr rather than in the GP since microperfusion of naltrindole in the SNr but not the GP or the striatum prevents the antiakinetic effect of SNC80 (Mabrouk et al. 2008). In the striatum, the antiparkinsonian effects of DOR with SCN80 require the concurrent blockade of nociceptin/orphanin FQ receptor (Mabrouk et al. 2014). Thus more investigations are required to clarify the role of DOR in the motor activity of PD.

L-DOPA-induced dyskinesia (LID) is a form of dyskinesia associated with L-DOPA used to treat PD. It often involves hyperkinetic movements, including chorea, dystonia, and athetosis, and may be accompanied by a worsening of motor performance (Gerlach et al. 2011). Once established, LID is difficult to treat (Thanvi et al. 2007). It has been suggested that the non-physiological pulsatile stimulation of the post-synaptic receptors by dopamine (produced by intermittent doses via L-DOPA or dopamine agonists) ultimately creates alterations in the striatum of the basal ganglia, causing it to release abnormal motor outputs that result in the involuntary movements (Calon et al. 2000; Olanow et al. 2006; Fabbrini et al. 2007). In MPTP-lesioned squirrel monkeys treated with L-DOPA, DOR and MOR-mediated G-protein activation is consistently enhanced in the striatum and this enhancement in the dorsal putamen and caudate is positively correlated with LID (Chen et al. 2005a). Non-subtype-selective opioid receptor antagonists (naloxone, naltrexone), via either i.p./s.c. injection or i.v. infusion, produce either an increase (Samadi et al. 2003) or decrease (Henry et al. 2001; Fox et al. 2004) in LID in the non-human primate PD model and PD patients. Thus the opposing conclusions of the role of opioid system in LID are present. Studies using selective DOR antagonist naltridine (i.p. or i.c.v) indicated a reduced LID response in both rodent and primate PD models, suggesting a pro-LID effect of DOR (Henry et al. 2001; Billet et al. 2012). Further investigation with [^{3}H]-DADPE binding to striatal membrane indicated a significant increase in density of DOR in the striatum of dyskinetic rats as compared to non-dyskinetic animals, and this difference was abolished by the corticostriatal deafferentation. This suggests that δ-opioid transmission modulates the expression of LIDs in rodents, and DORs involved in this effect are located on terminals of corticostriatal neurons (Billet et al. 2012). However, in a review paper, Samadi et al. (2006) proposed that enhanced opioid transmission after striatal DA depletion in PD and repeated L-DOPA treatment is an adaptive response that attenuates the short- and long-term changes responsible for the development of parkinsonian syndrome and the involuntary movements induced by the non-physiological stimulation of dopamine receptors in reorganized basal ganglia circuits. A recent study indicated that in reserpine-induced rat akinesia, treatment with apomorphine, a non-selective DA agonist which activates both D1-like and D2-like receptors, with some preference for the latter subtypes (Millan et al. 2002), not only reverses akinesia but also induces hyperkinesia, while opioid peptide MMP-2200 blocks the apomorphine-induced hyperkinesia but doesn't prevent the reserpine-induced akinesia (Yue et al. 2011). The effect of MMP-2200 is prevented by pre-administration of naloxone and partially blocked by DOR antagonist naltrindole, suggesting a potent effect of DOR on movements related to dopaminergic hyper-stimulation following striatal dopamine depletion (Yue et al. 2011). Thus further investigations are required to clarify the role of δ-opioid system in LID. It is worthy to note that the effects of opioid system on LID and PD are not always identical to each other. The caveats should be taken for the explanation of the effects of opioid system on PD based on the effects on LID. In this context, though we do not know whether DOR has such an effect, it has been shown that KOR agonist U50,488 reduces LID (abnormal involuntary movements, including axial, orolingual and forelimb dyskinesias contralateral to the lesion) in a dose-dependent fashion, but the anti-parkinsonian effect of L-DOPA is similarly reduced. Side effects were also developed, including sedation and vomiting in non-human primate PD model (Cox et al. 2007). That means KOR activation reduces LID but worsens parkinsonism. Thus the authors suggested that KOR agonists such as U50,488 may not be clinically useful antidyskinetic agents because they also reverse the anti-parkinsonian effect of L-DOPA (Cox et al. 2007).

3.5 Effects of DOR Activation on Dopaminergic Neuron Injury

DOR are distributed in the SN (Tempel and Zukin 1987; Mansour et al. 1994; Jenab et al. 1995; Gray et al. 2006; Léna et al. 2008) while PENK mRNA expression is very high in the human SN (Noda et al. 1982; Nisbet et al. 1995; Brana et al. 1995; Hurd 1996) but not found in the rat SN (Wamsley et al. 1980; Harlan et al. 1987;

Mansour et al. 1993; Winzer-Serhan et al. 2003). Therefore, enkephalins that activate DOR in the SN might originate from the release from the afferent terminals (e.g., stratonigrol projection). Thus the δ-opioid system may have relevance in the physiological and pathophysiological process in the SN (Saha et al. 2003; Méndez et al. 2004). Indeed, some studies have shown such a role (Borlongan et al. 2000; Lu et al. 2000; Sun et al. 2011).

In an in vivo neuroinflammation-induced SN neuronal death model, either a continuous two-week SN or systemic infusion of opioid antagonist, naloxone, significantly prevents neuronal death, with 85–94 % of the neurons survival (Lu et al. 2000). The authors suggested that naloxone may prevent death of nigrostriatal neurons by reducing release of cytotoxins from activated microglia or by blocking opioid receptor binding on dopaminergic and/or GABAergic neurons (Lu et al. 2000). This result seems against the idea of DOR neuroprotection, and supporting a detrimental role of opioid receptors. However, the caveats for this observation must be stated. Firstly, naloxone is a broad-spectrum opioid receptor antagonist for all three classical receptor subtypes (DOR, MOR, and KOR). Therefore, it cannot differentiate the roles of receptor subtypes in neuroinflammation-induced SN neuronal death; secondly, in the SN, not only is there DOR, but also other opioid receptors such as MOR and KOR (Mansour et al. 1994; Winzer-Serhan et al. 2003; Gray et al. 2006), and the receptor subtypes are not homologously abundant. For example, DOR mRNA is in low levels, whereas a higher number of cells express MOR mRNA in the SNc (Mansour et al. 1994). The intervention of different receptor subtypes to the same degree with the same ligand may have different, even opposing effects (Chao et al. 2007). Indeed, activation of DOR produces neuroprotection against insults such as excitatoxicity and hypoxia/ischemia (Chao and Xia 2010; He et al. 2013), while MOR activation has been shown to actually aggravate the neurotoxic effects, induce neuronal apoptosis, and inhibit cell survival (Kofke et al. 1999; Ammon-Treiber et al. 2005; Hu et al. 2002; Hauser et al. 1994; Yin et al. 1999), while blocking MOR activation is beneficial in cerebral ischemia/reperfusion injury (Liao et al. 2003; Chen et al. 2005b). Thus the result of Lu et al. (2000) is not surprising. Direct evidence of DOR protection against SN dopaminergic neuron degeneration has been emerging (Borlongan et al. 2000; Sun et al. 2011). Borlongan et al. (2000) first presented the evidence for this. They showed that pretreatment with DOR agonist DADLE (0.0025, 0.005, 0.01 g/ml) dose-dependently enhances cell viability of cultured primary rat fetal mesencephalic cells. In addition, DADLE administration in adult rats (4 mg/kg every 2 h, 4 injections, i.p.) prior to 6-OHDA lesions of the medial forebrain bundle, significantly reduces the severity of loss of tyrosine hydroxylase-immunoreactive neurons in the substantia nigra a month post-lesion (Borlongan et al. 2000). In a recently study by Sun et al. (2011), cultured dopaminergic SN neurons suffer denegration by chelating iron that is highly concentrated and has essential roles to SN neurons (Götz et al. 1990). Pre-administration of the DADLE significantly protects the SN cells from damage by iron deficiency. In their discussion, the authors (Sun et al. 2011) addressed the relevance of DOR protection of SN neurons from degenerative death in restless legs syndrome and PD, two disorders that can co-exist, and the former when severe may be an early feature of PD

(Rijsman et al. 2014; Wong et al. 2014). In a recent study, it was shown that neither subthreshold dose of nociceptin/orphanin FQ receptor antagonist J-113397, nor DOR agonist SCN80 alone is effective in reducing immobility time and improving rotarod performance in MPTP-induced rodent PD; however, the repeated administration of the combination indeed does help, and also improves stepping activity. Meanwhile striatal tyrosine hydroxylase immunohistochemistry does not reveal any sparing of striatal DA terminals after each drug alone or in combination (Mabrouk et al. 2014). This study suggests that repeated administration of the combination of subthreshold doses of J-113397 and SNC80 could provide prolonged symptomatic relief and protect DA terminals (which largely originate from the SNc) from MPTP-induced degeneration (Mabrouk et al. 2014).

Our most recent work in a cellular model of PD in HEK293 cells indicated that prolonged and severe hypoxia, similar to MPP^+, enhances α-synuclein expression and its oligomer formation leading to the collapse of mitochondrial membrane potential and cell damage (Chen et al. 2014). On the other hand, relatively short-term and less severe hypoxia enhances α-synuclein expression, but it does not cause oligomer formation and cellular injury. DOR activation with UFP 512 promotes CREB phosphorylation, markedly attenuates the hypoxia-induced α-synuclein dysregulation, and prevents the collapse of mitochondrial membrane potential and cell injury. This effect is completely reversed by DOR antagonism or knock-down of DOR expression with naltrindole or siRNA (Chen et al. 2014). Our data shines a novel light on DOR inhibition of α-synuclein dysregulation and its potential as a target for PD therapy/prevention.

The motor symptoms of PD result from the death of dopaminergic neurons in the SNc. Thus it is reasonable that the reagents that rescue SNc dopaminergic neurons from death may be beneficial to PD. Indeed, analyzing our data (Chen et al. 2014) and that of others' (Borlongan et al. 2000; Sun et al. 2011) together makes the beneficial effect of DOR against SN dopaminergic neuron death obvious. In this context, DOR may contribute to the improvement of the motor symptoms of PD. If this is completely verified and validated in PD animal models and patients, then it will break the conventional treatments in which obvious side-effects have been found unavoidable. It will provide a fundamental therapy rather than a symptom treatment for PD, and will have a revolutionary impact on the treatment of PD.

4 Is DOR Beneficial or Detrimental in Parkinson's Disease?

Although the answer is not very sure, the current evidence shows a prudent tendency toward beneficent.

Firstly, the complexity of the distribution and action of δ-opioid system (including DOR and its endogenous ligands) itself is still a barrier to finding a clear answer. As has been discussed earlier, δ-opioid system is distributed widely but also heterogeneously through the basal ganglia, and its effects are multiple and very complicated. The action of δ-opioid for certain activities is largely determined by the location of DOR. Thus, it is completely possible that the effects of activation of

DOR in the corticostriatal terminals are very different from those in the striatopallidal terminals. Therefore, the net output of the basal ganglia is dependent on the integration of DOR action within the basal ganglia nuclei and their current states. For the latter, it has been shown in the limbic system that opiates may indeed induce network shifts (state changes), which contribute to brain rewarding (Dejean et al. 2013). The extreme complexity of connectivity and functions of the basal ganglia nuclei makes the DOR effect more complicated. Secondly, the existing studies for the PD behavior observation gave the opposite results of both anti-parkinsonism and worsening of PD by DOR activation. DOR-deficient animals showed significant hyperlocomotion (Filliol et al. 2000), which suggests that DORs tonically inhibit locomotion. It is still a puzzle whether the increased activity of δ-opioid system in the basal ganglia (e.g., an increased expression of PPE-A and PENK mRNA in the striatum) is related to the bradykinesia in parkinsonian patients. Thirdly, though both the activity and expression of δ-opioid system are greatly changed, the solid causality between δ-opioid system and PD pathogenesis is still lacking. It is unknown whether the changes in δ-opioid system are causative factors for PD or if it is just a consequent result due to the loss of DA terminals where DOR coexist. These changes could even be just accompanying phenomena without any relations to PD pathogenesis. Fourthly, the proposed compensatory improvement of LID by DOR activation (Samadi et al. 2006) is sometimes used to support the beneficialness of δ-opioid system on PD. However, LID and PD are two different movement diseases (Sgroi et al. 2014) though they have intimate relations (Thanvi et al. 2007; Gerlach et al. 2011). Some opioids have been shown to reduce LID but worsen parkinsonism (Cox et al. 2007).

Even from the pure neuroprotection viewpoint, DOR protection from neuron death should be cautiously used to support the beneficialness of δ-opioid system on PD. As we have discussed earlier, the motor symptoms of PD result from the death of dopaminergic neurons in SNc, presumably resulted from extremely high vulnerability to multiple insults such as excitotoxicity, lysosomal and proteasomal dysfunction and the consequent neurotoxicity to misfolded proteins (α-synuclein), mitochondrial dysfunction and the consequent oxidative stress, and probably neuroinflammation (Dauer and Przedborski 2003; Subramaniam and Chesselet 2013; Antony et al. 2013; Stayte and Vissel 2014). Although DOR activation has been demonstrated to protect SN cell degeneration from neurotoxicity, oxidative and neuroimflammative insults (Borlongan et al. 2000; Sun et al. 2011), and also protects the cell line from hypoxic/MPP^+ insults and the consequent α-synuclein toxicity and mitochondrial dysfunction (Chen et al. 2014), the caveats for the beneficialness on PD cannot be ignored. Firstly, we need to consider the integrative effect of DOR action on the activity of other basal ganglia nuclei and their current states together, since as discussed above, DOR activation may produce different effects in different nuclei within the basal ganglia. Secondly, numerous strategies have been developed to prevent neuronal death from various conditions, but none of these strategies have been proven to effectively cure PD to date. The only proven and most efficacious pharmacological treatment is DA replacement with L-DOPA, even though its treatment is only symptomatic, rather than the fundamental therapy, and has the obviously severe side-effects for the long-term user (Olanow et al. 2006;

Brichta et al. 2013; Stayte and Vissel 2014). Even though the emerging non-dopaminergic treatments for PD as an alternative treatment to get rid of the side-effect of L-DOPA are becoming a scheduling program for the pipe-lines of drug development and clinical trials, DOR (the whole opioid family to be precise) and its endogenous agonists are still not in the tables of FDA approved non-dopaminergic trials (Olanow and Schapira 2013; Brichta et al. 2013; Stayte and Vissel 2014; Connolly and Lang 2014), suggesting the current state of uncertainty for DOR treatment of PD.

However, the uncertainty based on the previous research on the effects of DOR on PD does not in any way mean that we don't have a standpoint. We have recently found that in *parkin*-mutant transgenic mice, DOR expression is normal during early development but largely decreased in adulthood, especially when PD symptoms appear (Jiang et al. data to be published). Moreover, we found that in an in-vitro PD model with MPP^+ exposure, DOR activation protects cells against severe hypoxia and MPP+ insult via attenuating α-synuclein expression and aggregation (Chen et al. 2014). Our new data suggest that DOR may play a role in neuroprotection against PD pathophysiology. We therefore believe that δ-opioid system may be beneficial on PD. However, more investigations are required to validate this stance.

5 Conclusion Remarks

There are still numerous unsolved puzzles on the anatomical and functional organization and the role of δ-opioid system within the basal ganglia in PD. Existing evidence is full of contradictions and needs to be validated. Nevertheless, increasing evidence tends to support the beneficial effect of δ-opioid system. In this chapter, based on the retrospect of the studies on δ-opioid system within basal ganglia in health and PD, we believe that δ-opioid system is a beneficial, not detrimental factor for motor complications in PD. A simple fact is that DOR, highly distributed in the cortical region and basal ganglia, is neuroprotective. Several lines of evidence have shown that DOR density decreased in the brain of PD patients or PD animal model (Jiang et al., unpublished data). Such a pathological decrease undoubtly causes neuronal dysfunction and injury in these brain regions. However, more in-depth investigations are expected to provide solid and direct evidence, and the development of treatment strategies targeting to δ-opioid system is also expected, especially the roles of DOR in the functional regulation of basal ganglia and broard cortical regions. In regard to this aspect, some open questions are listed below:

1. What is the causative relationship between δ-opioid system and PD?
2. How does δ-opioid system modify the nuclei activity (firing rate, pattern) within the basal ganglia in PD?
3. How does δ-opioid system modify the output of the basal ganglia in PD?
4. What are roles of δ-opioid system in the other brain regions, including the cerebral cortex, thalamus, and several brain stem nuclei, which form more compli-

cated networks with both open- and closed-loop circuitry of the control of motor activity (Kreitzer 2009; Nelson and Kreitzer 2014) in PD?

5. How does δ-opioid system influence the basal ganglia via modifying the activity of the other brain regions with the motor circuitry?

Answering these questions will advance our understanding of δ-opioid system in PD, and help us develop novel therapeutic strategies for the treatment of PD and related motor complications.

Acknowledgements This work was supported by NIH (AT-004422, and HD-034852) and Vivian L Smith Neurologic Foundation.

References

Abou-Khalil B, Young AB, Penney JB (1984) Evidence for the presynaptic localization of opiate binding sites on striatal efferent fibers. Brain Res 323:21–29

Aizman O, Brismar H, Uhlén P, Zettergren E, Levey AI, Forssberg H, Greengard P, Aperia A (2000) Anatomical and physiological evidence for D1 and D2 dopamine receptor colocalization in neostriatal neurons. Nat Neurosci 3:226–230

Albin RL, Young AB, Penney JB (1989) The functional anatomy of basal ganglia disorders. Trends Neurosci 12:366–375

Alexander GE, Crutcher MD (1990) Functional architecture of basal ganglia circuits: neural substrates of parallel processing. Trends Neurosci 13:266–271

Ambrose LM, Gallagher SM, Unterwald EM, Van Bockstaele EJ (2006) Dopamine-D1 and delta-opioid receptors co-exist in rat striatal neurons. Neurosci Lett 399:191–196

Ammon-Treiber S, Stolze D, Schröder H, Loh H, Höllt V (2005) Effects of opioid antagonists and morphine in a hippocampal hypoxia/hypoglycemia model. Neuropharmacology 49:1160–1169

Anderson ME, Postupna N, Ruffo M (2003) Effects of high-frequency stimulation in the internal globus pallidus on the activity of thalamic neurons in the awake monkey. J Neurophysiol 89:1150–1160

Antony PM, Diederich NJ, Krüger R, Balling R (2013) The hallmarks of Parkinson's disease. FEBS J 280:5981–5993

Arenas E, Marsal J, Alberch J (1990) GABAA and GABAB antagonists prevent the opioid inhibition of endogenous acetylcholine release evoked by glutamate from rat neostriatal slices. Neurosci Lett 120:201–204

Aronin N, Chase K, DiFiglia M (1986) Glutamic acid decarboxylase and enkephalin immunoreactive axon terminals in the rat neostriatum synapse with striatonigral neurons. Brain Res 365:151–158

Aronin N, Difiglia M, Graveland GA, Schwartz WJ, Wu JY (1984) Localization of immunoreactive enkephalins in GABA synthesizing neurons of the rat neostriatum. Brain Res 300:376–380

Aubert I, Guigoni C, Li Q, Dovero S, Bioulac BH, Gross CE, Crossman AR, Bloch B, Bezard E (2007) Enhanced preproenkephalin-B-derived opioid transmission in striatum and subthalamic nucleus converges upon globus pallidus internalis in L-3,4-dihydroxyphenylalanine-induced dyskinesia. Biol Psychiatry 61:836–844

Ballion B, Mallet N, Bézard E, Lanciego JL, Gonon F (2008) Intratelencephalic corticostriatal neurons equally excite striatonigral and striatopallidal neurons and their discharge activity is selectively reduced in experimental parkinsonism. Eur J Neurosci 27:2313–2321

Baunez C, Yelnik J, Mallet L (2011) Six questions on the subthalamic nucleus: lessons from animal models and from stimulated patients. Neuroscience 198:193–204

Bausch SB, Patterson TA, Appleyard SM, Chavkin C (1995) Immunocytochemical localization of delta opioid receptors in mouse brain. J Chem Neuroanat 8:175–189

Bayon A, Shoemaker WJ, Bloom FE, Mauss A, Guillemin R (1979) Perinatal development of the endorphin- and enkephalin-containing systems in the rat brain. Brain Res 179:93–101

Beani L, Bianchi C, Siniscalchi A (1982) The effect of naloxone on opioid-induced inhibition and facilitation of acetylcholine release in brain slices. Br J Pharmacol 76:393–401

Benazzouz A, Mamad O, Abedi P, Bouali-Benazzouz R, Chetrit J (2014) Involvement of dopamine loss in extrastriatal basal ganglia nuclei in the pathophysiology of Parkinson's disease. Front Aging Neurosci 6:87. doi:10.3389/fnagi.2014.00087

Benhamou L, Cohen D (2014) Electrophysiological characterization of entopeduncular nucleus neurons in anesthetized and freely moving rats. Front Syst Neurosci 8:7. doi:10.3389/fnsys.2014.00007

Bertran-Gonzalez J, Bosch C, Maroteaux M, Matamales M, Hervé D, Valjent E, Girault JA (2008) Opposing patterns of signaling activation in dopamine D1 and D2 receptor-expressing striatal neurons in response to cocaine and haloperidol. J Neurosci 28:5671–5685

Bertran-Gonzalez J, Laurent V, Chieng BC, Christie MJ, Balleine BW (2013) Learning-related translocation of δ-opioid receptors on ventral striatal cholinergic interneurons mediates choice between goal-directed action. J Neurosci 33:16060–16071

Billet F, Costentin J, Dourmap N (2007) Influence of glial cells in the dopamine releasing effect resulting from the stimulation of striatal δ-opioid receptors. Neuroscience 150:131–143

Billet F, Costentin J, Dourmap N (2012) Influence of corticostriatal δ-opioid receptors on abnormal involuntary movements induced by L-DOPA in hemiparkinsonian rats. Exp Neurol 236:339–350

Billet F, Dourmap N, Costentin J (2004) Involvement of corticostriatal glutamatergic terminals in striatal dopamine release elicited by stimulation of δ-opioid receptors. Eur J Neurosci 20:2629–2638

Blume LC, Bass CE, Childers SR, Dalton GD, Roberts DC, Richardson JM, Xiao R, Selley DE, Howlett AC (2013) Striatal CB1 and D2 receptors regulate expression of each other, CRIP1A and δ opioid systems. J Neurochem 124:808–820

Borlongan CV, Su TP, Wang Y (2000) Treatment with delta opioid peptide enhances in vitro and in vivo survival of rat dopaminergic neurons. Neuroreport 11:923–926

Bosch C, Degos B, Deniau JM, Venance L (2011) Subthalamic nucleus high-frequency stimulation generates a concomitant synaptic excitation-inhibition in substantia nigra pars reticulata. J Physiol 589:4189–4207

Bosse KE, Jutkiewicz EM, Gnegy ME, Traynor JR (2008) The selective delta opioid agonist SNC80 enhances amphetamine-mediated efflux of dopamine from rat striatum. Neuropharmacology 55:755–762

Bourdenx M, Nilsson A, Wadensten H, Fälth M, Li Q, Crossman AR, Andrén PE, Bezard E (2014) Abnormal structure-specific peptide transmission and processing in a primate model of Parkinson's disease and l-DOPA-induced dyskinesia. Neurobiol Dis 62:307–312

Brana C, Charron G, Aubert I, Carles D, Martin-Negrier ML, Trouette H, Fournier MC, Vital C, Bloch B (1995) Ontogeny of the striatal neurons expressing neuropeptide genes in the human fetus and neonate. J Comp Neurol 360:488–505

Brann MR, Emson PC (1980) Microiontophoretic injection of fluorescent tracer combined with simultaneous immunofluorescent histochemistry for the demonstration of efferents from the caudate-putamen projecting to the globus pallidus. Neurosci Lett 16:61–65

Brennan MJ, Cantrill RC, Wylie BA (1980) Modulation of synaptosomal GABA release by enkephalin. Life Sci 27:1097–1101

Brichta L, Greengard P, Flajolet M (2013) Advances in the pharmacological treatment of Parkinson's disease: targeting neurotransmitter systems. Trends Neurosci 36:543–554

Britt JP, McGehee DS (2008) Presynaptic opioid and nicotinic receptor modulation of dopamine overflow in the nucleus accumbens. J Neurosci 28:1672–1681

Cachope R, Cheer JF (2014) Local control of striatal dopamine release. Front Behav Neurosci 8:188

Cai Z, Ratka A (2012) Opioid system and Alzheimer's disease. Neuromol Med 14:91–111
Calabresi P, Picconi B, Tozzi A, Ghiglieri V, Di Filippo M (2014) Direct and indirect pathways of basal ganglia: a critical reappraisal. Nat Neurosci 17:1022–1030
Calì T, Ottolini D, Brini M (2014) Calcium signaling in Parkinson's disease. Cell Tissue Res 357:439–454
Calon F, Birdi S, Rajput AH, Hornykiewicz O, Bédard PJ, Di Paolo T (2002) Increase of preproenkephalin mRNA levels in the putamen of Parkinson disease patients with levodopa-induced dyskinesias. J Neuropathol Exp Neurol 61:186–196
Calon F, Grondin R, Morissette M, Goulet M, Blanchet PJ, Di Paolo T, Bédard PJ (2000) Molecular basis of levodopa-induced dyskinesias. Ann Neurol 47:S70–S78
Cannon JR, Greenamyre JT (2013) Gene-environment interactions in Parkinson's disease: specific evidence in humans and mammalian models. Neurobiol Dis 57:38–46
Castrioto A, Lhommée E, Moro E, Krack P (2014) Mood and behavioural effects of subthalamic stimulation in Parkinson's disease. Lancet Neurol 13:287–305
Cazorla M, de Carvalho FD, Chohan MO, Shegda M, Chuhma N, Rayport S, Ahmari SE, Moore H, Kellendonk C (2014) Dopamine D2 receptors regulate the anatomical and functional balance of basal ganglia circuitry. Neuron 81:153–164
Chan V, Starr PA, Turner RS (2011) Bursts and oscillations as independent properties of neural activity in the parkinsonian globus pallidus internus. Neurobiol Dis 41:2–10
Chao D, Bazzy-Asaad A, Balboni G, Xia Y (2007) Delta-, but not mu-, opioid receptor stabilizes K^+ homeostasis by reducing Ca^{2+} influx in the cortex during acute hypoxia. J Cell Physiol 212:60–67
Chao D, Shen X, Xia Y (2013) From acupuncture to interaction between δ-opioid receptors and Na^+ channels: a potential pathway to inhibit epileptic hyperexcitability. Evid Base Compl Alternative Med 2013:216016. doi:10.1155/2013/216016
Chao D, Xia Y (2010) Ionic storm in hypoxic/ischemic stress: can opioid receptors subside it? Prog Neurobiol 90:439–470
Chen L, Togasaki DM, Langston JW, Di Monte DA, Quik M (2005a) Enhanced striatal opioid receptor-mediated G-protein activation in L-DOPA-treated dyskinetic monkeys. Neuroscience 132:409–420
Chen T, Li J, Chao D, Sandhu HK, Liao X, Zhao J, Wen G, Xia Y (2014) δ-Opioid receptor activation reduces α-synuclein overexpression and oligomer formation induced by MPP(+) and/or hypoxia. Exp Neurol 255:127–136
Chen ZC, Kuo JR, Huang YP, Lin MT (2005b) μ-opioid receptor blockade protects against circulatory shock and cerebral ischemia during heatstroke. J Cardiovasc Pharmacol 46:754–760
Chesselet MF, Chéramy A, Reisine TD, Glowinski J (1981) Morphine and delta-opiate agonists locally stimulate in vivo dopamine release in cat caudate nucleus. Nature 291:320–322
Chesselet MF, Chéramy A, Reisine TD, Lubetzki C, Desban M, Glowinski J (1983) Local and distal effects induced by unilateral striatal application of opiates in the absence or in the presence of naloxone on the release of dopamine in both caudate nuclei and substantiae nigrae of the cat. Brain Res 258:229–242
Chiken S, Nambu A (2013) High-frequency pallidal stimulation disrupts information flow through the pallidum by GABAergic inhibition. J Neurosci 33:2268–2280
Chiken S, Nambu A (2014) Disrupting neuronal transmission: mechanism of DBS? Front Syst Neurosci 8:33. doi:10.3389/fnsys.2014.00033
Chu Sin Chung P, Kieffer BL (2013) Delta opioid receptors in brain function and diseases. Pharmacol Ther 140:112–120
Cleary DR, Raslan AM, Rubin JE, Bahgat D, Viswanathan A, Heinricher MM, Burchiel KJ (2013) Deep brain stimulation entrains local neuronal firing in human globus pallidus internus. J Neurophysiol 109:978–987
Comb M, Seeburg PH, Adelman J, Eiden L, Herbert E (1982) Primary structure of the human Met- and Leu-enkephalin precursor and its mRNA. Nature 295:663–666
Connolly BS, Lang AE (2014) Pharmacological treatment of Parkinson disease: a review. JAMA 311:1670–1683

Corti O, Lesage S, Brice A (2011) What genetics tells us about the causes and mechanisms of Parkinson's disease. Physiol Rev 91:1161–1218

Cotzias GC, Papavasiliou PS, Gellene R (1969) L-dopa in Parkinson's syndrome. N Engl J Med 281:272

Cox H, Togasaki DM, Chen L, Langston JW, Di Monte DA, Quik M (2007) The selective kappa-opioid receptor agonist U50,488 reduces L-dopa-induced dyskinesias but worsens parkinsonism in MPTP-treated primates. Exp Neurol 205:101–107

Cuello AC, Paxinos G (1978) Evidence for a long Leu-enkephalin striopallidal pathway in rat brain. Nature 271:178–180

Cui G, Jun SB, Jin X, Pham MD, Vogel SS, Lovinger DM, Costa RM (2013) Concurrent activation of striatal direct and indirect pathways during action initiation. Nature 494:238–242

Dardiotis E, Xiromerisiou G, Hadjichristodoulou C, Tsatsakis AM, Wilks MF, Hadjigeorgiou GM (2013) The interplay between environmental and genetic factors in Parkinson's disease susceptibility: the evidence for pesticides. Toxicology 307:17–23

de Lau LM, Breteler MM (2006) Epidemiology of Parkinson's disease. Lancet Neurol 5:525–535

Dauer W, Przedborski S (2003) Parkinson's disease: mechanisms and models. Neuron 39:889–909

Dejean C, Boraud T, Le Moine C (2013) Opiate dependence induces network state shifts in the limic system. Neurobiol Dis 59:220–229

Dejean C, Gross CE, Bioulac B, Boraud T (2008) Dynamic changes in the cortex-basal ganglia network after dopamine depletion in the rat. J Neurophysiol 100:385–396

Dejean C, Nadjar A, Le Moine C, Bioulac B, Gross CE, Boraud T (2012) Evolution of the dynamic properties of the cortex-basal ganglia network after dopaminergic depletion in rats. Neurobiol Dis 46:402–413

Del Fiacco M, Paxinos G, Cuello AC (1982) Neostriatal enkephalin-immunoreactive neurones project to the globus pallidus. Brain Res 231:1–17

Delay-Goyet P, Zajac JM, Javoy-Agid F, Agid Y, Roques BP (1987) Regional distribution of mu, delta and kappa opioid receptors in human brains from controls and parkinsonian subjects. Brain Res 414:8–14

DeLong M, Wichmann T (2009) Update on models of basal ganglia function and dysfunction. Parkinsonism Relat Disord 15(Suppl 3):S237–S240

DeLong MR (1971) Activity of pallidal neurons during movement. J Neurophysiol 34:414–427

Delong MR (1990) Primate models of movement disorders of basal ganglia origin. Trends Neurosci 13:281–285

Deniau JM, Mailly P, Maurice N, Charpier S (2007) The pars reticulata of the substantia nigra: a window to basal ganglia output. Prog Brain Res 160:151–172

Dorval AD, Panjwani N, Qi RY, Grill WM (2009) Deep brain stimulation that abolishes Parkinsonian activity in basal ganglia improves thalamic relay fidelity in a computational circuit. Conf Proc IEEE Eng Med Biol Soc 2009:4230–4233

Dourmap N, Costentin J (1994) Involvement of glutamate receptors in the striatal encephalin-induced dopamine release. Eur J Pharmacol 253:R9–R11

Dvorzhak A, Gertler C, Harnack D, Grantyn R (2013) High frequency stimulation of the subthalamic nucleus leads to presynaptic GABA(B)-dependent depression of subthalamo-nigral afferents. PLoS One 8, e82191. doi:10.1371/journal.pone.0082191

Eghbali M, Santoro C, Paredes W, Gardner EL, Zukin RS (1987) Visualization of multiple opioid-receptor types in rat striatum after specific mesencephalic lesions. Proc Natl Acad Sci U S A 84:6582–6586

Ellens DJ, Leventhal DK (2013) Electrophysiology of basal ganglia and cortex in models of Parkinson disease. J Parkinsons Dis 3:241–254

Erbs E, Faget L, Scherrer G, Kessler P, Hentsch D, Vonesch JL, Matifas A, Kieffer BL, Massotte D (2012) Distribution of delta opioid receptor-expressing neurons in the mouse hippocampus. Neuroscience 221:203–213

Fabbrini G, Brotchie JM, Grandas F, Nomoto M, Goetz CG (2007) Levodopa-induced dyskinesias. Mov Disord 22:1379–1389

Feng Y, He X, Yang Y, Chao D, Lazarus LH, Xia Y (2012) Current research on opioid receptor function. Curr Drug Targets 13:230–246
Fernandez A, de Ceballos ML, Jenner P, Marsden CD (1992) Striatal neuropeptide levels in Parkinson's disease patients. Neurosci Lett 145:171–174
Fernandez A, de Ceballos ML, Jenner P, Marsden CD (1994) Neurotensin, substance P, delta and mu opioid receptors are decreased in basal ganglia of Parkinson's disease patients. Neuroscience 61:73–79
Fernandez A, de Ceballos ML, Rose S, Jenner P, Marsden CD (1996) Alterations in peptide levels in Parkinson's disease and incidental Lewy body disease. Brain 119:823–830
Filion M, Tremblay L (1991) Abnormal spontaneous activity of globus pallidus neurons in monkeys with MPTP-induced parkinsonism. Brain Res 547:142–151
Filliol D, Ghozland S, Chluba J, Martin M, Matthes HW, Simonin F, Befort K, Gavériaux-Ruff C, Dierich A, LeMeur M, Valverde O, Maldonado R, Kieffer BL (2000) Mice deficient for delta- and mu-opioid receptors exhibit opposing alterations of emotional responses. Nat Genet 25:195–200
Fox S, Silverdale M, Kellett M, Davies R, Steiger M, Fletcher N, Crossman A, Brotchie J (2004) Non-subtype-selective opioid receptor antagonism in treatment of levodopa-induced motor complications in Parkinson's disease. Mov Disord 19:554–560
Freeze BS, Kravitz AV, Hammack N, Berke JD, Kreitzer AC (2013) Control of basal ganglia output by direct and indirect pathway projection neurons. J Neurosci 33:18531–18539
Gao HM, Hong JS (2011) Gene-environment interactions: key to unraveling the mystery of Parkinson's disease. Prog Neurobiol 94:1–19
Ge S, Yang C, Li M, Li J, Chang X, Fu J, Chen L, Chang C, Wang X, Zhu J, Gao G (2012) Dopamine depletion increases the power and coherence of high-voltage spindles in the globus pallidus and motor cortex of freely moving rats. Brain Res 1465:66–79
Georges F, Normand E, Bloch B, Le Moine C (1998) Opioid receptor gene expression in the rat brain during ontogeny, with special reference to the mesostriatal system: an in situ hybridization study. Brain Res Dev Brain Res 109:187–199
Gerfen CR, Engber TM, Mahan LC, Susel Z, Chase TN, Monsma FJ Jr, Sibley DR (1990) D1 and D2 dopamine receptor-regulated gene expression of striatonigral and striatopallidal neurons. Science 250:1429–1432
Gerfen CR, McGinty JF, Young WS III (1991) Dopamine differentially regulates dynorphin, substance P, and enkephalin expression in striatal neurons: in situ hybridization histochemical analysis. J Neurosci 11:1016–1031
Gerfen CR, Young WS III (1988) Distribution of striatonigral and striatopallidal peptidergic neurons in both patch and matrix compartments: an in situ hybridization histochemistry and fluorescent retrograde tracing study. Brain Res 460:161–167
Gerlach M, Riederer P, Scheller D (2011) Mechanisms underlying and medical management of L-Dope-associated motor complications. J Neural Transm 118:1659–1660
Gittis AH, Kreitzer AC (2012) Striatal microcircuitry and movement disorders. Trends Neurosci 35:557–564
Goldberg JA, Bergman H (2011) Computational physiology of the neural networks of the primate globus pallidus: function and dysfunction. Neuroscience 198:171–192
Goldman JG, Postuma R (2014) Premotor and nonmotor features of Parkinson's disease. Curr Opin Neurol 27:434–441
Goldman SM (2014) Environmental toxins and Parkinson's disease. Annu Rev Pharmacol Toxicol 54:141–164
Götz ME, Freyberger A, Riederer P (1990) Oxidative stress: a role in the pathogenesis of Parkinson's disease. J Neural Transm Suppl 29:241–249
Gradinaru V, Mogri M, Thompson KR, Henderson JM, Deisseroth K (2009) Optical deconstruction of parkinsonian neural circuitry. Science 324:354–359
Gray AC, Coupar IM, White PJ (2006) Comparison of opioid receptor distributions in the rat central nervous system. Life Sci 79:674–685
Graybiel AM (1990) Neurotransmitters and neuromodulators in the basal ganglia. Trends Neurosci 13:244–254

Gross A, Sims RE, Swinny JD, Sieghart W, Bolam JP, Stanford IM (2011) Differential localization of GABA(A) receptor subunits in relation to rat striatopallidal and pallidopallidal synapses. Eur J Neurosci 33:868–878
Haber SN, Watson SJ (1985) The comparative distribution of enkephalin, dynorphin and substance P in the human globus pallidus and basal forebrain. Neuroscience 14:1011–1024
Hallett PJ, Brotchie JM (2007) Striatal delta opioid receptor binding in experimental models of Parkinson's disease and dyskinesia. Mov Disord 22:28–40
Hamani C, Saint-Cyr JA, Fraser J, Kaplitt M, Lozano AM (2004) The subthalamic nucleus in the context of movement disorders. Brain 127:4–20
Hammond C, Bergman H, Brown P (2007) Pathological synchronization in Parkinson's disease: networks, models and treatments. Trends Neurosci 30:357–364
Harlan RE, Shivers BD, Romano GJ, Howells RD, Pfaff DW (1987) Localization of preproenkephalin mRNA in the rat brain and spinal cord by in situ hybridization. J Comp Neurol 258:159–184
Hasbi A, Fan T, Alijaniaram M, Nguyen T, Perreault ML, O'Dowd BF, George SR (2009) Calcium signaling cascade links dopamine D1-D2 receptor heteromer to striatal BDNF production and neuronal growth. Proc Natl Acad Sci U S A 106:21377–21382
Hauser KF, Gurwell JA, Turbek CS (1994) Morphine inhibits Purkinje cell survival and dendritic differentiation in organotypic cultures of the mouse cerebellum. Exp Neurol 130:95–105
Häusser M (2014) Optogenetics: the age of light. Nat Methods 11:1012–1014
He X, Sandhu HK, Yang Y, Hua F, Belser N, Kim DH, Xia Y (2013) Neuroprotection against hypoxia/ischemia: δ-opioid receptor-mediated cellular/molecular events. Cell Mol Life Sci 70:2291–2303
Hegarty SV, Sullivan AM, O'Keeffe GW (2013) Midbrain dopaminergic neurons: a review of the molecular circuitry that regulates their development. Dev Biol 379:123–138
Henry B, Crossman AR, Brotchie JM (1999) Effect of repeated L-DOPA, bromocriptine, or lisuride administration on preproenkephalin-A and preproenkephalin-B mRNA levels in the striatum of the 6-hydroxydopamine-lesioned rat. Exp Neurol 155:204–220
Henry B, Fox SH, Crossman AR, Brotchie JM (2001) Mu- and delta-opioid receptor antagonists reduce levodopa-induced dyskinesia in the MPTP-lesioned primate model of Parkinson's disease. Exp Neurol 171:139–146
Hill MP, Hille CJ, Brotchie JM (2000) Delta-opioid receptor agonists as a therapeutic approach in Parkinson's disease. Drug News Perspect 13:261–268
Hille CJ, Fox SH, Maneuf YP, Crossman AR, Brotchie JM (2001) Antiparkinsonian action of a delta opioid agonist in rodent and primate models of Parkinson's disease. Exp Neurol 172:189–198
Hipólito L, Sánchez-Catalán MJ, Zanolini I, Polache A, Granero L (2008) Shell/core differences in mu- and delta-opioid receptor modulation of dopamine efflux in nucleus accumbens. Neuropharmacology 55:183–189
Homberg JR, Mul JD, De Wit E, Cuppen E (2009) Complete knockout of the nociception/orphanin FQ receptor in the rat does not induce compensatory changes in μ, δ, and κ opioid receptors. Neuroscience 163:308–315
Hong JS, Yang HY, Fratta W, Costa E (1977) Determination of methionine enkephalin in discrete regions of rat brain. Brain Res 134:383–386
Hong S, Hikosaka O (2008) The globus pallidus sends reward-related signals to the lateral habenula. Neuron 60:720–729
Hu S, Sheng WS, Lokensgard JR, Peterson PK (2002) Morphine induces apoptosis of human microglia and neurons. Neuropharmacology 42:829–836
Hurd YL (1996) Differential messenger RNA expression of prodynorphin and proenkephalin in the human brain. Neuroscience 72:767–783
Hutchison WD, Allan RJ, Opitz H, Levy R, Dostrovsky JO, Lang AE, Lozano AM (1998) Neurophysiological identification of the subthalamic nucleus in surgery for Parkinson's disease. Ann Neurol 44:622–628

Isomura Y, Takekawa T, Harukuni R, Handa T, Aizawa H, Takada M, Fukai T (2013) Reward-modulated motor information in identified striatum neurons. J Neurosci 33:10209–10220
Jaeger D, Kita H (2011) Functional connectivity and integrative properties of globus pallidus neurons. Neuroscience 198:44–53
Jenab S, Kest B, Franklin SO, Inturrisi CE (1995) Quantitative distribution of the delta opioid receptor mRNA in the mouse and rat CNS. Life Sci 56:2343–2355
Jiang ZG, North RA (1992) Pre- and postsynaptic inhibition by opioid in rat striatum. J Neurosci 12:356–361
Johansson PA, Andersson M, Andersson KE, Cenci MA (2001) Alterations in cortical and basal ganglia levels of opioid receptor binding in a rat model of l-DOPA-induced dyskinesia. Neurobiol Dis 8:220–239
Johnson MD, Vitek JL, McIntyre CC (2009) Pallidal stimulation that improves parkinsonian motor symptoms also modulates neuronal firing patterns in primary motor cortex in the MPTP-treated monkey. Exp Neurol 219:359–362
Jose X, Pineda JC, Rodriguez C, Mendoza E, Galarraga E, Bargas J, Barral J (2007) Delta opioids reduce the neurotransmitter release probability by enhancing transient (KV4) K^+-currents in corticostriatal synapses as evaluated by the paired pulse protocol. Neurosci Lett 414:150–154
Kalia SK, Sankar T, Lozano AM (2013) Deep brain stimulation for Parkinson's disease and other movement disorders. Curr Opin Neurol 26:374–380
Kita H, Chiken S, Tachibana Y, Nambu A (2006) Origins of GABA(A) and GABA(B) receptor-mediated responses of globus pallidus induced after stimulation of the putamen in the monkey. J Neurosci 26:6554–6562
Kliem MA, Pare JF, Khan ZU, Wichmann T, Smith Y (2010) Ultrastructural localization and function of dopamine D1-like receptors in the substantia nigra pars reticulata and the internal segment of the globus pallidus of parkinsonian monkeys. Eur J Neurosci 31:836–851
Kofke WA, Garman RH, Garman R, Rose ME (1999) Opioid neurotoxicity: fentanyl-induced exacerbation of cerebral ischemia in rats. Brain Res 818:326–334
Koizumi H, Morigaki R, Okita S, Nagahiro S, Kaji R, Nakagawa M, Goto S (2013) Response of striosomal opioid signaling to dopamine depletion in 6-hydroxydopamine-lesioned rat model of Parkinson's disease: a potential compensatory role. Front Cell Neurosci 7:74. doi:10.3389/fncel.2013.00074
Kornblum HI, Hurlbut DE, Leslie FM (1987) Postnatal development of multiple opioid receptors in rat brain. Del Brain Res 37:21–41
Kravitz AV, Freeze BS, Parker PR, Kay K, Thwin MT, Deisseroth K, Kreitzer AC (2010) Regulation of parkinsonian motor behaviours by optogenetic control of basal ganglia circuitry. Nature 466:622–626
Kreitzer AC (2009) Physiology and pharmacology of striatal neurons. Annu Rev Neurosci 32:127–147
Kreitzer AC, Malenka RC (2008) Striatal plasticity and basal ganglia circuit function. Neuron 60:543–554
Kress GJ, Yamawaki N, Wokosin DL, Wickersham IR, Shepherd GM, Surmeier DJ (2013) Convergent cortical innervation of striatal projection neurons. Nat Neurosci 16:665–667
Langston JW, Ballard P, Tetrud JW, Irwin I (1983) Chronic Parkinsonism in humans due to a product of meperidine-analog synthesis. Science 219:979–980
Le Moine C, Fauchey V, Jaber M (2002) Opioid receptor gene expression in dopamine transporter knock-out mice in adult and during development. Neuroscience 112:131–139
Le Moine C, Kieffer B, Gaveriaux-Ruff C, Befort K, Bloch B (1994) Delta-opioid receptor gene expression in the mouse forebrain: localization in cholinergic neurons of the striatum. Neuroscience 62:635–640
Leblois A, Meissner W, Bioulac B, Gross CE, Hansel D, Boraud T (2007) Late emergence of synchronized oscillatory activity in the pallidum during progressive Parkinsonism. Eur J Neurosci 26:1701–1713
Lees AJ, Hardy J, Revesz T (2009) Parkinson's disease. Lancet 373:2055–2066

Lei W, Jiao Y, Del Mar N, Reiner A (2004) Evidence for differential cortical input to direct pathway versus indirect pathway striatal projection neurons in rats. J Neurosci 24:8289–8299
Léna I, Bradshaw S, Pintar J, Kitchen I (2008) Adaptive changes in the expression of central opioid receptors in mice lacking the dopamine D2 receptor gene. Neuroscience 153:773–788
Liao SL, Chen WY, Raung SL, Chen CJ (2003) Neuroprotection of naloxone against ischemic injury in rats: role of mu receptor antagonism. Neurosci Lett 345:169–172
Liu LD, Prescott IA, Dostrovsky JO, Hodaie M, Lozano AM, Hutchison WD (2012) Frequency-dependent effects of electrical stimulation in the globus pallidus of dystonia patients. J Neurophysiol 108:5–17
Llorens-Cortes C, Javoy-Agid F, Agid Y, Taquet H, Schwartz JC (1984) Enkephalinergic markers in substantia nigra and caudate nucleus from Parkinsonian subjects. J Neurochem 43:874–877
Longoni R, Spina L, Mulas A, Carboni E, Garau L, Melchiorri P, Di Chiara G (1991) (D-Ala2) deltorphin II: D1-dependent stereotypies and stimulation of dopamine release in the nucleus accumbens. J Neurosci 11:1565–1576
Lord JA, Waterfield AA, Hughes J, Kosterlitz HW (1977) Endogenous opioid peptides: multiple agonists and receptors. Nature 267:495–499
Lu X, Bing G, Hagg T (2000) Naloxone prevents microglia-induced degeneration of dopaminergic substantia nigra neurons in adult rats. Neuroscience 97:285–291
Lucido AL, Morinville A, Gendron L, Stroh T, Beaudet A (2005) Prolonged morphine treatment selectively increases membrane recruitment of delta-opioid receptors in mouse basal ganglia. J Mol Neurosci 25:207–214
Mabrouk OS, Marti M, Salvadori S, Morari M (2009) The novel delta opioid receptor agonist UFP-512 dually modulates motor activity in hemiparkinsonian rats via control of the nigro-thalamic pathway. Neuroscience 164:360–369
Mabrouk OS, Viaro R, Volta M, Ledonne A, Mercuri N, Morari M (2014) Stimulation of δ opioid receptor and blockade of nociceptin/orphanin FQ receptor synergistically attenuate parkinsonism. J Neurosci 34:12953–12962
Mabrouk OS, Volta M, Marti M, Morari M (2008) Stimulation of delta opioid receptors located in substantia nigra reticulata but not globus pallidus or striatum restores motor activity in 6-hydroxydopamine lesioned rats: new insights into the role of delta receptors in parkinsonism. J Neurochem 107:1647–1659
Madar I, Lever JR, Kinter CM, Scheffel U, Ravert HT, Musachio JL, Mathews WB, Dannals RF, Frost JJ (1996) Imaging of delta opioid receptors in human brain by N1'-([11C]methyl)naltrindole and PET. Synapse 24:19–28
Mansour A, Fox CA, Burke S, Meng F, Thompson RC, Akil H, Watson SJ (1994) Mu, delta, and kappa opioid receptor mRNA expression in the rat CNS: an in situ hybridization study. J Comp Neurol 350:412–438
Mansour A, Thompson RC, Akil H, Watson SJ (1993) Delta opioid receptor mRNA distribution in the brain: comparison to delta receptor binding and proenkephalin mRNA. J Chem Neuroanat 6:351–362
McCormick SE, Stoessl AJ (2002) Blockade of nigral and pallidal opioid receptors suppresses vacuous chewing movements in a rodent model of tardive dyskinesia. Neuroscience 112:851–859
McCracken CB, Kiss ZH (2014) Time and frequency-dependent modulation of local field potential synchronization by deep brain stimulation. PLoS One 9, e102576. doi:10.1371/journal.pone.0102576
Méndez M, Morales-Mulia M, Leriche M (2004) [3H]DPDPE binding to delta opioid receptors in the rat mesocorticolimbic and nigrostriatal pathways is transiently increased by acute ethanol administration. Brain Res 1028:180–190
Millan MJ, Maiofiss L, Cussac D, Audinot V, Boutin JA, Newman-Tancredi A (2002) Differential actions of antiparkinson agents at multiple classes of monoaminergic receptor. I. A multivariate analysis of the binding profiles of 14 drugs at 21 native and cloned human receptor subtypes. J Pharmacol Exp Ther 303:791–804
Mink JW (1996) The basal ganglia: focused selection and inhibition of competing motor programs. Prog Neurobiol 50:381–425

Montgomery EB Jr (2006) Effects of GPi stimulation on human thalamic neuronal activity. Clin Neurophysiol 117:2691–2702

Montgomery EB Jr (2007) Basal ganglia physiology and pathophysiology: a reappraisal. Parkinsonism Relat Disord 13:455–465

Mulder AH, Wardeh G, Hogenboom F, Frankhuyzen AL (1984) Kappa- and delta-opioid receptor agonists differentially inhibit striatal dopamine and acetylcholine release. Nature 308:278–280

Nambu A (2007) Globus pallidus internal segment. Prog Brain Res 160:135–150

Nambu A (2008) Seven problems on the basal ganglia. Curr Opin Neurobiol 18:595–604

Nambu A, Tachibana Y (2014) Mechanism of parkinsonian neuronal oscillations in the primate basal ganglia: some considerations based on our recent work. Front Syst Neurosci 8:74. doi:10.3389/fnsys.2014.00074

Nambu A, Tokuno H, Takada M (2002) Functional significance of the cortico-subthalamo-pallidal 'hyperdirect' pathway. Neurosci Res 43:111–117

Nelson AB, Hammack N, Yang CF, Shah NM, Seal RP, Kreitzer AC (2014) Striatal cholinergic interneurons Drive GABA release from dopamine terminals. Neuron 82:63–70

Nelson AB, Kreitzer AC (2014) Reassessing models of basal ganglia function and dysfunction. Annu Rev Neurosci 37:117–135

Nisbet AP, Foster OJ, Kingsbury A, Eve DJ, Daniel SE, Marsden CD, Lees AJ (1995) Preproenkephalin and preprotachykinin messenger RNA expression in normal human basal ganglia and in Parkinson's disease. Neuroscience 66:361–376

Nisenbaum LK, Crowley WR, Kitai ST (1996) Partial striatal dopamine depletion differentially affects striatal substance P and enkephalin messenger RNA expression. Brain Res Mol Brain Res 37:209–216

Nisenbaum LK, Kitai ST, Crowley WR, Gerfen CR (1994) Temporal dissociation between changes in striatal enkephalin and substance P messenger RNAs following striatal dopamine depletion. Neuroscience 60:927–937

Noda M, Teranishi Y, Takahashi H, Toyosato M, Notake M, Nakanishi S, Numa S (1982) Isolation and structural organization of the human preproenkephalin gene. Nature 297:431–434

Obeso JA, Jahanshahi M, Alvarez L, Macias R, Pedroso I, Wilkinson L, Pavon N, Day B, Pinto S, Rodríguez-Oroz MC, Tejeiro J, Artieda J, Talelli P, Swayne O, Rodríguez R, Bhatia K, Rodriguez-Diaz M, Lopez G, Guridi J, Rothwell JC (2009) What can man do without basal ganglia motor output? The effect of combined unilateral subthalamotomy and pallidotomy in a patient with Parkinson's disease. Exp Neurol 220:283–292

Obeso JA, Marin C, Rodriguez-Oroz C, Blesa J, Benitez-Temiño B, Mena-Segovia J, Rodríguez M, Olanow CW (2008) The basal ganglia in Parkinson's disease: current concepts and unexplained observations. Ann Neurol 64(Suppl):S30–S46

Obeso JA, Rodriguez-Oroz MC, Blesa FJ, Guridi J (2006) The globus pallidus pars externa and Parkinson's disease. Ready for prime time? Exp Neurol 202:1–7

Olanow CW, Obeso JA, Stocchi F (2006) Continuous dopamine-receptor treatment of Parkinson's disease: scientific rationale and clinical implications. Lancet Neurol 5:677–687

Olanow CW, Schapira AH (2013) Therapeutic prospects for Parkinson disease. Ann Neurol 74:337–347

Olive MF, Anton B, Micevych P, Evans CJ, Maidment NT (1997) Presynaptic versus postsynaptic localization of mu and delta opioid receptors in dorsal and ventral striatopallidal pathways. J Neurosci 17:7471–7479

Ostermeier AM, Schlosser B, Schwender D, Sutor B (2000) Activation of mu- and delta-opioid receptors causes presynaptic inhibition of glutamatergic excitation in neocortical neurons. Anesthesiology 93:1053–1063

Palkovits M, Epelbaum J, Gros C (1981) Met-enkephalin concentrations in individual brain nuclei of ansa lenticularis and stria terminalis transected rats. Brain Res 216:203–209

Pan WX, Mao T, Dudman JT (2010) Inputs to the dorsal striatum of the mouse reflect the parallel circuit architecture of the forebrain. Front Neuroanat 4:147. doi:10.3389/fnana.2010.00147

Parent A, Hazrati LN (1995a) Functional anatomy of the basal ganglia. I. The cortico-basal ganglia-thalamo-cortical loop. Brain Res Brain Res Rev 20:91–127

Parent A, Hazrati LN (1995b) Functional anatomy of the basal ganglia. II. The place of subthalamic nucleus and external pallidum in basal ganglia circuitry. Brain Res Brain Res Rev 20:128–154

Peckys D, Landwehrmeyer GB (1999) Expression of mu, kappa, and delta opioid receptor messenger RNA in the human CNS: a 33P in situ hybridization study. Neuroscience 88: 1093–1135

Pentney RJW, Gratton A (1991) Effects of local delta and mu opioid receptor activation on basal and stimulated dopamine release in striatum and nucleus accumbens of rat: an in vivo electrochemical study. Neuroscience 45:95–102

Pérez-Navarro E, Alberch J, Marsal J (1993) Postnatal development of functional dopamine, opioid and tachykinin receptors that regulate acetylcholine release from rat neostriatal slices. Effect of 6-hydroxydopamine lesion. Int J Dev Neurosci 11:701–708

Perreault ML, Hasbi A, O'Dowd BF, George SR (2011) The dopamine d1-d2 receptor heteromer in striatal medium spiny neurons: evidence for a third distinct neuronal pathway in Basal Ganglia. Front Neuroanat 5:31. doi:10.3389/fnana.2011.00031

Pinna A, Di Chiara G (1998) Dopamine-dependent behavioural stimulation by non-peptide delta opioids BW 373U86 and SNC 80: 3. Facilitation of D1 and D2 responses in unilaterally 6-hydroxydopamine-lesioned rats. Behav Pharmacol 9:15–21

Pioro EP, Hughes JT, Cuello AC (1984) Loss of substance P and enkephalin immunoreactivity in the human substantia nigra after striato-pallidal infarction. Brain Res 292:339–347

Pollok B, Krause V, Martsch W, Wach C, Schnitzler A, Südmeyer M (2012) Motor-cortical oscillations in early stages of Parkinson's disease. J Physiol 590:3203–3212

Polymeropoulos MH, Lavedan C, Leroy E, Ide SE, Dehejia A, Dutra A, Pike B, Root H, Rubenstein J, Boyer R, Stenroos ES, Chandrasekharappa S, Athanassiadou A, Papapetropoulos T, Johnson WG, Lazzarini AM, Duvoisin RC, Di Iorio G, Golbe LI, Nussbaum RL (1997) Mutation in the alpha-synuclein gene identified in families with Parkinson's disease. Science 276:2045–2047

Pradhan AA, Befort K, Nozaki C, Gavériaux-Ruff C, Kieffer BL (2011) The delta opioid receptor: an evolving target for the treatment of brain disorders. Trends Pharmacol Sci 32:581–590

Pringsheim T, Jette N, Frolkis A, Steeves TD (2014) The prevalence of Parkinson's disease: a systematic review and meta-analysis. Mov Disord 29(13):1583–1590

Quik M, Police S, Langston JW, Di Monte DA (2002) Increases in striatal preproenkephalin gene expression are associated with nigrostriatal damage but not L-DOPA-induced dyskinesias in the squirrel monkey. Neuroscience 113:213–220

Ravenscroft P, Chalon S, Brotchie JM, Crossman AR (2004) Ropinirole versus L-DOPA effects on striatal opioid peptide precursors in a rodent model of Parkinson's disease: implications for dyskinesia. Exp Neurol 185:36–46

Rawls SM, McGinty JF (2000) Delta opioid receptors regulate calcium-dependent, amphetamine-evoked glutamate levels in the rat striatum: an in vivo microdialysis study. Brain Res 861:296–304

Reeve A, Simcox E, Turnbull D (2014) Ageing and Parkinson's disease: why is advancing age the biggest risk factor? Ageing Res Rev 14:19–30

Reimer S, Sirinathsinghji DJ, Nikolorakis KE, Höllt V (1992) Differential dopaminergic regulation of proenkephalin and prodynorphin mRNAs in the basal ganglia of rats. Brain Res Mol Brain Res 12:259–266

Reiner A, Hart NM, Lei W, Deng Y (2010) Corticostriatal projection neurons—dichotomous types and dichotomous functions. Front Neuroanat 4:142. doi:10.3389/fnana.2010.00142

Rijsman RM, Schoolderman LF, Rundervoort RS, Louter M (2014) Restless legs syndrome in Parkinson's disease. Parkinsonism Relat Disord 20(Suppl 1):S5–S9

Rivero-Ríos P, Gómez-Suaga P, Fdez E, Hilfiker S (2014) Upstream deregulation of calcium signaling in Parkinson's disease. Front Mol Neurosci 7:53. doi:10.3389/fnmol.2014.00053

Rivlin-Etzion M, Marmor O, Heimer G, Raz A, Nini A, Bergman H (2006) Basal ganglia oscillation and pathophysiology of movement disorders. Curr Opin Neurobiol 16:629–637

Rubchinsky LL, Park C, Worth RM (2012) Intermittent neural synchronization in Parkinson's disease. Nonlinear Dyn 68:329–346

Rubin JE, McIntyre CC, Turner RS, Wichmann T (2012) Basal ganglia activity patterns in parkinsonism and computational modeling of their downstream effects. Eur J Neurosci 36: 2213–2228
Saha SN, Bhatia SC, Nayar U (2003) Enkephalinergic involvement in substantia nigra in the modulation of hypothalamically-induced predatory attack behavior. Indian J Physiol Pharmacol 47:311–317
Samadi P, Bédard PJ, Rouillard C (2006) Opioids and motor complications in Parkinson's disease. Trends Pharmacol Sci 27:512–517
Samadi P, Grégoire L, Bédard PJ (2003) Opioid antagonists increase the dyskinetic response to dopaminergic agents in parkinsonian monkeys: interaction between dopamine and opioid systems. Neuropharmacology 45:954–963
Samii A, Nutt JG, Ransom BR (2004) Parkinson's disease. Lancet 363:1783–1793
Sandor NT, Lendvai B, Vizi ES (1992) Effect of selective opiate antagonists on striatal acetylchonline and dopamine release. Brain Res Bull 29:369–373
Sano H, Chiken S, Hikida T, Kobayashi K, Nambu A (2013) Signals through the striatopallidal indirect pathway stop movements by phasic excitation in the substantia nigra. J Neurosci 33:7583–7594
Scherrer G, Tryoen-Tóth P, Filliol D, Matifas A, Laustriat D, Cao YQ, Basbaum AI, Dierich A, Vonesh JL, Gavériaux-Ruff C, Kieffer BL (2006) Knockin mice expressing fluorescent delta-opioid receptors uncover G protein-coupled receptor dynamics in vivo. Proc Natl Acad Sci U S A 103:9691–9696
Schmidt R, Leventhal DK, Mallet N, Chen F, Berke JD (2013) Canceling actions involves a race between basal ganglia pathways. Nat Neurosci 16:1118–1124
Schwab BC, Heida T, Zhao Y, Marani E, van Gils SA, van Wezel RJ (2013) Synchrony in Parkinson's disease: importance of intrinsic properties of the external globus pallidus. Front Syst Neurosci 7:60. doi:10.3389/fnsys.2013.00060
Schwartz JP, Simantov R (1988) Developmental expression of proenkephalin mRNA in rat striatum and in striatal cultures. Brain Res 468:311–314
Sgroi S, Kaelin-Lang A, Capper-Loup C (2014) Spontaneous locomotor activity and L-DOPA-induced dyskinesia are not linked in 6-OHDA parkinsonian rats. Front Behav Neurosci 8:331. doi:10.3389/fnbeh.2014.00331
Shen KZ, Johnson SW (2002) Presynaptic modulation of synaptic transmission by opioid receptor in rat subthalamic nucleus in vitro. J Physiol 541:219–230
Shepherd GMG (2013) Corticostriatal connectivity and its role in disease. Nat Rev Neurosci 14:278–291
Shivers BD, Harlan RE, Romano GJ, Howells RD, Pfaff DW (1986) Cellular localization of proenkephalin mRNA in rat brain: gene expression in the caudate-putamen and cerebellar cortex. Proc Natl Acad Sci U S A 83:6221–6225
Sims RE, Woodhall GL, Wilson CL, Stanford IM (2008) Functional characterization of GABAergic pallidopallidal and striatopallidal synapses in the rat globus pallidus in vitro. Eur J Neurosci 28:2401–2408
Singleton AB, Farrer MJ, Bonifati V (2013) The genetics of Parkinson's disease: progress and therapeutic implications. Mov Disord 28:14–23
Smith JA, Leslie FM, Broide RS, Loughlin SE (1993) Long-term changes in striatal opioid systems after 6-hydroxydopamine lesion of rat substantia nigra. Neuroscience 55:935–951
Smith Y, Villalba R (2008) Striatal and extrastriatal dopamine in the basal ganglia: an overview of its anatomical organization in normal and Parkinsonian brains. Mov Dis 23(Suppl 3):S534–S547
Smith Y, Galvan A, Ellender TJ, Doig N, Villalba RM, Huerta-Ocampo I, Wichmann T, Bolam JP (2014) The thalamostriatal system in normal and diseased states. Front Syst Neurosci 8:5. doi:10.3389/fnsys.2014.00005
Spillantini MG, Schmidt ML, Lee VM, Trojanowski JQ, Jakes R, Goedert M (1997) Alpha-synuclein in Lewy bodies. Nature 388:839–840
Standford IM, Cooper AJ (1999) Presynaptic μ and δ opioid receptor modulation of GABAA IPSCs in the rat globus pallidus in vitro. J Neurosci 19:4796–4803

Stayte S, Vissel B (2014) Advances in non-dopaminergic treatments for Parkinson's disease. Front Neurosci 8:113. doi:10.3389/fnins.2014.00113

Steece KA, DeLeon-Jones FA, Meyerson LR, Lee JM, Fields JZ, Ritzmann RF (1986) In vivo down-regulation of rat striatal opioid receptors by chronic enkephalin. Brain Res Bull 17:255–257

Steiner H, Gerfen CR (1998) Role of dynorphin and encephalin in the regulation of striatal output pathways and behavior. Exp Brain Res 123:60–67

Subramaniam SR, Chesselet MF (2013) Mitochondrial dysfunction and oxidative stress in Parkinson's disease. Prog Neurobiol 106–107:17–32

Sun YM, Hoang T, Neubauer JA, Walters AS (2011) Opioids protect against substantia nigra cell degeneration under conditions of iron deprivation: a mechanism of possible relevance to the Restless Legs Syndrome (RLS) and Parkinson's disease. J Neurol Sci 304:93–101

Surmeier DJ, Ding J, Day M, Wang ZSM, Shen W (2007) D1 and D2 dopaminergic-receptor modulation of striatal glutamatergic signaling in striatal medium spiny neurons. Trends Neurosci 30:228–235

Svingos AL, Clarke CL, Pickel VM (1998) Cellular sites for activation of delta-opioid receptors in the rat nucleus accumbens shell: relationship with Met5-enkephalin. J Neurosci 18:1923–1933

Svingos AL, Clarke CL, Pickel VM (1999) Localization of the delta-opioid receptor and dopamine transporter in the nucleus accumbens shell: implications for opiate and psychostimulant cross-sensitization. Synapse 34:1–10

Tachibana Y, Kita H, Chiken S, Takada M, Nambu A (2008) Motor cortical control of internal pallidal activity through glutamatergic and GABAergic inputs in awake monkeys. Eur J Neurosci 27:238–253

Tamim MK, Samadi P, Morissette M, Grégoire L, Ouattara B, Lévesque D, Rouillard C, Di Paolo T (2010) Effect of non-dopaminergic drug treatment on Levodopa induced dyskinesias in MPTP monkeys: common implication of striatal neuropeptides. Neuropharmacology 58:286–296

Tanahashi S, Ueda Y, Nakajima A, Yamamura S, Nagase H, Okada M (2012) Novel δ1-receptor agonist KNT-127 increases the release of dopamine and L-glutamate in the striatum, nucleus accumbens and median pre-frontal cortex. Neuropharmacology 62:2057–2067

Tanaka E, North RA (1994) Opioid actions on rat anterior cingulated cortex neurons in vitro. J Neurosci 14:1106–1113

Tanibuchi I, Kitano H, Jinnai K (2009a) Substantia nigra output to prefrontal cortex via thalamus in monkeys. I Electrophysiological identification of thalamic relay neurons. J Neurophysiol 102:2933–2945

Tanibuchi I, Kitano H, Jinnai K (2009b) Substantia nigra output to prefrontal cortex via thalamus in monkeys. II Activity of thalamic relay neurons in delayed conditional go/no-go discrimination task. J Neurophysiol 102:2946–2954

Taquet H, Javoy-Agid F, Hamon M, Legrand JC, Agid Y, Cesselin F (1983) Parkinson's disease affects differently Met5- and Leu5-enkephalin in the human brain. Brain Res 280:379–382

Tel BC, Zeng BY, Cannizzaro C, Pearce RK, Rose S, Jenner P (2002) Alterations in striatal neuropeptide mRNA produced by repeated administration of L-DOPA, ropinirole or bromocriptine correlate with dyskinesia induction in MPTP-treated common marmosets. Neuroscience 115:1047–1058

Temel Y, Blokland A, Steinbusch HW, Visser-Vandewalle V (2005) The functional role of the subthalamic nucleus in cognitive and limbic circuits. Prog Neurobiol 76:393–413

Tempel A, Zukin RS (1987) Neuroanatomical patterns of the mu, delta, and kappa opioid receptors of rat brain as determined by quantitative in vitro autoradiography. Proc Natl Acad Sci U S A 84:4308–4312

Tepper JM, Tecuapetla F, Koós T, Ibáñez-Sandoval O (2010) Heterogeneity and diversity of striatal GABAergic interneurons. Front Neuroanat 4:150. doi:10.3389/fnana.2010.00150

Thanvi B, Lo N, Robinson T (2007) Levodopa-induced dyskinesia in Parkinson's disease: clinical features, pathogenesis, prevention and treatment. Postgrad Med J 83:384–388

Thibault D, Loustalot F, Fortin GM, Bourque MJ, Trudeau LÉ (2013) Evaluation of D1 and D2 dopamine receptor segregation in the developing striatum using BAC transgenic mice. PLoS One 8(7), e67219. doi:10.1371/journal.pone.0067219

Threlfell S, Lalic T, Platt NJ, Jennings KA, Deisseroth K, Cragg SJ (2012) Striatal dopamine release is triggered by synchronized activity in cholinergic interneurons. Neuron 75:58–64

Trinh J, Farrer M (2013) Advances in the genetics of Parkinson disease. Nat Rev Neurol 9:445–454

Tritsch NX, Sabatini BL (2012) Dopaminergic modulation of synaptic transmission in the cortex and striatum. Neuron 76:33–50

Tsao L-I, Cadet JL, Su T-P (1999) Reversal by peptide [D-Ala2, D-Leu5]enkephalin of the dopamine transporter loss caused by methamphetamine. Eur J Pharmacol 372:R5–R7

Tsao L-I, Ladenheim B, Andrews AM, Chiueh CC, Cadet JL, Su T-P (1998) Delta opioid peptide [D-Ala2, D-Leu5]enkephalin blocks the long-term loss of dopamine transporters induced by multiple administration of methamphetamine: involvement of opioid receptors and reactive oxygen species. J Pharmacol Exp Ther 287:322–331

Tsiokos C, Hu X, Pouratian N (2013) 200-300Hz movement modulated oscillations in the internal globus pallidus of patients with Parkinson's disease. Neurobiol Dis 54:464–474

Turner RS, Anderson ME (2005) Context-dependent modulation of movement-related discharge in the primate globus pallidus. J Neurosci 25:2965–2976

Ubeda-Bañon I, Novejarque A, Mohedano-Moriano A, Pro-Sistiaga P, de la Rosa-Prieto C, Insausti R, Martinez-Garcia F, Lanuza E, Martinez-Marcos A (2007) Projections from the posterolateral olfactory amygdala to the ventral striatum: neural basis for reinforcing properties of chemical stimuli. BMC Neurosci 8:103

Urigüen L, Berrendero F, Ledent C, Maldonado R, Manzanares J (2005) Kappa- and delta-opioid receptor functional activities are increased in the caudate putamen of cannabinoid CB1 receptor knockout mice. Eur J Neurosci 22:2106–2110

Utter AA, Basso MA (2008) The basal ganglia: an overview of circuits and function. Neurosci Biobehav Rev 32:333–342

Valjent E, Bertran-Gonzalez J, Hervé D, Fisone G, Girault JA (2009) Looking BAC at striatal signaling: cell-specific analysis in new transgenic mice. Trends Neurosci 32:538–547

Vance JM, Ali S, Bradley WG, Singer C, Di Monte DA (2010) Gene-environment interactions in Parkinson's disease and other forms of parkinsonism. Neurotoxicology 31:598–602

Vo A, Hiebert NM, Seergobin KN, Solcz S, Partridge A, MacDonald PA (2014) Dopaminergic medication impairs feedback-based stimulus-response learning but not response selection in Parkinson's disease. Front Hum Neurosci 8:784. doi:10.3389/fnhum.2014.00784

Waldhoer M, Bartlett SE, Whistler JL (2004) Opioid receptors. Annu Rev Biochem 73:953–990

Wall NR, De La Parra M, Callaway EM, Kreitzer AC (2013) Differential innervation of direct- and indirect-pathway striatal projection neurons. Neuron 79:347–360

Wamsley JK, Young WS III, Kuhar MJ (1980) Immunohistochemical localization of enkephalin in rat forebrain. Brain Res 190:153–174

Wang H, Cuzon VC, Pickel VM (2003) Ultrastructural localization of delta-opioid receptors in the rat caudate-putamen nucleus during postnatal development: relation to synaptogenesis. J Comp Neurol 467:343–353

Wang H, Pickel VM (2001) Preferential cytoplasmic localization of delta-opioid receptors in rat striatal patches: comparison with plasmalemmal mu-opioid receptors. J Neurosci 21:3242–3250

Wang HB, Deng YP, Reiner A (2007) In situ hybridization histochemical and immunohistochemical evidence that striatal projection neurons co-containing substance P and enkephaline are overrepresented in the striosomal compartment of striatum in rats. Neurosci Lett 425:195–199

Wang Y, Qu L, Wang XL, Gao L, Li ZZ, Gao GD, Yang Q (2014) Firing pattern modulation through SK channel current increase underlies neuronal survival in an organotypic slice model of Parkinson's disease. Mol Neurobiol 51(1):424–436

Westin JE, Andersson M, Lundblad M, Cenci MA (2001) Persistent changes in striatal gene expression induced by long-term L-DOPA treatment in a rat model of Parkinson's disease. Eur J Neurosci 14:1171–1176

Wichmann T, Bergman H, Starr PA, Subramanian T, Watts RL, DeLong MR (1999) Comparison of MPTP-induced changes in spontaneous neuronal discharge in the internal pallidal segment and in the substantia nigra pars reticulata in primates. Exp Brain Res 125:397–409

Wichmann T, Soares J (2006) Neuronal firing before and after burst discharges in the monkey basal ganglia is predictably patterned in the normal state and altered in Parkinsonism. J Neurophysiol 95:2120–2133

Winzer-Serhan UH, Chen Y, Leslie FM (2003) Expression of opioid peptides and receptors in striatum and substantia nigra during rat brain development. J Chem Neuroanat 26:17–36

Wong JC, Li Y, Schwarzschild MA, Ascherio A, Gao X (2014) Restless legs syndrome: an early clinical feature of Parkinson disease in men. Sleep 37:369–372

Xia Y, Haddad GG (1991) Ontogeny and distribution of opioid receptors in the rat brainstem. Brain Res 549:181–193

Xia Y, Haddad GG (2001) Major difference in the expression of μ- and δ-opioid receptors between turtle and rat brain. J Comp Neurol 436:202–210

Yamamura S, Ohoyama K, Nagase H, Okada M (2009) Zonisamide enhances delta receptor-associated neurotransmitter release in striato-pallidal pathway. Neuropharmacology 57:322–331

Yin D, Mufson RA, Wang R, Shi Y (1999) Fas-mediated cell death promoted by opioids. Nature 397:218

Yizhar O, Fenno LE, Davidson TJ, Mogri M, Deisseroth K (2011) Optogenetics in neural systems. Neuron 71:9–34

Yue X, Falk T, Zuniga LA, Szabò L, Porreca F, Polt R, Sherman SJ (2011) Effects of the novel glycopeptide opioid agonist MMP-22'00 in preclinical models of Parkinson's disease. Brain Res 1413:72–83

Zhou FM, Lee CR (2011) Intrinsic and integrative properties of substantia nigra pars reticulata neurons. Neuroscience 198:69–94

Zhu Y, Hsu MS, Pintar JE (1998) Developmental expression of the mu, kappa, and delta opioid receptor mRNAs in mouse. J Neurosci 18:2538–2549

Chapter 15
Are δ-Opioid Receptors Involved in Deep Brain Stimulation?

Albert J. Fenoy and Ying Xia

Abstract Despite its widespread use, the underlying mechanism of deep brain stimulation (DBS) remains unknown. Once thought to impart a "functional inactivation", there is now increasing evidence showing that DBS actually can both inhibit neurons and activate axons, generating a wide range of effects. This implies that the mechanisms that underlie DBS work not only locally but also at the network level. Therefore, not only may DBS induce membrane or synaptic plastic changes in neurons over a wide network, but it may also trigger cellular and molecular changes in other cells, especially astrocytes, where together the glial-neuronal interactions may explain effects that are not clearly rationalized by simple activation/inhibition theories alone. Recent studies suggest that (1) High frequency stimulation (HFS) activates astrocytes and leads to the release of gliotransmitters that can regulate surrounding neurons at the synapse; (2) Activated astrocytes modulate synaptic activity and increase axonal activation; (3) Activated astrocytes can signal further astrocytes across large networks, contributing to observed network effects induced by DBS; (4) Activated astrocytes can help explain the disparate effects of activation and inhibition induced by HFS at different sites; and (5) Astrocytes contribute to synaptic plasticity through long-term potentiation (LTP) and depression (LTD), possibly helping to mediate the long term effects of DBS. More recently, we found that δ-Opioid receptor (DOR) activation reduces α-synuclein overexpression and oligomer formation, thus attenuating cellular injury after MPP^+ exposure. Indeed, DOR activation protects both glial and neuronal cells against injury in our *in-vitro* studies. Furthermore, we demonstrate that PD pathophysiology is critically attributed to DOR impairment in the brain of a mouse PD model. Moreover, recent evidence suggests that DOR activation promotes the release of dopamine in the striatum. Since we have previously shown that brain stimulation increased the

A.J. Fenoy, M.D. (✉)
Department of Neurosurgery, University of Texas Medical School at Houston, Houston, TX 77030, USA
e-mail: albert.j.fenoy@uth.tmc.edu

Y. Xia, M.D., Ph.D.
University of Texas Medical School, Houston, TX, USA
e-mail: ying.xia@uth.tmc.edu; y55738088@gmail.com

Y. Xia (ed.), *Neural Functions of the Delta-Opioid Receptor*,
DOI 10.1007/978-3-319-25495-1_15

release of encephalin, an endogenous DOR agonist, in the brain, the mechanism of DBS may be, at least partially, attributed to DOR activity that confers protection in both neurons and astrocytes. Together, the plastic changes in these glial-neuronal interactions network-wide likely underlie the range of effects seen, from the variable temporal latencies to observed effect to global activation patterns. This chapter provides a broad review of our understanding of DBS mechanisms, and then presents recent research progress in the literature on how astrocytes play a key role in DBS efficacy and our novel observations on a potential DOR mechanism for DBS.

Keywords DBS • Astrocytes • Neurons • Plasticity • Network • δ-Opioid receptors

Abbreviations

AD	Alzheimer's disease
AMPA	α-Amino-3-hydroxy-5-methyl-4-isoxazolepropionic acid
AP	Action potential
ATP	Adenosine triphosphate
BA	Brodmann area
BDNF	Brain derived natriuretic peptide
BG	Basal ganglia
BOLD	Blood oxygen level dependent
CBGTC	Cortico-basal-ganglia-thalamo-cortical
CHR2	Channel rhodopsin 2
CSF	Cerebral spinal fluid
CT	Computed tomography
cZi	Caudal zona incerta
DBS	Deep brain stimulation
DOR	Delta opiod receptor
DRT	Dentato-rubral-thalamic tract
dTi	Diffusion tensor imaging
EAAT	Excitatory amino acid transporter
ET	Essential tremor
fMRI	Functional magnetic resonance imaging
FOG	Freezing of gait
GABA	γ-Aminobutyric acid
GDNF	Glial cell-derived neurotrophic factor
GFAP	Glial fibrillary acidic protein
GPe	Globus pallidus externus
GPi	Globus pallidus internus
GTS	Gilles de la Tourettes syndrome
HFS	High frequency stimulation
IL-1beta	Interleukin 1 beta
LFS	Low frequency stimulation
LTD	Long term depression

LTP	Long term potentiation
MAPK	Mitogen activated protein kinase
MDD	Major depressive disorder
MEG	Magnetoencephalography
mGluR	Metabotropic glutamate receptor
MPTP	1-Methyl-4-phenyl-1,2,3,6-tetrahydropyridine
MRI	Magnetic resonance imaging
MSN	Medium spiny neurons
NGF	Nerve growth factor
NKCC	Na-K-Cl co-transporter
NMDA	N-methyl-D-aspartate
OCD	Obsessive compulsive disorder
OGD	Oxygen glucose deprivation
PAS	Paired associative stimulation
PD	Parkinson's disease
PET	Positron-emission tomography
PKC	Protein kinase C
PPN	Pedunculopontine nucleus
PVG/PAG	Periventricular grey/periaqueductal grey
rCBF	Regional cerebral blood flow
SIC	Slow inward current
SICI	Short interval intracortical inhibition
sLORETA	Standardized low resolution brain electromagnetic tomography
SNc	Substantia nigra, pars compacta
SNr	Substantia nigra, pars reticulata
STN	Subthalamic nucleus
TNF-a	Tumor necrosis factor alpha
TRD	Treatment resistant depression
TTX	Tetrodotoxin
VA	Ventral anterior nucleus (thalamus)
Vim	Ventral intermediate nucleus (thalamus)
VLo	Ventral lateral oralis (thalamus)
Vop	Ventralis oralis posterior (thalamus)
VPL	Ventral posterior lateral (thalamus)
VPM	Ventral posterior medial (thalamus)

1 Introduction

Ever since its use to refine brain lesioning procedures over 50 years ago to the more current practice launched by Benabid et al. in (1987), deep brain stimulation (DBS) has become a widely recognized technique for reversible modulation of brain function that is adjunctive to medical management of neurological disorders. The proven efficacy in Parkinson's Disease (PD) (Benabid et al. 1987; DBS Group 2001; Krack et al. 2003; Rodriguez-Oroz et al. 2005) has led to its worldwide application in a spectrum of hyperkinetic diseases, including essential tremor (ET) (Benabid et al. 1991; Hubble et al. 1996; Ondo et al. 1998; Koller et al. 2001) dystonia (Cif et al. 2003; Yianni et al. 2003; Krauss et al. 2004; Vidailhet et al. 2005), cerebellar outflow tremor (Schulder et al. 1999; Berk et al. 2002; Hooper et al. 2002; Bittar et al. 2005), Gilles de la Tourette Syndrome (GTS) (Temel and Visser-Vandewalle 2004; Diederich et al. 2005; Houeto et al. 2005) as well as a growing number of other medically intractable disorders, including pain, obsessive-compulsive disorder (OCD) (Greenberg et al. 2006; Mallet et al. 2008a; Denys et al. 2010); major depressive disorder (MDD), (Mayberg et al. 2005; Lozano et al. 2008; Malone et al. 2009), cluster headache (Sillay et al. 2010) and epilepsy (Pereira et al. 2012). The spectrum of applications continues to grow as new targets are identified and as more treatment-resistant diseases seek other options for therapy.

Interestingly, despite its widespread use, the underlying mechanism of DBS for its therapeutic efficacy remains unknown. Although there once was a divergence in opinions on the primary action of DBS, ranging from a "functional inactivation" due to local depolarization block to cell activation, there is now increasing evidence that DBS actually can both inhibit neurons and activate axons, generating a wide range of effects. This implies that the mechanisms that underlie DBS work not only locally but at the network level.

Support for fiber activation as a critical mechanism of DBS is provided by the clinical beneficial effects obtained by stimulation of fiber paths in a wide range of disorders, such as the dentate-rubro-thalamic tract in tremor, as well as the internal capsule (Greenberg et al. 2010) in OCD and subgenual cingulate white matter in MDD (Mayberg et al. 2005). Yet, stimulation of subcortical nuclear structures such as the subthalamic nucleus (STN) and globus pallidus internus (GPi) in PD clearly cause widespread effect, as illustrated by cerebral blood flow changes in the cortex (Limousin et al. 1997; Haegelen et al. 2010). It seems that the mechanism cannot embody solely one hypothesis or the other, but more of a combination of the two.

When scouring the mechanisms hypothesized for DBS efficacy, in terms of both local and network effects, one cannot dismiss the fact that a non-neuronal factor, i.e., the glial interaction with neurons, is important in mediating cellular signaling. Astrocytes, a major type of glia, contain numerous processes that extend hundreds of millimeters and connect to those of other astrocytes through gap junctions, and surround and are intertwined in neuronal synapses forming what is termed the 'tripartate synapse' (Araque et al. 1999; Ventura and Harris 1999; Takano et al. 2006). This position makes them well poised to function actively in the regulation of neuronal communication. Indeed, they are active players in neuronal signaling. If modulation of axonal flow—and hence synaptic transmission—is considered a key element underlying the mechanism of DBS, then quite likely both neurons and astrocytes contribute to the DBS therapeutic effect. In addition, not only may DBS electronically stimulate the brain, but also chemically regulate the integrative function of the local environment and network via regulation of neurotransmitter/modulator release that may affect both neurons and astrocytes. Therefore, DBS-induced regulation of astrocytes, especially its interaction with neurons, quite likely contributes to the beneficial effect of DBS. In fact, the contribution from the astrocytes has been the subject of a few recent reviews (Vedam-Mai et al. 2012; Jantz and Watanabe

2013). However, these reviews have not been inclusive of all mechanisms by which astrocytic—neuronal interactions may contribute to the mechanisms of action of DBS. Our recent investigations on how astrocytes react to applied hypoxia or simulated ischemia (oxygen and glucose deprivation, OGD) gives us a new insight into their under-recognized importance in neuroprotection in disease states. Recent evidence suggests that activation of the delta-opioid receptor (DOR) promotes the release of dopamine in the striatum (Tanahashi et al. 2012) and protects against neuronal injury in PD states (also refer to Chap. 14). We also noticed that applied electrical stimulation to the hypothalamus evokes the release of enkephalin in the brain (Xia et al. 1988), which potentially promotes signaling cascades of DOR to confer neuroprotection (He et al. 2013). This is of direct significance to how DBS may influence networks far more than just the local milieu. These findings are discussed below in the context of the role of DOR activation in increasing astrocytic glutamate transport and other neuroprotective effects.

What follows is a unifying discussion reviewing the literature and current mindset of mechanisms underlying DBS and how astrocytes may be considered key players in promoting the wide range of observed DBS effects. Through these neuronal-glial interactions both locally and network wide, we hypothesize that perhaps the DBS activation of DOR can potentially be a key contributor to specific effects, such as neuroprotection.

2 DBS Induced Actions

The proposed mechanisms behind how DBS exerts its powerful effects have been the subject of debate over the years since its inception. Below is a synopsis of current hypotheses.

2.1 Background

Once Benabid et al. (1987) observed that high frequency stimulation (HFS) of the ventral intermediate (Vim) thalamic nucleus alleviates tremor, there has been increasing interest in deep brain stimulation as a therapeutic tool to treat a wide range of movement disorders. However, there has been a divergence of opinions on the action of DBS, ever since the original proposition by Benabid et al. as an intervention into a pathologically activated network, akin to "jamming" (Benabid et al. 1996).

Other early hypotheses on the mechanism of DBS were centered on the idea of reversible functional inactivation of the target structure, producing a lesion-like effect, achieved through mechanisms including local depolarization block, inactivation of neuronal voltage-dependent channels, functional deafferentation (Anderson et al. 2006; Benazzouz et al. 1993, 1995; Benazzouz and Hallett 2000; Beurrier et al. 2001; Magarinos-Ascone et al. 2002) and possibly inhibiting axon flow (Zheng et al. 2011).

This idea was based on observations that stimulation effects of the STN, GPi or the Vim on parkinsonian motor signs, essential tremor and dystonia were similar to lesions in these structures, which was further maintained by recording suppressed neuronal activity in these structures during stimulation (Dostrovsky et al. 2000; Wu et al. 2001).

Further non-synaptic mechanisms were proposed over subsequent years to account for functional silencing of pathological activity and hence silencing of target efferent activity, and associated aberrant network oscillations, including axonal conduction block through neurotransmitter depletion (Iremonger et al. 2006; Jensen and Durand 2009), antidromic effects (Li et al. 2007; Liu et al. 2012), mixed effects on cell somata (inhibition of intrinsic firing) and axons (excitation of target regions) (McIntyre et al. 2004a, b, c, d), and adenosinergic inhibition (Bekar et al. 2008).

Currently, there is increasing electrophysiological evidence that DBS actually inhibits neurons and activates axons, generating a wide range of effects not only locally but at the systemic level: there is less support of synaptic inhibition (Iremonger et al. 2006) than excitation and in general more reports are in favor of both orthodromic and antidromic activation as well as activating fibers of passage (Holsheimer et al. 2000; Dostrovsky and Lozano 2002; Maurice et al. 2003; McIntyre et al. 2004a, b, c, d; Perlmutter and Mink 2006; Chang et al. 2007; Kringelbach et al. 2007a, b; Johnson et al. 2008; Hammond et al. 2008; Montgomery and Gale 2008; Xu et al. 2008; Gubellini et al. 2009; McIntyre and Hahn 2010; Montgomery and Hilker 2009; Yousif et al. 2008; Gradinaru et al. 2009; McCairn and Turner 2009; Carlson et al. 2010; Hashimoto et al. 2003; Bar-Gad et al. 2004).

2.1.1 Electrical Stimulation In Situ: Basic Principles

Detailed reviews over the last years have discussed and described the wide range of both observational and theoretical data that could explain the varied effects of DBS clinically (McIntyre et al. 2004a, c, d; Gubellini et al. 2009), focusing on the potential effects of application of HFS on the structural differences of the neuron, within or without the context of a greater neuronal network.

To summarize, there are different electrical properties of each of the neuronal components—myelinated axons, cell bodies (somata), and dendrites—(Kringelbach et al. 2010) which directly translate into their excitability due to applied stimulation, and this varies in the normal as compared to the diseased brain. One also must not forget the neighboring glial structures that envelop the neuronal synapses, forming what is termed the 'tripartite synapse', are undoubtedly modulated by electrical stimulation (Araque et al. 1999; Ventura and Harris 1999; Takano et al. 2006).

The 'rheobase', or amplitude threshold, is the minimal current necessary to stimulate a neural element with a long stimulus duration; 'chronaxie' time refers to the minimum time interval required to excite a neural element using twice the intensity needed to elicit a threshold response. Myelinated axons are most excitable with chronaxies of about 30–200 μs, while those of cell bodies and dendrites are 1–10 ms (Holsheimer et al. 2000), indicating that most of the postsynaptic responses elicited from HFS are the result of axonal conduction due to their much increased excitability (Ranck 1975; Nowak and Bullier 1998a, b; McIntyre et al. 2004a), where larger diameter axons prioritize smaller diameter axons, followed by dendrites and somata (Carron et al. 2012).

Other complicating factors include the distance and orientation of the axons and cell body to a voltage gradient (Ranck 1975), properties of synaptic transmission on the target neurons, the variability in synaptic propagation both in terms of postsynaptic neuronal geometry and kinetics of the postsynaptic receptors (Anderson et al. 2004, 2006).

Stimulation parameters are also critical to understand in the overall mechanism of DBS. Current parameters are the result of trial and error to cause therapeutic effect, and vary by target, most of which are first witnessed almost immediately upon turning stimulation on (Volkmann et al. 2002, personal observation). Larger current density spheres (Butson and McIntyre 2006; Butson et al. 2007) likely modulate more fibers/neurons, and thus may cause a larger or more varied effect than smaller spheres centered at the same location. The parameter sets are somewhat arbitrary but remain within certain boundaries: high frequency stimulation between 130 and 180 Hz is most commonly applied to most targets of implantation (e.g., GPi, STN, and Vim for dystonia, PD, and ET), but some white matter targets (pedunculopontine nucleus, PPN; periventricular/periaqueductal grey, PVG/PAG) respond to low frequency stimulation (10–50 Hz). Changing frequency can have dramatic differences in clinical response, especially when placed within the context of a diseased network, which will be discussed in more detail below.

2.2 *Local Effects Due to HFS: Electrophysiology*

High frequency stimulation has been thought to produce its effects through a variety of mechanisms, varying from inhibition to activation of target structure neurons or passing axons, or a combination of both, which is discussed below.

2.2.1 Inhibition: Synaptic vs. Depolarization Block?

Local effects of DBS have historically been focused on inhibition of the target nucleus due to the replication of therapeutic effect seen with pallidotomy and lesions of the STN in PD. Traditionally, looking within the context of the classical pathophysiological model of PD (Albin et al. 1989; DeLong 1990), where dopaminergic neuronal loss in the substantia nigra pars compacta (SNc) ultimately manifests as hyperactivity seen in the STN and basal ganglia (BG) output structures, the substantia nigra pars reticulata (SNr) and GPi, application of HFS to the STN or GPi was thought to induce a functional inactivation by local depolarization block, as observed *in vitro* through inactivation of neuronal voltage-dependent sodium channels (Benazzouz et al. 1993, 1995; Benazzouz and Hallett 2000; Beurrier et al. 2001; Magarinos-Ascone et al. 2002; Anderson et al. 2006). Inhibition (including silencing and decreased activity) of cell activity is by far the most commonly observed result electrophysiologically. Most of the cells inhibited, though, due to HFS still demonstrate spontaneous activity; this observation combined with the fact that they are not preceded by an increased discharge, and that inhibition is not absolute (Hashimoto et al. 2003; Tai et al. 2003; Filali et al. 2004; Welter et al. 2004; Meissner et al. 2005; Carlson et al. 2010; Liu et al. 2012), makes a theory of complete depolarization block untenable.

As proposed, likely synaptic mechanisms are the origin of this inhibition due to activation of gamma-aminobutyric acid (GABA) axons, the repetition of this process at high frequency leading to a decrease in the mean firing rate (Meissner et al. 2005; Degos et al. 2005). In vitro, this was demonstrated by that STN stimulation evokes locally inhibitory postsynaptic potentials that are blocked by GABA antagonists, as well as in the rat during pallidal HFS where there was a blockade of inhibitory effect with GABA-A receptor antagonists and synaptic blockers (Chin and Hutchison 2008).

2.2.2 GPi and STN HFS Induced-Electrophysiologic Effects

Many groups (Dostrovsky et al. 2000; Wu et al. 2001; Filali et al. 2004; Liu et al. 2012) have reported that focal, therapeutic HFS of the human pallidum leads to inhibition of neuronal firing, and the hypothesized mechanism was synaptic GABA release from the afferent terminals. In the intraoperative GPi microstimulation and MER analysis by Liu et al. (2012), stimulation frequencies greater than 30 Hz led to cell firing declines and ultimate silencing at 50 Hz, whereas the field evoked potential (fEP) increased significantly until maximizing at 20–30 Hz, after which it declined. This is suggestive that there was a maximal amount of pre-synaptic GABA in this frequency range, declining at higher levels, perhaps due to saturation of reuptake or synaptic fatigue due to continued high release of neurotransmitter over long periods of time (Urbano et al. 2002); this could cause first an acute decrease in neuronal firing and then a plateau of effect after neurotransmitter depletion, as seen by others (Iremonger et al. 2006). In hippocampal in vitro preparations, GABAergic inhibitory post-synaptic potentials (IPSPs) were considered to be "labile" and attenuate after a period of repetitive stimulation due to reduction of driving force (McCarren and Alger 1985; Huguenard and Alger 1986) and decrease in synaptic release (Thompson and Gahwiler 1989).

It is unknown, as proffered by Liu et al. (2012), whether prolonged inhibition during long trains involves a continuous GABA release that sustains a postsynaptic inhibition. Microdialysis studies measuring GABA release per unit time in patients during HFS have found a high initial release of GABA followed by a continuous stable GABA release (Kilpatrick et al. 2010), suggesting that even if stimulation depressed some synapses, surrounding efferent fibers could be driven by HFS to sustain a postsynaptic inhibition of downstream targets via continuous GABA release. A sustained, regular firing at 100 Hz is considered physiological for basal ganglia output neurons (Liu et al. 2012). Looking at the classical model (Albin et al. 1989), excitation of GABAergic afferents such as those from striatum and GPe would inhibit GPi neurons, as has been consistently observed in previous studies (Anderson and Mullins 2003; Moran et al. 2011; Rosin et al. 2011), in agreement with the above findings. Together, these observations support a clear role of (high) frequency dependence on stimulation-induced effects.

Most commonly, though, HFS has not resulted in complete silencing of neurons. Rather, in the STN, some neurons have been shown to present an increased discharge rate following STN-HFS (Tai et al. 2003), possibly due to the ability of HFS to activate glutamatergic excitatory inputs, while other in vitro recordings have reported the triggering of spike discharges at the frequency of stimulation (Garcia et al. 2003). Again, this is likely due to axonal activation with back firing of the soma (Garcia et al. 2003); this is in accordance with other model observations of HFS—induced antidromic activation of nuclei which are connected to the target structure (in this case, the STN) (Maurice et al. 2003; Li et al. 2007). Noteworthy, besides the spikes evoked after each stimulation pulse, the spontaneous activity of the back-fired neurons was abolished, suggesting an inability of the activated neurons to further process incoming inputs, a mechanism combining functional inhibition of the soma and activation of the output of the neuron (McIntyre et al. 2004a; Deniau et al. 2010). Liu et al. (2012) also observed neurons following HFS at short latencies and low intensities manifesting collisions with preceding spontaneous activity indicative of antidromic activations; the fact that the cells were not completely silenced supported their conclusion that stimulation can bring about focal inhibition of neuronal spiking while exciting axonal fibers, which may in turn cause synaptic inhibition of the target.

In summary, HFS induced local effects can be explained, from an electrophysiological point of view, by the following: (1) HFS does not completely silence cellular activity; (2) Inhibition of cell bodies is most common, and may occur synaptically due to activation of GABAergic axons; (3) HFS promotes simultaneous inhibition of the soma and activation of axons; and (4) Some axons may be selectively activated/inhibited.

2.3 *Network Effects of HFS: Electrophysiology*

The above examples illustrating the acute local effects of somatic or axonal activation—chiefly in terms of firing rate or periodicity changes—of HFS on a target nucleus implicate the role of a larger network that is modulated in each disease process.

2.3.1 Orthodromic/Antidromic Fiber Activation

Review of the literature gives tremendous support for fiber activation based on recordings performed in all components of the basal ganglia, in nuclei both upstream and downstream from the site of stimulation—in the form of evoked synaptic effects, antidromic activations and changes in neurotransmitter release in the structures connected to the stimulated nucleus. As Parkinson's disease is the most common affliction for which DBS is implemented, and STN the most common target, most of the literature has explored mechanisms of action, and hence electrophysiology, in the context of STN-HFS, which is summarized below.

Evoked responses have been recorded following STN-HFS in **SNr** in the rat (Burbaud et al. 1994; Benazzouz et al. 1995; Windels et al. 2000, 2003; Maurice et al. 2003; Tai et al. 2003; Degos et al. 2005; Shi et al. 2006) and human (Maltete et al. 2007; Galati et al. 2006), **GPi** in the rat (Burbaud et al. 1994; Benazzouz et al. 1995; Windels et al. 2000, 2003; Shi et al. 2006), primate (Hashimoto et al. 2003; Dorval et al. 2008; Hahn et al. 2008), and human (Reese et al. 2008); **thalamus** in the rat (Benazzouz et al. 2000a) and primate (Dorval et al. 2008; Xu et al. 2008), **striatum** in the rat (Meissner et al. 2002; Lee et al. 2006; Gubellini et al. 2006; Shi et al. 2006), **contralateral STN** in the human (Novak et al. 2009), **cortex** in the rat (Li et al. 2007; Lehmkuhle et al. 2009) and human (Jech et al. 2006; Fraix et al. 2008; Potter-Nerger et al. 2008; Eusebio and Brown 2009; Eusebio et al. 2009; Kuriakose et al. 2009) and **PPN** in the rat (Florio et al. 2007).

In the case of STN-HFS, it seems that there is downstream excitation when glutamatergic STN axons are activated; inhibition, when GABAergic GPi efferents are activated; or a combination of effects influenced by both direct excitatory and indirect inhibitory projections. As McIntyre and Hahn describe (2010), STN-HFS would incur in the GPi a mixed effect due to excitatory STN-GPi projections and an inhibitory STN-GPe-GPi pathway, but one would also have to account for the role of antidromic activation of afferent cortical projections to STN (Li et al. 2007; Gradinaru et al. 2009), as well as activation of globus pallidus externus (GPe) to STN projections which pass through GPi (Kita et al. 2005).

With respect to SNr responses due to STN-HFS in PD models and in humans, the main effect of a decrease in cell firing—with some cells activated—goes in line with the hypothesis that there is fiber activation rather than from a simple deactivation due to inhibition of the STN cells. In their review, Deniau et al. (2010) posit that this inhibitory effect is due to activation of GABAergic synaptic inputs, as such inhibition is blocked by intranigral application of a GABA-A receptor antagonist (Maurice et al. 2003) and observation of STN-HFS-induced increase of GABA release in the SNR (Windels et al. 2000, 2003). The GPe is believed to be the source of such GABA in line with its varied connections (Kita 2007). Deniau et al. (2010) further elaborate that STN-HFS-induced activation of the pallidal-nigral pathway could be due to synaptic activation of glutamatergic subthalamo-pallidal projections, an axon reflex resulting from antidromic activation of pallidal neurons projecting to both the STN and the SNr, or direct stimulation of the pallido-nigral axons at the level of the

STN; the excitatory effects seen on SNr due to STN-HFS probably results from the activation of the glutamatergic subthalamo-nigral projections. In reality, there seems to be a complex interaction of stimulation intensity on the main effect of activation and inhibition (Maurice et al. 2003; Boulet et al. 2006; Deniau et al. 2010), largely resulting from a combination of the weight of multiple synaptic inputs on the cells recorded (Deniau et al. 2010).

In STN-HFS, there are wide effects due to orthodromic but also anitdromic fiber activation at synaptically connected structures, not only in the basal ganglia but over a wide network, including the cortex. Antidromic effects of STN-HFS in the cortex are likely due to axon activation and generation of synaptic effects in deep cortical layers, which has been associated with changes in excitability and dampening of oscillations in local field potentials (Jech et al. 2006; Li et al. 2007; Fraix et al. 2008; Potter-Nerger et al. 2008; Eusebio et al. 2009; Kuriakose et al. 2009; Lehmkuhle et al. 2009). Thalamo-STN neurons may also be antidromically activated and impact thalamo-cortical oscillators (Sugimoto et al. 1983). The complete picture is far from being resolved, but it is evident that multiple axonal pathways are influenced in complex manners.

2.3.2 Modulation of Network Activity: Firing Pattern vs. Rate

The above examples show that fiber activation may cause inhibition/activation of downstream structures, but it appears that there is a key element of oscillatory entrainment. This has been extensively investigated during MER in vivo.

In the 1-methyl-4-phenyl-1,2,3,6-tetrahydropyridine (MPTP) treated primate, the findings of increased GPi and GPe activity (Hashimoto et al. 2003) as well as a decrease in firing in the pallidal receiving activity of the ventral anterior/ventral lateralis pars oralis (VA/VLo) thalamus (Xu et al. 2008) support that STN-HFS increased the output of the nucleus. Moreover, in both the pallidum and VA/VLo, there was seen a change from burstiness to more periodic and regular firing, organized by the stimulus timing (Hahn et al. 2008; Dorval et al. 2008) with a shift in oscillatory activity from low to high frequencies. Clinically ineffective STN-HFS did not cause significant changes in cellular firing (Xu et al. 2008). These results from the Vitek group suggest that therapeutic STN-HFS activates subthalamo-pallidal projections and changes the discharge patterns of the GP neurons (McIntyre and Hahn 2010); the finding of activation of thalamic cerebellar receiving area ventroposterolateral (VPL) neurons suggest activation of adjacent cerebello-thalamic fiber bundles (Xu et al. 2008). Such findings corroborate more recent MRI studies (to be discussed below).

During GPi-HFS in the dystonic patient it was recently seen by Liu et al. (2012) that clinically effective stimulation caused a decrease in STN firing to 1 % of baseline, with clinically ineffective stimulation (such as at frequencies less than 100 Hz) having no effect on STN firing (Liu et al. 2012). This was thought to be due to the driving of fibers of passage from GPe to STN (Hazrati and Parent 1992; Smith et al. 1994), which pass thorough GPi, as predicted by modeling studies (McIntyre and

Hahn 2010) and tract tracing studies (Hazrati and Parent 1992; Smith et al. 1994). Simultaneous activation of GABAergic afferents caused inhibition of GPi neurons (Liu et al. 2012), as has been consistently observed in previous studies (Anderson and Mullins 2003; Moran et al. 2011; Rosin et al. 2011). Also, orthodromic activation of GPi output was seen following GPi-HFS by Montgomery (2006) who reported inhibited thalamic neuronal activity following a stimulus pulse, again likely due to GABAergic activation.

In the dystonic hamster model, Leblois et al. (2010) showed that pallidal HFS decreased the regularity of dystonic pallidal cell firing without a change in firing rate. A similar decrease in firing regularity was also observed in the SNr, as were multiphasic responses locked to stimulus pulses in both the pallidum and SNr suggestive of fiber activation: while no direct connection exists, striato-nigral GABAergic fibers and glutamatergic efferents of the STN are known to send collateral projections to the EP and SNr (Kita and Kitai 1987; Deniau et al. 1996). Similar findings of stimulus time locking onto firing in the GPi were found in parkinsonian non-human primates during GPi-HFS (Bar-Gad et al. 2004; McCairn and Turner 2009). This could be due to a direct action on the membrane of neurons surrounding the stimulation electrode, as seen in the entrainment of STN neurons by HFS in vitro by Garcia et al. (2003), or as this was also observed in SNr, could be due to simultaneous synaptic release in the SNr through antidromic activation of subthalamic, pallidal and/or striatal fibers.

The above studies suggest that modulation of firing pattern may be just as (or more) important as modulation of firing rate, which concurs with network activity. The dystonic GPi has a decreased firing rate from normal, similar to in similar to in GTS (Zhang and Barres 2010), whereas in PD the GPi has an increased firing rate from normal (Vitek et al. 1999; Hutchison et al. 2003; Starr et al. 2005; Tang et al. 2007). The effectiveness of GPi-HFS across each of these diseases with different characteristic firing rates—to break pathological oscillatory activity within the basal ganglia circuit—seems more critical and perhaps the key aspect to DBS efficacy here. This therefore again highly implicates axonal activation in the mechanism of DBS efficacy.

Thus, when intensity and frequency are increased to therapeutic efficacy, resulting in silencing of neuronal activity, or changes in pathological firing patterns, an implied mechanism again is synaptic and neurotransmitter modulation. As will be discussed in the modeling section below, the regularity of stimulation is as important as frequency, as temporally irregular stimulation at high frequencies is not clinically effective (Birdno et al. 2008; Kuncel et al. 2012).

In summary, HFS induced network effects by altering neuronal firing in several ways: (1) HFS incurs Antidromic and Orthodromic activation of axons; (2) Activation of axonal projections contribute to network wide effects, and can cause either inhibition or activation of target end structures due to the combined effect of synaptic inputs; (3) HFS causes modulation of firing pattern (disruption of pathologic burstiness) as well as changes firing rate through axonal and trans-synaptic activation in the basal ganglia.

2.3.3 Modulation of Network Activity: Frequency-Dependent Effects

Although the above evidence argues mechanistically for why high frequency (>100 Hz) stimulation is considered to be therapeutic, causing activation of axonal flow, not all targets to which DBS is applied respond equally. As mentioned, perhaps the high 'clinically therapeutic' frequency corresponds to the optimal physiologic functioning of the basal ganglia system's output neurons (Liu et al. 2012), a hypothesis that could be extended to other locations targeted.

Perhaps this can be reasoned to why low frequency stimulation (LFS, <50 Hz) is considered optimal for targets such as the PVG/PAG, PPN, and the ventromedial hypothalamus. Local field potential recordings have elucidated that there are enhanced 8–12 Hz power spectra in both the PVG/PAG and ventral posterior lateral and medial (VPL/VPM) thalamus of patients when in pain (Green et al. 2009; Kringelbach et al. 2010).

Similarly, PPN cells were found to fire slowly (10–11 Hz) in naïve animals while in those cells of PD models fire irregularly, and in bursts, at 18–20 Hz (Breit et al. 2001). Piallat et al. (2009) observed increased firing rate in PPN of PD patients during stepping relative to at rest, with most cells irregular and bursty (14.8 ± 10.9 to 21.0 ± 12.0 spike/s, $p < 0.05$). Such low inherent cell firing is likely due to the type of cell present, which at the PPN are predominantly both glutamate and acetylcholine cell bodies (Hamani et al. 2011; Stefani et al. 2007). When tried empirically, HFS applied to the PPN has been observed to lead to a decrease in locomotor activity (Nandi et al. 2002, 2008), whereas PPN-LFS improves locomotion in parkinsonian animals, and in the few human series where this improves freezing of gait effective frequency has been less than 80 Hz, most averaging 25 Hz (Stefani et al. 2007; Ferraye et al. 2010; Moro et al. 2010).

Perhaps also it is key to understand the exact structures that lie in proximity to the electrode, which can weigh into the frequency chosen. At stimulation under 10 Hz at the PPN site, Ferraye et al. (2010) witnessed oscillopsia (as believed to be due to proximity of lateral and caudal oculomotor fibers) and limb myoclonus (from the modulation of the thalamic projections (Bejjani et al. 2000) whereas paresthesias limited use of stimulation above 50 Hz, likely due to involvement of the medial lemniscus system. Possibly, stimulation at a frequency greater than the average cellular firing rate recorded is sufficient to override the pathologic bursty and irregular activity, to drive more normal synaptic transmission.

An additional argument is that perhaps the reason why LFS is effective in these locations is the same way how surrounding structures—astroglial bodies in the neuropil—can be differentially affected by DBS according to their relative distance from target, with a dissipated intensity delivered at higher radii (Melega et al. 2012). Somata and axons may be activated by lower frequencies by enhancing their firing and responsivity to other neural inputs (Vitek 2002). This may factor into play when different areas of the hypothalamus have been a target of DBS: HFS is effective in the posterior hypothalamus for cluster headache, (Sillay et al. 2010), whereas LFS (50 Hz) of the ventromedial hypothalamus in the animal (Melega et al. 2012), and human (Hamani et al. 2008) morbid obesity studies have resulted in modulation of weight gain, with no effects seen at 130 Hz. This might be suggestive of an induced increase in metabolic rate (Melega et al. 2012).

Looking more broadly at DBS in general, perhaps this could explain why HFS is effective at both purely white matter targets (ventral capsule) as well as at subcortical gray matter targets (STN, Vim, GPi); stimulation of the latter likely invokes variable activation of not only local synapses but larger axonal tracts (e.g., lenticular fasciulus, thalamic fasciulus, ansa lenticularis, dentato-rubro-thalamic tract). As we only have little empirical evidence, a more systematic analysis including electrophysiology is clearly warranted before any clear supposition can be extended.

In summary, DBS induced network effects are frequency-dependent and are of the following characteristics: (1) Both HFS and LFS can activate axons; (2) HFS and LFS are differentially effective when applied to specific systems, possibly due to the optimal physiologic functioning of the target; (3) HFS in basal ganglia nuclei can cause an inhibition of cellular firing, with clinical improvement, whereas LFS has no effect; (4) LFS outside the basal ganglia is variably effective, influenced by target anatomical and cellular composition; and (5) The finding that both HFS and LFS congruently activate fibers whereas only HFS can functionally inhibit a nuclear target is paradoxical and needs to be further explored.

2.4 Regularity of Stimulation is Essential in an Abnormally Oscillating System

Modulation of firing rate in the context of a larger closed loop network including all the upstream and downstream connections of BG structures, including cortex and thalamus, has been considered more recently by McIntyre and Hahn (2010) to play a large role in the mechanism of DBS. It is known that there are several parallel loops within the cortico-basal ganglia-thalamo-cortical (CBGTC) regions that are organized into motor program components (Mink 1996; Rubchinsky et al. 2003) with adjacent parallel limbic and associative loops (Parent and Hazrati 1995a; Haber et al. 2000).

In the low dopamine state, neuronal discharges in the STN and the basal ganglia output nuclei become irregular with burst of spikes and pauses, and there is a loss of segregation between CBGTC loops (Bergman et al. 1994; Pessiglione et al. 2005). Increasingly, it is believed that in these pathological states, it is the changes in the patterns of discharges themselves that are more significant than the absolute firing rates, with neurons exhibiting an abnormal tendency to synchronize, modifying the resonance properties of the cortico-basal ganglia loop circuits (Foffani et al. 2003; Degos et al. 2005; Heimer et al. 2006; Leblois et al. 2006; Braak and Del Tredici 2008; Gale et al. 2008; Eusebio and Brown 2009; Eusebio et al. 2009).

In PD models and patients, such synchronization is manifested by the transmission of increased low frequency oscillations in the beta band (13–30 Hz) through the BG circuits from the cortex (Bevan et al. 2002; Brown 2006; Mallet et al.

2008b). The presence of such oscillations in the striatum in the low dopamine state suggests that it cannot properly filter out the cortical input at the beta frequency (Bergman et al. 1994; Courtemanche et al. 2003; Gatev and Wichmann 2008). This frequency is considered to contribute to the akinetic clinical manifestations of PD (Hutchison et al. 2004), which is further supported by the correlation of increased movement and the dissolution of such synchronization thorough treatments (Kuhn et al. 2008; Ray et al. 2008; Bronte-Stewart et al. 2009).

The importance of oscillations as a biomarker of network dysfunction could also be transposed to non-motor disease; indeed, it appears that pathological oscillations in the non motor BG network due to the appearance of pathological bursts in the STN, mainly left sided, have been reported in other diseases such as OCD and could also explain the therapeutic effect of STN-DBS to treat OCD (Mallet et al. 2008a). This so-called 'network dysrhythmia', as coined by Llinás et al. (1999), explains how such pathologic resonant interaction between cortical and subcortical structures is a root cause of neurologic disorder, and helps to tie in the concept that not one single nucleus is the implicating factor (McIntyre and Hahn 2010). This network concept of a disease state (DeLong and Wichmann 2007) helps explain how a lesion or HFS to one of several nodes can disrupt aberrant network dynamics and provide similar therapeutic benefit (McIntyre and Hahn 2010), illustrated by the disparate firing rate changes observed in GPi via STN-HFS or lesioning (Montgomery 2007).

Computational studies over the years have provided insights into how models can be used to further address the network mechanisms of DBS, (Montgomery and Baker 2000; Rubin and Terman 2004; Grill et al. 2004; Feng et al. 2007; Shils et al. 2008; So et al. 2012) for instance, concluding how population -based models more appropriately explain the widespread effects of STN-HFS than those based only on simple firing rates. Both simplified neural models (Montgomery and Baker 2000) and more detailed theoretical analyses (Rubin and Terman 2004; Guo et al. 2008) have offered that HFS represents a noise source that disrupts pathological burstiness of the basal ganglia structures in PD, enabling interaction and hence information flow between multiple loops within a larger CBGTC network. Such modeling studies have examined how clinically therapeutic DBS, in contrast to irregular, non-therapeutic DBS trains, can regularize basal ganglia disordered activity and thereby improve thalamic relay fidelity (Dorval et al. 2010). The STN is one such location where multiple loops (direct, indirect, hyperdirect pathways) converge- through STN-HFS, orthodromic effects to the BG output nuclei and connected structures are manifested as a regularization in neuronal firing, (Degos et al. 2005; Hammond et al. 2007; Maltete et al. 2007; Dorval et al. 2008), where downstream synaptic effects at these multiple different levels may contribute to decrease the noisy signals sent to the premotor thalamo-cortical and brainstem network.

Antidromic activation of cortical afferents via the hyperdirect loop (Li et al. 2007; Gradinaru et al. 2009) has shown the possibility of STN-HFS to specifically interface with diffuse cortical circuits, and that together with regularized BG synaptic effects can modify the oscillatory properties of the CBGTC network and prevent the genesis and propagation of pathological synchronized oscillations (Devos et al. 2004; Li et al. 2007; Montgomery and Gale 2008). Thus, it seems that in regard to a global

CBGTC network, the effects of HFS result from a complex heterogeneous integration of the subtle local effects of inhibition of the soma (at target) and increased excitation of axons, which ripple downstream (and upstream) differently through synaptic activation of structures that either may show varying patterns of inhibition or excitation. The sum effect of these alterations, or corrections in aberrant discharges, induced by HFS is an improvement in the symptoms of the disease.

McIntyre and Hahn (2010), in reviewing the concept of a network mechanism of DBS action, offer how optimal DBS frequencies might correspond to resonant frequencies of a healthy CBGTC oscillating loop, where the inter-pulse interval of therapeutic DBS frequencies represents the characteristic period of the network (Grill et al. 2004; Montgomery 2007; Montgomery and Gale 2008). The role of entering periodicity into an irregularly oscillating circuit has been examined through experimental and computational modeling: it has been generally accepted that not only is frequency above 100 Hz important for therapeutic effect (Lozano et al. 2002), but so is the regularity of interpulse DBS trains (Birdno et al. 2007, 2008). This hypothesis may explain why HFS works at the BG circuit, and in areas where resonant frequencies are lower, why LFS may instead be more efficacious. Temporally irregular high-frequency DBS has been observed to be less effective than regular DBS with the same average frequency (Birdno et al. 2008; Dorval et al. 2010). Regular temporal spacing of DBS trains is thought to override the aberrancy in the target nucleus and network; the converse was observed when Ma and Wichmann (2004) stimulated STN of normal monkey with bursty spike trains, it developed parkinsonian symptoms.

This was recently revisited in a clinical and computational study of Vim DBS patients by Kuncel et al. (2012), where cycling with stimulation on for at least 60 % of the time was as effective as continuous 125 Hz, correlated with a reduction in firing pattern entropy—manifested as burstiness—of model neurons from baseline levels. Tremor and neural firing pattern entropy decreased as the percent on-stimulation time increased, minimizing the number of pauses (which ranged from 100 to 500 ms) that could allow the generation of rebound bursts (Person and Perkel 2005; Wichmann and Soares 2006), possibly by masking the intrinsic activity either through synaptically mediated inhibition (Dostrovsky et al. 2000) or regular firing synchronized with the stimulation (Grill et al. 2004). This exploration into a pathologic circuit let the authors posit that cycling with a sufficiently high percent on time resulted in fewer pauses and thus fewer opportunities for rebound burst to occur.

It is thus becoming clear that HFS imposition of regularity into a pathologically oscillating network is brought upon by a complex combination of both inhibition and excitation, not fully understood, originating at the level of the stimulated nucleus where a decreased excitability of the soma can be associated with an increased output due to direct activation of the axon (McIntyre et al. 2004c).

The emergence of biophysical modeling of neural activity, whereby multi-scale spatio-temporal patterns of neuronal activity in relation to the physical properties of brain tissue will be modeled mathematically, will likely orient the direction of future DBS innovation as we continue to learn more about these complex interactions, as reviewed by Modolo et al. (2011). Some computational models have helped tried to reconcile the temporal differences in clinical effect seen after DBS, as discussed above in neuroimaging studies, in relation to delayed non-motor changes (Le Jeune et al. 2009; Haegelen et al. 2010). In PD, there seems to be immediate improvement of tremor and rigidity while axial symptoms take time to improve; dystonic posturing takes months to improve after DBS onset. In a modeling study by Beuter and Modolo (2009), it is suggested that each set of symptoms are improved by different network mechanisms operating at multiple time scales, where immediate tremor abolishment is due to sudden desynchronization of subcortical neural activity while locomotion improvement may involve a delayed plastic reorganization. These delayed effects are seen in multiple diseases treated with DBS, and quite possibly involves synaptic plasticity.

In summary, it is important to keep three points in mind in terms of the regularization of firing for DBS: (1) Temporal Regularity in cell firing induced by DBS is essential for clinical improvement; (2) Irregular stimulation sequences are less effective clinically; and (3) Regular impulses of DBS causing disruption of pathologic burstiness ripple network-wide through axonal activation.

3 DBS Mechanistic Analysis: Beyond Electrophysiology

Electrophysiological and neurochemical/microdialysis studies have supported the hypothesis that HFS activates neurotransmitter output from both the target site and from up- and downstream locations.

3.1 Support for Fiber Activation: Neuro-chemical Studies

Following stimulation, the activation of efferent and afferent axons can give rise to sustained neurotransmitter release (Windels et al. 2000, 2003) modifying local, up- and downstream neuronal activity. However, this release lasts many minutes and does not follow the time scale of clinical effect, which is obvious after several seconds of stimulation at least for STN and VIM in PD and ET, and stops immediately when the stimulation is turned off. During STN-HFS, stimulation activates output from the target site (Windels et al. 2000; Hashimoto et al. 2003; Maurice et al. 2003; Galati et al. 2006). Glutamate levels were found to increase in the vicinity of the electrode site as well as to persist after termination of stimulation, implying other mediators, such as the astrocyte contribution (see discussion below) (Lee et al. 2007). STN-HFS has also been reported to increase striatal dopamine efflux (Bruet et al. 2001; Meissner et al. 2003; Lee et al. 2006; Lacombe et al. 2007; Walker et al. 2009; Shon et al. 2010); although the significance of this is uncertain, it may implicate STN-HFS in indirect dopaminergic effects on non-motor symptoms, as mediated through associative and limbic connectivity (Limousin et al. 1997; Haegelen et al. 2010), which is discussed more below.

3.2 *Support for Fiber Activation: Functional Imaging Studies*

Perhaps what could be considered the best proof that DBS modulates networks in various disease states via axonal activation are the consistent observations that cerebral metabolism is altered across the disease spectrum following intervention, with a clinically significant change in symptoms. Positron-emission tomography (PET) has been used for over a decade to evaluate regional differences in cerebral metabolism in disease, and similarly, the effect of DBS on brain metabolism. More recently, magnetic resonance (MR) diffusion tensor imaging (dTi) has given us the ability to visualize axonal tracts in the brain subserving cortical-cortical and cortical-subcortical connections, and can give clues to perhaps how such metabolic changes result following DBS. Very recently, some centers have begun more methodical protocols. Superposition of a pre-operative dTi map on a post-op helical computed tomography (CT) and identification of the active electrode contact allows detection of which fiber pathways are contacted, and potentially modulated (activated), allowing identification of terminal cortical field involvement and potential explanations for symptoms resulting. Our center has begun performing this protocol in order to help explain improvements in cognitive parameters as well as deteriorations in speech following DBS in PD, as shown in Fig. 15.1.

The use of imaging in this context will be especially important—especially when certain trends in metabolism are consistent- to broaden the spectrum of disease applications for DBS. What is potentially most indicative of DBS activation of axonal flow is when the electrode is implanted into a white matter target to cause effect—such as PPN, ventral internal capsule, PAG, and now the fornix.

Let us explore each disease process amenable to DBS intervention, and the results favoring network modulation via axonal activation.

3.2.1 Positron-Emission Tomography: PD

Multiple functional imaging experiments have explored and confirmed, beginning in 1997 (Limousin et al. 1997), that therapeutic HFS generates metabolic and blood flow changes in cortical and subcortical regions (Perlmutter and Mink 2006). PET studies of STN-HFS have confirmed that HFS drives STN output, inducing an increase in regional cerebral blood flow (rCBF) in thalamus and midbrain while decreasing rCBF in frontal cortical areas (Hershey et al. 2003; Payoux et al. 2004; Grafton et al. 2006; Karimi et al. 2008; Hilke et al. 2008). These studies are intriguing when one views them in the context of specific cognitive changes that occur in STN DBS patients following surgery in PD.

Delayed cognitive effects of DBS can be seen most clearly when one explores the functional imaging literature in STN-HFS in PD. Le Jeune et al. have used PET imaging in PD patients both 3 months before and after onset of STN-HFS to clinically correlate new symptoms of apathy (Le Jeune et al. 2009), as well as impaired facial emotion recognition (Le Jeune et al. 2008; Haegelen et al. 2010), with glucose

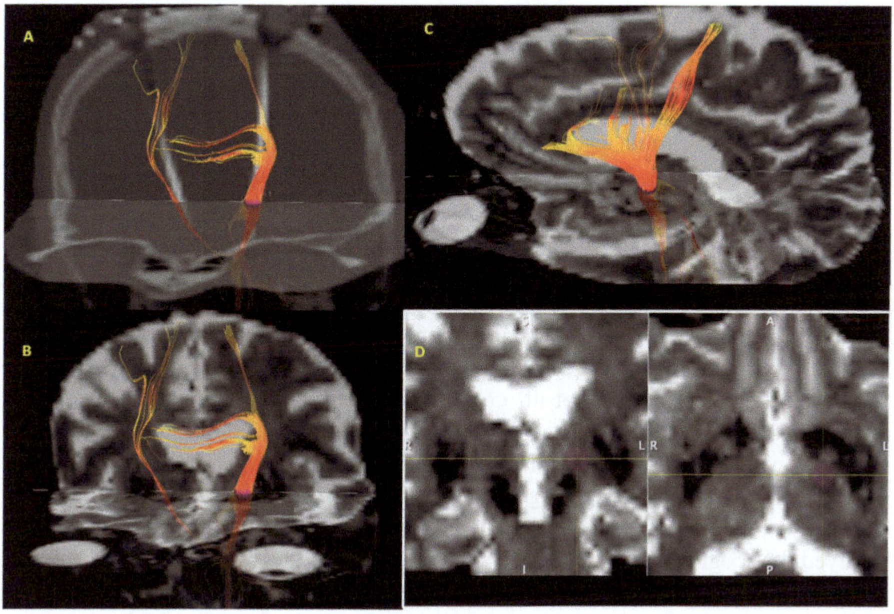

Fig. 15.1 Pre-operative MR dTi image super-imposed on post-operative 3D CT, s/p bilateral STN-DBS electrode placement, using CT to define contact positions, to identify involved fiber tracts. Illustrated are fiber pathways that intercepted sphere of tissue activation (*pink sphere*) at active contact on L DBS electrode for clinical efficacy based on stable 3-month parameter settings, assuming a basic isotropic model (Butson et al. 2007), plotted relative to post-operative CT (**a**) and pre-operative MRI (**b**, **c**). Depicted are fibers ascending to sensorimotor cortices and across the corpus callosum (**b**, **c**). Coronal and Axial slices (**d**) of active contact and sphere of activation showing position in dorsolateral STN. Only active contact 2 on *left* electrode depicted for clarity. Parameters for L STN electrode: C+, 2−; amplitude 1.7 v, pulse width 60 μs, 160 Hz. *Figures drawn in collaboration with Tim Ellmore, PhD, U T Houston HSC*

metabolism changes in prefrontal and cingulate cortices without any attendant perfusion changes. These findings substantiate empirical evidence of non-motor cognitive, motivational and emotional changes following STN-HFS by elucidating the involvement of cortical substrates (Dujardin et al. 2001, 2004; Biseul et al. 2005; Funkiewiez et al. 2004; Schroeder et al. 2004; Vicente et al. 2009; Le Jeune et al. 2010).

Both metabolic increases (Hilker et al. 2002, 2003) and conversely, decreases (Adolphs et al. 1996; Schroeder et al. 2002, 2003; Le Jeune et al. 2008, 2009) in the same anatomic areas under the same conditions let us posit that STN-HFS modifies medial prefrontal function through a disruption of the integration and transfer of limbic information from the cortex within the STN (Haegelen et al. 2010), manifested in alteration of the normal (or abnormal) function of a limbic circuit. This becomes more tenable when viewed in light of the changes in cerebral metabolism in right superior and middle frontal gyri and the anterior cingulate gyrus following STN-HFS in OCD responders (Le Jeune et al. 2010). Similar changes are also seen in PD following STN DBS (Le Jeune et al. 2009).

As it is known through experimental modeling and tract tracing that adjacent parallel circuits exist for motor, associative, and limbic functions (Alexander et al. 1986; Parent and Hazrati 1995a, b; Haber et al. 2000), the involvement of the anteromedial STN and its role in limbic function (Maurice et al. 1998a, b, 1999) upon dorsolateral STN stimulation is not impossible due to current diffusion into these loosely segregated sub-locations due to the sphere of stimulation density. Quite likely, reiterated here, it is the activation of the efferent fibers at this node within a circuit that manifests in metabolic changes concordant with symptom improvement. These findings further exemplify the local effect of HFS inhibition of soma and activation of axons, but add complexity as the clinical correlations are not immediate, and likely underlie some mechanism of induced neural plasticity from various new subcortical-cortical connectivity changes.

It must be realized that not all cognitive effects in PD, specifically executive dysfunction, observed following STN DBS are directly attributed to the effects of stimulation itself, and must factor changes in exogenous dopamine administration. There likely is some confounding effect of both pre-operative dopaminergic medication (Heo et al. 2008) as well as acute dopamine reduction following DBS, as it is believed that cognitive function may be influenced by dopamine level (Cools et al. 2002). Such a correlation with dopamine medication reduction has been correlated with temporary changes in executive function at 1 month after STN DBS (Zangaglia et al. 2009; Yamanaka et al. 2012) which was largely gone by 12 months.

It has been demonstrated by some that apathy, as well as depression and anxiety, that may occur after STN-DBS in PD is due to post operative reduction of dopamine which is allowed by the dramatic effect of STN-HFS (Volkmann et al. 2002; Thobois et al. 2010); others, however, did not find such an association between apathy trajectory and dopamine reduction (Kirsch-Darrow et al. 2011).

3.2.2 Positron-Emission Tomography: TRD, OCD, AD

PET studies of DBS for neuropsychiatric disorders, such as treatment refractory depression (TRD), show network wide changes in the cortico-striato-thalamocortical circuit. HFS of the white matter near the subgenual cingulate (Brodmann Area 25, [BA] 25), an area seen by PET to be metabolically hyperactive with relative other prefrontal hypoactivity in TRD patients (Drevets et al. 1997), has been seen to have significant deactivation 3 and 6 months post HFS in clinical responders, with concomitant increases in brainstem and prefrontal structures (Mayberg et al. 2005; Holtzheimer et al. 2012). Interestingly, such PET changes were similarly found upon HFS of the ventral capsule- nucleus accumbens at 6 months following simulation onset in TRD patient responders, implicating again that MDD is a network disease, and modulation of various axonal pathways subserving limbic structures—in the ventral capsule, anterior cingulum—has effect. Not surprisingly, PET studies in OCD after stimulation of the ventral caudate/ventral striatum have seen modulation of activity in the orbitofrontal and anterior cingulate cortices (Rauch et al. 2006).

A further use of PET imaging in a disease process to illustrate network-wide activations—and hence clinical effects seen—is the experimental use of DBS in Alzheimer's Disease (AD). The Toronto group (Laxton et al. 2010) has applied HFS to the fornix within the hippocampus in a phase 1 trial to evaluate incurred memory changes in AD patients at 1 year, correlated with PET metabolism changes seen at 1 month and again at 1 year. Increases were not only seen in temporal and parietal cortices, brain areas demonstrating large accumulations of amyloid deposits and the greatest impairment in glucose utilization in AD patients (Buckner et al. 2005), but also in the posterior cingulate and precuneus, parts of the more widespread brain default mode network that is affected early in the course of AD (Buckner et al. 2008; Seeley et al. 2009; Sperling et al. 2010; Laxton et al. 2010).

What is more exciting is this group's use of standardized low-resolution electromagnetic tomography (sLORETA) to explore the temporal and spatial propagation of the effects of stimulation to bring about these widespread activations: Short latency evoked responses were first localized to ipsilateral hippocampus and parahippocampal gyrus, whereas at longer latencies after stimulation, there was significant activation of the middle and posterior cingulate gyrus and precuneus (Laxton et al. 2010). The novel use of sLORETA here corroborates the PET data and supports the direct and trans-synaptic sequential activation of downstream targets connected to the DBS application site at the fornix—chiefly the hippocampus and the Papez circuit which is implicated in memory (Villain et al. 2008). It is the activation of not only this circuit but the more polysynaptically connected nodes within the brain default circuit (seen at increased latencies) that makes the authors conclude that DBS directly drives axonal flow. However, regarding the mechanisms of actions, it is important to highlight that in this case, DBS was applied on fibers (the fornix) and not on a grey nucleus.

3.2.3 Magnetoencephalography

An imaging modality that holds promise in elucidating whole-brain neural mechanisms due to DBS is Magnetoencephalography (MEG), due to its non-invasive ability to measure oscillatory communication between brain regions (Schnitzler and Gross 2005) and both excellent spatial and temporal resolution. Its use post-DBS is relatively new and quite underused. The first MEG study of DBS was performed in a patient under low frequency stimulation of the periventricular/periaqueductal grey (PVG/PAG) for severe phantom limb pain (Kringelbach et al. 2006, 2007a); with subjective improvement in pain there were corresponding widespread significant changes in the elicited power of neural activity including the mid-anterior orbitofrontal and subgenual cingulate cortices, areas known to be involved in pain relief (Kringelbach 2005). Similarly, MEG of high-frequency hypothalamic stimulation for cluster headache showed a similar pattern of changes in neural activity in a widespread cortical and sub-cortical network, including the orbitofrontal cortex (Ray et al. 2007).

3.2.4 MR Diffusion Tensor Imaging

The elucidation of fiber pathways through dTi, in seeking answers to terminally stimulated structures, can conversely raise questions as to exactly what is being stimulated to bring upon therapeutic effect of DBS. For instance, tremor control, which is immediate upon turning on stimulation, can occur via electrode placement in the Vim, ventralis oralis posterior nucleus of the thalamus (VOp), STN, the caudal zona incerta (cZi), or the posterior subthalamic region. Some centers choose cZi over Vim or STN (Plaha et al. 2004, 2008). What is common to all these structures is the nearby presence or traverse of the dentatorubrothalamic (DRT) tract: axons from dentate, globose and emboliform nuclei form the superior cerebellar peduncle, pass around and through the red nucleus and terminate in VOp and Vim thalamus. DTi visualizes fiber tracts according to the anisotropy of water diffusion and cannot differentiate afferent and efferent fibers or synapses; as such the DRT is seen to penetrate the thalamus, although synapses in the thalamus are known to exist in this functional projection (Hassler et al. 1979). Use of preoperative dTi to directly target the DRT directly has had great success in controlling head, appendicular, and MS tremor (Coenen et al. 2011, personal observation) using 180 Hz stimulation, rather than indirectly targeting the Vim. It is not proven whether stimulating these tracts leads to activation of the involved structures, but the involvement of these seems to be positively correlated with therapeutic effect.

Does this imply that really white matter tracts—e.g., DRT—are the true target, and by stimulating subcortical grey matter nuclei (e.g., Vim) the effect is produced via multi-synaptic routes to activate these fiber bundles? This remains unclear. In a more recent study of bilateral Vim DBS, report that tracts traversing the volume of activation of therapeutic electrode contact locations, obtained by using pre-operative dTi, showed a highly reproducible network of structural connectivity comprising motor cortical, subcortical, and cerebellar sites and the brainstem, which was not seen when the same analysis was performed using clinically ineffective contact sites. The Vim is an integrative area of BG efferents and the DRT en route to M1, and decades of stimulation here causing tremor suppression corroborates with the literature (Morel et al. 1997; Percheron et al. 1996; Deuschl et al. 2001; Nowinski et al. 2005) that it is indeed a key node within a spatially distributed tremor network. This view point also support why the posterior subthalamic area and cZI, as well as the DRT itself, provide the same tremor relief—and provide further evidence that HFS is effectual via axonal tract mediation.

As above, dTi can also be used to elucidate which nuclei are activated via stimulation of white matter, as explored by Pereira et al. (2010) in their study of circuitry involved in mediating DBS effects when targeting the PVG/PAG. They explored different mechanisms of pain relief due to activation via LFS in either the ventral or dorsal PAG –ventral PAG connectivity included the amygdala, nucleus accumbens, anterior cingulate and ventromedial prefrontal cortex, invoking parasympathetic activity, whereas prominent dorsal PAG connections included ventral posterior thalamus and primary somatosensory cortex, which when viewed in light of concordant changes in arterial blood pressure (Green et al. 2006; Basnayake et al. 2011) assume a sympathetic mechanism.

Imaging studies of other white matter targets support the concept of activation of axonal tracts due to DBS. Schweder et al. (2010) performed a post-DBS connectivity analysis using dTi 12 months after therapeutically effective bilateral LFS to the PPN, to improve freezing of gait (FOG) in PD. They found a normalization of pathological PPN connectivity, with new connections to cerebellum from PPN, before which were absent, and a regularization of a pre-op 'super-connectivity' to cortical/subcortical areas from PPN. Such findings 12 months after onset show that reorganization is apparent and due to stimulation; when viewed in light of cerebral metabolic changes by Ballanger et al. (2009) there is even further support for axonal modulation. PET imaging in patients with advanced PD who underwent bilateral PPN placement revealed blood flow increases on-stimulation (relative to off) not only in the mesencephalic locomotor region (PPN), but the thalamus, midbrain, and cerebellum as well as medial sensorimotor and caudal supplementary motor cortices, all of which are implicated in balance and motor control (Ballanger et al. 2009). Although not proof of effect, the combination of studies supports each finding, and gives a plausible mechanism to groups reporting significant symptom improvement in FOG of PD (Ferraye et al. 2010; Moro et al. 2010; Thevathasan et al. 2011).

In summary, we should realize the following three points in functional imaging support for fiber activation: (1) The use of metabolic imaging has shown HFS induced changes in rCBF throughout the brain across diseases, illustrating wide spread network modulation due to axonal activation; (2) Differential widespread activations/deactivations of end structures likely results from the sum of synaptic inputs of activated fibers; and (3) MR dTi illustrates fiber activation and structural involvement induced by DBS, providing clues for observed clinical effects.

4 Neuroplasticity Induced by DBS

Clinically, the application of GPi-HFS in dystonia allows us some insight into the mechanism of how benefits derived are not immediate but delayed. In dystonia, it may take several weeks or months to achieve maximum clinical benefit in patients treated not only by DBS (Yianni et al. 2003; Coubes et al. 2004; Vidailhet et al. 2005; Tisch et al. 2006a, b; Ruge et al. 2011a, b) but also following pallidotomy (Lozano et al. 1997). Similarly, there is often not an observed quick loss of effect once DBS is turned off in dystonia, taking again weeks for pre-surgical symptoms to re-emerge. It is contended that such a slowly progressive clinical time course, as well as concomitant commensurate normalization of electrophysiological measures is due to a process of neural reorganization.

Several authors have reported that short-interval intracortical inhibition (SICI), a measure of excitability of GABAA-ergic corticocortical connections, is less than normal in patients with dystonia (Ridding et al. 1995; Sommer et al. 2002; Huang et al. 2010), and could be contributing to muscle hyperactivity in this disease.

In contrast, paired associate plasticity (PAS) response, a test of long-term potentiation-like synaptic plasticity in cortical circuits (Stefan et al. 2000), has been reported to be more responsive in dystonia than in healthy individuals (Weise et al. 2006; Quartarone et al. 2009; Schwingenschuh et al. 2010).

Increased plasticity in dystonia could lead to the formation and maintenance of aberrant circuits within the motor system, ultimately manifesting as excessive involuntary movement that then becomes permanent. Ruge et al. (2009, 2011a, b) in their studies corroborate the view that increased plasticity is present in dystonic patients prior to surgery as compared to healthy controls (Weise et al. 2006; Quartarone et al. 2009; Schwingenschuh et al. 2010); this PAS response was nearly absent 1 month after GPi-HFS was begun, and then over the following months slowly increased towards levels observed in healthy individuals. Ruge et al. postulate that the quick loss in PAS response was due to disruption of aberrant basal ganglia signals, followed by the storage of more normal patterns of movement. In their series of ten patients with GPi-HFS for at least 4.5 years, when DBS was shut off for 2 days, there was insignificant change in both SICI and PAS responses and no change in clinical symptoms (Ruge et al. 2011a, b).

The authors reported that the lower the PAS response in the 'on' DBS state was, the more worsening of symptoms occurred when DBS was turned off. Patients with the highest PAS response were those after years of DBS and who experienced the least decline in clinical symptoms on stopping DBS for days, hypothesized to be due to "hard re-wiring of normal motor memories" (Ruge et al. 2011a, b). In this same population, it was seen that the patients with the least relapse of symptoms had good PAS responses and had the least current drain chronically, suggesting a correlation between DBS parameters and the degree of long-term potentiation-like plasticity, as seen in the cortex (Ziemann et al. 2004). This is not seen uniformly, however, as the opposite was demonstrated in a different dystonia patient study of GPi-HFS for at least 3 years where turning stimulation off resulted in a quick decline of symptoms within hours to presurgical levels, followed by a quick return of clinical improvement following restimulation (Grabli et al. 2009). Interestingly, these changes occurred rather fast in contrast to the initial delay in benefit at the start of the treatment.

Overall, it seems that if these 'new' representations are adequately robust, there is less chance that temporarily discontinuing stimulation allows aberrant basal ganglia output back into the system, where the return of symptoms is resisted by the stored patterns of movements that are still present in the motor system (Ruge et al. 2011a, b). As the authors suggest, initial clinical benefit is delayed because engrams of abnormal movement persist and take time to normalize (Ruge et al. 2011a).

Likely, a similar mechanism of plasticity is occurring in most other diseases with delayed improvement once HFS is begun, although this is not quantified. In the recent study of subgenual cingulate white matter HFS in TRD, increased symptom improvement was correlated with longer stimulation times, where at 24 weeks' time post-stimulation-onset there were only 41 % responders compared to a 92 % response rate at 2 years (Holtzheimer et al. 2012). When HFS was turned off for a 2 week period in three patients, there was a quick relapse of symptoms in all; once reinitiated, improvement occurred in all three of these individuals, but notably more gradually than with initial stimulation (Holtzheimer et al. 2012).

It is likely that such neuronal plasticity is fundamental to the efficacy of DBS in each of these scenarios, but its variability and complexity in extent and time make the mechanisms invoked remain fascinatingly unclear despite years of research. Synaptic modification is an important mechanism for altering neural circuit function. The striatum is the location and source of control of inhibitory inputs within the BG circuit (Tepper et al. 2004; Gustafson et al. 2006; Kreitzer and Malenka 2008), and thus modulation via synaptic plasticity within the striatum can regulate striatal output. This can have major effects in behavior, as elucidated in an eloquent study by whereby optogenetic stimulation of direct and indirect pathway medium spiny neurons (MSNs) caused either persistent reinforcement or transient punishment, respectively. Striatal plasticity has been the subject of a number of reviews in normal and in disease states (Ding et al. 2008; Kreitzer and Malenka 2008; Picconi et al. 2012). Activity-dependent modifications in synaptic efficacy, such as long-term depression (LTD) and long-term potentiation (LTP), represent key cellular substrates for adaptive motor control and procedural memory (Picconi et al. 2012); each has been observed to be induced at MSN synapses by HFS (Calabresi et al. 1992; Lovinger et al. 1993; Charpier and Deniau 1997).

As we will discuss below, in place of former views of their passivity, astrocytes are currently considered to play an intimate and active role at the neuronal synapse, forming what is referred to the 'tripartate synapse' (Ventura and Harris 1999; Araque et al. 1999; Takano et al. 2006); as such, they contribute to such modulations in neuronal communication in normal and disease states. In PD, the loss of dopamine afferents induces a compensatory over-activity and structural remodeling of the corticostriatal system, as seen in rodent and non-human primate models of parkinsonism, to continue information relay despite a reduction in corticostriatal glutamatergic synapses (Galarraga et al. 1987; Marti et al. 1999; Gubellini et al. 2002; Liang et al. 2009; Villalba and Smith 2011). Commensurate with neuronal terminal reshaping, perisynaptic astrocytes have been found to be increased in number, and undergo morphological changes, where the appositions between the axo-spinous complex and the astroglial processes were much tighter and continuous in histological studies of the MPTP primate striatum (Villalba and Smith 2011). Such structural changes in the tripartite synapse in disease states, whereby astrocytes contribute to the modulation of processing and integration of synaptic signaling through the formation of complex neuron–glia networks, allow them to be just as well-poised to be modulated by deep brain stimulation as neurons.

Astrocytes are the topic of the next section, where we will discuss how they are hypothesized to contribute to mediating the network effect of deep brain stimulation.

In summary, the key points of DBS-induced neuroplasticity are as follows: (1) HFS disrupts aberrantly formed circuits in certain disease states and normalizes functioning; (2) Temporally disparate HFS induced effects likely result from plastic changes due to synaptic modulation; and (3) Neuroplasticity induced by HFS involves astrocytic contribution at the synapse.

5 Role of Astrocytes in DBS

In terms of both local and network DBS effects, one cannot dismiss the fact that the 'other' factor in mediating cellular signaling is the contribution from the glia, namely the astrocytes. If modulation of axonal flow—and hence synaptic transmission—is considered a key element underlying the mechanism of DBS, then quite likely both neurons and astrocytes contribute to the DBS therapeutic effect. Astrocytes are active players in neuronal signaling, as they contain numerous processes that extend hundreds of millimeters and connect to those of other astrocytes through gap junctions, and surround and are intertwined in neuronal synapses forming what is termed the 'tripartate synapse' (Araque et al. 1999; Ventura and Harris 1999; Takano et al. 2006). This position makes them well poised to function actively in the regulation of neuronal communication. Until recently, this topic has not been considered as relevant in DBS mechanistic theory, and the potential that DOR activation may play a role is rarely discussed. We have recently reviewed research progress in this filed (Fenoy et al. 2014). In the following, we will take major contents from our recent publication (Fenoy et al. 2014, with permission) to discuss the role of astrocytes in mediating the effects of DBS in neurodegenerative disorders such as PD.

5.1 *Astrocytes and PD*

Astrocytes are the predominant glial cell in the central nervous system, and not only serve to regulate the external chemical environment of neurons by removing excess ions and recycle neurotransmitters released during synaptic transmission, but also signal to each other and to neurons. They, as well as microglia, react to and communicate changes detected within their immediate external milieu both locally and over large networks.

It has recently been suggested that a glial reaction and inflammatory processes participate in the cascade of events leading to neuronal degeneration in PD (Mena and Garcia de Yebenes 2008). Neuropathological examination of the substantia nigra in PD has shown increased glial fibrillary acidic protein (GFAP), confirming astrocytic reaction, and the presence of activated microglia has consistently been reported. In PD models and other neurodegenerative disease states, it seems that astrocytes play a conflicting role by releasing toxic cytokines, such as tumor necrosis factor-alpha (TNF-α), leading to neurodegeneration, but also by eliminating glutamate from the extracellular space, conferring neuroprotection.

Also, as it has been recently confirmed that astrocyte-derived glial cell line-derived neurotrophic factor (GDNF) is a potent inhibitor of microglial activation, it appears that astrocyte activation may be directly involved in midbrain protection from neurodegeneration through the inhibition of nigrostriatal neuroinflammation. In addition, we have recent evidence to suggest that DOR activation on the astrocyte may inhibit inflammatory cytokines and upregulate GDNF, leading to protection

(Tian et al. 2013a, b; He et al. 2013). This balance between promoting neurodegeneration and protection seems to be on a fine line, and likely there are other unidentified contributors that help to sway to one side over another. Potentially, it seems that DBS may be such an exogenous contributor, as discussed below.

5.2 Signaling Mechanisms Between Astrocytes and Neurons

Astrocytes, once activated in disease states such as PD, are known to display excitability due to a change in intracellular Ca^{2+} concentration, which can cause the extracellular release of gliotransmitters—glutamate, adenosine (from adenosine triphosphate (ATP) and D-serine—that act on neighboring neuronal pre-and post-synaptic receptors and can modulate synaptic transmission (Hamilton and Attwell 2010). Increased Ca^{2+} in astrocytes evokes post synaptic slow inward currents (SICs) in nearby neurons through astrocytic glutamate release, which binds to the NR2B subunit of extrasynaptic N-methyl-D-aspartate receptors (NMDA-R) (Araque et al. 1998a; Kang et al. 1998; Angulo et al. 2004; Fellin et al. 2004; Fiacco and McCarthy 2004; Tian et al. 2005; Jourdain et al. 2007; Perea and Araque 2007; Hamilton and Attwell 2010; Sasaki et al. 2011a); this increases neuronal excitability and increases probability of neurotransmitter release from neuronal pre-synaptic terminals at excitatory (Jourdain et al. 2007; Perea and Araque 2007) and inhibitory synapses (Araque et al. 1998b; Kang et al. 1998). Such an SIC is generated in many nearby neurons, and can promote synchrony of neuronal firing (Hamilton and Attwell 2010).

Astrocytes are also known to communicate with each other through Ca^{2+} waves. Possible mechanisms of Ca^{2+} wave propagation include the presence of hemichannels on astrocytes which are permeable to ATP, glutamate and glucose, (Cotrina et al. 2000; Ye et al. 2003; Retamal et al. 2007; Steinhäuser et al. 2012) and can mediate non-synaptic neuronal synchronization via glutamate release (Fellin et al. 2004; Tian et al. 2005); or gap junctions which can also mediate glucose uptake and trafficking from blood vessels and clear extracellular K^+ and glutamate released from neurons (Kofuji and Newman 2004; Rouach et al. 2008; Steinhäuser et al. 2012). In the seizure model, increased astrocytic coupling through these structures may facilitate the propagation of Ca^{2+} waves, leading to hypersynchronization and spread of ictal activity (Gomez-Gonzalo et al. 2010). It is unclear how these may function to cause widespread communication in other disease states or due to exogenous stimulation. Such astrocytic Ca^{2+} waves can either be initiated in response to neuronal activation via Gq G protein-coupled receptor signaling on astrocytic micro-domains (Agulhon et al. 2008), forming networks of cells, or can occur spontaneously even independent of neuronal activity (Cornell-Bell et al. 1990; Charles et al. 1991; Araque et al. 1998b), exhibiting so-called ‘pairwise synchronization’ with a nearly zero time lag in neighboring astrocyte activity (Hirase et al. 2004; Kuchibhotla et al. 2009; Sasaki et al. 2011a). Even sensory stimulation, as reported by Wang et al. (2006) via whisker stimulation in vivo can evoke calcium activity in the astrocytes in the barrel cortex.

Sasaki et al. (2011a) employed in vivo calcium imaging to reveal that neighboring astrocytes spontaneously exhibited synchronous Ca^{2+} elevations and formed locally correlated cell groups of 2–5 cells, which is believed to be due to release of glutamate upon astrocyte activation and subsequent activation of other astrocytes through metabotropic glutamate receptor type 1 activity, independent of neuronal activity (Sasaki et al. 2011a). These clusters were found to enhance the local excitability of neuronal networks by inducing tonic depolarization of neurons, whereas sporadic activity of single astrocytes—as elicited by Ca^{2+} uncaging by photostimulation—could not induce an action potential (AP) and was speculated to lower the threshold of synaptic plasticity by enhancing NMDA-R activity and SICs.

Kuga et al. (2011) monitored spontaneous calcium activity simultaneously from hundreds of mouse hippocampal and neocortical astrocytes *in vivo* and found participation of most in the propagation of regenerative waves from cell to cell—referred to as *glissandi*—which, as they were blocked by sodium channel blocker tetrodotoxin (TTX), are believed to emerge in response to changes in coordinated intrinsic neuronal network oscillations, or state changes. Neuronal activity, ATP, and gap junctions were found to mediate such waves (Kuga et al. 2011). As the generation of such waves were associated with a decrease in the power of infraslow field oscillations, which is correlated with the blood oxygen level-dependent (BOLD) signal in functional magnetic resonance imaging (fMRI), (Hinterberger et al. 2003; Nagai et al. 2004; Heo et al. 2008; Khader et al. 2008), it is speculated that these waves mediate the activity-dependent large scale modulations of cerebral blood flow (CBF). Astrocytes extend several processes, or endfeet, around the cerebral microvasculature (Takano et al. 2006). In response to intracellular Ca^{2+} influxes that cause astrocyte activation, Ca^{2+}-gated K^+ channels in these endfeet are also activated, resulting in increased extracellular K^+. It is this increased extracellular K^+ that causes the dilation of arterioles (Filosa et al. 2006; Hamilton and Attwell 2010). By being strategically positioned between synapses and blood vessels, astrocytes can indeed act as mediators of neurovascular coupling and may modulate CBF through such highly organized large-scale astrocyte dynamics (Zonta et al. 2003; Filosa et al. 2004; Mulligan and MacVicar 2004; Metea and Newman 2006; Takano et al. 2006; Halassa and Haydon 2010).

Interestingly, although short electrical microstimulation did activate astrocytes at the tip of the electrode, it did not induce such broad glissandi, which is expected in keeping with our network theory of cellular modulation. Such artificial activation likely disrupts the intrinsic nature of ongoing network activity, and when applied transiently, cannot induce a persistent shift in the global network state (Sasaki et al. 2007).

The ability of astrocytes to modulate synaptic transmission—directly on the AP—has been evidenced *in vitro* in hippocampal slices via the use of calcium imaging again, where uncaging of Ca^{2+} in peri-neuronal astrocytes was found to cause AP broadening through ionotropic axonal glutamate (2-amino-3-(5-methyl-3-oxo-1,2-oxazol-4-yl)propanoic acid) (AMPA) receptor activation (Sasaki et al. 2011b). Broadened APs triggered larger Ca^{2+} elevations in presynaptic boutons and facilitated synaptic transmission to postsynaptic neurons. Such broadening of APs occurred in response to the local application of glutamate and adenosine A1 antagonist to axon shafts within 400 μm of the axon hillock, making the suggestion that adenosine tonically inhibits AP (via A1 receptors) and that glutamate released from astrocytes mediates neuronal axonal effects—AP—via extra-synaptic modulation at non-NMDA receptors (Sasaki et al. 2011b).

These are yet other ways astrocytes are believed to influence neuronal signaling, and although this was only seen *in vitro* on unmyelinated axons in hippocampal slices, the potential for these mechanisms to exist elsewhere in the brain at subcortical or cortical locations is highly possible and exciting, as it builds yet new layers of complexity into circuit modulation. In disease states such as PD, the astroglial contribution has a contradictory role of both promoting neurodegeneration as well as conferring neuroprotection. The balance is fine, and can be modulated by exogenous stimuli such as HFS. Specifically, this above work could give explanations for the mechanisms for synaptic modulation underlying the beneficial effects seen upon stimulating specific nodes in a dysregulated circuit in certain neurodegenerative diseases. So there is evidence in vivo that neurons can activate astrocytes, as well as conversely that astrocytes can activate neurons. This could fit into the DBS mechanism theory a bit more readily, as quite possibly HFS can activate astrocytes locally, which can then cause multiple effects to modulate neuronal activity leading to the 'summed final' network effect seen. Evidence of HFS activation of astrocytic activity is detailed below to help gain focus on this still blurry concept.

5.3 Astrocytes and DBS

HFS, a popular strategy of DBS, induces a sudden change, or interruption of a network status—either physiological or pathological. As detailed above, it has been observed that HFS modulates CBF (Perlmutter et al. 2002; Kefalopoulou et al. 2010; Wyckhuys et al. 2010) associated with an increase or decrease in neuronal activity (Attwell et al. 2010). Possibly, after a neuronal network re-organization occurs, such as in dystonia or PD after years of HFS, such glissandi are representative of the new current steady-state of neuronal activity. It is thus also suspected that the observed DBS-induced modulation in CBF is due to large scale network astrocytic activation representative of such a new global status.

HFS has been known to directly activate astrocytes, and their impact on neuronal signaling has been observed for over a decade (Kang et al. 1998). Bekar et al. (2008) applied HFS to the ventrolateral thalamus to the mouse cortex in a tremor model in vivo and observed increased adenosine levels which they showed were Ca^{2+}—independent, and thus believed to be from astrocyte origin. ATP released from astrocytes is broken down in extracellular space by ecto-ATPase to adenosine; adenosine was seen to increase in the extracellular space around the electrode and is thought to be a key mediator in promoting the efficacy of HFS in suppressing tremor. When an adenosine A1-receptor antagonist or ecto-ATPase inhibitor was used, HFS-induced suppression of thalamic neuronal activity was abolished. Also,

HFS induced was seen to induce astrocytic Ca^{2+} waves that propagated away from the site of stimulation in a frequency and amplitude dependent manner.

These findings corroborate the later study of Tawfik et al. (2010), where the application of HFS to the thalamic slice resulted in abolishment of oscillations concomitant with the release of glutamate and adenosine. This was interpreted to be due to astrocyte activation, as the glutamate persisted despite TTX application—which would inhibit axonal synaptic release—but was eliminated with a vesicular H^+-ATPase inhibitor. This group also saw HFS induced triggering of Ca^{2+} waves and release of glutamate in astrocyte cultures.

The actual insertion of the DBS electrode into the Vim, STN, or GPi during surgery has indeed been seen to sometimes cause a reduction or abolishment of tremor in ET or PD patients (Mann et al. 2009; Morishita et al. 2010; personal observation), but also rigidity when surgery is performed under local anesthesia (personal observation). This so-called 'microthalamotomy effect' is believed to be mediated by the release of glutamate and adenosine (Chang et al. 2009; Shon et al. 2010); such mechanical stimulation can induce activation of calcium signaling and ATP release in cultured astrocytes (Newman and Zahs 1998; Kozlov et al. 2006), and thus it is possible that the adenosine and glutamate observed could have astrocyte origins.

The presence of mild reactive gliosis—defined by hypertrophy of astrocytes and increase in the production of the GFAP among other proteins (Pekny and Nilsson 2005)—around the implanted electrode as seen in several histological studies (Stock et al. 1979; Haberler et al. 2000; Pilitsis et al. 2008; Sun et al. 2008; DiLorenzo et al. 2010) have also implicated HFS-induced astrocytic activation. After DBS electrode implantation, regional neuro-inflammation in the rat was observed with concomitant memory impairment (Hirshler et al. 2010); in the macaque, reactive gliosis was seen after 3 months and 3 years (Griffith and Humphrey 2006); and in the human, reactive gliosis was observed in a post-mortem study 12 years after implantation in a ET patient with good control (DiLorenzo et al. 2010).

Induction of reactive astrocytosis may modulate local as well as widespread effects in neuronal circuits. It has been shown by Ortinski et al. (2010) that reactive astrocytosis mediated deficits in inhibitory synaptic currents via a down-regulation of glutamine synthetase and consequently GABA, triggering reversible hyperexcitability in hippocampal circuits. Implication of astrocyte activation by DBS was seen by Gradinaru et al. (2009) in their optogenetic study on PD circuitry, where channelrhodopsin 2 (CHR2) was introduced specifically into astroglia of 6-hydroxydopamine lesioned-rodents through a lentiviral vector driven by a GFAP-promoter. Optical stimulation of CHR2 activated astrocytes and triggered astrocytic Ca^{2+} waves, which resulted in an inhibition of neuronal firing in STN, though not sufficient enough to cause even trace responses in motor pathology (Gradinaru et al. 2009). Although this group did not use DBS, it can be inferred that activated astrocytes alone are likely not responsible for mediating all the effects of DBS, but only together in a complex manner with neuronal modulation do they contribute to efficacy.

Glutamate plays a large role in astrocyte activation, and the evidence that it can be induced by DBS in mounting. Dyskinesias in the naïve rat brought on by STN-HFS were found to be correlated with increased phosphorylation of the NR2B subunit of the NMDA-R in neurons of the STN, entopeduncular nucleus (correlate of the human GPi) and other locations (Quintana et al. 2010); blockade of NR2B/NMDA-Rs, found chiefly peri- or extraspynaptically (Chen and Roche 2007), was found to suppress STN-HFS induced dyskinesias, whereas blockade of non-competitive type of NMDA-R at the synapse was found to potentiate STN-HFS induced dyskinesias (Quintana et al. 2010; Sgambato-Faure and Cenci 2012).

Such dyskinesias were also associated with altered expression of glutamate transporters in SNr and striatum, along with increased extracellular concentration of glutamate (Dupre et al. 2011; Mela et al. 2012; Seifert and Steinhauser 2013; Sgambato-Faure and Cenci 2012), which has been found to persist after STN-HFS ceases (Lee et al. 2007). The authors make the suggestion that extra-synaptic NR2B/NMDA-Rs play a key role in STN-HFS-induced dyskinesias, which corroborates with findings that stimulation frequencies of 100–200 Hz were found to maximally activate extrasynaptic NMDA-Rs in hippocampal slices (Harris and Pettit 2008). This implies that not only are extra-synaptic NMDA-Rs preferentially activated by HFS and are critically linked to the production of results, but one can also contend that excessive extracellular glutamate plays a role in causing dyskinesias. Please refer to Fig. 15.2 for an illustration of the HFS impact on astrocytic interaction with neurons.

How HFS impacts the role of astrocytes in regulating glutamate release and transport is indeed complex. We know that the excessive excitatory stimulation of NMDA, AMPA and other glutamate receptors is neurotoxic. Astrocytic glutamate transporters play a critical role in cleaning up the released excitatory transmitters in extracellular space (Murugan et al. 2013; Stobart and Anderson 2013; Lopez-Bayghen and Ortega 2011; Kim et al. 2011). Our recent findings show that activation of DOR increases the expression and function of astrocytic glutamate transporters (Mena and Garcia de Yebenes 2008) suggesting that DOR may promote neuroprotection (Tian et al. 2013a, b; He et al. 2013; Yang et al. 2012).

These findings relate directly to support recent theories that HFS may confer neuroprotection (Spieles-Engemann et al. 2010). In our studies involving electrical stimulation of the hypothalamus in the rabbit (Xia et al. 1987, 1988), we observed an increased release of enkephalin, which is the only endogenous ligand for DOR. Such electrical stimulation likely caused DOR activation and increased astrocytic glutamate transport from the extracellular space. In support of DOR's beneficial effect against PD, a recent study shows that a novel DOR agonist, KNT-127, increases the release of dopamine in the striatum, nucleus accumbens and median pre-frontal cortex (Tanahashi et al. 2012). Very recent clinical observations that STN-DBS improves survival in severe PD (Ngoga et al. 2014) may likely be due to astrocyte activation and an increased role of astrocytes in removing extracellular oxidative toxins.

5.4 Astrocytes as Players at the Tripartite Synapse with DBS

As their processes are ideally located in apposition with pre and postsynaptic neuronal elements of synapses throughout the CNS (Ventura and Harris 1999; Witcher et al. 2007; Panatier et al. 2011), astrocytes have been implicated to be exquisitely

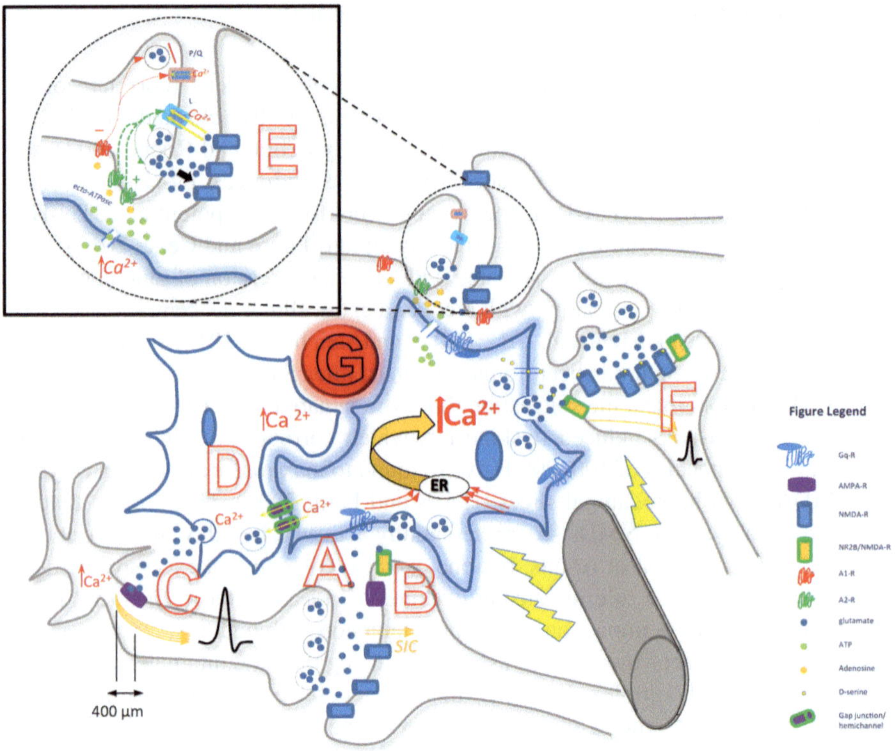

Fig. 15.2 Summary diagram depicting the proposed simultaneity of DBS effects on axons, astrocyte-astrocyte and astrocyte-neuronal interactions. DBS electrode is here activating one local astrocyte (*blue*), leading to increased intracelluar Ca^{2+}, causing hypermorphism and release of gliotransmitters, resulting in increased axonal flow and synaptic transmission. This astrocyte has multiple processes, each intimately participating at specific neuronal synapses; activation at one process/micro-domain may or may not spread to others. Multiple processes are likely occurring simultaneously. (*A*) Secondary activation by DBS: Increased synaptic activity between axon and dendrite, with release of neurotransmitters (e.g., glutamate), is sensed by the astrocytic Gq protein-coupled receptors on a process at the tripartite synapse (such as mGluR5), resulting in an increase in intracelluar Ca^{2+}. This activation can be confined to one particular micro-domain, or can be distributed throughout the astrocyte to all its processes. Each micro-domain can express different gliotransmitters particular to the synapse. (*B*) Increased Ca^{2+} in astrocytes evokes post synaptic slow inward currents (SICs) in nearby neurons through astrocytic glutamate release, which binds to the NR2B subunit of extrasynaptic NMDA-R and increases neuronal excitability and increases probability of neurotransmitter release from neuronal pre-synaptic terminals. (*C*) Ca^{2+} activation in astrocyte causes release of extra-synaptic glutamate; glutamate binds to peri-synaptic AMPA-R on pre-synaptic neuronal processes within 400 μm of axon hillock, causes influx of Ca^{2+} which broadens action potential (AP). (*D*) Astrocytes in clusters, co-activated and synchronized via gap junctions/hemichannels. (*E*) Astrocytic processes detect local synaptic activity—glutamate release—which causes a local Ca^{2+} evoked response mediated through mGluR5 activation; this leads to release of glial ATP which is converted by ecto-ATPase into adenosine. Adenosine then acts at pre-synaptic A1 coupled P/Q-type calcium channels to incur synaptic depression while it also acts at pre-synaptic A2 coupled L-type calcium channels to incur synaptic potentiation. Relative amounts of synaptic adenosine as well as interactions between each receptor determine the balance; here potentiation is depicted. Also, activation of pre-synaptic adenosine A2A receptors increases basal synaptic transmission. (*F*) Different gliotransmitters can be released at other processes due to Ca^{2+} activation in astrocyte. D-serine and glutamate are released by the activated astrocyte, modulate NMDA-R activity and synaptic plasticity. Glutamate released due to DBS astrocytic activation acts on neuronal extrasynaptic NR2B/NMDA-R, causing increased neuronal excitability and probability of downstream synaptic transmission. (*G*) Blood vessel proximity to astrocytes—increased CBF linked to state changes of astrocyte activity; DBS induced changes in CBF and interaction with astrocytes not resolved

involved in the regulation of basal synaptic transmission and plasticity in multiple brain locations (Matsui and Jahr 2004; Matsui et al. 2005; Panatier et al. 2006; Gordon et al. 2009; Gourine et al. 2010). This was eloquently investigated by Panatier et al. (2011) in the CA1 region of the hippocampus, where it was observed that astrocytic processes detect local synaptic activity—neurotransmitter release—which causes a local Ca^{2+} evoked response mediated through metabotropic glutamate subtype 5 receptor (mGluR5) activation. Such activated astrocytes were then found to increase the efficiency of basal synaptic transmission in CA1 pyramidal cells through activation of presynaptic adenosine A2A receptors (Panatier et al. 2011).

These authors contend that such glial modulation may serve to adjust synaptic efficacy so that it can conform to various plasticity events that occur at specific regions in the brain, as adenosine release can either increase transmitter release, via presynaptic adenosine A2A receptor activation—or decrease it via adenosine A1 receptor activation (Pascual et al. 2005; Serrano et al. 2006; Todd et al. 2010).

Such differential modulation of glutamate and GABA release by astrocyte-derived adenosine is hypothesized to contribute to the differential effects of DBS in the GPi and GPe. It is known that the GPi is inhibited by afferent GABAergic projections from the striatum and the GPe, and it is activated by a glutamatergic projection from the STN. A recent critique by Jantz and Watanabe (2013) on the experimental findings of DBS induced inhibition of the GPi and excitation of the GPe in the normal monkey by Chiken and Nambu (2013) highlights this point. Not only might the differences in the balance of GABAergic and glutamatergic inputs to these structures be a factor (Chiken and Nambu 2013), but also the relative activation of A2A receptors, which are known to be expressed at higher levels in the GPe and striatum than in the GPi, should certainly be considered. As A1 receptors are expressed widely in both the GPi and GPe, Jantz and Watanabe contend that the variation in adenosine receptor concentration can explain the different effects of DBS.

It is likely that both A1 and A2A receptors are also co-modulated to differentially effect synaptic plasticity, as seen at the neuromuscular junction (Todd et al. 2010), and that such regulation may vary with long-term potentiation (LTP) or depression (LTD) (more below). Group 1 mGluR (mGluR5) are the major subgroup of receptors implicated in the detection of glutamate by astrocytes in response to sustained neuronal activity, (Fellin et al. 2004; D'Ascenzo et al. 2007) and are most expressed in the striatum (Testa et al. 1994). Given their primarily peri- or extra-synaptic location, the activation of these astrocytic mGluR5, resulting in astrocyte activation and release of gliotransmitters, is thought to be a key player in the expression of postsynaptic neuronal NMDA-R and more permanent plastic effects.

It is yet unproven whether the local synaptic modulation brought upon by calcium signaling in an astrocytic process may extend to incur activation of the whole astrocyte, whereby multiple synapses in different spatial relations can be modulated, either uniformally or differentially, allowing an astrocyte to function as a 'global network switchboard operator. As astrocytes have such a particularly complex spatial relationship with adjacent neurons, maintaining several types of synapses (Bushong et al. 2002; Halassa et al. 2007), it is likely that they interact with and differentially modulate many synapses according to the specific conditions at each, and thus use different transmitters to effect different forms of plasticity.

5.5 Astrocytes in Synaptic Plasticity After DBS

Neuroplasticity is believed to be fundamental to DBS efficacy. How else can the disparate temporal effects of efficacy in tremor vs. those in dystonia be explained?

In dystonia, it may take several weeks or months to achieve maximum clinical benefit in patients treated not only by DBS (Yianni et al. 2003; Coubes et al. 2004; Vidailhet et al. 2005; Tisch et al. 2006a, b; Ruge et al. 2011a, b) but also following pallidotomy (Lozano et al. 1997). Similarly, there is often not an observed quick loss of effect once DBS is turned off in dystonia, taking again weeks for pre-surgical symptoms to re-emerge (Ruge et al. 2011a, b). It is contended that such a slowly progressive clinical time course, as well as concomitant commensurate normalization of electrophysiological measures is due to a LTP synaptic plasticity (Tisch et al. 2006a, b).

The striatum is the source of control of inhibitory inputs within the BG circuit (Tepper et al. 2004; Gustafson et al. 2006; Kreitzer and Malenka 2008), and thus modulation via synaptic plasticity within the striatum can regulate striatal output (for reviews in normal and in disease states see: Ding et al. 2008; Kreitzer and Malenka 2008; Picconi et al. 2012). Activity-dependent modifications in synaptic efficacy, such as LTP and LTD, represent key cellular substrates for adaptive motor control and procedural memory (Picconi et al. 2012); each has been observed to be induced at MSN synapses by HFS (Calabresi et al. 1992; Lovinger et al. 1993; Charpier and Deniau 1997).

In PD, the loss of dopamine afferents induces a compensatory over-activity and structural remodeling of the corticostriatal system, as seen in rodent and non-human primate models of parkinsonism, to continue information relay despite a reduction in corticostriatal glutamatergic synapses (Galarraga et al. 1987; Marti et al. 1999; Gubellini et al. 2002; Liang et al. 2009; Villalba and Smith 2011). Commensurate with neuronal terminal reshaping, perisynaptic astrocytes have been found to be increased in number, and undergo morphological changes, where the appositions between the axo-spinous complex and the astroglial processes were much tighter and continuous in histological studies of the MPTP primate striatum (Villalba and Smith 2011). Such structural changes in the tripartite synapse in disease states, whereby astrocytes contribute to the modulation of processing and integration of synaptic signaling through the formation of complex neuron–glia networks, allow them to be just as well-poised to be modulated by deep brain stimulation as neurons.

Modulation of the NMDA-R is directly involved in these processes, and has been observed to occur on two time scales, and differentially involves particular gliotransmitters (Deng et al. 2011) Astrocytes incur short time scale modulation of NMDA-R by releasing glutamate and by providing the NMDA-R coagonist D-serine, which like glycine, acts at separate and distinct NMDA-R populations (Deng et al. 2011). Astrocyte-derived D-serine modulates the NMDA receptor current and is necessary for the induction of synaptic plasticity (Schell et al. 1995, 1997; Mothet et al. 2000; Stevens et al. 2003; Yang et al. 2003; Panatier et al. 2006; Henneberger et al. 2010), as LTP was observed to be absent at NMDA-R without astrocytic presence (Yang et al. 2003) but resumed upon exogenous D-serine (Panatier et al. 2006).

D-serine is likely released by a Ca^{2+} trigger, possible via vesicular routes, (Henneberger et al. 2010, 2012) and can incur dynamic changes in NMDA-R mediated currents. It is believed that D-serine can be delivered to distinct populations of NMDA-Rs (e.g. synaptic, extrasynaptic) characterized by different subunit compositions and with variable affinity to co-agonists; different micro-domains on distinct astrocyte processes may differentially release D-Serine to specific NMDARs, which may in turn trigger different forms of plasticity (Liu et al. 2004; Henneberger et al. 2012). Because D-serine is known to cause internalization of NMDA-R subunits (Nong et al. 2003), the authors infer that the release of D-serine can be dynamic and provides the potential for second to second changes in NMDA-R mediated currents.

Other longer time scale modulation of synaptic plasticity at the NMDA-R by astrocytes has also recently been observed to involve the release of adenosine. Deng et al. (2011) used astrocyte specific inducible transgenic mice (dnSNARE) brain slices to demonstrate that activation of adenosine A1R in cortex can affect neuronal excitability by regulating the surface expression of NMDA-R through the activation of specific intracellular signaling pathways. Specifically, astrocytic activation of neuronal A1R led to downstream phosphorylation of the NR2A and NR2B subunits of the NMDA-R, which in turn decreased the rate of its endocytosis (Deng et al. 2011). Due to such involvement in the trafficking of the NMDA-R, adenosine is thus linked to a slower time scale of synaptic modulation, such as sleep wake cycles and memory (Deng et al. 2011; Florian et al. 2011). On the other hand, other global plastic changes can be implicated and might very well use this mechanism.

Combined with the differential activation of presynaptic receptors on causing potentiation or depression of neurotransmitter release, as influenced by its relative extra-synaptic concentration (at least at the neuromuscular junction (Todd et al. 2010), it appears that there are many ways astrocyte-derived adenosine can ultimately modulate synaptic transmission. Very likely, these mechanisms contribute to some of the temporal variability in effects seen after DBS. Together with the other gliotransmitters released upon activation, the astrocyte is well positioned to effect dynamic or delayed effects on neighboring neurons, which has implications on the overall function of the network (Vedam-Mai et al. 2012), especially after an exogenous stimulus is applied.

6 Potential Role of DOR in DBS

Our serial studies have shown that DOR activation is neuroprotective in many neurodegenerative conditions such as hypoxic/ischemic injury, excitatory insult. More recently, we found a clue showing that DOR may play a role in neuroprotection against PD (Chen et al. 2013. Moreover, there is evidence showing that DOR activation promotes the release of dopamine in the striatum (Tanahashi et al. 2012). Since we previously observed that brain stimulation could promote the release of enkephalin (Xia et al. 1988), an endogenous DOR ligand in the brain, it is our belief that DBS may yield benefits to the PD brain through the DOR mechanism, at least partially.

6.1 DOR and Neuroprotection

It has been generally established that DOR confers neuroprotection against hypoxic/ischemic neuronal insult (Zhang et al. 1999, 2002; Ma et al. 2005; Chao et al. 2007a, b, 2008; Chao and Xia 2010; He et al. 2013). The mechanisms behind such neuroprotection are varied at multiple levels. As discussed in our recent review (He et al. 2013), we can summarize the mechanisms of DOR neuroprotection as: (1) stabilization of the ionic homeostasis; (2) inhibition of excitatory transmitter release; (3) attenuation of disrupted neuronal transmission; (4) increase in antioxidant capacity; (5) regulation of intracellular pathways—inhibition of apoptotic signals and activation of pro-survival signaling; (6) regulation of specific gene and protein expression; and (7) up-regulation of endogenous opioid release and/or DOR expression. Of these, some occur immediately in response to acute hypoxic stress, while other changes appear gradually in response to a prolonged hypoxic/ischemic insult. For example, blockade of anoxic K^+ leakage has been shown to attenuate hypoxia/ischemia-induced neuronal death (Liu et al. 2003; Wei et al. 2003). Our results show that activation of DOR markedly reduces K^+ leakage, suggesting that DOR signals help stabilize ionic homeostasis (Chao et al. 2007a, b, 2008, 2009), which is apparently neuroprotective. Ca^{2+} influx is involved in anoxic K^+ leakage (Chao et al. 2007a, b). The blockade of K^+ efflux is partially aided by inhibition of Ca^{2+} influx because DOR activation causes a decrease in Ca^{2+}-dependent K^+ efflux via BK channels (Chao et al. 2007a, b). Another major mechanism underlying DOR neuroprotection against anoxic disruption of K^+ homeostasis is the inhibition of Na^+ entry through voltage-gated channels because the DOR effect disappeared in low Na^+ perfusion or with Na^+ channel blocker TTX (He et al. 2013; Chao et al. 2008, 2012; Kang et al. 2009).

DOR activation also leads to presynaptic inhibition of glutamate release (Chao and Xia 2010; Johnson and Turner 2010; Turner and Johnson 2011; Ostermeier et al. 2000; Tanaka and North 1994). Electrophysiological evidence suggests that DOR activation can prevent the release of glutamate from presynaptic vesicles (Ostermeier et al. 2000; Tanaka and North 1994), which may prevent K^+ efflux and Na^+ influx due to a reduction in NMDA receptor-mediated Na^+ influx (Chao and Xia 2010; Ostermeier et al. 2000). Recently, another association was highlighted as a potential aspect of DOR neuroprotection at the presynaptic level, i.e., DOR activation down-regulates Na–K-2Cl co-transporter (NKCC) expression that is associated with neuronal injury secondary to severe hypoxic/ischemic insult (Kahle et al. 2009; Yang et al. 2011). DOR activation inhibits glutamate release that in turn decreases NKCC activity thereby reducing cerebral edema and in the process protecting against neuronal damage (Yang et al. 2011).

As prolonged stress can potentially alter cell structure, macromolecules and gene expression, and signal transduction (Chao and Xia 2010; Zhang et al. 2006; Ma et al. 2005; Peng et al. 2009; Nandhu et al. 2010; Brouwer et al. 2004; Bindra et al. 2005; Appenzeller et al. 2006; Storey 2004; Lin et al. 2011), it is likely that DOR activation can mediate neuroprotection at multiple levels in a very intricate interplay of complex mechanisms in recognized neuroprotective signaling cascades. We found that the activation of DOR stimulates protein kinase C (PKC) and mitogen-activated protein kinase (MAPK)-ERK1/2 and eventually prevents phosphorylation of p38, thereby preventing cell death (Ma et al. 2005; Peng et al. 2009).

We also demonstrated an increase in antioxidant capacity following DOR activation in the ischemic rat brain after middle cerebral artery occlusion. Both superoxide dismutase and glutathione peroxidase activities increased and free radical and nitric oxide levels decreased following DOR activation that resulted in reduced caspase 3 expression in the ischemic regions and thereby prevented cell death (Yang et al. 2009).

6.2 *DOR and Astrocytes*

Astrocytes are intimately involved in the regulation and maintenance of their neuronal neighbors. Our recent findings show that DOR activation promotes the expression of excitatory amino acids transporters (EAATs) (Liang et al. 2014) and neurotrophic factors (Tian et al. 2013a, b) and attenuated TNF-α release/expression in the astrocytes, suggesting that DOR activation protects the brain through an astrocytic mechanism via upregulating EAATs-mediated capacity of uptaking excitatory amino acids and neurotrophic factors-mediated neuroprotection and down-regulating TNF-α mediated inflammatory action.

DOR expression on astrocytes may confer neuroprotection via the increased expression of EAATs and altered glutamate uptake. In our recent study, we measured mRNA and protein expression of EAATs in cultured mouse astrocytes exposed to a specific DOR agonist (UFP-512) with or without a DOR-antagonist or several inhibitors of downstream cellular signaling cascades (Liang et al. 2014). DOR activation increased mRNA and protein expression of EAAT1 and EAAT2, but not EAAT3, leading to increased glutamate uptake by astrocytes, which was largely reversed by DOR antagonist naltrindole and inhibitors of MEK, ERK, and p38. There was no effect by inhibitors of PKA, PKC, or PI3K. Thus we show that

DOR activation upregulates astrocytic EAATs via MEK-ERK-p38 signaling, suggesting a critical role for DOR in the regulation of astrocytic EAATs and protection against neuroexcitotoxicity and neurodegenerative diseases.

Neurotrophic factors such as brain-derived neurotrophic factor (BDNF), nerve growth factor (NGF), and glial cell line-derived neurotrophic factor (GDNF) enhance the growth, functional maintenance and phenotypic development of neurons. They are expressed in both neurons and astrocytes and play a crucial role in brain protection against hypoxic/ischemic encephalopathy and neurodegenerative diseases (Almeida et al. 2005; Chen et al. 2013; Abe et al. 1997; Kitagawa et al. 1998). DOR is expressed in astrocytes. Our studies have shown that astrocyte-specific pathways are also involved in DOR protection. Using a pure mouse brain astrocyte culture, we found that a potent DOR agonist UFP-512 enhanced mRNA expression of both astrocytic BDNF and NGF, but not GDNF, which was differentially suppressed by protein kinase A and C inhibitors.

Further evidence shows that DOR neuroprotective against stress involves the regulation of pro-inflammatory cytokines (Wang et al. 2014). Hypoxia largely increased the level of cellular pro-inflammatory cytokine TNF-α in the astrocytes but reduced it in neuron-like PC-12 cells. An exogenous addition of TNF-α to the culture media further worsened the hypoxic damage in the PC-12 cells. Hypoxia was found to increase the DOR expression in astrocytes but not in PC-12 cells. DOR protein expression activation reversed the hypoxic changes in TNF-α in both types of cells. In contrast, however, the interleukin-1 beta (IL-1β) levels did not significantly change in either cell when exposed to hypoxia or following DOR activation. Thus here our study suggests that DOR activation inhibits TNF-α mediated inflammatory responses following exposure to severe hypoxia with a differential regulation between glial and neuron-like cells.

6.3 DOR-Mediated Attenuation of PD Pathophysiology

We have recently found that in a PD animal model, DOR expression is normal during early development but largely decreased in adulthood, especially when PD symptoms appear, suggesting that DOR may play a role in neuroprotection against PD, while its impairment may contribute to the ontogenesis of PD. More recently, we obtained more direct evidence in an in vitro PD model with MPP^+ exposure showing that DOR is indeed protective against PD pathophysiology (Chen et al. 2013).

The pathophysiology of PD may potentially be caused by hypoxic/ischemic injury or MPP^+ with α-synuclein playing a key role. In our recent study, we observed that both hypoxia and MPP^+ exposure enhanced α-synuclein expression and oligomer formation in HEK293 cells (Chen et al. 2013). UFP-512 mediated DOR activation markedly attenuated both the hypoxic and MPP^+ induced cell injury and α-synuclein overexpression, which was largely reversed by DOR antagonism with naltrindole. Also, to further elucidate the mechanism behind the DOR-mediated cytoprotection, hypoxia/MPP^+ induced changes in mitochondrial membrane potential were measured since dysfunction of mitochondria plays a critical role in cell damage under stress. DOR activation was found to prevent mitochondrial membrane collapse by promoting transcription factor CREB phosphorylation, which likely was the cause of markedly attenuating the alpha-synuclein dysregulation and cell injury.

6.4 Increased Enkephalin by Brain Stimulation

We have previously observed that electrical stimulation of the brain induces an increase in leu-enkephalin, an endogenous DOR agonist, in the CSF, suggesting an increased release of enkephalin in the brain (Xia et al. 1988). Since DOR is neuroprotective against brain injury including PD pathophysiology through neuronal and astrocytic regulation as state above and DOR activation promotes the release of dopamine in the striatum (Tanahashi et al. 2012), DBS may promote brain enkephalin-DOR activity, thereby regulating astrocytic and neuronal function and generate a beneficial effect on the PD brain. Moreover, a DOR-induced upregulation of BDNF in astrocytes could also partially contribute to the therapeutic efficacy of DBS, and ultimately underlie the observations of increased neuronal (Spieles-Engemann et al. 2010) and patient survival in PD (Ngoga et al. 2014) after DBS. Indeed, BDNF was seen to increase in M1, GPi and SN with STN DBS in a rat model of PD (Spieles-Engemann et al. 2010). On the other hand, BDNF and other neurotrophins may play a role in neurogenesis that could play a role in restoration of behavior and increased survival, as it has been seen in the animal model of various pathologies where DBS has been instituted, including striatal stimulation in a rat chronic ischemia model and entorhinal stimulation in mice. Potential roles of DOR in DBS are summarized in Fig. 15.3.

7 Concluding Remarks

The currently available literature suggests that the efficacy of DBS depends upon the combined support of the complex neuronal-glial infrastructure. Whether or not this partnership is essential to DBS needs to be further proven, but mounting evidence supporting an astrocytic component cannot be excluded.

If we consider that not only may DBS induce axonal activation but also astrocytic activation, it becomes perhaps more plausible how synaptic plastic changes can occur over a wide network, which cannot be clearly rationalized by simple neuronal activation/inhibition theories alone.

It remains quite fascinating that such layers of complexity exist in these networks, can change in response to an electrical stimulus, and then produce such marked clinical amelioration of symptoms across a wide range of diseases. If one looks at the mechanisms of DBS in the context of a neurodegenerative/chronic inflammatory process such as PD, then it is plausible that the observed increased

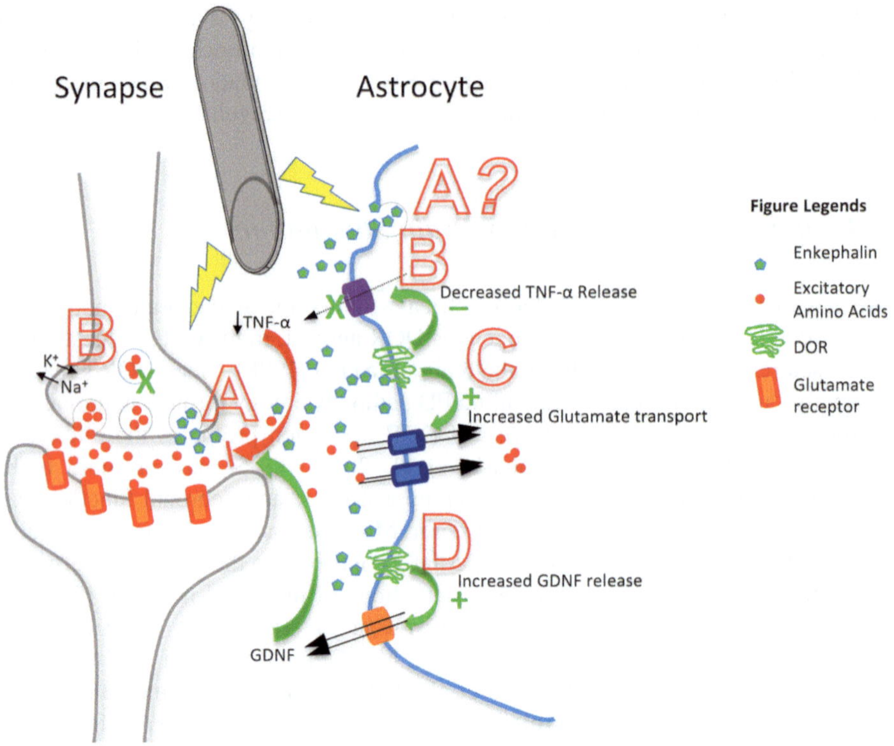

Fig. 15.3 Summary diagram depicting the proposed role of astrocytic DOR in DBS. Stimulation increases leu-enkephalin release into the synapse, which then activates delta-opioid receptors (DOR) located on the astrocytes at the tripartite synapse. DOR activation promotes neuroprotection here through three illustrated mechanisms: (*A*) Increased enkephalin release by neurons and possibly by astrocytes; (*B*) Decreased TNF-α release, leading to decreased neuronal dysfunction and injury, possibly through stabilized ionic homeostasis and decreased excessive excitatory neurotransmission (187); (*C*) Increased expression and function of glutamate transporters to reduce excessive excitatory amino acids, thus attenuating neurotoxicity; (*D*) Increased release of astrocyte-derived GDNF, which inhibits microglial activation and leads to decreased neuroinflammation

survival of DA neurons in the substantia nigra and increased clinical survival of PD patients is due to the conferred astrocytic neuroprotection via increased glutamate transport and decreased release of inflammatory cytokines, possibly through an upregulation of DOR. Conversely, if the mechanisms behind DBS efficacy are examined in tremor or other non-degenerative disease states, then astrocytic activation and release of gliotransmitters with resultant improved synaptic transmission and plasticity may be one of the dominant factors.

Our novel findings show that DOR activation is neuroprotective in a variety of ways, including attenuation of inflammatory cytokines and increased EAA uptake. Astrocytes are likely activated by DBS, which may in turn lead to increased DOR activation in astrocytes. Such increased DOR activation has the potential to prevent ongoing cellular injury taking place in neurodegenerative diseases such as PD, and possibly even neuronal regeneration. Could neuroprotection and/or neurogenesis be occurring in all DBS cases, but only verified in animal models? Quite possibly, and this potentiality remains very exciting, especially as it could be directly related to DOR activation.

Together, the plastic changes in these glial-neuronal interactions network-wide likely underlie the range of effects seen (Fenoy et al. 2014). Much more needs to be explored in vivo in both electrophysiological and imaging domains, but the evidence is mounting.

Acknowledgements YX was partially supported by NIH (AT-004422, and HD-034852) and Vivian L Smith Neurologic Foundation.

References

Abe K, Hayashi T, Itoyama Y (1997) Amelioration of brain edema by topical application of glial cell line-derived neurotrophic factor in reperfused rat brain. Neurosci Lett 231:37–40

Adolphs R, Damasio H, Tranel D, Damasio AR (1996) Cortical systems for the recognition of emotion in facial expressions. J Neurosci 16:7678–7687

Agulhon C, Petravicz J, McMullen AB, Sweger EJ, Minton SK, Taves SR et al (2008) What is the role of astrocyte calcium in neurophysiology? Neuron 59:932–946

Albin RL, Young AB, Penney JB (1989) The functional anatomy of basal ganglia disorders. Trends Neurosci 12:366–375

Alexander GE, DeLong MR, Strick PL (1986) Parallel organization of functionally segregated circuits linking basal ganglia and cortex. Annu Rev Neurosci 9:357–381

Almeida RD, Manadas BJ, Melo CV, Gomes JR, Mendes CS, Grãos MM et al (2005) Neuroprotection by BDNF against glutamate-induced apoptotic cell death is mediated by ERK and PI3-kinase pathways. Cell Death Differ 12(10):1329–1343

Anderson KE, Mullins J (2003) Behavioral changes associated with deep brain stimulation surgery for Parkinson's disease. Curr Neurol Neurosci Rep 3:306–313

Anderson T, Hu B, Pittman Q, Kiss ZH (2004) Mechanisms of deep brain stimulation: an intracellular study in rat thalamus. J Physiol 559:301–313

Anderson TR, Hu B, Iremonger K, Kiss ZH (2006) Selective attenuation of afferent synaptic transmission as a mechanism of thalamic deep brain stimulation-induced tremor arrest. J Neurosci 26:841–850

Angulo MC, Kozlov AS, Charpak S, Audinat E (2004) Glutamate released from glial cells synchronizes neuronal activity in the hippocampus. J Neurosci 24:6920–6927

Appenzeller O, Minko T, Qualls C, Pozharov V, Gamboa J, Gamboa A et al (2006) Gene expression, autonomic function and chronic hypoxia: lessons from the Andes. Clin Auton Res 16:217–222

Araque A, Parpura V, Sanzgiri RP, Haydon PG (1998a) Glutamate-dependent astrocyte modulation of synaptic transmission between cultured hippocampal neurons. Eur J Neurosci 10:2129–2142

Araque A, Sanzgiri RP, Parpura V, Haydon PG (1998b) Calcium elevation in astrocytes causes an NMDA receptor-dependent increase in the frequency of miniature synaptic currents in cultured hippocampal neurons. J Neurosci 18:6822–6829

Araque A, Parpura V, Sanzgiri RP, Haydon PG (1999) Tripartite synapses: glia, the unacknowledged partner. Trends Neurosci 22:208–215

Attwell D, Buchan AM, Charpak S, Lauritzen M, Macvicar BA, Newman EA (2010) Glial and neuronal control of brain blood flow. Nature 468:232–243

Ballanger B, Lozano AM, Moro E, van Eimeren T, Hamani C, Chen R, Cilia R, Houle S, Poon YY, Lang AE, Strafella AP (2009) Cerebral blood flow changes induced by pedunculopontine nucleus stimulation in patients with advanced Parkinson's disease: a [(15)O] H20 PET study. Hum Brain Mapp 30(12):3901–3909

Bar-Gad S, Elias S, Vaadia E, Bergman H (2004) Complex locking rather than complete cessation of neuronal activity in the globus pallidus of a 1-methyl-4-phenyl-1, 2, 3, 6-tetrahydropyridine-treated primate in response to pallidal microstimulation. J Neurosci 24:7410–7419

Basnayake SD, Hyam JA, Pereira EA, Schweder PM, Brittain JS, Aziz TZ, Green AL, Paterson DJ (2011) Identifying cardiovascular neurocircuitry involved in the exercise pressor reflex in humans using functional neurosurgery. J Appl Physiol 110(4):881–891

Bejjani BP, Arnulf I, Vidailhet M, Pidoux B, Damier P, Papadopoulos S et al (2000) Irregular jerky tremor, myoclonus, and thalamus: a study using low-frequency stimulation. Mov Disord 15:919–924

Bekar L, Libionka W, Tian GF, Xu Q, Torres A, Wang X et al (2008) Adenosine is crucial for deep brain stimulation-mediated attenuation of tremor. Nat Med 14:75–80

Benabid AL, Pollak P, Louveau A, Henry S, de Rougemont J (1987) Combined (thalamotomy and stimulation) stereotactic surgery of the VIM thalamic nucleus for bilateral Parkinson disease. Appl Neurophysiol 50:344–346

Benabid AL, Pollak P, Gervason C, Hoffmann D, Gao DM, Hommel M et al (1991) Long-term suppression of tremor by chronic stimulation of the ventral intermediate thalamic nucleus. Lancet 337:403–406

Benabid AL, Pollak P, Gao D, Hoffman D, Limousin P, Gay E, Payen I, Benazzouz A (1996) Chronic electrical stimulation of the ventralis intermedius nucleus of the thalamus as a treatment of movement disorders. J Neurosurg 84:203–214

Benazzouz A, Gross C, Feger J, Boraud T, Bioulac B (1993) Reversal of rigidity and improvement in motor performance by subthalamic high-frequency stimulation in MPTP-treated monkeys. Eur J Neurosci 5:382–389

Benazzouz A, Piallat B, Pollak P, Benabid AL (1995) Responses of substantia nigra pars reticulata and globus pallidus complex to high frequency stimulation of the subthalamic nucleus in rats: electrophysiological data. Neurosci Lett 189:77–80

Benazzouz A, Gao DM, Ni ZG, Piallat B, Bouali-Benazzouz R, Benabid AL (2000a) Effect of high-frequency stimulation of the subthalamic nucleus on the neuronal activities of the substantia nigra pars reticulate and ventrolateral nucleus of the thalamus in the rat. Neuroscience 99:289–295

Benazzouz A, Gao DM, Ni ZG, Benabid AL (2000b) High frequency stimulation of the STN influences the activity of dopamine neurons in the rat. Neuroreport 11:1593–1596

Bergman H, Wichmann T, Karmon B, DeLong MR (1994) The primate subthalamic nucleus. II. Neuronal activity in the MPTP model of parkinsonism. J Neurophysiol 72:507–520

Berk C, Carr J, Sinden M, Martzke J, Honey CR (2002) Thalamic deep brain stimulation for the treatment of tremor due to multiple sclerosis: a prospective study of tremor and quality of life. J Neurosurg 97:815–820

Beurrier C, Bioulac B, Audin J, Hammond C (2001) High-frequency stimulation produces a transient blockade of voltage-gated currents in subthalamic neurons. J Neurophysiol 85: 1351–1356

Beuter A, Modolo J (2009) Delayed and lasting effects of deep brain stimulation on locomotion in Parkinson's disease. Chaos 2:2–14

Bevan MD, Magill PJ, Terman D, Bolam JP, Wilson CJ (2002) Move to the rhythm: oscillations in the subthalamic nucleus-external globus pallidus network. Trends Neurosci 25:525–531

Bindra RS, Schaffer PJ, Meng A, Woo J, Maseide K, Roth ME et al (2005) Alterations in DNA repair gene expression under hypoxia: elucidating the mechanisms of hypoxia-induced genetic instability. Ann N Y Acad Sci 1059:184–195

Birdno MJ, Cooper SE, Rezai AR, Grill WM (2007) Pulse-to-pulse changes in the frequency of deep brain stimulation affect tremor and modeled neuronal activity. J Neurophysiol 98:1675–1684

Birdno MJ, Kuncel AM, Dorval AD, Turner DA, Grill WM (2008) Tremor varies as a function of the temporal regularity of deep brain stimulation. Neuroreport 19:599–602

Biseul I, Sauleau P, Haegelen C, Trebon P, Drapier D, Raoul S et al (2005) Fear recognition is impaired by subthalamic nucleus stimulation in Parkinson's disease. Neuropsychologia 43:1054–1059

Bittar RG, Hyam J, Nandi D, Wang S, Liu X, Joint C et al (2005) Thalamotomy versus thalamic stimulation for multiple sclerosis tremor. J Clin Neurosci 12:638–642

Boulet S, Lacombe E, Carcenac C, Feuerstein C, Sgambato-Faure V, Poupard A, Savasta M (2006) Subthalamic stimulation-induced forelimb dyskinesias are linked to an increase in glutamate levels in the substantia nigra pars reticulata. J Neurosci 26(42):10768–10776

Braak H, Del Tredici K (2008) Cortico-basal ganglia-cortical circuitry in Parkinson's disease reconsidered. Exp Neurol 212:226–229

Breit S, Bouali-Benazzouz R, Benabid AL, Benazzouz A (2001) Unilateral lesion of the nigrostriatal pathway induces an increase of neuronal activity of the pedunculopontine nucleus, which is reversed by the lesion of the subthalamic nucleus in the rat. Eur J Neurosci 14(11):1833–1842

Bronte-Stewart H, Barberini C, Koop MM, Hill BC, Henderson JM, Wingeier B (2009) The STN beta-band profile in Parkinson's disease is stationary and shows prolonged attenuation after deep brain stimulation. Exp Neurol 215:20–28

Brouwer M, Larkin P, Brown-Peterson N, King C, Manning S, Denslow N (2004) Effects of hypoxia on gene and protein expression in the blue crab, Callinectes sapidus. Mar Environ Res 58:787–792

Brown P (2006) Bad oscillations in Parkinson's disease. J Neural Transm Suppl 70:27–30

Bruet N, Windels F, Bertrand A, Feuerstein C, Poupard A, Savasta M (2001) High frequency stimulation of the subthalamic nucleus increases the extracellular contents of striatal dopamine in normal and partially dopaminergic denervated rats. J Neuropathol Exp Neurol 60:15–24

Buckner RL, Snyder AZ, Shannon BJ et al (2005) Molecular, structural, and functional characterization of Alzheimer's disease: evidence for a relationship between default activity, amyloid, and memory. J Neurosci 25:7709–7917

Buckner RL, Andrews-Hanna JR, Schacter DL (2008) The brain's default network: anatomy, function, and relevance to disease [review]. Ann N Y Acad Sci 1124:1–38

Burbaud P, Gross C, Bioulac B (1994) Effect of subthalamic high frequency stimulation on substantia nigra pars reticulata and globus pallidus neurons in normal rats. J Physiol 88:359–361

Bushong EA, Martone ME, Jones YZ, Ellisman MH (2002) Protoplasmic astrocytes in CA1 stratum radiatum occupy separate anatomical domains. J Neurosci 22:183–192

Butson CR, McIntyre CC (2006) Role of electrode design on the volume of tissue activated during deep brain stimulation. J Neural Eng 3:1–8

Butson CR, Cooper SE, Henderson JM, McIntyre CC (2007) Patient-specific analysis of the volume of tissue activated during deep brain stimulation. Neuroimage 34(2):661–670

Calabresi P, Maj R, Pisani A, Mercuri NB, Bernardi G (1992) Long-term synaptic depression in the striatum: physiological and pharmacological characterization. J Neurosci 12:4224–4233

Carlson JD, Clearly DR, Cetas JS, Heinricer MM, Burchiel KJ (2010) Deep brain stimulation does not silence neurons in subthalamic nucleus in Parkinson's patients. J Neurophysiol 103:962–967

Carron R, Chabardès S, Hammond C (2012) Mechanisms of action of high-frequency deep brain stimulation. A review of the literature and current concepts. Neurochirurgie 30:382–391

Chang JY, Shi LH, Luo F, Zhang WM, Woodward DJ (2007) Studies of the neural mechanisms of deep brain stimulation in rodent models of Parkinson's disease. Neurosci Biobehav Rev 31:643–657

Chang SY, Shon YM, Agnesi F, Lee KH (2009) Microthalamotomy effect during deep brain stimulation: potential involvement of adenosine and glutamate efflux. Conf Proc IEEE Eng Med Biol Soc 2009:3294–3297

Chao D, Xia Y (2010) Ionic storm in hypoxic/ischemic stress: can opioid receptors subside it? Prog Neurobiol 90:439–470

Chao D, Donnelly DF, Feng Y, Bazzy-Asaad A, Xia Y (2007a) Cortical delta-opioid receptors potentiate K^+ homeostasis during anoxia and oxygen-glucose deprivation. J Cereb Blood Flow Metab 27:356–368

Chao D, Bazzy-Asaad A, Balboni G, Xia Y (2007b) delta-, but not mu-, opioid receptor stabilizes K(+) homeostasis by reducing Ca(2+) influx in the cortex during acute hypoxia. J Cell Physiol 212:60–67

Chao D, Bazzy-Asaad A, Balboni G, Salvadori S, Xia Y (2008) Activation of DOR attenuates anoxic K? derangement via inhibition of Na? entry in mouse cortex. Cereb Cortex 18: 2217–2227

Chao D, He X, Yang Y, Bazzy-Asaad A, Lazarus LH, Balboni G, Kim DH, Xia Y (2012) DOR activation inhibits anoxic/ischemic Na? influx through Na? channels via PKC mechanisms in the cortex. Exp Neurol 236:228–239

Charles AC, Merrill JE, Dirksen ER, Sanderson MJ (1991) Intercellular signaling in glial cells: calcium waves and oscillations in response to mechanical stimulation and glutamate. Neuron 6:983–992

Charpier S, Deniau JM (1997) In vivo activity-dependent plasticity at cortico-striatal connections: evidence for physiological long-term potentiation. Proc Natl Acad Sci U S A 94:7036–7040

Chen BS, Roche KW (2007) Regulation of NMDA receptors by phosphorylation. Neuropharmacology 53(3):362–368

Chen A, Xiong LJ, Tong Y, Mao M (2013) The neuroprotective roles of BDNF in hypoxic ischemic brain injury. Biomed Rep 1(2):167–176

Chiken S, Nambu A (2013) High-frequency pallidal stimulation disrupts information flow through the pallidum by GABAergic inhibition. J Neurosci 33:2268–2280

Chin GC, Hutchison WD (2008) Effects of cobalt and bicuculline on focal microstimulation of rat pallidal neurons in vivo. Brain Stimul 1:134–150

Cif L, El Fertit H, Vayssiere N, Hemm S, Hardouin E, Gannau A et al (2003) Treatment of dystonic syndromes by chronic electrical stimulation of the internal globus pallidus. J Neurosurg Sci 47:52–55

Coenen V, Allert N, Madler B (2011) A role of diffusion tensor imaging fiber tracking in deep brain stimulation surgery: DBS of the dentato-rubro-thalamic tract (drt) for the treatment of therapy-refractory tremor. Acta Neurochir 153:1579–1585

Cools R, Stefanova E, Barker RA, Robbins TW, Owen AM (2002) Dopaminergic modulation of high-level cognition in Parkinson's disease: the role of the prefrontal cortex revealed by PET. Brain 125(pt 3):584–594

Cornell-Bell AH, Finkbeiner SM, Cooper MS, Smith SJ (1990) Glutamate induces calcium waves in cultured astrocytes: long-range glial signaling. Science 247:470–473

Cotrina ML, Lin JHC, Lopez-Garcıa JC, Naus CCG, Nedergaard M (2000) ATP-mediated glia signaling. J Neurosci 20:2835–2844

Coubes P, Cif L, El Fertit H, Hemm S, Vayssiere N, Serrat S et al (2004) Electrical stimulation of the globus pallidus internus in patients with primary generalized dystonia: long-term results. J Neurosurg 101:189–194

Courtemanche R, Fujii N, Graybiel AM (2003) Synchronous, focally modulated beta-band oscillations characterize local field potential activity in the striatum of awake behaving monkeys. J Neurosci 23:11741–11752

D'Ascenzo M, Fellin T, Terunuma M, Revilla-Sanchez R, Meaney DF, Auberson YP, Moss SJ, Haydon PG (2007) mGluR5 stimulates gliotransmission in the nucleus accumbens. Proc Natl Acad Sci U S A 104:1995–2000

Deep-Brain Stimulation For Parkinson's Disease Study Group (2001) Deep-brain stimulation of the subthalamic nucleus or the pars interna of the globus pallidus in Parkinson's disease. N Engl J Med 345:956–963

Degos B, Deniau JM, Thierry AM, Glowinski J, Pezard L, Maurice N (2005) Neuroleptic-induced catalepsy: electrophysiological mechanisms of functional recovery induced by high-frequency stimulation of the subthalamic nucleus. J Neurosci 25:7687–7696

DeLong MR (1990) Primate models of movement disorders of basal ganglia origin. Trends Neurosci 13:281–285

DeLong MR, Wichmann T (2007) Circuits and circuit disorders of the basal ganglia. Arch Neurol 64:20–24

Deng Q, Terunuma M, Fellin T, Moss SJ, Haydon PG (2011) Astrocytic activation of A1 receptors regulates the surface expression of NMDA receptors through a Src kinase dependent pathway. Glia 59(7):1084–1093

Deniau JM, Menetrey A, Charpier S (1996) The lamellar organization of the rat substantia nigra pars reticulata: segregated patterns of striatal afferents and relationship to the topography of corticostriatal projections. Neuroscience 73:761–781

Deniau JM, Degos B, Bosch C, Maurice N (2010) Deep brain stimulation mechanisms: beyond the concept of local functional inhibition. Eur J Neurosci 32(7):1080–1091

Denys D, Mantione M, Figee M, van den Munckhof P, Koerselman F, Westenberg H (2010) Deep brain stimulation of the nucleus accumbens for treatment-refractory obsessive-compulsive disorder. Arch Gen Psychiatry 67(10):1061–1068

Deuschl G, Raethjen J, Lindemann M, Krack P (2001) The pathophysiology of tremor. Muscle Nerve 24:716–735

Devos D, Labyt E, Derambure P, Bourriez JL, Cassim F, Reyns N et al (2004) Subthalamic nucleus stimulation modulates motor cortex oscillatory activity in Parkinson's disease. Brain 127:408–419

Diederich NJ, Kalteis K, Stamenkovic M, Pieri V, Alesch F (2005) Efficient internal pallidal stimulation in Gilles de la Tourette syndrome: a case report. Mov Disord 20:1496–1499

DiLorenzo DJ, Jankovic J, Simpson RK, Takei H, Powell SZ (2010) Long-term deep brain stimulation for essential tremor: 12-year clinicopathologic follow-up. Mov Disord 25:232–238

Ding J, Peterson JD, Surmeier DJ (2008) Corticostriatal and thalamostriatal synapses have distinctive properties. J Neurosci 28:6483–6492

Dorval AD, Russo GS, Hashimoto T, Xu W, Grill WM, Vitek JL (2008) Deep brain stimulation reduces neuronal entropy in the MPTP-primate model of Parkinson's disease. J Neurophysiol 100:2807–2818

Dorval AD, Kuncel AM, Birdno MJ, Turner DA, Grill WM (2010) Deep brain stimulation alleviates parkinsonian bradykinesia by regularizing pallidal activity. J Neurophysiol 104(2):911–921

Dostrovsky JO, Lozano AM (2002) Mechanisms of deep brain stimulation. Mov Disord 17:S63–S68

Dostrovsky JO, Levy R, Wu JP, Hutchison WD, Tasker RR, Lozano AM (2000) Microstimulation-induced inhibition of neuronal firing in human globus pallidus. J Neurophysiol 84:570–574

Drevets WC, Price JL, Simpson JR Jr, Todd RD, Reich T, Vannier M, Raichle ME (1997) Subgenual prefrontal cortex abnormalities in mood disorders. Nature 386:824–827

Dujardin K, Defebvre L, Krystkowiak P, Blond S, Destée A (2001) Influence of chronic bilateral stimulation of the subthalamic nucleus on cognitive function in Parkinson's disease. J Neurol 248:603–611

Dujardin K, Blairy S, Defebvre L, Duhem S, Noël Y, Hess U et al (2004) Deficits in decoding emotional facial expressions in Parkinson's disease. Neuropsychologia 42:239–250

Dupre KB, Ostock CY, Eskow Jaunarajs KL, Button T, Savage LM, Wolf W, Bishop C (2011) Local modulation of striatal glutamate efflux by serotonin 1A receptor stimulation in dyskinetic, hemiparkinsonian rats. Exp Neurol 229(2):288–299

Eusebio A, Brown P (2009) Synchronisation in the beta frequency-band—the bad boy of parkinsonism or an innocent bystander? Exp Neurol 217:1–3

Eusebio A, Pogosyan A, Wang S, Averbeck B, Gaynor LD, Cantiniaux S et al (2009) Resonance in subthalamo-cortical circuits in Parkinson's disease. Brain 132:2139–2150

Fellin T, Pascual O, Gobbo S, Pozzan T, Haydon PG, Carmignoto G (2004) Neuronal synchrony mediated by astrocytic glutamate through activation of extrasynaptic NMDA receptors. Neuron 443:729–743

Feng XJ, Greenwald B, Rabitz H, Shea-Brown E, Kosut R (2007) Toward closed-loop optimization of deep brain stimulation for Parkinson's disease: concepts and lessons from a computational model. J Neural Eng 4:14–21

Fenoy AJ, Goetz L, Chabardès S, Xia Y (2014) Deep brain stimulation: are astrocytes a key driver behind the scene? CNS Neurosci Ther 20(3):191–201

Ferraye MU, Debu B, Fraix V, Goetz L, Ardouin C, Yelnik J et al (2010) Effects of pedunculopontine nucleus area stimulation on gait disorders in Parkinson's disease. Brain 133:205–214

Fiacco TA, McCarthy KD (2004) Intracellular astrocyte calcium waves in situ increase the frequency of spontaneous AMPA receptor currents in CA1 pyramidal neurons. J Neurosci 24:722–732

Filali M, Hutchison WD, Palter VN, Lozano AM, Dostrovsky JO (2004) Stimulation-induced inhibition of neuronal firing in human subthalamic nucleus. Exp Brain Res 156:274–281

Filosa JA, Bonev AD, Nelson MT (2004) Calcium dynamics in cortical astrocytes and arterioles during neurovascular coupling. Circ Res 95:73–81

Filosa JA, Bonev AD, Straub SV, Meredith AL, Wilkerson MK, Aldrich RW, Nelson MT (2006) Local potassium signaling couples neuronal activity to vasodilation in the brain. Nat Neurosci 9(11):1397–1403

Florian C, Vecsey CG, Halassa MM, Haydon PG, Abel T (2011) Astrocyte-derived adenosine and A1 receptor activity contribute to sleep loss-induced deficits in hippocampal synaptic plasticity and memory in mice. J Neurosci 31(19):6956–6962

Florio T, Scarnati E, Confalone G, Minchella D, Galati S, Stanzione P et al (2007) High-frequency stimulation of the subthalamic nucleus modulates the activity of pedunculopontine neurons through direct activation of excitatory fibres as well as through indirect activation of inhibitory pallidal fibres in the rat. Eur J Neurosci 25:1174–1186

Foffani G, Priori A, Egidi M, Rampini P, Tamma F, Caputo E et al (2003) 300-Hz subthalamic oscillations in Parkinson's disease. Brain 126:2153–2163

Fraix V, Pollak P, Vercueil L, Benabid AL, Mauguiere F (2008) Effects of subthalamic nucleus stimulation on motor cortex excitability in Parkinson's disease. Clin Neurophysiol 119:2513–2518

Funkiewiez A, Ardouin C, Caputo E, Krack P, Fraix V, Klinger H et al (2004) Long term effects of bilateral subthalamic nucleus stimulation on cognitive function, mood, and behaviour in Parkinson's disease. J Neurol Neurosurg Psychiatry 75:834–839

Galarraga E, Bargas J, Martinez-Fong D, Aceves J (1987) Spontaneous synaptic potentials in dopamine-denervated neostriatal neurons. Neurosci Lett 81:351–355

Galati S, Mazzone P, Fedele E, Pisani A, Peppe A, Pierantozzi M et al (2006) Biochemical and electrophysiological changes of substantia nigra pars reticulata driven by subthalamic stimulation in patients with Parkinson's disease. Eur J Neurosci 23:2923–2928

Gale JT, Amirnovin R, Williams ZM, Flaherty AW, Eskandar EN (2008) From symphony to cacophony: pathophysiology of the human basal ganglia in Parkinson disease. Neurosci Biobehav Rev 32:378–387

Garcia L, Audin J, D'Alessandro G, Bioulac B, Hammond C (2003) Dual effect of high-frequency stimulation on subthalamic neuron activity. J Neurosci 23:8743–8751

Gatev P, Wichmann T (2008) Interactions between cortical rhythms and spiking activity of single basal ganglia neurons in the normal and parkinsonian state. Cereb Cortex 19:1330–1344

Gomez-Gonzalo M, Losi G, Chiavegato A, Zonta M, Cammarota M, Brondi M et al (2010) An excitatory loop with astrocytes contributes to drive neurons to seizure threshold. PLoS Biol 8, e1000352

Gordon GRJ, Iremonger KJ, Kantevari S, Ellis-Davies GCR, MacVicar BA, Bains JS (2009) Astrocyte-mediated distributed plasticity at hypothalamic glutamate synapses. Neuron 64:391–403
Gourine AV, Kasymov V, Marina N, Tang F, Figueiredo MF, Lane S et al (2010) Astrocytes control breathing through pH-dependent release of ATP. Science 329:571–575
Grabli D, Ewenczyk C, Coelho-Braga MC et al (2009) Interruption of deep brain stimulation of the globus pallidus in primary generalized dystonia. Mov Disord 24:2363–2369
Gradinaru V, Mogri M, Thompson KR, Henderson JM, Deisseroth K (2009) Optical deconstruction of parkinsonian neural circuitry. Science 324:354–359
Grafton ST, Turner RS, Desmurget M, Bakay R, Delong M, Vitek J et al (2006) Normalizing motor-related brain activity: subthalamic nucleus stimulation in Parkinson disease. Neurology 66:1192–1199
Green AL, Wang S, Owen SL, Xie K, Bittar RG, Stein JF, Paterson DJ, Aziz TZ (2006) Stimulating the human midbrain to reveal the link between pain and blood pressure. Pain 124(3):349–359
Green AL, Wang S, Stein JF, Pereira EA, Kringelbach ML, Liu X et al (2009) Neural signatures in patients with neuropathic pain. Neurology 72:569–571
Greenberg BD, Malone DA, Friehs GM, Rezai AR, Kubu CS, Malloy PF et al (2006) Three-year outcomes in deep brain stimulation for highly resistant obsessive-compulsive disorder. Neuropsychopharmacology 31(11):2384–2393
Greenberg BD, Gabriëls LA, Malone DA Jr, Rezai AR, Friehs GM, Okun MS et al (2010) Deep brain stimulation of the ventral internal capsule/ventral striatum for obsessive-compulsive disorder: worldwide experience. Mol Psychiatry 15(1):64–79
Griffith RW, Humphrey DR (2006) Long-term gliosis around chronically implanted platinum electrodes in the Rhesus macaque motor cortex. Neurosci Lett 406:81–86
Grill WM, Snyder AN, Miocinovic S (2004) Deep brain stimulation creates an informational lesion of the stimulated nucleus. Neuroreport 15:1137–1140
Gubellini P, Picconi B, Bari M, Battista N, Calabresi P, Centonze D et al (2002) Experimental parkinsonism alters endocannabinoid degradation: implications for striatal glutamatergic transmission. J Neurosci 22:6900–6907
Gubellini P, Eusebio A, Oueslati A, Melon C, Kerkerian-Le Goff L, Salin P (2006) Chronic high-frequency stimulation of the subthalamic nucleus and L-DOPA treatment in experimental parkinsonism: effects on motor behaviour and striatal glutamate transmission. Eur J Neurosci 24:1802–1814
Gubellini P, Salin P, Kerkerian-Le GL, Baunez C (2009) Deep brain stimulation in neurological diseases and experimental models: from molecule to complex behavior. Prog Neurobiol 89:79–123
Guo Y, Rubin JE, McIntyre CC, Vitek JL, Terman D (2008) Thalamocortical relay fidelity varies across subthalamic nucleus deep brain stimulation protocols in a data-driven computational model. J Neurophysiol 99:1477–1492
Gustafson N, Gireesh-Dharmaraj E, Czubayko U, Blackwell KT, Plenz D (2006) A comparative voltage and current-clamp analysis of feedback and feedforward synaptic transmission in the striatal microcircuit in vitro. J Neurophysiol 95:737–752
Haber SN, Fudge JL, McFarland NR (2000) Striatonigrostriatal pathways in primates form an ascending spiral from the shell to the dorsolateral striatum. J Neurosci 20:2369–2382
Haberler C, Alesch F, Mazal PR, Pilz P, Jellinger K, Pinter MM et al (2000) No tissue damage by chronic deep brain stimulation in Parkinson's disease. Ann Neurol 48:372–376
Haegelen C, García-Lorenzo D, Le Jeune F, Péron J, Gibaud B, Riffaud L et al (2010) SPECT and PET analysis of subthalamic stimulation in Parkinson's disease: analysis using a manual segmentation. J Neurol 257(3):375–382
Hahn PJ, Russo GS, Hashimoto T, Miocinovic S, Xu W, McIntyre CC et al (2008) Pallidal burst activity during therapeutic deep brain stimulation. Exp Neurol 211:243–251
Halassa MM, Haydon PG (2010) Integrated brain circuits: astrocytic networks modulate neuronal activity and behavior. Annu Rev Physiol 72:335–355

Halassa MM, Fellin T, Takano H, Dong J-H, Haydon PG (2007) Synaptic islands defined by the territory of a single astrocyte. J Neurosci 27:6473–6477
Hamani C, McAndrews MP, Cohn M, Oh M, Zumsteg D et al (2008) Memory enhancement induced by hypothalamic/fornix deep brain stimulation. Ann Neurol 63:119–123
Hamani C, Moro E, Lozano AM (2011) The pedunculopontine nucleus as a target for deep brain stimulation. J Neural Transm 118(10):1461–1468
Hamilton NB, Attwell D (2010) Do astrocytes really exocytose neurotransmitters? Nat Rev Neurosci 11:227–238
Hammond C, Bergman H, Brown P (2007) Pathological synchronization in Parkinson's disease: networks, models and treatments. Trends Neurosci 30:357–364
Hammond C, Ammari R, Bioulac B, Garcia L (2008) Latest view on the mechanism of action of deep brain stimulation. Mov Disord 23:2111–2121
Harris AZ, Pettit DL (2008) Recruiting extrasynaptic NMDA receptors augments synaptic signaling. J Neurophysiol 99(2):524–533
Hashimoto T, Elder CM, Okun MS, Patrick SK, Vitek JL (2003) Stimulation of the subthalamic nucleus changes the firing pattern of pallidal neurons. J Neurosci 23:1916–1923
Hassler R, Mundinger F, Riechert T (1979) Stereotaxis in Parkinson syndrome. Springer, Berlin, pp 1–336
Hazrati LN, Parent A (1992) Differential patterns of arborization of striatal and subthalamic fibers in the two pallidal segments in primates. Brain Res 598:311–315
He XZ, Sandhu HK, Yang YL, Hua F, Belser N, Kim DH, Xia Y (2013) Neuroprotection against hypoxia/ischemia: δ-opioid receptor-mediated cellular/molecular event. Cell Mol Life Sci 70:2291–2303
Heimer G, Rivlin M, Israel Z, Bergman H (2006) Synchronizing activity of basal ganglia and pathophysiology of Parkinson's disease. J Neural Transm Suppl 70:17–20
Henneberger C, Papouin T, Oliet SHR, Rusakov DA (2010) Long-term potentiation depends on release of D-serine from astrocytes. Nature 463:232–236
Henneberger C, Bard L, Rusakov DA (2012) D-Serine: a key to synaptic plasticity? Int J Biochem Cell Biol 44(4):587–590
Heo JH, Lee KM, Paek SH, Kim MJ, Lee JY, Kim JY et al (2008) The effects of bilateral subthalamic nucleus deep brain stimulation (STN DBS) on cognition in Parkinson disease. J Neurol Sci 273(1–2):19–24
Hershey T, Revilla FJ, Wernle AR, McGee-Minnich L, Antenr JV, Videen TO et al (2003) Cortical and subcortical blood flow effects of subthalamic nucleus stimulation in PD. Neurology 61:816–821
Hilke R, Voges J, Weber T, Kracht LW, Roggendorf J, Baudrexel S et al (2008) STN-DBS activates the target area in Parkinson disease: an FDG-PET study. Neurology 71:708–713
Hilker R, Voges J, Thiel A, Ghaemi M, Kerholz K, Sturm V et al (2002) Deep brain stimulation of the subthalamic nucleus versus levodopa challenge in Parkinson's disease: measuring the on- and off-conditions with FDG-PET. J Neural Transm 109:1257–1264
Hilker R, Voges J, Weisenbach S, Kalbe E, Burghaus L, Ghaemi M et al (2003) Subthalamic nucleus stimulation restores glucose metabolism in associative and limbic cortices and in cerebellum: evidence from a FDG-PET study in advanced Parkinson's disease. J Cereb Blood Flow Metab 24:7–16
Hinterberger T, Veit R, Strehl U, Trevorrow T, Erb M, Kotchoubey B et al (2003) Brain areas activated in fMRI during self-regulation of slow cortical potentials (SCPs). Exp Brain Res 152:113–122
Hirase H, Qian L, Bartho P, Buzsaki G (2004) Calcium dynamics of cortical astrocytic networks in vivo. PLoS Biol 2, e96
Hirshler YK, Polat U, Biegon A (2010) Intracranial electrode implantation produces regional neuroinflammation and memory deficits in rats. Exp Neurol 222(42–50):254–267
Holsheimer J, Demeulemeester H, Nuttin B, de Sutter P (2000) Identification of the target neuronal elements in electrical deep brain stimulation. Eur J Neurosci 12:4573–4577

Holtzheimer PE, Kelley ME, Gross RE, Filkowski MM, Garlow SJ, Barrocas A et al (2012) Subcallosal cingulate deep brain stimulation for treatment-resistant unipolar and bipolar depression. Arch Gen Psychiatry 69(2):150–158

Hooper J, Taylor R, Pentland B, Whittle IR (2002) A prospective study of thalamic deep brain stimulation for the treatment of movement disorders in multiple sclerosis. Br J Neurosurg 16:102–109

Houeto JL, Karachi C, Mallet L, Pillon B, Yelnik J, Mesnage V et al (2005) Tourette's syndrome and deep brain stimulation. J Neurol Neurosurg Psychiatry 76:992–995

Huang YZ, Rothwell JC, Lu CS, Wang J, Chen RS (2010) Restoration of motor inhibition through an abnormal premotor-motor connection in dystonia. Mov Disord 25:689–696

Hubble JP, Busenbark KL, Wilkinson S, Penn RD, Lyons K, Koller WC (1996) Deep brain stimulation for essential tremor. Neurology 46:1150–1153

Huguenard JR, Alger BE (1986) Whole-cell voltage-clamp study of the fading of GABA activated currents in acutely dissociated hippocampal neurons. J Neurophysiol 56:1–18

Hutchison WD, Lang AE, Dostrovsky JO, Lozano A (2003) Pallidal neuronal activity: implications for models of dystonia. Ann Neurol 53:480–488

Hutchison WD, Dostrovsky JO, Walters JR, Courtemanche R, Boraud T, Goldberg J et al (2004) Neuronal oscillations in the basal ganglia and movement disorders: evidence from whole animal and human recordings. J Neurosci 24:9240–9243

Iremonger KJ, Anderson TR, Hu B, Kiss ZH (2006) Cellular mechanisms preventing sustained activation of cortex during subcortical high-frequency stimulation. J Neurophysiol 96: 613–621

Jantz JJ, Watanabe M (2013) Pallidal deep brain stimulation modulates afferent fibers, efferent fibers, and glia. J Neurosci 33:9873–9875

Jech R, Ruzicka E, Urgosık D, Serranova T, Volfova M, Novakova O et al (2006) Deep brain stimulation of the subthalamic nucleus affects resting EEG and visual evoked potentials in Parkinson's disease. Clin Neurophysiol 117:1017–1028

Jensen AL, Durand DM (2009) High frequency stimulation can block axonal conduction. Exp Neurol 220:57–70

Johnson SM, Turner SMF (2010) Protecting motor networks during perinatal ischemia: the case for delta-opioid receptors. Ann N Y Acad Sci 1198:260–270

Johnson MD, Miocinovic S, McIntyre CC, Vitek JL (2008) Mechanisms and targets of deep brain stimulation in movement disorders. Neurotherapeutics 5:294–308

Jourdain P, Bergersen LH, Bhaukaurally K, Bezzi P, Santello M, Domercq M et al (2007) Glutamate exocytosis from astrocytes controls synaptic strength. Nat Neurosci 10:331–339

Kahle KT, Simard JM, Staley KJ, Nahed BV, Jones PS, Sun D (2009) Molecular mechanisms of ischemic cerebral edema: role of electroneutral ion transport. Physiology (Bethesda) 24: 257–265

Kang J, Jiang L, Goldman SA, Nedergaard M (1998) Astrocyte-mediated potentiation of inhibitory synaptic transmission. Nat Neurosci 1(8):683–692

Kang X, Chao D, Gu Q, Ding G, Wang Y, Balboni G et al (2009) Delta-opioid receptors protect from anoxic disruption of Na homeostasis via Na channel regulation. Cell Mol Life Sci 66(21):3505–3516

Karimi M, Golchin N, Tabbal SD, Hershey T, Videen TO, Wu J et al (2008) Subthalamic nucleus stimulation-induced regional blood flow responses correlate with improvement of motor signs in Parkinson disease. Brain 131:2710–2719

Kefalopoulou Z, Paschali A, Markaki E, Ellul J, Chroni E, Vassilakos P et al (2010) Regional cerebral blood flow changes induced by deep brain stimulation in secondary dystonia. Acta Neurochir 152:1007–1014

Khader P, Schicke T, Roder B, Rosler F (2008) On the relationship between slow cortical potentials and BOLD signal changes in humans. Int J Psychophysiol 67:252–261

Kilpatrick M, Church E, Danish S, Stiefel M, Jaggi J, Halpern C et al (2010) Intracerebral microdialysis during deep brain stimulation surgery. J Neurosci Methods 190:106–111

Kim K, Lee SG, Kegelman TP, Su ZZ, Das SK, Dash R et al (2011) Role of excitatory amino acid transporter-2 (EAAT2) and glutamate in neurodegeneration: opportunities for developing novel therapeutics. J Cell Physiol 226:2484–2493

Kirsch-Darrow L, Zahodne LB, Marsiske M, Okun MS, Foote KD, Bowers D (2011) The trajectory of apathy after deep brain stimulation: from pre-surgery to 6 months post-surgery in Parkinson's disease. Parkinsonism Relat Disord 17(3):182–188

Kita H (2007) Globus pallidus external segment. Prog Brain Res 160:111–133

Kita H, Kitai ST (1987) Efferent projections of the subthalamic nucleus in the rat: light and electron microscopic analysis with the PHA-L method. J Comp Neurol 260:435–452

Kita H, Tachibana Y, Nambu A, Chiken S (2005) Balance of monosynaptic excitatory and disynaptic inhibitory responses of the globus pallidus induced after stimulation of the subthalamic nucleus in the monkey. J Neurosci 25:8611–8619

Kitagawa H, Hayashi T, Mitsumoto Y, Koga N, Itoyama Y, Abe K (1998) Reduction of ischemic brain injury by topical application of glial cell line-derived neurotrophic factor after permanent middle cerebral artery occlusion in rats. Stroke 29:1417–1422

Kofuji P, Newman EA (2004) Potassium buffering in the central nervous system. Neuroscience 129:1045–1056

Koller WC, Lyons KE, Wilkinson SB, Troster AI, Pahwa R (2001) Long-term safety and efficacy of unilateral deep brain stimulation of the thalamus in essential tremor. Mov Disord 16:464–468

Kozlov AS, Angulo MC, Audinat E, Charpak S (2006) Target cell-specific modulation of neuronal activity by astrocytes. Proc Natl Acad Sci U S A 103:10058–10063

Krack P, Batir A, Van Blercom N, Chabardes S, Fraix V, Ardouin C et al (2003) Five-year follow-up of bilateral stimulation of the subthalamic nucleus in advanced Parkinson's disease. N Engl J Med 349:1925–1934

Krauss JK, Yianni J, Loher TJ, Aziz TZ (2004) Deep brain stimulation for dystonia. J Clin Neurophysiol 21:18–30

Kreitzer AC, Malenka RC (2008) Striatal plasticity and basal ganglia circuit function. Neuron 60(4):543–554

Kringelbach ML (2005) The orbitofrontal cortex: linking reward to hedonic experience. Nat Rev Neurosci 6:691–702

Kringelbach ML, Jenkinson N, Green A, Hansen PC, Cornelissen PL, Holliday IE et al (2006) Deep brain stimulation and chronic pain mapped with MEG. Soc Neurosci 782:1

Kringelbach ML, Jenkinson N, Green A, Owen SLF, Hansen PC, Cornelissen PL et al (2007a) Deep brain stimulation for chronic pain investigated with magnetoencephalography. Neuroreport 18:223–228

Kringelbach ML, Jenkinson N, Owen SL, Aziz TZ (2007b) Translational principles of deep brain stimulation. Nat Rev Neurosci 8:623–635

Kringelbach ML, Green AL, Owen SL, Schweder PM, Aziz TZ (2010) Sing the mind electric—principles of deep brain stimulation. Eur J Neurosci 32(7):1070–1079

Kuchibhotla KV, Lattarulo CR, Hyman BT, Bacskai BJ (2009) Synchronous hyperactivity and intercellular calcium waves in astrocytes in Alzheimer mice. Science 323:1211–1215

Kuga N, Sasaki T, Takahara Y, Matsuki N, Ikegaya Y (2011) Large-scale calcium waves traveling through astrocytic networks in vivo. J Neurosci 31(7):2607–2614

Kuhn AA, Kempf F, Brucke C, Gaynor Doyle L, Martinez-Torres I, Pogosyan A et al (2008) High-frequency stimulation of the subthalamic nucleus suppresses oscillatory beta activity in patients with Parkinson's disease in parallel with improvement in motor performance. J Neurosci 28:6165–6173

Kuncel AM, Birdno MJ, Swan BD, Grill WM (2012) Tremor reduction and modeled neural activity during cycling thalamic deep brain stimulation. Clin Neurophysiol 123(5):1044–1052

Kuriakose R, Saha U, Castillo G, Udupa K, Ni Z, Gunraj C et al (2009) The nature and time course of cortical activation following subthalamic stimulation in Parkinson's disease. Cereb Cortex 20:1926–1936

Lacombe E, Carcenac C, Boulet S, Feuerstein C, Bertrand A, Poupard A et al (2007) High-frequency stimulation of the subthalamic nucleus prolongs the increase in striatal dopamine induced by acute l-3,4-dihydroxyphenylalanine in dopaminergic denervated rats. Eur J Neurosci 26:1670–1680

Laxton AW, Tang-Wai DF, McAndrews MP, Zumsteg D, Wennberg R, Keren R et al (2010) A phase I trial of deep brain stimulation of memory circuits in Alzheimer's disease. Ann Neurol 68(4):521–534

Le Jeune F, Péron J, Biseul I, Fournier S, Sauleau P, Drapier S et al (2008) Subthalamic nucleus stimulation affects orbitofrontal cortex in facial emotion recognition: a pet study. Brain 131:1599–1608

Le Jeune F, Drapier D, Bourguignon A, Péron J, Mesbah H, Drapier S et al (2009) Subthalamic nucleus stimulation in Parkinson disease induces apathy: a PET study. Neurology 73:1746–1751

Le Jeune F, Péron J, Grandjean D, Drapier S, Haegelen C, Garin E et al (2010) Subthalamic nucleus stimulation affects limbic and associative circuits: a PET study. Eur J Nucl Med Mol Imaging 37(8):1512–1520

Leblois A, Boraud T, Meissner W, Bergman H, Hansel D (2006) Competition between feedback loops underlies normal and pathological dynamics in the basal ganglia. J Neurosci 26:3567–3583

Leblois A, Reese R, Labarre D, Hamann M, Richter A, Boraud T et al (2010) Deep brain stimulation changes basal ganglia output nuclei firing pattern in the dystonic hamster. Neurobiol Dis 38(2):288–298

Lee KH, Blaha CD, Harris BT, Cooper S, Hitti FL, Leiter JC, Roberts DW, Kim U (2006) Dopamine efflux in the rat striatum evoked by electrical stimulation of the subthalamic nucleus: potential mechanism of action in Parkinson's disease. Eur J Neurosci 23:1005–1014

Lee KH, Kristic K, van Hoff R, Hitti FL, Blaha C, Harris B et al (2007) High-frequency stimulation of the subthalamic nucleus increases glutamate in the subthalamic nucleus of rats as demonstrated by in vivo enzyme-linked glutamate sensor. Brain Res 1162:121–129

Lehmkuhle MJ, Bhangoo SS, Kipke DR (2009) The electrocorticogram signal can be modulated with deep brain stimulation of the subthalamic nucleus in the hemiparkinsonian rat. J Neurophysiol 102:1811–1820

Li S, Arbuthnott GW, Jutras MJ, Goldberg JA, Jaeger D (2007) Resonant antidromic cortical circuit activation as a consequence of high-frequency subthalamic deep-brain stimulation. J Neurophysiol 98:3525–3537

Liang L, DeLong MR, Papa SM (2009) Inversion of dopamine responses in striatal medium spiny neurons and involuntary movements. J Neurosci 28:7537–7547

Limousin P, Greene J, Pollak P, Rothwell J, Benabid AL, Frackowiak R (1997) Changes in cerebral activity pattern due to subthalamic nucleus or internal pallidum stimulation in Parkinson's disease. Ann Neurol 42(3):283–291

Lin HW, Thompson JW, Morris KC, Perez-Pinzon MA (2011) Signal transducers and activators of transcription: STATs-mediated mitochondrial neuroprotection. Antioxid Redox Signal 14:1853–1861

Liu D, Slevin JR, Lu C, Chan SL, Hansson M, Elmer E, Mattson MP (2003) Involvement of mitochondrial K^+ release and cellular efflux in ischemic and apoptotic neuronal death. J Neurochem 86:966–979

Liu L, Wong TP, Pozza MF, Lingenhoehl K, Wang Y, Sheng M et al (2004) Role of NMDA receptor subtypes in governing the direction of hippocampal synaptic plasticity. Science 304:1021–1024

Liu LD, Prescott IA, Dostrovsky JO, Hodaie M, Lozano AM, Hutchison WD (2012) Frequency-dependent effects of electrical stimulation in the globus pallidus of dystonia patients. J Neurophysiol 108(1):5–17

Llinás RR, Ribary U, Jeanmonod D, Kronberg E, Mitra PP (1999) Thalamocortical dysrhythmia: a neurological and neuropsychiatric syndrome characterized by magnetoencephalography. Proc Natl Acad Sci U S A 96:15222–15227

Lopez-Bayghen E, Ortega A (2011) Glial glutamate transporters: new actors in brain signaling. IUBMB Life 63:816–823
Lovinger DM, Tyler EC, Merritt A (1993) Short- and long-term synaptic depression in rat neostriatum. J Neurophysiol 70:1937–1949
Lozano AM, Kumar R, Gross RE, Giladi N, Hutchison WD, Dostrovsky JO et al (1997) Globus pallidus internus pallidotomy for generalized dystonia. Mov Disord 12:865–870
Lozano AM, Dostrovsky J, Chen R, Ashby P (2002) Deep brain stimulation for Parkinson's disease: disrupting the disruption. Lancet Neurol 1:225–231
Lozano AM, Mayberg HS, Giacobbe P, Hamani C, Craddock RC, Kennedy SH (2008) Subcallosal cingulate gyrus deep brain stimulation for treatment-resistant depression. Biol Psychiatry 64:461–467
Ma Y, Wichmann T (2004) Disruption of motor performance by basal ganglia stimulation. Soc Neurosci Abstract 416.2, San Diego, CA
Ma MC, Qian H, Ghassemi F, Zhao P, Xia Y (2005) Oxygen-sensitive {delta}-opioid receptor-regulated survival and death signals: novel insights into neuronal preconditioning and protection. J Biol Chem 280:16208–16218
Magarinos-Ascone C, Pazo JH, Macada R, Buno W (2002) High frequency stimulation of the subthalamic nucleus silences subthalamic neurons: a possible cellular mechanism in Parkinson's disease. Neuroscience 115:1109–1117
Mallet L, Polosan M, Jaafari N, Baup N, Welter ML, Fontaine D et al (2008a) Subthalamic nucleus stimulation in severe obsessive-compulsive disorder. N Engl J Med 359(20):2121–2134
Mallet N, Pogosyan A, Sharott A, Csicsvari J, Bolam JP, Brown P et al (2008b) Disrupted dopamine transmission and the emergence of exaggerated beta oscillations in subthalamic nucleus and cerebral cortex. J Neurosci 28:4795–4806
Malone DA Jr, Dougherty DD, Rezai AR, Carpenter LL, Friehs GM, Esander EN et al (2009) Deep brain stimulation of the ventral capsule/ventral striatum for treatment-resistant depression. Biol Psychiatry 65:267–275
Maltete D, Jodoin N, Karchi C, Houeto JL, Navarro S, Cornu P et al (2007) Subthalamic stimulation and neuronal activity in the substantia nigra in Parkinson's disease. J Neurophysiol 97:4017–4022
Mann JM, Foote KD, Garvan CW, Fernandez HH, Jacobson CE, Rodriguez RL et al (2009) Brain penetration effects of microelectrodes and DBS leads in STN or GPi. J Neurol Neurosurg Psychiatry 80:794–797
Marti M, Sbrenna S, Fuxe K, Bianchi C, Beani L, Morari M (1999) In vitro evidence for increased facilitation of striatal acetylcholine release via pre- and postsynaptic NMDA receptors in hemiparkinsonian rats. J Neurochem 72:875–878
Matsui K, Jahr CE (2004) Differential control of synaptic and ectopic vesicular release of glutamate. J Neurosci 24:8932–8939
Matsui K, Jahr CE, Rubio ME (2005) High-concentration rapid transients of glutamate mediate neural-glial communication via ectopic release. J Neurosci 25:7538–7547
Maurice N, Deniau JM, Glowinski J, Thierry AM (1998a) Relationships between the prefrontal cortex and the basal ganglia in the rat: physiology of the corticosubthalamic circuits. J Neurosci 18:9539–9546
Maurice N, Deniau JM, Menetrey A, Glowinski J, Thierry AM (1998b) Prefrontal cortex-basal ganglia circuits in the rat: involvement of ventral pallidum and subthalamic nucleus. Synapse 29:363–370
Maurice N, Deniau JM, Glowinski J, Thierry AM (1999) Relationships between the prefrontal cortex and the basal ganglia in the rat: physiology of the corticonigral circuits. J Neurosci 19:4671–4681
Maurice N, Thierry AM, Glowinski J, Deniau JM (2003) Spontaneous and evoked activity of substantia nigra pars reticulata neurons during high-frequency stimulation of the subthalamic nucleus. J Neurosci 23:9929–9936
Mayberg HS, Lozano AM, Voon V, McNeely HE, Seminowicz D, Hamani C et al (2005) Deep brain stimulation for treatment-resistant depression. Neuron 45:651–660

McCairn KW, Turner RS (2009) Deep brain stimulation of the globus pallidus internus in the parkinsonian primate: local entrainment and suppression of low-frequency oscillations. J Neurophysiol 101(4):1941–1960
McCarren M, Alger BE (1985) Use-dependent depression of IPSPs in rat hippocampal pyramidal cells in vitro. J Neurophysiol 53:557–571
McIntyre CC, Hahn PJ (2010) Network perspectives on the mechanisms of deep brain stimulation. Neurobiol Dis 38:329–337
McIntyre CC, Grill WM, Sherman DL, Thakor NV (2004a) Cellular effects of deep brain stimulation: model-based analysis of activation and inhibition. J Neurophysiol 91:1457–1469
McIntyre CC, Mori S, Sherman DL, Thakor NV, Vitek JL (2004b) Electric field and stimulating influence generated by deep brain stimulation of the subthalamic nucleus. Clin Neurophysiol 115:589–595
McIntyre CC, Savasta M, Kerkerian-Le GL, Vitek JL (2004c) Uncovering the mechanism(s) of action of deep brain stimulation: activation, inhibition, or both. Clin Neurophysiol 115:1239–1248
McIntyre CC, Savasta M, Walter BL, Vitek JL (2004d) How does deep brain stimulation work? Present understanding and future questions. J Clin Neurophysiol 21:40–50
Meissner W, Harnack D, Paul G, Reum T, Sohr R, Morgenstern R et al (2002) Deep brain stimulation of subthalamic neurons increases striatal dopamine metabolism and induces contralateral circling in freely moving 6-hydroxydopamine-lesioned rats. Neurosci Lett 328:105–108
Meissner W, Harnack D, Reese R, Paul G, Reum T, Ansorge M et al (2003) High-frequency stimulation of the subthalamic nucleus enhances striatal dopamine release and metabolism in rats. J Neurochem 85:601–609
Meissner W, Leblois A, Hansel D, Bioulac B, Gross CE, Benazzouz A et al (2005) Subthalamic high frequency stimulation resets subthalamic firing and reduces abnormal oscillations. Brain 128(pt 10):2372–2382
Mela F, Marti M, Bido S, Cenci MA, Morari M (2012) In vivo evidence for a differential contribution of striatal and nigral D1 and D2 receptors to l-DOPA induced dyskinesia and the accompanying surge of nigral amino acid levels. Neurobiol Dis 45(1):573–582
Melega WP, Lacan G, Gorgulho AA, Behnke EJ, De Salles AA (2012) Hypothalamic deep brain stimulation reduces weight gain in an obesity-animal model. PLoS One 7(1), e30672
Mena MA, Garcia de Yebenes J (2008) Glial cells as players in parkinsonism: the "good", the "bad", and the "mysterious" glia. Neuroscientist 14:544–560
Metea MR, Newman EA (2006) Glial cells dilate and constrict blood vessels: a mechanism of neurovascular coupling. J Neurosci 26:2862–2870
Mink JW (1996) The basal ganglia: focused selection and inhibition of competing motor programs. Prog Neurobiol 50:381–425
Modolo J, Legros A, Thomas AW, Beuter A (2011) Model-driven therapeutic treatment of neurological disorders: reshaping brain rhythms with neuromodulation. Interface Focus 1(1):61–74
Montgomery EB Jr (2006) Effects of GPi stimulation on human thalamic neuronal activity. Clin Neurophysiol 117:2691–2702
Montgomery EB Jr (2007) Basal ganglia physiology and pathophysiology: a reappraisal. Parkinsonism Relat Disord 13:455–465
Montgomery EB Jr, Baker KB (2000) Mechanisms of deep brain stimulation and future technical developments. Neurol Res 22:259–266
Montgomery EB Jr, Gale JT (2008) Mechanisms of action of deep brain stimulation (DBS). Neurosci Biobehav Rev 32:388–407
Montgomery EB Jr, Hilker R (2009) STN-DBS activates the target area in Parkinson disease. Neurology 73:327
Moran A, Stein E, Tischler H, Belelovsky K, Bar-Gad I (2011) Dynamic stereotypic responses of Basal Ganglia neurons to subthalamic nucleus high-frequency stimulation in the parkinsonian primate. Front Syst Neurosci 5:21
Morel A, Magnin M, Jeanmonod D (1997) Multiarchitectonic and stereotactic atlas of the human thalamus. J Comp Neurol 387:588–630

Morishita T, Foote KD, Wu SS, Jacobson CE, Rodriguez RL, Haq IU et al (2010) Brain penetration effects of microelectrodes and deep brain stimulation leads in ventral intermediate nucleus stimulation for essential tremor. J Neurosurg 112:491–496

Moro E, Hamani C, Poon YY, Al-Khairallah T, Dostrovsky JO, Hutchison WD et al (2010) Unilateral pedunculopontine stimulation improves falls in Parkinson's disease. Brain 133:215–224

Mothet JP, Parent AT, Wolosker H, Brady RO, Linden DJ, Ferris CD et al (2000) D-serine is an endogenous ligand for the glycine site of the N-methyl-D-aspartate receptor. Proc Natl Acad Sci U S A 97:4926–4931

Mulligan SJ, MacVicar BA (2004) Calcium transients in astrocyte endfeet cause cerebrovascular constrictions. Nature 431:195–199

Murugan M, Ling EA, Kaur C (2013) Dysregulated glutamate uptake by astrocytes causes oligodendroglia death in hypoxic perventricular white matter damage. Mol Cell Neurosci 56C: 342–354

Nagai Y, Critchley HD, Featherstone E, Fenwick PB, Trimble MR, Dolan RJ (2004) Brain activity relating to the contingent negative variation: an fMRI investigation. Neuroimage 10:1232–1241

Nandhu MS, Naijil G, Smijin S, Jayanarayanan S, Paulose CS (2010) Opioid system functional regulation in neurological disease management. J Neurosci Res 88:3215–3221

Nandi D, Liu X, Winter JL, Aziz TZ, Stein JF (2002) Deep brain stimulation of the pedunculopontine region in the normal nonhuman primate. J Clin Neurosci 9:170–174

Nandi D, Jenkinson N, Stein J, Aziz T (2008) The pedunculopontine nucleus in Parkinson's disease: primate studies. Br J Neurosurg 22(suppl 1):S4–S8

Newman EA, Zahs KR (1998) Modulation of neuronal activity by glial cells in the retina. J Neurosci 18:4022–4028

Ngoga D, Mitchell R, Kausar J, Hodson J, Harries A, Pall H (2014) Deep brain stimulation improves survival in severe Parkinson's disease. J Neurol Neurosurg Psychiatry 85:17–22

Nong Y, Huang YQ, Ju W, Kalia LV, Ahmadian G, Wang YT et al (2003) Glycine binding primes NMDA receptor internalization. Nature 422:302–307

Novak P, Klemp JA, Ridings LW, Lyons KE, Pahwa R, Nazarro JM (2009) Effect of deep brain stimulation of the subthalamic nucleus upon the contralateral subthalamic nucleus in Parkinson disease. Neurosci Lett 463:12–16

Nowak LG, Bullier J (1998a) Axons, but not cell bodies, are activated by electrical stimulation in cortical gray matter. I. Evidence from chronaxie measurements. Exp Brain Res 118:477–488

Nowak LG, Bullier J (1998b) Axons, but not cell bodies, are activated by electrical stimulation in cortical gray matter. II. Evidence from selective inactivation of cell bodies and axon initial segments. Exp Brain Res 118:489–500

Nowinski WL, Belov D, Thirunavuukarasuu A, Benabid AL (2005) A probabilistic functional atlas of the VIM nucleus constructed from pre-, intra- and postoperative electrophysiological and neuroimaging data acquired during the surgical treatment of Parkinson's disease patients. Stereotact Funct Neurosurg 83(5–6):190–196

Ondo W, Jankovic J, Schwartz K, Almaguer M, Simpson RK (1998) Unilateral thalamic deep brain stimulation for refractory essential tremor and Parkinson's disease tremor. Neurology 51:1063–1069

Ortinski PI, Dong J, Mungenast A, Yue C, Takano H, Watson DJ et al (2010) Selective induction of astrocytic gliosis generates deficits in neuronal inhibition. Nat Neurosci 13:584–591

Ostermeier AM, Schlosser B, Schwender D, Sutor B (2000) Activation of mu and delta-opioid receptors causes presynaptic inhibition of glutamatergic excitation in neocortical neurons. Anesthesiology 93:1053–1063

Panatier A, Theodosis DT, Mothet J-P, Touquet B, Pollegioni L, Poulain DA et al (2006) Glia-derived D-serine controls NMDA receptor activity and synaptic memory. Cell 125:775–784

Panatier A, Vallée J, Haber M, Murai KK, Lacaille JC, Robitaille R (2011) Astrocytes are endogenous regulators of basal transmission at central synapses. Cell 146(5):785–798

Parent A, Hazrati LN (1995a) Functional anatomy of the basal ganglia. I. The corticobasal ganglia-thalamo-cortical loop. Brain Res Rev 20:91–127
Parent A, Hazrati LN (1995b) Functional anatomy of the basal ganglia. II. The place of subthalamic nucleus and external pallidum in basal ganglia circuitry. Brain Res Rev 20:128–154
Pascual O, Casper KB, Kubera C, Zhang J, Revilla-Sanchez R, Sul J-Y et al (2005) Astrocytic purinergic signaling coordinates synaptic networks. Science 310:113–116
Payoux P, Remy P, Damier P, Miloudi M, Loubinoux I, Pidoux B et al (2004) Subthalamic nucleus stimulation reduces abnormal motor cortical overactivity in Parkinson disease. Arch Neurol 61:1307–1313
Pekny M, Nilsson M (2005) Astrocyte activation and reactive gliosis. Glia 50:427–434
Peng P-HH, Huang H-SS, Lee Y-JJ, Chen Y-SS, Ma M-C (2009) Novel role for the delta-opioid receptor in hypoxic pre-conditioning in rat retinas. J Neurochem 108:741–754
Percheron G, Francois C, Talbi B, Yelnik J, Fenelon G (1996) The primate motor thalamus. Brain Res Brain Res Rev 22:93–181
Perea G, Araque A (2007) Astrocytes potentiate transmitter release at single hippocampal synapses. Science 317:1083–1086
Pereira EA, Lu G, Wang S, Schweder PM, Hyam JA, Stein JF et al (2010) Ventral periaqueductal grey stimulation alters heart rate variability in humans with chronic pain. Exp Neurol 223(2):574–581
Pereira EA, Green AL, Stacey RJ, Aziz TZ (2012) Refractory epilepsy and deep brain stimulation. J Clin Neurosci 19(1):27–33
Perlmutter JS, Mink JW (2006) Deep brain stimulation. Annu Rev Neurosci 29:229–257
Perlmutter JS, Mink JW, Bastian AJ, Zackowski K, Hershey T, Miyawaki E et al (2002) Blood flow responses to deep brain stimulation of thalamus. Neurology 58:1388–1394
Person AL, Perkel DJ (2005) Unitary IPSPs drive precise thalamic spiking in a circuit required for learning. Neuron 46:129–140
Pessiglione M, Guehl D, Rolland AS, Francois C, Hirsch EC, Feger J et al (2005) Thalamic neuronal activity in dopamine-depleted primates: evidence for a loss of functional segregation within basal ganglia circuits. J Neurosci 25:1523–1531
Piallat B, Chabardès S, Torres N, Fraix V, Goetz L, Seigneuret E et al (2009) Gait is associated with an increase in tonic firing of the sub-cuneiform nucleus neurons. Neuroscience 158(4):1201–1205
Picconi B, Piccoli G, Calabresi P (2012) Synaptic dysfunction in Parkinson's disease. Adv Exp Med Biol 970:553–572
Pilitsis JG, Metman LV, Toleikis JR, Hughes LE, Sani SB, Bakay RA (2008) Factors involved in long-term efficacy of deep brain stimulation of the thalamus for essential tremor. J Neurosurg 109:640–646
Plaha P, Patel NK, Gill SS (2004) Stimulation of the subthalamic region for essential tremor. J Neurosurg 101:48–54
Plaha P, Khan S, Gill SS (2008) Bilateral stimulation of the caudal zona incerta nucleus for tremor control. J Neurol Neurosurg Psychiatry 79:504–513
Potter-Nerger M, Ilic TV, Siebner HR, Deuschl G, Volkmann J (2008) Subthalamic nucleus stimulation restores corticospinal facilitation in Parkinson's disease. Mov Disord 23:2210–2215
Quartarone A, Rizzo V, Terranova C et al (2009) Abnormal sensorimotor plasticity in organic but not in psychogenic dystonia. Brain 132:2871–2877
Quintana A, Melon C, Kerkerian-Le Goff L, Salin P, Savasta M, Sgambato-Faure V (2010) Forelimb dyskinesia mediated by high-frequency stimulation of the subthalamic nucleus is linked to rapid activation of the NR2B subunit of N-methyl-D-aspartate receptors. Eur J Neurosci 32(3):423–434
Ranck JB Jr (1975) Which elements are excited in electrical stimulation of mammalian central nervous system: a review. Brain Res 98:417–440
Rauch SL, Dougherty DD, Malone D, Rezai A, Friehs G, Fischman AJ et al (2006) A functional neuroimaging investigation of deep brain stimulation in patients with obsessive-compulsive disorder. J Neurosurg 104(4):558–565

Ray NJ, Kringelbach ML, Jenkinson N, Owen SLF, Davies P, Wang S et al (2007) Using magnetoencephalography to investigate deep brain stimulation for cluster headache. Biomed Imaging Interv J 3, e25

Ray NJ, Jenkinson N, Wang S, Holland P, Brittain JS, Joint C et al (2008) Local field potential beta activity in the subthalamic nucleus of patients with Parkinson's disease is associated with improvements in bradykinesia after dopamine and deep brain stimulation. Exp Neurol 213:108–113

Reese R, Stiegerwald F, Potter M, Herzog J, Deuschl G, Volkmann J et al (2008) High-frequency stimulation of the subthalamic nucleus increases pallidal neuronal firing rate in a patient with Parkinson's disease. Mov Disord 23:1945–1947

Retamal MA, Froger N, Palacios-Prado N, Ezan P, Saez PJ, Saez JC, Giaume C et al (2007) Cx43 hemichannels and gap junction channels in astrocytes are regulated oppositely by proinflammatory cytokines released from activated microglia. J Neurosci 27:13781–13792

Ridding MC, Sheean G, Rothwell JC, Inzelberg R, Kujirai T et al (1995) Changes in the balance between motor cortical excitation and inhibition in focal, task specific dystonia. J Neurol Neurosurg Psychiatry 59:493–498

Rodriguez-Oroz MC, Obeso JA, Lang AE, Houeto JL, Pollak P, Rehncrona S et al (2005) Bilateral deep brain stimulation in Parkinson's disease: a multicentre study with 4 years follow-up. Brain 128:2240–2249

Rosin B, Slovik M, Mitelman R, Rivlin-Etzion M, Haber SN, Israel Z et al (2011) Closed-loop deep brain stimulation is superior in ameliorating parkinsonism. Neuron 72:370–384

Rouach N, Koulakoff A, Abudara V, Willecke K, Giaume C (2008) Astroglial metabolic networks sustain hippocampal synaptic transmission. Science 322:1551–1555

Rubchinsky LL, Kopell N, Sigvardt KA (2003) Modeling facilitation and inhibition of competing motor programs in basal ganglia subthalamic nucleus-pallidal circuits. Proc Natl Acad Sci U S A 100:14427–14432

Rubin JE, Terman D (2004) High frequency stimulation of the subthalamic nucleus eliminates pathological thalamic rhythmicity in a computational model. J Comput Neurosci 16:211–235

Ruge D, Tisch S, Limousin P, Hariz MI, Bhatia KP, Quinn N et al (2009) Longitudinal effects of deep brain stimulation in the globus pallidus on intracortical GABAergic inhibition and LTP-like plasticity in dystonia. Mov Disord 24:105–106

Ruge D, Tisch S, Hariz MI, Zrinzo L, Bhatia KP, Quinn NP et al (2011a) Deep brain stimulation effects in dystonia: time course of electrophysiological changes in early treatment. Mov Disord 26(10):1913–1921

Ruge D, Cif L, Limousin P, Gonzalez V, Vasques X, Hariz MI et al (2011b) Shaping reversibility? Long-term deep brain stimulation in dystonia: the relationship between effects on electrophysiology and clinical symptoms. Brain 134(pt 7):2106–2115

Sasaki T, Matsuki N, Ikegaya Y (2007) Metastability of active CA3 networks. J Neurosci 27:517–528

Sasaki T, Kuga N, Namiki S, Matsuki N, Ikegaya Y (2011a) Locally synchronized astrocytes. Cereb Cortex 21(8):1889–1900

Sasaki T, Matsuki N, Ikegaya Y (2011b) Action-potential modulation during axonal conduction. Science 331(6017):599–601

Schell MJ, Molliver ME, Snyder SH (1995) D-serine, an endogenous synaptic modulator—localization to astrocytes and glutamate-stimulated release. Proc Natl Acad Sci U S A 92: 3948–3952

Schell MJ, Brady RO, Molliver ME, Snyder SH (1997) D-Serine as a neuromodulator: regional and developmental localizations in rat brain glia resemble NMDA receptors. J Neurosci 17:1604–1615

Schnitzler A, Gross J (2005) Normal and pathological oscillatory communication in the brain. Nat Rev Neurosci 6:285–296

Schroeder U, Kuehler A, Haslinger B, Erhard P, Fogel W, Tronnier VM et al (2002) Subthalamic nucleus stimulation affects striato-anterior cingulate cortex circuit in a response conflict task: a PET study. Brain 125:1995–2004

Schroeder U, Kuehler A, Lange KW, Haslinger B, Tronnier VM, Krause M et al (2003) Subthalamic nucleus stimulation affects a frontotemporal network: a PET study. Ann Neurol 54:445–450
Schroeder U, Kuehler A, Hennenlotter A, Haslinger B, Tronnier VM, Krause M et al (2004) Facial expression recognition and subthalamic nucleus stimulation. J Neurol Neurosurg Psychiatry 75:648–650
Schulder M, Sernas T, Mahalick D, Adler R, Cook S (1999) Thalamic stimulation in patients with multiple sclerosis. Stereotact Funct Neurosurg 72:196–201
Schweder PM, Joint C, Hansen PC, Green AL, Quaghebeur G, Aziz TZ (2010) Chronic pedunculopontine nucleus stimulation restores functional connectivity. Neuroreport 21(17):1065–1068
Schwingenschuh P, Ruge D, Edwards MJ, Terranova C, Katschnig P, Carrillo F et al (2010) Distinguishing SWEDDs patients with asymmetric resting tremor from Parkinson's disease: a clinical and electrophysiological study. Mov Disord 25:560–569
Seeley WW, Crawford RK, Zhou J, Miller BL, Greicius MD (2009) Neurodegenerative diseases target large-scale human brain networks. Neuron 62:42–52
Seifert G, Steinhauser C (2013) Neuron-astrocyte signaling and epilepsy. Exp Neurol 244:4–10
Serrano A, Haddjeri N, Lacaille J-C, Robitaille R (2006) GABAergic network activation of glial cells underlies hippocampal heterosynaptic depression. J Neurosci 26:5370–5382
Sgambato-Faure V, Cenci MA (2012) Glutamatergic mechanisms in the dyskinesias induced by pharmacological dopamine replacement and deep brain stimulation for the treatment of Parkinson's disease. Prog Neurobiol 96:69–86
Shi LH, Luo F, Woodward DJ, Chang JY (2006) Basal ganglia neural responses during behaviorally effective deep brain stimulation of the subthalamic nucleus in rats performing a treadmill locomotion test. Synapse 59:445–457
Shils JL, Mei LZ, Arle JE (2008) Modeling parkinsonian circuitry and the DBS electrode. II. Evaluation of a computer simulation model of the basal ganglia with and without subthalamic nucleus stimulation. Stereotact Funct Neurosurg 86:16–29
Shon YM, Lee KH, Goerss SJ, Kim IY, Kimble C, Van Gompel JJ et al (2010) High frequency stimulation of the subthalamic nucleus evokes striatal dopamine release in a large animal model of human DBS neurosurgery. Neurosci Lett 475:136–140
Sillay KA, Sani S, Starr PA (2010) Deep brain stimulation for medically intractable cluster headache. Neurobiol Dis 38(3):361–368
Smith Y, Wichmann T, DeLong MR (1994) Synaptic innervation of neurones in the internal pallidal segment by the subthalamic nucleus and the external pallidum in monkeys. J Comp Neurol 343:297–318
So RQ, Kent AR, Grill WM (2012) Relative contributions of local cell and passing fiber activation and silencing to changes in thalamic fidelity during deep brain stimulation and lesioning: a computational modeling study. J Comput Neurosci 32(3):499–519
Sommer M, Ruge D, Tergau F, Beuche W, Altenmuller E, Paulus W (2002) Intracortical excitability in the hand motor representation in hand dystonia and blepharospasm. Mov Disord 17:1017–1025
Sperling RA, Dickerson BC, Pihlajamaki M, Vannini P, LaViolette PS, Vitolo OV et al (2010) Functional alterations in memory networks in early Alzheimer's disease. Neuromolecular Med 12:27–43
Spieles-Engemann AL, Behbehani MM, Collier TJ, Wohlgenant SL, Steece-Collier K, Paumier K et al (2010) Stimulation of the rat subthalamic nucleus is neuroprotective following significant nigral dopamine neuron loss. Neurobiol Dis 39:105–115
Starr PA, Rau GM, Davis V, Marks WJ Jr, Ostrem JL, Simmons D et al (2005) Spontaneous pallidal neuronal activity in human dystonia: comparison with Parkinson's disease and normal macaque. J Neurophysiol 93:3165–3176
Stefan K, Kunesch E, Cohen LG, Benecke R, Classen J (2000) Induction of plasticity in the human motor cortex by paired associative stimulation. Brain 123(pt 3):572–584
Stefani A, Lozano AM, Peppe A, Stanzione P, Galati S, Tropepi D et al (2007) Bilateral deep brain stimulation of the pedunculopontine and subthalamic nuclei in severe Parkinson's disease. Brain 130:1596–1607

Steinhäuser C, Seifert G, Bedner P (2012) Astrocyte dysfunction in temporal lobe epilepsy: K(+) channels and gap junction coupling. Glia 60(8):1192–1202

Stevens ER, Esguerra M, Kim PM, Newman EA, Snyder SH, Zahs KR et al (2003) D-serine and serine racemase are present in the vertebrate retina and contribute to the physiological activation of NMDA receptors. Proc Natl Acad Sci U S A 100:6789–6794

Stobart JL, Anderson CM (2013) Multifunctional role of astrocytes as gatekeeps of neuronal energy supply. Front Cell Neurosci 7:38

Stock G, Sturm V, Schmitt HP, Schlor KH (1979) The influence of chronic deep brain stimulation on excitability and morphology of the stimulated tissue. Acta Neurochir 47:123–129

Storey KB (2004) Adventures in oxygen metabolism. Comp Biochem Physiol B Biochem Mol Biol 139:359–369

Sugimoto T, Hattori T, Mizuno N, Itoh K, Sato M (1983) Direct projections from the centre median-parafascicular complex to the subthalamic nucleus in the cat and rat. J Comp Neurol 214:209–216

Sun DA, Yu H, Spooner J, Tatsas AD, Davis T, Abel TW et al (2008) Postmortem analysis following 71 months of deep brain stimulation of the subthalamic nucleus for Parkinson disease. J Neurosurg 109:325–329

Tai CH, Boraud T, Bezard E, Bioulac B, Gross C, Benazzouz A (2003) Electrophysiological and metabolic evidence that high-frequency stimulation of the subthalamic nucleus bridles neuronal activity in the subthalamic nucleus and the substantia nigra reticulata. FASEB J 17: 1820–1830

Takano T, Tian GF, Peng W, Lou N, Libionka W, Han X et al (2006) Astrocyte-mediated control of cerebral blood flow. Nat Neurosci 9:260–267

Tanahashi S, Ueda Y, Nakajima A, Yamamura S, Nagase H, Okada M (2012) Novel δ1-receptor agonist KNT-127 increases the release of dopamine and L-glutamate in the striatum, nucleus accumbens and median pre-frontal cortex. Neuropharmacology 62:2057–2067

Tanaka E, North RA (1994) Opioid actions on rat anterior cingulate cortex neurons in vitro. J Neurosci 14:1106–1113

Tang JK, Moro E, Mahant N, Hutchison WD, Lang AE, Lozano AM et al (2007) Neuronal firing rates and patterns in the globus pallidus internus of patients with cervical dystonia differ from those with Parkinson's disease. J Neurophysiol 98:720–729

Tawfik VL, Chang SY, Hitti FL, Roberts DW, Leiter JC, Jovanovic S et al (2010) Deep brain stimulation results in local glutamate and adenosine release: investigation into the role of astrocytes. Neurosurgery 67:367–375

Temel Y, Visser-Vandewalle V (2004) Surgery in Tourette syndrome. Mov Disord 19:3–14

Tepper JM, Koos T, Wilson CJ (2004) GABAergic microcircuits in the neostriatum. Trends Neurosci 27:662–669

Testa CM, Standaert DG, Young AB, Penney JB Jr (1994) Metabotropic glutamate receptor mRNA expression in the basal ganglia of the rat. J Neurosci 14(5 pt 2):3005–3018

Thevathasan W, Coyne TJ, Hyam JA, Kerr G, Jenkinson N, Aziz TZ et al (2011) Pedunculopontine nucleus stimulation improves gait freezing in Parkinson disease. Neurosurgery 69(6): 1248–1253

Thobois S, Ardouin C, Lhommée E, Klinger H, Lagrange C, Xie J et al (2010) Non-motor dopamine withdrawal syndrome after surgery for Parkinson's disease: predictors and underlying mesolimbic denervation. Brain 133(pt 4):1111–1127

Thompson SM, Gahwiler BH (1989) Activity-dependent disinhibition. I. Repetitive stimulation reduces IPSP driving force and conductance in the hippocampus in vitro. J Neurophysiol 61:501–511

Tian GF, Azmi H, Takano T, Xu Q, Peng W, Lin J et al (2005) An astrocytic basis of epilepsy. Nat Med 11:973–981

Tian X, Guo J, Zhu M, Li M, Wu G, Xia Y (2013a) d-Opioid receptor activation rescues the functional TrkB receptor and protects the brain from ischemia-reperfusion injury in the rat. PLoS One 8:e69252. doi:10.1371/journal.pone.0069252

Tian X, Hua F, Sandhu HK, Chao D, Balboni G, Salvadori S et al (2013b) Effect of d-opioid receptor activation on BDNF-TrkB vs. TNF-a in the mouse cortex exposed to prolonged hypoxia. Int J Mol Sci 14:15959–15976

Tisch S, Limousin P, Rothwell JC, Asselman P, Zrinzo L, Jahanshahi M et al (2006a) Changes in forearm reciprocal inhibition following pallidal stimulation for dystonia. Neurology 66: 1091–1093

Tisch S, Limousin P, Rothwell JC, Asselman P, Quinn N, Jahanshahi M et al (2006b) Changes in blink reflex excitability after globus pallidus internus stimulation for dystonia. Mov Disord 21:1650–1655

Todd KJ, Darabid H, Robitaille R (2010) Perisynaptic glia discriminate patterns of motor nerve activity and influence plasticity at the neuromuscular junction. J Neurosci 30:11870–11882

Turner SMF, Johnson SM (2011) Delta-opioid receptor activation prolongs respiratory motor output during oxygen-glucose deprivation in neonatal rat spinal cord in vitro. Neuroscience 187:70–83

Urbano FJ, Leznik E, Ilinas RR (2002) Cortical activation patterns evoked by afferent axons stimuli at different frequencies: an in vitro voltage-sensitive dye imaging study. Thalamus Rel Syst 1:371–378

Vedam-Mai V, van Battum EY, Kamphuis W, Feenstra MG, Denys D, Reynolds BA et al (2012) Deep brain stimulation and the role of astrocytes. Mol Psychiatry 17(2):124–131

Ventura R, Harris KM (1999) Three-dimensional relationships between hippocampal synapses and astrocytes. J Neurosci 19:6897–6906

Vicente S, Biseul I, Péron J, Philippot P, Drapier S, Drapier D et al (2009) Subthalamic nucleus stimulation affects subjective emotional experience in Parkinson's disease patients. Neuropsychologia 47:1928–1937

Vidailhet M, Vercueil L, Houeto JL, Krystkowiak P, Benabid AL, Cornu P et al (2005) Bilateral deep-brain stimulation of the globuspallidus in primary generalized dystonia. N Engl J Med 352:459–467

Villain N, Desgranges B, Viader F, de la Sayette V, Mézenge F, Landeau B et al (2008) Relationships between hippocampal atrophy, white matter disruption, and gray matter hypometabolism in Alzheimer's disease. J Neurosci 28(24):6174–6181

Villalba RM, Smith Y (2011) Neuroglial plasticity at the striatal glutamatergic synapses in Parkinson's disease. Front Syst Neurosci 5:68–75

Vitek JL (2002) Commentary: mechanisms of deep brain stimulation: excitation or inhibition. Mov Disord 17(S3):S69–S72

Vitek JL, Chockkan V, Zhang JY, Kaneoke Y, Evatt M, DeLong MR et al (1999) Neuronal activity in the basal ganglia in patients with generalized dystonia and hemiballismus. Ann Neurol 46:22–35

Volkmann J, Herzog J, Kopper F, Deuschl G (2002) Introduction to the programming of deep brain stimulators. Mov Disord 17:S181–S187

Walker RH, Koch RJ, Moore C, Meshul CK (2009) Subthalamic nucleus stimulation and lesioning have distinct state-dependent effects upon striatal dopamine metabolism. Synapse 63: 136–146

Wang X, Lou N, Xu Q, Tian GF, Peng WG, Han X et al (2006) Astrocytic Ca^{2+} signaling evoked by sensory stimulation in vivo. Nat Neurosci 9:816–823

Wei L, Yu SP, Gottron F, Snider BJ, Zipfel GJ, Choi DW (2003) Potassium channel blockers attenuate hypoxia and ischemia-induced neuronal death in vitro and in vivo. Stroke 34:1281–1286

Weise D, Schramm A, Stefan K, Wolters A, Reiners K, Naumann M et al (2006) The two sides of associative plasticity in writer's cramp. Brain 129:2709–2721

Welter ML, Houeto JL, Bonnet AM, Bejjani PB, Mesnage V, Dormont D et al (2004) Effects of high-frequency stimulation on subthalamic neuronal activity in Parkinsonian patients. Arch Neurol 61:89–96

Wichmann T, Soares J (2006) Neuronal firing before and after burst discharges in the monkey basal ganglia is predictably patterned in the normal state and altered in parkinsonism. J Neurophysiol 95:2120–2133

Windels F, Bruet N, Poupard A, Urbain N, Chouvet G, Feuerstein C et al (2000) Effects of high frequency stimulation of subthalamic nucleus on extracellular glutamate and GABA in substantia nigra and globus pallidus in the normal rat. Eur J Neurosci 12:4141–4146

Windels F, Bruet N, Poupard A, Feuerstein C, Bertrand A, Savasta M (2003) Influence of the frequency parameter on extracellular glutamate and gamma-aminobutyric acid in substantia nigra and globus pallidus during electrical stimulation of subthalamic nucleus in rats. J Neurosci Res 72:259–267

Witcher MR, Kirov SA, Harris KM (2007) Plasticity of perisynaptic astroglia during synaptogenesis in the mature rat hippocampus. Glia 55:13–23

Wu YR, Levy R, Ashby P, Tasker RR, Dostrovsky JO (2001) Does stimulation of the GPi control dyskinesia by activating inhibitory axons? Mov Disord 16:208–216

Wyckhuys T, Staelens S, Van NB, Deleye S, Hallez H, Vonck K et al (2010) Hippocampal deep brain stimulation induces decreased rCBF in the hippocampal formation of the rat. Neuroimage 52:55–61

Xia Y, Zhang AZ, Cao XD, Tang QM, Xu XR (1987) Inhibitory effect of analogous electroacupuncture on sympathetic cardiovascular response to stimulation of hypothalamic defence area in rabbits. J Tradit Chin Med 7:211–214

Xia Y, Zhang AZ, Cao XD (1988) Increased leu-enkephalin immunoreactivity in cerebrospinal fluid during stimulation of hypothalamic defence area in rabbits. Sheng Li Xue Bao 40:379–381

Xu W, Russo GS, Hashimoto T, Zhang J, Vitek JL (2008) Subthalamic nucleus stimulation modulates thalamic neuronal activity. J Neurosci 28:11916–11924

Yamanaka T, Ishii F, Umemura A, Miyata M, Horiba M, Oka Y et al (2012) Deterioration of executive function after subtlamaic deep brain stimulation in Parkinson's disease. Clin Neurol Neurosurg 114(4):347–351

Yang Y, Ge W, Chen Y, Zhang Z, Shen W, Wu C et al (2003) Contribution of astrocytes to hippocampal long-term potentiation through release of D-serine. Proc Natl Acad Sci U S A 100:15194–15199

Yang Y, Xia X, Zhang Y, Wang Q, Li L, Luo G, Xia Y (2009) d-Opioid receptor activation attenuates oxidative injury in the ischemic rat brain. BMC Biol 7:55

Yang L, Wang H, Shah K, Karamyan VT, Abbruscato TJ (2011) Opioid receptor agonists reduce brain edema in stroke. Brain Res 1383:307–316

Yang Y, Zhi F, He X, Moore ML, Kang X, Chao D (2012) d-opioid receptor activation and microRNA expression of the rat cortex in hypoxia. PLoS One 7:e51524

Ye ZC, Wyeth MS, Baltan-Tekkok S, Ransom BR (2003) Functional hemicannels in astrocytes: a novel mechanism of glutamate release. J Neurosci 23:3588–3596

Yianni J, Bain P, Giladi N, Auca M, Gregory R, Joint C et al (2003) Globus pallidus internus deep brain stimulation for dystonic conditions: a prospective audit. Mov Disord 18:436–442

Yousif N, Bayford R, Liu X (2008) The influence of reactivity of the electrode-brain interface on the crossing electric current in therapeutic deep brain stimulation. Neuroscience 156:597–606

Zangaglia R, Pacchetti C, Pasotti C, Mancini F, Servello D, Sinforiani E et al (2009) Deep brain stimulation and cognitive functions in Parkinson's disease: a three-year controlled study. Mov Disord 24(11):1621–1628

Zhang Y, Barres BA (2010) Astrocyte heterogeneity: an underappreciated topic in neurobiology. Curr Opin Neurobiol 20:588–594

Zhang JH, Xia Y, Haddad GG (1999) Activation of d-opioid receptors protects cortical neurons from glutamate excitotoxic injury. Soc Neurosci Abstract 25:736

Zhang J, Gibney GT, Zhao P, Xia Y (2002) Neuroprotective role of delta-opioid receptors in cortical neurons. Am J Physiol Cell Physiol 282:C1225–C1234

Zhang J, Qian H, Zhao P, Hong S–S, Xia Y (2006) Rapid hypoxia preconditioning protects cortical neurons from glutamate toxicity through delta-opioid receptor. Stroke 37:1094–1099

Zheng F, Lammert K, Nixdorf-Bergweiler BE, Steigerwald F, Volkmann J, Alzheimer C (2011) Axonal failure during high frequency stimulation of rat subthalamic nucleus. J Physiol 589:2781–2793

Ziemann U, Ilic TV, Pauli C, Meintzschel F, Ruge D (2004) Learning modifies subsequent induction of long-term potentiation-like and long-term depression-like plasticity in human motor cortex. J Neurosci 24:1666–1672
Zonta M, Angulo MC, Gobbo S, Rosengarten B, Hossmann KA, Pozzan T et al (2003) Neuron-to-astrocyte signaling is central to the dynamic control of brain microcirculation. Nat Neurosci 6:43–50

Chapter 16
Acupuncture, Opioid Receptors and Na^+ Channels: A Novel Insight into Inhibition of Epileptic Hyperexcitability

Dongman Chao and Ying Xia

Abstract Epilepsy is one of the most common neurological disorders, affecting about 1 % of the population. Although the precise mechanism of its pathophysiological changes in the brain is unknown, epilepsy has been recognized as a disorder of brain excitability characterized by recurrent unprovoked seizures that result from the abnormal, excessive, and synchronous activity of clusters of nerve cells in the brain. Currently available therapies, including medical, surgical, and other strategies such as ketogenic diet and vagus nerve stimulation, are all used for the control of epileptic symptoms with their own limitations and complications. Seeking new strategies to cure this serious disorder still poses a big challenge in the field of medicine. Recent studies suggest that acupuncture may exert its antiepileptic effects by normalizing the disrupted neuronal and network excitability through several mechanisms, including lowering the overexcited neuronal activity, enhancing the inhibitory system, and attenuating the excitatory system in the brain via regulation of the interaction between δ-opioid receptors (DOR) and Na^+ channels. This chapter reviews the progress in this field and summarizes new knowledge based on our work and the work of others.

Keywords Acupuncture • Epilepsy • Excitability • Neurons • δ-Opioid receptors • Na^+ channels

This chapter is a modified version of article "From Acupuncture to Interaction between δ-Opioid Receptors and Na^+ Channels: A Potential Pathway to Inhibit Epileptic Hyperexcitability" published in Evidence Based Complement Alternative Medicine. 2013: 216016 (doi: 10.1155/2013/216016) with the permission of Hindawi Publishing Corporation for republishing the contents.

D. Chao, Ph.D.
Department of Neurosurgery, The University of Texas Medical School at Houston, Houston, TX 77030, USA
e-mail: chaodm21@yahoo.com

Y. Xia, M.D., Ph.D. (✉)
University of Texas Medical School, Houston, TX, USA
e-mail: ying.xia@uth.tmc.edu; y55738088@gmail.com

Y. Xia (ed.), *Neural Functions of the Delta-Opioid Receptor*,
DOI 10.1007/978-3-319-25495-1_16

Abbreviations

BFNIS	Benign familial neonatal-infantile seizures
DOR	δ-Opioid receptors
fMRI	Functional magnetic resonance imaging
GEFS+	Generalized epilepsy with febrile seizure plus
KOR	κ-Opioid receptor
MOR	μ-Opioid receptor
SMEI	Severe myoclonic epilepsy of infancy
TTX	Tetrodotoxin

1 Introduction

Epilepsy is one of the most common neurological disorders, affecting about 1 % of the population. Currently available therapies, including medical treatment, surgical treatment, and other treatment strategies such as ketogenic diet and vagus nerve stimulation are symptomatic with their own limitations and complications (Chao and Xia 2012). Understanding the cellular and molecular mechanisms of epilepsy and seeking new strategies to cure this disorder still pose a big challenge.

Epilepsy has been recognized as a disorder of brain excitability characterized by recurrent unprovoked seizures that result from the abnormal, excessive, and synchronous activity of clusters of nerve cells in the brain. About 40 % of epilepsies are mainly caused by genetic factors, while the other 60 % are acquired and etiological epilepsies. Irrespective of the type of epilepsy, the changes in neuronal excitability that underlie the pathogenesis of epilepsy occur in a complex process that induces abnormal activity, not only in individual neurons, but also in a population of hyperexcitable neurons in highly synchronous activities that are propagated through normal or pathological pathways.

Epileptic hyperexcitability results from multiple factors such as innate ability of neurons in the cortex and hippocampus, alterations in membrane properties, imbalance between excitatory and inhibitory transmission, alterations in existing synaptic contacts/circuits and/or the establishment of new synaptic contacts/circuits, and the inability of glial cells to maintain glutamate and K^+ homeostasis (Chao and Xia 2012). Among these factors, the most important determinant is the intrinsic electrogenic property of each neuron that depends on the function of ion channels like Na^+, K^+, and Ca^{2+} channels in the membrane (Errington et al. 2005). In particular, Na^+ channels are responsible for the initiation and propagation of action potential, and are critical determinants of intrinsic neuronal excitability.

Acupuncture is one of the oriental medical therapeutic techniques that involves insertion of fine needles into specific body areas (acupoints or meridian points) and swift manual manipulation (e.g., rotating, lifting, thrusting, retaining etc.) that results in the manifestation of the *Qi* or *De Qi* (acquisition of energy) phenomenon that refers to a mixed sensation of soreness, numbness, swelling, sinking, and heat that appears in the acupoints. *De Qi* is an important component of the therapeutic effect and may be necessary for the clinical efficacy of acupuncture (Kong et al. 2007; Mao et al. 2007; Lin 2008). In modern practice, electrical stimulation (vs. manual manipulation) of acupoints, i.e., electroacupuncture, is increasingly gaining popularity among various acupunctural modalities. Both animal and clinical studies indicate that acupuncture is effective in certain kinds of epilepsy. Compared to the conventional Western medicine and surgical treatments, significant advantages of acupuncture treatment include its safety, convenience, and minimal side effects (NIH Consensus Development panel on Acupuncture 1998; Jindal et al. 2008; Wu et al. 2008).

Acupuncture exerts its antiepileptic effects through normalization of the disrupted neuronal excitability (Chao and Xia 2012). Some of the acupuncture-induced effects involve the activation of opioid system (Chao et al. 1999; Han 2003; Wang et al. 2008; Wen et al. 2010a, b; Liang and Xia 2012). We have previously demonstrated that electroacupuncture has a neuroprotective role in the brain against ischemic injury via the δ-opioid system (Zhao et al. 2002; Tian et al. 2008). DOR expression/activation inhibits Na^+ channel activity (Kang et al. 2009; Chattopadhyay et al. 2008), and thus attenuates the disruption of Na^+-K^+ homeostasis in the cortex under normoxic/hypoxic conditions (Kang et al. 2009; Chao et al. 2007a, b, 2008, 2009, 2012). Since the electrolyte homeostasis (e.g., Na^+ Ca^{2+}, K^+, and Cl^-) gets disrupted between the intra- and extra-cellular space during epileptic seizures, and since Na^+ channel up-regulation participates in the pathological changes of several neurological disorders such as hypoxic/ischemic injury and epilepsy (Xia et al. 2000, 2003; Stafstrom 2007; Mantegazza et al. 2010), acupuncture may attenuate epileptic seizures through the DOR-mediated inhibition of Na^+ channels (Chao and Xia 2012). Research in this new field may help us in the pursuit of novel therapeutic solutions for epileptic hyperexcitability and seizures.

2 Pathological Genesis of Epileptic Brain Hyperexcitability

The most prominent feature of epilepsy is brain hyperexcitability. At least two major factors contribute to hyperexcitability in the epileptic brain. The first and also the most significant contributor is the altered intrinsic electrogenic properties of neurons (neuronal excitability), which depend on the functioning of membrane ion channels, namely Na^+, K^+, and Ca^{2+} channels. The second factor is the synaptic imbalance that involves disruption of chemical transmission from the neighboring cells within the network via ligand gated ion channels and G-protein coupled metabotropic receptors, as well as rapid electrical transmission via gap junctions (network excitability) (Errington et al. 2005). For both major factors, epilepsy results from the abnormal activity in the neuronal networks. However, hyperexcitability due to altered ion channel function contributes to the seizure-prone state (Stafstrom 2007).

Numerous studies have shown the existence of neuronal networks of epilepsy in epileptic patients with the aid of EEG-functional magnetic resonance imaging (EEG-fMRI) (Aghakhani et al. 2004; Berman et al. 2010; Moeller et al. 2010; Szaflarski et al. 2010; Zhang et al. 2010; Chao and Xia 2012). In epilepsy, synaptic input from neighboring cells within the network is disrupted, which results in a progressive increase in excitability and epileptogenesis. This imbalanced neurotransmission between excitatory and inhibitory activities is the most prominent feature. Exaggerated glutamatergic excitatory transmission, or decreased GABAergic inhibitory transmission between the inhibitory and the excitatory systems, or a combination of the two, can lead to overexcitation of the neurons. A causal association between such imbalances in neurotransmitter activity has been causatively linked to seizure activity and the development of chronic epilepsy (Chao and Xia 2012). In addition, aberrant fiber sprouting and robust synaptic reorganization, due to the neuronal injury/loss associated with brain damage under conditions such as status epilepticus, stroke, brain trauma, and developmental malformation or deafferentation often observed in the hippocampus and cortex, also instigate a recurrent excitatory circuit by forming synapses on the dendrites of neighboring neurons (e.g., granule cells in hippocampus). A small cluster of pathologically interconnected neurons in this aberrant recurrent network are capable of generating powerful hypersynchronous bursts of discharges, initiating epileptogenesis via a kindling effect and development of epileptic discharges that spread throughout the limbic system, and eventually resulting in temporal lobe epilepsy (Bragin et al. 2000; Williams et al. 2004; Morgan and Soltesz 2008; Chao and Xia 2012).

Ion channels in the neuronal membrane have a critical impact on neuronal excitability. Ion channels, especially voltage-gated Na^+, K^+ and Ca^{2+} channels, are functional proteins embedded in the plasma membrane. They are critical determinants of neuronal excitability as they influence the generation and propagation of action potentials in neurons. Action potential is the cellular language by which neurons communicate with one another. During the firing of action potential, the voltage-dependent Na^+ channels dominate the explosive, regenerative activation of inward currents during the rising phase, while a fraction of the voltage-dependent K^+ channels chiefly contribute to the repolarization and hyperpolarizing overshoot phase of the action potential. Voltage-dependent Ca^{2+} channels generally make little contribution to the rising phase of action potential because their activation kinetics is slower than that of Na^+ channels. However, Ca^{2+} entry through voltage-gated Ca^{2+} channels activates Ca^{2+}-activated K^+ channels that indirectly contributes to the late repolarization and after-hyperpolarization, which follows the rising phase of action potential (Bean 2007). Basically, voltage-gated Na^+ channels along with K^+ and Ca^{2+} channels, determine the capacity of action potential generation, the shape of action potential, and the firing rate, and therefore have a great impact on neuronal excitability. Na^+ channel states (opening, closing, and inactivation) make a yes-or-no decision regarding the firing of an action potential, and the channel kinetics (activation and deactivation) and current density play an essential role in the amplitude and duration of an action potential. Therefore, abnormities of the function (opening, closing, and inactivation), expression, or structure (e.g., mutation) of ion channels may be responsible for the hyperexcitability of neurons and contribute to the consequent epilepsy.

3 Clinical Practice of Acupuncture Therapy for Epilepsy

Acupuncture treatment for epilepsy has been employed for thousands of years through all the dynasties in China (Yang and Cheng 2010). With the aid of advanced techniques, many modern scientific researchers have demonstrated the anti-epileptic effects of acupuncture in animal studies (Chao et al. 2001; Guo et al. 2008b, c; Kim et al. 2008; Zhang et al. 2008a, b; Goiz-Marquez et al 2009; Kang et al. 2013). For example, we found that electroacupunctural stimulation of Jinsuo (GV-8) and Fengfu (GV-16) significantly improves epileptic EEG and seizure behaviors through the increase in the endogenous melatonin levels in a penicillin-induced rat model (Chao et al. 2001). Furthermore, we found in a kainic acid-induced seizure model that electroacupuncture attenuates epileptic seizures, which is relatively specific to stimulation parameters and acupoints (Kang et al. 2013). Our findings show that (1) low- or high-frequency electroacupuncture at different acupoints, e.g., Renzhong (GV-26) plus Dazhui (GV-14), Jinsuo (GV-8) plus Yaoqi (EXB-9), Neiguan (PC-6) plus Quchi (LI-11), and Fenglong (ST-40) plus Yongquan (KI-1), reduced epileptic seizures with an exception of low-frequency electroacupuncture at Neiguan (PC-6) and Quchi (LI-11); (2) low-frequency electroacupuncture induced a better effect at Fenglong (ST-40) plus Yongquan (KI-1) than at other acupoints; (3) there was no significant difference in effects of high-frequency electroacupuncture at these acupoints; and (4) high-frequency electroacupuncture elicited a greater effect than low-frequency electroacupuncture, with an exception of Jinsuo (GV-8)+Yaoqi (EXB-9). The electroacupuncture-induced attenuation appeared to last for 1–1.5 h after electroacupuncture, with no appreciable effect in either EEG or behavioral tests during the first hour after electroacupuncture (Kang et al. 2013).

Despite substantial evidence from animal research in support of antiepileptic effects of acupuncture, a large number of clinical studies have also shown that patients have benefited from acupuncture for control of their seizures, especially in cases with refractory epilepsy (Chen et al. 1983; Shi et al. 1987; Yang 1990; Wen 2000; Deng et al. 2001; Lin et al. 2001; Wang 2001; Ma et al. 2001; Mao and Guo 2005; Zhang et al. 2006; Zhuang et al. 2006; Ren 2006; Xu 2006; Song et al. 2006), though results from one study beg to differ (Kloster et al. 1999).

These are extremely heterogenous clinical cases (Chen et al. 1983; Shi et al. 1987; Yang 1990; Wen 2000; Deng et al. 2001; Lin et al. 2001; Wang 2001; Ma et al. 2001; Mao and Guo 2005; Zhang et al. 2006; Zhuang et al. 2006; Ren 2006; Xu 2006; Song et al. 2006). In these cases, the patients treated varied over a range of ages from infants to elderly, and the type of epilepsy from absence seizure, febrile convulsion, generalized clonic-tonic seizure, to even status epilepticus. The case numbers reported also vary from a few to over 100, many of which were resistant to antiepileptic drugs. The acupoints and acupuncture methods used in these reports were highly heterogeneous as well. The overall treatment effects of acupuncture are principally manifested by the reduction in symptoms (e.g., a reduction of seizure frequency, shortness of episodes etc.), functional recovery (e.g., smoothed breathing from short and quick breathing with the occasional inability to breath, recovery from unconsciousness etc.), improved EEG (e.g., reduction of spike wave, desychronization etc.), and/or decreased epilepsy scores. The outcome of acupuncture therapy in these reports is relatively significant, and the total effective rate of treatment is as high as up to 98 %.

For the report that was unable to prove a beneficial effect of acupuncture in chronic intractable epilepsy (Kloster et al. 1999), as the authors acknowledged, the negative results could in part be ascribed to the small sample size for this observation (Kloster et al. 1999). Also, the selection of acupoints and manipulation/stimulation parameters (e.g., the rate, intensity, duration of swift rotation, lifting, thrusting, retaining) may also be responsible for this observation. In this respect, as has been mentioned earlier in this section, electroacupunctural attenuation of epileptic seizures is relatively specific to stimulation parameters and acupoints in our animal studies (Kang et al. 2013).

4 Potential Mechanisms of Acupuncture Inhibition of Neuronal Hyper-Excitability

4.1 Acupuncture and Opioid System

Classic opioid systems include endogenous peptides and δ-, μ- and κ-opioid receptors (DOR, MOR and KOR, respectively). Endogenous opioid peptides enkephalin, β-endorphin, and dynorphin in the brain preferentially bind to DOR, MOR and KOR respectively under physiological concentrations, and have multiple functions in the brain. Acupuncture and electroacupuncture have been well recognized to regulate endogenous opioid systems in the central nervous system (Chao et al. 1999; Wen et al. 2010a, b; Liang and Xia 2012).

Along with other biomediators (neurotransmitters, neuromodulators and/or signaling molecules) (Chao and Xia 2012), opioid system is also involved in the anticonvulsant effect of acupuncture, though the role of different opioid systems in the acupuncture suppression of epilepsy appears to be very complex.

Some studies show that the blockade of opioid system in the brain can diminish inhibitory effects of acupuncture on epilepsy, while its activation enhances it. In this regard, Wu and his coworkers reported that in a rat model of penicillin-induced seizures, the inhibitory effect of electroacupuncture on cortical epileptiform discharges could be reversed by the injection of a broad spectrum opioid receptor antagonist, naloxone, via various routes including intraperitoneal, intracerebroventricular, intranucleus (periaqueductal gray matter of midbrain, accumbens), or even directly into the cortical site where penicillin was applied. Naloxone injection into the reticular formation of midbrain and the reticular nucleus of thalamus had little effect on acupuncture attenuation of epilepsy, and injection of naloxone into the above-mentioned sites without application of penicillin on the cortex did not induce epileptic burst (Wu et al. 1986; Wu 1992). These findings suggest a specific activation of opioid system by electroacupuncture in the epileptic brain. In the hippocampus, activation of KOR with U50488H enhances the antiepileptic effect of electroacupuncture in electroconvulsive rats, while blockade of KOR with antagonist MR2266 or anti-dynorphin serum abates it (Gao and Cheng 1998). Therefore, the authors concluded that endogenously released opioid peptides mediate the suppressive effect of electroacupuncture on seizures (Wu et al. 1986; Wu 1992).

In support of the view of Wu et al., enhanced biosynthesis of central enkephalin by application of electroacupuncture has been found in the rat brain (Yuan and Han 1985). In the rat brain of experimental seizures model, hippocampal dynorphin concentration and dynorphin immunoreactivity in hippocampal mossy fiber and hilus are increased by electroacupuncture (Wang and Cheng 1992, 1994, 1995). Also, the seizure-associated increase in leu-enkephalin and β-endorphin levels in the hippocampus and augmented expression of preproenkephalin mRNA in several brain regions (entorhinal cortex, subiculum, hippocampal CA1 area, amygdaloidal nucleus, and piriform cortex) in rats, are suppressed by electroacupuncture (He et al. 1989; He and Cao 1989; Wang et al. 1994; Wang and Cheng 1994). Therefore, acupuncture can suppress epilepsy by regulating the synthesis and release of endogenous opioids.

Acupuncture can also regulate the expression and activity of opioid receptors. Radioreceptor-binding assay, combined with autoradiography, revealed that repeated electroconvulsive shock resulted in epileptiform EEG, behavioral seizures, and an accompanied increase in the opioid receptor densities in several brain regions (caudate nucleus, hippocampus, habenula nucleus, and amygdale) in rats. Electroacupuncture at Fengfu (GV-16) and Jingsuo (GV-8) was found to suppress seizure activities and significantly decrease the opioid receptor densities in these brain regions except the caudate nucleus (He et al. 1990).

Electroacupuncture also can up-regulate DOR expression in the ischemic brain, and thus protect the brain from ischemic injury (Tian et al. 2008) that can cause epilepsy secondary to cerebral infarction (Camilo and Goldstein 2004). Most interestingly, direct sub-acute high frequency electrical stimulation of parahippocampal cortex in the patients with intractable medial temporal lobe epilepsy also produces an antiepileptic effect, which is associated with reduced opioid peptide binding including ^{3}H-DAMGO, ^{3}H-DPDPE and ^{3}H-nociceptin (the ligand for MOR, DOR, and a fourth opioid, nociception receptor, respectively) in the same brain regions in patients with epilepsy (Rocha et al. 2007).

Therefore, it is evident in both animals and patients with epilepsy that acupuncture can regulate the activities of endogenous opioids and their receptors in the brain, thereby exerting its antiepileptic effects.

4.2 Opioid Receptors and Na^+ Channels

Opioid receptors are members of the G protein-coupled receptor superfamily (Raynor et al. 1994), and can functionally couple with ion channels, including Na^+ channels (Witkowski and Szulczyk 2006; Chattopadhyay et al. 2008; Kang et al. 2009).

Our previous data implied an interaction between opioid receptors and Na^+ channels. For example, we observed that DOR down-regulation (Zhao et al. 2005) is associated with an up-regulation of voltage-gated Na^+ channels in a mutant brain with epileptic seizures (Xia et al. 2003). An increased Na^+ channel density (Xia et al. 2000) along with decreased DOR density (Xia et al. 2001) occurred in hypoxia-exposed brain. Activation of presynaptic DOR by enkephalin prevents the increase in neuronal Nav1.7 in the dorsal root ganglia, which relieves pain in diabetic neuropathy (Chattopadhyay et al. 2008). More recently, our studies indicated that DOR activation attenuates hypoxic disruption of K^+-Na^+ homeostasis, which largely relies on DOR inhibition on Na^+ influx through Na^+ channels (Chao et al. 2007a, b, 2008, 2009, 2012; Kang et al. 2009). These results suggest that under pathological conditions DOR could mediate an inhibitory regulation of Na^+ channels in the brain.

Furthermore, our direct evidence gained from electrophysiological studies shows that activation of DOR indeed inhibits Na^+ channel activity (Kang et al. 2009). Remy et al. (2004) observed that SNC80 (1–1000 μM), a putative DOR agonist, reduced the maximal Na^+ current amplitude in a dose-dependent manner, and selectively prolonged the course of recovery from slow inactivation without effects on fast inactivation processes (Remy et al. 2004). However, the authors concluded that this effect was opioid receptor-independent since the effects of SNC80 were not mimicked by another DOR agonist DPDPE (10 μM), and were not inhibited by high concentrations of opioid receptor antagonists, naloxone (50–300 μM) and naltrindole (10 and 100 μM) (Remy et al. 2004). This conclusion is debatable due to several important issues (e.g., experimental procedures, specificity and dose of drugs used etc.) (see review by Chao and Xia 2010). We recently took the advantages of *Xenopus* oocytes with co-expressed DOR and Na^+ channels to explore a "pure" interaction between DOR and Na^+ channels, and found: (1) Nav1.2 expression induced tetrodotoxin (TTX)-sensitive inward currents; (2) DOR expression reduced the inward currents; (3) Activation of DOR reduced the amplitude of the current and rightward shifted the activation curve of the current in the oocytes with both Nav1.2 and DOR, but these effects were not observed in oocytes with Nav1.2 alone; (4) The DOR agonist-induced inhibition of Nav1.2 currents was dose-dependent and saturable; and (5) The selective DOR agonist had no effect on naive oocytes. These findings present the first demonstration that activation of DOR inhibits Na^+ channel function by decreasing the amplitude of Na^+ currents and increasing its threshold for activation (Kang et al. 2009). Similar inhibitory effects of Na^+ channels are also found on MOR and KOR. For example, in acutely isolated cortical neurons, the application of 1 μM of DAMGO, a specific MOR agonist, caused a decrease in the Na^+ current amplitude to approximately 79 % of the controls. Moreover, DAMGO decreased the maximum current activation rate, prolonged its time-dependent inactivation, shifted the half inactivation voltage from −63.4 to −71.5 mV, and prolonged the time constant of recovery from inactivation from 5.4 to 7.4 ms (Witkowski and Szulczyk 2006). DAMGO also inhibited TTX-resistant voltage-dependent Na^+ current in dorsal root ganglion neurons (Gold and Levine 1996). U50488, a KOR agonist, decreases voltage-activated Na^+ currents in colon sensory neurons (Su et al. 2002). Therefore, it seems that inhibition of Na^+ channel activities, which depends on signal molecules such as protein kinases (Witkowski and Szulczyk 2006; Chattopadhyay et al. 2008; Chao et al. 2012), is one of the common characteristics of opioid receptors.

4.3 Na^+ Channels and Epileptic Hyper-Excitability

As shown by previous reviews (Catterall 2000; Goldin 2001), brain Na^+ channels consist of a 260 kDa α subunit with two auxiliary β subunits. The α subunit forms the ion-conducting pore and the activation and inactivation gates that regulate voltage-dependent sodium flux across the plasma membrane, and β subunit modifies the kinetics and voltage dependence of the gates. By now at least nine α subtypes (Nav1.1–Nav1.9) and four β subtypes (β1–β4) have been found to express in the excitable cells (Table 16.1). Nav1.1, Nav1.2, Nav1.3 and Nav1.6 are the primary subtypes in the central nervous system (Catterall 2000; Goldin 2001; Ragsdale 2008; Escayg and Goldin 2010; Mantegazza et al. 2010). The developmental expression and cellular localization of these subtypes are different in the brain. Nav1.1 expression is first detectable at postnatal day 7, increases during the third postnatal week, and peaks at the end of the first postnatal month, after which levels decrease by about 50 % in the adult. Nav1.2 expression also increases during the third postnatal week and continues to increase thereafter, until the maximal level is reached in adulthood. In rodents, Nav1.3 channels are highly expressed in the brain during the embryonic period, then peak at birth, and finally decline after birth as

Table 16.1 Voltage-gated sodium channels

Subunit	α	β
Subtypes	Nav1.1–Nav1.9	β1–β4
Location	Prevalent in the CNS: Nav1.1, Nav1.2, Nav1.3 and Nav1.6	Two β subunits associated with an α subunit
	Abundant in muscle: Nav1.4, Nav1.5	
	Primarily in the peripheral nervous system: Nav1.7, Nav1.8 and Nav1.9	
Cellular distribution	Nav1.1 and Nav1.3: primary localized in cell body	Expressed in a complementary fashion (either β1 or β3, and β2 or β4) with α subunit
	Nav1.2: high expressed in unmyelinated or pre-myelinated axons and dendrites	
	Nav1.6: expressed at nodes of Ranvier and axon initial segments as well as in the somata and dendrites of many projection neurons	
Function	Forms the ion-conducting pore and the activation and inactivation gates	Modify the kinetics and voltage dependence of gating
		Serve as cell adhesion molecules for integrating the channels into the appropriate subcellular domains

Nav1.1 and Nav1.2 channels take over. However, Nav1.3 remains detectable at a lower level during adulthood. Comparatively higher levels of Nav1.3 exist in the human adult brain. Nav1.1 and Nav1.3 are primary localized in the cell body and are preferentially expressed in the GABAergic neurons. Nav1.2 is particularly highly expressed in the unmyelinated or pre-myelinated axons and dendrites. Nav1.6 subtype is expressed at the nodes of Ranvier and the initial segments of axon, as well as in the somata and dendrites of many projection neurons (Catterall 2000; Goldin 2001; Ragsdale 2008; Escayg and Goldin 2010; Mantegazza et al. 2010). Na^+ channels are responsible for the initiation and propagation of action potentials, and influence the subthreshold electrophysiology. Therefore, they are crucial determinants of intrinsic neuronal excitability (Bean 2007; Stafstrom 2007; Ragsdale 2008). Since epilepsy is regarded as an "electrical storm" of brain hyperexcitability (Ragsdale 2008), altered density or biophysical properties of Na^+ channels may have important consequences on the neuronal excitability, and may contribute to the pathophysiology of brain diseases associated with altered excitability, such as epilepsy (Ragsdale 2008; Catterall et al. 2010; Mantegazza et al. 2010; Meisler et al. 2010; Cheah et al. 2012). The importance of Na^+ channels in brain hyperexcitability and consequent epilepsy is further supported by the fact that a large number of antiepileptic drugs exert their antiepileptic effect by interacting with Na^+ channels (Armijo et al. 2005; Errington et al. 2005).

4.3.1 Expressional and Functional Alterations in Na^+ Channels during Epilepsy

Accumulating evidence indicates that altered expression and functional regulation of Na^+ channels in the neurons play important roles in brain hyperexcitability and epileptic phenotype in both acquired and inherited epilepsy. In the cortex of a genetically seizure susceptible E1 mouse brain and spontaneously epileptic rat hippocampus, a significant increase in total Na^+ channel mRNA and protein, as well as in Nav1.1, Nav1.3 and β1 subunits, was observed to contribute to the generation of epileptiform activity and the observed seizure phenotypes (Sashihara et al. 1992; Guo et al. 2008a). The expression of Nav1.1 in hippocampal CA1, and of Nav1.2 in cortical neurons, was found to be significantly increased, which was accompanied with increased neuronal excitability and spontaneous epileptic seizures in the Na^+/H^+ exchanger null mutant mouse (Gu et al. 2001; Xia et al. 2003). In a kindling seizure model, selectively increased expression of Nav1.6 mRNA and proteins in hippocampal CA3 neurons and Nav1.6 immunoactivity in the medial entorhinal neurons resulted in hyperexcitability in these brain regions (Blumenfeld et al. 2009; Hargus et al. 2011). In acute status epilepticus models, a transient up-regulation of Na^+ channel mRNAs encoding Nav1.2 and Nav1.3 subunits was observed in hippocampal neurons (Bartolomei et al. 1997; Aronica et al. 2001). In contrast, the rats with status epilepticus and chronically developed spontaneous epileptic seizures showed a selectively persistent down-regulation of Nav1.2, Nav1.6, and β1 subunits, as well as short-term down-regulation of β2 subunit. In addition, an increased excitability, manifested by the augmented window current due to the significant positive shift of inactivation potential and negative shift of activation potential and the resultant increased overlap between the activation and inactivation curve, in these neurons in hippocampal dentate gyrus was observed (Ellerkmann et al. 2003). This phenomenon may be caused by the down-regulation of β subunit expression, since β1/β2 subunit, if co-expressed with α subunits, favors inactivation, accelerates recovery of Na^+ currents (Toib et al. 1998; Chen et al. 2002; Aman et al. 2009), which increases the number of Na^+ channels available to be activated, and thus increases the firing rate. In the human brain with temporal lobe epilepsy, Nav1.3 mRNA in the pyramidal cells of hippocampal CA4 area is significantly up-regulated, and Nav1.2 mRNA in the remaining pyramidal cells of hippocampal CA1, CA2 and CA3 areas is largely down-regulated, while Nav1.1 and Nav1.6 do not show any differences in their expression in the hippocampus (Whitaker et al. 2001). *SCN7A* gene-encoding atypical Na^+ channels, Na_X, were recently reported to be increasingly and persistently expressed in the pyramidal neurons and astrocytes of the hippocampal CA1 and CA3 areas in patients with drug-resistant temporal lobe epilepsy and epileptic rats, and are possibly responsible for the enhanced brain excitability and epileptogenesis (Gorter et al. 2010). In summary, the increased amplitude and density of the voltage-dependent Na^+ currents, shortened phase of inactivation, enhanced window currents due to shift towards depolarization of inactivation currents, and the more negative activation, have been associated with neuronal hyperexcitability and the development of some types of epilepsy (Vreugdenhil et al. 1998; Gu et al. 2001; Ketelaars et al. 2001; Ellerkmann et al. 2003; Xia et al. 2003). Na^+ channels, even the few that fail to inactivate, carry the persistent fraction of Na^+ currents (that are not sensitive to TTX). Though small, they may drive the membrane towards the firing threshold. Especially under physiological conditions, the persistent Na^+ currents serve to amplify or spatially integrate synaptic potentials, allow excitable cells to generate subthreshold oscillations, and reduce the threshold for repetitive action potential firing. Therefore, they increase excitability of neurons associated with epilepsy (Rhodes et al. 2004; Stafstrom 2007; Blumenfeld et al. 2009; Epsztein et al. 2010; Hargus et al. 2011). Despite the differences in epilepsy types/models and Na^+ channel subtypes in these reports, these findings suggest altered expression and functional regulation of Na^+ channels (that carry Na^+ currents including both TTX-sensitive and -resistant one) are critically involved in brain hyperexcitability and the pathology of epilepsy.

4.3.2 Insights from Genetic Epilepsy Model

Genetic factors are important for the intrinsic excitability of neurons. Studies on Na^+ channel mutation in the genetic epilepsy model broaden the view on the roles and the underlying mechanisms of Na^+ channels in brain hyperexcitability and the pathophysiology of epilepsy. Gene mutations that result in channel dysfunction (channelopathies) play an essential role in neuronal excitability, leading to the development of a variety of epilepsy syndromes. The most convincing data on the role of Na^+ channels in brain hyperexcitability and epileptogenesis come from the identification of several hundred mutations of Na^+ channels which lead to inherited epileptic syndromes. These syndromes range in severity from relatively mild disorders such as benign familial neonatal-infantile seizures (BFNIS), simple febrile seizures, and generalized epilepsy with febrile seizure plus (GEFS+), to severe epileptic encephalopathy, such as severe myoclonic epilepsy of infancy (SMEI, also called Dravet's Syndrome), SMEI borderline, intractable childhood epilepsy with generalized tonic-clonic seizures (See reviews in Ragsdale 2008; Stafstrom 2007; Catterall et al. 2010; Escayg and Goldin 2010; Meisler et al. 2010).

So far, several hundred mutations of many Na^+ channel subtypes, including Nav1.1 (*SCN1A*), Nav1.2 (*SCN2A*), Nav1.3 (*SCN3A*), Nav1.6 (*SCN8A*), and even Nav1.7 (*SCN9A*) (which is predominantly expressed in peripheral nervous system), as well as β1 subunit (*SCN1B*), have been causally linked to a variety of genetic epilepsies (Meisler et al. 2010). However, the genotype-phenotype correlations for Na^+ channel epilepsy are very complicated with high heterogeneity. This heterogeneity of genotype-phenotype for Na^+ channel epilepsy is reflected in the observation that the same gene mutation may result in different phenotypes of epilepsy, and in turn, a single phenotype may be a result of different gene mutations of Na^+ channels. For example, mutations in *SCN1A* have been reported to cause epilepsy with the symptoms ranging from febrile seizures and GEFS+ to SMEI, and mutations in *SCN2A* are identified to cause BFNIS, GEFS+, SMEI, and intractable epilepsy with mental decline (Ragsdale 2008; Catterall et al. 2010; Escayg and Goldin 2010; Meisler et al. 2010). On the other hand, Dravet's Syndrome is reported to have mutations in *SCN1A*, *SCN2A* and *SCN1B*, and the mutations in both *SCN1A* and *SCN2A* lead to GEFS+ (Escayg and Goldin 2010; Meisler et al. 2010). Among all the mutations, NaV1.1 mutations account for the majority of established epilepsy syndromes in children (Ragsdale 2008; Lossin 2009; Catterall et al. 2010; Escayg and Goldin 2010; Meisler et al. 2010). Approximately 30 *SCN1A* mutations have been identified to account for GEFS+ and all of them are missense mutations. Many other missense mutations in *SCN1A* also contribute to Dravet's Syndrome, but most of the Dravet's Syndrome results from *SCN1A* mutations caused by frame shift, nonsense, and splice-site mutations, which lead to a truncated protein and haploinsufficiency of *SCN1A* (Ragsdale 2008; Lossin 2009; Catterall et al. 2010; Escayg and Goldin 2010; Meisler et al. 2010; Bechi et al. 2012; Cheah et al. 2012). Unlike *SCN1A*-GEFS+ mutations that show a Mendelian inheritance pattern within affected families, most *SCN1A*-SMEI mutations occur *de novo* in the affected child.

Na^+ channel mutations may result in either gain-of-function or loss-of-function, which has been proposed to respectively increase and decrease neuronal excitability (Escayg and Goldin 2010). The gain-of-function of Na^+ channels is manifested by the increased current density, negative shift of activation, positive shift of inactivation, enhanced persistent Na^+ currents, or mixed effects on channel kinetics but with the net effect of an increase in activity. The loss-of-function of Na^+ channels is reflected by the changes that are opposite to those seen with gain-of-function. However, both gain-of-function and loss-of-function in Nav1.1 can predispose the brain to abnormal excitability, i.e., brain hyperexcitability and consequent epilepsy syndrome. The increase in neuronal excitability due to gain-of-function is quite straightforward to be understood and is further supported by the fact that many antiepileptic drugs developed are Na$^+$ channel blockers (Armijo et al. 2005; Errington et al. 2005). It is very surprising, however, that "loss-of-function" also makes the brain very hyperexcitable, although it decreases the neuronal excitability. Fortunately, this mystery was unveiled recently with targeted knock-out mice and rats carrying mutated *SCN1A*. These mice and rats developed epileptic seizures (SMEI and GEFS+) and sporadically died within the first postnatal month for the SMEI mice (Yu et al. 2006; Kalume et al. 2007; Ogiwara et al. 2007; Ohno et al. 2010; Cheah et al. 2012). Immunohistochemical analysis revealed that in a developing rodent brain, Nav1.1 was predominantly expressed in the inhibitory GABAergic interneurons of the neocortex and hippocampus, as well as the cerebellar GABAergic Purkinje cells that serve as the output pathway for information on movement, coordination, and balance from the cerebellar cortex (Yu et al. 2006; Kalume et al. 2007; Ogiwara et al. 2007; Ohno et al. 2010). Mutations or deletion of Nav1.1 lead to the loss of sustained high-frequency firing of action potential and excitability in the hippocampal and cortical inhibitory interneurons and Purkinje neurons, which allows hyperexcitability of principal neurons (e.g., dentate granule cells, pyramidal neurons) in the neuronal networks, thus leading to brain hyperexcitability, and subsequent epilepsy. A unified loss-of-function hypothesis for Nav1.1 genetic epilepsies has been proposed (Ragsdale 2008; Catterall et al. 2010). According to this hypothesis, the severity of epileptic phenotypes of Nav1.1 mutations is dependent on the extent of Nav1.1 functional damage due to mutations, as for instance, the spectrum of severity of Nav1.1-associated forms of epilepsy results from an increasing severity of loss-of-function mutations in Nav1.1 channels and increasing impairment of action potential firing of the GABAergic inhibitory neurons (Catterall et al. 2010). As mentioned earlier, mild impairment of Nav1.1 channel function causes febrile seizures, and moderate to severe impairment of Nav1.1 function due to missense or nonsense mutations causes GEFS+ and SMEI (Ragsdale 2008; Catterall et al. 2010).

4.3.3 Na$^+$ Channel Based Neuronal and Network Excitability

The above discussion on altered expression, functions, and mutations in Na$^+$ channels strongly supports the belief that abnormal Na$^+$ channel activities are critically involved in neuronal hyperexcitability and the pathology of epilepsy. However, it should be pointed out that emphasizing the important roles of Na$^+$ channels in intrinsic neuronal excitability and consequent epilepsy does not mean that Na$^+$ channels hold any less importance in network excitability. As we elaborated earlier, both neuronal excitability and network excitability contribute to the hyperexcitability in the epileptic brain. Since epileptic seizures are regarded as an "electrical storm" of brain hyperexcitability (Ragsdale 2008), they involve not only intrinsic neuronal properties, but also a cluster of anatomically and functionally associated neurons that form epileptic networks and propagate the extremely synchronized "electrical storm" within the networks (Chao and Xia 2012). Therefore, it is important to realize that Na$^+$ channels regulate brain excitability by altering both intrinsic neuronal and network excitability. The genetic models of SMEI show that Nav1.1

mutation or deletion affects the inhibitory interneurons in the hippocampus and cortex (Yu et al. 2006; Ogiwara et al. 2007; Cheah et al. 2012). Even though the excitatory transmission is not affected, the balance between inhibition and excitation within the networks is severely disrupted (Yu et al. 2006; Ogiwara et al. 2007). In this model, though the embryonic Nav1.3 channels were found to be up-regulated, probably as a partial compensatory response to the impaired inhibition and disrupted balance between inhibitory and excitatory transmission (Yu et al. 2006), the network excitability is greatly enhanced due to Nav1.1 mutation that leads to a loss of sustained high-frequency firing of action potential in the hippocampal and cortical inhibitory interneurons and a limited compensatory capacity of Nav1.3 channels, thereby, making the brain very hyperexcitable. Therefore, Na^+ channels play a central role in epileptogenesis, which involve interplay of both intrinsic neuronal properties and network activities.

5 Concluding Remarks

Na^+ channel expressional and functional up-regulation has been demonstrated to be critical to epileptic hyperexcitability and seizures, while the inhibitory regulation of Na^+ channels by DOR (Kang et al. 2009; Chao et al. 2008, 2009, 2012) may contribute to the proper control of neuronal excitability. An impairment of such a balance mechanism, e.g., Na^+ channel up-regulation and/or DOR down-regulation in genetic or acquired conditions, may lead to neuronal dysfunction and eventually neurological diseases, especially epileptic seizures. Indeed, Na^+ channel mutations have been causally linked to human epilepsy. There is also abundant supporting evidence from epileptic animals. For example, in the mutant brain exhibiting spontaneous epilepsy, Na^+ channel was up-regulated (Xia et al. 2003), while DOR was down-regulated in the same brain (Zhao et al. 2005), suggesting a potential role of DOR impairment in the pathophysiology of epilepsy associated with genetic abnormality.

Since Na^+ channel up-regulation contributes greatly to some kinds of epileptic hyperexcitability that lead to epilepsies, the DOR-mediated inhibition of Na^+ channels could provide a novel clue to open a vast potential of solutions to epileptic seizures. In fact, many antiepileptic drugs are actually inhibitors of Na^+ channels (Armijo et al. 2005; Errington et al. 2005). Moreover, acupuncture can regulate the activities of endogenous opioids and their receptors in both animal brains and in patients with epilepsy, thus exerting its antiepileptic effects. Therefore, stimulating appropriate acupoints with suitable manipulations may be a useful strategy for the treatment of epilepsy. Figure 16.1 presents a schematic demonstration regarding the interaction between acupuncture, opioids, and Na^+ channels in the regulation of hyper-excitability and epileptic seizures in the brain.

However, some issues need attention with regard to the association between acupuncture therapy for epilepsy and the role of opioids and Na^+ channels. As previously discussed, loss-of-function of Na^+ channels in the inhibitory interneurons can cause brain hyperexcitability and epilepsy. Therefore, it is possible that activa-

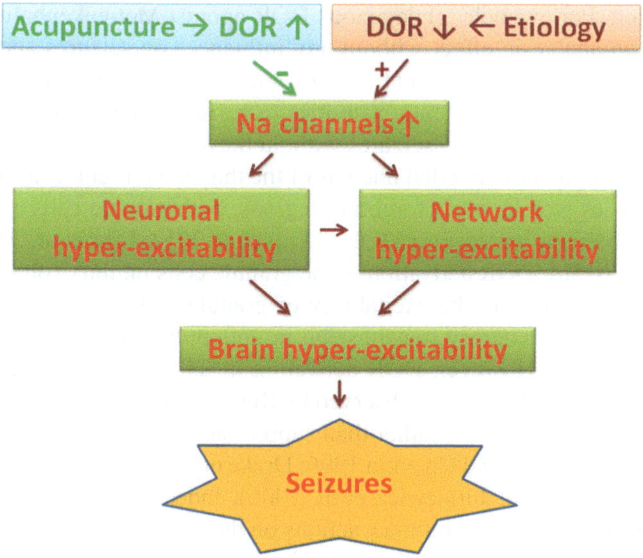

Fig. 16.1 Schematic demonstration of the potential relation between acupuncture, opioid, and Na^+ channels in the regulation of brain hyperexcitability and epileptic seizures. Acupuncture can regulate the levels of endogenous opioids and their receptors in the brain. The released opioids activate δ-opioid receptors, and Na^+ channels are inhibited by activated δ-opioid receptors via signaling molecules such as PKC. Thus the neuronal discharges are inhibited and the over-excited brain is "cooled," leading to the termination of epilepsy

tion of opioid system by acupuncture causes inhibition of Na^+ channel activity in inhibitory interneurons, which may further aggravate the symptoms of epilepsy. Despite the demonstration of an antiepileptic effect following DOR activation in some studies (Tortella et al. 1988; Tortella and Long 1985; Koide et al. 1992; Madar et al. 1997), several other reports showed opposite results (Snead 1986; Stutzmann et al. 1986; Negus et al. 1998; Grecksch et al. 1999; Yajima et al. 2000; Danielsson et al. 2006; Jutkiewicz et al. 2005, 2006; Chu Sin Chung et al. 2015a). For example, as early as the 1970s, two groups independently observed the dose-related, naloxone-reversible epileptic properties of the endogenous DOR agonist Met-enkephalin when it was intracerebroventricularly injected to study its analgesic and catatonic properties (Bloom et al. 1976; Urca et al. 1977). DOR agonists SNC80 and TAN-67 are also pro-convulsive (Yajima et al. 2000; Danielsson et al. 2006; Jutkiewicz et al. 2006; Chu Sin Chung et al. 2015a) though it is reported that SNC80 inhibits Na^+ channel activity (Remy et al. 2004). The reasons for the complex and mixed effects of δ-opioid system on seizures are not well clarified yet, but could be partially related to multiple factors like animal species (e.g., pro-convulsive in rats but has little effect in rhesus monkeys) (Negus et al. 1998; Jutkiewicz et al. 2006; Danielsson et al. 2006), seizure types, the methods of drug administration (Jutkiewicz et al. 2005), dose used (Bausch et al. 2005), and target neurons etc. (also see Chao and Xia 2012). Among these factors, the location of DOR on the target neuron seems critical and important. For example, endogenous DOR agonist Met-enkephalin was found to produce an analgesic effect without any influences on seizures and other EEG changes when injected into the ventral, caudal midbrain periaqueductal gray matter. However, it did generate epileptic activities (e.g., intermittent myoclonic twitches and wet-dog shakes) and EEG seizures without the analgesic effect when given into or near the forebrain dorsomedial nucleus of the thalamus (Frenk et al. 1978). In the hippocampus, as well as other forebrain regions, both granule cells and inhibitory GABAergic interneurons express DOR (Rezaï et al. 2012; Erbs et al. 2012; Chu Sin Chung et al. 2015b). DOR activation in the granule cells inhibits voltage-gated Na^+ channels, and thus lowers the excitability of granule neurons (Remy et al. 2004), which reduces the excitatory transmission in epileptic network, and subsequently suppresses seizures. However, DOR activation in inhibitory interneurons leads to the inhibition of Na^+ channels as observed by Remy et al. (2004) in granule neurons, and may result in facilitation, rather than suppression, of seizures via post-synaptic de-inhibition (Simmons and Chavkin 1996; Drake et al. 2007), as has been observed (Danielsson et al. 2006; Jutkiewicz et al. 2006). Indeed, DOR activation in these interneurons does produce inhibitory actions on their own activity (Zieglgänsberger et al. 1979; Madison and Nicoll 1988; Neumaier et al. 1988; Piskorowski and Chevaleyre 2013; Rezaï et al. 2013), but this inhibition of inhibitory interneurons may lead to neurons in the network to become excited via post-synaptic de-inhibition (Zieglgänsberger et al. 1979; Madison and Nicoll 1988; Neumaier et al. 1988). Thus, facilitation, rather than suppression of seizures may occur. This view is further supported by a recent study since dose-dependent seizure events produced by SNC80 are no longer detected in a conditional DOR knockout mouse line which targets the receptor gene specifically in GABAergic neurons of the forebrain (Chu Sin Chung et al. 2015a). Therefore, selective activation/inhibition of opioid system in certain locations of epileptic networks with acupuncture therapy is critical to the therapeutic outcomes, but is also challengeable for acuptuncture operators. It has been shown that in seizure rat models, acupuncture increases dynorphin synthesis and release, and enhances the activity of KOR (which is abundantly distributed in hippocampal granule cell and perforant path) (Simmons and Chavkin 1996; Drake et al. 2007) in the hippocampus, and decreases DOR activity, as well as synthesis and release of enkephalin (which mainly influences inhibitory interneuron activity) (Rezaï et al. 2012; Erbs et al. 2012). Therefore, acupuncture balances the inhibition and excitation in the network, and thus suppresses seizures (Chao and Xia 2012). In this way, acupuncture exerts its antiepileptic effects by normalizing the disrupted neuronal and network excitability (by lowering the overexcited neuronal activity through multiple strategies like enhancing the inhibitory system and attenuating the excitatory one in the brain via regulation of the activities of opioid-Na^+ channels). We believe that further clarification on the correlations between acupuncture, opioids, and Na^+ channels can better help us explore the mystery of acupuncture therapy on epilepsy.

Acknowledgements This work was supported by NIH (AT-004422, and HD-034852) and Vivian L Smith Neurologic Foundation.

References

Aghakhani Y, Bagshaw AP, Bénar CG, Hawco C, Andermann F, Dubeau F, Gotman J (2004) fMRI activation during spike and wave discharges in idiopathic generalized epilepsy. Brain 127:1127–1144

Aman TK, Grieco-Calub TM, Chen C, Rusconi R, Slat EA, Isom LL, Roman IM (2009) Regulation of persistent Na current by interactions between β subunits of voltage-gated Na channel. J Neurosci 29:2027–2042

Armijo JA, Shushtarian M, Valdizan EM, Cuadrado A, de las Cuevas I, Adin J (2005) Ion channels and epilepsy. Curr Pharm Des 11:1975–2003

Aronica E, Yankaya B, Troost D, van Vliet EA, Lopes da Silva FH, Gorter JA (2001) Induction of neonatal sodium channel II and III α-isoform mRNAs in neurons and microglia after status epilepticus in the rat hippocampus. Eur J Neurosci 13:1261–1266

Bartolomei F, Gastaldi M, Massacrier A, Planells R, Nicolas S, Cau P (1997) Changes in the mRNAs encoding subtypes I, II and III sodium channel alpha subunits following kainite-induced seizures in rat brain. J Neurocytol 26:667–678

Bausch SB, Garland JP, Yamada J (2005) The delta opioid receptor agonist, SNC80, has complex, dose-dependent effects on pilocarpine-induced seizures in Sprague-Dawley rats. Brain Res 1045:38–44

Bean BP (2007) The action potential in mammalian central neurons. Nat Rev Neurosci 8:451–465

Bechi G, Scalmani P, Schiavon E, Rusconi R, Franceschetti S, Mantegazza M (2012) Pure haploinsufficiency for Dravet syndrome Na(V)1.1 (SCN1A) sodium channel truncating mutations. Epilepsia 53:87–100

Berman R, Negishi M, Vestal M, Spann M, Chung MH, Bai X, Purcaro M, Motelow JE, Danielson N, Dix-Cooper L, Enev M, Novotny EJ, Constable RT, Blumenfeld H (2010) Simultaneous EEG, fMRI, and behavior in typical childhood absence seizures. Epilepsia 51:2011–2022

Bloom F, Segal D, Ling N, Guillemin R (1976) Endorphins: profound behavioral effects in rats suggest new etiological factors in mental illness. Science 194:630–632

Blumenfeld H, Lampert A, Klein JP, Mission J, Chen MC, Riera M, Dib-Hajj S, Brennan AR, Hains BC, Waxman SG (2009) Role of hippocampal sodium channel Nav1.6 in kindling epileptogenesis. Epilepsia 50:44–55

Bragin A, Wilson CL, Engel J Jr (2000) Chronic epileptogenesis requires development of a network of pathologically interconnected neuron clusters: a hypothesis. Epilepsia 42(suppl 6):S144–S152

Camilo O, Goldstein LB (2004) Seizures and epilepsy after ischemic stroke. Stroke 35: 1769–1775

Catterall WA (2000) From ionic currents to molecular review mechanisms: the structure and function of voltage-gated sodium channels. Neuron 26:13–25

Catterall WA, Kalume F, Oakley JC (2010) Nav1.1 channels and epilepsy. J Physiol 588: 1849–1859

Chao D, Balboni G, Lazarus LH, Salvadori S, Xia Y (2009) Na^+ mechanism of DOR protection against anoxic K^+ derangement in the cortex. Cell Mol Life Sci 66:1105–1115

Chao D, Bazzy-Asaad A, Balboni G, Salvadori S, Xia Y (2008) Activation of DOR attenuates anoxic K^+ derangement via inhibition of Na^+ entry in mouse cortex. Cereb Cortex 18:2217–2227

Chao D, Bazzy-Asaad A, Balboni G, Xia Y (2007a) Delta-, but not mu-, opioid receptor stabilizes K^+ homeostasis by reducing Ca^{2+} influx in the cortex during acute hypoxia. J Cell Physiol 212:60–67

Chao D, Donnelly DF, Feng Y, Bazzy-Asaad A, Xia Y (2007b) Cortical delta-opioid receptors potentiate K^+ homeostasis during anoxia and oxygen-glucose deprivation. J Cereb Blood Flow Metab 27:356–368

Chao D, He X, Yang Y, Bazzy-Asaad A, Lazarus LH, Balboni G, Kim DH, Xia Y (2012) DOR activation inhibits anoxic/ischemic Na^+ influx through Na^+ channels via PKC mechanisms in the cortex. Exp Neurol 236:228–239

Chao D, Xia Y (2010) Ionic storm in hypoxic/ischemic stress: can opioid receptors subside it? Prog Neurobiol 90:439–470

Chao D, Xia Y (2012) Acupuncture treatment of epilepsy. In: Xia Y, Ding G, Wu G (eds) Current research in acupuncture. Springer, New York, pp 129–214

Chao DM, Chen G, Cheng JS (2001) Melatonin might be one possible medium of electroacupuncture anti-seizures. Acupunct Electrother Res 26:39–48

Chao DM, Shen LL, Tjen-A-Looi S, Pitsillides KF, Li P, Longhurst JC (1999) Naloxone reverses inhibitory effect of electroacupuncture on sympathetic cardiovascular reflex responses. Am J Physiol Heart Circ Physiol 276:H2127–H2134

Chattopadhyay M, Mata M, Fink DJ (2008) Continuous δ-opioid receptor activation reduces neuronal voltage-gated sodium channel (Nav1.7) levels through activation of protein kinase C in painful diabetic neuropathy. J Neurosci 28:6625–6658

Cheah CS, Yu FH, Westenbrook RE, Kalume FK, Oakley JC, Potter GB, Rubenstein JL, Catterall WA (2012) Specific deletion of Nav1.1 sodium channels in inhibitory interneurons causes seizures and premature death in a mouse model of Dravet syndrome. Proc Natl Acad Sci U S A 109:14646–14651

Chen C, Bharucha V, Chen Y, Westenbrock RE, Brown A, Malhotra JD, Jones D, Avery C, Gillespic PJ III, Kazen-Gillespie KA, Kazarinova-Noyes K, Shrager P, Saunders TL, Macdonald RL, Ransom BR, Scheuer T, Catterall WA, Isom LL (2002) Reduced sodium channel density, altered voltage dependence of inactivation, and increased susceptibility to seizures in mice lacking sodium channel β2-subunits. Proc Natl Acad Sci U S A 99:17072–17077

Chen K, Chen G, Feng X (1983) Observation of immediate effect of acupuncture on electroencephalograms in epileptic patients. J Tradit Chin Med 3:121–124

Chu Sin Chung P, Boehrer A, Stephan A, Matifas A, Scherrer G, Darcq E, Befort K, Kieffer BL (2015a) Delta opioid receptors expressed in forebrain GABAergic neurons are responsible for SNC80-induced seizures. Behav Brain Res 278:429–434

Chu Sin Chung P, Keyworth HL, Martin-Garcia E, Charbogne P, Darcq E, Bailey A, Filliol D, Matifas A, Scherrer G, Ouagazzal AM, Gaveriaux-Ruff C, Befort K, Maldonado R, Kitchen I, Kieffer BL (2015b) A novel anxiogenic role for the delta opioid receptor expressed in GABAergic forebrain neurons. Biol Psychiatry 77:404–415

Danielsson I, Gasior M, Stevenson GW, Folk JE, Rice KC, Negus SS (2006) Electroencephalographic and convulsant effects of the delta opioid agonist SNC80 in rhesus monkeys. Pharmacol Biochem Behav 85:428–434

Deng YJ, Wang JJ, Lin YP, Liu WY, Wang LH (2001) Clinical observation on treatment of epilepsy general tonic-clonic attack with catgut implantation at acupoint plus antiepileptic Western Medicine of small dose. Zhongguo Zhen Jiu 21:271–273

Drake CT, Chavkin C, Milner TA (2007) Opioid systems in the dentate gyrus. Prog Brain Res 163:245–263

Ellerkmann RK, Remy S, Chen J, Sochivko D, Elger CE, Urban BW, Becker A, Beck H (2003) Molecular and functional changes in voltage-dependent Na^+ channels following pilocarpine-induced status epilepticus in rat dentate granule cells. Neuroscience 119:323–333

Epsztein J, Sola E, Represa A, Ben-Ari Y, Crépel V (2010) A selective interplay between aberrant EPSPKA and INaP reduces spike time precision in dentate granule cells of epileptic rats. Cereb Cortex 20:898–911

Erbs E, Faget L, Scherrer G, Kessler P, Hentsch D, Vonesch JL, Matifas A, Kieffer BL, Massotte D (2012) Distribution of delta opioid receptor-expressing neurons in the mouse hippocampus. Neuroscience 221:203–213

Errington AC, Stöhr T, Lees G (2005) Voltage gated ion channels: targets for anticonvulsant drugs. Curr Top Med Chem 5:15–30

Escayg A, Goldin AL (2010) Sodium channel SCN1A and epilepsy: mutations and mechanisms. Epilepsia 51:1650–1658

Frenk H, McCarty BC, Liebeskind JC (1978) Different brain areas mediate the analgesic and epileptic properties of enkephalin. Science 200:335–337

Gao H, Cheng J (1998) Role of intrahippocampal kappa opioid receptors in inhibiting the seizure by electroacupuncture (EA) in rats. Zhen Ci Yan Jiu 23:122–125

Goiz-Marquez G, Caballero S, Solis H, Rodriguez C, Sumano H (2009) Electroencephalographic evaluation of gold wire implants inserted in acupuncture points in dogs with epileptic seizures. Res Vet Sci 86:152–161

Gold MS, Levine JD (1996) DAMGO inhibits prostaglandin E2-induced potentiation of a TTX-resistant Na^+ current in rat sensory neurons in vitro. Neurosci Lett 212:83–86

Goldin A (2001) Resurgence of sodium channel research. Annu Rev Physiol 63:871–894

Gorter JA, Zurolo E, Lyer A, Fluiter K, van Vliet EA, Baayen JC, Aronica E (2010) Induction of sodium channel NaX (SCN7A) expression in rat and human hippocampus in temporal lobe epilepsy. Epilepsia 51:1791–1800

Grecksch G, Becker A, Schroeder H, Höllt V (1999) Involvement of delta-opioid receptors in pentylenetetrazol kindling development and kindling-related processes in rats. Naunyn Schmiedebergs Arch Pharmacol 360:151–156

Gu XQ, Yao H, Haddad GG (2001) Increased neuronal excitability and seizures in the Na^+/H^+ exchanger null mutant mouse. Am J Physiol Cell Physiol 281:C498–C503

Guo F, Yu N, Cai JQ, Quinn T, Zong ZH, Zeng YJ, Hao LY (2008a) Voltage-gated sodium channel Nav1.1, Nav1.3 and β1 subunit were up-regulated in the hippocampus of spontaneously epileptic rat. Brain Res Bull 75:179–187

Guo J, Liu J, Fu W, Ma W, Xu Z, Yuan M, Song J, Hu J (2008b) The effect of electroacupuncture on spontaneous recurrent seizure and expression of GAD67 mRNA in dentate gyrus in a rat model of epilepsy. Brain Res 1188:165–172

Guo J, Liu J, Fu W, Ma W, Xu Z, Yuan M, Zhou X, Hu J (2008c) Effect of electroacupuncture stimulation of hindlimb on seizure incidence and supragranular mossy fiber sprouting in a rat model of epilepsy. J Physiol Sci 58:309–315

Han JS (2003) Acupuncture: neuropeptide release produced by electrical stimulation of different frequencies. Trends Neurosci 26:17–22

Hargus NJ, Merrick EC, Nigam A, Kalmar CL, Baheti AR, Bertram EH III, Patel MK (2011) Temporal lobe epilepsy induces intrinsic alterations in Na channel gating in layer II medial entorhinal cortex neurons. Neurobiol Dis 41:361–376

He XP, Cao XD (1989) Effects of intrahippocampal delta-receptors on inhibition of electroconvulsive shock by electro-acupuncture. Zhongguo Yao Li Xue Bao 10:197–201

He XP, Chen BY, Zhu JM, Cao XD (1989) Change of Leu-enkephalin- and B-endorphin-like immunoreactivity in the hippocampus after electroconvulsive shock and electroacupuncture. Acupunct Electrother Res 14:131–139

He XP, Zhu JM, Huang DK, Li KY, Cao XD (1990) Effect of electroacupuncture on electroconvulsive shock—an autoradiographic study for opioid receptors. Sheng Li Xue Bao 42:149–154

Jindal V, Ge A, Mansky PJ (2008) Safety and efficacy of acupuncture in children: a review of the evidence. J Pediatr Hematol Oncol 30:431–442

Jutkiewicz EM, Baladi MG, Folk JE, Rice KC, Woods JH (2006) The convulsive and electroencephalographic changes produced by nonpeptidic delta-opioid agonists in rats: comparison with pentylenetetrazol. J Pharmacol Exp Ther 317:1337–1348

Jutkiewicz EM, Rice KC, Traynor JR, Woods JH (2005) Separation of the convulsions and antidepressant-like effects produced by the delta-opioid agonist SNC80 in rats. Psychopharmacology (Berl) 182:588–596

Kalume F, Yu FH, Westenbroek RE, Scheuer T, Catterall WA (2007) Reduced sodium current in Purkinje neurons from Nav1.1 mutant mice: implications of ataxia in severe myoclonic epilepsy in infancy. J Neurosci 27:11065–11074

Kang X, Chao D, Gu Q, Ding G, Wang Y, Balboni G, Lazarus LH, Xia Y (2009) δ-Opioid receptors protect from anoxic disruption of Na^+ homeostasis via Na^+ channel regulation. Cell Mol Life Sci 66:3505–3516

Kang X, Shen X, Xia Y (2013) Electroacupuncture-induced attenuation of experimental epilepsy: a comparative evaluation of acupoints and stimulation parameters. Evid Based Complement Alternat Med 2013, 149612. doi:10.1155/2013/149612
Ketelaars SOM, Gorter JA, van Vliet EA, Lopes da Silva FH, Wadman WJ (2001) Sodium currents in isolated rat CA1 pyramidal and dentate granule neurons in the post-status epilepticus model of epilepsy. Neuroscience 105:109–120
Kim ST, Jeon S, Park HJ, Hong MS, Jeong WB, Kim JH, Kim Y, Lee HJ, Park HJ, Chung JH (2008) Acupuncture inhibits kainic acid-induce hippocampal cell death in mice. J Physiol Sci 58:31–38
Kloster R, Larsson PG, Lossius R, Nakken KO, Dahl R, Xue XL, Zhou WX, Kinge E, Røssberg E (1999) The effect of acupuncture in chronic intractable epilepsy. Seizure 8:170–174
Koide S, Onishi H, Yamagami S, Kawakita Y (1992) Effects of morphine and D-Ala2-Leu5-enkephalin in the seizure-sensitive E1 mouse. Neurochem Res 17:779–783
Kong J, Gollub R, Huang T, Polich G, Napadow V, Hui K, Vangel M, Rosen B, Kaptchuk TJ (2007) Acupuncture de qi, from qualitative history to quantitative measurement. J Altern Complement Med 13:1059–1070
Liang JF, Xia Y (2012) Acupuncture modulation of central neurotransmitters. In: Xia Y, Ding G, Wu G (eds) Current research in acupuncture. Springer, New York, pp 1–36
Lin J, Deng QP, Zhang JW (2001) Observation on therapeutic effect of 160 cases of epilepsy treated with acupoint catgut embedding therapy. Zhongguo Zhen Jiu 21(11):14–15
Lin ZN (2008) What is the primary entity of Meridian? J Accord Integr Med 4:243–247
Lossin C (2009) A catalog of SCN1A variants. Brain Dev 31:114–130
Ma R, Zhang X, Liu Y, Li X, Yang C, Xiong J (2001) Clinical observation on treatment of tono-clonic attack of infantile epilepsy with acupuncture plus Xi Feng capsule. J Tradit Chin Med 42:276–278
Madar I, Lesser RP, Krauss G, Zubieta JK, Lever JR, Kinter CM, Ravert HT, Musachoi JL, Mathews WB, Dannals RF, Frost JJ (1997) Imaging of delta- and mu-opioid receptors in temporal lobe epilepsy by positron emission tomography. Ann Neurol 41:358–367
Madison DV, Nicoll RA (1988) Enkephalin hyperpolarizes interneurones in the rat hippocampus. J Physiol 398:123–130
Mantegazza M, Curia G, Biagini G, Ragsdale DS, Avoli M (2010) Voltage-gated sodium channels as therapeutic targets in epilepsy and other neurological disorders. Lancet Neurol 9:413–424
Mao JJ, Farrar JT, Armstrong K, Donahue A, Ngo J, Bowman MA (2007) De qi: Chinese acupuncture patients' experiences and beliefs regarding acupuncture needling sensation—an exploratory survey. Acupunct Med 25:158–165
Mao KY, Guo Z (2005) Observations on the efficacy of combined acupuncture and medicine for treating epilepsy secondary to cerebral infarction. Shanghai Zhen Jiu Za Zhi 24:17–18
Meisler MH, O'Brien JE, Sharkey LM (2010) Sodium channel gene family: epilepsy mutations, gene interactions and modifier effects. J Physiol 588:1841–1848
Moeller F, Muhle H, Wiegand G, Wolff S, Stephani U, Siniatchkin M (2010) EEG-fMRI study of generalized spike and wave discharges without transitory cognitive impairment. Epilepsy Behav 18:313–316
Morgan RJ, Soltesz I (2008) Nonrandom connectivity of the epileptic dentate gyrus predicts a major role for neuronal hubs in seizures. Proc Natl Acad Sci U S A 105:6179–6184
Negus SS, Gatch MB, Mello NK, Zhang X, Rice K (1998) Behavioral effects of the delta-selective opioid agonist SNC80 and related compounds in rhesus monkeys. J Pharmacol Exp Ther 286:362–375
Neumaier JF, Mailheau S, Chavkin C (1988) Opioid receptor-mediated responses in the dentate gyrus and CA1 region of the rat hippocampus. J Pharmacol Exp Ther 244:564–570
NIH consensus development panel on acupuncture (1998) Acupuncture. JAMA 280:1518–1524
Ogiwara I, Miyamoto H, Morita N, Atapour N, Mazaki E, Inoue I, Takeuchi T, Itohara S, Yanagawa Y, Obata K, Furuichi T, Hensch TK, Yamakawa K (2007) Nav1.1 localizes to axons of parvalbumin-positive inhibitory interneurons: a circuit basis for epileptic seizures in mice carry an Scn1a gene mutation. J Neurosci 27:5903–5914

Ohno Y, Sofue N, Ishihara S, Mashimo T, Sasa M, Serikawa T (2010) Scn1a missense mutation impairs GABAA receptor-mediated synaptic transmission in the hippocampus. Biochem Biophys Res Commun 400:117–122

Piskorowski RA, Chevaleyre V (2013) Delta-opioid receptors mediate unique plasticity onto parvalbumin-expressing interneurons in area CA2 of the hippocampus. J Neurosci 33: 14567–14578

Ragsdale DS (2008) How do mutant Nav1.1 sodium channels cause epilepsy. Brain Res Rev 58:149–159

Raynor K, Kong H, Chen Y, Yasuda K, Yu L, Bell GI, Reisine T (1994) Pharmacological characterization of the cloned κ-, δ-, and μ-opioid receptors. Mol Pharmacol 45:330–334

Remy C, Remy S, Beck H, Swandulla D, Hans D (2004) Modulation of voltage-dependent sodium channels by the δ-agonist SNC80 in acutely isolated rat hippocampal neurons. Neuropharmacology 47:1102–1112

Ren Y (2006) Acupuncture treatment of Jacksonian epilepsy-a report of 98 cases. J Tradit Chin Med 26:177–178

Rezaï X, Faget L, Bednarek E, Schwab Y, Kieffer BL, Massotte D (2012) Mouse delta opioid receptors are located on presynaptic afferents to hippocampal pyramidal cells. Cell Mol Neurobiol 32:509–516

Rezaï X, Kieffer BL, Roux MJ, Massotte D (2013) Delta opioid receptors regulate temporoammonic-activated feedforward inhibition to the mouse CA1 hippocampus. PLoS One 8, e79081

Rhodes TH, Lossin C, Vanoye CG, Wang DW, George AL Jr (2004) Noninactivating voltage-gated sodium channels in severe myoclonic epilepsy of infancy. Proc Natl Acad Sci U S A 101: 11147–11152

Rocha L, Cuellar-Herrera M, Velasco M, Velasco F, Velasco AL, Jiménez F, Orozco-Suarez S, Borsodi A (2007) Opioid receptor binding in parahippocampus of patients with temporal lobe epilepsy: its association with the antiepileptic effects of subacute electrical stimulation. Seizure 16:645–652

Sashihara S, Yanagihara N, Kobayashi H, Izumi F, Tsuji S, Murai Y, Mita T (1992) Overproduction of voltage-dependent Na^+ channels in the developing brain of genetically seizure-susceptible E1 mice. Neuroscience 48:285–291

Shi Z, Gong B, Jia Y, Huo Z (1987) The efficacy of electro-acupuncture on 98 cases of epilepsy. J Tradit Chin Med 7:21–22

Simmons ML, Chavkin C (1996) Endogenous opioid regulation of hippocampal function. Int Rev Neurobiol 39:145–196

Snead OC III (1986) Opiate-induced seizures: a study of mu and delta specific mechanisms. Exp Neurol 93:348–358

Song YP, Yang W, Guo HM, Han YY (2006) Clinical observation on acupuncture combined with medicine for treatment of infantile febrile convulsion. Zhongguo Zhen Jiu 26:561–562

Stafstrom CE (2007) Persistent sodium current and its role in epilepsy. Epilepsy Curr 7:15–22

Stutzmann JM, Böhme GA, Roques BP, Blanchard JC (1986) Differential electrographic patterns for specific mu- and delta-opioid peptides in rats. Eur J Pharmacol 123:53–59

Su X, Joshi SK, Kardos S, Gebhart GF (2002) Sodium channel blocking actions of the κ-opioid receptor agonist U50,488 contribute to its visceral antinociceptive effects. J Neurophysiol 87:1271–1279

Szaflarski JP, Difrancesco M, Hirschauer T, Banks C, Privitera MD, Gotman J, Holland SK (2010) Cortical and subcortical contributions to absence seizure onset examined with EEG/fMRI. Epilepsy Behav 18:404–413

Tian XS, Zhou F, Yang R, Xia Y, Wu GC, Guo JC (2008) Electroacupuncture protects the brain against acute ischemic injury via up-regulation of delta-opioid receptor in rats. Zhong Xi Yi Jie He Xue Bao 6:632–638

Toib A, Lyakhov V, Marom S (1998) Interaction between duration of activity and time course of recovery from slow inactivation in mammalian brain Na^+ channels. J Neurosci 18:1893–1903

Tortella FC, Echevarria E, Robles L, Mosberg HI, Holaday JW (1988) Anticonvulsant effects of mu (DAGO) and delta (DPDPE) enkephalins in rats. Peptides 9:1177–1181

Tortella FC, Long JB (1985) Endogenous anticonvulsant substance in rat cerebrospinal fluid after a generalized seizure. Science 228:1106–1108
Urca G, Frenk H, Liebeskind JC, Taylor AN (1977) Morphine and enkephalin: analgesic and epileptic properties. Science 197:83–86
Vreugdenhil M, Faas GC, Wadman WJ (1998) Sodium current in isolated rat CA1 neurons after kindling epileptogenesis. Neuroscience 86:99–107
Wang BE, Cheng JS (1992) The relationship between dynorphon and acupuncture anticonvulsion in hippocampus. Chin Sci Bull 37:1321–1323
Wang BE, Cheng JS (1994) Alteration of dynorphin1-8 and leu-enkephalin in rat hippocampus during seizure and electroacupuncture. Zhongguo Yao Li Xue Bao 15:155–157
Wang BE, Cheng JS (1995) Release of hippocampal dynorphin during electro-stimulated seizure and acupuncture anti-convulsion. Shanghai Zhen Jiu Za Zhi 14:33–34
Wang BE, Yang R, Cheng JS (1994) Effect of electroacupuncture on the level of preproenkephalin mRNA in rat during penicillin-induced epilepsy. Acupunct Electrother Res 19:129–140
Wang JC (2001) One hundred and twenty cases of epilepsy treated by three acupoints on back. Shanghai Zhen Jiu Za Zhi 20:20
Wang SM, Kain ZN, White P (2008) Acupuncture analgesia: I. The scientific basis. Anesth Analg 106:602–610
Wen GQ, He XZ, Lu Y, Xia Y (2010a) Effect of acupuncture on neurotransmitters/modulators. In: Xia Y, Cao XD, Wu GC, Cheng JS (eds) Acupuncture therapy for neurological diseases: a neurobiological view. Springer, Heidelberg, pp 120–142
Wen GQ, Yang YL, Lu Y, Xia Y (2010b) Acupuncture-induced activation of endogenous opioid system. In: Xia Y, Cao XD, Wu GC, Cheng JS (eds) Acupuncture therapy for neurological diseases: a neurobiological view. Springer, Heidelberg, pp 104–119
Wen T (2000) Treatment of 78 cases of epilepsy in narcotic abstaining. Zhongguo Zhen Jiu 20(4):34
Whitaker WRJ, Faull RLM, Dragunow M, Mee EW, Emson PC, Clare JJ (2001) Changes in the mRNAs encoding voltage-gated sodium channel type II and III in human epileptic hippocampus. Neuroscience 106:275–285
Williams P, Dou P, Dudek FE (2004) Epilepsy and synaptic reorganization in a perinatal rat model of hypoxia-ischemia. Epilepsia 45:1210–1218
Witkowski G, Szulczyk P (2006) Opioid μ receptor activation inhibits sodium currents in prefrontal cortical neurons via a protein kinase A- and C-dependent mechanism. Brain Res 1094:92–106
Wu D (1992) Mechanism of acupuncture in suppressing epileptic seizures. J Tradit Chin Med 12:187–192
Wu DZ, Ma JY, Li WJ (1986) Inhibitory effect of electro-acupuncture on penicillin-induced cortical epileptiform discharges. Sheng Li Xue Bao 38:325–331
Wu Y, Zou LP, Han TL, Zheng H, Caspi O, Wong V, Su Y, Shen KL (2008) Randomized controlled trial of traditional Chinese medicine (acupuncture and tuina) in cerebral palsy: part 1-any increase in seizure in integrated acupuncture and rehabilitation group versus rehabilitation group? J Altern Complement Med 14:1005–1009
Xia Y, Cao H, Zhang JH, Chen NY, Siegel K, Agulnik M, Haddad GG (2001) Effect of δ-opioid receptor activation on Na^+ channel expression in cortical neurons subjected to prolonged hypoxia in culture. Soc Neurosci Abstr Program No. 740.6
Xia Y, Fung ML, O'Reilly JP, Haddad GG (2000) Increased neuronal excitability after long-term O2 deprivation is mediated mainly by sodium channels. Brain Res Mol Brain Res 76:211–219
Xia Y, Zhao P, Xue J, Gu XQ, Sun X, Yao H, Haddad GG (2003) Na^+ channel expression and neuronal function in the Na^+/H^+ exchanger 1 null mutant mouse. J Neurophysiol 89:229–236
Xu ZF (2006) Clinical observation on treatment of epileptic seizure by combined catgut embedding and herbal medicine. Shanghai Zhen Jiu Za Zhi 25:13–14

Yajima Y, Narita M, Takahashi-Nakano Y, Misawa M, Nagase H, Mizoguchi H, Tseng LF, Suzuki T (2000) Effects of differential modulation of mu-, delta- and kappa-opioid systems on bicuculline-induced convulsions in the mouse. Brain Res 862:120–126

Yang J (1990) Treatment of status epilepticus with acupuncture. J Tradit Chin Med 10:101–102

Yang R, Cheng J (2010) Effect of acupuncture on epilepsy. In: Xia Y, Cao X, Wu G, Cheng J (eds) Acupuncture therapy for neurological diseases. Springer, Heidelberg, pp 326–364

Yu FH, Mantegazza M, Westenbroek RE, Robbins CA, Kalume F, Burton KA, Spain WJ, McKnight GS, Scheuer T, Catterall WA (2006) Reduced sodium current in GABAergic interneurons in a mouse model of severe myoclonic epilepsy in infancy. Nat Neurosci 9:1142–1149

Yuan H, Han JS (1985) Electroacupuncture accelerates the biogenesis of central enkephalins in the rat. Sheng Li Xue Bao 37:265–273

Zhang J, Li YZ, Zhuang LX (2006) Observation on therapeutic effect of 90 tonic-clonic epilepsy patients treated by catgut implantation therapy. Zhen Jiu Lin Chuang Za Zhi 22:8–10

Zhang JL, Zhang SP, Zhang HQ (2008a) Antiepileptic effect of electroacupuncture vs. vagus nerve stimulation in the rat thalamus. Neurosci Lett 441:183–187

Zhang JL, Zhang SP, Zhang HQ (2008b) Antiepileptic effect of electroacupuncture vs. vagus nerve stimulation on cortical epileptiform activities. J Neurol Sci 270:114–121

Zhang Z, Lu G, Zhong Y, Tan Q, Chen H, Liao W, Tian L, Li Z, Shi J, Liu Y (2010) fMRI study of mesial temporal lobe epilepsy using amplitude of low-frequency fluctuation analysis. Hum Brain Mapp 31:1851–1861

Zhao P, Guo JC, Hong SS, Bazzy-Asaad A, Cheng JS, Xia Y (2002) Electro-acupuncture and brain protection from cerebral ischemia: the role of delta-opioid receptor. Soc Neurosci Abstr 28:736

Zhao P, Ma MC, Qian H, Xia Y (2005) Down-regulation of delta-opioid receptors in Na^+/H^+ exchanger 1 null mutant mouse brain with epilepsy. Neurosci Res 53:442–446

Zhuang LX, Zhang J, Li YZ (2006) Clinical observation on catgut implantation at acupoint for treatment of general paroxysmal epilepsy. Zhongguo Zhen Jiu 26:611–613

Zieglgänsberger W, French ED, Siggins GR, Bloom FE (1979) Opioid peptides may excite hippocampal pyramidal neurons by inhibiting adjacent inhibitory interneurons. Science 205:415–417

Chapter 17
Delta Opioids in Protection of the Heart and Brain

Jan M. Schilling, Daniel N. Deussen, David M. Roth, and Hemal H. Patel

Abstract Delta opioids have actions both centrally and peripherally to produce cell protection. This chapter will focus on the role of delta opioid receptors in protection of the heart and brain, the classic and non-classic molecular pathways involved, and considerations to clinical translation. We will consider trigger, mediator and end effectors involved in acute protective pathways in heart and brain. Non-classic pathways will assess the role of delta opioid in remote and postconditioning, hibernation, sustained ligand activated protection, flavonols as opioid like ligands, and membrane microdomains in modulating delta opioid protective pathways.

Keywords Protection • Preconditioning • Postconditioning • Remote conditioning • Caveolae • Neuronal • Heart

Abbreviations

AKT	Protein kinase B
AV	Atrioventricular
BDNF	Brain derived neurotrophic factor
CABG	Coronary artery bypass graft
CV	Cranioventricular
Cx43	Connexin 43
DADLE	D-Ala^2, D-Leu^5-enkephalin
DOR	Delta opioid receptors
GSK3β	Glycogen synthase kinase 3β
HSP	Heat shock protein
IPC	Ischemic preconditioning
JNK-c	Jun N terminal kinase
MAPK	Mitogen activated protein kinase

J.M. Schilling, M.D. • D.N. Deussen, M.S. • D.M. Roth, M.D., Ph.D. • H.H. Patel Ph.D. (✉)
VA San Diego Healthcare System, 3350 La Jolla Village Drive, San Diego, CA 92161, USA

Department of Anesthesiology, University of California, San Diego, CA 92093, USA
e-mail: hepatel@ucsd.edu

Y. Xia (ed.), *Neural Functions of the Delta-Opioid Receptor*,
DOI 10.1007/978-3-319-25495-1_17

MCAO	Middle cerebral artery occlusion
mitoK_{ATP}	Mitochondrial ATP dependent potassium channel
mPTP	Mitochondrial permeability transition pore
NGF	Nerve growth factor
NOS	Nitric oxide synthase
PI3K	Phosphatidylinositol-4,5-bisphosphate 3-kinase
PKA	Protein kinase A
PKC	Protein kinase C
PKG	Protein kinase G
ppENk	Prepro encephalin
SA	Sinoatrial
SLP	Sustained ligand activated protection

1 Introduction

The heart and brain seem as though they are very different: the heart is a muscle in constant motion and utilizes fat as its primary fuel source whereas the brain is largely fixed in a hard shell, is made up largely of lipids and predominantly uses glucose as a fuel source. Though different, there are striking similarities between these two organs that echo redundant pathways of metabolism and function, each is essentially a bag of terminally differentiated cells that form interconnected networks with others of their type to perform coordinated functions, these cells in turn are supported by surrounding proliferative cells (i.e., fibroblasts or glia) that provide a scaffold for growth and function, are highly structured, sensitive to injury (especially oxygen deprivation), undergo aging, and have high mitochondrial metabolic activity. Given these similarities, it is likely that many adaptive and maladaptive signaling pathways are highly redundant in the heart and brain. This chapter will focus on the protective role of δ-opioid receptor (DOR) signaling in the heart and brain, assess common and divergent mechanisms, provide a description of novel and less well studied mechanisms of protection, and offer perspective for the future.

2 Ischemic Preconditioning and Organ Protection

In 1986 Murry et al. (1986) discovered the phenomenon of ischemic preconditioning (IPC) where brief periods of ischemia protected from subsequent lethal ischemia-reperfusion injury. Protection from myocardial infarction via IPC is a biphasic event: the acute phase occurs immediately after the IPC stimulus, lasts 1–3 h and is transient (Kuzuya et al. 1993). The delayed phase of protection is seen 12–24 h after the initial stimulus and lasts up to 72 h (Marber et al. 1993). IPC is a promiscuous stimulus that involves the initiation of many shared and interconnected signaling pathways (Hausenloy et al. 2005; Hausenloy and Yellon 2006). Importantly, this phenomenon is ubiquitous and occurs in virtually every organ tested including the brain. The protective pathways involve three key steps: trigger, mediator, and end effector phases. As will be discussed below, the δ-opioid receptor is a clear trigger in heart and brain for protection, and it recruits a series of mediators that converge upon end effectors to affect protection.

Since the original discovery of this endogenous protection nearly three decades ago, a myriad of studies have been undertaken to define the molecular mechanisms responsible and the subsequent identification of a therapeutic target that can be exploited clinically. Though there is no shortage of therapeutic targets, none have been clinically translated; this is likely due to many complicating factors leading to a disconnect between the preclinical models where robust protection is observed and the reality of the clinical setting where modifiers of protection (i.e., age, sex, disease, etc.) work to undermine the robustness of the uncomplicated conditions observed in many preclinical models.

3 The Classic Pathways of Protection

3.1 Triggers of Cardiac Protection: The δ-Opioid Receptor as a Therapeutic Target

The quest for a therapeutic agent must primarily draw from an understanding and exploitation of processes that are ongoing normally either in the presence or absence of disease. That is, a therapeutic, and exogenously administered substance must take the role of or modulate one of the players in the natural internal process to produce a beneficial effect. It is therefore essential to approach the systematic characterization of a disease and potential therapeutic initially with a clear understanding of the players in the internal milieu that have the potential to be exploited.

Opioids are predominantly thought of as mediators of analgesia and having an origin in the central nervous system. However, there is a large body of literature that describes the existence of opioid systems in peripheral organs including the heart (Howells et al. 1986; Mantelli et al. 1987; Springhorn and Claycomb 1989; Ventura et al. 1998; Weil et al. 1998) and gastrointestinal tract (Fickel et al. 1997; Tang et al. 1982). In addition to precursor molecules being found in the heart, opioid receptors are also present and stimulation by exogenously administrated opioid agonists has been shown to be cardioprotective (Fryer et al. 2000; Schultz et al. 1996, 1998a, b). In terms of endogenous opioids, it has been shown that the heart is predominantly a storage site for the precursor molecule, preproenkephalin (ppENK), which modulates the activity of δ-opioid receptors. The heart contains a greater amount of preproenkephalin mRNA than the brain, and this preproenkephalin mRNA was shown to be contained predominantly in the ventricles (Howells et al. 1986). Though the message level is high, it appears that the actual peptide level is very low.

In addition, the heart has also been shown to contain κ-opioid receptors and the myocyte is capable of expressing, synthesizing and releasing dynorphin, an endogenous activator of κ-opioid receptors (Ventura et al. 1994). Therefore, the myocardium is an organ capable of modulating its own activity or the activity of other organ systems by using opioids in an autocrine, paracrine, or endocrine fashion.

It has been argued that stresses such as ischemia increase both the expression of ppENK mRNA (Paradis et al. 1992) and the content of opioid peptides (Maslov and Lishmanov 1995). It is suggested that enkephalins are rapidly inactivated by several different peptidases (i.e., enkephalinases), which would account for their very short half-life and possible inactivity in the setting of acute protection (Bausback and Ward 1986; Llorens-Cortes et al. 1992; Schwartz et al. 1981). The ppENK gene encodes numerous peptides that include four Met^5-enkephalin sequences, one Leu^5-enkephalin, and two extended sequences for Met^5-enkephalin-Arg-Phe, and Met^5-enkephalin-Arg6-Gly^7-Leu^8. The predominant peptide found in the ventricles is Met^5-enkephalin-Arg-Phe by immunohistochemistry (Barron et al. 1995). It has been proposed that cardioprotection is mediated by δ-opioid receptor stimulation in the rat (Schultz et al. 1998a, b). Exogenous non-peptide opioid agonists (Fryer et al. 2000; Schultz et al. 1998a, b) and metabolically stable peptide agonists (Karck et al. 2001) have been shown to protect the heart from acute ischemia. This suggests that opioids have direct actions on the myocardium to induce protective signals. It has been shown that exogenous administration of endogenous opioid peptides can induce protective signals. Takasaki et al. (1999) showed that in adult isolated rabbit cardiomyocytes that Met^5-enkephalin, Leu^5-enkephalin, and Met^5-enkephalin-Arg-Phe produced a protective effect against simulated ischemia. In addition, Dickson et al. (2001) showed in isolated rabbit hearts that a concentrate obtained from a globally preconditioned heart is capable of protecting a virgin myocardium not undergoing preconditioning and that this effect is sensitive to blockade by naloxone suggesting the concentrate is partly composed of endogenous opioids (discussed later as "remote protection").

It is evident that opioids have protective potential. The remainder of the chapter will focus on classic and novel mechanisms of this protection and attempt to compare similarities in cardiac and neuronal protection induced by δ opioid receptor signaling.

3.2 Mediators of Cardiac Protection: Cytoplasmic Associated Signaling Events

The rapid nature of acute protection implies that post-translational modification of proteins (Fryer et al. 2001a, b) (i.e., phosphorylation or dephosphorylation) that alters protein localization and interactions may be key features of protection. Downstream signal cascades involved include: tyrosine kinases (Src), prosurvival kinases including phosphatidylinositol-3-kinase (PI3K) and protein kinase B (Akt) with downstream phosphorylation/inactivation of glycogen synthase kinase 3β (GSK3β), activation of several protein kinases including protein kinase G (PKG), protein kinase A (PKA), and protein kinase C (PKC) isoforms, nitric oxide synthase (NOS), and mitogen activated protein kinases (MAPK) (Hausenloy and Yellon 2006; Juhaszova et al. 2004).

The early events of cardiac protective signal transduction appear to involve the activation of Src kinases (Hattori et al. 2001; Maulik et al. 1996) which function to phosphorylate tyrosine residues on a number of downstream mediator and effector molecules. PKC has long been implicated in preconditioning (Armstrong et al. 1994). More recent studies have determined isoform selective (Fryer et al. 2001a, b; Inagaki et al. 2003) and membrane compartment (Hassouna et al. 2004; Ludwig et al. 2003) specific roles for PKC in regulating and potentially serving as a convergence point for cardiac protective signaling. PKC-ε has come to the forefront as the likely candidate for downstream transduction of signaling from membrane associated events to mitochondria where it may modulate mitochondrial ATP dependent potassium (mitoK_{ATP}) channels (Costa et al. 2005) and the mitochondrial permeability transition pore (mPTP) (Baines et al. 2003). Other protein kinases such as PKG (Costa et al. 2005) and PKA (Sanada et al. 2004) have also been implicated in preconditioning though the exact mechanisms are unknown.

Downstream kinase activation, as a result of Src or PKC, has been extensively studied in cardiac protection. Mitogen activated protein kinases (MAPK) are comprised of a number of kinases which are involved in cardiac protection and include extracellular signal-regulated kinase (ERK1/2) (Ping et al. 1999, p 38; (Weinbrenner et al. 1997), c-Jun N terminal kinase (JNK) (Haq et al. 1998), and big MAPK 1 (BMK1) (Cameron et al. 2004). A second survival kinase pathway activated by cardiac protective stimuli involves PI3K-Akt (Tong et al. 2000). Akt has been shown to activate a number of pro-survival/anti-apoptotic pathways (Hausenloy et al. 2005) with an important downstream target being GSK3β, whose phosphorylation keeps the mPTP from opening (Juhaszova et al. 2004). In the aging heart, a loss of δ-opioid receptor mediated cardioprotection was suggested to be dependent on failure to activate p38 MAPK and HSP27 (Peart et al. 2007). An additional molecule proposed to play a role in the protection is membrane and mitochondrial localized gap junction protein connexin 43 (Cx43). δ-receptor agonism with DADLE induces phosphorylation of Akt and GSK-3β in a H9c2 cell model and this phosphorylation is reduced by knockdown of Cx43 as is the protection (Ishikawa et al. 2012). It was suggested that Cx43 phosphorylation plays a crucial role in δ-opioid-induced suppression of gap junction permeability (Miura et al. 2007).

Opioids also produce a profound second window (delayed) of protection in the heart that is dependent on the induction of various mediators. The trigger events appear to involve not only direct engagement of opioid receptors (Macedo et al. 2014), but may involve a biphasic response of opioid receptor stimulation. At higher doses, classical signaling is engaged, whereas at low doses the delayed protection appears to be opioid receptor independent and may involve direct effects on mitochondrial reactive oxygen species generation (Patel et al. 2001, 2004a, b). The opioid induced protection was sensitive to free radical scavengers but not opioid receptor antagonism. Data also suggest that sarcolemmal KATP channels may be involved as additional triggers of delayed protection induced by opioids (Patel et al. 2005). Though the triggering events are divergent, the mediators induced appear to be common and include 12-lipooxygenase (Patel et al. 2003; Tsutsumi et al. 2006), cyclooxygenase-2 (Hausenloy and Yellon 2010; Patel et al. 2004a, b), and inducible nitric oxide synthase (Patel et al. 2004a, b). Other mediators of delayed protection include heme oxygenase, heat shock proteins, and superoxide dismutase (Baxter et al. 1994; Hausenloy and Yellon 2010).

As seen from the description above, a long list of varied mediators exist in survival signaling and offer a myriad of therapeutic targets.

3.3 End Effectors of Cardiac Protection: Mitochondria as a Target

Mitochondria are not only the source of energy for living organisms, but serve a role in determining the life/death fate of individual cells, whole organs, and ultimately the entire organism. Three main mitochondrial phenomenon: (1) opening of mito-K_{ATP} channels, (2) generation of a small ROS burst, and (3) maintenance of the mPTP have been linked to cardiac protective effects and they are likely not mutually exclusive. The first evidence for the involvement of mitochondria in preconditioning centered around the ability of virtually all preconditioning stimuli to be blocked by treatment with mitoK_{ATP} channel inhibitors, a concept first suggested by Gross and Auchampach (Auchampach et al. 1992; Gross and Auchampach 1992). The protective role of mitoK_{ATP} has been attributed to both triggering and mediator mechanisms. Treatment with the mitoK_{ATP} channel opener, diazoxide, produces cardiac protection and this has been tied to generation of ROS (Pain et al. 2000). Paradoxically, the generation of ROS in cells has been linked both to cellular damage and a more controlled, small burst to cellular signaling during preconditioning and decreased oxidant injury (Becker et al. 1999; VandenHoek et al. 1998, 2000; Yao et al. 1999). The source of this ROS burst is debatable but clearly involves some aspect of mitochondrial function. The mediator function of mitoK_{ATP} has been linked to preservation of the integrity of the mPTP (Akao et al. 2003). Mitochondrial permeability transition pore opening leads to induction of apoptosis as solutes enter the mitochondria and effectively shut down the electrochemical gradient necessary for energy generation and cytochrome C is released from the intermembrane space to the cytosol where it can activate caspases (Kroemer et al. 1998). Additionally, mitoK_{ATP} channel function has been linked to mitochondrial calcium homeostasis (Wang and Ashraf 1999) and bioenergetics (Kowaltowski et al. 2001) likely as a secondary function of maintaining mPTP integrity. Therapies that target mPTP opening protect the heart from ischemic injury, and it appears that the various pathway components described above converge upon this phenomenon as a likely end effector of cardiac protection (Murphy and Steenbergen 2007).

4 Connecting the Brain and the Heart

δ-Opioid receptor signaling can also influence the brain, and it is clear that the protective mechanisms in brain and heart are connected. There have been a lot of experiments to evaluate the connection between the central nervous system and the effects on the heart. Blinder et al. (2005) have shown that anatomically, vagal preganglionic neurons to the heart originate in the ventrolateral part of the nucleus ambiguous and project to postganglionic neurons in intracardiac ganglia, including the sinoatrial (SA), atrioventricular (AV) and cranioventricular (CV) ganglia. They furthermore demonstrate that these pathways selectively modulate heart rate, AV conduction and left ventricular contractility. Through retrograde labeling, it was shown that enkephalinergic terminals not only modulate heart rate, but also atrioventricular conduction and left ventricular contractility (Blinder et al. 2005). One experimental approach has been demonstrated in cats that received electroacupuncture resulting in activation of preganglionic parasympathetic neurons in the nucleus ambiguous likely through an enkephalinergic mechanism (Guo et al. 2012), which may bind δ- and μ-opioid receptors to induce tolerance in a focal cerebral ischemia model (Xiong et al. 2007).

Additionally, enkephanlins have hypotensive effects during high vagal activity, but produce hypertension during high sympathetic activity (Caffrey et al. 1985). The inhibiting effects of morphine in a model of right vagal nerve stimulation was observed by Kostelitz as early as 1959 (Kosterlitz and Taylor 1959). Rat and rabbit preparations showed responsiveness to opioid treatment and heart rate and blood pressure were decreased through acute stimulation of the right vagus nerve. Later experiments showed that intrinsic enkephalins are able to inhibit vagal control of heart rate, contractile function, and coronary blood flow in canine models (Caffrey 1999). It was suggested that the mechanism includes opioid receptors pre-junctionally on vagal nerve terminals or at a location close to the parasympathetic ganglia. In experimental injections of opioid antagonists into the ventrolateral periaqueductal grey matter, the only opioid receptor antagonist that was able to inhibit hypotension in response to hemorrhage in a significant manner was a δ opioid receptor antagonist (naltrindole) (Cavun et al. 2001).

5 Neuronal Protection

5.1 δ-Opioid Receptors and Neuronal Protection

δ-opioid receptor mediated protection is robust in the brain. Xia et al. have published a series of papers that demonstrate the neuroprotective role of DOR in hypoxic/excitotoxic stress (Ma et al. 2005; Yang et al. 2009; Zhang et al. 2000, 2001, 2002, 2006). The classical mechanism for neuronal protection through δ-opioid signaling has been recently summarized in a review by Feng et al. (2012) where DOR signaling triggers (depending on stress duration and severity) multiple independent mechanisms leading to: (1) preservation of neuronal survival, (2) stabilization of ionic homeostasis, (3) an increase in pro-survival signaling, and (4) increased antioxidative capacity (Feng et al. 2009).

Both peptide and non-peptide agonists of DOR have been shown to produce neuroprotection. Kao et al. demonstrated that BW373U86 (non-peptide DOR agonist) and dynorphin A 1–13 (κ-agonist) were able to attenuate cerebral ischemic injury in a focal cerebral I/R rat model (Kao et al. 2008). Furthermore, BW373U86 was shown to increase survival time of mice during lethal hypoxia (Bofetiado et al. 1996), and may involve cAMP response element-binding protein (CREB) in mediating these neuroprotective effects (Yang et al. 2014). D-Ala^2, D-Leu^5-enkephalin (DADLE, a peptide DOR agonist) produced postconditioning (Wang et al. 2011), as well as protection from mild forebrain ischemia (Iwata et al. 2007). In these protective paradigms, naltrindole or the Akt antagonist LY294002 abolished protection suggesting activation of similar signaling cascades as the heart. DADLE was also shown to improve behavioral deficits as assessed by the Morris water maze and tests for motor function likely via preservation of CA1 neurons (Su et al. 2007). In a middle cerebral artery occlusion model, DADLE showed decreased infarct size, near complete preservation of the striata, blunted p-53 mRNA expression, and an increase in neurotrophic factors (Borlongan et al. 2009).

After focal cerebral ischemia, different sublocalization of receptors was noticed pre- and postsynaptically. Boutin et al. showed that μ- and δ-receptors were localized postsynaptically on cortical interneurons, while κ-receptors were found presynaptically on cortical afferent pathways (Boutin et al. 1999). It has been observed that a postsynaptic decrease of K^+ efflux stabilizes the membrane and is neuroprotective suggesting receptor localization may dictate specific events in the brain that lead to neuroproteciton (Chao et al. 2007a, b, 2008, 2009). Also, NMDA receptors have been shown to be involved in protecting the brain through a reduction in Na^+ and Ca^{2+} entering the cell (Chao et al. 2009; Chao and Xia 2010). Neuronal δ-opioid receptors, but not μ- and κ-opioid receptor activation, has been shown to protect neocortical neurons from glutamate excitotoxicity (Zhang et al. 2002, 2006). The dimeric enkephalin peptide, biphalin, has also been shown to reduce glutamate induced neuronal damage and this protection is abolished by naltrexone (a non-selective opioid receptor antagonist), while reduced cell death is seen even after delayed biphalin administration after glutamate challenge (Kawalec et al. 2011).

Protective mechanisms previously outlined in the heart, can similarly be found in the brain. In vitro studies in CHO/DOR cells have shown that activating Gi/Go-coupled DOR by norclozapine, an atypical antipsychotic drug and potent DOR agonist, results in activation of AKT likely via PI3Kα activation, regulated through Src-dependent transactivation of the IGF-I receptor (Olianas et al. 2011). Other studies utilizing PC12 cells have found PI3K/AKT/NF-κB signaling as one of the main downstream signaling pathways in nerve growth factor (NGF)-induced temporal expression of the DOR gene (Chen et al. 2008). In NGF-responsive rodent neuronal cells, DADLE acts through the NGF-induced G protein-coupled Oprd1 receptors and mediates neuroprotection and differentiation partly through MAPK and PI3K/AKT signaling (Sen et al. 2013). Furthermore, cross-talk between ERK and p38 MAPKs under the control of the DOR-G protein-protein kinase C pathway has been suggested (Ma et al. 2005). Additionally, attenuation of ERK 1/2 activity and TNF-alpha production has been shown to be neuroprotective in the rat brain after treatment with the ultra short acting synthetic opioid remifentanil (Jeong et al. 2012). Chao et al. suggested that activation of DOR protects the cortex against anoxia- or oxygen glucose deprivation-induced derangement of potassium homeostasis, and that this protection occurs via a PKC-dependent and PKA-independent pathway (Chao et al. 2007a, b). The role for mitochondrial KATP channels and free radical production has also been suggested to be neuroprotective and DOR dependent in the brain (Lim et al. 2004). In the brain, Cx43 and the mPTP are also proposed as targets to prevent injury (Azarashvili et al. 2011; Boltze et al. 2011). Lastly, Zhu et al. suggested that DOR activation plays an important role in neuroprotection against mitochondrial respiratory chain injury where in a sodium azide model of neuronal damage, δ- but not μ-opioid receptor activation was protective (Zhu et al. 2009). More recently the same group described the mechanism of DADLE induced protection against mitochondrial dysfunction from acute sodium azide involves PKC-ERK and mitochondrial ERK activation (Zhu et al. 2011). Additional mechanisms may involve increased antioxidant capacity upon DOR activation where an increase in the antioxidant enzymes glutathione peroxidase and superoxide dismutase was observed (Yang et al. 2009). Such similarities suggest much overlap in the signaling mechanism leading to protection in the heart and brain and suggest that some of these may be common ubiquitous mechanisms operative in many different organs and likely represent robust therapeutic targets.

5.2 *Brain Specific Mechanisms of Protection*

Mechanisms specific to the brain involve the brain-derived neurotrophic factor (BDNF)-TrkB pathways. Narita et al. suggested that DOR play a crucial role in neurogenesis and neuroprotection, mainly through the activation of Trk-dependent tyrosine kinase, which could link to specific protective pathways (i.e., PI3K, PKC, CaMKII and MEK) (Narita et al. 2006). Other studies looking at TrkB signaling found DOR rescues full length TrkB to limit cerebral ischemia (Tian et al. 2013a, b). Furthermore, it was shown that prolonged hypoxia accompanies a down-regulation of BDNF-TrkB signaling which may lead to an increase in TNF-alpha in the cortex; however, when stimulated with a DOR agonist, BDNF-TrkB signaling was up-regulated and decreased TNF-alpha levels in the hypoxic cortex were observed (Tian et al. 2013a, b).

6 The Non-classic Pathways of Ischemic Protection

6.1 *Alternatives to Preconditioning: Role of Opioids in Remote Protection and Postconditioning*

The major limitation of ischemic preconditioning both in the acute and delayed setting is lack of clinical applicability. In order to be effective, the ischemic event would need to be predicted and the patient undergo prophylactic intervention. To overcome these limitations two alternatives were proposed that have much clearer clinical applicability: remote protection and postconditioning.

Remote protection shows that non-lethal injury/ischemia to a vascular bed that is "remote" to the organ of interest can lead to the subsequent release upon reperfusion of humoral factors that produce organ protection. In the early 1990s Przyklenk and colleagues detected a transfer of the protective effect of preconditioning in one coronary bed to another remote coronary bed after reperfusion (Przyklenk et al. 1993). It was initially observed in a rat model, that mesenteric artery occlusion-reperfusion led to protection of the heart from ischemia-reperfusion injury and this protection could be blocked by treatment with naloxone, a non selective opioid receptor antagonist (Patel et al. 2002). The initial observation was later expanded to suggest δ-opioid receptor specific effects (Weinbrenner et al. 2004). Interestingly, it was shown that dialyzed blood samples from human individuals exposed to short bouts of high intensity exercise or multiple cycles of upper limb ischemia could produce subsequent protection of rabbit myocardium exposed to ischemia-reperfusion injury and the protection was attenuated by naloxone treatment (Michelsen et al. 2012). Such data suggest the peripheral organs release opioids in response to stress and these opioids then act upon the myocardium to produce protection. The findings were later expanded to suggest that remote protection not only involved activation of δ-opioid receptors but also required interaction with adenosine receptors (Surendra et al. 2013). Ultimately, this phenomenon was found to be ubiquitous and functional in brain ischemia as well (Zhou et al. 2011).

Postconditioning involves multiple periods of reperfusion alternated with brief periods of ischemia immediately following the index ischemic event (Vinten-Johansen and Shi 2013). These intermittent restraints of blood flow during early reperfusion have proven their protective potential in clinical trials in the heart, brain, spinal cord, and other organs (Zhao 2010). Zhao et al. showed that postconditioning, with three cycles of 30 s reperfusion alternating with 30 s of LAD occlusion, is as effective as pre-conditioning for 5 min (Zhao et al. 2003). The main advantage of postconditioning is the ease of transfer into the clinical setting, because patients are usually already hospitalized with acute symptoms. Postconditioning has probable applications in bypass surgery, coronary surgery, transplantation and revascularization.

Postconditioning can also be induced pharmacologically with treatment with various agonists at the time of reperfusion. δ-opioid receptors have been shown to be involved in postconditioning where treatment with agonists induces postconditioning and antagonists inhibit postconditioning (Jang et al. 2008; Tsutsumi et al. 2007). This effect induced by opioids as well as other agents appears to involve generating signaling bursts of reactive oxygen species (Penna et al. 2006; Tsutsumi et al. 2007). Postconditioning with remifentanil leads to similar protection when compared to ischemic postconditioning and involves both δ- and κ-opioid receptor activation (Guo et al. 2011; Wong et al. 2010). These studies provided evidence that pre- and postconditioning are at least partly dependent on opioid receptors and both strategies have justification in clinical settings of cardiac protection. Whether postconditioning offers additional protection to preconditioning, has been tested in an open-chest canine model by Halkos and colleagues where the effect of preconditioning was not augmented by a subsequent postconditioning stimulus (Halkos et al. 2004).

6.2 *Opioids and Hibernation*

Hibernation is a phenomenon in which an organism undergoes of profound state of inactivity and suppression of metabolism for an extended period of time to then emerge almost instantly from this state unharmed and ready to maintain waking activity and metabolism. A DOR agonist like compound has been suggested to be the trigger for hibernation (Horton et al. 1998; Oeltgen et al. 1982, 1988). It was shown, that a δ-agonist (Delt-D(var)) as well as plasma from the hibernating woodchuck, but not summer activated woodchuck plasma, protects the brain in a model of middle cerebral artery occlusion (MCAO) (Govindaswami et al. 2008). For further reading about this concept, a review by Borlogan et al. links hibernation and neuroprotection (Borlongan et al. 2004).

6.3 *Flavonoids as Opioid Agonists and Protective Molecules*

Several studies show a protective role of flavonoids at the membrane surface through interactions with both lipid and protein components of these membranes (Hendrich 2006; Oteiza et al. 2005). Previous studies showed flavonoid binding with various receptors: androgen and/or estrogen, insulin-like growth factor-1, epidermal growth factor, aryl hydrocarbon, Fas, vascular endothelial growth factor, and laminin (Patra et al. 2008). Flavonoids possess both opioid receptor antagonist and agonist activity. Structure-activity relationships of flavonoids with opioid receptor ligands show binding activity *in vitro* (Katavic et al. 2007) and flavonoids reduce morphine-induced opiate withdrawal (Capasso 2007; Capasso et al. 1998, 2000; Trang et al. 2003). Other groups have concluded that antinociception induced by flavonoids underlies the activation of the opioid system since this observed analgesic effect was reversed by naloxone (Maleki-Dizaji et al. 2007). Collectively, such studies provide evidence for the interaction of flavonoids with opioid receptors. We have previously shown that the flavonoid epicatechin has DOR binding activity and promotes cardiac protection (Panneerselvam et al. 2010). Furthermore, we recently published that epicatechin induced cardioprotection is mediated through mitochondrial targeted effects where the opioid specific effect of epicatechin impacted mitochondrial structure and function (Panneerselvam et al. 2013). Such data suggest two important possibilities: (1) there may be natural sources of compounds that have DOR activity that may be of benefit to the heart thereby limiting the side effects of direct peptide and non-peptide agonists and (2) mitochondrial actions of opioids suggest that there may be intracellular sites where opioid receptors may be localized providing new modalities for modulating cell physiology and function and providing novel therapeutic targets for disease.

6.4 Sustained Ligand-Activated Preconditioning

Marmor et al. (2004) observed patients in methadone programs in the context of cardiovascular disease. Their epidemiological observations implicated a cardioprotective effect of long-term opioid exposure, with significantly fewer occurrences of severe coronary artery disease compared to frequency-matched decedents. Morphine is a frequently used opioid for pain modulation and substance abuse and is among the best-studied opioid receptor agonist. Though classified as a μ-opioid receptor agonist, morphine has strong affinity for δ- and κ-receptors, and as such it is a non-specific opioid agonist. Morphine injections during reperfusion in anesthetized open-chest Wistar rats produce reductions in infarct-size that are abolished by δ-opioid receptor antagonism (Schultz et al. 1997). Peart and colleagues have shown through a series of elegant studies that chronic exposure to morphine produced a profound protected state in the heart (Peart and Gross 2004a, b, 2006; Peart et al. 2011). These protective effects were not blocked by classic inhibitors of acute protection but were rather shown to involve PKA activation (Peart and Gross 2006). Interestingly, it was observed that the effects of chronic opioid stimulation resulted in tolerance to central analgesic effects but not the cardioprotective effects (Peart and Gross 2004a, b). It is possible to speculate that these divergent central and peripheral effects of chronic opioids may be related to differential membrane receptor localization to specific membrane microdomains or different kinetics and signaling related to opioid receptor desensitization in the heart vs. brain.

6.5 δ-Opioid Protection and Plasma Membrane Microdomains

An emerging idea in signal transduction suggests signaling molecules exist as dynamic, spatially organized multi-protein complexes in lipid-rich microdomains of the plasma membrane continuously forming and dissociating under basal or stimulated conditions (Okamoto et al. 1998). Caveolae are cholesterol and sphingolipid-enriched microenvironments that form microscopically distinct flask-like invaginations of the plasma membrane (Palade 1953). Our laboratory as well as others, have shown that the caveolar structural proteins, caveolins, act as scaffolding molecules to aid in localization and regulation of receptors (Insel et al. 2005) and signaling molecules to facilitate coordinated, precise and rapid regulation of cell function (Lisanti et al. 1994). Caveolins function as chaperones and scaffolds recruiting signaling molecules to caveolae to provide direct temporal and spatial regulation of signal transduction (Shaul and Anderson 1998; Williams and Lisanti 2004). Caveolins can inhibit activity of signaling proteins by interaction of the caveolin scaffolding domain with a caveolin binding motif present in many proteins found in caveolae including eNOS and ERK1/2 (Engelman et al. 1998; Feron et al. 1998; Kamoun et al. 2006). Alternatively, caveolins can promote signaling via enhanced receptor-effector coupling or enhanced receptor affinity when caveolins are up regulated or overexpressed (Feron and Balligand 2006; Raikar et al. 2006). This has led to the concept of a "caveolar paradox" in which caveolins may produce direct allosteric inhibition of molecules such as eNOS under basal conditions but facilitate increased signaling upon agonist stimulation through compartmentalization (Feron and Balligand 2006; Feron and Kelly 2001). Recent data suggest that Cav-1 is phosphorylated producing physiological consequences (Swaney et al. 2006). No studies have established the actual phosphorylated state of Cav-3 though putative phosphorylation sites exist. Caveolins exist as oligomeric complexes that allow them to form high molecular weight scaffolds (Monier et al. 1995; Woodman et al. 2004). Importantly, Cav-3 mutations have revealed that an inability to form oligomeric complexes leads to skeletal muscle defects (Woodman et al. 2004). Studies in failing hearts suggest that caveolins dissociate from caveolae (Ratajczak et al. 2003) presumably as the oligomer is broken down.

We have shown that caveolae and caveolins are critical for protecting the heart from ischemia-reperfusion injury (Tsutsumi et al. 2008, 2010a, b). Furthermore, we have shown a connection with opioid signaling and caveolin (Head et al. 2005; Patel et al. 2006; Peart et al. 2014; Tsutsumi et al. 2010a, b, 2014; Zhao et al. 2006). We initially showed that opioid receptors localized to caveolar microdomains with DOR being the predominant receptor in the cardiac myocyte (Head et al. 2005; Patel et al. 2006). We further showed that DOR mediated protection via SNC-121 (a non-peptide δ opioid agonist) was abolished by methyl-β-cyclodextrin, a compound that disrupts caveolae, suggesting that DOR localized to caveolae were critical to cardiac protection (Patel et al. 2006). In a follow-up study we showed that caveolin-3 knockout mice that lack cardiac myocyte caveolae were unable to be protected by SNC-121 and the endogenous IPC-like protection observed in heart specific Cav-3 overexpressing mice could be attenuated by naloxone treatment (Tsutsumi et al. 2010a, b). Such data further support the notion that DOR localization to caveolae is critical to cardiac protection. Though classic modes of protection induced by opioids are dependent on caveolae, non conventional modes such as sustained ligand-activated preconditioning (SLP) appear to be caveolin-independent (See Hoe et al. 2014). Additionally, it appears that caveolae may be a critical feature of loss of protection in situations such as the aged or diseased myocardium where protective pathways are compromised (Peart et al. 2014). Though caveolin proteins have been described in neurons and their role in neuronal protection appears to be critical, the link to opioid receptor signaling and caveolin in neuronal protection has yet to be established (Head et al. 2008, 2010). Such data suggest that caveolins may be important targets for modulation of opioid receptor signaling in various organ systems and may be key control points for protective signaling.

7 From Bench to Bedside: Clinical Studies and Implications

Immunohistological investigations in human (Sobanski et al. 2014) as well as porcine (Theisen et al. 2014) heart tissue show the existence of δ-, κ- and μ-opioid receptors. Messenger RNA encoding the δ-OR has been identified in human ventricles and atria (Bell et al. 2000). When isolated human right atrial trabeculae were subjected to hypoxia, opioid agonists remifentanil and sufentanil produced myocardial protection, measured as sustained force of contraction (Lemoine et al. 2011). Similar effects were observed with morphine though the protection was lost in aged human hearts (Peart et al. 2014).

Promising pilot studies of ischemic postconditioning in the clinical setting utilizing balloon angioplasty (Laskey 2005), support clinical relevance and the need for further research into underlying mechanisms. Endogenous opioids are essential for the myocardial protection in humans. When naloxone was administered intravenously during balloon inflation, the protective effects were significantly reduced compared to placebo treatment (Tomai et al. 1999). In human patients treatment with 40 mg of morphine before cardiopulmonary bypass improved global myocardial function, compared to administration of 1000 μg fentanyl (Murphy et al. 2006). Morphine also suppressed the inflammatory response (IL-6, CD 11b, CD 18, postoperative hyperthermia) when compared to fentanyl when used during anesthesia for cardiopulmonary bypass surgery (Murphy et al. 2007). Remifentanil, a short-acting opioid receptor agonist, reduces the hemodynamic and metabolic responses in coronary artery bypass graft (CABG) surgery patients, when used as an addition to fentanyl/isoflourane based anesthesia (Pleym et al. 2004). Xu et al. also investigated the cardioprotective properties of remifentanil in off-pump CABG surgery and found a positive benefit (Xu et al. 2009).

8 Concluding Remarks

The heart and brain contain overlapping δ-opioid receptor mediated pathways likely as a consequence of highly conserved signaling that preserve terminally differentiated, highly structured, metabolic, and stress sensitive cells (i.e., cardiac myocytes and neurons) for vital organ and organism preservation. Though classic mechanisms have been described over nearly the past three decades the clinical reality is that no particular compound or targeted approach has been shown to be of rigorous clinical benefit. The non-classic pathways and approaches therefore define a logical basis and path for continuing the quest to define clinically relevant therapeutic interventions for organ protection. The δ-opioid receptor system is poised to be not only a model system for such interventions but also a likely endpoint for clinical translation.

References

Akao M, O'Rourke B, Kusuoka H, Teshima Y, Jones SP, Marban E (2003) Differential actions of cardioprotective agents on the mitochondrial death pathway. Circ Res 92:195–202

Armstrong S, Downey JM, Ganote CE (1994) Preconditioning of isolated rabbit cardiomyocytes: induction by metabolic stress and blockade by the adenosine antagonist SPT and calphostin C, a protein kinase C inhibitor. Cardiovasc Res 28:72–77

Auchampach JA, Grover GJ, Gross GJ (1992) Blockade of ischaemic preconditioning in dogs by the novel ATP dependent potassium channel antagonist sodium 5-hydroxydecanoate. Cardiovasc Res 26:1054–1062

Azarashvili T, Baburina Y, Grachev D, Krestinina O, Evtodienko Y, Stricker R, Reiser G (2011) Calcium-induced permeability transition in rat brain mitochondria is promoted by carbenoxolone through targeting connexin43. Am J Physiol Cell Physiol 300:C707–C720

Baines CP, Song CX, Zheng YT, Wang GW, Zhang J, Wang OL, Guo Y, Bolli R, Cardwell EM, Ping P (2003) Protein kinase cepsilon interacts with and inhibits the permeability transition pore in cardiac mitochondria. Circ Res 92:873–880

Barron BA, Oakford LX, Gaugl JF, Caffrey JL (1995) Methionine-enkephalin-Arg-Phe immunoreactivity in heart tissue. Peptides 16:1221–1227

Bausback HH, Ward PE (1986) Degradation of low-molecular-weight opioid peptides by vascular plasma membrane aminopeptidase. Biochim Biophys Acta 882:437–444

Baxter GF, Marber MS, Patel VC, Yellon DM (1994) Adenosine receptor involvement in a delayed phase of myocardial protection 24 hours after ischemic preconditioning. Circulation 90:2993–3000

Becker LB, VandenHoek TL, Shao Z-H, Li C-Q, Schumacker PT (1999) Generation of superoxide in cardiomyocytes during ischemia before reperfusion. Am J Physiol Heart Circ Physiol 277:H2240–H2246

Bell SP, Sack MN, Patel A, Opie LH, Yellon DM (2000) Delta opioid receptor stimulation mimics ischemic preconditioning in human heart muscle. J Am Coll Cardiol 36:2296–2302

Blinder KJ, Johnson TA, Massari VJ (2005) Enkephalins and functionally specific vagal preganglionic neurons to the heart: ultrastructural studies in the cat. Auton Neurosci 120:52–61

Bofetiado DM, Mayfield KP, D'Alecy LG (1996) Alkaloid delta agonist BW373U86 increases hypoxic tolerance. Anesth Analg 82:1237–1241

Boltze J, Kranz A, Wagner DC, Reymann K, Reiser G, Hess DC (2011) Recent advances in basic and translational stroke research. Expert Rev Neurother 11:199–202

Borlongan CV, Wang Y, Su TP (2004) Delta opioid peptide (D-Ala 2, D-Leu 5) enkephalin: linking hibernation and neuroprotection. Front Biosci 9:3392–3398

Borlongan CV, Hayashi T, Oeltgen PR, Su TP, Wang Y (2009) Hibernation-like state induced by an opioid peptide protects against experimental stroke. BMC Biol 7:31

Boutin H, Dauphin F, MacKenzie ET, Jauzac P (1999) Differential time-course decreases in nonselective, mu-, delta-, and kappa-opioid receptors after focal cerebral ischemia in mice. Stroke 30:1271–1277

Caffrey JL (1999) Enkephalin inhibits vagal control of heart rate, contractile force and coronary blood flow in the canine heart in vivo. J Auton Nerv Syst 76:75–82

Caffrey JL, Wooldridge CB, Gaugl JF (1985) The interaction of endogenous opiates with autonomic circulatory control in the dog. Circ Shock 17:233–242

Cameron SJ, Itoh S, Baines CP, Zhang C, Ohta S, Che W, Glassman M, Lee JD, Yan C, Yang J, Abe J (2004) Activation of big MAP kinase 1 (BMK1/ERK5) inhibits cardiac injury after myocardial ischemia and reperfusion. FEBS Lett 566:255–260

Capasso A (2007) The effect of flavonol glycosides on opiate withdrawal. Med Chem 3:327–331

Capasso A, Piacente S, Pizza C, Sorrentino L (1998) Flavonoids reduce morphine withdrawal in-vitro. J Pharm Pharmacol 50:561–564

Capasso A, Saturnino P, Simone FD, Aquino R (2000) Flavonol glycosides from Aristeguietia discolor reduce morphine withdrawal in vitro. Phytother Res 14:538–540

Cavun S, Resch GE, Evec AD, Rapacon-Baker MM, Millington WR (2001) Blockade of delta opioid receptors in the ventrolateral periaqueductal gray region inhibits the fall in arterial pressure evoked by hemorrhage. J Pharmacol Exp Ther 297:612–619

Chao D, Xia Y (2010) Ionic storm in hypoxic/ischemic stress: can opioid receptors subside it? Prog Neurobiol 90:439–470

Chao D, Bazzy-Asaad A, Balboni G, Xia Y (2007a) Delta-, but not mu-, opioid receptor stabilizes K(+) homeostasis by reducing Ca(2+) influx in the cortex during acute hypoxia. J Cell Physiol 212:60–67

Chao D, Donnelly DF, Feng Y, Bazzy-Asaad A, Xia Y (2007b) Cortical delta-opioid receptors potentiate K^+ homeostasis during anoxia and oxygen-glucose deprivation. J Cereb Blood Flow Metab 27:356–368

Chao D, Bazzy-Asaad A, Balboni G, Salvadori S, Xia Y (2008) Activation of DOR attenuates anoxic K^+ derangement via inhibition of Na^+ entry in mouse cortex. Cereb Cortex 18:2217–2227

Chao D, Balboni G, Lazarus LH, Salvadori S, Xia Y (2009) Na^+ mechanism of delta-opioid receptor induced protection from anoxic K^+ leakage in the cortex. Cell Mol Life Sci 66:1105–1115

Chen YL, Law PY, Loh HH (2008) NGF/PI3K signaling-mediated epigenetic regulation of delta opioid receptor gene expression. Biochem Biophys Res Commun 368:755–760

Costa AD, Garlid KD, West IC, Lincoln TM, Downey JM, Cohen MV, Critz SD (2005) Protein kinase G transmits the cardioprotective signal from cytosol to mitochondria. Circ Res 97:329–336

Dickson EW, Blehar DJ, Carraway RE, Heard SO, Steinberg G, Przyklenk K (2001) Naloxone blocks transferred preconditioning in isolated rabbit hearts. J Mol Cell Cardiol 33:1751–1756

Engelman JA, Chu C, Lin A, Jo H, Ikezu T, Okamoto T, Kohtz DS, Lisanti MP (1998) Caveolin-mediated regulation of signaling along the p42/44 MAP kinase cascade in vivo. A role for the caveolin-scaffolding domain. FEBS Lett 428:205–211

Feng Y, Chao D, He X, Yang Y, Kang X, HL L, Xia Y (2009) A novel insight into neuroprotection against hypoxic/ischemic stress. Sheng Li Xue Bao 61:585–592

Feng Y, He X, Yang Y, Chao D, Lazarus LH, Xia Y (2012) Current research on opioid receptor function. Curr Drug Targets 13:230–246

Feron O, Balligand J-L (2006) Caveolins and the regulation of endothelial nitric oxide synthase in the heart. Cardiovasc Res 69:788–797

Feron O, Kelly RA (2001) The Caveolar paradox: suppressing, inducing, and terminating eNOS signaling. Circ Res 88:129–131

Feron O, Dessy C, Opel DJ, Arstall MA, Kelly RA, Michel T (1998) Modulation of the endothelial nitric-oxide synthase-caveolin interaction in cardiac myocytes. Implications for the autonomic regulation of heart rate. J Biol Chem 273:30249–30254

Fickel J, Bagnol D, Watson SJ, Akil H (1997) Opioid receptor expression in the rat gastrointestinal tract: a quantitative study with comparison to the brain. Mol Brain Res 46:1–8

Fryer RM, Hsu AK, Nagase H, Gross GJ (2000) Opioid-induced cardioprotection against myocardial infarction and arrhythmias: mitochondrial versus sarcolemmal ATP-sensitive potassium channels. J Pharmacol Exp Ther 294:451–457

Fryer RM, Patel HH, Hsu AK, Gross GJ (2001a) Stress-activated protein kinase phosphorylation during cardioprotection in the ischemic myocardium. Am J Physiol Heart Circ Physiol 281:H1184–H1192

Fryer RM, Wang Y, Hsu AK, Gross GJ (2001b) Essential activation of PKC-delta in opioid-initiated cardioprotection. Am J Physiol Heart Circ Physiol 280:H1346–H1353

Govindaswami M, Brown SA, Yu J, Zhu H, Bishop PD, Kindy MS, Oeltgen PR (2008) Delta 2-specific opioid receptor agonist and hibernating woodchuck plasma fraction provide ischemic neuroprotection. Acad Emerg Med 15:250–257

Gross G, Auchampach J (1992) Blockade of ATP-sensitive potassium channels prevents myocardial preconditioning in dogs. Circ Res 70:223–233

Guo HT, Zhang RH, Zhang Y, Zhang LJ, Li J, Shi QX, Wang YM, Fan R, Bi H, Yin W, Pei JM (2011) Endogenous kappa-opioid peptide mediates the cardioprotection induced by ischemic postconditioning. J Cardiovasc Pharmacol 58:207–215

Guo ZL, Li M, Longhurst JC (2012) Nucleus ambiguus cholinergic neurons activated by acupuncture: relation to enkephalin. Brain Res 1442:25–35

Halkos ME, Kerendi F, Corvera JS, Wang NP, Kin H, Payne CS, Sun HY, Guyton RA, Vinten-Johansen J, Zhao ZQ (2004) Myocardial protection with postconditioning is not enhanced by ischemic preconditioning. Ann Thorac Surg 78:961–969

Haq SE, Clerk A, Sugden PH (1998) Activation of mitogen-activated protein kinases (p38-MAPKs, SAPKs/JNKs and ERKs) by adenosine in the perfused rat heart. FEBS Lett 434:305–308

Hassouna A, Matata BM, Galinanes M (2004) PKC {epsilon} is upstream and PKC {alpha} is downstream of MitoKATP channels in the signal transduction pathway of ischemic preconditioning of human myocardium. Am J Physiol Cell Physiol 287:1418–1425

Hattori R, Otani H, Uchiyama T, Imamura H, Cui J, Maulik N, Cordis GA, Zhu L, Das DK (2001) Src tyrosine kinase is the trigger but not the mediator of ischemic preconditioning. Am J Physiol Heart Circ Physiol 281:H1066–H1074

Hausenloy DJ, Yellon DM (2006) Survival kinases in ischemic preconditioning and postconditioning. Cardiovasc Res 70:240–253

Hausenloy DJ, Yellon DM (2010) The second window of preconditioning (SWOP) where are we now? Cardiovasc Drugs Ther 24:235–254

Hausenloy DJ, Tsang A, Yellon DM (2005) The reperfusion injury salvage kinase pathway: a common target for both ischemic preconditioning and postconditioning. Trends Cardiovasc Med 15:69–75

Head BP, Patel HH, Roth DM, Lai NC, Niesman IR, Farquhar MG, Insel PA (2005) G-protein-coupled receptor signaling components localize in both sarcolemmal and intracellular caveolin-3-associated microdomains in adult cardiac myocytes. J Biol Chem 280:31036–31044

Head BP, Patel HH, Tsutsumi YM, Hu Y, Mejia T, Mora RC, Insel PA, Roth DM, Drummond JC, Patel PM (2008) Caveolin-1 expression is essential for N-methyl-D-aspartate receptor-mediated Src and extracellular signal-regulated kinase 1/2 activation and protection of primary neurons from ischemic cell death. FASEB J 22:828–840

Head BP, Peart JN, Panneerselvam M, Yokoyama T, Pearn ML, Niesman IR, Bonds JA, Schilling JM, Miyanohara A, Headrick J, Ali SS, Roth DM, Patel PM, Patel HH (2010) Loss of caveolin-1 accelerates neurodegeneration and aging. PLoS One 5:e15697

Hendrich AB (2006) Flavonoid-membrane interactions: possible consequences for biological effects of some polyphenolic compounds. Acta Pharmacol Sin 27:27–40

Horton ND, Kaftani DJ, Bruce DS, Bailey EC, Krober AS, Jones JR, Turker M, Khattar N, Su TP, Bolling SF, Oeltgen PR (1998) Isolation and partial characterization of an opioid-like 88 kDa hibernation-related protein. Comp Biochem Physiol B Biochem Mol Biol 119:787–805

Howells RD, Kilpatrick DL, Bailey LC, Noe M, Udenfriend S (1986) Proenkephalin mRNA in rat heart. Proc Natl Acad Sci U S A 83:1960–1963

Inagaki K, Hahn HS, Dorn GW II, Mochly-Rosen D (2003) Additive protection of the ischemic heart ex vivo by combined treatment with delta-protein kinase C inhibitor and epsilon-protein kinase C activator. Circulation 108:869–875

Insel PA, Head BP, Ostrom RS, Patel HH, Swaney JS, Tang CM, Roth DM (2005) Caveolae and lipid rafts: G protein-coupled receptor signaling microdomains in cardiac myocytes. Ann N Y Acad Sci 1047:166–172

Ishikawa S, Kuno A, Tanno M, Miki T, Kouzu H, Itoh T, Sato T, Sunaga D, Murase H, Miura T (2012) Role of connexin-43 in protective PI3K-Akt-GSK-3beta signaling in cardiomyocytes. Am J Physiol Heart Circ Physiol 302:H2536–H2544

Iwata M, Inoue S, Kawaguchi M, Nakamura M, Konishi N, Furuya H (2007) Effects of delta-opioid receptor stimulation and inhibition on hippocampal survival in a rat model of forebrain ischaemia. Br J Anaesth 99:538–546

Jang Y, Xi J, Wang H, Mueller RA, Norfleet EA, Xu Z (2008) Postconditioning prevents reperfusion injury by activating delta-opioid receptors. Anesthesiology 108:243–250

Jeong S, Kim SJ, Jeong C, Lee S, Jeong H, Lee J, Yoo KY (2012) Neuroprotective effects of remifentanil against transient focal cerebral ischemia in rats. J Neurosurg Anesthesiol 24:51–57

Juhaszova M, Zorov DB, Kim SH, Pepe S, Fu Q, Fishbein KW, Ziman BD, Wang S, Ytrehus K, Antos CL, Olson EN, Sollott SJ (2004) Glycogen synthase kinase-3beta mediates convergence of protection signaling to inhibit the mitochondrial permeability transition pore. J Clin Invest 113:1535–1549

Kamoun WS, Karaa A, Kresge N, Merkel SM, Korneszczuk K, Clemens MG (2006) LPS inhibits endothelin-1-induced endothelial NOS activation in hepatic sinusoidal cells through a negative feedback involving caveolin-1. Hepatology 43:182–190

Kao TK, Ou YC, Liao SL, Chen WY, Wang CC, Chen SY, Chiang AN, Chen CJ (2008) Opioids modulate post-ischemic progression in a rat model of stroke. Neurochem Int 52:1256–1265

Karck M, Tanaka S, Bolling SF, Simon A, Su TP, Oeltgen PR, Haverich A (2001) Myocardial protection by ischemic preconditioning and delta-opioid receptor activation in the isolated working rat heart. J Thorac Cardiovasc Surg 122:986–992

Katavic PL, Lamb K, Navarro H, Prisinzano TE (2007) Flavonoids as opioid receptor ligands: identification and preliminary structure-activity relationships. J Nat Prod 70:1278–1282

Kawalec M, Kowalczyk JE, Beresewicz M, Lipkowski AW, Zablocka B (2011) Neuroprotective potential of biphalin, multireceptor opioid peptide, against excitotoxic injury in hippocampal organotypic culture. Neurochem Res 36:2091–2095

Kosterlitz HW, Taylor DW (1959) The effect of morphine on vagal inhibition of the heart. Br J Pharmacol Chemother 14:209–214

Kowaltowski AJ, Seetharaman S, Paucek P, Garlid KD (2001) Bioenergetic consequences of opening the ATP-sensitive K^+ channel of heart mitochondria. Am J Physiol Heart Circ Physiol 280:H649–H657

Kroemer G, Dallaporta B, Resche-Rigon M (1998) The mitochondrial death/life regulator in apoptosis and necrosis. Annu Rev Physiol 60:619–642

Kuzuya T, Hoshida S, Yamashita N, Fuji H, Oe H, Hori M, Kamada T, Tada M (1993) Delayed effects of sublethal ischemia on the acquisition of tolerance to ischemia. Circ Res 72:1293–1299

Laskey WK (2005) Brief repetitive balloon occlusions enhance reperfusion during percutaneous coronary intervention for acute myocardial infarction: a pilot study. Catheter Cardiovasc Interv 65:361–367

Lemoine S, Zhu L, Massetti M, Gerard JL, Hanouz JL (2011) Continuous administration of remifentanil and sufentanil induces cardioprotection in human myocardium, in vitro. Acta Anaesthesiol Scand 55:758–764

Lim YJ, Zheng S, Zuo Z (2004) Morphine preconditions Purkinje cells against cell death under in vitro simulated ischemia-reperfusion conditions. Anesthesiology 100:562–568

Lisanti MP, Scherer PE, Tang Z, Sargiacomo M (1994) Caveolae, caveolin and caveolin-rich membrane domains: a signalling hypothesis. Trends Cell Biol 4:231–235

Llorens-Cortes C, Huang H, Vicart P, Gasc JM, Paulin D, Corvol P (1992) Identification and characterization of neutral endopeptidase in endothelial cells from venous or arterial origins. J Biol Chem 267:14012–14018

Ludwig LM, Patel HH, Gross GJ, Kersten JR, Pagel P, Warltier DC (2003) Morphine enhances pharmacological preconditioning by isoflurane: role of mitochondrial KATP channels and opioid receptors. Anesthesiology 98:705–711

Ma MC, Qian H, Ghassemi F, Zhao P, Xia Y (2005) Oxygen-sensitive {delta}-opioid receptor-regulated survival and death signals: novel insights into neuronal preconditioning and protection. J Biol Chem 280:16208–16218

Macedo AF, Taylor FC, Casas JP, Adler A, Prieto-Merino D, Ebrahim S (2014) Unintended effects of statins from observational studies in the general population: systematic review and meta-analysis. BMC Med 12:51

Maleki-Dizaji N, Fathiazad F, Garjani A (2007) Antinociceptive properties of extracts and two flavonoids isolated from leaves of Danae racemosa. Arch Pharm Res 30:1536–1542

Mantelli L, Corti V, Ledda F (1987) On the presence of opioid receptors in guinea-pig ventricular tissue. Gen Pharmacol 18:309–313

Marber MS, Latchman DS, Walker JM, Yellon DM (1993) Cardiac stress protein elevation 24 hours after brief ischemia or heat stress is associated with resistance to myocardial infarction. Circulation 88:1264–1272

Marmor M, Penn A, Widmer K, Levin RI, Maslansky R (2004) Coronary artery disease and opioid use. Am J Cardiol 93:1295–1297

Maslov LN, Lishmanov YB (1995) Change in opioid peptide level in the heart and blood plasma during acute myocardial ischemia complicated by ventricular fibrillation. Clin Exp Pharmacol Physiol 22:812–816

Maulik N, Watanabe M, Zu YL, Huang CK, Cordis GA, Schley JA, Das DK (1996) Ischemic preconditioning triggers the activation of MAP kinases and MAPKAP kinase 2 in rat hearts. FEBS Lett 396:233–237

Michelsen MM, Stottrup NB, Schmidt MR, Lofgren B, Jensen RV, Tropak M, St-Michel EJ, Redington AN, Botker HE (2012) Exercise-induced cardioprotection is mediated by a bloodborne, transferable factor. Basic Res Cardiol 107:260

Miura T, Yano T, Naitoh K, Nishihara M, Miki T, Tanno M, Shimamoto K (2007) Delta-opioid receptor activation before ischemia reduces gap junction permeability in ischemic myocardium by PKC-epsilon-mediated phosphorylation of connexin 43. Am J Physiol Heart Circ Physiol 293:H1425–H1431

Monier S, Parton RG, Vogel F, Behlke J, Henske A, Kurzchalia TV (1995) VIP21-caveolin, a membrane protein constituent of the caveolar coat, oligomerizes in vivo and in vitro. Mol Biol Cell 6:911–927

Murphy E, Steenbergen C (2007) Preconditioning: the mitochondrial connection. Annu Rev Physiol 69:51–67

Murphy GS, Szokol JW, Marymont JH, Avram MJ, Vender JS (2006) Opioids and cardioprotection: the impact of morphine and fentanyl on recovery of ventricular function after cardiopulmonary bypass. J Cardiothorac Vasc Anesth 20:493–502

Murphy GS, Szokol JW, Marymont JH, Avram MJ, Vender JS (2007) The effects of morphine and fentanyl on the inflammatory response to cardiopulmonary bypass in patients undergoing elective coronary artery bypass graft surgery. Anesth Analg 104:1334–1342, table of contents

Murry CE, Jennings RB, Reimer KA (1986) Preconditioning with ischemia: a delay of lethal cell injury in ischemic myocardium. Circulation 74:1124–1136

Narita M, Kuzumaki N, Miyatake M, Sato F, Wachi H, Seyama Y, Suzuki T (2006) Role of delta-opioid receptor function in neurogenesis and neuroprotection. J Neurochem 97:1494–1505

Oeltgen PR, Walsh JW, Hamann SR, Randall DC, Spurrier WA, Myers RD (1982) Hibernation "trigger": opioid-like inhibitory action on brain function of the monkey. Pharmacol Biochem Behav 17:1271–1274

Oeltgen PR, Nilekani SP, Nuchols PA, Spurrier WA, Su TP (1988) Further studies on opioids and hibernation: delta opioid receptor ligand selectively induced hibernation in summer-active ground squirrels. Life Sci 43:1565–1574

Okamoto T, Schlegel A, Scherer PE, Lisanti MP (1998) Caveolins, a family of scaffolding proteins for organizing "preassembled signaling complexes" at the plasma membrane. J Biol Chem 273:5419–5422

Olianas MC, Dedoni S, Onali P (2011) Regulation of PI3K/Akt signaling by N-desmethyl clozapine through activation of δ-opioid receptor. Eur J Pharmacol 660:341–350

Oteiza PI, Erlejman AG, Verstraeten SV, Keen CL, Fraga CG (2005) Flavonoid-membrane interactions: a protective role of flavonoids at the membrane surface? Clin Dev Immunol 12:19–25

Pain T, Yang X-M, Critz SD, Yue Y, Nakano A, Liu GS, Heusch G, Cohen MV, Downey JM (2000) Opening of mitochondrial K_{ATP} channels triggers the preconditioned state by generating free radicals. Circ Res 87:460–466

Palade G (1953) Fine structure of blood capillaries. J Appl Phys 24:1424

Panneerselvam M, Tsutsumi YM, Bonds JA, Horikawa YT, Saldana M, Dalton ND, Head BP, Patel PM, Roth DM, Patel HH (2010) Dark chocolate receptors: epicatechin-induced cardiac protection is dependent on delta-opioid receptor stimulation. Am J Physiol Heart Circ Physiol 299:H1604–H1609

Panneerselvam M, Ali SS, Finley JC, Kellerhals SE, Migita MY, Head BP, Patel PM, Roth DM, Patel HH (2013) Epicatechin regulation of mitochondrial structure and function is opioid receptor dependent. Mol Nutr Food Res 57:1007–1014

Paradis P, Dumont M, Bélichard P, Rouleau J-L, Lemaire S, Brakier-Gingras L (1992) Increased preproenkephalin A gene expression in the rat heart after induction of myocardial infarction. Biochem Cell Biol 70:593–598

Patel HH, Hsu A, Moore J, Gross GJ (2001) BW373U86, a delta opioid agonist, partially mediates delayed cardioprotection via a free radical mechanism that is independent of opioid receptor stimulation. J Mol Cell Cardiol 33:1455–1465

Patel HH, Moore J, Hsu AK, Gross GJ (2002) Cardioprotection at a distance: mesenteric artery occlusion protects the myocardium via an opioid sensitive mechanism. J Mol Cell Cardiol 34:1317–1323

Patel HH, Fryer RM, Gross ER, Bundey RA, Hsu AK, Isbell M, Eusebi LO, Jensen RV, Gullans SR, Insel PA, Nithipatikom K, Gross GJ (2003) 12-Lipoxygenase in opioid-induced delayed cardioprotection: gene array, mass spectrometric, and pharmacological analyses. Circ Res 92:676–682

Patel HH, Hsu A, Gross GJ (2004a) Delayed cardioprotection is mediated via a non-peptide delta opioid agonist, SNC-121, independent of opioid receptor stimulation. Basic Res Cardiol 99:38–45

Patel HH, Hsu AK, Gross GJ (2004b) COX-2 and iNOS in opioid-induced delayed cardioprotection in the intact rat. Life Sci 75:129–140

Patel HH, Gross ER, Peart JN, Hsu AK, Gross GJ (2005) Sarcolemmal KATP channel triggers delayed ischemic preconditioning in rats. Am J Physiol Heart Circ Physiol 288:H445–H447

Patel HH, Head BP, Petersen HN, Niesman IR, Huang D, Gross GJ, Insel PA, Roth DM (2006) Protection of adult rat cardiac myocytes from ischemic cell death: role of caveolar microdomains and delta-opioid receptors. Am J Physiol Heart Circ Physiol 291:H344–H350

Patra SK, Rizzi F, Silva A, Rugina DO, Bettuzzi S (2008) Molecular targets of (–)-epigallocatechin-3-gallate (EGCG): specificity and interaction with membrane lipid rafts. J Physiol Pharmacol 9:217–235

Peart JN, Gross GJ (2004a) Chronic exposure to morphine produces a marked cardioprotective phenotype in aged mouse hearts. Exp Gerontol 39:1021–1026

Peart JN, Gross GJ (2004b) Morphine-tolerant mice exhibit a profound and persistent cardioprotective phenotype. Circulation 109:1219–1222

Peart JN, Gross GJ (2006) Cardioprotective effects of acute and chronic opioid treatment are mediated via different signaling pathways. Am J Physiol Heart Circ Physiol 291:H1746–H1753

Peart JN, Gross ER, Headrick JP, Gross GJ (2007) Impaired p38 MAPK/HSP27 signaling underlies aging-related failure in opioid-mediated cardioprotection. J Mol Cell Cardiol 42:972–980

Peart JN, Hoe LES, Gross GJ, Headrick JP (2011) Sustained ligand-activated preconditioning via δ-opioid receptors. J Pharmacol Exp Ther 336:274–281

Peart JN, Pepe S, Reichelt ME, Beckett N, See Hoe L, Ozberk V, Niesman IR, Patel HH, Headrick JP (2014) Dysfunctional survival-signaling and stress-intolerance in aged murine and human myocardium. Exp Gerontol 50:72–81

Penna C, Rastaldo R, Mancardi D, Raimondo S, Cappello S, Gattullo D, Losano G, Pagliaro P (2006) Post-conditioning induced cardioprotection requires signaling through a redox-sensitive mechanism, mitochondrial ATP-sensitive K^+ channel and protein kinase C activation. Basic Res Cardiol 101:180–189

Ping P, Zhang J, Cao X, Li RC, Kong D, Tang XL, Qiu Y, Manchikalapudi S, Auchampach JA, Black RG, Bolli R (1999) PKC-dependent activation of p44/p42 MAPKs during myocardial ischemia-reperfusion in conscious rabbits. Am J Physiol 276:H1468–H1481

Pleym H, Stenseth R, Wiseth R, Karevold A, Dale O (2004) Supplemental remifentanil during coronary artery bypass grafting is followed by a transient postoperative cardiac depression. Acta Anaesthesiol Scand 48:1155–1162

Przyklenk K, Bauer B, Ovize M, Kloner RA, Whittaker P (1993) Regional ischemic 'preconditioning' protects remote virgin myocardium from subsequent sustained coronary occlusion. Circulation 87:893–899

Raikar LS, Vallejo J, Lloyd PG, Hardin CD (2006) Overexpression of caveolin-1 results in increased plasma membrane targeting of glycolytic enzymes: the structural basis for a membrane associated metabolic compartment. J Cell Biochem 98:861–871

Ratajczak P, Damy T, Heymes C, Oliviero P, Marotte F, Robidel E, Sercombe R, Boczkowski J, Rappaport L, Samuel JL (2003) Caveolin-1 and -3 dissociations from caveolae to cytosol in the heart during aging and after myocardial infarction in rat. Cardiovasc Res 57:358–369

Sanada S, Asanuma H, Tsukamoto O, Minamino T, Node K, Takashima S, Fukushima T, Ogai A, Shinozaki Y, Fujita M, Hirata A, Okuda H, Shimokawa H, Tomoike H, Hori M, Kitakaze M (2004) Protein kinase A as another mediator of ischemic preconditioning independent of protein kinase C. Circulation 110:51–57

Schultz JEJ, Hsu AK, Gross GJ (1996) Morphine mimics the cardioprotective effect of ischemic preconditioning via a glibenclamide-sensitive mechanism in the rat heart. Circ Res 78:1100–1104

Schultz JJ, Hsu AK, Gross GJ (1997) Ischemic preconditioning and morphine-induced cardioprotection involve the delta (delta)-opioid receptor in the intact rat heart. J Mol Cell Cardiol 29:2187–2195

Schultz JE, Hsu AK, Gross GJ (1998a) Ischemic preconditioning in the intact rat heart is mediated by delta1- but not mu- or kappa-opioid receptors. Circulation 97:1282–1289

Schultz J-J, Hsu AK, Nagase H, Gross GJ (1998b) TAN-67, a delta 1-opioid receptor agonist, reduces infarct size via activation of Gi/o proteins and KATP channels. Am J Physiol Heart Circ Physiol 274:H909–H914

Schwartz J-C, Malfroy B, Sdl B (1981) Biological inactivation of enkephalins and the role of enkephalin-dipeptidyl-carboxypeptidase (enkephalinase) as neuropeptidase. Life Sci 29:1715–1740

See Hoe LE, Schilling JM, Tarbit E, Kiessling CJ, Busija AR, Niesman IR, Du Toit EF, Ashton KJ, Roth DM, Headrick JP, Patel HH, Peart JN (2014) Sarcolemmal cholesterol and caveolin-3 dependence of cardiac function, ischemic tolerance and opioidergic cardioprotection. Am J Physiol Heart Circ Physiol 307(6):H895–H903

Sen D, Huchital M, Chen YL (2013) Crosstalk between delta opioid receptor and nerve growth factor signaling modulates neuroprotection and differentiation in rodent cell models. Int J Mol Sci 14:21114–21139

Shaul PW, Anderson RG (1998) Role of plasmalemmal caveolae in signal transduction. Am J Physiol 275:L843–L851

Sobanski P, Krajnik M, Shaqura M, Bloch-Boguslawska E, Schafer M, Mousa SA (2014) The presence of mu-, delta-, and kappa-opioid receptors in human heart tissue. Heart Vessels 29(6):855–863

Springhorn JP, Claycomb WC (1989) Preproenkephalin mRNA expression in developing rat hart and in cultured ventricular cardiac muscle cells. Biochem J 258:73–78

Su DS, Wang ZH, Zheng YJ, Zhao YH, Wang XR (2007) Dose-dependent neuroprotection of delta opioid peptide [D-Ala2, D-Leu5] enkephalin in neuronal death and retarded behavior induced by forebrain ischemia in rats. Neurosci Lett 423:113–117

Surendra H, Diaz RJ, Harvey K, Tropak M, Callahan J, Hinek A, Hossain T, Redington A, Wilson GJ (2013) Interaction of delta and kappa opioid receptors with adenosine A1 receptors mediates cardioprotection by remote ischemic preconditioning. J Mol Cell Cardiol 60:142–150

Swaney JS, Patel HH, Yokoyama U, Head BP, Roth DM, Insel PA (2006) Focal adhesions in (myo) fibroblasts scaffold adenylyl cyclase with phosphorylated caveolin. J Biol Chem 281: 17173–17179

Takasaki Y, Wolff RA, Chien GL, Winkle DMV (1999) Met^5-enkephalin protects isolated adult rabbit cardiomyocytes via a d-opioid receptors. Am J Physiol Heart Circ Physiol 277: H2442–H2450

Tang J, Yang HYT, Costa E (1982) Distribution of met-enkephalin-arg^6-phe^7 in various tissues of rats and guinea pigs. Neuropharmacology 21:595–600

Theisen MM, Schlottmann S, August C, Herzog C, Theilmeier G, Maas M, Blumenstiel JM, Weber TP, Van Aken HK, Kaerlein KT (2014) Detection and distribution of opioid peptide receptors in porcine myocardial tissue. Pharmacol Res 84C:45–49

Tian X, Guo J, Zhu M, Li M, Wu G, Xia Y (2013a) Delta-opioid receptor activation rescues the functional TrkB receptor and protects the brain from ischemia-reperfusion injury in the rat. PLoS One 8:e69252

Tian X, Hua F, Sandhu HK, Chao D, Balboni G, Salvadori S, He X, Xia Y (2013b) Effect of delta-opioid receptor activation on BDNF-TrkB vs. TNF-alpha in the mouse cortex exposed to prolonged hypoxia. Int J Mol Sci 14:15959–15976

Tomai F, Crea F, Gaspardone A, Versaci F, Ghini AS, Ferri C, Desideri G, Chiariello L, Gioffre PA (1999) Effects of naloxone on myocardial ischemic preconditioning in humans. J Am Coll Cardiol 33:1863–1869

Tong H, Chen W, Steenbergen C, Murphy E (2000) Ischemic preconditioning activates phosphatidylinositol-3-kinase upstream of protein kinase C. Circ Res 87:309–315

Trang T, Sutak M, Quirion R, Jhamandas K (2003) Spinal administration of lipoxygenase inhibitors suppresses behavioural and neurochemical manifestations of naloxone-precipitated opioid withdrawal. Br J Pharmacol 140:295–304

Tsutsumi YM, Patel HH, Huang D, Roth DM (2006) Role of 12-lipoxygenase in volatile anesthetic-induced delayed preconditioning in mice. Am J Physiol Heart Circ Physiol 291:H979–H983

Tsutsumi YM, Yokoyama T, Horikawa Y, Roth DM, Patel HH (2007) Reactive oxygen species trigger ischemic and pharmacological postconditioning: in vivo and in vitro characterization. Life Sci 81:1223–1227

Tsutsumi YM, Horikawa YT, Jennings MM, Kidd MW, Niesman IR, Yokoyama U, Head BP, Hagiwara Y, Ishikawa Y, Miyanohara A, Patel PM, Insel PA, Patel HH, Roth DM (2008) Cardiac-specific overexpression of caveolin-3 induces endogenous cardiac protection by mimicking ischemic preconditioning. Circulation 118:1979–1988

Tsutsumi YM, Kawaraguchi Y, Horikawa YT, Niesman IR, Kidd MW, Chin-Lee B, Head BP, Patel PM, Roth DM, Patel HH (2010a) Role of caveolin-3 and glucose transporter-4 in isoflurane-induced delayed cardiac protection. Anesthesiology 112:1136–1145

Tsutsumi YM, Kawaraguchi Y, Niesman IR, Patel HH, Roth DM (2010b) Opioid-induced preconditioning is dependent on caveolin-3 expression. Anesth Analg 111:1117–1121

Tsutsumi YM, Tsutsumi R, Horikawa YT, Sakai Y, Hamaguchi E, Ishikawa Y, Yokoyama U, Kasai A, Kambe N, Tanaka K (2014) Geranylgeranylacetone protects the heart via caveolae and caveolin-3. Life Sci 101:43–48

VandenHoek TL, Becker LB, Shao Z, Li C, Schumacker PT (1998) Reactive oxygen species released from mitochondria during brief hypoxia induce preconditioning in cardiomyocytes. J Biol Chem 273:18092–18098
VandenHoek TL, Becker LB, Shao Z-H, Li C-Q, Schumacker PT (2000) Preconditioning in cardiomyocytes protects by attenuating oxidant stress at reperfusion. Circ Res 86:541–548
Ventura C, Guarnieri C, Vaona I, Campana G, Pintus G, Spampinato S (1994) Dynorphin gene expression and release in the myocardial cell. J Biol Chem 269:5384–5386
Ventura C, Pintus G, Tadolini B (1998) Opioid peptide gene expression in the myocardial cell. Trends Cardiovasc Med 8:102–110
Vinten-Johansen J, Shi W (2013) The science and clinical translation of remote postconditioning. J Cardiovasc Med (Hagerstown) 14:206–213
Wang Y, Ashraf M (1999) Role of protein kinase C in mitochondrial KATP channel-mediated protection against Ca^{2+} overload injury in rat myocardium. Circ Res 84:1156–1165
Wang S, Duan Y, Su D, Li W, Tan J, Yang D, Wang W, Zhao Z, Wang X (2011) Delta opioid peptide [D-Ala2, D-Leu5] enkephalin (DADLE) triggers postconditioning against transient forebrain ischemia. Eur J Pharmacol 658:140–144
Weil J, Eschenhagen T, Fleige G, Mittmann C, Orthey E, Scholz H (1998) Localization of preproenkephalin mRNA in rat heart: selective gene expression in left ventricular myocardium. Am J Physiol Heart Circ Physiol 275:H378–H384
Weinbrenner C, Liu GS, Cohen MV, Downey JM (1997) Phosphorylation of tyrosine 182 of p38 mitogen-activated protein kinase correlates with the protection of preconditioning in the rabbit heart. J Mol Cell Cardiol 29:2383–2391
Weinbrenner C, Schulze F, Sarvary L, Strasser RH (2004) Remote preconditioning by infrarenal aortic occlusion is operative via delta1-opioid receptors and free radicals in vivo in the rat heart. Cardiovasc Res 61:591–599
Williams TM, Lisanti MP (2004) The caveolin proteins. Genome Biol 5:214
Wong GT, Li R, Jiang LL, Irwin MG (2010) Remifentanil post-conditioning attenuates cardiac ischemia-reperfusion injury via kappa or delta opioid receptor activation. Acta Anaesthesiol Scand 54:510–518
Woodman SE, Sotgia F, Galbiati F, Minetti C, Lisanti MP (2004) Caveolinopathies: mutations in caveolin-3 cause four distinct autosomal dominant muscle diseases. Neurology 62:538–543
Xiong LZ, Yang J, Wang Q, Lu ZH (2007) Involvement of delta-and mu-opioid receptors in the delayed cerebral ischemic tolerance induced by repeated electroacupuncture preconditioning in rats. Chin Med J (Engl) 120:394–399
Xu ZD, Jin M, He WX, Xia SX, Zhao YF, He B, Cao DX, Peng SL, Li J, Cao MH (2009) [Remifentanil preconditioning lowers cardiac troponin I levels in patients undergoing off-pump coronary artery bypass graft surgery]. Nan Fang Yi Ke Da Xue Xue Bao 29:1554–1556
Yang Y, Xia X, Zhang Y, Wang Q, Li L, Luo G, Xia Y (2009) Delta-opioid receptor activation attenuates oxidative injury in the ischemic rat brain. BMC Biol 7:55
Yang L, Zhao X, Sun M, Sun X, Yao L, Yu D, Ding Q, Gao C, Chai W (2014) Delta opioid receptor agonist BW373U86 attenuates post-resuscitation brain injury in a rat model of asphyxial cardiac arrest. Resuscitation 85:299–305
Yao Z, Tong J, Tan X, Li C, Shao Z, Kim WC, VandenHoek TL, Becker LB, Head CA, Schumacker PT (1999) Role of reactive oxygen species in acetylcholine-induced preconditioning in cardiomyocytes. Am J Physiol Heart Circ Physiol 277:H2504–H2509
Zhang J, Haddad GG, Xia Y (2000) Delta-, but not mu- and kappa-, opioid receptor activation protects neocortical neurons from glutamate-induced excitotoxic injury. Brain Res 885: 143–153
Zhang JH, Gibney GT, Xia Y (2001) Effect of prolonged hypoxia on Na^{+} channel mRNA subtypes in the developing rat cortex. Brain Res Mol Brain Res 91:154–158
Zhang J, Gibney GT, Zhao P, Xia Y (2002) Neuroprotective role of delta-opioid receptors in cortical neurons. Am J Physiol Cell Physiol 282:C1225–C1234

Zhang J, Qian H, Zhao P, Hong SS, Xia Y (2006) Rapid hypoxia preconditioning protects cortical neurons from glutamate toxicity through delta-opioid receptor. Stroke 37:1094–1099

Zhao ZQ (2010) Postconditioning in reperfusion injury: a status report. Cardiovasc Drugs Ther 24:265–279

Zhao ZQ, Corvera JS, Halkos ME, Kerendi F, Wang NP, Guyton RA, Vinten-Johansen J (2003) Inhibition of myocardial injury by ischemic postconditioning during reperfusion: comparison with ischemic preconditioning. Am J Physiol Heart Circ Physiol 285:H579–H588

Zhao H, Loh HH, Law PY (2006) Adenylyl cyclase superactivation induced by long-term treatment with opioid agonist is dependent on receptor localized within lipid rafts and is independent of receptor internalization. Mol Pharmacol 69:1421–1432

Zhou Y, Fathali N, Lekic T, Ostrowski RP, Chen C, Martin RD, Tang J, Zhang JH (2011) Remote limb ischemic postconditioning protects against neonatal hypoxic-ischemic brain injury in rat pups by the opioid receptor/Akt pathway. Stroke 42:439–444

Zhu M, Li MW, Tian XS, Ou XM, Zhu CQ, Guo JC (2009) Neuroprotective role of delta-opioid receptors against mitochondrial respiratory chain injury. Brain Res 1252:183–191

Zhu M, Li M, Yang F, Ou X, Ren Q, Gao H, Zhu C, Guo J (2011) Mitochondrial ERK plays a key role in delta-opioid receptor neuroprotection against acute mitochondrial dysfunction. Neurochem Int 59:739–748

Chapter 18
The Various Functions of Opioids in Pathophysiological Conditions

Huimin Gao, Dongman Chao, Sebastian J. Villarreal, Daniel Yiu, Guoqiang Wen, Tao Chen, Yong Xia, Fei Yao, Qing Wang, and Ying Xia

Abstract Opioids and their analogues have been used as clinical analgesics for several centuries, with reports dating back to as early as 4000 B.C. Their anti-pain action is primarily accomplished through three major opioid receptor subtypes: μ-opioid receptor (MOR), δ-opioid receptor (DOR), and κ-opioid receptor (KOR). Although these opioid receptors have been cloned in many species, their complex functions, other than pain modulation, are still largely unknown in both physiological and pathophysiological conditions. However, recent studies have uncovered fresh insights into opioid receptor-mediated functions and their underlying mechanisms. Studies suggest that these opioid receptors participate in various physiological activities. Therefore, their dysfunction may play a role in numerous diseases, including hypoxic/ischemic injury in brain, epileptic seizures, Parkinson's disease, Multiple Sclerosis, cardiovascular disease, respiratory depression, pulmonary artery hypertension, malignancies, diabetic cutaneous wounds, immune disease, chronic

H. Gao, M.D., Ph.D.
Department of Neurosurgery, The University of Texas Medical School at Houston, Houston, TX 77030, USA

Department of Neurology, The Third Affiliated Hospital of Sun-Sen University, Guangzhou, Guangdong 510060, China

D. Chao, Ph.D. • S.J. Villarreal, M.D. • D. Yiu, B.A. • Y. Xia, M.D., Ph.D. • F. Yao, M.D., Ph.D.
Department of Neurosurgery, The University of Texas Medical School at Houston, Houston, TX 77030, USA

Y. Xia, M.D., Ph.D. (✉)
University of Texas Medical School, Houston, TX, USA
e-mail: ying.xia@uth.tmc.edu; y55738088@gmail.com

G. Wen, M.D.
Department of Neurology, Hainan General Hospital, Haikou, Hainan 570311, China

T. Chen, M.D., Ph.D.
Department of Neurosurgery, The University of Texas Medical School at Houston, Houston, TX 77030, USA

Department of Neurology, Hainan General Hospital, Haikou, Hainan 570311, China

Q. Wang, M.D., Ph.D.
Department of Neurology, The Third Affiliated Hospital of Sun-Sen University, Guangzhou, Guangdong 510060, China

Y. Xia (ed.), *Neural Functions of the Delta-Opioid Receptor*,
DOI 10.1007/978-3-319-25495-1_18

kidney disease, uremic pruritus, renal ischemia-reperfusion injury, and pre-eclampsia. There is also strong evidence showing that opioid receptors are involved in the regulation of stress, feeding and obesity, and play an essential role in hibernation in some species. One of the most exciting findings in the past decade is DOR-mediated neuroprotection and cardioprotection. The upregulation of DOR expression and activity increases the neuronal and myocardial tolerance to hypoxic and/or ischemic stress. The DOR signal triggers different mechanisms at multiple levels, depending on stress duration and severity, to preserve neuronal and myocardial survival, including the stabilization of homeostasis and increase in pro-survival signaling. DOR-mediated neuroprotection and cardioprotection may potentially influence clinical pharmacology in terms of prevention and treatment of life-threatening conditions like stroke and myocardial infarction. The main purpose of this chapter is to update recent research on opioids and their various functions, particularly in pathophysiological conditions. This chapter aims to deliver an informative reference for better understanding the opioid system from a/the pathophysiological point of view.

Keywords Opioids • Opioid receptors • Brain • Heart • Lung • Kidney • Pain • Protection • Hibernation • Hypoxia • Ischemia diseases

Abbreviations

AHR	Airway hyperresponsiveness
BMDCs	Bone marrow dendritic cells
BNTX	7-Benzylidenenaltrexone
BUN	Blood urea nitrogen
CKD-aP	Chronic kidney disease-associated pruritus
DG	Dentate gyrus
DOR	δ-Opioid receptor
EMT	Epithelial mesenchymal transition
GBM	Glioblastoma
HCC	Hepatocellular carcinoma
IRI	Renal ischaemia-reperfusion
KO	Knockout
KOR	κ-Opioid receptor
L-NAME	N-nitro-L-arginine methyl ester
LPS	Lipopolysaccharide
MS	Multiple Sclerosis
MOR	μ-Opioid receptor
NB	Neuroblastoma
NPC	Neural progenitor cells
NTX	Naltrexone
OGF	Opioid growth factor

ORL	Opioid receptor-like orphan receptor
PAH	Pulmonary artery hypertension
PASMCs	Pulmonary arterial smooth muscle cells
PD	Parkinson's disease
PDYN	Pro-dynorphin
PENK	Pro-enkephalin
RIPC	Remote ischaemic preconditioning
PIS	Penicillin-induced seizures
PKC	Protein kinase C
POMC	Pro-opiomelanocortin
RR-EAE	Relapse-remitting experimental autoimmune encephalomyelitis
RUPP	Uteroplacental perfusion
TK	Tyrosine kinase
WT	Wild-type

1 Introduction

The term "opioid" refers to a compound and chemical that shares the characteristics with opium (lachrymal papaveris, poppy tears). It has been used for medicinal purposes for many centuries dating back to 4000 B.C. The first know medical use of opioids was primarily for their analgesic properties. Generally, opioids exert their pharmacological actions by binding to specific opioid receptors. One exception to this is the "non-opioid" effect that occurs when administration leads to excessively high concentrations of opioids. Throughout history, clinicians and scientists around the world have mainly investigated the anti-pain function of opioids and the various opioid receptors. More recently, multiple studies have provided us with significant evidence illuminating the complex and numerous functions and effects of opioids throughout the body.

Besides pain modulation, opioid receptors play an important role in various physiological functions. Therefore, their dysfunctions may lead to pathophysiological conditions involved in various diseases. However, the complex functions of opioid receptors, as well the underlying mechanisms, have not been well fully understood. The limited availability of data, controversies, and diametrically opposite results are inevitable. For example, in vivo studies (Hosobuchi et al. 1982; Adams et al. 1986; Skarphedinsson and Thoren 1988; Olinger et al. 1990) showed that opioid receptor inhibition by opioid antagonists, such as the non-selective morphinan naloxone, protected the brain from ischemia-induced injury, while other in vivo studies (Hayward et al. 1993; Mayfield and D'Alecy 1994; Endoh et al. 1999) suggested that opioid receptor activation with opioid agonists were protective or extended animal survival time during severe hypoxia. Moreover, scientists have not reached an agreement on opioid involvement in epileptogenesis, and whether it is anticonvulsant and proconvulsant anticonvulsant or proconvulsant is still controversial.

Several previous reviews have addressed the history of opioid research, classification of opioid receptors, signal transduction properties of these receptors with regards to G-protein coupling, adenylyl cyclase, and cAMP, as well as early studies of opioid receptor function, especially on opioid-induced analgesia and tolerance/dependence (Law and Loh 1999; Wiesenfeld et al. 1999; Pasternak 2004; Waldhoer et al. 2004; Barry and Zuo 2005; Bohn and Rachal 2006; Ananthan 2006; Lambert 2008; Kieffer and Evans 2009; Kreek et al. 2009; Feng et al. 2009; Zhao et al. 2002; Chao and Xia 2010; He et al. 2013). Recent studies utilizing modern approaches have produced new information on opioid receptor-mediated function and the underlying mechanisms. For instance, our extensive studies have shown a specific role of DOR in neuroprotection against hypoxic/ischemic injury (Zhang et al. 1999, 2000, 2001, 2002, 2006; Ma et al. 2005; Tian et al. 2008a, b, c; Yang et al. 2009; Chao and Xia 2010; Feng et al. 2011; Chen et al. 2014). Many studies from other laboratories have also demonstrated that DOR activation is indeed neuroprotective against hypoxic/ischemic stress (Lim et al. 2004; Horiuchi et al. 2004, 2008; Borlongan et al. 2009; Zhao et al. 2006; Su et al. 2007; Iwata et al. 2007; Charron et al. 2008; Kao et al. 2008; Govindaswami et al. 2008; Pamenter et al. 2008; Peng et al. 2009a, b; Zhu et al. 2009). Opioid receptor activation mediates cellular/molecular events by regulating ionic homeostasis (refer to Chaps. 5 and 6) and multiple signal pathways (refer to Chaps. 1, 16 and 17).

The main purpose of this chapter is to update recent advances in opioid receptor research, based on our previous review (Feng et al. 2012), in terms of their functions in pathophysiological conditions. For the reader's convenience, we have briefly summarized the background information on endogenous opioids and opioid receptors, and listed the commonly used opioid ligands in research. Despite extensive research, controversies still linger. We have attempted to present a comprehensive and objective overview on this topic, though we may not necessarily agree with all the conclusions proposed by the original articles. In this way, we believe, this chapter shall provide an informative, non-biased reference for better understanding the opioid processes within the body and further elucidate on the opioid receptor function in physiological, pathophysiological and pharmacological views.

2 Opioids and Opioid Receptors

The term "opioids" is composed of a broad range of compounds that have intrinsic actions similar to morphine but with distinct and unique chemical structures that include an array of molecules from alkaloids to peptides (Martin 1967). Endogenous opioid peptides are small molecules that are naturally produced in the central nervous system (CNS) and in various glands throughout the body. Opioids exert their action by acting as ligands that bind to opioid receptors, which are a diverse group of G-protein coupled transmembrane cytoplasmic receptors, and play an important function in the neurohormonal system.

2.1 Endogenous and Exogenous Opioids

Throughout human history, opioids have been widely used as analgesics to combat pain or induce ecstasy in medical and non-medical situations. Despite its long history of usage, the underlying mechanisms of opioid action remain a topic of research. In 1960s, Tsou and Jang (Tsou and Jang 1964) performed groundbreaking studies by direct microinjection of morphine into the brain to produce analgesia. This provided the impetus for future studies on the role of opioids on brain function. Unfortunately, due to the ongoing political unrest in China at that time, their noteworthy work was never formally published in any scientific or medical journal. By the early 1970s, Pert and Yaksh (Pert and Yaksh 1974, 1975) published their data delineating sites of morphine-induced analgesia in the primate brain. Furthermore, Liebeskind and colleagues observed that brain stimulation in certain areas, particularly the periaqueductal gray region, produced pronounced analgesia (Mayer and Liebeskind 1974) that was blocked by the opioid antagonist, naloxone (Akil et al. 1976). This observation strongly suggested the existence of endogenous opioids in the brain. Shortly thereafter, two groups independently reported that brain extracts mimicked the ability of morphine to inhibit electrically induced contractions of the mouse vas deferens, which was blocked by naloxone (Hughes et al. 1975; Snuder and Matthysse 1975). These discoveries gave rise to the identification and isolation of two enkephalin pentapeptides, namely [Met5] enkephalin (H-Tyr-Gly-GlyPhe-Met-OH) and [Leu5] enkephalin (H-Tyr-Gly-Gly-PheLeu-OH) (Hughes et al. 1975), followed closely by the identification of the opioid receptor-like orphan receptor (ORL) which was isolated concurrently by two independent groups, each named differently, ORL-1 by Mollereau et al. (1994) and Meunier et al. (1995), and orphanin FQ by Reinscheid et al. (1995). But it is commonly referred to as nociceptin. Two years later, Zadina et al. (1997) isolated endomorphin-1 and -2, endogenous MOR agonists with very high affinity and selectivity for MOR in both bovine and human brain extracts (Zadina et al. 1997; Hackler et al. 1997). So far, at least ten endogenous opioids have been identified in the brain, which belong to distinct families of compounds.

Endogenous opioids can originate from the same or different precursor proteins. All mammalian opioid peptides are derived from three precursors, i.e., pro-opiomelanocortin (POMC), pro-enkephalin (PENK) and pro-dynorphin (PDYN), which are translated from separate genes (Höllt 1992). Although the sequence of [Met5] enkephalin is contained in the sequence of POMC (Li et al. 1976), POMC is not the source of [Met5] enkephalin peptide. Instead, [Met5] enkephalin is encoded by the preproenkephalin gene. A 31-amino-acid portion of β-lipotropin was isolated and named as β-endorphin (Li et al. 1976). It possessed an affinity for μ-opioid receptors. Subsequently, a series of polypeptides containing the N-terminal [Leu5] enkephalin sequence were also isolated and designated as another opioid group called dynorphins (Goldstein et al. 1979). Cloning of the peptide precursors demonstrated that one precursor contained both [Met5] enkephalin and [Leu5] enkephalin transcripts, a second precursor existed for the dynorphin peptides, and a third for β-endorphin and other pro-opiomelanocortin peptides (adrenocorticotropin and melanoctye-stimulating hormone) (Comb et al. 1982). Many C-terminal extended enkephalins were also identified and demonstrated a distinct spectrum of bioactivities (Barr and Van Stipdonk 2002). To date, however, the precursor for endomorphins remains to be discovered.

Upon identification of the unique properties of opioids, especially their effectiveness in producing analgesia, many exogenous opioids were synthetically developed to enhance their properties for an eventual application in clinical research. However, among the vast plethora of compounds recognized in the literature during the last several decades, morphine still remains the most popularly used exogenous opioid for pain relief, since it is the quintessential MOR agonist and exhibits higher affinity for MOR than for either DOR or KOR (Martin 1983; Paterson et al. 1983). Although many exogenous opioids with pharmacological properties similar to morphine have been used both clinically and in the laboratory, they differ widely with respect to their efficacies, potency, side effects, clearance, and passage through the blood–brain barrier (Zöllner and Schäfer 2008; Zöllner and Stein 2007).

Ongoing studies of functions of opioids have been broadly extended to various physiological and pathophysiological conditions. For example, methionine encephalin ([Met5] enkephalin) is a potent endogenous opioid peptide and plays a role in the regulation of cell division and cell growth during development. In order to distinguish its growth modulating characteristics from neurotransmission, this 5-amino acid neuropeptide is also referred to as an/the opioid growth factor (OGF) that interacts with the so-called OGF receptor (OGFr) (Zagon and McLaughlin 1983a, b, c; Zagon et al. 1988, 1990a, b, 1994; Hauser et al. 1987; McLaughlin and Zagon 1987, 2012; Guo et al. 2013).

2.2 *Opioid Receptors*

The action of opioid compounds is mediated through activation of specific opioid receptors. Opioid receptors exist not only in the nervous system (Mansour et al. 1995; Vogt et al. 1995; Xia and Haddad 1991, 2001; Abbadic et al. 2004; Beadles-Bohling and Wiren 2005), but also in peripheral organs, such as the heart, lungs, kidney, liver, gastrointestinal, and reproductive systems (Wittert et al. 1996; Villemagne et al. 2002; Peng et al. 2012). However, the expression and distribution of these receptors vary significantly among different organs as well as among different animal species (Barry and Zuo 2005; also refer to Chaps. 3 and 4).

2.2.1 Major Opioid Receptor Categories

Three major opioid receptor families, the μ-, κ-, and δ-opioid receptors (MOR, DOR, and KOR) were cloned in early 1990s (Evans et al. 1992; Zastawny et al. 1994; Kieffer et al. 1992; Law and Loh 1999; Pasternak 2004), and a fourth member of the opioid receptor family, nociceptin or orphanin FQ receptor (NOP) or the opioid receptor-like orphan receptor (ORL), was also added to the list in 1994 (Mollercau et al. 1994). Opioid receptors have high sequence homology and belong to the large family of seven-transmembrane G protein-coupled receptors. Because of the homologies between ORL/NOP and other opioid receptors, the function of ORL/NOP has also been focused primarily on pain and analgesia (Darland et al. 1998; Meis 2003; Heinricher 2005; Rizzi et al. 2006; Reiss et al. 2008; Scoto et al. 2007; Chen and Somme 2007). Other functions of ORL/NOP receptors include dynamic homeostatic mechanisms, and the action is not localized in one particular body region since the receptors circulate all throughout the body (Lambert 2008; Stevens et al. 2007; Doggrell 2007).

In general, endogenous opioid peptides are not highly selective or specific to one particular type of opioid receptors, except the endomorphins (Zadina et al. 1997). This is due to three major factors: (1) all peptide ligands contain the N-terminal residue Tyr (except Phe in the case of nociceptin), comparable to a functional hydroxyl group in the morphinans, which is a requirement for interaction within the ligand-bind domain of opioid receptors; (2) MOR, DOR, and KOR have many common similarities in their primary structures, in addition to function and intracellular signaling mechanisms; and (3) the formation of homomeric and heteromeric complexes between opioid receptors and nonopioid receptors modify their response to a given opioid ligand (Law and Loh 1999; Pasternak 2004; Waldhoer et al. 2004; Barry and Zuo 2005; Ananthan 2006; Zhang et al. 2009). In other words, the outcome of opioid action could be the result of an interaction with different opioid receptor complexes. However, synthetic opioid peptides and alkaloids have a relatively high selectivity to MOR, DOR, and KOR and have been widely used to define the pharmacological properties of opioid receptors.

Studies suggest that the physical interactions between opioid receptors play a major role in defining their pharmacological and physiologically behaviors (Garzon et al. 1982; Demoliou-Mason and Barnard 1986). For example, δ-agonists were shown to enhance the analgesic potency and efficacy of μ-agonists, whereas δ-antagonists prevented or reduced the development of tolerance and physical dependence by μ-agonists (Ananthan 2006). Another example is LY255582, an opioid antagonist, which produced effects on feeding (or appetite) and body weight gain through a heteromeric combination of MOR, DOR, and/or KOR (Gackenheimer et al. 2005). Therefore, new ligands with dual μ-/δ-agonism or μ-/δ-antagonism, or mixed μ-agonism/δ-antagonism or vice versa are emerging as promising unique approaches to the development of analgesic drugs (Ananthan 2006; Fujita et al. 2004; Balboni et al. 2006).

2.2.2 Opioid Receptor Subtypes

The development of novel opioid analogs that elicited different biological effects gave way to the detection of pharmacological subtypes of MOR, DOR, and KOR, although the molecular basis of these subtypes has not yet been well established

(Traynor 1989; Traynor and Elliot 1993; Borsodi and Tóth 1995; Zaki et al. 1996). Of the three opioid receptor families, MOR subtypes have been most extensively studied due to their role in mediating the actions of morphine and other clinically relevant analgesic agents, as well as drugs of abuse. Although only one μ-opioid receptor gene (Oprm) has been isolated to date, pharmacological observations have given rise to the concept of multiple μ-receptors along with the cross-tolerance to different μ-opioids (Wolozin and Pasternak 1981). Genetic and physiological variations influence the sensitivity of MOR towards different μ-opioid ligands and differences in their potency across various species and animal strains (Pasternak 2004), which led to the proposal that receptors could be classified as MOR1, MOR2, and MOR3. Similarly, KOR multiplicity is also a complex issue under debate. Clark et al. (Clark et al. 1989) presented evidence from binding studies in the guinea-pig cerebellum that suggested the existence of four KOR subtypes, KOR1a, KOR1b, KOR2, and KOR3, classified by pharmacological studies (Butelman et al. 1998; Caudle et al. 1998). In addition, DOR has been demonstrated to exist in at least two subtypes, DOR1 and DOR2, according to their pharmacological attributes (Traynor and Elliot 1993; Zaki et al. 1996; Jiang et al. 1991; Sofuoglu et al. 1991, 1993; Mattia et al. 1991). Rothman first proposed the existence of DOR subtypes in a study (Rothman et al. 1984, 1985) in which a δ-site associated with the μ- δ-opioid receptor complex was termed the δcx binding site, while non-associated or non-complexed δ-site as the δncx site, corresponded to DOR1 (Rothman et al. 1992). However, this complex terminology is not used in the opioid literature. Although many studies have been conducted on opioid receptor subtypes, an important question that hovers over the issue of various subtypes is the lack of a specific cDNA corresponding to a receptor subtype, e.g., only one receptor gene has been isolated for each opioid receptor in spite of the existence of the multiple subtypes (Giros et al. 1995; Liang et al. 1995; Jordan et al. 2000). DOR has been cloned in several mammalian species including mice (Evans et al. 1992; Kieffer et al. 1992; Augustin et al. 1995), rats (Chen et al. 1993; Abood et al. 1994), and humans (Simonin et al. 1994), yet no cDNA corresponding to DOR1 and DOR2 has been found.

The emerging notion on the occurrence of pharmacological subtypes of opioid receptors as a result of receptor heteromerization that leads to new pharmacological properties is largely being accepted. The properties of these heteromerized receptors are often found to be similar to those of the previously described subtypes (Zhang et al. 2009; Jordan et al. 2000). The subtypes of opioid receptors may also be considered as splice variants of the same gene, since six new splice variants from the human μ-opioid receptor gene have been identified and characterized, yielding a total of ten human splice variants of MOR1 that differed in both potency and efficacy (Pan et al. 2005). Nonetheless, all variants were equally selective for μ-opioid ligands, thereby, confirming their classification as μ-opioid receptors (Pastermak 2005).

2.3 *Commonly Employed Opioid Agonist and Antagonist Ligands*

Some common opioid agonist and antagonist ligands are listed as followed, and are mostly used for the purpose of research. MOR selective agonists include cUENK6, DALDA, DAMGO, DIPP-$NH_2\Psi$, Endomorphin-1, -2, Fentanyl, Methadone, Remifentanil, Sufentanil, Oxymorphone, Morphiceptins, Dermorphin, c [YGwFG], CYT-1010, and Ac-D-Trp-Phe-GlyNH2. DOR selective agonists are HIT (Hibernation Induction Trigger), JOM-13, SNC80, SNC-121, UFP-502, UFP-512, BW3733U86, Deltorphins I, II, and III. Moreover, in studies of pharmalogical properties, TAN-67 was shown to have hδ1-receptor affinity. CP-OH has δcx receptor affinity. DSLET has δ2 and δcx receptors affinity. DPDPE has δ1, δcx, and δnex receptors affinity. Main KOR selective agonists are Bremazocine, BRL 52537, CI-977, Dynorphin A, ICI-197067, Ketocyclazocine, PD-117302, Salvinorin A, U-62066, U-50488H, and U-69593. In addition, Pentazocine has μ-, κ-receptors affinity. EKC and Etorphine have μ-, δ-, κ-receptors. DADLE has δ, δcx, δnex, and μ-receptors affinity. On the other hand, MOR selective antagonists include β-FNA, Naltrexone, CTAP, CTOP, Cyprodime, Naloxonazine, Naloxone, ADL5510, ADC5510, and [N-allyl-Dmt[1]] endomorphin-1,-2. Common DOR selective antagonists are Naltrindole, DIPP-$NH_2\Psi$, ICI-174864, N,N$(Me)_2$-Dmt-Tic-OH, TIPP,UFP-501, BNTX, and Naltriben. DALCE is an antagonist to δ1 and δnex. KOR selective antagonists are GNTI, TRK-820, 5-AMN, 5-MABN, Nor-BNI, U50488H. MZ has affinity to μ-, δ-receptors. Diprenorphine and LY255582 have selectivity to all three classical receptors (also refer to Chap. 2).

However, this list is by no means complete, as it would be quite extensive and beyond the scope of this chapter. It should be noted that the selectivity mentioned above is based on the predominant spectrum of activity.

3 Role of Opioid Receptors in Pain Modulation and Drug Abuse

A majority of studies on opioids are associated with analgesia (Kieffer and Gavériaux-Ruff 2002; Eisinger and Schulz 2005; Hanlon et al. 2011). Many acute insults result in the release of endogenous opioids and increased opioid levels in the blood, which counteracts the noxious stimulus and pain, i.e., stress induced analgesia by opioids. Circulating levels of β-endorphin are significantly elevated following muscle injury, fixed-pressure hemorrhagic shock, and lipopolysaccharide (LPS) administration in animal models, which serves as an index of endogenous opioid system activation (Molina 2002). Similar stress-induced elevations in the production and release of endogenous opioids have been demonstrated in clinical studies, oral surgery (Teoulloa et al. 1997), as well as gynaecological and abdominal surgeries, (Kho et al. 1993; Ozarda et al. 2002) that resulted in increased β-endorphin levels in patients. Stress-induced analgesia can be partially reversed by the broad-spectrum opioid antagonist naloxone (Yamada and Nabeshima 1995), further emphasizing the involvement of endogenous opioids in this process. It is generally accepted that endogenous opioids tonically regulate nociceptive information, and that all naturally occurring exogenous opioid agonists produce analgesia. Today, opiate drugs are widely used in the treatment of various pain including injury pain, inflammatory pain, neuropathic pain, cancer pain, and post-operative pain (Smeester et al. 2014; Santos et al. 2014; Shaqura et al. 2014; Kouchek et al. 2013).

Opioid knockout receptor models provide a powerful tool for studying analgesia and other physiological effects attributed to the opioid system (Kieffer and Gavériaux-Ruff 2002). Strong evidence shows that MOR plays a central role in analgesia. Mice lacking MOR have increased sensitivity to heat, suggesting the existence of a MOR-mediated tone in thermal nociception (Kieffer and Gavériaux-Ruff 2002) and a reduction in stress-induced analgesia (LaBuda et al. 2000). In contrast, DOR- and KOR- deficient mice did not exhibit any alteration in heat perception. Many of the currently available clinical opioid analgesics exert their effects primarily through MOR (Ananthan 2006). After knocking out MOR, the analgesic effects of δ- agonists are either unchanged or diminished (Gavériaux-Ruff and Kieffer 2002), while MOR agonists fail to exhibit analgesia (Matthes et al. 1996; Yamada et al. 2006). However, DOR agonists can enhance the analgesic potency and efficacy of MOR agonists, and the antagonists can prevent or reduce the development of tolerance and physical dependence to MOR agonists, suggesting that DOR and MOR interaction contributes to MOR tolerance (Ananthan 2006; Zhang et al. 2013). All of these observations indicate a vital role of MOR in pain modulation.

Immunohistochemical and ultrastructural analysis revealed that DOR exists in peripheral nerves in the molar dental pulp (Fristad et al. 2006), and the bradykinin, an inflammatory mediator, causes a reaction in trigeminal ganglia sensory neurons, which results in a rapid increase in significant accumulation of DOR in the plasma membrane, thus contributing to peripheral opioid analgesia (Fristad et al. 2006; Patwardhan et al. 2005). Peripheral DOR attenuates mechanical hypersensitivity via activation of peripheral G protein-coupled inwardly rectifying potassium channels (Chung et al. 2014), while at central terminals of myelinated mechanoreceptors, activated DOR inhibits voltage-gated calcium channels and synaptic input to the spinal dorsal horn (Bardoni et al. 2014). It was observed in rodent models that DOR1 and DOR2 antagonists respectively blocked the supraspinal anti-nociceptive activation by DOR agonists DPDPE and deltorphin (Jiang et al. 1991; Sofuoglu et al. 1991, 1993; Mattia et al. 1991). In DOR knockout animals, DOR agonist-induced analgesia, was either abolished, reduced, or unchanged, depending on the nociceptive assay and route of administration (Zhu et al. 1999). The recorded analgesia could be a result of the cross-reactivity of DOR agonists at MOR in vivo (Gavériaux-Ruff and Kieffer 2002). Contrary to the prevailing view, Scherrer et al. (2009) demonstrated that, under resting conditions, DOR is trafficked to the cell surface independent of substance P and internalized following activation by DOR agonists. There is also evidence showing that the segregated DOR and MOR distribution is paralleled by a remarkably selective functional contribution of the two receptors to the control of mechanical and thermal pain, respectively. Intrathecal SNC80, a selective nonpeptidic DOR agonist, attenuated carrageenan-induced tactile allodynia via blocking formalin-induced substance P release, but it did not change acute thermal latencies or carrageenan-induced thermal hyperalgesia (Kouchek et al. 2013).

KOR is also known to play a role in pain modulation. Even though KOR deficient and KOR agonist did not exhibit any alteration in the perception of thermal or mechanical pain (Ga Lawson et al. 2010), the KOR mutant mice did have an enhanced pain response to the peritoneal acetic acid injection (Gavériaux-Ruff and Kieffer 2002). Furthermore, dynorphin-deficient mice displayed a mild increase in pain induced by inflammation (Wang et al. 2001). Additionally, KOR agonists have been shown to mitigate hyperalgesia under persistent inflammatory pain states (Auh and Ro 2012) by means of the PI3K/AKT/nNOS/NO pathway (Cunha et al. 2012). Interestingly, KOR has been shown to not only attenuate pain, but to also relieve pain-related depression (Leitl et al. 2014). Moreover, compared to MOR and DOR, KOR chiefly mediates analgesia to visceral pain (Gebhart 2000).

Remarkably, the anti-noceptive effects of opioids have been shown to be sex-dependent specific. For example, in female rats a tenfold higher dose of DPDPE, a DOR agonist, is required in order to produce the same level of anti-hyperalgesia in comparison to male rats (Saloman et al. 2011). Additionally for KOR, the difference of analgesia between males and females becomes even more evident when the dose of KOR agonists has higher concentration levels (Auh and Ro 2012; Liu et al. 2013). The central point here is that activation of KOR in the spinal cord produces an estrogen-dependent modulation of acute and inflammation-induced chronic nociception (Lawson et al. 2010). Moreover, the predominant endogenous MOR ligand, endomorphin 2 (EM2) induced antinociception are sex-dependent and stage of estrus cycle-dependent (Liu et al. 2013). Furthermore, Zhang et al. (2014) revealed that testosterone plays a key role in the regulation of MOR in inflammatory conditions, involving the trigeminal ganglia. Theses examples of sex-specific opioid analgesia indicated that even spinal cord MOR/KOR heterodimers could serves as pharmacological targets for female-specific control (Chakrabarti et al. 2010).

Although opioids have an array of different mechanisms, the combination of different opioids is known to produce synergism. For example, the co-administration of two opioids increases its analgesia effects, the magnitude of which is dependent on the intensity of pain stimulus (Miranda et al. 2014; Borroto-Escuela et al. 2013; Rech et al. 2012). Moreover, opioid and a2-adrenergic receptor (a2AR) agonists have also been show to have synergistic effects for which the DOR is sufficient for such spinal opioid-adrenergic interactions to occur (Giovannitti et al. 2015). As our understanding of anti-nociception has advanced, scientists have developed various opioid-peptides that can target different types of pain. One such example is crotalphine, DEL-6 and DK-4. In the cancer pain model, crotalphine produces its antinociceptive effects via interactions with KOR and DOR (Brigatte et al. 2013).

Likewise, DEL-6 and DK-4 have been shown to recruit DOR and MOR in order to attenuate the nociceptive responses to acute thermal stimuli (Kotlinska et al. 2013).

Multiple studies from the Caron and Lefkowitz laboratories used β- arrestin knockout mice to show that β- arrestins regulate opioid receptor-mediated signal transduction, and therefore play an important role in opioid-induced analgesia along with tolerance and dependence (Bohn et al. 1999, 2000, 2002, 2003). These findings have been confirmed by studies carried by Pei and Ma (Chen et al. 2001; Xiang et al. 2001; Kang et al. 2005; Zhang et al. 2005). Evidence has shown that MOR-mediated pain modulation in association with NF-kB and MEK1/2 pathways. Specifically, NF-kB inhibitor and MEK1/2 inhibitor have been shown to synergize morphine's analgesia along with up-regulate the expression of MOR and DOR mRNA (Popiolek-Barczyk et al. 2014). In addition, the heme oxygenase 1 (HO-1) inducer and cobalt protoporhyrin IX (CoPP) also increases the anti-nociceptive effects of morphine and peripheral expression of MOR and DOR in inflammatory pain conditions (Carcolé et al. 2014). Interestingly, in the dorsal root ganglion, increased methylation of the MOR gene proximal promoter reduces morphine's analgesic effects in neuropathic pain (Zhou et al. 2014). In conditions of neuropathic pain, MOR and KOR analgesia are dependent on modulations and interactions with the neuro-immune systems, while DOR analgesia is independent of microglial activation that is likely due to the fact that microglial cells lack DOR (Mika et al. 2014).

Excessive use of opioids leads to opioid addiction. Drug abuse and addiction are now considered a neurological pathology due to their effects on brain function. Neurological impairments that were in addiction reflect drug-induced neuronal dysfunction and neurotoxicity. The abuse of drugs directly or indirectly affect various neurotransmitter systems, particularly dopaminergic and glutamatergic systems (Cunha-Oliveira et al. 2008; LaLumiere and Kalivas 2008; Vander et al. 2014). The treatment of drug abuse and addiction is the most expensive of all the neuropsychiatric disorders, which may be associated with personality disorders and a diminished quality of life (Cunha-Oliveira et al. 2008; Brinbaum et al. 2006; Karrow et al. 2008; Gastfriend 2014). Behavioral patterns of addiction include compulsive drug-seeking, persistent abuse of substances, and the high probability of relapse even after prolonged drug-free periods (Ivanov et al. 2006).

The mechanisms underlying neurological disorders caused by opioid addiction are not fully understood yet. The neurological basis of positive reinforcement relies on the function of dopaminergic neurons, which result in an increased dopamine release in the mesolimbic brain structures. Certain aspects of opioid dependence and withdrawal syndrome are also related to noradrenergic and serotonergic systems, as well as both the excitatory and inhibitory amino acid systems, and peptidergic systems. An important role in neurochemical mechanisms of opioid reward, dependence, and vulnerability to addiction has been ascribed to the activation of endogenous opioid peptides, particularly those acting via the MOR and KOR.

Opioid abuse leads to opioid tolerance in the nervous system. Receptor tolerance and/or adaptation involve complex mechanisms of receptor regulation, including desensitization and internalization (Koch and Höllt 2008). Several important processes have been identified, including upregulation of cAMP/PKA and cAMP response element-binding signaling, and, perhaps, MAPK cascades in opioid sensitive neurons, which might not only influence tolerance and withdrawal, but also synaptic plasticity during the cycles between intoxication and withdrawal (Christie 2008). Intracellular molecules of signal transmission are also involved in opioid tolerance and dependence, including G proteins, cyclic cAMP, MAP kinases, and some transcription factors. The latter link in this chain of reactions modifies the expression of target genes, and it may be responsible for the long-lasting neural plasticity induced by opioids (Christie 2008; Przewlocki 2004). In general, drug abuse induces adaptive changes in opioid receptors that occur following acute (e.g., desensitization and/or internalization) and chronic (e.g., adaptive tolerance and down-regulation) opioid administration (Harrison et al. 1998; Kreek et al. 2004; Mayer and Höllt 2006; Iskandar et al. 2007). Future work on prescribed drugs of abuse will include identification of clinical practices that minimize the risks of addiction, the development of guidelines for early detection and management of addiction, and the development of clinically effective agents that minimize the risks of abuse (Compton and Volkow 2006).

In addition, opioid abuse may predispose an individual to opportunistic infections, such as hepatitis, bacterial pneumonia, tuberculosis, abscesses, CNS infections, endocarditis, and even AIDS (Quaglio et al. 2003; Ordaz-Sanchez et al. 2003; Korolenko et al. 2007; Hauser and Knapp 2014). One of the explanations for this phenomenon is that opioids suppress immune function (also refer to 4.9 below), including antimicrobial resistance, antibody production, monocyte-mediated phagocytosis, and neutrophil and monocyte chemotaxis (Ordaz-Sanchez et al. 2003).

4 Opioid Receptors and Neurological Disorders

The opioid receptor in the CNS represents one of the most extensive and diverse peptidergic neurotransmission systems, and is widely involved in various pleiotropic functions. Opioids and their receptors are essential for the proper function of various physiological processes in the CNS. Several devastating neurological disorders have been associated either directly or indirectly with opioid receptor dysregulation. Therefore, the expression of opioid receptor and regulation of opioid receptor activity may have a profound effect on these neurological disorders.

4.1 Hypoxic/Ischemic Injury

In the past decade, one of the most exciting findings is the DOR-mediated neuroprotection against hypoxic/ischemic injury. Historically, the controversies over the role of opioids in the neuronal responses to hypoxic/ischemic injuries have not reached an agreement. Some scholars (Mayfield and D'Alecy 1994; Endoh et al. 1999; Hayward et al. 1993; Summers et al. 2003) have shown that opioid receptor activation, by opioid agonists, protects brain from ischemia and prolongs the animal survival time during severe hypoxia, while others (Hosobuchi et al. 1982; Adams et al. 1986; Skarphedinsson and Thoren 1988; Olinger et al. 1990; Chen et al. 2000, 2001) demonstrated opioid receptor inhibition with intravenous injection of high dose naloxone protects the brain from ischemia-induced injury via interacting with DOR, MOR, and KOR.

Our serial publications since 1999 have demonstrated that DOR is neuroprotective against hypoxic/ischemic/excitotoxic injury in the brain, especially in the cortex (Zhang et al. 1999, 2000, 2001, 2002, 2006; Ma et al. 2005; Tian et al. 2008a, b, c; Yang et al. 2009; Zhao et al. 2002; Chao et al. 2006), and have clarified the historical controversies on the role of opioids and their receptors in hypoxic/ischemic neuronal injury (Chao and Xia 2010; Feng et al. 2009; He et al. 2013). Furthermore, we have found that the mechanisms for the DOR neuroprotection involves the stabilization of ionic homeostasis (Zhao et al. 2002; Chao et al. 2006, 2007a, b, 2008, 2009), increases in pro-survival signals (Ma et al. 2005; Chao et al. 2007b), and attenuation of oxidative injury (Yang et al. 2009), upregulation of excitatory amino acid transporters (Liang et al. 2014), as well as regulation of microRNA expression (Yang et al. 2012; He et al. 2013).

We applied glutamate to cultured cortical neurons to mimic neuro-excitotoxicity, as it is a key mediator of ischemic/hypoxic injury. The addition of glutamate induced neuronal injury and cell death in a dose-dependent manner with a major difference between the immature and mature neurons. Activation of DOR as the result of 10 μM DADLE reduced glutamate-induced injury by almost half, and naltrindole (a DOR antagonist, 10 μM) completely blocked this protective effect. In sharp contrast, administration of MOR and KOR agonists (DAMGO and U50, 488H respectively, 5–10 μM) did not induce appreciable neuro-protection, and MOR or KOR antagonists had no significant effect on glutamate-induced injury. These data demonstrate that the activation of DOR, but not MOR and KOR, protects neocortical neurons from exerts neuroprotection against glutamate excitotoxicity (Zhang et al. 1999, 2000). DOR neuroprotection against hypoxic stress, ischemic injury, or excitotoxic injury was shown in our series of studies (Zhang et al. 1999, 2000, 2001, 2002, 2006; Ma et al. 2005; Tian et al. 2008a, b, c; Yang et al. 2009, 2012; Zhao et al. 2002; Liang et al. 2014; Chen et al. 2014; Tian et al. 2013a, b; Zhou et al. 2013; Chao et al. 2013; He et al. 2012), as well as in many studies conducted by other researchers (Lim et al. 2004; Horiuchi et al. 2004, 2008; Borlongan and Wang 2004; Borlongan et al. 2009; Zhao et al. 2006; Su et al. 2007; Iwata et al. 2007; Charron et al. 2008; Kao et al. 2008; Govindaswami et al. 2008; Pamenter and Buck 2008; Peng et al. 2009a, b; Zhu et al. 2009).

4.2 *Epileptic Seizures*

Epilepsy is defined as a disorder in which a person has repeated seizures over a period time. Seizures are episodes of disturbed brain activity that cause changes in attention or behavior of the individual afflicted. Epileptic seizures can vary widely in their physical manifestations, but are generally characterized by recurrent, unprovoked excitability of neurons within a region of the brain. The biologic bases of epilepsy are variable and complex. There are many potential factors that contribute to the development of epilepsy, such as genetic abnormality, trama, hypoxic/ischemic brain injury and abnormal function of neurotransmitters/neuromodulators.

Endogenous opioids, play various roles in epileptogenesis (Chao and Xia 2012; Chao et al. 2013; Hammers et al. 2007). For example, as early as the 1970s, endogenous DOR agonist, Met-enkephalin, was observed to have dose-related naloxone-reversible epileptic seizures (Tortella et al. 1988). Despite their variance, these studies demonstrate a close relationship between the opioid systems and seizures.

Nevertheless, the role of opioids in epileptogenesis and epileptic seizure remains controversial. Both anticonvulsant and proconvulsant effects of the three major opioid systems (δ-, μ-, and κ-), in addition to nociceptin opioid peptides, have been proposed in the literature (Pinsky et al. 1986; Tortella et al. 1985, 1988; Tortella and Long 1985; Grecksch et al. 1999; Yajinma et al. 2000; Bregola et al. 2002; Feng et al. 2004; Bausch et al. 2005; Danielsson et al. 2006; Solbrig et al. 2006; Sabory et al. 2007; Zhang et al. 2009; Scwarzer 2009). The role of opioids on epiletogenesis and epileptic seizures depends on the selective characters of the opioid receptors (Solbrig et al. 2006; Scwarzer 2009), the dose of the opioids (Bausch et al. 2005; Yajinma et al. 2000; Danielsson et al. 2006; Negus et al. 1998; Jutkiewicz et al. 2006), and the specific types of seizure (Bregola et al. 2002; Feng et al. 2004). For example, a putative DOR agonist, SNC80, has a dose-dependent effect on seizures, with low-doses inhibiting seizures and high-doses promoting them. Furthermore, nociceptin/orphanin FQ may play a facilitatory role in epileptogenesis in kainite and kindling models of epilepsy in rats (Bregola et al. 2002), although they have a receptor-specific inhibitory effect on penicillin-induced seizures (PIS) in rats (Feng et al. 2004).

The action of δ- opioid system is complicated in terms of its role in epilepsy. Opioids have been noted to contain both anticonvulsant and pro-convulsant properties in various forms of epilepsy (Chao and Xia 2012). The reasons for the complex and diverse effects of the δ- opioid system on seizures are not well understood, but could partially be related to the several factors including animal species, seizure types, the methods for drug administration, dose used, and target neurons, etc. (Chao and Xia 2012). Endogenous opioids generally play an inhibitory role in modulation of neuronal excitability, which provide a straightforward explanation to their anticonvulsant effect. However, if this inhibition occurs on inhibitory interneurons, the opioids will facilitate, rather than suppress seizures via post-synaptic disinhibition (Chao and Xia 2012; Chao et al. 2013). Therefore, more delicate studies are needed to increase familiarity with the effects of opioid action on epilepsy.

Refer to Chap. 16 for potential roles of DOR in epilepsy.

4.3 *Parkinson's Disease*

Parkinson's disease (PD) is the second most common neurodegenerative disease that results from the loss of dopaminergic neurons in the substantia nigra pars compacta. This results in a consequential striatal dopamine deficiency, which leads to a cascade of functional changes in basal ganglia circuitry and ultimately manifests itself as a movement disorder. Opioid receptors (DOR, MOR, KOR), particularly DOR, and their cognate endogenous opioids (enkephalin, dynorphin) are widely distributed throughout the central nervous system. They have been demonstrated to be involved in the pathophysiology of PD; however, there has been no solid conclusion drawn. Nevertheless, activation of KOR effectively ameliorates L-DOPA-induced dyskinesia symptoms in the rat model of Parkinson's disease (Ikeda et al. 2009). Furthermore, abundant clinical evidence shows that various movement disorders, such as dystonia, can emerge from general anesthesia and can dramatically reverse with the administration of naloxone (Iselin-Chaves et al. 2009). A recent study shows that a selective MOR antagonist, such as ADC5510, can reduce L-DOPA-induced dyskinesias in the Parkinson's disease model (Koprich et al. 2011). Moreover, it is apparent that MOR has a very different role compared to that of DOR and KOR in Parkinson's disease. However, there is a lack of in-depth and mechanistic reports for most functions in which they are involved.

In regards to the δ-opioid system, both the δ-opioid peptide precursor proenkephalin, as well as DOR, are distributed throughout the basal ganglia, and help regulate the activity of basal ganglia nuclei. The changes in DOR expression in the PD brain are still full of ambiguity, even though the changes in endogenous enkephalins and their precursors are relatively explicit in PD. Moreover, the role of DOR systems in the PD is controversial in the literature. Activation of DOR in the basal ganglia may exert dual effects, either anti-akinetic or pro-akinetic, (Mabrouk et al. 2009). For example, several lines of evidence suggest that DOR agonists via systemic administration have a therapeutic effect on Parkinson's disease (Hill et al. 2000; Mabrouk et al. 2008; Chen et al. 2014), whereas some reports suggested a pro-Parkinsonian role of delta opioid systems (Pinna and Di Chiara 1998; McCormick and Stoessl 2002). Since DOR is present at both the presynaptic and postsynaptic sites in addition to its heterogeneous distribution throughout various brain regions, more investigation with better technology and novel approaches are needed to further assess DOR's effects on PD. Toward this aim, our recent studies suggest that δ-opioid receptor activation reduces α-synuclein overexpression and oligomer formation induced by MPP^+ and/or hypoxia DOR activation ameliorates MPP^+- and/or hypoxia-induced α-synuclein, suggesting a protective role of DOR in the brain against PD pathophysiology (Chen et al. 2014). For more detailed information regarding the role of DOR in PD, readers are recommended to refer to Chaps. 14 and 15.

4.4 *Multiple Sclerosis*

Multiple Sclerosis (MS) is an autoimmune inflammatory disease which causes damage to the insulating sheathes around neuronal cells in the CNS. MS is characterized with a female predominance and a wide array of signs and symptoms that include vision changes, sensory or motor deficits, and psychiatric problems, along with a relapsing or progressive course. MS has a wide range of signs and symptoms, pain being a common symptom. Substantial data has shown that opioids are involved in the onset and progression of experimental autoimmune encephalomyelitis (EAE) in the MS animal model (Zagon et al. 2010; Campbell et al. 2012). Opioid growth factor (OGF) and low dose naltrexone are inhibitory to the onset and progression of EAE, but a high dose of naltrexone does not have the same effect. Campbell et al. (2012) showed that on the sixth day of OGF treatment, those EAE mice models had the decreased behavioral scores about 45 % relative to saline EAE control groups. Moreover, during the 40-day period, the OGF-treated group demonstrated attenuated behavior, manifesting as limp tails and wobbly gaits, while the saline-treated EAE mice displayed continued paralysis of one or more limbs (Campbell et al. 2012). Furthermore, Rahn et al. (2011) showed low-dose naltrexone and OGF treated EAE mice display behavioral symptoms of EAE, 68 % and 63 %, respectively. This is in sharp contrast to 100 % of the control EAE mice, which displayed EAE symptoms. Additionally, Zagon et al. (2009) showed that there were no differences in incidence, severity, disease index, the number of activated astrocytes, demyelination, and neuronal injury between high-dose treated EAE mice and saline treated EAE mice. However, over 33 % of the low dose naltrexone EAE mice did not exhibit behavioral signs of disease and they had reduced severity and disease index.

In the reaction of specific rosette formation of lymphocytes, morphine was shown to eliminate the number of rosette-forming cells (Loseva and Koliaskin 1981). It was discovered that blood lymphocytes from MS patients had enhanced adhesive activity to opioids, which are specific to morphine. Therefore, the morphine test became an immunopharmacologic diagnosis method via the reaction of specific rosette formation (Shutov et al. 1999).

Interestingly, there is a probable link between opioid peptide β-endorphin and the heterogeneity of the clinical course seen in MS patients. Peripheral blood mononuclear cells (PBMC) concentrations of β-endorphin have been found to be highest in the patients with relapsing remitting MS, and lowest in patients with the progressive forms. This suggests that β-endorphin is involved in suppressing inflammatory process (Gironi et al. 2003). Nonetheless, the regulation of inflammatory process is the key function of opioids, specifically OGF in MS. For example, low-dose naltrexone (LDN) operates as a novel anti-inflammatory agent via action on microglial cells in MS (Younger et al. 2014). Naltrexone is one of the molecular approaches to modulation of the OGF-OGFr axis. Other pharmacological and molecular methods to modulation of the OGF-OGFr axis are genetic manipulations of OGFr expression and antibody neutralization of OGF (McLaughlin and Zagon 2012). OGF via OGF-OGFr axis inhibits DNA synthesis to prevent T lymphocytes, microglia/macrophages, and astrocytes from proliferation (Hammer et al. 2013; Campbell et al. 2012, 2013). By means of the proliferation mechanism, OGF prevents relapses in relapse remitting experiment autoimmune encephalomyelitis (RR-EAE). Comparing OGF-treated RR-EAE animals and saline treated control groups, OGF-treated RR-EAE animals showed less severe clinical disease, lower median cumulative disease scores, longer periods of remission, a decrease in frequency of relapses, shorter duration of relapses, and also showed significantly lower numbers of T-lymphocytes, microglia/macrophages, and activated astrocytes (Hammer et al. 2013). Campbell et al. (2012, 2013) provide further insights into how the mechanisms of OGF-OGFr axis regulate astrocyte proliferation.

Through knockdown of the classical opioid receptors and OGFr, only OGFr changes the number of astrocyte cells with OGF treatment, instead of MOR, DOR, and KOR. Additionally, OGF treatment can only suppress the cell cycle of activated, cultured astrocytes via suppressing its DNA synthesis, but it cannot alter astrocyte migration, apoptosis, and necrosis.

In addition to the studies of EAE animal model of MS, clinical researches also proved evidence that opioids are involved in the MS disease process. In 2008, a 6-month, phase II multicenter-pilot trial with low-dose naltrexone for primary progressive multiple sclerosis (PPMS) indicated that LDN is safe, tolerable, and can improve spasticity, pain, fatigue, and depression, there by improving the patients' quality of life (Gironi et al. 2008)

In the clinical arena, Kalman et al. (2002) found that MS patients are relatively insensitive to opiate treatment. Consequently, some studies do not support the routine administration of opioids for MS patients with central pain (CP) (Kalman et al. 2006). In a study examining 14 opioid-free patients with constant, non-fluctuating, long-lasting CP due to MS, only four of the patients were found to be "opioid responders" with high-dose morphine, while the other ten patients had similar responsiveness to placebo and low-dose morphine. In order to further investigate the underlying reason for loss of opiate analgesia in MS patients, Lynch et al. (2008) utilized the Theiler's murine encephalomyelitis virus (TMEV) model of MS to explore μ-, δ-, and κ-opioid receptors in mRNA expression. They found that all of the receptors decreased, which would explain the increased nociception in TMEV mice and MS patients. Although MS patients are less sensitive to opiate treatment, when MS patients are treated for pain by intrathecal morphine, morphine overdose can still cause severe complications. Yilmaz et al. (2003) described a case of a 47-year-old woman with secondary progressive multiple sclerosis that was treated with continuous infusion of intrathecal morphine. The patient subsequently suffered a status epilepticus attack after an overdosed bolus injection of morphine. Fortunately, after treatment with continuous intravenous naloxone infusion and medication to control hypertension, the seizure activity resolved, the patient eventually recovered. Such a case gives rise to concerns and caution whenever administrating opioid medications.

4.5 *Amyotrophic Lateral Sclerosis*

Amyotrophic Lateral Sclerosis (ALS) is an idiopathic disorder that affects both upper and lower motor neurons, causing muscle atrophy, weight loss, and respiratory failure. ALS pathogenesis includes inflammation, oxidative stress, glutamate excitotoxicity, apoptosis, mitochondrial dysfunction, protein aggregation, and cell death (Zhu et al. 2014). Studies have found that the levels of glutamate in the serum and spinal fluid of patients with ALS are higher than healthy people (Al-Chalabi and Leigh 2000), and glutamate excitotoxicity seems to be the core cause (Blasco et al. 2014). Currently in clinical practice, riluzole is the only treatment that may slow down the progression of the disease, which is thought to work by inhibiting glutamate release. Excitatory amino acid transporters (EAATs) regulate extracellular glutamate concentrations, thereby preventing excitatory amino accumulation and neuroexcitotoxicity. Our studies (Zhang et al. 2000, 2002) have proved that DOR has a neuroprotection against glutamate-induced excitotoxicity neuronal injury. In our recent studies (Liang et al. 2014), by exposing mouse astrocytes to a specific DOR agonist, UFP-512 and/or DOR antagonist, naltrindole, we found that DOR up-regulates astrocytic EAATs, EAAT1 and EAAT2, both in mRNA and protein levels via the MEK-ERK-p38 signaling pathway. Moreover, DOR significantly increases glutamate uptake of astrocytes, which is apparently a benefit for relieving glutamate-induced excitotoxicity. Our novel findings imply that DOR may enhance glutamate uptake through an upregulation of EAATs in the astrocytes.

4.6 Stress

The role of the opioid function in controlling pain, reward, and addiction is well established, but its role in regulating other emotional responses is poorly documented (Filliol et al. 2000). Recent studies have shown that endogenous opioid systems are indeed associated with the regulation of emotional responses. In particular, it has been reported that δ-opioid receptors act as natural inhibitors of stress and anxiety (Saitoh et al. 2005). DOR agonists have been shown to produce anti-depressant and anxiolytic effects in rodent models. In addition, DOR agonists have been shown to increase expression of brain-derived neurotrophic factor (BDNF) mRNA, a side-effect of some antidepressants, which may be important to research for the clinical efficacy of anti-depressant drugs (Torregrossa et al. 2006; Naidu et al. 2007). Zhang et al. (2007) reported that centrally administered nor-BNI, like most clinically used anti-depressants, upregulated BDNF mRNA expression in the rat hippocampus. These findings further imply that central KOR mediates anti-depressant-like effects by BDNF gene upregulation. Some data demonstrate dynamic changes in MOR neurotransmission in response to an experimentally induced negative affective state (Zubieta et al. 2003). In addition, opioids alter the expression of opioid receptors in the condition of stress. In the mice exposed to stress, for example, morphine induced a region specific increase in KOR (Taracha et al. 2009; Vien et al. 2009).

5 Opioid Receptors and Cardiovascular Conditions

Opioid peptides, including β-endorphin (Forman et al. 1989), enkephalins (Lang et al. 1983; Weihe et al. 1983, 1985; Barron et al. 1992), and dynorphins/dynorphin-like peptides (Weihe et al. 1985; Spampinato and Goldstein 1983) are present in the heart. In addition to myocardial cells, δ- and κ-opioid peptides have been shown to exist in sympathetic nerve fibers and ganglion cells (Weihe et al. 1983; Steele et al. 1996).

Myocardial cells are one of the known sites of opioid peptide synthesis, storage, and release (Barron et al. 1995).

5.1 Role of Opioid Receptors in the Regulation of Blood Pressure and Heart Rate

Short-term cholestasis is associated with resistance against ischemia/reperfusion-induced arrhythmia, which depends on availability of endogenous opioids (Hajrasouliha et al. 2005).

Both the KOR agonists, ethylketocyclazocine and enadoline, elicited an increased heart rate with little effect on blood pressure through the combined action on central and peripheral receptors in conscious squirrel monkeys that could be blocked by the opioid antagonist naltrexone. This effect appeared to be specific for KOR, since the MOR agonist morphine did not mimic the effects of the KOR agonists (Schindler et al. 2007). Micro-injections of DOR antagonist naltrindole into the ventrolateral periaqueductal gray, a region importantly involved in opioid analgesia, inhibited hemorrhagic hypotension significantly in both conscious and anesthetized rats, while MOR and KOR antagonists were ineffective (Cayun et al. 2001). However, it is reported that after hemorrhaging and during the re-compensatory period, stimulation of δ-opioid receptors led to an improved mean arterial pressure in conscious, freely moving male rats (McBride et al. 2005). Similar effects of enhanced mean arterial blood pressure were also observed in shock patients undergoing treatment with opioid antagonists (Boeuf et al. 2003).

MOR has been shown to be involved in the regulation of cardiovascular function. For example, cyclo [Nε, Nβ-carbonyl-D-Lys2, Dap5] enkephalinamide (cUENK6), a MOR agonist, was found to stimulate excretion of urine, sodium, potassium, and cGMP to suppress the stress-induced elevation in blood pressure and heart rate (Gutkowska et al. 2004). Intra-thecal injection of endomorphins, elicited a decrease in systemic arterial pressure and heart rate in a dose-dependent manner (Wang et al. 2007). As naloxone significantly antagonized the effects of vaso-depression and bradycardia, it is possible that they were elicited by the activation of MOR in the rat spinal cord (Wang et al. 2007). In contrast, Mao and Wang (2005) reported that micro-injections of endomorphin-1 (EM-1) or nociceptin into the nucleus tractus solitarii increased blood pressure and heart rate. These contrary results indicate that the cardiovascular effects are likely to be dependent on the site of action of opioid peptides.

5.2 Opioid Receptor Medicated Cardioprotection

Myocardial opioid peptide levels are elevated during episodes of stress, such as ischemia (Peart et al. 2005; Eliasson et al. 1998). Opioid receptors, especially DOR and KOR, are also involved in cardio-protection against ischemic injury. Gross et al. (Gross 2003; Gross et al. 2004; Peart et al. 2003) observed that opioid receptors are important in mediating ischemic pre-conditioning (IPC) in the heart of dogs. In addition, both DOR (Wang et al. 2001; Schultz et al. 1997) and KOR (Wang et al. 2001; Wu et al. 1999) have been shown to mediate cardio-protection by preconditioning myocardial ischemia and metabolic inhibition. Animal experiments have demonstrated that the opioid system can modulate hemodynamic and cardiovascular activity. In an animal model of hemorrhagic shock, opioid receptor antagonists ameliorated hemodynamic instability (Faden and Holaday 1979; Holaday 1984).

DOR stimulation with selective agonists could mimic the cardio-protective effects of IPC (Peart et al. 2003). In an animal model of severe hemorrhagic shock, following DADLE treatment improved hemodynamic stability and a prolonged survival was observed (Summers et al. 2003). Although some investigators believe that DOR1 plays an important role in the DOR mediated cardiac protection, others maintain an opposite view. For instance, Lasukova et al. (2004) showed that DOR agonist, DPDPE, induced anti-arrhythmic and cardio-protective effects in coronary artery occlusion and reperfusion, and concluded that this protection manifested secondary to stimulation of cardiac DOR1. On the other hand, Maslov et al. (2009) believed that DOR2 plays the vital role in cardio-protection. They reported that a pre-treatment with peripheral and DOR2 antagonists completely abolished the cardio-protective effects of deltorphin II, while DOR1 antagonist 7-benzylidenenaltrexone (BNTX) had no effect. In addition, protein kinase C (PKC) inhibitor chelerythrine and the NO-synthase (NOS) inhibitor L-NAME (N-nitro-L-arginine methyl ester) also reversed the effects of deltorphin II. The nonselective ATP-sensitive K+ (KATP) channel inhibitor glibenclamide, and the selective mitochondrial KATP channel inhibitor, 5-hydroxydecanoic acids, abolished the infarct-sparing effect of deltorphin II. Inhibition of tyrosine kinase (TK) with genistein, the ganglion blocker, hexamethonium, and the depletion of endogenous catecholamine storage with guanethidine, reversed the antia-rrhythmic action of deltorphin II, but did not change its infarct-sparing action. Based on all these findings, they concluded that the cardio-protective mechanism of deltorphin II is mediated via stimulation of DOR-2. PKC and NOS are involved in both its infarctsparing and anti-arrhythmic effects. Infarct-sparing effect is dependent upon mitochondrial KATP channel activation, while the anti-arrhythmic effect is dependent upon TK activation. Endogenous catecholamine depletion reduced antiarrhythmic effects, but did not alter the infarct-sparing effect of deltorphin II. This controversy provides additional evidence on the uncertainty over the existence of DOR subtypes.

In studies on KOR, Wu et al. (1999) showed that KOR plays a role in cardioprotection during the delayed phase of protection, while Wang et al. (2001) demonstrated that KOR activation mediated both the anti-arrhythmic and infarct limiting effects of IPC. Furthermore, Rong et al. (2009) found the KOR activation attenuated myocardial apoptosis and infarction after ischemia/reperfusion.

6 Opioid Receptors and Respiratory Disorders

Further evaluation found that opioids and the opioidergic systems extend to the lungs, specifically the pulmonary neuroendocrine cells (PNECs) and sensory C-fibers within the bronchial epithelium and pulmonary vasculature. In terms of respiratory system, opioids and their receptors are well known, as their respiratory depression via its centrally mediated actions following their analgesic effects. The peripheral mechanism of ventilation depression has been elucidated by studies in both animals and humans (Krajnik et al. 2014; Trescot et al. 2008; Dahan et al. 2010; Yeadon and Kitchen 1989; Shook et al. 1990; Young et al. 2013). One other function of opioids in respiratory systems is the regulation of hypoxic pulmonary hypertension (Zhang et al. 2013; Wu et al. 2013; Peng et al. 2009a, b; Li et al. 2009).

6.1 Respiration Depression

Opioids affect ventilation, leading to respiratory depression in both animals and humans (Yeadon and Kitchen 1989; Pattinson 2008). This effect occurs in part due to a direct action of opioids on respiratory generating structures in the brain (Pattinson 2008). It has been known that there are high densities of opioid receptors in the brain areas related to respiration (Mansour et al. 1993; Mutolo et al. 2007) and local application of opioid agonists to these areas depressed the activity of respiratory related neurons (Mutolo et al. 2007; Haji et al. 2000). Opioid receptors show a heterogeneous variation in the depression of respiration by opioids, as MOR activation shows a marked decrease in respiratory frequency and complete apnea, less pronounced respiratory depressant effect by DOR, and the absence of an effect on respiration upon KOR activation (Mutolo et al. 2007). Also, there appears to be crosstalk among opioid receptors regarding their depression on respiratory response. For example, it has been shown that respiratory depression produced by DOR agonists is suppressed, while the respiratory response to the activation of KOR is maintained in MOR knockout mice (Matthes et al. 1998). The phenomenon could be a result of μ-/δ-receptor heteromeric interactions, which exhibit regional selectivity across different areas of the brain.

Owing to an extensive clinical use of opioid drugs for pain relief (e.g., cancer-induced or post-operative), the accompanying undesirable side effect of opioid-induced respiratory depression has become an important concern. Though associated mortality is low, respiratory depression poses a significant clinical problem for patients treated with opioids, especially in the post-operative period (Pattinson 2008). Opioid drugs depress the rate and depth of breathing, blunt respiratory responsiveness to CO2 and hypoxia, increase upper airway resistance, and reduce pulmonary compliance. Opiate respiratory is mainly due to the agonist activation of μ- and δ-receptor subtypes, and involves specific types of respiratory-related neurons in the ventro lateral medulla and the dorsolateral pons (Xia 1989; Lalley 2008).

Under certain conditions, this adverse effect may be lethal. For instance, severe stress (hemorrhagic shock, trauma, bacterial infection) resulting in a massive release of endogenous opioids, or an overdose of opioid analogues (drug abuse, addiction, and uncontrolled use for pain relief), can cause a severe respiratory depression, hypotension, and even death.

6.2 Hypoxia Pulmonary Hypertension (HPH)

Opioid receptor, specifically κ-opioid receptor, is involved in the pathophysiology of hypoxic pulmonary artery hypertension. During episodes of hypoxia, pulmonary arterial smooth muscle cells (PASMCs) proliferate and the expression of κ-opioid receptors decreases. This results in a decrease in both mean artery pressure and right ventricular pressure in HPH rats, and suppressed the proliferation of the PASMCs in a dose-dependent manner. All of these effects can be reversed by nor-BNI, a highly selective κ-opioid receptor antagonist. U50 and 488H stimulate and up-regulate κ-opioid receptor expression in order to attenuate HPH (Zhang et al. 2013). The administration of U50 and 588H has a variety of effects, which include endothelial relaxation, an increased responsiveness to acetylcholine (ACh), as well as increases in levels of serum NO, eNOS phosphorylation, and T-AOC. Conversely, a decreased iNOS activity was seen in the pulmonary artery of chronically hypoxic rats, as well as attenuation of HPH- induced enhancement of gp91 (phox) expression and nitrotyrosine content (Wu et al. 2013).

7 Opioid Receptors and Nephrological Disease

In the kidneys, the expression of different kinds of opioid receptors is highly varied. In practicality, the kidney is a type of DOR organ. For example, in kidneys, DOR has been noted to be almost 45-fold higher in number than KOR (Peng et al. 2012). Several independent studies from our laboratory have identified DOR expression in HEK293t cell line, a specific cell line that is originally derived from human embryonic kidney cells (Chen et al. 2014; Shan et al. 2014). Opioids role in the pathological conditions affecting the renal system is complex and remarkable. To illustrate further, the DOR agonist, dalargin, can protect kidney from gentamicin-induced nephrotoxicity, while morphine, MOR agonist, can further aggravate chronic kidney disease (Weber et al. 2012). Moreover, chronic kidney disease-associated pruritus can be treated by KOR agonist nalfurafine (Wikstrom et al. 2005). Furthermore, studies have shown that opioid pathway is activated in renal ischaemia-reperfusion injury (Wever et al. 2013).

7.1 Nephrological Protection

Different kinds of opioid agonists have diverse effects on nephrological protection. Studies demonstrated that δ-opioid receptor agonist protected the kidney from toxicity, while morphine could induce renal disease (Plotnikov et al. 2013; Weber et al. 2012, 2013). Through comparing the effects of the lithium ions and δ-opioid receptors agonist, dalargin on gentamicin-induced nephrotoxicity, Plotnikow found out both of them reduced renal tubular cell death and diminished the damage. The mechanism is to promote glycogen synthase kinase 3β(GSK-3β) phosphorylation and to induce the nephroprotective signaling pathways (Plotnikov et al. 2013). On the other hand, morphine stimulates mitogenic signaling promoting mesangial cell proliferation and renal dysfunction and renal pathology in sickle mice, corresponding to clinically morphine-induced renal disease (Weber et al. 2012, 2013). Morphine mediates the egolmerular filtration barrier by compromising the integrity of podocytes followed by inducing albuminuria and accelerates the progression of chronic kidney disease (Lan et al. 2013).

7.2 *Uremic Pruritus*

An opioid receptor agonist, nalfurafine effectively reduced itching and itching intensity in uremic pruritus and chronic kidney disease-associated pruritus (CKD-aP) (Wikstrom et al. 2005; Mettang et al. 2015). Nalfurafine, a promising treatment for acute itch, as well as ongoing itch of dry skin, suppresses both pruritogen- and touch-evoked scratching behavior in acute and chronic itch mice (Akiyama et al. 2014). Clinically, κ-opioid receptor agonist nalfurafine is an alternative treatment for patient suffering from chronic kidney disease-associated pruritus or intractable pruritus under dialysis. It has shown a significant amelioration of pruritus and acceptable adverse effects (Xander et al. 2013; Kumagai et al. 2012).

7.3 *Renal Ischaemia-reperfusion Injury (IRI)*

Renal ischaemia-reperfusion injury is a severe problem causing high mortality rates. Remote ischaemic preconditioning (RIPC) is an alternative strategy to relieve the injury for brief ischaemia of a limb protecting the kidney against a prolonged ischaemia insult. The opioid receptor antagonist naloxone completely blocked this protective effect of RIPC. RIPC by brief hind limb ischaemia is probably the consequence of endorphin release from the hind limb (Wever et al. 2013). The underlying mechanism is obscure and this is a promising opioid signaling in renal RIPC. However, the blockade of central opioid receptor by naloxone can attenuate the ischemic acute kidney injury in the mice via the suppression of renal sympatho-excitation (Mutoh et al. 2013).

8 Opioid Receptors and Conditions of Obstetrics and Gynecology

Opioid receptors also been noted to play a role in obstetrics and gynecologic conditions including pre-eclampsia, breast cancer, and ovarian cancer.

8.1 Pre-eclampsia

Pre-eclampsia is a disorder, which occurs in the second or third trimester in pregnant females that is characterized by high blood pressure, thrombocytopenia, increased protein in the urine, and impaired liver function. Precelampsia is associated with multiple maternal and fetal adverse effects, such as encephaledema, renal injury, hepatic dysfunction, heart failure. If left untreated, preeclampsia can lead to eclampsia, which is preeclampsia with the addition of life-threatening seizures. Therefore, the effective prevention and treatment of preeclampsia is associated with a reduction of multiple maternal and fetal potential adverse effects. Opioids play an important role here. For example, in the reduced uteroplacental perfusion (RUPP) model of preeclampsia, morphine decreased systolic blood pressure and improved overall renal function via decreasing blood urea nitrogen (BUN), creatinine, urinary protein, and by increasing urinary nitrite/nitrate excretion and creatinine clearance (Javadian et al. 2013).

8.2 Gynecological Oncology

Activation of SSTR2 and opioid receptors can inhibit the tumor promoting signals in human breast cancer cells (Kharmate et al. 2013). Endogenous opioid and opioid growth factors suppress women's cancers, such as triple-negative breast and ovarian cancer. OGF inhibits DNA synthesis by upregulation of cyclin-dependent inhibitory kinases, without disrupting cell migration, differentiation, or apoptosis. OGF receptor is a determinant in the proliferation of triple-negative breast cancer and ovarian cancer, and can be genetically modified to alter neoplastic cell replication in vitro and in nude mice. Preclinical studies warrant the use of OGF alone, or in combination, for treatment of triple-negative breast and ovarian cancer patients, and so far the data have supported the transition of these novel therapies for treatment of human ovarian cancer from the bench to bedside in order to provide additional targets for treatment of this devastating disease (McLaughlin and Zagon 2014; Zagon et al. 2013).

9 Opioid Receptors and Reproductive Function

As previously described, opioid receptors exist in various organs and tissues of reproductive tracts (Soverchia et al. 2006; Fabbri et al. 1988; Tsong 1982; Wittert et al. 1996). Endogenous opioid peptides (EOP) can also be synthesized in the testis and different compartments of the male genital tract. LH and CRF stimulate the production of EOPs in Leydig cells. Germ cell factors regulate opioid gene expression in sertoli cells (Fujisawa et al. 1993). Additionally, opioid recursors are also expressed in somatic and germ cells of the testes (Fabbri et al. 1989; Subiran et al. 2011).

Genesis and expression of opioid precursors and opioids suggest that opioids are implicated in the reproductive function. Opioids regulate reproductive physiology at multiple sites, serving as a multi-messenger. Opioids and opioid receptors regulate the reproduction mainly at two levels, the levels of the central nervous system and the levels of testes and sperm. The opioid system is involved in the control of hormonal releasing (Lei and Rao 2001; Heckert and Griswold 2002; Cicero et al. 1979), regulation of testicular function and the sperm fertility, as well as sperm motility (Subiran et al. 2011). A series of studies showed that opioids could suppress the secretion of gonadotropin, LH and the sex hormonal steroids including testosterone, and estradiol through inhibition of the release of GnRH or directly action on specific binding sites on the gonadotrophs of the pituitary. On one hand, endogenous opioid peptides are synthesized in the Leydig cells and Sertoli cell. The other, opioid exerts an inhibitory paracrine effect on Sertoli cell function, which the effects can be reversed by naloxone. Sperm motility is one of the key factors for reproduction (Quill et al. 2006) and opioid involvement in such activity. Some experiments (Ahmed and Kurkar 2014; Albrizio et al. 2014; Xu et al. 2013) show that opioid decreases sperm motility. In this aspect, Xu (2011, 2012) described that remifentanil impairment on sperm motility has short-term effect and is reversed in 15–30 min in vitro.

10 Roles of Opioid Receptors in Cell Proliferation and Wound Repair

Cell proliferation presents a double-edged sword. On one hand, cell proliferation promotes much-needed repair and regeneration that is required for life. However, on the other hand, if cell proliferation occurs without proper regulation, it can lead to a variety of disease states including cancer. Not surprisingly, opioid receptors are involved in the regulation of cell proliferation and cell growth. It is for this reason that a proper understanding of the processes, controls, and regulations involved in cell proliferation is of vital importance.

10.1 Cell Proliferation

It has been well established that opioid receptors do play a role in cell proliferation. A literature review of reports on opioid receptors and their role in cell proliferation reveals that there are two main opioid receptors involved in proliferation, MOR and DOR. As Malendowicz et al. (2005) observed μ- and δ-receptor activation inhibits the growth of immature adrenal cells, while stimulating adrenal regeneration without affecting the proliferation of cultured adrenocortical cells. Furthermore, MOR regulates neurogenesis with strain specificity. MOR knockout (KO) mice on two background strains, both C57BL/6 and 129S6, enhanced cell proliferation in the dentate gyrus (DG) when compared to wild-type (WT) mice. However, C57BL/6 MOR KO mice increased survival rate of new neurons and decreased the number of apoptotic cells, while 129S6 MOR KO mice exerted little effects on survival rate and had a larger number of apoptotic cells in the region. Collectively, MOR KO increased the granule cell number in the DG of C57BL/6 mice and did not alter the amount of the cell in the same region of 129S6 mice (Cominski et al. 2014). However, in term of neural progenitor cells (NPC), mu receptors have different effects on proliferation and differentiation. Short-term morphine exposure suppressed NPC proliferation and promoted apoptosis by activation caspase-3 expression in a dose-dependent manner and induced differentiation. And all these alterations were reversed with the addition of the opioid antagonist naloxone (Willner et al. 2014).

On the other hand, δ-opioid receptor agonist SNC80 promotes neural differentiation from multipotent neural stem cells obtained from embryonic C3H mouse forebrains. δ-opioid receptor activation leads to neurite outgrowth and neuronal differentiation (Georganta et al. 2013). In addition, inhibition of δ-opioid receptors employing naltrindole, inhibit brain glioma cell proliferation through regulating the expression levels of the cellular internal phosphorylated PKC proteins. Naltrindole promotes mitochondrion to release cytochrome to cytoplasm and activated caspase-9 and -3 leading to apoptosis (Zhou et al. 2013). Theses findings have mooted the concept that δ-opioid receptors play a crucial role in neurogenesis and neuroprotection.

10.2 Wound Repair

Selective blockade of the OGF-OGFr axis by naltrexone (NTX) accelerates fibroblast proliferation and wound healing, and enhance cutaneous wound closure in diabetic and nondiabetic rats alike. Topical naltrexone as treatment benefits type 2 diabetic cutaneous wounds, epithelial-related complications of T2D, reducing the residual wound size by 13–30 % between days 8 and 14 relative to diabetic mice receiving saline (Immonen et al. 2013; Immonen et al. 2014a, b). In addition, topical naltrexone promotes full-thickness wound closure in type 1 diabetic rats by stimulating angiogenesis (McLaughlin and Zagon 2012, 2014).

11 Opioid Receptor and Cancer

Cancer is a diseases state traditionally characterized by unregulated cell growth. There are several types of cancers that are categorized by their cell-type of origin, such as neuroblastoma, hepatocellular carcinoma, renal cell carcinoma. Opioid receptors play an important role in cell proliferation and cell growth, and are therefore a novel strategy for targeting cancer treatments. In the past, there have been many scientists and physicians that have devoted their time and efforts towards investigating the relationship between opioid receptors and cancer-states. Despite recent epidemiologic findings that show that cancer recurrence may be based on specific anesthetic regimens (Maher et al. 2014; Lennon et al. 2012a), a serial of conflicting thoughts about tumor genesis and treatments remain. For example, KOR has been reported to suppress tumor angiogenesis by inhibiting VEGF signaling (Kuzumaki et al. 2012; Yamamizu et al. 2013, 2014). Additionally, morphine's effects on tumors, whether it promotes the tumor growth by in inhibiting apoptosis, promoting angiogenesis and migration of tumor cells, or whether it inhibits tumor growth by exerting proapoptotic and antiangiogenic effects are very much associated with the different doses of morphine administrated and varies with different types cancers (Fiore et al. 2013; Mathew et al. 2011). Moreover, some recent studies have shown that the OGF-OGFr (opioid growth factor-opioid growth factor receptor) system is an intrinsic biological regulator of cell proliferation in some cancers including ovarian cancer, hepatocellular cancer, to name a few (Donahue et al. 2009; Avella et al. 2010). There is also a recent study which indicates that morphine improved the antitumor effects on MCF-7 cells in combination with 5-Fluorouracil (Ge et al. 2014). However, more studies are needed for the validation of this remarkable observation and its underlying mechanisms.

11.1 Neuroblastoma

Neuroblastoma (NB) is the most common extracranial solid cancer seen in childhood and is associated with a high mortality rate. A cancer cell differentiation state is one of the significant factors that must be taken into consideration when planning a treatment regimen. Different differentiation states have varied responsiveness even to same medications. The same drug at different dosages and different concentrations can have two very different effects. Opioids, specifically morphine, have been observed to modify neuroblastoma progression. For example, morphine induced differentiated SHSY-5Y human neuroblastoma cultured cells into apoptosis through the c-Jun N-terminal kinase (JNK)/caspase pathway in a dose-dependent manner. While simultaneously, morphine stimulated undifferentiated cell survival via activating the mitogen-activated protein kinase (MAPK)/extracellular signal-regulated kinase (ERK) pathway (Fiore et al. 2013). However, Gonzalez-Nunez et al. (2014) described that the high concentration of 1 μM of morphine inhibited cell proliferation in undifferentiated SH-SY5Y. Morphine induces cell death through Belcin 1- and ATG5-dependent autophagy (Zhao et al. 2010). Gonzalez's study, including exploring miRNA levels, raises concerns about the joint regulation of opioid and miRNA networks contributing a novel strategy for tumor growth. Morphine, in a low concentration of 10 nM, promoted cell proliferation in undifferentiated cells and decreased the miR133b expression, while 1 μM of morphine inhibited cell proliferation and miR133b and miR128 expression without triggering apoptosis. With the addition of naloxone, the classical antagonist, miR128 expression remained unchanged suggesting the effects were mediated by non-opioid receptor mechanisms (Gonzalez-Nunez et al. 2014).

11.2 Glioblastoma Multiforme

Glioblastoma multifore (GBM) is highly malignant tumor that arises from astrocytes and has a rich supply of blood vessels. It has been verified that human gliomas contain opioid and opioid receptors (Pan et al. 1998; Olsen et al. 2005). In fact, KOR binding density is one method for stratification of grades of gliomas. The KOR binding densities of tissues arranged from highest to lowest is as follows: non-neoplastic brain tissue, Grade 2 and 3 astrocytoma, and glioblastoma multiforme. Through the density of KOR binding, we may distinguish GBM from less aggressive, lower grade astrocytomas (Pan et al. 1998). Opioid receptor activation using D, L-methadone, a MOR agonist, triggers the down-regulation of cAMP, inhibiting tumor growth via activation of the XIAP/Bel-xL apoptosis pathways, thereby further sensitizing GBM cells, along with their untreatable GBM stem cells for doxorubin-induced apoptosis. Opioid receptor activation depressed cell growth in GBMs and improved the effectiveness of anti-cancer drugs (Fricsen et al. 2014).

11.3 Lung Cancer

A retrospective analysis demonstrated that a higher dose of opioids during the initial 96-h postoperative period was followed by a higher recurrence rate of non-small-cell lung cancer (Maher et al. 2014). Conversely, another analysis determined that opioid administration was a risk factor for overall survival in patients with stage I cancer, but not stage II and III disease (Cata et al. 2014). Many studies lead us to believe that MOR can influence cancer progression. For example, it has been verified that over-expression of the MOR in human non-small cell lung cancer promotes tumor proliferation, migration, and epithelial mesenchymal transition (EMT) (Mathew et al. 2011; Lennon et al. 2012a, b, 2014; Singleton et al. 2014). Interestingly, MOR has been noted to regulate opioids and growth factor-induced EGF receptor signaling which, in turn, stimulates Src, Gab-1, PI3K, Akt, STAT3, and mTOR activation, tumor proliferation, metastasis, and EMT during lung cancer progression. On the other hand, Koodie et al. (2014) revealed that morphine inhibited the migration of tumor-infiltrating leukocytes and angiogenesis in long-term cultures of mouse Lewis lung carcinoma cells compared with a placebo. At present, a consensus as to whether morphine has overall beneficial or harmful effects in cancer treatments has not been reached in agreement, and is a field that requires further investigations.

11.4 Gastrointestinal Cancer

In 1992, Iishi et al. (1992) proved that administration of Met-ENK and Leu-ENK significantly promoted gastric carcinogenesis induced by N-methyl-N′-nitro-N-nitrosoguanidine. In contrast, exogenous morphine inhibits human gastric cancer

MGC-803 cell growth by cell cycle arrest in the G2/M phase and apoptosis induction through activating caspase-9 and caspase-3, and inhibiting survival and NF-κB (Qin et al. 2012a). Also, exogenous fentanyl inhibits progression of human gastric cancer MGC-803 cells by downregulating NF-κB and upregulating PTEN in vitro (Qin et al. 2012b). Morphine can enhance human colon cancer HCT116 cells proliferation under morphine exposure for 6 h in the concentration 5–50 ng/mL, but fentanyl cannot alter the proliferation under various concentrations (1–10 ng/mL), neither exposure for 6 h nor 24 h. However, both morphine and fentanyl do not affect 5-fluorouracil sensitivity in human colon cancer HCT 116 cells (Nomura et al. 2014). Although opioids as a risk factor for gastrointestinal cancer in humans have been shown by a population-based case–control study (Naghibzadeh Tahami et al. 2014), the specific effects of opioids and their opioid receptors on gastrointestinal cancer requires additional research and investigation before we arrive at a conclusion.

11.5 Liver Cancer

MOR and DOR are highly expressed in hepatocellular carcinoma (HCC) (Lu et al. 2013; Tang et al. 2013). Studies indicate that MOR's down-regulation of human hepato-cellular carcinoma cells inhibited the proliferation of cancer and increased their apoptotic rate by promoting phosphorylation of the MKK 7, and thus activating JNK. The tumor cell cycle was blocked in G0/G1 phase, and the tumor growth in the mice studies was arrested (Lu et al. 2013). Furthermore, activation of δ-opioid receptor by DADLE inhibited the apoptosis of human liver cancer HepG2 cells by activation of PKC pathway and enhanced the sensitivity to cis-diammine dichloridoplatium. DADLE can thereby improve the chemosensitivity of HepG2 cells to CDDP (Tang et al. 2012). Additionally, an up-regulation of DOR in liver cancer promotes progression, both in vitro and in vivo studies (Tang et al. 2013).

Nevertheless, there are still some opioids that act as negative regulators of cellular processes. For example, the opioid ethylketocyclazocine (EKC) inhibits cell proliferation and induces apoptosis. This inhibitory effect is related to the interaction of opioids with somatostatin receptors, instead of opioid receptors, found on hepatocellular carcinoma cells (Notas et al. 2007). Therefore, it appears that the major function of opioids with opioid receptors on hepatocellular carcinoma is to enhance proliferation and inhibit the apoptosis, while simultaneously those same opioids may have an entirely different function through its interactions with other receptors, such as the somatostatin receptors.

11.6 Pancreatic Cancer

Opioid growth factor (OGF) has been used in the naming of met-enkephalin, a DOR ligand, because of its growth modulating properties. Many studies suggest that OGF and its receptors are present in human pancreatic cells (Bostwick et al. 1987;

Zagon et al. 1997, 1999, 2000, 2007, 2008; Donahue et al. 2011; Zagon and MsLaughlin 2012, 2014). Interestingly, the level of opioid growth factor in the plasma of patients with pancreatic cancer is significantly higher than age-matched control patients without cancer. Furthermore, the sensitivity and specificity of OGF for pancreatic cancer have been found to be greater than either CA19-9 or CEA (Smith et al. 2000). Additionally, it has been noted that the overexpression of the opioid growth factor receptors can inhibit and delay human pancreatic cancer progression (Zagon et al. 2007, 2008; Zagon and McLaughlin 2014) It has been postulated that this occurs via the OGF-OGFr axis, utilizing the p21 pathway (Cheng et al. 2008). Clinically, pancreatic cancer patients subjected to administration of opioid growth factor biotherapy have benefited from benefit from the stabilizing of their disease and the slowing of its progression (Smith et al. 2004, 2010).

12 Opioid Receptors and Immune Function

In general, opioids assist in the modulation of the immune response through their interaction with receptors found in both central and peripheral neuro-humoral systems (Sheng et al. 1997; Welters 2003; Vallejo et al. 2004). For example, acute and chronic opioid administration is known to have inhibitory effects on both humoral and cellular immune responses, which include antibody production, natural killer cell activity, cytokine expression, and phagocytic activity (Welters 2003; Vallejo et al. 2004; Eisenstein et al. 2006; Zou et al. 2007; Jamali et al. 2007). In vivo and in vitro studies indicated that opioid receptor stimulation exerts suppression of multiple components of the immune defense response including natural killer (NK) cell activity (Carr et al. 1993), neutrophil complement and immunoglobulin receptor expression (Welters et al. 2000), chemokine-induced chemotaxis (Grimm et al. 1998), and phagocytosis (Menzebach et al. 2004). Bai et al. (1999) found that after intracerebroventricular injection of β-endorphin in rats, the mitogen-induced spleen lymphocyte DNA synthesis, hemolysin formation, and IgG production were reduced significantly, suggesting a role of the central opioid system in immune regulation. Pretreatment with U50 488, a κ-opioid ligand, significantly reduced lipopolysaccharide (LPS)-stimulated interleukin-6 (IL-6) production. This effect was mediated by the KOR through inhibition by KOR antagonist-nor-binaltorphimine (nor-BNI) (Parkhill and Bidlack 2006). Treatment of normal human astrocytes with morphine leads to significant down-regulation of gene expression for β-chemokines, MCP-1, and MIP-1β, while reciprocal up-regulation of the expression of their specific receptors, CCR2b, CCR3, and CCR5, respectively, and reversed by naloxone (Mahajan et al. 2006). In addition, acute administration of morphine significantly decreased NK cell cytotoxicity and interferon-gamma mRNA levels, and increased the mortality rate of mice infected with herpes simplex virus 1 (Jamali et al. 2007). DOR agonist, DPDPE, triggers monocyte adhesion (Pello et al. 2006). Ordaz-Sanchez et al. (2003) observed that SNC80, another DOR agonist, significantly stimulated rat thymic and human leukocyte chemotaxis. The effects of SNC80 on the chemotaxis of rat and human leukocytes were antagonized by naloxone, indicating that the modulation of chemotaxis via opioid receptors. It has been suggested that activation of the KOR induces an anti-inflammatory response through the down-regulation of cytokine, chemokine, and chemokine receptor expression, while activation of MOR favors a proinflammatory response (Finley et al. 2008). Investigation into the ORL1/nociceptin system demonstrated a down-regulation of immune function secondary to the activation of this receptor (Lambert 2008).

In summary, opioid modulation of the immune response in animals is mediated through the direct interaction with opioid receptors expressed by immune cells. The influence of the opioids on the immune response in vivo is likely to be a result of involvement of both the central nervous system and the hypothalamic-pituitary-adrenal axis (McCarthy et al. 2011). Sharp et al. (1985) demonstrated that the β-endorphinand dynorphin stimulate the superoxide production by neutrophils and peritoneal macrophages. Inhibition of superoxide production by morphine is likely to be mediated via soluble factors, such as growth factor β, produced by the lymphoid cells present in the peripheral blood (Chao et al. 1992).

Pharmacological evidence has demonstrated the presence of MOR, KOR, and DOR, as well as non-classical opioid-like receptors, in cells of the immune system. Studies have shown that the phagocytic activity of non-elicited macrophages is inhibited by in vitro administration with μ-, κ-, or δ-opioid agonists (Szabo et al. 1993). An antagonist blocks the inhibitory activity of each agonist with the appropriate opioid receptor selectivity.

TNF-α and IL-6 are the two of the major cytokines secreted by activated macrophages. One study (Roy et al. 1998) showed that morphine differentially modulates LPS induced expression of IL-6 and TNF-α, and Greeneltch et al. (2004) found that naltrexone was capable of preventing LPS-induced septic shock mortality by indirect inhibition of TNF-α production in vivo. Although Alicea et al. (1996) reported that macrophage function is significantly modulated following activation of the KOR, morphine and the endogenous opioids (β-endorphin, [Met5] enkephalin, and dynorphin) also induce chemotaxis of human monocytes and neutrophils (Grimm et al. 1998; Van Epps and Saland 1984; Makman et al. 1995; Ruff et al. 1985). Morphine modulation of several immune functions, including macrophage phagocytosis and macrophage secretion of TNF-α, was not observed in the MOR knockout animals suggesting that these functions are mediated by the classical MOR receptor (Roy et al. 1998).

That fact that opioids and opioid receptors are involved in immune activity suggests that opioid receptors could be a potential target in some immune-related disease. Interestingly, low-dose naltrexone performs to promote maturation of bone marrow dendritic cells (BMDCs), thereby giving rise to novel immune-regulation strategies for host immunity in cancer therapy, and perhaps even DC-based vaccines for a large number of diseases (Meng et al. 2013). Moreover, nociceptin/orphanin FQ receptor activation is seen to decrease the airway hyper-responsiveness (AHR) induced by allergens in sensitized mice (Sullo et al. 2013). Furthermore, data from our work indicates that DOR safeguards against hypoxic injury of astrocytes and PC12 cells by regulating inflammatory cytokines (Wang et al. 2014).

13 Role of Opioid Receptors in Feeding and Obesity

Opioid receptors also participate in the regulation of feeding. Opioid agonists and antagonists have corresponding stimulatory and inhibitory effects on feeding. In addition to studies aimed at identifying the relevant receptor subtypes and sites of action in brain, there has been an increasing interest in the role of opioids on diet/taste preferences, food reward, and the overlap of food reward with others types of reward. Some data suggest a role for opioids in the control of appetite for specific macronutrients, but there is also evidence for their role in the stimulation of intake based on already-existing diet or taste preferences and in controlling intake motivated by hedonics rather than by energy needs (Gosnell and Levine 2009).

Agonist stimulation of opioid receptors increases feeding in rodent, while opioid antagonists inhibit food intake and weight gain in ob/ob mice (Marczak et al. 2009). The pan-opioid antagonist, LY255582, produces a sustained reduction in food intake and body weight in rodent models of obesity. The study on mice indicated that LY2555823 produces its effects on feeding and body weight gain, likely through a combination of μ-, δ- and κ-receptor activity (Gackenheimer et al. 2005). The κ-antagonist, nor-binaltorphimine, significantly increased food intake, while naltrexone (general antagonist) and naloxonazine (μ-antagonist) both reduced feeding (Cooper et al. 2010). Research suggested that syndyphalin-33 (SD33), a μ-opioid receptor ligand, increases food intake in sheep after intravenous injection, and its effects are mediated via opioid receptors (Obese et al. 2007).

Opioid receptors as obesity-related factors affect body weight and obesity. Studies showed that stimulation of μ-opioid receptors preferentially increases the intake of a high fat diet. The increased levels of hypothalamic μ-opioid receptors in Osborne-Mendel rats could contribute to their preference for a high fat diet and increased susceptibility to obesity (Barnes et al. 2006; Zuberi et al. 2008). The μ-opioid receptor signaling in the nucleus accumbens core and shell is necessary for palatable diet-induced hyperphagia and obesity to fully develop in rats (Lenard et al. 2010). Czyzyk et al. (2010) fed KOR-knockout (KO) and wild-type (WT) mice with high-energy diet (HED) for 16 weeks. They found that the KO mice had 28 % lower body weight and 45 % lower fat mass when compared to WT mice fed an HED. Furthermore, KOR deficiency led to an attenuation of triglyceride synthesis in the liver. This study provides the evidence that KOR plays an essential physiological role in the control of hepatic lipid metabolism, and KOR activation is a permissive signal toward fat storage. Marczak et al. (2009) administered H-Dmt-Tic-Lys-NH-CH2-Ph (MZ-2), a potent μ-/δ-opioid receptor antagonist, in mice and found that MZ-2 exhibited the following: (1) reduction in weight gain in sedentary obese mice that persisted beyond the treatment period without any significant effection leaner mice; (2) stimulation of voluntary running behavior on exercise wheels of both groups of mice; (3) decrease in fat content, enhanced bone mineral density (BMD), and decreased serum insulin and glucose levels in obese mice; and (4) MZ-2 (30 μM) increased BMD in human osteoblast cells (MG-63) comparable to naltrexone, while morphine inhibited mineral nodule formation. Thus, MZ-2 offers potential for application in the clinical management of obesity, insulin and glucose levels, and osteoporosis.

14 Concluding Remarks

The opioid system is considered as one of the most complex and interwoven neurotransmitter systems in the body. The functional effects of opioids are mediated by three distinct opioid receptors with over-lapping distribution and functions in the brain and other major organs. Although opioids are predominantly used for therapeutic pain relief in a clinical setting, other potential roles of the opioid system throughout the body, especially in the brain, are coming into focus with an emergence of new and specific analogues that have brought forth their subtle differences and enabled the investigators to unravel this issue. Since the cloning of the three major opioid receptors, many studies have been performed to elucidate their function and show that besides the involvement of pain modulation, opioid receptors are widely involved in various pleiotropic functions throughout the body, including ionic homeostasis, cell proliferation, emotional response, epileptic seizures, immune function, feeding, obesity, respiratory, and cardiovascular control. They also play a critical role in hibernation. One of the most exciting findings in the past decade is neuroprotection and cardioprotection mediated by opioid receptors, especially DOR. The underlying mechanisms involve the stabilization of ionic homeostasis in acute stage of hypoxic/ischemic stress, up-regulation of survival molecules and anti-oxidative capacity, and the down-regulation of apoptosis during the long-term stress of hypoxia/ischemia. These recent findings can have significant clinical implications with regards to the role of opioids, either endogenous or exogenous, in limiting the pathogenesis of brain and heart from hypoxic/ischemic damage.

Currently, studies investigating the role of opioid activity in cancer cells are numerous, yet a clear function of opioids on cancer cells remains controversial. Whether opioids promote tumor progress or suppresses it depends on several factors including the specific type of cancer, cancer progression, and on which specific opioid pathways are activated in greater proportions. For example, in different types and stages of cancer, the opioids that may be expressed can vary tremendously in both the types of receptors present and their quantity. In some tumor cells, opioids may activate a greater number of opioid receptors, thereby promoting proliferation. Contrariwise, opioids may act upon other receptors in greater numbers, leading to an inhibition of cellular proliferation. Generally, DOR activation is known to protect cells and inhibit apoptosis, thereby promoting tumor genesis. However, specific roles of various opioid receptors in cancer cells may vary depending on the types of cancers, their stages, and many other factors, which are incompletely understood yet at present.

Many important issues, such as the difference between MOR and DOR, despite structural similarity among them, in terms of neuroprotection still remain to be clarified. Morphine protects purkinje cells against cell death under in vitro simulated ischemia-reperfusion conditions, which is mediated by DOR, but not MOR (Lim et al. 2004). In rat hippocampal slices, however, exposure to morphine aggravates the neurotoxic effects of a subsequent hypoxia/hypoglycemia in a concentration-dependent manner (Ammon et al. 2007a, b). It seems that opioid receptors have differential roles in neuronal responses to hypoxic/ischemic stress across different brain regions/cells (e.g., cortical vs. hippocampal cells). MOR may induce a toxic effect on neurons in some brain regions underhypoxic/ischemic condition. It is still unclear whether this is a reason why some investigators observed neuroprotection induced by opioid blockers, such as naltrexone under certain conditions (San-Emeterio and Hurlé 2006). More intricate and in-depth studies are needed to address many complex issues like this. Although the role of DOR in neuroprotection is relatively clear, it is important to develop more specific and potent DOR-agonists with lesser side effects to strengthen their neuroprotection as they might prove useful for patients with stroke, cardiac ischemia, and hypoxic encephalopathy.

Acknowledgements The work was supported by NIH (HD-034852 and AT-004422) and Vivian L Smith Neurologic Foundation. GW and TC were supported by the National Natural Science Foundation of China (81260204; 81060096).

References

Abbadic C, Pan YX, Pastemak GW (2004) Immunohistochemical study of the expression of exon11-containing mu opioid receptor variants in mouse brain. Neuroscience 127:419–430

Abood ME, Noel MA, Farnsworth JS, Tao Q (1994) Molecular cloning and expression of a delta-opioid receptor from rat brain. J Neurosci Res 37:714–719

Adams HP Jr, Olinger CP, Barsan WG, Butler MJ, Graff-Radford NR, Brott TG, Biller J, Damasio H, Tomsick T, Goldberg M (1986) A dose-escalation study of large doses of naloxone for treatment of patients with acute cerebral ischemia. Stroke 17:404–409

Ahmed MA, Kurkar A (2014) Effects of opioid (tramadol) treatment on testicular functions in adult male rats: the role of nitric oxidative stress. Clin Exp Pharmacol Physiol 41:317–323

Akil H, Mayer DJ, Liebeskind JC (1976) Antagonism of stimulation-produced analgesia by naloxone, a narcotic antagonist. Science 191:961–962

Akiyama T, Carsten MI, Piecha D, Steppan S, Carstens E (2014) Nalfurafine suppresses pruritogen- and touch-evoked scratching behavior in models of acute and chronic itch in mice. Acta Derm Venereol 95:147–150

Albrizio M, Guaricci AC, Milano S, Macri F, Aiudi G (2014) Mu opioid receptor in spermatozoa, eggs and larvae of gilthead sea bream (Sparus Aurata) and its involvement in stress related to aquaculture. Fish Physiol Biochem 40:997–1009

Al-Chalabi A, Leigh PN (2000) Recent advances in amyotrophic lateral sclerosis. Curr Opin Neurol 13:397–405

Alicea C, Belkowski S, Eisenstein TK, Adler MW, Rogers TI (1996) Inhibition of primary murine macrophage cytokine production in vitro following treatment with the kappa-opioid agonist U50, 488H. J Neuroimmunol 64:83–90

Ammon-Treiber S, Stolze D, Höllt V (2007a) Differential effects of mu-opioid receptor agonists in a hippocampal hypoxia/hypoglycemia model. Brain Res 1183:60–65

Ammon-Treiber S, Stolze D, Schröder H, Loh H, Höllt V (2007b) Effects of opioid antagonists and morphine in a hippocampal hypoxia/hypoglycemia model. Neuropharmacology 49: 1160–1169

Ananthan S (2006) Opioid ligands with mixed mu/delta opioid receptor interactions: an emerging approach to novel analgesics. AAPS J 8:118–125

Augustin LB, Felsheim RF, Min BH, Fuchs SM, Fuchs JA, Loh HH (1995) Genomic structure of the mouse delta opioid receptor gene. Biochem Biophys Res Commun 207:111–119

Auh QS, Ro JY (2012) Effects of peripheral κ opioid receptor activation on inflammatory mechanical hyperalgesia in male and female rats. Neurosci Lett 524:111–115

Avella DM, Kimchi ET, Donahue RN, Taqaram HR, McLquqhlin PJ, Zaqon IS, Staveley-O'Carroll KF (2010) The opioid growth factor-opioid growth factor receptor axis regulates cell proliferation of human hepatocellular cancer. Am J Physiol Regul Integr Comp Physiol 298:459–466

Bai HB, Du JZ, Zheng XX (1999) Beta-endorphin involved in the regulation of humoral immune function of rats during acute hypoxia. Sheng Li Xue Bao 51:258–262

Balboni G, Onnis V, Conqie C, Zotti M, Sasaki Y, Ambo A, Bryant SD, Jinsmaa Y, Lazarus LH, Trapella C, Salvadori S (2006) Effect of lysine at C-terminus of the Dmt-Tic opioid pharmacophore. J Med Chem 49:5610–5617

Bardoni R, Tawfik VL, Wang D, François A, Solorzano C, Shuster SA, Choudhury P, Betelli C, Cassidy C, Smith K, de Nooij JC, Mennicken F, O'Donnell D, Kieffer BL, Woodbury CJ, Basbaum AI, MacDermott AB, Scherrer G (2014) Delta opioid receptors presynaptically regulate cutaneous mechanosensory neuron input to the spinal cord dorsal horn. Neuron 81:1312–1327

Barnes MJ, Holmes G, Primeaux SD, York DA, Bray GA (2006) Increased expression of mu opioid receptors in animals susceptible to diet-induced obesity. Peptides 27:3292–3298

Barr JM, Van Stipdonk MJ (2002) Multi-stage tandem mass spectrometry of metal cationized leucine enkephalin and leucine enkephalin amide. Rapid Commun Mass Spectrom 16:566–578

Barron BA, Gu H, Gaugl JF, Caffrey JL (1992) Screening for opioids in dog heart. J Mol Cell Cardiol 24:67–77

Barron BA, Jones CE, Caffrey JL (1995) Pericardial repair depresses canine cardiac catecholamines and met-enkephalin. Regul Pept 59:313–320

Barry U, Zuo Z (2005) Opioids: old drugs for potential new applications. Curr Pharm Des 11:1343–1350

Bausch B, Garland JP, Yamada J (2005) The delta opioid receptor agonist, SNC80, has complex, dose-dependent effects on pilocarpine-induced seizures in Sprague–Dawley rats. Brain Res 1045:38–44

Beadles-Bohling AS, Wiren KM (2005) Alteration of kappa-opioid receptor system expression in distinct brain regions of a genetic model of enhanced ethanol withdrawal severity. Brain Res 1045:77–89

Blasco H, Mavel S, Corcia P, Gordon PH (2014) The glutamate hypothesis in ALS: pathophysiology and drug development. Curr Med Chem 21:3551–3575

Boeuf B, Poirier V, Gauvin F, Guerguerian AM, Roy C, Farrell CA, Lacroix J (2003) Naloxone for shock. Cochrane Database Syst Rev 4, CD004443

Bohn LM, Rachal KM (2006) Opioid receptor signaling: relevance for gastrointestinal therapy. Curr Opin Pharmacol 6:559–563

Bohn LM, Lefkowitz RJ, Gainetdinov RR, Pepp lK, Caron MG, Lin FT (1999) Enhanced morphine analgesia in mice lacking beta-arrestin 2. Science 286:2495–2498

Bohn LM, Gainetdinov RR, Lin FT, Lefkowitz RJ, Caron MG (2000) Mu-opioid receptor desensitization by beta-arrestin-2 determines morphine to tolerance but not dependence. Nature 408:720–723

Bohn LM, Lefkowitz RJ, Caron MGJ (2002) Differential mechanisms of morphine antinociceptive tolerance revealed in (beta) arrestin-2 knock-out mice. J Neurosci 22:10494–10500

Bohn LM, Gainetdinov RR, Sotnikova TD, Medvedev IO, Lefkowitz RJ, Dykstra LA, Caron MG (2003) Enhanced rewarding properties of morphine, but not cocaine, in beta (arrestin)-2 knock-out mice. J Neurosci 23:1026–1073

Borlongan CV, WangY STP (2004) Delta opioid peptide (D-Ala2, DLeu5) encephalin: linking hibernation and neuroprotection. Front Biosci 9:3392–3398

Borlongan CV, Hayashi T, Oeltgen PR, Su TP, Wang Y (2009) Hibernation-like state induced by an opioid peptide protects against experimental stroke. BMC Biol 7:31

Borroto-Escuela DO, Romero-Fernandez W, Rivera A, Van Craenenbroeck K, Tarakanov AO, Agnati LF, Fuxe K (2013) On the g-protein-coupled receptor heteromers and their allosteric receptor-receptor interactions in the central nervous system: focus on their role in pain modulation. Evid Based Complement Alternat Med 2013:563716

Borsodi A, Tóth G (1995) Characterization of opioid receptor types and subtypes with new ligands. Ann N Y Acad Sci 757:339–352

Bostwick DG, Null WE, Holmes D, Weber E, Barchas JD, Bensch KG (1987) Expression of opioid peptides in tumors. N Engl J Med 317:1439–1443

Bregola G, Zucchini S, Frigati L, Candeletti S, Romualdi P, Reinscheid R, Simonato M (2002) Involvement of the neuropeptide orphanin FQ/nociception in kainate and kindling seizures and epileptogenesis. Epilepsia 43:18–19

Brigatte P, Konno K, Gutierrez VP, Sampaio SC, Zambelli VO, Picolo G, Curi R, Cury Y (2013) Peripheral kappa and delta opioid receptors are involved in the antinociceptive effect of crotalphine in a rat model of cancer pain. Pharmacol Biochem Behav 109:1–7

Brinbaum HG, White AG, Reynolds JL, Greenberg PE, Zhang M, Vallow S, Schein JR, Katz NP (2006) Estimated costs of prescription opioid analgesic abuse in the United States in 2001: a societal perspective. Clin J Pain 22:667–676

Butelman ER, Ko MC, Sobczyk-Kojiro K, Mosberg HI, Van Bemmel B, Zernig G, Wods JH (1998) Kappa-opioid receptor binding populations in rhesus monkey brain: relationship to an assay of thermal antinociception. J Pharmacol Exp Ther 285:595–601

Campbell AM, Zagon IS, McLaughlin PJ (2012) Opioid growth factor arrests the progression of clinical disease and spinal cord pathology in established experimental autoimmune encephalomyelitis. Brain Res 1472:138–148

Cao S, Chao D, Zhou H, Balboni G, Xia Y (2014) A novel mechanism for cytoprotection against hypoxic injury: δ-opioid receptor-mediated increase in Nrf2 translocation. Br J Pharmacol 172:1869–1881

Carcolé M, Castany S, Leánez S, Pol O (2014) Treatment with a heme oxygenase 1 inducer enhances the antinociceptive effects of μ-opioid, δ-opioid, and cannabinoid 2 receptors during inflammatory pain. J Pharmacol Exp Ther 351:224–232

Carr D, Gebhart B, Paul D (1993) Alpha adrenergic and mu-2 opioid receptors are involved in morphine-induced suppression of splenocyte natural killer activity. J Pharmacol Exp Ther 264:1179–1186

Cata JP, Keerty V, Keerty D, Feng L, Norman PH, Gottumukkala V, Mehran JR, Engle M (2014) A retrospective analysis of the effect of intraoperative opioid dose on cancer recurrence after non-small cell lung cancer resection. Cancer Med 3:900–908

Caudle RM, Finegold AA, Mannes AJ, Tobias MD, Kenshalo DR, Iadarola MJ (1998) Spinal kappa 1 and kappa 2 opioid binding sites in rats, guinea pigs, monkeys and humans. Neuroreport 9:2523–2525

Cayun S, Resch GF, Evee AD, Rapacon-Baker MM, Millipgton WR (2001) Blockade of delta opioid receptors in the ventrolateral periaqueductal gray region inhibits the fall in arterial pressure evoked by hemorrhage. J Pharmacol Exp Ther 297:612–619

Chakrabarti S, Liu NJ, Gintzler AR (2010) Formation of mu-/kappa-opioid receptor heterodimer is sex-dependent and mediates female-specific opioid analgesia. Proc Natl Acad Sci U S A 107:20115–20119

Chao D, Xia Y (2010) Ionic storm in hypoxic/ischemic stress: can opioid receptors subside it? Prog Neurobiol 90:439–470

Chao D, Xia Y (2012) Acupuncture treatment of epilepsy. In: Xia Y, Ding G, Wu G (eds) Current research in acupuncture. Springer, New York, pp 129–214

Chao CC, Hu S, Molitor TW, Zhou Y, Murtaugh MP, Tsang M, Peterson PK (1992) Morphine potentiates transforming growth factor-beta release from human peripheral blood mononuclear cell cultures. J Pharmacol Exp Ther 262:19–24

Chao D, Qian H, Ghassemi F, Chen JS, Xia Y (2006) Transgenic overexpression of δ-opioid receptors protects the cortex from anoxic disruption of ionic homeostasis. Soc Neurosci Abstract Program, No. 87:19/MM68

Chao D, Bazzy-Asaad A, Balboni G, Xia Y (2007a) δ-, but not μ-, opioid receptor stabilizes K+ homoeostasis by reducing Ca2+ influx in the cortex during acute hypoxia. J Cell Physiol 2012:60–67

Chao D, Donnelly DF, Feng Y, Bazzy-Asaad A, Xia Y (2007b) Cortical delta-opioid receptors potentiate K+ homeostasis during anoxia and oxygen-glucose deprivation. J Cereb Blood Flow Metab 27:356–368

Chao D, Bazzy-Asaad A, Balboni G, Salvadori S, Xia Y (2008) Activation of DOR attenuates anoxic K+ derangement via inhibition of Na+ entry in mouse cortex. Cereb Cortex 18:2217–2227

Chao D, Balboni G, Lazarus LH, Salvadori S, Xia Y (2009) Na+ mechanism of δ-opioid receptor induced protection from anoxic K+ leakage in the cortex. Cell Mol Life Sci 66:1105–1115

Chao D, Shen X, Xia Y (2013) From acupuncture to interaction between δ-opioid receptors and Na (+) channels: a potential pathway to inhibit epileptic hyperexcitability. Evid Based Complement Alternat Med 2013:216016

Charron C, Messier C, Plamondon H (2008) Neuroprotection and functional recovery conferred by administration of kappa- and delta-opioid agonists in a rat model of global ischemia. Physiol Behav 939:502–511

Chen CJ, Cheng FC, Liao SL, Chen WY, Lin NN, Kuo JS (2000) Effects of naloxone on lactate, pyruvate metabolism and antioxidant enzyme activity in rat cerebralischemia/reperfusion. Neurosci Lett 287:113–116

Chen T, Li J, Chao D, Sandhu HK, Liao X, Zhao J, Wen G, Xia Y (2014) δ-Opioid receptor activation reduces α-synuclein overexpression and oligomer formation induced by MPP(+) and/or hypoxia. Exp Neurol 255:127–136

Chen Y, Somme C (2007) Activation of the nociception opioid system in rat sensory neurons produces antinociceptive effects in inflammatory pain: involvement of inflammatory mediators. J Neurosci Res 85:1478–1488

Chen Y, Mestek A, Liu J, Yu L (1993) Molecular cloning of a rat κ-opioid receptor reveals sequence similarities to the m and d opioid receptors. Biochem J 295:625–628

Chen CJ, Liao SL, Chen WY, Hong JS, Kuo JS (2001) Cerebral ischemia/reperfusion injury in rat brain: effects of naloxone. Neuroreport 12:1245–1249

Chen CJ, Cheng FC, Liao SL, Chen WY, Lin NN, Kuo JS (2000) Effects of naloxone on lactate, pyruvate metabolism and antioxidant enzyme activity in rat cerebral ischemia/reperfusion. Neurosci Lett 287:113–116

Chen T, Li J, Chao D, Sandhu HK, Liao X, Zhao J, Wen G, Xia Y (2014) δ-Opioid receptor activation reduced α-synuclein overexpression and oligomer formation induced by MPP (+) and/or hypoxia. Exp Neurol 255:127–136

Cheng F, McLaughlin PJ, Verderame MF, Zagon IS (2008) The OGF-OGFr axis utilizes the p21 pathway to restrict progression of human pancreatic cancer. Mol Cancer 7:5

Christie MJ (2008) Cellular neuroadaptations to chronic opioids: tolerance, withdrawal and addiction. Br J Pharmacol 154:384–396.

Chung MK, Cho YS, Bae YC, Lee J, Zhang X, Ro JY (2014) Peripheral G protein-coupled inwardly rectifying potassium channels are involved in δ-opioid receptor-mediated anti-hyperalgesia in rat masseter muscle. Eur J Pain 18:29–38

Cicero TJ, Schainker BA, Meyer ER (1979) Endogenous opioids participate in the regulation of the hypothalamus-pituitary-luteinizing hormone axis and testosterone's negative feedback control of luteinizing hormone. Endocrinology 104:1286–1291

Clark JA, Liu L, Price M, Hersh B, Edelson M, Pasternak GW (1989) Kappa opiate receptor multiplicity: evidence for two U50, 488-sensitive kappa 1 subtypes and a novel kappa 3 subtype. J Pharmacol Exp Ther 251:461–468

Comb M, Seeburg PH, Adelman J, Eiden L, Herbert E (1982) Primary structure of the human met- and leu-enkephalin precursor and its mRNA. Nature 295:663–666

Cominski TP, Ansonoff MA, Turchin CE, Pintar JE (2014) Loss of the mu opioid receptor induces strain-specific alterations in hippocampal neurogenesis and spatial learning. Neuroscience 278:11–19

Compton W, Volkow ND (2006) Abuse of prescription drugs and the risk of addiction. Drug Alcohol Depend 83:4–7

Cooper PD, Dennis SR, Woodman JD, Cowlings A, Donnelly C (2010) Effect of opioid compounds on feeding and activity of the cockroach, Periplaneta Americana. Comp Biochem Physiol C Toxicol Pharmacol 151:298–302
Cunha TM, Souza GR, Domingues AC, Carreira EU, Lotufo CM, Funez MI, Verri WA Jr, Cunha FQ, Ferreira SH (2012) Stimulation of peripheral kappa opioid receptors inhibits inflammatory hyperalgesia via activation of the PI3Kγ/AKT/nNOS/NO signaling pathway. Mol Pain 8:10
Cunha-Oliveira T, Rego AC, Oliveira CR (2008) Cellular and molecular mechanisms involved in the neurotoxicity of opioid and psychostimulant drugs. Brain Res Rev 58:192–208
Czyzyk TA, Nogueiras R, Lockwood JF, McLomzie JH, Coskun T, Pintar JE, Hammond C, Tschop MH, Stantnick MA (2010) kappa-Opioid receptors control the metabolic response to a high-energy diet in mice. FASEB J 24:1151–1159
Dahan A, Aarts L, Smith TW (2010) Incidence, reversal, and prevention of opioid-induced respiratory depression. Anesthesiology 112:226–238
Danielsson I, Gasior M, Stevenson GW, Folk JE, Rice KC, Negus SS (2006) Electroencephalographic and convulsant effects of the delta opioid agonist SNC80 in rhesus monkeys. Pharmacol Biochem Behav 85:428–434
Darland T, Heinricher MM, Grandy DK (1998) Orphanin FQ/nociception: a role in pain and analgesia, but so much more. Trends Neurosci 21:215–221
Demoliou-Mason CD, Barnard EA (1986) Distinct subtypes of the opioid receptor with allosteric interactions in brain membranes. J Neurochem 46:1118–1128
Doggrell SA (2007) Cardiovascular and renal effects of nociception/orphanin FQ: a new mediator to target? Curr Opin Investig Drugs 8:742–749
Donahue LA, Kulik L, Baker T, Ganger DR, Gupta R, Memon K, Abecassis MM, Salem R, Lewandowski RJ (2013) Yttrium-90 radioembolization for the treatment of unresectable hepatocellular carcinoma in patients with transjugular intrahepatic portosystemic shunts. J Vasc Interv Radiol 24(1):74–80
Donahue RN, McLaughlin PJ, Zagon IS (2009) Cell proliferation of human ovarian cancer is regulated by the opioid growth factor opioid growth factor receptor axis. Am J Physiol Regul Integr Comp Physiol 296:1716–1725
Donahue RN, McLaughlin PJ, Zagon IS (2011) Low-dose naltrexone targets the opioid growth factor-opioid growth factor receptor pathway to inhibit cell proliferation: mechanistic evidence from a tissue culture model. Exp Biol Med (Maywood) 236:1036–1050
Eisenstein TK, Rahim RT, Feng P, Thingalaya N, Meissler JJ (2006) Effects of opioid tolerance and withdrawal on the immune system. J Neuroimmune Pharmacol 1:237–249
Eisinger DA, Schulz R (2005) Mechanism and consequences of delta opioid receptor internalization. Crit Rev Neurobiol 17:1–26
Eliasson T, Mannheimer C, Waagstein F, Andersson B, Berqh CH, Auqustinsson LE, Hedner T, Larson G (1998) Myocardial turnover of endogenous opioid and calcitonin-gene-related peptide in the human heart and the effects of spinal cord stimulation on pacing-induced angina pectoris. Cardiology 89:170–177
Endoh H, Taga K, Yamakura T, Sato K, Watanabe I, Fukuda S, Shimoji K (1999) Effects of naloxone and morphine on acute hypoxic survival in mice. Crit Care Med 27:1929–1933
Evans CJ, Keith DE, Morrison H, Magendzo K, Edwards RH (1992) Cloning of a delta opioid receptor functional expression. Science 258:1952–1955
Fabbri A, Knox G, Buczko E, Dufau ML (1988) Beta-endorphin production by the fetal Leydig cell: regulation and implication for paracrine control of Sertoli cell function. Endocrinology 122:749–755
Fabbri A, Jannini EA, Gnessi L, Ulisse S, Moretti C, Isidori A (1989) Neuroendocrine control of male reproductive function: the opioid system as a model of control at multiple sites. J Steroid Biochem 32:145–150
Faden AI, Holaday JW (1979) Opiate antagonists: a role in the treatment of hypovolemic shock. Science 205:317–318
Feng Y, Chao DM, Li WM, Cao YX, Wang YQ, Wu GC (2004) Inhibition of nociception/orphanin FQ on penicillin-induced seizures in rats. Brain Res 1020:214–219

Feng Y, Chao D, He X, Yang Y, Kang X, Lazarus LH, Xia Y (2009) A novel insight into neuroprotection against hypoxic/ischemic stress. Sheng Li Xue Bao 61:585–592

Feng Y, He X, Yang Y, Chen J, Yin K, Xia Y (2011) Effect of delta opioid receptor over-expression on cortical expression of GABAA receptor alpha1-subunit in hypoxia. Chin J Physiol 54:118–123

Feng Y, He X, Yang Y, Chao D, Lazarus LH, Xia Y (2012) Current research on opioid receptor function. Curr Drug Targets 13:230–246

Filliol D, Ghozland S, Chluba J, Martin M, Matthes HW, Simonin F, Befort K, Gaveriaux-Ruff C, Dierich A, LeMeur M, Valverde O, Maldonado R, Kieffer BL (2000) Mice deficient for delta- and mu-opioid receptors exhibit opposing alterations of emotional responses. Nat Genet 25:195–200

Finley MJ, Happel CM, Kaminsky DE, Rogers TJ (2008) Opioid and nociception receptors regulate cytokine and cytokine receptor expression. Cell Immunol 252:146–154

Fiore G, Ghelardini C, Bruni G, Guarna M, Bianchi E (2013) Differentiation state affects morphine induced cell regulation in neuroblastoma cultured cells. Neurosci Lett 555:51–56

Forman LJ, Estilow S, Hock CE (1989) Localization of beta-endorphin in the rat heart and modulation by testosterone. Pro Soc Exp Biol Med 190:240–245

Fricsen C, Hormann I, Roscher M, Fichtner I, Alt A, Hilger R, Debatin KM, Miltner E (2014) Opioid receptor activation triggering downregulation of cAMP improves effectiveness of anti-cancer drugs in treatment of glioblastoma. Cell Cycle 13:156–170

Fristad I, Berggreen E, Haug SR (2006) Delta opioid receptors in small and medium-sized trigeminal neurons supporting the dental pulp of rats. Arch Oral Biol 51:273–281

Fujisawa M, Bardin CW, Morris PL (1993) Germ cell factor(s) regulates opioid gene expression in Sertoli cells. Recent Prog Horm Res 48:497–503

Fujita Y, Tsuda Y, Li T, Motoyama Takahashi M, Shimizu Y, Yokoi T, Sasaki Y, Ambo A, Kita A, Jinsmaa Y, Bryant SD, Lazarus LH, Okada Y (2004) Development of potent bifunctional endomorphin-2 analogues with mixed mu-/delta-opioid agonist and delta-opioid antagonist properties. J Med Chem 47(14):3591–3599

Gackenheimer SL, Suter TM, Pintar JE, Quimby SJ, Wheeler WJ, Mitch CH, Gehlert DR, Statnick MA (2005) Localization of opioid receptor antagonist [3H]-LY255582 binding sites in mouse brain: comparison with the distribution of mu, delta and kappa binding sites. Neuropeptides 39:559–567

Garzon J, Jen M, Sanchez-Blaquez P, Lee N (1982) Dynorphin (1–13), a long-lasting inhibitor of opiate receptor binding in vitro. Life Sci 31:1789–1792

Gastfriend DR (2014) A pharmaceutical industry perspective on the economics of treatments for alcohol and opioid use disorders. Ann N Y Acad Sci 1327:112–130

Gavériaux-Ruff C, Kieffer BL (2002) Opioid receptor genes inactivated in mice: the highlights. Neuropeptides 36:62–71

Ge ZH, Wang ZX, Yu TL, Yang N, Sun Y, Hao CL, Sun LX (2014) Morphine improved the antitumor effects on MCF-7 cells in combination with 5-Fluorouracil. Biomed Pharmacother 68:299–305

Gebhart GF (2000) J.J. Bonica lecture-2000: Physiology, pathophysiology, and pharmacology of visceral pain. Reg Anesth Pain Med 25:632–638

Georganta EM, Tsoutsi L, Gaitanou M, Georgoussi Z (2013) δ-opioid receptor activation leads to neurite outgrowth and neuronal differentiation via a STAT5B-Gαi/o pathway. J Neurochem 127:329–341

Giovannitti JA Jr, Thoms SM, Crawford JJ (2015) Alpha-2 adrenergic receptor agonists: a review of current clinical applications. Anesth Prog 62:31–39

Gironi M, Furlan R, Rovaris M, Comi G, Filippi M, Panerai AE, Sacerdote P (2003) Beta endorphin concentrations in PBMC of patients with different clinical phenotypes of multiple sclerosis. J Neurol Neurosurg Psychiatry 74:495–497

Gironi M, Martinelli-Boneschi F, Sacerdote P, Solaro C, Zaffaroni M, Cavarretta R, Moiola L, Bucello S, Radaelli M, Pilato V, Rodegher M, Cursi M, Franchi S, Martinelli V, Nemni R, Comi G, Martino G (2008) A pilot trial of low-dose naltrexone in primary progression multiple sclerosis. Mult Scler 14:1076–1083

Giros B, Pohl M, Rochelle JM, Seldin MF (1995) Chromosomal localization of opioid peptic and receptor genes in the mouse. Life Sci 56:360–375

Goldstein A, Tachibana S, Lowney LI, Hunkapiller M, Hood L (1979) Dynorphin-(1-13), an extraordinarily potent opioid peptide. Proc Natl Acad Sci U S A 76:6666–6670

Gonzalez-Nunez V, Noriega-Prieto JA, Rodrlfuez RE (2014) Morphine modulates cell proliferation through mir133b & mir128 in the neuroblastoma SH-SY5Y cell line. Biochim Biophys Acta 1842:566–572

Gosnell BA, Levine AS (2009) Reward systems and food intake: role of opioids. Int J Obes (Lond) 33:54–58

Govindaswami M, Brown SA, Yu J, Zhu H, Bishop PD, Kindy MS, Oeltgen PR (2008) Delta 2-specific opioid receptor agonist and hibernating woodchuck plasma fraction provide ischemic neuroprotection. Acad Emerg Med 15:250–257

Grecksch G, Becker A, Schroeder H, Höllt V (1999) Involvement of delta-opioid receptors in pentylenetetrazol kindling development and kindling-related processes in rats. Naunyn Schmiedebergs Arch Pharmacol 360:151–156

Greeneltch KM, Haudenschild CC, Keegan AD, Shi Y (2004) The opioid antagonist naltrexone blocks acute endotoxic shock by inhibiting tumor necrosis factor-alpha production. Brain Behav Immun 18:476–484

Grimm MC, Ben-Baruch A, Taub DD, Howard OM, Resau JH, Wang JM, Ali H, Richardson R, Snyderman R, Oppenheim JJ (1998) Opiates transdeactivate chemokine receptors: delta and mu opiate-receptor mediated heterologous desensitization. J Exp Med 188:317–325

Gross GJ (2003) Role of opioids in acute and delayed preconditioning. J Mol Cell Cardiol 35:709–718

Gross ER, Hsu AK, Gross GJ (2004) Opioid-induced cardioprotection occurs via glycogen synthase kinase beta inhibition during reperfusion in intact rat hearts. Circ Res 94:960–966

Guo Y, Wang L, Zhou Z, Wang M, Liu R, Wang L, Jiang Q, Song L (2013) An opioid growth factor receptor (OGFR) for [Met5]-enkephalin in Chlamys farreri. Fish Shellfish Immunol 34:1228–1235

Gutkowska J, Jankowski M, Pawlak D, Mukaddam-Daher S, Izdebski J (2004) The cardiovascular and renal effects of a highly potent mu-opioid receptor agonist, cyclo [Ne, Nb-carbonyl-D-Lys2, Dap5]enkephalinamide. Eur J Pharmacol 496:167–174

Hackler L, Zadina JE, Ge L-J, Kastin AJ (1997) Isolationo of relatively large amounts of endomorphin-1 and endomorphin-2 from human brain cortex. Peptides 18:1635–1639

Haji A, Takeda R, Okazaki M (2000) Neuropharmacology of control of respiratory rhythm and pattern in mature mammals. Pharmacol Ther 86:277–304

Hajrasouliha AR, Tavakoli S, Jabehdar-Maralani P, Ebrahimi F, Shafaroodi H, Mirkhani SH, Amanpour S, Dehpour AR (2005) Cholestatic liver disease modulates susceptibility to ischemia/reperfusion- induced arrhythmia, but not necrosis and hemodynamic instability: the role of endogenous opioid peptides. J Hepatol 43:491–498

Hammer LA, Zagon IS, McLaughlin PJ (2013) Treatment of a relapse-remitting model of multiple sclerosis with opioid growth factor. Brain Res Bull 98:122–131

Hammers A, Asselin M, Hinz R, Kitchen I, Brools DJ, Duncan JS, Koepp MJ (2007) Upregulation of opioid receptor binding following spontaneous epileptic seizures. Brain 130:1009–1016

Hanlon KE, Herman DS, Agnes RS, Largent-Milnes TM, Kumarasinghe IR, Ma SW, Guo W, Lee YS, Ossipov MH, Hruby VJ, Lai J, Porreca F, Vanderah TW (2011) Novel peptide ligands with dual acting pharmacophores designed for the pathophysiology of neuropathic pain. Brain Res 1395:1–11

Harrison LM, Kastin AL, Zadina JE (1998) Opiate tolerance and dependence: receptors, G-proteins, and antiopiates. Peptides 19:1603–1630

Hauser KF, Knapp PE (2014) Interactions of HIV and drugs of abuse: the importance of glia, neural progenitors, and host genetic factors. Int Rev Neurobiol 118:231–313

Hauser KF, McLaughlin PJ, Zagon IS (1987) Endogenous opioid regulate dendritic growth and spine formation in developing rat brain. Brain Res 416:157–161

Hayward NJ, McKnight AT, Woodruff GN (1993) Neuroprotective effect of the kappa-agonist enadoline (CI-997) in rat models of focal cerebral ischaemia. Eur J Neurosci 5:961–967

He A, Hersh EV (2012) A review of intranasal ketorolac tromethamine for the short-term management of moderate to moderately severe pain that requires analgesia at the opioid level. Curr Med Res Opin 28:1873–1880

He BJ, Tong PJ, Li J, Jing HT, Yao XM (2013) Auricular acupressure for analgesia in perioperative period of total knee arthroplasty. Pain Med 14:1608–1613

Heckert LL, Griswold MD (2002) The expression of the follicle-stimulating hormone receptor in spermatogenesis. Recent Prog Horm Res 57:127–148

Heinricher MM (2005) Nociceptin/orphanin FQ: pain, stress and neural circuits. Life Sci 77:3127–3132

Hill MP, Hille CJ, Brotchie JM (2000) Delta-opioid receptor agonists as a therapeutic approach in Parkinson's disease. Drug News Perspect 13:261–268

Holaday JW (1984) Opiate antagonists in shock and trauma. Am J Emerg Med 2:8–12

Höllt V (1992) Precursors of opioid peptides and their gene regulation. Clin Neuropharmacol 15:50–51

Horiuchi T, Kawaguchi M, Sakamoto T, Kurita N, Inoue S, Nakamura M, Konishi N, Furuya H (2004) The effects of the delta-opioid agonist SNC80 on hind-limb motor function and neuronal injury after spinal cord ischmia in rats. Anesth Analg 99:235–240

Horiuchi T, Kawaguchi M, KuritaN IS, Sakamoto T, Nakamura M, Konishi N, Furuya H (2008) Effects of delta-opioid agonist SNC80 on white matter injury following spinal cord ischemia in normothermic and mildly hypothermic rats. J Anesth 22:32–37

Hosobuchi Y, Baskin DS, Woo SK (1982) Reversal of induced ischemic neurologic deficit in gerbils by the opiate antagonist naloxone. Science 215:69–71

Hughes J, Smith TW, Kosterlitz HW, Fothergill LA, Morgan BA, Morris HR (1975) Identification of two related pentapeptides from the brain with potent opiate agonist activity. Nature 258:577–579

Iishi H, Tatsuta M, Baba M, Okuda S, Taniguchi H (1992) Enhancement by methionine- and leucine-enkephalin of gastric carcinogenesis induced by N-methyl-N′-nitro-N-nitrosoguanidine in Wistar rats. Oncology 49:407–410

Ikeda K, Yoshikawa S, Kurokawa T, Yuzawa N, Nakao K, Mochizuki H (2009) TRK-820, a selective kappa opioid receptor agonist, could effectively ameliorate L-DOPA-induced dyskinesia symptoms in a rat model of Parkinson's disease. Eur J Pharmacol 620:42–48

Immonen JA, Zagon IS, Lewis GS, McLaughlin PJ (2013) Topical treatment with the opioid antagonist naltrexone accelerates the remodeling phase of full-thickness wound healing in type 1 diabetic rats. Exp Biol Med (Maywood) 238:1127–1135

Immonen JA, Zagon IS, McLaughlin PJ (2014a) Selective blockade of the OGF-OGFr pathway by naltrexone accelerates fibroblast proliferation and wound healing. Exp Bliol Med (Maywood) 239:1300–1309

Immonen JA, Zagon IS, McLaughlin PJ (2014b) Topical naltrexone as treatment for Type 2 diabetic cutancous wounds. Adv Wound Care (New Rochelle) 3:419–427

Iselin-Chaves IA, Grötzsch H, Besson M, Burkhard PR, Savoldelli GL (2009) Naloxone-responsive acute dystonia and parkinsonism following general anaesthesia. Anaesthesia 64:1359–1362

Iskandar SB, Abi-Saleh BS, Mechleb BK, Fahrig SA (2007) Methadone and torsade de pointes: case report and review of the literature. Tenn Med 100:35–37

Ivanov IS, Schulz KP, Palmero RC, Newcorn JH (2006) Neurorobiology and evidence-based biological treatments for substance abuse disorders. CNS Spectr 11:864–877

Iwata M, Inoue S, Kawaguchi M, Nakamura M, Konishi N, Furuya H (2007) Effects of delta-opioid receptor stimulation and inhibition on hippocampal survival in a rat model of forebrain ischaemia. Br J Anaesth 99:538–546

Jamali A, Bamdad T, Soleimanjahi H, Pakdel FG, Arefian E (2007) Acute morphine administration reduces white blood cells' capability to induce innate resistance against HSV-1 infection in BALB/c mice. Neuroimmunomodulation 14:16–23

Javadian P, Salmanian B, Javadi-Paydar M, Shamshirsaz AA, Ejtemaci Mehr S, Gharedaghi MH, Dehpour AR (2013) Effect of morphine on the reduced uteroplacental perfusion model of preeclampsia in rats. Eur J Obstet Gynecol Reprod Biol 168:161–166

Jiang Q, Takemori AE, Sultana M, Portoqhese PS, Bowen WD, Mosberg HI, Porreca F (1991) Differential antagonism of opioid delta antinociception by [D-Ala2, Leu5, Cys6]enkephalin and naltrindole 5′-isothiocyanate: evidence for delta receptor subtypes. J Pharmacol Exp Ther 257:1069–1075

Jordan BA, Cyejie S, Devi LA (2000) Opioids and their complicated receptor complexes. Neuropsychopharmacology 23:5–18

Jutkiewicz EM, Baladi MG, Folk JE, Rice KC, Woods JH (2006) The convulsive and electroencephalographic changes produced by nonpeptidic delta-opioid agonists in rats: comparison with pentylenetetrazol. J Pharmacol Exp Ther 317:1337–1348

Kalman S, Osterberg A, Sörensen J, Boivie J, Bertler A (2002) Morphine responsiveness in a group of well-defined multiple sclerosis patients: a study with i.v. morphine. Eur J Pain 6:69–80

Kang J, Shi Y, Xiang B, Qu B, Su W, Zhu M, Zhang M, Bao G, Wang F, Zhang X, Yang R, Fan F, Chen X, Pei G, Ma L (2005) A nuclear function of beta-arrestin 1 in GPCR signaling: regulation of histone acetylation and gene transcription. Cell 123:833–847

Kao TK, Ou YC, Liao SL, Chen WY, Wang CC, Chen SY, Chiang AN, Chen CJ (2008) Opioids modulate post-ischemic progression in a rat model of stroke. Neurochem Int 52:1256–1265

Karrow A, Verthein U, Krausz M, Schafer I (2008) Association of personality disorders, family conflicts and treatment with quality of life in opiate addiction. Eur Addict Res 14:38–46

Kharmate G, Rajput PS, Lin YC, Kumar U (2013) Inhibition of tumor promoting signals by activation of SSTR 2 and opioid receptors in human breast cancer cells. Cancer Cell Int 13:93

Kho HG, Kloppenborg PW, van Egmond J (1993) Effects of acupuncture and transcutaneous stimulation analgesia on plasma hormone levels during and after major abdominal surgery. Eur J Anaesthesiol 10:197–208

Kieffer BL, Evans CJ (2009) Opioid receptors: from binding sites to visible molecules in vivo. Neuropharmacology 56:205–212

Kieffer BL, Gavériaux-Ruff C (2002) Exploring the opioid system by gene knockout. Prog Neurobiol 66:285–306

Kieffer BL, Befort K, Gaveiau-Ruff C, Hirth CG (1992) The delta opioid receptor: isolation of a cDNA clone by expression cloning and pharmacological characterization. Proc Natl Acad Sci U S A 89:12048–12052

Koch T, Höllt V (2008) Role of receptor internalization in opioid tolerance and dependence. Pharmacol Ther 117:199–206

Koodie L, Yuan H, Pumper JA, Yu H, Charboneau R, Ramkrishnan S, Roy S (2014) Morphine inhibits migration of tumor-infiltrating leukocytes and suppresses angiogenesis associated with tumor growth in mice. Am J Pathol 184:1073–1084

Koprich JB, Fox SH, Johnston TH, Goodman A, Le Bourdonnec B, Dolle RE, DeHaven RN, DeHaven RN, DeHaven-Hudkins DL, Little PJ, Brotchie JM (2011) The selective mu-opioid receptor antagonist, ADC5510, reduces L-DODA induced dyskinesia in the MPTP macaque model of Parkinson's disease. Mov Disord 26:1225–1233

Korolenko TA, Goncharova IA, Anterejkina LI, Levina OA, Korolenko CP (2007) Influence of opiate addiction on liver cell damage of patients with viral hepatitis C. Alaska Med 49:75–78

Kotlinska JH, Gibula-Bruzda E, Witkowska E, Chung NN, Schiller PW, Izdebski J (2013) Antinociceptive effects of two deltorphins analogs in the tail-immersion test in rats. Peptides 39:103–110

Kouchek M, Takasusuki T, Terashima T, Yaksh TL, Xu Q (2013) Effects of intrathecal SNC80, a delta receptor ligand, on nociceptive threshold and dorsal horn substance p release. J Pharmacol Exp Ther 347:258–264

Krajnik M, Jassem E, Sobanski P (2014) Opioid receptor bronchial tree: current science. Curr Opin Support Palliat Care 8:191–199

Kreek MJ, Nielsen DA, LaForge KS (2004) Genes associated with addiction: alcoholism, opiate, and cocaine addiction. Neuromolecular Med 5:85–108

Kreek MJ, Zhou Y, Butelman ER, Levran O (2009) Opiate and cocaine addiction: from bench to clinic and back to the bench. Curr Opin Pharmacol 9:74–80

Kumagai H, Ebata T, Takamori K, Miyasato K, Muramatsu T, Nakamoto H, Kurihara M, Yanagita T, Suzuki H (2012) Efficacy and safety of a novel κ-agonist for managing intractable pruritus in dialysis patients. Am J Nephrol 36:175–183

Kuzumaki N, Suzuki A, Narita M, Hosoya T, Nagasawa A, Imai S, Yamamizu K, Morita H, Nagase H, Okada Y, Okano HJ, Yamshita JK, Okano H, Suzuki T, Narita M (2012) Effect of κ- opioid receptor agonist on the growth of non-small cell lung cancer (NSCLC) cells. Br J Cancer 106:1148–1152

LaBuda CJ, Sora I, Uhl GR, Fuchs PN (2000) Stress-induced analgesia in mu-opioid receptor knockout mice reveals normal function of the delta-opioid receptor system. Brain Res 869:1–5

Lalley PM (2008) Opioidergic and dopaminergic modulation of respiration. Respir Physiol Neurobiol 164:160–167

LaLumiere RT, Kalivas PW (2008) Glutamate release in the nucleus accumbens core is necessary for heroin seeking. J Neurosci 28:3170–3177

Lambert DG (2008) The nociception/orphanin FQ receptor: a target with broad therapeutic potential. Nat Rev Drug Discov 7:694–711

Lan X, Chandel N, Cheng K, Lederman R, Saleem MA, Mathieson PW, Husain M, Crosson JT, Gupta K, Malhotra A, Singhal PC (2013) Morphine induces albuminuria by compromising podocyte integrity. PLoS One 8:e55748

Lang RE, Hermann K, Dietz R, Gaida W, Ganten D, Kraft K, Unger T (1983) Evidence for the presence of enkephalins in the heart. Life Sci 32:399–406

Lasukova TV, Krylatov AV, Maslov LN, Lishmanov lB, Gross GJ, Podoksenov A (2004) Antiarrhythmic and cardioprotective effect of stimulation of delta 1-opiate receptors in myocardial ischemia and reperfusion. Patol Fiziol Eksp Ter 3:12–15

Law PY, Loh HH (1999) Regulation of opioid receptor activities. J Pharmacol Exp Ther 289:607–624

Lawson KP, Nag S, Thompson AD, Mokha SS (2010) Sex-specificity and estrogen-dependence of kappa opioid receptor-mediated antinociception and antihyperalgesia. Pain 151:806–815

Lei ZM, Rao CV (2001) Neural actions of luteinizing hormone and human chorionic gonadotropin. Semin Reprod Med 19:103–109

Leitl MD, Onvani S, Bowers MS, Cheng K, Rice KC, Carlezon WA Jr, Banks ML, Negus SS (2014) Pain-related depression of the mesolimbic dopamine system in rats: expression, blockade by analgesics, and role of endogenous κ-opioids. Neuropsychopharmacology 39:614–624

Lenard NR, Zheng H, Berthoud HR (2010) Chronic suppression of mu opioid receptor signaling in the nucleus accumbens attenuates development of diet-induced obesity in rats. Int J Obes (Lond) 34:1001–1010

Lennon FE, Mirzapoiazova T, Mambetsariev B, Salgia R, Moss J, Singleton PA (2012a) Overexpression of the μ-opioid receptor in human non-small cell lung cancer promotes Akt and mTOR activation, tumor growth, and metastasis. Anesthesiology 116:857–867

Lennon FE, Moss J, Singleton PA (2012b) The μ-opioid receptor in cancer progression: is there a direct effect? Anesthesiology 116:940–945

Lennon FE, Mirzapoiazova T, Mambetsariev B, Poroyko VA, Salgia R, Moss J, Singletion PA (2014) The Mu opioid receptor promotes opioid and growth factor-induced proliferation, migration and Epithelial Mesenchymal Transition (EMT) in human lung cancer. PLoS One 9:91577

Li CH, Chung D, Doneen BA (1976) Isolation, characterization and opiate activity of beta-endorphin from human pituitary glands. Biochem Biophys Res Commun 72:1542–1547

Li J, Zhang P, Zhang QY, Zhang SM, Guo HT, Bi H, Wang YM, Sun X, Liu JC, Cheng L, Cui Q, Yu SQ, Kaye AD, Yi DH, Pei JM (2009) Effects of U50, 488H on hypoxia pulmonary hypertension and its underlying mechanism. Vascul Pharmacol 51:72–77

Liang Y, Mestek A, Yu L, Carr LG (1995) Cloning and characterization of the promoter region of the mouse mu opioid receptor gene. Brain Res 679:82–88

Liang J, Chao D, Sandhu HK, Yu Y, Zhang L, Balboni G, Kim DH, Xia Y (2014) δ-Opioid receptors up-regulate excitatory amino acid transporters in mouse astrocytes. Br J Pharmacol 171:5417–5430
Lim YJ, Zheng S, Zuo Z (2004) Morphine preconditions purkinje cells against cell death under in vitro simulated ischemia-reperfusion conditions. Anesthesiology 100:562–568
Liu NJ, Schnell S, Wessendorf MW, Gintzler AR (2013) Sex, pain, and opioids: interdependent influences of sex and pain modality on dynorphin-mediated antinociception in rats. J Pharmacol Exp Ther 344:522–530
Loseva TM, Koliaskin GI (1981) Opiate receptors of peripheral blood lymphocytes (effect of morphine and naloxone on human T-lymphocytes. Zh Nevropatol Psikhiatr Im S S Korsakova 81:1006–1008
Lu J, Liu Z, Zhao L, Tian H, Liu X, Yuan C (2013) In vivo and in vitro inhibition of human liver cancer progress by downregulation of the μ-opioid receptor and relevant mechanisms. Oncol Rep 30:1731–1738
Lynch JL, Alley JF, Wellman L, Beitz AJ (2008) Decreased spinal cord opioid receptor mRNA expression and antinociception in a Theiler's murine encephalomyelitis virus model of multiple sclerosis. Brain Res 1191:180–191
Ma MC, Qian H, Ghassemi F, Zhao P, Xia Y (2005) Oxygen-sensitive delta-opioid receptor-regulated survival and death signals: novel insights into neuronal preconditioning and protection. J Biol Chem 280:16208–16218
Mabrouk OS, Volta M, Marti M, Morari M (2008) Stimulation of delta opioid receptors located in substantia nigra reticulate but not globus pallidus or striatum restores motor activity in 6-hydroxydopamine lesioned rats: new insights into the role of delta receptors in parkinsonism. J Neurochem 107:1647–1659
Mabrouk OS, Marti M, Salvadori S, Morari M (2009) The novel delta opioid receptor agonist UFP-512 dually modulates motor activity in hemiparkinsonian rats via control of the nigro-thalamic pathway. Neuroscience 164:360–369
Mahajan SD, Aalinkeel R, Reynolds JL, Nair BB, Fernandez SF, Schwartz SA, Nair MP (2006) Morphine exacerbates HIV-1 viral protein gp120 induced modulation of chemokine gene expression in U373 astrocytoma cells. Curr HIV Res 3:277–288
Maher DP, Wong W, White PF, McKenna R Jr, Rosner H, Shamloo B, Louy C, Wender R, Yumul R, Zhang V (2014) Association of increased postoperative opioid administration with non-small-cell lung cancer recurrence: a retrospective analysis. Br J Anaesth 113:88–94
Makman MH, Dvorkin B, Stefano GB (1995) Murine macrophage cell lines contain mu 3-opiate receptors. Eur J Pharmacol 273:5–6
Malendowicz LK, Rebuffat P, Tortorella C, Nussdorfer GG, Ziolkowska A, Hochol A (2005) Effects of met-enkephalin on cell proliferation in different models of adrenocortical-cell growth. Int J Mol Med 15:841–845
Mansour A, Fox CA, Akil H, Watson SJ (1995) Opioid-receptor mRNA expression in the rat CNS: anatomical and functional implications. Trends Neurosci 18:22–29
Mansour A, Thompson RC, Akil H, Watson SJ (1993) Delta opioid receptor mRNA distribution in the brain: comparison to delta receptor binding and proenkephalin mRNA. J Chem Neuroanat 6:351–362
Mao L, Wang JQ (2005) Cardiovascular responses to microinjection of nociceptin and endomorphin-1 into the nucleus tractus solitarii in conscious rats. Neuroscience 132:1009–1015
Marczak ED, Jinsmaa Y, Myers PH, Blankenship T, Wilson R, Balboni G, Salvadori S, Lazarus LH (2009) Orally administered H-Dmt-Tic-Lys-NH-CH2-Ph (MZ-2), a potent mu/delta-opioid receptor antagonist, regulates obese-related factors in mice. Eur J Pharmacol 616:115–121
Martin W (1967) Opioid antagonists. Pharmacol Rev 19:283–323
Martin WR (1983) Pharmacology of opioids. Pharmacol Rev 35:283–323
Maslov LN, Lishmanov YB, Oeltgen PR, Barzakh EI, Krylatov AV, Govindaswami M, Brown SA (2009) Activation of peripheral delta2 opioid receptors increases cardiac tolerance to ischemia/reperfusion injury Involvement of protein kinase C, NO-synthase, KATP channels and the autonomic nervous system. Life Sci 84:657–663

Mathew B, Lennon FE, Siegler J, Mirzapoiazova T, Mambetsariev N, Sammani S, Gerhold LM, LaRiviere PJ, Chen CT, Garcia JG, Salgia R, Moss J, Singleton PA (2011) The novel role of the mu opioid receptor in lung cancer progression: a laboratory investigation. Anesth Analg 112:558–567

Matthes HW, Maldonado R, Simonin F, Valverde O, Slowe S, Kitchen I, Befort K, Dierich A, Le Meur M, Dolle P, Tzavara E, Hanone J, Roques BP, Kieffer BL (1996) Loss of morphine-induced analgesia, reward effect and withdrawal symptoms in mice lacing the mu-opioid-receptor gene. Nature 383:819–823

Matthes HW, Smadja C, Valverde O, Vonesch JL, Foutz AS, Boudinot E, Denavit-Saubié M, Severini C, Negri L, Roques BP, Maldonado R, Kieffer BL (1998) Activity of the delta-opioid receptor is partially reduced, whereas activity of the kappa-receptor is maintained in mice lacking the mu-receptor. J Neurosci 18:7285–7295

Mattia A, Vanderah T, Mosberg HI, Porreca F (1991) Lack of antinociceptive cross-tolerance between [D-Pen2, D-Pen5]enkephalin and [D-Ala2]deltorphin II in mice: evidence for delta receptor subtypes. J Pharmacol Exp Ther 258:583–587

Mayer P, Höllt V (2006) Pharmacogenetics of opioid receptors and addiction. Pharmacogenet Genomics 16:1–7

Mayer D, Liebeskind JC (1974) Pain reduction by focal electrical stimulation of the brain: an anatomical and behavioral analysis. Brain Res 68:73–93

Mayfield KP, D'Alecy LG (1994) Delta-1 opioid agonist acutely increases hypoxic tolerance. J Pharmacol Exp Ther 268:683–688

McBride SM, Smith-Sonneborn J, Oeltgen P, Flynn FW (2005) Delta2 opioid receptor agonist facilitates mean arterial pressure recovery after hemorrhage in conscious rats. Shock 23:264–268

McCarthy LE, Rogers TJ (2011) Alteration of early T cell development by opioid and superantigen stimulation. Adv Exp Med Biol 493:163–167

McCormick SE, Stoessl AJ (2002) Blockade of nigral and pallidal opioid receptors suppresses vacuous chewing movements in a rodent model of tardive dyskinesia. Neuroscience 112:851–859

McLaughlin PJ, Zagon IS (1987) Modulation of human neuroblastoma transplanted into nude mice by endogenous opioid systems. Life Sci 41:1465–1472

McLaughlin PJ, Zagon IS (2012) The opioid growth factor-opioid growth factor receptor axis: homeostatic regulator of cell proliferation and its implications for health and disease. Biochem Pharmacol 86:746–755

McLaughlin PJ, Zagon IS (2014) Novel treatment for triple-negative breast and ovarian cancer: endogenous opioid suppression of women's cancers. Expert Rev Anticancer Ther 14:247–250

Meis S (2003) Nociceptin/orphanin FQ: actions within the brain. Neuroscientist 9:158–168

Meng J, Meng Y, Plotnikoff NP, Youkilis G, Griffin N, Shan FInt (2013) Low dose naltrexone (LDN) enhances maturation of bone marrow dendritic cells (BMDCs). Immunopharmacol. 17:1084–1089

Menzebach A, Hirsch J, Nost R, Mogk M, Hempelmann G, Welters ID (2004) Morphine inhibits complement receptor expression, phagocytosis and oxidative burst by a nitric oxide dependent mechanism. Anasthesiol Intensivmed Notfallmed Schmerzther 39:204–211

Mettang T, Kremer AE (2015) Uremic pruritus. Kidney Int 87:685–691

Meunier JC, Mollereau C, Toll L, Suaudeau C, Moisand C, Alvinerie P, Butour JL, Guillemot JC, Ferrara P, Monsarrat B (1995) Isolation and structure of the endogenous agonist of opioid receptor-like ORL1 receptor. Nature 377:532–535

Mika J, Popiolek-Barczyk K, Rojewska E, Makuch W, Starowicz K, Przewlocka B (2014) Delta-opioid receptor analgesia is independent of microglial activation in a rat model of neuropathic pain. PLoS One 9:104420

Miranda HF, Noriega V, Zanetta P, Prieto J, Prieto-Rayo J, Aranda N, Sierralta F (2014) Isobolographic analysis of the opioid-opioid interactions in a tonic and a phasic mouse model of induced nociceptive pain. J Biomed Sci 21:62

Molina PE (2002) Stress-specific opioid modulation of haemodynamic counter-regulation. Clin Exp Pharmacol Physiol 29:248–253

Mollercau C, Parmentier M, Mailleux P, Butour JL, Moisand C, Chalon P, Caput D, Vassart G, Meunier JC (1994) ORL1, a novel member of the opioid receptor family: cloning, functional expression and localization. FEBS Lett 341:33–38

Mutoh J, Ohsawa M, Hisa H (2013) Effect of naloxone on ischemic acute kidney injury in the mouse. Neuropharmacology 71:10–18

Mutolo D, Bongianni F, Einum J, Dubuc R, Pantaleo T (2007) Opioid-induced depression in the lamprey respiratory network. Neuroscience 150:720–729

Naghibzadeh Tahami A, Khanjani N, Yazdi Feyzabadi V, Varzandeh M, Haghdoost AA (2014) Opium as a risk factor for upper gastrointestinal cancers: a population-based case–control study in Iran. Arch Iran Med 17:2–6

Naidu PS, Lichtman AH, Archer CC, May EL, Harris LS, Aceto MD (2007) NIH 11082 produces anti-depressant-like activity in the mouse tail-suspension test through a delta-opioid receptor mechanism of action. Eur J Pharmacol 566:132–136

Negus SS, Gatch MB, Mello NK, Zhang X, Rice K (1998) Behavioral effects of the delta-selective opioid agonist SNC80 and related compounds in rhesus monkeys. J Pharmacol Exp Ther 286:362–375

Nomura Y, Kawaraguchi Y, Sugimoto H, Furuya H, Kawaguchi M (2014) Effects of morphine and fentanyl on 5-fluorouracil sensitivity in human colon cancer HCT116 cells. J Anesth 28:298–301

Notas G, Kampa M, Nifli AP, Xiadakis K, Papasava D, Thermos K, Kouroumalis E, Castanas E (2007) The inhibitory effect of opioids on HepG2 cells is mediated via interaction with somatostatin receptors. Eur J Pharmacol 555:1–7

Obese FY, Whitlock BK, Steele BP, Buonomo FC, Sartin JL (2007) Long-term feed intake regulation in sheep is mediated by opioid receptors. J Anim Sci 85:11–17

Olinger CP, Adams HP Jr, Brott TG, Biller J, Barsan WG, Toffol GJ, Eberle RW, Marler JR (1990) High-dose intravenous naloxone for the treatment of acute ischemic stroke. Stroke 21:721–725

Olsen P, Rasmussen M, Zhu W, Tonnesen E, Stefano GB (2005) Human gliomas contain morphine. Med Sci Monit 11:18–21

Ordaz-Sanchez I, Weber RJ, Rice KC, Zhang X, Rodriquez-Padilla C, Tamez-Guerra R, Mendez-Vazquez JL, Gomez-Flores R (2003) Chemotaxis of human and rat leukocytes by the delta-selective non-peptidic opioid SNC 80. Rev Latinoam Microbiol 45:16–23

Ozarda Ilcol Y, Ozyurt G, Kilicturgay S, Uncu G, Ulus IH (2002) The decline in serum choline concentration in humans during and after surgery is associated with the elevation of cortisol, adrenocorticotropic hormone, prolactin and beta-endorphin concentrations. Neurosci Lett 324:41–44

Pamenter ME, Buck LT (2008) δ-Opioid receptor antagonism induces NMDA receptor-dependent excitotoxicity in anoxic turtle cortex. J Exp Biol 21:3512–3517

Pan EC, Bohn LM, Belcheva MM, Thomas GE, Manepalli AN, Mamone JY, Johnson FE, Coscia CJ (1998) Kappa-opioid receptor binding varies inversely with tumor grade in human gliomas. Cancer 83:2561–2566

Pan L, Xu J, Yu R, Xu MM, Pan YX, Pasternak GW (2005) Identification and characterization of six new alternatively spliced variants of the human mu opioid receptor gene. Neuroscience 133:209–220

Parkhill AL, Bidlack JM (2006) Reduction of lipopolysaccharide-induced interleukin-6 production by the kappa opioid U50, 488 in a mouse monocyte-like cell line. Int Immunopharmacol 6:1013–1019

Pastermak GW (2005) Molecular biology of opioid analgesia. J Pain Symptom Manage 29:2–9

Pasternak GW (2004) Multiple opiate receptors: déjà vu all over again. Neuropharmacology 47:312–323

Paterson SJ, Robson LE, Kosterlitz HW (1983) Classification of opioid receptors. Br Med Bull 39:31–36

Pattinson KT (2008) Opioids and the control of respiration. Br J Anaesth 100:747–758

Patwardhan AM, Berg KA, Akopain AN, Jeske NA, Gamper N, Clarke WP, Harqreaves KM (2005) Bradykinin-Induced functional competence and trafficking of the δ-opioid receptor in trigeminal nociceptors. J Neurosci 25:8825–8832

Peart JN, Patel HH, Gross GJ (2003) Delta-opioid receptor activation mimics ischemic preconditioning in the canine heart. J Cardiovasc Pharmacol 42:78–81

Peart JN, Gross ER, Gross GJ (2005) Opioid-induced preconditioning: recent advances and future perspectives. Vascul Pharmacol 42:211–218

Pello OM, Duthey B, García-Bernal D, Rodríguez-Frade JM, Stein JV, Teixido J, Martínez C, Mellado M (2006) Opioids trigger alpha 5 beta 1 integrin-mediated monocyte adhesion. J Immunol 176:1675–1685

Peng PH, Huang HS, Lee YJ, Chen YS, Ma MC (2009a) Novel role for the δ-opioid receptor in hypoxic preconditioning in rat retinas. J Neurochem 108:741–754

Peng P, Huang LY, Li J, Fan R, Zhang SM, Wang YM, Hu YZ, Sun X, Kaye AD, Pei JM (2009b) Distribution of kappa-opioid receptor in the pulmonary artery and its changes during hypoxia. Anat Rec 292:1062–1067

Peng J, Sarkara S, Chang S (2012) Opioid receptor expression in human brain and peripheral tissues using absolute quantitative real-time RT-PCR. Drug Alcohol Depend 124:223–228

Pert A, Yaksh T (1974) Sites of morphine induced analgesia in the primate brain: relation to pain pathway. Brain Res 80:135–140

Pert A, Yaksh T (1975) Localization of the antinociceptive action of morphine in primate brain. Pharmacol Biochem Behav 3:133–138

Pinna A, Di Chiara G (1998) Dopamine-dependent behavioural stimulation by non-peptide delta opioids BW 373U86 and SNC 80: 3. Facilitation of D1 and D2 responses in unilaterally 6-hydroxydopamine-lesioned rats. Behav Pharmacol 9:15–21

Pinsky C, Bose R, Frederickson RC (1986) Mu- and delta-opioid modulation of electrically-induced epileptic seizures in mice. NIDA Res Monogr 75:534–536

Plotnikov EY, Grebenchikov OA, Babenko VA, Pevzner IB, Zorova LD, Likhvantse VV, Zorov DB (2013) Nephroprotective effect of GSK-3β inhibition by lithium ions and δ-opioid receptor agonist dalargin on gentamicin-induced nephrotoxicity. Toxicol Lett 220:303–308

Popiolek-Barczyk K, Makuch W, Rojewska E, Pilat D, Mika J (2014) Inhibition of intracellular signaling pathways NF-κB and MEK1/2 attenuates neuropathic pain development and enhances morphine analgesia. Pharmacol Rep 66:845–851

Przewlocki R (2004) Opioid abuse and brain gene expression. Eur J Pharmacol 500:331–349

Qin Y, Chen J, Li L, Liao CJ, Liang YB, Guan EJ, Xie YB (2012a) Exogenous morphine inhibits human gastric cancer MGC-803 cell growth by cell cycle arrest and apoptosis induction. Asian Pac J Cancer Prev 13:1377–1382

Qin Y, Li L, Chen J, Tang X, Liao C, Xie Y, Xiao Q (2012b) Fentanyl inhibits progression of human gastric cancer MGC-803 cells by NF-kappaB downregulation and PTEN upregulation in vitro. Oncol Res 20:61–69

Quaglio GL, Lugoboni F, Pajusco B, Sarti M, Talamini G, Mezzelani P, Des Jarlais DC, GICS (2003) Hepatitis C virus infection: prevalence, predictor variables and prevention opportunities among drug users in Italy. J Viral Hepat 10:394–400

Quill TA, Wang D, Garbers DL (2006) Insights into sperm cell motility signaling through sNHE and the CatSpers. Mol Cell Endocrinol 250:84–92

Rahn KA, McLaughlin PJ, Zagon IS (2011) Prevention and diminished expression of experimental autoimmune encephalomyelitis by low dose naltrexone (LDN) or opioid growth factor (OGF) for an extended period: Therapeutic implications for multiple sclerosis. Brain Res 1381:243–253

Rech RH, Mokler DJ, Briggs SL (2012) Effects of combined opioids on pain and mood in mammals. Pain Res Treat 2012:145965

Reinscheid RK, Nothacker HP, Bourson A (1995) Orphanin FQ: a neuropeptide that activates an opioid like G protein-coupled receptor. Science 270:792–794

Reiss D, Wichmann J, Tekeshima H, Kieffer BL, Ouagazzal AM (2008) Effects of nociceptin/orphanin FQ receptor (NOP) agonist, Ro64-6198, on reactivity to acute pain in mice: comparison to morphine. Eur J Pharmacol 579:141–148

Rizzi A, Nazzaro C, Marzola GG, Zucchini S, Trapella C, Guerrini R, Zeilhofer HU, Reqoli D, Calo' G (2006) Endogenous nociceptin/orphanin FQ signaling produces opposite spinal antinociceptive and supraspinal pronociceptive effects in the mouse formalin test: pharmacological and genetic evidences. Pain 124:100–108

Rong F, Peng Z, Ye MX, Zhang QY, Zhao Y, Zhang SM, Guo HT, Hui B, Wang YM, Liang C, Cui Q, Yu SQ, Yi DH, Pei JM (2009) Myocardial apoptosis and infarction after ischemia/reperfusion are attenuated by kappa-opioid receptor agonist. Arch Med Res 40:227–234

Rothman RB, Bowen WD, Bykov V, Schumacher UK, Pert CB, Jacobson AE, Burke TR Jr, Rice KC (1984) Preparation of rat brain membranes greatly enriched with either type-I-delta or type-II-delta opiate binding sites using site directed alkylating agents: evidence for a two-site allosteric model. Neuropeptides 4:201–215

Rothman RB, Bykov V, Danks JA, Jacobson AE, Burke TR Jr, Rice KC, Herkenham M (1985) Preparation of rat brain membranes highly enriched with opiate kappa binding sites using site-directed acylating agents: optimization of assay conditions. Neuropeptides 6:503–516

Rothman RB, Bykov V, JacobsonAE RKC, Long JE, Bowen WD (1992) A study of the effect of the irreversible delta receptor antagonist [D-Ala2, Leu5, Cys6]-enkephalin on delta cx and delta ncx opioid binding sites in vitro and in vivo. Peptides 13:691–694

Roy S, Cain KJ, Chapin RB, Charbonecau RG, Barke RA (1998) Morphine modulates NF kappa B activation in macrophages. Biochem Biophys Res Commun 245:392–396

Ruff MR, Wahl SM, Mergenhagen S, Pert CB (1985) Opiate receptor mediated chemotaxis of human monocytes. Neuropeptides 5:363–366

Sabory E, Derchansky M, Ismaili M, Jahromi SS, Brull R, Carlen PL, El Beheiry H (2007) Mechanisms of morphine enhancement of spontaneous seizure activity. Anesth Analg 105:1729–1735

Saitoh A, Yoshikawa Y, Onodera K, Kamei J (2005) Role of delta-opioid receptor subtypes in anxiety-related behaviors in the elevated plus-maze in rats. Psychopharmacology (Berl) 182:327–334

Saloman JL, Niu KY, Ro JY (2011) Activation of peripheral delta-opioid receptors leads to anti-hyperalgesic responses in the masseter muscle of male and female rats. Neuroscience 190:379–385

San-Emeterio EP, Hurlé MA (2006) Modulation of brain apoptosis-related proteins by the opioid antagonist naltrexone in mice. Neurosci Lett 403:276–279

Santos FM, Grecco LH, Pereira MG, Oliveira ME, Rocha PA, Silva JT, Martins DO, Miyabara EH, Chacur M (2014) The neural mobilization technique modulates the expression of endogenous opioids in the periaqueductal gray and improves muscle strength and mobility in rats with neuropathic pain. Behav Brain Funct 10:19

Scherrer G, Imamachi N, Gao YQ, Contet C, Mennicken F, O'Donnell D, Kieffer BL, Basbaum A (2009) Dissociation of the opioid receptor mechanisms that control mechanical and the pain. Cell 137:1148–1159

Schindler CW, Graczyk Z, Gilman JP, Nequs SS, Berqman J, Mello NK, Goldberg SR (2007) Effects of kappa opioid agonists alone and in combination with cocaine on heart rate and blood pressure in conscious squirrel monkeys. Eur J Pharmacol 576:107–113

Schultz JJ, Hsu AK, Gross GJ (1997) Ischaemic preconditioning and morphine-induced cardioprotection involve the delta (δ)-opioid receptor in the intact rat heart. J Mol Cell Cardiol 29:2187–2195

Scoto GM, Aricò G, Ronsisvalle S, Parenti C (2007) Blockade of the nociception/orphanin FQ/NOP receptor system in the rat ventrolateral periaqueductal gray potentiates DAMGO analgesia. Peptides 28:1441–1446

Scwarzer S (2009) 30 Years of Dynorphins-new insights on their functions in neuropsychiatric disease. Pharmacol Ther 123:353–370

Shaqura M, Khalefa BI, Shakibaei M, Zöllner C, Al-Khrasani M, Fürst S, Schäfer M, Mousa SA (2014) New insights into mechanisms of opioid inhibitory effects on capsaicin-induced TRPV1 activity during painful diabetic neuropathy. Neuropharmacology 85:142–150

Sharp BM, Keane WF, Suh HJ, Gekker G, Tsukayama D, Peterson PK (1985) Opioid peptides rapidly stimulate superoxide production by human polymorphonuclear leukocytes and macrophages. Endocrinology 117:793–795

Sheng WS, Hu S, Gekker G, Zhu S, Peterson PK, Chao CC (1997) Immunomodulatory role of opioids in the central nervous system. Arch Immunol Ther Exp (Warsz) 45:359–366

Shook JE, Watkins WD, Camporesi EM (1990) Differential roles of opioid receptors in respiration, respiratory disease, and opiate-induced respiratory depression. Am Rev Respir Dis 142:895–909

Shutov AA, Bykova AA, Volodina EV (1999) Immunopharmacological method for the diagnosis of multiple sclerosis. Zh Nevrol Psikhiatr Im S S Korsakova 99:35–37

Simonin F, Befort K, Gaveriaux-Ruff C et al (1994) The human delta opioid receptor: genomic organization, cDNA cloning, functional expression, and distribution in human brain. Mol Pharmacol 46:1015–1021

Singleton PA, Mirzapoiazova T, Hasina R, Salgia R, Moss J (2014) Increased μ-opioid receptor expression in metastatic lung cancer. Br J Anaesth 113:103–108

Skarphedinsson JO, Thoren P (1988) Endorphin mechanisms are responsible for the beneficial effects of opioid antagonists on cerebral function during relative cerebral ischaemia in rats. Acta Physiol Scand 132:281–288

Smeester BA, Lunzer MM, Akgün E, Beitz AJ, Portoghese PS (2014) Targeting putative mu opioid/metabotropic glutamate receptor-5 heteromers produces potent antinociception in a chronic murine bone cancer model. Eur J Pharmacol 734:48–52

Smith JP, Conter RL, Demers TM, McLaughlin PJ, Zagon IS (2000) Elevated levels of opioid growth factor in the plasma of patients with pancreatic cancer. Pancreas 21:158–164

Smith JP, Conter RL, Bingaman SI, Harvey HA, Mauger DT, Ahmad M, Demers LM, Stanley WB, McLaughlin PJ, Zagon IS (2004) Treatment of advanced pancreatic cancer with opioid growth factor: phase I. Anticancer Drugs 15:203–209

Smith JP, Bingaman SI, Mauger DT, Harvey HH, Demers LM, Zagon IS (2010) Opioid growth factor improves clinical benefit and survival in patients with advanced pancreatic cancer. Open Access J Clin Trials 2010:37–48

Snuder SH, Matthysse S (1975) Opiate receptor mechanisms. Neurosci Res Program Bull 13:1–66

Sofuoglu M, Portoghese PS, Takemori AE (1991) Differential antagonism of delta opioid agonists by naltrindole and its benzofuran analog (NTB) in mice: evidence for delta opioid receptor subtypes. J Pharmacol Exp Ther 257:676–680

Sofuoglu M, Portoghese PS, Takemori AE (1993) 7-Benzylidenenaltrexone (BNTX): a selective delta 1 opioid receptor antagonist in the mouse spinal cord. Life Sci 52:769–775

Solbrig MV, Adrian R, Baratta J, Lauterborn JC, Koob GF (2006) Kappa opioid control of seizures produced by a virus in an animal model. Brain 129:642–654

Soverchia L, Mosconi G, Ruqqeri B, Ballarini P, Catone G, Deql'lnnocenti S, Nabissi M, Polzonetti-Maqni AM (2006) Proopiomelanocortin gene expression and beta-endorphin localization in the pituitary, testis, and epididymis of stallion. Mol Reprod Dev 73:1–8

Spampinato S, Goldstein A (1983) Immunoreactive dynorphin in rat tissues and plasma. Neuropeptides 3:193–212

Steele PA, Aromataris EC, Riederer BM (1996) Endogenous opioid peptides in parasympathetic, sympathetic and sensory nerves in the guinea-pig heart. Cell Tissue Res 284:331–339

Stevens CW, Brasel CM, Mohan S (2007) Cloning and bioinformatics of amphibian mu, delta, kappa, and nociception opioid receptors expressed in brain tissue: evidence for opioid receptor of divergence in mammals. Neurosci Lett 419:189–194

Su DS, Wang ZH, Zheng YJ, Zhao YH, Wang XR (2007) Dose-dependent neuroprotection of delta opioid peptide [D-Ala2, DLeu5] encephalin in neuronal death and retarded behavior induced by forebrain ischemia in rats. Neurosci Lett 423:113–117

Subirán N, Casis L, Irazusta J (2011) Regulation of male fertility by the opioid system. Mol Med 17:846–853

Sullo N, Roviezzo F, Matteis M, Lanaro A, Calo G, Guerrini R, De Gruttloa L, Spaziano G, Cirino G, Rossi F, DAgostino B (2013) Nociceptin/orphanin FQ receptor activation decreases the airway hyperresponsiveness induced by allergen in sensitized mice. Am J Physiol Lung Cell Mol Physiol 304:657–664

Summers RL, Li Z, Hildebrandt D (2003) Effect of a delta receptor agonist on duration of survival during hemorrhagic shock. Acad Emerg Med 10:587–593

Szabo I, Rojavin M, Bussiere JL, Eisenstein TK, Adler MW, Rogers TJ (1993) Suppression of peritoneal macrophage phagocytosis of Candida albicans by opioid. J Pharmacol Exp Ther 267:703–706

Tang B, DU J, Cao ZM, Liang R, Sun DG, Jin XL, Wang LM (2012) DADLE suppresses the proliferation of human liver cancer HepG2 cells by activation of PKC pathway and elevates the sensitivity to cis-diammine dichloridoplatium. Zhonghua Zhong Liu Za Zhi 34:425–429

Tang B, Li Y, Yuan S, Tomlinson S, He S (2013) Upregulation of the δ-opioid receptor in liver cancer promotes liver cancer progression both in vitro and in vivo. Int J Oncol 43:1281–1290

Taracha E, Mierzejewski P, Lehner M (2009) Stress-opioid interactions: a comparison of morphine and methadone. Pharmacol Rep 61:424–435

Teoulloa E, Hargreaves KM, Dionne RA (1997) Ibuprofen elevates immunoreactive beta-endorphin levels in humans during surgical stress. Clin Pharmacol Ther 62:74–81

Tian XS, Zhou F, Yang R, Xia Y, Wu GC, Guo JC (2008a) Electroacupuncture protects the brain against acute ischemic injury via up-regulation of delta-opioid receptor in rats. J Chin Integr Med 6:632–638

Tian XS, Zhou F, Yang R, Xia Y, Wu GC, Guo JC (2008b) Role of δ-opioid receptors in cumulative electro-acupuncture induced protection from ischemic injury in the rat brain. Shanghai J TCM 42:71–74

Tian XS, Zhou F, Yang R, Xia Y, Wu GC, Guo JC (2008c) Effects of intracerebroventricular injection of delta-opioid receptor agonist TAN-67 or antagonist naltrindole on acute cerebral ischemia in rat. Sheng Li Xue Bao 60:475–484

Tian X, Hua F, Sandhu HK, Chao D, Balboni G, Salvadori S, He X, Xia Y (2013a) Effect of δ-opioid receptor activation on BDNF-TrkB vs. TNF-α in the mouse cortex exposed to prolonged hypoxia. Int J Mol Sci 14:15959–15976

Tian X, Guo J, Zhu M, Li M, Wu G, Xia Y (2013b) δ-Opioid receptor activation rescues the functional TrkB receptor and protects the brain from ischemia-reperfusion injury in the rat. PLoS One 8:e69252

Torregrossa MM, Jutkiewicz EM, Mosberg HI, Balboni G, Watson SJ, Woods JH (2006) Peptidic delta opioid receptor agonists produce antidepressant-like effects in the forced swim test and regulate BDNF mRNA expression in rats. Brain Res 1069:172–181

Tortella FC, Long JB (1998) Characterization of opioid peptide-like anticonvulsant activity in rat cerebrospinal fluid. Brain Res 456:139–146

Tortella FC, Robles L, Holaday JW (1985) The anticonvulsant effects of DADLE are primarily mediated by activation of delta opioid receptors: interactions between delta and mu receptor antagonists. Life Sci 37:497–503

Tortella FC, Echevarria E, Robles L, Mosberg HI, Holaday JW (1988) Anticonvulsant effects of mu (DAGO) and delta (DPDPE) enkephalins in rats. Peptides 9:1177–1181

Traynor J (1989) Subtypes of κ-opioid receptor: fact or fiction? Trends Pharmacol Sci 10:52–53

Traynor J, Elliot J (1993) δ-opioid receptor subtypes and cross talk with μ-receptors. Trends Pharmacol Sci 14:84–85

Trescot AM, Datta S, Lee M, Hansen H (2008) Opioid pharmacology. Pain Physician 11:133–153

Tsong SD (1982) ACTH and beta-endorphin-related peptides are present in multiple sites in the reproductive tract of the male rat. Endocrinology 1109:2204–2206

Tsou K, Jang CS (1964) Studies on the site of analgesic action of morphine by intracerebral micro-injection. Sci Sin 13:1099–1109

Vallejo R, de Leon-Cassola O, Benyamin R (2004) Opioid therapy and immunosuppression: a review. Am J Ther 11:354–365

Van Epps DE, Saland L (1984) Beta-endorphin and met-enkephalin stimulate human peripheral blood mononuclear cell chemotaxis. J Immunol 132:3046–3053

Vander Weele CM, Porter-Stransky KA, Mabrouk OS, Lovic V, Singer BF, Kennedy RT, Aragona BJ (2014) Rapid dopamine transmission within the nucleus accumbens: dramatic difference between morphine and oxycodone delivery. Eur J Neurosci 40:3041–3054

Vien N, Gleason CA, Hays SL, McPherson RJ, Chavkin C, Juul SE (2009) Effects of neonatal stress and morphine on kappa opioid receptor signaling. Neonatology 96:235–243

Villemagne PS, Dannals RF, Ravert HT, Frost JJ (2002) PET imaging of human cardiac opioid receptors. Eur J Nucl Med Mol Imaging 29:1385–1388

Vogt BA, Wiley RG, Jensen EL (1995) Localization of mu and delta opioid receptors to anterior cingulated afferents and projection neurons and input/output model of mu regulation. Exp Neurol 135:83–92

Waldhoer M, Bartlett SE, Whistler JL (2004) Opioid receptors. Annu Rev Biochem 73:953–990

Wang GY, Wu S, Pei JM, Yu XC, Wong TM (2001) Kappa-but not delta-opioid receptors mediate effects of ischemic preconditioning on both infarct and arrhythmia in rats. Am J Physiol Heart Circ Physiol 280:384–391

Wang J, Gao Q, Shen J, Ye TM, Xia Q (2007) Kappa-opioid receptor mediates the cardioprotective effect of ischemic postconditioning. Zhejiang Da Xue Xue Bao Yi Xue Ban 36:41–47

Wang Q, Chao D, Chen T, Sandhu H, Xia Y (2014) δ-Opioid receptors and inflammatory cytokines in hypoxia: differential regulation between glial and neuron-like cells. Transl Stroke Res 5:476–483

Weber ML, Vang D, Velho PE, Gupta P, Crosson JT, Hebbel RP, Gupta K (2012) Morphine promotes renal pathology in sickle mice. Int J Nephrol Renovasc Dis 2012:109–118

Weber ML, Chen C, Li Y, Faroqui M, Nguyen J, Poonawala T, Hebbel RP, Gupta K (2013) Morphine stimulates platelet-derived growth factor receptor-β signalling in mesangial cells in vitro and transgenic sickle mouse kidney in vivo. Br J Anaesth 111:1004–1012

Weihe E, McKnight AT, Corbett AD, Kosterlitz HW (1985) Proenkephalin- and prodynorphin-derived opioid peptides in guinea-pig heart. Neuropeptides 5:453–456

Weihe E, McKnight AT, Corbett AD, Hartschuh W, Reinecke M, Kosterliz HW (1983) Characterization of opioid peptides in guinea-pig heart. Neuropeptides 5:453–456

Welters ID (2003) Is immunomodulation by opioid drugs of clinical relevance? Curr Opin Anaesthesiol 16:509–513

Welters ID, Menzebach A, Goumon Y, Langegeld TW, Teschemacher H, Hempelmann G, Stefano GB (2000) Morphine suppresses complement receptor expression, phagocytosis, and respiratory burst in neutrophils by a nitric oxide and mu (3) opiate receptor dependent mechanism. J Neuroimmunol 111:139–145

Wever KE, Mascreeuw R, Wagener FA, Verweij VG, Peters JG, Petijis JC, Van der Vilet JA, Warle MC, Rongen GA (2013) Humoral signaling compounds in remote ischaemic preconditioning of the kidney, a role for the opioid receptor. Nephrol Dial Transplant 28:1721–1732

Wiesenfeld-Hallin Z, de Araúja LG, Alster P, Xu XJ, Hökfelt T (1999) Cholecystokinin/opioid interactions. Brain Res 848:78–89

Wikstrom B, Gellert R, Ladefoged SD, Danda Y, Akai M, Ide K, Ogasawara M, Kawashima Y, Ueno K, Mori A, Ueno Y (2005) Kappa-opioid system in uremic pruritus: multicenter, randomized, double-blind, placebo-controlled clinical studies. J Am Soc Nephrol 16:3742–3747

Willner D, Cohen-Yeshurun A, Avidan A, Ozersky V, Shohami E, Leker RR (2014) Short term morphine exposure in vitro alters proliferation and differentiation of neural progenitor cells and promotes apoptosis via mu receptors. PLoS One 29:9

Wittert G, Hope P, Pyle D (1996) Tissue distribution of opioid receptor gene expression in the rat. Biochem Biophys Res Commun 218:877–8781

Wolozin BL, Pasternak GW (1981) Classification of multiple morphine and encephalin binding sites in the central nervous system. Proc Natl Acad Sci U S A 78:6181–6185

Wu S, Li HY, Wong TM (1999) Cardioprotection of preconditioning by metabolic inhibition in the rat ventricular myocyte. Involvement of kappa-opioid receptor. Circ Res 84:1388–1395

Wu Q, Wang HY, Li J, Zhou P, Wang QL, Zhao L, Fan R, Wang YM, Xu XZ, Yi DH, Yu SQ, Pei JM (2013) K-opioid receptor stimulation improves endothelial function in hypoxic pulmonary hypertension. PLoS One 8:60850

Xander C, Meerpohl JJ, Galandi D, Buroh S, Schwarzer G, Antes G, Becker G (2013) Pharmacological interventions for pruritus in adult palliative care patients. Cochrane Database Syst Rev 6, CD008320
Xia Y (1989) Pressor effect of naloxone in acute experimental rabbits. Chin Pharmacol Bull 5:162–164
Xia Y, Haddad GG (1991) Ontogeny and distribution of opioid receptors in the rat brainstem. Brain Res 549:181–193
Xia Y, Haddad GG (2001) Major different in the expression of delta- and mu-opioid receptors between turtle and rat brain. J Comp Neurol 436:202–210
Xiang B, Yu GH, Guo J, Chen L, Hu W, Pei G, Ma L (2001) Heterologous activation of protein kinase C stimulates phosphorylation of delta-opioid receptor at serine 344, resulting in beta-arrestin-and clathrin-mediated receptor internalization. J Biol Chem 276:4709–4716
Xu B (2011) Effects of remifentanil combined with naloxone on human sperm motility. Zhonghua Nan Ke Xue 17:926–929
Xu B (2012) Remifentanil impairs sperm motility reversibly in vitro. J Biochem Mol Toxicol 26:176–177
Xu B, Wang ZP, Wang YJ, Lu PH, Wang XH (2013) The toxic effect of opioid analgesics on human sperm motility in vitro. Drug Chem Toxicol 36:205–208
Yajinma Y, Narita M, Takahashi-Nakano Y, Misawa M, Nagase H, Mizoquchi H, Tseng LF, Suzuki T (2000) Effects of differential modulation of mu-, delta- and kappa-opioid systems on bicuculline-induced convulsions in the mouse. Brain Res 862:120–126
Yamada K, Nabeshima T (1995) Stress-induced behavioral responses and multiple opioid systems in the brain. Behav Brain Res 67:133–145
Yamada H, Shimoyama N, Sora I, Uhl GR, Fukuda Y, Moriya H, Shimoyama M (2006) Morphine can produce analgesia via spinal kappa opioid receptors in the absence of mu opioid receptors. Brain Res 1083:61–69
Yamamizu K, Furuta S, Hamada Y, Yamashita A, Kuzumaki N, Narita M, Doi K, Katayama S, Nagase H, Yamasita JK, Narita M (2013) κ Opioids inhibit tumor angiogenesis by suppressing VEGF signaling. Sci Rep 3:3213
Yamamizu K, Hamada Y, Narita M (2014) κ Opioid receptor ligands regulate angiogenesis in development and in tumours. Br J Pharmacol 172:268–276
Yang Y, Xia X, Zhang Y, Wang Q, Li L, Luo G, Xia Y (2009) Delta-opioid receptor activation attenuates oxidative injury in the ischemic rat brain. BMC Biol 7:55
Yang Y, Zhi F, He X, Moore M, Kang X, Chao D, Wang R, Kim DH, Xia Y (2012) δ-opioid receptor activation and microRNA expression of the rat cortex in hypoxia. PLoS One 7:51524
Yeadon M, Kitchen I (1989) Opioids and respiration. Prog Neurobiol 33:1–16
Yilmaz A, Sogut A, Kilinc M, Sogut AG (2003) Successful treatment of intrathecal morphine overdose. Neurol India 51:410–411
Young AP, Gruber RB, Discala JF, May WJ, Palmer LA, Lewis SJ (2013) Co-activation of μ- and δ-opioid receptors elicits tolerance to morphine-induced ventilatory peroxynitrite. Respir Physiol Neurobiol 186:255–264
Younger J, Parkitny L, McLain D (2014) The use of low-dose naltrexone (LDN) as a novel anti-inflammatory treatment for chronic pain. Clin Rheumatol 33:451–459
Zadina JE, Hackler L, Ge L-J, Kastin AJ (1997) A potent and selective endogenous agonist for the mu-opiate receptor. Nature 386:499–502
Zagon IS, McLaughlin PJ (1983a) Naltrexone modulates tumor response in mice with neuroblastoma. Science 221:671–673
Zagon IS, McLaughlin PJ (1983b) Opioid antagonists inhibit the growth of metastatic murine neuroblastoma. Cancer Lett 21:89–94
Zagon IS, McLaughlin PJ (1983c) Increased brain size and cellular content in infant rats treated with an opiate antagonist. Science 221:1179–1180
Zagon IS, McLaughlin PJ (2014) Opioid growth factor and the treatment of human pancreatic cancer: a review. World J Gastroenterol 20:2218–2223
Zagon IS, MsLaughlin PJ (2012) Targeting opioidergic pathways as a novel biological treatment for advanced pancreatic cancer. Expert Rev Gastroenterol Hepatol 6:133–135

Zagon IS, Goodman SR, McLaughlin PJ (1988) Characterization of opioid binding sites in murine neuroblastoma. Brain Res 499:80–88

Zagon IS, Gibo D, McLaughlin PJ (1990a) Expression of zeta (zeta), a growth-related opioid receptor, in metastatic adenocarcinoma of the human cerebellum. J Natl Cancer Inst 82:325–327

Zagon IS, Goodman SR, MeLaughlin PJ (1990b) Demonstration and characterization of zeta (zeta), a growth-related opioid receptor, in a neuroblastoma cell line. Brain Res 511:181–186

Zagon IS, Wu Y, McLaughlin PJ (1994) Opioid growth factor inhibits DNA synthesis in mouse tongue epithelium in a circadian rhythm-dependent manner. Am J Physiol 267:645–652

Zagon IS, Hytrek SD, Smith JP, McLaughlin PJ (1997) Opioid growth factor (OGF) inhibits human pancreatic cancer transplanted into nude mice. Cancer Lett 112:167–175

Zagon IS, Smith JP, McLaughlin PJ (1999) Human pancreatic cancer cell proliferation in tissue culture is tonically inhibited by opioid growth factor. Int J Oncol 14:577–584

Zagon IS, Smith JP, Conter R, McLaughlin PJ (2000) Identification and characterization of opioid growth factor receptor in human pancreatic adenocarcinoma. Int J Mol Med 5:77–84

Zagon IS, Kreiner S, Heslop JJ, Conway AB, Morgan CR, MaLaughlin PJ (2008) Prevention and delay in progression of human pancreatic cancer by stable overexpression of the opioid growth factor receptor. Int J Oncol 33:317–323

Zagon IS, Rahn KA, Turel AP, McLaughlin PJ (2009) Endogenous opioids regulate expression of experimental autoimmune encephalomyelitis: a new paradigm for the treatment of multiple sclerosis. Exp Biol Med (Maywood) 234:1383–1392

Zagon IS, Rahn KA, Bonneau RH, Turel AP, McLaughlin PJ (2010) Opioid growth factor suppresses expression of experimental autoimmune encephalomyelitis. Brain Res 1310:154–161

Zagon IS, Donahue R, McLaughlin PJ (2013) Targeting the opioid growth factor: opioid growth factor receptor axis for treatment of human ovarian cancer. Exp Biol Med (Maywood) 238:579–587

Zagon IS, Verderame MF, Hankins J, McLaughlin PJ (2007) Overexpression of the opioid growth factor receptor potentiates growth inhibition in humanpancreatic cancer cells. Int J Oncol 30:775–783

Zaki PA, Bilsky EJ, Vanderah TW, Lai J, Evans CJ, Porreca F (1996) Opioid receptor types and subtypes: the delta receptor as a model. Annu Rev Pharmacol Toxicol 36:379–401

Zastawny RL, George SR, Nguyen T, Cheng R, Tsatsos J, Briones-Urbina R, O'Dowd BF (1994) Cloning, characterization, and distribution of a mu-opioid receptor in rat brain. J Neurochem 62:2009–2105

Zhang JH, Xia Y, Haddad GG (1999) Activation of delta-opioid receptors protects cortical neurons from glutamate excitotoxic injury. Soc Neurosci 28:736

Zhang JH, Haddad GG, Xia Y (2000) delta- but not mu-and kappa-opioid receptor activation protects neocortical neuron from glutamate-induced excitotoxic injury. Brain Res 885:143–153

Zhang JH, Gibney GT, Xia Y (2001) Effect of prolonged hypoxia on expression of NA+ channel mRNA subtypes in the developing rat cortex. Brain Res Mol Brain Res 91:154–158

Zhang JH, Gibney GT, Xia Y (2002) Neuroprotective role of delta-opioid receptors in cortical neurons. Am J Physiol Cell Physiol 282:C1225–C3124

Zhang X, Wang F, Chen X, Li J, Xiang B, Zhang YQ, Li BM, Ma L (2005) Beta-arrestin1 and beta-arrestin2 are differentially required for phosphorylation-dependent and -independent internalization of delta-opioid receptors. J Neurochem 95:169–178

Zhang J, Qian H, Zhao P, Hong SS, Xia Y (2006) Rapid hypoxia preconditioning protects cortical neurons from glutamate toxicity through delta-opioid receptor. Stroke 37:1094–1099

Zhang H, Shi YG, Woods JH, Watson SJ, Ko MC (2007) Central kappa opioid receptor-mediated antidepressant-like effects of nor-Binaltorphimine: behavioral and BDNF mRNA expression studies. Eur J Pharmacol 570:89–96

Zhang S, Yekkirala A, Tang Y, Portoghese PS (2009) A bivalent ligand (KMN-21) antagonist for mu/kappa heterodimeric opioid receptors. Bioorg Med Chem Lett 19:6978–6980

Zhang L, Li J, Shi Q, Fan R, Kaye AJ, Wang Y, Sun X, Rivera FB, Kaye AD, Pei J (2013) Role of κ-opioid receptor in hypoxic pulmonary artery hypertension and its underlying mechanism. Am J Ther 20:329–336

Zhang X, Zhang Y, Asgar J, Niu KY, Lee J, Lee KS, Schneider M, Ro JY (2014) Sex differences in μ-opioid receptor expression in trigeminal ganglia under a myositis condition in rats. Eur J Pain 8:151–161

Zhao P, Guo JC, Xia Y (2002) Electro-acupuncture and brain protection from cerebral ischemia: the role of delta-opioid receptor. Soc Neurosci Abstract Program, No. 490.13

Zhao P, Huang Y, Zuo Z (2006) Opioid preconditioning induces opioid receptor-dependent delayed neuroprotection against ischemia in rats. J Neuropathol Exp Neurol 65:945–952

Zhao L, Zhu Y, Wang D, Chen M, Gao P, Xiao W, Rao G, Wang X, Jin H, Xu L, Sui N, Chen Q (2010) Morphine induces Beclin 1- and ATG5-dependent autophagy in human neuroblastoma SH-SY5Y cells and in the rat hippocampus. Autophagy 6:386–394

Zhou F, Guo J, Cheng J, Wu G, Xia Y (2013) Effect of electroacupuncture on rat ischemic brain injury: importance of stimulation duration. Evid Based Complement Alternat Med 2013:87852

Zhou XL, Yu LN, Wang Y, Tang LH, Peng YN, Cao JL, Yan M (2014) Increased methylation of the MOR gene proximal promoter in primary sensory neurons plays a crucial role in the decreased analgesic effect of opioids in neuropathic pain. Mol Pain 10:51

Zhu Y, King MA, Schuller AGP, Nitsche JF, Reidl M, Elde RP, Unterwald E, Paseternak GW, Pintar JE (1999) Retention of supraspinal delta-like analgesia and loss of morphine tolerance in delta opioid receptor knockout mice. Neuron 24:243–252

Zhu M, Li M, Tian X, Ou X, Zhu C, Guo J (2009) Neuroprotective role of δ-opioid receptors against mitochondrial respiratory chain injury. Brain Res 1252:183–191

Zhu Y, Fotinos A, Mao LL, Atassi N, Zhou EW, Ahmad S, Guan Y, Berry JD, Cudkowicz ME, Wang X (2014) Neuroprotective agents target molecular mechanisms of disease in ALS. Drug Discov Today 6446:343

Zöllner C, Schäfer M (2008) Opioids in anesthesia. Anaesthesist 57:729–740

Zöllner C, Stein C (2007) Opioids. Handb Exp Pharmacol 2007:31–63

Zou W, Guo Q, Wang E, Cai J, Cheng Z (2007) Intrathecal morphine suppresses immune function in rats with inflammatory-induced pain. J Int Med Res 36:626–636

Zuberi AR, Townsend L, Patterson L, Zheng H, Berthoud HR (2008) Increased adiposity on normal diet, but decreased susceptibility to diet-induced obesity in mu-opioid receptor-deficient mice. Eur J Pharmacol 585:14–23

Zubieta J-K, Ketter TA, Bueller JA, Xu Y, Kilbourn MR, Yong EA, Koeppe RA (2003) Regulation of human affective responses by anterior cingulated and limbic μ-opioid neurotransmission. Arch Gen Psychiatry 60:1145–1153

MIX
Papier aus verantwortungsvollen Quellen
Paper from responsible sources
FSC® C105338

FSC
www.fsc.org

If you have any concerns about our products,
you can contact us on
ProductSafety@springernature.com

In case Publisher is established outside the EU,
the EU authorized representative is:
Springer Nature Customer Service Center GmbH
Europaplatz 3, 69115 Heidelberg, Germany

Printed by Libri Plureos GmbH
in Hamburg, Germany